MW01202271

VOLVO PENTA STERN DRIVES 1968-91 REPAIR MANUAL

SELOC™

President Dean F. Morgantini, S.A.E.
Vice President–Finance Barry L. Beck
Vice President–Sales Glenn D. Potere

Executive Editor Kevin M. G. Maher, A.S.E.

Manager–Marine/Recreation James R. Marotta, A.S.E.
Manager–Consumer Automotive Richard Schwartz, A.S.E.

Production Specialists Brian Hollingsworth, Melinda Possinger
Project Managers Will Kessler, A.S.E., S.A.E., Thomas A. Mellon, A.S.E., S.A.E., Richard Rivele, Todd W. Stidham, A.S.E., Ron Webb

Authors Joan and Clarence Coles

Brought to you by

CHILTON™ MARINE

Manufactured in USA

1020 Andrew Drive
West Chester, PA 19380
ISBN 0-89330-011-X
3456789012 8765432109

SAFETY NOTICE

Proper service and repair procedures are vital to the safe, reliable operation of all marine engines, as well as the personal safety of those performing repairs. This manual outlines procedures for servicing and repairing stern drives using safe, effective methods. The procedures contain many NOTES, CAUTIONS and WARNINGS which should be followed, along with standard procedures, to eliminate the possibility of personal injury or improper service which could damage the vessel or compromise its safety.

It is important to note that repair procedures and techniques, tools and parts for servicing marine engines, as well as the skill and experience of the individual performing the work, vary widely. It is not possible to anticipate all of the conceivable ways or conditions under which these engines may be serviced, or to provide cautions as to all possible hazards that may result. Standard and accepted safety precautions and equipment should be used during cutting, grinding, chiseling, prying, or any other process that can cause material removal or projectiles.

Some procedures require the use of tools specially designed for a specific purpose. Before substituting another tool or procedure, you must be completely satisfied that neither your personal safety, nor the performance of the marine engine, will be compromised.

Although information in this manual is based on industry sources and is complete as possible at the time of publication, the possibility exists that some vehicle manufacturers made later changes which could not be included here. While striving for total accuracy, Chilton Marine cannot assume responsibility for any errors, changes or omissions that may occur in the compilation of this data.

PART NUMBERS

Part numbers listed in this reference are not recommendations by Chilton Marine for any product brand name. They are references that can be used with interchange manuals and aftermarket supplier catalogs to locate each brand supplier's discrete part number.

SPECIAL TOOLS

Special tools are recommended by the marine manufacturer to perform a specific task. Use has been kept to a minimum, but, where absolutely necessary, they are referred to in the text by the part number of the tool manufacturer. These tools can be purchased, under the appropriate part number, from your local dealer or regional distributor, or an equivalent tool can be purchased locally from a tool supplier or parts outlet. Before substituting any tool for the one recommended, read the SAFETY NOTICE at the top of this page.

ACKNOWLEDGMENTS

Chilton Marine expresses sincere appreciation to Volvo Penta of the Americas Inc. for their assistance in the production of this manual.

ALL RIGHTS RESERVED

FOREWORD

This is a comprehensive tune-up and repair manual for Volvo Penta stern drives manufactured between 1968 and 1991. Competition, high-performance, and commercial units (including aftermarket equipment), are not covered. The book has been designed and written for the professional mechanic, the do-it-yourselfer, and the student developing his mechanical skills.

Professional Mechanics will find it to be an additional tool for use in their daily work on outboard units because of the many special techniques described.

Boating enthusiasts interested in performing their own work and in keeping their unit operating in the most efficient manner will find the step-by-step illustrated procedures used throughout the manual extremely valuable. In fact, many have said this book almost equals an experienced mechanic looking over their shoulder giving them advice.

Students and Instructors have found the chapters divided into practical areas of interest and work. Technical trade schools, from Florida to Michigan and west to California, as well as the U.S. Navy and Coast Guard, have adopted these manuals as a standard classroom text.

Troubleshooting sections have been included in many chapters to assist the individual performing the work in quickly and accurately isolating problems to a specific area without unnecessary expense and time-consuming work. As an added aid and one of the unique features of this book, many worn parts are illustrated to identify and clarify when an item should be replaced.

Illustrations and procedural steps are so closely related and identified with matching numbers that, in most cases, captions are not used. Exploded drawings show internal parts and their interrelationship with the major component.

TABLE OF CONTENTS

5 IGNITION (Continued)

6 ELECTRICAL

APPENDIX

1
SAFETY

1-1 INTRODUCTION

Your boat probably represents a sizeable investment for you. In order to protect this investment and to receive the maximum amount of enjoyment from your boat it must be cared for properly while being used and when it is out of the water. Always store your boat with the bow higher than the stern and be sure to remove the transom drain plug and the inner hull drain plugs. If you use any type of cover to protect your boat, plastic, canvas, whatever, be sure to allow for some movement of air through the hull.

Proper ventilation will assure evaporation of any condensation that may form due to changes in temperature and humidity.

1-2 CLEANING, WAXING, AND POLISHING

Any boat should be washed with clear water after each use to remove surface dirt and any salt deposits from use in salt water. Regular rinsing will extend the time between waxing and polishing. It will also give you "pride of ownership", by having a sharp looking piece of equipment. Elbow grease, a mild detergent, and a brush will be required to remove stubborn dirt, oil, and other unsightly deposits.

Stay away from harsh abrasives or strong chemical cleaners. A white buffing compound can be used to restore the original gloss to a scratched, dull, or faded area. The finish of your boat should be thoroughly cleaned, buffed, and polished at least once

*The **Spirit of Western Engine**, powered by a modified Volvo Penta AQ290, 350 CID engine and equipped with a Volvo stock Model 280 stern drive, attained 110 mph, when driven by Mike Basso during endurance competition. The boat was clocked by radar on a measured course. The only change to the lower unit was slight streamlining to the leading edge. Safety is a prime consideration for racing teams competing in such high-performance events. The **"Spirit"** is owned by Western Engine Distributors, Santa Ana, California, Volvo Penta distributors for Southern California and neighboring areas until mid 1987, when Volvo eliminated the "three tier" distribution system.*

Pride of Ownership *This well-maintained boat, engine, stern drive, and trailer, reflect the owner's pride in his/her equipment. The rewards are efficiency, economy, enjoyment, and safe boating.*

each season. Take care when buffing or polishing with a marine cleaner not to overheat the surface you are working, because you will burn it.

1-3 CONTROLLING CORROSION

Since man first started out on the water, corrosion on his craft has been his enemy. The first form was merely rot in the wood and then it was rust, followed by other forms of destructive corrosion in the more modern materials. One defense against corrosion is to use similar metals throughout

Zincs used with the Volvo Penta stern drive units. The bar attaches to the underside of the transom shield and the ring is secured to the bearing housing ahead of the propeller.

A neglected boat and stern drive. Such corrosion and marine growth will be costly to the owner and greatly reduce his boating enjoyment through poor performance. What a contrast with the equipment shown in the left column.

the boat. Even though this is difficult to do in designing a new boat, particularly the undersides, similar metals should be used whenever and wherever possible.

A second defense against corrosion is to insulate dissimilar metals. This can be done by using an exterior coating of Sea Skin or by insulating them with plastic or rubber gaskets.

Using Zinc

The proper amount of zinc attached to a boat is extremely important. The use of too much zinc can cause wood burning by placing the metals close together and they become "hot". On the other hand, using too small a zinc plate will cause more rapid deterioration of the metal you are trying to protect. If in doubt, consider the fact that it is far better to replace the zincs than to replace planking or other expensive metal parts from having an excess of zinc.

Two proper size zincs may be purchased from the local Volvo Penta marine dealer, as shown in the accompanying illustration.

A partially corroded zinc bar from the transom shield. An excellent example of how the inexpensive zincs are sacrificed to save more costly parts. The zincs should be checked regularly and changed when they show this much deterioration.

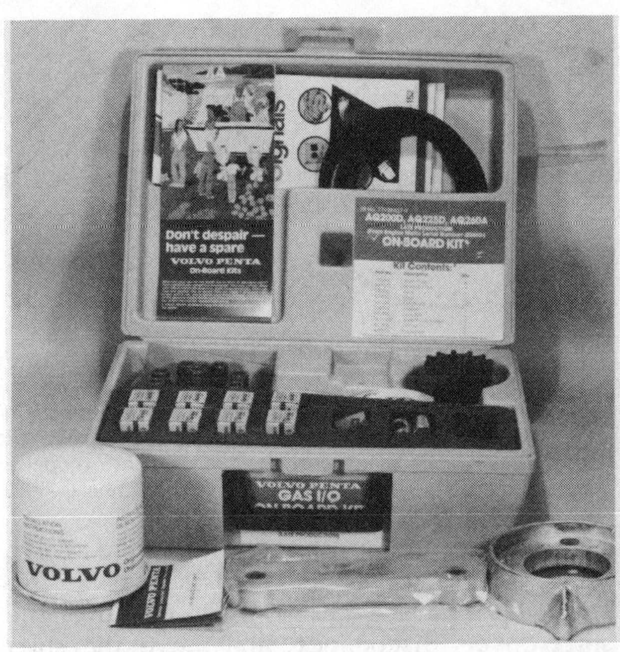

A Volvo Penta spare parts kit includes a raw water pump impeller, set of spark plugs, condenser, V-belt for the alternator, and a fuse. Such a kit, sold at modest cost, could prove most valuable in an emergency situation. A zinc bar and ring (lower right), and a spare oil filter (left), might be added to the kit by the prudent boat owner.

The bar attaches to the underside of the transom shield and the ring is secured to the double bearing housing ahead of the propeller.

The other illustration on the previous page is an excellent example of how the inexpensive zincs are sacrificed to save more costly parts. The zincs should be checked regularly and changed when they show signs of deteriorating.

1-4 PROPELLERS

As you know, the propeller is actually what moves the boat through the water. This is how it is done. The propeller operates in water in much the manner as a wood screw does in wood. The propeller "bites" into the water as it rotates. Water passes between the blades and out to the rear in

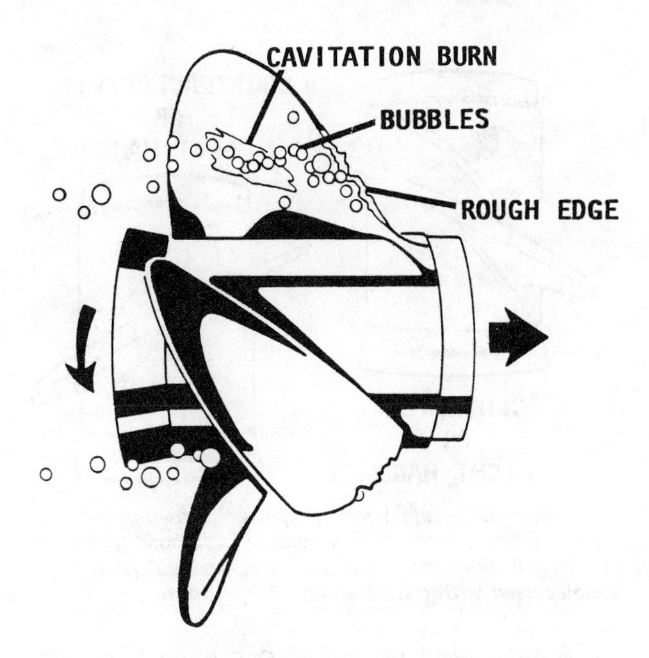

Cavitation air bubbles formed at the propeller. Manufacturers are constantly fighting this problem, as explained in the text.

the shape of a cone. The propeller "biting" through the water in much the same manner as a wood auger is what propels the boat.

Ventilation

Ventilation is the forming of voids in the water just ahead of the propeller blades. Marine propulsion designers are constantly fighting the battle against the formation of these voids due to excessive blade tip speed and engine wear. The voids may be filled with air or water vapor, or they may actually be a partial vacuum. Ventilation may be caused by installing a piece of equipment too close to the lower unit, such as the knot indicator pickup, depth sounder, or bait tank pickup.

Rotation

Propellers are manufactured as right-hand rotation (RH), and as left-hand rotation (LH). The standard propeller for Volvo Penta stern drive units is L.H. rotation.

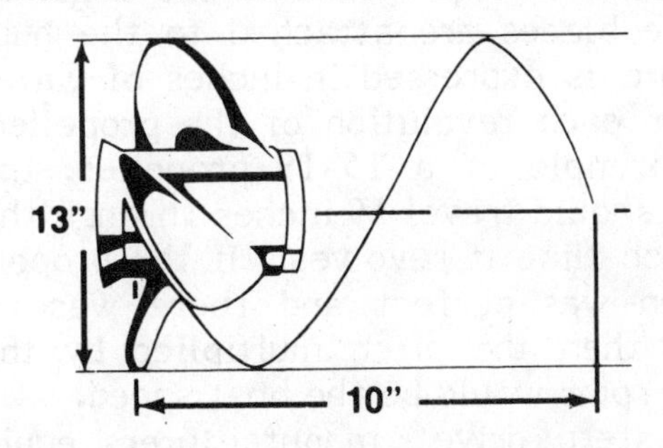

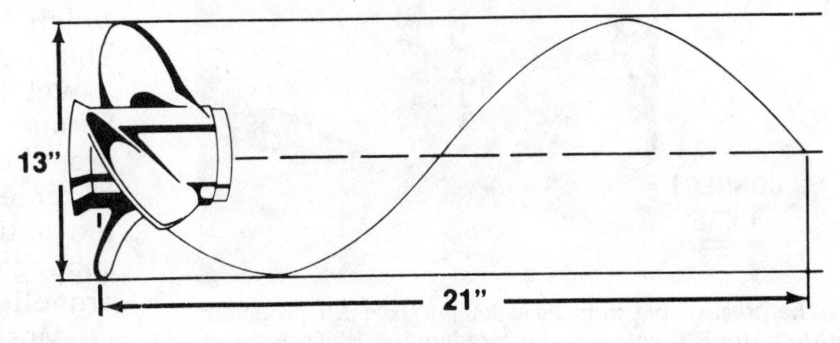

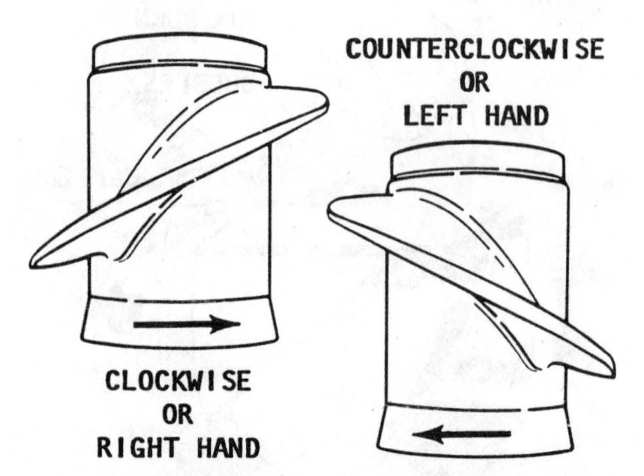

Right- and left-hand propeller showing how the angle of the blades is reversed. Left-hand propellers are by far the most popular. A right-hand propeller is usually used with a dual engine installation.

A left-hand propeller can easily be identified by observing it as shown in the accompanying illustration. Observe how the blade of the left-hand propeller slants from upper left to lower right, as shown. The right-hand propeller slants in the opposite direction, from lower left to upper right.

When the LH propeller is observed rotating from astern the boat, it will be rotating counterclockwise when the engine is in forward gear. The right-hand propeller will rotate clockwise.

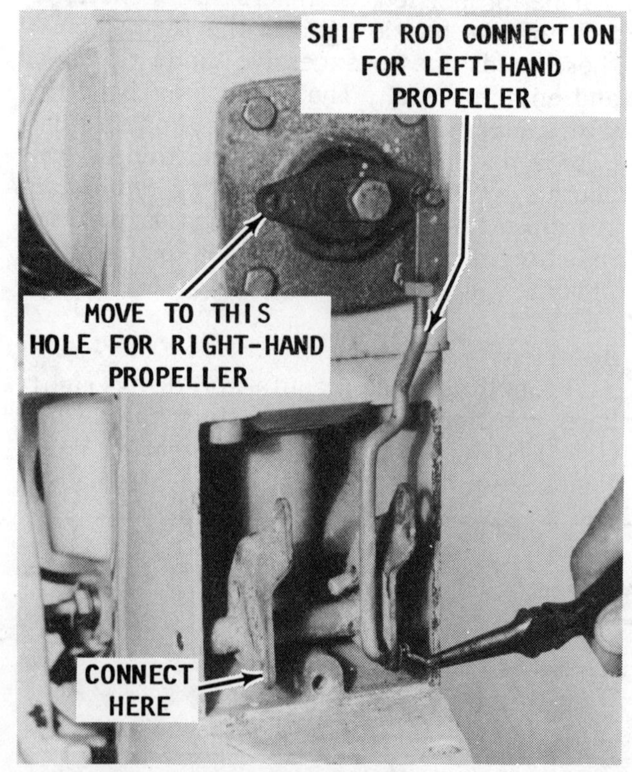

The stern drive may be changed from LH propeller rotation to RH rotation by moving the shift rod, as described in the text.

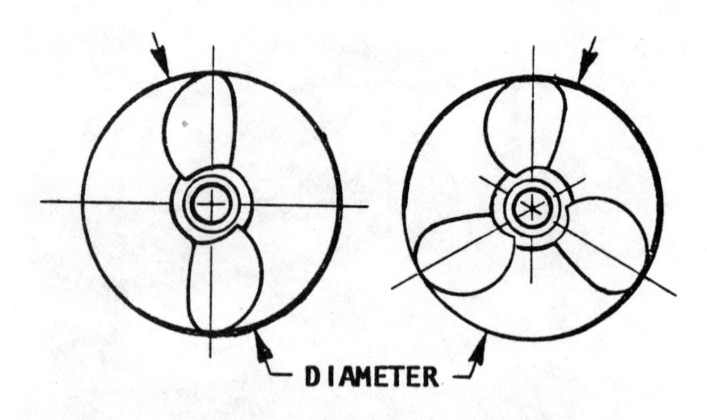

Diameter and pitch are the two basic dimensions of a propeller. The diameter is measured across the circumference of a circle scribed by the propeller blades, as shown.

Changing propeller rotation on a Volvo Penta stern drive is a simple task. No modifications are necessary. Remove the shift cover, disconnect, and move the shift rod to the opposite end of the eccentric piston arm, as indicated. Connect the rod to the opposite hole in the yoke and secure it in place with the washer and cotter key. **DO NOT** rotate the rod when making the new connection. The bend in the rod **MUST** face the same direction for both installations.

If an adjustment is necessary, see Chapter 8, Retaining Pawl Adjustment, Page 8-5.

Diameter and Pitch

Only two dimensions of the propeller are of real interest to the boat owner: the diameter and the pitch. These two dimensions are stamped on the propeller hub and always appear in the same order: the diameter first and then the pitch. For instance, the number 15-19 stamped on the hub, would mean the propeller had a diameter of 15 inches with a pitch of 19.

The diameter is the measured distance from the tip of one blade to the tip of the other as shown in the accompanying illustration.

The pitch of a propeller is the angle at which the blades are attached to the hub. This figure is expressed in inches of water travel for each revolution of the propeller. In our example of a 15-19 propeller, the propeller should travel 19 inches through the water each time it revolves. If the propeller action was perfect and there was no slippage, then the pitch multiplied by the propeller rpms would be the boat speed.

Most stern drive manufacturers equip their units with a standard propeller with a diameter and pitch they consider to be best suited to the engine and the boat. Such a propeller allows the engine to run as near to the rated rpm and horsepower (at full throttle) as possible for the boat design.

The blade area of the propeller determines its load-carrying capacity. A two-blade propeller is used for high-speed running under very light loads.

A four-blade propeller is installed in boats intended to operate at low speeds under very heavy loads such as tugs, barges, or large houseboats. The three-blade propeller is the happy medium covering the wide range between the high performance units and the load carrying workhorses.

Progressive Pitch

Progressive pitch is a blade design innovation that improves performance when forward and rotational speed is high and/or the propeller breaks the surface of the water.

Progressive pitch starts low at the leading edge and progressively increases to the trailing edge, as shown in the accompanying illustration. The average pitch over the entire blade is the number assigned to that propeller. In the illustration of the progressive pitch, the average pitch assigned to the propeller would be 21.

Propeller Rake

If a propeller blade is examined on a cut extending directly through the center of the hub, and if the blade is set vertical to the propeller hub, as shown in the accompanying illustration, the propeller is said to have a zero degree (0°) rake. As the blade slants back, the rake increases. Standard propellers have a rake angle from 0° to 15°.

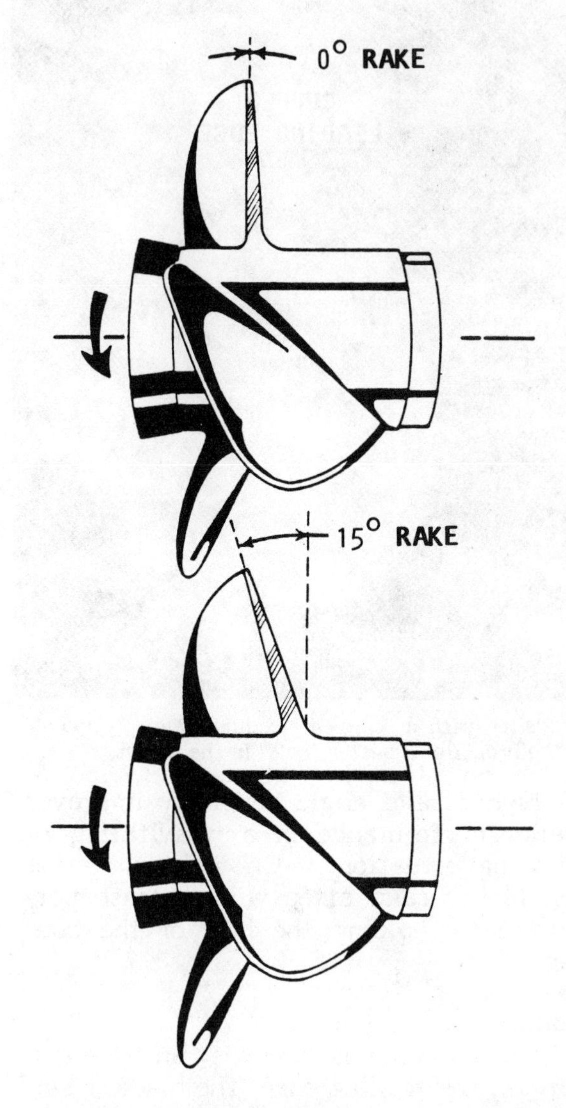

Illustration depicting the rake of a propeller, as explained in the text.

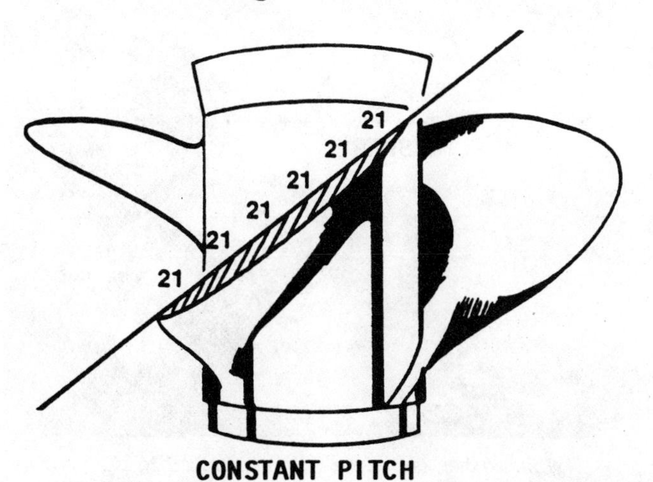

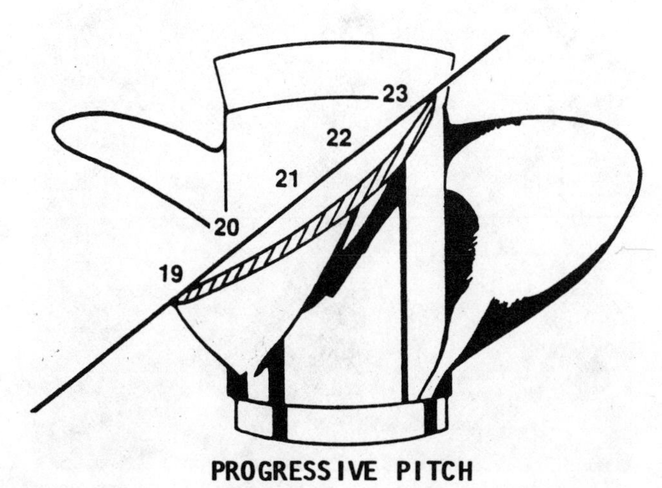

Comparison of a constant and progressive pitch propeller. Notice how the pitch of the progressive pitch propeller, right, changes to give the blade more thrust and therefore, the boat more speed.

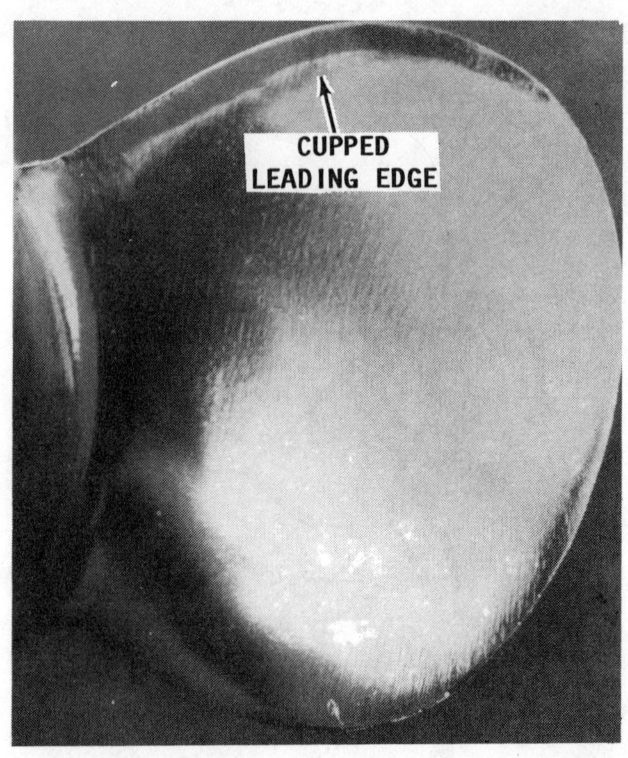

Propeller with a "cupped" leading edge. "Cupping" gives the propeller a better "hold" in the water.

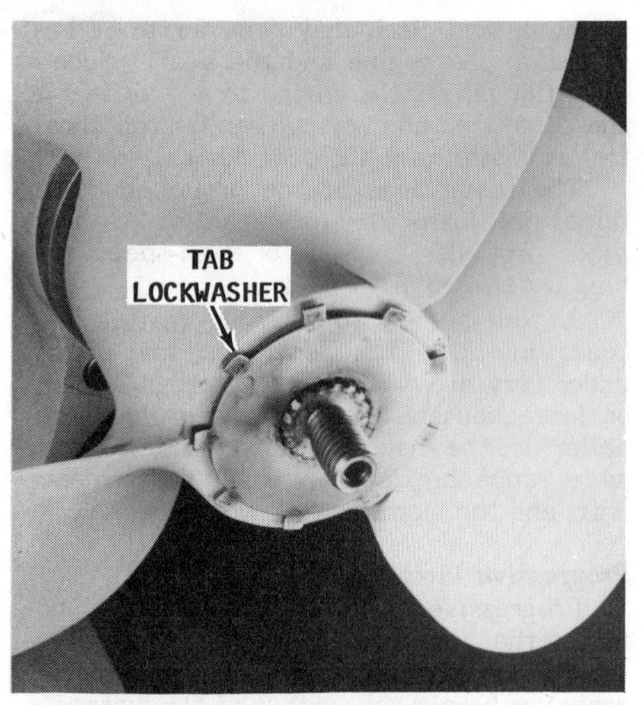

Tab lockwasher used on early model propeller installations. The tabs are bent down into slots in the propeller. The cone is installed after the propeller and tightened with a tool inserted into a hole in the cone.

A higher rake angle generally improves propeller performance in a cavitating or ventilating situation. On lighter, faster boats, higher rake often will increase performance by holding the bow of the boat higher.

Cupping

If the propeller is cast with an edge curl inward on the trailing edge, the blade is said to have a cup. In most cases, cupped blades improve performance. The cup helps the blades to **"HOLD"** and not break loose, when operating in a cavitating or ventilating situation. This action permits the engine to be trimmed out further, or to be mounted higher on the transom. This is especially true on high-performance boats. Either of these two adjustments will usually add to higher speed.

The cup has the effect of adding to the propeller pitch. Cupping usually will reduce

Propeller installation on a Duoprop lower unit. The propellers are a "matched" set with size and identification number stamped on the side of the hub.

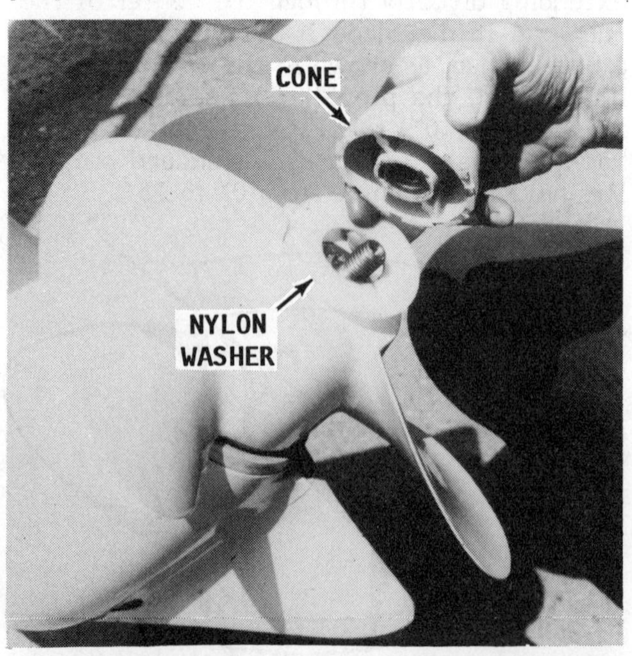

Nylon washer and cone being installed to the propeller shaft behind a late model propeller. The cone is secured in place with an Allen-head screw in the end of the shaft.

full-throttle engine speed about 150 to 300 rpm below the same pitch propeller without a cup to the blade. A propeller repair shop is able to increase or decrease the cup on the blades. This change, as explained, will alter engine rpm to meet specific operating demands. Cups are rapidly becoming standard on propellers.

In order for a cup to be the most effective, the cup should be completely concave (hollowed) and finished with a sharp corner. If the cup has any convex rounding, the effectiveness of the cup will be reduced.

Vibration

Your propeller should be checked regularly to be sure all blades are in good condition. If any of the blades become bent or nicked, this condition will set up vibrations in the drive unit and the motor. If the vibration becomes very serious it will cause a loss of power, efficiency, and boat performance. If the vibration is allowed to continue over a period of time it can have a damaging effect on many of the operating parts.

Vibration in boats can never be completely eliminated, but it can be reduced by keeping all parts in good working condition and through proper maintenance and lubrication. Vibration can also be reduced in some cases by increasing the number of blades. For this reason, many racers use two-blade props and luxury cruisers have four- and five-blade props installed.

Shock Absorbers/Cushion Hubs

The shock absorber, sometimes referred to cushion hub, in the propeller plays a very important role in protecting the shafting, gears, and engine against the shock of a blow, should the propeller strike an underwater object. The shock absorber allows the propeller to stop rotating at the instant of impact while the power train continues turning.

How much impact the propeller is able to withstand, before causing the shock absorber to slip, is calculated to be more than the force needed to propel the boat, but less than the amount that could damage any part of the power train. Under normal propulsion loads of moving the boat through the water, the hub will not slip. However, it will slip if the propeller strikes an object with a force that would be great enough to stop any part of the power train.

If the power train was to absorb an impact great enough to stop rotation, even for an instant, something would have to give and be damaged. If a propeller is subjected to repeated striking of underwater objects, it would eventually slip on its clutch hub under normal loads. If the propeller should start to slip, a new shock absorber/cushion hub would have to be installed.

Propeller Selection

There is no standard propeller that will do the proper job in very many cases. The list of sizes and weights of boats is almost endless. This fact coupled with the many boat-engine combinations makes the propeller selection for a specific purpose a difficult job. In fact, in many cases the propeller is changed after a few test runs. Proper selection is aided through the use of charts set up for various engines and boats. These charts should be studied and understood when buying a propeller. However, bear in mind, the charts are based on average boats with average loads, therefore, it may be necessary to make a change in size or pitch, in order to obtain the desired results for the hull design or load condition.

Propellers are available with a wide range of pitch. Remember, a low pitch takes a smaller bite of the water than the high pitch propeller. This means the low pitch propeller will travel less distance through the water per revolution. The low pitch will require less horsepower and will allow the engine to run faster.

All engine manufacturers design their units to operate with full throttle at, or slightly above, the rated rpm. If you run your engine at the rated rpm, you will increase spark plug life, receive better fuel economy, and obtain the best performance from your boat and engine. Therefore, take time to make the proper propeller selection for the rated rpm of your engine at full throttle with what you consider to be an average load. Your boat will then be correctly balanced between engine and propeller throughout the entire speed range.

A reliable tachometer must be used to measure engine speed at full throttle to ensure the engine will achieve full horsepower and operate efficiently and safely. To test for the correct propeller, make your run in a body of smooth water with the lower unit in forward gear at full throttle.

If the reading is above the manufacturer's recommended operating range, you must try propellers of greater pitch, until you find the one that allows the engine to operate continually within the recommended full throttle range.

If the engine is unable to deliver top performance and you feel it is properly tuned, then the propeller may not be to blame. Operating conditions have a marked effect on performance. For instance, an engine will lose rpm when run in very cold water. It will also lose rpm when run in salt water as compared with fresh water. A hot, low-barometer day will also cause your engine to lose power.

1-5 FUEL SYSTEM

With Built-in Fuel Tank

All parts of the fuel system should be selected and installed to provide maximum service and protection against leakage. Reinforced flexible sections should be installed in fuel lines where there is a lot of motion, such as at the engine connection. The flaring of copper tubing should be annealed after it is formed as a protection against hardening.

CAUTION: Compression fittings should **NOT** be used because they are so easily overtightened, which places them under a strain and subjects them to fatigue. Such conditions will cause the fitting to leak after it is connected a second time.

The capacity of the fuel filter must be large enough to handle the demands of the engine as specified by the engine manufacturer.

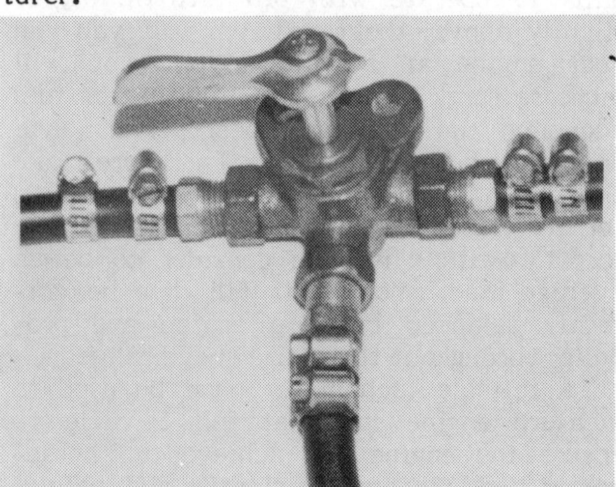

A three-position valve permits fuel to be drawn from either tank or to be shut off completely. Such an arrangement prevents accidental siphoning of fuel from the tank.

All fittings and outlets must come out the top of the tank. An anti-siphon device should be installed close to the tank. This spring-loaded valve will automatically prevent fuel from being siphoned out of the tank if the line from the tank to the fuel pump should be damaged and begin to leak.

A manually-operated valve should be installed if anti-siphon protection is not provided. This valve should be installed in the fuel line as close to the gas tank as possible. Such a valve will maintain anti-siphon protection between the tank and the engine.

Fuel tanks should be mounted in dry, well ventilated places. Ideally, the fuel tanks should be installed above the cockpit floors, where any leakage will be quickly detected.

In order to obtain maximum circulation of air around fuel tanks, the tank should not come in contact with the boat hull except through the necessary supports. The supporting surfaces and hold-downs must fasten the tank firmly and they should be insulated from the tank surfaces. This insulation material should be non-abrasive and non-absorbent material. Fuel tanks installed in the forward portion of the boat should be

Taking on fuel at an automobile service station. Safety grounding is accomplished through the hose fixtures. The fuel tank should be kept full to prevent condensation.

especially well secured and protected because shock loads in this area can be as high as 20 to 25 g's ("g" equals force of gravity).

Engine Compartment Ventilation

All motorboats built after April 25, 1940 and before August 1, 1980, powered by a gasoline engine or by fuels having a flashpoint of 110° F. or less **MUST** have the following, which is quoted from a Coast Guard publication dated 1984:

At least two ventilation ducts fitted with cowls or their equivalent for the purpose of properly and efficiently ventilating the bilges of every engine and fuel tank compartment. There shall be at least one exhaust duct installed so as to extend to the lower portion of the bilge and at least one intake duct installed so as to extend to a point at least midway to the bilge or at least below the level of the carburetor air intake.

All boats built after July 31, 1978 but prior to August 1, 1980, the requirement for ventilation of the fuel tank compartment can be omitted if there is no electrical source of ignition in the fuel tank compartment and if the fuel tank vents to the outside of the boat. After August 1, 1980, all boats with gasoline engines must be built with ventilation systems which comply with Coast Guard standards. The standard requires the engine compartment to be equipped with a blower. Now, if the blower and ventilation system do not meet Coast Guard regulations, the owner can recall the manufacturer to make it right. Owners can not recall the manufacturer for boats built prior to August 1, 1980. The operator is required to keep the system in proper operating condition.

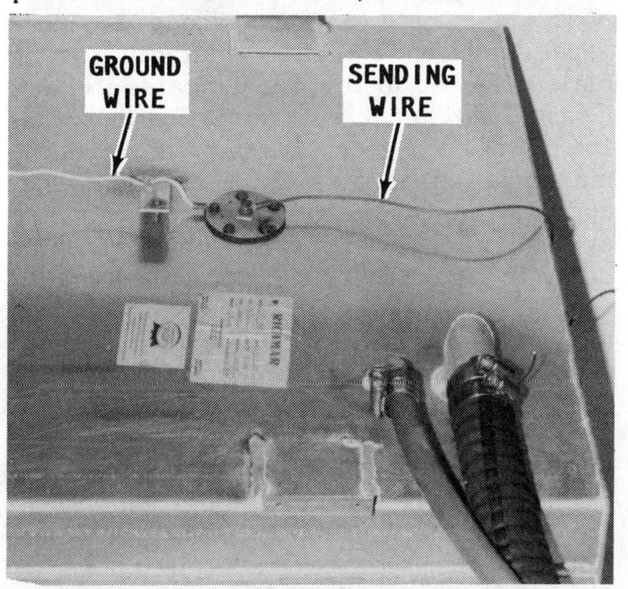

A fuel tank properly grounded to prevent static electricity. Static electricity could be extremely dangerous when taking on fuel.

In addition to the blower requirement for the engine compartment, a tag **MUST** be affixed to the dashboard stating the blower must be operated for at least four minutes before an attempt is made to start the engine.

This tag requirement is made to the boat manufacturer and the boat owner is responsible for keeping it in place and in readable condition. The tag must be yellow and able to withstand salt spray.

Citation

An owner may receive a citation --ticket and fine -- if the ventilation system is not Coast Guard Approved, in good operating order, and the blower tag is not affixed to the dashboard.

Flame Arrestors

A gasoline engine installed in a motorboat or motor vessel after April 25, 1940, except outboard motors, must have a Coast Guard Approved flame arrestor fitted to the carburetor. This requirement applies if the engine is enclosed or if the carburetor is below the gunwale of the boat.

Automotive Replacement Parts

When replacing fuel, electrical, and other parts, check to be sure they are marine type and Coast Guard Approved. Automotive parts are not made to the high standards of marine parts. The carburetors must **NOT** leak fuel; alternators, generators, and voltage regulators **MUST** be able to operate in a gas fume enclosed area without exploding; etc. Automotive parts could cause a fire endangering the crew and the craft. The part may look the same and even have a similar number, but if it is not **MARINE** it is not safe to use on the boat and will not pass Coast Guard inspection.

Taking On Fuel

The fuel tank of the boat should be kept full to prevent water from entering the system through condensation caused by temperature changes. Water droplets forming is one of the greatest enemies of the fuel system. By keeping the tank full, the air space in the tank is kept to an absolute minimum and there is no room for moisture to form. It is a good practice not to store fuel in the tank over an extended period, say for six months. Today, fuels contain ingredients that change into gums when stored for any length of time. These gums and varnish products will cause carburetor problems and poor spark plug performance. An additive (Sta-Bil) is available and can be used to prevent gums and varnish from forming.

Static Electricity

In very simple terms, static electricity is called frictional electricity. It is generated by two dissimilar materials moving over each other. One form is gasoline flowing through a pipe or into the air. Another form is when you brush your hair or walk across a synthetic carpet and then touch a metal object. All of these actions cause an electrical charge. In most cases, static electricity is generated during very dry weather conditions, but when you are filling the fuel tank on a boat it can happen at any time.

Fuel Tank Grounding

One area of protection against the build-up of static electricity is to have the fuel tank properly grounded (also known as bonding). A direct metal-to-metal contact from the fuel hose nozzle to the water in which the boat is floating. If the fill pipe is made of metal, and the fuel nozzle makes a good contact with the deck plate, then a good ground is made.

As an economy measure, some boats use rubber or plastic filler pipes because of compound bends in the pipe. Such a fill line does not give any kind of ground and if your boat has this type of installation and you do not want to replace the filler pipe with a metal one, then it is possible to connect the deck fitting to the tank with a copper wire. The wire should be 8 gauge or larger.

The fuel line from the tank to the engine should provide a continuous metal-to-metal contact for proper grounding. If any part of this line is plastic or other non-metallic material, then a copper wire must be connected to bridge the non-metal material. The power train provides a ground through the engine and drive shaft, to the propeller in the water.

Fiberglass fuel tanks pose problems of their own. Fortunately, this material has almost totally disappeared as a suitable substance for fuel tanks. If, however, the boat you are servicing, does have a fiberglass tank, or one is being installed, or repaired, it is almost mandatory that you check with the Coast Guard Recreational Boating Standards Office in your district before proceeding with any work. The new standards are very specific and the Coast Guard is extremely rigid about enforcing the regulations.

Anything you can feel as a "shock" is enough to set off an explosion. Did you know that under certain atmospheric conditions you can cause a static explosion yourself, particularly if you are wearing synthetic clothing. It is almost a certainty you could cause a static spark if you are **NOT** wearing insulated rubber-soled shoes.

As soon as the deck fitting is opened, fumes are released to the air. Therefore, to be safe you should ground yourself before opening the fill pipe deck fitting. One way to ground yourself is to dip your hand in the water overside to discharge the electricity in your body before opening the filler cap. Another method is to touch the engine block or any metal fitting on the dock which goes down into the water.

1-6 LOADING

In order to receive maximum enjoyment, with safety and performance, from your boat, take care not to exceed the load capacity given by the manufacturer. A plate attached to the hull indicates the U.S. Coast Guard capacity information in pounds for persons and gear. If the plate states the maximum person capacity to be 750 pounds and you assume each person to weigh an average of 150 lbs., then the boat could carry five persons safely. If you add another 250 lbs. for motor and gear, and the maximum weight capacity for persons and gear is 1,000 lbs. or more, then the five persons and gear would be within the limit.

Try to load the boat evenly port and starboard. If you place more weight on one side than on the other, the boat will list to the heavy side and make steering difficult. You will also get better performance by placing heavy supplies aft of the center to keep the bow light for more efficient planing.

U.S. COAST GUARD
MAXIMUM CAPACITIES
PERSONS OR LBS
LBS PERSONS,MOTOR,GEAR
H.P. MOTOR
THIS BOAT COMPLIES WITH U.S. COAST GUARD SAFETY STANDARDS IN EFFECT ON THE DATE OF CERTIFICATION
RECREATIONAL ENTERPRISES
RIALTO, CALIF.

U.S. Coast Guard plate affixed to all new boats. When the blanks are filled in, the plate will indicate the Coast Guard's recommendations for persons, gear, and horsepower to ensure safe operation of the boat. These recommendations should not be exceeded, as explained in the text.

Clarification

Much confusion arises from the terms, certification, requirements, approval, regulations, etc. Perhaps the following may clarify a couple of these points.

1- The Coast Guard does not approve boats in the same manner as they "Approve" life jackets. The Coast Guard applies a formula to inform the public of what is safe for a particular craft.

2- If a boat has to meet a particular regulation, it must have a Coast Guard certification plate. The public has been led to believe this indicates approval of the Coast Guard. Not so.

3- The certification plate means a willingness of the manufacturer to meet the Coast Guard regulations for that particular craft. The manufacturer may recall a boat if it fails to meet the Coast Guard requirements.

4- The Coast Guard certification plate, see accompanying illustration, may or may not be metal. The plate is a regulation for the manufacturer. It is only a warning plate and the public does not have to adhere to the restrictions set forth on it. Again, the plate sets forth information as to the Coast Guard's opinion for safety on that particular boat.

5- Coast Guard Approved equipment is equipment which has been approved by the Commandant of the U.S. Coast Guard and has been determined to be in compliance with Coast Guard specifications and regulations relating to the materials, construction, and performance of such equipment.

1-7 HORSEPOWER

The maximum horsepower engine for each individual boat should not be increased by any great amount without checking requirements from the Coast Guard in your area. The Coast Guard determines horsepower requirements based on the length, beam, and depth of the hull. **TAKE CARE NOT** to exceed the maximum horsepower listed on the plate or the warranty and possibly the insurance on the boat may become void.

1-8 FLOTATION

If your boat is less than 20 ft. overall, a Coast Guard or BIA (Boating Industry of America) now changed to NMMA (National Marine Manufacturers Association) requirement is that the boat must have buoyant material built into the hull (usually foam) to keep it from sinking if it should become swamped. Coast Guard requirements are mandatory but the NMMA is voluntary.

"Kept from sinking" is defined as the ability of the flotation material to keep the boat from sinking when filled with water and with passengers clinging to the hull. One restriction is that the total weight of the motor, passengers, and equipment aboard does not exceed the maximum load capacity listed on the plate.

Life Preservers —Personal Flotation Devices (PFDs)

The Coast Guard requires at least one Coast Guard approved life-saving device be carried on board all motorboats for each person on board. Devices approved are identified by a tag indicating Coast Guard approval. Such devices may be life preservers, buoyant vests, ring buoys, or buoyant cushions. Cushions used for seating are serviceable if air cannot be squeezed out of it. Once air is released when the cushion is squeezed, it is no longer fit as a flotation device. New foam cushions dipped in a rubberized material are almost indestructible.

Life preservers have been classified by the Coast Guard into five type categories. All PFDs presently acceptable on recreational

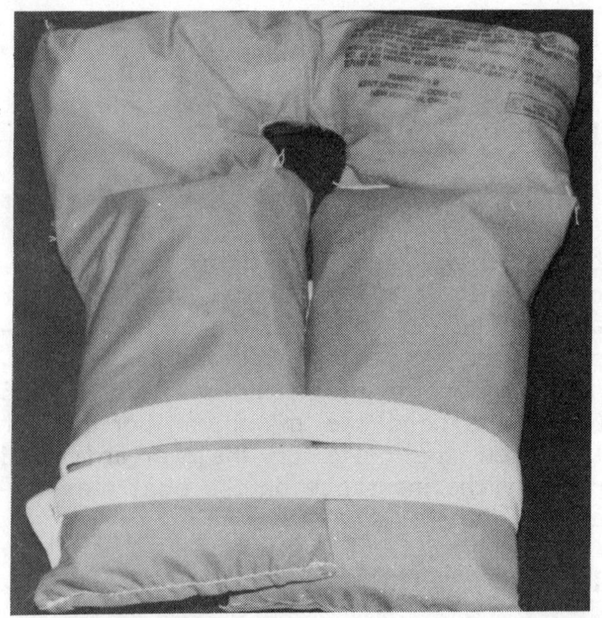

*Type I PFD Coast Guard Approved life jacket. This type flotation device provides the greatest amount of buoyancy. **NEVER** use them for cushions or other purposes.*

boats fall into one of these five designations. All PFDs **MUST** be U.S. Coast Guard approved, in good and serviceable condition, and of an appropriate size for the persons who intend to wear them. Wearable PFDs **MUST** be readily accessible and throwable devices **MUST** be immediately available for use.

Type I PFD has the greatest required buoyancy and is designed to turn most **UNCONSCIOUS** persons in the water from a face down position to a vertical or slightly backward position. The adult size device provides a minimum buoyancy of 22 pounds and the child size provides a minimum buoyancy of 11 pounds. The Type I PFD provides the greatest protection to its wearer and is most effective for all waters and conditions.

A Type IV PFD cushion device intended to be thrown to a person in the water. If air can be squeezed out of the cushion it is no longer fit for service as a PFD.

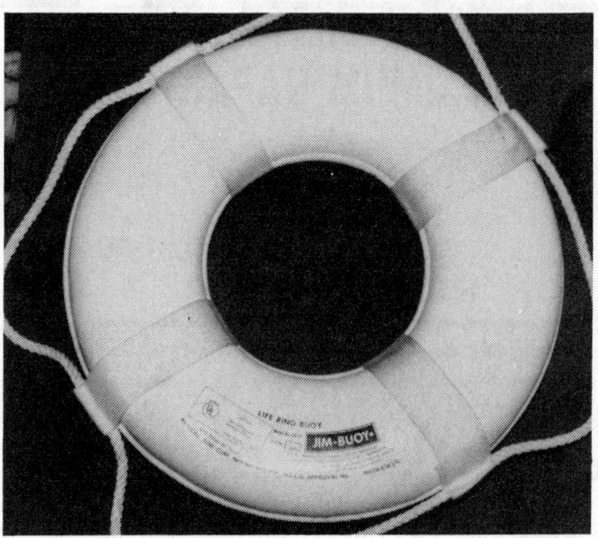

Type IV ring buoy also designed to be thrown to a person in the water. On ocean cruisers, this type device usually has a weighted pole with flag and light attached to the buoy.

Type II PFD is designed to turn its wearer in a vertical or slightly backward position in the water. The turning action is not as pronounced as with a Type I. The device will not turn as many different type persons under the same conditions as the Type I. An adult size device provides a minimum buoyancy of 15½ pounds, the medium child size provides a minimum of 11 pounds, and the infant and small child sizes provide a minimum buoyancy of 7 pounds.

Type III PFD is designed to permit the wearer to place himself (herself) in a vertical or slightly backward position. The Type III device has the same buoyancy as the Type II PFD but it has little or no turning ability. Many of the Type III PFD are designed to be particularly useful when water skiing, sailing, hunting, fishing, or engaging in other water sports. Several of this type will also provide increased hypothermia protection.

Type IV PFD is designed to be thrown to a person in the water and grasped and held by the user until rescued. It is **NOT** designed to be worn. The most common Type IV PFD is a ring buoy or a buoyant cushion.

Type V PFD is any PFD approved for restricted use.

Coast Guard regulations state, in general terms, that on all boats less than 16 ft. overall, one Type I, II, III, or IV device shall be carried on board for each person in the boat. On boats over 26 ft., one Type I, II, or III device shall be carried on board for each person in the boat **plus** one Type IV device.

It is an accepted fact that most boating people own life preservers, but too few actually wear them. There is little or no excuse for not wearing one because the modern comfortable designs available today do not subtract from an individual's boating pleasure. Make a life jacket available to your crew and advise each member to wear it. If you are a crew member ask your skipper to issue you one, especially when boating in rough weather, cold water, or when running at high speed. Naturally, a life jacket should be a must for non-swimmers any time they are out on the water in a boat.

1-9 EMERGENCY EQUIPMENT

Visual Distress Signals
The Regulation

Since January 1, 1981, Coast Guard Regulations require all recreation boats when used on coastal waters, which includes the Great Lakes, the territorial seas and those waters directly connected to the Great Lakes and the territorial seas, up to a point where the waters are less than two miles wide, and boats owned in the United States when operating on the high seas to be equipped with visual distress signals.

The only exceptions are during daytime (sunrise to sunset) for:

Recreational boats less than 16 ft. (5 meters) in length.

Boats participating in organized events such as races, regattas or marine parades.

Open sailboats not equipped with propulsion machinery and less than 26 ft. (8 meters) in length.

Manually propelled boats.

The above listed boats need to carry night signals when used on these waters at night.

Pyrotechnic visual distress signaling devices **MUST** be Coast Guard Approved, in serviceable condition and stowed to be readily accessible. If they are marked with a date showing the serviceable life, this date must not have passed. Launchers, produced before Jan. 1, 1981, intended for use with approved signals are not required to be Coast Guard Approved.

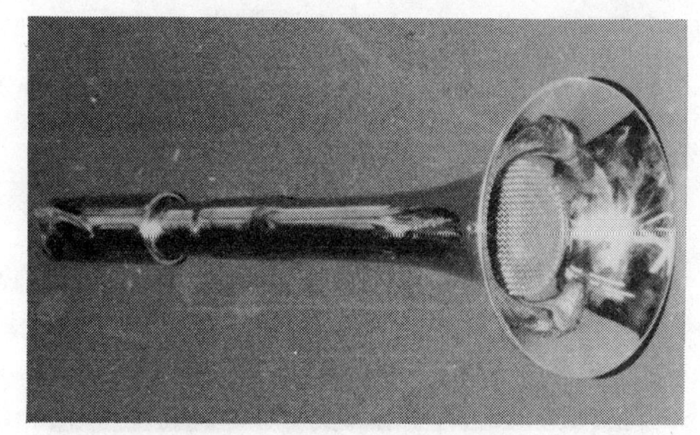

A sounding device should be mounted close to the helmsman for use in sounding an emergency alarm.

USCG Approved pyrotechnic visual distress signals and associated devices include:

Pyrotechnic red flares, hand held or aerial.

Pyrotechnic orange smoke, hand held or floating.

Launchers for aerial red meteors or parachute flares.

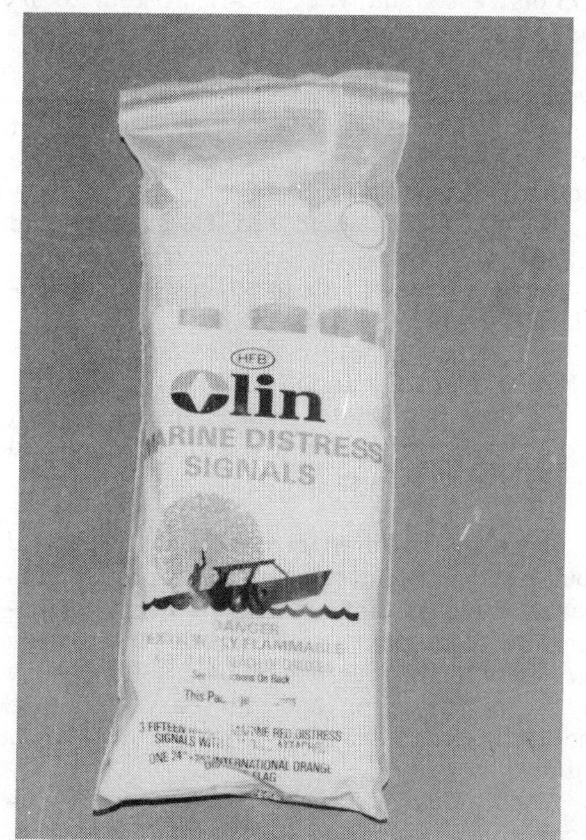

Moisture protected flares should be carried on board for use as a distress signal in an emergency.

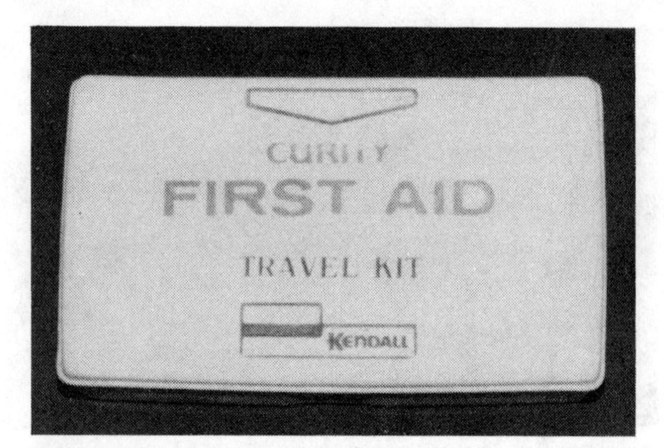

An adequately stocked first-aid kit should be on board for the safety of crew and guests.

Non-pyrotechnic visual distress signaling devices must carry the manufacturer's certification that they meet Coast Guard requirements. They must be in serviceable condition and stowed so as to be readily accessible.

This group includes:

Orange distress flag at least 3 x 3 feet with a black square and ball on an orange background.

Electric distress light -- not a flashlight but an approved electric distress light which **MUST** automatically flash the international **SOS** distress signal (. . . - - - - . . .) four to six times each minute.

Types and Quantities

The following variety and combination of devices may be carried in order to meet the requirements.

1- Three hand-held red flares (day and night).

2- One electric distress light (night only).

3- One hand-held red flare and two parachute flares (day and night).

4- One hand-held orange smoke signal, two floating orange smoke signals (day) and one electric distress light (day and night).

If young children are frequently aboard your boat, careful selection and proper stowage of visual distress signals becomes especially important. If you elect to carry pyrotechnic devices, you should select those in tough packaging and not easy to ignite should the devices fall into the hands of children.

Coast Guard Approved pyrotechnic devices carry an expiration date. This date can **NOT** exceed 42 months from the date of manufacture and at such time the device can no longer be counted toward the minimum requirements.

SPECIAL WORDS

In some states the launchers for meteors and parachute flares may be considered a firearm. Therefore, check with your state authorities before acquiring such a launcher.

First Aid Kits

The first-aid kit is similar to an insurance policy or life jacket. You hope you don't have to use it but if needed, you want it there. It is only natural to overlook this essential item because, let's face it, who likes to think of unpleasantness when planning to have only a good time. However, the prudent skipper is prepared ahead of time, and is thus able to handle the emergency without a lot of fuss.

Good commercial first-aid kits are available such as the Johnson and Johnson "Marine First-Aid Kit". With a very modest expenditure, a well-stocked and adequate kit can be prepared at home.

Any kit should include instruments supplies, and a set of instructions for their use. Instruments should be protected in a watertight case and should include: scissors, tweezers, tourniquet, thermometer, safety pins, eye-washing cup, and a hot water bottle. The supplies in the kit should include: assorted bandages in addition to the various sizes of "band-aids", adhesive tape, absorbent cotton, applicators, petroleum jelly, antiseptic (liquid and ointment), local ointment, aspirin, eye ointment, antihistamine, ammonia inhalent, sea-sickness pills, antacid pills, and a laxative. You may want to consult your family physician about including antibiotics. Be sure your kit contains a first-aid manual because even though you have taken the Red Cross course, you may be the patient and have to rely on an untrained crew for care.

Fire Extinguishers

All fire extinguishers must bear Underwriters Laboratory (UL) "Marine Type" approved labels. With the UL certification, the extinguisher does not have to have a Coast Guard approval number. The Coast Guard classifies fire extinguishers according to their size and type.

Type B-I or B-II Designed for extinguishing flammable liquids. Required on all motorboats.

The Coast Guard considers a boat having one or more of the following conditions as a "boat of closed construction" subject to fire extinguisher regulations.

1- Inboard engine or engines.

2- Closed compartments under thwarts and seats wherein portable fuel tanks may be stored.

3- Double bottoms not sealed to the hull or which are not completely filled with flotation materials.

4- Closed living spaces.

5- Closed stowage compartments in which combustible or flammable material is stored.

6- Permanently installed fuel tanks.

Detailed classification of fire extinguishers is by agent and size:

B-I contains 1-1/4 gallons foam, 4 pounds carbon dioxide, 2 pounds dry chemical, and 2-1/2 pounds freon.

B-II contains 2-1/2 gallons foam, 15 pounds carbon dioxide, and 10 pounds dry chemical.

The class of motorboat dictates how many fire extinguishers are required on board. One B-II unit can be substituted for two B-I extinguishers. When the engine compartment of a motorboat is equipped with a fixed (built-in) extinguishing system, one less portable B-I unit is required.

Dry chemical fire extinguishers without gauges or indicating devices must be weighed and tagged every 6 months. If the gross weight of a carbon dioxide (CO_2) fire extinguisher is reduced by more than 10% of the net weight, the extinguisher is not acceptable and must be recharged.

READ labels on fire extinguishers. If the extinguisher is U.L. listed, it is approved for marine use.

DOUBLE the number of fire extinguishers recommended by the Coast Guard, because their requirements are a bare **MINIMUM** for safe operation. Your boat, family, and crew, must certainly be worth much more than "bare minimum".

1-10 COMPASS

Selection

The safety of the boat and her crew may depend on her compass. In many areas weather conditions can change so rapidly that within minutes a skipper may find himself "socked-in" by a fog bank, a rain squall,

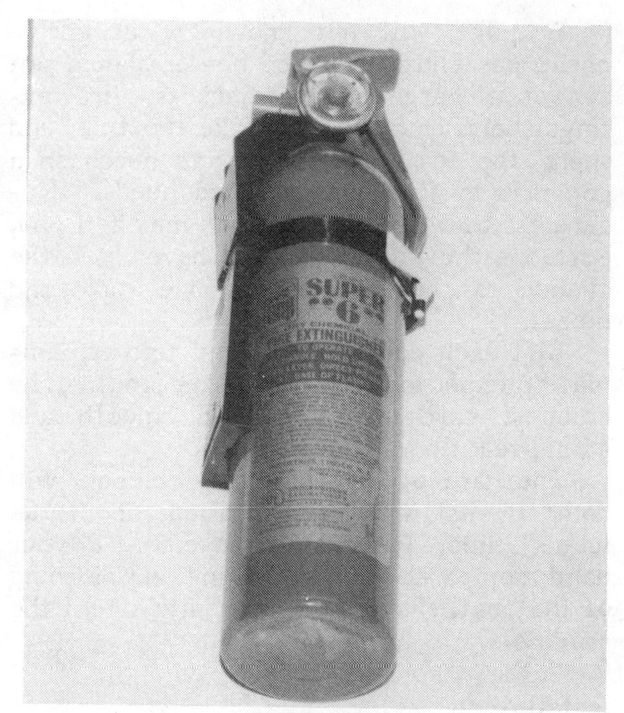

A suitable fire extinguisher should be mounted close to the helmsman for emergency use.

or just poor visibility. Under these conditions, he may have no other means of keeping to his desired course except with the compass. When crossing an open body of water, his compass may be the only means of making an accurate landfall.

During thick weather when you can neither see nor hear the expected aids to navigation, attempting to run out the time on a given course can disrupt the pleasure of the cruise. The skipper gains little comfort in a chain of soundings that does not match those given on the chart for the expected area. Any stranding, even for a short time, can be an unnerving experience.

Do not hesitate to spend a few extra dollars for a good reliable compass. If in doubt, seek advice from fellow boaters.

A pilot will not knowingly accept a cheap parachute. A good boater should not accept a bargain in lifejackets, fire extinguishers, or compass. Take the time and spend the few extra dollars to purchase a compass to fit your expected needs. Regardless of what the salesman may tell you, postpone buying until you have had the chance to check more than one make and model.

Lift each compass, tilt and turn it, simulating expected motions of the boat. The compass card should have a smooth and stable reaction.

The card of a good quality compass will come to rest without oscillations about the lubber's line. Reasonable movement in your hand, comparable to the rolling and pitching of the boat, should not materially affect the reading.

Installation

Proper installation of the compass does not happen by accident. Make a critical check of the proposed location to be sure compass placement will permit the helmsman to use it with comfort and accuracy. First, the compass should be placed directly in front of the helmsman and in such a position that it can be viewed without body stress as he sits or stands in a posture of relaxed alertness. The compass should be in the helmsman's zone of comfort. If the compass is too far away, he may have to bend forward to watch it; too close and he must rear backward for relief.

The compass is a delicate instrument and deserves respect. It should be mounted securely and in position where it can be easily observed by the helmsman.

Second, give some thought to comfort in heavy weather and poor visibilty conditions during the day and night. In some cases, the compass position may be partially determined by the location of the wheel, shift lever, and throttle handle.

Third, inspect the compass site to be sure the instrument will be at least two feet from any engine indicators, bilge vapor detectors, magnetic instruments, or any steel or iron objects. If the compass cannot be placed at least two feet (six feet would be better) from one of these influences, then either the compass or the other object must be moved, if first order accuracy is to be expected.

Once the compass location appears to be satisfactory, give the compass a test before installation. Hidden influences may be concealed under the cabin top, forward of the cabin aft bulkhead, within the cockpit ceiling, or in a wood-covered stanchion.

Move the compass around in the area of the proposed location. Keep an eye on the card. A magnetic influence is the only thing that will make the card turn. You can quickly find any such influence with the compass. If the influence can not be moved away or replaced by one of non-magnetic material, test to determine whether it is merely magnetic, a small piece of iron or steel, or some magnetized steel. Bring the north pole of the compass near the object, then shift and bring the south pole near it. Both the north and south poles will be attracted if the compass is demagnetized. If the object attracts one pole and repels the other, then the compass is magnetized. If your compass needs to be demagnetized, take it to a shop equipped to do the job **PROPERLY.**

After you have moved the compass around in the proposed mounting area, hold it down or tape it in position. Test everything you feel might affect the compass and cause a deviation from a true reading. Rotate the wheel from hard over to hard over. Switch on and off all the lights, radios, radio direction finder, radio telephone, depth finder and the shipboard intercom, if one is installed. Sound the electric whistle, turn on the windshield wipers, start the engine (with water circulating through the engine), work the throttle, and move the gear shift lever.

If the boat has an auxiliary generator, start it.

If the card moves during any one of these tests, the compass should be relocated. Naturally, if something like the windshield wipers cause a slight deviation, it may be necessary for you to make a different deviation table to use only when certain pieces of equipment is operating. Bear in mind, following a course that is only off a degree or two for several hours can make considerable difference at the end, putting you on a reef, rock, or shoal.

Check to be sure the intended compass site is solid. Vibration will increase pivot wear.

Now, you are ready to mount the compass. To prevent an error on all courses, the line through the lubber line and the compass card pivot must be exactly parallel to the keel of the boat. You can establish the fore-and-aft line of the boat with a stout cord or string. Use care to transfer this line to the compass site. If necessary, shim the base of the compass until the stile-type lubber line (the one affixed to the case and not gimbaled) is vertical when the boat is on an even keel. Drill the holes and mount the compass.

Magnetic Items After Installation

Many times an owner will install an expensive stereo system in the cabin of his boat. It is not uncommon for the speakers to be mounted on the aft bulkhead up against the overhead (ceiling). In almost every case, this position places one of the speakers in very close proximity to the compass, mounted above the ceiling.

As we all know, a magnet is used in the operation of the speaker. Therefore, it is very likely that the speaker, mounted almost under the compass in the cabin will have a very pronounced affect on the compass accuracy.

Consider the following test and the accompanying photographs as prove of the statements made.

First, the compass was read as 190 degrees while the boat was secure in her slip.

Next a full can of diet coke in an **aluminum** can was placed on one side and the compass read as 204 degrees, a good 14 degrees off.

Next, the full can was moved to the opposite side of the compass and again a reading was observed. This time as 189 degrees, 11 degrees off from the original reading.

Finally the contents of the can were consumed, the can placed on both sides of the compass with **NO** affect on the compass reading.

Two very important conclusions can be drawn from these tests.

1- Something must have been in the contents of the can to affect the compass so drastically.

"Innocent" objects close to the compass, such as diet coke in an aluminum can, may cause serious problems and lead to disaster, as these three photos and the accompanying text prove.

2- Keep even "innocent" things clear of the compass to avoid any possible error in the boat's heading.

REMEMBER, a boat moving through the water at 10 knots on a compass error of just 5 degrees will be almost 1.5 miles off course in only **ONE** hour. At night, or in thick weather, this could very possibly put the boat on a reef, rock, or shoal, with disastrous results.

1-11 POWER STEERING SYSTEM

USCG or BIA certification of a steering system means that all materials, equipment, and installation of the steering parts meet or exceed specific standards for strength, type, and maneuverability. Service procedures for the power steering system which may be installed as optional equipment are given in Chapter 10.

1-12 ANCHORS

One of the most important pieces of equipment in the boat next to the power plant is the ground tackle carried. The engine makes the boat go and the anchor and its line are what hold it in place when the boat is not secured to a dock or on the beach.

The anchor must be of suitable size, type, and weight to give the skipper peace of mind when his boat is at anchor. Under certain conditions, a second, smaller, lighter anchor may help to keep the boat in a favorable position during a non-emergency daytime situation.

*The weight of the anchor **MUST** be adequate to secure the boat without dragging.*

In order for the anchor to hold properly, a piece of chain must be attached to the anchor and then the nylon anchor line attached to the chain. The amount of chain should equal or exceed the length of the boat. Such a piece of chain will ensure that the anchor stock will lay in an approximate horizontal position and permit the flutes to dig into the bottom and hold.

1-13 MISCELLANEOUS EQUIPMENT

In addition to the equipment you are legally required to carry in the boat and those previously mentioned, some extra items will add to your boating pleasure and safety. Practical suggestions would include: a bailing device (bucket, pump, etc.), boat hook, fenders, spare propeller, spare engine parts, tools, an auxiliary means of propulsion (paddle or oars), spare can of gasoline, flashlight, and extra warm clothing. The area of your boating activity, weather conditions, length of stay aboard your boat, and the specific purpose will all contribute to the kind and amount of stores you put aboard. When it comes to personal gear, heed the advice of veteran boaters who say, "Decide on how little you think you can get by with, then cut it in half".

Bilge Pumps

Automatic bilge pumps should be equipped with an overriding manual switch. They should also have an indicator in the operator's position to advise the helmsman when the pump is operating. Select a pump that will stabilize its temperature within the manufacturer's specified limits when it is operated continuously. The pump motor should be a sealed or arcless type, suitable for a marine atmosphere. Place the bilge pump inlets so excess bilge water can be removed at all normal boat trims. The intakes should be properly screened to prevent the pump from sucking up debris from the bilge. Intake tubing should be of a high quality and stiff enough to resist kinking and not collapse under maximum pump suction condition if the intake becomes blocked.

To test operation of the bilge pump, operate the pump switch. If the motor does not run, disconnect the leads to the motor. Connect a voltmeter to the leads and see if voltage is indicated. If voltage is not indicated, then the problem must be in a blown fuse, defective switch, or some other area of the electrical system.

If the meter indicates voltage is present at the leads, then remove, disassemble, and inspect the bilge pump. Clean it, reassemble, connect the leads, and operate the switch again. If the motor still fails to run, the pump must be replaced.

To test the bilge pump switch, first disconnect the leads from the pump and connect them to a test light or ohmmeter. Next, hold the switch firmly against the mounting location in order to make a good ground. Now, tilt the opposite end of the switch upward until it is activated as indicated by the test light coming on or the ohmmeter showing continuity. Finally, lower the switch slowly toward the mounting position until it is deactivated. Measure the distance between the point the switch was activated and the point it was deactivated. For proper service, the switch should deactivate between 1/2-inch and 1/4-inch from the planned mounting position. **CAUTION: The switch must never be mounted lower than the bilge pump pickup.**

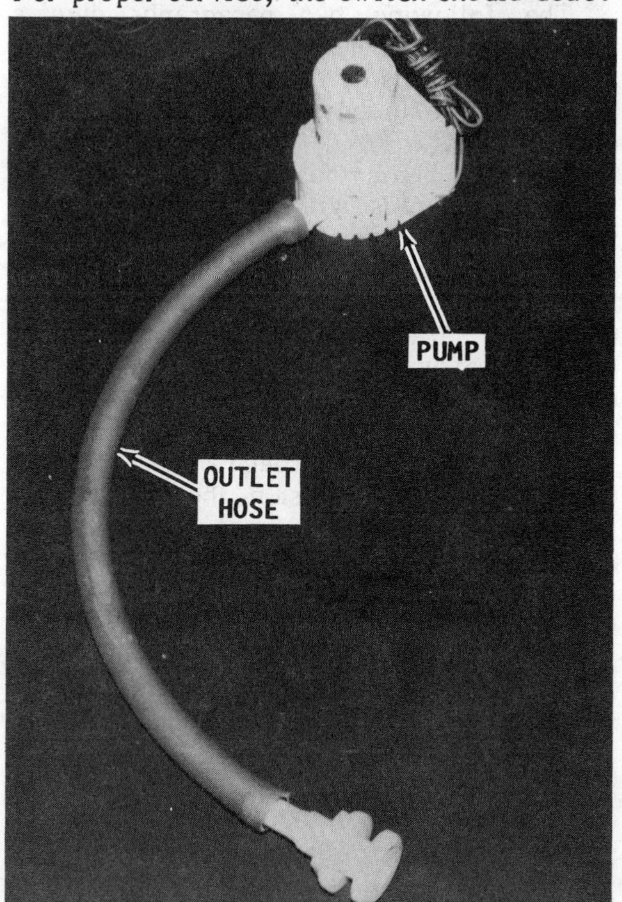

The bilge pump line must be cleaned frequently to ensure the entire bilge pump system will function properly when a demand is made during an emergency.

1-14 BOATING ACCIDENT REPORTS

New federal and state regulations require an accident report to be filed with the nearest State boating authority within 48 hours if a person is lost, disappears, or is injured to the degree of needing medical treatment beyond first aid.

Accidents involving only property or equipment damage **MUST** be reported within 10 days if the damage is in excess of $200. Some States require reporting of accidents with property damage less than $200 or total boat loss.

A **$500 PENALTY** may be asessed for failure to submit the report.

WORD OF ADVICE

Take time to make a copy of the report to keep for your records or for the insurance company. Once the report is filed, the Coast Guard will not give out a copy, even to the person who filed the report.

The report must give details of the accident and include:

1- The date, time, and exact location of the occurrence.

2- The name of each person who died, was lost, or injured.

3- The number and name of the vessel.

4- The names and addresses of the owner and operator.

If the operator cannot file the report for any reason, each person on aboard **MUST** notify the authorities, or determine that the report has been filed.

1-15 NAVIGATION

Buoys

In the United States, a buoyage system is used as an assist to all boaters of all size craft to navigate our coastal waters and our navigable rivers in safety. When properly read and understood, these buoys and markers will permit the boater to cruise with comparative confidence that he will be able to avoid reefs, rocks, shoals, and other hazards.

In the spring of 1983, the Coast Guard began making modifications to U.S. aids to navigation in support of an agreement spon-

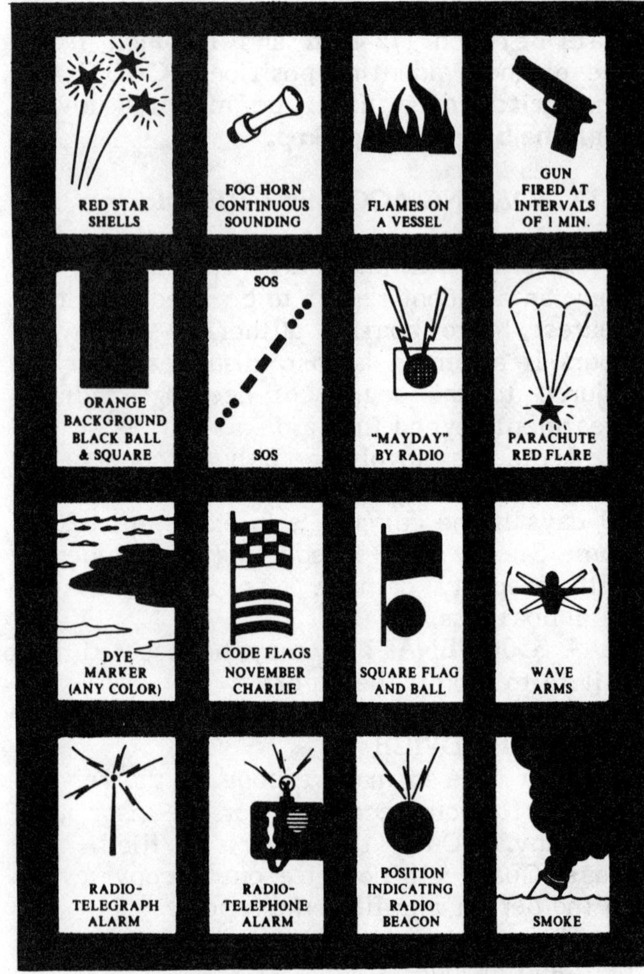

Internationally accepted distress signals.

sored by the International Associaiton of Lighthouse Authorities (IALA) and signed by representatives from most of the maritime nations of the world. The primary purpose of the modifications is to improve safety by making buoyage systems around the world more alike and less confusing.

The modifications should be completed by the end of 1989.

Lights

The following information regarding lights required on boats between sunset and sunrise or during restricted visibility is taken directly from a U.S. Coast Guard publication dated 1984.

The terms **"PORT"** and **"STARBOARD"** are used to refer to the left and right side of the boat, when looking forward. One easy way to remember this basic fundamental is to consider the words "port" and "left" both have four letters and go together.

Waterway Rules

On the water, certain basic safe-operating practices must be followed. You should learn and practice them, for to **know**, is to be able to handle your boat with confidence and safety. Knowledge of what to do, and not do, will add a great deal to the enjoyment you will receive from your boating investment.

Rules of the Road

The best advice possible and a Coast Guard requirement for boats over 39' 4" (12 meters) since 1981, is to obtain an official copy of the "Rules of the Road", which includes Inland Waterways, Western Rivers, and the Great Lakes for study and ready reference.

The following two paragraphs give a **VERY** brief condensed and abbreviated -- almost a synopsis of the rules and should not be considered in any way as covering the entire subject.

Powered boats must yield the right-of-way to all boats without motors, except when being overtaken. When meeting another boat head-on, keep to starboard, unless you are too far to port to make this practical. When overtaking another boat, the right-of-way belongs to the boat being overtaken. If your boat is being passed, you must maintain course and speed.

When two boats approach at an angle and there is danger of collision, the boat to port must give way to the boat to starboard. Always keep to starboard in a narrow channel or canal. Boats underway must stay clear of vessels fishing with nets, lines, or trawls. (Fishing boats are not allowed to fish in channels or to obstruct navigation.)

2
TUNING

2-1 INTRODUCTION

The efficiency, reliability, fuel economy and enjoyment you receive from the engine's performance are all directly dependent on having it tuned properly. The importance of performing service work in the sequence detailed in this chapter cannot be overemphasized. Before making any adjustments, check the specifications in the Appendix. Do not rely on your memory.

Before beginning to tune any engine, check to be sure the engine cylinders have satisfactory compression. An engine with worn piston rings, burnt valves, or a blown gasket cannot be made to perform properly no matter how much time and expense is spent on the tune-up. Poor compression must be corrected or the tune-up will not give the desired results.

2-2 TUNING FOR PERFORMANCE

First, check the battery to be sure it will deliver enough energy. The battery must not only crank the engine rapidly enough to draw in the proper amount of air-fuel mixture and compress it, but the battery must also have enough energy left to energize the ignition system for a hot spark to ignite the mixture.

Mechanical checks and service includes adjusting the fan belt, checking the compression, tightening the cylinder head and manifold bolts, and adjusting the valves.

Ignition system service includes replacing and adjusting the contact points, checking the ignition advance, and adjusting the timing.

Fuel system service and carburetor adjustments are made after all ignition checks and adjustments have been completed. It is very important to complete the ignition work before moving to the fuel system because the ignition adjustments have so much affect on the carburetion.

2-3 MECHANICAL TASKS

Drive Belt Adjustment

Check the drive belt at least once a year or during each tune-up. Replace the belt if there is any evidence of a crack or tear on the under surface.

Adjust the drive belt until the belt will depress approximately 1/4" (6.35 mm) under finger pressure.

If it is necessary to install a new belt, be sure to use only the size and type recommended by the engine manufacturer.

Adjust the belt tension by first loosening the brace and pivot bolts, and then pivoting the alternator away from the engine until the belt deflection is 1/4 inch (6.35mm) when you exert a downward pressure against the belt. Make the belt deflection measurement midway between the circulating pump pulley and the alternator pulley. Tighten the alternator bracket and pivot bolts securely.

If the battery has a record of not holding a full charge, check the tension of the drive belt. The failure to hold a full charge may also be caused by a faulty alternator or a defective voltage regulator. The cause of a battery using too much water is usually attributed to the voltage regulator being set too high. In this case, replace the regulator. Refer to Chapter 6 for service details of the electrical system.

Battery Service

Check the battery, cables, and surrounding area for signs of corrosion. Look for loose or broken carriers; cracked or bulged cases; dirt and acid; electrolyte leakage; and low electrolyte level. Add distilled water to the cells to bring them up to the proper level. Keep the top of the battery clean and be sure the battery is securely fastened in position.

When cleaning the battery **BE SURE** the vent plugs are tight to prevent any of the cleaning solution from entering the cells. First, wash the battery with a diluted ammonia or soda solution to neutralize any acid, and then flush the solution off with clean water.

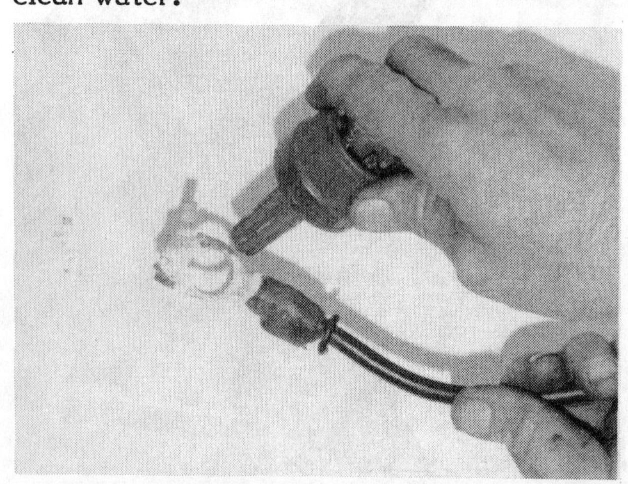

An inexpensive brush is the best tool for cleaning the inside diameter of battery cable connectors.

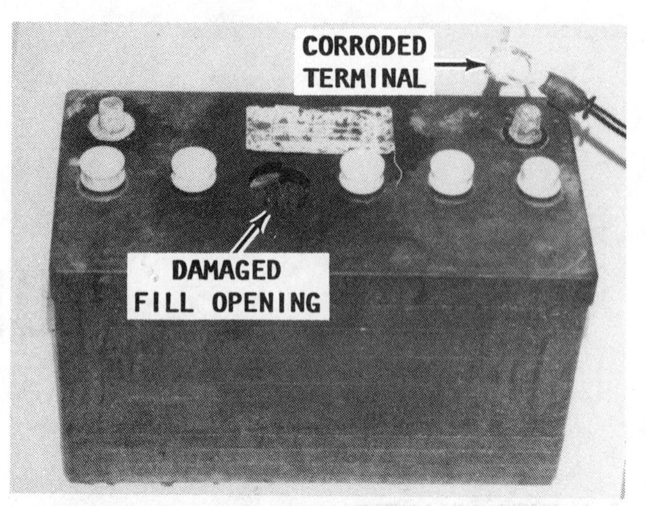

Many electrical problems can be traced to corroded battery terminals. The most serious damage could be a burned-out alternator.

Keep the battery hold-down device tight enough to prevent any movement but not so tight that it puts the battery case under a strain.

Tighten the battery cables on the battery posts to ensure a good contact. Clean the posts or cable terminals with a wire brush if they have become corroded. After the posts and terminals are clean, apply a thin coating of Multipurpose Lubricant to both as a preventative measure against corrosion forming.

Using a special tool to clean the battery terminals. After cleaning, and the cables have been connected, coat the terminals and connectors with lubricant to prevent further corrosion.

Test the battery electrolyte at regular intervals.

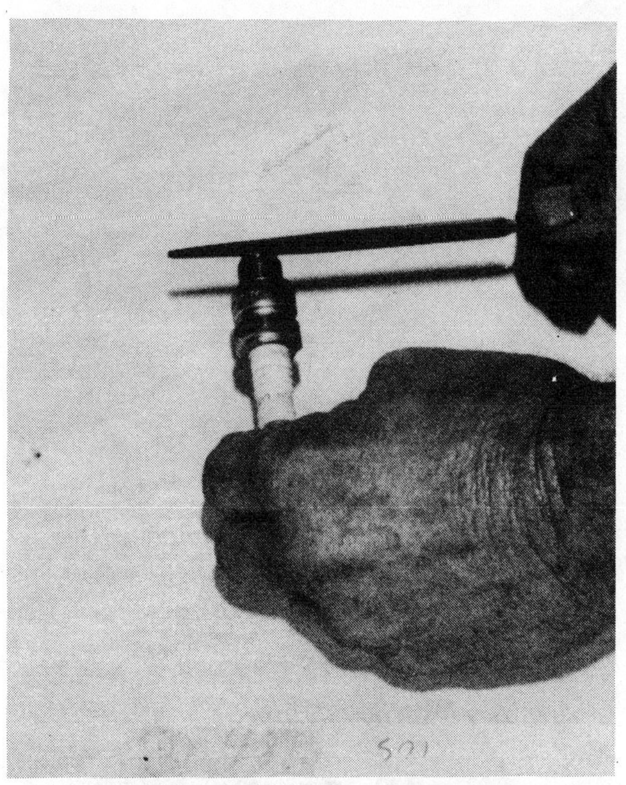

Engine misfire and poor fuel economy can be caused by corroded spark plug electrodes. To regain performance, file and reset each spark plug, if they are the least bit corroded.

Spark Plug Cleaning

Grasp the molded cap, then twist slightly and pull it loose from the plug. Do not pull on the wire, or the connection inside the cap may become separated or the boot may be damaged. Remove the spark plugs and be careful not to tilt the socket, to prevent cracking the insulator. Compare the spark plugs with the illustration on this page to determine how the engine has been running. Clean and gap the spark plugs. Use the specifications in the Appendix for the proper gap dimension.

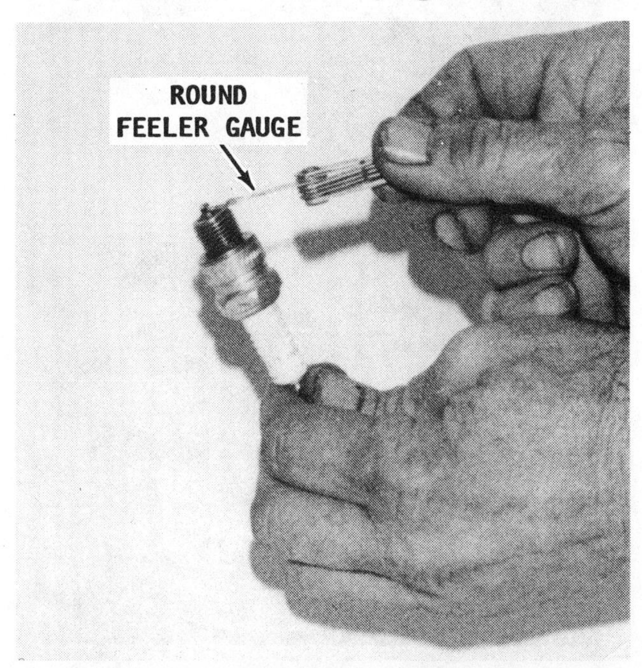

***ALWAYS** use a round gauge to measure spark plug gap. A flat gauge will not give an accurate reading. **NEVER** use a gasket on a spark plug with a tapered seat.*

Spark plug fouled by oil. The engine may need an overhaul.

Excessive overheating, heavy load. Use a spark plug with a lower heat rating.

Powdery deposits have melted and shorted-out this spark plug.

Red, brown, or yellow deposits, are by-products of combustion from fuel and lubricating oil.

Cylinder and Manifold Bolt Tightening

Tighten the cylinder head and manifold bolts in the sequence detailed by the diagrams in Chapter 3, Engine Service, and according to the torque specification given in the Appendix.

Compression Testing

Insert a compression gauge in each spark plug opening one-at-a-time; crank the engine and check the compression. A significant variation between cylinders is far more important than the actual reading of each one individually. A difference of over 20 psi, indicates a ring or valve problem. To determine which needs attention, insert a teaspoonful of oil into the spark plug opening of the low reading cylinder, and then crank the engine a few times to distribute the oil. Now, check the compression again to see if inserting the oil caused a change. If the reading went up, then the compression loss is due to worn rings. If the reading remained the same, the loss is due to a burned valve.

A compression gauge inserted into the spark plug opening for a compression check. A variation between cylinders is more important than the actual individual readings.

Valve Adjusting

Proper valve adjustment allows the hydraulic lifters to operate in the center of their designed travel. Valve adjusting is not a simple one-two operation. Therefore, this procedure is covered in detail in Chapter 3 under the appropriate heading:

Volvo Penta In-Line Engines
Volvo Penta OHC Engines
GMC V6 and V8 Engines
Ford V8 Engines

2-4 IGNITION SERVICE

It is not possible to do a good job of replacing the contact points with the distributor in the engine. To remove the distributor, turn the crankshaft until the rotor points to the No. 1 cylinder position. Most distributor caps have a "1" stamped on the cap. Remove the hold-down bolt. Remove the distributor slowly and you will notice the rotor turn. When the rotor stops turning, the distributor gear is free of the camshaft gear. At this point scribe a mark on the distributor housing in line with the edge of the rotor as an aid to installation.

At the time of installation, align the rotor on the mark you scribed on the housing, then lower the distributor slowly and you will see the rotor turn back until it is pointing to the No. 1 cylinder position.

Use a couple squirts of engine oil into the cylinder to determine if the compression loss is a burned valve or worn piston rings, as explained in the text.

Replace a pitted rotor.

A crack on the inside surface of the distributor cap can cause hard starting and misfire.

Always replace the contact points rather than filing them. It is seldom necessary to replace the condenser.

Adjust the point gap according to the specifications listed in the Appendix. Keep

A crack on the outside of the distributor cap may also be the cause of hard starting and misfire, especially at high speed.

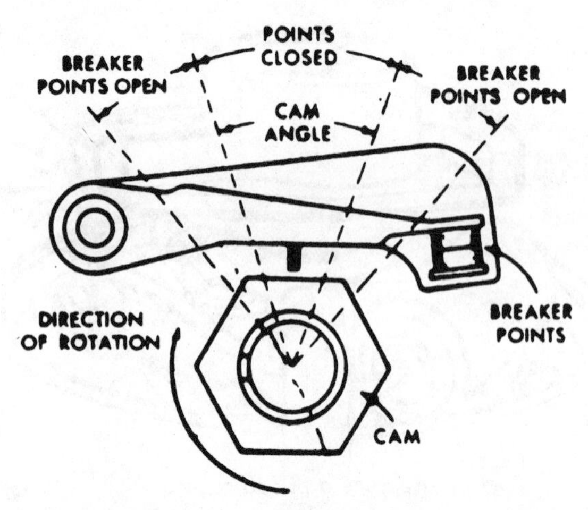

The dwell angle is directly related to the point gap.

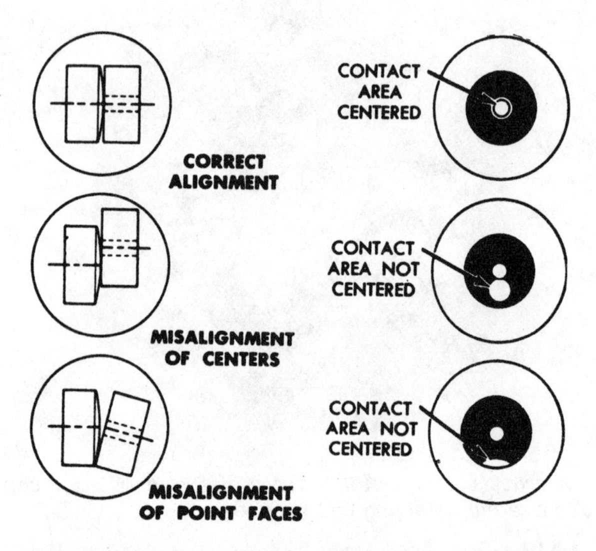

Breaker point alignment guide. Always bend the stationary point for aligning the contact point faces; otherwise, misalignment may result between the rubbing block and the cam.

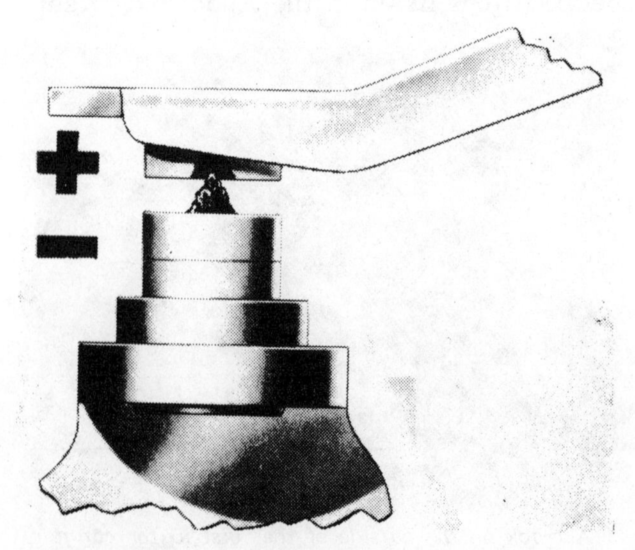

Check the contact points for abnormal wear, burning, or pitting.

the feeler gauge blade clean, because the slightest amount of oil film will cause trouble when it oxidizes.

It is best to add 0.003" to the clearance specification when installing a new set of contact points to compensate for initial rubbing block wear. **ALWAYS** keep the contact point retaining screw **SNUG** during the adjustment to prevent the gap from changing when it is finally tightened.

After the proper gap adjustment has been made, apply a light layer of heavy grease to coat the distributor cam. Turn the distributor shaft in the normal direction of rotation so the lubricant is wiped off against the back of the rubbing block. The grease will remain on the rubbing block as a reservoir to supply lubricant as the block wears. Wipe off any excess lubricant. Leave only the grease stored on the rubbing block.

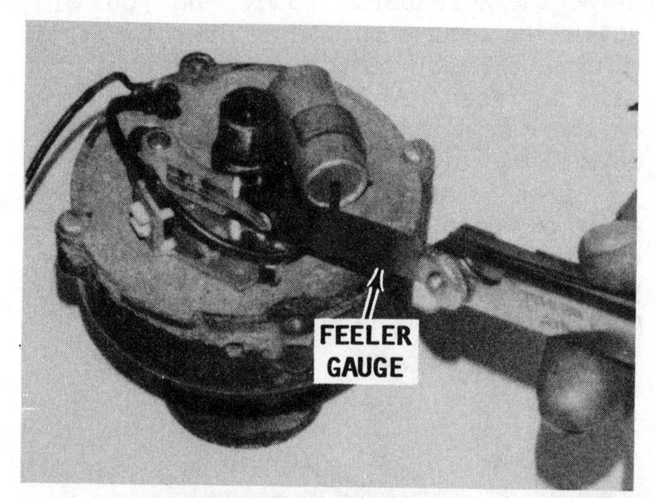

Adjusting the point gap.

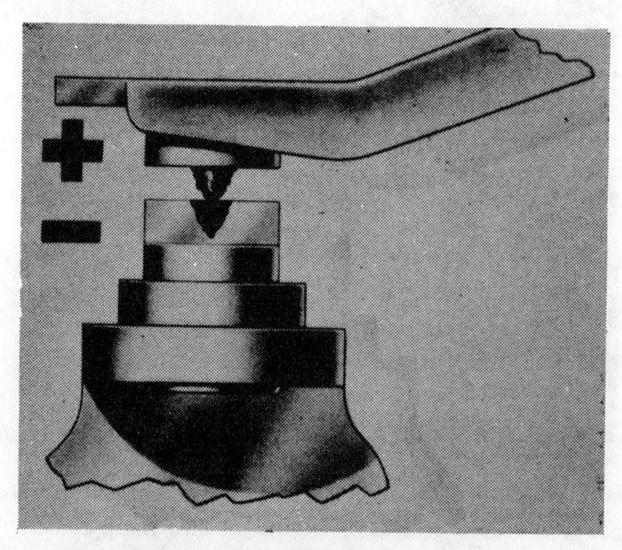

A faulty condenser will cause abnormal contact point wear and loss of engine performance.

Replace the distributor in the engine, with the rotor pointing toward the mark you scribed on the housing. Tighten the hold-down bolt. The ignition timing will be adjusted after the engine is running, as described in the following paragraphs.

Ignition Timing

Connect a tachometer to the coil leads. Connect a power timing light to the No. 1 spark plug. Do not puncture the wire or boot when you install the timing light because a puncture could start a voltage leak and lead to problems later.

Start the engine and adjust the speed and timing to the specifications listed in the Appendix.

CAUTION: Water must circulate through the lower unit to the engine any time the engine is run to prevent damage to the water pump mounted on the engine.

If a point adjustment is required, stop the engine, remove the cap, and make the adjustment. Remember, the dwell setting (gap) of the contact points affects the ignition timing; therefore, it is essential the dwell be set before adjusting the ignition timing.

Idle the engine at 600 rpm or less to be sure no centrifugal advance is taking place. Adjust the ignition timing by rotating the distributor housing. Tighten the hold-down bolt.

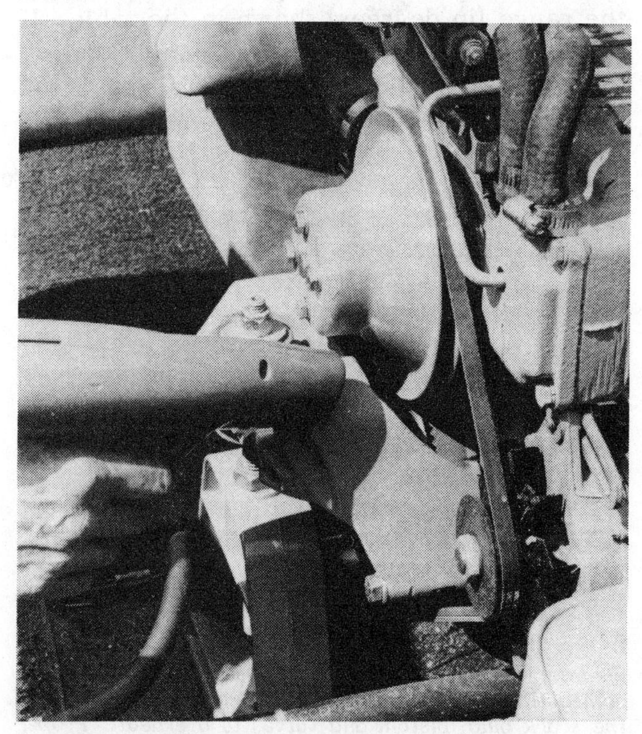

Using a timing light connected to the No. 1 spark plug lead to adjust the timing.

Timing light aimed at the timing marks on the damper and on the timing chain cover — AQ200, AQ225, AQ260, AQ280, and AQ290 engines.

Timing light aimed at the timing pointer and the mark on the damper — AQ130 and AQ170 engines.

Timing pointer and timing notch on the crankshaft pulley -- AQ 120, AQ 125, AQ 140, and AQ 145 engines.

Timing dot on the accessory drive pulley, the mark on the belt, and the notch on the cover plate -- AQ 120, AQ 125, AQ 140, and AQ 145 engines.

Twist the rotor and check the automatic advance weights for good movement. The rotor must feel springy in one direction and solid in the opposite direction. Engine performance will suffer if the rotor turns sluggishly.

Checking Ignition Advance

Ignition timing is varied according to engine speed by means of a centrifugal advance unit. During tune-up, it is essential to check the operation of this unit. An accurate check can be made using a timing light with a timing gauge and an advance control knob. If you do not have this piece of test equipment, a rough check can be made to be sure the system is functioning properly.

With the ignition timing properly adjusted and the timing light connected to No. 1 spark plug wire, increase engine speed to the rpm given in the "Strobe Time." column of the Tune-up Adjustments in the Appendix. The ignition timing must advance to the range (in degrees) shown in the table. If the speed fails to advance to the required range, remove the distributor and check the advance mechanism under the breaker plate.

Damage to this piston was caused by abnormal combustion because the fuel exploded violently, causing the spark plug, piston, and valves to overheat. Proper adjustment of the spark advance or a change in the fuel to a higher octane rating may correct the problem.

2-5 FUEL SYSTEM SERVICE

General Practices

Change the fuel filter in the base of the carburetor at least once a year. When the filter is changed, be sure to use a new gasket under the nut to prevent fuel from leaking out of the filter.

Change the water separator every year. Clean the flame arrestor after every 100 hours of operation.

Carburetor Adjustment

Because the carburetor is required to accurately control and mix the air and fuel quantities entering the combustion chamber, proper adjustments are critical to efficient engine operation. Dirt and gum in the passages restrict the flow of air or fuel causing a lean operating condition; hesitation on acceleration; and lack of power on demand.

The carburetor control linkage is subject to wear which will change the synchronization and fuel mixture. These changes will affect engine performance and fuel economy. Therefore, accurate fine carburetor adjustments can hardly be made or expected if the carburetor is not in satisfactory condition. If considerable difficulty is encountered in making the adjustments, the remedy may be to take time for a carburetor overhaul, see Chapter 4.

To make a preliminary adjustment, turn the idle mixture adjusting needles inward until they **BARELY** make contact with their seats, then back the needle out the specified number of turns.

Idle speed adjusting screw on GMC engines. AQ200, AQ225, AQ260, AQ280, and AQ290 engines.

NEVER turn the idle mixture screws **TIGHTLY** against their seats or they will be **DAMAGED.** Disconnect the throttle cable. Start the engine and run it at idle speed.

CAUTION: Water must circulate through the lower unit to the engine any time the engine is run to prevent damage to the water pump mounted on the engine.

Air/Fuel mixture adjusting screw location on the Solex downdraft carburetor — AQ120, AQ125, AQ140, and AQ145 engines.

Idle speed adjustment screw location on the Solex downdraft carburetor — AQ120, AQ125, AQ140, and AQ145 engines.

Adjust each idle mixture needle to obtain the highest and steadiest manifold vacuum reading. If a vacuum gauge is not available, obtain the smoothest running, maximum idle speed by turning one of the idle adjusting needles in until engine speed begins to fall off, then back the needle off over the "high spot" until the engine rpm again drops off. Set the idle adjusting needle halfway between the two points for an acceptable idle mixture setting. Repeat this procedure with the other needle.

If these adjustments result in an increase in idle rpm, reset the idle speed adjusting screw to obtain the specified idle rpm and again adjust the idle mixture adjusting needles.

Shift the unit into forward gear and readjust the idle speed screw to obtain the recommended idle speed as given in the Appendix.

Stop the engine and install the throttle cable. Check to be sure the throttle valves are in the full open position when the remote-control is in the full forward position. On the 120 hp to 165 hp units, with the throttle valves fully open, turn the wide-open throttle stop adjusting screw clockwise until the screw just touches the throttle lever. Tighten the set nut securely to prevent the adjustment screw from turning. Return the shift lever to the neutral gear and idle position. The idle-stop screw should be against its stop.

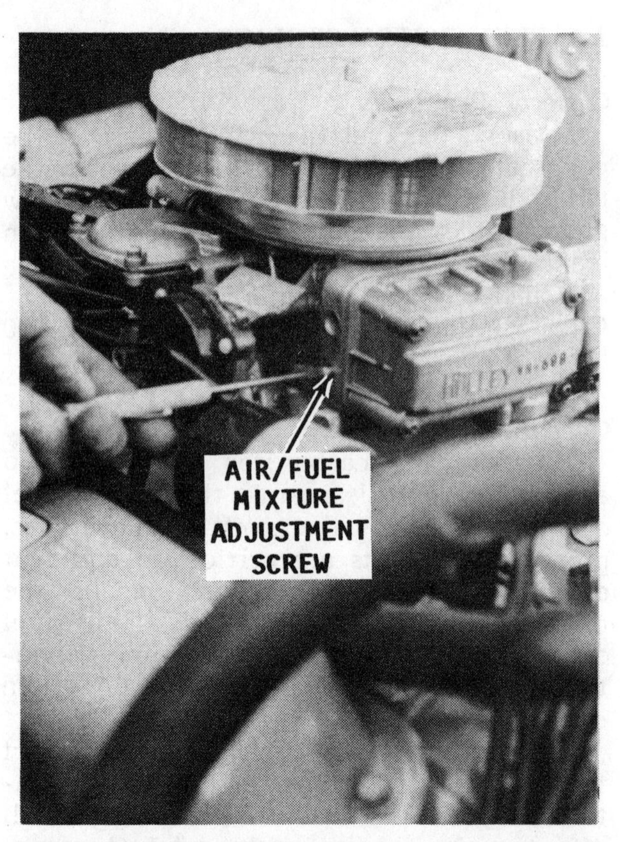

Air/Fuel mixture adjustment screw location on the early model AQ190 and AQ240 engines. A second adjustment screw is located on the opposite side of the carburetor.

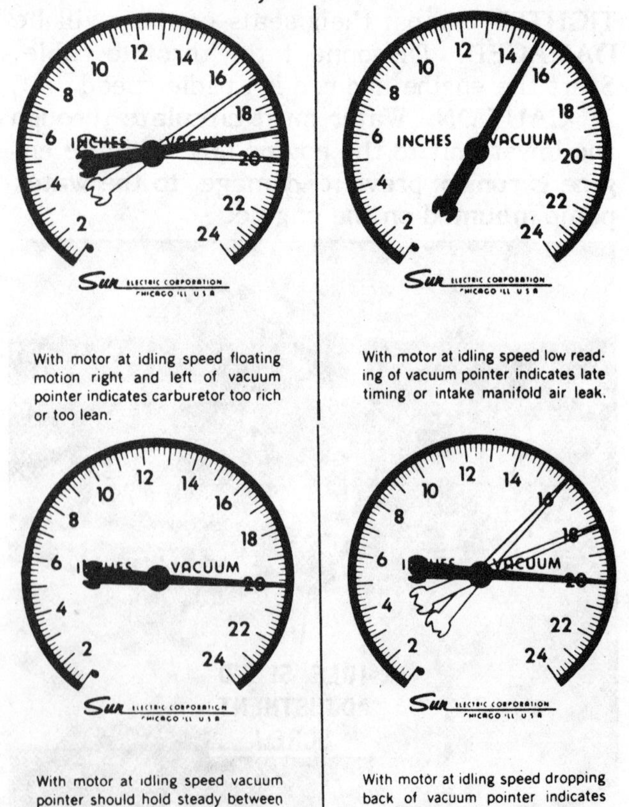

A vacuum gauge is a most useful tool when tuning an engine. The readings reveal a great deal concerning carburetor and valve performance.

Calibration for High Altitudes

Increased spark advance or carburetor recalibration have very little affect on high-altitude performance. Tests have proven this statement to be true. However, a marked increase in performance can be obtained by changing propellers for high-altitude operation.

Changing the prop should be the only modification considered to obtain the rated rpm. Any other recommendation should be considered a special case and should be referred to the factory branch or area distributor for specific jet sizes and spark timing settings.

Changing jet sizes and spark timing settings will: Cause engine failure if operated at lower elevations; result in increased fuel economy but will not have any significant affect on performance; and may cause added problems at much lower elevations.

3
ENGINE

3-1 GENERAL PRINCIPLES

Engine specification charts are located in the Appendix. These charts can be used to determine the engine type, size, and specifications.

All engines used to power inboard-outboard boats, except some of the two-cycle engines operate on the four-stroke cycle principle. During this cycle, the piston travels the length of its stroke four times. As the piston travels up and down, the crankshaft is rotated halfway (180 degrees). To complete one full cycle, the crankshaft rotates two complete turns; the camshaft, which controls the valves, is driven by the crankshaft at half crankshaft speed. Valve action, intake and exhaust, occurs once in each four-stroke cycle, and the piston acts as an air pump during the two remaining strokes.

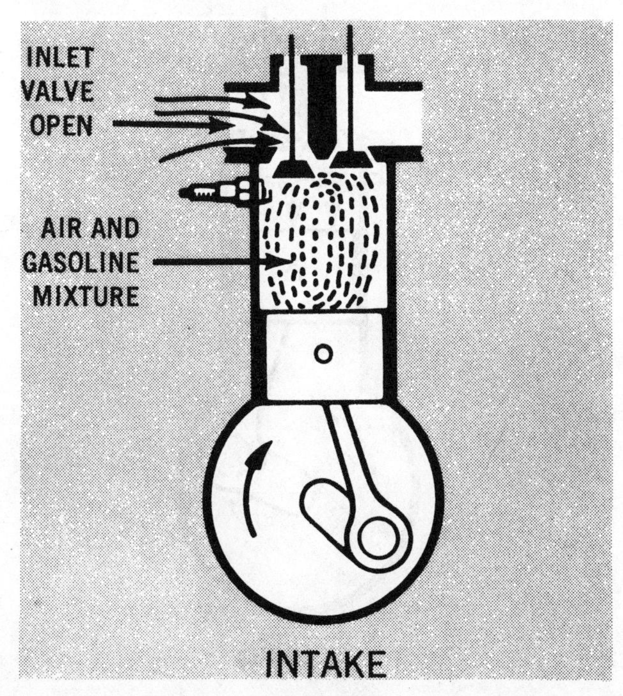

Intake Stroke

The intake valve is opened as the piston moves down the cylinder, and this creates an area of pressure lower than the surrounding atmosphere. Atmospheric pressure will cause air to flow into this low-pressure area. By directing the air flow through the carburetor, a measured amount of vaporized fuel is added. When the piston reaches the bottom of the intake stroke, the cylinder is filled with air and vaporized fuel. The exhaust valve is closed during the intake stroke.

Compression Stroke

When the piston starts to move upward, the compression stroke begins. The intake valve closes, trapping the air-fuel mixture in the cylinder. The upward movement of

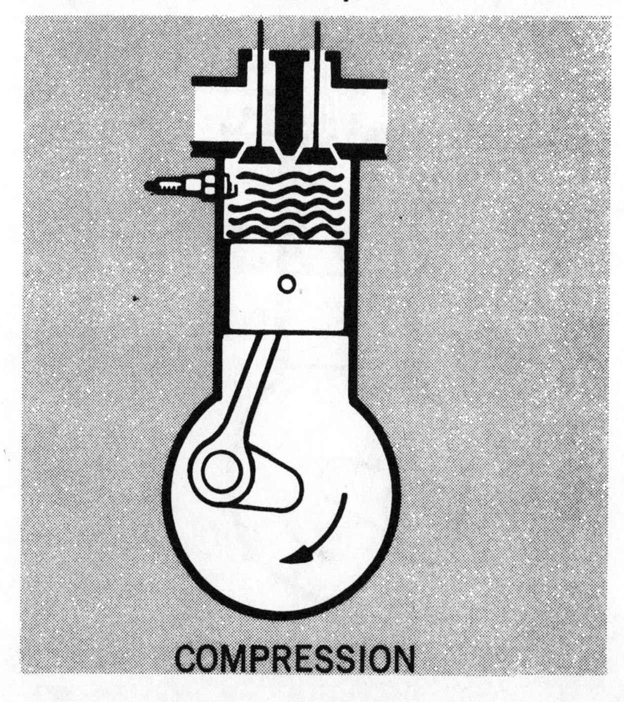

the piston compresses the mixture to a fraction of its original volume; exact pressure depends principally on the compression ratio of the engine.

Power Stroke

The power stroke is produced by igniting the compressed air-fuel mixture. When the spark plug arcs, the mixture ignites and burns very rapidly during the power stroke. The resulting high temperature expands the gases, creating very high pressure on top of the piston, which drives the piston down. This downward motion of the piston is transmitted through the connecting rod and is converted into rotary motion by the crankshaft. Both the intake and exhaust valves are closed during the power stroke.

Exhaust Stroke

The exhaust valve opens just before the piston completes the power stroke. Pressure in the cylinder at this time causes the exhaust gas to rush into the exhaust manifold (blowdown). The upward movement of the piston on its exhaust stroke expels most of the remaining exhaust gas.

As the piston pauses momentarily at the top of the exhaust stroke, the inertia of the exhausting gas tends to remove any remaining gas in the combustion chamber; however, a small amount always remains to dilute the incoming mixture. This unexpelled gas is captured in the clearance area between the piston and the cylinder head.

Combustion

The power delivered from the piston to the crankshaft is the result of a pressure increase in the gas mixture above the piston. This pressure increase occurs as the mixture is heated, first by compression, and then (on the down stroke) by burning. The burning fuel supplies heat that raises temperature and, at the same time, raises pressure. Actually, about 75 percent of the mixture in the cylinder is composed of nitrogen gas that does not burn but expands when heated by the burning of the combustible elements, and it is this expanding nitrogen that supplies most of the pressure on the piston.

The fuel and oxygen must burn smoothly within the combustion chamber to take full advantage of this heating effect. Maximum power would not be delivered to the piston if an explosion took place, because the entire force would be spent in one sharp hammer-like blow, occurring too fast for the piston to follow.

Instead, burning takes place evenly as the flame moves across the combustion chamber. Burning must be completed by the time the piston is about half-way down so maximum pressure will be developed in the cylinder at the time the piston applies its greatest force to the crankshaft. This will be when the mechanical advantage of the connecting rod and crankshaft is at a maximum.

At the beginning of the power stroke (as

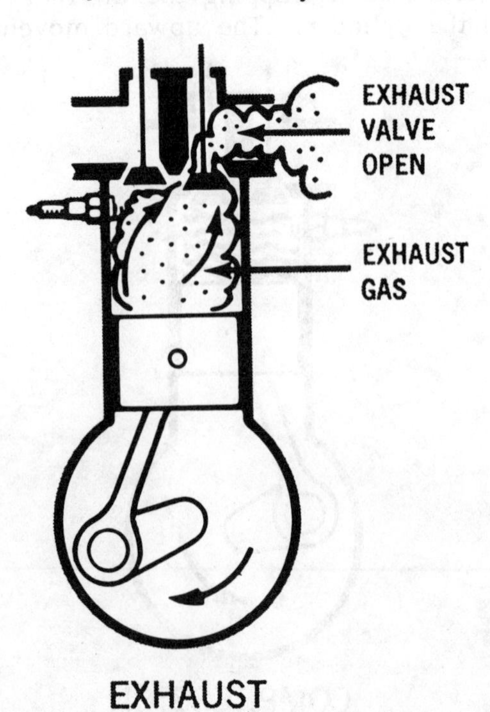

EXHAUST

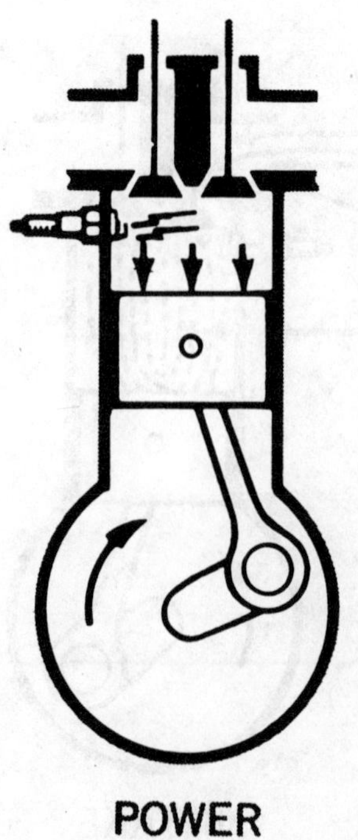

POWER

the piston is driven down by the pressure), the volume above the piston increases, which would normally allow the pressure in the cylinder to drop. However, combustion is still in progress, and this continues to raise the temperature of the gases, expanding them and maintaining a continuous pressure on the piston as it travels downward. This provides a smooth application of power throughout the effective part of the power stroke to make the most efficient use of the energy released by the burning fuel.

VALVE TIMING

On the power stroke, the exhaust valve opens before bottom dead center in order to get the exhaust gases started out of the combustion chamber under the remaining pressure (blowdown). At the end of the exhaust stroke, the intake valve opens before top dead center in order to start the air-fuel mixture moving into the combustion chamber. These processes are functions of camshaft design and valve timing.

Valves always open and close at the same time in the cycle; the timing is not variable with speed and load as is ignition timing. There is, however, one particular speed for each given engine at which the air-fuel mixture will pack itself into the combustion chambers most effectively. This is the speed at which the engine puts out its peak torque. At low engine speeds, compression is somewhat suppressed due to the slight reverse flow of gases through the valves just as they open or close when the mixture is not moving fast enough to take advantage of the time lag. At high speeds, the valve timing does not allow enough time during the valve opening and closing periods for effective packing of the air-fuel mixture into the cylinders.

ENGINE TYPES

The following engine manufacturers and the engine sizes or designations covered in this manual are as follows:

Volvo Penta In-Line

Pushrod engines -- 4-cylinder
AQ105, 115, & 130 Series (121 CID)
Pushrod engines -- 6-cylinder
AQ165 & 170 Series (182 CID)
Single overhead cam **(OHC)** engines
AQ120, 125, 131, 140, 145, & 151 Series (130, & 141 CID)
Double overhead cam **(DOHC)** engines
AQ171 Series (152 CID)

GMC V6

AQ175 & 205 Series and Model 431 (229, & 262 CID)

GMC V8

AQ200, 211, 225, 231, 240, 255, 260, 271, 280, 290, 311, & 740 Series (305, 307, 350, & 454 CID)

Ford V8

AQ190 & 240 Series (302 & 351 CID)

CHAPTER ORGANIZATION

Many tasks can be performed while the engine remains in the boat, because the engine shroud can be dismantled for access.

THEREFORE, this chapter is divided into general areas by section as follows:

3-2 Troubleshooting
3-3 Engine removal
3-4 Engine installation
3-5 thru **3-8** Intake & exhaust manifolds
3-9 Rocker arm service
3-10 thru **3-13** Cylinder head R & I
3-14 Cylinder head reconditioning
3-15 thru **3-18** Camshaft service
3-19 thru **3-23** Valve lash adjustment
3-24 thru **3-26** Timing belt or chain
3-27 Piston, ring, & rod service
3-28 & **3-29** Crankshaft & main bearings
3-30 Oil pump service

To prevent constant repetition of procedures, the above section numbers are referenced throughout the chapter.

3-2 TROUBLESHOOTING ALL UNITS

Troubleshooting must be a well thought-out procedure. To be successful, start by accurately determining the problem; then use a logical approach to arrive at the proper solution. The common phrase, "shotgun approach" only leads to wasted time, money, and frustration. Obviously, if the instructions are to be of maximum benefit, they must be followed exactly.

When an engine does not start, the trouble must be localized to one of four general areas; starting system, ignition, fuel, or compression. Each of these areas must be systematically inspected until the trouble is located in one of them, and then detailed tests of that system must be made to isolate the part causing the starting problem.

TROUBLESHOOTING CHECK

When using this Troubleshooting Check, proceed sequentially through each of the tests until a defect is uncovered. Then skip to the detailed testing procedure and check for that system. For example, if, when using the Troubleshooting Check procedure, the first two systems, starter and ignition test OK, but the third test shows there is trouble in the fuel system, proceed to the detailed test under the Fuel System Troubleshooting Check in Chapter 4.

CRANKING SYSTEM TEST

1- Turn the ignition switch to the **START** position, and the starter should crank the engine at a normal rate of speed.

If the starter cranks the engine slowly or doesn't crank it at all, the trouble is in the cranking system, and you should proceed to the Cranking System in Chapter 6 for the detailed testing procedure that will help you uncover the starting problem.

IGNITION SYSTEM TEST

2- Disconnect a spark plug wire and hold it about 1/4 inch from a spark plug or ground. Crank the engine with the ignition switch turned ON.

If there is no spark or if the spark is very weak, the trouble is in the ignition system, and you should proceed to the Ignition Troubleshooting in Chapter 5 for the detailed testing procedure that will help you to uncover the ignition system problem.

FUEL SYSTEM TEST

3- This test is to determine whether or not there is fuel in the carburetor. Remove the flame arrestor, and look down into the throat of the carburetor. Open and close the throttle several times to see if fuel is squirting out of the pump jets as shown in the accompanying illustration. *Note: The top of the carburetor has been removed in this illustration for photographic clarity.*

1

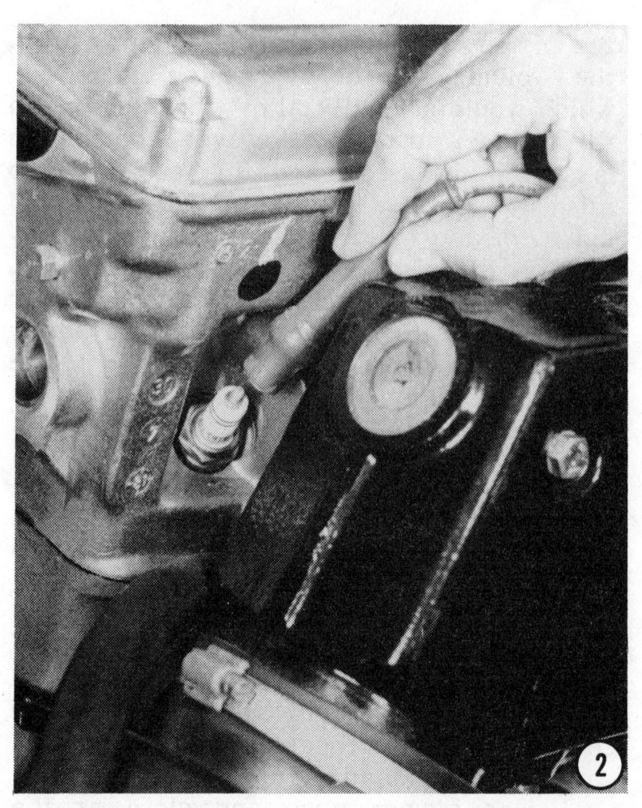

2

3

Consider whether the fuel in the tank is fresh or several months old. Many fuels tend to "sour" in three to four months. Others take longer but in no case should you be attempting to start the engine if the fuel has been in the tank over a year.

If no fuel is discharged from the pump jets, then the trouble is in the fuel system, and you should proceed to the Fuel Troubleshooting in Chapter 4 to isolate the problem.

COMPRESSION TEST

4- Good compression is the key to engine performance. An engine with worn piston rings, burned valves, or blown gaskets cannot be made to perform satisfactorily until the mechanical defects are corrected. Generally, a compression gauge is used to determine the cranking pressure within each cylinder. However, today's large displacement engines generally have considerable valve overlap, and the resulting compression reading may be much lower than the manufacturer's specifications of about 150-170 psi. It is entirely possible to obtain a reading as low as 120 psi on a modern engine which is in good mechanical condition. Such an engine is said to "exhale" at cranking speed, even though everything is perfectly normal at operating speeds.

To make a compression test, remove the spark plugs and lay them out in the order of removal. This is extremely important so you can "read" the firing end of each spark plug. After the spark plugs are removed, insert the rubber adapter of the compression gauge into one cylinder and have a helper crank the engine.

GROUND the **PRIMARY SIDE** of the coil to prevent **DAMAGE** to it. The throttle valve and choke **MUST** be in the **WIDE-OPEN** position in order to obtain maximum readings. Crank the engine through several revolutions to obtain the highest reading on the compression gauge or record an equal number of pulses for each cylinder.

The significance in a compression test is the variation in pressure readings between cylinders. As long as this variation is within 20-30 psi, the engine is normal. If a greater variation exists, then the low-reading cylinder should be checked by making a cylinder

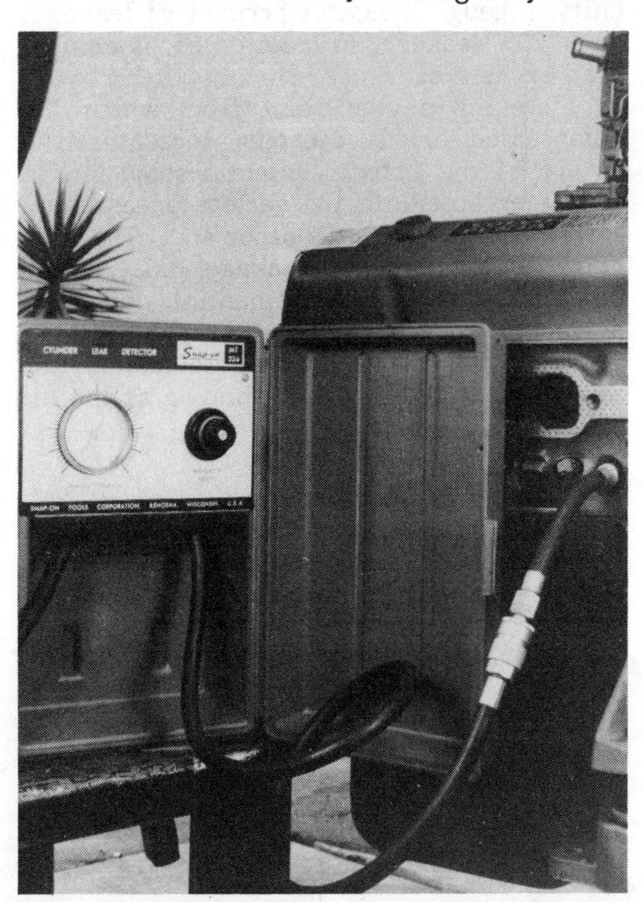

Use compressed air in a cylinder when the piston is at TDC, firing position and listen carefully. The source of leaking air will indicate the source of the problem. A hissing sound through the carburetor indicates a leaking intake valve.

leak test. A simple leak test can be made by first inserting a teaspoonful of oil into the spark plug opening of the low reading cylinder, and then cranking the engine a few times to distribute the oil. Check the compression again to see if inserting the oil caused a change. If the reading went up, then the compression loss is probably due to worn rings. If the reading remained the same, the loss may be due to a burned valve.

TROUBLESHOOTING MECHANICAL ENGINE PROBLEMS

LEAK TEST

A more definite and scientific method of performing a leak test and determining the problem area than the simple method described in the previous paragraph is to use compressed air.

Install an air hose adapter into the spark plug port. With the piston at top dead center, firing position, apply 60-70 psi of air. On commercially built air adapter units, a gauge indicates percent of leakage. Over 20% leakage, in most cases, is considered excessive.

Listening at the point from which the compressed air is escaping indicates the nature of the defect. Insert a short length of heater hose into the various areas being tested and listen at the other end. The hose helps to amplify the leakage noise. Air hissing from the exhaust manifold indicates a leaking exhaust valve. Air heard in the carburetor air horn indicates a leaking intake valve. If you hear air hissing at the oil filler pipe, the rings are worn.

VACUUM GAUGE

A vacuum gauge is a relatively inexpensive piece of test equipment that can be very handy in isolating mechanical troubles in an internal-combustion engine. As with a compression gauge, you cannot rely soley on the actual numerical reading of the vacuum gauge. Instead, relative readings and typical actions of the needle provide clues to some types of troubles.

Normal idle vacuum in the intake manifold ranges from 15 to 22" Hg. On later-model engines, lower and less steady intake manifold vacuum readings are becoming increasingly common because of the greater use of high-lift cams and the increase in the amount of valve overlap.

In addition to these factors, altitude affects a vacuum gauge reading. That's right. At high elevations, a vacuum gauge will read about one inch lower for each 1,000 ft. of elevation above sea level. Another outside influence on the vacuum gauge is a change in barometric pressure. Because of these factors, which are not determined by the condition of the engine, it is much more important for you to watch the action of the needle than its actual reading. After you have worked with a vacuum gauge just a few times you will be able to recognize such problems areas as sticking valves, a tight valve lash adjustment, or a restriction in the exhaust system.

DYNAMOMETER TESTING

An inexpensive method of making a dynamometer-type of engine test, is to use a vacuum gauge and a set of shorting wires. With these two items, it is possible to isolate mechanical troubles to one or more cylinders. When making this type of test, the spark plugs are shorted out one-at-a-time until the engine is running on one

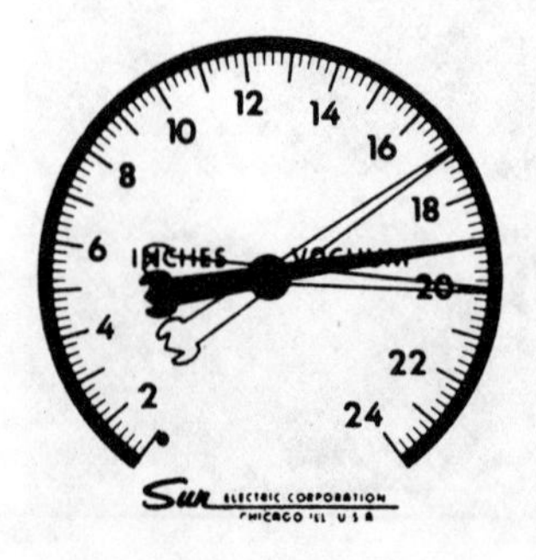

With motor at idling speed floating motion right and left of vacuum pointer indicates carburetor too rich or too lean.

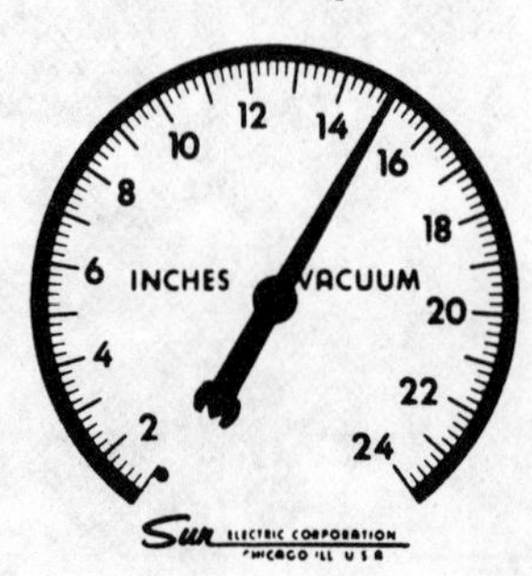

With motor at idling speed low reading of vacuum pointer indicates late timing or intake manifold air leak.

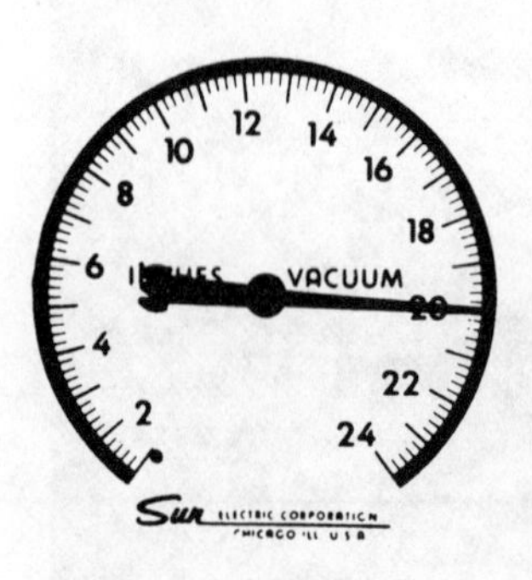

With motor at idling speed vacuum pointer should hold steady between 15 and 21.

With motor at idling speed dropping back of vacuum pointer indicates sticky valves.

cylinder. Now, if a vacuum gauge is attached to the intake manifold, a reading can be obtained to compare the efficiency of each cylinder. A low-reading cylinder can be the result of inefficiency in the ignition, fuel, or compression system.

To make this test the ground clip of the shorting wires must be attached to a **good ground.**

Start the engine.

CAUTION: Water must circulate through the lower unit to the engine any time the engine is run to prevent damage to the water pump mounted on the engine.

Have a helper advance the throttle as you successively short out all of the cylinders until the engine is running on one cylinder. You will discover it will be necessary to open the throttle wide in order to keep the engine running on one cylinder at a time.

Observe the vacuum gauge reading, and then move the shorting clip from one of the spark plugs to the spark plug of the cylinder you just tested. Observe the vacuum gauge reading, and then test each of the other cylinders in turn by running the engine on that one spark plug. A weak cylinder or a cylinder that is not firing is easily determined, but the most important part of this test is the ability to compare the relative power (vacuum) of each firing stroke.

EXCESSIVE OIL CONSUMPTION

High oil comsumption can usually be traced to one of four general areas:

1- A clogged positive crankcase ventilation system.

2- Piston rings not sealing.

3- Excessive valve stem-to-guide clearance.

4- Cracked intake manifold (the type that serves as a valve chamber cover).

High oil consumption complaints many times are the result of oil leaks rather than actual consumption by the engine. For this reason, before assuming an engine is burning oil, examine the exterior for evidence of oil leaks. In analyzing the problem, consideration must be given to the fact that oil has only three routes available to enter the combustion chambers. The oil may get past the piston rings, enter through the valve guides, or seaps in through the intake manifold. Two very definite clues will indicate the engine is actually burning excessive oil.

Spark plug fouled by oil. The engine may need an overhaul.

One is carbon deposits in the exhaust outlet and the other is oil-fouled spark plugs.

ENGINE NOISES

Engine noises can be generally classified as knocks, slaps, clicks, or squeaks. These noises are usually caused by loose bearings, sloppy pistons, worn gears, or other moving parts of the engine. Most common types of noises are either synchronized to engine speed or to one-half engine speed. Noises that are timed to engine speed are sounds that are related to the crankshaft, rods, pistons, and pins. The cause of noises that seem to be timed to one-half engine speed are usually in the valve-train. To determine whether the noise is timed to engine speed or one-half engine speed, operate the engine at a slow idle and observe whether the noise is synchronized with the flashes of a timing light.

A main bearing knock is usually identified by a dull thud that is noticeable when the engine is under a load. Attempting to move the boat under power with it tied to the dock will bring out a main bearing knock. If you pull the spark plug wire from one plug at-a-time and the noise disappears when a particular wire is removed, then the noise is probably coming from that cylinder, either the rod bearing, piston pin, or the piston. If a rod bearing is loose, the noise will be loudest when the engine is decelerating. Piston pin noise and piston slap are generally louder when a cold engine is first started.

Many times the source of an unusual sound may be isolated by using a stethoscope or other listening device. One such device is a long shank screwdriver. Allow your hand to extend over the end of the handle, press an ear to your hand, and then probe with the other end of the screwdriver by touching it firmly against the block at each cylinder or noise area. Take care and use good judgment when using this method of attempting to detect the cause of a problem because the noise will travel through other metallic parts of the engine and could lead to a false interpretation of what you are hearing.

Carbon build-up in the combustion chamber can cause interference with a piston. Fuel pumps can knock, belts can be noisy, distributors can emit clicking noises, and alternators can contribute to unusual sounds. Flywheels, water pumps, and loose manifolds can also cause noise problems.

3-3 ENGINE REMOVAL

GENERAL INFORMATION

Three methods are available for separating the engine from the Volvo Penta stern drive and removal from the boat. One method is to disconnect the stern drive and pull the power plant, leaving the bell housing connected to the engine.

The second method is to remove the engine from the boat leaving the bell housing connected to the stern drive. This method is not an easy task because the bell housing is very difficult to separate from the engine while still mounted in the boat. Three bolts securing the bell housing to the engine are located down under the housing. These three bolts are extremely difficult to remove and install due to the small space between the housing and the engine and the awkward position required to reach them from the bilge of the boat. These three bolts are pointed out in disassembling section when the bell housing is separated from the engine block.

A third method is to "pull" only the upper gear housing, and then to remove the engine. This is by far the most popular and the procedures in this section follow this method. However, if the lower unit requires service, the entire stern drive may be removed by first driving out the hinge pins, and then removing related hardware to "pull" the stern drive.

The following procedures cover the removal of a 4-cylinder push rod AQ130 engine from a boat equipped with a Model 280 stern drive. The removal of other power plant installations is almost identical, especially disconnecting the bell housing from the stern drive. The major differences for removal of an OHC, V6, or V8 engine is in disconnecting water, fuel, and accessories prior to lifting the engine out of the boat.

Procedures to cover every possible installation and combination of accessories would not be practical in this type of publication. Therefore, follow the procedures closely, but bear-in-mind, the installation you are servicing may have added items requiring attention.

REMOVAL

1- Remove the plug and drain the lubricant from the stern drive. As the lubricant drains, check its color. Any whitish, milkish color is an indication water has entered the stern drive. From time-to-time, rub some of the lubricant between your fingers. Any metal particles in the lubricant will thus be very evident.

2- Remove the shift cover. Disconnect the shift linkage at the intermediate housing by first pulling the cotter pin and washer, and then sliding the rod free.

SPECIAL WORDS

When the hose clamps are loosened, a socket wrench with a swivel joint from a 1/4" drive set should be used. If the socket set is not available use extreme **CAUTION**

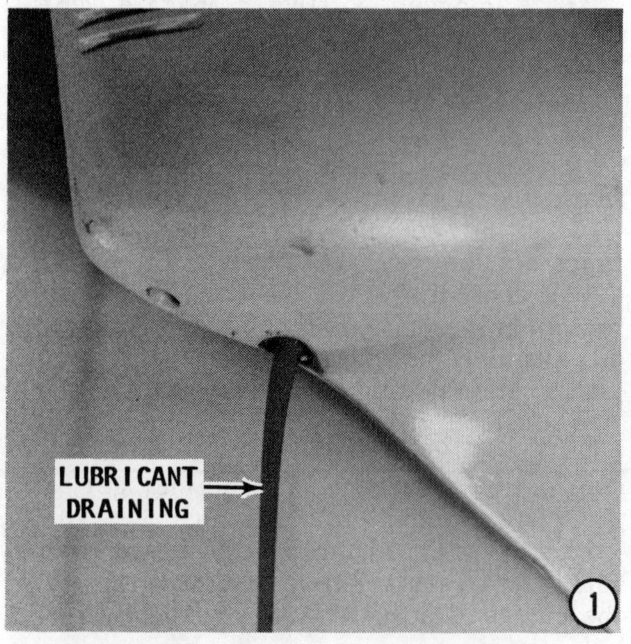

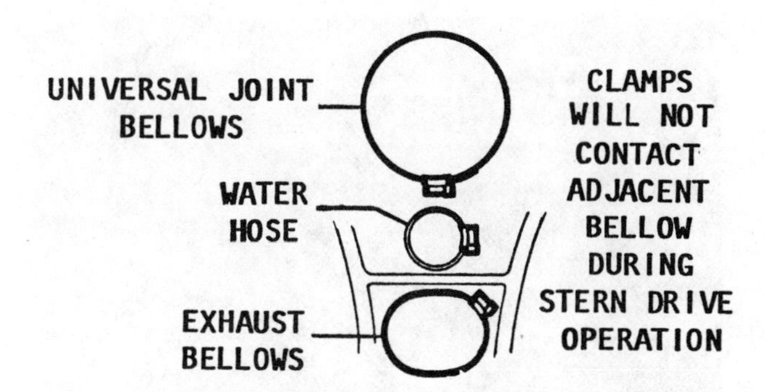

Proper arrangement of the bellows and hose clamps. In this position the tightening device on one clamp will not chafe against another bellow when the stern drive is in motion.

when loosening the clamp to prevent the screwdriver from slipping off the fastener and puncturing a hole in one of the rubber boots. Also, the clamps **MUST** be positioned as shown in the accompanying drawing to prevent the clamp ring tightening hardware from coming in contact with another rubber boot when the stern drive is in motion. Such action could eventually cause the boot to leak, causing extensive damage to expensive parts.

3- Remove the steering helmet attachment bolt. Slide the steering helmet forward.

4- Loosen the universal joint bellows clamp.

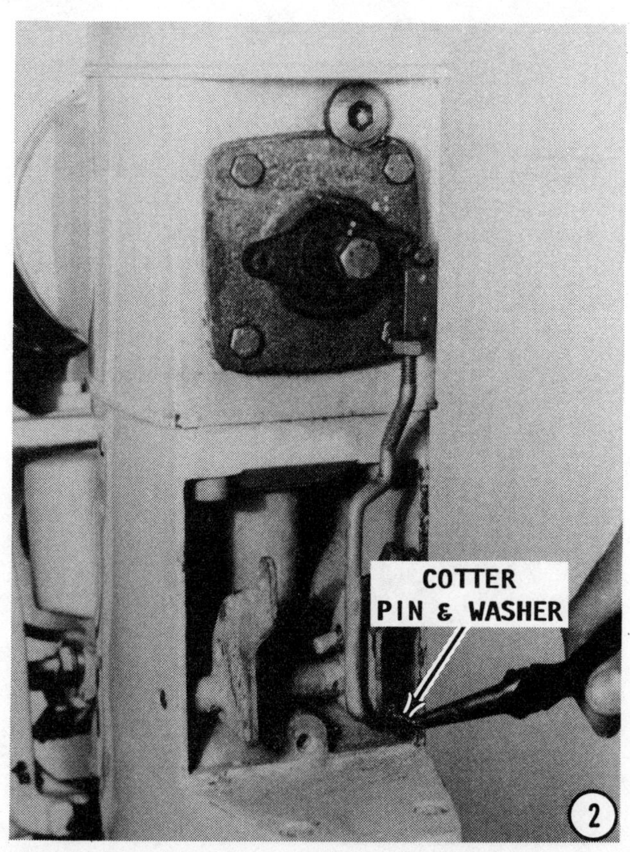

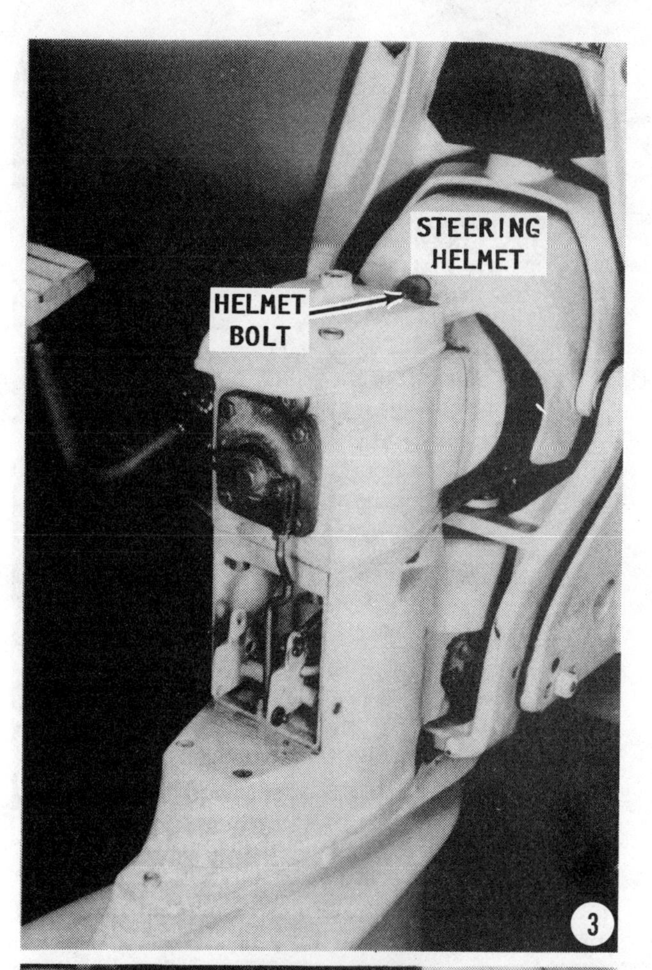

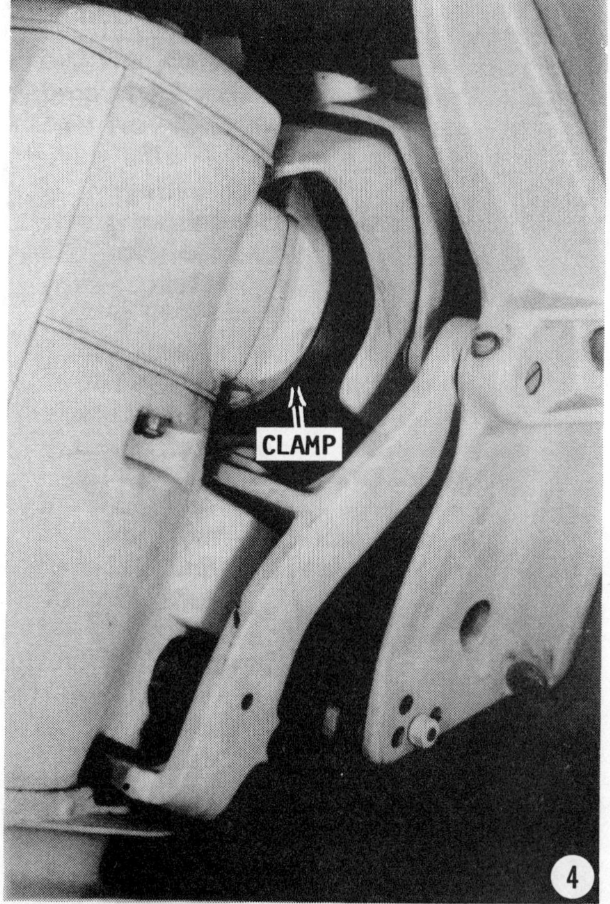

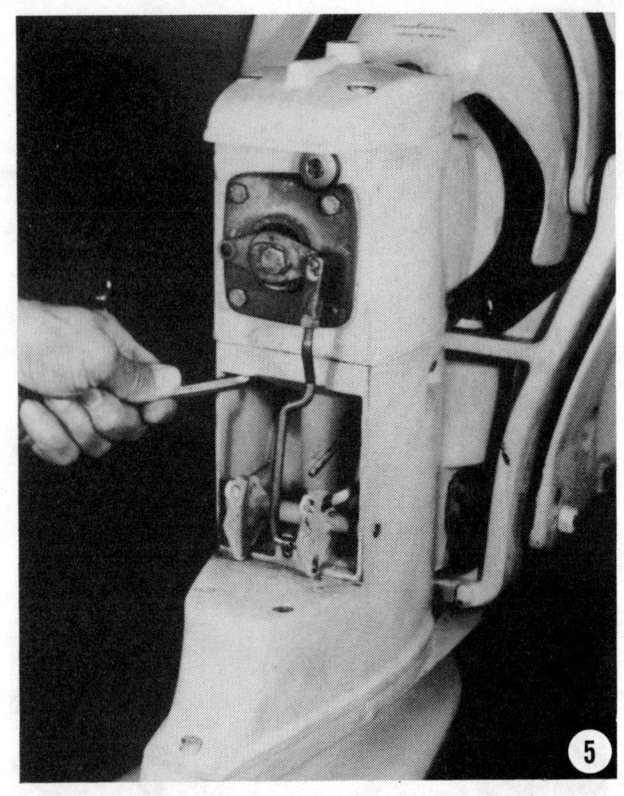
5

7

5- Remove the four bolts securing the upper gear assembly to the intermediate housing. These bolts are accessible from the intermediate housing, as shown.

6- Lift the upper gear assembly straight up off the intermediate housing, and then work it free of the universal joint bellows.

7- Loosen the clamp, and then remove the universal joint bellows.

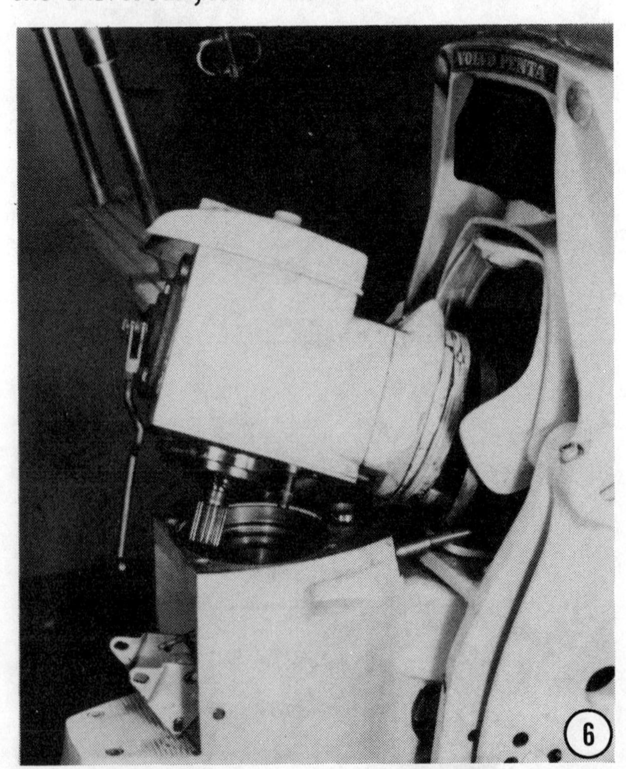

6

8

Disconnect the fuel line from the fuel pump, Loosen the clamp around the water intake hose. Loosen the exhaust boot clamp. Disconnect all electrical, accessory, and control wiring and linkage. These items will vary greatly depending on the power plant installation. Therefore, check a second time to be sure all have been disconnected and the engine is ready for removal. Most installation will have a master "Cannon-type" plug in the wiring harness approximately 4-6 ft. from the engine.

SPECIAL WORDS

A few power plants for the Volvo Penta stern drive units are cantilevered out from the transom by the engine mounts. Most engine installations have front engine mounts. If the unit being serviced does not have the front engine mounts, **NOW** is the time to secure a chain, web strapping, or other suitable device to the engine lifting brackets. One bracket is located on the port side forward and the other on the starboard side aft. If the unit has front engine mounts, back out the lag screws.

8- Secure the lifting device to the engine. Take up the slack to put just a small strain on the lifting device.

9- Cover the opening in the intermediate housing to prevent any foreign material from falling in. Remove the six bolts in the clamp ring.

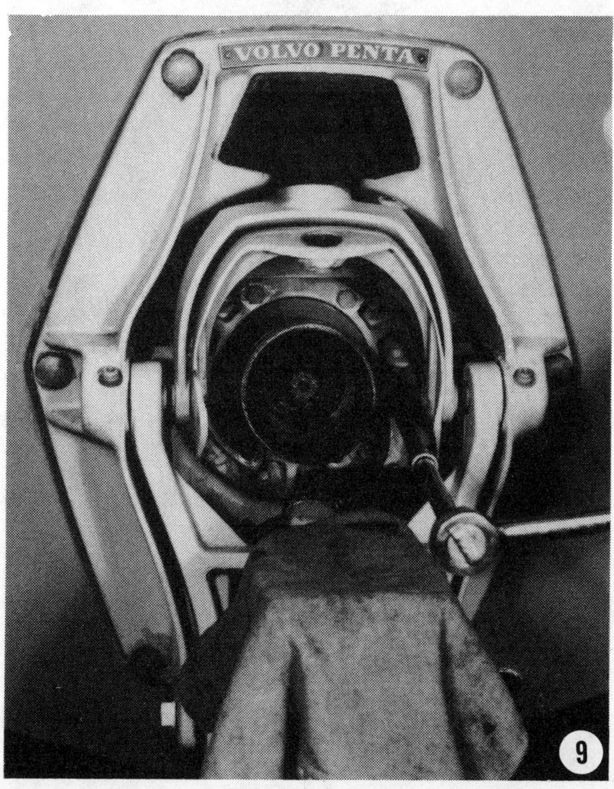

9

10- Lift the engine slightly. **CAREFULLY** work a screwdriver around the exhaust hose to loosen it from the elbow. Rock the engine, and at the same time attempt to move it forward to work the exhaust hose free of the fitting. Working this hose free of the elbow may require a little time and patience. If the task cannot be accomplished, it may be necessary to cut the hose free, and then to install a new hose and clamps during installation. **CHECK** to be sure a new hose is available before cutting the old one. Once the hose is free, slide the engine forward to allow the bell housing to clear the boat, and then lift the engine from the boat. Secure the engine in an upright position on the shop deck with blocks, or secure it in an engine work stand.

GOOD WORDS

To service the various parts of the block, see the appropriate section of this chapter. Detailed procedures are presented covering almost every task to completely recondition the block. Therefore, only those areas requiring attention need to be worked. To service components attached to the block, refer to other chapters.

10

3-4 ENGINE INSTALLATION

FIRST, THESE WORDS

The following procedures cover the installation of a 4-cylinder push rod AQ130 engine into a boat equipped with a Model 280 stern drive. The installation of other power plants is almost identical, especially connecting the bell housing to the stern drive. The major differences for installation of an OHC, V6, or V8 engine is in connecting water, fuel, and accessories after the engine is secured in the boat.

As with the removal instructions, step-by-step procedures to cover every possible installation and combination of accessories would not be practical in this type of publication. Therefore, follow the procedures closely, but bear-in-mind, the power plant you are installing may have added items requiring attention.

The instructions pickup the work after the engine has been reconditioned and is ready for installation into the boat.

INSTALLATION

1- Secure the lifting device to the engine brackets. One is located on the port side forward and the other on the starboard side aft. Lift the engine into the boat and begin to lower it into position.

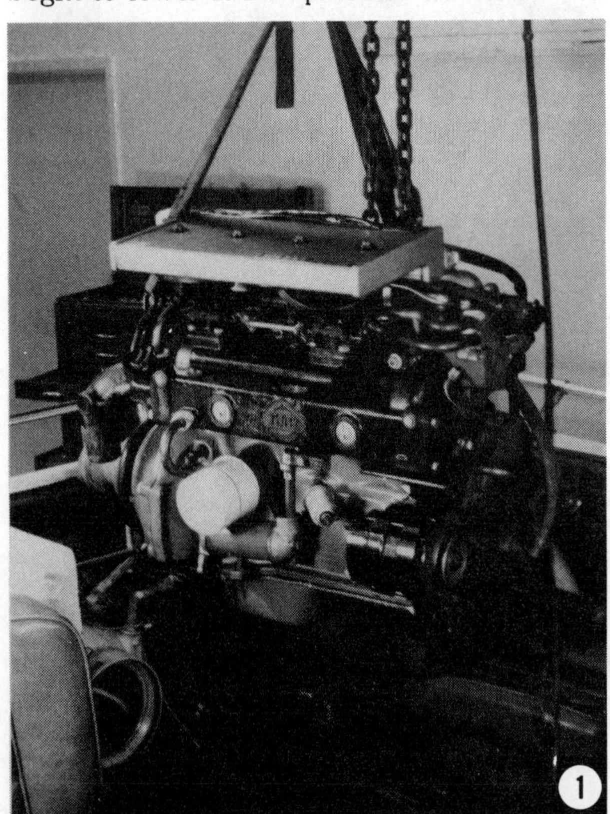

1

2

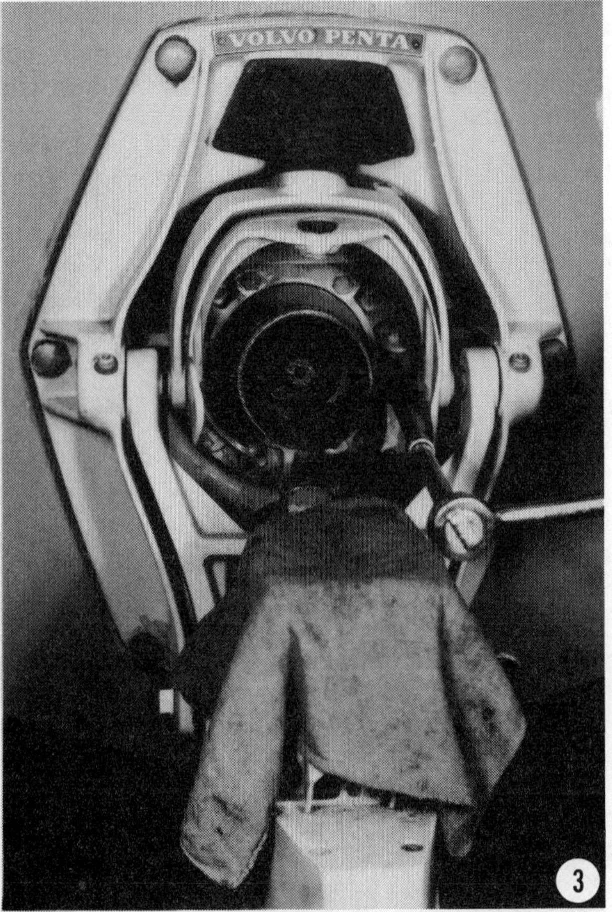

3

2- CAREFULLY lower the engine and begin to move it aft with the bell housing sliding through the transom shield. At the same time start to slide the exhaust hose up onto the exhaust manifold. Work the engine aft and down into position.

3- From outside the boat, install the clamp ring and start the six attaching bolts. **DO NOT** tighten the bolts at this time. Inside the boat, secure the engine mounts in position. After the engine mounts are secure, tighten the clamp ring bolts.

SPECIAL WORDS

To ensure proper engine/stern drive alignment, an even gap **MUST** exist between the bell housing and the mating transom shield. Take several measurements between the bell housing and the transom shield. These measurements should be within 1/8". If the engine is cocked to one side, the measurements will indicate this fact. If the mounts have not been disturbed, this alignment should not have changed since the engine was removed. The measurements will also reveal if the front of the engine is high or low. Adjust the studs at the front engine mounts to obtain satisfactory measurements. Connect the fuel, electrical, water, accessories, and other items disconnected during removal.

4- Slide the universal joint bellows onto the bell housing. Secure the bellows with the clamp.

SPECIAL WORDS

This is an excellent time to install the drain plug in the bottom of the lower unit and pour in lubricating oil. Use a funnel inserted into the oil distribution tube and fill the unit until the lubricant level is up to the bottom of the vertical shaft coupling. Much time can be saved by filling the lower unit in this manner.

5- Check the mating surfaces of the upper gear housing and the intermediate housing to be sure they are clean. Install a **NEW** O-ring around the bearing housing of the upper gear assembly. Install a **NEW** O-ring around the oil distribution tube. Coat the splines in the bell housing with a thin coating of water-resistant grease. This grease will aid installation and will assist in preventing the universal joint splines from seizing with the splines in the bell housing. Hold the upper gear assembly on its side with one hand and grasp the universal joint

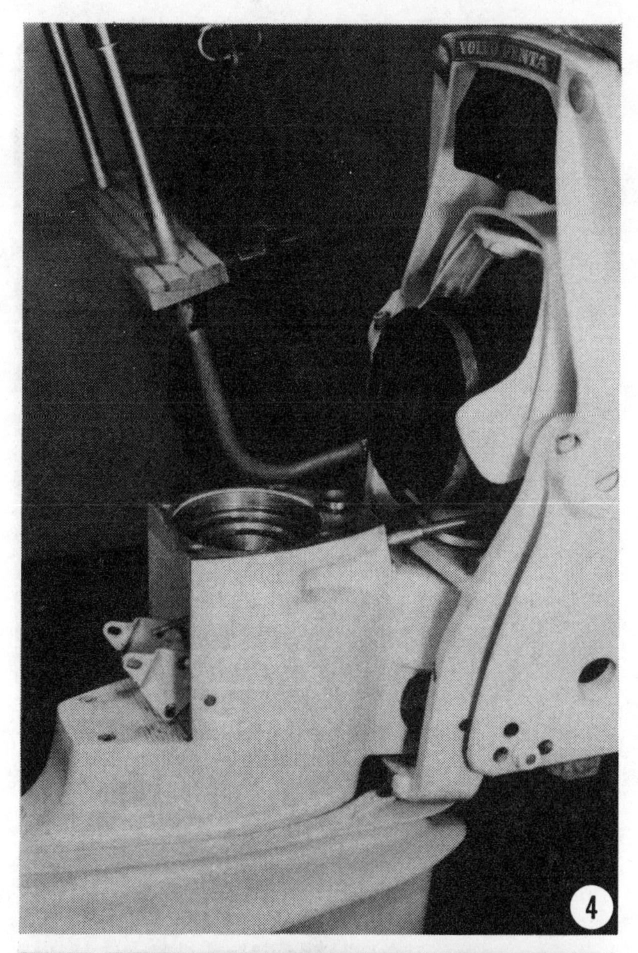

4

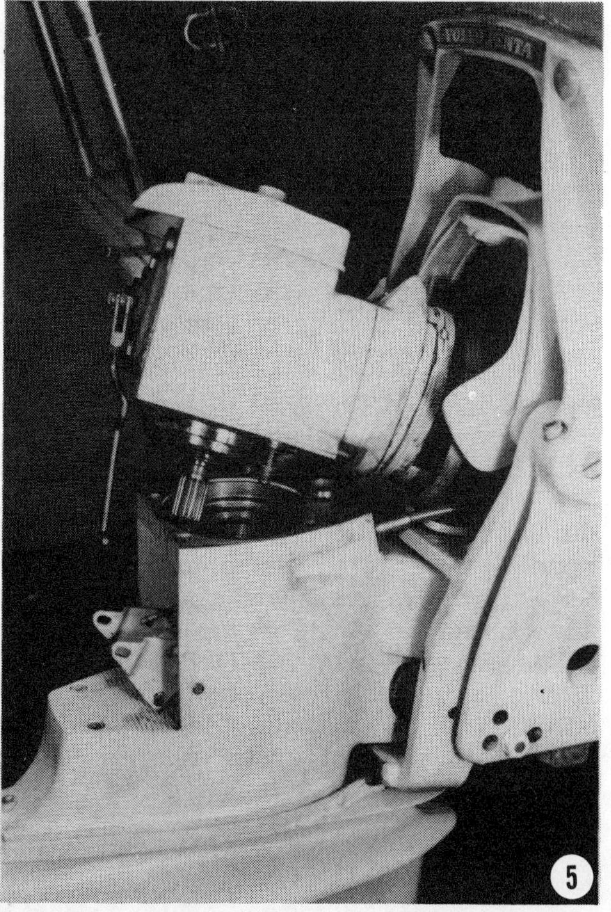

5

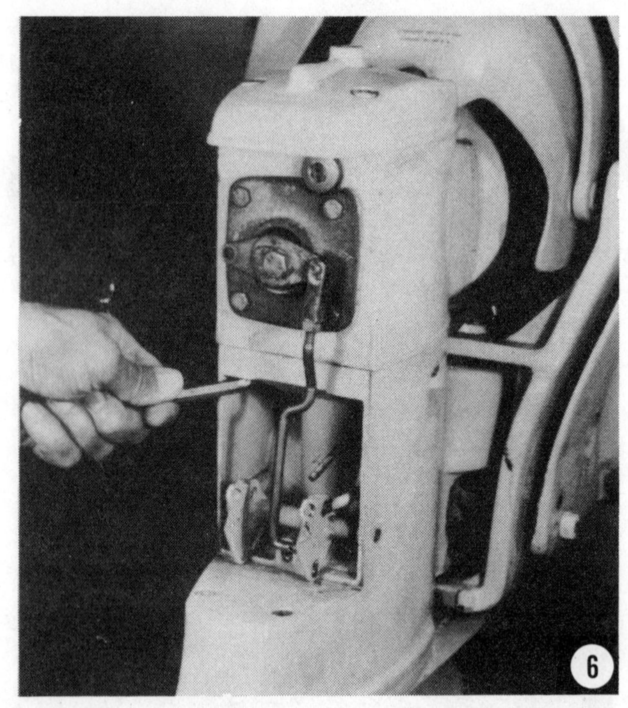

assembly with the other hand. **CAREFULLY** index the splines of the universal joint into the splines in the bell housing. Slide the upper gear housing forward slightly with the universal joint splines sliding and indexing with the splines in the bell housing and at the same time rotate the upper gear housing to an upright position. Now, lower the upper gear housing down onto the intermediate housing with the splines on the vertical shaft indexing with the splines in the coupling.

6- Secure the upper gear housing to the intermediate housing with the two bolts and the two nuts. Tighten the hardware alternately and evenly in a criss-cross pattern.

7- Slide the universal joint bellows into position on the upper gear assembly. Secure the bellows with the clamp.

CRITICAL WORDS

When the hose clamps are tightened, a socket wrench with a swivel joint from a 1/4" drive set should be used. If the socket set is not available use extreme **CAUTION** when tightening the clamp to prevent the screwdriver from slipping off the fastener and puncturing a hole in one of the rubber boots. Also, the clamps **MUST** be positioned as shown in the accompanying drawing to prevent the clamp ring tightening hardware from coming in contact with another rubber boot when the stern drive is in motion. Such action could eventually cause the boot to leak, causing extensive damage to expensive parts.

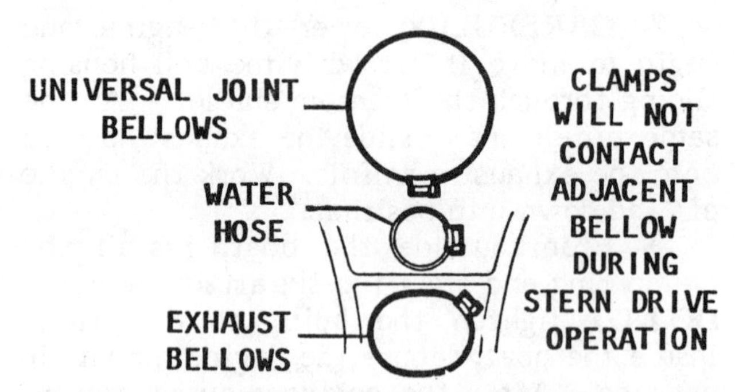

Care must be exercised when installing the hose clamps to be sure they are positioned as shown. The tightening device must not chafe against an adjacent bellow when the stern drive is in motion.

8- Place the steering helmet in position and secure it with the attaching bolt.

9- Connect the shift linkage from the upper gear housing to the intermediate housing and secure it with the stainless steel washer and cotter pin.

10- Install the shift cover and secure it in place with the two screws.

11- Remove the special lubricant fill bolt on the upper gear housing cover and fill the stern drive with the same oil used in the power plant.

12- Fill the stern drive with lubricant until the level reaches the top of the flat section of the dip stick, as shown.

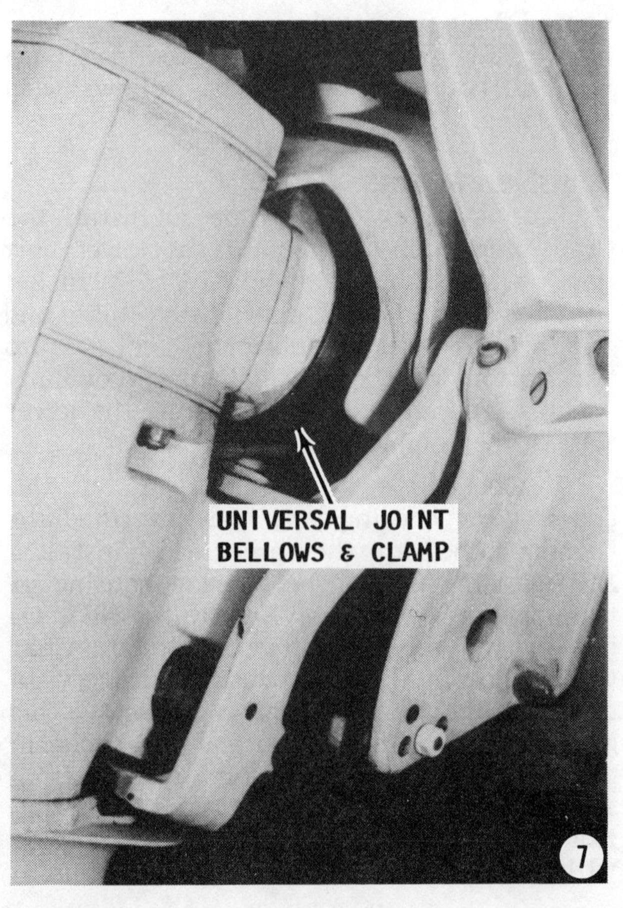

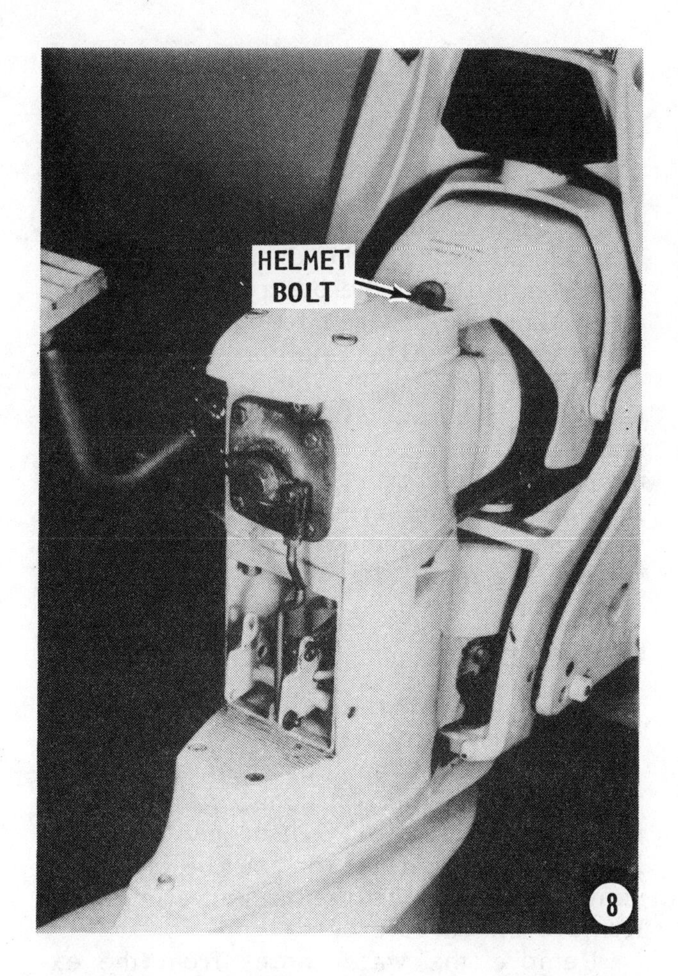
HELMET
BOLT
8

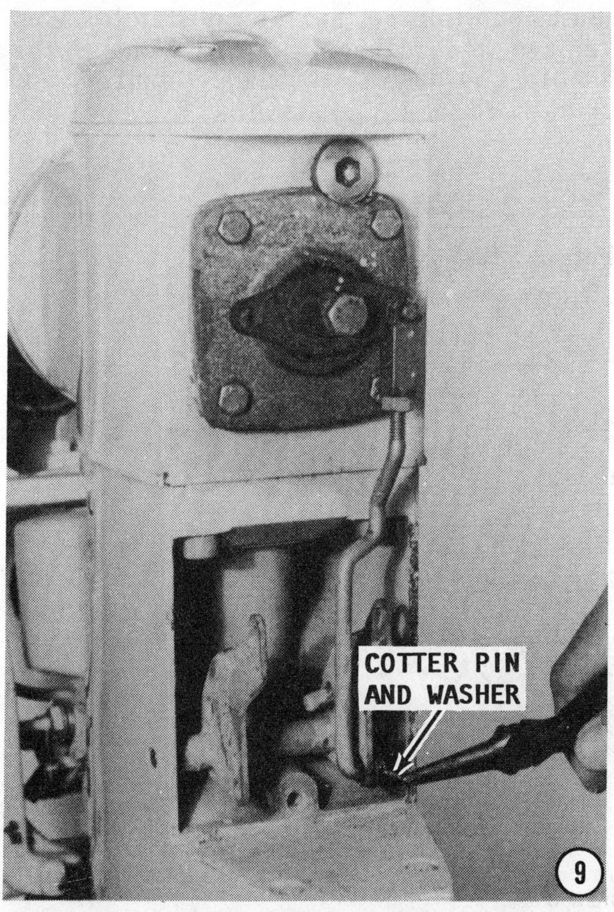
COTTER PIN
AND WASHER
9

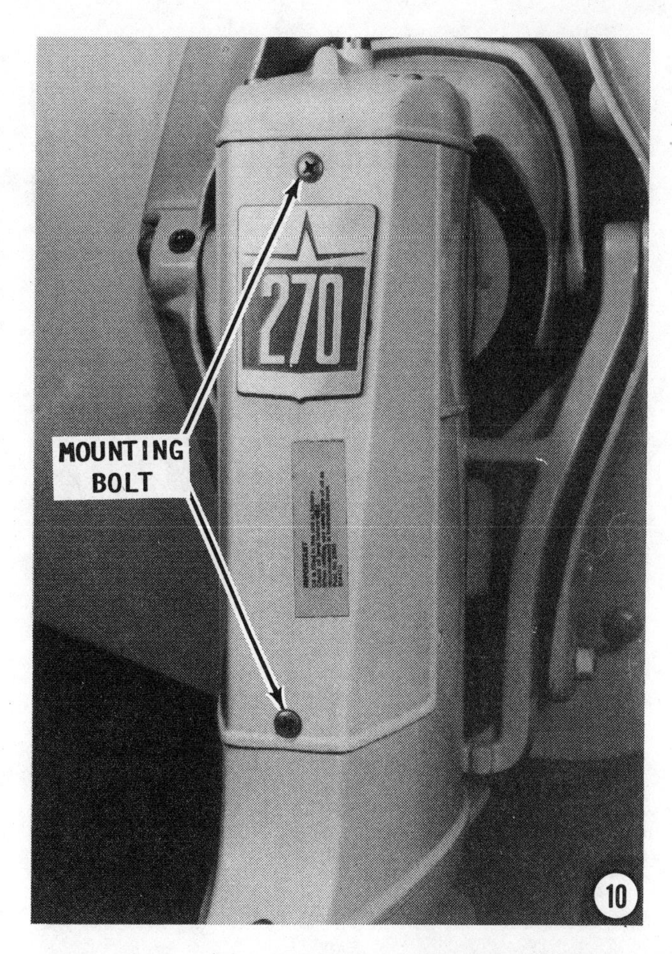
270
MOUNTING
BOLT
10

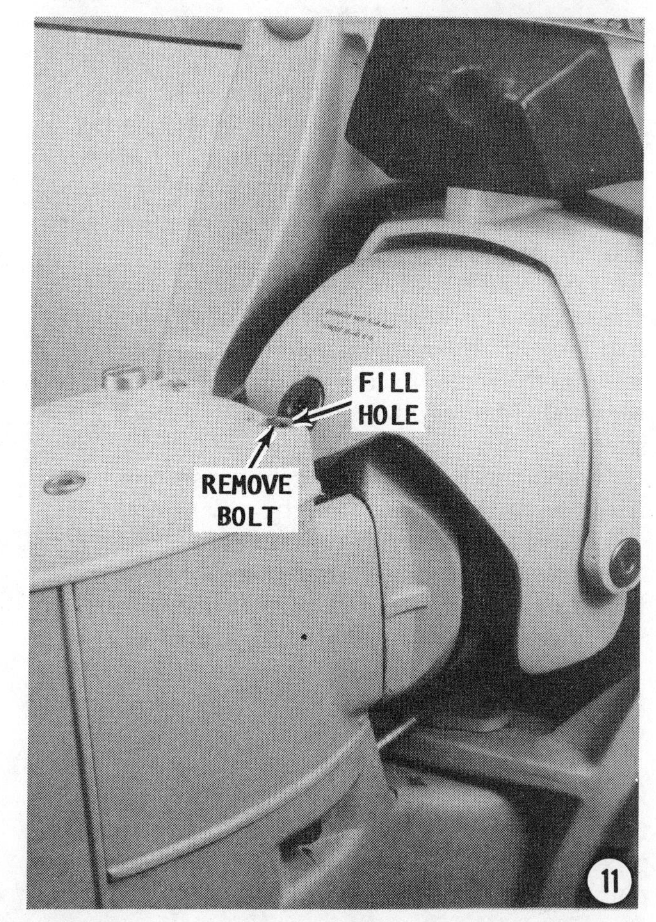
FILL
HOLE
REMOVE
BOLT
11

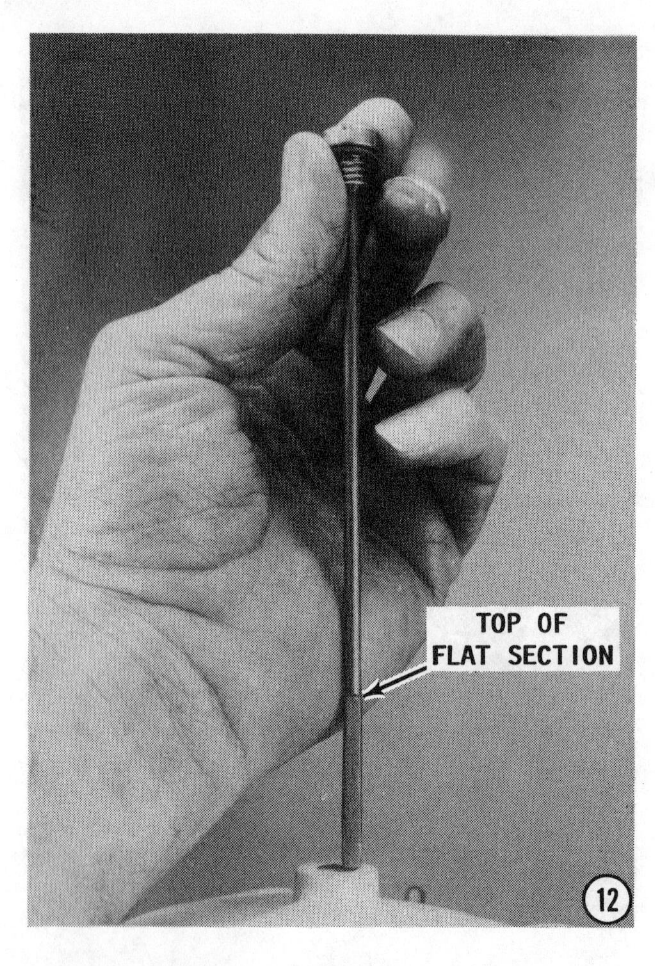

SPECIAL WORDS

When filling the stern drive, pour in a few ounces and then wait about 10 seconds for the lubricant to trickle down through the unit before adding more lubricant.

Allow enough time for the lubricant to stabilize before taking a measurement with the dip stick.

USE EXTREME CAUTION not to overfill the unit. The lubricant **MUST** have room for expansion. If an excess amount of lubricant is added, seals in the stern drive may be seriously damaged from the increased pressure.

Check to be sure the crankcase has been filled with the proper weight oil.

Close the water drain valves.

Move the boat into a body of water, a test tank, or connect a flush attachment to the lower unit.

Start the engine, run it at reduced speed for the first hour. Check the completed work.

CAUTION: Water must circulate through the lower unit to the engine any time the engine is run to prevent damage to the water pump mounted on the engine.

3-5 INTAKE/EXHAUST MANIFOLD IN-LINE OHV AND SINGLE OHC ENGINES (Not DOHC Engines)

On Volvo Penta in-line pushrod and single OHC engines, the intake manifold is cast aluminum and the exhaust manifold a cast iron unit. On the pushrod engines, both assemblies are secured to the engine block using the same attaching hardware. Therefore, when one is removed, the other is also removed. On an OHC engine the intake manifold is on the port side and the exhaust manifold on the starboard side. These assemblies seldom create problems. This short section is included to enable you to remove the manifold in order to accomplish other work.

REMOVAL

Open the drain valves and drain the water from the engine block.

Remove the flame arrestor screws and the flame arrestor. Disconnect the throttle rod at the bellcrank. Disconnect the fuel lines from the carburetor and the crankcase ventilation hose at the rocker arm cover. Remove the carburetor.

Remove the water hoses from the exhaust manifold to the thermostat housing. Remove the exhaust hose. Remove the manifold-to-head attaching bolts and clamps. Remove the manifold.

On an in-line pushrod engine, the same attaching hardware is used to secure the intake and the exhaust manifolds.

Removal/installation of the exhaust manifold on an in-line pushrod engine.

CLEANING

Scrape all gasket material from the gasket surface. Remove any deposits in the combustion chambers with a wire brush. Take care not to damage the gasket surface. Inspect the exhaust manifold water passages to be sure they are not corroded closed or clogged with foreign material. A small sharp pointed tool may be used to clean the passageways. A drill bit may also be used and the corrosion drilled out.

A cast iron exhaust manifold may be dipped in an acid solution similar to that used in a radiator shop.

If the manifold is badly corroded and cannot be cleaned to open all passageways, it **MUST** be replaced.

INSTALLATION

Apply a light coating of Permatex, Form-A-Gasket, or equivalent, to both sides of the manifold gasket or gaskets. The OHC engine of course has a separate gasket for each manifold. Place the gasket over the manifold end studs and carefully install the manifold. The Form-A-Gasket will hold the gasket in proper place. Install the bolts and clamps with one hand while holding the manifold in place with the other. Tighten the manifold bolts **ALTERNATELY** and **EVENLY** to a torque value of 14.8 ft-lb (20 Nm). Connect the exhaust hoses to the manifold. Replace the water hoses from the thermostat-to-exhaust manifold.

Install the carburetor, flame arrestor, and the crankcase ventilation hose. Connect the throttle rod. Replace the flame arrestor.

Close the water drain valves.

SPECIAL WORDS

If the engine installation is equipped with a fresh water cooling package, fill the unit with a 50% water and 50% antifreeze solution. The water is added through the heat exchanger fill cap.

Start the engine and check for leaks.

CAUTION: Water must circulate through the lower unit to the engine any time the engine is run to prevent damage to the water pump mounted on the engine.

3-6 INTAKE AND EXHAUST MANIFOLDS DOHC ENGINE

FIRST, THESE WORDS

Usually the exhaust manifold is removed as a preliminary task prior to performing other work on the block. **HOWEVER,** if an exhaust leak has developed at the manifold gasket or if there is a crack in the manifold, a very loud noise will be noticed when the engine is operating. In most cases a white powdery substance will be noticed in the area of the leak. In either case the manifold would be removed.

In almost all cases, the intake manifold is only removed prior to performing service work on the head.

The carburetors may remain attached to the intake manifold when the manifold is removed. Naturally, if the manifold requires service, the carburetors would then be removed.

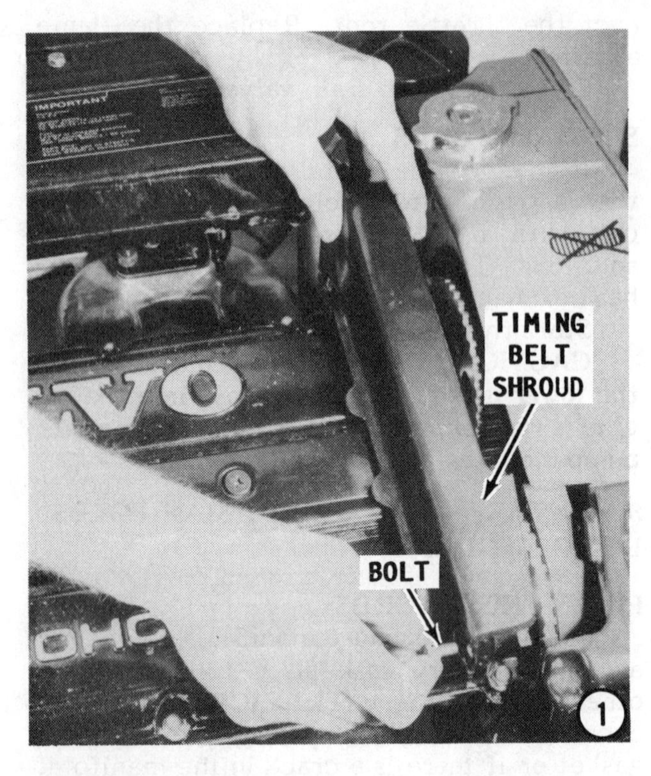

AUTHORS' WORDS

If the design of the engine compartment and the shrouds surrounding the engine permit, the intake and exhaust manifolds may be removed with the engine mounted in the boat. If the engine remains in the boat, drain the water from the block and the heat exchanger. Procedures to remove the engine from the boat are outlined in Section 3-3.

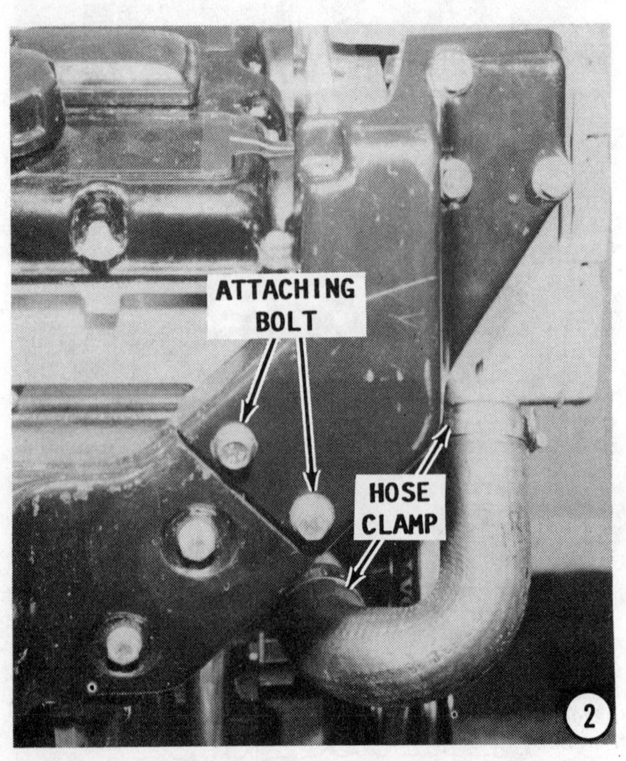

The following illustrations accompanying this section were taken with the engine removed from the boat for photographic clarity.

Exhaust Manifold
Removal

1- Disconnect the oil hose at the flame arrestor. Thread a bolt or other type plug into the end of the hose to prevent oil from draining from the oil trap. Remove the flame arrestor from the carburetors. Remove the two bolts securing the timing belt shroud to the heat exchanger. Lift the shroud clear of the heat exchanger.

2- Remove the two bolts on the starboard side heat exchanger bracket. Loosen the screw on the upper coolant hose clamp and pry the hose from the heat exchanger. The hose need not be removed from the block.

3- Remove the two bolts securing the thermostat housing to the heat exchanger. The housing will stay in place when the exchanger is removed.

4- Remove the two bolts on the flange securing the copper cooling water pipe to the exhaust manifold. This pipe must be moved aside to permit the heat exchanger to be lifted free.

5- Grasp the heat exchanger unit with both hands and lift it clear of the engine block.

6- Remove a total of nine bolts securing the exhaust manifold to the cylinder head. Two of these bolts are shorter than the

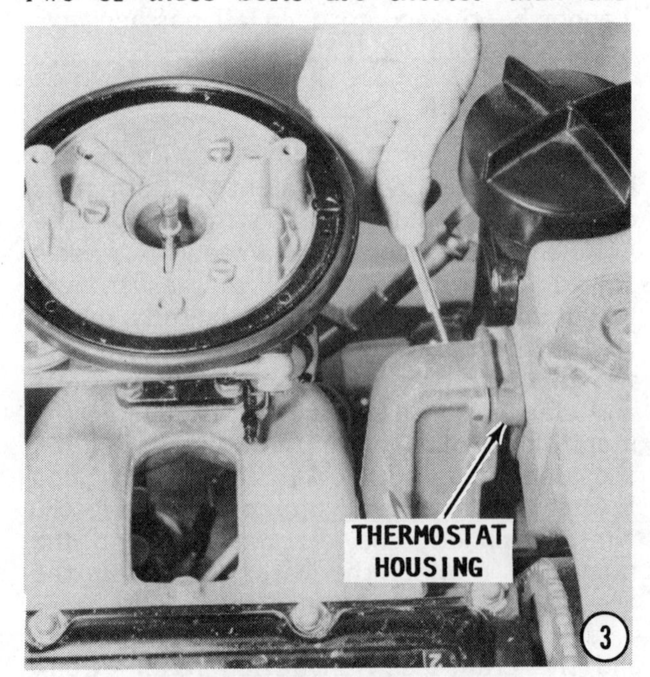

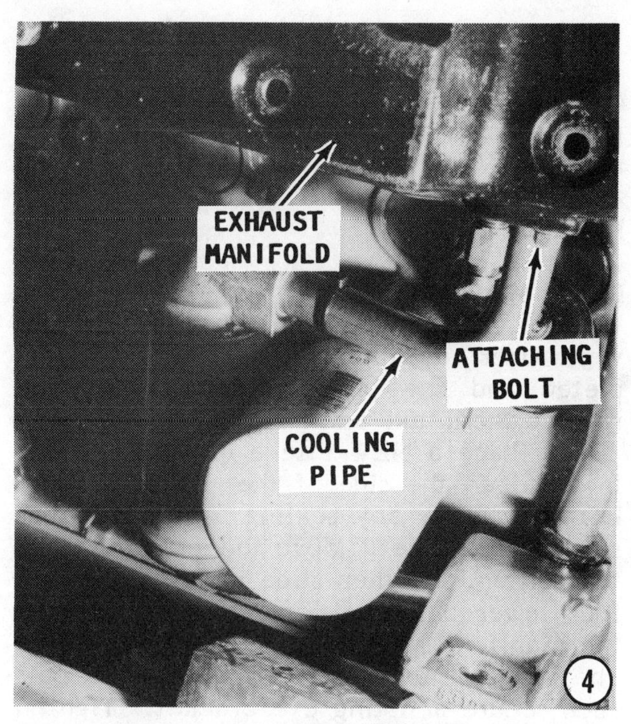

other seven. The two short bolts are the two most forward ones. Save the two center bolts until last and be prepared to take the weight of the manifold when they are removed. Lift the manifold clear. Remove and discard the manifold gasket.

Intake Manifold
Removal

7- Remove the carburetor covers, air filters, and rubber seals from the carburetors. Disconnect the fuel line at the fuel pump and the Red vacuum line leading to the intake manifold. Disconnect the yellow hose to the fuel enrichment device. Disconnect the throttle linkage. The two carburetors will remain in place on the intake manifold. Remove the oil dipstick from the block. Remove the bolt securing the dipstick bracket to the manifold and remove the bracket.

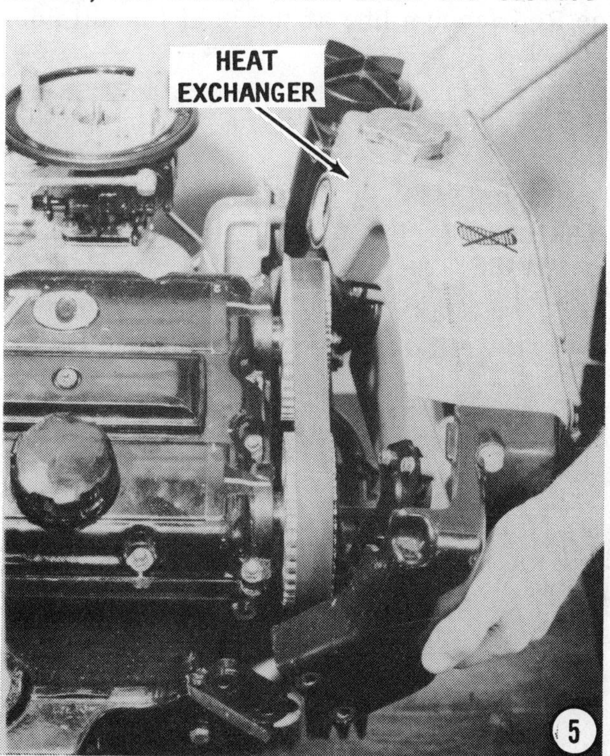

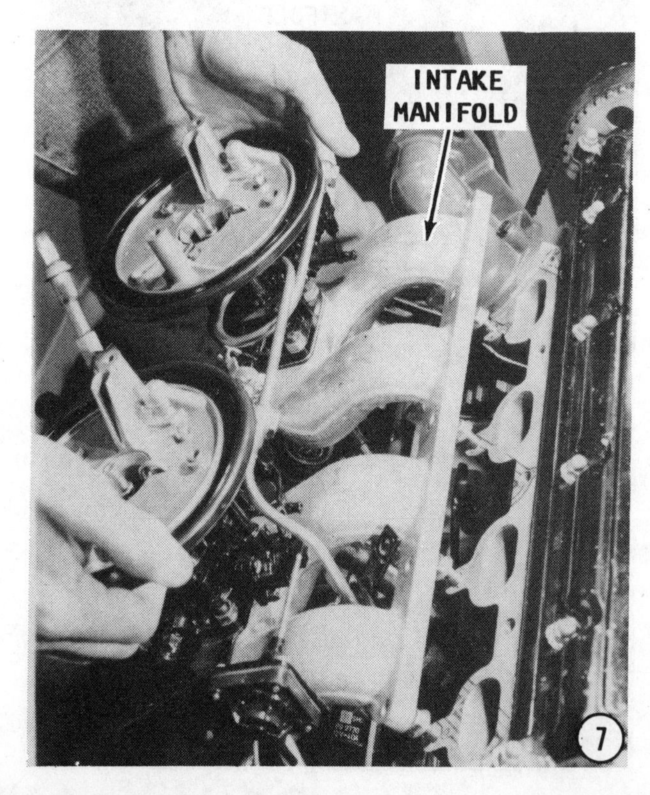

Remove the upper five bolts securing the manifold to the head. The forward and aft bolts secure lifting eye brackets. Loosen, but **DO NOT** remove the lower four bolts. Slots in the manifold permit the manifold to slide free without removing the bolts. After the lower bolts are loosened, slide the manifold upward a bit for the slots to clear the lower four bolts, and then remove the manifold.

CLEANING AND INSPECTING

Scrape all gasket material from the gasket surface. Clean away the gasket material from the mating surface on the block. Remove any deposits in the passageways with a wire brush. Take care not to damage the gasket surface. Inspect the exhaust manifold water passages to be sure they are not corroded, closed, or clogged with foreign material. A small sharp pointed tool may be used to clean the passageways. A drill bit may also be used and the corrosion drilled out.

A cast iron exhaust manifold may be dipped in an acid solution similar to that used in a radiator shop.

Carefully inspect the manifold for any sign of a crack. Many times a white powdery substance will be discovered in the area of a crack. If the manifold is badly corroded and cannot be cleaned to open all passageways, or if a crack is discovered the manifold **MUST** be replaced.

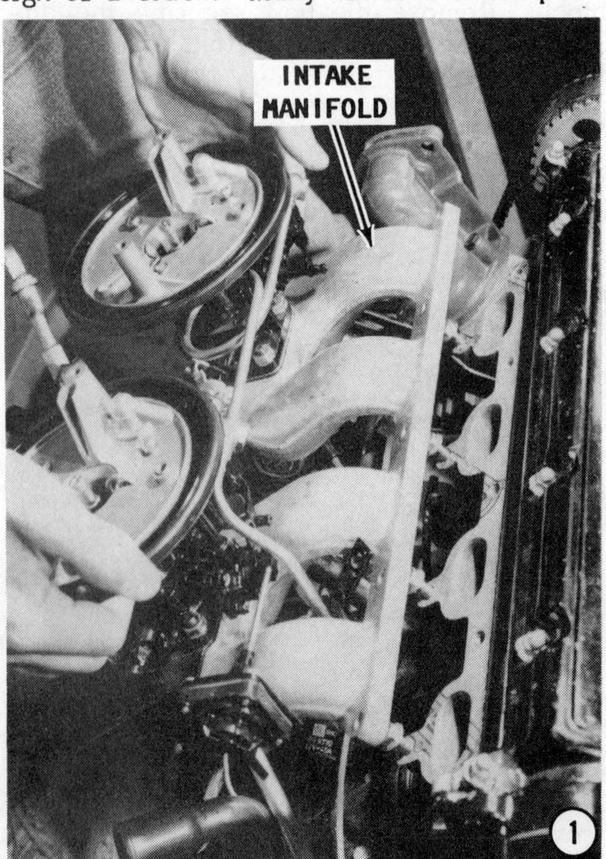

INSTALLATION

Intake Manifold

The following procedures pickup the work after any other tasks have been completed and the power plant is ready to receive the intake and exhaust manifolds.

1- Place a **NEW** intake manifold gasket in place on the side of the cylinder head. **DO NOT** apply any sealing agent to either side of the gasket. Move the manifold close to position, and then slide it down with the four lower bolts indexing into the slots in the manifold. Install the upper five bolts. The forward most bolt and the aft most bolt each secure a lifting eye bracket. Tighten all nine bolts evenly and alternately starting in the center and working toward both ends in a criss cross pattern. Install and secure the dipstick bracket to the manifold. Slide the dipstick down the tube and into the pan.

Connect the throttle linkage to both carburetors.

2- Connect the Yellow hose to the fuel enrichment device. Install the rubber seals, air filters, and covers onto the carburetors. Connect the fuel line at the fuel pump and the Red vacuum line at the intake manifold.

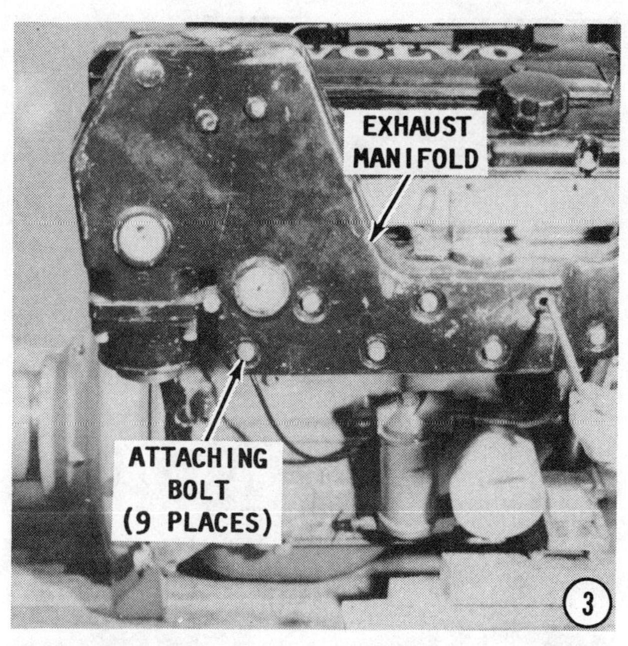

Exhaust Manifold

3- Place a **NEW** exhaust manifold gasket in position on the head. **DO NOT** apply any sealing agent to either side of the gasket. The gasket will only fit one way. Move the manifold into place and start a couple bolts to take the weight. Secure the manifold with a total of nine bolts. Two bolts are shorter than the other seven. The short bolts are the two most forward. Tighten the bolts alternately and evenly from the center working toward both ends in a criss cross pattern.

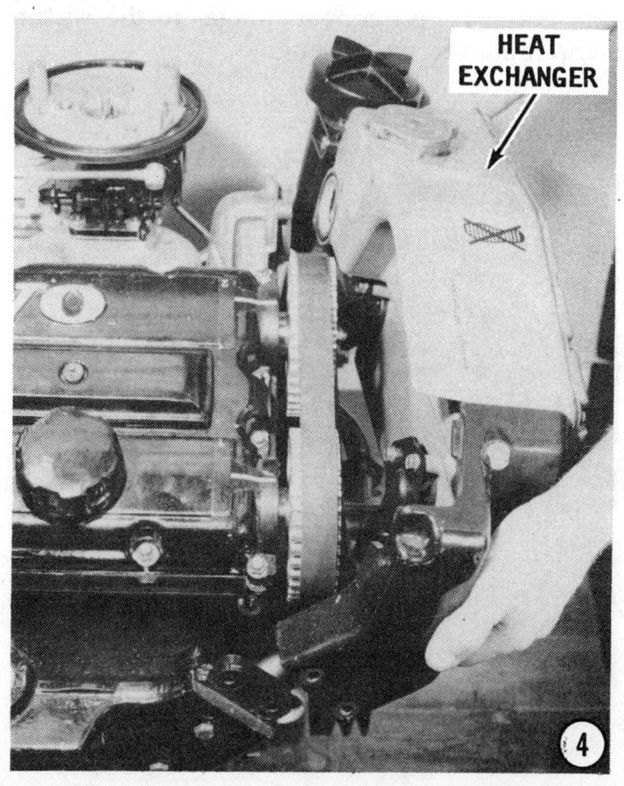

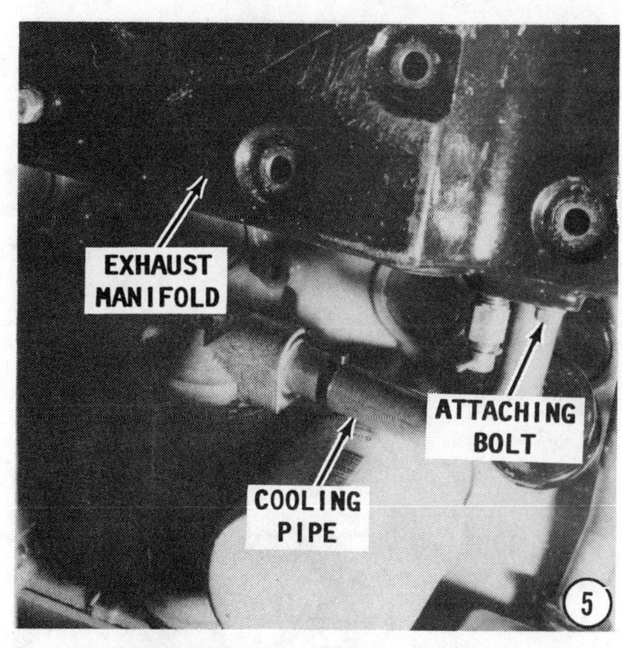

4- Move the heat exchanger up to the engine and rest it in place.

GOOD WORDS

For ease of installation, the exchanger should be in place, but not secured, before the cooling pipe is installed in the following step.

5- Connect the pipe from the oil cooler to the exhaust manifold with the two bolts through the pipe flange and into the manifold.

6- Coat both sides of the thermostat housing gasket with Permatex (non-drying), or equivalent. Move the heat exchanger up to the thermostat housing and secure them together with the two attaching bolts.

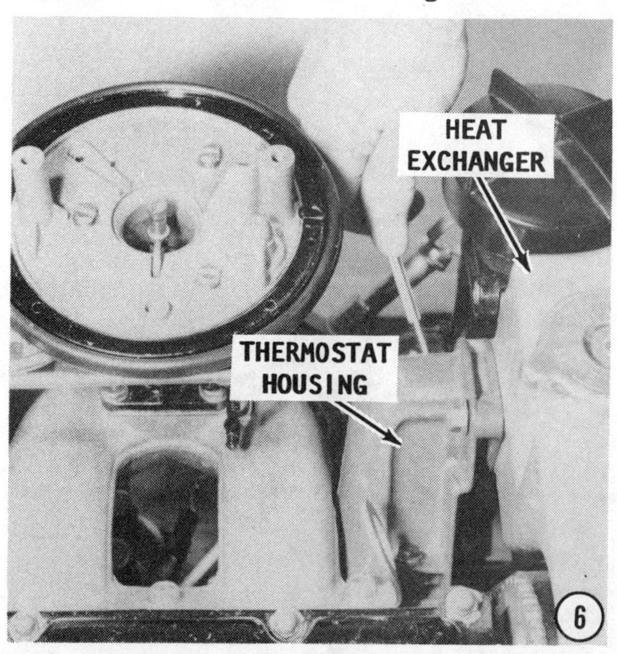

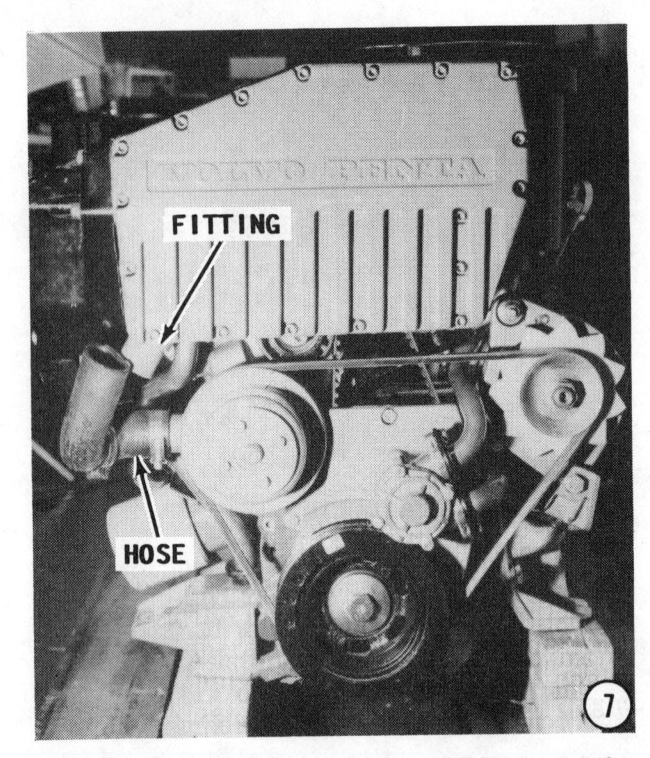

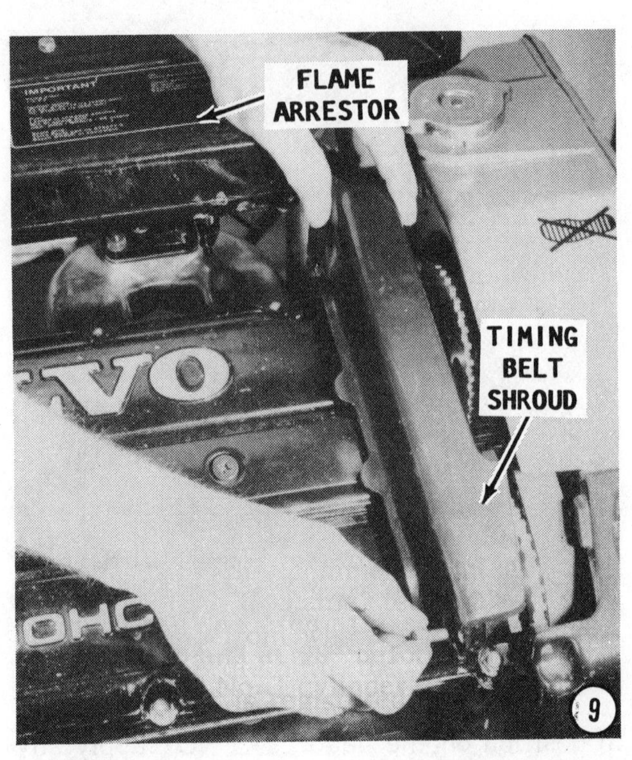

7- Slide the main coolant hose onto the heat exchanger fitting. If the hose is stubborn, coat the inside surface with some soap. Secure the hose in place with the clamp.

8- Install and tighten the two bolts securing the heat exchanger to the exhaust manifold.

9- Install the timing belt shroud over the timing belt and secure it in place with the two bolts and washers. Install the flame arrestor over the carburetors.

10- Close the water drain valves. If the engine was removed from the boat, install the engine, see Section 3-4. Fill the cooling system.

Start the engine and check the completed work.

CAUTION: Water must circulate through the lower unit to the engine any time the engine is run to prevent damage to the water pump mounted on the engine.

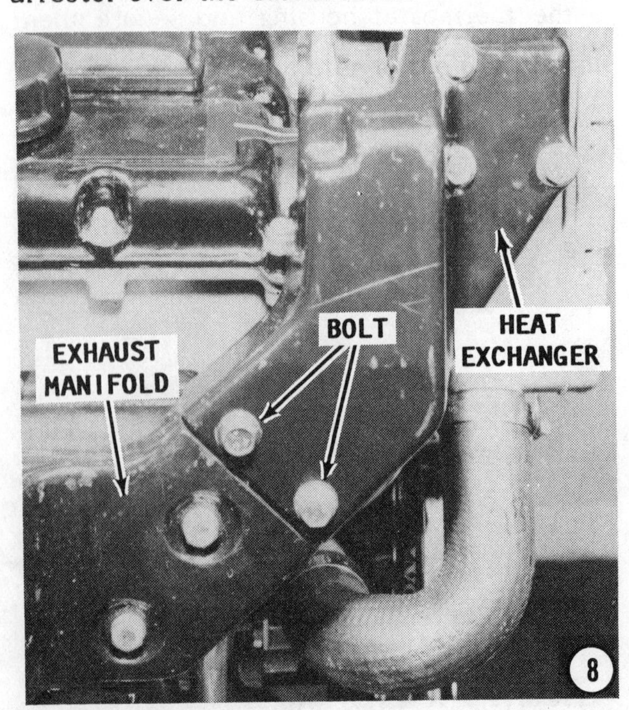

3-7 INTAKE MANIFOLD V6 OR V8 ENGINE

REMOVAL

Drain the water from the block by the drain plugs from the block -- port and starboard just above the crankcase pan line. Remove the flame arrestor from the carburetor. Disconnect the battery cables at the battery. Disconnect the hoses, fuel line, and throttle linkage at the carburetor. Disconnect the wires to the coil and temperature sending switch.

Remove the distributor cap and mark the distributor housing indicating the position of the rotor. Remove the distributor clamp and then pull the distributor out of the block. Move the distributor cap out of the way. Remove the coil and bracket.

Remove the bolts attaching the intake manifold to the head. **TAKE NOTE** that the bolts on the intake manifold of a **V6** engine may vary in length. Therefore, identify them in some way with the holes from which they are removed, as an aid during manifold installation.

Lift the intake manifold, with the carburetor attached, from the engine. **Discard** all gaskets and seals.

CLEANING

Clean the gasket surfaces of the intake manifold and the heads. Check the old gaskets to determine if there has been any exhaust leakage. Any evidence of exhaust leakage would indicate a crack in the head. Any sign of water in an intake manifold port would indicate either a crack in the manifold or a crack in the head.

The carburetor/s may be left in place when removing the intake manifold on a V6 or V8 engine.

INSTALLATION

Place new rubber intake manifold seals in position at the front and rear rails of the cylinder block. **BE SURE** the pointed ends of the seal fit snugly against the block and head.

Apply a light coating of Permatex, Form-A-Gasket, or equivalent, to the area between the water passages on the head and the manifold. Apply just a "dab" of the silicone Permatex onto the block at the corners where the flat gasket on the cylinder heads meet the front and rear seals in the valley, as shown. Install the intake manifold gaskets onto the heads. Set the intake manifold in place carefully and start the two guide bolts on each side.

Lift the manifold slightly and slip the gaskets into position as shown. **TAKE CARE** to align the three intake manifold holes in the gasket with the matching ports in the head and the manifold. The gasket should be installed as shown for the left side and reversed for the right side installation.

Install a manifold attaching bolt in the open bolt hole. The open bolt hole is held to close tolerances and the bolt in this location

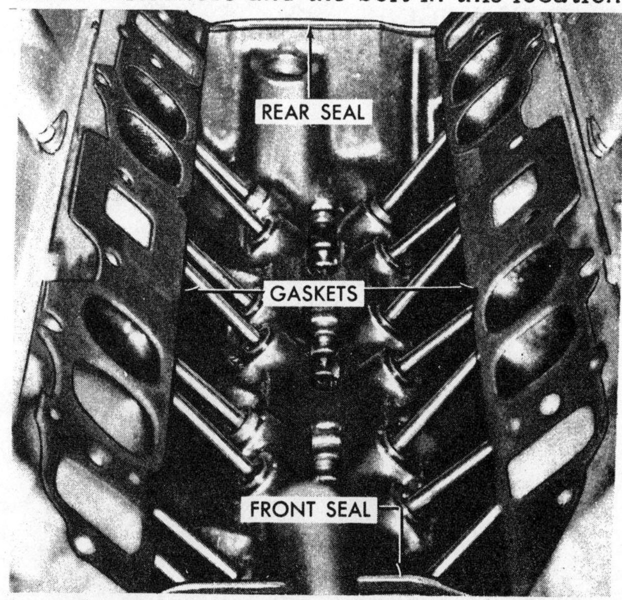

Placement of manifold gaskets and seals prior to installing the manifold. Just a "dab" of Permatex onto the block at the corners, as shown, is an aid during installation.

serves to locate the intake manifold perfectly fore-and-aft. Install the remaining manifold-to-cylinder head bolts with the longer bolts at the forward location. Tighten the bolts in the sequence shown in the accompanying illustration, and to a torque value of 25-35 ft-lbs (33.9-47.5 Nm).

Install the coil. Slide the distributor into place with the rotor pointing to the mark you made prior to removing the distributor. Snap the distributor cap in place. If the crankshaft was rotated while the distributor was out of the block, the engine will have to be timed. See Chapter 5 for detailed procedures to time the engine properly.

Connect the battery cables at the battery; the hoses to the thermostat housing; the throttle linkage and fuel line at the carburetor; and the wires to the coil and temperature sending switch.

Install the water drain plugs.

SPECIAL WORDS

If the engine installation is equipped with a fresh water cooling package, fill the unit with a 50% water and 50% antifreeze solution. The water is added through the heat exchanger fill cap.

Start the engine and check for leaks.

CAUTION: Water must circulate through the lower unit to the engine any time the engine is run to prevent damage to the water pump mounted on the engine.

3-8 EXHAUST MANIFOLD V6 OR V8 ENGINE

REMOVAL

Remove all water hoses and exhaust hoses from the elbow to the side of the exhaust housing.

Remove the attaching bolts, except one, preferably the **CENTER** bolt. Support the manifold in a convenient manner. Now, remove the last bolt, and then lift the manifold free of the head.

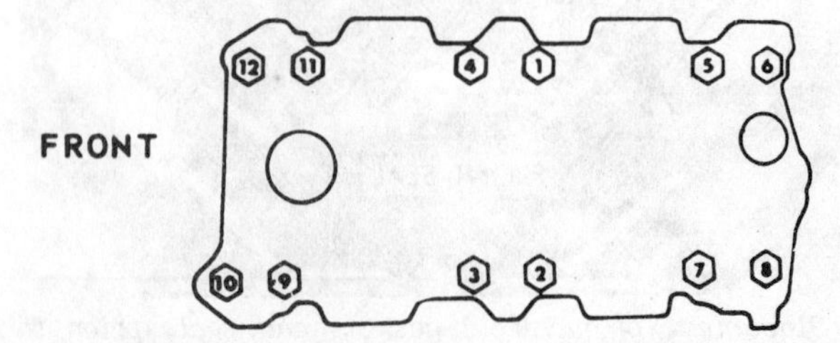

Tightening sequence for the intake manifold bolts on a V8 engine. On a V6 engine, start with the center bolts and work in both directions, as on the V8.

CLEANING AND INSPECTING

Scrape all gasket material from the gasket surface. Remove any deposits in the combustion chambers with a wire brush. Take care not to damage the gasket surface. Inspect the exhaust manifold water passages to be sure they are not corroded closed or clogged with foreign material. A small sharp pointed tool may be used to clean the passageways. A drill bit may also be used and the corrosion drilled out.

A cast iron exhaust manifold may be dipped in an acid solution similar to that used in a radiator shop.

If the manifold is badly corroded and cannot be cleaned to open all passageways, it **MUST** be replaced.

On all V6 models and on late model V8 engines, the exhaust manifolds have a removeable riser. The riser should be removed, the passageways inspected, and cleaned to ensure proper water circulation.

INSTALLATION

Install the manifolds to the cylinder head and start the bolts into the cylinder head. Tighten the bolts **ALTERNATELY** and **EVENLY**. The manufacturer does not always install a gasket between the exhaust manifold and the engine block, however, if any evidence indicates an exhaust leak, a new gasket should be installed. They are readily available at the nearest automotive parts dealer. A substitute for a new gasket

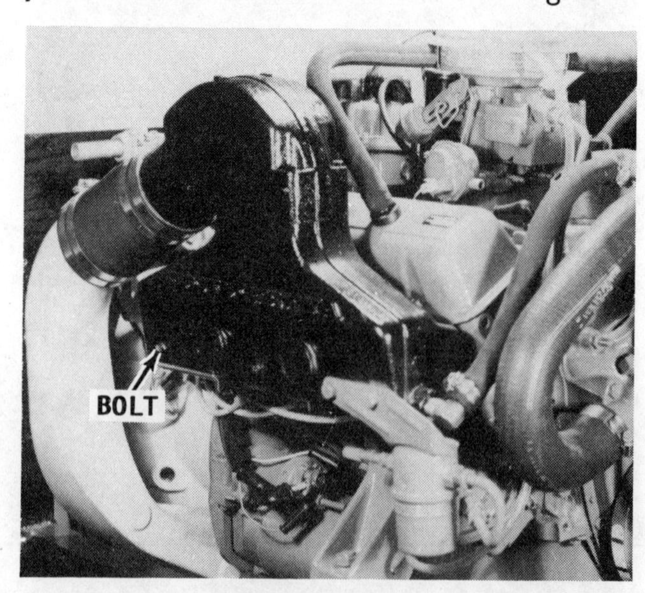

The exhaust manifold on a V6 or V8 engine is quite heavy. Therefore, be sure to support it before the last (center) bolt is removed to avoid personal injury.

would be to coat the surfaces of the manifolds and the matching surfaces on the cylinder head with Permatex "Form-A-Gasket" or equivalent.

Install the water hoses and exhaust hoses to the manifold.

Start the engine and check for leaks.

CAUTION: Water must circulate through the lower unit to the engine any time the engine is run to prevent damage to the water pump mounted on the engine.

3-9 ROCKER ARM SERVICE

REMOVAL — IN-LINE PUSHROD ENGINE

Remove the flame arrestor. Disconnect the crankcase ventilation hoses at the rocker arm cover. Disconnect all wires from the rocker arm cover clips. Remove the rocker arm cover.

Remove the rocker arm bolts. Remove the rocker arm shaft assembly. Withdraw the pushrods. Take time and **CARE** to place the pushrods in a rack so each can be reinstalled into the same position from which it was removed.

INSTALLATION — IN-LINE ENGINE

Install the push rods with the cup end up. **BE SURE** each rod is installed into the same location from which it was removed.

Lubricate the complete rocker arm assembly with standard engine oil.

Install the rocker arm and shaft assembly. Secure it in place with the bolts.

Adjust the valve lash according to the procedures outlined in Section 3-19.

Removing the rocker arm on an in-line pushrod engine. Notice the manifolds and thermostat have already been removed.

REMOVAL — V6 AND V8 ENGINE

Remove the rocker arm covers and gaskets. Remove the rocker arm nuts, rocker arm balls, and rocker arms.

Withdraw the pushrods. Take time and **CARE** to place the pushrods in a rack so each can be reinstalled into the same position from which it was removed.

GOOD WORDS

If the lifters are **NOT** to be serviced, protect the lifters and camshaft by covering the area with suitable cloths.

If the lifters **ARE** to be serviced, then set up some kind of system to hold the push rods. Remove each one in order so they will be installed back in the identical hole from which they were removed. Remove the lifters and keep them in order similar to the push rods. Each **MUST** be installed into the hole from which it was removed.

ROCKER ARM STUD REPLACEMENT

Replace any rocker arm studs with damaged threads with standard studs. If the studs are loose in the head, oversize studs are available in 0.003" or 0.010" oversize. These oversize studs can be installed after reaming the holes with the proper size tool for 0.003" or 0.010" oversize in the follow-

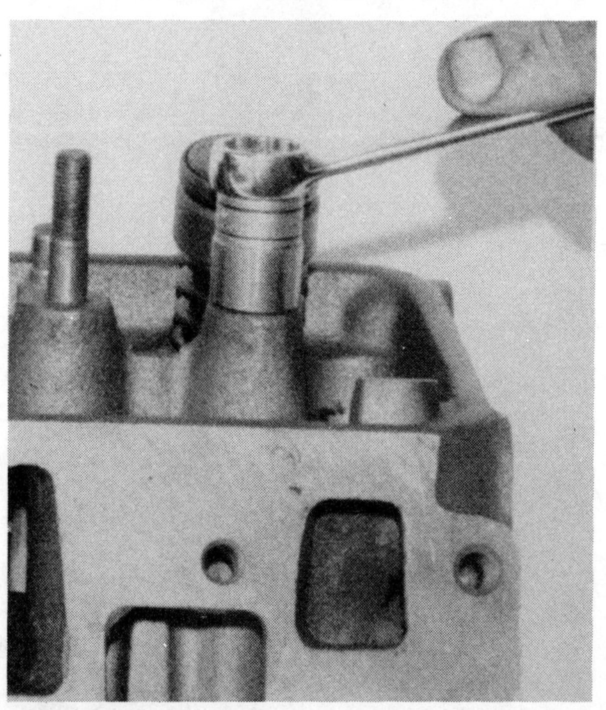

Rocker arm studs are pressed in. Therefore, a damaged rocker arm stud may be removed, using a stud puller.

ing manner. First, remove the old stud by placing a stud removal tool over the stud. Next, install the nut and flat washer and remove the stud by turning the nut. Now, ream the hole for an oversize stud using the proper size tool. After the hole has been properly reamed, coat the press-fit area of the stud with hypoid axle lubricant. Finally, install the new stud using a stud installation tool. The tool **MUST** bottom on the head.

CLEANING AND INSPECTING

Clean and inspect all parts. Pay particular attention to cleaning all of the oil holes. Replace any part excessively worn.

ROCKER ARM INSTALLATION V6 AND V8 ENGINE

Slide the pushrods down into the same hole from which they were removed.

Install the rocker arm assemblies onto the same stud from which they were removed.

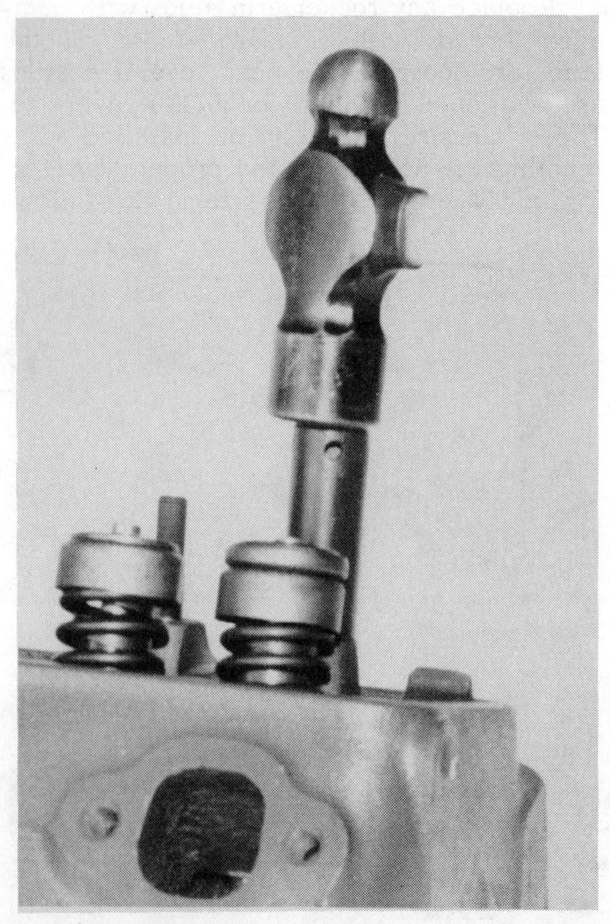

Before driving the new rocker arm stud into place with an installer tool, always coat the parts with hypoid axle lubricant.

Adjust the valve lash according to the procedures outlined in Section 3-22 for GMC V6 and V8 engines; Section 3-23 for Ford V8.

Install the rocker arm covers.

Connect the wires to the rocker arm cover clips.

Connect the crankcase ventilation hose to the rocker arm cover.

Install the flame arrestor.

3-10 CYLINDER HEAD IN-LINE PUSHROD ENGINE

REMOVAL

Remove the flame arrestor. Disconnect the crankcase ventilation hoses at the rocker arm cover. Disconnect all wires from the rocker arm cover clips. Remove the rocker arm cover.

Drain the block, and then disconnect the water hoses. Disconnect the spark plug wires. Remove the spark plugs and take care not to tilt the spark plug socket. If the socket is not kept straight with the plug, the insulator may be cracked.

Remove the intake and exhaust manifold assemblies; see Section 3-5.

Remove the rocker arm bolts, rocker arm assembly, and pushrods. Take time and care to place the pushrods in a rack so each can be reinstalled into the same position from which it was removed.

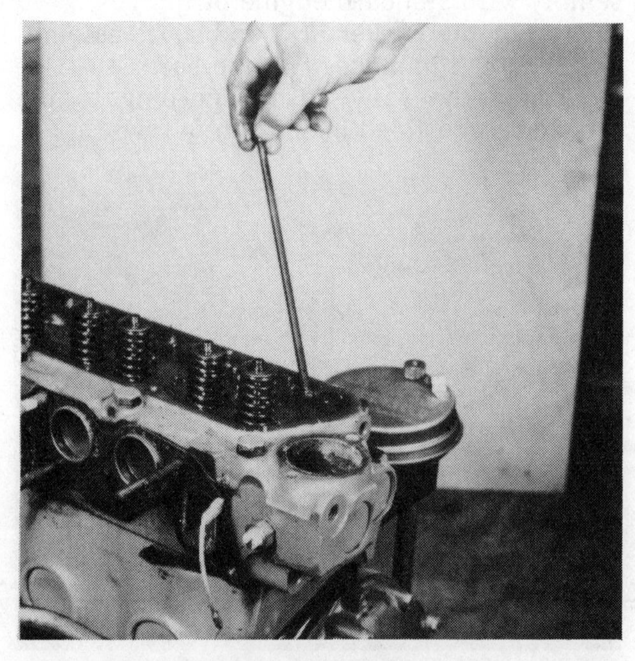

*Removing the push rods from an in-line pushrod engine. Each rod **MUST** be installed back into the same position from which it was removed.*

Disconnect the fuel lines from the retaining clip, and then disconnect the wires from the temperature sending units. Disconnect the battery ground strap at the cylinder head. Remove the ignition coil. Loosen and remove the cylinder head bolts, take off the head, and **DISCARD** the gasket.

To recondition the head, valves, and related parts, see Section 3-14.

INSTALLATION
IN-LINE PUSHROD ENGINE

The following procedures pickup the work after the head, valves, and related parts have been reconditioned and assembled according to the instructions outlined in Section 3-14.

The gasket surfaces on both the head and the block must be clean of any foreign material and free of nicks or heavy scratches. Bolt threads in the block and threads on the cylinder head bolts must be clean. Dirt on the threads will affect bolt torque values. **DO NOT** use any kind of gasket sealer on a composition steel-asbestos gasket. **Always** use a marine head gasket due to the possibility of the boat being used in salt water.

Place the gasket in position over the dowel pins, with the bead facing **UP**. Carefully guide the cylinder head into place over the dowel pins and gasket. Coat the threads of the cylinder head bolts with sealing compound and install them fingertight.

Tighten the bolts **ALTERNATELY** and **EVENLY** in the sequence shown in the accompanying illustration. **BEAR IN MIND**, uneven tightening of the head bolts can distort the cylinder bores, causing compression loss and excessive oil consumption.

Tighten the bolts in three rounds to a torque value of 66 ft-lb (90 Nm). On the first round, tighten the bolts to 1/2 the total torque value; on the second round to 3/4 the total torque value; and to the final torque value on the third and final round.

After the engine has been started and warmed to operating temperature, allow it to cool for about 30 minutes. Loosen each bolt just a wee bit, and then tighten it again to the required final torque value. Loosening the bolts just a bit will overcome static friction.

Install the push rods and rocker arms. Install the manifold assembly according to the procedures outlined in Section 3-4.

Install the coil, spark plugs, and high-tension wires. Connect the upper water hose and the engine ground strap. Connect the temperature sending unit wires. Connect the fuel lines in the clip at the water outlet.

Removing the head from an in-line pushrod engine. A new head gasket should always be used whenever the head is removed for any reason.

Starting the head bolts on a 4-cylinder in-line pushrod engine. Next, a torque wrench will be used to tighten the bolts in the sequence indicated on the next page.

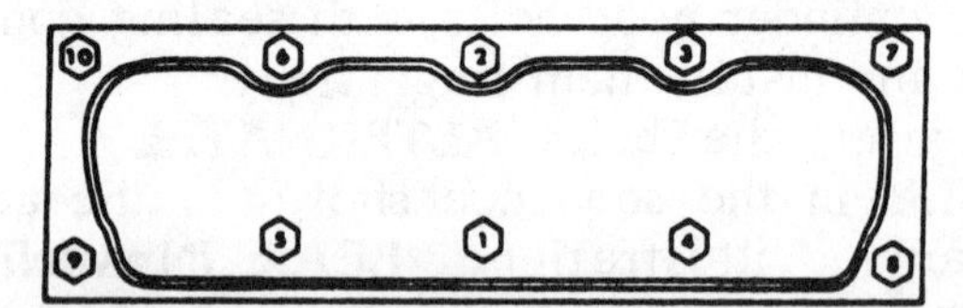

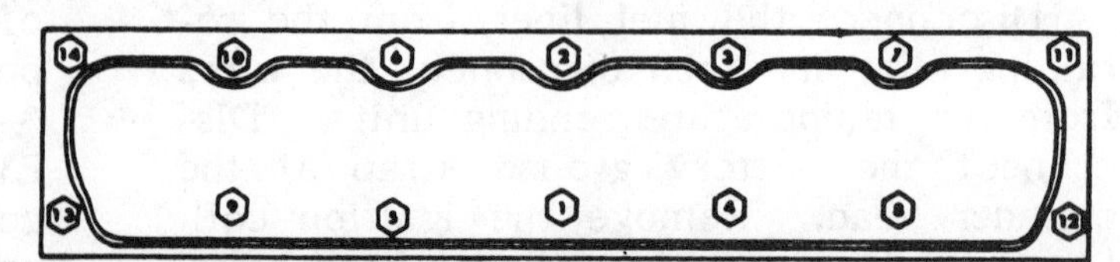

Cylinder head bolt tightening sequence for a 4-cylinder (left), and a 6-cylinder (right), in-line pushrod engine. It is a simple pattern, starting in the center and working toward both ends alternately, one set of bolts at-a-time.

3-11 HEAD SERVICE SINGLE OHC ENGINE

REMOVAL

Rotate the crankshaft until No. 1 cylinder is at TDC (top dead center), on the compression stroke. In this position the timing notch on the crankshaft pulley will be aligned with the **"0"** mark on the timing belt cover; the timing notch on the camshaft pulley will be aligned with the mark on the valve cover; and the timing dot on the accessory drive pulley will be aligned with the notch on the accessory drive backing plate. If the crankshaft and the accessory drive pulley are not rotated while the cylinder head is being serviced, installation will be greatly simplified.

Remove the intake and exhaust manifolds according to the procedures outlined in Section 3-5.

Remove the valve cover, gasket and the crescent shaped rubber seal.

Remove the bolt securing the drive pulley to the end of the camshaft. Remove the shroud from the camshaft pulley. Loosen the nut on the timing belt tensioner. Using a pair of channel-lock pliers, compress the spring. While the spring is compressed, slide the timing belt free of the camshaft drive pulley.

SPECIAL WORDS

It is not necessary to remove the camshaft in order to remove the head. If the head is being removed in order to simply replace a blown head gasket, the camshaft need not be disturbed. However, if the camshaft assembly or related items require service, then the camshaft would be removed at this time. See Section 3-16 to remove and service the camshaft.

The underneath side of a head from a 4-cylinder OHC engine. The head has been reconditioned as described in this chapter and is ready for installation.

Remove the Allen-head screws securing the cylinder head to the block. Lift the head clear of the block. Remove and discard the gasket. Remove the spark plugs.

To recondition the head, valves, and related parts, see Section 3-14.

HEAD INSTALLATION SINGLE OHC ENGINE

The following procedures pickup the work after the head, valves, and related parts have been reconditioned and assembled according to the instructions outlined in Section 3-14.

An OHC engine with the head installed and ready for the camshaft installation.

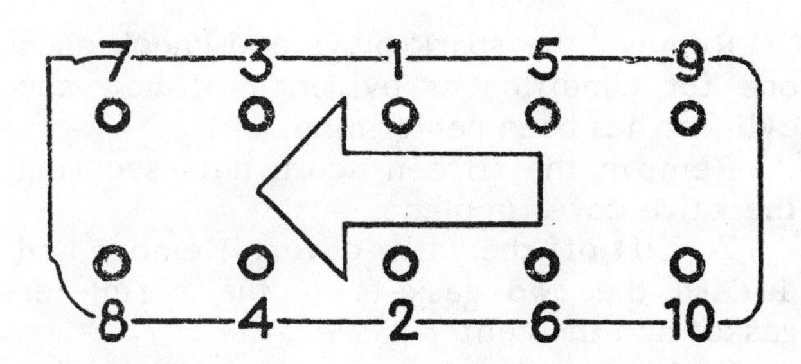

Cylinder head bolt tightening sequence for a 4-cylinder OHC engine. The bolts should be tightened in three rounds, as explained in the text.

Clean the engine block gasket surfaces. Check carefully to be sure no foreign material has fallen into the cylinder bores or bolt holes. Clean out all bolt holes with compressed air. One further step for a good job, is to run the correct size tap down each bolt hole.

Lay a **NEW** gasket in position on the block. Lower the cylinder head onto the block with the holes in the head aligned with the matching holes in the block. Lubricate the threads of the head bolts, and then install them finger tight.

Tighten the bolts **ALTERNATELY** and **EVENLY** in the sequence shown in the accompanying illustration. **BEAR IN MIND,** uneven tightening of the head bolts can distort the cylinder bores, causing compression loss and excessive oil consumption.

Tighten the bolts in three rounds to a torque value of 66 ft-lb (90 Nm). On the first round tighten the bolts to 1/2 the total torque value; on the second round to 3/4 the total torque value; and to the final torque value on the third and final round.

After the engine has been started and warmed to operating temperature, allow it cool for about 30 minutes. Loosen each bolt just a wee bit, and then tighten it again to the required final torque value. Loosening the bolts just a bit will overcome static friction.

If the camshaft was removed, install the camshaft according to the procedures outlined in Section 3-16.

Adjust the valves according to the procedures outlined in Section 3-14. Install a new cresent shaped seal and a **NEW** valve cover gasket. Position the valve cover in place and secure it with the attaching hardware.

Install the intake and exhaust manifolds according to the procedures outlined in Section 3-5.

3-12 CYLINDER HEAD DOUBLE OHC ENGINE

The following procedures pickup the work after the heat exchanger and the intake and exhaust manifolds have been removed, see Section 3-6.

AUTHORS' WORDS

If the design of the engine compartment and the shrouds surrounding the engine permit, the intake and exhaust manifolds, then the head, may be removed with the engine mounted in the boat.

The following illustrations accompanying this section were taken with the engine removed from the boat for photographic clarity.

HEAD REMOVAL

1- Rotate the crankshaft with the proper size socket on the crankshaft nut until the two marks on the timing belt are aligned with the two marks on the valve cover, one mark for each camshaft. The marks are decals on the valve cover. If the decals are no longer visible, rotate the crankshaft until the notch on each camshaft sprocket is

straight up. In this position, a vertical line would pass from the notch down through the center of the camshaft sprocket nut.

WORDS OF ADVICE

If the crankshaft and the idler pulley are not rotated while the cylinder head is being serviced, installation will be greatly simplified.

Remove the two bolts securing the spark plug cover, and then remove the cover from the head. Take time to identify each spark plug lead to ensure it will be installed to the correct spark plug. The original equipment wires are marked 1, 2, 3, and 4. These numbers may still be legible.

Twist and pull the high tension spark leads free of the spark plugs. Move the leads aft out of the way.

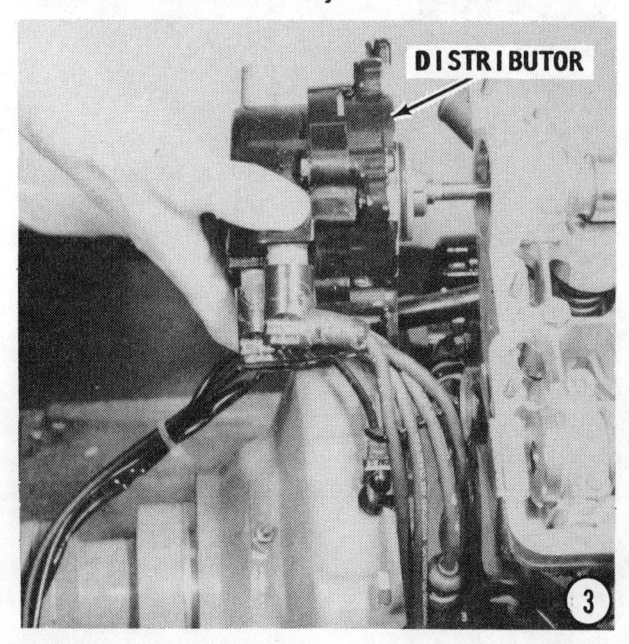

Remove the spark plugs and check each one for condition as evidence of how the cylinder has been performing.

Remove the fifteen acorn nuts securing the valve cover in place.

2- Lift off the valve cover. Remove and discard the two gaskets -- the perimeter gasket and the center gasket.

3- Loosen the center bolt on the idler pulley to allow the wheel to rotate. This wheel is pivoted on an eccentric to provide slack in the timing belt. As mentioned earlier, if the crankshaft is not rotated, installation will go quickly and easily.

Remove the two bolts securing the distributor in place. Gently pull straight back and remove the distributor from the head. **DO NOT** attempt to rotate the distributor shaft.

SPECIAL WORDS

When the camshaft carrier is removed in the next step, some of the hydraulic lifters will fall free onto the surface of the head. Make an effort to turn the carrier over rather quickly to prevent all of the lifters falling free. Take extra precautions to prevent the lifters from falling to the floor and being damaged. Any damaged lifters must be replaced.

If at all possible, keep track of the lifters in an effort to install them back into the same position from which they were removed. The wear pattern for each lifter will be slightly different from its neighbor.

4- Remove the outside perimeter nuts and the center row of nuts securing the camshaft carrier to the head. Lift the camshaft carrier from the head and turn it

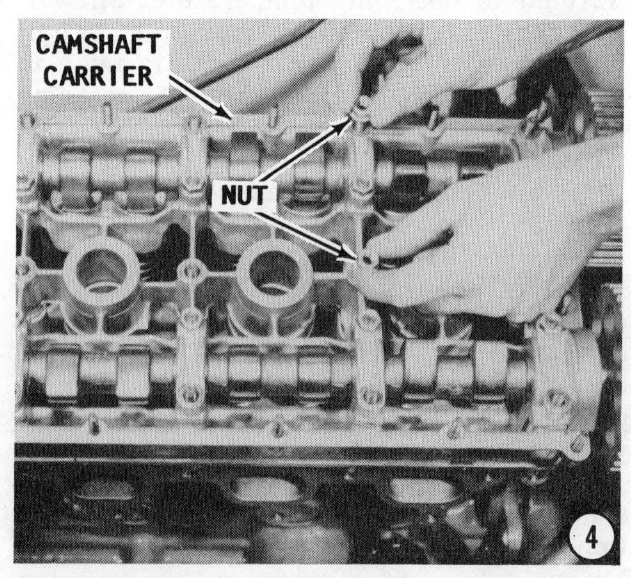

over rather quickly to prevent as many lifters from falling free as possible.

5- Place the camshaft carrier on a suitable work surface in its normal upright position and allow the hydraulic lifters to slowly fall free from their bores, one by one.

If the lifters are to be installed for further use, place them in a shallow pan in the same order from which they were removed. Add clean engine oil to the pan until the oil level is above the small hole in the lifter to prevent complete bleed down.

CRITICAL WORDS

It is important to keep the lifters in order because if the lifters and the camshafts are fit for continued use, a wear pattern is established between each lifter and its matching cam lobe. If the lifters are mixed during installation, the camshaft wear is greatly accelerated.

Further removal and service of the camshafts is covered in Section 3-17.

6- Remove the two bolts securing the thermostat housing to the head, and then remove the housing. Lift the thermostat out of the head, if it did not come free with the housing.

7- Pry off the hoses leading to the fuel enrichment vacuum valve. Use a pair of channel lock pliers and remove the fuel enrichment vacuum control valve. Disconnect the temperature sender wire at the connector on the head.

8- A loosening sequence, as indicated in the accompanying illustration, is given for the aluminum head of this engine to prevent distortion. Remove the ten head bolts following the prescribed sequence. Notice the difference in head bolt length. Remember where the short bolts are installed.

9- Lift off the head and place it on a suitable clean wooden work surface or on a

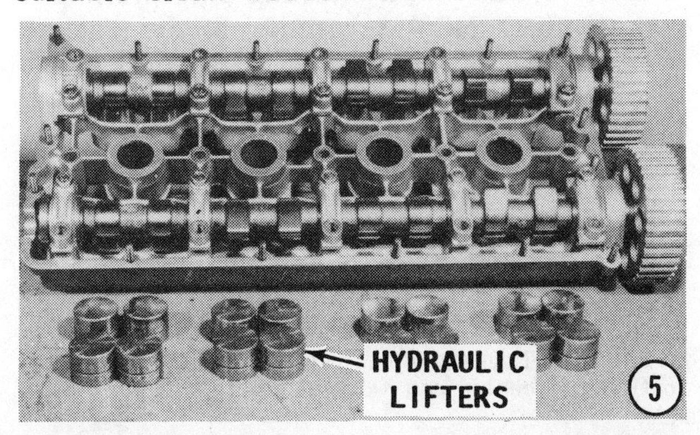

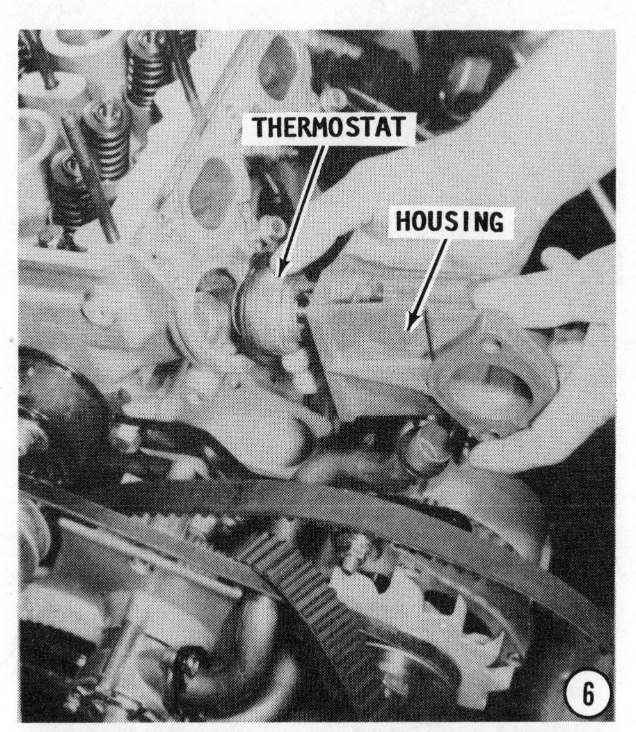

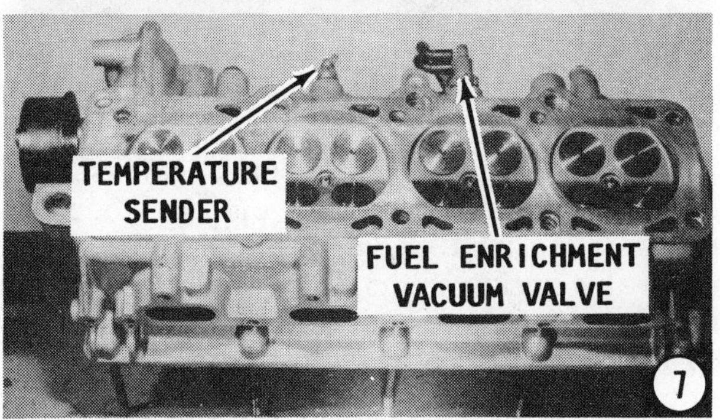

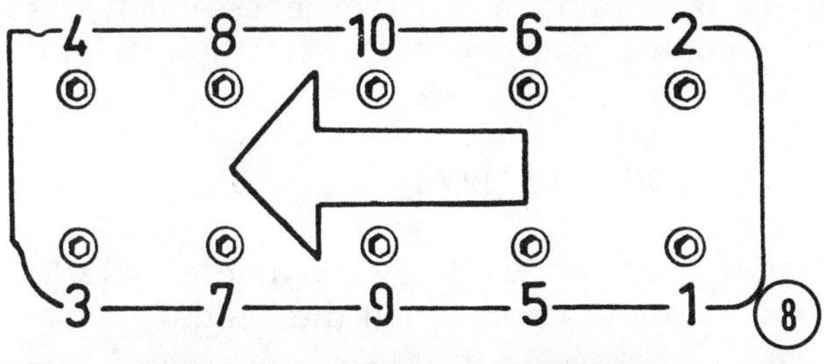

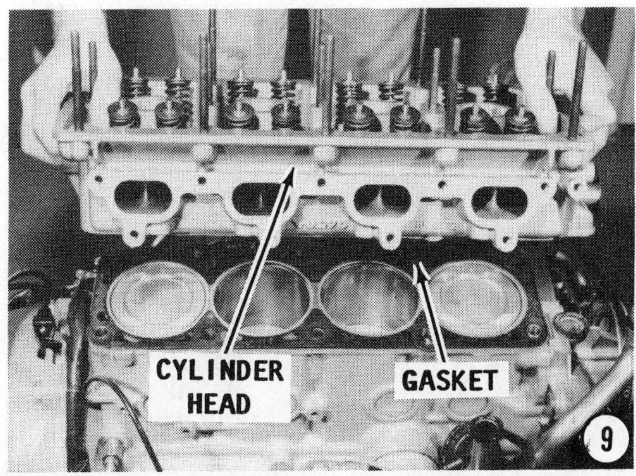

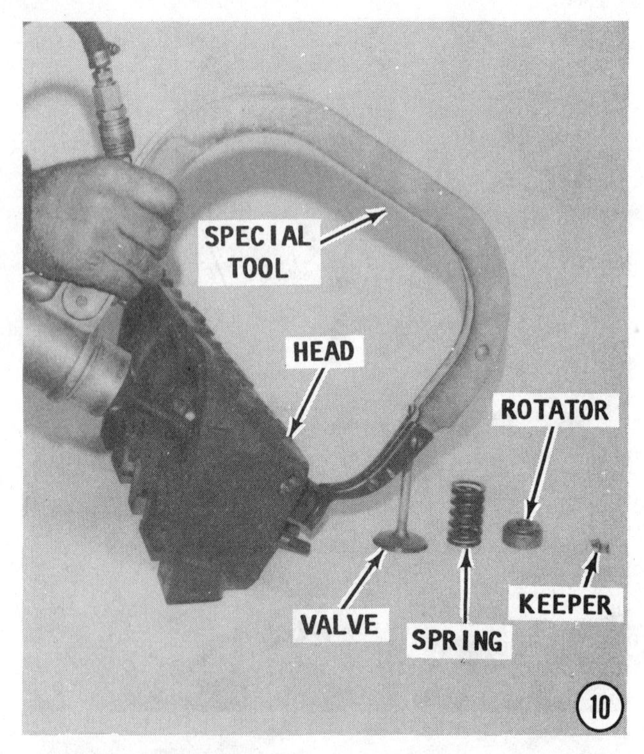

padded metal surface to prevent scratching the gasket sealing surfaces.

10- Use a tool to compress the valve springs to remove the keepers. Release the compressor, and then take off the keepers, rotators, springs, spring dampers, oil seals, and valve spring shims. Pay attention to the number of shims under each of the valve springs. The number of shims is very important during assembly. Remove the valves from the cylinder head and place them in a rack in their proper sequence so they can be installed back into their original positions.

SERVICING THE HEAD

See Section 3-14 for complete detailed procedures to recondition the head.

HEAD INSTALLATION

The following procedures pickup the work after the head has been completely reconditioned and assembled with valves in place.

1- Position the head gasket in place on the block with the word **"VOLVO"** facing **UP** and **FORWARD.** The manufacturer states **NO** sealant is to be used on either side of the gasket. Lower the head onto the block with the holes in the head indexing with the pin in the block.

2- Thread the ten head bolts through the head and into the block. Three of the head bolts are shorter in length than the other seven. Install these three short bolts on the port side between the forward and aft bolts, as shown in the accompanying illustration.

Tighten the ten bolts in three stages and in the sequence indicated in the accompanying illustration.

On the first stage, tighten to 14.5 ft lbs (20Nm); on the second stage, tighten to 29 ft lbs (40Nm).

3- For the third and final stage, proceed as follows: Obtain a marking pen and make a mark (**DO NOT** scratch) on the gasket

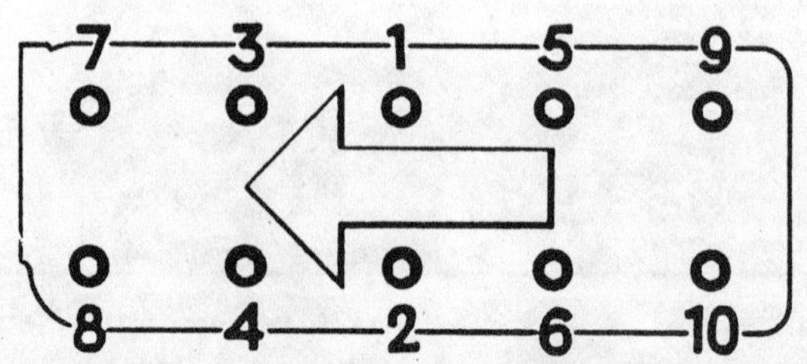

Tightening sequence for the cylinder head bolts on a 4-cylinder in-line engine. Notice how the pattern is: start in the center and work alternately toward both ends.

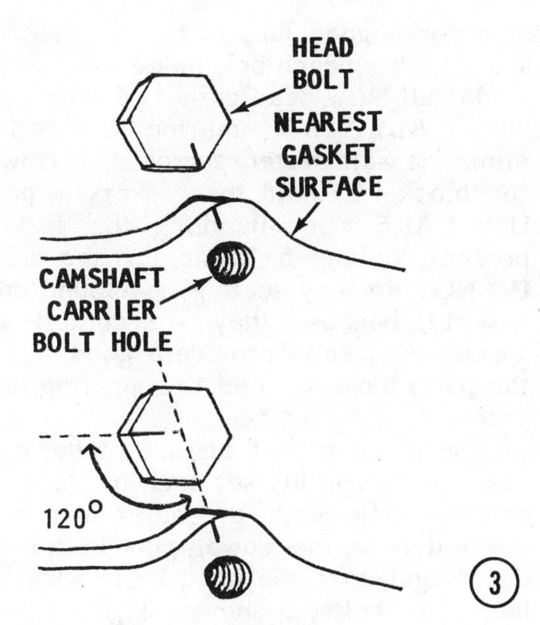

surface and a matching mark on the bolt head aligned with the first mark.

Now, for the third stage, tighten the bolt through 120° arc (1/3 of circle).

CRITICAL WORDS

DO NOT tighten the head bolts any further than stated, 120° after the 29 ft lbs (40Nm) has been reached. The third stage angular tightening closely approaches the elastic limit of the bolt.

4- Apply a thin coat of non-drying Permatex, or equivalent, sealant to **BOTH** sides of a new thermostat gasket. Place the gasket in position on the head. Install the thermostat with the larger portion facing **OUT**. Install the thermostat housing and secure it in place with the two attaching bolts.

Install the camshafts and related parts, see Section 3-17.

3-13 CYLINDER HEAD V6 AND V8 ENGINES

REMOVAL

Drain the water from the block by removing the drain plugs -- port and starboard just above the crankcase pan line.

Remove the intake manifold, see Section 3-7.

Remove the rocker arm covers and gaskets. Remove the rocker arms. **BE SURE** to identify the rocker arms to ensure they will be installed on the same side from which they were removed.

Withdraw the pushrods. Take time and care to place the pushrods in a rack so each can be reinstalled into the same position from which it was removed.

If the lifters are **NOT** to be serviced, protect the lifters and camshaft by covering the area with suitable cloths.

If the lifters **ARE** to be serviced, remove each one in order so they will be installed back in the identical hole from which they were removed. Each **MUST** be installed into the hole from which it was removed. A magnet is a handy tool for removing the lifters.

Remove the alternator mounting bracket and brace attaching bolts. Move the alternator out of the way.

Remove the exhaust manifolds; see Section 3-8.

Remove the cylinder head bolts, and then lift the cylinder heads free of the block.

INSTALLATION V6 AND V8 ENGINES

The following procedures pickup the work after the head, valves, and related

Location of the three different size head bolts on a V6 engine.

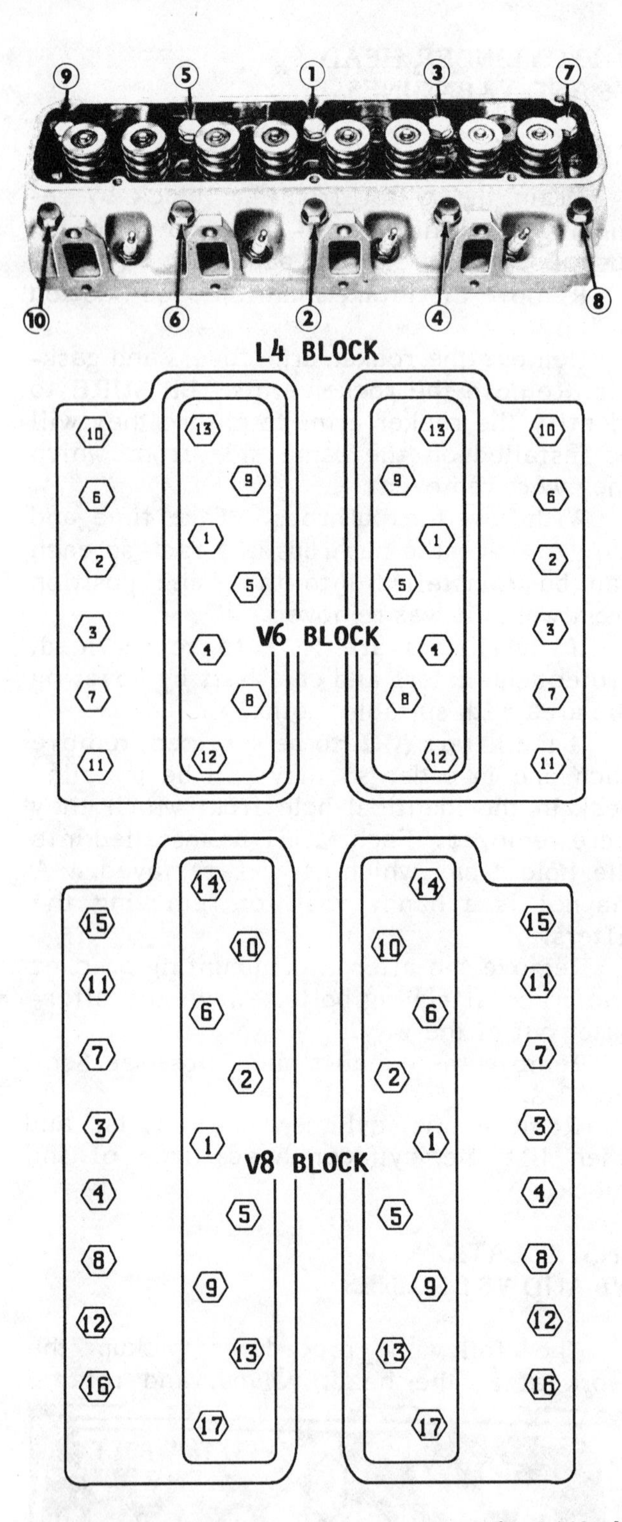

Cylinder head bolt tightening sequence for L4 blocks (top), V6 blocks (center), and V8 blocks (lower). Torque values are listed in the Appendix.

parts have been reconditioned and assembled according to the instructions outlined in Section 3-14.

Clean the engine block gasket surfaces. Check carefully to be sure no foreign material has fallen into the cylinder bores, bolt holes, or valve lifter area. Clean out all bolt holes with compressed air. One further step for a good job, is to run the correct size tap down each bolt hole.

Install **NEW** head gaskets on the cylinder block. **ALWAYS** use marine head gaskets to minimize salt water corrosion. Dowels in the block will hold the gaskets in position. **USE CARE** when handling the gaskets to prevent kinking or damaging the surfaces. **DO NOT** use any sealing compound on head gaskets, because they are coated with a special lacquer to provide a good seal, once the parts have warmed to operating temperature.

Clean the gasket surface of the cylinder heads and carefully set each in place on the engine block, with the holes in the heads indexed over the dowel pins in the block. Clean and coat the head bolts with sealer. Install the bolts, as shown. Tighten the head bolts in the sequence shown, making three rounds to a final torque value of 68 ft-lb (92.2 Nm). **BEAR IN MIND,** uneven tightening of the head bolts can distort the cylinder bores, causing compression loss and excessive oil consumption. On the first round tighten the bolts to 1/2 the total torque value; on the second round to 3/4 the total torque value; and to the final torque value on the third and final round.

Install the exhaust manifold to each cylinder head, see Section 3-8. Install the exhaust hoses to the exhaust housing.

Install the intake manifold according to the procedures outlined in Section 3-7.

Install the water drain plugs.

3-14 HEAD RECONDITIONING

HEAD DISASSEMBLING

Use a tool to compress the valve springs to remove the keys. Release the compres-

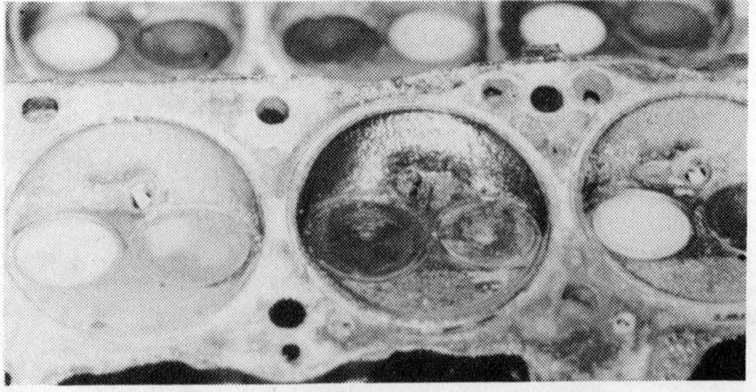

Notice the oil around the intake valve and the darker coloring of the exhaust valve in the right combustion chamber. The compression in the right cylinder was lower than in the left. Also, the intake valve guide and seal are defective allowing oil to leak into the combustion chamber.

sor, and then take off the spring caps, spring shields, springs, spring dampers, oil seals, and valve spring shims. Pay attention to the number of shims under each of the valve springs. This will be very important during assembly. Remove the valves from the cylinder head and place them in a rack in their proper sequence so they can be installed back into their original positions.

CLEANING AND INSPECTING

Clean all carbon from the combustion chambers and valve ports. Use lacquer thinner to cut the gum on the valve guides. It is this gum that causes sticky valves. Clean all carbon and sludge from the push rods, rocker arms, and pushrod guides. Clean all carbon deposits from the head gasket mating surface. **TAKE CARE** not to damage the gasket surface.

Check the flatness of the gasket surface with a straightedge and a feeler gauge. Surface irregularity must not exceed 0.003" in any six-inch space, and the total must not exceed 0.007" for the entire length of the head. If necessary, the cylinder head gasket surface can be machined.

CAUTION: Do not remove more than 0.010" of stock.

Clean the inside of the valve guides with a wire brush and lacquer thinner to remove all gum and carbon deposits. Any gum or deposits could prevent the valve from closing properly.

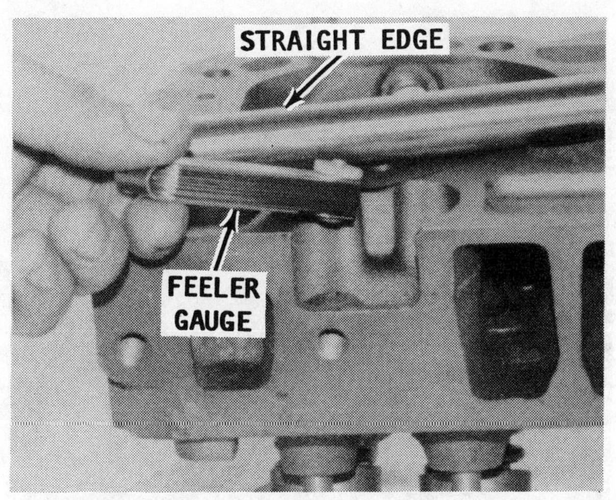

Checking the flatness of the head gasket surface with a straight edge and feeler gauge. Surface irregularities must not exceed 0.003" (0.08 mm) in any 6-inch space.

Inspect the cylinder head for cracks in the exhaust ports, combustion chambers, or external cracks to the water chamber. Inspect the valves for burned heads, cracked faces, or damaged stems. Bear-in-mind, excessive valve stem-to-bore clearance will increase oil consumption and may cause valve breakage. Insufficient clearance will

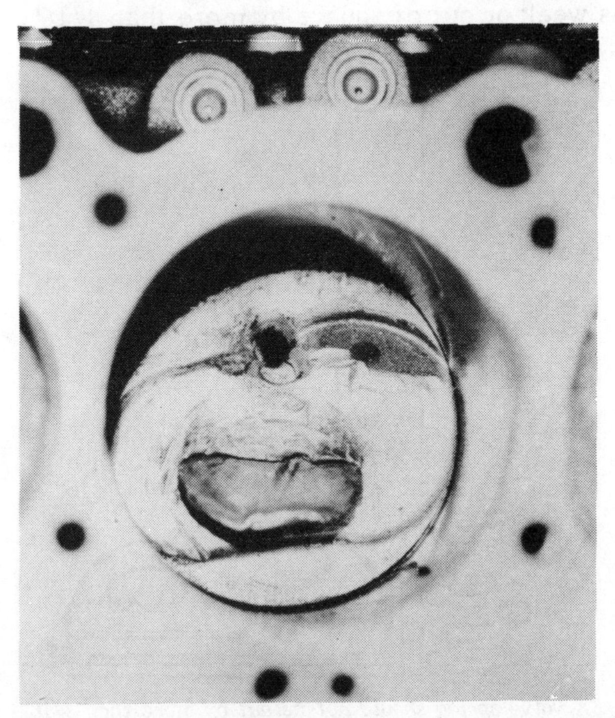

Improper installation techniques caused this valve to drop out and strike the piston.

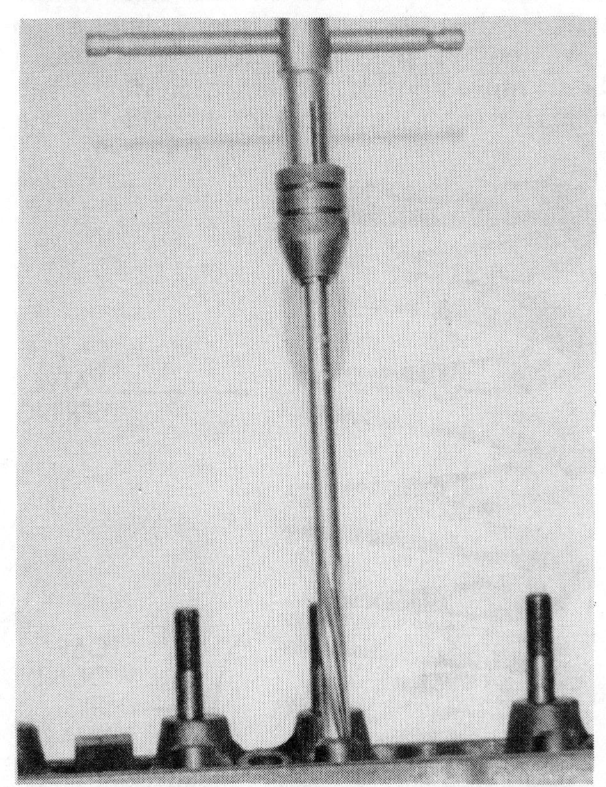

To restore production clearances, worn valve guides can be reamed oversize, and then valves with oversize stems installed.

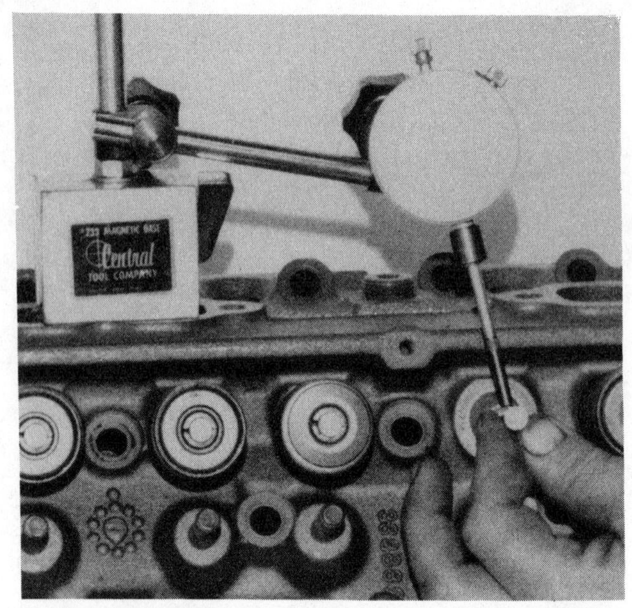

The wear limit for intake valve guides is 0.001" (0.03 mm) and for exhaust guides, 0.002" (0.05 mm). Valve guide wear can be accurately measured by checking the deflection of a new valve stem with a dial indicator.

result in noisy and sticky valves and also disturb engine smoothness.

VALVE GUIDES

Measure the valve stem clearance with a dial indicator clamped on one side of the cylinder head. Locate the indicator so movement of the valve stem will cause a direct movement of the indicator stem. The indicator stem must contact the side of the valve stem just above the valve guide. With a new valve and the head dropped about 1/16" off the valve seat, move the stem of the valve from side-to-side, with a light pressure to obtain a clearance reading. If the clearance exceeds 0.001" for the intake or 0.002" for the exhaust, it will be necessary to ream the valve guides for oversize valve stems.

Valves with oversize stems are available for inlet and exhaust valves in the following sizes: 0.003", 0.015", and 0.030". Use the proper size reamer to obtain satisfactory clearance between the valve stem and the valve guide.

No matter what type of equipment is used, it is essential that the valve guides are free from carbon, gum, or dirt. The valve **MUST** be clean in order for the pilot to center properly in the guide.

VALVE SPRINGS

Check the valve springs for the correct tension against the Specifications in the Appendix. A quick check can be made by laying all of the springs on a flat surface and comparing the heights, which **must** be even. Also, the ends **must be square** or the spring will tend to cock the valve stem. Weak valve springs cause poor engine performance; therefore replace any spring that is weak or out of square by more than 1/16".

VALVE SPRING

CLOSE WOUND COILS TOWARD HEAD

Enlargment of a valve coil spring. Notice the close wound coils at the bottom (toward the head).

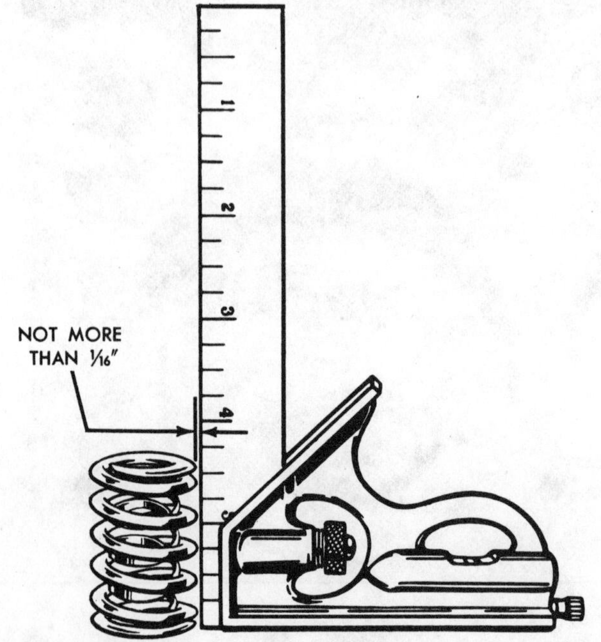

A valve spring should not be out by more than 1/16" (1.59 mm) when it is rotated against a square on a flat surface, as shown. Replace defective springs.

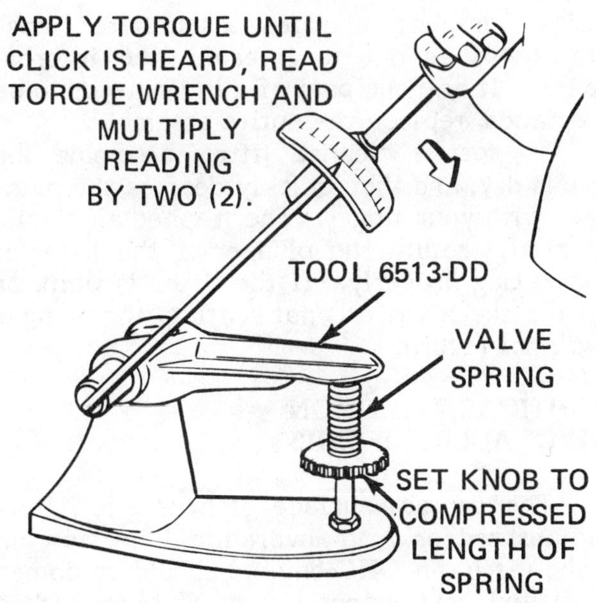

The valve spring tension should be checked with an accurate tester and a torque wrench.

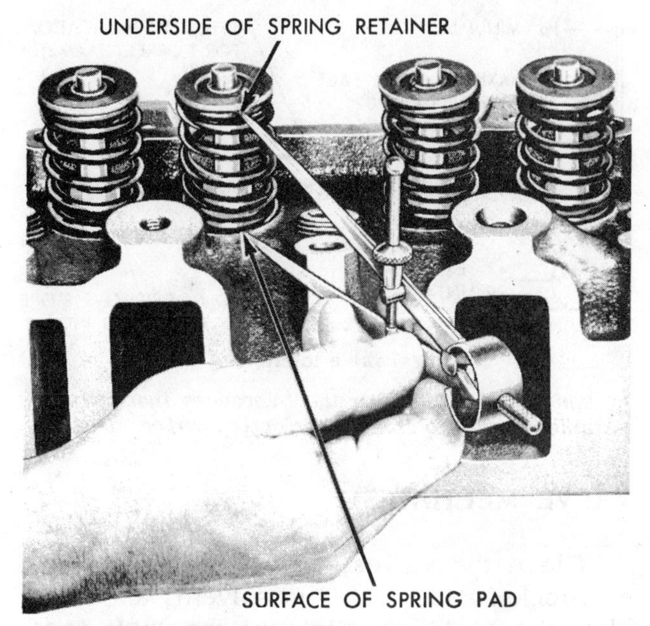

Measuring the height of the installed valve springs. Compare this measurement with those given in the Specifications in the Appendix.

VALVES

Grind the 45° valve face to 44° angle for a 1° interference angle. Remove only enough stock to correct run-out or to dress off the pits and grooves. If the edge of the valve head is less than 1/32" after grinding, replace the valve as it will run too hot in the engine. Lightly lap the valves into their seats with fine grinding compound.

Measure the valve seat widths, which should be as specified in the Appendix. The seats can be narrowed by removing stock from the top and bottom edges by using a 30° stone and a 60° stone.

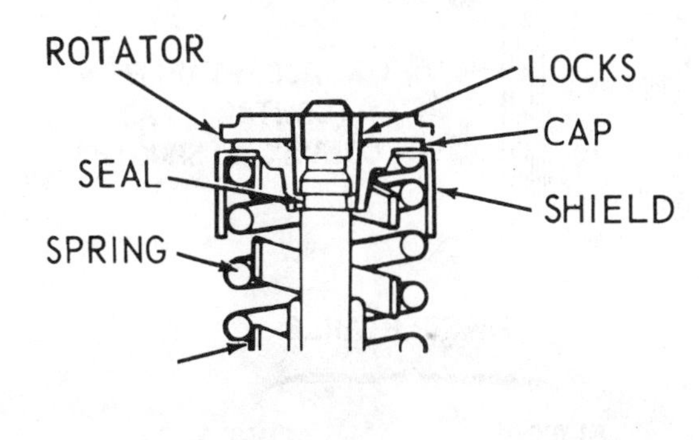

Typical arrangement of the valve mechanism parts. In addition to the parts shown, some late model units have a valve rotator, and some models do not have a damper spring.

The finished seat should contact the approximate center of the valve face. To determine the position of the seat on the valve face, coat the seat with Prussian blue, and then rotate the valve in place with a light pressure. The blue pattern on the valve face will show the position of the seat.

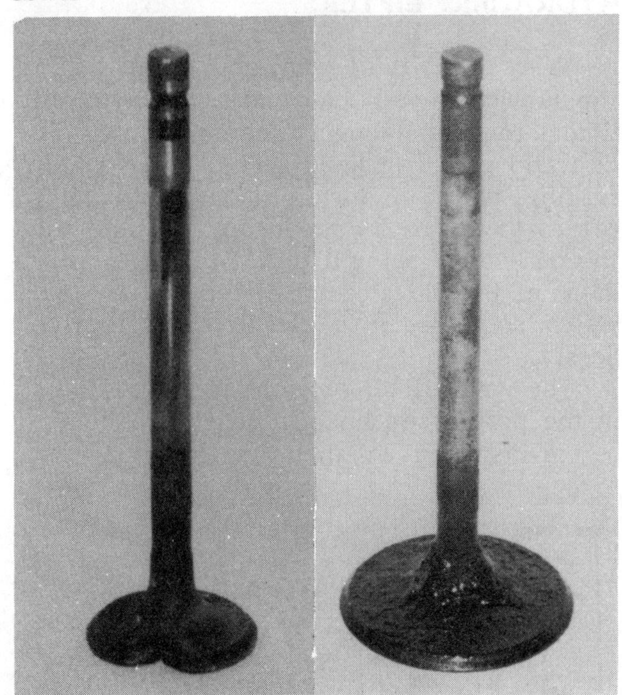

A severely burned exhuast valve (left). The valve was sticking in the guide as evidenced by the gum on the neck of the stem. The wet oily condition of the valve on the right shows excessive lubricant was leaking past the seal and valve guide. Take time to clean the valve guides thoroughly.

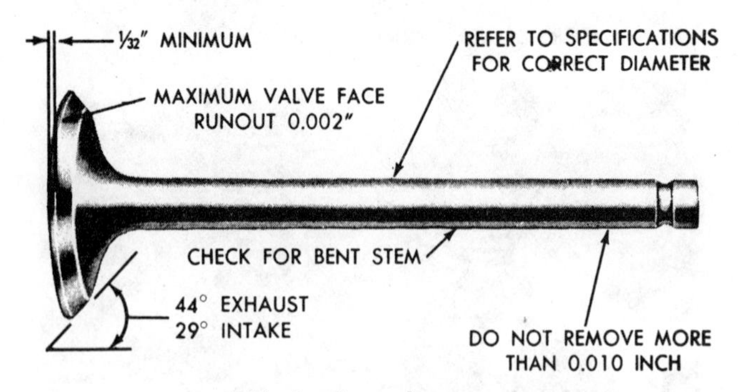

Critical valve tolerances.

Typical valve showing the tolerances that are acceptable in order to obtain satisfactory performance.

VALVE MECHANISM

Clean the valves, springs, spring retainers, locks, and sleeves in solvent, and then blow the parts dry. Inspect the valve face and the head for pits, grooves, and scores. Inspect the stem for wear and the end for grooves. The face must be trued on a valve grinding machine, which will remove minor pits and grooves. Valves with serious defects, or those having heads with a knife edge, must be replaced.

HYDRAULIC LIFTERS

Dirt, deposits of gum, and air bubbles in the lubricating oil can cause the hydraulic lifters to wear enough to cause failure. The dirt and gum can keep a check valve from seating, which in turn will cause the oil to return to the reservoir during the time that the pushrod is being lifted. Excessive movement of the parts of the lifter causes wear, which soon destroys the lifters effectiveness.

The valve lifter assemblies must be kept in the proper sequence in order for them to be re-installed in their original position.

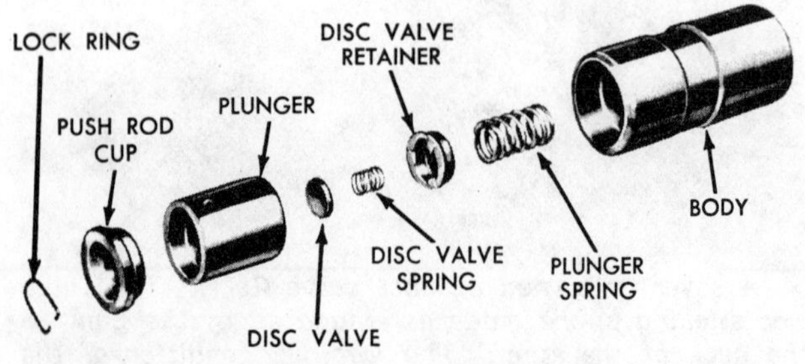

*Typical hydraulic valve lifter parts arranged in order of assembly. The lifter **MUST** be free of dirt and gum for maximum performance.*

Clean, inspect, and test each lifter separately so as not to intermix the internal parts. If any one part of a lifter needs to be replaced, replace the entire assembly.

To test a cleaned lifter, assemble the parts dry, and then quickly depress the plunger with your finger. The trapped air should partially return the plunger if the lifter is operating properly. If the lifter is worn, or if the check valve is not seating, the plunger will not return.

CRITICAL WORDS ON HYDRAULIC LIFTERS

The contact surface of hydrualic lifter-to-pushrod or from hydraulic lifter-to-cam lobe (as in an OHC engine) should be domed (convex) and smooth. A flat or dished (concave) surface indicates excessive wear and the lifter should be replaced.

Hydraulic lifters should be primed (pumped up) and lightly coated with engine oil **PRIOR** to installation A hydraulic lifter which is not adequately primed prior to installation may never pump up once the engine is operating. Such a condition could lead to internal scoring and incorrect operation, commonly termed "sticky lifters".

VALVE SEATS

Reconditioning the valve seats is very important, because the seating of the valves must be perfect for the engine to deliver

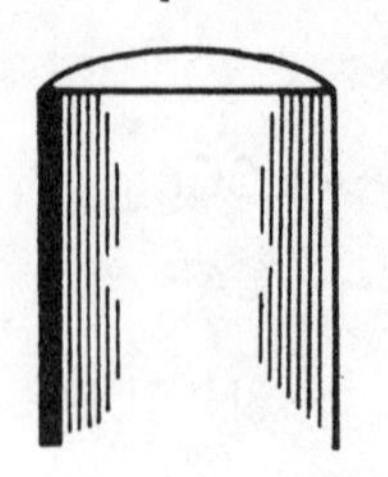

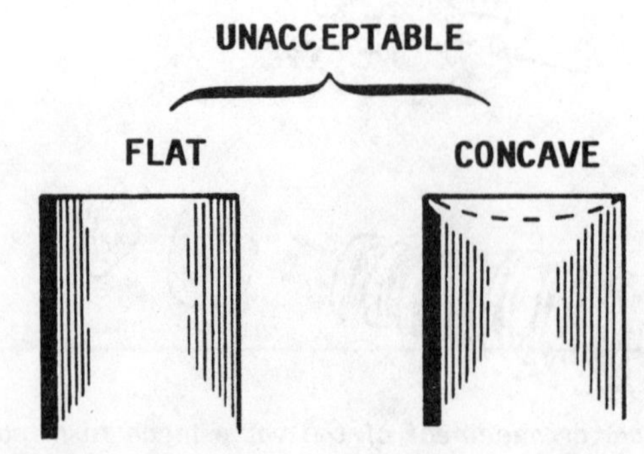

The top surface of the lifter should be inspected for a wear pattern and replaced if unacceptable.

Refacing the valve seat using a cutting tool. ***ALWAYS*** *hold the tool firmly and rotate it clockwise. Take care to hold the tool at a right angle to the valve pocket, as shown.*

Clean all carbon from the cylinder head with a wire brush in a drill. The carbon ***MUST*** *be removed before the valve seat can be ground.*

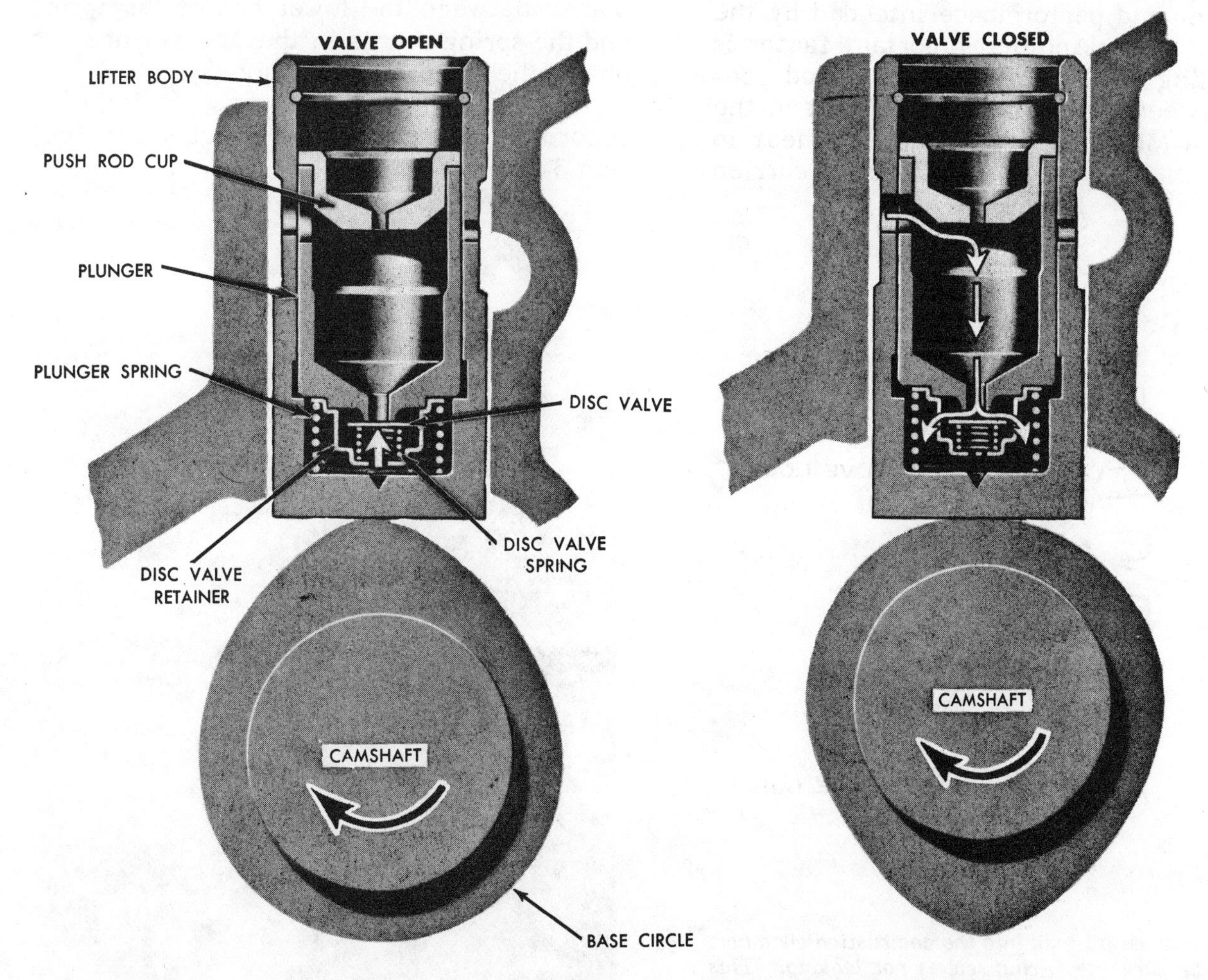

Relationship of the cam lobe and the valve lifter in the valve open position (left) and in the the valve closed position (right).

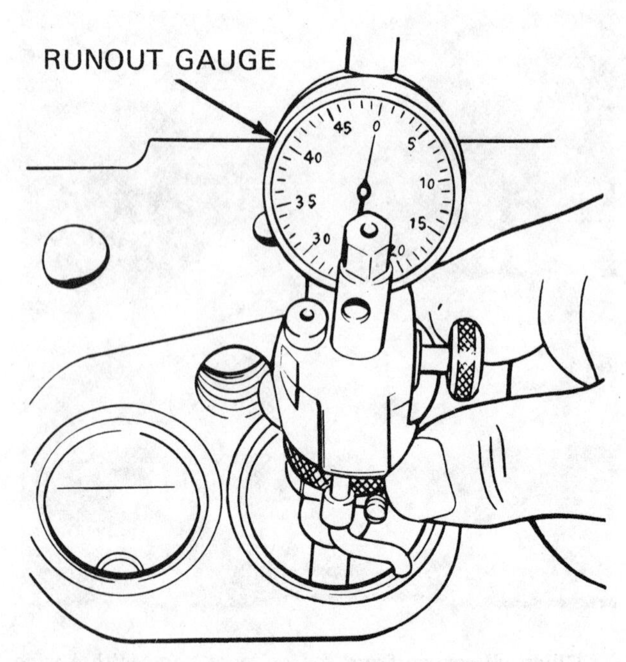

After the seat has been reground, check to be sure it is concentric with the guide using a dial indicator. The runout should not exceed 0.002" (0.05 mm).

the power and performance intended by the manufacturer. Another important factor is the cooling of the valve heads. Good contact between each valve and its seat in the head is a **MUST** to ensure that the heat in the valve head will be properly carried away.

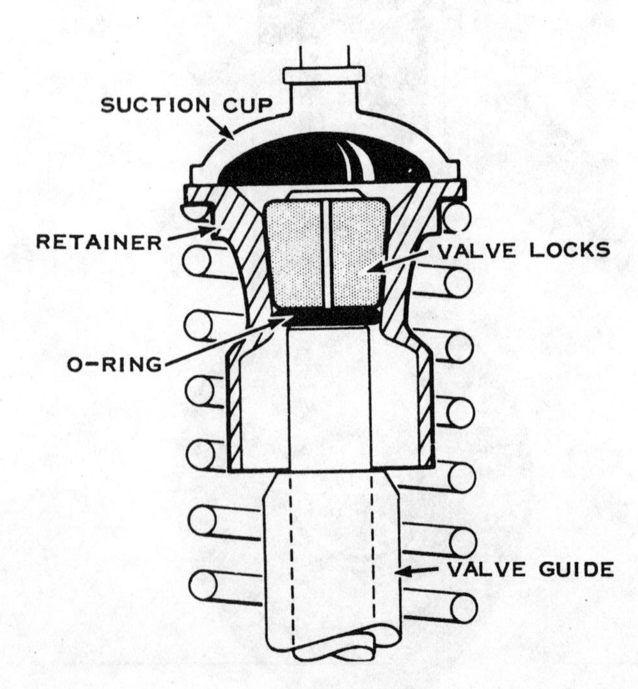

To prevent an oil leak into the combustion chamber, check to be sure the O-ring seal is not leaking. This can be done by applying vacuum to the assembly with a suction cup.

CYLINDER HEAD ASSEMBLING

Start with No. 1 cylinder. Lubricate the exhaust valve with a generous amount of oil. Place the well-lubricated valve in the port and the valve spring and cap in position at the same time. The spring end with the closed coil **must** be against the cylinder head. Place the spring and rotator on the exhaust valves, and then compress the spring and install the oil seal and valve keys. Pay special attention that the seal is **flat** and not twisted in the valve stem groove and that the keys seat properly.

Assemble the remaining valves, valve springs, rotators, spring caps, oil seals, and valve keys in the cylinder head in the same manner as the first. Check the seals by placing a vacuum cup over the valve stem and cap; squeeze the vacuum cup to make sure there is no leak past the oil seals.

Measure and compare the valve spring compressed height with the specifications in the Appendix. If necessary, shims can be placed between the lower end of the spring and the spring recess in the cylinder head to obtain the required dimension.

The head is now ready for installation according to the procedures oulined in Section 3-13.

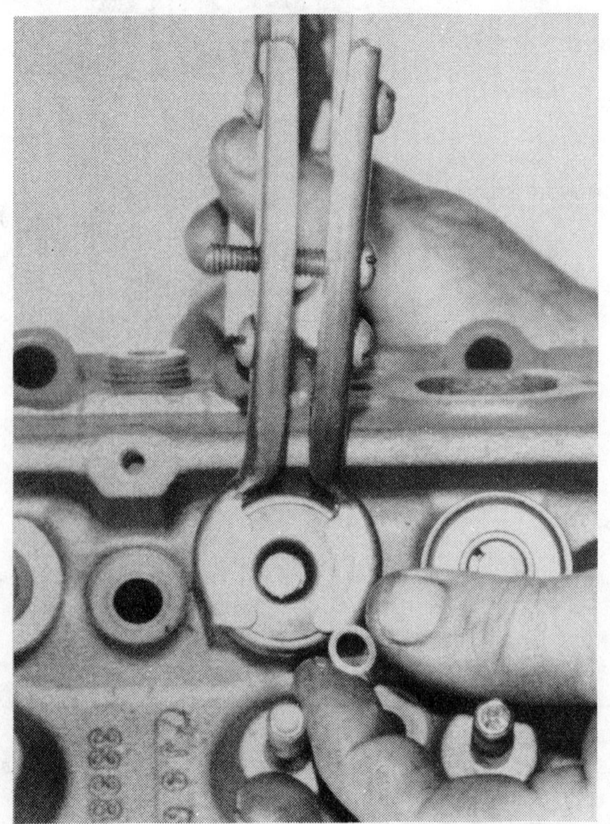

Installing the valves into the cylinder head.

3-15 CAMSHAFT SERVICE IN-LINE PUSHROD ENGINE

If the engine is to perform properly, the valves must open and close at a predetermined precise moment for maximum efficiency. On the power stroke, the exhaust valve must open just before bottom dead center in order to permit the exhaust gases to leave the combustion chamber under the remaining pressure (blowdown). On the exhaust stroke, the intake valve must open just before top dead center in order to permit the air-fuel mixture to enter the combustion chamber. The movement of the valves are functions of the camshaft design and the valve timing. Therefore, excessive wear of any camshaft part will affect engine performance.

CAMSHAFT LIFT MEASUREMENT

If improper valve operation is indicated, measure the lift of each pushrod in consecutive order and record the readings. Remove the valve mechanism. Position a dial indicator with a ball socket adapter on the pushrod. Rotate the crankshaft slowly in the operating direction until the lifter is on the heel of the cam lobe. At this point, the pushrod will be in its lowest position.

Set the dial indicator on zero, then rotate the crankshaft slowly, or attach an auxiliary starter switch and "bump" the engine over, until the pushrod is in the fully raised position.

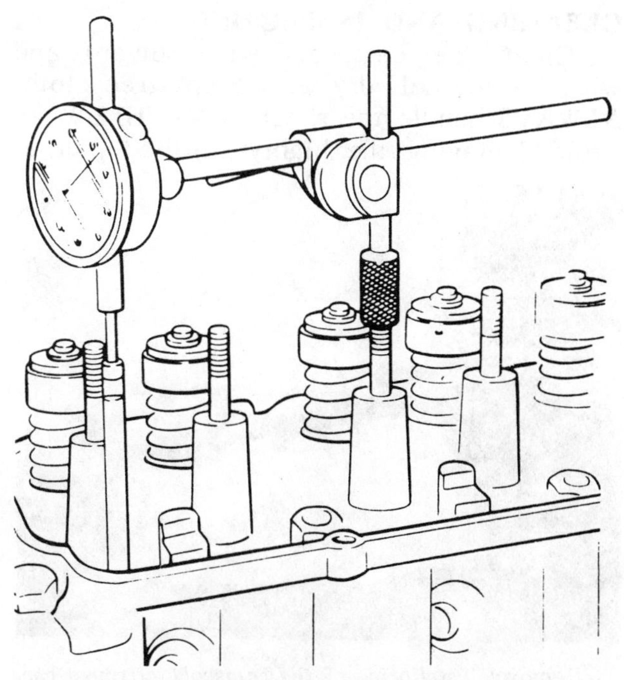

Measure the lift of each pushrod in consecutive order and record the readings.

The distributor primary lead **MUST** be disconnected from the negative post on the coil and the ignition switch **MUST** be in the **ON** position, or the grounding circuit in the ignition switch will be **DAMAGED.**

Compare the total lift recorded from the dial indicator with the specifications listed in the Appendix.

Continue to rotate the camshaft until the indicator reads zero. This point will be a check on the accuracy of the original indicator reading.

If the camshaft readings for all lobes are within specifications, remove the dial indicator assembly.

CAMSHAFT REMOVAL

The engine does not have to be removed from the boat if the camshaft merely needs servicing. However, if the camshaft bearings require replacement, then considerable work must be done including removal of the engine. Removal and replacement of the camshaft bearings should be performed by qualified mechanics in a shop equipped to handle such work.

To remove the camshaft, begin by removing the valve cover and gasket. Next, loosen the valve rocker arm nuts until the pivot rocker arms clear the pushrods. Now, note the position of the distributor rotor, and then remove the distributor. Next,

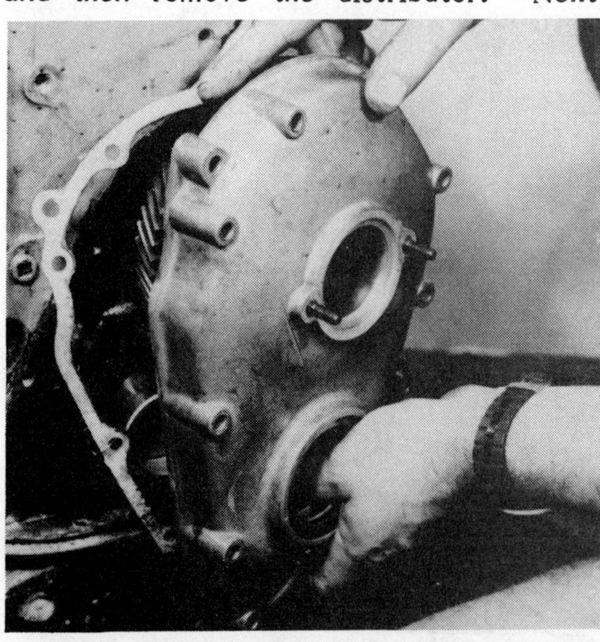

Removal/installation of the front timing gear cover.

Alignment of the timing mark on the crankshaft gear with the mark on the camshaft gear.

remove the ignition coil and side cover gasket.

TAKE TIME to set up a system to keep the pushrods and valve lifters in order to **ENSURE** each will be installed back into the exact location from which it was removed. Pull out the pushrods and valve lifters in order.

Remove the nut from the forward end of the crankshaft. Slide the crankshaft pulley free of the crankshaft.

Remove the attaching bolts and washer from the perimeter of the timing gear cover. Remove the cover. **DISCARD** the gasket.

Removal/installation of the camshaft gear.

Removal/installation of the camshaft retaining plate, after the gear has been removed.

Rotate the crankshaft until the No. 1 cylinder is at top dead center (TDC). The timing mark on the camshaft should be aligned with the timing mark on the crankshaft gear.

Remove the nut securing the camshaft gear to the end of the cam shaft. Slide the gear free of the camshaft.

Remove the two bolts securing the camshaft retaining plate. Remove the plate from the camshaft.

Withdraw the camshaft.

CLEANING AND INSPECTING

Clean the camshaft with solvent and wipe the journals dry with a lint-free cloth. **ALWAYS** handle the shaft **CAREFULLY** to avoid damaging the highly finished journal

*Removal/installation of the camshaft. Utmost care **MUST** be used not to damage the highly polished bearing surfaces or cam lobes.*

surfaces. Blow out all of the oil passages with compressed air.

Clean the gasket surfaces on the block and crankcase front cover. Cut the tabs from the new oil pan front seal. Use a sharp knife to ensure a clean cut.

Check the diameter of the three camshaft bearings with a micrometer for out-of-round condition. If the journals are out-of-round more than 0.001", the camshaft should be replaced. Compare your findings with the Specifications in the Appendix.

Check the camshaft for alignment. This is best done using "V" blocks and a dial indicator. The dial indicator will indicate the exact amount the camshaft is out of true. If it is out more than 0.002" dial indicator reading, the camshaft should be replaced. When using the dial indicator in this manner the high reading indicates the high point of the shaft. Examine the camshaft bearings and if any one bearing needs to be replaced, **ALL THREE** should be replaced.

CAMSHAFT INSTALLATION

Coat the cam lobes with Molykote lubricant, or equivalent to **ENSURE** proper lubrication during initial startup.

Install the camshaft assembly in the engine block. Take care to move the shaft straight into the block and not to damage the bearings or cams.

Measuring the camshaft bearing journals with a micrometer. The out-of-round should not exceed 0.001" (0.025mm).

Using a dial indicator to check the camshaft runout. Runout for the camshaft should not exceed 0.004" (0.102 mm). Runout for the crankshaft should not exceed 0.003" (0.076 mm).

Install the camshaft retaining plate and secure it in place with two attaching bolts. Tighten the bolts to a torque value 80 in-lbs (9 Nm).

With the No. 1 piston at TDC on the compression stroke, slide the timing gear onto the camshaft with the timing marks on the crankshaft gear and the camshaft gear aligned, as shown.

Secure the camshaft gear with the nut on the end of the camshaft. Tighten the nut to a torque value of 100 ft-lbs (136 Nm).

Check the camshaft and crankshaft gear runout with a dial indicator. The camshaft

Using a feeler gauge to check the backlash between the camshaft gear and the crankshaft gear.

gear runout should not exceed 0.004" and the crankshaft gear runout should not exceed 0.003".

If the gear runout is more than the 0.003" limit, the gear will have to be removed and any burrs cleaned from the shaft. If this does not reduce the runout to 0.003", then either the shaft or the gear must be replaced.

Check the backlash between the timing gear teeth with a narrow feeler gauge or dial indicator. The backlash should not be less than 0.004" nor more than 0.006".

Install the seal on the front cover, pressing the tips into the holes provided in the cover. Coat the gasket with sealer and place it in position on the cover.

Install the front cover-to-block attaching screws. Tighten all of the cover attaching screws to a torque value of 80 in-lbs (9 Nm). Install the crankshaft pulley.

Coat the oil seal contact area on the hub with engine oil. Position the hub over the crankshaft and key, then start the hub into position with a mallet. Drive the hub onto the crankshaft until the hub bottoms against the crankshaft gear. The crankshaft extends slightly through the hub, therefore a hollow tool **MUST** be used to drive the hub completely to the bottom position. Secure the hub with the center bolt. Tighten the center bolt to a torque value of 55 ft-lbs (75 Nm). Install the pulley onto the hub.

Timing notch on the camshaft pulley aligned with the mark on the valve cover. No. 1 piston is at TDC.

Install the valve lifters and pushrods. Install the rocker arms onto the pushrods and tighten the rocker arm bolts **ONLY** enough to hold the pushrod until the valves are adjusted. Replace the side cover using a new gasket. Install the ignition coil. Install the distributor rotor and cap.

Adjust the valve lash according to the procedures outlined in Section 3-19.

CAMSHAFT BEARING SERVICE

Removal and replacement of the camshaft bearings should be performed by qualified mechanics in a shop equipped to handle such work. However, in most cases the block bores for the bearings can be bored to a larger size and oversize bearings installed. **BE SURE** to check with your local marine dealer for oversize bearings available.

3-16 CAMSHAFT SERVICE SINGLE OHC ENGINE

The following procedures can be accomplished while the engine remains in the boat and without removing the head.

Front view of a 4-cylinder OHC engine. Items requiring removal before the camshaft can be serviced are identified.

PREPARATION

A few tasks must be performed before the camshaft can be serviced. Several items of exterior equipment on the engine must be removed to gain access for removal of the camshaft drive pulley and the camshaft.

Begin, by obtaining a suitable container, and draining the fresh water/antifreeze solution from the block.

Next, remove the raw water pump. If any problems arise, see Chapter 7.

Next, loosen the bolts on the alternator tensioning bracket, and then remove the alternator drive belt.

Remove the timing belt cover. The upper part of the cover is fitted with a guide. This guide requires the cover to be moved outward just a bit before it is free and can be removed.

Now, remove the crankshaft outer drive pulley, by removing the six attaching bolts.

Remove the fresh water tank. Again, if difficulties arise, see Chapter 7.

The front of the engine is now ready for removal of the timing belt, camshaft pulley, and the camshaft.

Rotate the crankshaft until No. 1 cylinder is at TDC (top dead center), on the compression stroke. In this position the timing notch on the crankshaft pulley will be aligned with the mark on the inner guide plate; the timing notch on the camshaft pulley will be aligned with the notch on the valve cover; and the timing dot on the accessory drive pulley will be aligned with the notch on the accessory driveshaft backing plate, as shown in the accompanying illustrations on the previous page and on this page.

Remove the flame arrestor. Remove the valve cover, gasket, and the crescent shaped seal.

If the crankshaft and the accessory drive pulley are not rotated while the camshaft is being serviced, installation will be greatly simplified.

Remove the shroud from the camshaft pulley. Loosen the nut on the timing belt tensioner. Using a pair of channel-lock pliers, compress the spring. While the spring is compressed, slide the timing belt free of the camshaft drive pulley.

Notice how the camshaft bearing caps are numbered from the front of the engine to the rear. These numbers will help to ensure the same bearing caps will be installed back into the same position from which they were removed.

CRITICAL WORDS

The camshaft bearing caps **MUST** be loosened alternately and evenly to prevent

Timing notch on the crankshaft pulley aligned with the mark on the inner guide plate. No. 1 piston is at TDC on the compression stroke.

Timing dot on the accessory drive pulley aligned with the notch on the backing plate. The stripe on a new timing belt is also aligned.

flexing the camshaft, which could cause distortion.

SPECIAL WORDS

Two methods are available to remove the camshaft. One involves the use of a special Volvo tool, and the other method removes the camshaft if the special tool is not available.

Tool Available

Obtain Volvo Penta tool No. 9995021. Remove the **CENTER** camshaft bearing cap. Spring washers are installed under the nuts. If necessary, use a blunt chisel or similar tool against the lug and **CAREFULLY** break the cap loose.

Install the special Volvo tool on the camshaft, using the nuts for the bearing cap. This tool will hold the camshaft in place while the other bearing caps are removed. Remove the other bearing caps. Remove the seal from the end of the camshaft. Remove the special tool, and then lift the camshaft free of the cylinder head.

Tool Not Available

Loosen the nuts on the center bearing cap just a wee bit. Loosen the other nuts on the bearing caps just a bit working **ALTERNATELY** and **EVENLY**. Work in both directions from the center. Continue to loosen the nuts, working in several rounds. After the nuts have all been loosened a little, use a blunt chisel and **CAREFULLY** tap the cap loose. Continue to loosen the nuts evenly, and then remove the rod caps.

Installation of special Volvo tool when removing the camshaft. The text gives an alternate method.

Lift the camshaft free of the cylinder head. The shim material and retainers will remain on the valve stems. When the camshaft is removed, the front oil seal will come off with it. Discard the seal.

Remove the nut from the end of the camshaft and remove the camshaft drive pulley.

CLEANING AND INSPECTING

Clean the camshaft with solvent and wipe the journals dry with a lint-free cloth. **ALWAYS** handle the shaft **CAREFULLY** to avoid damaging the highly finished journal surfaces. Blow out all of the oil passages with compressed air.

Check the circumference of the five camshaft bearing journals with a micrometer for out-of-round condition. If the journals are out-of-round more than 0.001", the camshaft should be replaced.

Mount the bearing caps in the same position from which they were removed. Tighten the nuts to a torque value of 15 ft-lbs (20 Nm). Now, measure the inside diameter of the bearings. The difference between the circumference of the camshaft journals and the inside diameter of the bearings must not exceed 0.001" to 0.003" (0.030 mm to 0.071 mm).

If the measurements are not within the required tolerance, the cylinder head should be replaced.

Measuring the camshaft bearing journals with a micrometer. The out-of-round should not exceed 0.001" (0.025mm).

INSTALLATION

Coat the camshaft with engine oil. If the camshaft is new, coat the bearing journals with some type of cam lubricant to assist the camshaft to "wear in".

GOOD WORDS

As with the removal procedures, there are two methods available to install the camshaft. One involves the use of a special Volvo tool, and the other method installs the camshaft if the special tool is not available.

Tool Available

Obtain Volvo Penta tool No. 9995021. Place the camshaft in position on the cylinder head, with the pin for the camshaft pulley facing upward. The end that is drilled and tapped for the camshaft pulley is the forward end. Lay the rear bearing cap in position. Install the special tool on the **CENTER** bearing journal and depress the camshaft.

Lubricate the rubber lip of the camshaft seal and then **CAREFULLY** install it onto the camshaft. **TAKE CARE** not to damage the rubber lip while installing it. If the old sealing ring is used, rotate the ring to obtain a new wear surface against the camshaft.

Lay the other four bearing caps in position and start the nuts. Coat the forward bearing cap surface that will make contact with the cylinder head with sealing compound before it is installed.

Check to be sure the seals are properly installed before the bearing cap nuts are tightened.

Remove the special tool. Oil and install the last bearing cap. Tighten the nuts **ALTERNATELY** and **EVENLY** just a little at-a-time to prevent distorting the camshaft. Work from the center bearing in both directions. Continue tightening the nuts to a torque value of 15 ft-lbs (20 Nm).

Tool Not Available

Place the camshaft in position on the cylinder head with the pin for the camshaft pulley facing upward. The end that is drilled and tapped for the camshaft pulley is the forward end. Install the bearing caps into the same position from which they were removed. Tighten the nuts just snug. **CAREFULLY** slide a new oil seal onto the front end of the camshaft.

Tighten the nuts **ALTERNATELY** and **EVENLY** just a little at-a-time to prevent distorting the camshaft. Work from the center bearing in both directions. Continue tightening the nuts to a torque value of 15 ft-lbs (20 Nm).

After Bearing Cap Installation

Slide the camshaft pulley onto the end of the camshaft. Install the belt guide onto the end of the camshaft with the notch indexed over the roll pin in the camshaft. Install the bolt and washer into the end of the camshaft.

Attach special Volvo Tool No. 9995034 to the camshaft pulley. If the special tool is not available, insert a couple of dowels into the two holes in the pulley, and then use a large screwdriver across the two dowels to hold the pulley while the nut is tightened with a torque wrench. Tighten the nut to a torque value of 35 ft-lbs (50 Nm).

Temporarily set the valve cover in place. It is not necessary to secure it in place at this time. Rotate the camshaft **CLOCKWISE** until the notch in the camshaft pulley is aligned with the notch on the valve cover.

SPECIAL WORDS

Check to be sure the crankshaft was not rotated while the camshaft was removed from the cylinder head. With the No. 1 cylinder at TDC on the compression stroke, the notch on the crankshaft pulley should be aligned with the mark on the inner guide plate and the dot on the accessory driveshaft pulley should be aligned with the notch on the accessory driveshaft backing plate, as shown in the illustrations on Page 3-44 and 3-45.

Compress the spring of the timing belt tensioner with a pair of channel-lock pliers,

Using a special Volvo tool to remove or install the camshaft. The text gives an alternate method.

and then insert a punch or pin through the hole in the tensioner shaft to hold the tensioner relaxed. Loosen the nut and slide the timing belt up over the camshaft pulley. If the marks on the belt are still visible, the belt can be installed with the marks on the belt aligned with the timing marks on the pulleys. However, if the marks are not visible, the belt can still be installed and the engine will be properly timed, because of the marks on the camshaft pulley, the accessory drive pulley, and the crankshaft pulley.

Compress the spring again with the channel-lock pliers and remove the punch or pin through the hole. Rotate the crankshaft in a **CLOCKWISE** direction with a socket wrench on the crankshaft pulley bolt. **DO NOT** rotate the crankshaft in the wrong direction or the belt will hop over teeth and the adjustment will be lost. Continue rotating the crankshaft through a couple revolutions. This action will adjust the tension on the belt by removing all the slack. Now, tighten the nut on the belt tensioner.

Illustration to show routing of the timing belt and a punch through the hole in the tensioner shaft, as described in the text.

Check the valve lash, and adjust if necessary, according to the procedures in Section 3-19.

COMPLETION

After the valve lash has been checked and adjusted, install and connect the external equipment removed from the front of the engine in preparation to servicing the camshaft.

Install the fresh water tank, see Chapter 7.

Install the crankshaft outer drive pulley with the six attaching bolts. Tighten the bolts securely.

Install the timing belt cover. The cover must be moved inward and then into place because of the guide. Secure the cover with the attaching hardware.

Install the raw water pump, see Chapter 7.

Install the alternator belt and then tighten the tensioning bracket until the belt can be depressed approximately 3/16" (5 mm) with finger pressure, as shown in the accompanying illustration.

Proper tension for the alternator drive belt is about 3/16" (4.8 mm) deflection with finger pressure.

Close the drain valve for the block, and then fill the fresh water tank with a 50% water and 50% anti-freeze solution.

Start the engine and check the completed work.

CAUTION: Water must circulate through the lower unit to the engine any time the engine is run to prevent damage to the water pump mounted on the engine.

To tune the engine, see Chapter 2.

3-17 CAMSHAFT SERVICE DOUBLE OHC ENGINE

FIRST, THESE WORDS

On a DOHC engine, both camshafts are contained in a "pan" called the camshaft carrier. The camshafts may be serviced by removing the camshaft carrier or by leaving the camshaft carrier attached to the head. If the head is to be serviced, the camshaft carrier must be removed.

Good shop practice dictates the head be serviced, if the camshaft requires work. Therefore, the following procedures include removal of the camshaft carrier and performing camshaft tasks on the workbench.

CRITICAL WORDS

The port and starboard camshafts and pulleys **MUST** be installed back into the same position from which they were removed.

PRELIMINARY TASKS

Several preliminary tasks must be performed before access is gained to the camshafts. In most cases, the engine does not have to be removed from the boat.

Begin by obtaining a suitable container and draining the fresh water/antifreeze solution from the block.

Remove the heat exchanger. Remove the timing belt shroud.

Rotate the crankshaft until the two marks on the timing belt align with the two marks on the valve cover. If the decal marks on the valve cover are no longer visible, rotate the crankshaft until the notch on both camshaft pulleys is straight up. In this position, an imaginary vertical line would pass from the notch down through the center camshaft pulley nut.

Loosen the center bolt on the idler pulley and move the idler to provide some slack on the timing belt. The pulley is on an eccentric and movement for slack is possible without removing the pulley. Slip the timing belt free of each pulley.

Remove the spark plug cover, and then the valve cover.

Remove the two bolts securing the distributor in place. Gently pull straight back and remove the distributor from the head. Make an effort not to rotate the distributor shaft while the distributor is out of the head. Taking this precaution will assist during installation by having confidence the rotor is pointing to the No. 1 contact.

SPECIAL WORDS

No special tools are required to remove or install the camshafts on a DOHC engine.

If the only work to be performed is on the camshafts and the hydraulic lifters, the intake and exhaust manifolds need not be removed. If service work is to be performed on the head, then the manifolds must be removed. See Section 3-6 to remove the manifolds and Section 3-12 to remove the head.

CAMSHAFT SERVICE

MORE SPECIAL WORDS

When the camshaft carrier is removed in the next step, some of the hydraulic lifters will fall free onto the surface of the head. Make an effort to turn the carrier over rather quickly to prevent all of the lifters falling free. Take extra precautions to prevent the lifters from falling to the floor and being damaged. Any damaged lifters must be replaced.

If at all possible, keep track of the lifters in an effort to install them back into

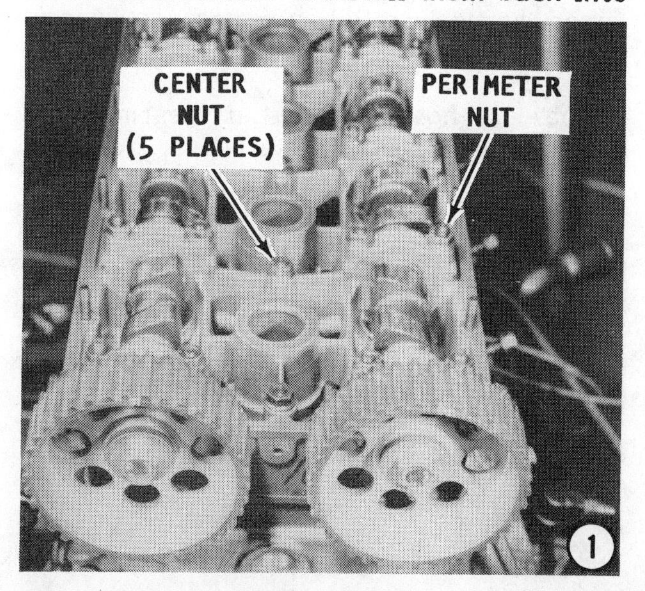

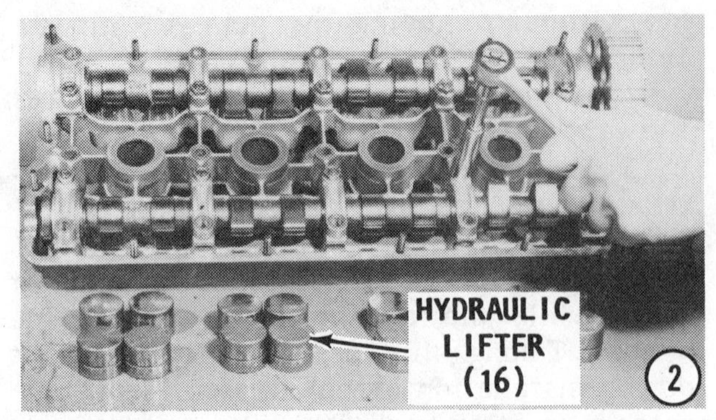

the same position from which they were removed. The wear pattern for each lifter will be slightly different from its neighbor.

1- Remove the outside perimeter nuts and the center row of nuts securing the camshaft carrier to the head. Lift the camshaft carrier from the head and turn it over rather quickly to prevent as many lifters from falling free as possible.

2- Place the camshaft carrier on a suitable work surface in its normal upright position. Now lift the carrier just a bit and allow the hydraulic lifters to slowly fall free from their bores, one by one.

If the lifters are to be installed for further use, place them in a shallow pan in the same order in which they were removed. Add clean engine oil to the pan until the oil level is above the small hole in the lifter to prevent complete bleed down.

CRITICAL WORDS

It is important to keep the lifters in order because if the lifters and the camshaft are fit for continued use, a wear pattern is established between each lifter and its matching cam lobe. If the lifters are mixed during installation, camshaft wear is greatly accelerated.

Observe how the camshaft bearing caps are numbered port and starboard, from the

forward end aft: 1 through 5 on the port camshaft and 6 through 0 (for 10) on the starboard camshaft. These numbers are to **ENSURE** each cap will be installed back into the same position from which it was removed. Bearing cap numbers 1, 5, and 6 can only be installed in one direction. The other caps could be installed backwards but **MUST** be installed to enable the numbers to be upright when viewed from the aft end of the camshaft carrier.

Remove the nuts securing the camshaft bearing caps to the carrier.

3- Lay out each bearing cap on the work surface in the order and direction from which the cap was removed.

4- Remove the "half moon" oil seal from the starboard side.

5- Lift out each camshaft with the pulleys still attached. If a camshaft/s or

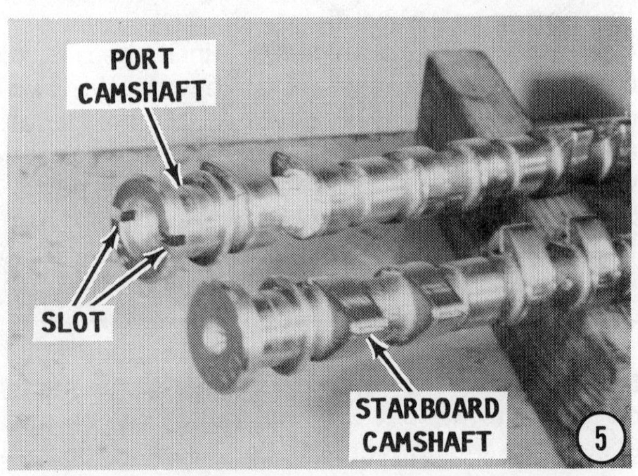

pulley/s or oil seals are to be replaced, the pulley must be removed from the camshaft. To prevent the pulley from rotating when the center nut is being loosened, obtain counterhold Volvo special tool No. 5034. If the special tool is not available, insert a wooden dowel into each of the two holes in the face of the pulley. Use a large screwdriver, or similar tool, across the two dowels while the nut is being loosened. Remove the center bolt and washer.

Push the pulley from the camshaft. Store the camshaft on two blocks of wood a third and two-thirds along their lengths to prevent distorting the camshaft and to prevent damage to the lobes and bearing surfaces.

SPECIAL WORDS

Observe the cutouts in the end of the port side camshaft. These cutouts are to accommodate the tangs on the distributor shaft. The cutouts also positively identify the camshaft as the port side camshaft.

CLEANING AND INSPECTING

Clean the camshafts with solvent and wipe the journals dry with a lint-free cloth. **ALWAYS** handle the shafts **CAREFULLY** to avoid damaging the highly finished journal surfaces. Blow out all of the oil passages with compressed air.

Measuring the camshaft bearing journals with a micrometer. The out-of-round should not exceed 0.001" (0.025mm).

Check the circumference of the five camshaft bearing journals with a micrometer for out-of-round condition. If the journals are out-of-round more than 0.001", the camshaft should be replaced.

Mount the bearing caps in the same position from which they were removed. Tighten the nuts to a torque value of 15 ft-lbs (20 Nm). Now, measure the inside diameter of the bearings. The difference between the circumference of the camshaft journals and the inside diameter of the bearings must not exceed 0.001" to 0.003" (0.030 mm to 0.071mm).

If the measurements are not within the required tolerance, the cylinder head should be replaced.

Good shop practice dictates the hydraulic lifters be replaced during an engine overhaul. It is **ESSENTIAL** to replace the complete set of lifters if the camshaft is replaced. Installing used lifters with a new camshaft will greatly accelerate lobe wear.

CAMSHAFT INSTALLATION DOHC ENGINE

1- If the camshaft oil seals were removed, install a new seal onto the forward end of each camshaft with the lip of the seal facing aft (the length of the camshaft) and the flat side of the seal facing forward. In this position, the seal will prevent oil in the carrier from escaping. Fit each pulley to the end of the camshaft. To prevent the pulley from rotating when the center nut is being tightened, obtain Volvo special tool No. 5034. If the special tool is not available, obtain two pieces of dowel. Insert the dowels into the two holes in the face of the pulley and use a large screwdriver, or similar tool, across the two dowels while the nut is being tightened. Tighten the bolt to a torque value of 36 ft lbs (50Nm).

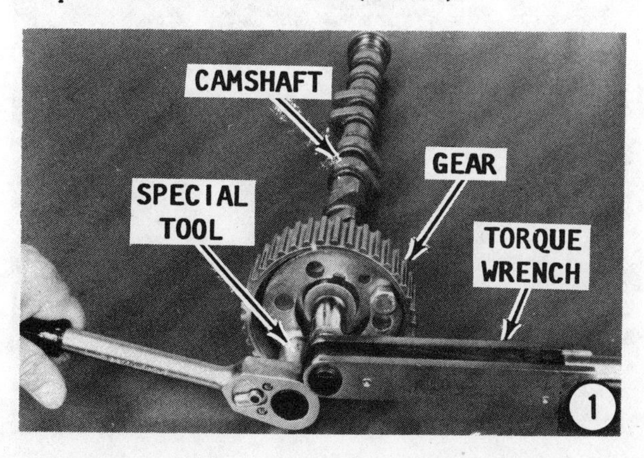

Repeat the procedure for the second camshaft.

2- Apply a narrow bead of Loctite 573, or equivalent to the upper gasket sealing surface of the head. Lower the camshaft carrier over the studs of the head. The carrier can only be installed one way with the semi-circular seal aft on the starboard side. Check to be sure the seal is in place. Thread the five acorn nuts onto the center line of studs through the carrier. Tighten these five nuts securely.

CRITICAL WORDS ON HYDRAULIC LIFTERS

The contact surface of each hydraulic lifter to cam lobe should be domed (convex) and smooth. A flat or dished (concave) surface indicates excessive wear and the lifter should be replaced.

Hydraulic lifters should be primed (pumped up) and lightly coated with engine oil **PRIOR** to installation A hydraulic lifter which is not adequately primed prior to installation may never pump up once the engine is operating. Such a condition could lead to internal scoring and incorrect operation, commonly termed "sticky lifters".

Slide each primed and lubricated hydraulic lifter into the same bore from which it was removed.

WORDS OF WISDOM

Read the remainder of this step carefully and understand what is to be accomplished and how before attempting to install the camshafts. Each camshaft **MUST** be placed, not rotated, into the correct position because the valves could strike the pistons if the camshaft is rotated.

Check to be sure the crankshaft has not been disturbed from the TDC position.

Place the camshaft with the slots in the aft end into the port side bearing races. When the camshaft is placed into the bearings the two most forward lobes **MUST** face toward the starboard camshaft at an oblique angle. Place the starboard side camshaft into the bearing races with the two most forward lobes facing the port side camshaft at an oblique angle. In this position, the timing notches in both pulleys will be pointing straight up, or very close to it. Actually, in most cases, the port side pulley will be slightly off, but when the bearing caps are tightened, the notch will come around to the straight up position.

3- Apply a small amount of sealing compound to the base of bearing caps No. 1,

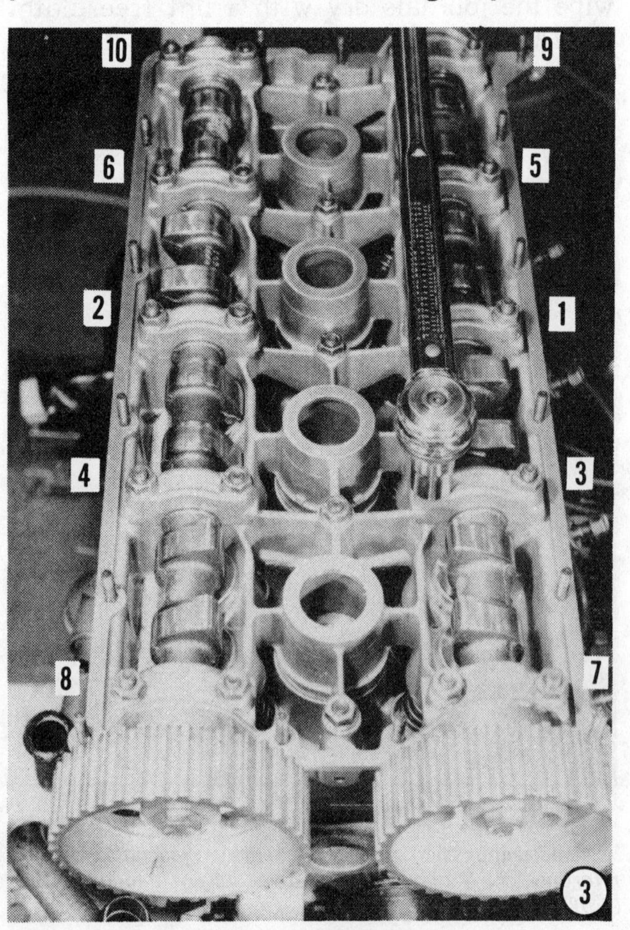

No. 5, and No. 6. Keep the sealer off the actual bearing surfaces. Install bearing caps No. 1 through No. 5 in order from forward working aft on the port side bearings. Install bearing caps No. 6 through No. 0 (10) onto the starboard bearings.

Thread the acorn nuts onto the bearing studs and tighten them **ONLY** "hand tight".

Now, tighten the bearing cap acorn nuts following the sequence indicated in the accompanying illustration to a torque value of 14.5 ft lbs (20Nm).

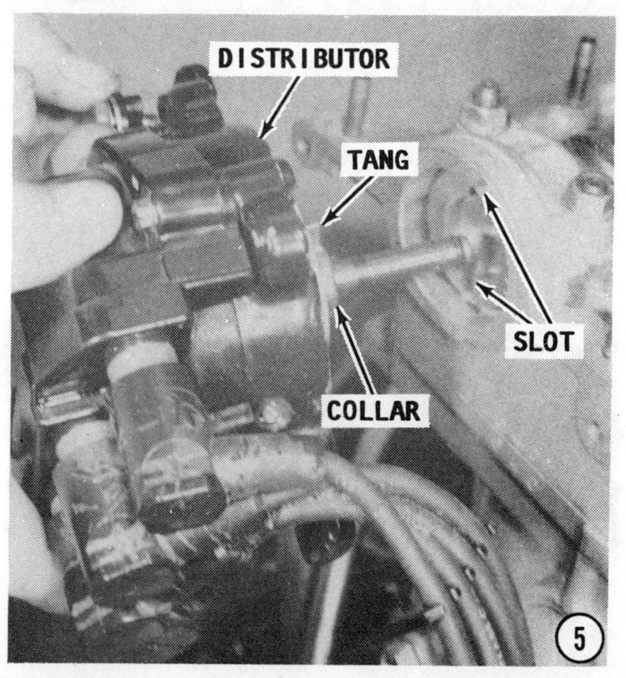

4- Place two **NEW** gaskets in position on the camshaft carrier, with the arrow on the center gasket facing **FORWARD. DO NOT** apply any sealing agent to either side of either gasket. The perimeter gasket can only be installed one way. Connect the hoses to the fuel enrichment vacuum valve and the wire connector to the temperature sending unit on the port side of the cylinder head. Install the valve cover and secure it in place with the acorn nuts. No torque value is given for these acorn nuts.

5- Apply a light coating of lubricant to the distributor shaft. Check the condition of the two O-rings used to seal the distributor to the cylinder head. If the seals appear damaged, they should be replaced. Install the distributor with the two tangs on the distributor shaft indexing with the two grooves in the end of the port side camshaft. The slots in the camshaft are offset. Therefore, it is impossible to install the distributor out of phase. Install the two bolts to secure the distributor in place.

Install the distributor cap and then tighten the two attaching bolts.

Install the timing belt, see Section 3-25.

Install the intake and exhaust manifolds, also the heat exchanger, see Section 3-6.

3-18 CAMSHAFT SERVICE V6 AND V8 ENGINE

If the engine is to perform properly, the valves must open and close at a predetermined precise moment. On the power stroke, the exhaust valve must open just before bottom dead center in order to permit the exhaust gases to leave the combustion chamber under the remaining pressure (blowdown). On the exhaust stroke, the intake valve must open just before top dead center in order to permit the air-fuel mixture to enter the combustion chamber. The movement of the valves are functions of the camshaft design and the valve timing. Therefore, excessive wear of any camshaft part will affect engine performance.

REMOVAL

Drain the water from the block by removing the drain plugs -- port and starboard just above the crankcase pan line.

Remove the intake manifold, see Section 3-7.

Remove the timing sprocket, see Section 3-26.

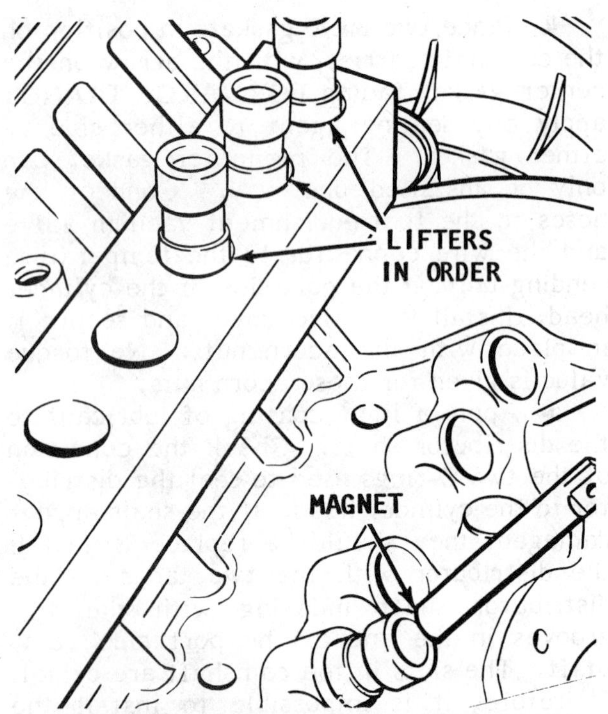

Withdraw the valve lifters with a magnet and keep them in order.

Remove the rocker arm covers. Loosen the rocker arm nuts, and then turn the rocker arms aside.

Set up some kind of system to hold the push rods, and then remove each one in order so they will be installed back in the identical hole from which they were removed.

TAKE CARE not to get any dirt into the engine particularly into the valve lifters. Use cloths and compressed air to clean the cylinder heads and adjacent parts.

Remove the valve lifters and keep them in order. Each **MUST** be installed into the same location from which it was removed.

Remove the fuel pump and push rod.

CAUTION: Be sure to plug the fuel line to prevent fuel from siphoning out of the fuel tank.

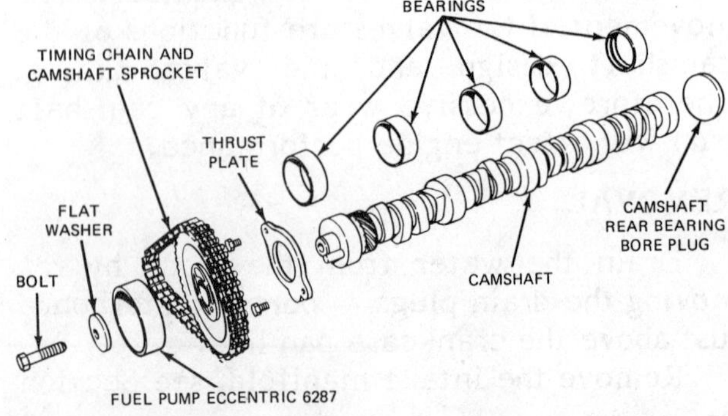

Typical arrangement of associated parts for the camshaft of a V8 engine. The V6 has four less cam lobes.

Thread two bolts about 4" in length into the end of the camshaft as an aid in removing the camshaft. Grasp the bolts firmly and **CAREFULLY** slide the camshaft out of the engine block.

CLEANING AND INSPECTING

Clean the camshaft with solvent and wipe the journals dry with a lint-free cloth. **ALWAYS** handle the shaft **CAREFULLY** to avoid damaging the highly finished journal surfaces. Blow out all of the oil passages with compressed air.

Clean the gasket surfaces on the block and crankcase front cover. Check the diameter of the five camshaft bearing surfaces with a micrometer for out-of-round condition, taper, and wear. If the journals are out-of-round more than 0.001", the camshaft should be replaced.

Check the camshaft for alignment. This is best done using **"V"** blocks and a dial indicator. The dial indicator will indicate the exact amount the camshaft is out of true. If it is out more than 0.002" dial indicator reading, the camshaft should be replaced. When using the dial indicator in this manner the high reading indicates the high point of the shaft. Examine the camshaft bearings and if any one bearing needs to be replaced, **ALL FIVE** should be replaced.

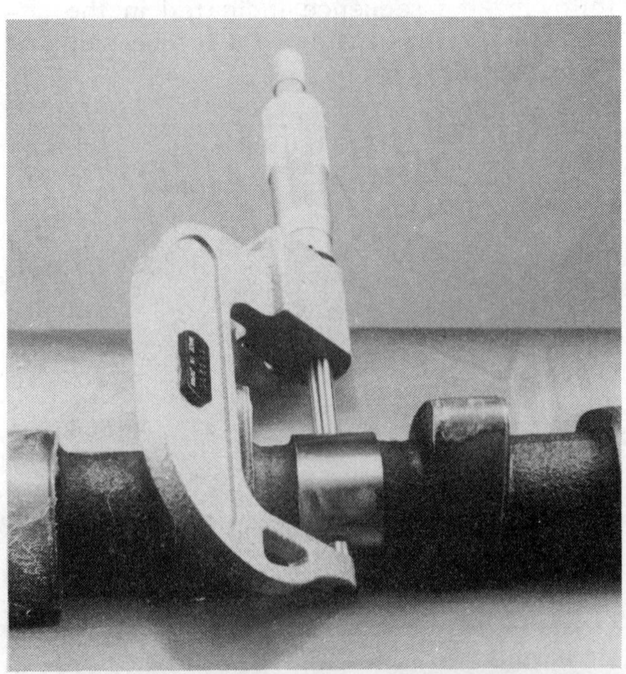
Measuring the camshaft bearing journals with a micrometer. The out-of-round should not exceed 0.001" (0.025mm).

CAMSHAFT BEARING REMOVAL

Removal and replacement of the camshaft bearings should be performed by qualified mechanics in a shop equipped to handle such work. However, in most cases the block bores for the bearings can be bored to a larger size and oversize bearings installed. **BE SURE** to check with your local marine dealer for oversize bearings available.

CAMSHAFT INSTALLATION

Lubricate the camshaft bearing journals with engine oil and the camshaft lobes with Molykote or equivalent.

Install the two 5/16"-18x4" bolts in the camshaft bolt holes. Install the camshaft into the engine block. **TAKE CARE** to move the shaft straight into the block and not to damage the bearings or cams.

Install the timing sprocket, timing chain, and cover, see Section 3-26.

Install the fuel pump. Install the valve lifters and push rods into the **SAME POSITION** from which they were removed. Position the rocker arm over the push rods and start the rocker arm nuts. Tighten the rocker arm nuts slowly and at the same time rotate the push rod. **STOP** tightening the nut when the push rod turns with difficulty, and move on to the next valve. Continue in a similar manner until all of the nuts have been tightened as just described. For the actual valve adjustment, see Section 3-19.

Install the intake manifold, see Section 3-5.

3-19 VALVE LASH ADJUSTMENT

IN-LINE PUSHROD ENGINE (Not OHC Engine)

Proper valve lash can be done with the engine hot or cold and **NOT RUNNING.** The clearance is the same for the intake and exhaust valves. While making the adjustment, use two feeler gauges, one **"GO"** gauge 0.020" (0.50 mm) thick and the other a **"NO GO"** gauge 0.022" (0.55 mm) thick. The clearance is adjusted so that the **"GO"** gauge can be inserted easily while the **"NO GO"** gauge can not enter.

Remove the rocker arm cover.

4-Cylinder Engine

When the No. 1 piston is at TDC on the compression stroke, adjust valves No. 1, 2, 3, and 5, counted from the front of the engine.

When the No. 4 piston is at TDC on the compression stroke, adjust valves No. 4, 6, 7, and 8.

6-Cylinder Engine

With the No. 1 piston at TDC on the compression stroke, adjust valves No. 1, 2, 3, 6, 7, and 10, counted from the front of the engine. With the No. 6 cylinder at TDC, adjust valves No. 4, 5, 8, 9, 11, and 12.

3-20 VALVE LASH ADJUSTMENT SINGLE OHC ENGINE

Remove the flame arrestor and the valve cover. Remove the No. 1 spark plug. Hold a thumb over the spark plug opening and at the same time have an assistant rotate the crankshaft with a socket wrench on the center bolt until the No. 1 piston is at TDC on the compression stroke. **ALWAYS** rotate the crankshaft in the normal rotation direction, **CLOCKWISE,** when facing the engine from the front. With No. 1 at TDC, both cam lobes for the No. 1 cylinder will point upward at equal angles. The timing notch on the crankshaft pulley will be at the Zero position.

SPECIAL WORDS

The valve adjustment may be made with the engine either hot or cold, **NOT** running.

Timing mark and numbers on the timing belt cover of an OHC engine.

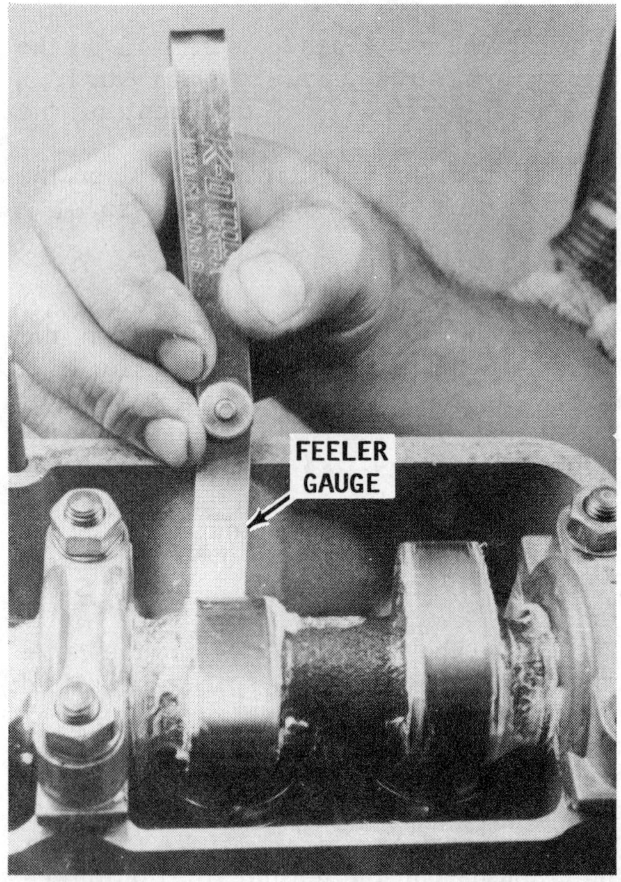

Checking the valve clearance on an OHC engine. "Checking" and "adjusting" clearances differ.

Changing shim material from the top of the pushers using special Volvo tools.

Engine Hot

Check the valve clearance by inserting a feeler gauge between the base circle of the camshaft and the shim on top of the valve stem. The clearance for the intake and exhaust valves should be 0.012" to 0.020" (0.30 mm to 0.50 mm).

Engine Cold

The valve clearance should be 0.010" to 0.018" (0.25 mm to 0.45 mm).

If the valve clearance is within the limits specified in the previous paragraphs, rotate the crankshaft until the next piston in firing order is at TDC, the No. 3, and repeat the check. Continue checking by bringing each piston in firing order to the TDC compression position.

ADJUSTING CLEARANCE

If any of the valve clearances are not within the specified tolerance range, the shim material on top of the valve pusher must be changed. When making an adjustment, the clearances are closer than when just checking.

Engine Hot

Valve clearance should be 0.016" to 0.018" (0.40 mm to 0.45 mm).

Engine Cold

Valve clearance should be 0.014" to 0.016" (0.35 mm to 0.40 mm).

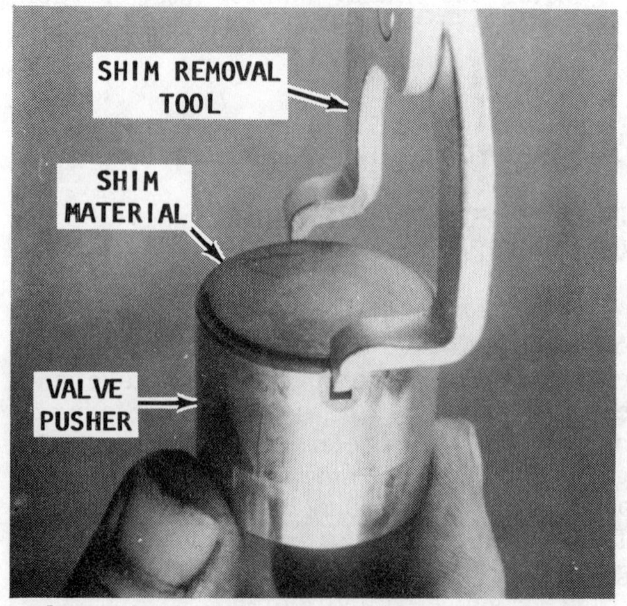

Instructional picture to show the relationship of the special tool, shim material, and the valve pusher.

To change the shim material, first rotate the valve shim retainer until the groove is at a right angle to the camshaft. Next, install special Volvo Tool No. 9995022. Depress the valve with the tool. Use special Volvo Shim Remover Tool No. 9995026 and remove the shim material.

Now, measure the thickness of the shim material. If the valve clearance was too great, which is usually the case, then purchase shim material from the local Volvo Penta dealer to obtain a satisfactory valve clearance measurement.

Shim material is available in different thickness between 0.140" to 0.166" (3.55 to 4.20mm) at intervals of 0.002" (0.05mm).

Coat new shim material with oil and install it with the marking facing down.

Rotate the crankshaft and camshaft through a couple complete revolutions, and then recheck the clearance.

Rotate the crankshaft until the next piston in firing order is at TDC, the No. 3 cylinder, and repeat the adjustment. Continue making the adjustment with new shim material, as required, by bringing each piston in firing order to the TDC compression position.

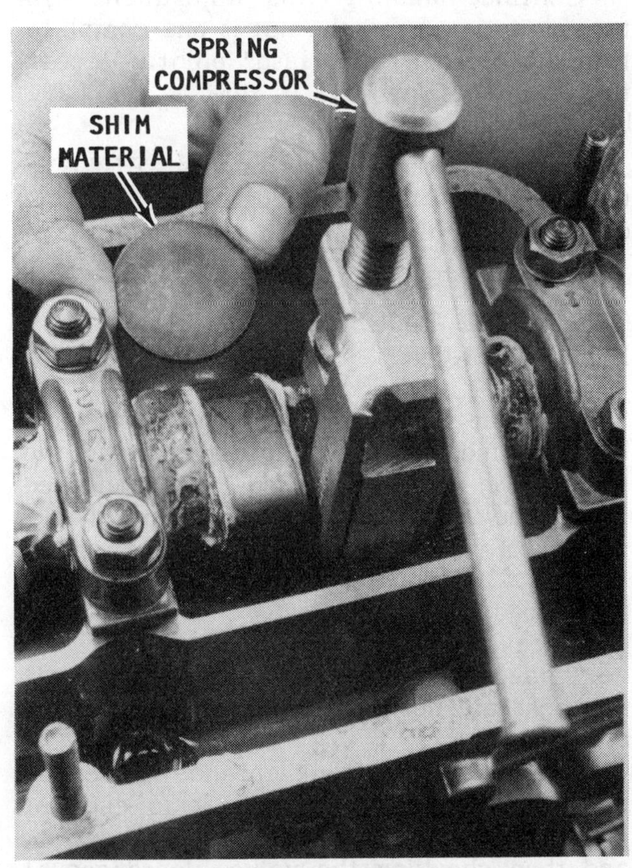

Shim material ready for installation. The special Volvo spring compressor is in place.

3-21 VALVE LASH ADJUSTMENT DOUBLE OHC ENGINE

The DOHC engine is equipped with hydraulic valve lifters. These lifters do not have a lash adjustment. Instead, the lifter provides zero valve clearance which results in very quiet operation. There is no tappet noise as the cam lobes move against the lifters. A variation due to wear of parts or temperature changes is compensated hydraulically.

3-22 VALVE LASH ADJUSTMENT GMC V6 AND GMC V8 ENGINES

The first procedure in this section covers: **EXHAUST VALVES 1, 3, 4, and 8; INTAKE VALVES 1, 2, 5, and 7.**

Crank the engine **SLOWLY** until two conditions exist: (1) the No. 1 piston is in the firing position, and (2) the mark on the harmonic balancer is aligned with the center or **"O"** mark on the timing tab fastened to the timing chain cover.

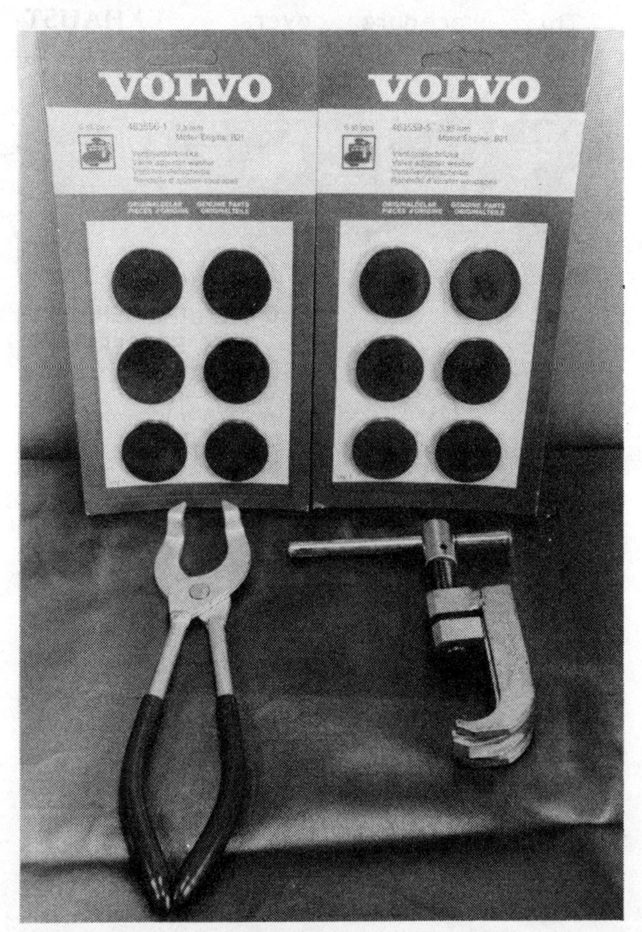

Sets of shim material along with the shim removal tool (left) and the spring compressor (right).

These two conditions may be determined by placing your fingers on the No.1 cylinder valves as the mark on the balancer comes near the "O" mark. If the valves are not moving, the engine is in the No. 1 firing position. If the valves move as the mark comes up to the timing tab, the engine is in the No. 6 firing position. This means the engine must be turned over one more time to reach the No. 1 position.

The actual valve lash adjustment is made by first, backing off the adjusting nut (rocker arm stud nut) until there is a small amount of play in the push rod. Next, tighten the adjusting nut to barely remove the clearance between the push rod and rocker arm. This position may be determined by rotating the push rod with your fingers while at the same time you tighten the nut, as shown in the illustration on Page 3-41.

At the point when you can not rotate the push rod easily, the clearance has been eliminated. Now, tighten the adjusting nut 3/4 turn more to place the hydraulic lifter plunger in the center of its travel. No further adjustment is required.

This procedure covers: **EXHAUST VALVES 2, 5, 6,** and **7; INTAKE VALVES 3, 4, 6,** and **8.**

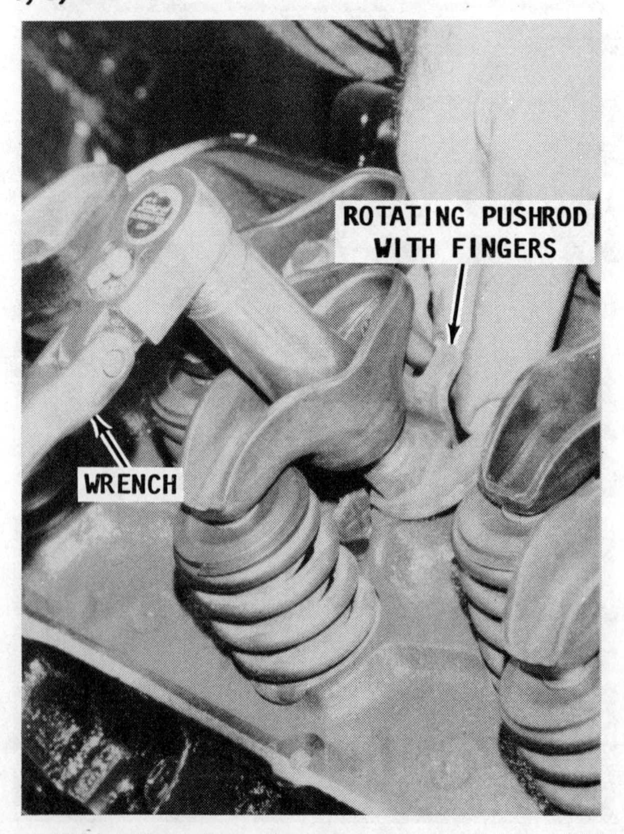

Making the valve adjustment on a GMC V6 or GMC V8 engine, as described in the text.

Crank the engine **SLOWLY** one revolution from its position in the previous procedure until the No. 6 piston is in the firing position and the mark on the harmonic balancer is again aligned with the center or "O" mark on the timing tab fastened to the timing chain cover.

Repeat the sequence given in the previous procedure until the valves listed have been adjusted.

Continue in the same manner for the same numbered valves on the other head.

For those who prefer to adjust the valve lash while the engine is running, the preferred method is to find the "zero lash" described in the first procedure of this section, and then to slowly turn the adjusting nut 1/4 turn and wait several engine revolutions for the lifter to bleed down before making a further adjustment. **DO NOT** attempt to turn the adjusting nut one full turn while the engine is operating. Adjustment in this manner will not allow the lifters to bleed down which could result in valve train damage, most likely a bent push rod or rods.

Continue making this adjustment 1/4 turn at-a-time until the nut is one complete turn down from "zero lash" point. Repeat the sequence until all of the valves have been properly adjusted.

Install the rocker arm covers. **ALWAYS** use new gaskets. Check to be sure the cover hole reinforcements are in place.

3-23 VALVE ADJUSTMENT FORD V8 ENGINE

Anytime the valves and the seats have been ground, or if you detect noise in the valve train which is not due to a collapsed lifter, the valve clearance **MUST** be checked. The valves are not adjustable. However, 0.030" shorter push rods, or 0.030" longer rods are available.

Check the accompanying diagram on the next page and notice how three crankshaft positions are designated by the leters **A, B,** and **C.**

The collapsed lifter clearance is checked by first turning the crankshaft to each of the three positions indicated in the diagram, and then checking the valve clearances at each position as follows:

Ford AQ 190A

Position A

With the No. 1 piston at TDC, at the end of the compression stroke, check the following valves:

No. 1 intake, No. 1 exhaust
No. 7 intake, No. 5 exhaust
No. 8 intake, No. 4 exhaust

Position B

Rotate the crankshaft to Position B, and then check the following valves:

No. 5 intake, No. 2 exhaust
No. 4 intake, No. 6 exhaust

Position C

Rotate the crankshaft to Position C, and then check the following valves:

No. 2 intake, No. 7 exhaust
No. 3 intake, No. 3 exhaust
No. 6 intake, No. 8 exhaust

Ford AQ 240A

Position A

With the No. 1 piston at TDC, at the end of the compression stroke, check the following valves:

No. 1 intake, No. 1 exhaust
No. 4 intake, No. 3 exhaust
No. 8 intake, No. 7 exhaust

Position B

Rotate the crankshaft to Position B, and then check the following valves:

No. 3 intake, No. 2 exhaust
No. 7 intake, No. 6 exhaust

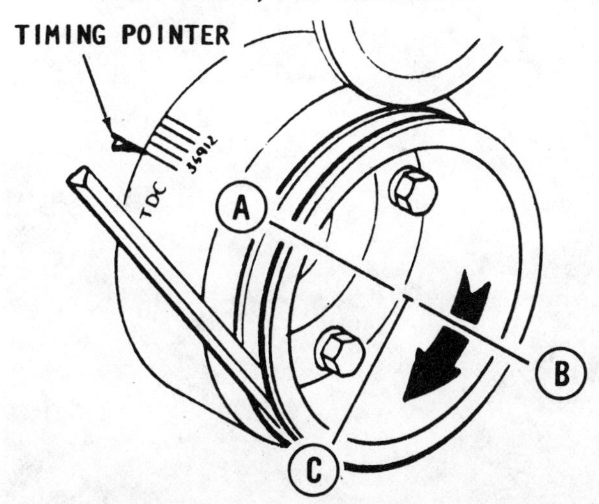

Crankshaft position for checking the valve lash. The No. 1 piston is at TDC at the end of the compression stroke for the "A" position. A chalk mark is then made 90° apart for points "B" and "C".

Position C

Rotate the crankshaft to Position C, and then check the following valves:

No. 2 intake, No. 4 exhaust
No. 6 intake, No. 8 exhaust
No. 5 intake, No. 5 exhaust

To make the actual valve lash clearance check, first apply pressure on the valve lifter with Tool T711A-6513AC to bleed down the lifter plunger until it is fully collapsed, and then check the clearance with a feeler gauge. The clearance should be 0.083" (2.11mm) to 0.183" (4.65mm). **BE SURE** the feeler gauge is no wider than 3/8" (9.52mm). If the clearance is not within the prescribed limits, replace the push rod with a longer or shorter one as required.

Install the rocker arm covers. **ALWAYS** use new gaskets. Check to be sure the cover hole reinforcements are in place. Tighten the screws to the torque value given in the Specifications in the Appendix.

Close the water drain valves.

Start the engine and check for leaks.

3-24 TIMING BELT SERVICE SINGLE OHC ENGINE

GOOD WORDS

The timing belt should be changed after about 500 hours of engine operation or:

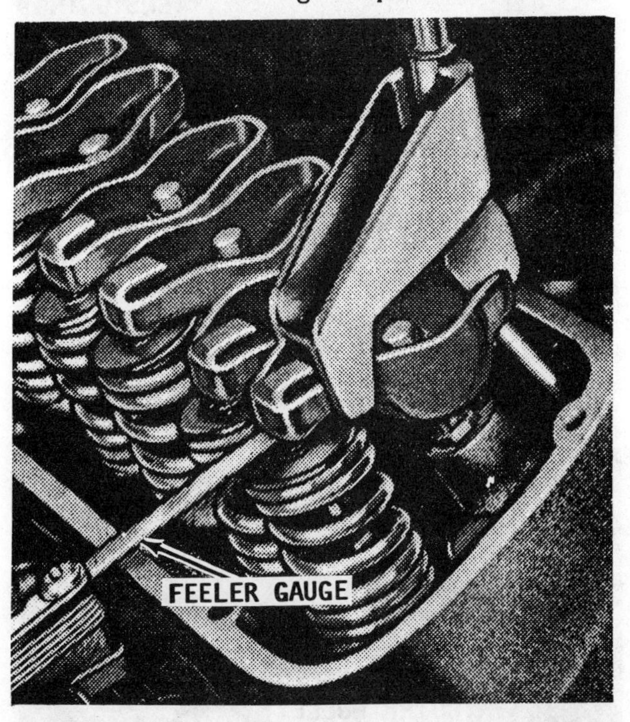

Using a feeler gauge to check the collapsed hydraulic lifter lash. To keep the clearance between 0.090" and 0.190", shorter or longer push rods may be installed.

a- the cord re-inforcement and the rubber begin to separate.

b- the tread on the timing belt shows signs of wear.

The following procedures provide detailed instructions to remove and install the timing belt.

The instructions begin from the absolute beginning, or worse condition, namely installing the timing belt because the old one broke.

PREPARATION

First, it will be necessary to remove some of the exterior equipment on the engine in order to gain access to the drive pulleys for removal of the old belt or the installation of a new one.

Begin, by obtaining a suitable container, and draining the fresh water/antifreeze solution from the block.

Next, remove the raw water pump. If any problems arise, see Chapter 7.

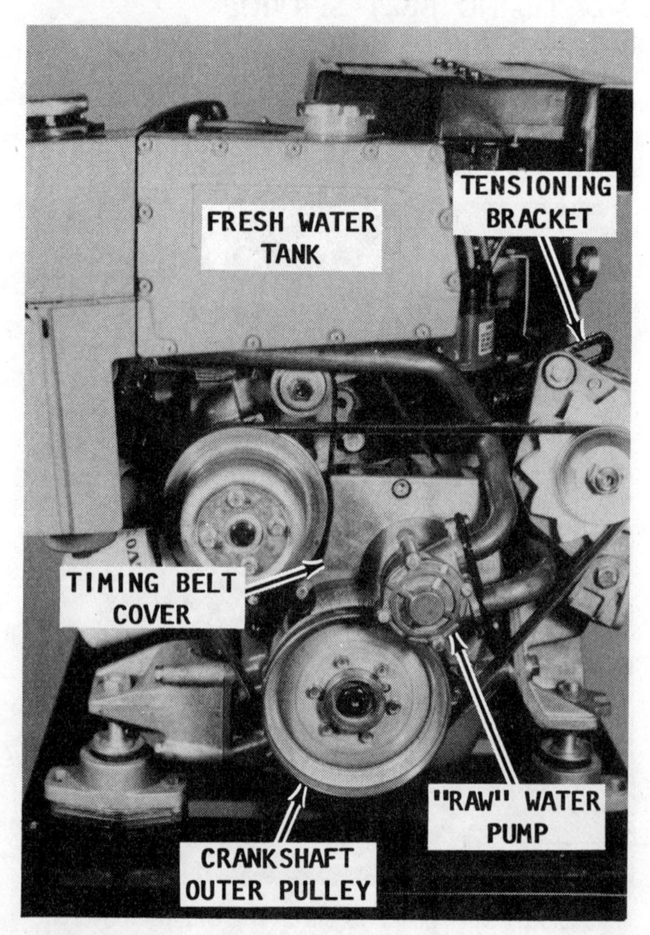

Items that must be removed from the front of an OHC engine before the timing belt can be replaced.

Next, loosen the bolts on the alternator tensioning bracket, and then remove the alternator drive pulley.

Remove the timing belt cover. The upper part of the cover is fitted with a guide. This guide requires the cover to be moved outward just a bit before it is free and can be removed.

Now, remove the crankshaft outer drive pulley, by removing the six attaching bolts.

Remove the fresh water tank. Again, if difficulties arise, see Chapter 7.

The front of the engine is now ready for removal of the old belt or installation of the new one.

REMOVAL

Rotate the crankshaft until No. 1 cylinder is at TDC (top dead center) on the compression stroke. In this position the timing notch on the crankshaft pulley will be aligned with the mark on the inner guide plate; the timing mark notch on the camshaft pulley will be aligned with the timing mark on the valve cover; and the timing mark dot on the accessory drive pulley will be aligned with the notch on the accessory drive backing plate.

If the crankshaft and the accessory drive pulley are not rotated while the timing belt is removed, installation will be simplified.

Timing notch on the crankshaft pulley aligned with the mark on the inner guide plate. No. 1 piston is at TDC on the compression stroke.

Loosen the nut on the belt tensioner. Compress the spring with a pair of channel-lock pliers, and then insert a punch or pin into the hole to hold the spring in the compressed position. While the spring is compressed, remove the timing belt.

INSTALLATION

GOOD WORDS

Proper installation of the timing belt is not a difficult task. The procedure simply involves having the crankshaft, the camshaft, and the accessory drive shaft, all in the correct position, when the timing belt is installed. Fortunately, all the pulleys have some type of timing mark that can be aligned with a matching mark on the engine when the No. 1 piston is at TDC on the compression stroke. If none of the shafts have been rotated while the timing belt was off, the task is even more simple.

The procedures pickup the work after the exterior equipment has been removed from the front of the engine, as outlined under **REMOVAL** of this section.

If the timing belt was broken, perform the instructions in this paragraph. If the belt was not broken, skip to removal of the No. 1 spark plug. Remove the valve cover. Rotate the camshaft until both valves for the No. 1 cylinder are fully closed. Temporarily place the valve cover in position. The timing mark on the camshaft pulley should be aligned with the mark on the valve cover.

Remove the No. 1 spark plug. Cover the spark plug opening with a thumb and have an assistant rotate the crankshaft with a socket wrench on the center bolt until the piston coming up on the compression stroke can be felt. Continue rotating the crankshaft until the No. 1 piston is at TDC on the compression stroke.

The timing mark notch on the crankshaft pulley will be aligned with the mark on the inner guide plate. Check the camshaft pulley timing mark notch to be sure it is aligned with the mark on the valve cover. If necessary, rotate the camshaft slightly until the marks are aligned. Finally, check the accessory drive pulley. The timing mark dot should be aligned with the notch on the accessory drive backing plate. If the timing belt was broken, it will be necessary to rotate the accessory drive pulley until the timing mark on the pulley is properly align-

Camshaft drive pulley timing notch aligned with the timing mark on the valve cover.

Timing dot on the accessory drive pulley aligned with the notch on the backing plate. The stripe on a new timing belt is also aligned.

ed with the notch on the block. Check the distributor to be sure the rotor is in position for No. 1 spark plug to fire.

The tensioner spring should still be compressed with the punch or pin in the shaft hole, which was inserted during removal. While the spring is still compressed, install the timing belt onto the pulley.

If a new belt is being installed, stripes on the belt can be positioned with the timing marks on the pulley. The stripes may still show on an old belt, but the stripes are not essential for proper timing. The timing belt mark with two stripes is to be aligned with the timing mark notch on the crankshaft pulley. In this position, one of the marks with a single stripe will be aligned with the dot on the accessory drive pulley, and the other mark on the timing belt (also with a single stripe), will be aligned with the timing mark notch on the camshaft pulley.

Again, use the channel-lock pliers to hold the tensioner spring, and then remove the punch or pin. Release the spring. Leave the belt tensioner nut loose.

With a socket wrench on the center nut of the crankshaft, rotate the crankshaft **CLOCKWISE** through a couple revolutions to take up the backlash between the pulleys. **DO NOT** rotate the crankshaft in the wrong direction or the belt will hop over teeth and the adjustment will be lost.

Illustration to show routing of the timing belt and a punch through the hole in the tensioner shaft, as described in the text.

Check to be sure the belt tensioner can move around its shaft. Tighten the nut to a torque value of 37 ft-lbs (50 Nm).

Again, rotate the crankshaft until the timing mark notch on the crankshaft pulley is aligned with the mark on the belt cover, when the cover is temporarily held in place. Check to be sure the timing mark notch on the camshaft pulley is aligned with the mark on the valve cover and the timing mark dot on the accessory drive pulley is aligned with the notch on the accessory drive backing plate.

COMPLETION

Install the fresh water tank, see Chapter 7.

Install the crankshaft outer drive pulley with the six attaching bolts. Tighten the bolts securely.

Install the timing belt cover. The cover must be moved inward and then into place because of the guide. Secure the cover with the attaching hardware.

Install the alternator pulley, and then tighten tensioning bracket until the belt can be depressed approximately 3/16" (5 mm) with finger pressure.

*Checking the completed work by backing the boat into a test tank. If major work was performed, run the engine at **REDUCED** speed for the first hour.*

Install the raw water pump, see Chapter 7.

Close the drain valve for the block, and then fill the fresh water tank with a 50% water and 50% anti-freeze solution.

Start the engine and check the completed work.

CAUTION: Water must circulate through the lower unit to the engine any time the engine is run to prevent damage to the water pump mounted on the engine.

To tune the engine, see Chapter 2.

3-25 TIMING BELT SERVICE DOUBLE OHC ENGINE

MOST CRITICAL WORDS

If a timing belt should break on a DOHC engine, while the engine is operating, such action will almost certainly cause severe damage to the valve train, due to the extremely close proximity of the pistons and the valves. In fact, the pistons and valves are so close, the piston crowns have grooves called "valve reliefs" to accommodate the valve heads at TDC. If the crown did not have these "reliefs" the piston would strike the valve head and bend the valve stem.

A **BROKEN** timing belt, while the engine was operating, **DEMANDS** the valve carrier and cylinder head be removed and a thorough inspection of the damage made.

As a result of the foregoing information, common **SENSE** dictates it is most important to periodically check the condition of the timing belt. Replace the belt as soon as there is evidence of noticeable wear, any sign of stretching (looseness), any fraying (loose fibers), etc.

REMOVAL

A broken or frayed timing belt may be easily removed. Simply pull out the broken belt. If the belt is worn, cut the belt and pull it out.

HOWEVER, several pieces of equipment must be removed to gain access for installation of a new timing belt.

Begin by obtaining a suitable container and draining the raw water.

Loosen, but do not remove, the top and bottom alternator bolts. Move the alternator toward the block to obtain slack in the drive belt. Remove the drive belt.

Disconnect the two hoses from the raw water pump. Remove the raw water pump. If any problems arise, see Chapter 7.

Remove the timing gear cover.

Remove the intake and exhaust manifolds and the heat exchanger, see Section 3-6.

Remove the cylinder head, see Section 3-12. This task is required to inspect and assess the damage only if the belt broke while the engine was operating.

The engine is now ready to receive a new timing belt.

INSTALLATION

The following procedures pickup the work after inspection and replacement of any damaged parts to the head or block have been accomplished **AND** the head and camshaft carrier have been installed, per Sections 3-12 and 3-17.

If a worn belt is being replaced, of course the head and camshaft carrier would not have been removed.

Temporarily install the timing gear cover, just bring the atttaching bolts up snug. Now, rotate the crankshaft pulley with the proper size socket until the timing marks on the crankshaft pulleys align with the 0° mark on the timing gear cover. Remove the timing gear cover.

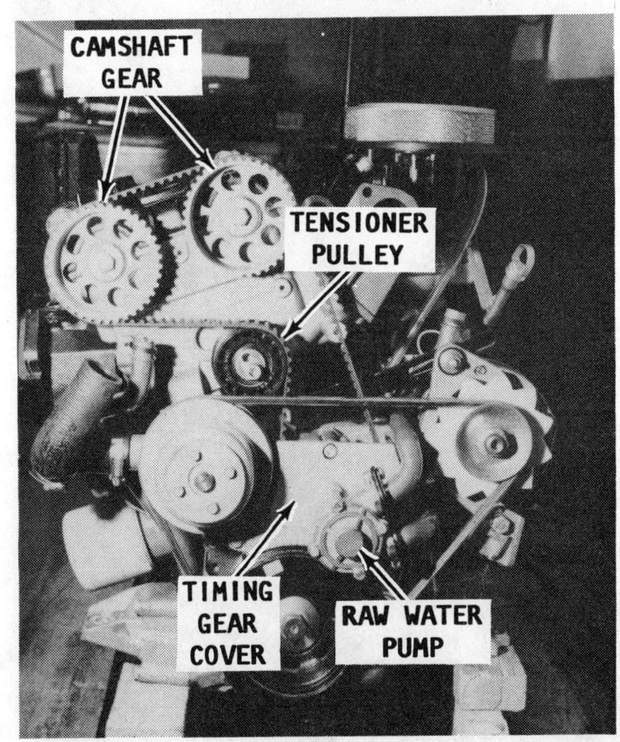

Several pieces of equipment must be removed from the front of the engine to gain access for installation of a new timing belt.

SPECIAL WORDS

If the camshaft carrier was installed properly, and the crankshaft has not been disturbed, the timing marks on the crankshaft pulley should align and the notches on each camshaft gear should point straight **UP.**

Identify the two yellow stripes on the back side of the new timing belt. These two marks will be aligned with the timing marks on the valve cover.

CRITICAL WORDS

The timing belt **MUST** installed in only one exact position without disturbing the position of the crankshaft pulley or the camshaft gears.

Loosen the tensioner gear. This is accomplished by preventing the center bolt from rotating with the proper size socket wrench and at the same time loosening the Allen-head screw with the proper size Allen wrench. Once the bolts are loosened, the tenioner gear is on an eccentric and may be moved to the left to provide slack for the timing belt.

Loosening (or tightening), the tensioner pulley on a DOHC engine.

Next, attempt to estimate the location of the timing belt on its route with the timing marks roughly aligned with the marks on the valve cover. Slip the lower end of the belt around the crankshaft gear, then up around the intermediate gear, around the back side of the tenioner pulley, and observe if the marks will align when the belt is installed around the camshaft gear. If an adjustment is required, move the belt downward and slip a tooth or two around the crankshaft gear, without disturbing its position, and then try the alignment marks again.

Now, install the timing belt between the rails of the tensioner pulley and onto the timing gears with the timing marks aligned with the timing marks on the valve cover. If the decal timing marks on the valve cover are no longer visible, an imaginary **VERTICAL** line should pass from each timing mark on the belt down through the notch on each camshaft gear and on through the center of the gear attaching bolt.

Move the tensioner pulley to the right to place tension on the belt. Hold this position and tighten the center bolt and the Allen-

Testing the belt tension on a DOHC engine, as described in the text.

head screw. Hold the center nut with the correct size socket and tighten the Allen-head screw with an 8mm Allen wrench.

To check for correct tension, first rotate the crankshaft through two or three complete revolutions. Next, attempt to twist the vertical part of the belt. The attempt should only allow the belt to be twisted 90°, until the flat side is facing forward as shown in the accompanying illustration. Adust the tensioner pulley, as required for proper belt tension.

Install the intake and exhaust manifolds and the heat exchanger, see Section 3-6.

Install the timing gear cover. Install the raw water pump. If any problems arise, see Chapter 7.

Determine if a new alternator drive belt is needed. Install the new alternator drive belt and tighten the belt to the proper tension by moving the alternator away from the block and then tightening the upper bolt. Attempt to depress the belt in the center horizontal portion with finger pressure. The attempt should allow the belt to be depressed approximately 3/16" (5mm). Tighten the upper and lower alternator bolts.

Close the drain valves and fill with coolant.

Start the engine and check the completed work.

CAUTION: Water must circulate through the lower unit to the engine any time the engine is run to prevent damage to the water pump mounted on the engine.

3-26 TIMING CHAIN SERVICE V6 AND V8 ENGINE

The camshaft is driven at half engine speed by a sprocket secured to the forward end of the shaft and a single-row chain encircling a sprocket on the forward end of the cranksahft. Timing chains are subject to stretching after prolonged use, and therefore, require replacement. Access to the chain is gained by first removing the timing chain cover. The work can be performed while the engine is mounted in the boat.

PREPARATION

Several pieces of external equipment must be removed from the front of the engine before the timing chain cover and timing chain can be removed.

First, drain the water from the block by removing the drain plugs -- port and starboard just above the crankcase pan line.

Next, slacken the belt tension on the alternator bracket. Remove the raw water pump. If the boat is equipped with power steering, remove the power steering pump.

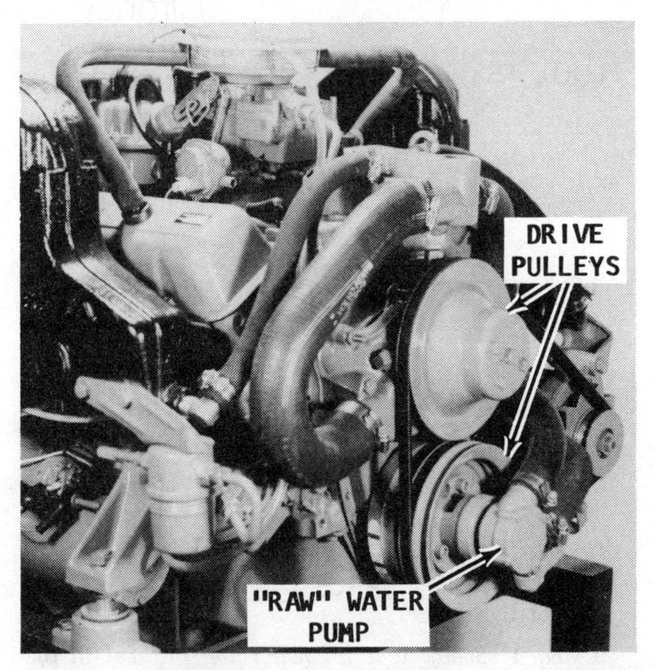

Items to be removed from the front of a V6 engine before the timing chain can be replaced.

Items to be removed from a V8 engine before the timing chain can be replaced.

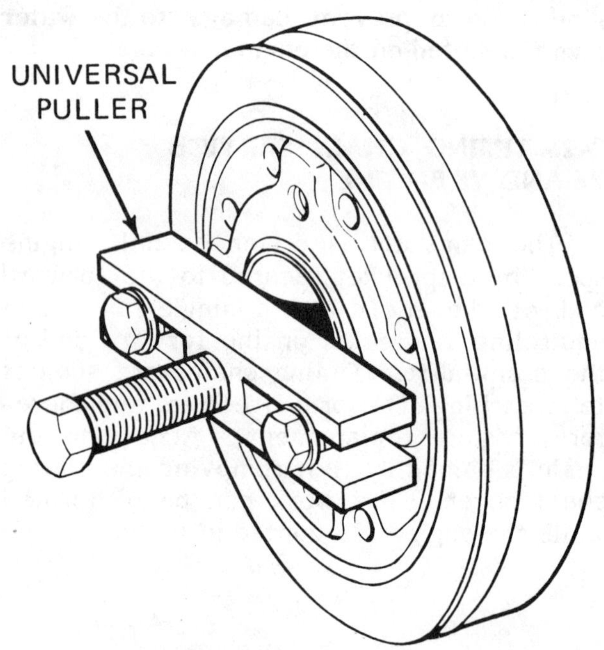

Universal puller used to remove the torsional damper from the crankshaft.

Remove the accessory drive pulleys. Remove the water circulation pump.

REMOVAL

Remove the torsional damper retaining bolt. Install a universal puller and remove the damper.

Remove the two oil pan-to-front cover attaching bolts. Remove the front cover-to-block attaching bolts. On some "V" block engines, it may be necessary to move the cover off just a bit and then to cut the oil pan gasket flush with the block face with a sharp knife, **BEFORE** removing the cover from the block. Cutting the pan gasket at this time means it will not be necessary to remove the pan in order to replace the front part of the pan gasket when installing the timing chain cover. Remove the cover and discard the gasket.

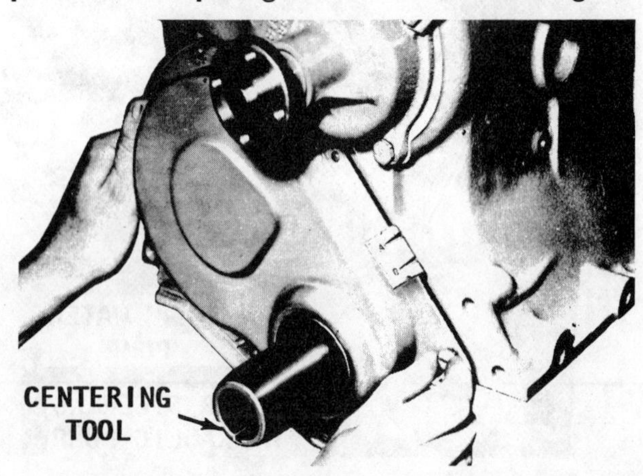

Using a centering tool to install the timing chain cover. Exercise care not to damage the seal.

Rotate the crankshaft until the timing mark on the camshaft sprocket is aligned with the timing mark on the crankshaft sprocket, as shown.

Remove the three camshaft sprocket-to-camshaft bolts. The camshaft sprocket is a light press-fit onto the camshaft. Tap the sprocket lightly on the lower edge of the camshaft to dislodge it.

Remove the camshaft sprocket and the timing chain together. If the crankshaft sprocket is to be replaced, remove the sprocket. The key will probably stay on the crankshaft.

TIMING CHAIN INSTALLATION

Position the key in the keyway of the crankshaft. Align the keyway of the crankshaft sprocket with the key, and then install the sprocket. Engage the timing chain onto the camshaft sprocket.

Hold the sprocket in the vertical position, with the chain hanging down, and then align the timing mark on the camshaft with the timing mark on the crankshaft sprocket. Now, align the dowel in the camshaft with the dowel hole in the camshaft sprocket, and then slide the sprocket onto the camshaft. **NEVER** try to drive the sprocket onto the camshaft because the **WELSH PLUG** at the rear of the engine will be dislodged.

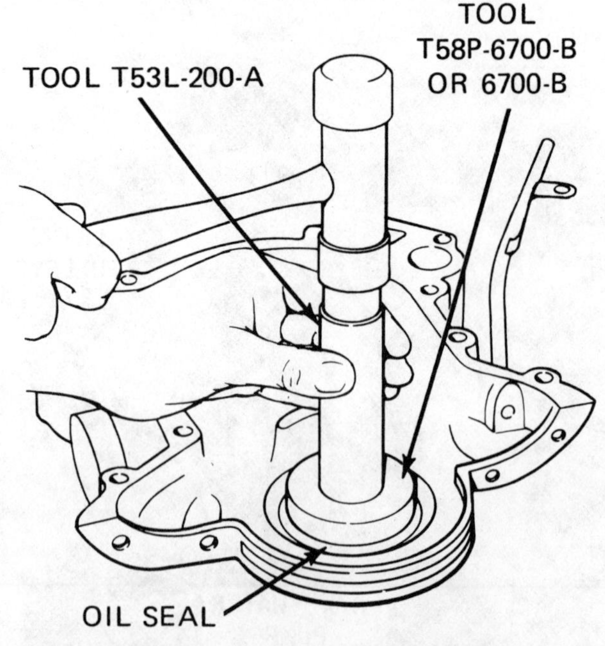

Using a special tool to install the oil seal in the cover. A hollow piece of tubing can be substituted for the tool.

Take up on the three mounting bolts to draw the camshaft sprocket onto the shaft. Check to be sure the dowel in the shaft indexes with the hole in the sprocket. Tighten the bolts to the value given in the Specifications in the Appendix. Coat the timing chain with engine oil.

While the timing chain cover is removed would be an excellent time to check the end play of the camshaft. This can be accomplished using a dial indicator, as shown. The end play should not exceed 0.0015" (0.038 mm).

ALWAYS intall a new oil seal in the timing chain cover or an oil leak at this point is very likely to develop. Use special tools or fabricate an alternate method of driving the seal into place without damage to the seal. Use care not to distort the cover while driving the seal.

Coat the cover gasket with sealer and place it in position over the dowel pins in the cylinder block. Coat the timing cover seal lip with engine oil and place the cover in position over the dowel pins.

If a timing chain cover alignment tool is available, install the tool. If the tool is not available, use some type of alternate tool to **ENSURE** the seal makes even contact all around the crankshaft when the cover is installed.

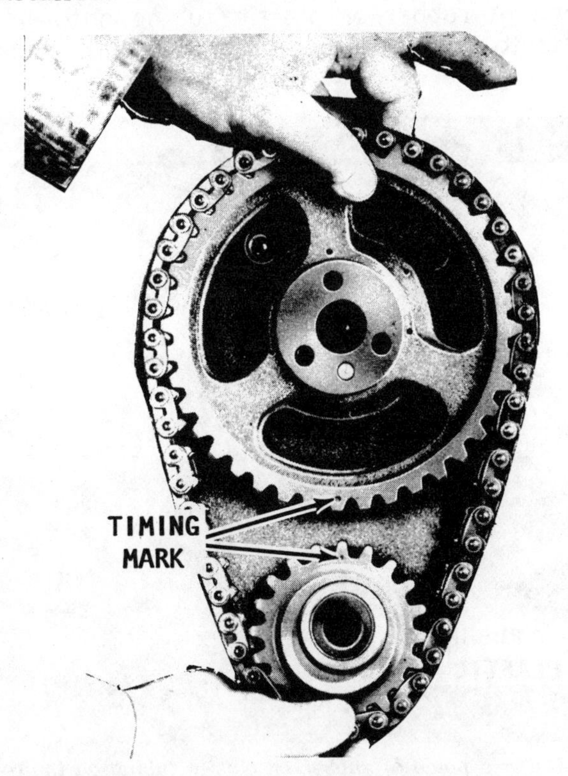

Installing the timing chain and camshaft sprocket, at the same time.

Push downward on the cover, and while holding the cover in this position, install the pan-to-cover attaching bolts.

Install the attaching bolts and tighten them to the torque value given in the Specifications in the Appendix.

Coat the timing chain cover seal area on the harmonic balancer hub with engine oil. Align the keyway and start the balancer onto the crankshaft. Seat the harmonic balancer onto the crankshaft using a large washer and a 7/16"x18x4" bolt.

Remove the bolt and washer. Install the belt pulley and tighten the screws to the torque value given in the Specifications in the Appendix.

Thread a 7/16"x20x2" bolt and thick washer into the crankshaft and tighten it to the torque value given in the Specifications.

COMPLETION

Install the accessory drive pulleys. If the boat is equipped with power steering, install the power steering pump. Install the raw water pump. Install the drive belts, and then tighten them. Secure the alternator bracket to hold the tension on the alternator drive belt.

Install the water drain plugs.

Move the boat into a body of water, a test tank, or connect a flush attachment to the lower unit.

Start the engine and check for leaks.

CAUTION: Water must circulate through the lower unit to the engine any time the engine is run to prevent damage to the water pump mounted on the engine.

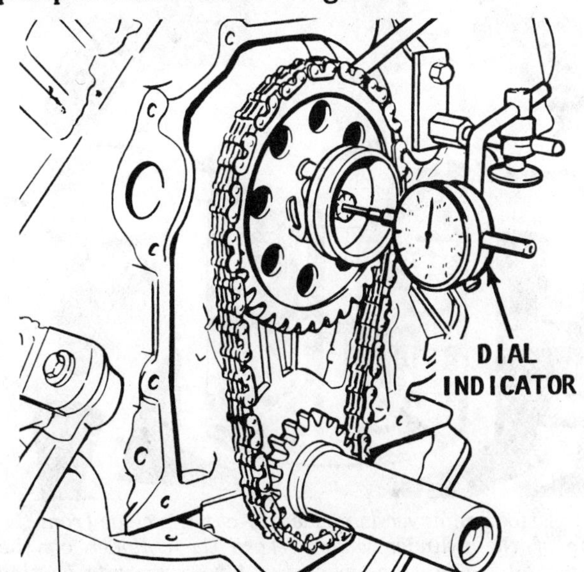

Using a dial indicator mounted on a jig to check the camshaft end "play".

3-27 PISTON, RING, & ROD SERVICE

The following procedures pickup the work after the engine has been removed from the boat, see Section 3-2, the manifolds have been removed, see Section 3-4 or Section 3-5 and 3-6, the rocker arms have been removed, see Section 3-7, the head or heads have been removed, see Section 3-8. Also other accessories not covered in the previous mentioned sections have been removed, such as the water circulation pump, oil cooler, etc. and the oil drained from the crankcase.

With the engine secured in a work stand or sitting on suitable blocks of wood on the deck with the pan facing upward, remove the pan.

SPECIAL WORDS

If a shoulder or ridge exists in the cylinder bores above the ring travel, they **MUST** be removed with a ridge reamer or the rings may be damaged or the ring lands (the area between the grooves) may be damaged during piston removal.

Before using the reamer, turn the crankshaft until the piston is at the bottom of its stroke, and then place a cloth on top of the piston to collect the cuttings. After the ridge is removed, turn the crankshaft until the next piston is at the top of its stroke to remove the cloth and cuttings.

All parts **MUST** be kept together because if the old pistons are serviceable, they **MUST** be installed on the rods from which they were removed and installed in the same bore. Identify each piston, connecting rod, and cap, with a quick drying paint or silver pencil to ensure each part will be replaced in the exact position from which it was removed.

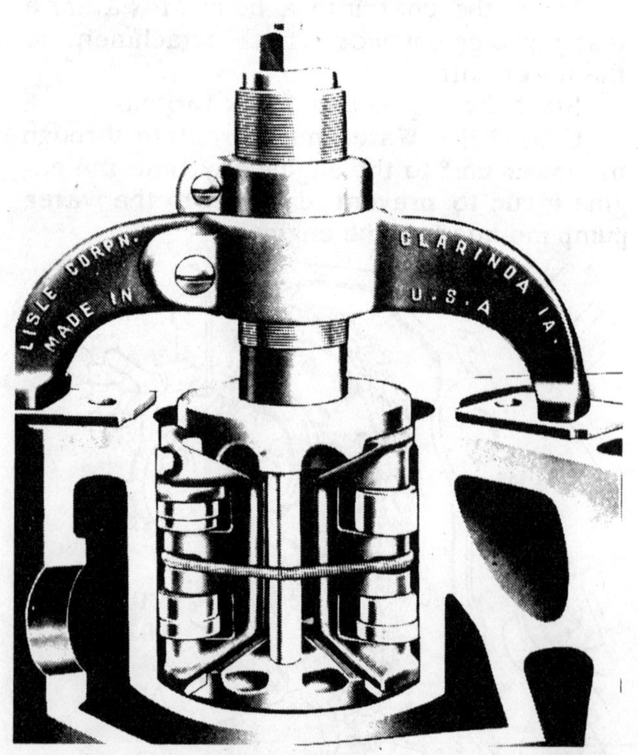

A ridge remover is needed to cut the ridge from the top of the cylinder walls, before the cylinder can be removed. The stop under the blade prevents cutting into the walls too deeply. Do not cut more than 1/32" (0.79 mm) below the bottom of the ridge.

Numbering System

Beginning at the forward end of the crankcase, the cylinders of an in-line engine are numbered from the front of the block, 1-2-3-4. In a GMC V6 engine, the cylinders in the port bank are numbered 1-3-5 and in the starboard bank 2-4-6. In a GMC V8 engine, the cylinders in the port bank are numbered 1-3-5-7 and in the starboard bank, 2-4-6-8. In a Ford V8 engine, the cylinders in the port bank are numbered 5, 6, 7, 8 and in the starboard bank, 1, 2, 3, 4.

Remove the cap and bearing shell from the No. 1 connecting rod. Install a short piece of rubber or plastic tubing onto the bolts to hold the upper half of the bearing

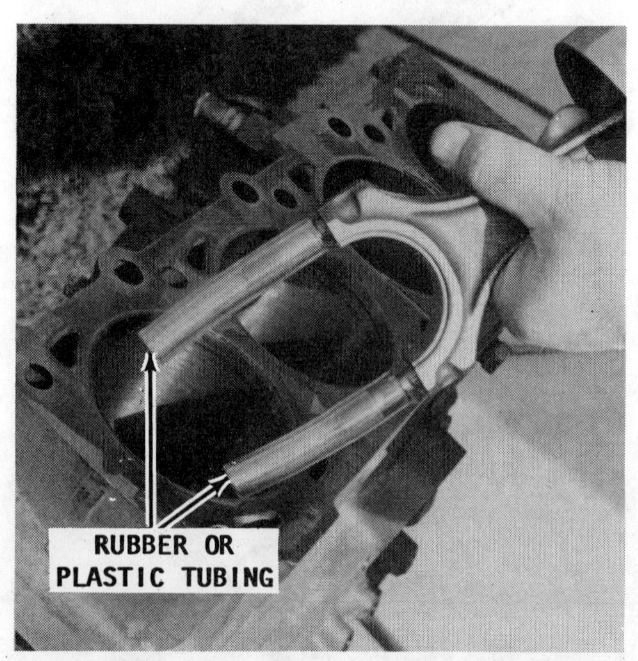

Using a piece of rubber or plastic tubing on the rod threads to protect the cylinder walls and the crankshaft journals during removal and installation of the piston.

Preignition caused this piston crown to melt. Such damage in one cylinder, means other piston assemblies need service. Proper timing is critical.

shell in place. These pieces of hose will also protect the cylinder walls and the crankshaft journals, as the piston and rod are removed.

Push the piston and rod assembly up and out of the cylinder. As each piston comes free of the cylinder block, remove the pieces of rubber hose, or plastic, and install the bearing shells and cap on the rod. It will be necessary to turn the crankshaft slightly to disconnect some of the connecting rod and piston assemblies and to push them out of their cylinders.

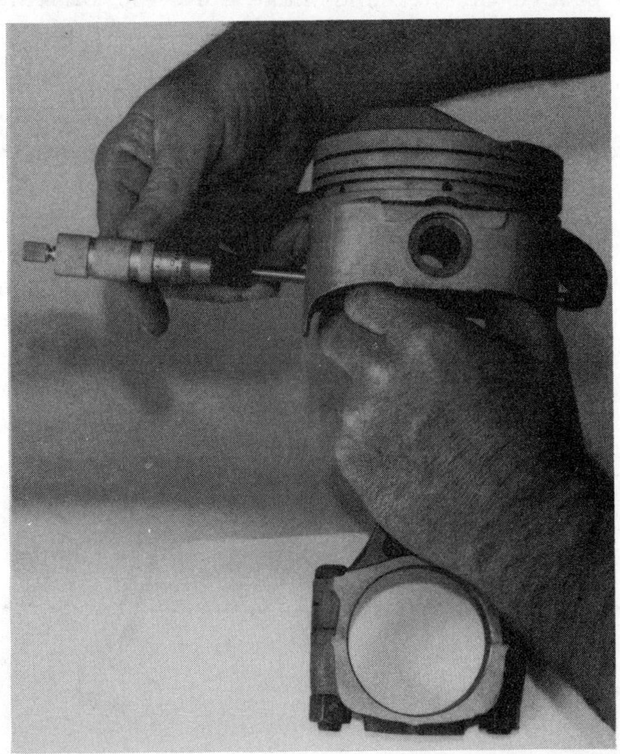

Using a micrometer to measure the piston circumference.

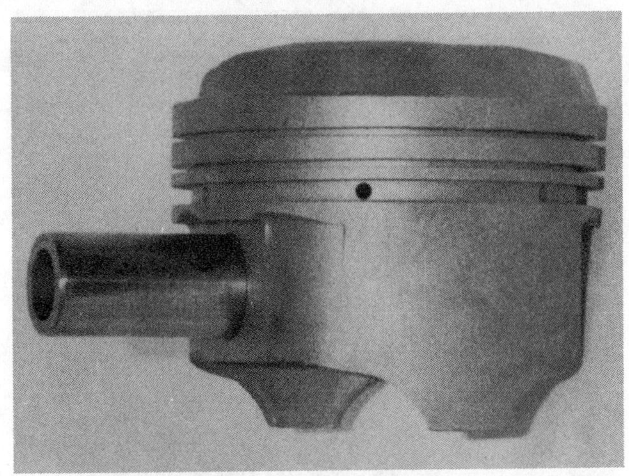

A properly fitted pin should support its own weight in either pin boss when it has been coated with light engine oil at room temperature.

Continue in a similar manner until all the piston and rod assemblies have been removed, partially reassembled to keep the parts matched, and the assemblies hung or placed in order.

CLEANING AND INSPECTING

If the engine has over 750 hours of service, or if the piston and rod assembly has been removed, it is considered good shop practice to replace the piston pins. Loose piston pins, coupled with tight rings, will

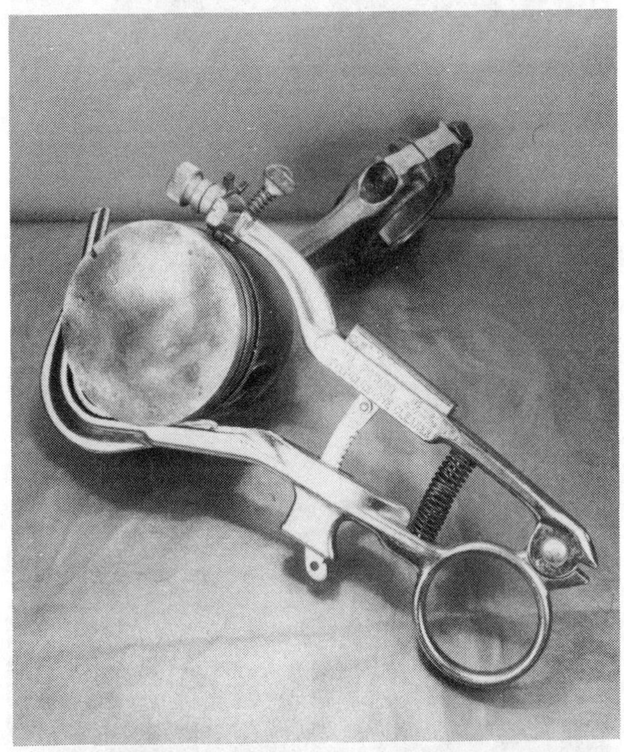

Reconditioning piston ring grooves using a special tool. This task can also be accomplished using the smooth factory end of a broken ring.

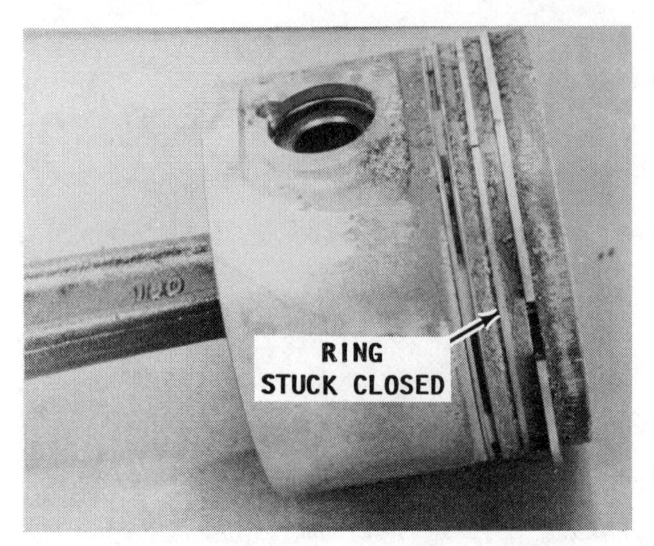

Used piston with corrosion and carbon build-up. The second ring appears stuck closed in the groove.

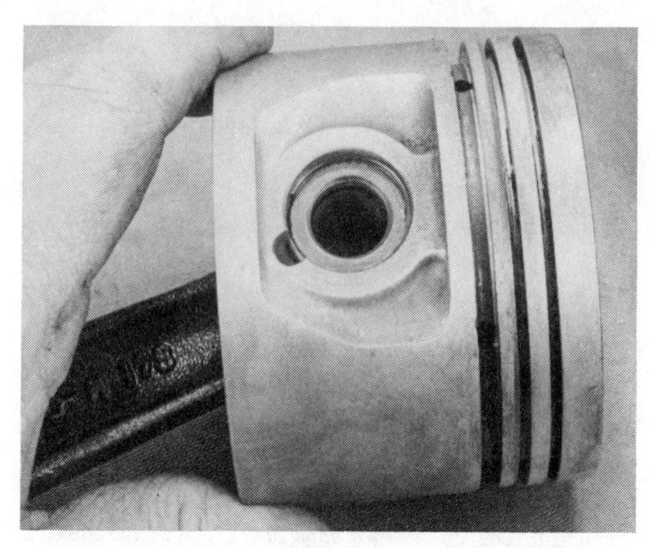

A cleaned piston, ready for new rings and installation into the block.

cause piston pin noises. These noises may disappear as the engine loosens.

Most mechanics have the piston pin work done by a machine shop with the necessary equipment and trained people to perform a precision job. If done in the machine shop, the connecting rods will be aligned so the pistons and rings will run true with the cylinder walls.

An arbor press and a special tool should be used to remove pressed-in piston pins.

PISTONS AND PINS

Remove the compression rings with an expander. Remove the oil ring by removing the two rails and the spacer-expander, which are separate pieces in each piston's third groove.

Use a cleaning solvent to remove the varnish from the piston skirts and pins. **NEVER** use a wire-brush on any part of the piston. Clean the ring grooves with a groove cleaner and make sure the oil ring holes and slots are clean. Inspect the piston ring lands (the material between the ring grooves) the skirts, and pin bosses for cracks, or other damage. Check the top of the piston for any sign of "piston burn" caused by pre-ignition, poor quality gasoline, improper timing, or water in the gas.

Replace any piston that is damaged or shows signs of excessive wear. Inspect the grooves for nicks or burrs that might cause

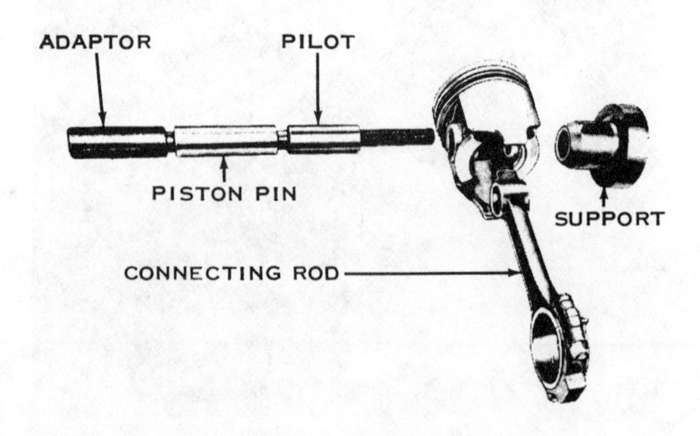

Arrangement of special tools required to properly install a piston pin through the piston.

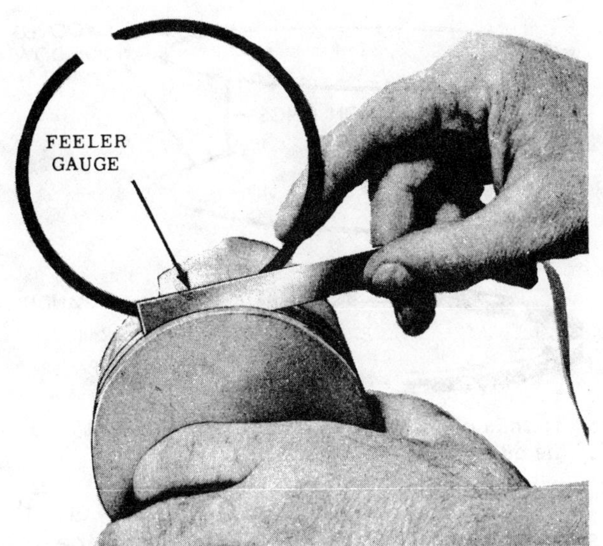

Checking the ring side clearance in the piston groove. The side clearance of used and new rings should not exceed those given in the text.

the rings to hang-up. Measure the piston skirt (across the centerline of the piston pin) and check the clearance in the cylinder bore. The clearance should not be greater than 0.0025".

After the rings are installed on the piston, the clearance in the grooves needs to be checked with a feeler gauge. The recommended clearance between the ring and the upper land is 0.006". Ring wear forms a step at the inner portion of the upper land. If the piston grooves have worn to the extent to cause high steps on the upper land, the step will interfere with the operation of new rings and the ring clearance will be too much. Therefore, if steps exist in any of the upper lands the piston should be replaced.

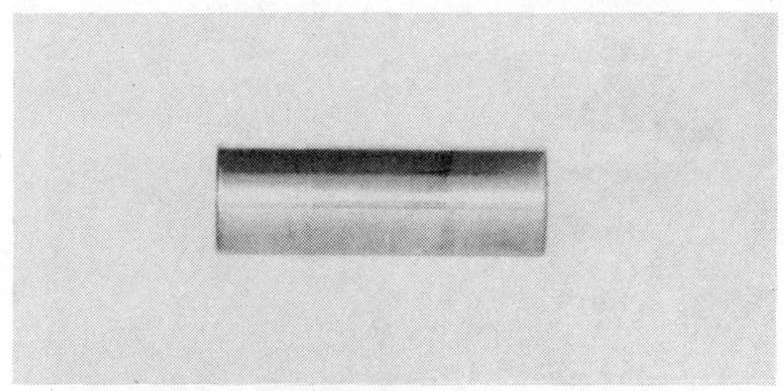

Three measurements with a micrometer should be made to determine piston pin wear. One, on the unworn section (center), the second and third, on both ends. Differences will indicate the amount of wear.

New compression rings on new pistons should have a side clearance of 0.0012" to 0.0032". The oil ring should have a side clearance of 0.0005" to 0.0065".

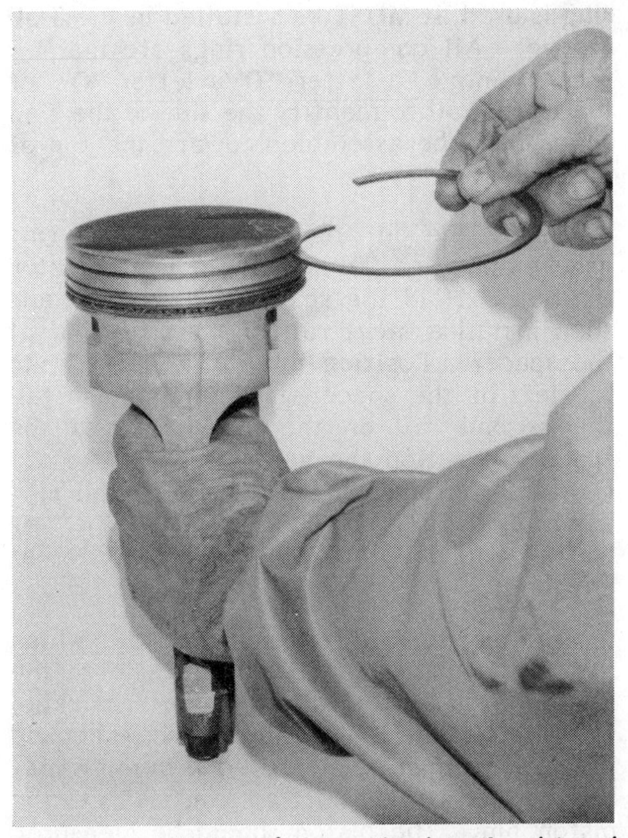

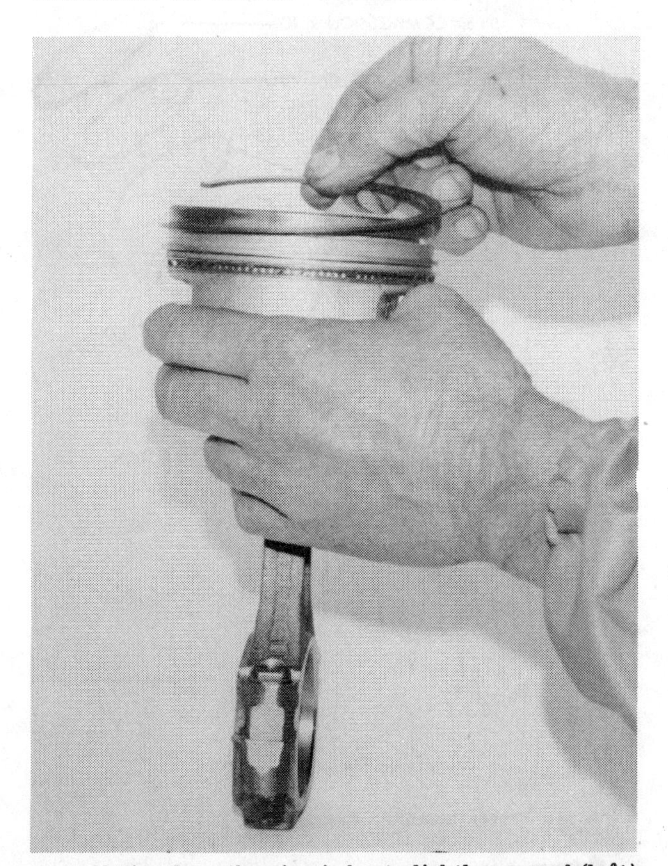

Proper installation of a new ring into the piston ring groove. Notice how the ring is bent slightly upward (left). After the ring is started, the ring is rotated around and over the top of the piston (right). The ring will feed into the groove and the end will finally snap into place.

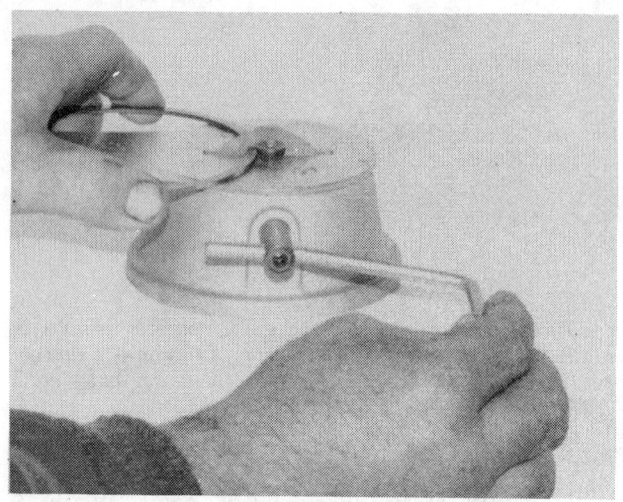

Grinding ring ends to obtain proper end gap. This task can also be accomplished with a thin fine file.

PISTON ASSEMBLING

Set up Tool BT-6408 and Adapter BT-6408-4, as shown. The piston and rod assembly must be mated as shown for right and left bank rods. The notch in the piston **MUST** face forward for the right bank cylinders, and face aft for the left bank cylinders.

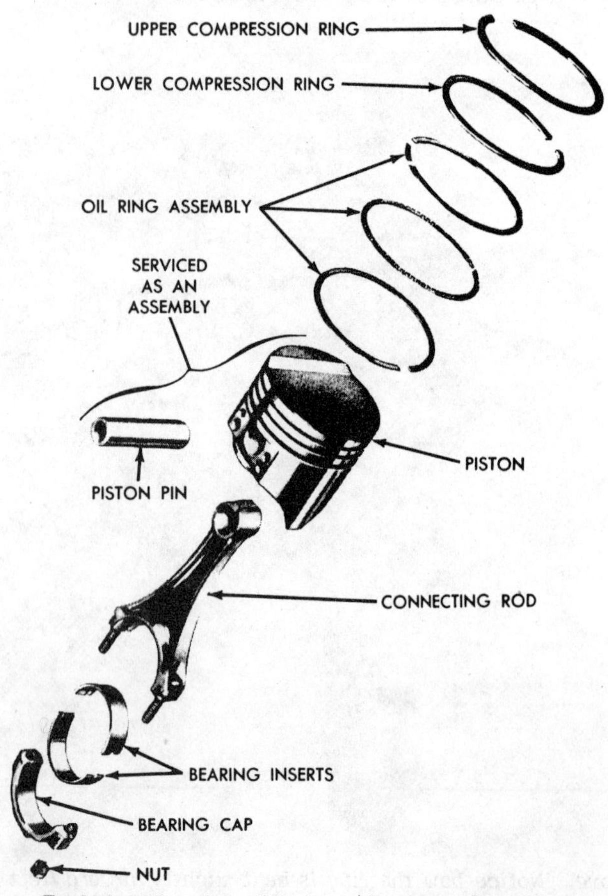

Exploded drawing showing all parts of the piston, ring, and rod parts.

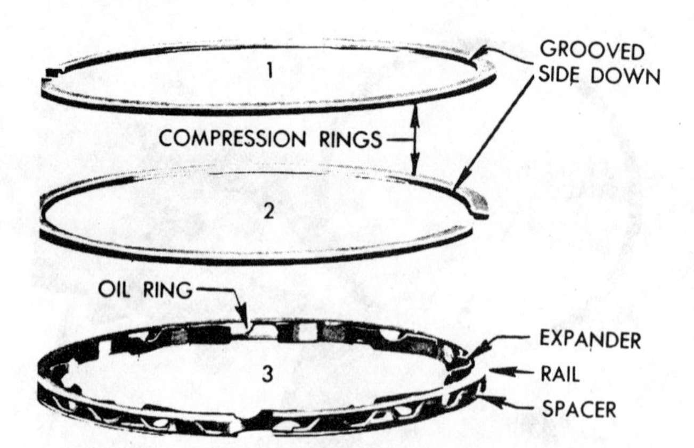

Details of the compression rings and the three parts of the oil ring.

Assemble the piston and rod on the spring-loaded guide pin. Lubricate the piston pin to avoid damage when it is pressed through the connecting rod. Install the drive pin in the upper end of the piston pin. Press on the drive pin until the piston pin bottoms. Remove the piston and rod assembly from the press.

Rotate the piston on the pin to be sure the pin was not damaged during the pressing operation.

If a single chrome plated compression ring is used, it **MUST** be installed in the **TOP** groove. All compression rings are marked with a dimple, a letter **"T"** a letter **"O"**, or the word **TOP** to identify the side of the ring which must be assembled toward the top of the piston.

Install the oil ring spacer in the oil ring groove and align the ring gap with the piston pin hole. Hold the spacer ends butted, and then install a steel rail on the top side of the spacer. Position the gap at least 1" to the left of the spacer gap, and then install the second rail on the lower side of the spacer. Position the gap at least 1" to the right of the spacer gap. Flex the oil ring assembly in its groove to make sure the ring is free and does not bind in the groove at any point.

The pistons are cam-ground, which means the diameter at a right angle to the piston pin is more than the diameter parallel to the piston pin. When a piston is checked for size, it must be measured with micrometers applied to the skirt at points 90° to the piston pin. The piston should be measured for fitting purposes 1/4" below the bottom of the oil ring groove.

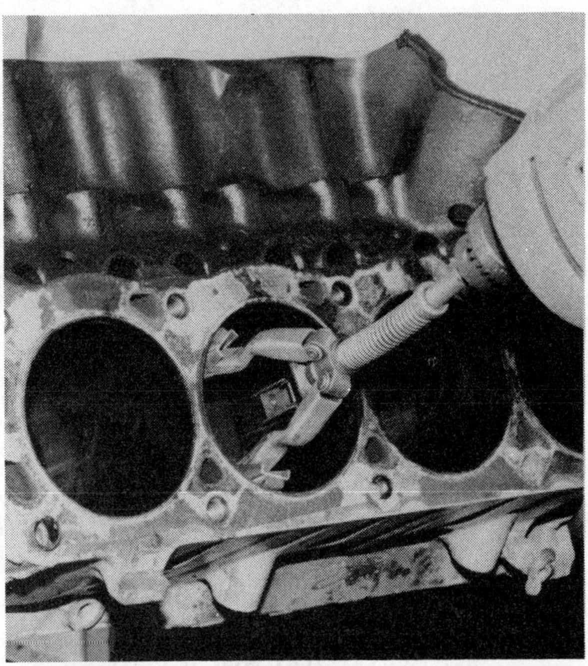

Refinishing the cylinder wall in a V8 block using one type of hone attached to an electric drill. ***ALWAYS*** *keep the tool moving in long even strokes over the entire depth of the cylinder.*

Inspect the piston pin bores and piston pins for wear. Piston pin bores and piston pins must be free of varnish or scuffing when being measured with a dial bore gauge or an inside micrometer. If the clearance is greater than 0.0001", the piston and/or piston pin should be replaced.

Inspect the bearing surfaces of the piston pins. Use a micrometer to check for wear by measuring the worn and unworn surfaces. Rough or worn pins **MUST** be replaced. Test the fit of each pin with its piston boss. If the boss is worn out-of-round, the piston and pin assembly **MUST** be replaced.

Refinishing the cylinder wall in an in-line engine using a different type of hone from the one shown in the left column.

The pistons are locked in the rod by a press fit and the pins turn in the pistons. Oversize pins are available in 0.0015", 0,003", 0.005", and 0.010" oversize.

CYLINDER BORES

Inspect the cylinder walls for scoring, roughness, or ridges which indicate excessive wear. Check the cylinder gauge at the top, middle, and bottom of the bore, both

Examples of various wear patterns on bearing halves, including possible reasons for the condition.

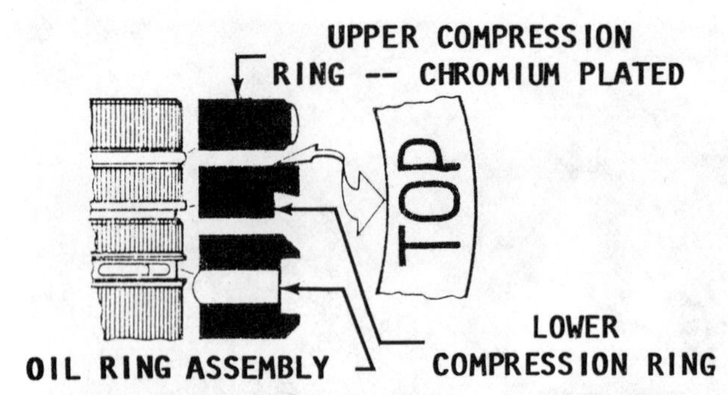

Ring arrangement on the piston for an OHC engine.

parallel and at right angles to the centerline of the engine.

A cylinder bore which is tapered 0.005" or more, or is out-of-round 0.003" or more, must be reconditioned.

ROD BOLTS AND NUTS

Check the rod bolts and nuts for defects in the threads. Inspect the inside of the rod bearing bore for evidence of galling, which indicates the insert is loose enough to move around. Check the parting cheeks to be sure the cap or rod has not been filed. Replace any defective rods.

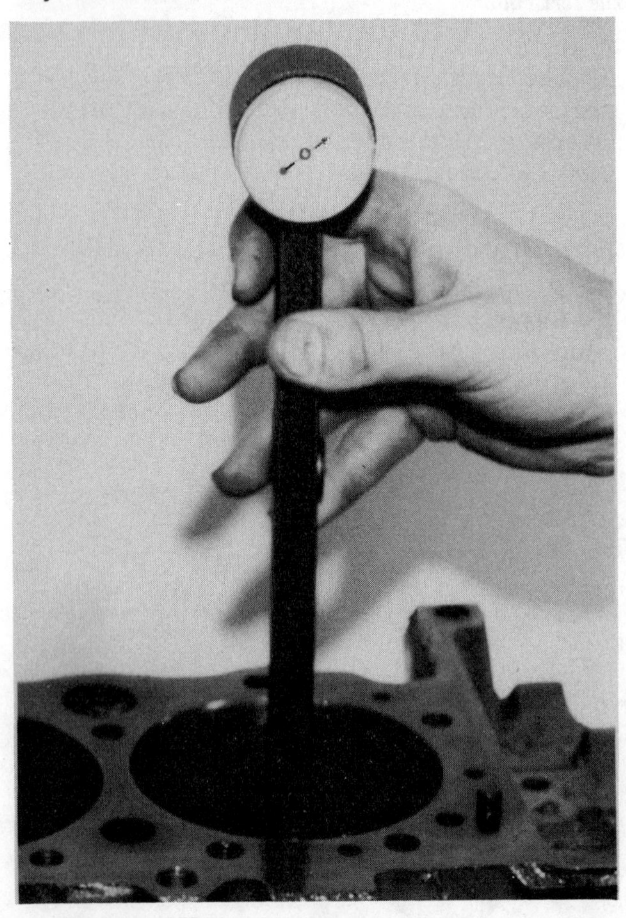

Cylinder wall taper and wear measured with a bore gauge indicator.

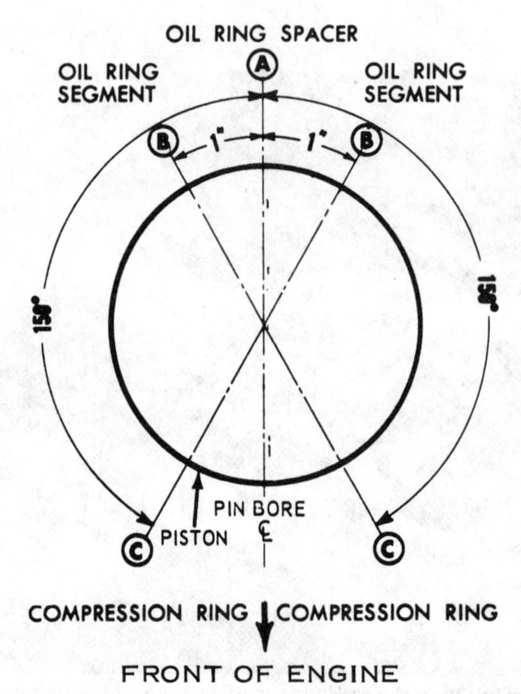

Typical ring spacing arrangement. The degrees are approximate, but follow the general pattern.

RINGS

The glazed cylinder walls should be slightly dulled, without increasing the bore diameter, if new piston rings are to be installed. This glazing is best accomplished using the finest grade of stones of a cylinder hone. The cylinder bores and piston grooves must be clean, dry, and free of any carbon deposits or burrs. New piston rings must be checked for clearance in the piston grooves and for end gap in the cylinder bores.

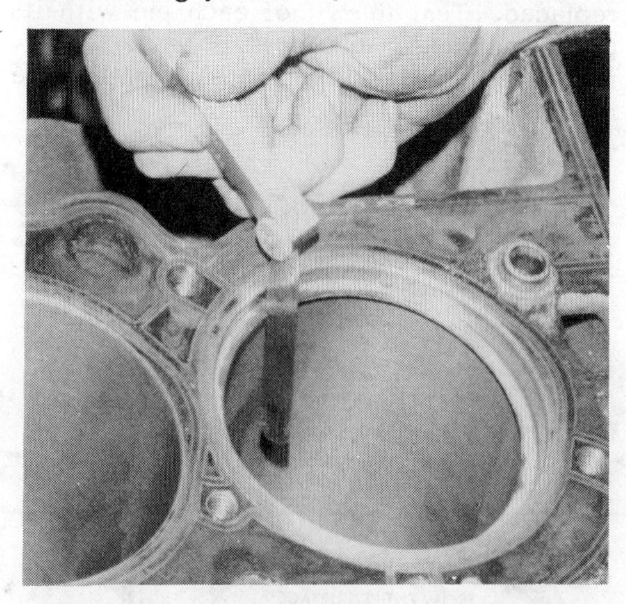

Measuring ring end gap in the cylinder with a feeler gauge. Push the ring down with an inverted piston.

The oil-control rings consist of two segments (rails) and a spacer. Piston rings are furnished in standard sizes and 0.020", 0.030", and 0.040" oversize. To check the end gap of compression rings, first place each ring in the cylinder into which it will be used. Next, square the ring in the bore using the upper end of a piston. Now measure the gap with a feeler gauge. End gap for the top ring should be 0.010" to 0.020"; for the second ring, 0.013" to 0.023". The oil ring rail gap should be 0.015" to 0.055". If the gap is less than specified, the ends of the ring can be filed carefully with a fine file until the proper gap is reached.

On the 4-cylinder single OHC engine and the double OHC engine, the upper ring is chromium-plated. The lower compression ring is marked **"TOP"**. Rotate the piston rings until the open gaps are as indicated in the accompanying illustration (about 120° from each other).

No. 1 piston with a ring compressor in place. The piston is ready to be forced into the cylinder.

The bore of the cylinders and the diameter of the pistons are very closely matched. Therefore, The "D" piston must be installed in the "D" cylinders and the "E" pistons in the "E" cylinders. The pistons **CANNOT** be exchanged.

PISTON CROWNS DOHC ENGINES ONLY

The pistons in a DOHC engine have two sets of "valve reliefs" cut in the crown to accommodate the opening of the valves at TDC. This engine block also has the letters "D" and "E" embossed on the starboard side of the head gasket surface.

The pistons also have either a "D" or an "E" embossed on the crown.

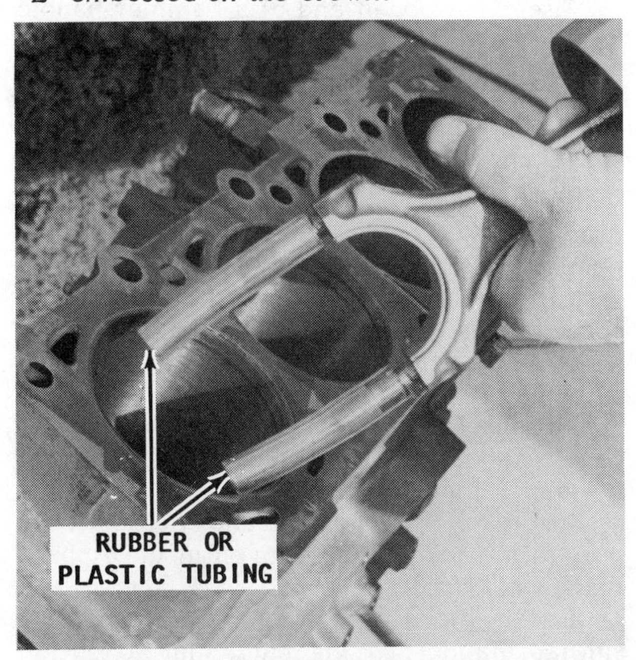

Using a piece of rubber or plastic tubing on the rod threads to protect the cylinder walls and the crankshaft journals during removal and installation of the piston.

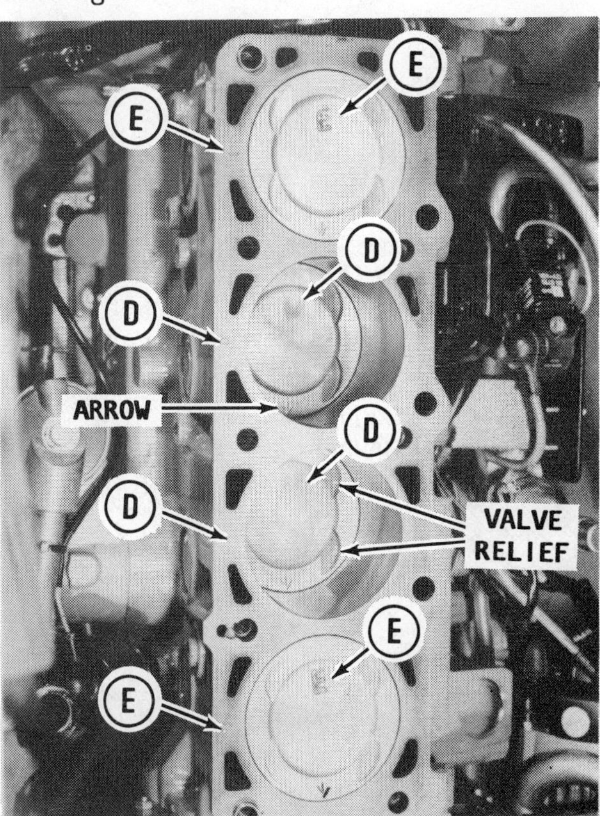

On a DOHC engine, letters are embossed on the block and piston to indicate position of installation. An arrow is also embossed on the piston crown to indicate "forward".

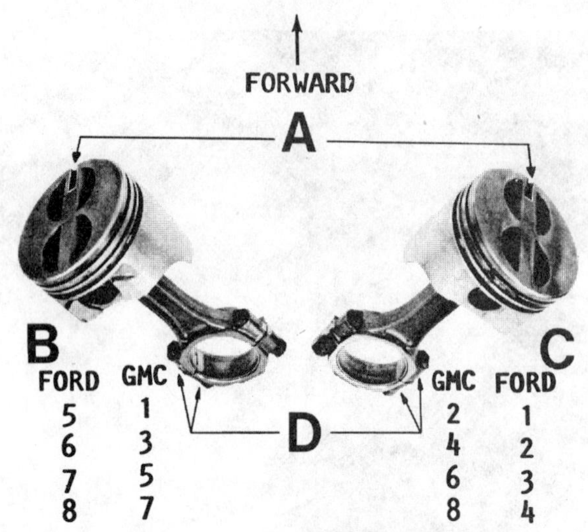

Piston and cylinder numbering system used on "V" engines. The number 7 & 8 are dropped for a V6.

The manufacturer has slightly oversize "standard" pistons available to replace original pistons even if the bore is not machined.

Each of the four pistons has an arrow embossed on the crown near one edge. This arrow **ABSOLUTELY MUST** face forward after the piston is installed in the block to accommodate the valve head at TDC.

ALL BLOCKS INCLUDING DOHC ENGINES

After you are satisfied the cylinder bores, pistons, connecting rod bearings, and crankshaft journals are as clean as possible, then coat all of the bearing surfaces with engine oil. Position the crankpin straight down. Remove the connecting rod cap and with the bearing upper shell seated in the rod, slide a piece of rubber or plastic tubing onto the rod bolts to hold the upper bearing shell in place and to prevent damage to the cylinder wall or the rod bearing journal on the crankshaft as the piston is installed.

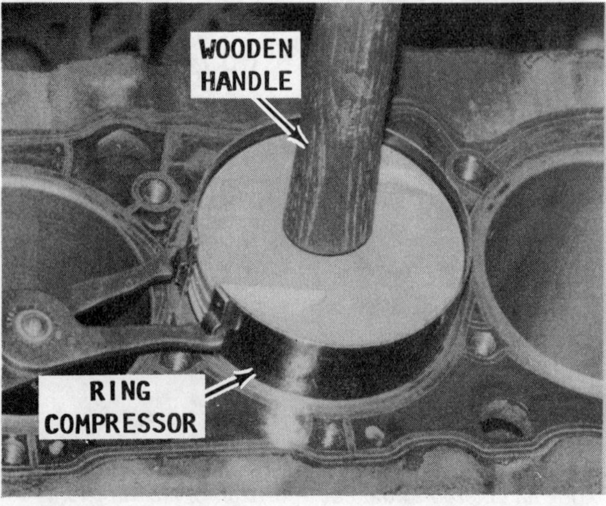

Using the wooden handle of a tool to force the piston into the cylinder.

Coat the piston and rings with engine oil, and then install the assembly into the proper numbered cylinder bore by compressing the rings with a ring compressor.

V6 and V8 Engines

On a V6 or V8 engine, the pistons are installed with the arrow or mark on the piston facing **FORWARD** and the numbered side of the rod facing outboard for both bank of cylinders. In the accompanying illustration, at the top of the previous column:

A Piston marking forward.
B Port bank, GMC 1,3,5,7.
B Port bank, Ford 5, 6, 7, 8
C Starboard bank, GMC 2, 4, 6, 8
C Starboard bank, Ford 1, 2, 3, 4
D Connecting rod number

For V6 engine, the number 7 and 8 cylinders are dropped.

On a 4-cylinder in-line engine, pistons and connecting rods are installed with the marks on the pistons pointing **FORWARD** and the marking on the connecting rod turned towards the starboard (the oil filter) side of the engine.

Use the wooden end of a hammer handle to push the piston down into the cylinder. **NEVER** hammer on the piston in an attempt to get it into the cylinder because a ring may be caught and you could snap it.

Place a piece of Plastigage the full width of the rod bearing journal on the crankshaft. Check to be sure the rod cap number or marking matches with the rod. Install the cap and tighten the cap to the torque value given in the Appendix.

Remove the rod cap and measure the flattened Plastigage with the scale on the side of the Plastigage package. The amount of clearance can be accurately determined by this method.

Install the cap again and tighten both nuts to the torque value given in the Appendix.

Install the other piston assemblies in a similar manner paying particular attention to install each used piston assembly into the same cylinder from which it was removed.

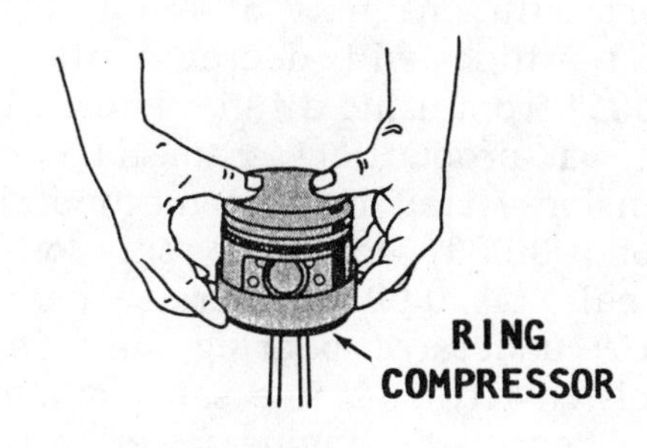

FIRST

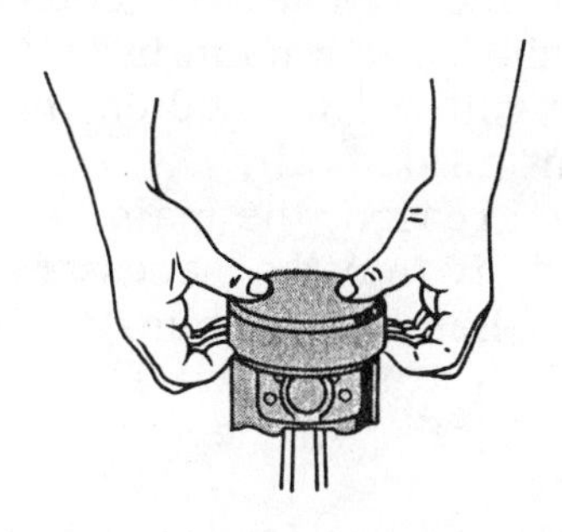

SECOND

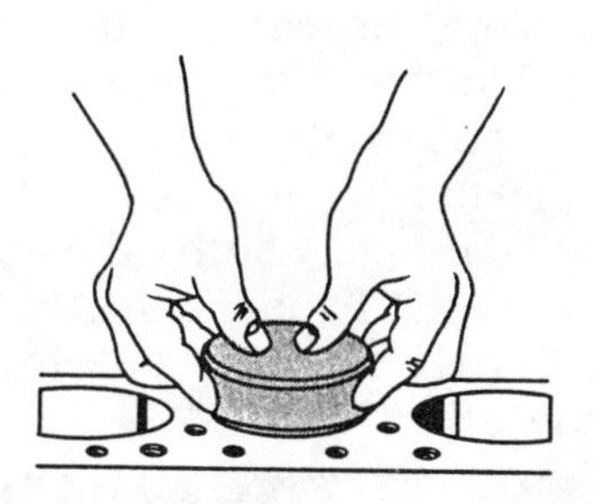

THIRD

LAST

The four stages of correct piston installation. If more than one "tap" with the mallet handle is required, it would be advisable to remove the assembly and check the rings for damage.

After all piston and rod assemblies are properly installed and the bearing cap nuts have been tightened to the proper torque value, check to be sure the oil spit holes in the connecting rods are facing the camshaft, and the edge of the rod cap is on the same side as the conical boss on the connecting rod web. These marks (the rib and boss) will be toward the other connecting rod on the same crankpin. Check the end clearance between the connecting rods on each crankpin. The clearance should be between 0.005" and 0.012".

Install the oil pump, see Section 3-30.

Install the oil pan gaskets, seals, and the oil pan.

The engine block is now ready for the head and accessories. See the appropriate section of this chapter and other chapters, as required.

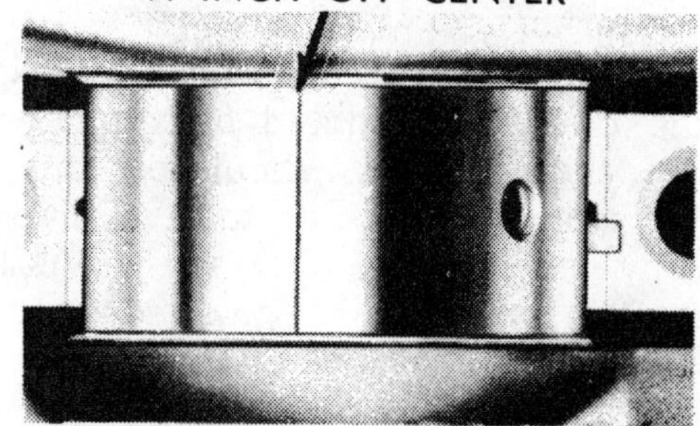

INSTALLING PLASTIGAGE

CHECK WIDTH OF PLASTIGAGE

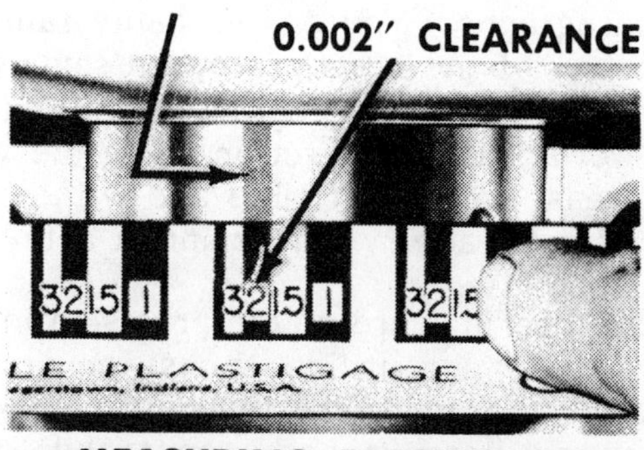

MEASURING PLASTIGAGE

After the connecting rod cap has been properly tightened, and then removed, the flattened Plastigage can be compared with the scale on the side of the package. The amount of clearance can accurately be determined in this manner.

3-28 CRANKSHAFT AND MAIN BEARINGS

ABOUT THIS SECTION

In most cases, the crankshaft is removed during an engine overhaul, when pistons, rods, rings, valves, and other service work is being performed. Only rarely, under very unusual circumstances, would the crankshaft and main bearings be removed for service without other tasks being accomplished on the engine. The difficulty and cost of performing the additional work while the crankshaft is being serviced would be extremely nominal.

Therefore, the following procedures realistically assume the head will be serviced, and the pistons, rings, and rods, will be removed, cleaned and checked, while the crankshaft is being removed.

Service work affecting the crankshaft or the crankshaft bearings may also affect the connecting rod bearings. Therefore, it is highly recommended that the rod bearings be carefully inspected as they are removed, see Section 3-27.

If the crankpins are worn to the extent that the crankshaft should be replaced or reground, then this work should be done in a qualified shop. Attempting to save time and money by merely replacing the crankshaft bearing inserts will not give satisfactory engine performance.

CRITICAL WORDS

BEFORE starting the actual removal of the crankshaft and main bearings, **TAKE TIME** to read the information in the next few paragraphs. You are probably familiar with many of the facts given, but some of it may be new for these engines and all of the good words will assist you in doing the work in the least amount of time and with assurance of satisfactory performance after the job is complete.

Main bearings are of the precision insert type and do not utilize shims for adjustment. If clearances are found to be excessive, a new bearing, both upper and lower halves must be installed. Bearings are available in standard size and as 0.001", 0.009", and 0.020" undersize.

In order to obtain close tolerances in production, selective fitting of both rod and main bearing inserts is necessary. For this reason you may find one half of a standard insert with one half of a 0.001" undersize insert which will decrease the clearance 0.0005" from using a full standard bearing.

If a production crankshaft cannot be precision fitted by this method, it is then ground 0.009" undersize only on the main journals. A 0.009" undersize bearing and a 0.010" undersize bearing may be used for precision fitting in the same manner as just described. Any engine fitted with a 0.009" undersize crankshaft will be identified by the following markings: **".009"** will be stamped on the crankshaft counterweight forward of the center main bearing journal. A figure **"9"** will be stamped on the block at the left front oil pan rail.

If, for any reason, the main bearing caps are replaced, it may be necessary to shim the bearing. Laminated shims for each cap are available for service. The amount of shimming required will depend on the bearing clearance.

In general, except for No. 1 bearing, the lower half of the bearing will show more wear and fatigue than the upper half. If you determine, from inspection, that the lower half is suitable for use, then it is safe to assume the upper half is also satisfactory. **HOWEVER**, if the lower half shows evidence of wear or damage, both the upper and lower halves **MUST** be replaced. **NEVER** replace one half of the bearing without replacing the other half.

Strip of Plastigage in place on the journal.

CHECKING MAIN BEARING CLEARANCE

Plastigage is soluble in oil, therefore, clean the crankshaft journal thoroughly of all traces of oil. With engine held in the bottom-up position, turn the crankshaft until the oil hole is down to avoid getting oil on the Plastigage. Whenever the crankshaft is rotated with the rear bearing cap removed, hold the oil seal to prevent it from rotating out of position in the crankcase. Place paper shims in the upper half of adjacent bearings and tighten the cap bolts to take the weight off of the shell of the bearing being checked.

Place a piece of Plastigage lengthwise along the bottom center of the lower bearing shell, and then install the cap with the shell and tighten the bolt nuts to a torque value given in the Specifications in the Appendix.

Remove the bearing cap and measure the flattened Plastigage at its widest point using the scale printed on the Plastigage envelope. The number within the graduation which most closely corresponds to the width of the Plastigage indicates the bearing clearance in thousandths of an inch.

Tightening the bearing cap with the Plastigage underneath to determine bearing clearance.

Under normal service conditions, main bearing journals will wear evenly and are seldom found to be out-of-round. However, if a bearing is being fitted to an out-of-round journal (0.001" maximum) be sure to fit the bearing to the maximum diameter of the journal.

If the bearing is fitted to the minimum diameter and the journal is out-of-round by 0.001", interference between the bearing and the journal will result in rapid bearing failure. If the flattened gauging plastic tapers toward the middle or ends, there is a difference in clearance indicating taper, low spot, or other irregularity of the bearing or journal. **BE SURE** to measure the journal with a micrometer if the flattened Plastigage indicates more than 0.001" difference.

If the bearing clearance is within specifications, the bearing insert is satisfactory. If the clearance is not within specifications, replace the insert. **ALWAYS** replace both the upper and lower insert together as a unit. **NEVER** just one half.

If a new bearing is being installed in the cap, and the clearance is less than 0.001", inspect the cap for burrs or nicks. If none

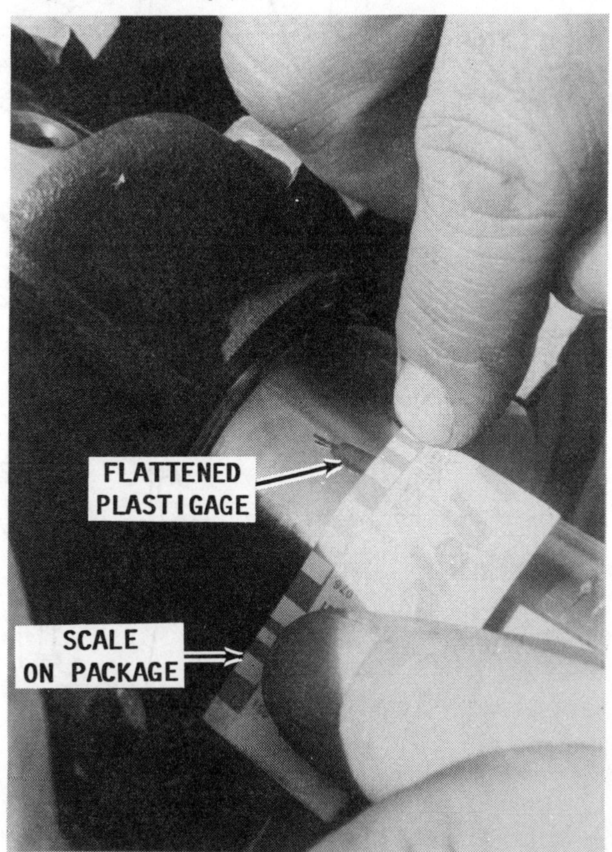

Comparing the flattened Plastigage with the scale on the side of the package.

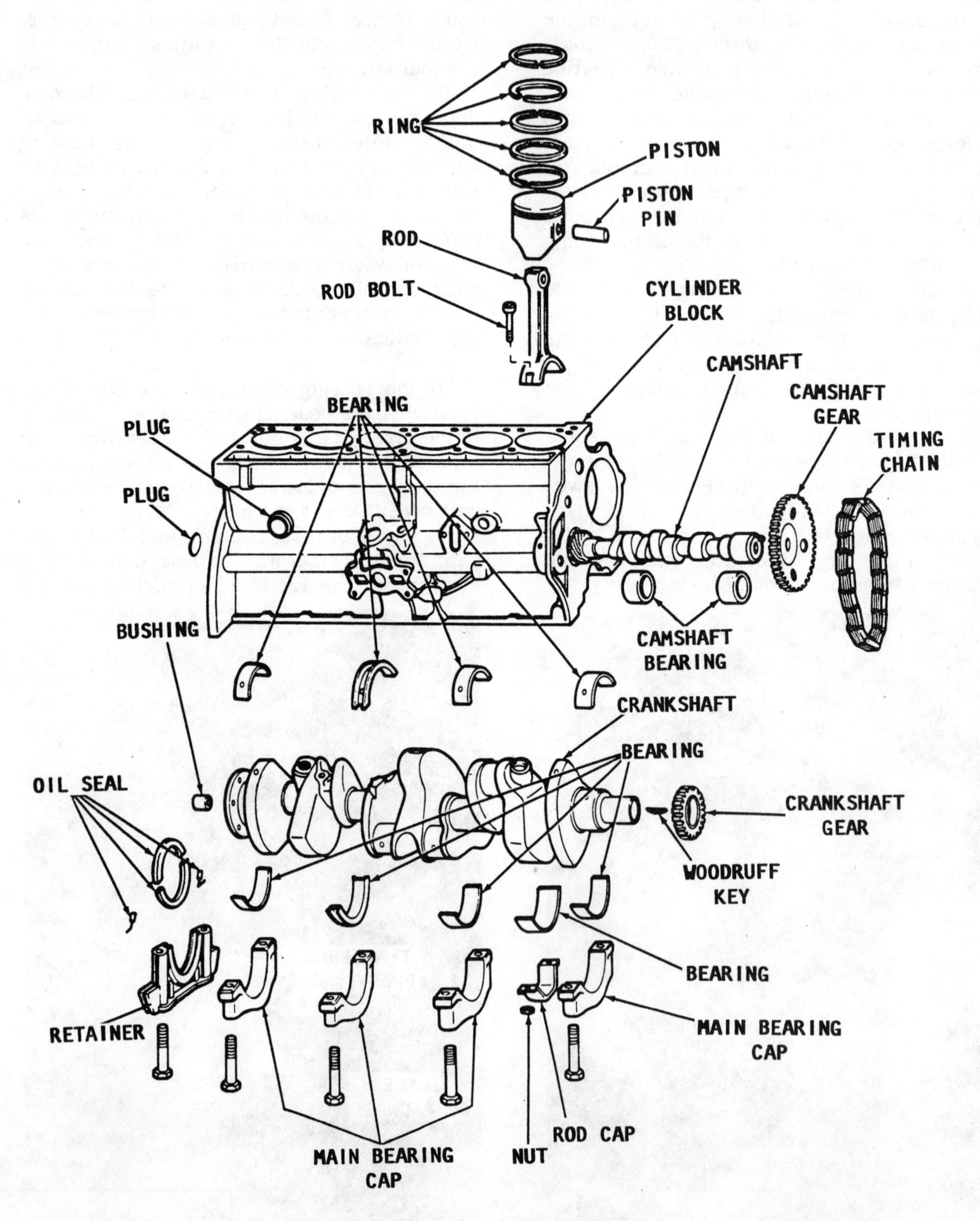

Exploded drawing of a Volvo Penta 6-cylinder in-line pushrod engine, with major parts identified.

are found, then install shims on the bearing shoulders as required.

A standard 0.001" or 0.002" undersize bearing may give the proper clearance. If the undersize bearing does not give proper clearance, the crankshaft journal will have to be reground for use with the next undersize bearing.

Continue the procedure until all of the bearings have been checked. Rotate the crankshaft to be sure there is an even drag for the complete turn without binding in any one spot.

Measure the crankshaft end play by first forcing the crankshaft to the extreme front position. Next measure the clearance at the front end of the rear main bearing with a feeler gauge and compare the results with the Specifications given in the Appendix.

CRANKSHAFT REMOVAL IN-LINE ENGINES

Remove the engine; see Section 3-3.

Remove the manifolds, see Section 3-6.

Remove the head, see Section 3-10.

Remove the timing belt on an OHC engine, see Section 3-24.

Pushrod and OHC Engines

At the rear of the engine, remove the six bolts easily reached and exposed securing the bell housing to the block.

Remove the three bolts from the underneath side of the bell housing, as shown. These three bolts are threaded in from the

Removing/installing the six exposed bolts securing the bell housing to the cylinder block.

REMEMBER *the three hidden bolts underneath threaded in from the forward side of the flange.*

forward side of the block flange. They are the three bolts that make separating the bell housing from the engine very difficult while still mounted in the boat.

Separate the bell housing from the engine block, by sliding it free of the dowels imbeded in the block. Disconnect and remove the starter motor.

Remove the six bolts attaching the vibration damper to the flywheel. Remove the damper.

Remove the eight bolts securing the flywheel to the crankshaft. Lift the flywheel free of the crankshaft.

Remove the two screws underneath securing the pan and the rear sealing flange together. Remove the bolts securing the

Removing/installing the attaching bolts securing the vibration damper to the flywheel.

Removal/installation of the bolts securing the flywheel to the crankshaft.

Removal/installation of the raw water pump attaching hardware.

flange to the block, and then slide the flange free of the crankshaft.

If the seal in the flange appears to be damaged, it can be removed and a new seal driven into place.

Disconnect and remove the wiring harness. Make a record of each connection to ensure they will be connected in the same manner from which they were removed. One method of recording is to wrap a piece of masking tape to each wire, and then to identify the wire with a pen or other suitable marker. Remove the cover for the raw water strainer. The screen should be removed and cleaned at regular intervals.

Pushrod Engine

At the front of the cylinder block, remove the alternator. Be sure to make a note of how the wires are connected to the terminals, as an aid during assembling. The manufacturer has helped a little by making each of the terminals a different size.

Remove the two attaching nuts, and then separate the raw water pump from the engine block. **DISCARD** the O-ring.

Remove the nut from the forward end of the crankshaft. Slide the crankshaft pulley free of the crankshaft.

Remove the attaching bolts and washers from the perimeter of the timing gear cov-

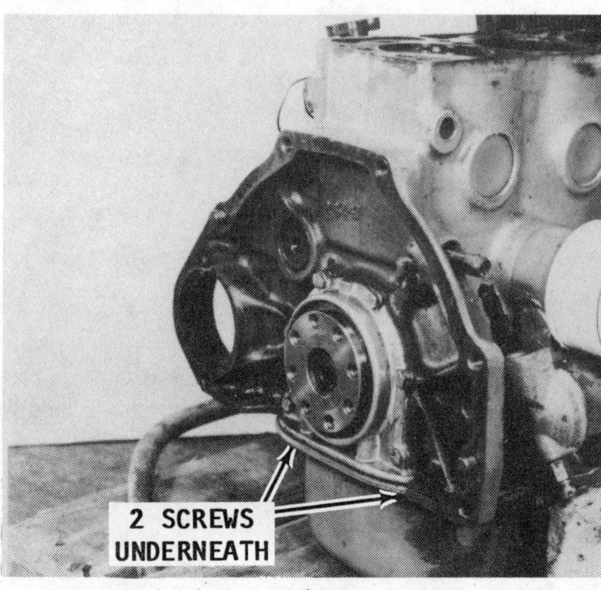

Removal/installation of the rear sealing ring.

Removal/installation of the crankshaft pulley.

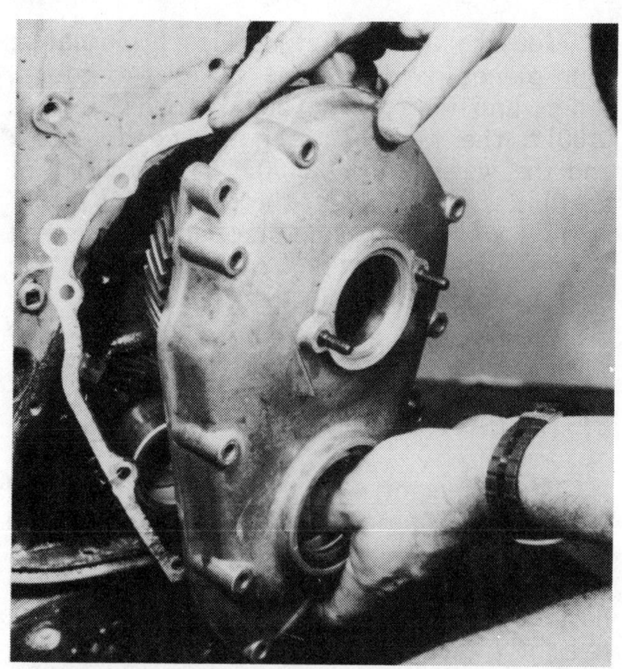

*Removal/installation of the rear cover. New seals should **ALWAYS** be installed during an overhaul.*

Using a puller to remove the crankshaft gear from the crankshaft.

er. Remove the cover. **DISCARD** the gasket.

Rotate the crankshaft until the No. 1 cylinder is at TDC. The timing mark on the camshaft should be aligned with the timing mark on the crankshaft gear. Use a puller and remove the hub.

Attach a "puller" to the drive gear on the end of the crankshaft. Take up on the "puller" until the gear is loose and can be removed from the crankshaft. Remove the oil nozzle for cleaning.

Drain the oil from the pan. If an engine work stand is not being used, and assuming the head has been removed, **CAREFULLY**

Alignment of the timing mark on the crankshaft gear and the mark on the camshaft gear.

lay the cylinder block down on suitable blocks of wood in the inverted position, with the pan facing upward. Remove the bolts securing the pan to the block. Remove the pan and **DISCARD** the gasket.

Remove the oil pump; see Section 3-30.

Remove the connecting rod caps. **TAKE TIME** to mark each bearing cap so you will be able to install it back onto the same rod from which it was removed. If the pistons are to be serviced, continue with that work, see Section 3-27.

BEFORE PROCEEDING

Before the main bearing caps are removed, perform a bearing check according to the procedures outlined at the beginning of this section, starting with the "Critical Words" on Page 3-78.

Using a bearing remover to pull the bearing from the end of the crankshaft.

Crankshaft cleaned, measured, and ready for installation.

Notice how the bearing caps are numbered from the front to the rear of the engine.

Remove the main bearing caps, and then lift the crankshaft out of the block. **ALWAYS** handle the shaft carefully to avoid damaging the highly finished journal surfaces.

The bearing in the end of the crankshaft may be removed by first releasing the lock-ring, and then working the bearing free with a bearing puller.

CLEANING

Clean the crankshaft with solvent and wipe the journals dry with a lint-free cloth. Blow out all of the oil passages with compressed air. Oil passageways lead from the rod journal to the main bearing journal. **BE CAREFUL** not to blow the dirt into the main bearing journal bore.

Support the crankshaft on **"V"** blocks at the No. 1 and No. 4 main bearing journals. Inspect each rod bearing journal surface. If the journal surface is scored or ridged, the crankshaft must be replaced or reground to ensure satisfactory performance with new bearings. A slight roughness may be polished out with fine-grit polishing cloth thoroughly wetted with engine oil. Burrs may be honed off with a fine stone.

Crankshaft in place ready for installation of the bearing caps and the Plastigage measurement.

Measure the diameter of each journal at four places to determine the out-of-round, taper, and wear. The out-of-round limit is 0.001"; the taper must not exceed 0.001"; and the wear limit is 0.0035". If any one of the limits is exceeded, the crankshaft **MUST** be reground to an undersize, and undersized bearing inserts must be installed. Check the Specifications listed in the Appendix.

Remove the old bearing shells and the rear oil seal.

INSTALLATION

If the bearing was removed from the end of the crankshaft, install the bearing and secure it in place with the lock ring.

Install new bearing shells. The upper half of the bearing shell has a hole. This half of the shell **MUST** be inserted between the crankshaft and the block. **CAREFULLY** place the crankshaft into the bearing halves on the block.

Clean the journals thoroughly of all traces of oil, and then place a piece of Plastigage on the bearing surface, the full width of the cap. Install the cap and torque the retaining bolts according to the specifications listed in the Appendix. **DO NOT** turn the crankshaft with the Plastigage in place or you will distort it and the reading will have no value.

Remove the bearing cap. Use the scale on the plastic strip to determine the clearance. The bearing journal is tapered and so is the Plastigage. Measure the Plastigage at the widest point and also at the narrowest point. These measurements will give you the minimum and maximum clearances.

Final tightening of the bearing cap after the Plastigage measurement indicated satisfactory clearance.

If the clearance exceeds 0.0025", a new insert should be installed. After you have installed the new insert make another Plastigage measurement. If the new clearance still does not give the proper clearance according to the specifications, then an undersized insert should be used.

Lubricate the rear bearing seal with oil. **DO NOT** allow any oil to get on the parting surface. Gradually push the seal with a hammer handle until the seal is rolled into place. To replace the upper half of the seal, use a small hammer and brass punch. Tap one end of the seal into the block groove, and then push the seal until it protrudes out the other side.

Notice how the bearing caps are numbered from the front to the rear of the engine. **TAKE TIME** to ensure each cap is installed in the correct position. Tighten the bolts just fingertight. Pry the crankshaft against the thrust surface of the upper half of the thrust bearing (the cap at the flywheel end). Hold the crankshaft in this position, and pry the thrust bearing cap in the opposite direction. This action will align the thrust surfaces of both halves of the bearing. Retain the pressure on the crankshaft against the thrust bearing, and tighten the cap bolts to the torque value of 92 ft-lbs (125 Nm).

Check the "end play" with a feeler gauge Acceptable "end play" is 0.0014" to 0.0058" (0.037 mm to 0.147 mm)

Install the rod caps, see Section 3-27.
Install the oil pump, see Section 3-30.
Install the oil pan.

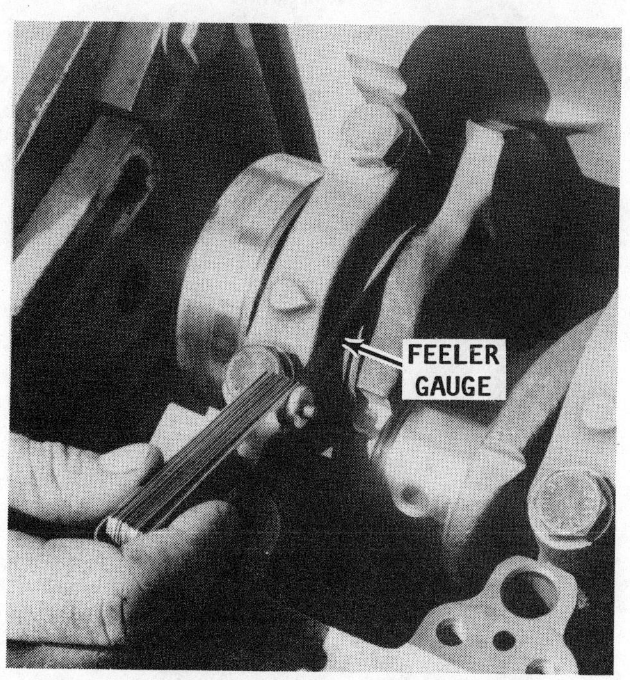

Checking the crankshaft "end play" with a feeler gauge.

Pushrod Engine

Rotate the crankshaft until the No. 1 cylinder is at TDC on the compression stroke. Set the key in the keyway of the crankshaft. A little grease will hold it in place. Install the drive gear onto the end of the crankshaft. The timing mark on the crankshaft gear should be aligned with the gear on the camshaft, as shown in the accompanying illustration. Coat the inside surface of the hub and the crankshaft with grease. Install the hub with the slot in the hub aligned with the key in the crankshaft.

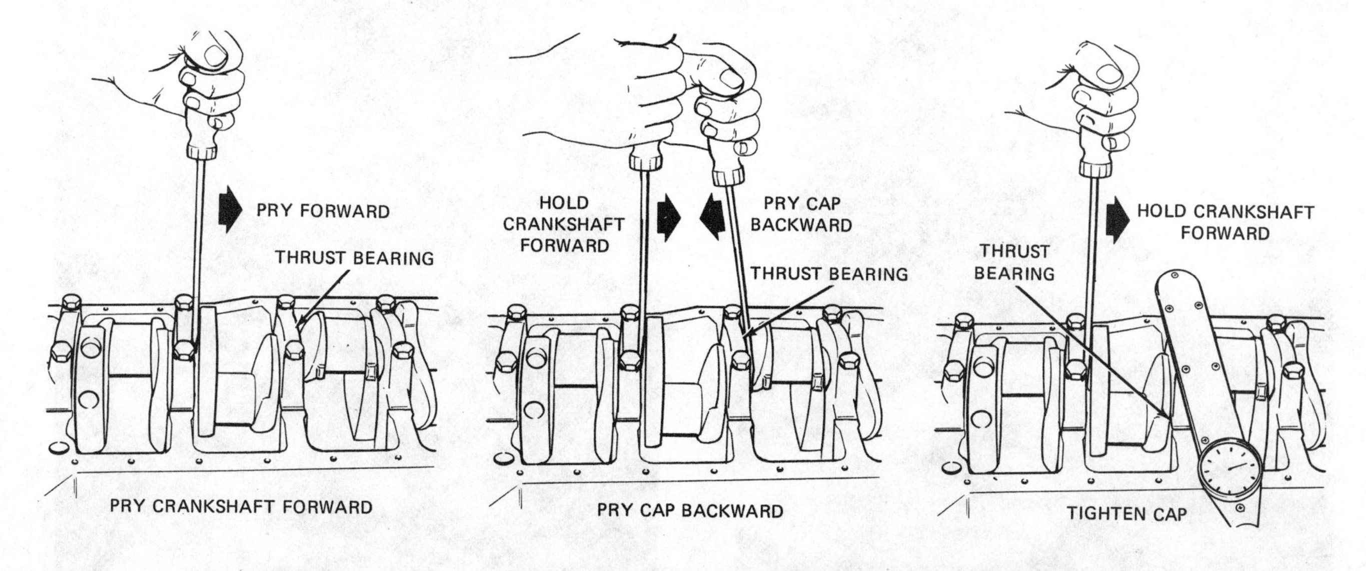

*The thrust bearing **MUST** be aligned properly before it is tightened. The thrust bearing is aligned by first prying the crankshaft forward, then prying the main bearing cap backward, and finally tightening the bolts to the torque value given in the Specifications in the Appendix.*

Timing marks on the crankshaft gear and the camshaft gear properly aligned -- pushrod engine.

Draw the hub into position with the crankshaft center bolt.

Install new seals into the timing gear cover. Install the cover and tighten the attaching bolts to a torque value of 8.0 ft-lbs (10 Nm).

Use a new O-ring and install the raw water pump.

Single & Double OHC Engine

Install a new crankshaft seal and a new accessory driveshaft seal into the front cover. Install the front cover using a **NEW** gasket. **REMEMBER** the two screws up

Installing the hub onto the crankshaft. The center crankshaft bolt can be used to "pull" the hub into place.

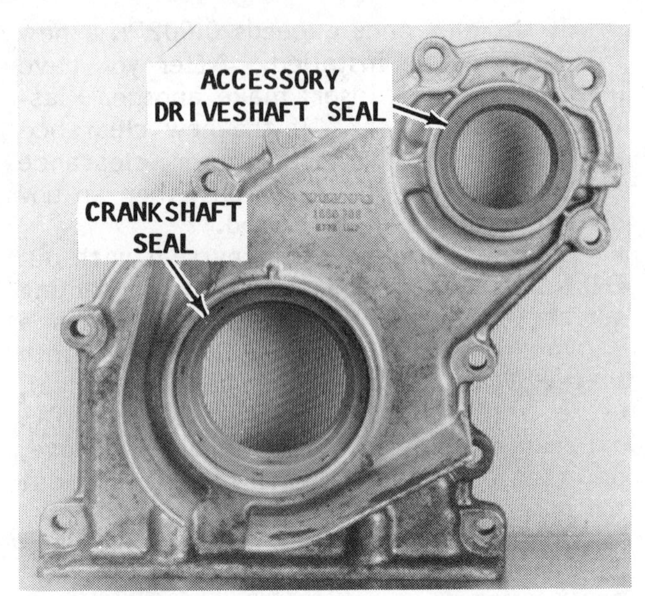

***ALWAYS** install new seals in the front cover during an overhaul — OHC engine.*

through the pan into the lower flange of the front cover. Tighten the attaching bolts to a torque value of 8.0 ft-lbs (10 Nm).

Rotate the crankshaft until the No. 1 cylinder is at TDC on the compression stroke. Place the key in the keyway. A dab of grease will hold in place. Install the inner guideplate and crankshaft drive pulley onto the end of the crankshaft, with the notch on the drive pulley aligned with the mark on the front cover. Install the key.

Install the timing belt, see Section 3-24

Securing the front cover with the attaching hardware on a pushrod engine.

Installing the crankshaft pulley after the front cover has been secured.

for single OHC engine and Section 3-25 for double OHC engine.

Pushrod and OHC Engines

At the rear of the engine, install a **NEW** sealing ring into the rear sealing flange. Lubricate the rubber lip of the seal and the mating surface on the crankshaft. Move the sealing flange into position against the engine block. **TAKE CARE** to prevent the rubber lip from being damaged by the edge of the crankshaft or being turned inside-out allowing the spring to pop out of position. Secure the ring in position with the attaching bolts. **REMEMBER** the two screws through the pan into the lower flange of the ring.

*A **NEW** O-ring should be installed any time the raw water pump is removed.*

Installing the eight flywheel bolts. Tighten the bolts to a torque value of 52 ft-lbs (70 Nm).

*A **NEW** seal should always be installed, if the rear sealing ring has been removed. **REMEMBER** the two screws to be installed up through the pan into the lower edge of the ring. Use care when installing the ring not to damage the seal, as described in the text.*

Moving the vibration damper into place onto the flywheel. Secure the damper with the attaching bolts.

*Installing the bell housing onto the cylinder block. **REMEMBER** the three bolts inserted underneath from the forward side of the cylinder block flange.*

Coat the inside surface of the flywheel with anti-sieze compound. Install the flywheel and tighten the eight attaching bolts to a torque value of 52 ft-lbs (70 Nm).

Install the vibration damper to the flywheel and secure it in place with the six attaching screws.

Place the bell housing in position, and then secure it with the **NINE** attaching bolts. **REMEMBER** the three bolts inserted underneath from the forward side of the cylinder block plate. Tighten the bolts securely.

If other service was performed on the engine, complete the assembling work by following the procedures in the appropriate sections of this chapter.

Install the engine, see Section 3-4.

Move the boat into a body of water, a test tank, or connect a flush attachment to the lower unit.

Start the engine and check the completed work. Operate the engine at **REDUCED** speed for the first hour.

*Checking the completed work by running the engine with the boat in a test tank. Operate the engine at **REDUCED** speed for the first hour.*

CAUTION: Water must circulate through the lower unit to the engine any time the engine is run to prevent damage to the water pump mounted on the engine.

3-29 CRANKSHAFT AND MAIN BEARINGS V6 AND V8 ENGINE

As explained in the introductory paragraphs to this section beginning on Page 3-78, the assumption is made that other service work will be performed on the engine while it is out of the boat and the crankshaft is removed.

Service work affecting the crankshaft or the crankshaft bearings may also affect the connecting rod bearings. Therefore, it is highly recommended that the rod bearings be carefully inspected as they are removed, see Section 3-27.

If the crankpins are worn to the extent that the crankshaft should be replaced or reground, then this work should be done in a qualified shop. Attempting to save time and money by merely replacing the crankshaft bearing inserts will not give satisfactory engine performance.

CRANKSHAFT REMOVAL

Drain the water from the block by opening all of the drain valves.

Remove the engine, Section 3-3.

Remove the oil pan.

Remove the oil pump.

Remove the flywheel and coupler.

Remove and identify the connecting rod caps, to ensure they will be installed back onto the same rods from which they were removed. Push the pistons to the top of the cylinders.

SPECIAL WORDS

The AQ271C and other new GMC 350 CID V8 engines have a new designed rear main seal. This new seal consists of a separate assembly which is bolted to the block and provides a better seal against oil leakage at this point.

Perform a main bearing clearance check as outlined in this section, beginning with the "Critical Words" on Page 3-78.

Remove and identify the main bearing caps. Inspect the caps and journals for signs of excessive wear. Remove the crankshaft from the block and support it on "V" blocks at the No. 1 and No. 4 main bearing journals. Inspect each rod bearing journal surface. If the journal surface is scored or ridged, the crankshaft must be replaced or reground to ensure satisfactory performance with new bearings. A slight roughness may be polished out with fine-grit polishing cloth thoroughly wetted with engine oil. Burrs may be honed off with a fine stone.

Use a dial indicator to check the runout at the main bearing journals. Total indicator readings at each journal should not exceed 0.003".

CLEANING AND INSPECTING

Clean the crankshaft with solvent and wipe the journals dry with a lint-free cloth. **ALWAYS** handle the shaft carefully to avoid damaging the highly finished journal surfaces. Blow out all of the oil passages with compressed air. Oil passageways lead from the rod to the main bearing journals. **TAKE CARE** not to blow dirt into the main bearing journal bore.

Measure the diameter of each journal at four places to determine the out-of-round, taper, and wear. The out-of-round limit is 0.001"; the taper must not exceed 0.001"; and the wear limit is 0.0035". If any of these limits is exceeded, the crankshaft must be reground to an undersize, and undersized bearing inserts must be installed.

While checking the runout at each journal take time to note the relation of the

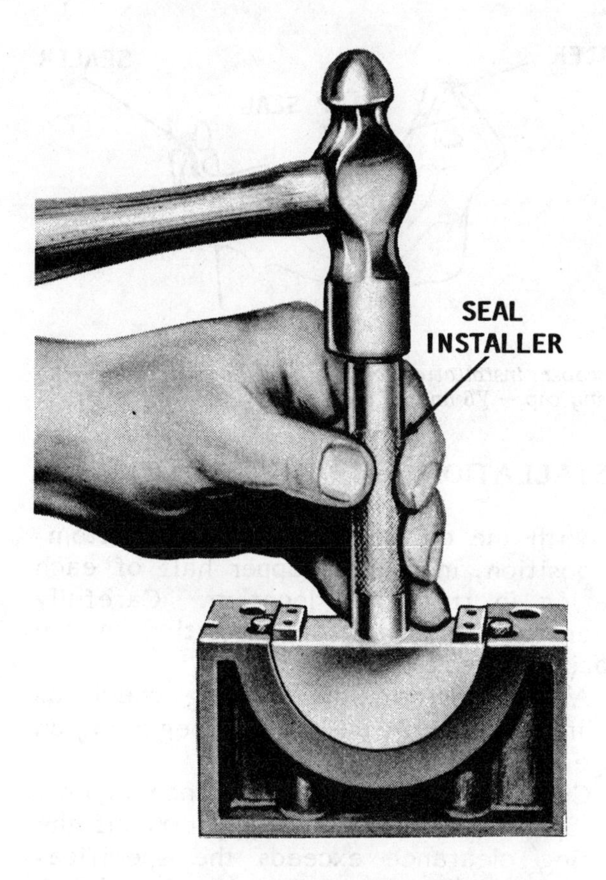

Use a rounded tool to roll the top oil seal into its groove after it has been well lubricated.

"high" spot (or maximum eccentricity) on each journal to the others. "High" spots on all journals should come at the same angular location. If the "high" spots do not appear to be in the same angular locations, the crankshaft has a "crook" or "dogleg" in it making it unsatisfactory for service.

Install a new rear main bearing oil seal in the cylinder block and main bearing cap.

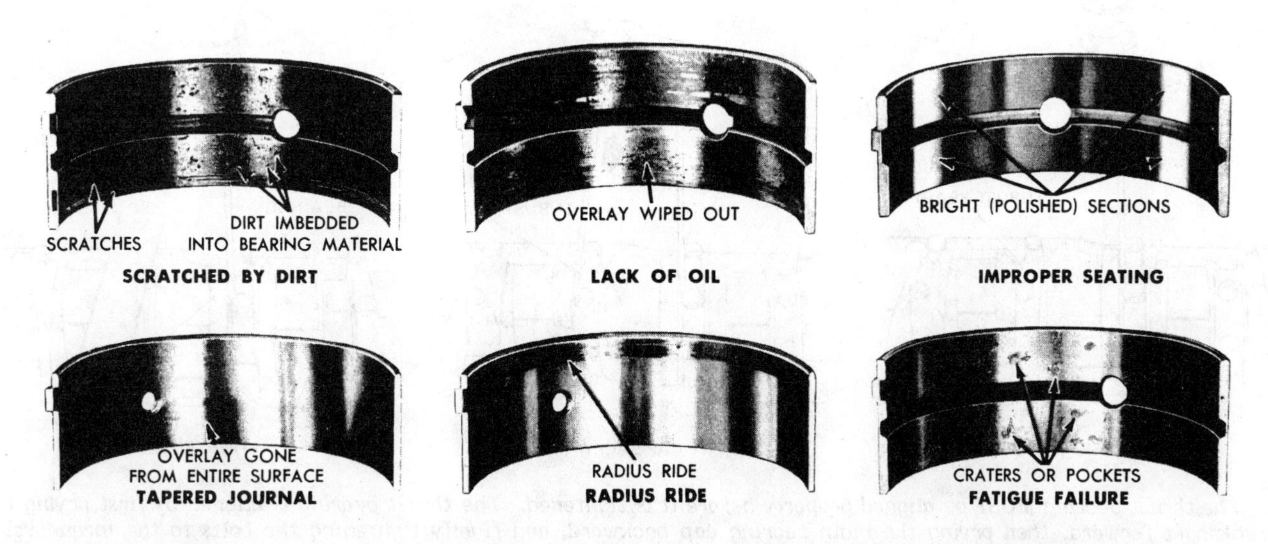

Examples of various wear patterns on bearing halves, including possible reasons for the condition.

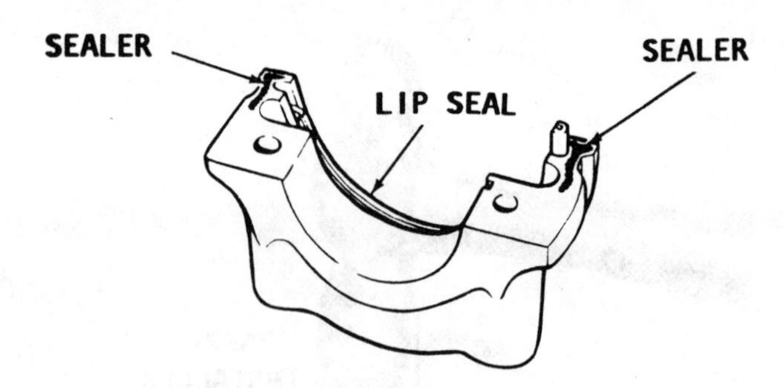

Proper installation of the seal in the rear main bearing cap -- V6 and V8 engine.

INSTALLATION

With the engine secured in the bottom-up position, install the upper half of each bearing in its proper location. Carefully lower the crankshaft into position in the bearing halves on the block.

Again, perform the bearing check as outlined earlier in this section beginning on Page 3-79

Compare the bearing clearance against the Specifications in the Appendix. If the bearing clearance exceeds the specifications, it is advisable to install a new bearing. However, if the bearing is in good condition and is not being checked because of bearing noise, it is not necessary to replace the bearing. After finishing the Plastigage check, install the bearing cap again and tighten the cap bolts to the torque value given in the Specifications in the Appendix, then loosen it 1/2 turn.

Continue making the Plastigage test at each bearing. When all bearings have been

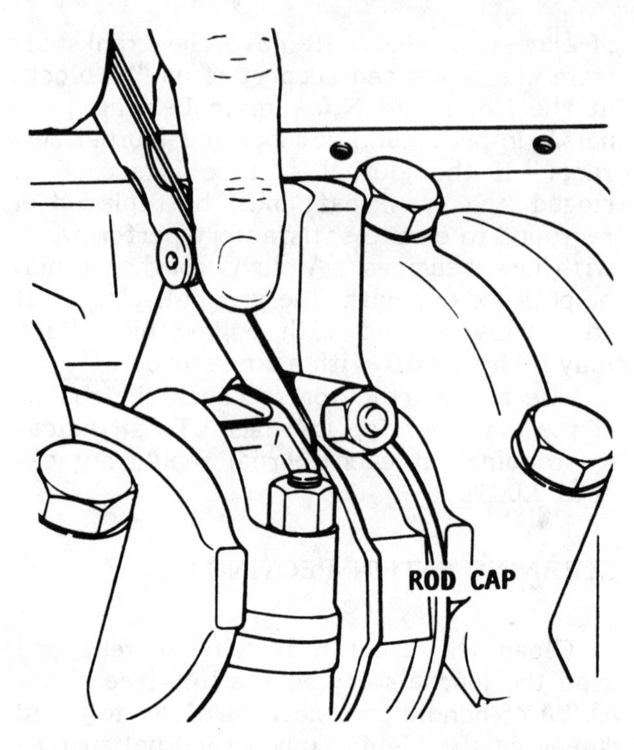

Checking the piston rod side clearance using a feeler gauge, as described in the text.

installed and tested, tighten the cap bolts again to the required torque valve.

Install the rod caps, Section 3-27.

Install the flywheel and coupler.

Install the oil pump.

Install the oil pan.

If other service was performed on the engine, complete the assembling work by following the procedures in the appropriate sections of this chapter.

Install the engine, Section 3-4.

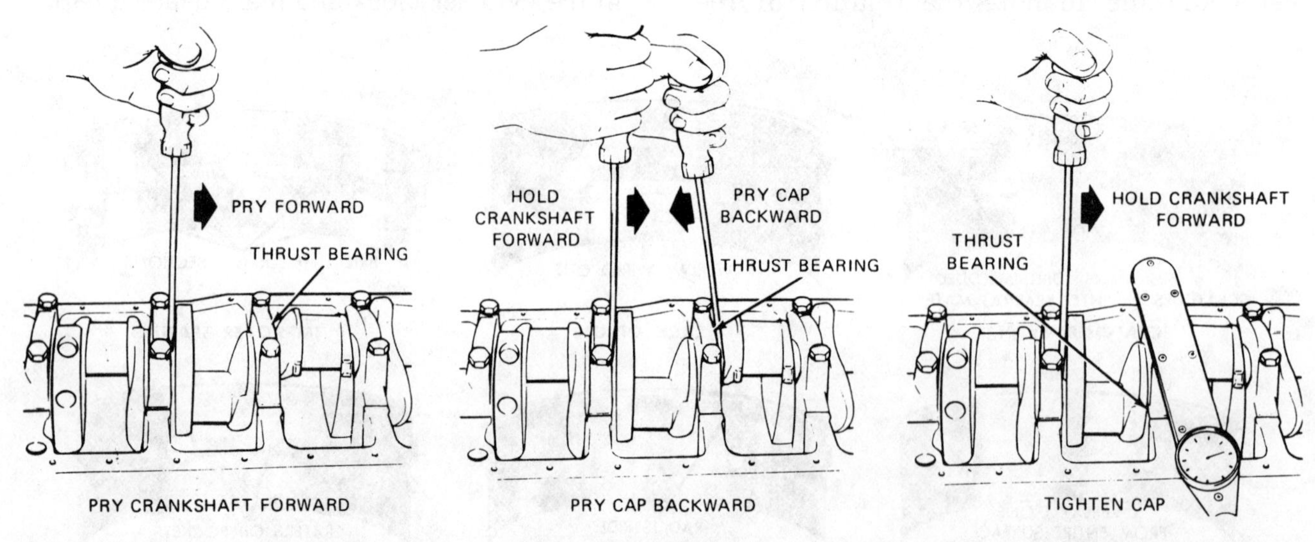

*The thrust bearing **MUST** be aligned properly before it is tightened. The thrust bearing is aligned by first prying the crankshaft forward, then prying the main bearing cap backward, and finally tightening the bolts to the torque value given in the Specifications in the Appendix.*

Fill the crankcase with the proper weight oil.

Close the water drain valves.

Move the boat into a body of water, a test tank, or connect a flush attachment to the lower unit.

Start the engine and check the completed work. Operate the engine at **REDUCED** speed for the first hour.

CAUTION: Water must circulate through the lower unit to the engine any time the engine is run to prevent damage to the water pump mounted on the engine.

3-30 OIL PUMP SERVICE

This section is divided into two parts. The first portion covers the in-line pushrod, single OHC, and double OHC engines. The same oil pump is used in these three cylinder blocks.

The second portion of this section covers the oil pumps installed on the GMC V6 and V8; and the Ford V8 engines. The oil pumps differ in GMC and Ford engines, but the procedures are very similar. Exploded drawings and differences are identified.

IN-LINE ENGINES PUSHROD, SINGLE OHC, AND DOHC

Description

The oil pump consists of two gears and a pressure relief valve, enclosed in a two-piece housing. The relief valve is a spring and ball arrangement. In a pushrod in-line engine, the oil pump is driven by the distributor shaft, which is driven by a helical gear on the camshaft. In an OHC engine, the oil pump is driven by the distributor shaft, which is driven by a helical gear on the accessory drive shaft.

From the pump, oil is forced through an oil cooler and conventional oil filter before circulating through the engine, as shown in the oil circulation routing drawing on this page.

REMOVAL

Remove the engine, see Section 3-3.

Remove the oil pan.

Remove the two oil pump flange mounting bolts. Remove the pump, screen, and delivery pipe as an assembly. Remove the

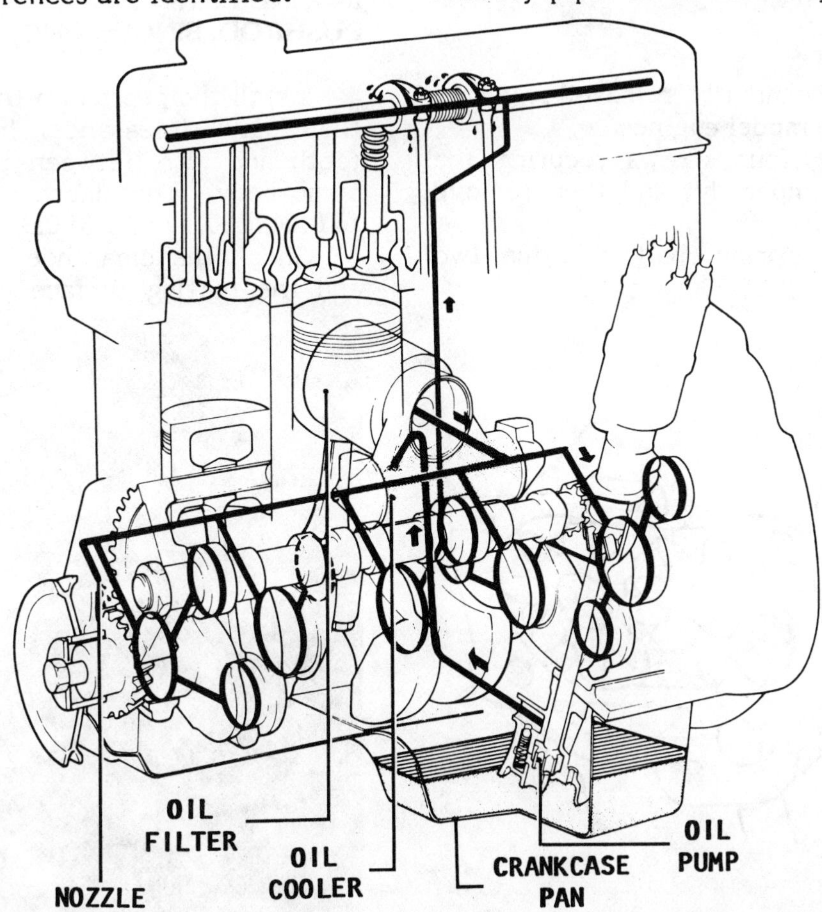

Oil circulation routing in an in-line pushrod engine. In an OHC engine, the circulation is almost identical except for the circulation for the overhead camshaft.

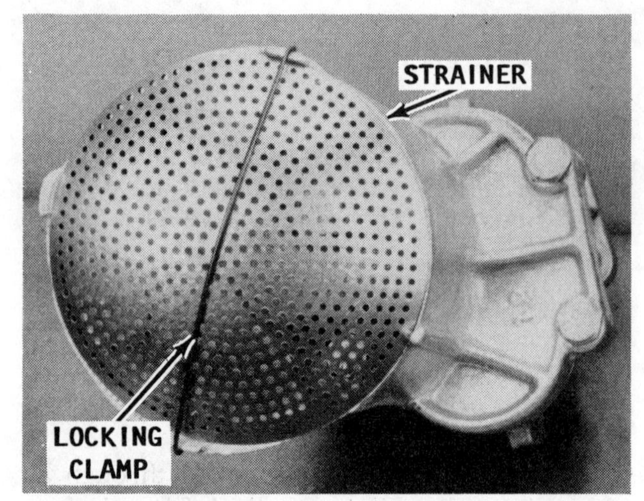

Strainer and locking clamp installed on early model oil pumps.

rubber ring for the oil pipe from the cylinder block. The ring may stay with the pipe when it is removed.

DISASSEMBLING AND CLEANING

Slide the delivery pipe out of the pump body. Remove the locking clamp, and then remove the strainer.

SPECIAL WORDS

The strainer cannot be removed from the oil pump on late model engines.

Remove the four screws securing the cover to the pump body, and then remove the cover.

Remove the spring, ball, and the two gears.

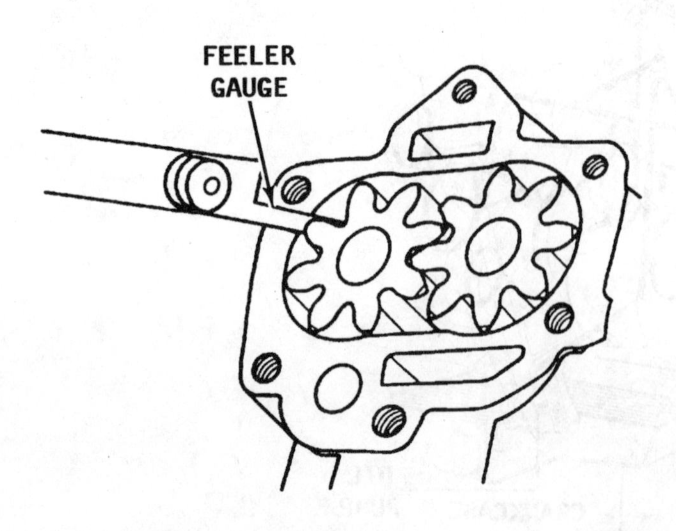

Measuring the clearance between the gear teeth and the pump body.

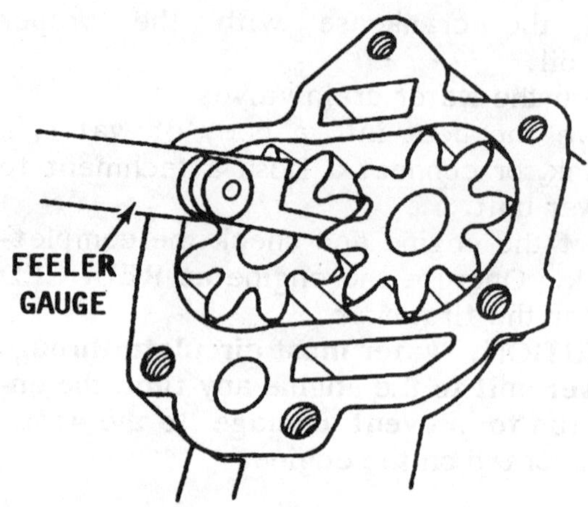

Checking the "play" (clearance) between the gear teeth with a feeler gauge.

Clean all the parts thoroughly and replace the gears if they appear to be worn. The gears **MUST** be replaced as a pair.

Inspect the sealing rings on both ends of the delivery pipe. These rings are made of a special rubber and are manufactured to very close tolerances. Therefore, use only genuine Volvo Penta sealing rings.

ASSEMBLING AND INSTALLATION IN-LINE ENGINES PUSHROD, SINGLE OHC, AND DOHC

Install the gears into the housing. Check the "play" (clearance) between the gear teeth and also between the teeth and the pump body. The clearance should be 0.006" to 0.014" (0.15 mm to 0.35 mm).

Check the clearance between the gears and the mating surface of the pump body

Oil pump cleaned and properly installed, with the delivery tube in place, ready for service.

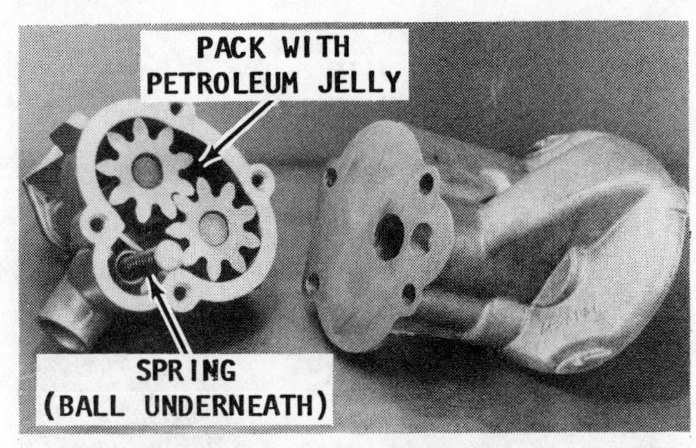

Pump ready to be packed with Petroleum jelly, prior to installation.

with a straight edge and feeler gauge, as shown. This clearance should be 0.005" to 0.008" (0.02 mm to 0.12 mm).

If the gear end clearance is satisfactory, remove the gears and pack the gear pocket full of petroleum jelly. **DO NOT** use chassis lube.

CAUTION: Unless the pocket is packed with petroleum jelly it may not prime itself when the engine is started.

Insert the ball into place, and then the spring. Install the pump cover and secure it in place with the attaching screws.

If the strainer was removed (early model units only) install the strainer and secure it in place with the locking clamp.

If new sealing rings are to be installed onto the ends of the delivery pipe, coat the rings with soapy water before installation. This simple task will enable the pipe to take up its proper position more easily. Hand press the delivery pipe into the pump housing.

Install the pump onto the cylinder block with the delivery pipe sliding into the hole in the block. The slot in the pump shaft must index over the tang on the end of the oil pump driveshaft in the block. No gasket is used between the pump and the block. Check to be sure the pump flange is completely against the mating surface of the block before installing the attaching screws. Install and tighten the screws.

Install the pan. Install the engine, see Section 3-4.

Start the engine and check that the oil pressure light goes out. Check for oil leaks.

CAUTION: Water must circulate through the lower unit to the engine any time the engine is run to prevent damage to the water pump mounted on the engine.

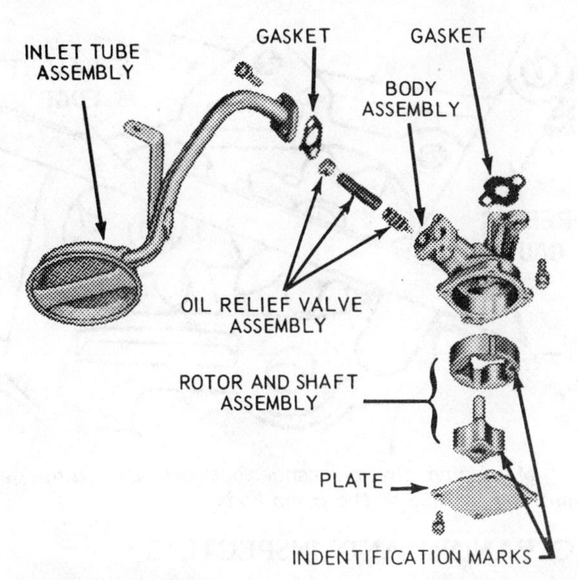

Exploded drawing of the oil pump used on AQ190 and AQ240 engines.

OIL PUMP SERVICE V6 AND V8 ENGINES

Description

The oil pump consists of two gears and a pressure regulator valve, enclosed in a two-piece housing. The oil pump is mounted on the cylinder block and is driven by the distributor shaft, which is driven by a helical gear on the camshaft.

REMOVAL

Remove the engine, see Section 3-3.

Remove the oil pan.

Remove the pump-to-rear main bearing cap bolt, and then remove the pump and extension shaft.

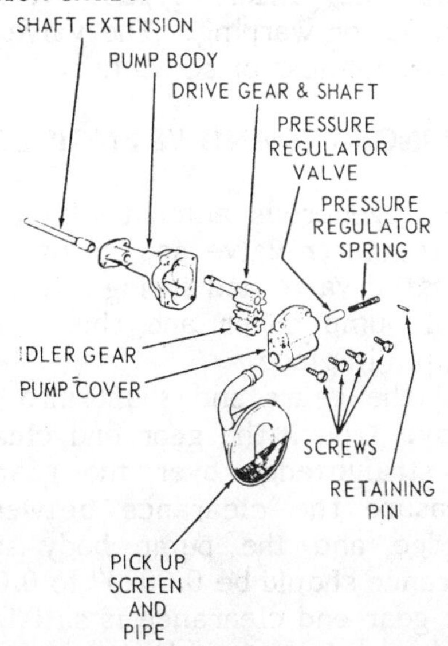

Exploded view of the oil pump installed on all V6 and V8 engines except those listed in the illustration above.

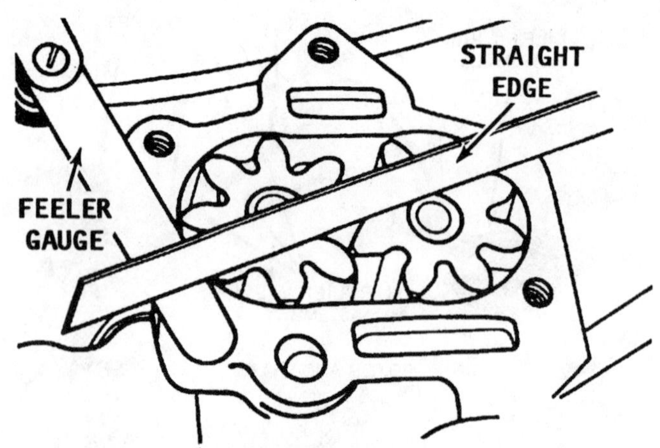

Measuring the clearance between the pump gears and the surface of the pump body.

CLEANING AND INSPECTING

Wash the gears thoroughly and inspect them for wear and scores. If either gear is defective, they must be replaced as a **PAIR.**

Remove the oil pressure regulator valve cap, spring, and valve. The oil filter bypass valve and spring **MUST NOT** be removed because they are staked in place.

Wash the parts removed and check each one carefully. Inspect the regulator valve for wear or scores. Check the regulator valve spring to be sure it is not worn on its side or has collapsed. If in doubt about the condition of the spring, install a new one. Clean the screen staked in the cover.

Check the regulator valve in its bore in the cover. The clearance for the valve should be only a slip fit. If any side clearance can be felt, the valve or the cover should be replaced.

Inspect the filter bypass valve for cracks, nicks, or warping. The valve should be flat with no nicks or scratches.

ASSEMBLING — V6 AND V8 ENGINES

Apply a generous amount of oil to the pressure regulator valve and spring. Install the lubricated valve and spring into the bore of the oil pump cover and then slide the retaining pin in place.

Install the gears and shaft into the oil pump body. Check the gear end clearance. Place a straightedge over the gears, and then measure the clearance between the straightedge and the pump body surface. The clearance should be 0.0023" to 0.0058".

If the gear end clearance is satisfactory, remove the gears and pack the gear pocket full of petroleum jelly. **DO NOT** use chassis lube.

Oil pump cleaned, installed, and ready to be packed with Petroleum jelly, as described in the text.

CAUTION: Unless the pocket is packed with petroleum jelly it may not prime itself when the engine is started.

Install the gears and shaft again, pushing the gears into the petroleum jelly. Place a new gasket in position, and then install the cover screws. Tighten the screws alternately and evenly to the torque value given in the Specifications in the Appendix.

If the oil pump pickup screen was removed, apply sealer to the mating surfaces before you drive the pipe into position. An **AIR LEAK** could cause a **LOSS** of **OIL PRESSURE.**

INSTALLATION — V6 AND V8 ENGINES

Assemble the pump and extension shaft to the rear main bearing cap with the slot in the top end of the extension shaft aligned with the drive tang on the lower end of the distributor driveshaft. Install the pump-to-rear bearing cap bolt. The correct position for the oil pump screen is with the bottom edge parallel to the oil pan rails.

Install the oil pan.

Install the engine, see Section 3-4.

Fill the crankcase with the proper oil.

Close the water drain valves.

Start the engine and check for leaks.

CAUTION: Water must circulate through the lower unit to the engine any time the engine is run to prevent damage to the water pump mounted on the engine.

4
FUEL

4-1 INTRODUCTION

The fuel system includes the fuel tank, fuel pump, fuel filters, carburetor, connecting lines, and the parts associated with these items. Regular maintenance of this system, as an aid to satisfactory performance from the engine, is limited to changing the fuel filter and cleaning the flame arrester at regular intervals.

If a sudden increase in gas consumption is noticed, or if the engine does not perform properly, and if the flame arrester has been checked and found to be clean, then an overhaul, including boil-out, of the carburetor, or replacement of the fuel pump may be in order.

SINGLE-DIAPHRAGM FUEL PUMP

The fuel pump sucks gasoline from the fuel tank and delivers it to the carburetor in sufficient quantities, under pressure, to satisfy engine demands under all operating conditions.

The pump is operated by a two-part rocker arm. The outer part rides on an eccentric on the camshaft and is held in constant contact with the camshaft by a strong return spring. The inner part is connected to the fuel pump diaphragm by a short connecting rod. As the camshaft rotates, the rocker arm moves up and down. As the outer part of the rocker arm moves downward, the inner part moves upward, pulling the fuel diaphragm upward. This upward movement compresses the diaphragm spring and creates a vacuum in the fuel chamber below the diaphragm. The vacuum causes the outlet valve to close and permits fuel from the gas tank to enter the chamber by way of the fuel filter and the inlet valve.

Now, as the eccentric on the camshaft allows the outer part of the rocker arm to move upward, the inner part moves downward, releasing its hold on the connecting rod. The compressed diaphragm spring then exerts pressure on the diaphragm, which closes the inlet valve and forces fuel out through the outlet valve to the carburetor.

DUAL-DIAPHRAGM FUEL PUMP

Some fuel pumps have two diaphragms and a sight bowl attached on the outside of the pump. The diaphragms are separated by a metal spacer. This type of fuel pump has four important safety features:

1- If the primary diaphragm fails, the pump continues to function with the second diaphragm.

2- Fuel cannot leak outward from the pump. The only possible place it can leak to is into the space between the diaphragms.

3- Fuel observed in the sight bowl indicates a faulty pump.

4- The possibility of both diaphragms failing at the same time is extremely remote because they are made of different materials and are shaped differently.

No maintenance is required or possible on the dual-diaphragm pump. If fuel is detected in the sight bowl replace the pump.

4-2 FUEL SYSTEM TROUBLESHOOTING

The following paragraphs provide an orderly sequence of tests to pinpoint problems in the system. It is very rare for the carburetor by itself to cause failure of the engine to start.

Many times fuel system troubles are caused by a plugged fuel filter, a defective fuel pump, or by a leak in the line from the fuel tank to the fuel pump. Would you believe, a good majority of starting troubles which are traced to the fuel system are the result of an empty fuel tank and to aged fuel.

Fuel will begin to sour in three to four months and will cause engine starting problems. A fuel additive such as Sta-Bil may be used to prevent gum from forming during storage or prolonged idle periods.

If the automatic choke should stick in the open position, the engine would have trouble starting. This condition can be quickly corrected by rapid movement of the accelerator which will discharge fuel into the intake manifold and the engine will start.

If the automatic choke should stick in the closed position, the engine will flood making it very difficult to start. To correct this condition, move the fast idle or warm up lever to the wide-open position as the engine is cranked. This action will activate the unloader linkage to open the choke valve and help the flooded engine to start.

Mike Basso, driving this boat equipped with a high performance engine and standard Volvo Penta stock Model 280 stern drive, averaged 110 mph (177 kph) for two runs over a measured course. Imagine the demand on the complete fuel system to deliver the quantity of fuel required by such performance.

When the engine is hot, the fuel system can cause starting problems. After a hot engine is shut down, the temperature inside the fuel bowl may rise to 200°F and cause the fuel to actually boil. All carburetors are vented to allow this pressure to escape to the atmosphere. However, some of the fuel may percolate over the high-speed nozzle and overflow into the intake manifold.

In order for this raw fuel to vaporize enough to burn, considerable air must be added to lean out the mixture. Therefore, the only remedy is to open the throttle as wide as possible and to crank the engine until enough air is drawn in to provide the proper mixture for the engine to start. **NEVER** move the throttle lever back-and-forth in an attempt to start a hot engine. This action will only compound the problem by adding more fuel to an already too-rich mixture.

If the needle valve and seat assembly is leaking, an excessive amount of fuel may enter the intake manifold in the following manner: After the engine is shut down, the pressure left in the fuel line will force fuel past the leaking needle valve. This extra fuel will raise the level in the fuel bowl and cause fuel to overflow into the intake manifold.

A continuous overflow of fuel into the intake manifold may be due to a sticking or defective float which would cause an extra high level of fuel in the bowl and overflow into the intake manifold.

ROUGH ENGINE IDLE

If an engine does not idle smoothly, the most reasonable approach to the problem is to perform a tune-up to eliminate such areas as: defective points; faulty spark plugs; and timing out of adjustment.

Other problems that can prevent an engine from running smoothly include: An air leak in the intake manifold; uneven compression between the cylinders; and sticky valves.

Of course any problem in the carburetor affecting the air-fuel mixture will also prevent the engine from operating smoothly at idle speed. These problems usually include: Too high a fuel level in the bowl; a heavy float; leaking needle valve and seat; a dirty flame arrestor; defective automatic choke; and improper adjustments for idle mixture or idle speed.

EXCESSIVE FUEL CONSUMPTION

Excessive fuel consumption can be the result of any one of three conditions, or a combination of all three.

1- Inefficient engine operation.

2- Faulty condition of the hull, including excessive marine growth.

3- Poor boating habits of the operator.

If the fuel consumption suddenly increases over what could be considered normal, then the cause can probably be attributed to the engine or boat and not the operator.

Marine growth on the hull can have a very marked effect on boat performance. This is why sail boats always try to have a haul out as close to race time as possible. While you are checking the bottom take note of the propeller condition. A bent blade or other damage will definitely cause poor boat performance.

If the hull and propeller are in good shape, then check the fuel system for a leak. Check the line between the fuel pump and the carburetor while the engine is running and the line between the fuel tank and the pump when the engine is not running. A leak between the tank and the pump many times will not appear when the engine is operating because the suction created by the pump sucking fuel will not allow the fuel to leak. Once the engine is turned off and the suction no longer exists, fuel may begin to leak.

If a minor tune-up has been performed and the spark plugs, points, and timing are properly adjusted, then the problem most likely is in the carburetor and an overhaul is in order. Check the power valve and the needle valve and seat for leaking. Use extra care when making any adjustments affecting the fuel consumption, such as the float level, automatic choke, vacuum-control, and the power valve. Any time the automatic choke is checked, **BE SURE** the heat tube is open and the vacuum system is operating properly.

The flame arrester should be cleaned at regular intervals.

ENGINE SURGE

If the engine operates as if the load on the boat were being constantly increased and decreased even though you are attempting to maintain a constant engine speed, the problem can most likely be attributed to the fuel pump.

The next few paragraphs briefly describe operation of the fuel pump. This description is followed by detailed procedures for testing the pressure; testing the volume; removing; and installing the fuel pump.

Because the fuel pump diaphragm is moved downward only by the diaphragm spring, the pump delivers fuel to the carburetor only when the pressure in the outlet line is less than the pressure exerted by the diaphragm spring. This lower pressure condition exists when the carburetor float needle valve is unseated and the fuel passages from the pump into the carburetor float chamber are open.

When the needle valve is closed and held in place by the pressure of the fuel on the float, the pump builds up pressure in the fuel chamber until it overcomes the pressure of the diaphragm spring. This pressure almost stops movement of the diaphragm until more fuel is needed in the carburetor float bowl.

From this description you can appreciate why the the fuel pump diaphragm and the carburetor float must be in good condition at all times for proper engine performance.

4-3 FUEL LINE AND FUEL PUMP TESTS

CAUTION: Gasoline will be flowing in the engine compartment during this test. Therefore, guard against fire by grounding the high-tension wire to prevent it from sparking.

The high tension wire between the coil and the distributor can be grounded by either pulling it out of the distributor cap and grounding it, or by connecting a jumper wire from the primary (distributor) side of the igniton coil to a good ground.

Disconnect the fuel line at the carburetor. Place a suitable container over the end of the fuel line to catch the fuel discharged, and then crank the engine. If the

Fuel pump installed on a 4-cylinder OHC engine.

fuel pump is operating properly, a healthy stream of fuel should pulse out of the line.

If the engine does not start even though there is adequate fuel flow from the fuel line, the fuel filter in the carburetor inlet may be plugged or the fuel inlet needle valve and the seat may be gummed together and prevent adequate fuel flow.

Continue cranking the engine and catching the fuel for about 15 pulses to determine if the amount of fuel decreases with each pulse or maintains a constant amount. A decrease in the discharge indicates a restriction in the line. If the fuel line is plugged, the fuel stream may stop. If there is fuel in the fuel tank but no fuel flows out of the fuel line while the engine is being cranked, the problem may be in one of three areas:

1- The line from the fuel pump to the carburetor may be plugged as already mentioned.

2- The fuel pump may be defective.

3- The line from the fuel tank to the fuel pump may be plugged or the line may be leaking air.

The following test explores these possibilities.

FUEL LINE TEST

Before spending time and effort checking the fuel system, make a careful survey to ensure some object such as a tackle box, etc. is not resting on the fuel line and restricting the flow of fuel.

The fuel line from the tank to the fuel filter can be quickly tested by disconnecting the existing fuel line at the fuel filter and connecting a six-gallon portable tank and fuel line. This simple substitution eliminates the fuel tank and fuel lines in the boat. Now, start the engine and check the performance.

If the problem has been corrected, the fuel system between the fuel pump inlet and the fuel tank, including the fuel tank is at fault. This area includes the fuel line, the fuel pickup in the tank, the fuel filter, anti-siphon valve, the fuel tank vent, excessive foreign matter in the fuel tank, and loose fuel fittings sucking air into the system. Improper size fuel fittings can also restrict fuel flow.

FUEL PUMP TESTING

Most fuel pumps on late-model engines are of the sealed-type and cannot be repaired. The cost of a new pump is nominal, therefore, even if the pump is not sealed, it is usually more practical to replace the pump instead of attempting to repair it. For your safety, the new pump must be a Coast Guard approved marine-type unit.

Proper fuel pressure and volume are both necessary for proper engine performance. If the pressure is not adequate, the fuel level in the float bowl of the carburetor will be low and result in a lean mixture and fuel starvation at high speeds. If the pressure is too high the fuel level in the float bowl will rise and result in a rich mixture and flooding.

If the volume is not adequate, the engine will be starved at high speeds.

Fuel pump testing consists of checking the output pressure of the pump and the

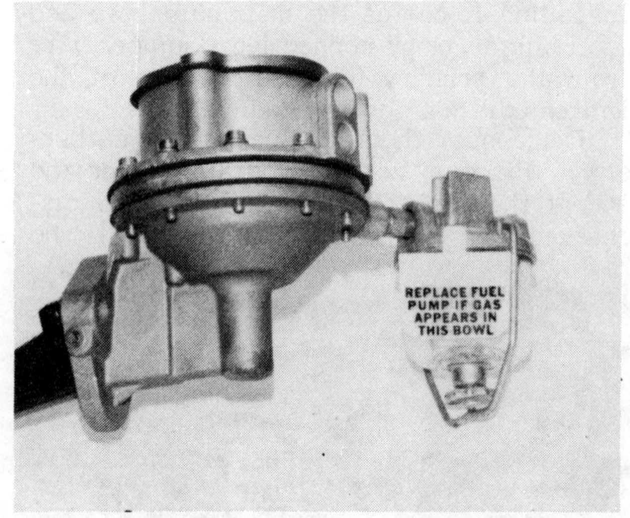

*Fuel in the sight glass indicates a ruptured diaphragm. The fuel pump **MUST** be replaced.*

The Volvo fuel pump is equipped with a manual lever which can be actuated to assist engine start.

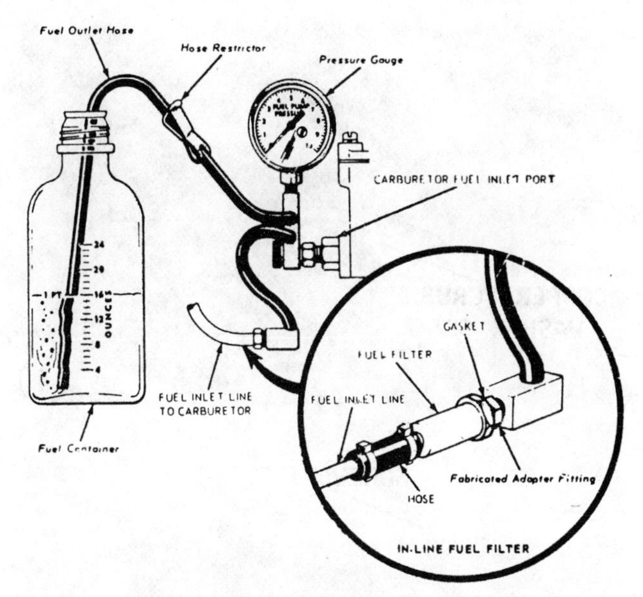

Arrangement of pressure gauge and container to test fuel pump pressure and volume.

volume delivered to the carburetor.

FUEL PUMP PRESSURE TEST

Remove the flame arrester. Disconnect the fuel line at the carburetor. **TAKE CARE** not to spill fuel on a hot engine because it may **IGNITE.**

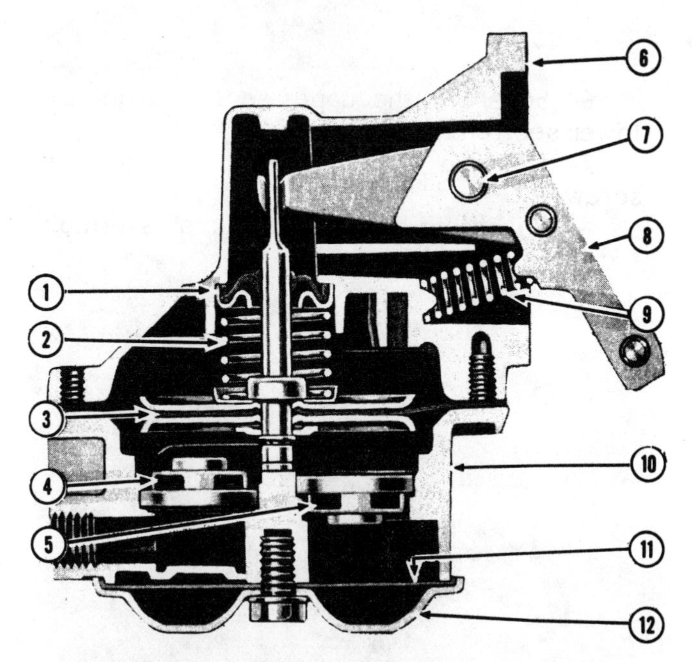

V-8

1	Oil Seal and Retainer	7	Pivot Pin
2	Diaphragm Spring	8	Rocker Arm and Lever Assembly
3	Diaphram Assembly	9	Rocker Arm Return Spring
4	Inlet Valve	10	Fuel Cover
5	Outlet Valve	11	Pulsator Diaphragm
6	Pump Body	12	Pulsator Cover

Cutaway view of the fuel pump used on V8 engines.

ALWAYS have a fire extinguisher handy when working on any part of the fuel system. **REMEMBER,** a very small amount of fuel vapor in the bilge, has the potential explosive power of one stick of **DYNAMITE.**

Connect a pressure gauge, restrictor, container, and flexible hose between the fuel filter and the carburetor. Start the engine. With the engine idling, vent the outlet hose into the container by opening the hose restrictor momentarily. Close the hose restrictor and allow the pressure to stabilize. The pressure reading should be between 3.5 and 5.5 psi.

FUEL PUMP VOLUME TEST

If the fuel pump pressure is within the 3.5 to 5.5 psi range, test the volume by opening the hose restrictor with the engine idling and collect the fuel discharged into the graduated container. The fuel pump should discharge a pint of fuel in 30 seconds for a six-cylinder engine and in 20 seconds for a V8 engine.

4-4 VOLVO FUEL PUMP SERVICE

INTRODUCTION

Two types of fuel pumps are used on the Volvo Penta 4- and 6-cylinder in-line engines. One is a sealed unit and cannot be repaired. If troubleshooting indicates the fuel pump is at fault, the pump must be replaced with a new unit.

The other type fuel pump may be serviced. Replacement parts are available at the local Volvo Penta marine dealer.

The fuel strainer may be removed, cleaned, and replaced, while the pump remains in place on the engine block.

The following procedures give detailed instructions to remove, service, and install the repairable fuel pump.

REMOVAL

1- Have a small container available to catch any residual fuel left in the fuel lines. Disconnect the fuel line to the pump from the fuel tank. Plug the line to prevent fuel from siphoning out of the tank. Disconnect the fuel line to the carburetors. Use two wrenches, as shown, to prevent damaging the fittings.

2- Remove the mounting bolts from the engine block. Pull the fuel pump, two gaskets and plastic insulator plate free. **DISCARD** the gaskets.

DISASSEMBLING

3- Remove the strainer from the pump body.

4- Remove the copper "crush" washer seal from the strainer, and then clean the strainer thoroughly.

5- Scribe a mark on the upper and lower sections of the pump as an aid during assembling. Remove the screws securing the upper and lower sections together.

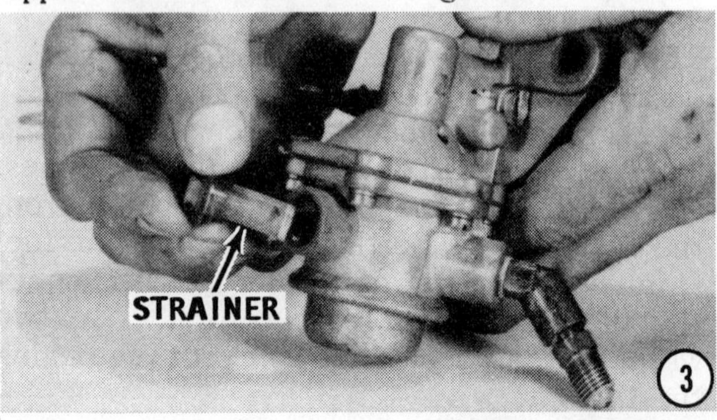

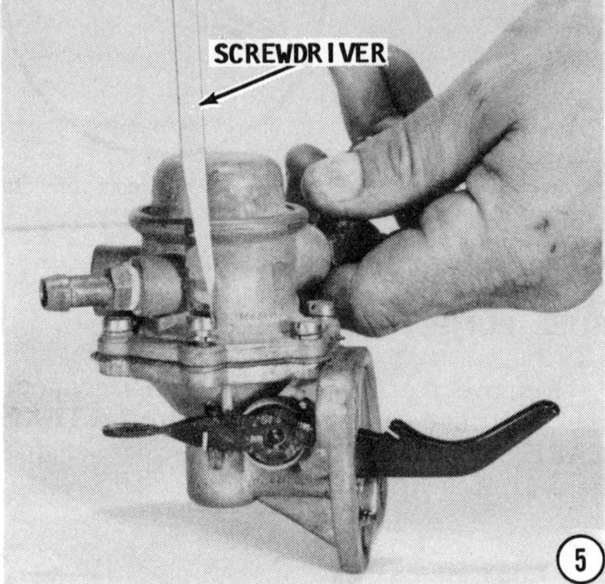

6- Separate the upper section from the lower section.

7- Remove the manual pump shaft set screw.

8- Withdraw the hand pump assembly from the pump body.

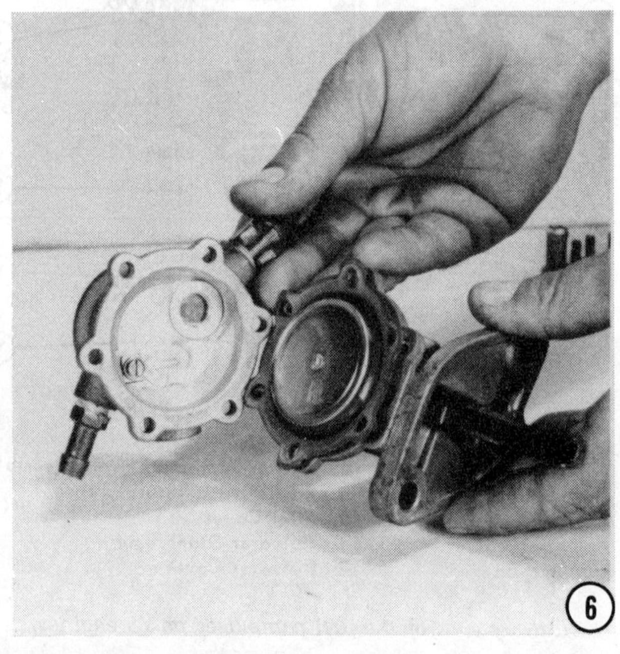

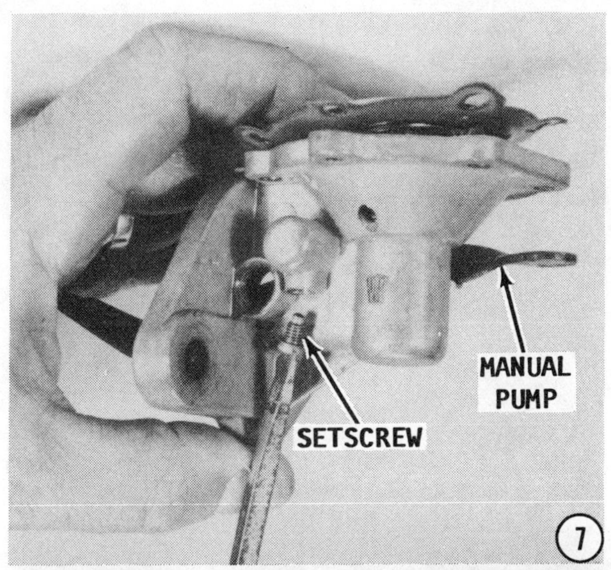

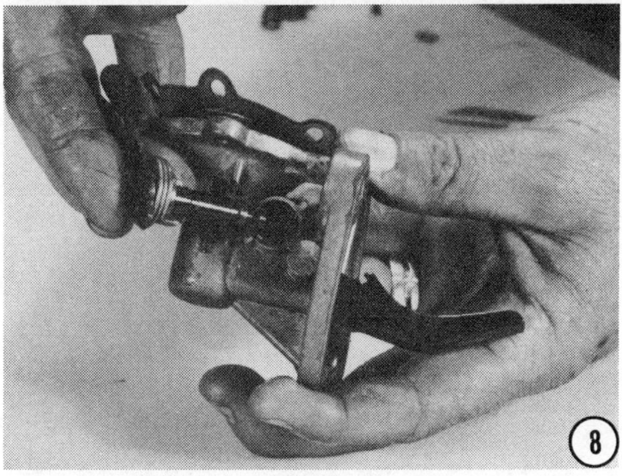

9- Remove the actuating lever pivot pin.

10- Unhook the actuating lever return spring, and then CAREFULLY pull the lever free of the diaphragm shaft. Pull the diaphragm out of the pump body.

11- Remove the holddown screw for the inlet check valve, and then remove the valve and piece of spring steel.

SPECIAL WORDS

Make a note of how the valve was installed. The inlet check valve **MUST** be installed properly to ensure it will close with the spring tension.

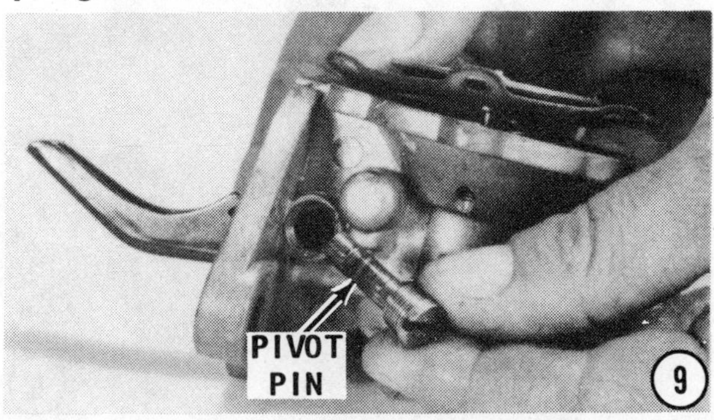

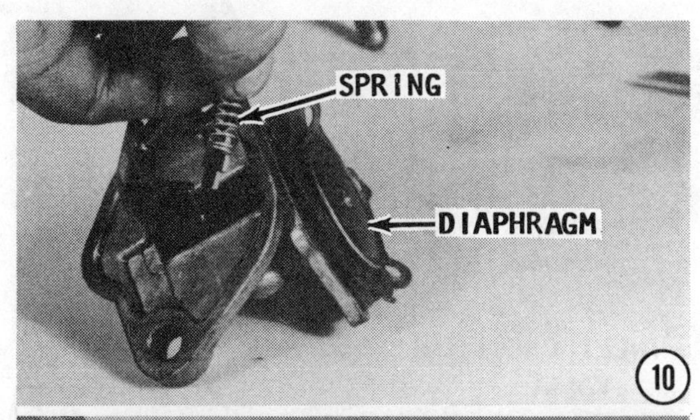

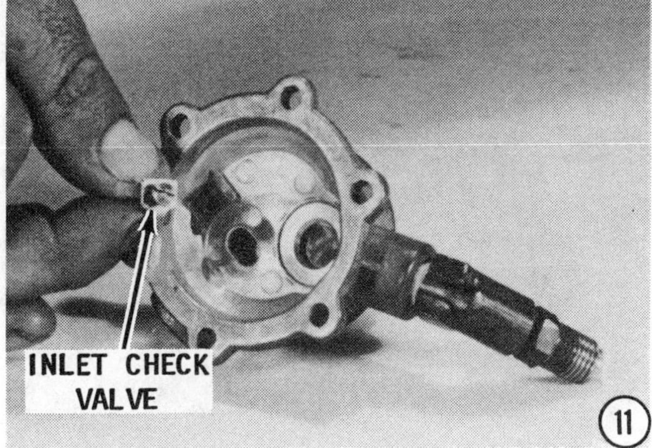

CLEANING AND INSPECTING

Wash all parts except the diaphragm in solvent and blow them dry with compressed air.

Check the diaphragm for holes or other damage.

Blow passageways with compressed air.

A pump repair kit is available from the local Volvo Penta dealer. The kit contains a new diaphragm, new gaskets, sealing washers, and all moving or wearing parts except the rocker arm.

ALWAYS use new seals and gaskets. If any of the other parts appear to be the least bit defective, purchase a kit and replace all the parts included in the package.

ASSEMBLING

1- Place the inlet check valve and piece of spring steel in the same position in the upper section of the pump body from which it was removed.

SPECIAL WORDS

The inlet check valve and piece of spring steel **MUST** be installed with the contour facing in such a way that the valve will close with the spring tension. Secure the valve in place with the holddown screw. Tighten the screw **JUST** snug.

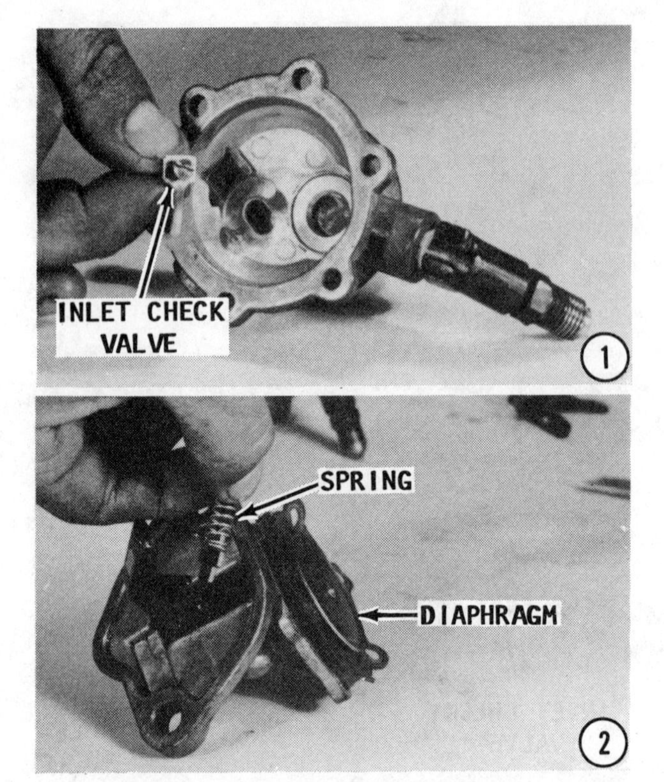

2- Insert the diaphragm in the lower section of the pump body. Rotate the diaphragm until the flat part on the end of the shaft is aligned to receive the little cutout in the end of the actuating lever. Slide the actuating lever into the lower section until the cutout engages with the flat part on the end of the diaphragm shaft. Slide the spring over the "T" shaped flange on the plunger, and then snap the other end into place in the pump body.

3- Slide the pivot pin through the body and actuating lever. Thread it into place. Tighten the pin snugly.

4- Insert the manual pump shaft into place. Move the handle until it extends straight out (90°) to the mating surface of the pump body.

5- Secure the manual pump in place with the set screw. The set screw will index into a groove in the manual pump shaft.

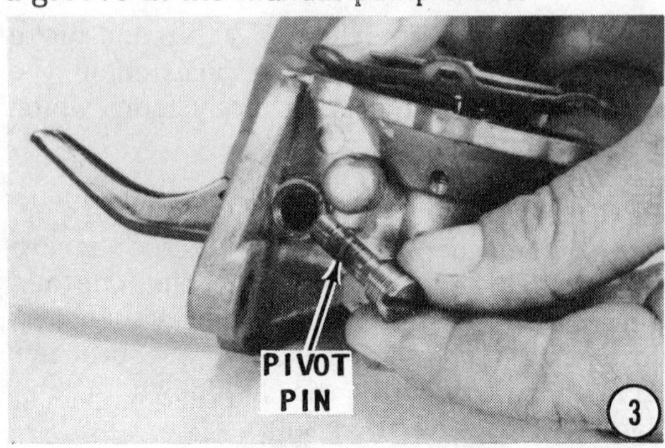

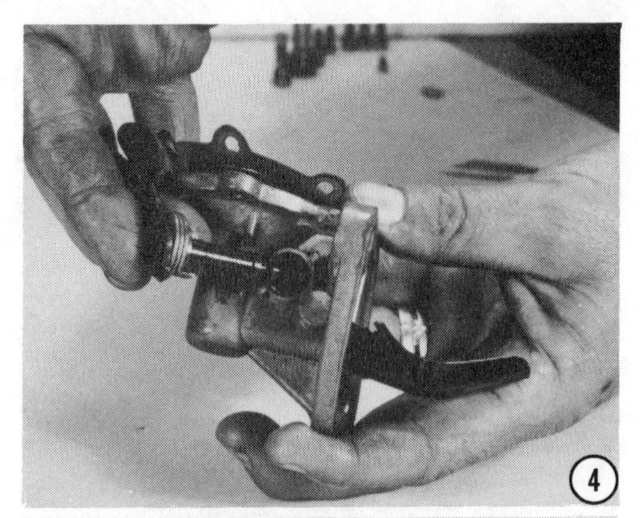

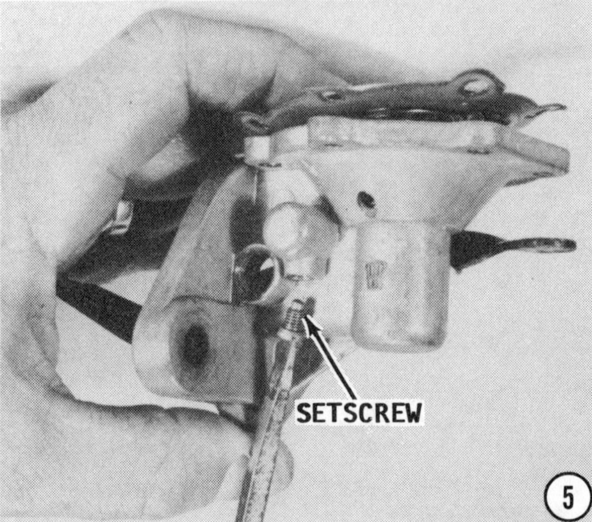

6- Check to be sure the mating surfaces of the upper and lower sections of the pump body are clean. Mate the upper and lower sections with the scribe marks made during disassembling aligned.

7- Secure the two sections together with the attaching screws. Tighten the screws alternately and evenly in a criss-cross pattern until they are all snug.

8- Install a **NEW** copper "crush" washer onto the strainer.

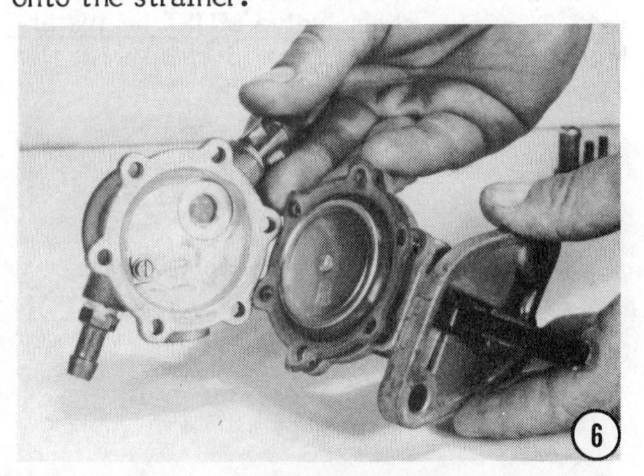

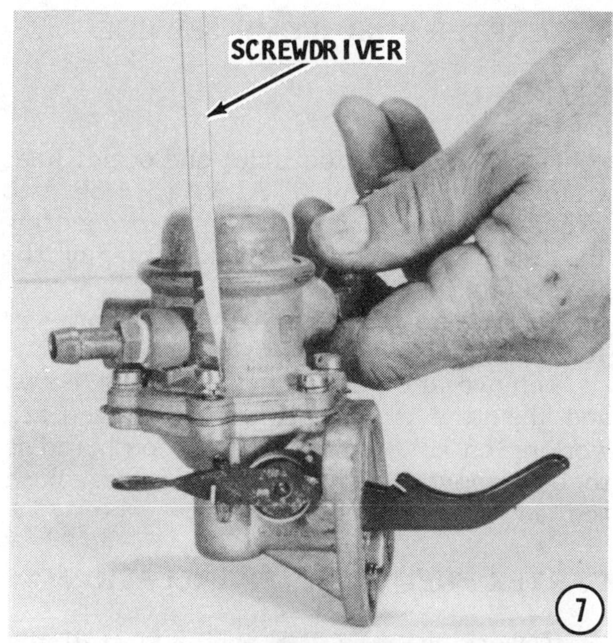

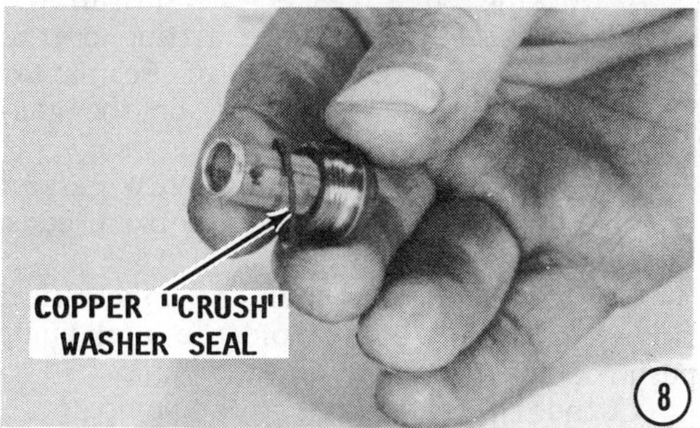

9- Thread the strainer into the pump body and tighten it snugly.

10- Check the action of the manual pump and the pump plunger.

11- Check the surface of the block to be sure it is clean and free of any old gasket material.

12- Install one gasket, the plastic insulator plate, another gasket, and then the pump, in that order to the engine block.

13- Secure the pump in place with the two attaching bolts. Tighten the bolts evenly and alternately.

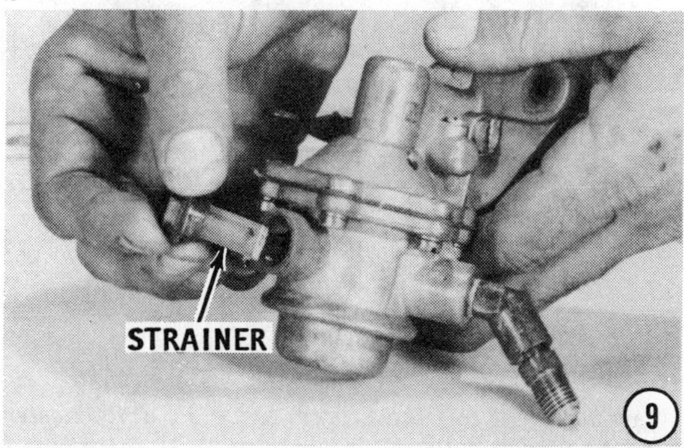

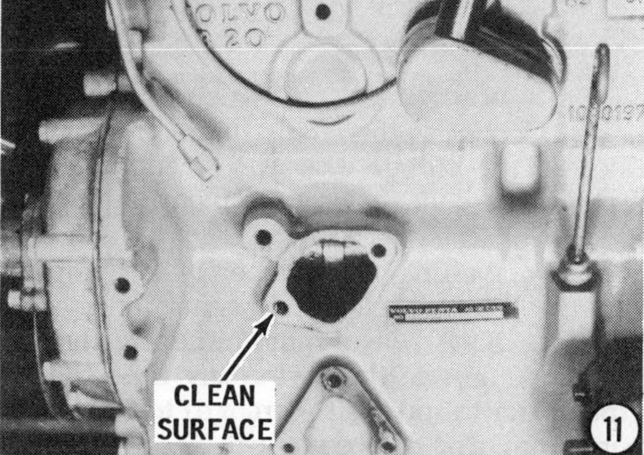

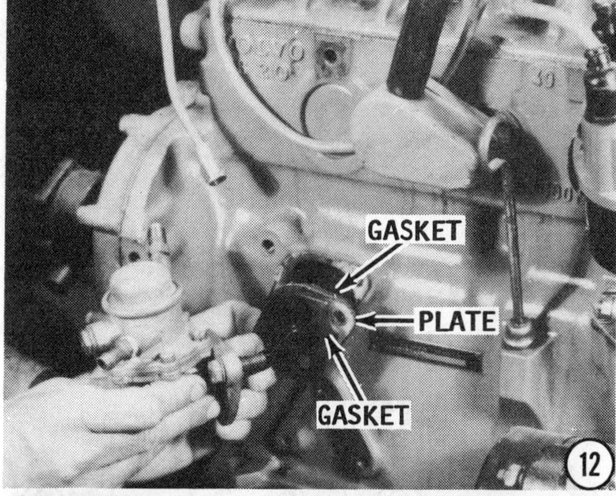

14- Connect the fuel inlet and outlet lines to the pump. Use two wrenches when making the connection to prevent damage to the fittings.

Start the engine and check the completed work for fuel leaks and performance.

CAUTION: Water must circulate through the lower unit to the engine any time the engine is run to prevent damage to the water pump mounted on the engine.

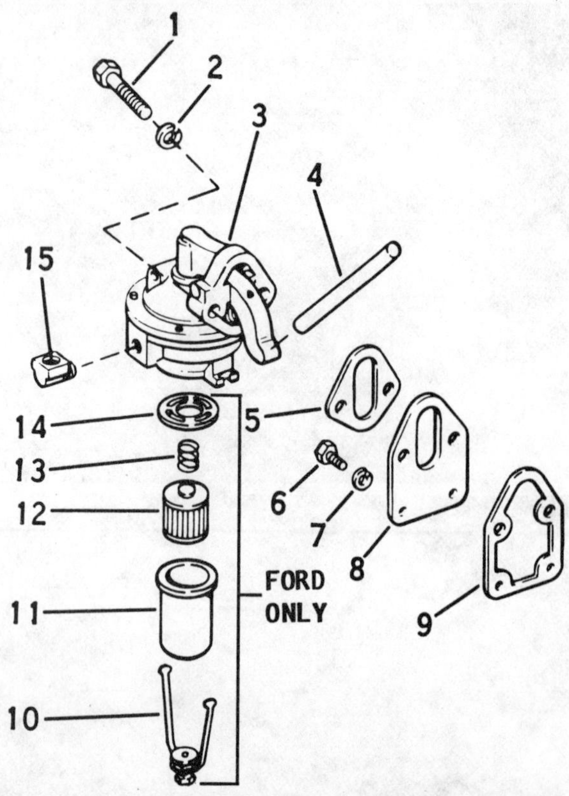

Exploded view of a Carter fuel pump. A fuel pump repair kit is available which consists of all moving or wearing parts except the rocker arm. If the rocker arm or a casting is damaged, it is advisable to replace the pump rather than attempting to repair it. (1) screw, (2) lockwasher, (3) fuel pump assembly, (4) push rod, (5) gasket, (6) screw, (7) lockwasher, (8) plate, (9) gasket, (10) yoke, (11) bowl, (12) filter element, (13) spring, (14) gasket, (15) fitting.

4-5 CARTER FUEL PUMP SERVICE

REMOVAL

Disconnect the fuel inlet and outlet lines from the fuel pump. **ALWAYS USE** two wrenches when disconnecting or connecting the outlet line fitting to avoid damaging the fuel pump.

Be sure to plug the fuel line to prevent fuel from siphoning out of the fuel tank.

Remove the fuel pump mounting bolts, and then the pump and gasket. If you are working on a V8 engine and the push rod is to be removed, remove the pipe plug, push rod, and the fuel pump adapter.

CARTER FUEL PUMP INSTALLATION

If working on a V-8 engine, install the fuel pump push rod and pipe fitting or the adapter. Lay down a bead of Permatex, Form-A-Gasket, or equivalent, on the gasket and pipe fitting.

Install the fuel pump and a **NEW** gasket. Use sealer on the mounting bolt threads. Tighten the bolts securely.

On V8 engines, a pair of mechanical fingers can be used to hold the fuel pump push rod up while installing the pump.

Connect the fuel lines to the pump.

Start the engine and check for leaks.

CAUTION: Water must circulate through the lower unit to the engine any time the engine is run to prevent damage to the water pump mounted on the engine.

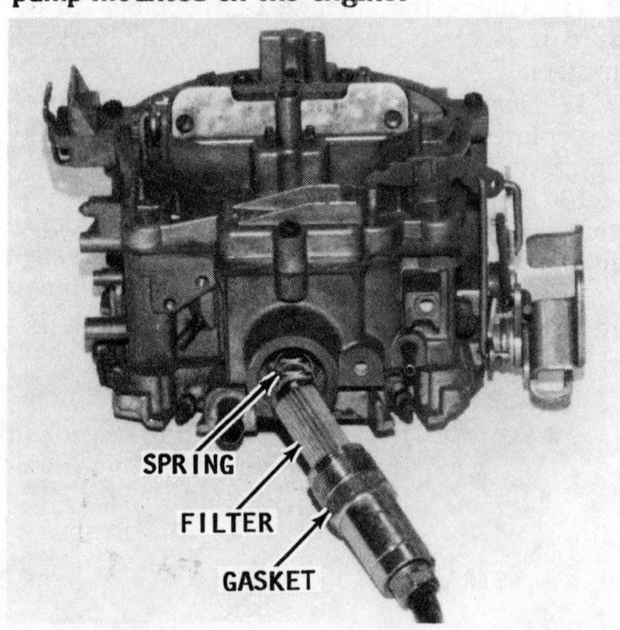

Arrangement of fuel filter parts on a Rochester four-barrel carburetor.

4-6 FUEL SYSTEM ACCESSORIES

FUEL FILTER REPLACEMENT

Most marine engines have some type of in-line fuel filter. This filter should be replaced every 100 hours of operation or sooner if you suspect it may be clogged.

To replace the in-line filter element, first remove the flame arrester. Next, loosen the retaining clamps securing the hoses to the fuel filter. Disconnect the fuel filter from the hoses and discard the retaining clamps.

Install **NEW** clamps on the hoses. Connect the hoses to the new filter. Tighten the filter, and then position the fuel line hose clamps in place and crimp them securely. Start the engine and check for leaks.

CAUTION: Water must circulate through the lower unit to the engine any time the engine is run to prevent damage to the water pump mounted on the engine.

FLAME ARRESTER CLEANING

The flame arrester should be removed and cleaned every 50 hours. It is not necessary to replace the arrester unless it is damaged and will not seat properly on the carburetor.

Remove and clean the arrester with solvent, and then blow it dry with compressed air. Replace it on the carburetor.

4-7 CARBURETORS

INTRODUCTION

In the simplest terms, a carburetor is merely a metering device which mixes the proper amount of fuel and air for delivery to the cylinders under all operating conditions.

Flame arrester mounted on a four-barrel carburetor on a V8 engine.

When the engine is idling, the mixture is roughly 10 parts air to 1 part fuel. At high speed or under heavy load, the mixture is about 12 parts air to 1 part fuel.

The fuel is held in reserve in the float chamber of the carburetor. A float valve in this chamber admits fuel from the fuel pump to replace the fuel leaving the chamber and burned by the engine. Metering jets extend from the fuel chamber into the carburetor throat.

The downward movement of a piston creates a suction that draws air into the carburetor throat. There is a restriction in the throat called a venturi. This venturi reduces air pressure at this point by increasing the air velocity.

The difference between the air pressure in the float chamber and the pressure in the carburetor throat causes the fuel to be forced through the metering jets and into the air stream. This mixture of fuel and air is then burned in the engine cylinders.

From this description, you can appreciate why the jets must be clean, free of gum, and adjusted properly for satisfactory engine performance.

The low-speed jet has an adjustable needle to compensate for changing atmospheric conditions. The high-speed jet is a fixed orifice.

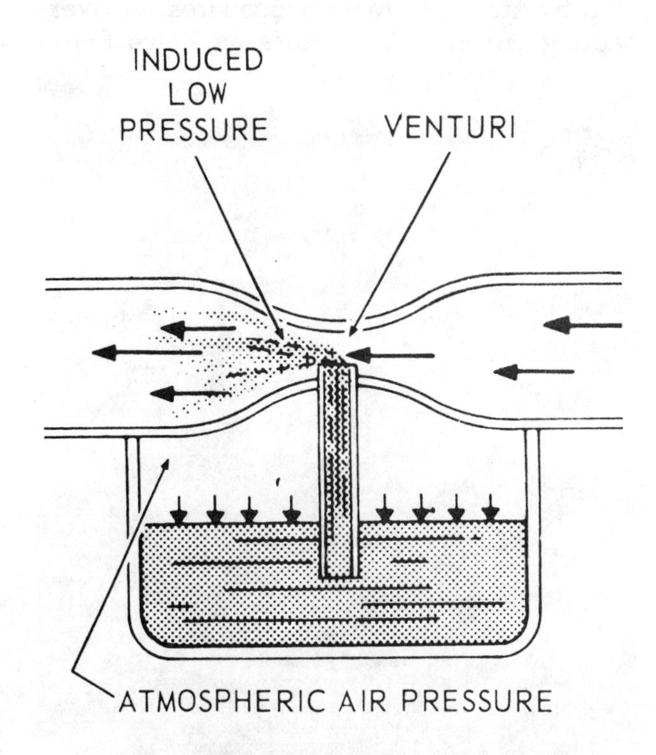

Air flow principle for a modern carburetor.

The volume of the fuel-air mixture drawn into the engine to regulate engine speed is controlled by a throttle valve. An extra amount of fuel is required to start a cold engine. This extra fuel is delivered by a choke valve installed ahead of the metering jets and the venturi or venturis as the case may be. After the engine starts and warms to operating temperature, the choke is gradually opened to restore the normal air-fuel mixture.

The throat of a carburetor is usually referred to as a "barrel". Single, double, or four barrel carburetors have a metering jet, needle valve, throttle, and choke plate for each barrel or pair of barrels.

"SOUR" FUEL

Even with the high price of fuel, removing gasoline that has been standing unused over a long period of time is still the easiest and least expensive preventative maintenance possible.

In most cases, this old gas can be used without harmful effects in an automobile using regular gasoline.

CARBURETOR SERVICE

The remainder of this chapter provides step-by-step illustrated procedures for overhauling current carburetors on Volvo Penta powerplants. Exploded drawings and bench adjustments necessary for proper performance are included.

The information is presented in a logical sequence as follows:

Carburetor	Section
Solex down-draft	4-8
Solex side-draft	4-9
Rochester 2GC and 2GV	4-10
Rochester 4-barrel 4MV	4-11
Holley 2-barrel	4-12
Holley 4-barrel 4150 and 4160	4-12
Holley 4-barrel 4011	4-13

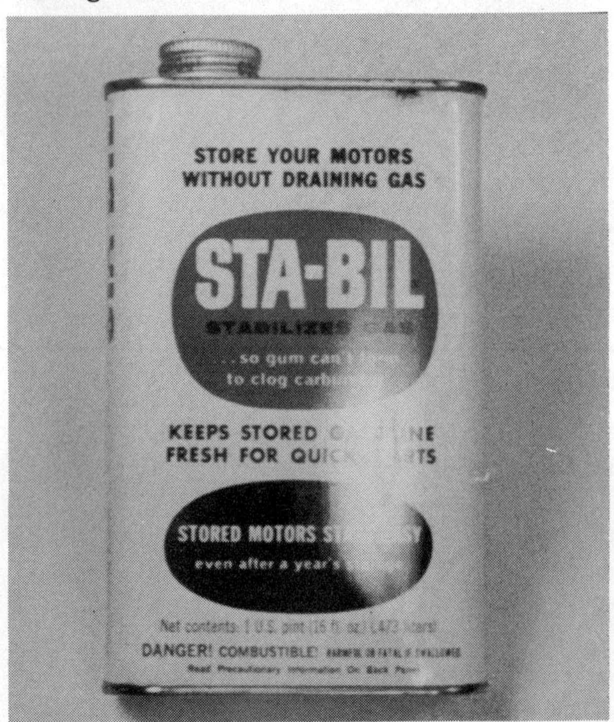

A fuel additive will keep fuel "fresh" for up to twelve full months.

4-8 SOLEX DOWNDRAFT CARBURETOR

DESCRIPTION AND OPERATION

The Solex downdraft carburetor is a very basic, simple unit. The main jet, idling jet, acceleration jet, and the emulsion jet are all fixed jets.

This carburetor has an acceleration pump which also aids during hard starting in cold weather. No other cold start device is used.

The float chamber is a part of the carburetor body. The chamber has internal ventilation. This built-in ventilation prevents external fuel flooding.

A single Solex downdraft carburetor may be installed on the AQ115A engine, two may be installed together as on the AQ130C engine, or three installed and connected in tandem as on the AQ170A engine.

Service procedures for the multiple installation are identical as for a single installation. Therefore, only one set of service instruction are given. If more than one carburetor is installed on the engine being serviced, simply repeat the listed procedures for each unit.

REMOVAL

1- Disconnect the battery leads at the battery terminals to prevent the possibility of a spark igniting fuel fumes trapped in the engine compartment. **TAKE CARE** that the engine compartment area is well ventilated to permit any accumulated fuel fumes to escape. Remove the hardware securing the flame arrester to the carburetor/s. Disconnect the linkage and the fuel line. Have a suitable container handy to catch any residual fuel in the line. Remove the four nuts attaching the carburetor to the intake manifold. Remove the carburetor, and then cover the intake manifold opening to prevent any foreign material from entering.

DISASSEMBLING

2- With the carburetor on the work bench, remove the four screws securing the float bowl cover, and then lift the cover free of the carburetor. Remove and **DISCARD** the gasket.

3- Use the proper size socket and remove the needle seat and float valve.

4- If servicing a carburetor from a late model AQ125, or AQ145, one of the multiple

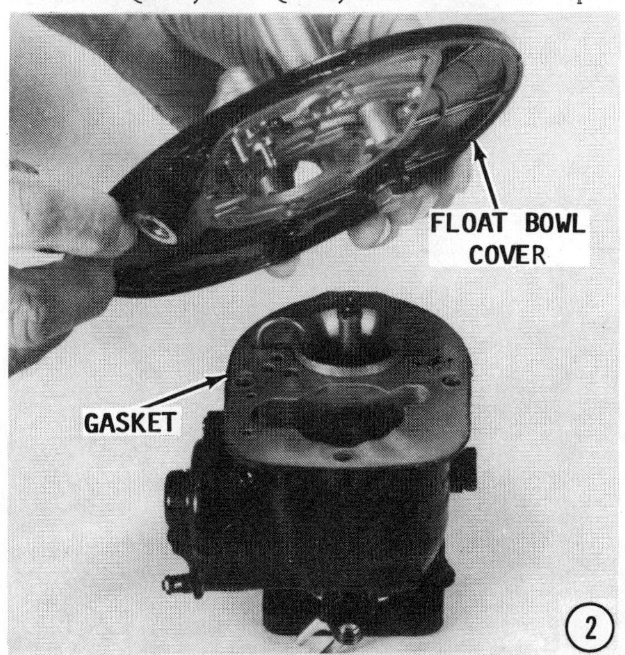

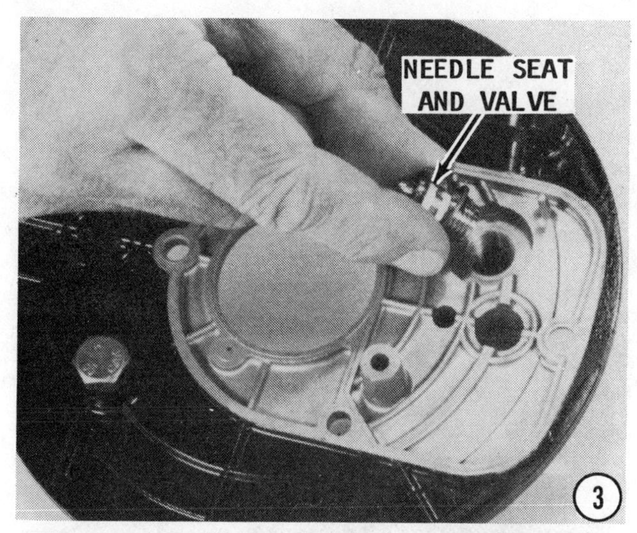

carburetors will have an air vent from the fuel pump to the bottom of the base plate. This air vent may be removed for cleaning, using the proper size socket wrench.

5- Lift the float hinge pin and float free of the slot in the carburetor body. Slide the pin out of the hinge.

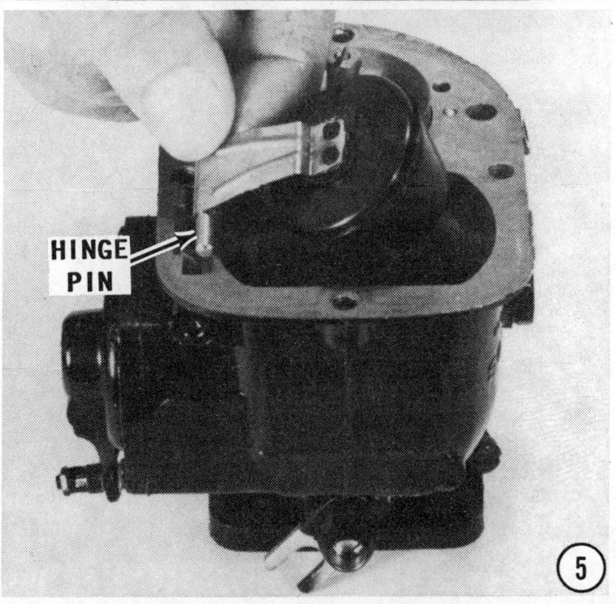

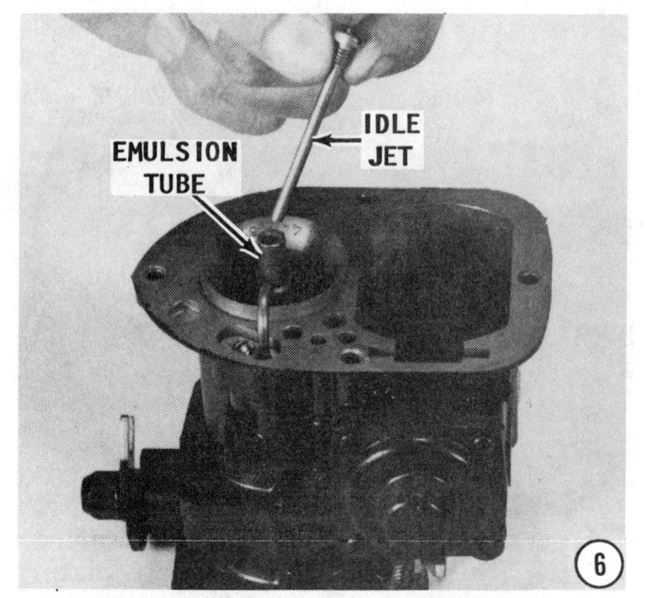

6- Remove the idle jet from the emulsion tube.

7- Lift out the accelerator pump nozzle.

8- Remove the gasket from the accelerator pump nozzle opening.

9- Remove the four screws securing the accelerator pump to the carburetor body.

GOOD WORDS

The cover does not have to be removed from the rod for normal carburetor cleaning and service. However, if the rod is damaged, or the cover needs to be removed for good reason, perform Step 10. If the cover is not to be removed, proceed directly to Step 11.

10- Use the proper size end wrench and remove the nut from the end of the accelerator pump rod. **COUNT** the number of turns required to remove the nut as an aid to

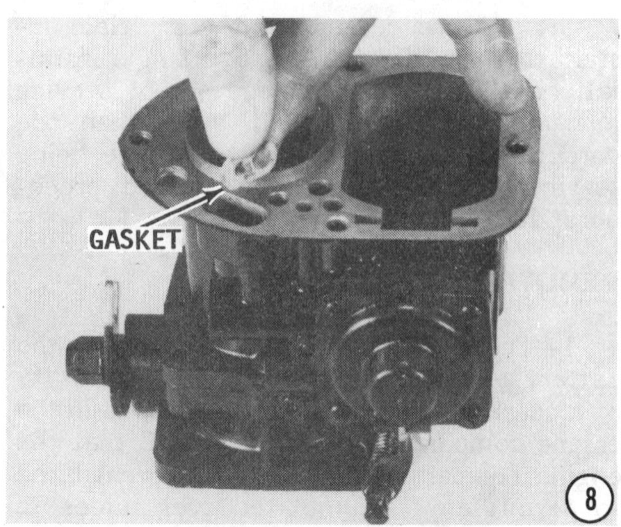

assembling. An alternate method is to measure the distance the rod extends through the nut before the nut is loosened. This nut provides the adjustment for proper operation of the acceleration pump. Slide the spring free of the pump rod. Separate the accelerator pump cover from the carburetor body.

11- Lift the diaphragm and spring out of the pump cavity.

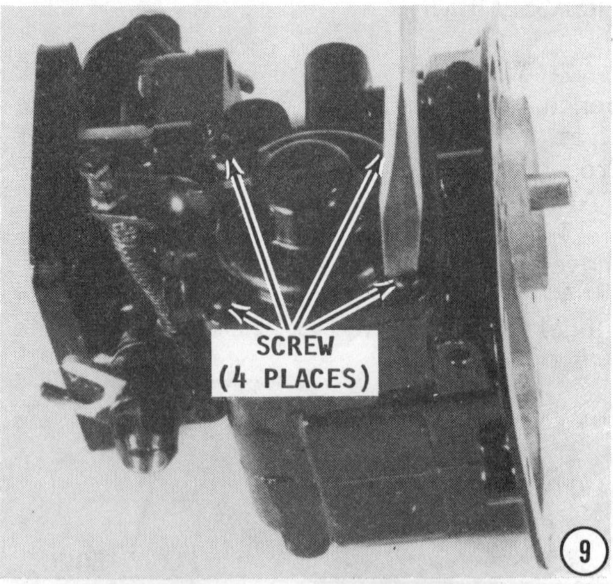

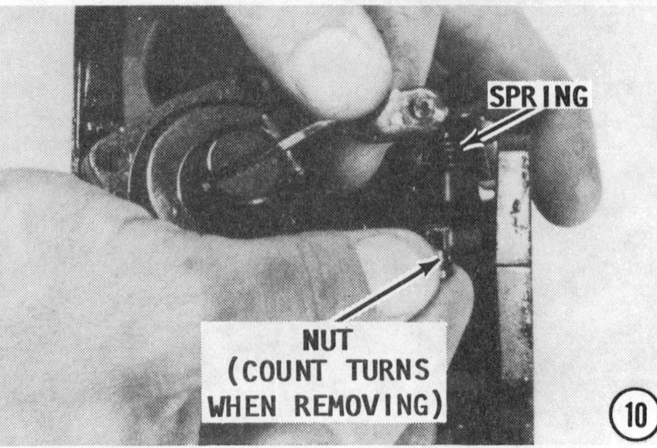

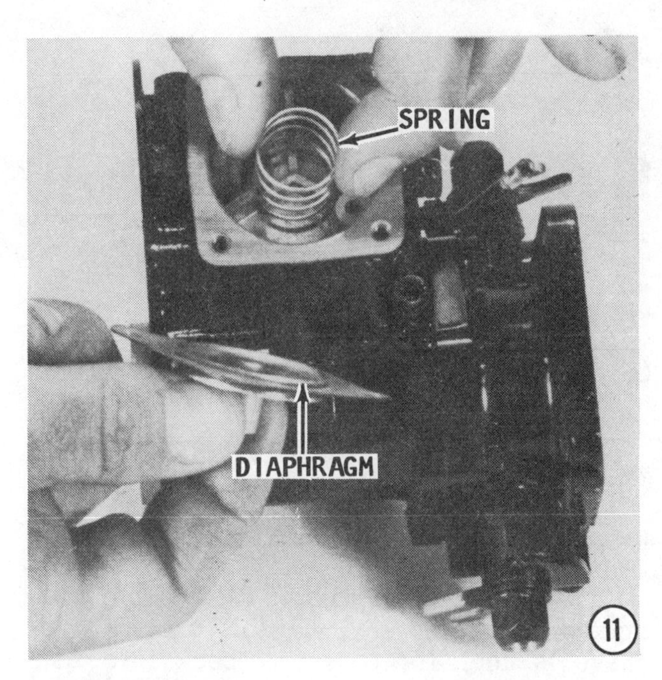

GOOD WORDS

The main jet is removable, but under normal maintenance, it is not necessary to remove it. This main jet is the one that would be changed for size if the unit was to be used at higher altitudes. Check with the local Volvo dealer for the size jets available for the unit being serviced.

12- Using the proper size end wrench, loosen, and then remove the main jet. Remove and **DISCARD** the copper washer.

13- Should the main jet need to be removed from the jet body, use the proper size box-end wrench and a screwdriver. Rotate the jet **COUNTERCLOCKWISE** until it is free of the jet body. If the main jet is

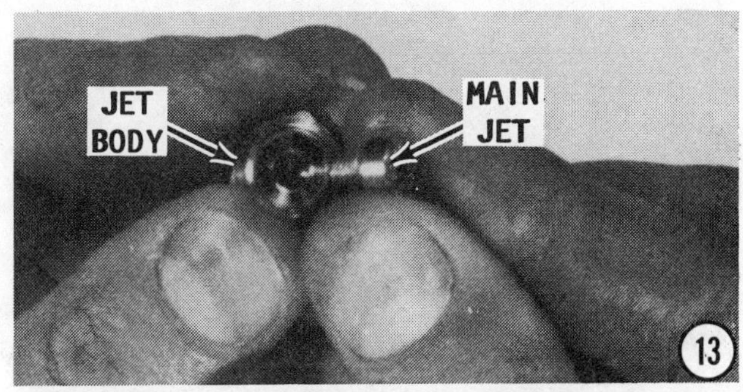

changed, the emulsion jet should also be changed. These two jets are sold and **MUST** be installed as a matched set.

14- Remove the air/fuel adjusting screw.

15- Remove the air jet from the carburetor body.

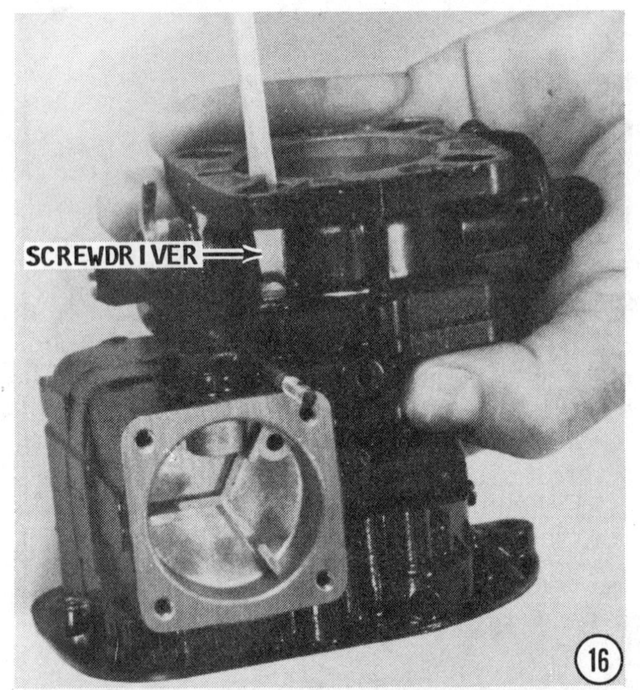

16- Scribe a small mark on the carburetor body and a matching mark on the base. These marks will guarantee the base is matched to the carburetor body properly during assembling. It is possible to mate the base and body 180° in error. Therefore, the requirement for the scribe marks. Remove the four attaching screws, and then separate the base from the main carburetor body.

17- Remove and **DISCARD** the gasket between the base and the main carburetor body.

18- Normally the butterfly and shaft do not have to be removed. However, if either of these items appear to be damaged, first remove the two Phillips head screws securing the butterfly to the shaft. Next, slide the butterfly free from the groove in the shaft. Remove the nut from the end of the shaft, and then pull the butterfly shaft free of the carburetor base.

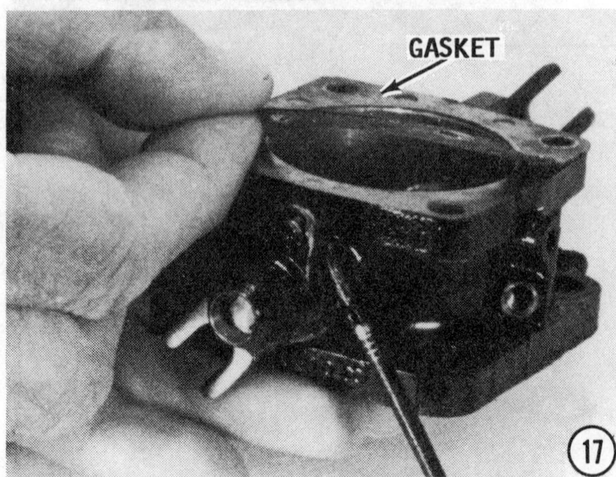

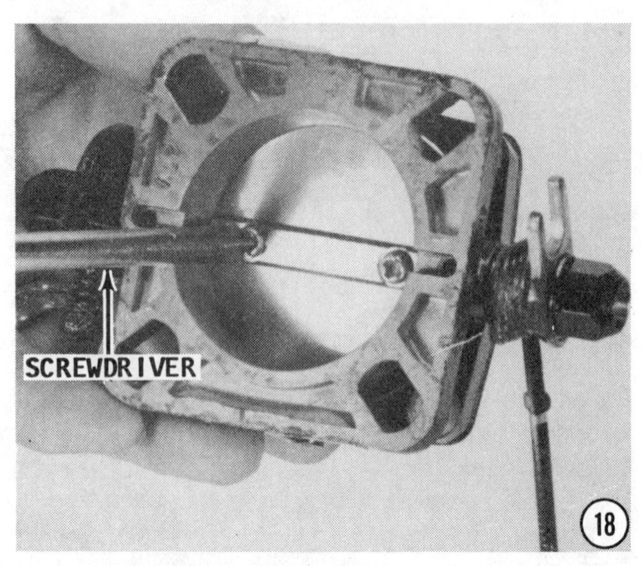

19- Remove the check valve from the accelerator pump cavity. This check valve prevents fuel from running back into the float bowl when the accelerator pump is operating.

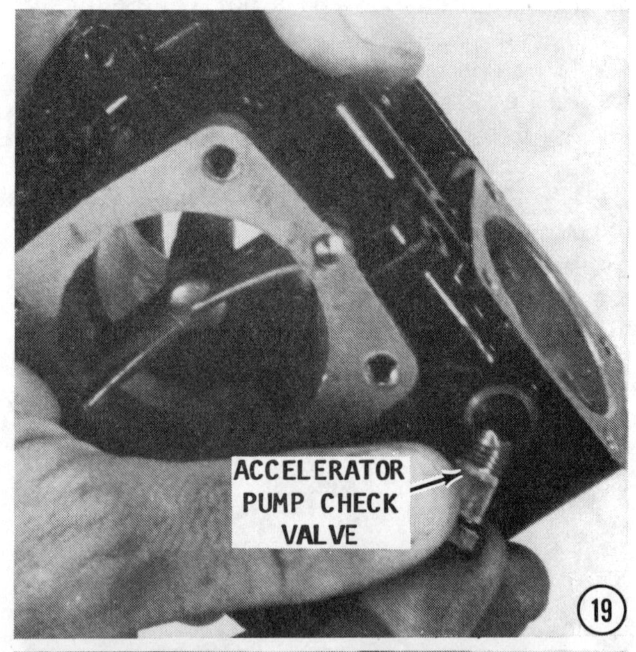

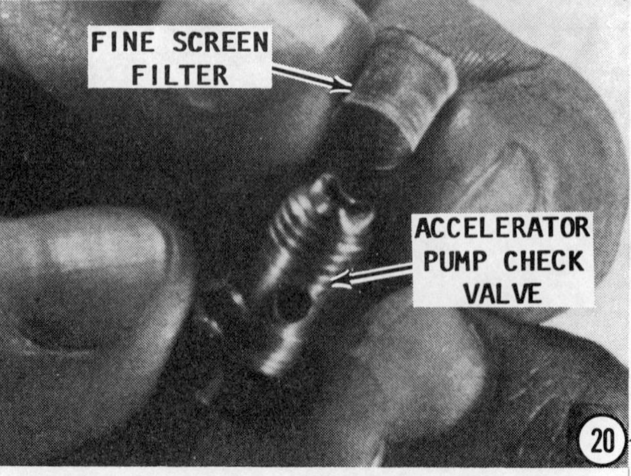

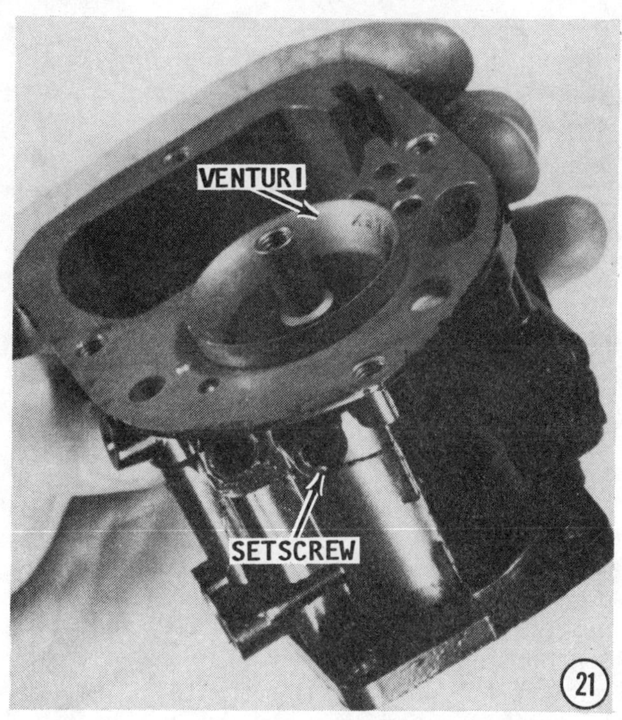

20- Slide the fine screen filter free of the check valve. This fine screen should be **CAREFULLY** cleaned each time the carburetor is serviced, because it is a source of carburetor problems. Its function is to prevent any foreign material from entering the accelerator pump cavity and passageways. This tiny fine screen does its job very well.

21- A small set screw secures the venturi in the carburetor body. Actually there is no reason to remove the venturi unless it is severely damaged. Simply remove the setscrew, and then pull the venturi free of the carburetor body.

CLEANING AND INSPECTING

NEVER dip rubber parts, plastic parts, diaphragms, or pump plungers in carburetor cleaner.

Place all metal parts in a screen-type tray and dip them in carburetor solvent until they appear completely clean, then blow them dry with compressed air.

Blow out all passages in the castings with compressed air. Check all parts and passages to be sure they are not clogged or contain any deposits. **NEVER** use a piece of wire or any type of pointed instrument to clean drilled passages or calibrated holes in a carburetor.

CAREFULLY clean the tiny fine screen from the accelerator pump check valve in carburetor cleaner and blow it dry with compressed air. Inspect it thoroughly to be sure it is clean and not damaged. This screen plays an important role in preventing any foreign material from entering the accelerator pump nozzle or passageways.

Inspect the idle mixture needle for any type of damage. Carefully check the float needle and seat assembly for wear. Good shop practice is to always install a new factory matched set to avoid internal leaks. Such leaks could cause rough idling and excessive fuel consumption.

Check the springs to be sure they have not become distorted or lost their tension. If in doubt, compare them with a new one, if possible.

Inspect the sealing surfaces of the casting for damage.

Most of the parts that should be replaced during a carburetor overhaul are included in an overhaul kit available from the local Volvo Penta marine dealer.

ASSEMBLING

1- If the venturi was removed because it was damaged, **CAREFULLY** slide the new one into the carburetor body. **TAKE CARE** not to crimp or bend the venturi lip. Secure the venturi in place with the setscrew.

2- Slip the sealing washer onto the check valve. This washer may be aluminum or copper. Slide the tiny fine screen onto the accelerator pump check valve with the shoulder of the screen towards the threaded

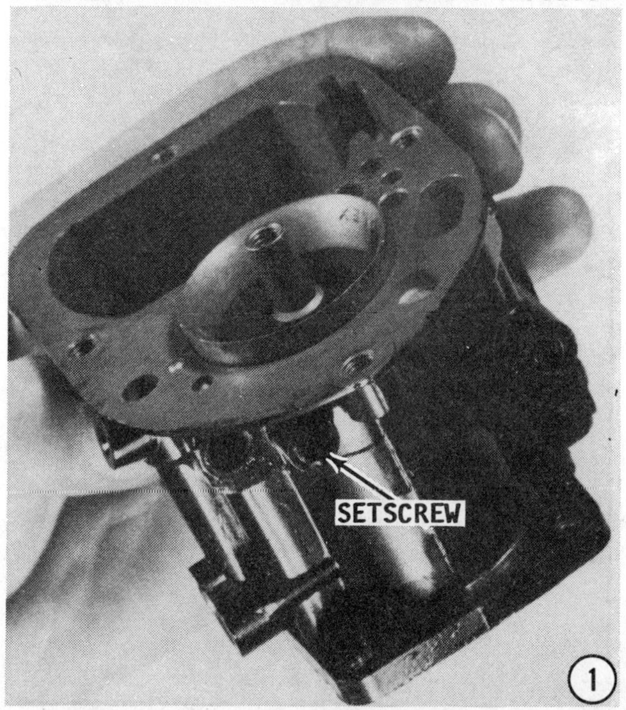

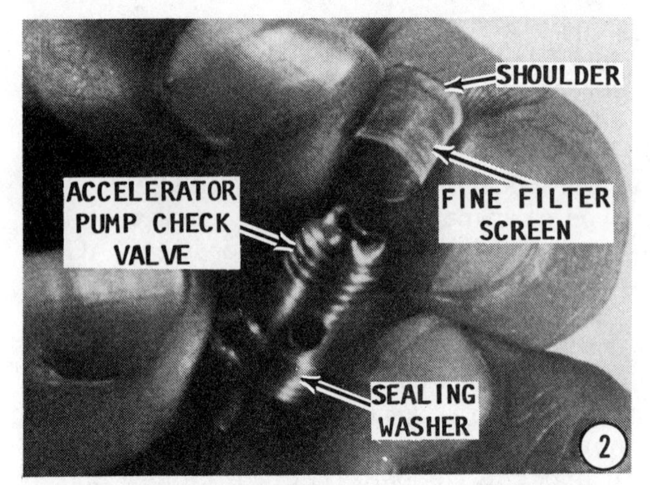

end of the valve, as shown.

3- Thread the accelerator pump check valve into the carburetor body. Tighten the check valve until it is just snug. **DO NOT** overtighten or damage may be caused to the sealing washer.

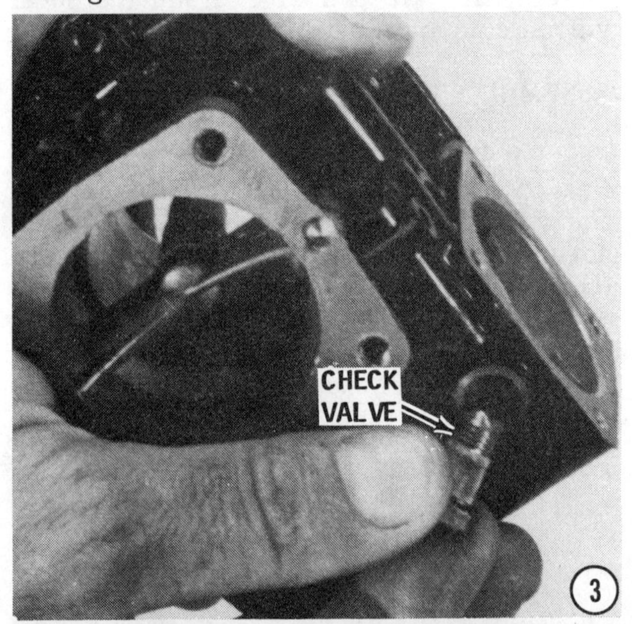

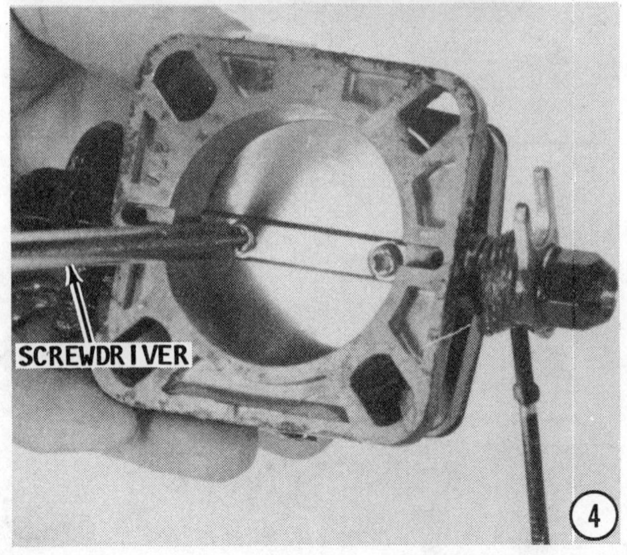

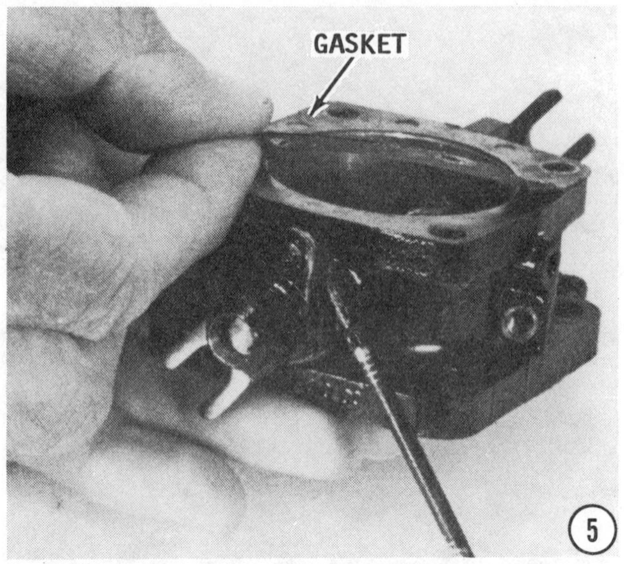

4- If the butterfly and shaft were removed, slide the shaft through the carburetor housing and secure it in place with the nut on the end of the shaft. Push the butterfly through the slot in the shaft and secure it in place with the two Phillips head screws. Tighten the screws securely. Check to be sure the butterfly rotates smoothly with no sign of binding.

5- Check to be sure the mating surfaces of the carburetor body and the base are clean and free of any old gasket material. Position a **NEW** gasket in place on the carburetor base with the holes in the gasket aligned with the holes in the base.

6- Mate the carburetor base to the body with the scribe marks made during disassembly aligned and matching holes aligned.

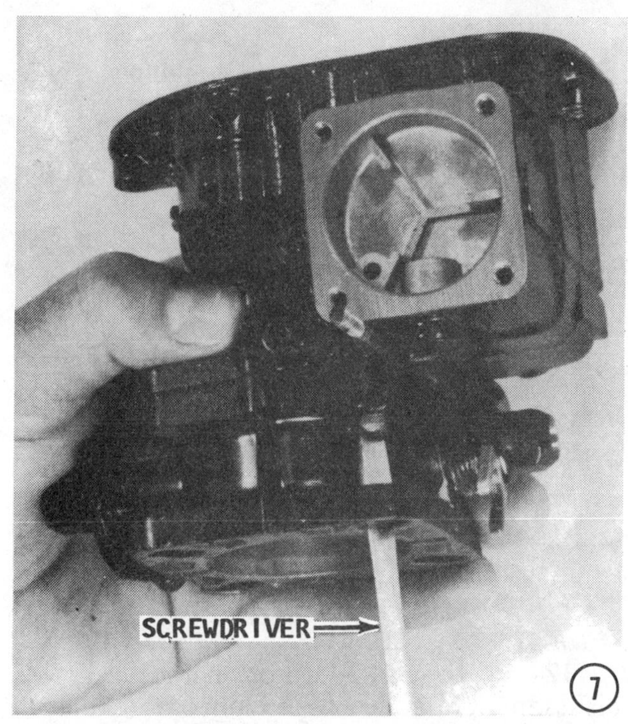

7- Secure the base to the carburetor body with the four attaching screws. Tighten the screws evenly and alternately in a criss-cross pattern.

8- Check the condition of the air/fuel mixture adjusting screw to be sure the tip has not been bent or damaged by overtightening. The tip **MUST** be in excellent condition to provide satisfactory service. Slide the spring onto the shaft of the screw and then thread the air/fuel mixture adjusting screw into the carburetor body. **CAREFULLY** tighten the screw until it **BARELY** seats and then back it out **1-1/4** complete turns as a preliminary adjustment at this time.

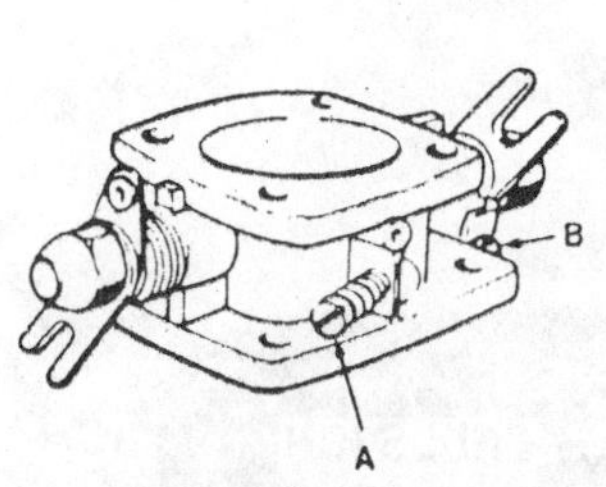

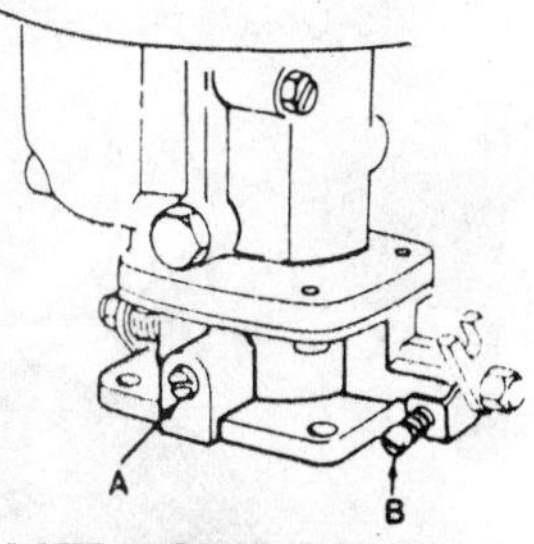

EARLY MODEL SOLEX **LATE MODEL SOLEX**

The Solex unit being serviced may be an early model or late model, as indicated in this illustration. Initial bench adjustment of the idle screws differ between the two models, as shown in the following charts.

Set Idle Screw "A"

Tighten the screw lightly against the seat, and then back it out, per the following chart.

Screw "A	Early Model	Late Model
AQ 131	2	9.5
AQ 151	2	8
AQ 171	2	10

Set Idle Screw "B"

Tighten the screw until it barely makes contact with the carburetor lever, and then tighten it further, per the following chart.

Screw "B	Early Model	Late Model
AQ131	2	2
AQ151	1-1/2	1
AQ171	1-1/2	1-1/4

9- Place a **NEW** copper washer into the main jet recess in the carburetor body. If the main jet was removed from the body during disassembling, thread the jet into the body. Hold the jet body securely with the proper size box-end wrench and tighten the jet with a screwdriver. Thread the assembled main jet and body into the carburetor body and tighten it with the proper size wrench until the copper washer is crushed **SLIGHTLY**. The crushed copper washer will provide an adequate seal.

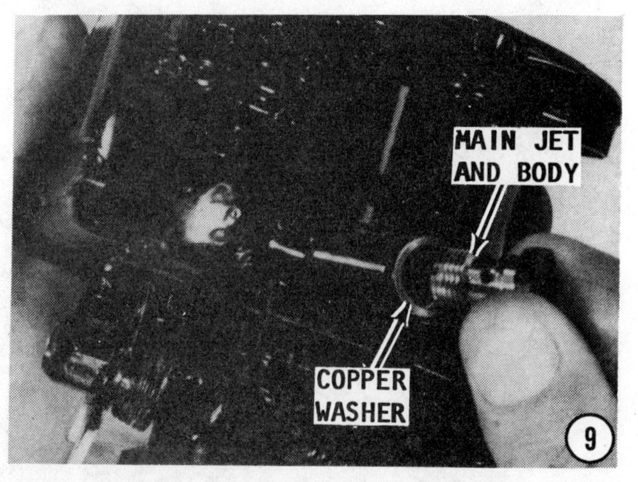

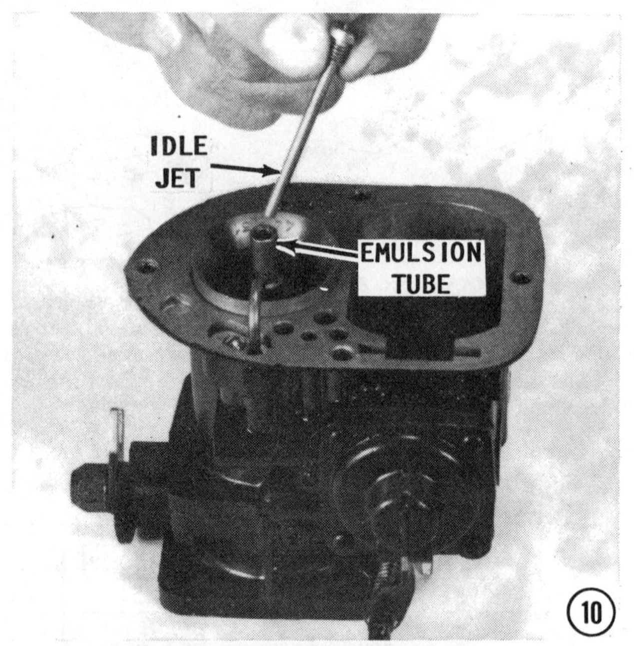

10- Thread the idle jet into the emulsion tube. **SLOWLY** tighten the jet until it is snug.

SPECIAL WORDS

If the main jet was replaced, the size changed for operation at a different altitude, the idle jet **MUST** also be changed. These two jets **MUST** be replaced as a set for maximum performance. The two are usually sold as a set at the local Volvo Penta marine dealer.

11- Place the spring onto the diaphragm post. Position the diaphragm onto the carburetor surface with the spring and diaphragm post facing upward, as indicated in the accompanying line drawing. Align the holes in the diaphragm with the matching holes in the carburetor body.

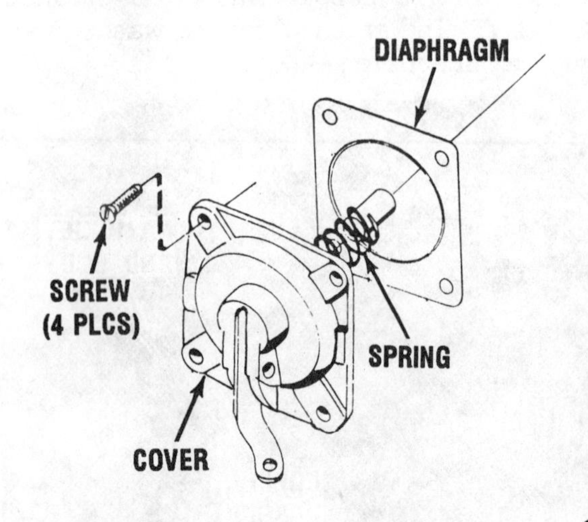

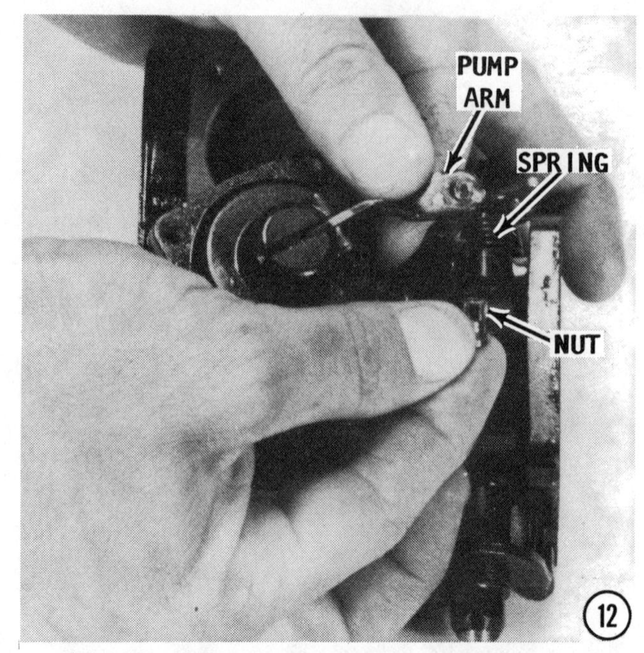

12- If the accelerator pump rod was disassembled, slide the spring onto the rod, then the pump arm. Thread the nut onto the pump rod and tighten it the **SAME** number of turns recorded during disassembling. An alternate method was to have measured the distance the rod extends through the nut before the nut was loosened. If the measurement was made, tighten the nut until this same measurement is obtained. This nut provides the adjustment to ensure proper operation of the acceleration pump.

13- Secure the accelerator pump cover to the carburetor body with the four attaching screws. Tighten the pump cover attaching screws alternately and evenly.

14- Insert a **NEW** gasket into the accelerator pump nozzle recess.

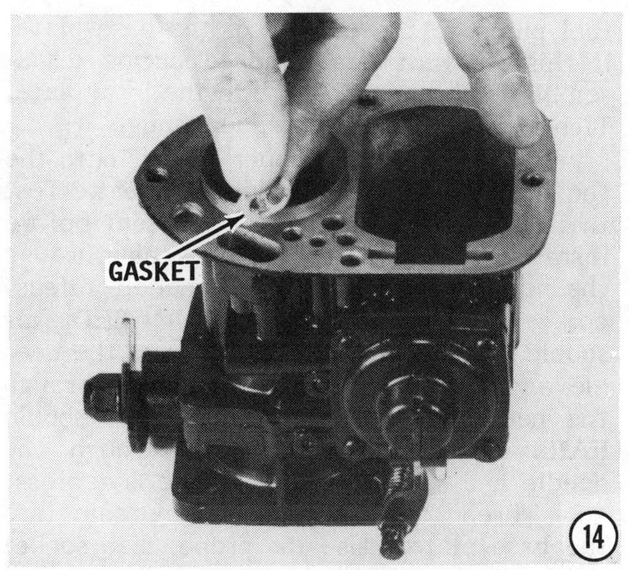

15- Slide the accelerator pump nozzle into its recess in the carburetor body. After the nozzle is in place, the curved end will point down into the venturi.

16- Push the float hinge pin through the hinge.

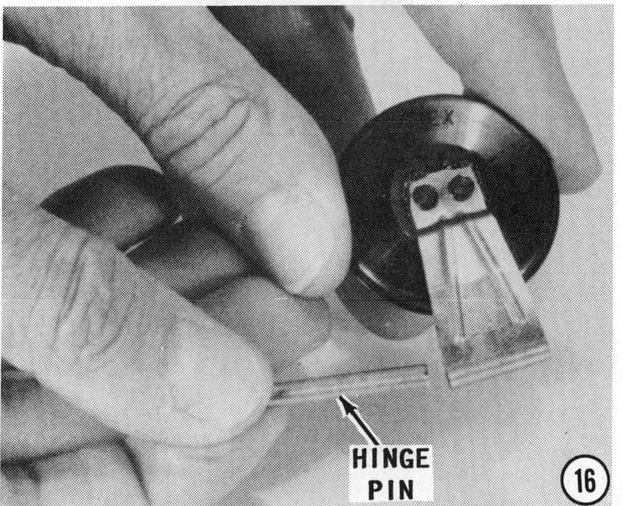

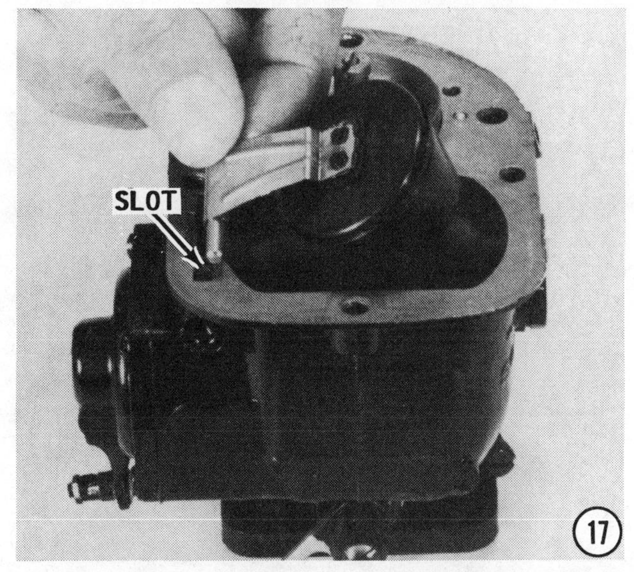

17- Slide the assembled float and hinge pin into the slots in the carburetor body with the float facing downward, as shown.

18- If servicing a carburetor from a late model AQ125, or AQ145, one of the multiple carburetors will have an air vent from the

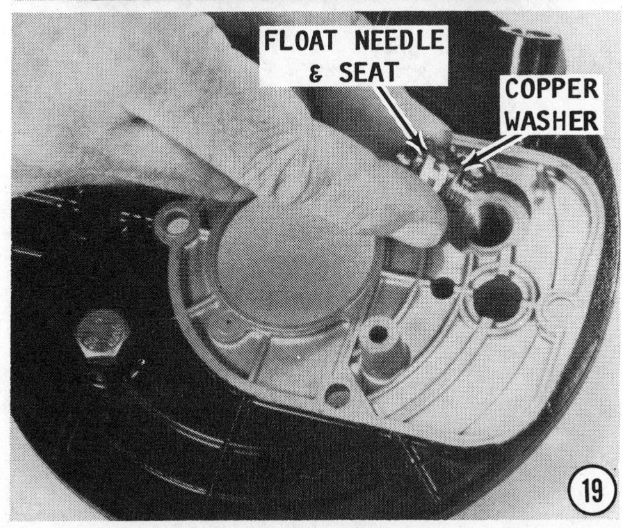

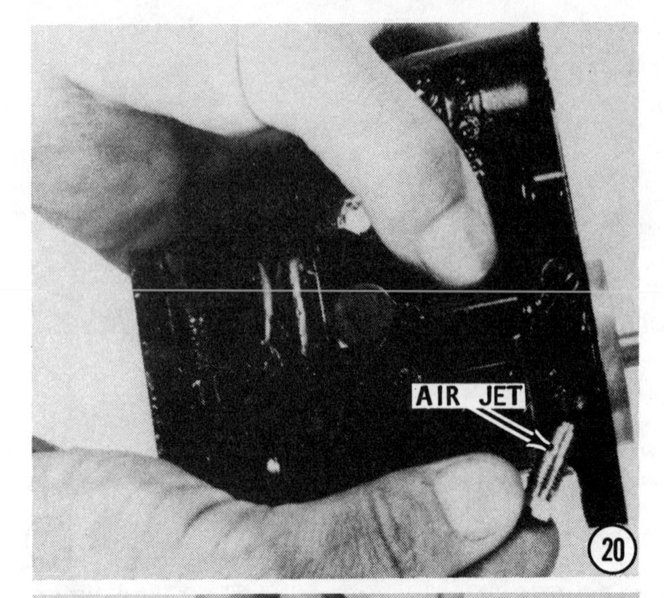

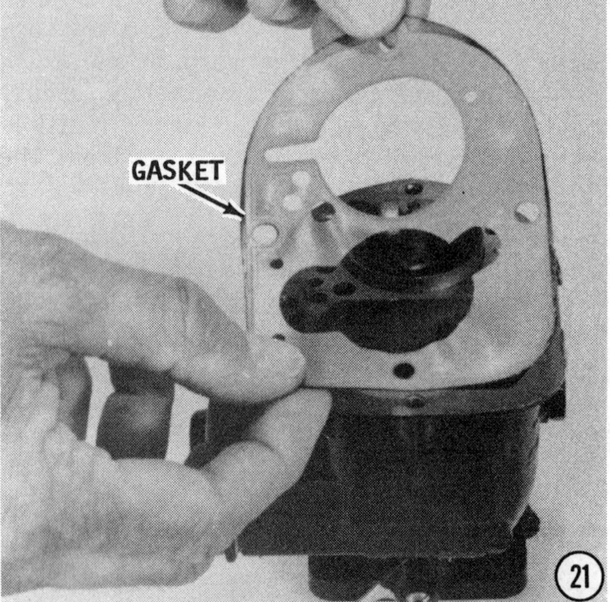

fuel pump to the bottom of the base plate. If this air vent was removed during disassembling, thread it back into the base plate. Tighten the vent until it is just snug.

19- Slide a **NEW** copper washer onto the the shaft of the float needle and seat. Test the operation of the needle and seat before installing it. Attempt to blow air through the needle and seat with the needle released. The attempt should **SUCCEED,** air should pass through. Now, depress the needle and again attempt to blow air through the needle and seat. The attempt should **FAIL.** Air should not pass through the needle and seat when the needle is depressed. Thread the float needle and seat into the base plate. Use the proper size socket wrench and tighten the seat until it is just snug. **DO NOT** overtighten the seat or the copper washer will be crushed and an effective seal will be lost.

20- Thread the air jet into the carburetor body. Tighten the jet until it is just snug.

21- Check to be sure the mating surface of the carburetor body is clean and free of any old gasket material. Lay a **NEW** gasket onto the carburetor surface with the holes in the gasket aligned with the holes in the carburetor.

22- Set the float bowl cover down over the carburetor body with the holes in the cover aligned with the holes in the gasket and carburetor body.

23- Secure the float bowl cover to the carburetor with the four attaching screws. Tighten the screws alternately and evenly.

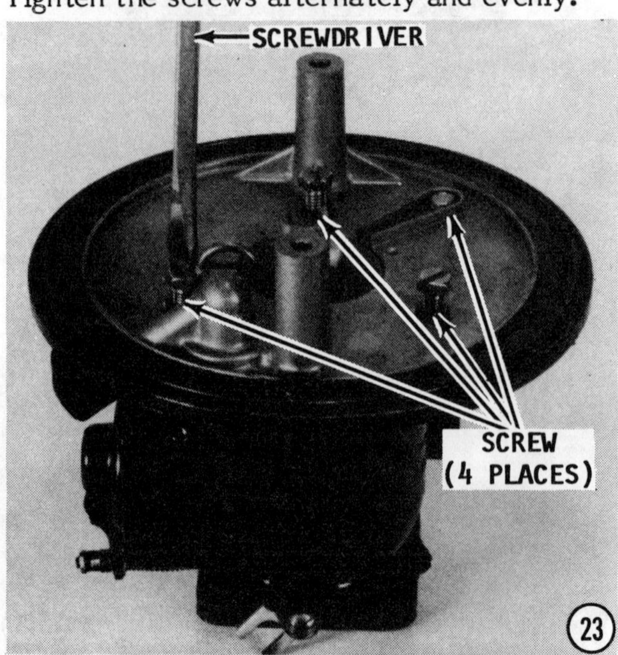

24- Remove the protective cover placed over the intake manifold opening. Check the mating surfaces of the intake manifold and the carburetor to be sure they are clean, smooth, and free of any old gasket material. Place a **NEW** gasket in position on the intake manifold. Mount the carburetor and secure it in place with the four attaching screws. Connect the fuel line. Use **NEW** copper washers if the old ones appear to be the least bit damaged. Connect the linkage. Do not install the flame arrestor until after the carburetor/s have been adjusted.

ADJUSTMENTS AQ115 SINGLE INSTALLATION

25- When the control lever is in the neutral position, the spring-loaded fitting on the end of the control rod is pushed downward, the movement should be about 5/64" (2.0 mm) (measurement **"A"** in the accompanying illustration. If an adjustment is required, loosen the locknut under the spring sleeve and then rotate the fitting until the correct measurement is obtained.

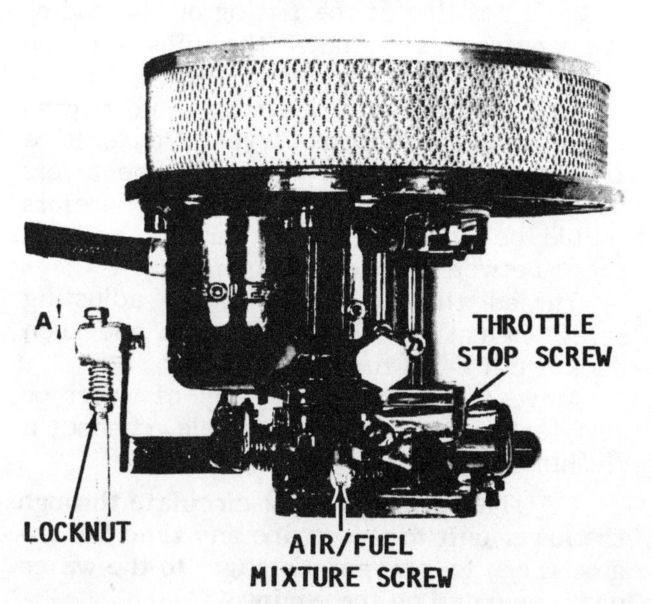

Move the boat into a body of water or test tank. If this is not possible, connect a flushing device to the lower unit.

CAUTION: Water must circulate through the lower unit to the engine any time the engine is run to prevent damage to the water pump mounted on the engine.

Start the engine and allow it to warm to operating temperature. Adjust the idle speed screw until engine rpm indicated on the tachometer is within the range given in the Appendix. If necessary, adjust the throttle stop screw until the engine is operating at the required rpm.

SLOWLY adjust the air/fuel adjusting screw until the highest steady idle speed is obtained. Under normal operating conditions, the final position of the air/fuel mixture adjusting screw will be approximately 1-1/4 turns out from the initial setting as given in Step 8 on Page 4-19.

ADJUSTMENTS AND SYNCHRONIZING DOUBLE CARBURETOR INSTALLATION

SPECIAL WORDS

The objective of the synchronizing procedure is to ensure that both carburetors open at the **SAME TIME.** Your goal is to be sure both carburetors are drawing air and fuel equally, at idle speed. While making the adjustments, **MAKE SURE** that when the throttle is advanced, both carburetors open at the same time.

Small boat, equipped with an AQ130 engine and Model 280 stern drive, operating in a test tank during final adjustments.

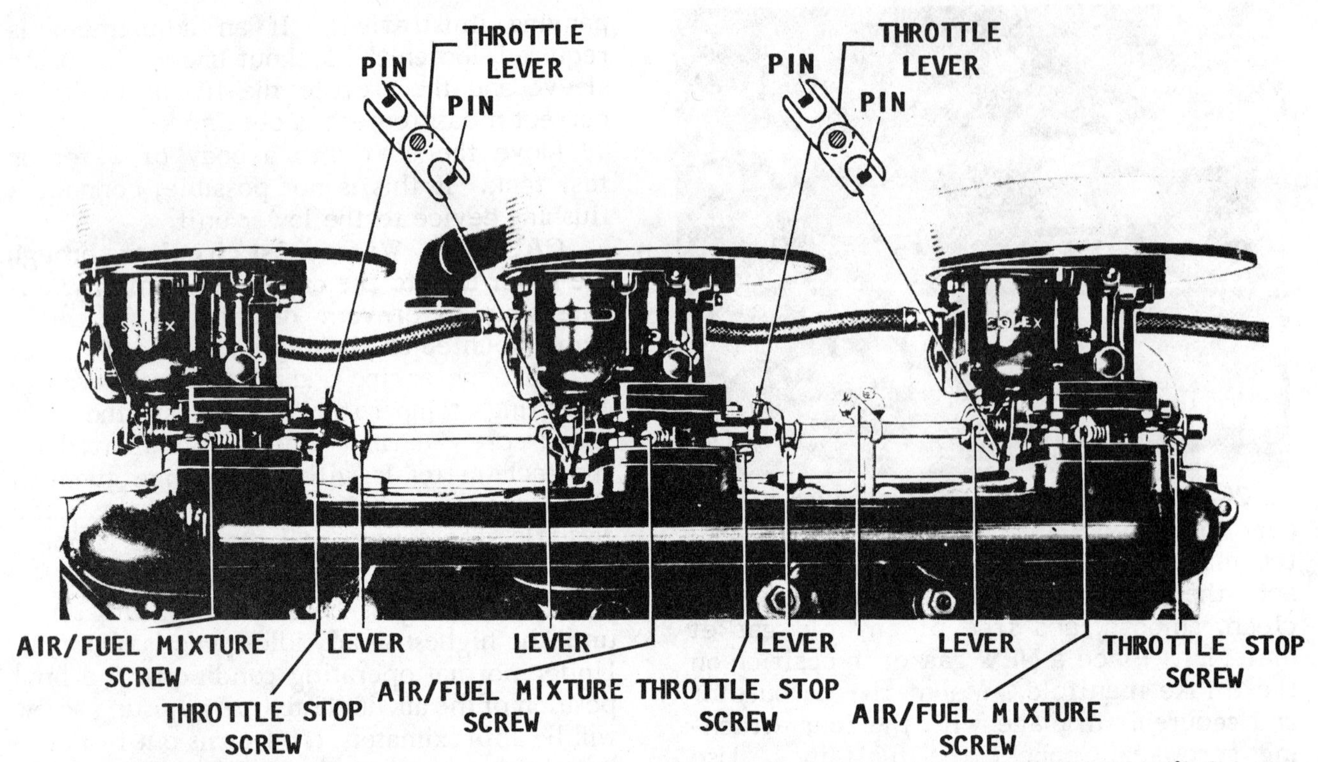

Multiple Solex down-draft carburetor installation. The synchronizing adjustment points are identified.

As an aid to adjusting multiple carburetors, a three-pound coffee container can be adapted to hold a Uni-Sync vacuum measuring device, as shown in the accompanying illustration. This arrangement will prove to be an excellent tool. The coffee can inverted will fit on the top base of each carburetor. The vacuum of the carburetor can be accurately measured and adjustments made until both carburetors are pulling equally.

UNI-SYNC

3-POUND
COFFEE CONTAINER

A coffee container and Uni-Sync vacuum measuring device used when synchronizing multiple carburetors.

Release the clamp nut for one of the levers. Back off all the throttle stop screws until they just make contact but do not press against the boss on the carburetor body. From this position, screw the throttle stop screws in exactly **ONE** full turn. Adjust and lock the lever so that both the throttle levers are actuated at the same time. Adjust the position of the fitting on the end of the control cable until the pins for the levers are in the middle of the gap on the throttle levers when the fitting is connected to the end of the control lever. Connect the fitting and lock it in position. The levers between the rear and forward carburetors should be adjusted until there is a small space between the levers and pins.

Thread the air/fuel mixture adjusting screw in until it **BARELY** seats and then back it out **1-1/4** turns.

Move the boat into a body of water or test tank. If this is not possible, connect a flushing device to the lower unit.

CAUTION: Water must circulate through the lower unit to the engine any time the engine is run to prevent damage to the water pump mounted on the engine.

Start the engine and allow it to warm to operating temperature. Check the rpm idle speed as indicated on the tachometer with the speed listed in the Appendix. If necessary, adjust the idle speed screw the same amount on each carburetor until engine rpm is within the range given in the Appendix.

ADJUSTMENTS AND SYNCHRONIZING TRIPLE CARBURETOR INSTALLATION

SPECIAL WORDS

When making the carburetor adjustments on the AQ170A, the intake silencer and the speed cable should be removed.

The objective of the synchronizing procedure is to ensure that all three carburetors open at the **SAME TIME.** Your goal is to be sure the three carburetors are drawing air and fuel equally, at idle idle speed. While making the adjustments, **MAKE SURE** that when the throttle is advanced, all carburetors open at the same time.

As an aid to adjusting multiple carburetors, a three-pound coffee container can be adapted to hold a Uni-Sync vacuum measuring device, as shown in the accompanying illustration. This simple arrangement will prove to be an **EXCELLENT** tool. The coffee can inverted will fit on the top base of each carburetor. The vacuum of the carburetor can be accurately measured and adjustments made until all three carburetors are pulling equally.

Loosen the clamp nut on one of the levers on the front and rear intermediate shaft. Back off all the throttle stop screws until they just make contact, but do not press against, the boss on the carburetor. From this position tighten all the throttle stop screws exactly **ONE** full turn.

Adjust and lock the lever on the front intermediate shaft so that both the throttle levers are actuated at the same time. Adjust the position of the fitting on the end of the control cable so that the pins for the levers are in the middle of the gap on the throttle levers when the fitting is connected to the control lever. Connect the fitting to the control lever for the front intermediate shaft and lock it.

Tighten the air/fuel mixture adjusting screw until it **BARELY** seats, and then back it out **1-1/4** turns.

Install the intake silencer.

Move the boat into a body of water or test tank. If this is not possible, connect a flushing device to the lower unit.

CAUTION: Water must circulate through the lower unit to the engine any time the engine is run to prevent damage to the water pump mounted on the engine.

Start the engine and allow it to warm to operating temperature. Check the rpm idle speed as indicated on the tachometer with the speed listed in the Appendix. If necessary, adjust the idle speed screw the same amount on each carburetor until engine rpm is within the range given in the Appendix.

4-9 SOLEX SIDE-DRAFT CARBURETOR

DESCRIPTION AND OPERATION

The Solex side-draft carburetor is a single-barrel (valve) unit. The valve is adjustable. A tapered metering needle is fixed in the air valve and its vertical location is determined by the existing vacuum in the valve. These carburetors may be installed as a single unit on the AQ120 or AQ140 OHC engines.

REMOVAL AND DISASSEMBLING

1- Disconnect the battery leads at the battery terminals to prevent the possibility of a spark igniting fuel fumes trapped in the engine compartment. **TAKE CARE** that the engine compartment area is well ventilated to permit any accumulated fuel fumes to escape. Remove the hardware attaching the carburetor to the intake manifold. Move the carburetor clear of the intake manifold and then disconnect the linkage and the fuel line. Catch any residual fuel in the line into a suitable container.

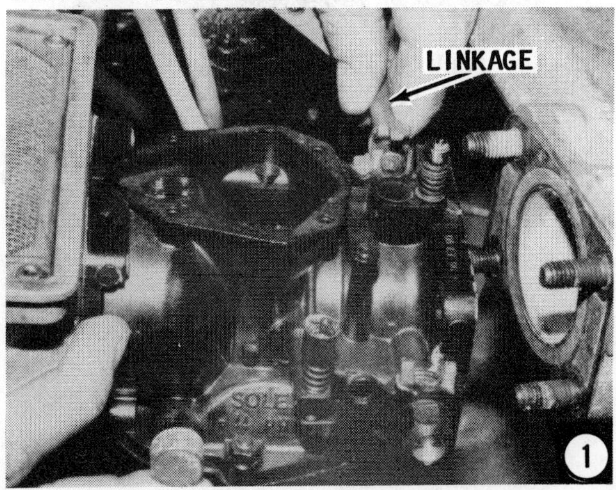

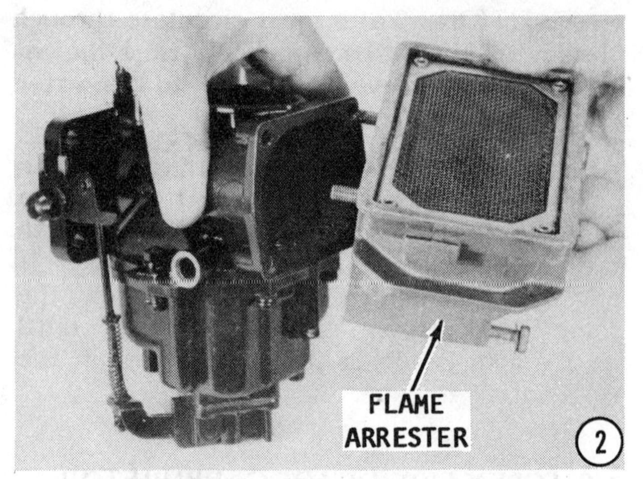

2- With the carburetor on the work bench, remove the four nuts securing the flame arrester to the carburetor body. Separate the flame arrestor from the carburetor.

SPECIAL WORDS

If the accelerator pump cover, accelerator pump rod, or spring, are not damaged, the cover may be left attached to the pump rod while other service work is being performed on the carburetor.

3- If the accelerator pump cover is to be removed, **COUNT** the number of turns required to back off and remove the nut on the end of the accelerator pump rod. An alternative to counting the number of turns for the nut, is to first measure the distance the pump rod extends through the nut before starting to remove the nut. Lift the pump cover off the pump rod, and then slide the spring free.

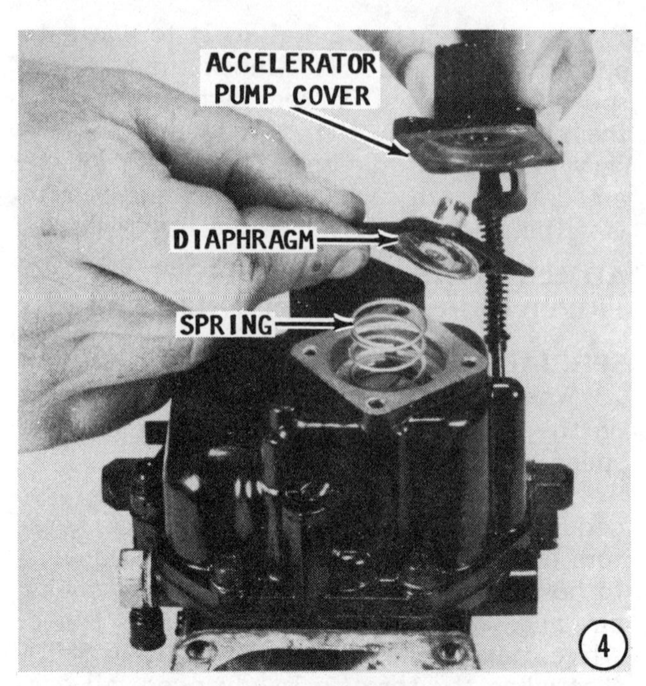

4- Remove the attaching screws, and then lift the accelerator pump cover free of the carburetor body. The pump diaphragm and spring can now be removed.

5- Remove the six screws securing the float bowl to the carburetor body. Separate the float bowl from the carburetor body. Remove and **DISCARD** the gasket.

6- Remove the idle jet, and then remove the main jet. It is not necessary to remove the body for the main jet.

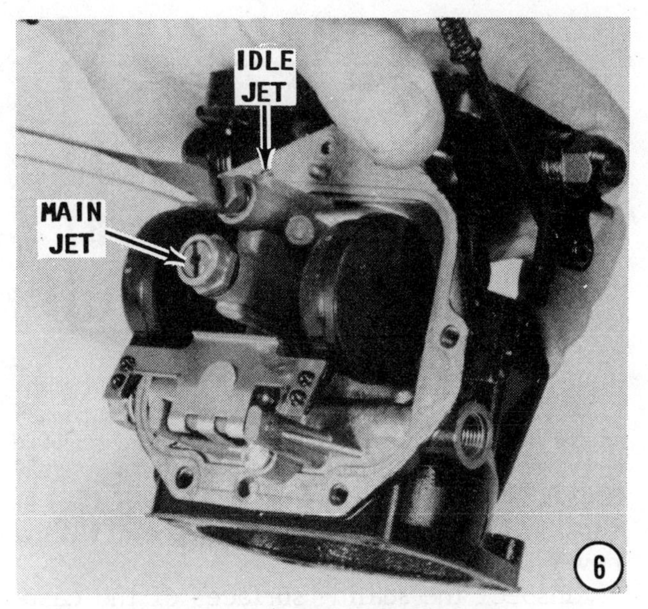

7- Remove the float hinge pin screw, and then lift the float assembly free.

8- Remove the float valve needle and seat.

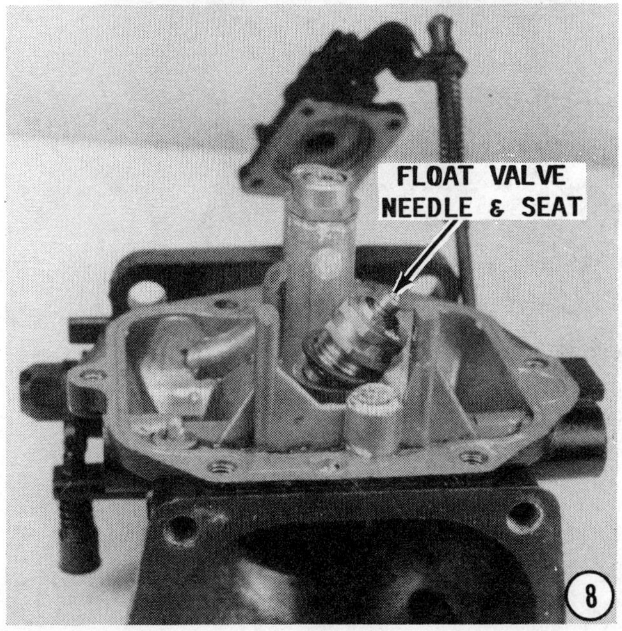

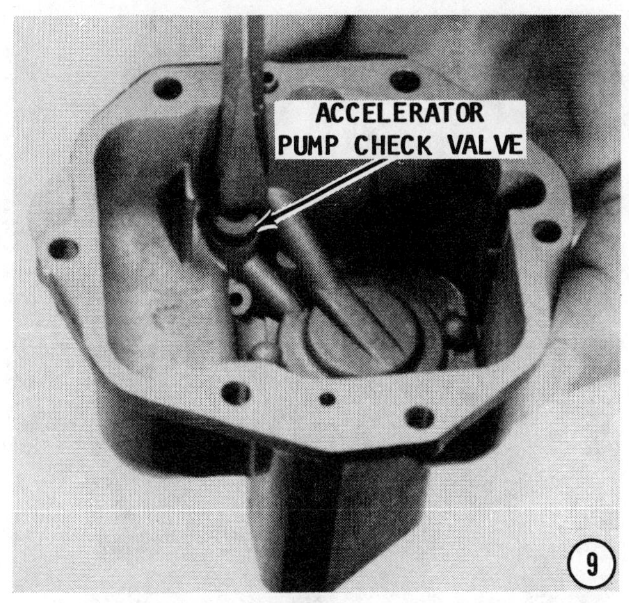

9- Remove the check valve for the accelerator pump.

10- Remove the air/fuel adjusting screw.

11- Remove the screw opening the pas-

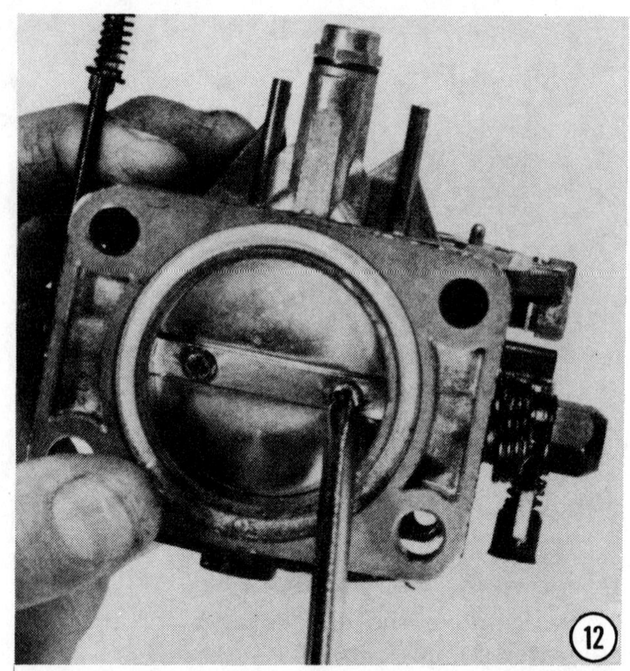

sageway for the accelerator pump nozzle. Removal of this screw permits blowing the passageway clean with compressed air.

12- Remove the two screws securing the butterfly to the shaft. Slide the butterfly out of the groove and free of the shaft. The butterfly is **ONLY** removed from the shaft because it has been damaged. If necessary, the nut can be removed from one end of the shaft, the opposite end from the throttle arm, and then the shaft pulled free from the throttle arm end.

CLEANING AND INSPECTING

NEVER dip rubber parts, plastic parts, diaphragms, or pump plungers in carburetor cleaner.

Place all metal parts in a screen-type tray and dip them in carburetor solvent until they appear completely clean, then blow them dry with compressed air.

Blow out all passages in the castings with compressed air. Check all parts and passages to be sure they are not clogged or contain any deposits. **NEVER** use a piece of wire or any type of pointed instrument to clean drilled passages or calibrated holes in a carburetor.

Inspect the idle mixture needle for any type of damage. Carefully check the float needle and seat assembly for wear. Good shop practice is to always install a new factory matched set to avoid leaks.

Check the springs to be sure they have not become distorted or have lost their

A clean flame arrester is essential for maximum engine performance. The arrester should be removed and cleaned at least once each season or following 50 hours of operation.

tension. If in doubt, compare them with a new one, if possible.

Inspect the sealing surfaces of the casting for damage.

Most of the parts that should be replaced during a carburetor overhaul are included in an overhaul kit available from the local Volvo Penta marine dealer.

ASSEMBLING

1- If the butterfly was disassembled and the shaft removed, slide the shaft through the carburetor body from the throttle arm side. Force the butterfly into the groove in the shaft, and then secure it to the shaft with the two Phillips head screws.

2- Thread the screw covering the accelerator pump nozzle passage into the carburetor body and tighten it just snugly.

3- Thread the air/fuel adjusting screw into place.

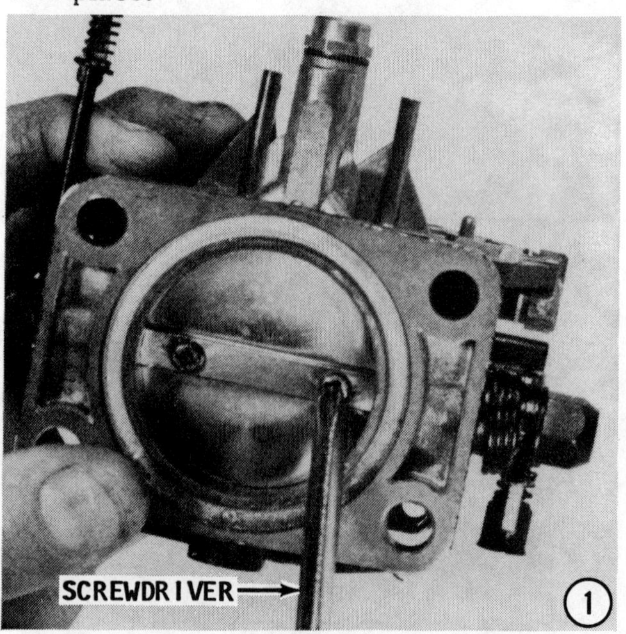

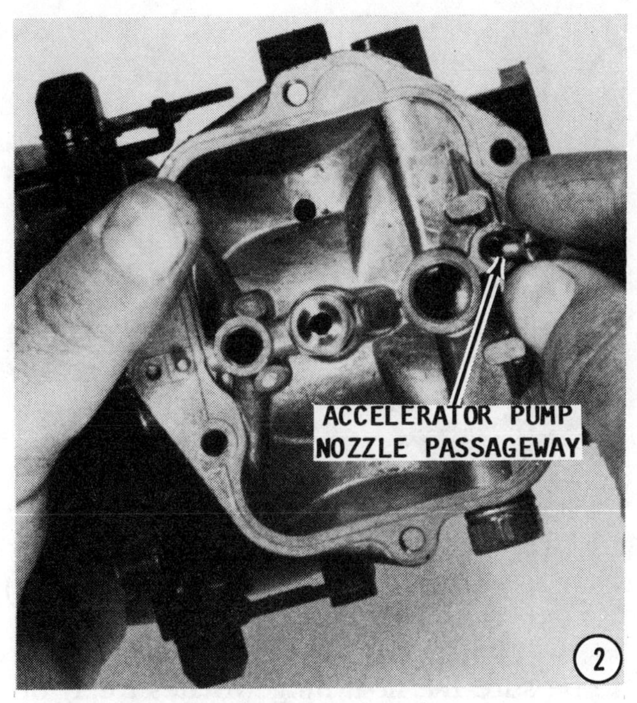

4- Slowly tighten the air/fuel adjusting screw until it **BARELY** seats, and then back it out **FOUR AND ONE-HALF** (4-1/2) complete turns as a preliminary adjustment at this time.

5- Install the check valve for the accelerator pump. Tighten the valve snugly.

6- Insert the float valve needle and seat into the carburetor body, and then tighten the needle securely.

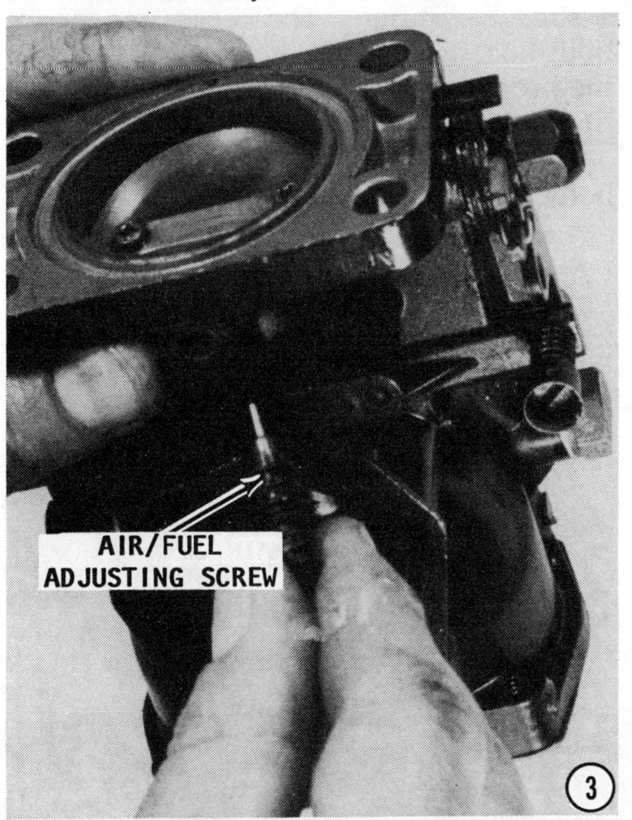

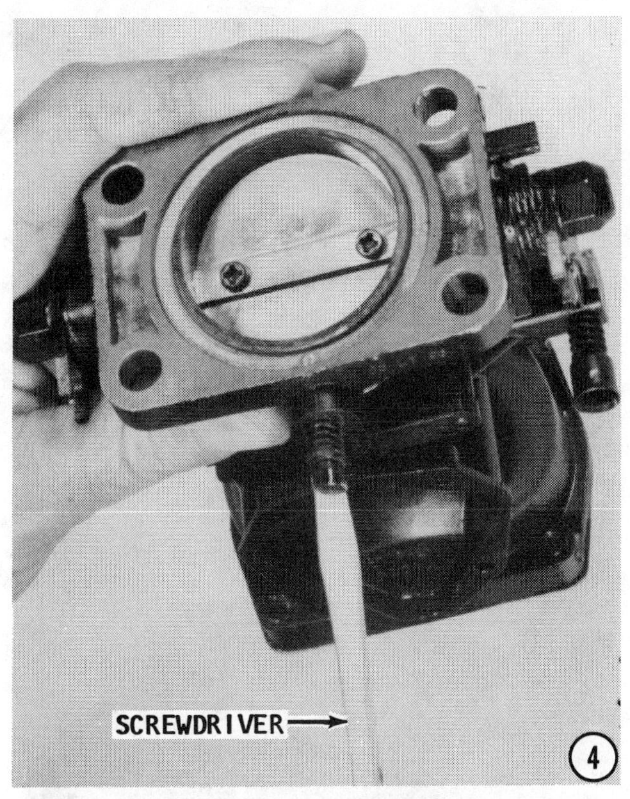

7- Place the float in position with the holes in the hinge aligned with the holes in the brackets. Hold the float in position and slide the hinge pin through the bracket without threads in the hole, then through the hinge, and then thread the pin into the other bracket.

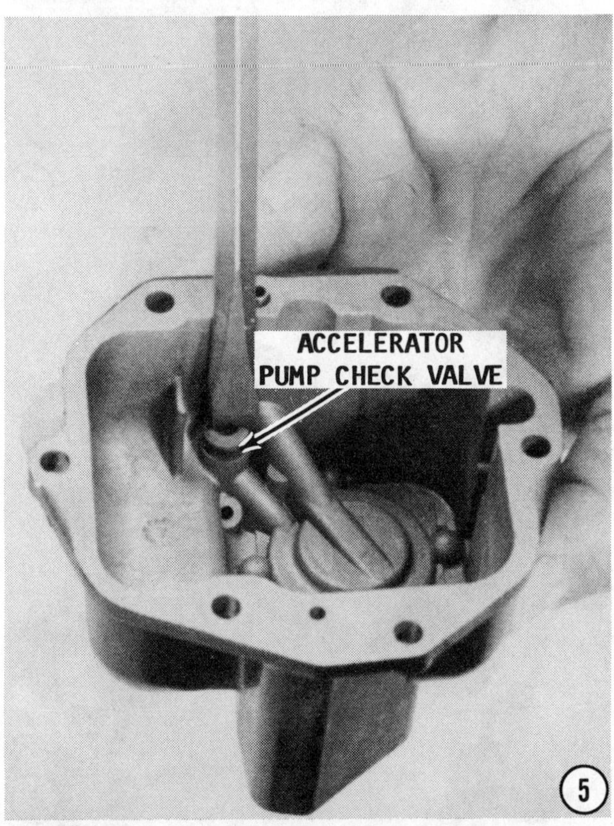

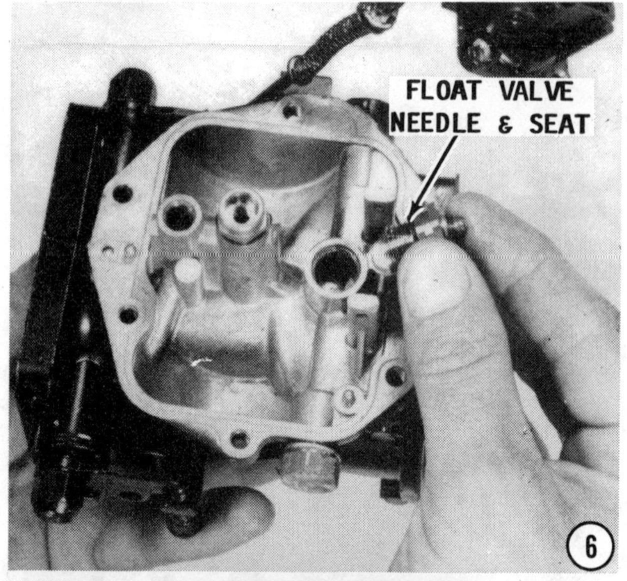

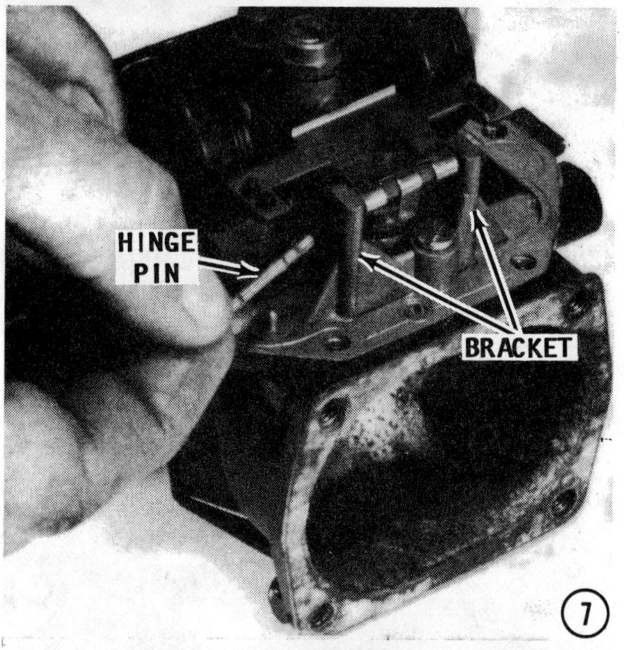

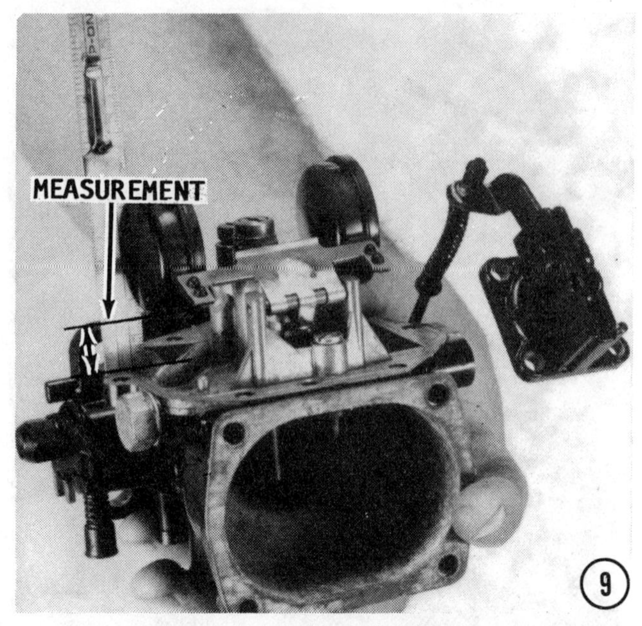

8- Tighten the hinge pin snugly. Check to be sure the float hinge rotates freely on the pin, with no sign of binding.

9- To adjust the float position, hold the carburetor in a level attitude, as shown, with the floats facing upward. Now, measure the distance from the machined surface of the carburetor to the bottom of one float, as indicated. This measurement should be a minimumn of 7/8" (22mm), and a maximum of 1-1/32" (26mm). Check the other float in a similar manner. The measurement for both floats should be the same. To obtain the proper measurement, **CAREFULLY** bend the hinge tab either downward or upward.

10- Thread the main jet into the main jet body. Tighten the jet securely.

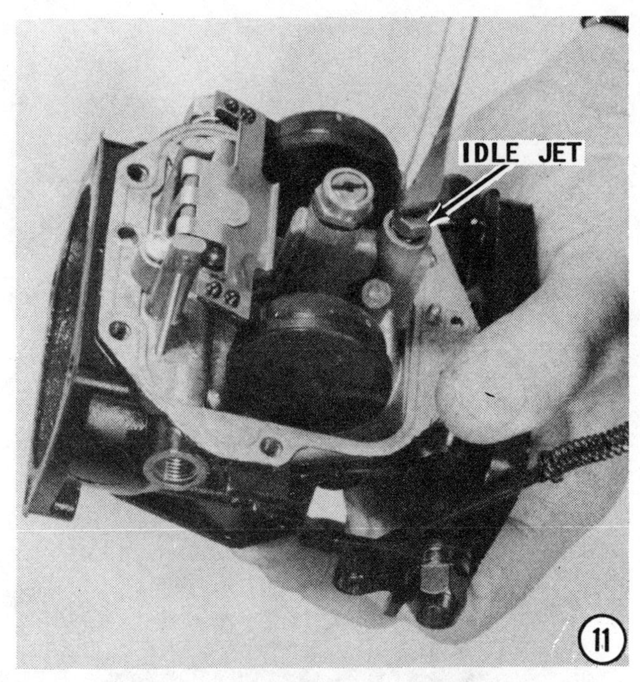

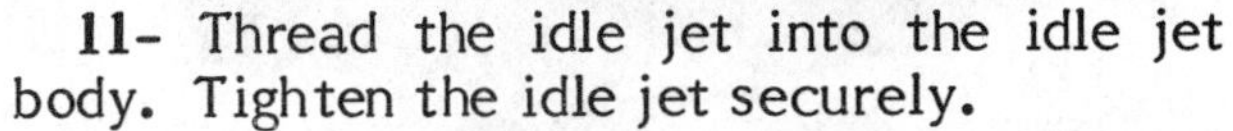

11- Thread the idle jet into the idle jet body. Tighten the idle jet securely.

12- Check to be sure the surface of the carburetor body is clean and completely free of any old gasket material. Lay a **NEW** gasket in place on the carburetor body with the holes in the gasket aligned with the matching holes in the body.

13- Check to be sure the matching surface of the float bowl is clean and free of any old gasket material.

14- Secure the float bowl to the carburetor body with the six attaching screws. Tighten the screws evenly and alternately to prevent warping the bowl.

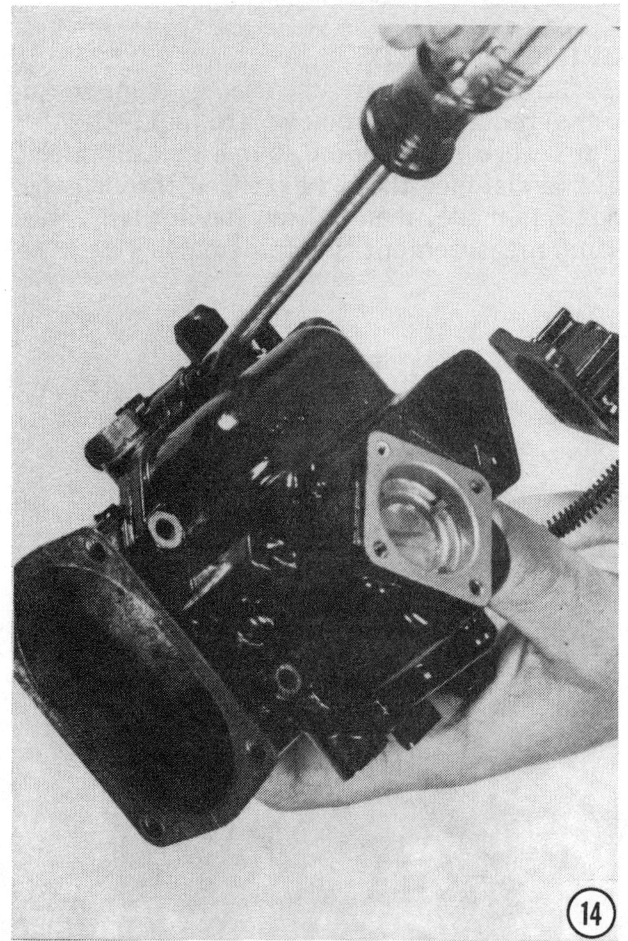

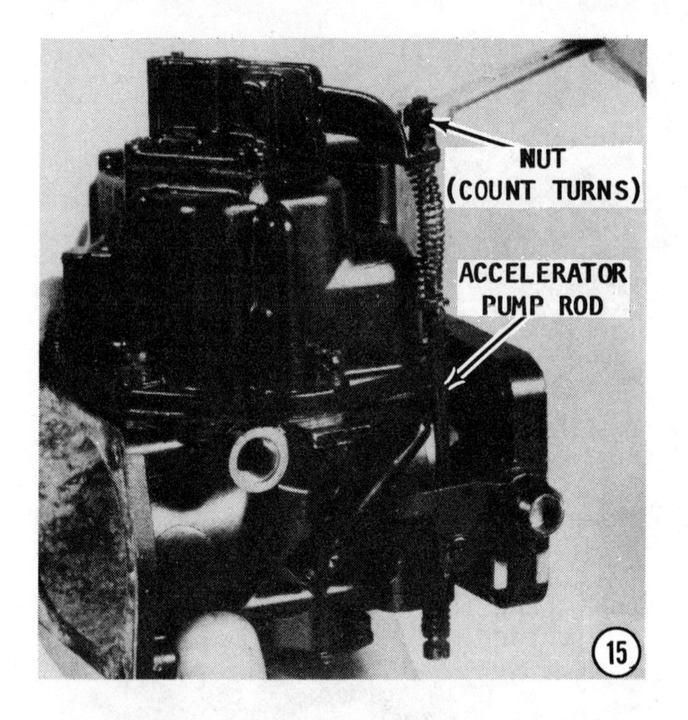

15- If the accelerator pump cover was removed from the accelerator pump rod, slide the spring onto the rod. Slide the arm of the pump cover onto the rod and then thread the nut onto the rod.

CRITICAL WORDS

Tighten the nut the same number of turns required to remove the nut. If the turns were not counted, but a measurement of the distance the rod extended through the nut was made, then tighten the nut until this same measurement is obtained.

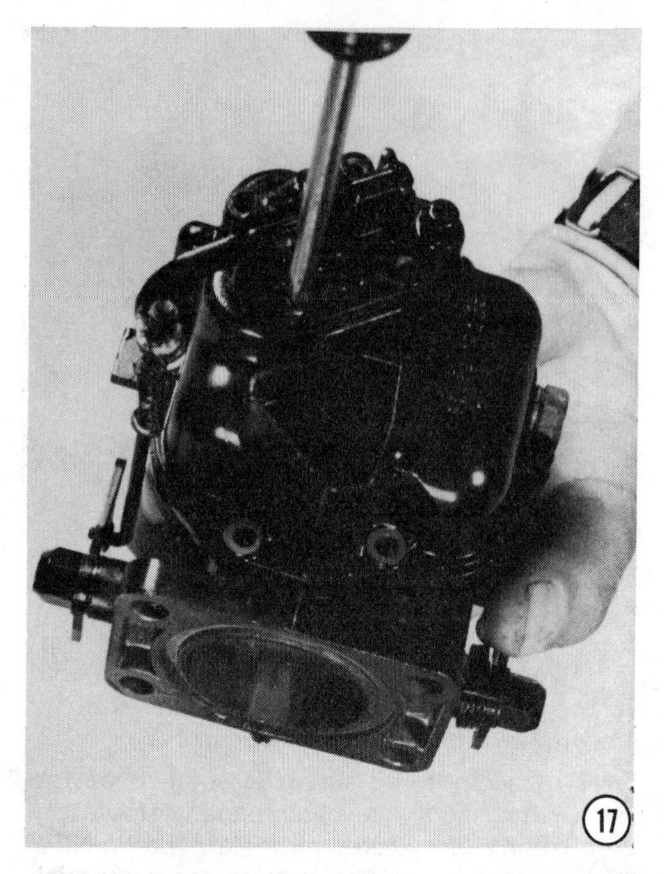

16- Set the spring into the pump recess, and then lay the diaphragm onto the spring with the screw holes in the diaphragm aligned with the matching holes in the carburetor.

17- Secure the accelerator pump cover in place with the attaching screws. Tighten the screws evenly and alternately.

18- Attach a **CLEAN** flame arrester to the carburetor and secure it in place with the four nuts. Tighten the nuts securely.

19- Hold the assembled carburetor in approximate position on the engine and attach the connecting linkage and the fuel line.

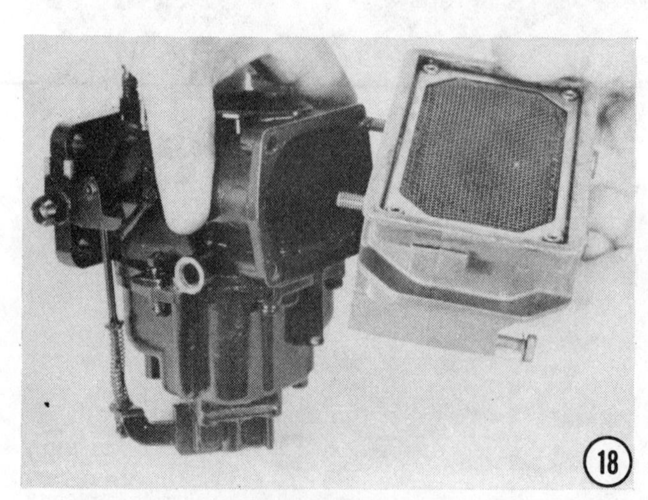

20- Secure the carburetor to the intake manifold with the attaching hardware. Connect the battery cables to the battery.

Move the boat into a body of water or test tank. If this is not possible, connect a flushing device to the lower unit.

CAUTION: Water must circulate through the lower unit to the engine any time the engine is run to prevent damage to the water pump mounted on the engine.

Start the engine and allow it to warm to operating temperature, until the choke is wide open. Adjust the idle speed screw until 500 rpm is indicated on the tachometer. **SLOWLY** adjust the air/fuel adjusting screw until the highest steady idle speed is obtained.

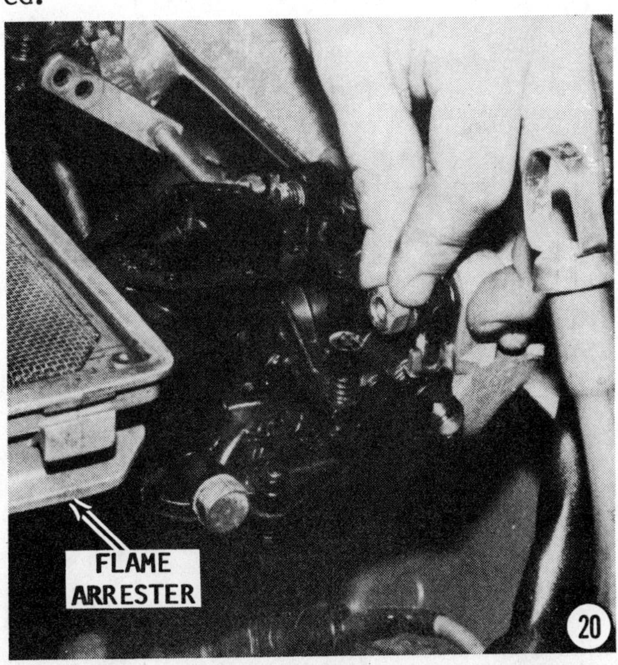

The Rochester 2GC (two-barrel) carburetor.

4-10 ROCHESTER CARBURETOR MODEL 2GC

This current two-barrel carburetor is available in two different models: 2GC and 2GV. The Model 2GV has an automatic choke with the thermostatic coil installed on the engine manifold. The coil is then connected to the choke valve through linkage. This model has a vacuum-break diaphragm on the carburetor instead of the conventional choke housing on the air horn. The Model 2GC has the automatic choke as a part of the carburetor.

DISASSEMBLING

1- Remove the idle vent valve and the pump link. Remove the screw at the end of the choke shaft and the fast-idle cam attaching screw, then remove the fast-idle linkage as an assembly. Remove the air horn attaching screws and lift the air horn straight up from the body.

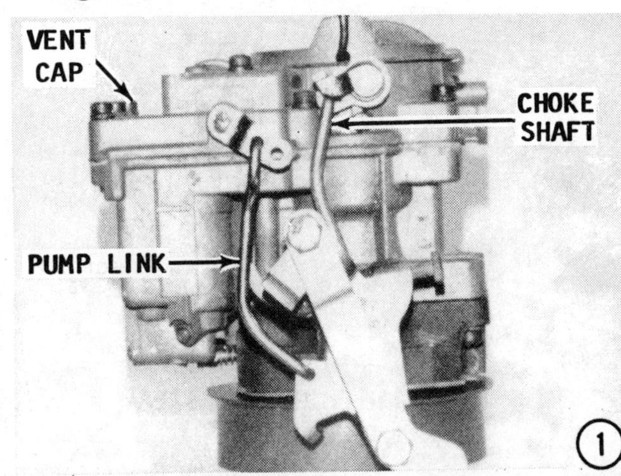

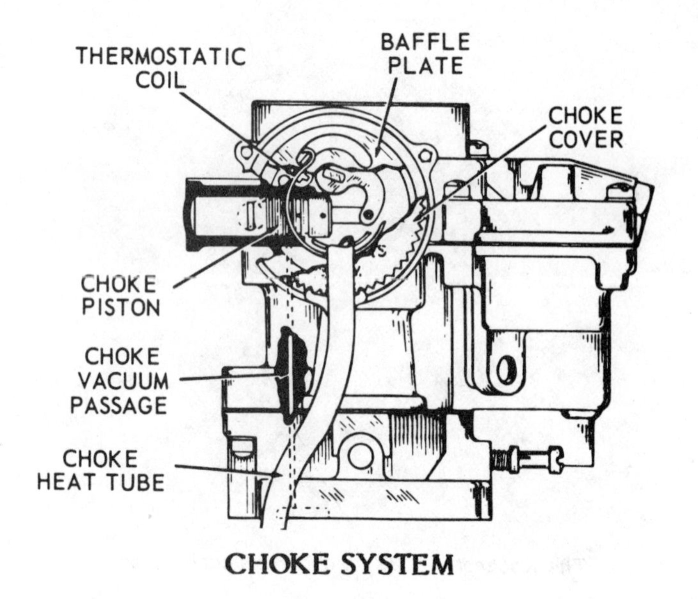

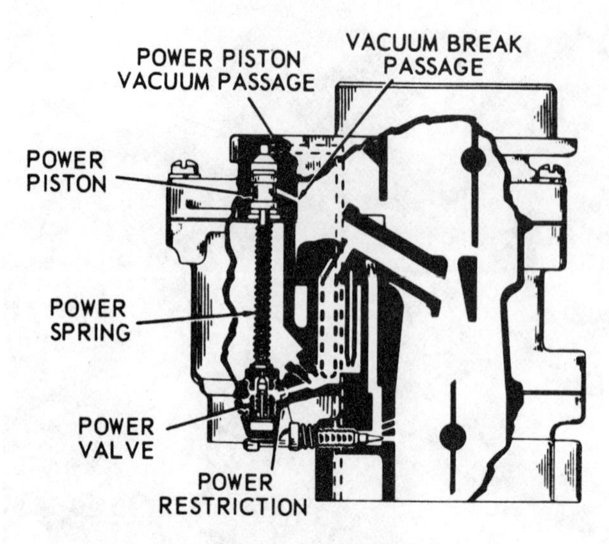

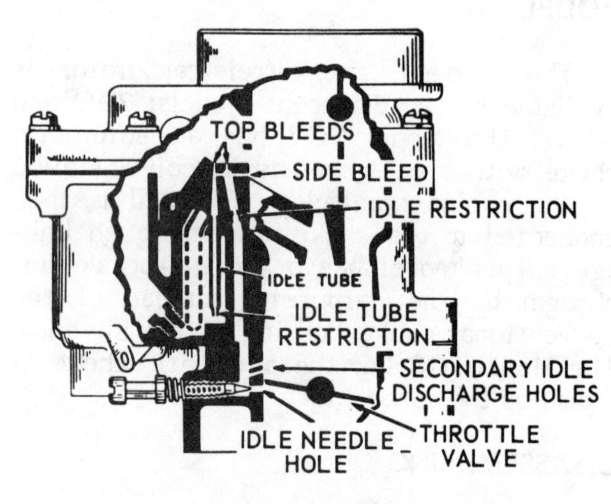

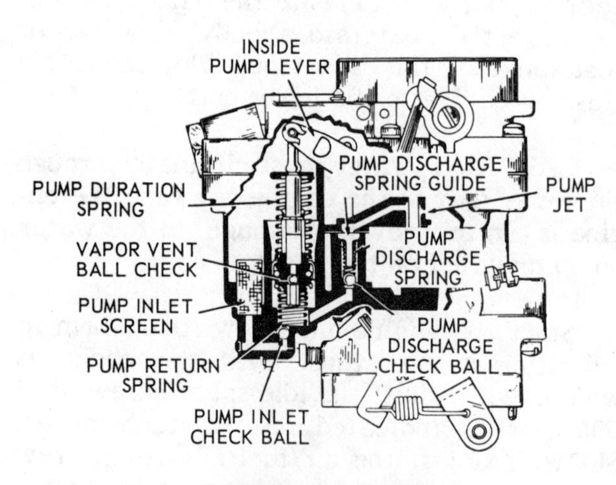

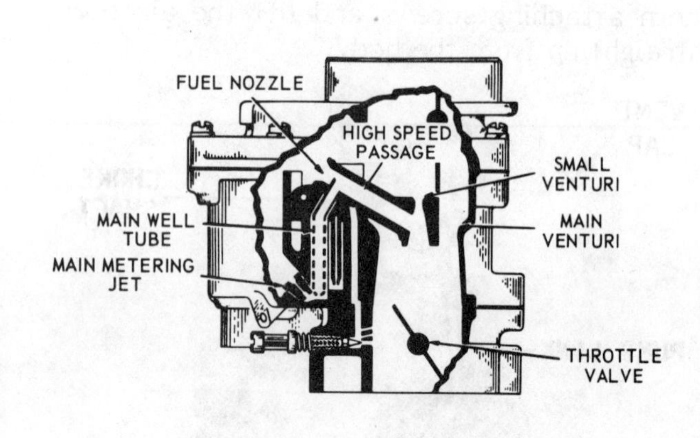

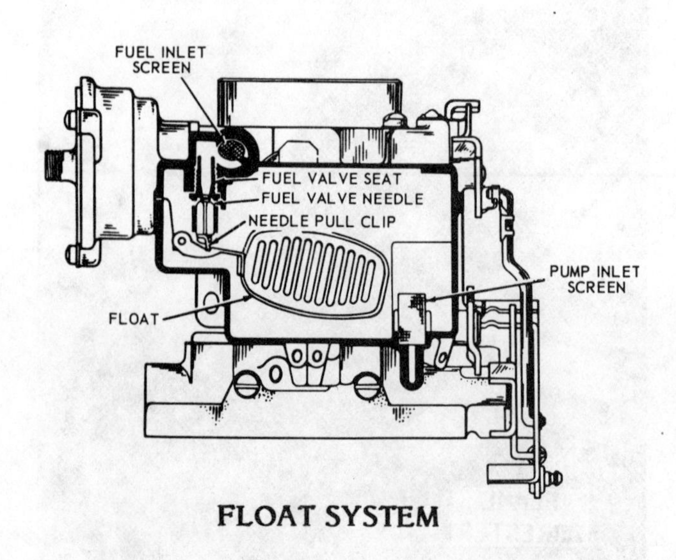

Details of the various systems of the Rochester 2GC carburetor.

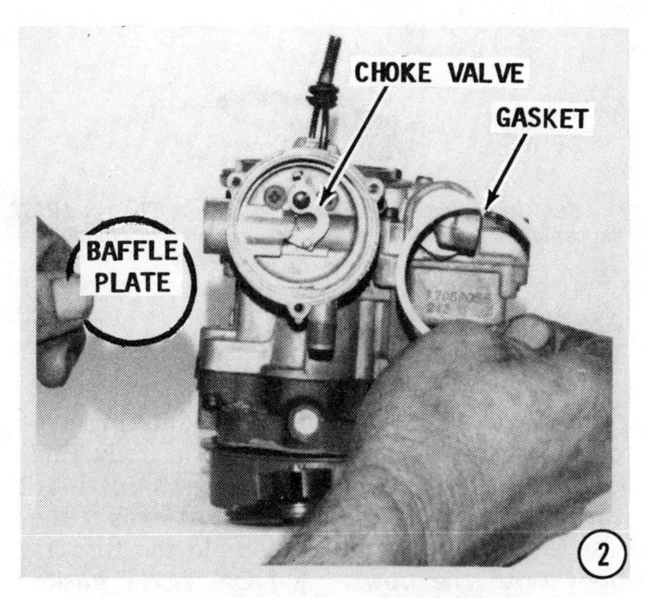

2- Disassemble the automatic choke by first removing the two choke valve retaining screws. However, under normal conditions, the choke shaft is not removed. You may have to file the staked ends of the screws in order to remove them. Lift out the choke valve. Remove the choke cover, with the thermostatic coil attached, the gasket and the baffle plate. Now, rotate the choke shaft counterclockwise to guide the piston from its bore. Remove the shaft and piston assembly.

3- Turn the air horn upside down and remove the float hinge pin. Lift off the float, and remove the float needle. Depress the power piston shaft which will allow the spring to snap and in turn will force the piston from the casting and allow the power piston to be removed. Remove the accelerator pump plunger and the pump lever and shaft assembly. Remove the two choke housing-to-air horn attaching screws and lift off the choke housing. Discard the gasket.

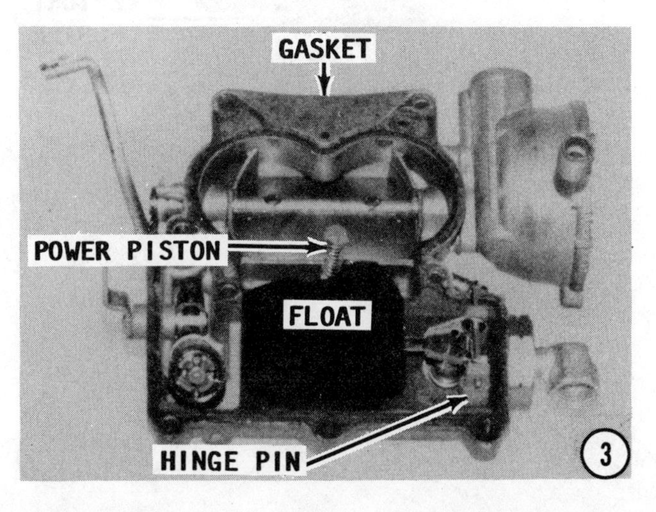

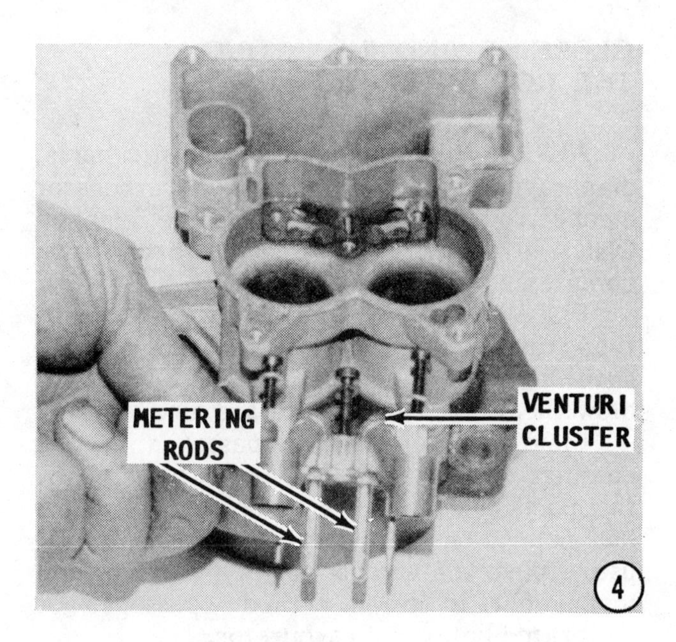

4- Remove the pump plunger, return spring, main metering jets, and power valve from the main body. Some models have an aluminum inlet ball in the bottom of the pump well. This ball will fall out when the carburetor is turned over. Remove the venturi cluster by removing the attaching screws.

5- Use a pair of needle-nosed pliers to remove the discharge ball spring T-shaped retainer. Now, take out the pump discharge spring and the steel discharge ball. Turn the carburetor over and remove the three throttle body-to-bowl attaching screws. Lift off the throttle body. **DO NOT** disassemble the throttle body. Replacement parts are **NOT** available because of the close relationships between the throttle plates and the idle ports.

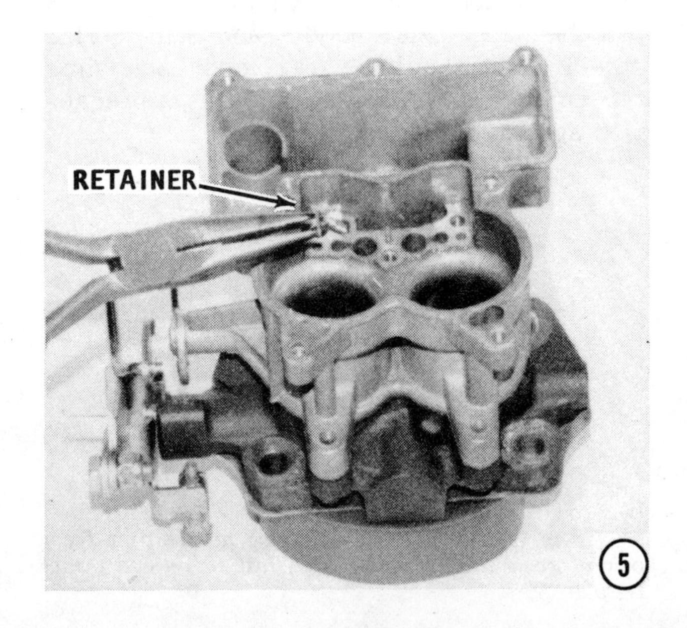

CLEANING AND INSPECTING THE ROCHESTER 2GC

NEVER dip rubber parts, plastic parts, diaphragms, or pump plungers in carburetor cleaner. These parts should be cleaned **ONLY** in solvent, and then blown dry with compressed air.

Place all of the metal parts in a screen-type tray and dip them in carburetor cleaner until they appear completely clean, then blow them dry with compressed air.

Blow out all of the passages in the castings with compressed air. Check all of the parts and passages to be sure they are not clogged or contain any deposits. **NEVER** use a piece of wire or any type of pointed instrument to clean drilled passages or calibrated holes in a carburetor.

Move the throttle shaft back-and-forth to check for wear. If the shaft appears to be too loose, replace the complete throttle body because individual replacement parts are not available.

Inspect the main body, air horn, and venturi cluster gasket surfaces for cracks and burrs which might cause a leak.

Shake the float to determine if there is any liquid inside, and if there is, replace the float. Check the float arm needle contacting surface and replace the float if this surface has a groove worn in it.

Inspect the tapered section of the idle adjusting needles and replace any that have developed a groove.

Most of the parts that should be replaced during a carburetor overhaul are included in an overhaul kit available from your local marine dealer. This kit will also contain a matched fuel inlet needle and seat. This combination should be replaced each time the carburetor is disassembled as a precaution against leakage.

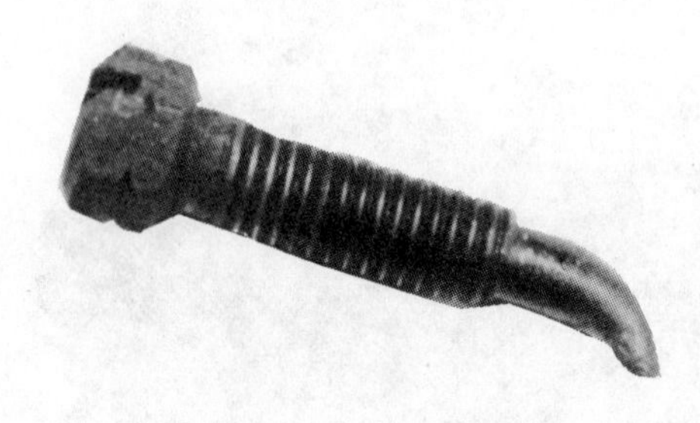

Tip of an idle mixture adjusting needle bent from being forced into its seat and against the closed throttle plate.

*Fuel inlet needle with a worn ridge. The set **MUST** be replaced to prevent a leak and carburetor flooding.*

ASSEMBLING THE ROCHESTER 2GC

1- Install the idle mixture adjusting needles and springs into the throttle body **FINGERTIGHT.** Now, back the screws out 1-1/2 turns as a rough adjustment at this time. Use a **NEW** gasket and assemble the throttle body onto the bowl. A **NON-VENT** gasket **MUST** be used because during hot-engine operation, the fuel in the carburetor tends to "percolate" due to engine heat. If a vented, automotive-type gasket is used, these fuel vapors will be vented directly into the atmosphere. A genuine replacement gasket will prevent these fuel vapors from venting. Insert the pump discharge check steel ball, spring, and T-shaped retainer into the top of the main body. Use a **NEW** gasket and install the venturi cluster. **BE SURE** the undercut screw has a gasket and is placed in the center hole. Install the main metering jets and the power valve, with a **NEW** gasket. Install the pump return spring in the pump well and the pump inlet screen in the bottom of the bowl.

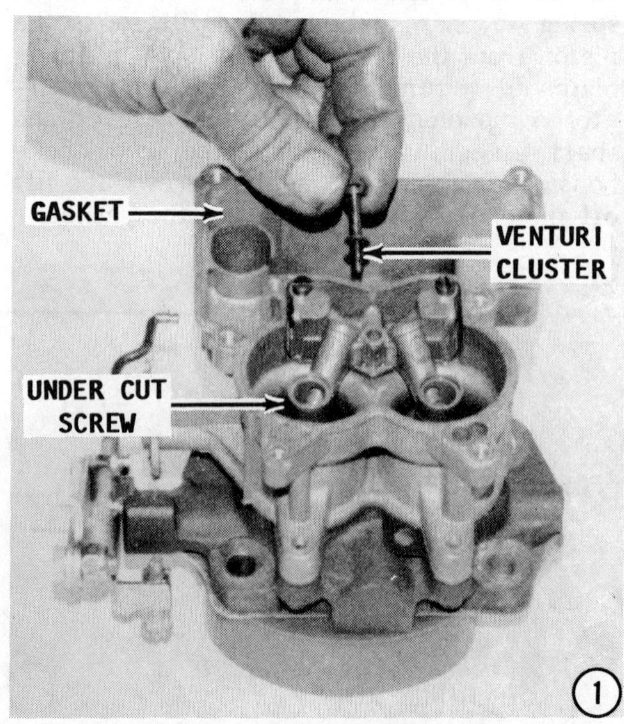

1. Idle Stop Lever Screw
2. Air Horn Gasket
3. Long Air Horn Screw
4. Short Air Horn Screw
5. Air Horn Lockwasher
6. Pump Lever
7. Pump Screw
8. Pump Clip
9. Pull Clip
10. Choke Gasket
11. Choke Valve
12. Pump Discharge Ball
13. Pump Discharge Spring
14. Pump Spring Guide
15. Venturi Gasket
16. Outer Venturi Screw
17. Center Venturi Screw
18. Outer Venturi Lockwasher
19. Center Venturi Gasket
20. Main Metering Jet
21. Power Valve Gasket
22. Choke Rod
23. Idle Speed Stop Lever
24. Throttle Body Gasket
25. Throttle Body Screw
26. Throttle Body Lockwasher
27. Pump Rod
28. Pump Rod Clip
29. Idle Stop Screw
30. Idle Adjusting Needle
31. Idle Needle Spring
32. Pump Shaft and Lever Assembly
33. Pump Assembly
34. Air Horn Assembly
35. Power Piston Assembly
36. Float Assembly
37. Needle and Seat Assembly
38. Needle Seat Gasket
39. Choke Housing Assembly
40. Thermostat Cover
41. Thermostat Cover Gasket
42. Choke Lever and Link Assembly
43. Choke Shaft Assembly
44. Choke Lever and Collar Assembly
45. Float Bowl Assembly
46. Venturi Cluster Assembly
47. Power Valve Assembly
48. Throttle Body Assembly
49. Float Hinge Pin
50. Choke Housing Screw
51. Baffle Plate
52. Thermostat Cover Retainer
53. Thermostat Cover Screw
54. Choke Lever Screw
55. Choke Valve Screw
56. Choke Piston
57. Choke Piston Pin
58. Lead Ball Plug
59. Expansion Plug
60. Pump Return Spring
61. Fuel Line Fitting

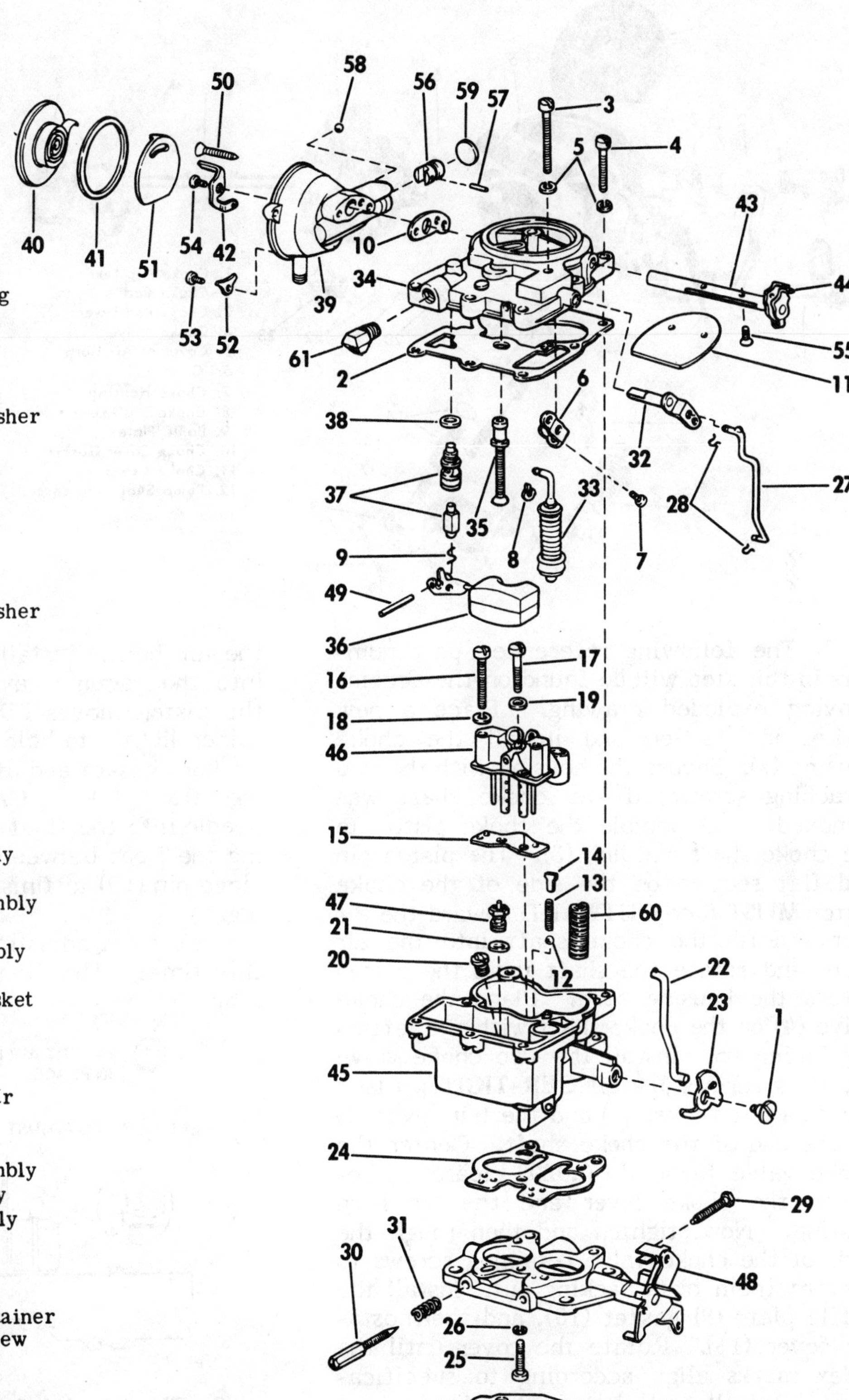

Exploded view of a Rochester 2GC carburetor.

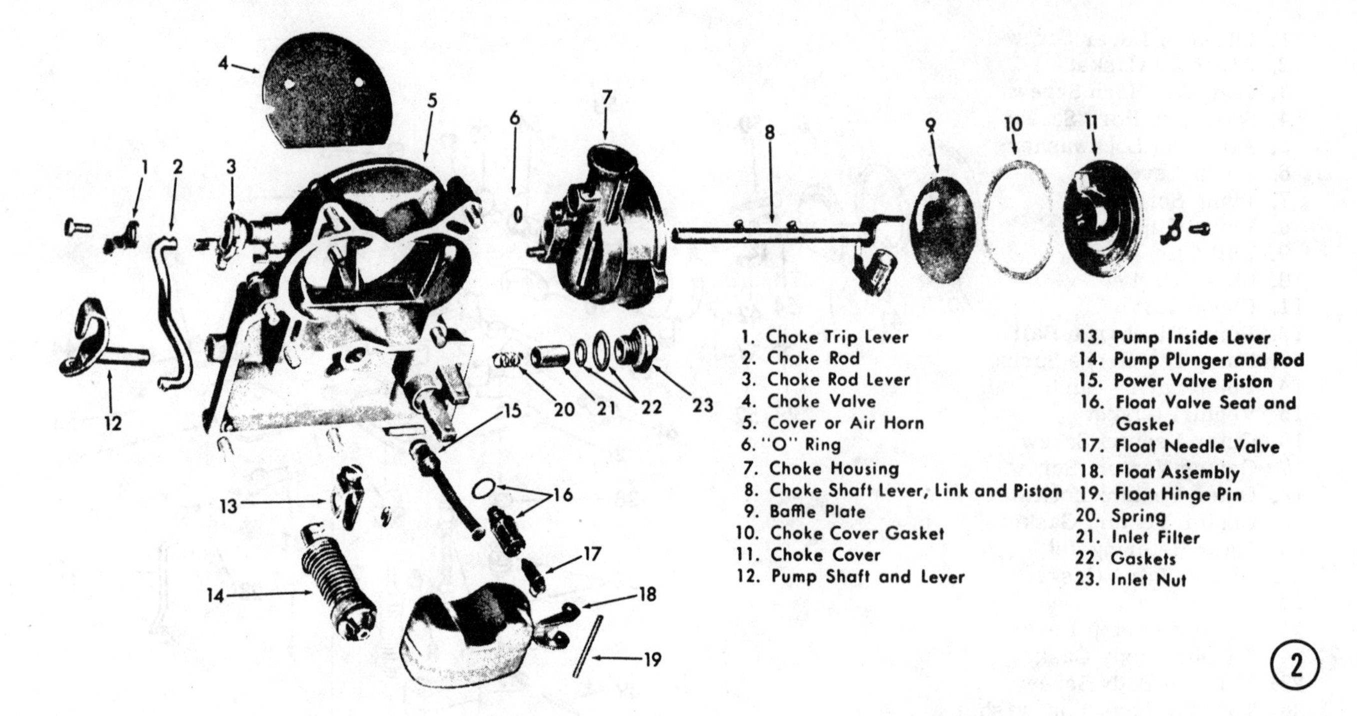

2- The following referenced part numbers in this step will be found on the accompanying exploded drawing. Place a new gasket in position and install the choke housing (7). Secure the housing with the two attaching screws, if the choke shaft was removed. Assemble the choke piston to the choke shaft and link (8). The piston pin and flat section on the side of the choke piston **MUST** face **OUTWARD** toward the air horn. Push the choke shaft into the air horn, and rotate the shaft until the piston enters the housing bore. Place the choke valve (4) on the choke shaft with the letters **RP** facing up. Install the two choke valve retaining screws just **FINGER-TIGHT.** Place the choke rod lever (3) and the trip lever (1) on the end of the choke shaft. Center the choke valve to obtain 0.020" clearance between the choke lever and the air horn casting. Now, tighten and then rough the ends of the choke valve retaining screws to prevent them from backing out. Install the baffle plate (9) gasket (10), and thermostatic cover (11). Rotate the cover until the index marks align according to specifications. Install and tighten the three cover retainers and retaining screws. Install the outer pump lever (12) in the air horn. Assemble the inner pump arm (13) and tighten the screw . Attach the pump plunger (14) to the inner arm (13), with the pump shaft pointing **INWARD.** Install the horseshoe retainer. Position the screen on the float needle seat (16) and screw the assembly into the air horn. Install the power piston (15) into the vacuum cavity. Check to be sure the piston moves **FREELY.** Stake the retainer lightly to hold it in place. Install the air horn gasket and attach the needle (17) to the float (18). **CAREFULLY** insert the needle into the float needle seat while guiding the float between the bosses. Insert the hinge pin (19) to finish the air horn assembling.

Two float adjustments **MUST** be made at this time: The float level, and the float drop.

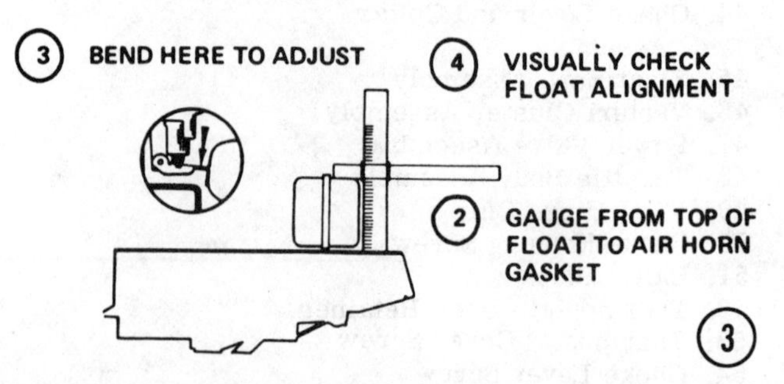

3- Float Level Adjustment: models with **VERTICAL** seam in the float. Measure the distance from the top of the float to the air horn, with the gasket in place. Compare your measurement with those given in the rebuild kit. **CAREFULLY** bend the float arm at the rear of the float as shown in the accompanying illustration, until the required measurement is reached.

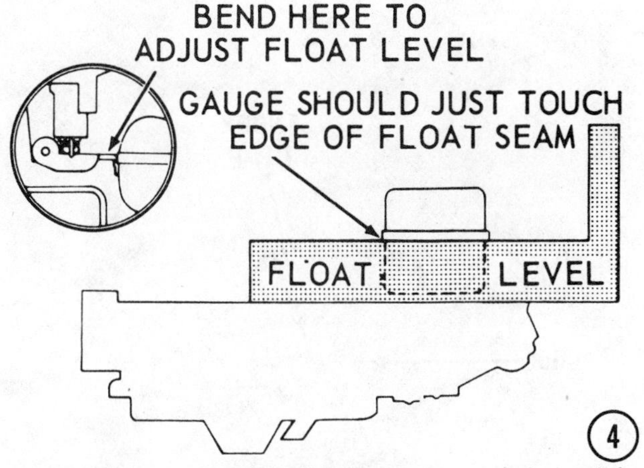

4

4- **Float Level Adjustment:** models with **HORIZONTAL** seam in the float. Measure the distance from the lower edge of the seam to the air horn, with the gasket in place. Compare your measurement with the those given in the rebuild kit. **CAREFULLY** bend the float arm at the rear of the float as shown in the accompanying illustration.

① INVERT AIR HORN WITH GASKET IN PLACE

② MEASURE FROM LIP AT TOE OF FLOAT TO AIR HORN GASKET

③ BEND HERE TO ADJUST

④ VISUALLY CHECK FLOAT ALIGNMENT

5

5- **Float Level Adjustment:** models with **NITROPHYL-TYPE** (hollow) float. Measure the distance from the lip at the toe of the float to the air horn, with the gasket in place. Compare your measurement with those given in the rebuild kit. **CAREFULLY** bend the float arm at the rear of the float, as shown in the accompanying illustration, until the required measurement is reached.

MEASURE SPECIFIED DISTANCE FROM GASKET SURFACE TO BOTTOM OF FLOAT

BEND FLOAT TANG TO ADJUST FOR PROPER SETTING

SCRIBED MARK ON GAUGE INDICATES PROPER SETTING

FLOAT DROP

6

6- **Float Drop Adjustment:** Turn the air horn right side up to allow the float to move to the wide-open position. Now, measure the distance from the air horn gasket to the bottom of the float. Compare your measurement with those given in the rebuild kit. **CAREFULLY** bend the float tang until the required measurement is reached, as shown in the accompanying illustration.

7- Replace the assembled air horn onto the bowl and guide the accelerator pump plunger into its well. Install and tighten the cover screws. Install the idle needle into the spring, then into the throttle body. Install the pump rod, fast-idle rod, the fast-idle cam, and screw.

BENCH ADJUSTMENTS FOR THE ROCHESTER 2GC

8- Back out the idle stop screw and completely close the throttle valves in their bores. Place a pump gauge across the top of the carburetor air horn ring, as shown, with the 1-5/32" leg of the gauge pointing downward towards the top of the pump rod. The lower edge of the gauge leg should just touch the top of the pump rod. **CAREFULLY** bend the pump rod, as required to obtain the proper setting.

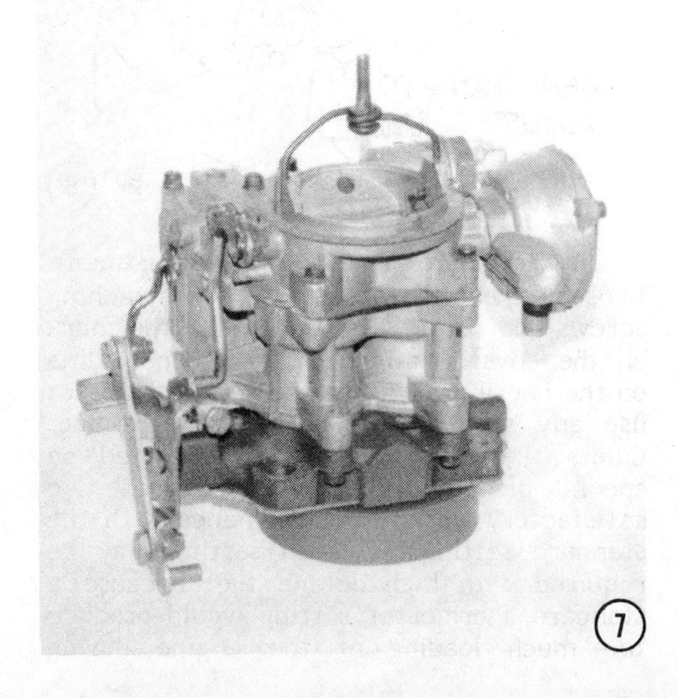

7

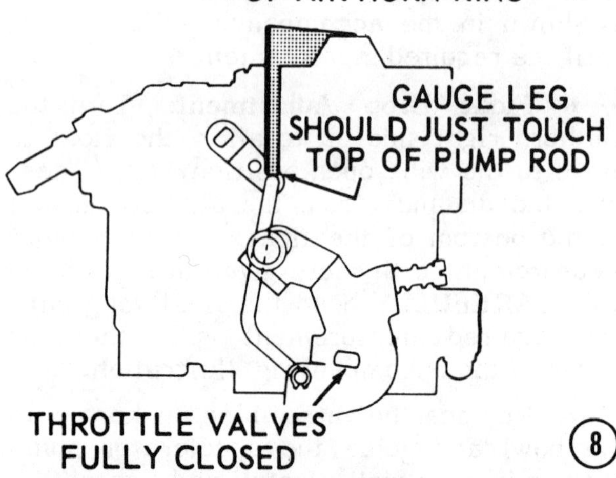

8

9- Unloader Adjustment: Open the throttle valves wide. The choke valve should open only enough to allow the specified gauge between the upper end of the valve and the inner air horn wall. **CAREFULLY** bend the tang on the throttle lever until the proper adjustment is reached.

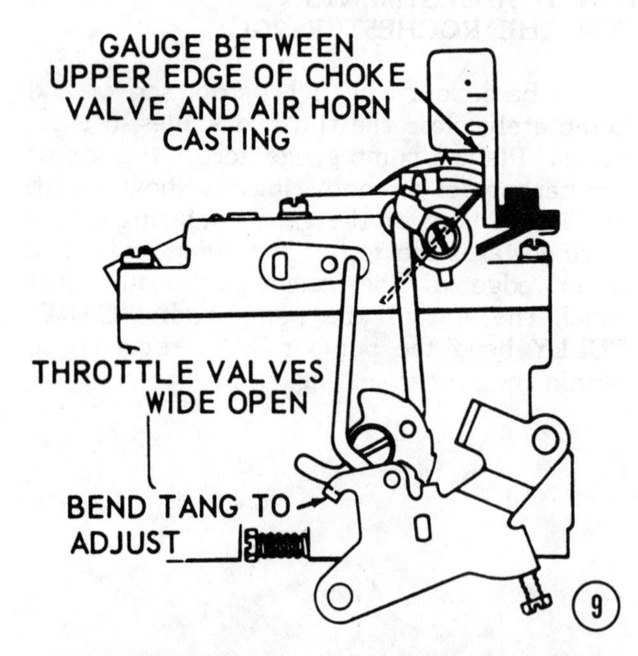

9

10- Automatic Choke Adjustment: Loosen the thermostat cover attaching screws and rotate the cover until the mark on the cover is aligned with the index line on the housing. Tighten the screws. Do not use any setting except the standard one, unless the engine is usually operated on special blends of fuel which do not give satisfactory warm-up performance with the standard setting. A lean setting may be required with high octane fuel because a standard thermostat setting would produce too much loading of the engine during

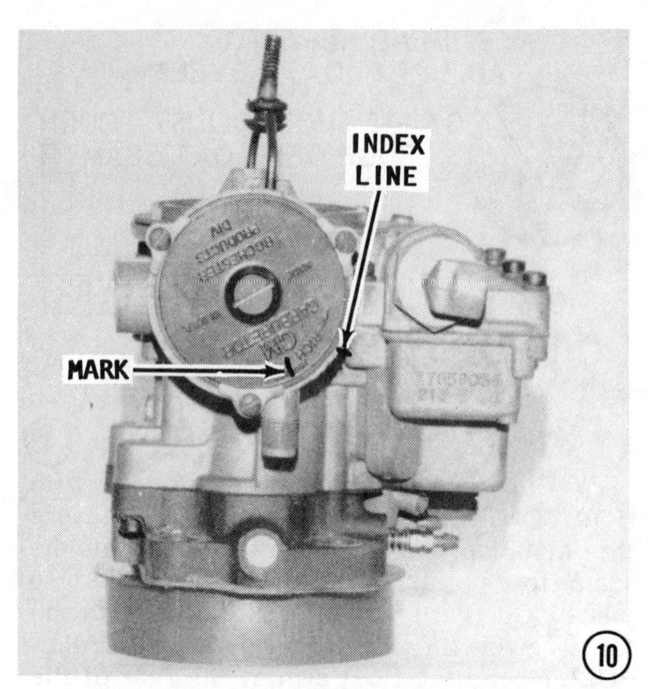

10

warm-up. A rich setting should be used only if a lot of spitting occurs during engine warm-up with the standard setting. Whenever the setting is changed for either richer or leaner operation, the cover should be moved just one point at-a-time, and the results tested with the engine cold.

11- Idle Speed and Mixture Adjustments: Install a clean flame arrestor. Start the engine and allow it to run until it has warmed to operating temperature and the choke has moved to the full open position.

CAUTION: Water must circulate through the lower unit to the engine any time the engine is run to prevent damage to the water pump mounted on the engine.

Stop the engine and disconnect the throttle cable from the throttle lever. Now, turn the idle mixture adjusting screws in until they just barely make contact with their seats, then back out 1-1/4 turns as a rough adjustment. **TAKE CARE** not to turn the screws in tightly against the seats or the needles and seats will be **DAMAGED.**

With water running through the engine and outdrive, start the engine again and shift the drive unit into forward gear and run at idle rpm. Adjust the idle mixture needle for the highest and steadiest manifold vacuum reading. If a vacuum gauge is not available, obtain the smoothest running, maximum idle speed by first turning the idle adjusting needle in until the engine rpm begins to drop slightly. From this point, back the needle out over the "high spot" until the engine rpm again begins to drop.

Now, set the idle adjusting needle halfway between the two points for a proper idle mixture. If these adjustments result in an increase in idle rpm, reset the idle speed adjusting screw to obtain the specified idle rpm and again adjust the idle mixture adjusting needle. Shift the drive unit into neutral.

Stop the engine and install the throttle cable. Check to be sure the throttle valves are fully open when the remote control is in the full forward position. **ALWAYS** have a helper turn the propeller when shifting without the engine running in order to engage the shift dog.

The engine **MUST NOT BE RUNNING** to make the following test. With the throttle valves fully open, turn the wide-open throttle stop adjusting screw clockwise until the screw just makes contact with the throttle lever. Tighten the set nut securely to prevent the adjustment screw from turning. Now, return the shift control lever to the neutral position, idle, and check to see if the idle stop screw is against the stop. Shift into forward gear. Readjust the idle speed screw until the recommended idle rpm is reached.

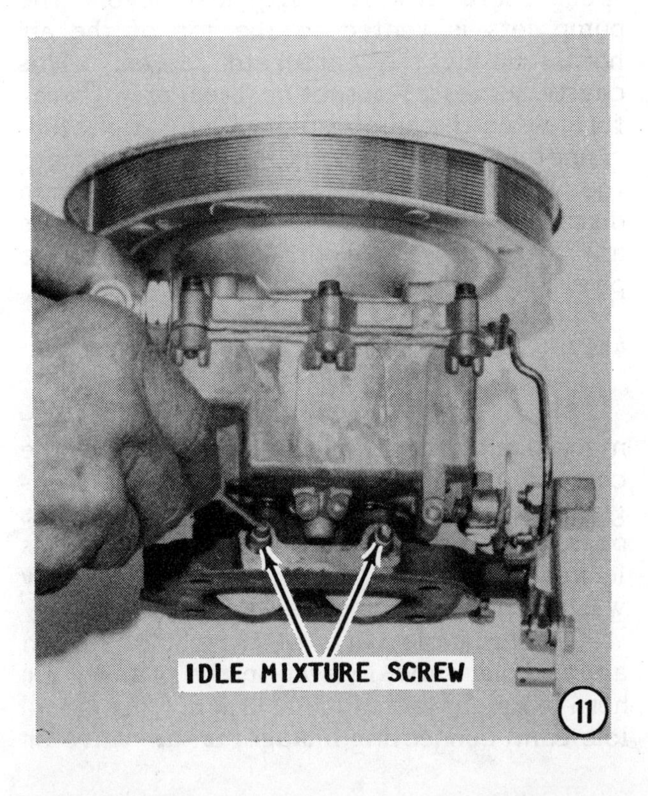

4-11 ROCHESTER CARBURETOR MODEL 4MV

DESCRIPTION

The 4MV Rochester carburetor is a four-barrel (Quadrajet) unit. The air-fuel mixture is controlled with a secondary-side air valve and tapered metering rods.

The Quadrajet carburetor has two stages. The primary (fuel inlet) side has small 1-3/8" bores with a triple venturi set-up equipped with plain-tube nozzles. The carburetor operates much the same as other carburetors using the venturi principle. The triple venturi, plus the small primary bores, makes for a more stable and finer fuel control during idle and partial throttle operation. When the throttle is partially open, the fuel metering is accomplished with tapered metering rods, positioned by a vacuum-responsive piston and operating in specially designed jets.

The secondary side has two large, 2-1/4", bores. These large bores, when added to the primary side bores, provide enough air capacity to meet most engine requirements. The air valve is used in the secondary side for metering control and backs-up the primary bores to meet air and fuel demands of the engine.

The secondary air valve operates the tapered metering rods. These rods move in orifice plates and thus control fuel flow from the secondary nozzles in direct relation to the air flowing through the secondary bores.

The float bowl is designed to avoid problems of fuel spillage during sharp turns of

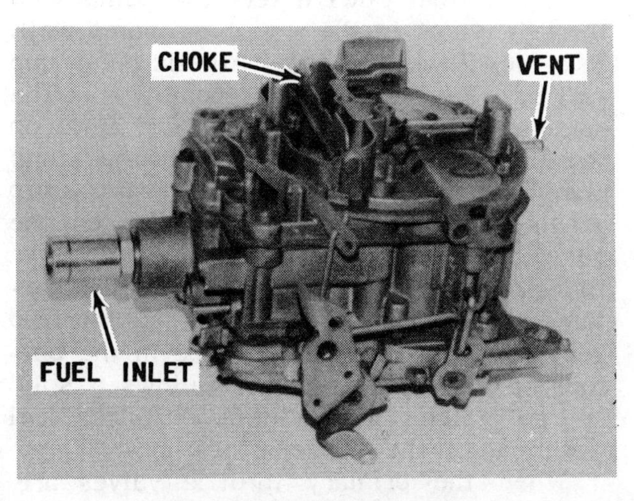

The Rochester 4MV carburetor.

the boat which could result in engine cut-out and delayed fuel flow. The bowl reservoir is smaller than most four-barrel carburetors to reduce fuel evaporation during hot engine shut-down.

The float system has one pontoon float and fuel valve which makes servicing much easier than on some other model carburetors. A fuel filter is located in the float bowl ahead of the float needle valve. This filter is easily removed for cleaning or replacement.

The throttle body is made of aluminum as part of a weight-reduction program and also to improve heat transfer away from the fuel bowl and prevent the fuel from "percolating" during hot engine shut-down. A heat insulator gasket is used between the throttle body and bowl to help prevent "percolating".

4MV ACCELERATING SYSTEM

When the throttle is opened suddenly during acceleration, the air flow and manifold vacuum change almost at the same time. The fuel, which is heavier, has a tendency to lag behind. This condition causes a lean mixture for just a moment. It is at this time, the accelerator pump comes into play by providing the extra fuel necessary to maintain the proper mixture.

The accelerator pump system is located in the primary side of the carburetor. The system consists of spring-loaded pump plunger and a return spring. The plunger is operated by a pump lever on the air horn which is connected to the throttle lever by a pump rod.

When the pump plunger moves upward in the pump well during throttle closing, fuel from the float bowl enters the pump well through a slot in the top of the pump well. This fuel flows past the synthetic pump cup seal into the bottom of the pump well. The pump cup floats and moves up and down on the pump plunger head. When the pump plunger is moved upward, the flat on the top of the cup unseats from the flat on the plunger head and allows fuel to move through the inside of the cup into the bottom of the pump well. This action also vents any vapors which may be in the bottom of the pump well and allows a solid charge of fuel to be maintained in the fuel well beneath the plunger head.

When the primary throttle valves are opened, the connecting linkage forces the

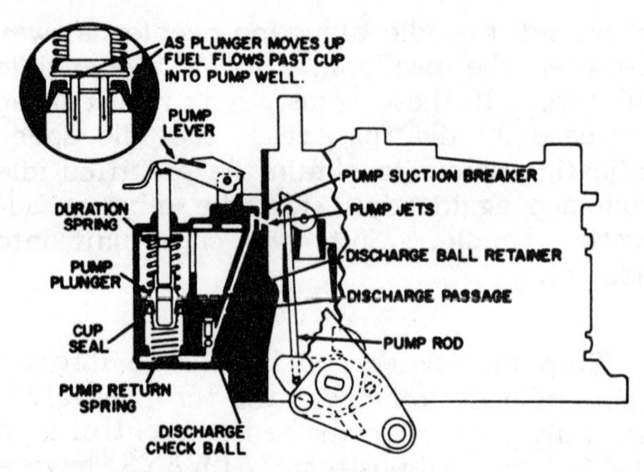

ACCELERATING PUMP SYSTEM

pump plunger downward. The pump cup seats at once, and fuel is forced through the pump discharge check ball and passes on through the passage to the pump jets located in the air horn. From these jets, the fuel sprays into the venturi area of each primary bore. As mentioned earlier, the pump plunger is spring-loaded. The upper time portion of the spring is balanced with the bottom pump return spring to permit a smooth sustained charge of fuel to be delivered during the acceleration period.

The pump discharge check ball seats in the pump discharge passage during upward motion of the pump plunger to prevent air from being drawn into the passage. Without this arrangement, there would be a momentary lag during acceleration.

A vacuum exists at the jets during high-speed operation. A cavity just beyond the pump jets is vented to the top of the air horn, outside the carburetor bores. This cavity serves as a suction breaker. Therefore, when the pump is not in operation, fuel cannot be pulled out of the pump jets into the venturi area, but ensures a full pump discharge when needed and still prevents any fuel spill over from the pump discharge passage.

4MV CHOKE SYSTEM

The choke valve is located in the primary side of the carburetor and provides the correct air-fuel mixture for quick cold-engine starting and until the engine reaches its operating temperature. The air valve is locked closed until the engine is completely warmed and the choke valve is wide open.

The principle parts of the choke system are a choke valve located in the primary air horn bore, a vacuum diaphragm unit, fast-idle cam, connecting linkage, an air valve or

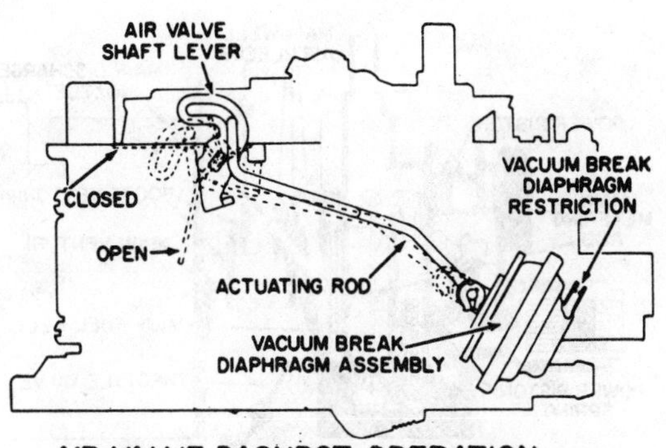

AIR VALVE DASHPOT OPERATION

secondary throttle valve lockout lever, and a thermostatic coil.

While the engine is being cranked, the tension of the thermostatic coil holds the choke valve closed. The closed choke valve restricts air flow through the carburetor to provide a richer mixture for starting.

After the engine starts, manifold vacuum applied to the vacuum diaphragm unit opens the choke valve the proper amount for the engine to run without loading or stalling.

Other conditions exist after the engine starts: The cold-enrichment feed holes are no longer in a low-pressure area so they cease to feed fuel. These holes are then used as secondary main well air bleeds. The fast-idle cam follower lever on the end of the primary throttle shaft drops from the highest step on the fast-idle cam to a lower step when the throttle is opened. The lever in this position gives the engine enough fast idle and a correct fuel mixture for smooth operation until full operating temperature is reached. Once the engine has warmed-up, the thermostatic coil heats and eases its tension and allows the choke valve to open farther due to the intake air pushing on the off-set choke valve. The choke valve continues to open until the thermostatic coil is completely relaxed and the choke is fully open.

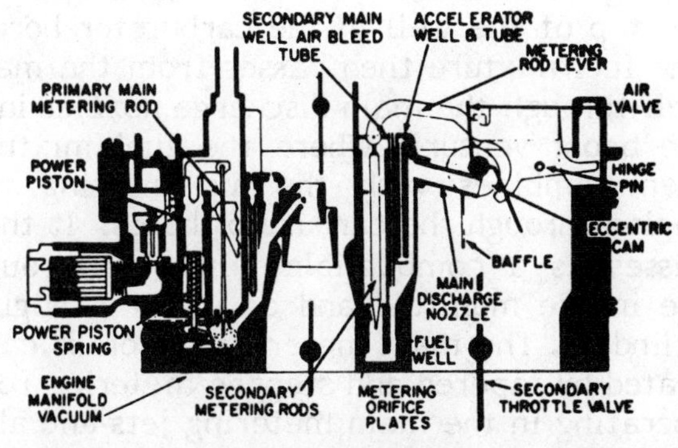

POWER SYSTEM

4MV POWER SYSTEM

The power system provides extra fuel to meet power demands during heavy engine loads and during high-speed operation. This richer mixture is supplied through the main metering systems in the primary and secondary sides of the carburetor.

The power system enriches the fuel mixture in the two primary bores. This system is made up of a vacuum-operated power piston and a spring located in a cylinder connected by a passage to intake manifold vacuum. The spring under the power piston operates against manifold vacuum and pushes the power piston upward.

During partial throttle and cruising ranges, the manifold vacuum is enough to hold the power piston down against spring tension so the larger diameter of the metering rod tip is held in the main metering jet orifice.

As engine load is increased until more fuel and a richer mixture is necessary, the power piston spring overcomes the vacuum pull on the power piston, and the tapered tips of the metering rods move upward in the main metering jet orifices. The smaller diameter of the metering rod tip permits more fuel to pass through the main metering jet and enriches the fuel mixture to meet the added power demands. When engine load decreases, the manifold vacuum rises and extra fuel and a richer mixture is no longer required. The higher vacuum pulls the power piston down against spring tension, and this action moves the larger diameter of the metering rod into the metering jet orifice, returning the fuel mixture to normal.

The primary side of the carburetor supplies enough air and fuel for low-speed operation. More air and fuel are required at higher speeds to meet engine demands and it is the secondary side of the carburetor that meets these requirements.

The secondary side of the 4MV has a separate and independent metering system. This system consists of two large throttle valves connected by a shaft and linkage to the primary throttle shaft. Fuel metering is controlled by spring-loaded air valves, metering orifice plates, secondary metering rods, main fuel wells with bleed tubes, fuel-discharge nozzles, accelerating wells, and tubes.

A lever on the primary throttle shaft, through a connecting link to the secondary throttle shaft, begins to open the secondary throttle valves when the engine reaches a point where the primary bores cannot deliver the quantity of air and fuel demanded by the engine. As the secondary throttle valves are opened, engine manifold vacuum (reduced pressure) is applied directly beneath the air valves. Atmospheric pressure on top of the air valves overcomes spring tension and forces them open, allowing metered air to pass through the secondary bores of the carburetor.

The secondary main discharge nozzles are located above the secondary throttle valves and just below the center of the air valves. There is one nozzle for each bore.

Because the valves are located in a low-pressure area, they feed fuel in the following manner:

When the secondary throttle valves are opened, atmospheric pressure opens the air valves. This action rotates a plastic cam attached to the center of the air-valve shaft. The cam movement lifts the secondary metering rods out of the secondary orifice plates through the metering rod lever. As the throttle valves are opened still farther and engine speed continues to increase, air flow through the secondary side increases and opens the air valve more. The opening valves lift the secondary metering rods farther out of the orifice plates. The metering rods are tapered. This design allows fuel flow through the secondary metering orifice plates in direct proportion to air flow through the secondary bores. This system allows correct air-fuel mixtures through the secondary bores to be controlled by the depth of the metering rods in the orifice plates.

The actual depth of the metering rods in the orifice plates is factory adjusted in relation to air-valve position and to meet air-fuel requirements for each engine. If an adjustment should become necessary due to replacement of parts, a service setting is possible.

4MV METERING SYSTEM

The main metering system supplies fuel to the engine from off-idle to wide-open throttle. The primary bores supply fuel and air during this range through plain-tube nozzles and the venturi principle. The main

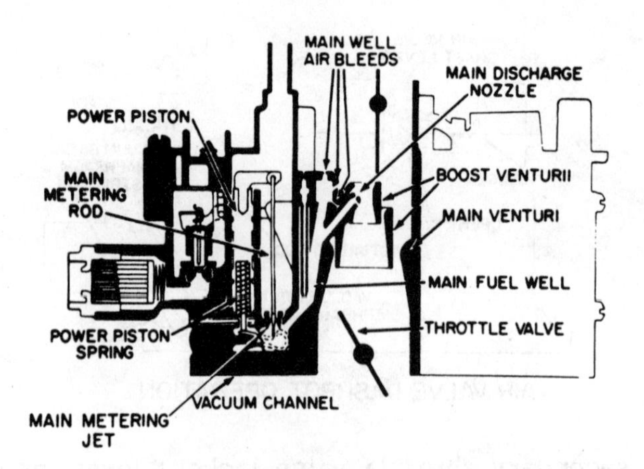

MAIN METERING SYSTEM

metering system starts to operate when air flow increases through the venturi system and more fuel is required to supply the correct air-fuel mixture to the engine. Fuel from the idle system gradually diminishes as the lower pressures are now in the venturi area.

The main metering system consists of the main metering jets, vacuum-operated metering rods, main fuel wells, main well air bleeds, fuel discharge nozzles, and triple venturis.

When the primary throttle valves open beyond the off-idle range and allow more air to enter the engine intake manifold, air velocity increases in the carburetor venturi. The increased velocity causes a drop in pressure in the large venturi, which increases many times in the boost venturi. Because the low pressure (vacuum) is now in the smallest boost venturi, fuel flows from the main discharge nozzle in the following manner:

Fuel from the float bowl flows through the main metering jets into the main fuel wells. It passes upward in the main well and is fed with air by an air bleed located at the top of the well. The fuel is further fed air through calibrated air bleeds located near the top of the well in the carburetor bores. The fuel mixture then passes from the main well through the main discharge nozzles into the boost venturis where the fuel mixture then combines with the air entering the engine through the carburetor bores. It then passes as a combustible mixture through the intake manifold and on into the engine cylinders. The main metering system is calibrated by tapered and stepped metering rods operating in the main metering jets and also through the main well air bleeds.

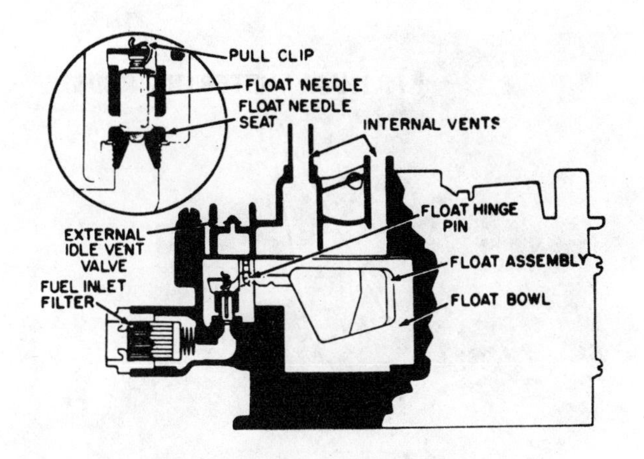

FLOAT SYSTEM

4MV FLOAT SYSTEM

The float bowl is located between the primary bores and adjacent to the secondary bores. This position assures an adequate fuel supply to all carburetor bores and does much to maintain excellent engine performance when the bow of the boat is high or during high-speed tight turns. The float pontoon is solid and made of a light plastic material. The combination of these two features gives added buoyancy to the float and allows the use of a single float to maintain a constant fuel level.

The parts of the float system include: The float bowl, a single pontoon float, float hinge pin-and-retainer combination, float valve and seat, and a slot valve pull clip. A plastic filler block is located in the top of the float chamber over the float valve to prevent fuel slosh into this area.

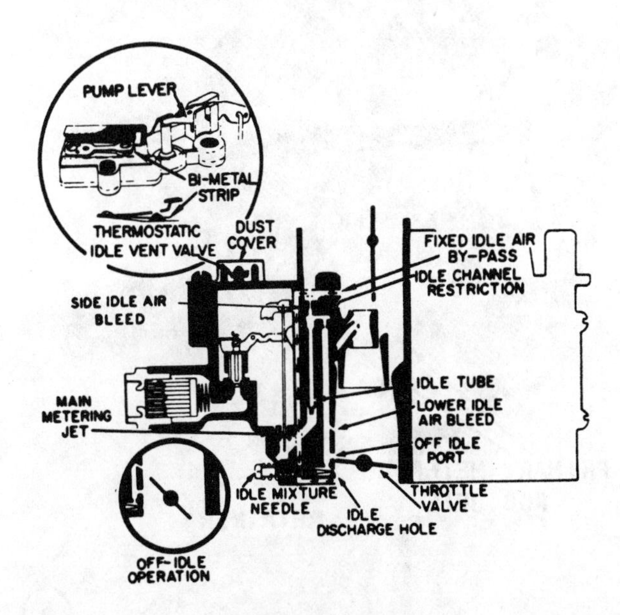

IDLE SYSTEM

4MV IDLE SYSTEM

The idle system is located on the fuel inlet (primary) side of the carburetor to supply the correct air-fuel mixture during idle and off-idle operation. The idle system is used during this period because air flow through the carburetor venturi is not great enough to obtain good metering from the main disharge nozzles.

DISASSEMBLING THE ROCHESTER 4MV

1- Place the carburetor on the work bench in the upright position. Remove the idle vent valve attaching screw, and then remove the idle vent valve assembly. Remove the clips from the upper end of the choke rod, disconnect the choke rod from the upper choke shaft lever, and then remove the choke rod from the bowl. Detach the spring clip from the upper end of the pump rod, and then disconnect the pump rod from the pump lever. Remove the nine air horn-to-bowl attaching screws. Two screws are located next to the primary venturi. Lift straight up on the air horn and remove it. **TAKE CARE** not to bend the accelerating well and air bleed tubes sticking out from the air horn.

2- Hold the air valve wide open and then remove the secondary metering rods by tilting and sliding the rods from the holes in the hanger. Remove the dashpot piston from the air-valve link by rotating the bend through the hole, and then remove the dashpot from the air horn by rotating the bend through the air horn. Further disassembly of the air horn is not necessary. **DO NOT** remove the air valves, air valve shaft, and

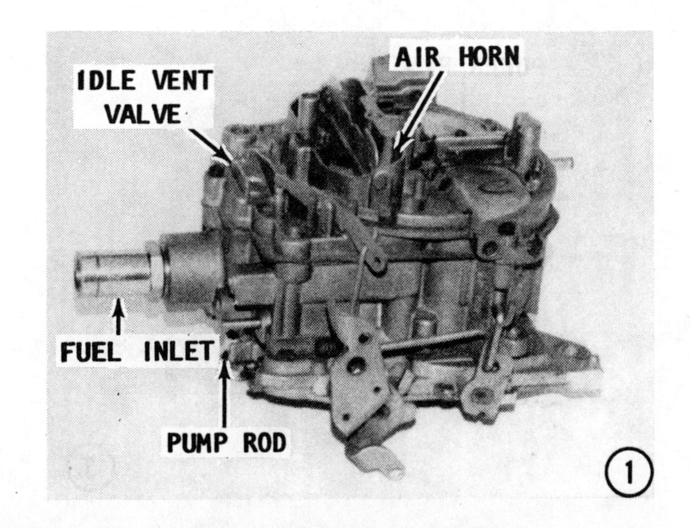

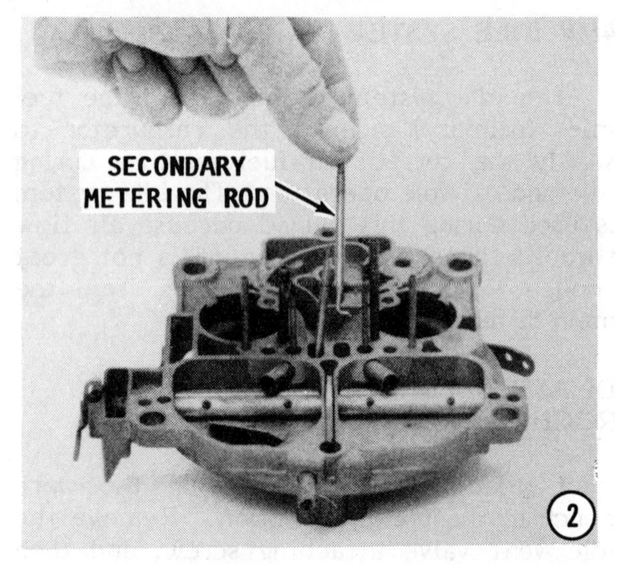

secondary metering rod hangers because they are calibrated. **DO NOT** attempt to remove the high-speed air bleeds and accelerating well tubes because they are pressed into position.

3- Remove the power piston from the pump well. Release the air horn gasket from the dowels on the secondary side of the bowl, and then pry the gasket from around the power piston and primary metering rods. Remove the pump return spring from the pump well.

4- Remove the plastic filler from over the float valve. Use a pair of needle-nosed pliers and pull straight up on the metering rod hanger directly over the power piston and remove the power piston and the primary metering rods. Disconnect the tension spring from the top of each rod and then

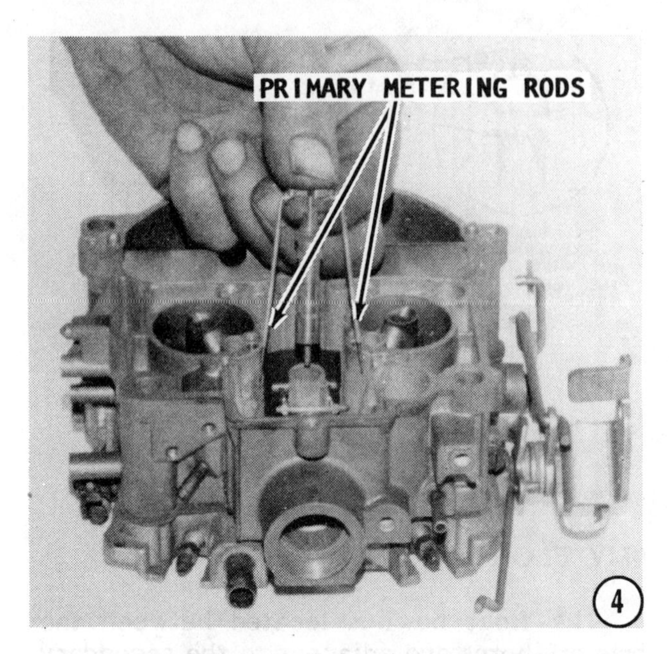

rotate the rod and remove the metering rods from the power piston. Pull up just a bit on the float assembly hinge pin until the pin can be removed by sliding it toward the pump well. Disengage the needle valve pull clip by sliding the float assembly toward the front of the bowl. **TAKE EXTRA CARE** not to distort the pull clip.

5- Remove the two screws from the float needle retainer, and then lift out the retainer and needle assembly. **NEVER** attempt to remove the needle seat because it is staked and tested at the factory. If the

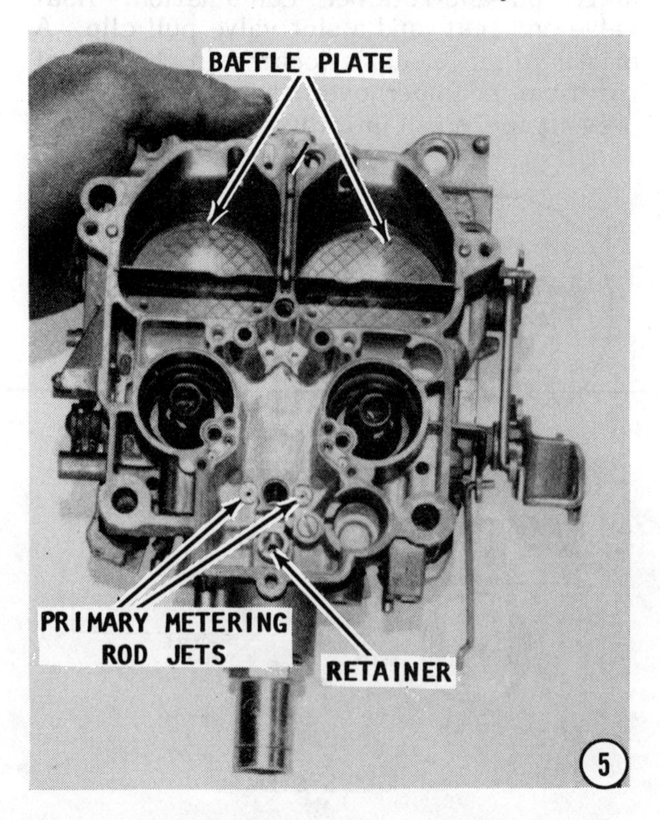

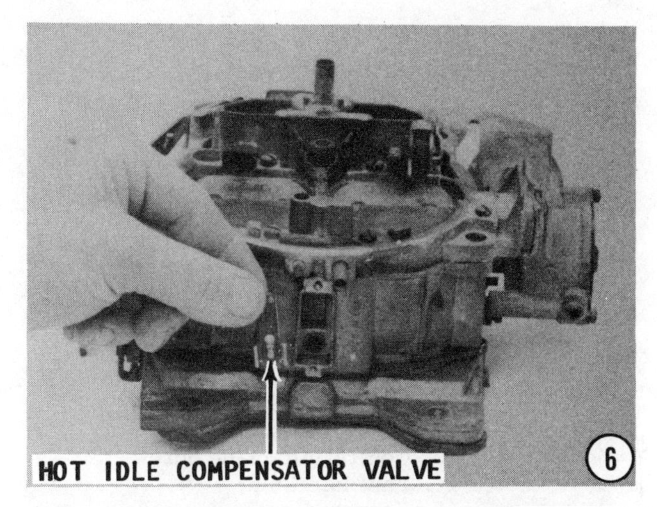

float assembly is damaged, replace the assembly. Remove both primary metering rod jets. Remove the pump discharge check ball retainer and the check ball. Remove the baffle plates from the secondary side of the bowl. Disconnect the vacuum hose from the tube connection on the bowl and from the vacuum break assembly. Remove the retaining screw, and then lift the assembly from the float bowl.

6- Remove the two screws from the hot-idle compensator cover. Lift the hot-idle compensator and O-ring from the float bowl. Remove the fuel inlet filter retaining nut, gasket, filter, and spring. Remove the throttle body by taking out the throttle body-to-bowl attaching screws, and then lift off the insulator gasket. Remove the idle mixture screws and springs. **TAKE CARE** not to damage the secondary throttle valves.

CLEANING AND INSPECTING THE ROCHESTER 4MV

NEVER dip rubber parts, plastic parts, diaphragms, pump plungers or the vacuum-break assembly in carburetor cleaner. Place all of the metal parts in a screen-type tray and dip them in carburetor solvent until they appear completely clean, then blow them dry with compressed air.

Blow out all of the passages in the castings with compressed air. Check all of the parts and passages to be sure they are

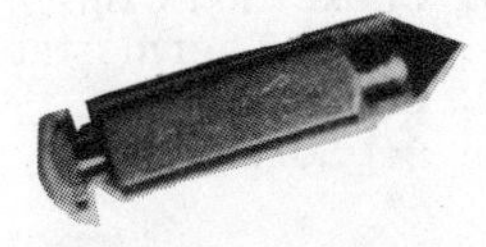

*Fuel inlet needle with a worn ridge. The set **MUST** be replaced to prevent a leak and carburetor flooding.*

Tip of an idle mixture adjusting needle bent from being forced into its seat and against the closed throttle plate.

not clogged or contain any deposits. **NEVER** use a piece of wire or any type of pointed instrument to clean drilled passages or calibrated holes in a carburetor.

Inspect the idle mixture needles for damage. Check the float needle and diaphragm for wear. Inspect the upper and lower surfaces of the carburetor castings for damage. Inspect the holes in the levers for being out-of-round. Check the fast-idle cam for wear or damage. Check the air valve for binding. If the air valve is damaged, the complete air horn assembly must be replaced.

Most of the parts that should be replaced during a carburetor overhaul, including the latest updated parts, are found in a carburetor kit available at your local marine dealer. In addition to the parts, most kits include the latest specifications which are so important when making bench adjustments.

ASSEMBLING THE 4MV

1- Turn the idle mixture adjusting screws in until they are barely seated, and then back them out two turns as a rough adjustment at this time. **NEVER** turn the adjusting screws down tight into their seats

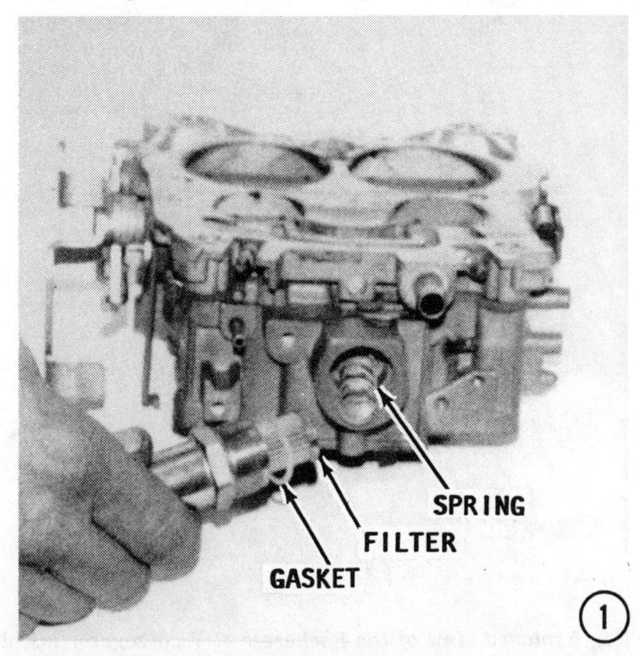

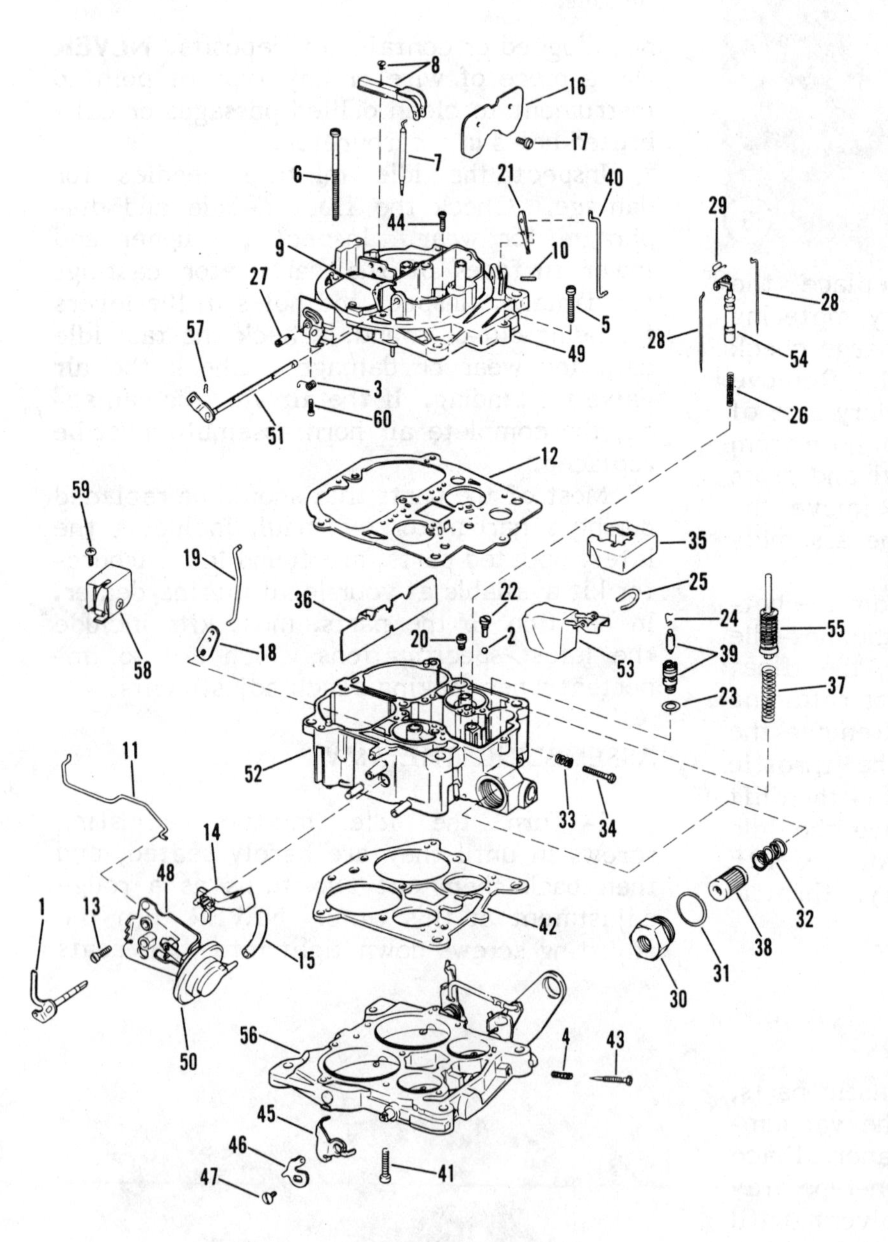

1. Vacuum break lever and shaft
2. Pump Discharge Ball
3. Secondary Air Valve Adjustment Spring
4. Idle Needle Spring
5. Air Horn Screw
6. Air Horn Screw
7. Secondary Metering Rod
8. Secondary Metering Rod Holder
9. Air Valve Lockout Lever
10. Roll Pin
11. Vacuum Break Rod
12. Air Horn Gasket
13. Control Attaching Screw
14. Fast Idle Cam
15. Vacuum Hose
16. Choke Valve
17. Choke Valve Screw
18. Intermediate Choke Lever
19. Choke Rod
20. Primary Jet
21. Pump Actuating Lever
22. Pump Discharge Ball Retainer
23. Needle Seat Gasket
24. Float Needle Pull Clip
25. Float Assembly Hinge Pin
26. Power Piston Spring
27. Secondary Air Valve Adjustment Shaft
28. Primary Metering Rod
29. Primary Metering Rod Spring
30. Fuel Inlet Filter Nut
31. Filter Nut Gasket
32. Fuel Filter Relief Spring
33. Idle Stop Screw Spring
34. Idle Screw
35. Float Bowl Insert
36. Secondary Baffle Plate
37. Pump Return Spring
38. Fuel Inlet Filter
39. Needle and Seat Assembly
40. Pump Rod
41. Throttle Body Screw
42. Throttle Body Gasket
43. Idle Needle
44. Air Horn Screw
45. Cam Lever
46. Fast Idle Lever
47. Idle Lever Screw
48. Vacuum Break Control Bracket
49. Air Horn Assembly
50. Vacuum Break Control Assembly
51. Choke Shaft & Lever Assembly
52. Float Bowl Assy
53. Float Assembly
54. Power Piston Assembly
55. Pump Assembly
56. Throttle Body Assembly
57. Choke Rod Clip
58. Choke Thermostat Assembly
59. Choke Screw
60. Set Screw

Exploded view of the Rochester 4MV carburetor installed on GMC V8 engines. Principle parts are identified.

or they will be damaged. Install the pump rod in the lower hole of the throttle lever by rotating the rod. Place a new insulator gasket on the bowl with the holes in the gasket indexed over the two dowels. Install, and then tighten the throttle body-to-bowl screws evenly. Install the fuel inlet filter spring, filter, new gasket, and inlet nut. Tighten the nut. Position a **NEW** hot-idle compensator O-ring seal in the recess in the bowl, and then install the compensator.

2- Install the U-bend end of the vacuum-break rod in the diaphragm link, with the end toward the bracket, and then slide the grooved end of the rod into the hole of the actuating lever. Install the spring clip to retain the rod in the vacuum-break rod. Install the fast-idle cam on the choke housing assembly. Check to be sure the fast-idle cam actuating pin on the middle choke shaft is located in the cutout area of the fast-idle cam. Connect the choke rod to the plain end of the choke rod actuating lever, and then hold the choke rod with the grooved end pointing inward and position the choke rod actuating lever in the well of the float bowl. Install the choke assembly, with the shaft engaged with the hole in the actuating lever. Install and tighten the retaining screw. Remove the choke rod from the lever. This rod will be installed later. Install and connect the vacuum hose.

3- Install the baffle plates in the secondary side of the bowl with the notches facing up. Install the primary main metering jets. Install a **NEW** float needle seat. Install the pump discharge ball check and retainer in the passage next to the pump well.

4- Install the pull clip on the needle with the open end toward the front of the bowl. Install the float by sliding the float lever

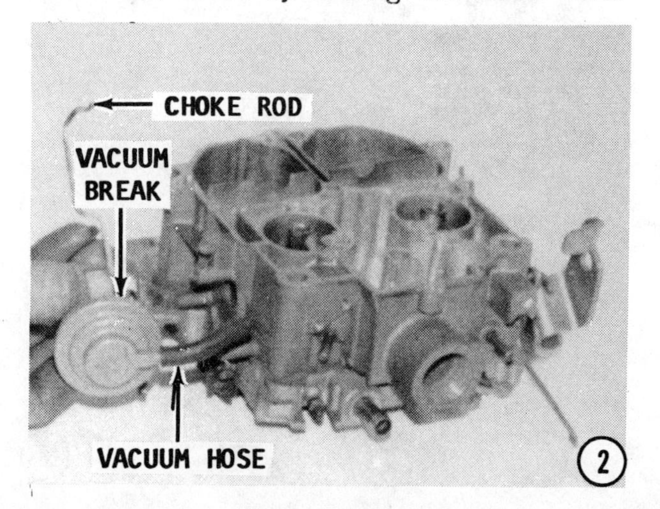

under the pull clip from the front to the back. Hold the float assembly by the toe and with the float lever in the pull clip, install the retaining pin from the pump-well side. **TAKE CARE** not to distort the pull clip.

5- Float Level Adjustment: Measure the distance from the top of the float bowl gasket surface, with the gasket removed, to the top of the float at the toe end. **CHECK TO BE SURE** the retaining pin is held firmly in place and the tang of the float is seated on the float needle when making the measurement. Check your measurement with the Specifications in the Appendix.

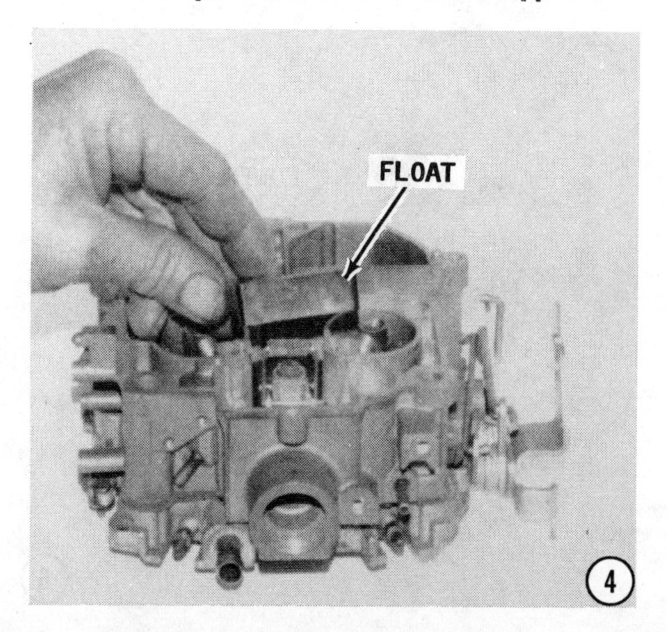

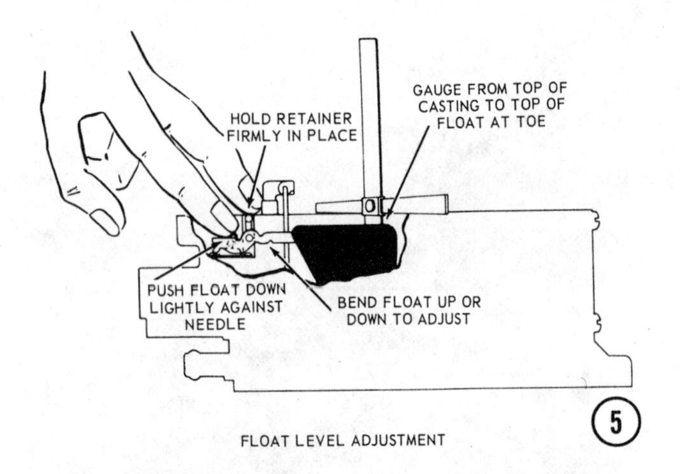

FLOAT LEVEL ADJUSTMENT

CAREFULLY bend the float up or down until the correct measurement is reached.

6- Install the power piston spring in the power piston well. Install the primary metering rods, if they were removed during disassembly. Be sure the tension spring is connected to the top of each metering rod. Install the power piston assembly in the well with the metering jets. On early models, it is necessary to press down on the power-piston to be sure the retaining pin is engaged with the hole in the throttle body gasket. On later models, a sleeve around the piston holds the piston in place during assembly.

7- Install the plastic filler over the float needle. Press it down firmly until it is seated.

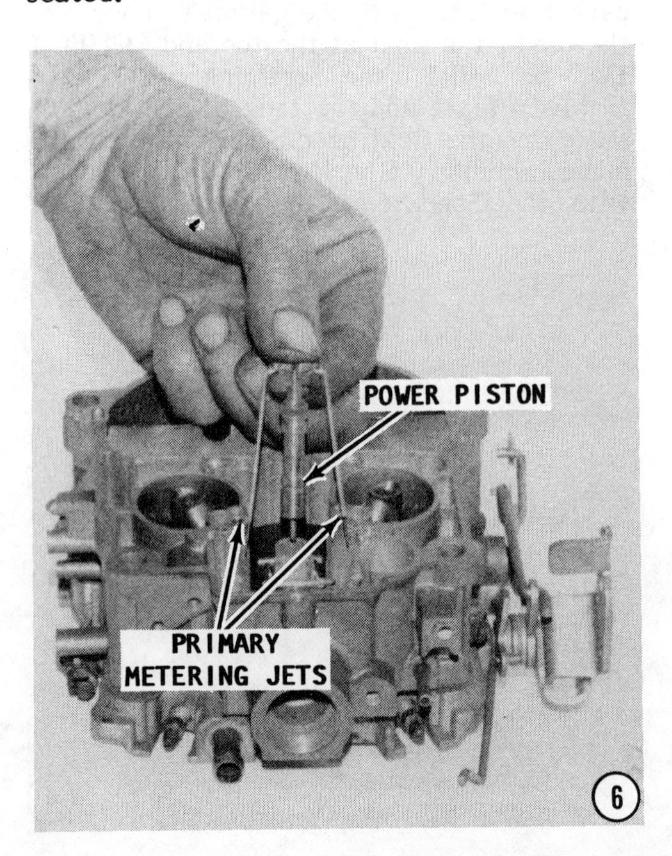

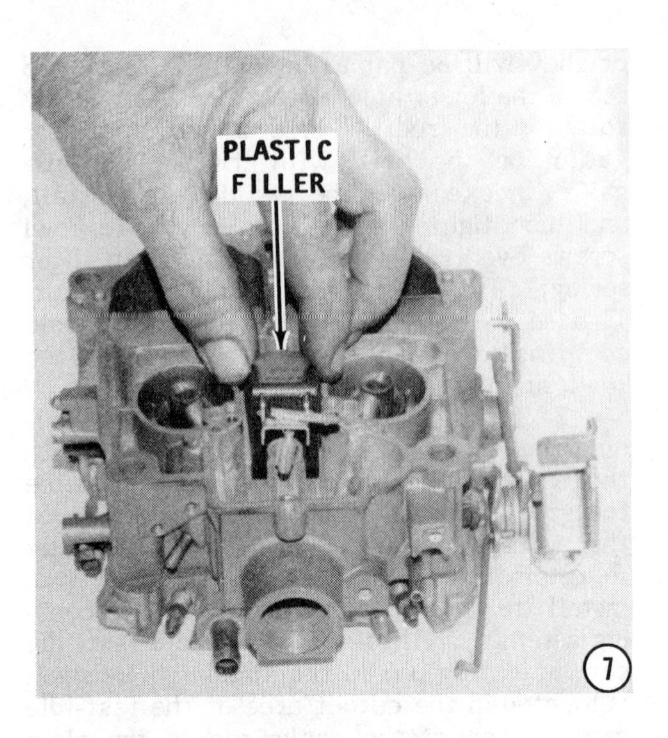

8- Place the pump return spring in the pump well. Install the air horn gasket around the primary metering rods and piston. Install the gasket on the secondary side of the bowl with the holes in the gasket indexed over the two dowels. Install the accelerating pump plunger in the pump well.

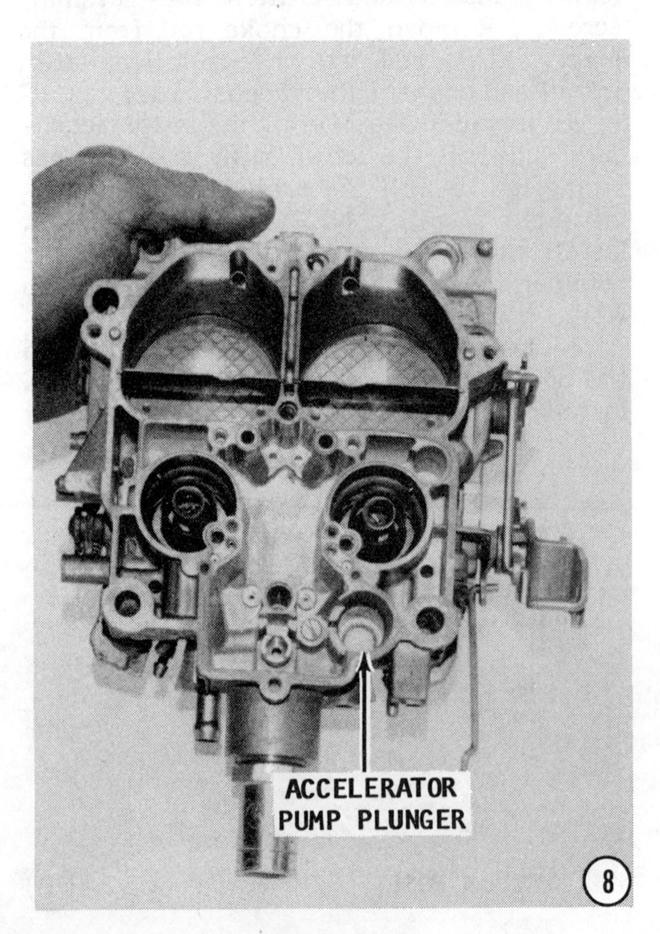

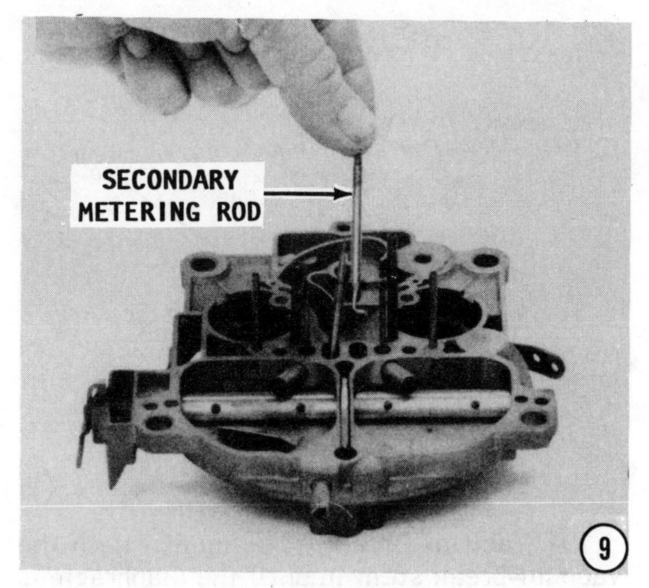

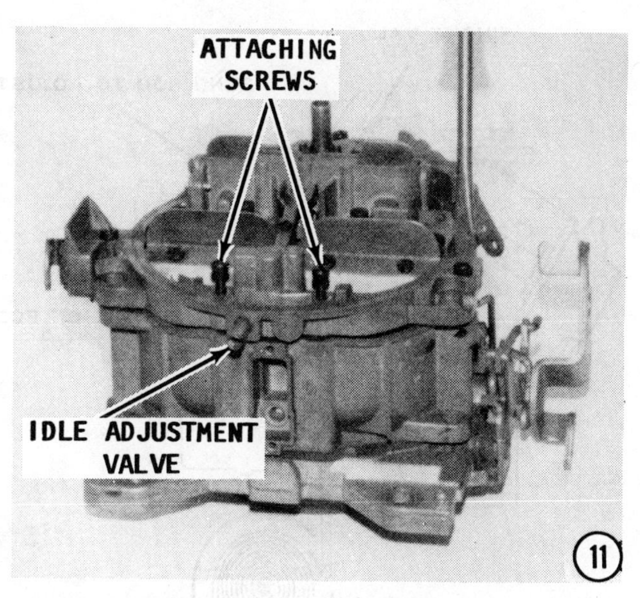

9- Install the secondary metering rods. Hold the air valve wide open and check to be sure the rods are positioned with their upper ends through the hanger holes and pointing toward each other.

10- Slowly position the air horn assembly on the bowl and **CAREFULLY** insert the secondary metering rods, the high-speed air bleeds, and the accelerating well tubes through the holes of the air horn gasket. **NEVER** force the air horn assembly onto the float bowl because you may distort the secondary metering plates. If you move the air horn assembly slightly sideways it will center the metering rods in the metering plates. Install the attaching screws as follows: The four long air horn screws around the secondary side; two short screws in the center section; one short screw above the fuel inlet; and the two countersunk screws in the primary venturi area. Tighten the screws evenly and in the sequence as shown in the accompanying illustration.

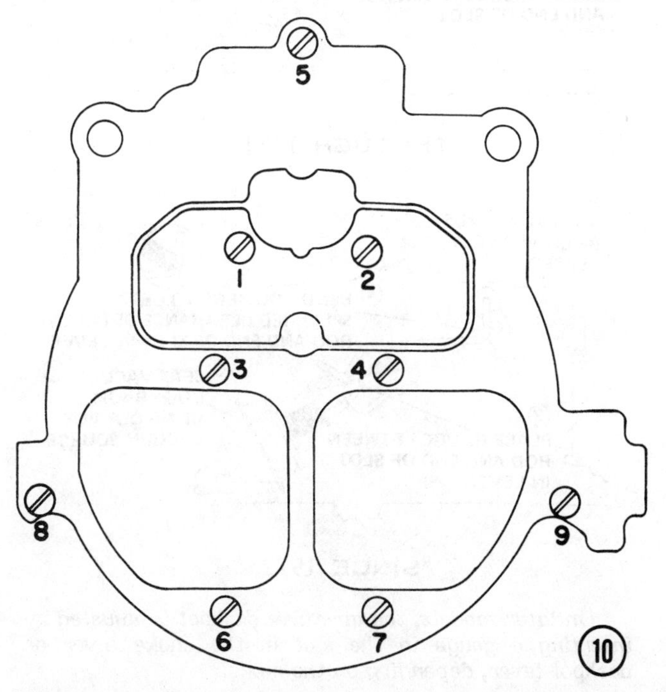

11- Install the idle vent actuating rod in the pump lever. Connect the pump rod to the inner hole of the pump lever and secure it with a spring clip. Connect the choke rod in the lower choke lever and secure it with a spring clip. Install the idle vent valve with the actuating rod engaged, and then tighten the attaching screws.

4MV BENCH ADJUSTMENTS

12- Pump Adjustment: Disconnect the secondary actuating link. Measure the distance from the top of the choke valve wall, next to the vent stack, to the top of the pump stem, with the throttle valves completely closed and the pump rod in the specified hole in the lever. Compare your

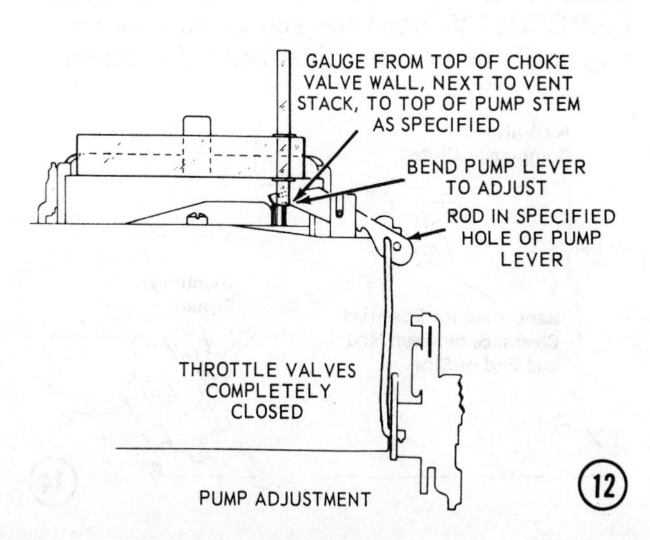

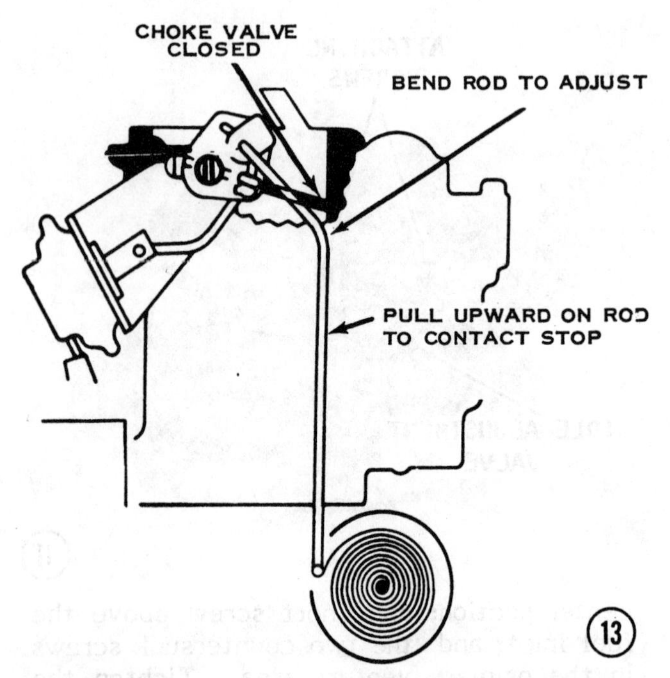

(13)

measurement with the Specifications. **CAREFULLY** bend the pump lever until the specified dimension is obtained. Connect the actuating link.

13- Choke Rod Adjustment: First, place the cam follower on the second step of the fast-idle cam and against the high step. Next, rotate the choke valve toward the closed position by pushing down **LIGHTLY** on the vacuum-break lever. Now, measure the distance between the lower edge of the choke valve and the air horn wall. Compare your measurement with the Specifications. **CAREFULLY** bend the choke rod until the proper measurement is obtained.

14-Air Valve Dashpot Adjustment: Push the vacuum-break stem in until the diaphragm is seated. Measure the distance between the dashpot rod and the end of the slot in the air valve lever. Compare your measurement with the Specifications. **CAREFULLY** bend the rod at the air valve end until the proper adjustment is reached.

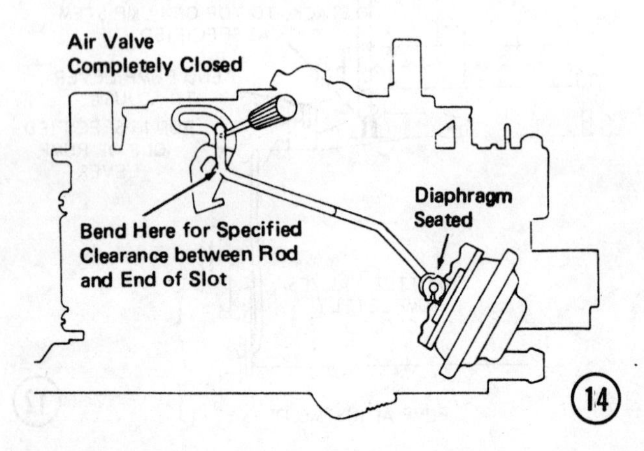

(14)

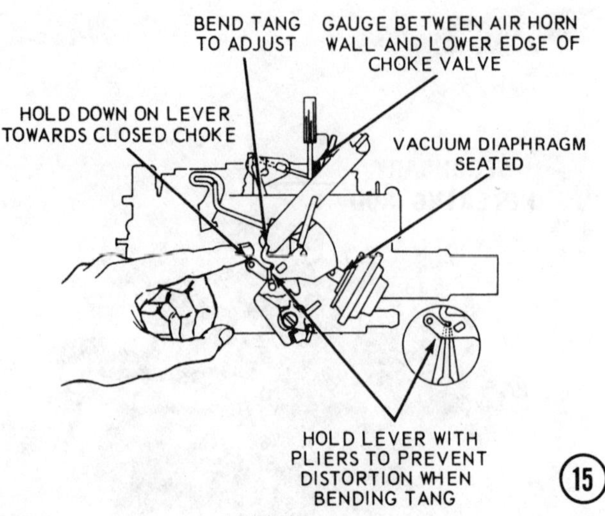

(15)

15- Vacuum-Break Adjustment: Push the vacuum-break stem in until the diaphragm is seated. At the same time, hold the choke valve toward the closed position and measure the distance between the lower edge of the choke valve and the air horn wall. Compare your measurement with the Specifications. **CAREFULLY** bend the vacuum-break tang until the proper adjustment is reached.

16- Unloader Adjustment: Secure a rubber band on the vacuum-break to hold

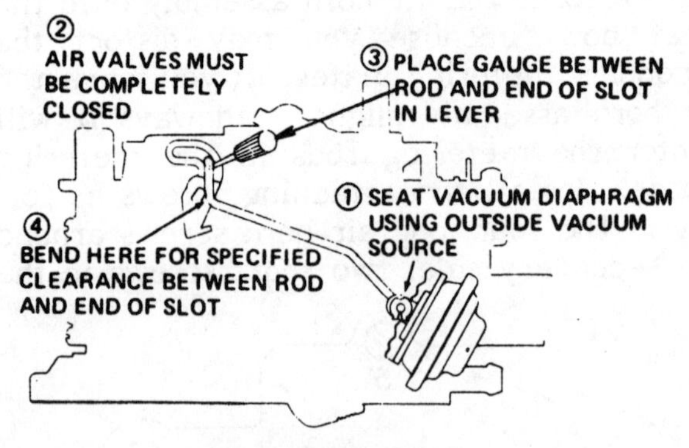

THROUGH 1971

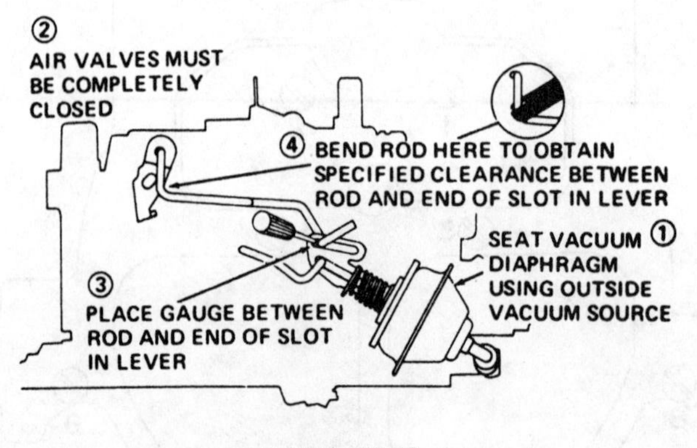

SINCE 1972

On later models, the air-valve dashpot is adjusted by inserting a gauge in the slot in the choke lever or dashpot lever, depending on the model.

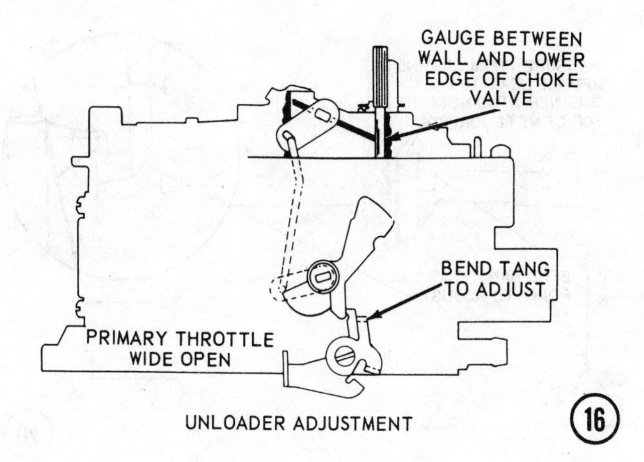

the choke valve in the closed position. Move the primary throttle valves to the wide-open position and at the same time measure the distance between the edge of the choke valve and the air horn wall. Compare your measurement with the Specifications. **CAREFULLY** bend the fast-idle lever tang until the proper adjustment is reached.

17- Air-Valve Lockout Adjustment: With the choke valve wide open, force the thermostatic spring tang to move the choke rod to the top of the slot in the choke lever. Now, move the air valve toward the open position. Measure the distance between the lockout tang and the front edge of the air valve, as shown in the accompanying illustration.

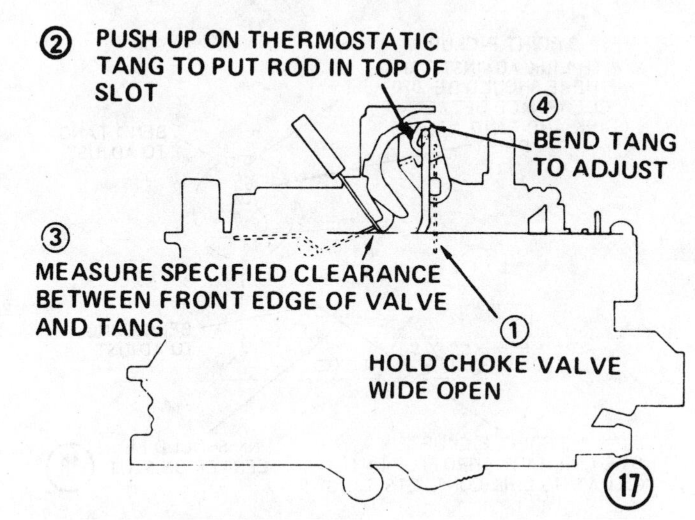

CAREFULLY bend the upper end of the air valve lockout lever tang until the required adjustment is reached. Finally, open the choke valve to its wide-open position by applying force to the underside of the choke valve. **BE SURE** the choke rod is in the bottom of the slot in the choke lever, the air valve lockout tang holds the air valve closed.

18- Secondary Opening Adjustment: Open the primary throttle valves until the actuating link contacts the tang on the secondary lever, and then measure the distance between the link and the tang on the secondary lever, which should be 0.070". **CAREFULLY** bend the tang on the secondary lever until the proper adjustment is reached.

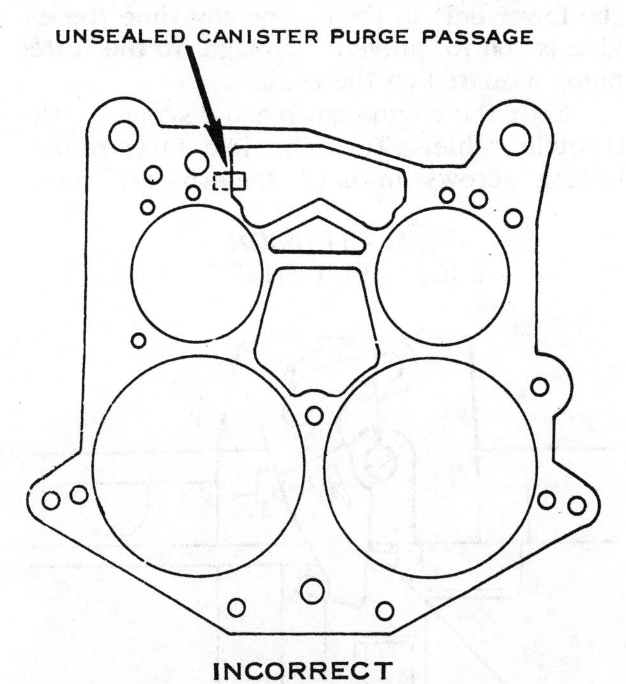

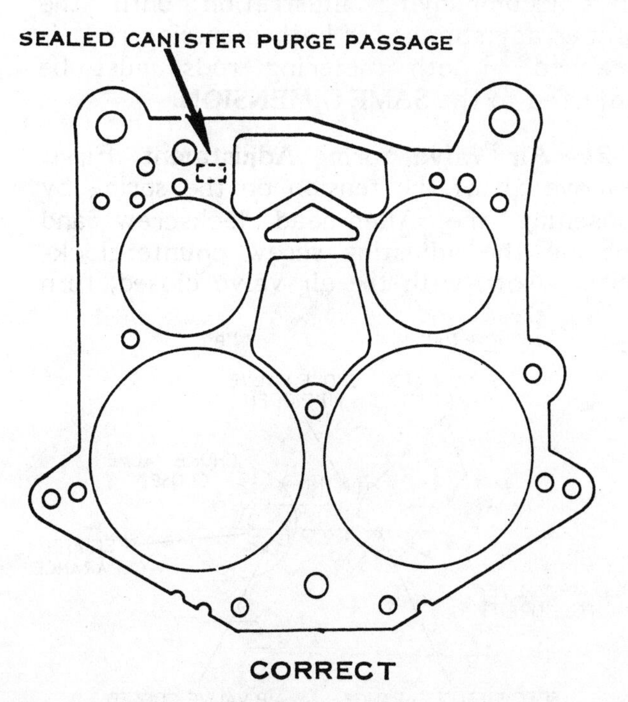

The addition of a charcoal canister purge port on models since 1970 has resulted in a change in the gaskets used between the float bowl and the throttle body. A vacuum leak will result if the wrong gasket is used because air will bypass the primary throttle valves through the canister purge passageway (left) and the engine will not idle smoothly.

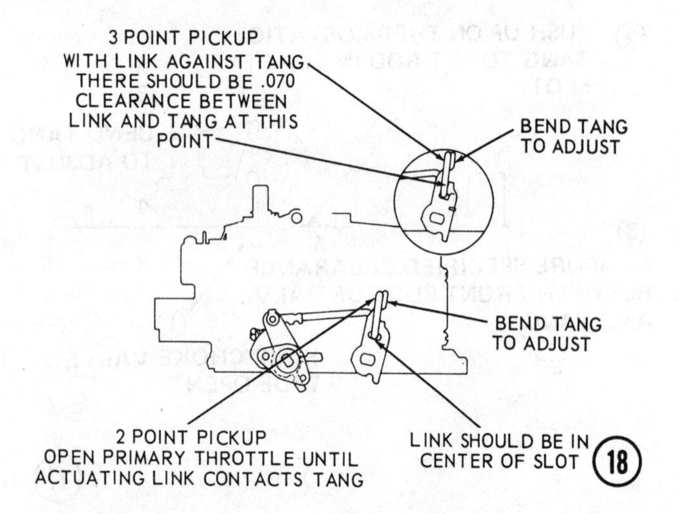

19- Secondary Closing Adjustment: After the idle speed has been adjusted and with the tang on the throttle lever against the actuating lever, measure the distance between the actuating link and the front of the slot in the secondary lever. Compare your measurement with the Specifications. **CAREFULLY** bend the tang on the throttle lever until the proper adjustment is reached.

20- Secondary Metering Rod Adjusment: Measure the distance from the top of the metering rod to the top of the air horn casting and also to the flame arrestor stud hole. Compare your measurement with the Specifications. **CAREFULLY** bend the metering rod hanger at the point shown in the accompanying illustration until the proper adjustment for both metering rods is reached. Both metering rods must be adjusted to the **SAME DIMENSION.**

21- Air Valve Spring Adjustment: First, remove all of the tension on the spring by loosening the Allen-head lockscrew and turning the adjusting screw counterclockwise. Now, with the air valve closed, turn

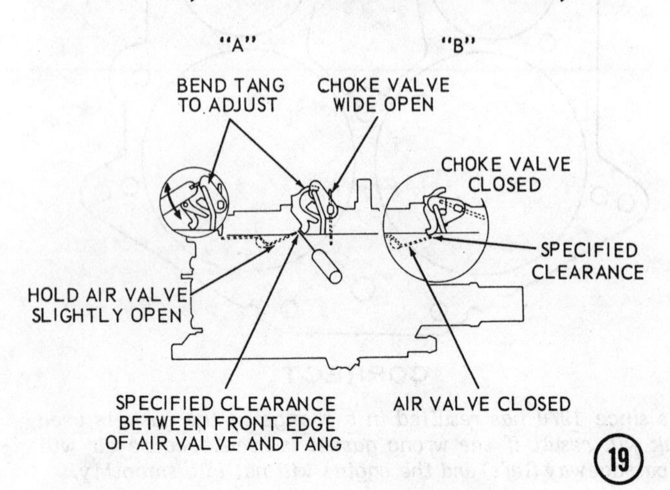

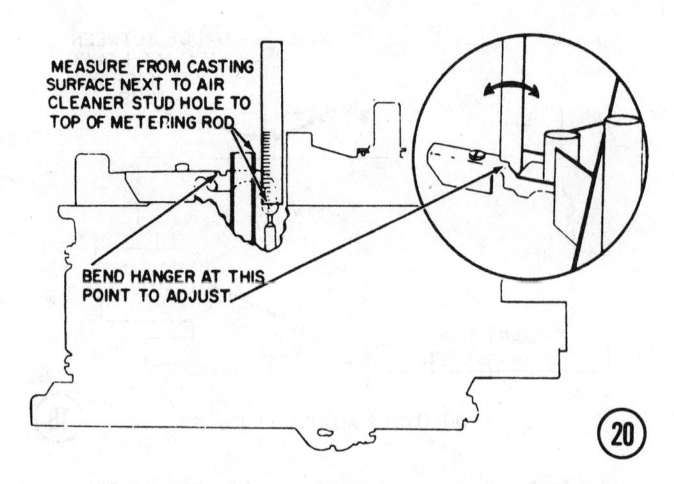

the adjusting screw clockwise until the torsion spring just contacts the pin on the shaft, and then 3/8 turn more. Hold the adjusting screw in this position, and tighten the lockscrew.

22- Choke Coil Rod Adjustment: Close the choke valve so the choke rod is at the bottom of the choke lever slot. Now, pull the choke coil rod up to the end of its travel. The top of the hole must be even with the bottom of the rod. **CAREFULLY** bend the choke coil rod until the proper adjustment is reached.

23- Idle Speed and Mixture Adjustments: Start the engine and allow it to warm to operating temperature with the flame arrestor in place.

CAUTION: Water must circulate through the lower unit to the engine any time the engine is run to prevent damage to the water pump mounted on the engine.

Shut the engine down and disconnect the throttle cable. Turn the idle mixture adjusting screws in until they barely touch

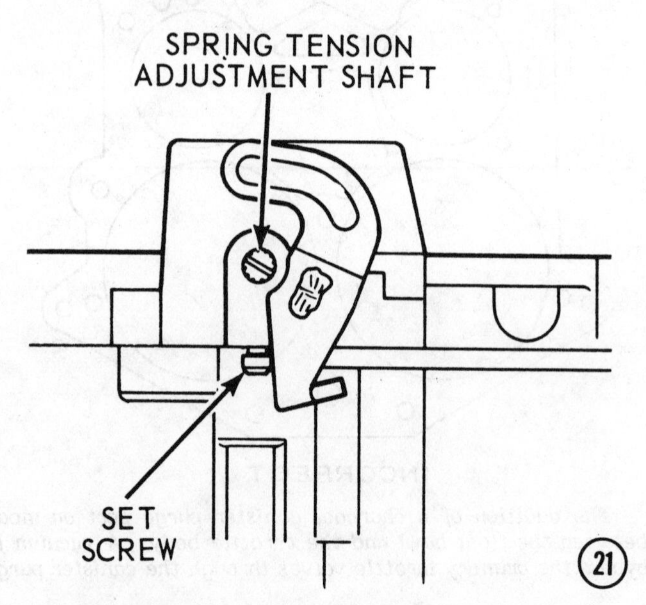

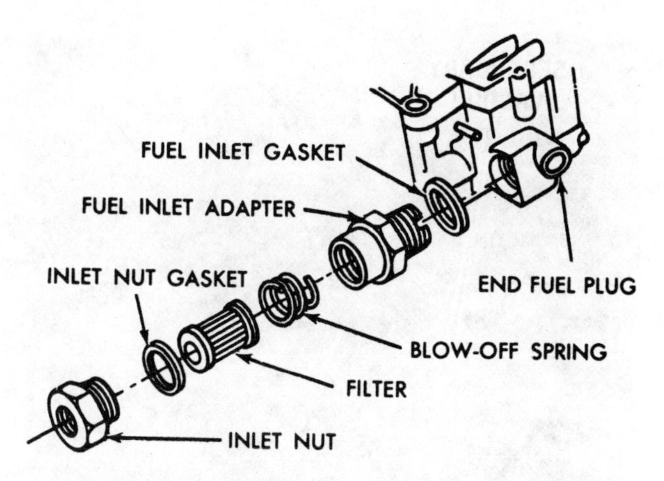

Arrangement of fuel filter parts including a self-tapping fuel inlet adapter which may be installed if the threads in the inlet hole of the carburetor are damaged. Use of the adapter will save considerable expense, because the carburetor does not have to be replaced.

their seats, and then back them out one full turn as a rough adjustment at this time. **NEVER** turn the adjusting needles in hard or the needle and seat will be **DAMAGED.** Start the engine and run it at idle speed. Adjust the idle mixture needle for the highest and steadiest manifold vacuum reading. If you do not have a vacuum gauge, obtain the smoothest running, maximum idle speed by turning the idle adjusting needle in until the engine rpm begins to drop off, then back the needle off over the "high spot" until the engine rpm again begins to drop off. Set the idle adjusting needle halfway between the two points as an idle speed setting. Repeat

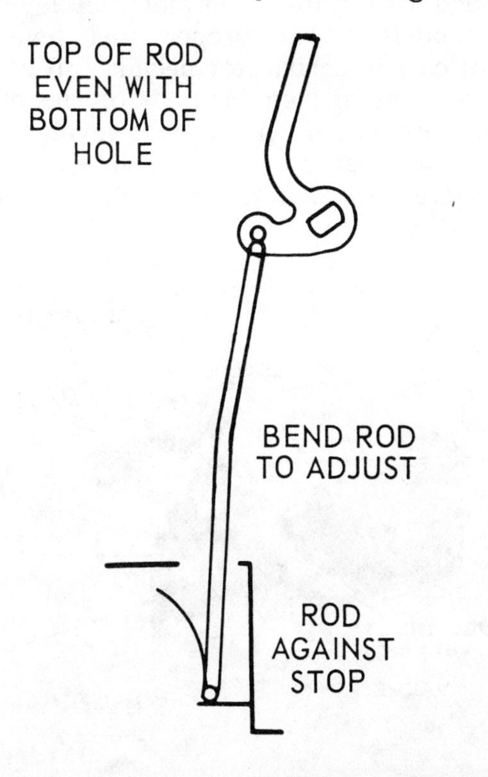

Tip of an idle mixture adjusting needle bent from being forced into its seat and against the closed throttle plate.

the procedure with the other needle. If the adjustments result in an increase in idle rpm, reset the idle speed adjusting screw to obtain between 550 and 600 rpm. Again adjust the idle mixture adjusting needles. Stop the engine and install the throttle cable. Check to be sure the throttle valves are fully open when the remote control is in the full forward position. Shift the unit into forward gear and re-adjust the idle speed screw until the rpm is between 550 and 600 rpm.

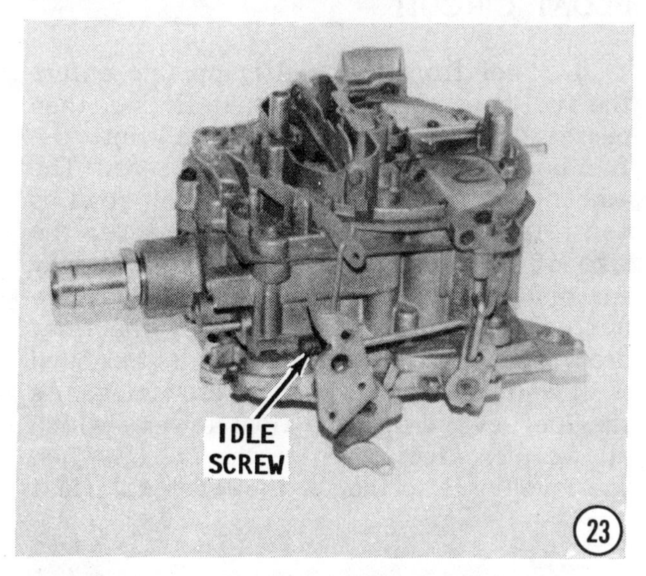

4-12 HOLLEY CARBURETORS MODEL 2300 - TWO BARREL AND MODEL 4150, 4160 & 4011 - FOUR BARREL

DESCRIPTION

The Holley two-barrel and four-barrel carburetors are designed and built very similarly. The four-barrel unit can be considered as dual two-barrel carburetors, mounted side-by-side, with each having its own fuel bowl and float system. Each side has its own metering body. The two primary bores have a single choke valve, and the primary side has a power valve, accelerating pump, and adjustable idle system. The throttles on the secondary side are controlled by a vacuum diaphragm. The secondary metering body has only fixed idle and high-speed metering systems.

The two-barrel carburetor is basically the primary side of the four-barrel. Therefore, the overhaul procedures are the same for the two-barrel model as those provided in the following section for the four-barrel.

The four-barrel carburetors have a dual, high-speed system made up of primary and secondary circuits. The primary circuits are composed of the float, idle, acceleration, main metering, power, and choke circuits. The secondary circuit becomes operational when the engine demands extra power. Each circuit will be described under separate headings.

The illustrations accompanying this section are keyed by number to the paragraphs which describe the parts being shown.

FLOAT CIRCUIT

1- Fuel from the fuel pump line enters the fuel bowl through the inlet fitting, then passes through a filter, and finally into the fuel bowl through the needle and seat. The amount of fuel entering the fuel bowl is controlled by the fuel pump pressure, the size of the hole in the needle seat, and by the distance the needle is permitted to rise out of the seat as determined by the float drop. Therefore, the fuel level in the bowl is determined by the float level setting. As the fuel level drops, the float lowers, which allows more fuel to enter the bowl. When the level rises to the setting level, the float

Holley two-barrel carburetor.

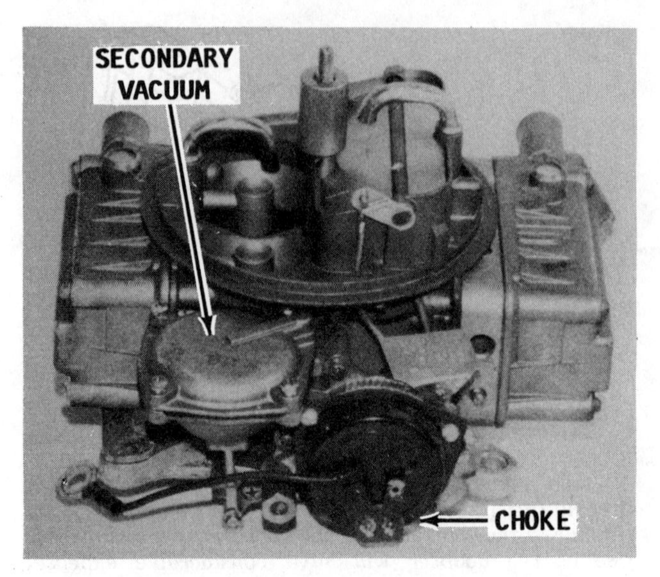

Holley four-barrel carburetor.

pushes the needle into the seat and thus shuts off the flow of fuel or at least restricts the amount entering the bowl. In this manner, the float rising and lowering allows only enough fuel to enter the bowl to replace the fuel used.

2- The Holley carburetors are equipped with one of three common types of needle and seat assemblies, as shown in the accompanying illustration: (1) spring loaded; (2) solid needle; and (3) externally adjustable.

To prevent dirt in the fuel from flooding, most of the needles have a special soft plastic tip. A spring under the float stabilizes and maintains a normal fuel level in most models. The proper fuel level is critical on any carburetor because all of the basic settings and calibrations of the other systems are based on the fuel level in the bowl. The float system is equipped with a vent valve in order to pressurize the fuel

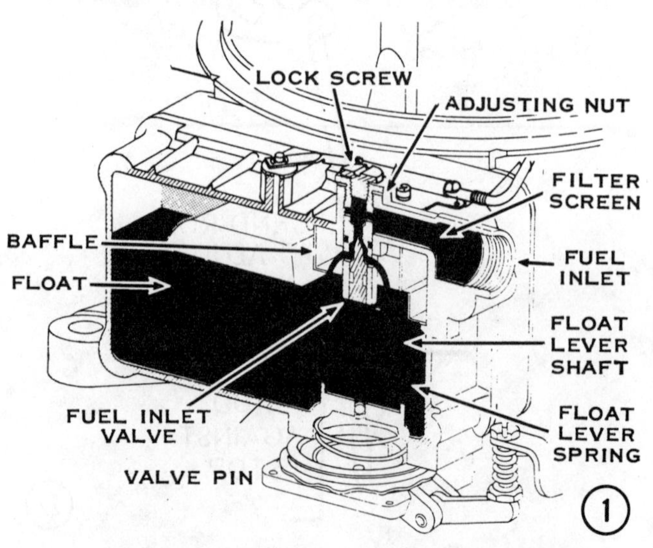

and to create a pressure differential. The reduced pressure in the venturi creates a pressure differential forcing the fuel to flow out of the fuel bowl to the discharge nozzle.

3- Fuel passes through the main metering jet to the vertical passageway. Near the top of this passageway is an idle feed restriction. This restriction performs the same function as an idle tube in other carburetors by metering the fuel for low-speed operation. From the restriction, the fuel moves horizontally across the carburetor to another vertical passageway, and then down the second passageway to the idle mixture screw port.

An idle air bleed is located in the low-speed fuel passageway, above the idle mixture adjustment screw. The idle air mixes with the fuel to form an emulsion that is easier to atomize when it is discharged. This air-fuel mixture moves much faster than solid fuel. The idle air bleed prevents fuel from syphoning from the fuel bowl when the engine is not running.

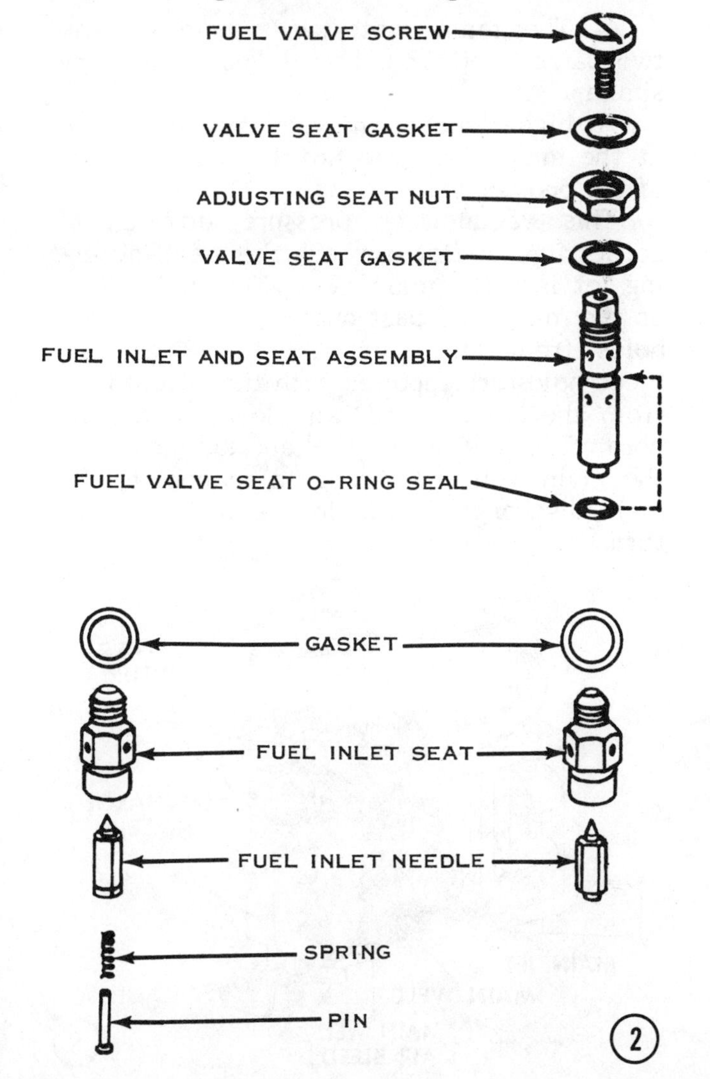

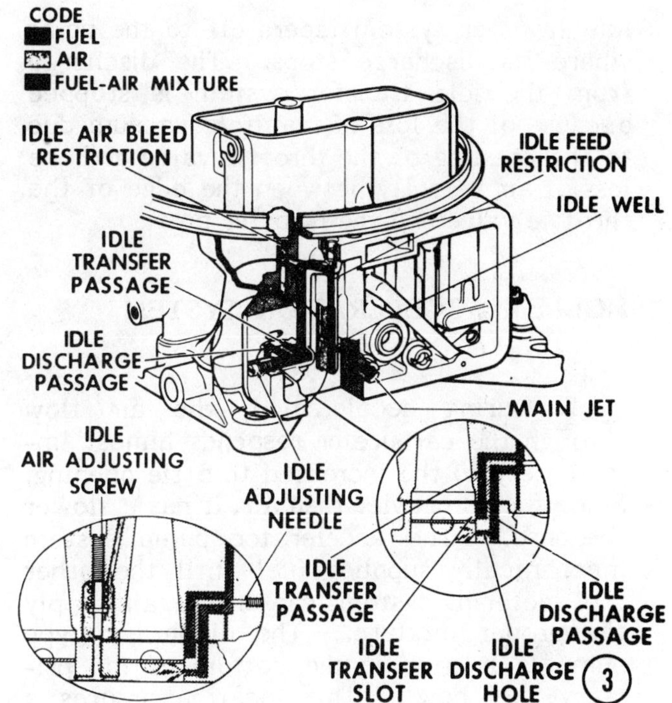

4- The fuel branches into two passageways at the idle mixture adjusting screw. One branch is controlled by the idle mixture needle. This branch exits below the throttle valve and supplies the fuel for the hot curb idle. The other branch exits above the throttle valve and supplies the fuel for the transfer from idle to the main metering stage. The mixture of air and fuel flows past the pointed tip of the adjusting needle. If the needle is backed out, the volume of the mixture is increased making it richer. If the needle is turned inward, the volume is decreased and the mixture becomes leaner. Fuel from the needle then moves through a passage and is discharged into the throttle bore below the throttle plate. When the throttle plates are opened above the idle position, more fuel is fed through the idle transfer passageway to supply the added demands of the engine.

If the throttle is opened wider, air speed in the venturi increases and the main metering system begins to function. As the flow increases in the main metering system, the

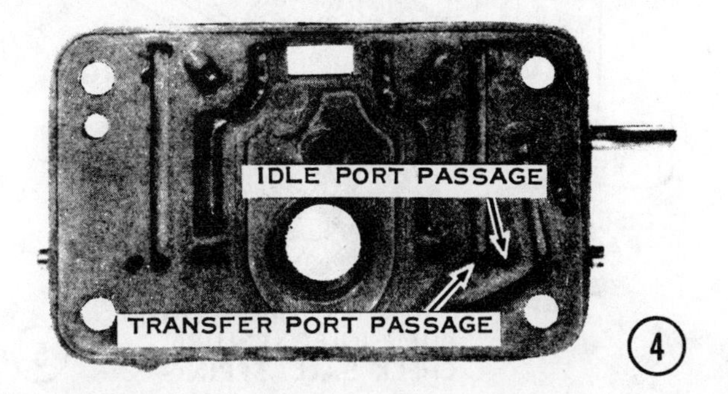

idle transfer system tapers off to the point where its discharge stops. The discharge from the idle transfer system is stopped because of the loss of manifold vacuum due to the opening of the throttle valve and the loss of air velocity between the edge of the throttle valve and the transfer port.

HOLLEY ACCELERATION SYSTEM

5- During acceleration, the air flow through the carburetor responds almost immediately to the increased throttle opening. Since fuel is heavier than air, it has a slower response. The accelerator pump system mechanically supplies fuel until the other fuel metering systems can once again supply the proper mixture. The diaphragm-type pump is located in the bottom of the primary fuel bowl. This location assures a more solid charge of fuel (fewer bubbles).

When the throttle is opened, the pump linkage, actuated by a cam on the throttle lever, forces the pump diaphragm up. As the diaphragm moves up, the pressure forces the pump inlet check ball or valve onto its seat, thereby preventing the fuel from flowing back into the fuel bowl.

The fuel passes through a short passage in the fuel bowl into the long diagonal passage in the metering body. It next flows into the main body passage and then into the pump discharge chamber. The pressure of the fuel causes the discharge valve to raise, and fuel is then discharged into the venturi.

The pump override spring is an important part of the accelerator system. When the accelerator is moved rapidly to the wide open position, the override spring is compressed and allows full pump travel. The spring applies pressure to maintain the pump discharge. Without the spring, the pump linkage would be bent or broken due to the resistance of the fuel which is not compressible.

As the throttle moves toward the closed position, the linkage returns to its original position and the diaphragm return spring forces the diaphragm down. The pump inlet check valve is moved off its seat and the diaphragm chamber is refilled with fuel from the fuel bowl.

The accelerator pump delivery rate is controlled by the pump cam, linkage, the override spring, and the size of the discharge holes.

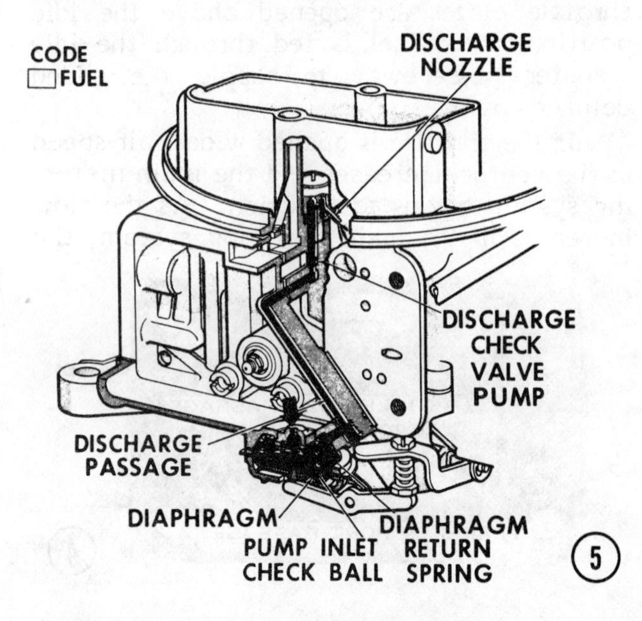

HOLLEY MAIN METERING

6- The main metering system on the two-barrel and four-barrel model is very similar.

At higher speeds the vacuum is increased at the main discharge nozzle in the center of the booster venturi.

This vacuum or pressure differential causes fuel to flow through the main metering jet into the main well. The fuel moves up the main well past one or more air bleed holes from the main airwell. These air bleed holes are supplied with the filtered air from the "high speed" air bleeds in the air horn. The mixture of fuel and air moves up the main well and through a channel to the main discharge nozzle in the booster venturi.

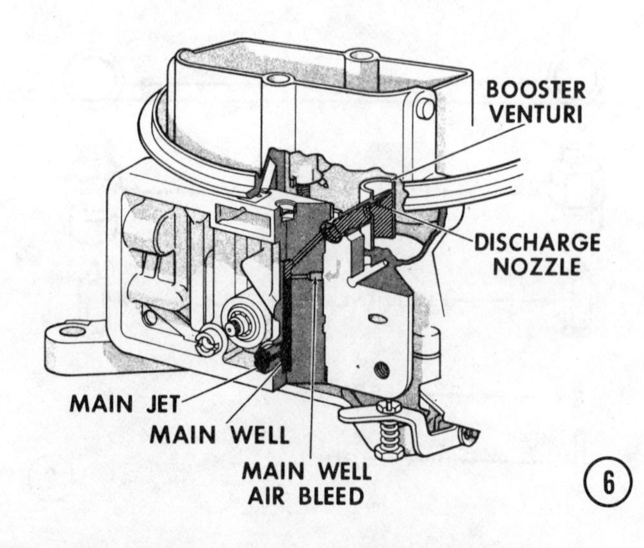

POWER ENRICHMENT SYSTEM

7- During high speed or heavy load operation, when manifold vacuum is low, the power system provides added fuel for power operation. A vacuum passage in the throttle body transmits vacuum to the power valve vacuum chamber in the main body. All of the power valves used in this series of carburetors are actuated by a vacuum diaphragm. Manifold vacuum is applied to the vacuum side of the diaphragm to hold it closed at idle and normal moderate load conditions.

When manifold vacuum drops below the power valves' calibration, the power valve spring opens the valve to admit additional fuel. This fuel is metered by the power valve channel restrictions into the main well and is added to the fuel flowing from the main metering jets.

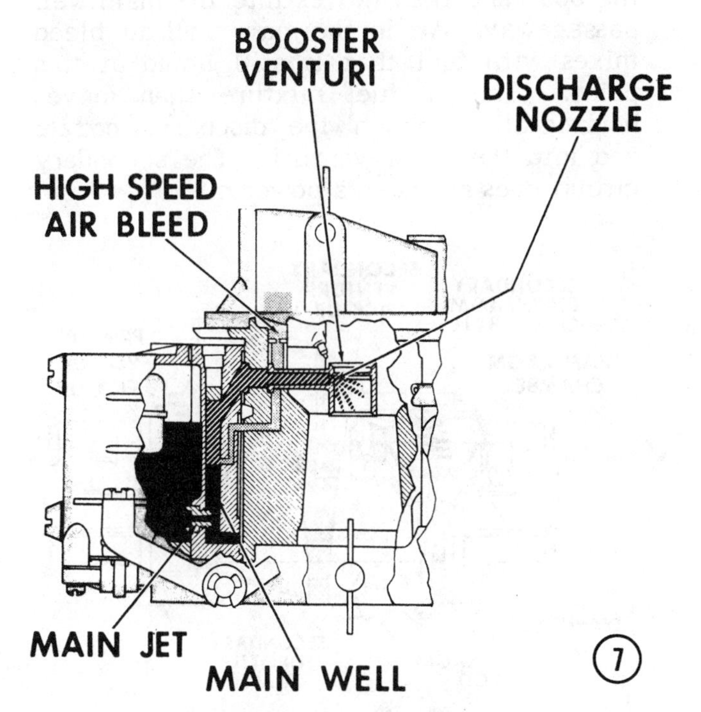

HOLLEY AUTOMATIC CHOKE

8- The choke system has a bi-metal spring. When this spring cools, it closes the choke. An electric heating coil is installed in the choke spring cover to cause the bi-metal spring to open the choke. This heating element substitutes for the normal heat tube of other carburetors. Correct wiring for the heating element is directly to the stator terminal of the alternator. The heater receives half of the charging voltage and does not receive any voltage until the engine is running and the alternator is producing current. If the heater is wired incorrectly -- directly to the ignition to receive a full 12-volts -- it will have a very short life. Once the heater coil burns out, warming of the bi-metal strip will be accomplished only after a very long time by the manifold heat. Such a condition wastes considerable fuel.

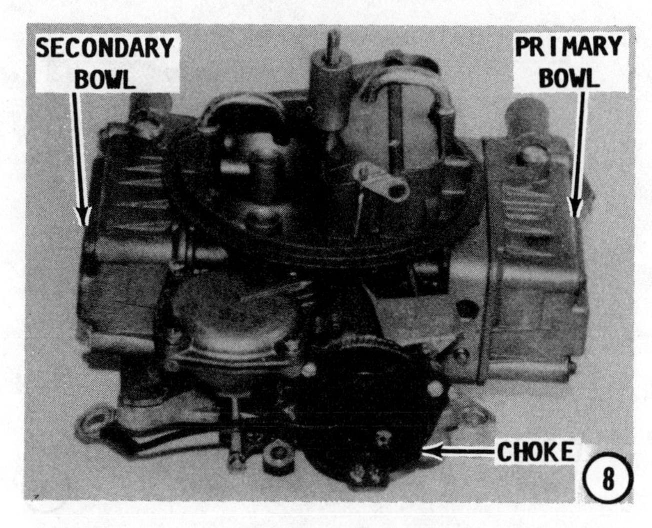

SECONDARY SYSTEMS

VACUUM SECONDARY OPERATION

9- At lower speeds the secondary throttle valves remain closed, allowing the engine to maintain proper air-fuel velocities and distribution for lower speed, light load operation. When engine demand increases to a point where additional breathing capacity is needed, the vacuum controlled secondary throttle valves begin to open automatically.

Vacuum from one of the primary venturi and one of the secondary venturi is channeled to the top of the secondary diaphragm. The bottom of the diaphragm is open to atmospheric pressure. At higher speeds and higher primary venturi vacuum, the diaphragm, operating through a rod and secondary throttle lever, will start to open the secondary throttle valves. This action will start to compress the secondary diaphragm spring. As the secondary throttle valves open further, a vacuum signal is created in the secondary venturi. This additional vacuum assists in opening the secondary throttle valves to the maximum designed opening. The secondary opening rate is controlled by the diaphragm spring and the size of the

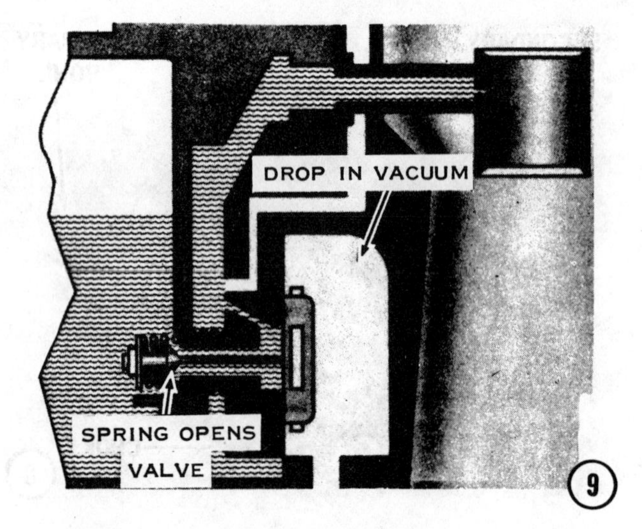

vacuum restrictions in the venturi.

When the engine speed is reduced, venturi vacuum decreases and the diaphragm spring starts to push the diaphragm down to start closing the secondaries. Closing the primary throttle valves moves the secondary throttle connecting link.

Most production model carburetors have a ball check and bypass bleed installed in the diaphragm passage. The ball permits a smooth, even opening of the secondaries, but lifts off the inlet bleed to cause rapid closing of the secondaries when the primary throttle valves are closed.

SECONDARY FUEL METERING

SECONDARY FLOAT CIRCUIT

10- The secondary system has a separate fuel bowl. Fuel is usually supplied to the secondary bowl by a transfer tube from the primary fuel inlet fitting.

The secondary fuel bowl is equipped with a fuel inlet valve and float assembly similar to the primary side. The specified fuel level on the secondary side is usually slightly lower than the primary side.

Many of the Holley carburetors use a fuel balance tube between the primary and secondary fuel bowls. This tube is made of brass and connects the bowls just above the fuel level to prevent flooding of the secondary fuel bowl.

SECONDARY IDLE CIRCUIT

11- The secondary side of the carburetor is seldom used. Therefore, the fuel in the secondary fuel bowl could become stale and develop gum and varnish. To avoid this problem, a fixed idle is designed into the secondary side. Anytime the engine is running, some fuel is used from the secondary bowl and fresh fuel flows in to replace it. The secondary idle mixture is not adjustable. The transfer operates in the same manner as on the primary side.

SECONDARY MAIN METERING CIRCUIT

12- The basic principles of the four-barrel carburetor are the same as on the two-barrel unit except for detailed construction differences. As the secondary throttle valves are opened further, air velocity increases in the boost venturi. This action allows fuel to flow through the main metering system.

Fuel from the secondary fuel bowl enters the lower main metering holes in the metering body and then moves into the main well passageway. Air in the main well air bleed mixes with fuel that is still liquid at this point. The air-fuel mixture then moves horizontally through the discharge nozzle and into the boost venturi. The secondary circuit does not have a power circuit.

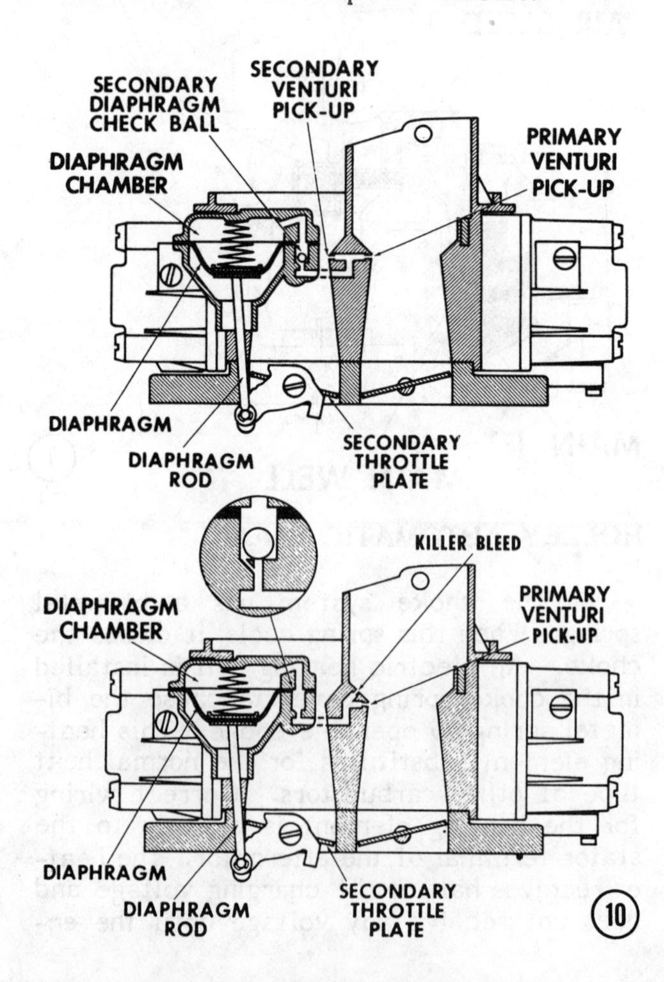

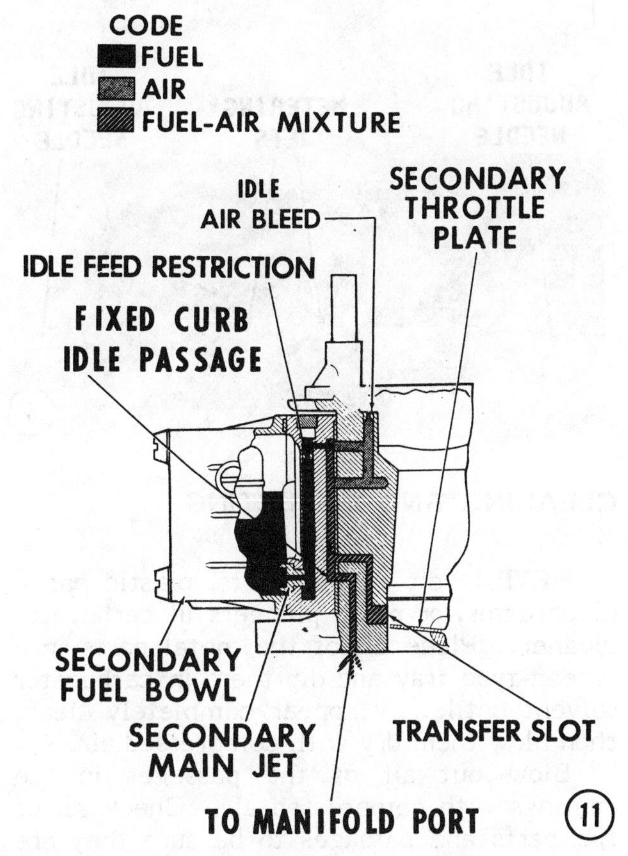

DISASSEMBLING HOLLEY CARBURETOR MODEL 2300 — TWO BARREL AND MODEL 4150 AND 4160 — FOUR BARREL (Model 4011 Covered in Next Section — 4-13)

1- Remove the four primary fuel bowl retaining screws. Slide the bowl straight off. Discard the gaskets. Pull off the fuel transfer tube and discard the O-ring seals.

2- Remove the secondary fuel bowl assembly. Detach the secondary metering body. **TAKE CARE** and remove the secondary metering body, plate, and gaskets. Slide the balance tube, washers, and O-rings out of the main body from either end.

TAKE NOTE of the position of the fast-

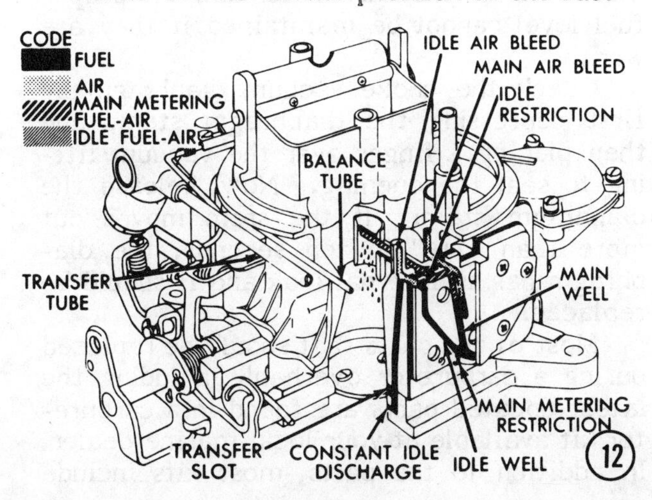

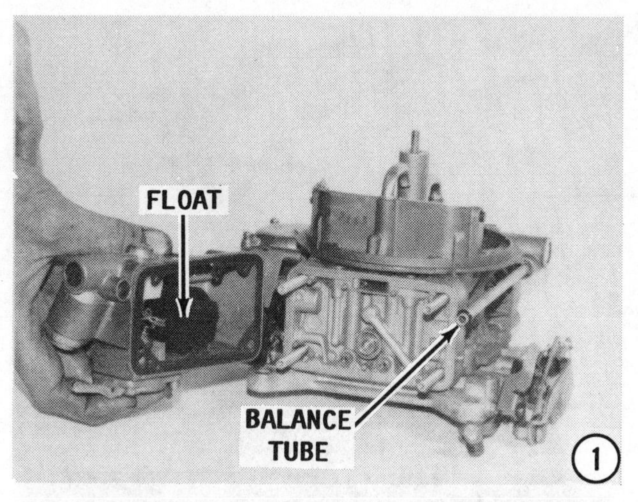

idle cam to the cam lever for proper assembly. Disconnect the link from the fast-idle cam lever and cam, and then slide the lever and cam off of the stub shaft, and at the same time, disengage the choke rod from the cam lever. Disconnect the secondary vacuum diaphragm assembly stem from the secondary stop lever, and then remove the assembly. Discard the gasket. Remove the choke plate retaining screws by first filing the backs of the screws to remove the stake marks. Remove the choke plate. Remove the choke shaft.

3- Remove the accelerator pump discharge nozzle by first removing the retaining screw. Discard both gaskets. Turn the carburetor over and catch the accelerating pump discharge needle as it falls out. Remove the six screws securing the throttle body and main body together and then separate the parts. Discard the gasket.

4- Remove the float retainer E-clip, and then slide out the float and spring. Remove the fuel inlet needle, and then take out the float baffle. Remove the fuel inlet needle seat and gasket. Discard the gasket. Re-

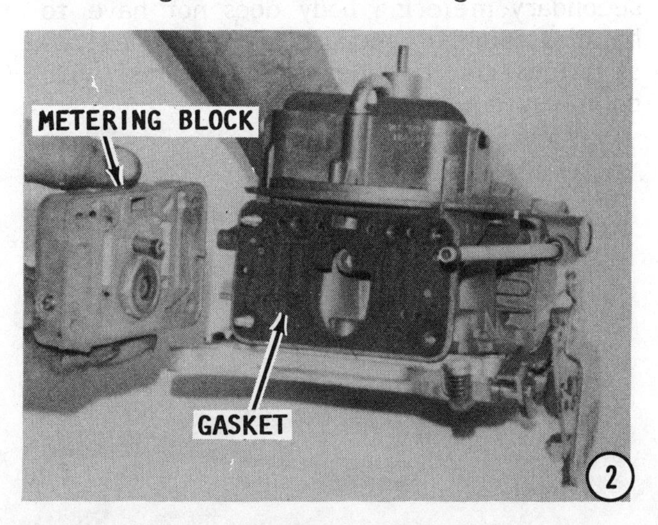

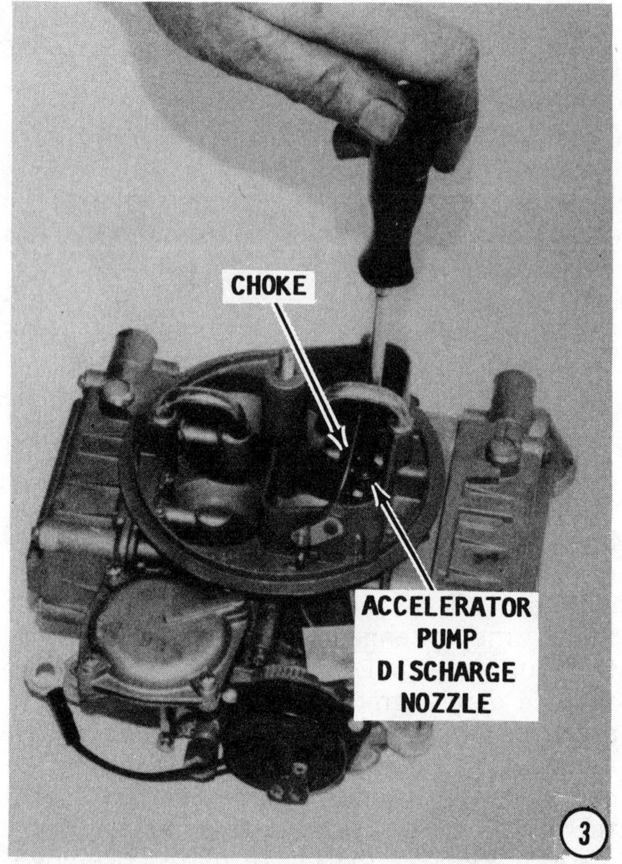

move the fuel inlet fitting and discard the gasket. Remove the bowl vent valve assembly and accelerator pump cover on the primary fuel bowl. Remove the diaphragm and spring.

5- The power valve assembly should be replaced each time the carburetor is overhauled. This assembly should be removed from the primary metering body, using Tool No. 3747. The metering jets do not have to be removed because they can be cleaned with compressed air while they are still in place in the metering body. Remove the idle adjusting needles and gaskets. The secondary metering body does not have to be disassembled because the metering restrictions can be adequately cleaned with compressed air.

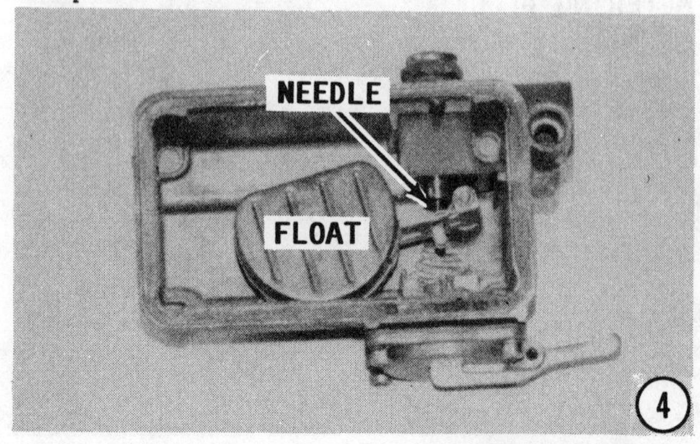

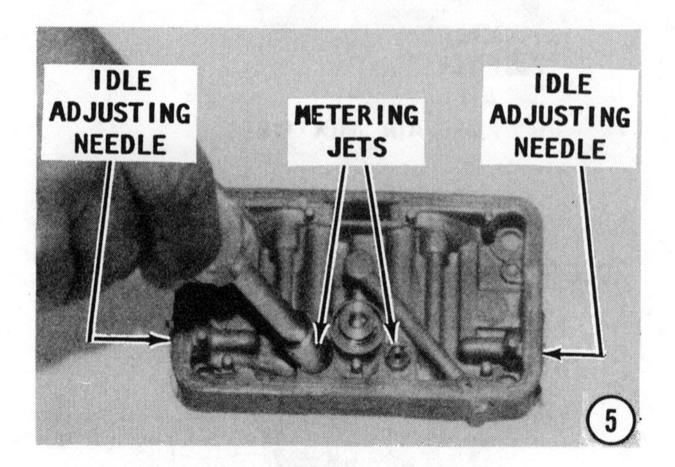

CLEANING AND INSPECTING

NEVER dip rubber parts, plastic parts, diaphragms, or pump plungers in carburetor cleaner. Place all of the metal parts in a screen-type tray and dip them in carburetor solvent until they appear completely clean, then blow them dry with compressed air.

Blow out all of the passages in the castings with compressed air. Check all of the parts and passages to be sure they are not clogged or contain any deposits. **NEVER** use a piece of wire or any type of pointed instrument to clean drilled passages or calibrated holes in a carburetor.

Inspect the choke and throttle shafts for excessive wear and replace the complete carburetor if the throttle shaft is worn.

Check the floats for leaking by shaking them and listening for fluid movement inside. If a float contains fluid, it must be replaced. Inspect the arm needle contact surface and the float shaft and replace them if either has any grooves worn in it.

ALWAYS replace the needle valve-and-seat assemblies. These parts receive the most wear in the carburetor and the proper fuel level cannot be maintained if they are worn.

Check the choke vacuum diaphragm by first depressing the diaphragm stem, and then placing a finger over the vacuum fitting to seal the opening. Now, release the diaphragm stem. If the stem moves out more than 1/16" in ten seconds, the diaphragm has an internal leak and it should be replaced.

Most of the parts that should be replaced during a carburetor overhaul, including the latest updated parts are found in a carburetor kit available at your local marine dealer. In addition to the parts, most kits include

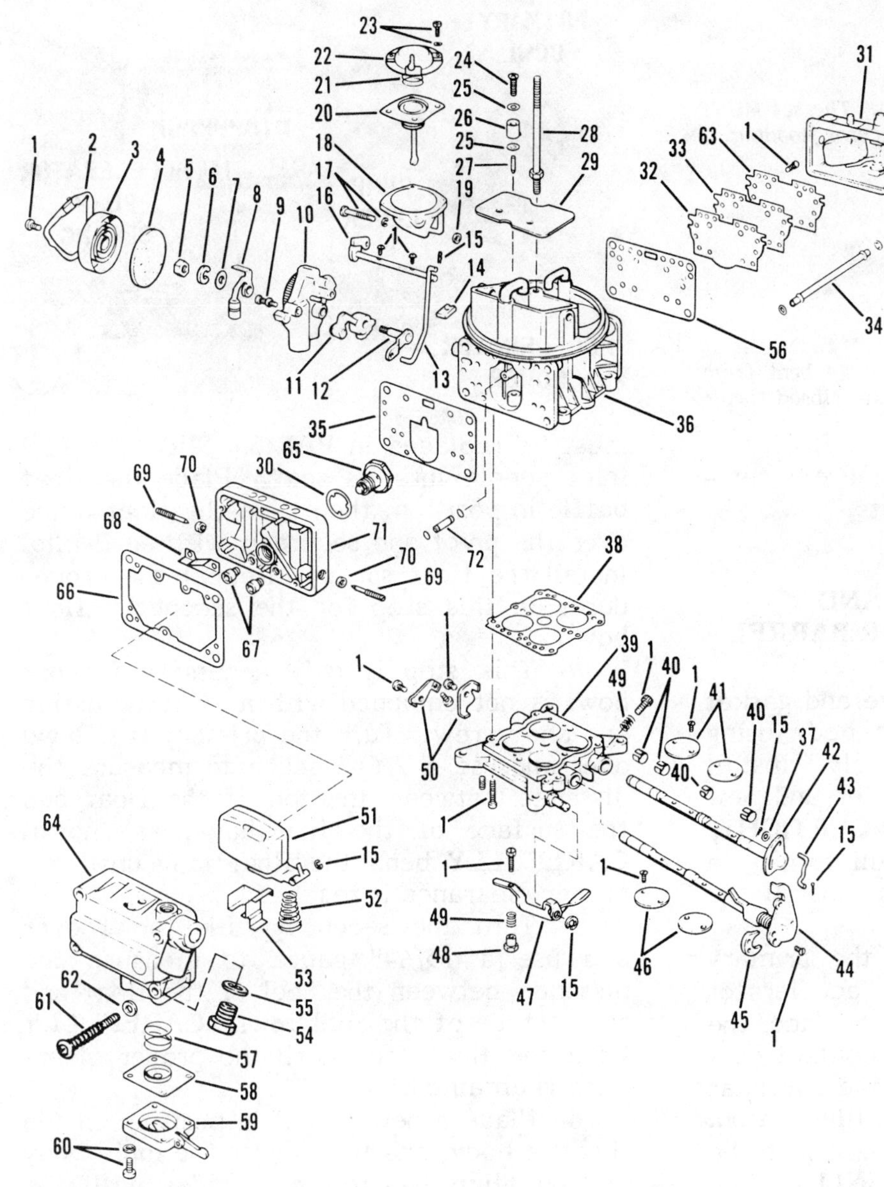

1. Screw
2. Choke Thermostat Housing Clamp
3. Choke Thermostat Housing and Spring
4. Choke Thermostat Housing Gasket
5. Nut
6. Lockwasher
7. Spacer
8. Choke Thermostat Lever Link and Piston Assembly
9. Screw and Washer
10. Choke Housing
11. Fast Idle Cam Assembly
12. Choke Housing Shaft and Lever
13. Choke Rod
14. Choke Rod Seal
15. Retainer
16. Choke Shaft
17. Screw and Washer
18. Secondary Housing
19. Gasket
20. Diaphragm Assembly
21. Diaphragm Spring
22. Cover
23. Screw and Washer
24. Screw
25. Gasket
26. Accelerating Pump Discharge Nozzle
27. Accelerating Pump Discharge Needle
28. Flame Arrestor Anchor Screw
29. Choke Plate
30. Power Valve Gasket
31. Secondary Fuel Bowl
32. Secondary Plate
33. Secondary Plate
34. Fuel Line
35. Primary Meter Block Gasket
36. Main Body
37. Washer
38. Throttle Body-to-main Body Gasket
39. Throttle Body
40. Shaft Bushings
41. Secondary Throttle Plates
42. Secondary Throttle Shaft
43. Throttle Connecting Rod
44. Primary Throttle Shaft Assembly
45. Accelerating Pump Cam
46. Primary Throttle Plates
47. Accelerating Pump Operating Lever
48. Sleeve Nut
49. Spring
50. Diaphragm Lever Assembly
51. Float
52. Float Spring
53. Baffle Plate
54. Fuel Inlet Fitting
55. Gasket
56. Gasket
57. Diaphragm Spring
58. Diaphragm Assembly
59. Accelerating Pump Cover
60. Retaining Screw and Lockwasher
61. Screw
62. Gasket
63. Secondary Metering Block
64. Primary Fuel Bowl
65. Power Valve
66. Primary Fuel Bowl Gasket
67. Main Jets
68. Baffle Plate
69. Idle Adjusting Needle
70. Seal
71. Primary Metering Block
72. Tube and O-Rings

Exploded drawing of the Holley four-barrel carburetor with most parts identified.

*Fuel inlet needle with a worn ridge. The set **MUST** be replaced to prevent a leak and carburetor flooding.*

Tip of an idle mixture adjusting needle bent from being forced into its seat and against the closed throttle plate.

the latest specifications which are so important when making adjustments.

ASSEMBLING THE HOLLEY MODEL 2300 — TWO BARREL AND MODEL 4150 AND 4160 — FOUR BARREL

1- Install a new power valve and gasket in the primary main metering body using Tool C-3747 to tighten it securely. Install the idle mixture adjusting needles and new gaskets. Tighten the needles **FINGERTIGHT**, then back each one out one full turn as a rough adjustment at this time. Install the main metering jets.

2- Assemble the parts to the primary fuel bowl by first placing the accelerator pump spring in position. Next place the diaphragm in position with the contact button facing the **PUMP LEVER** in the cover, as shown in the accompanying illustration. Place the cover over the diaphragm with the pump lever **FACING** the **FUEL INLET** side, as shown. Start the attaching screws, then depress the pump lever to center the diaphragm, then tighten the screws. Install a new gasket over the fuel inlet fitting, then the fitting and tighten it securely.

3- Install a **NEW** needle valve seat and gasket and tighten it securely. The needle valve and seat are a **MATCHED SET** and must be replaced in **PAIRS**. Slide the fuel inlet needle into its seat. Place the float baffle in position, then slide the float hinge over the pivot and secure it with an E-clip. Install the float spring. Repeat the procedures in this step for the secondary float bowl.

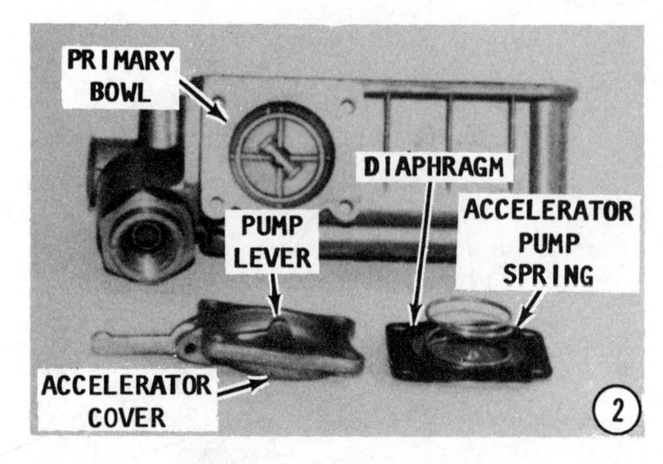

4- This step is only necessary if your bowl is not equipped with a float adjusting nut and screw. Turn the primary fuel bowl over and use a 7/64" gauge to measure the distance between the toe of the float and the surface of the fuel bowl, as shown. **CAREFULLY** bend the float tang until the proper clearance is reached.

5- Turn the secondary fuel bowl over and use a 15/64" gauge to measure the distance between the heel of the float and the surface of the fuel bowl. **CAREFULLY** bend the float tang until the proper clearance is obtained.

6- Place a new gasket in position on the throttle body, and then lower the main body as you align the roll pin guides with the openings in the main body. The primary bores of the throttle body **MUST** be on the **SAME** side as the primary venturi. Hold the

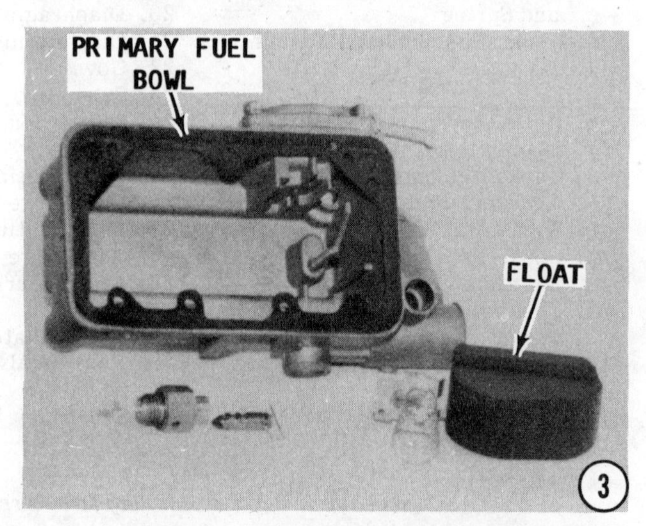

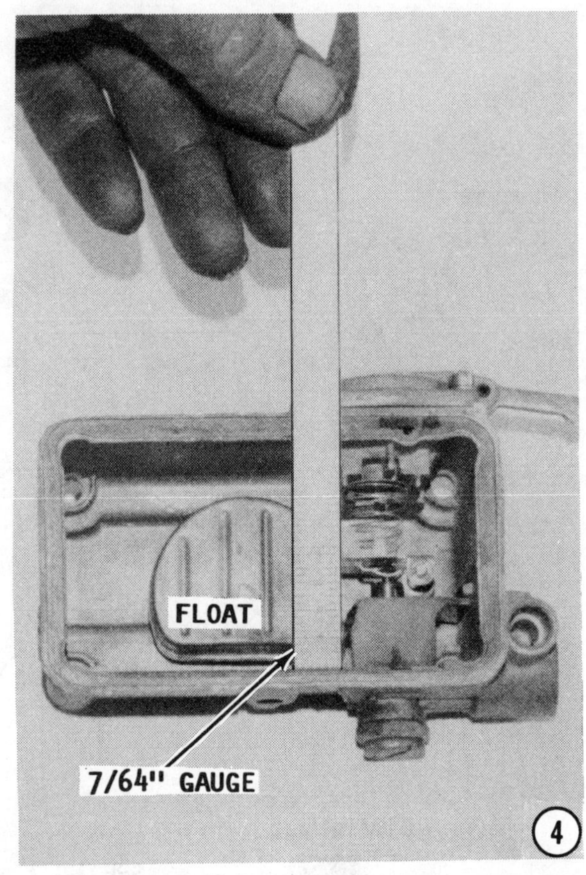

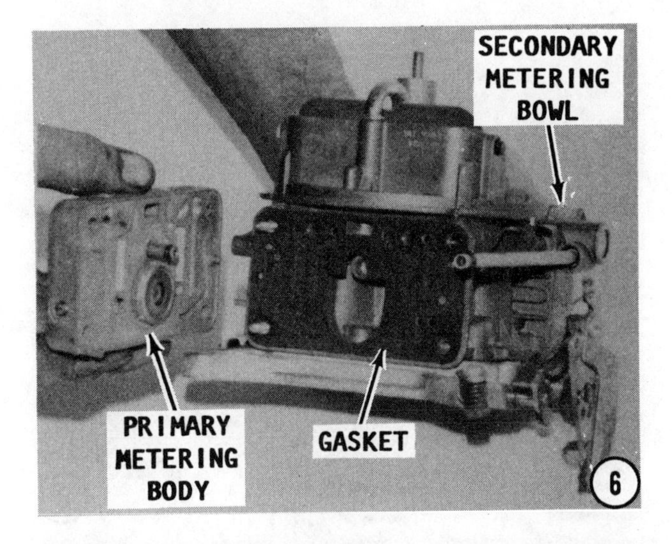

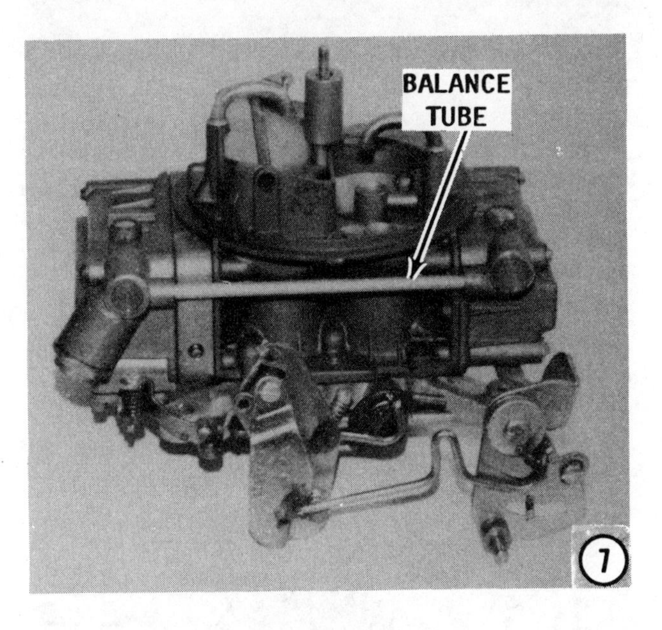

parts together, turn them over and install and tighten them securely. Push an O-ring seal into each recess, then install a flat washer at each end. Install a **NEW** gasket, then the plate, another **NEW** gasket, and the secondary metering body with the restrictions at the **BOTTOM.** Secure it all together with the attaching screws.

7- Position the balance tube, if one is used, so one inch extends beyond the secondary metering body, as shown.

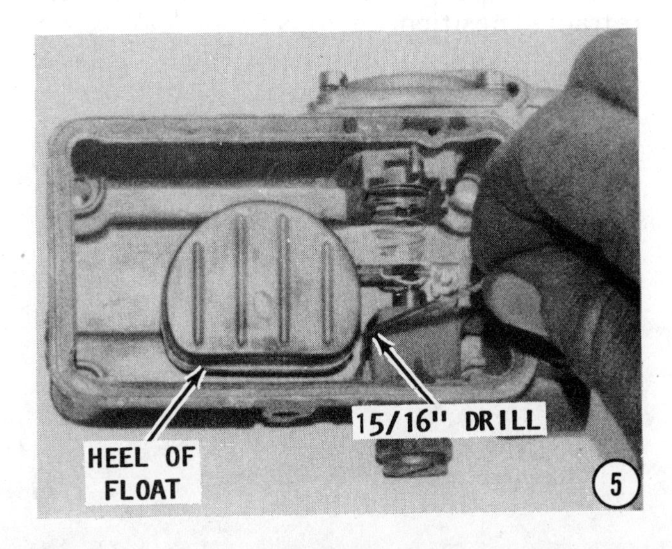

8- Slide the assembled secondary fuel bowl and a **NEW** gasket over the balance tube. Install and tighten the attaching screws. Slide a new O-ring seal on each end of the fuel transfer tube, and then press the tube into the opening in the secondary bowl. Place a **NEW** gasket over the primary metering body pin, then slide the body over the balance tube and into place on the main body. Slide a **NEW** gasket over the metering body, and then install the primary fuel bowl over the balance tube and down against the metering body. Slide **NEW** gaskets over the long fuel bowl mounting screws, then install and tighen them securely. **NEVER** use an old gasket because they will always leak following a second installation.

9- Install the accelerating pump discharge needle into the discharge passageway in the center of the primary venturi.

10- Install the pump discharge nozzle gas-

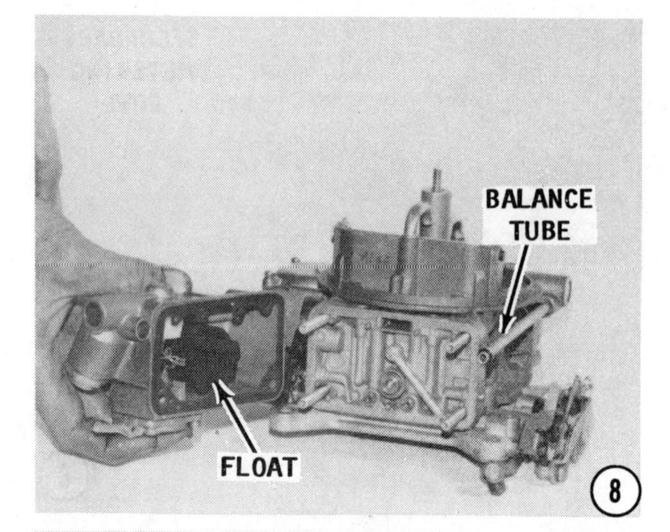

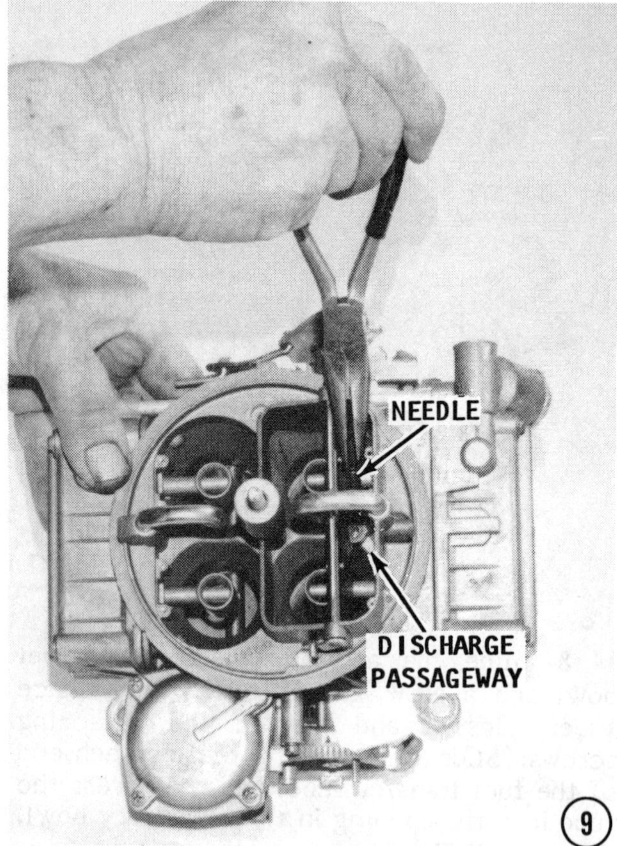

ket, nozzle, mounting screw, and gasket. Align the notch in the rear of the nozzle with the projection on the boss of the casting. Install the choke valve and shaft.

11–Assemble the secondary throttle-opening diaphragm by first sliding a new diaphragm into the housing. Next, place the diaphragm in position with the vacuum hole in the housing aligned with the vacuum hole in the diaphragm. Now, install the return spring with the coiled end snapped over the button in the cover. Lower the cover into position with the vacuum port in the cover aligned with the port in the housing. Keep

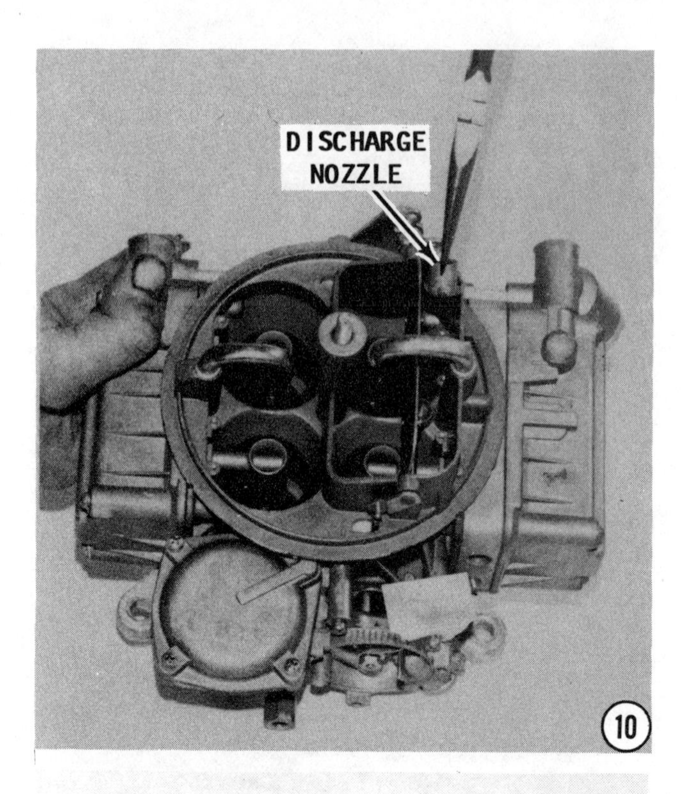

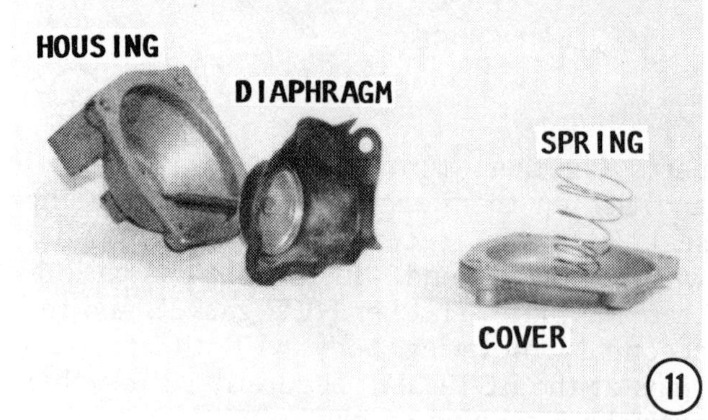

the diaphragm flat while the cover is being installed. Install and tighten the attaching screws. Check the assembly for air leaks by depressing the diaphragm stem and then placing your finger over the port to seal it. If the diaphragm does not remain in the retracted position, there is an air leak.

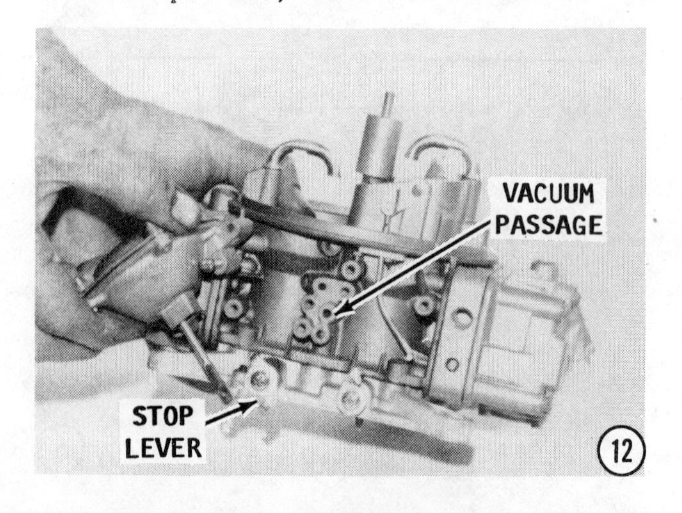

12- Place a **NEW** gasket in the vacuum passageway recess in the diaphragm housing, then install the secondary diaphragm assembly and at the same time, engage the stem with the secondary stop lever. Install and tighten the attaching screws.

CARBURETOR ADJUSTMENTS
MODEL 2300 -- TWO BARREL AND MODEL 4150 AND 4160 -- FOUR BARREL

Carburetor adjustments **MUST** be performed in the sequence outlined in the following steps.

For best performance and economy results, follow the specifications included in the carburetor repair kit.

1- **Automatic Choke Adjustment:** The automatic choke is set at the factory to give maximum performance under all weather conditions. If necessary, the choke can be returned to its original position by aligning the index mark with the proper specification.

2- A richer or leaner mixture can be obtained for the warm-up period, if desired, by rotating the thermostat cover. **NEVER** set the index mark on the cover more than two marks from the specified setting.

3- **Fuel Level Adjustment:** Loosen the lock screw and turn the adjusting nut clockwise to lower the fuel level and counterclockwise to raise the fuel level. A 1/16 turn of the nut will equal approximately 1/16 inch difference in fuel level. After the adjustment has been made, tighten the lock screw. On newer model carburetors, this adjustment can only be done with the bowl removed.

4- With the engine approximately level and the engine running, the fuel level in the sight plug must be in line with the threads at the bottom of the plug within 1/32". This

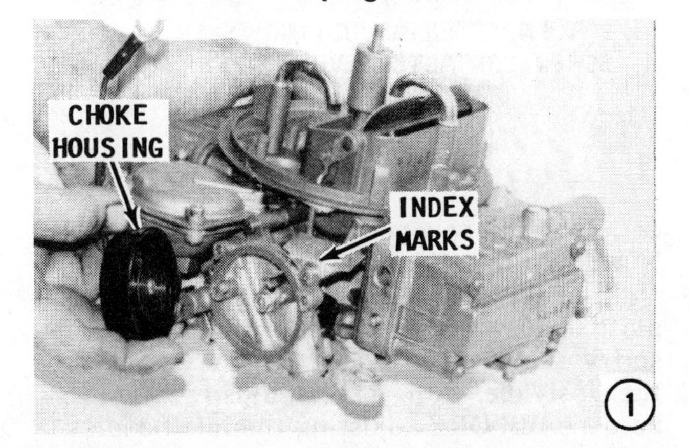

1

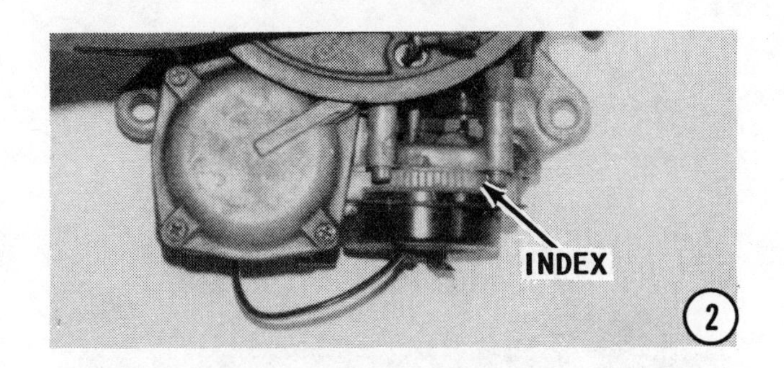

2

sight plug is not used on late model carburetors.

5- **Accelerator Pump Lever Clearance:** First, move the throttle to the wide-open position. Hold the pump lever down. You should be able to insert a 0.015" gauge between the adjusting nut and the lever. Rotate the pump override screw to obtain the correct clearance. There must be **NO FREE PLAY** with the pump lever at idle.

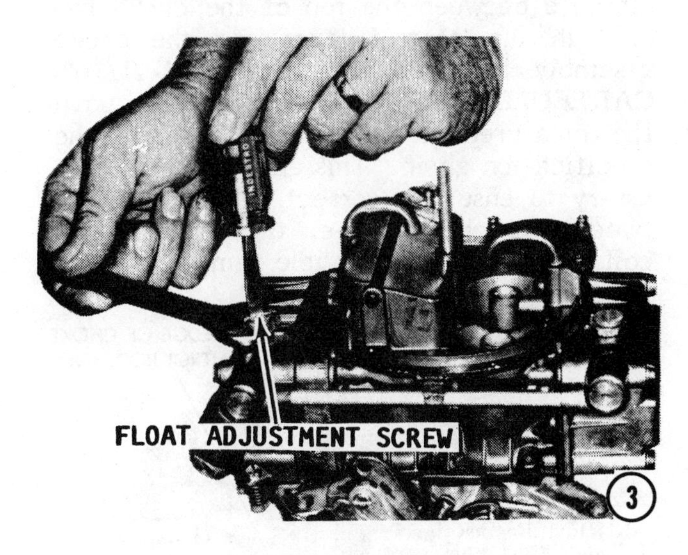

3

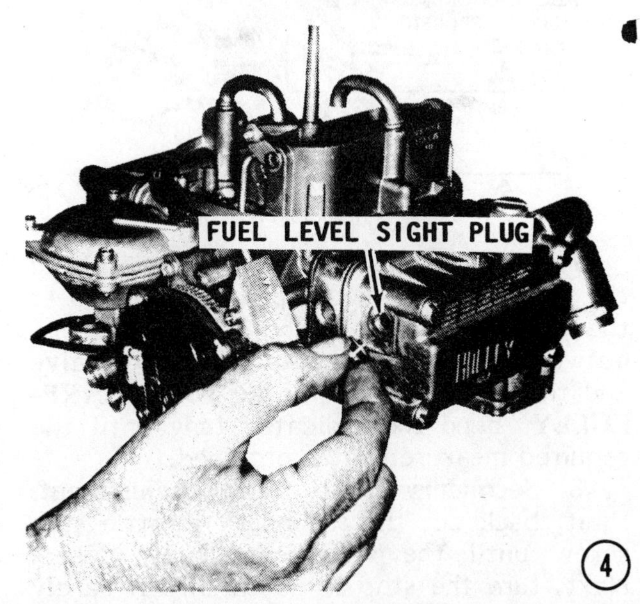

4

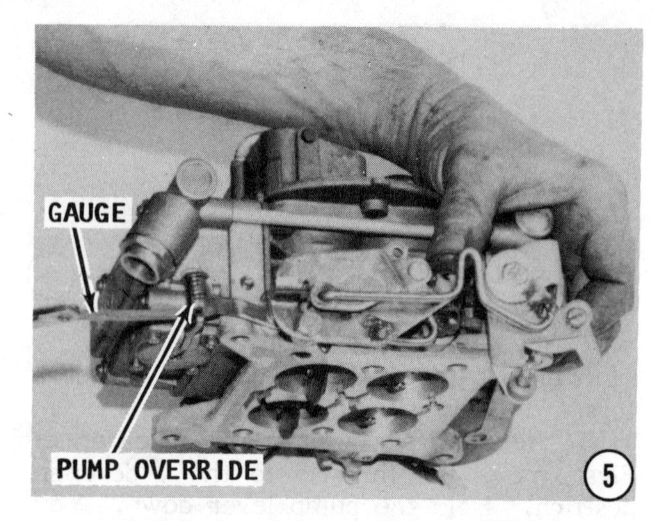

6- Choke Control Lever Adjustment: First, open the throttle to the midposition. Next, close the choke valve with a light pressure on the choke control lever. With the carburetor on the engine, measure the distance between the top of the choke rod hole in the control lever and the choke assembly and compare it with 1-11/16". **CAREFULLY** bend the choke shaft rod until the measurement is within 1/64" of the specification given. This adjustment is necessary to ensure a correct relationship between the choke valve, the thermostatic coil spring, and the fast-idle cam.

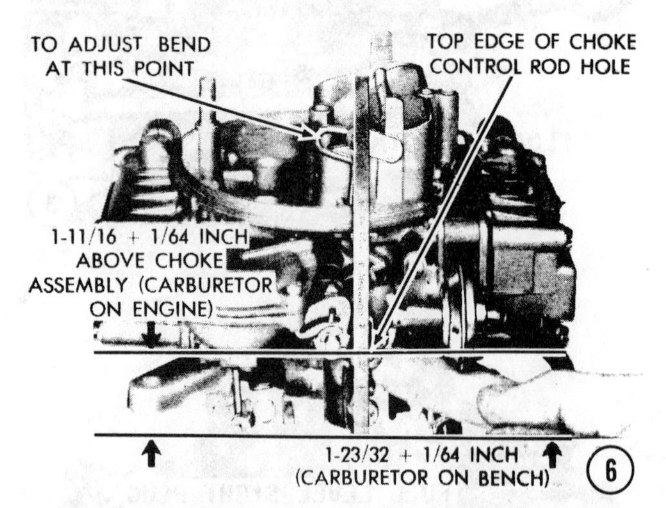

7- Choke Unloader Adjustment: Hold the throttle in the wide-open position and at the same time insert the specified gauge between the upper edge of the choke valve and the inner wall of the air horn. **CAREFULLY** bend the indicated tang until the required measurement is obtained.

8- Secondary Throttle Valve Adjustment: First, back out the secondary throttle stop screw until the valves are fully closed. Next, turn the stop screw in until it barely

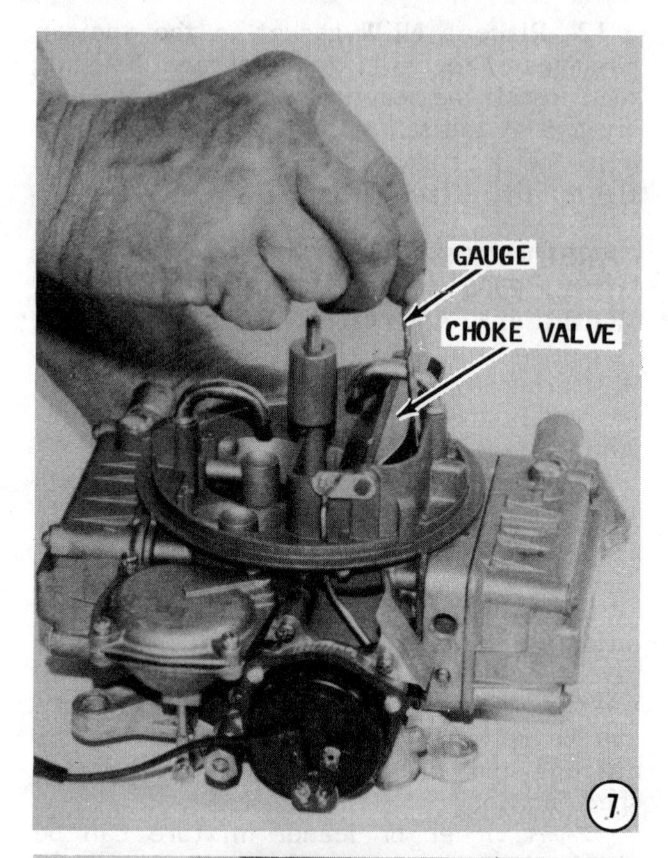

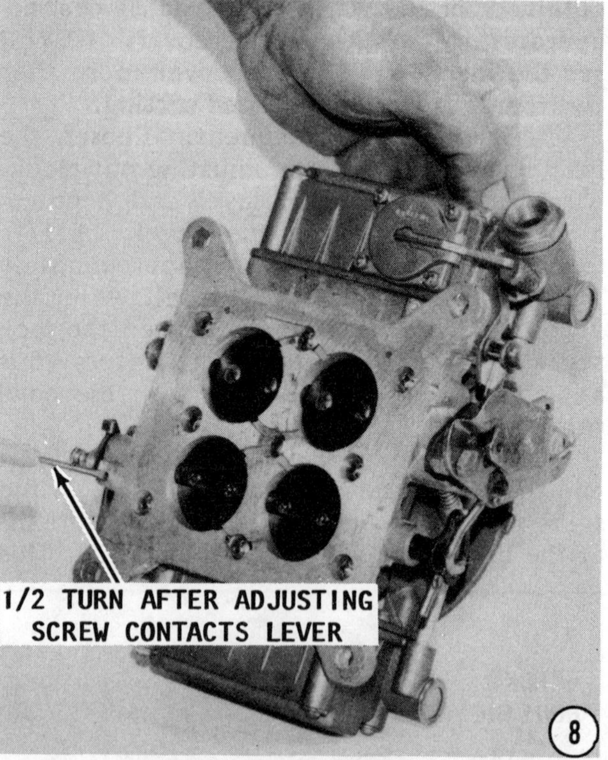

contacts the secondary throttle stop lever. Now, turn the adjusting screw in 1/2 turn.

9- Fast Idle Adjustment: With the choke plate wide open, the fast-idle adjusting screw should just touch the lowest step of the fast-idle cam. This adjustment MUST be done **BEFORE** adjusting the idle needles.

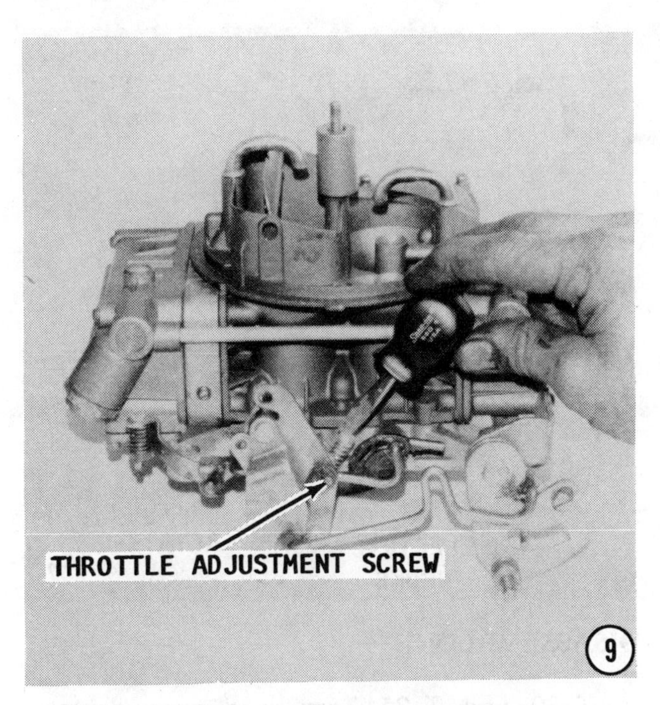

10- Idle Speed and Mixture Adjustment: Turn each idle speed needle in until it barely seats, and then back it out one full turn. **TAKE CARE** not to seat the needles tightly because it would groove the tip of the needle. If the needle should become grooved, a smooth idle cannot be obtained. Start the engine and allow it to warm to full operating temperature.

CAUTION: Water must circulate through the lower unit to the engine any time the engine is run to prevent damage to the water pump mounted on the engine.

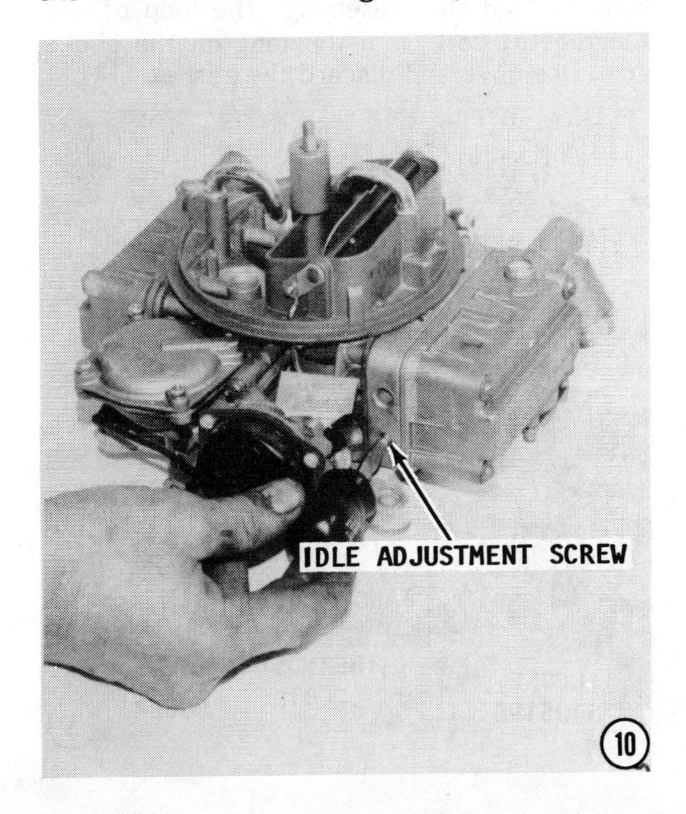

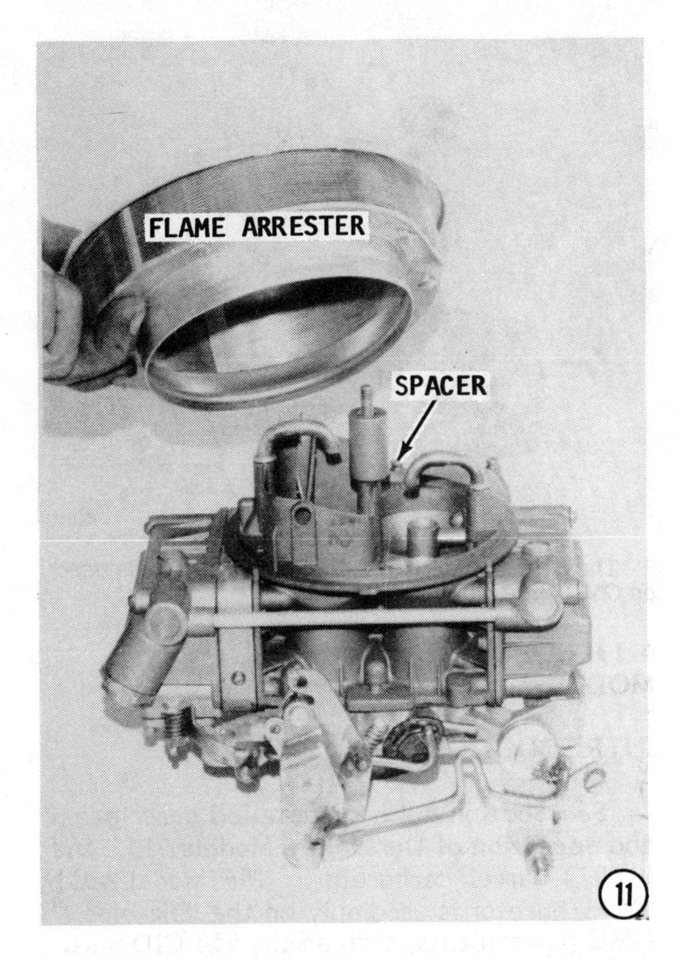

Adjust the idle-speed screw until the required rpm is reached as given in the Specifications in the Appendix. Now, set one of the two idle adjusting needles for the highest steady manifold vacuum reading. If a vacuum gauge is not available, turn the idle adjusting needle inward until the rpm begins to drop off, then back the needle out over the "highspot" until the rpm again drops off. Set the idle adjusting needle halfway between these two points as an idle speed setting. Repeat the procedure for the other needle. If these adjustments result in an increase in idle rpm, reset the idle speed screw to obtain the required idle rpm, and then adjust the idle adjusting needles again.

11- Position the sealing ring over the carburetor. The ring has three cutouts to make it fit easily. Install the spacer over the center screw, and then install the flame arrestor. **CHECK TO BE SURE** the flame arrestor is clean because a dirty one will consume too much fuel and cause the engine to run rough at high speeds.

The Holley Model 4011 4-barrel carburetor is used on GMC large block engines.

4-13 HOLLEY CARBURETOR MODEL 4011 4V

INTRODUCTION

See Section 4-12 for detailed description and operation of the Holley Models 4150 and 4160 4-barrel carburetor. The Model 4011 4V carburetor is used only on the "big block" GMC powerplants, such as the 454 CID unit.

The following procedures pickup the work after the flame arrestor, fuel lines, electric choke harness, and linkage has been disconnected and the carburetor removed from the manifold.

Use a clean suitable work surface, make an attempt to keep parts in a logical order. Be sure to separate parts removed from each side -- primary and secondary -- to ensure all components are returned to their original location.

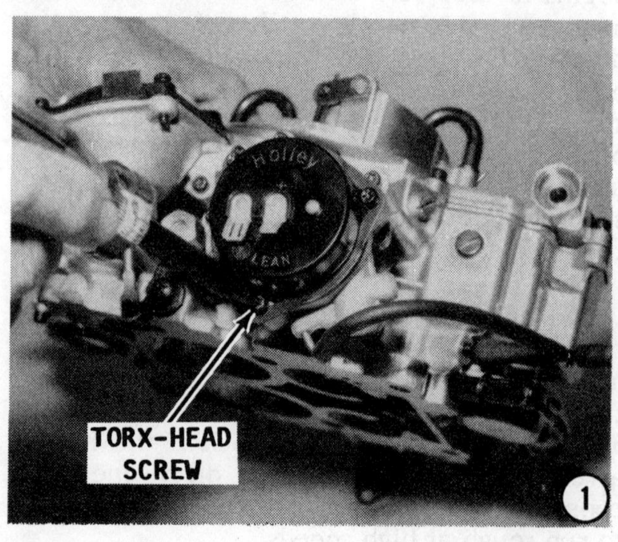

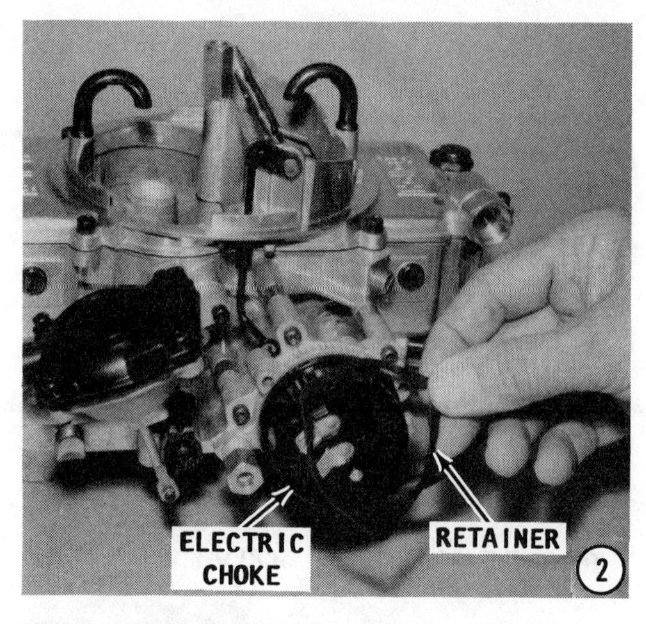

DISASSEMBLING

T-20 and T-25 Torx-head screwdrivers are absolutely necessary when working on this model carburetor.

DO NOT commence disassembling work on the carburetor until a rebuild kit including a full set of gaskets, diaphragms, O-rings and springs has been obtained and is at hand.

1- Remove the three Torx-head screws from the retainer around the electric choke.

2- Remove the retainer from the electric choke.

3- Pull the choke cover free of the housing, and then disengage the loop of the thermostat coil from the tang on the choke arm. Remove and discard the gasket.

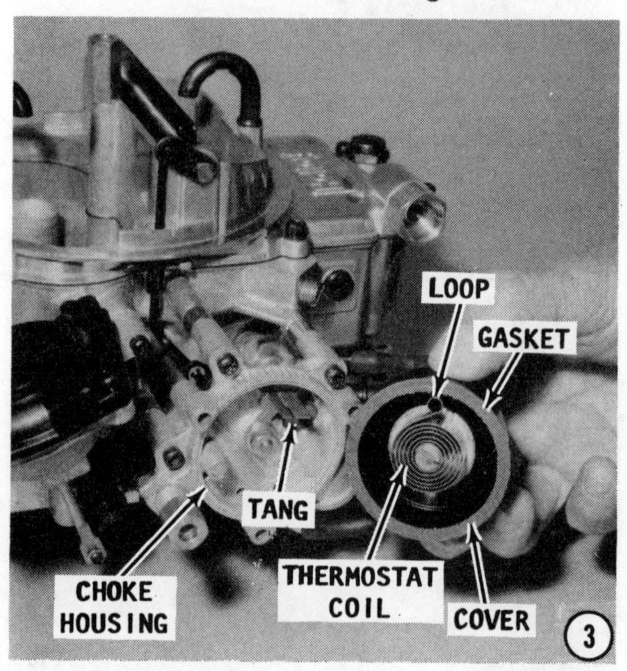

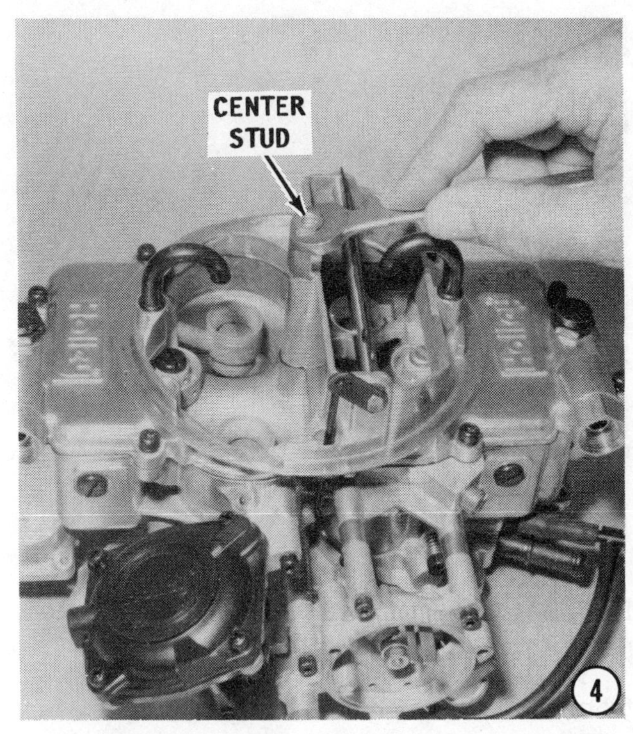

4- Remove the center stud from the carburetor body.

WARNING

The circlip to be removed in the next step is very small and under considerable tension. Therefore, use utmost caution and cover the circlip with one hand while removing the clip with a pair of needlenose pliers.

5- Using a pair of needlenose pliers, pry the circlip free from the lower end of choke link.

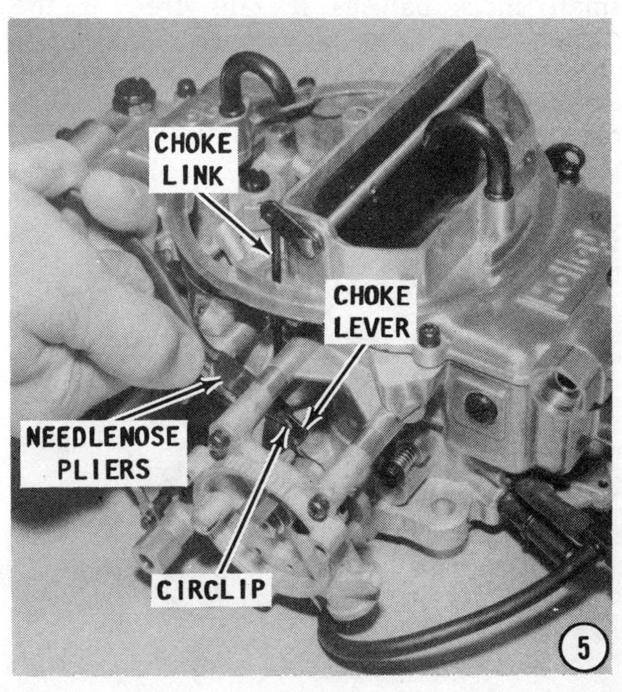

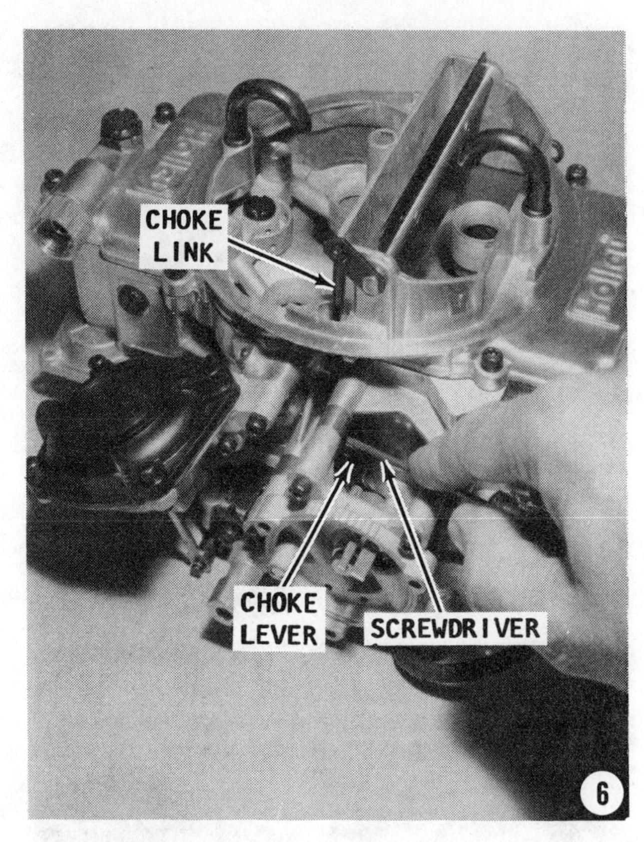

6- Insert a small screwdriver between the choke link and the choke lever, and then pry them apart.

7- Lift the airhorn assembly up and free of the carburetor body. The choke linkage and floats are attached and will come up with the airhorn. If necessary, rotate the free end of the choke link to allow the end to pass through the hole in the carburetor body.

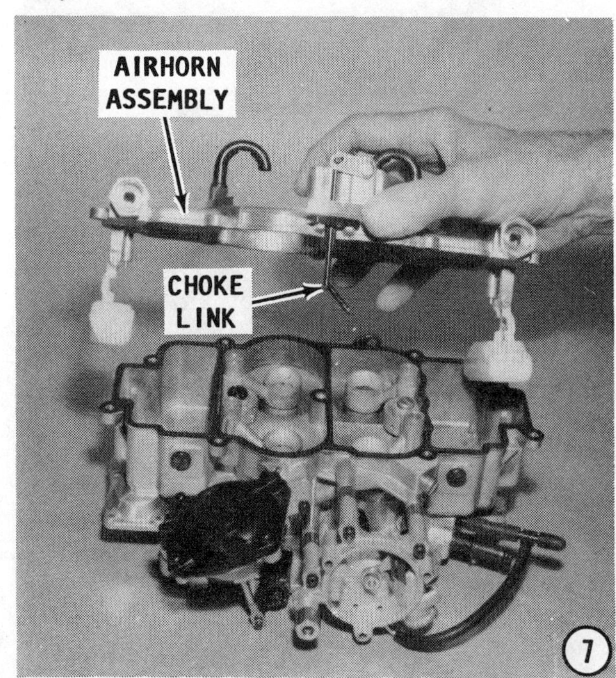

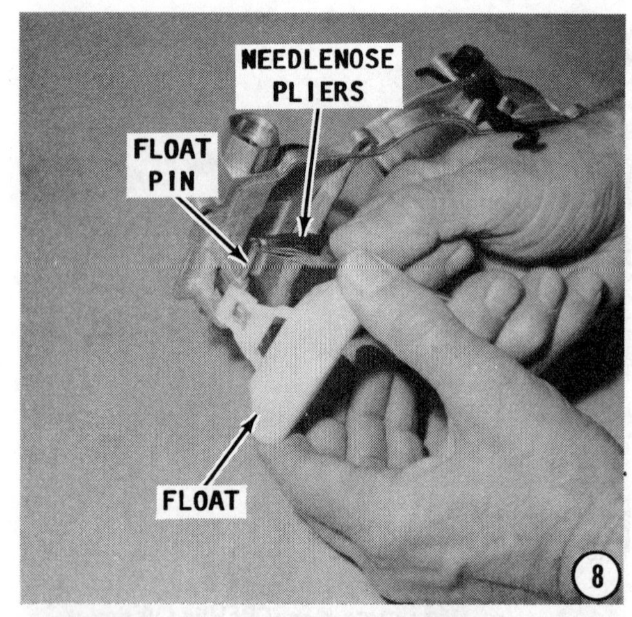

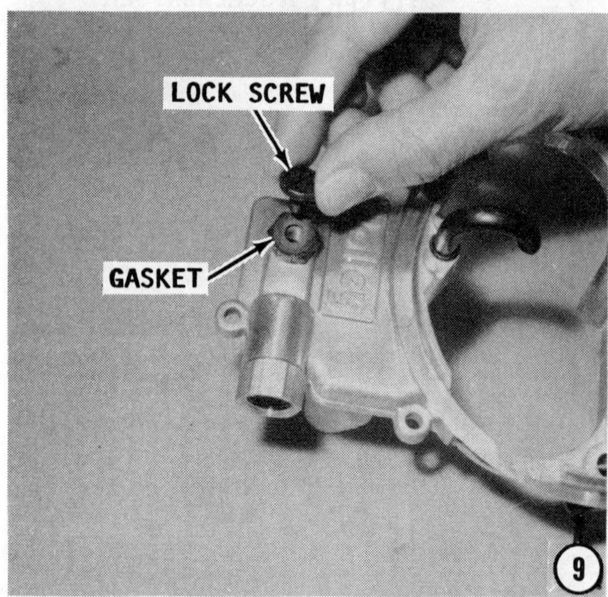

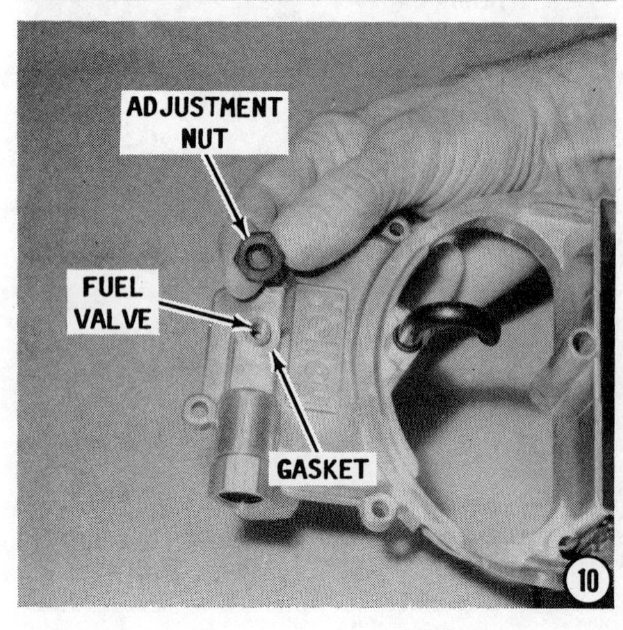

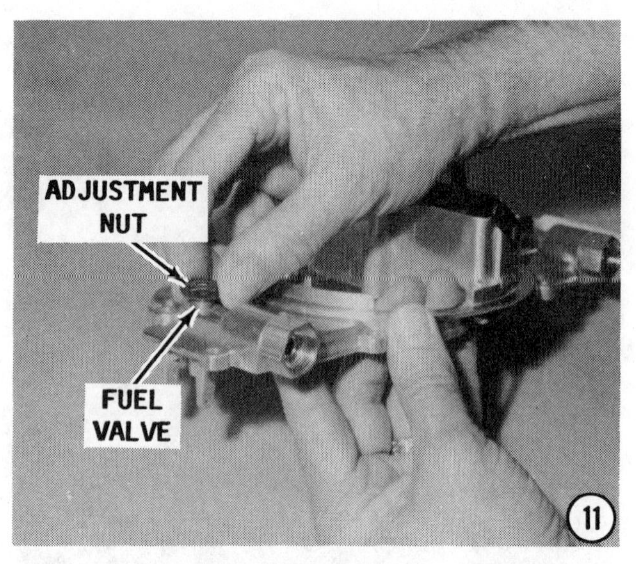

8- Push one of the float pins free of the float boss, and then remove the float. Remove the other float in the same manner.

9- Remove the lock screw and gasket from each side of the airhorn.

10- Remove the adjustment nut and gasket from each side of the airhorn.

11- Using the adjustment nut as a tool, unscrew both fuel valves from the airhorn.

12- Lift off and discard the carburetor main body gasket.

13- Remove the banjo bolt from the primary cluster.

14- Lift out the primary cluster. Remove and discard the gasket.

15- Lift out the weight rod from the center hole on the primary side. Turn the carburetor upside down and **CATCH** the very small check ball as it falls free of the center hole. Use the proper size screwdriver and remove the seat valve for the accelerator pump fuel channel, located under the check ball which was just removed.

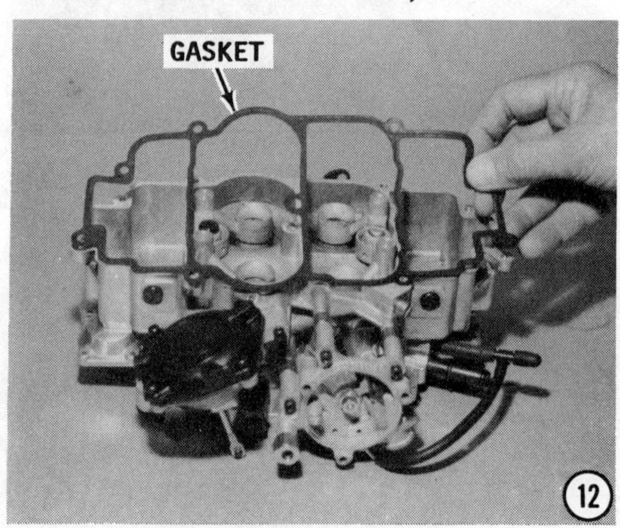

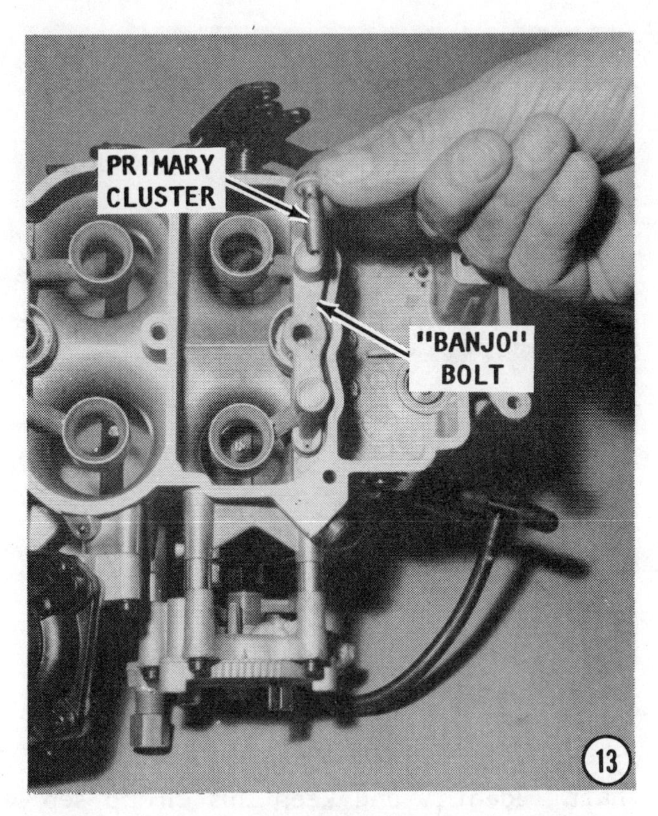

16- Remove the bolt from the secondary cluster. Lift out the cluster. Remove and discard the gasket. The secondary cluster **DOES NOT** have a weight rod and check ball.

17- Using the proper size slotted screwdriver, remove a total of four main jets

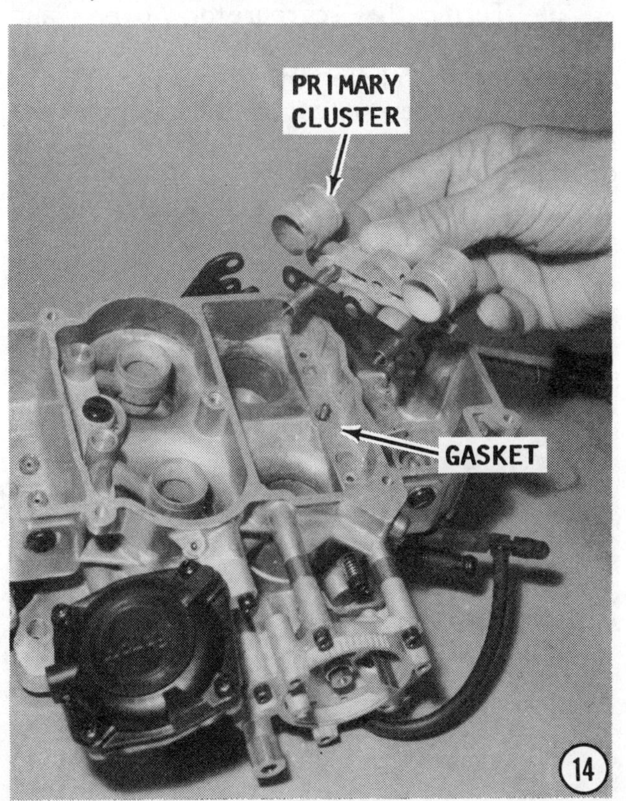

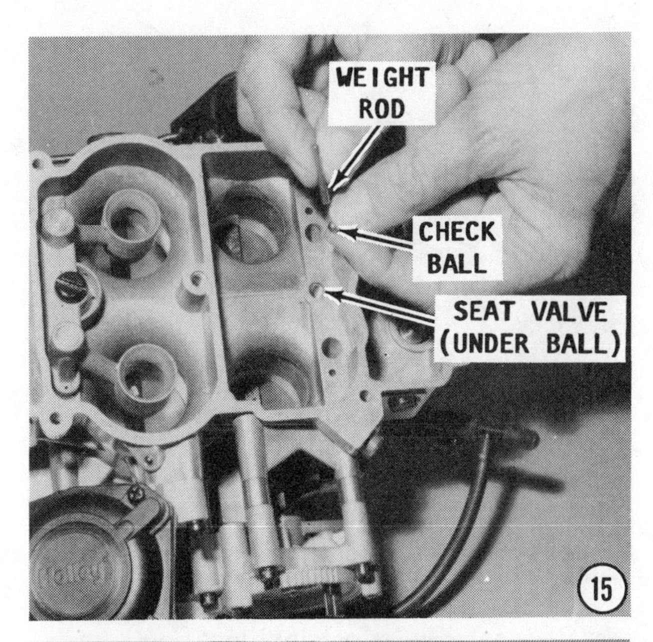

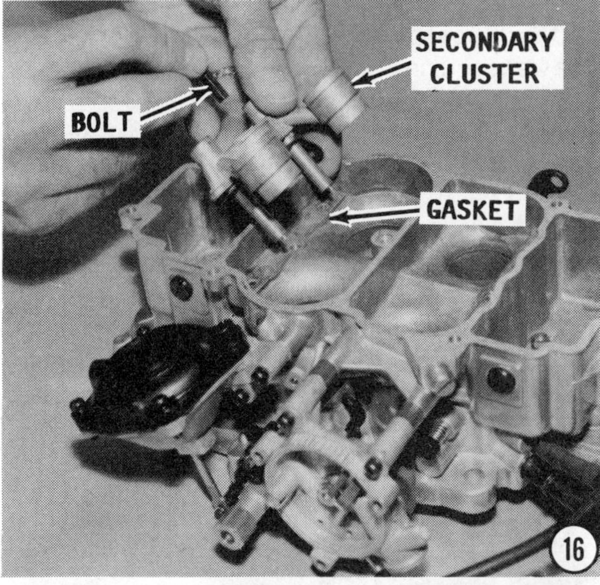

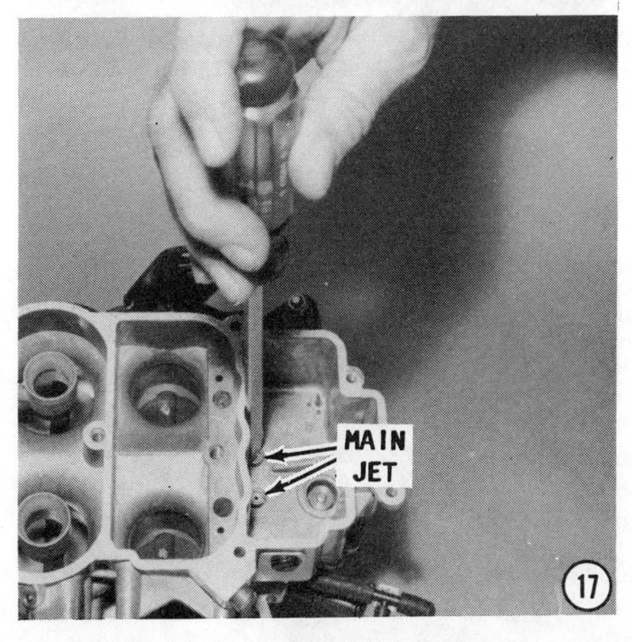

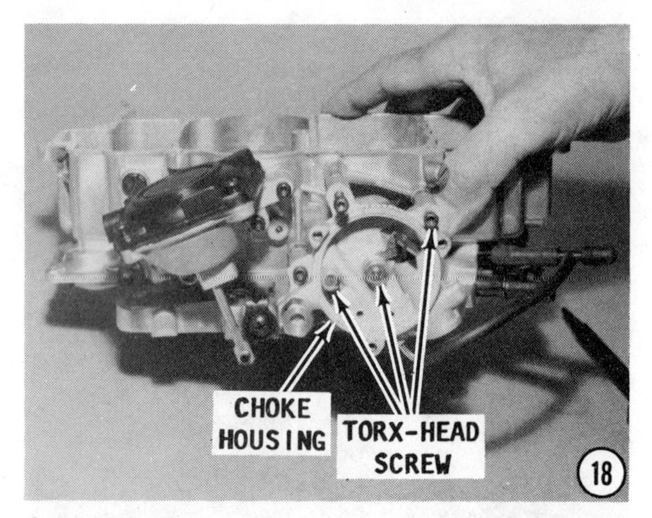

from the primary and secondary fuel bowls. Take **EXTRA** precaution to identify and keep the jets separate to ensure each is installed back in its original location.

18- Loosen the three long Torx-head screws securing the choke housing to the carburetor body.

19- Remove each of the Torx-head screws and be sure to **SAVE** the small spacer used on each bolt, as shown.

20- Disconnect the vacuum hose from the fitting under the primary fuel bowl. Set the choke housing with vacuum hose attached, to one side.

WARNING

The circlip to be removed in the next step is very small and under considerable tension. Therefore, use utmost caution and cover the circlip with one hand while removing the clip with a pair of needlenose pliers.

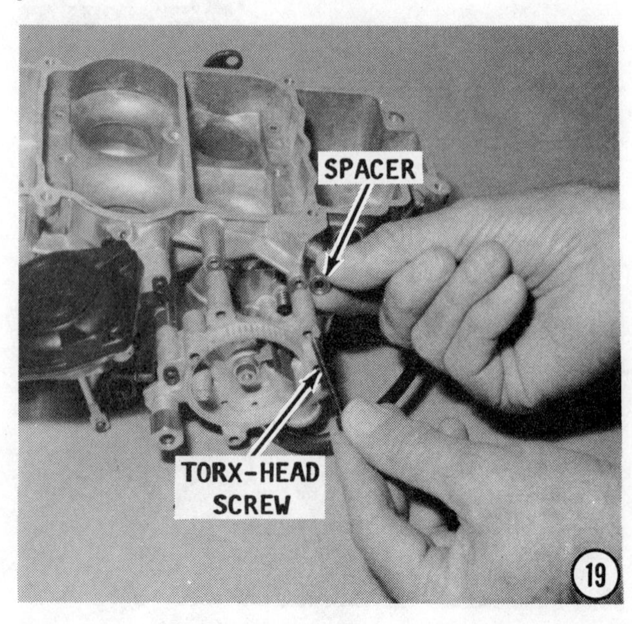

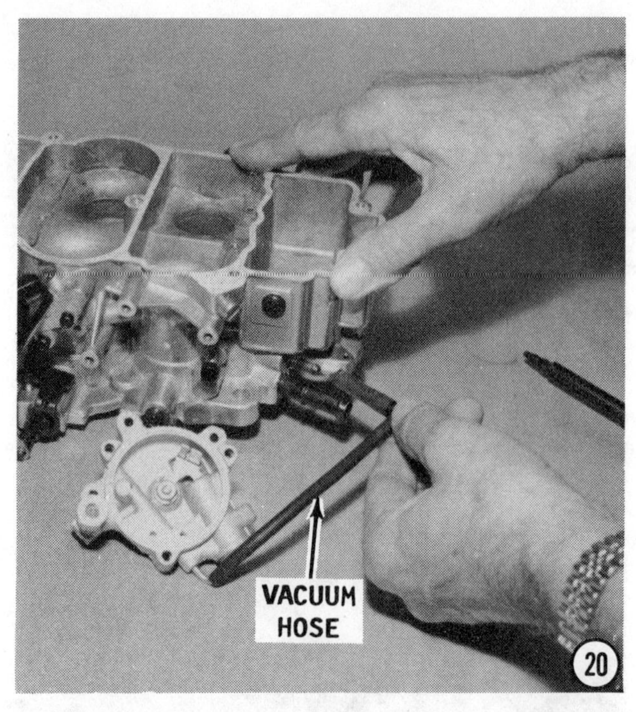

21- Using a small screwdriver, pry the circlip free of the secondary diaphragm shaft. Identify and keep this circlip separate from the clip removed earlier in Step 5.

22- Remove the four Torx-head screws securing the cover to the diaphragm housing. Lift off the cover and spring.

23- Lift out the secondary diaphragm from the housing.

24- Turn the carburetor over and **CATCH** the small check valve ball as it falls free of the fifth hole in the housing.

25- Remove one of the idle adjustment needles, and spring as shown, from the side of the primary fuel bowl.

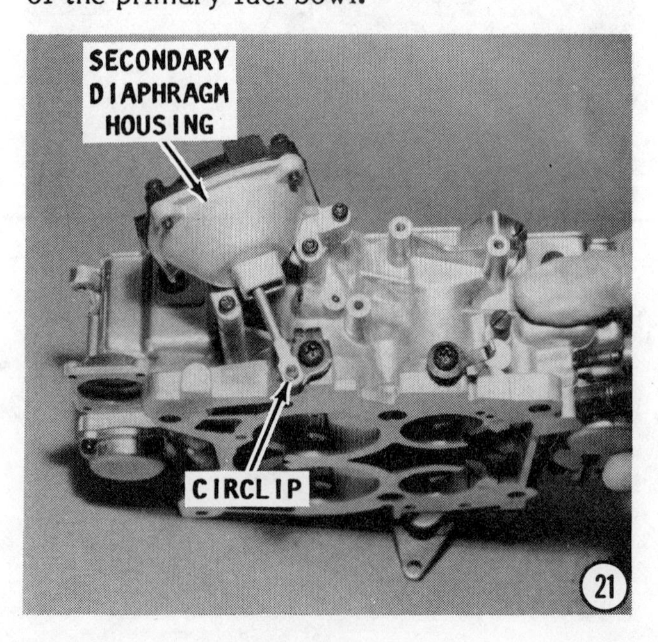

26- Remove the other idle adjustment needle and spring from the other side of the primary fuel bowl.

SPECIAL WORDS

The primary and secondary power valves are identical. Each is removed in the same manner as described in the next three steps.

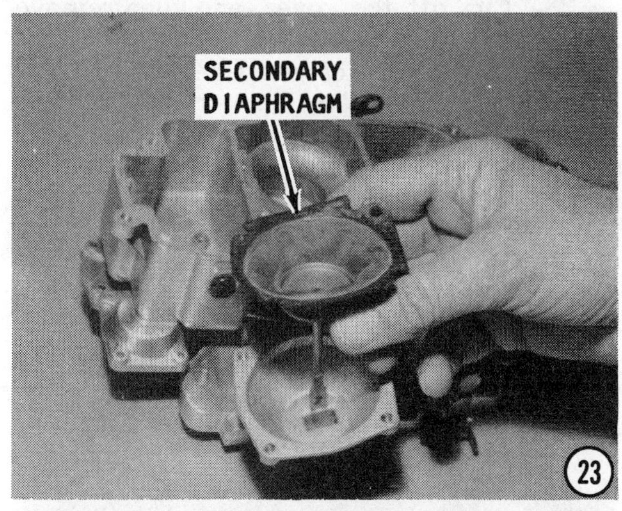

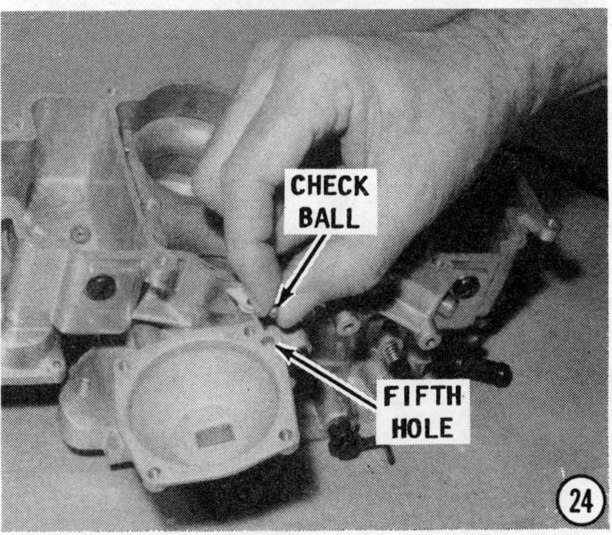

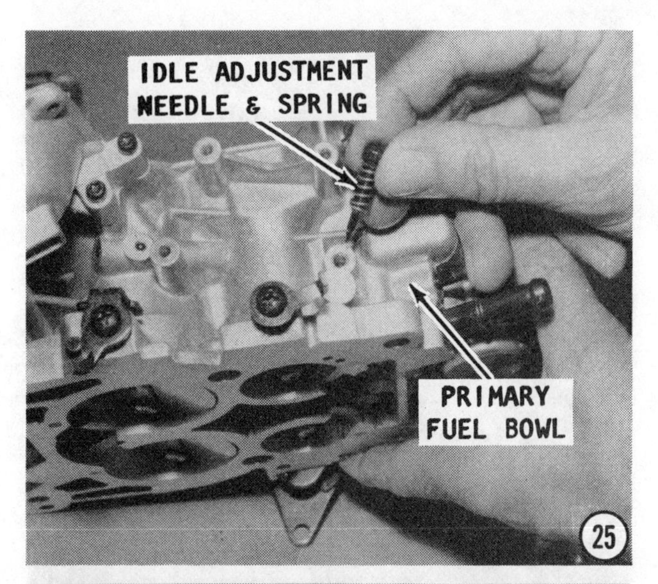

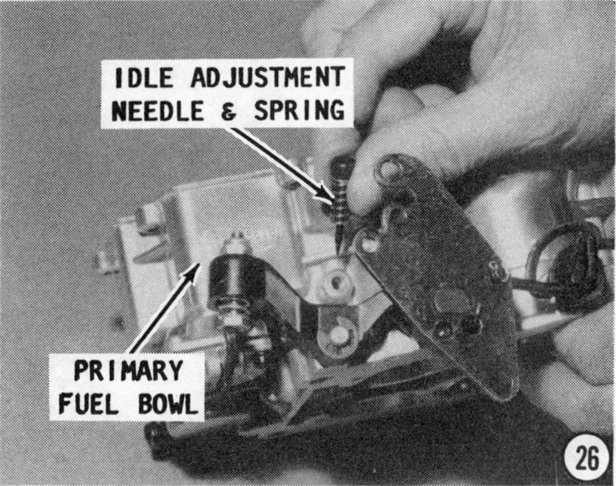

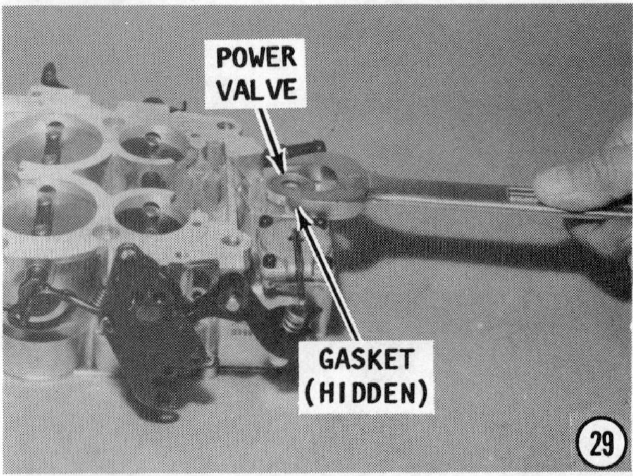

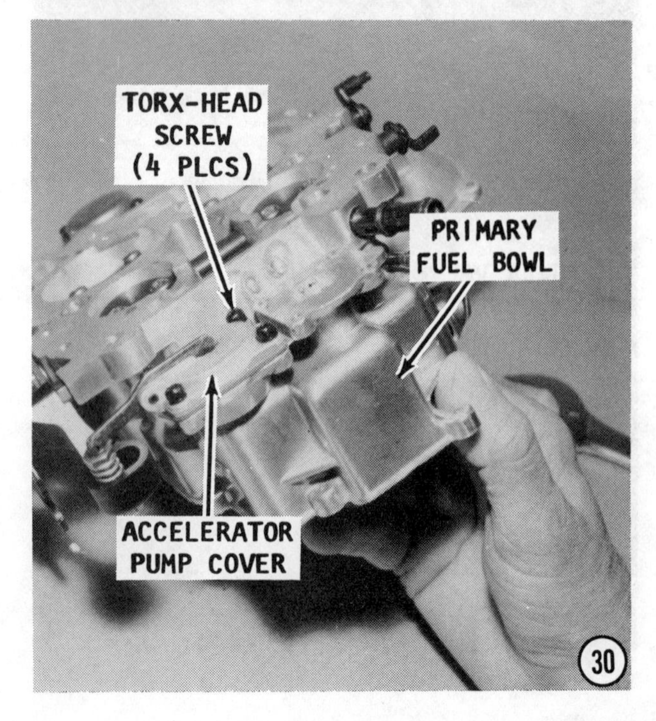

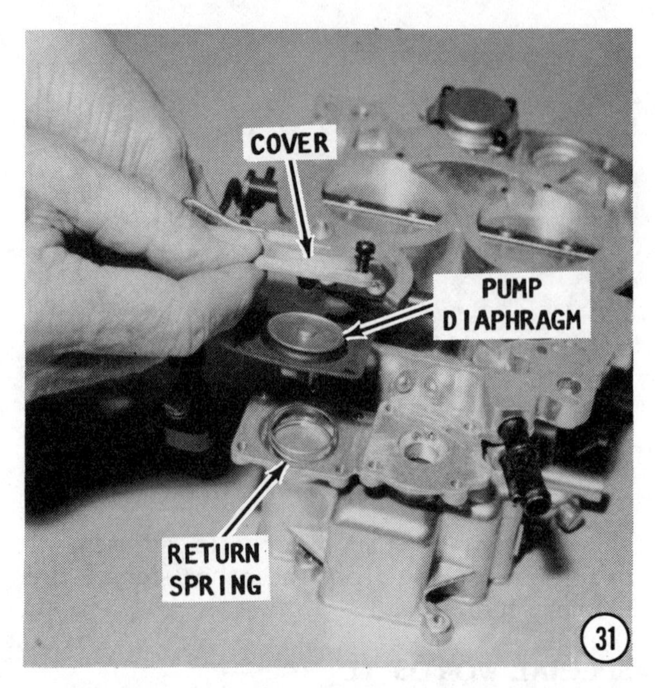

27- Remove the four Torx-head screws securing the power valve cover to the carburetor body.

28- Lift off the cover, and then remove and discard the gasket.

29- Using the correct size wrench, unscrew the power valve from the body. Remove and discard the gasket. The power valve **CANNOT** be disassembled even though diassembly appears possible. The valve is replaced as a complete unit.

30- Remove the four Torx-head screws securing the accelerator pump cover to the base of the primary fuel bowl.

31- Lift off the cover. Withdraw the pump diaphragm and return spring.

32- Using a small screwdriver, lift up an edge of the check valve diaphragm, and then

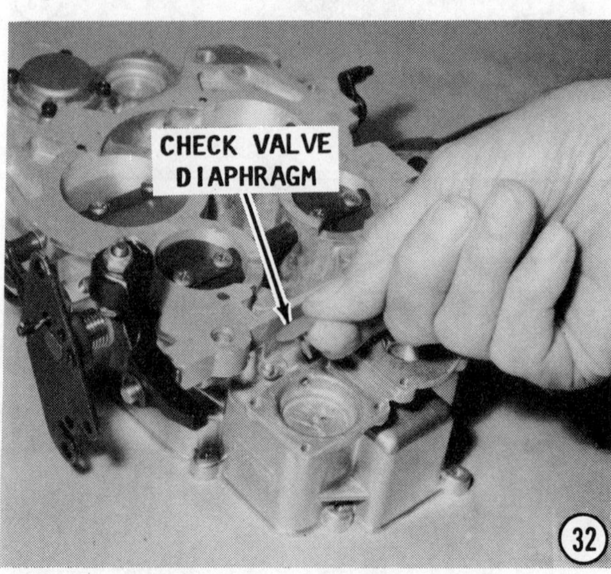

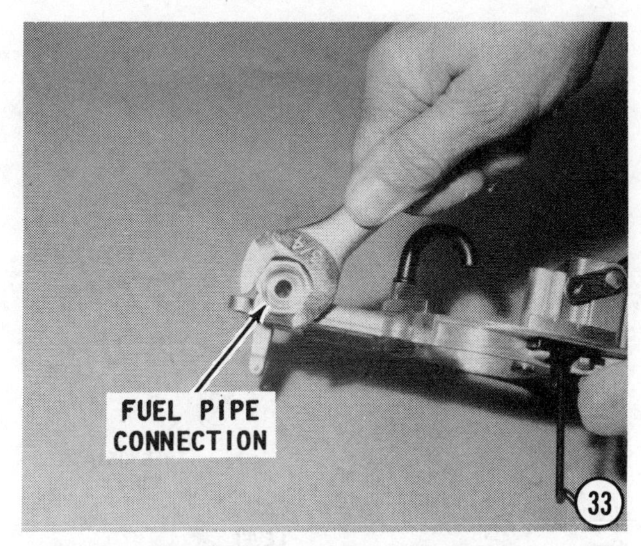

pull the diaphragm out by hand. Handle with **CARE**, because the diaphragm can be easily damaged.

33- Using a 3/4" open end wrench, remove the fuel pipe connections on the airhorn.

34- Pull out the gasket and filter screen from the primary inlet and the secondary inlet on the airhorn.

CLEANING AND INSPECTING

NEVER dip rubber parts, plastic parts, or diaphragms in carburetor cleaner. Place all of the metal parts in a screen-type tray and dip them in carburetor solvent until they appear completely clean, then blow them dry with compressed air.

Blow out all of the passages in the castings with compressed air. Check all of the parts and passages to be sure they are not clogged or contain any deposits. **NEVER**

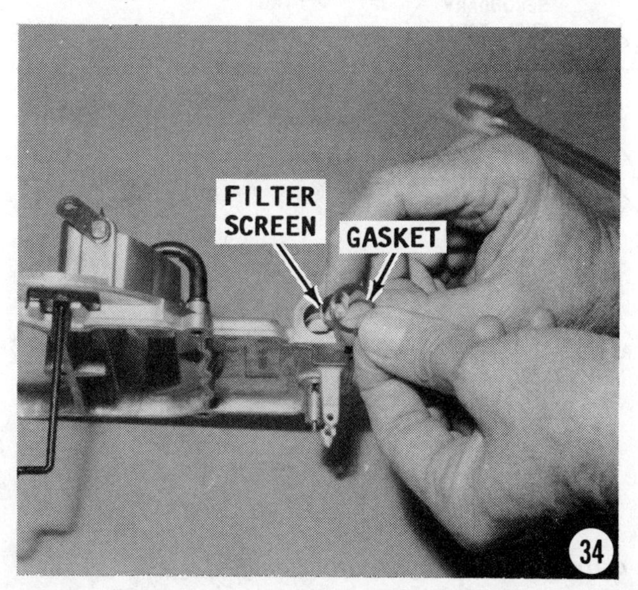

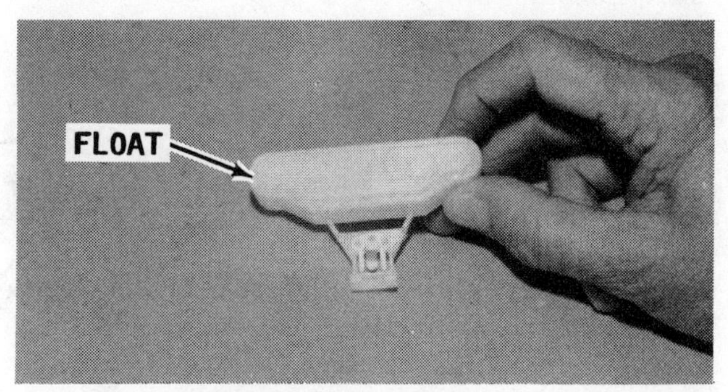

Check the float for any evidence of leakage by shaking it and listening for the sound of fluid movement inside.

use a piece of wire or any type of pointed instrument to clean drilled passages or calibrated holes in a carburetor.

Inspect the choke and throttle shafts for excessive wear and replace the complete carburetor if the throttle shaft is worn.

Check the floats for leaking by shaking them and listening for fluid movement inside. If a float contains fluid, it must be replaced. Inspect the seat valve contact surface and the float shaft and replace them if either has any grooves worn in it.

ALWAYS replace the valve seat assemblies. These parts receive the most wear in the carburetor and the proper fuel level cannot be maintained if they are worn.

Check the secondary vacuum diaphragm by first depressing the diaphragm stem, and then placing a finger over the vacuum fitting to seal the opening - the diaphragm housing must be removed from the carburetor to perform this test and completely assembled. Now, release the diaphragm stem. If the stem moves out more than 1/16" in ten seconds, the diaphragm has an internal leak and it should be replaced.

Most of the parts that should be replaced during a carburetor overhaul, including the latest updated parts are found in a carburetor kit available at your local marine dealer. In addition to the parts, most kits include the latest specifications which are so important when making adjustments.

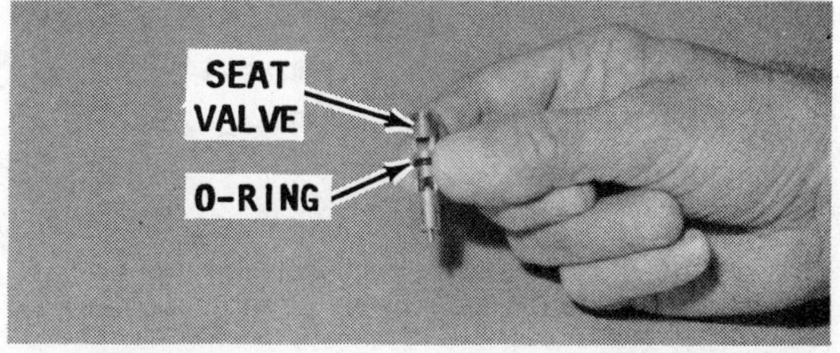

Proper fuel level in the fuel bowls cannot be maintained if the valve seat assemblies are worn.

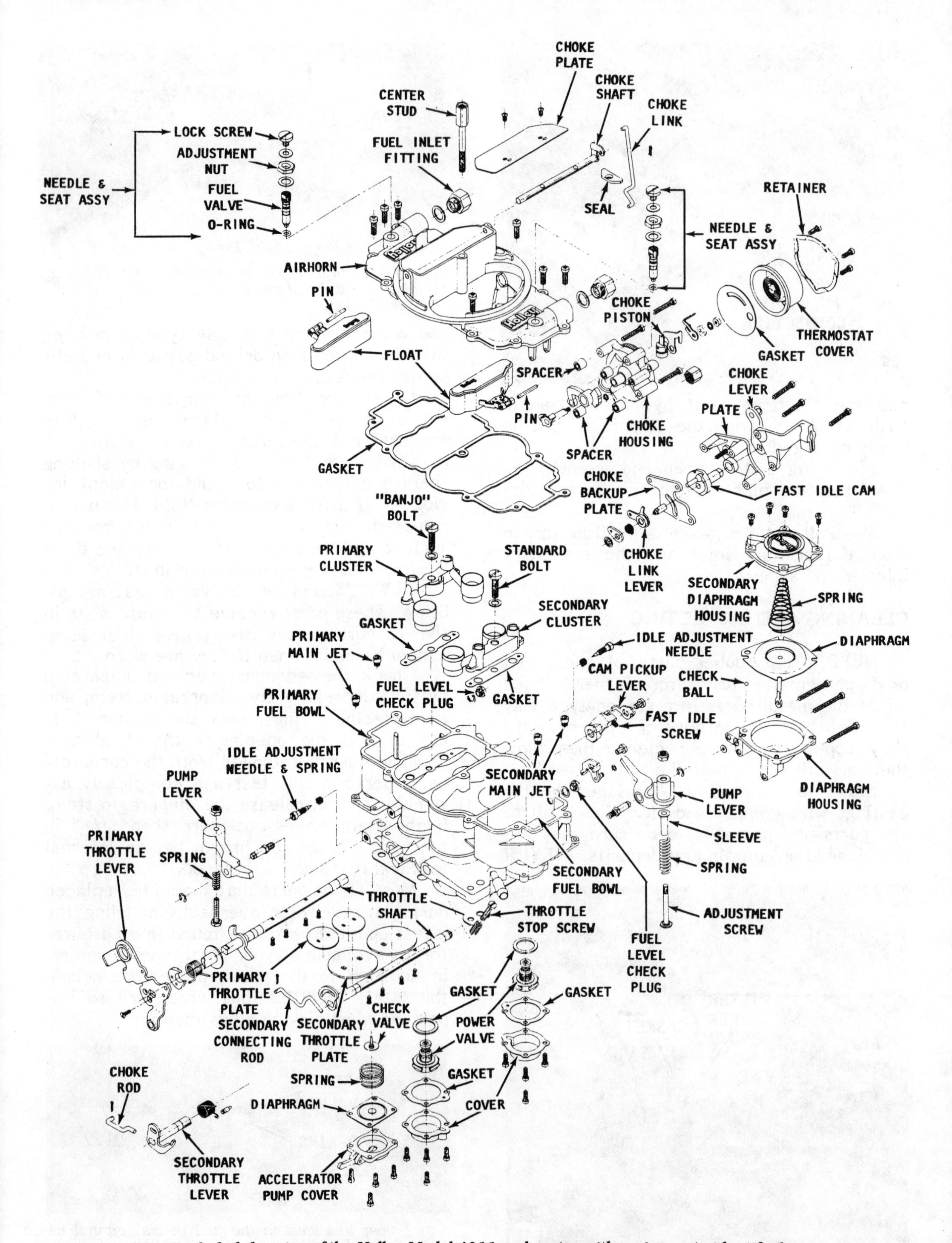

Exploded drawing of the Holley Model 4011 carburetor with major parts identified.

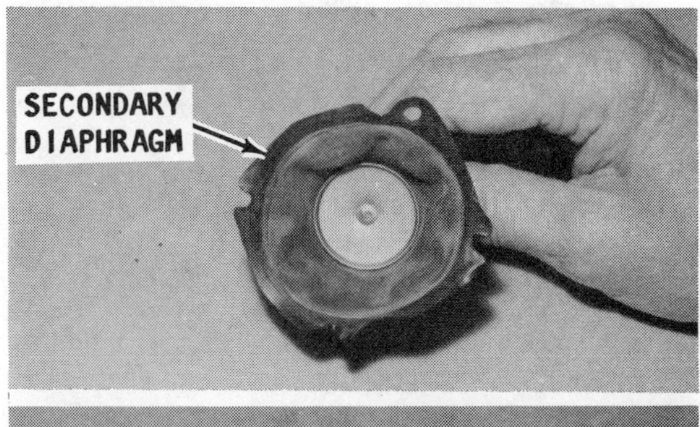

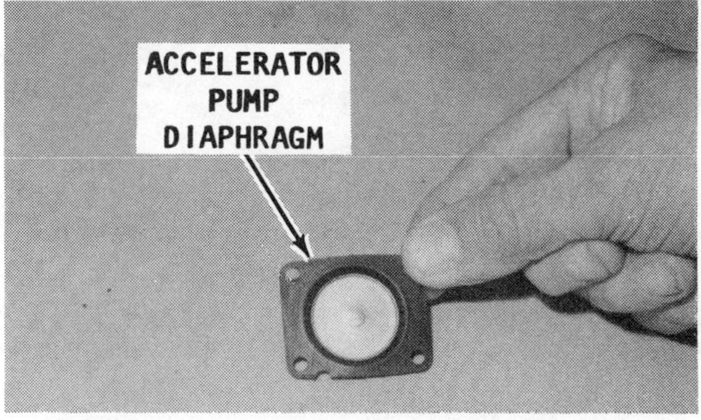

*Check the secondary diaphragm (above) and the accelerator pump diaphragm (below), for pin holes or tears. A damaged diaphragm **MUST** be replaced.*

ASSEMBLING HOLLEY 4011

1- Slide the primary filter screen into the airhorn opening. Position a **NEW** gasket in place over the opening. Install the secondary filter screen and gasket.

2- Install the two fuel pipe connections and tighten them securely.

3- Insert the end of the check valve diaphragm into the primary fuel bowl.

4- Hold the diaphragm in place with a finger and at the same time turn the carburetor body over. Now, using a pair of needlenose pliers, pull the protruding end of

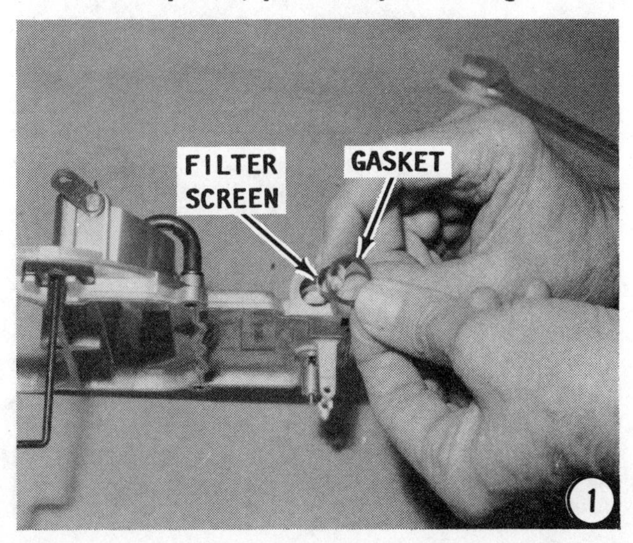

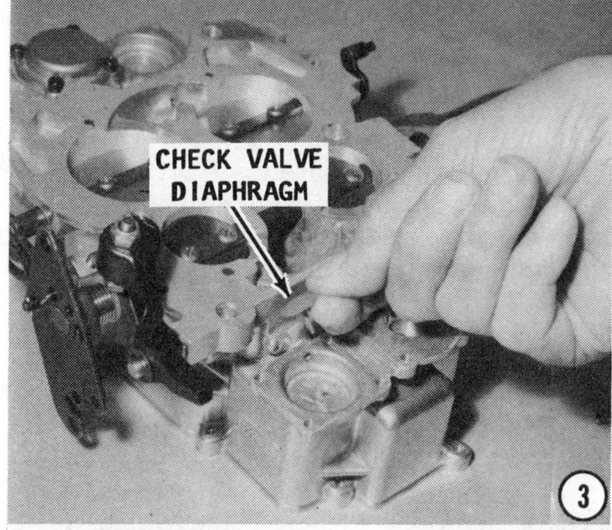

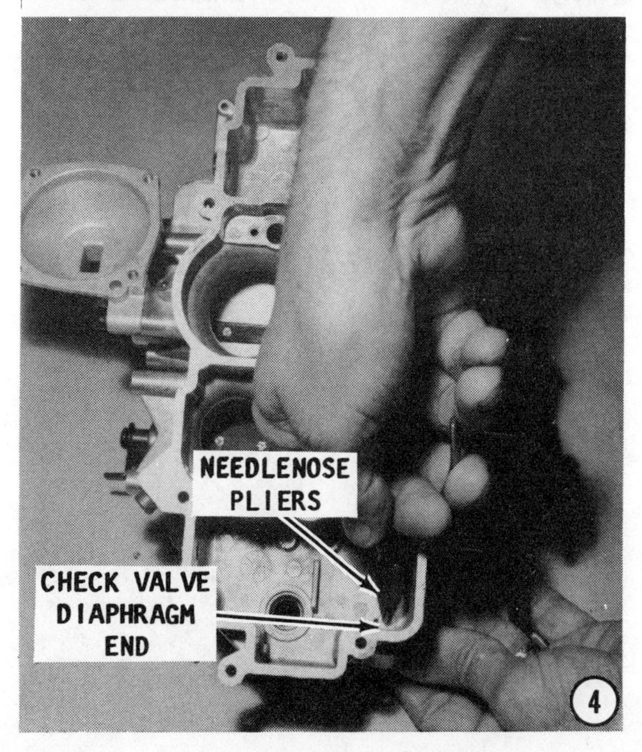

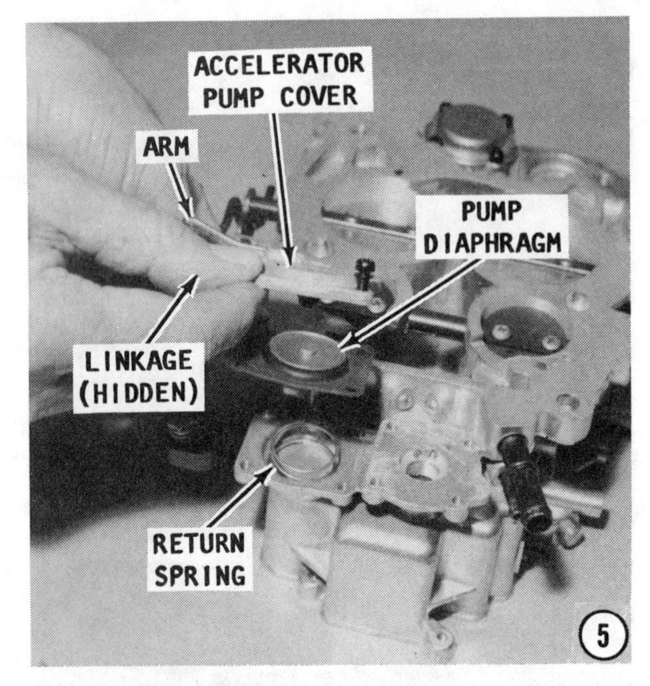

the check valve diaphragm up until the diaphragm snaps into place.

5- Position the return spring in place over the diaphragm. Place the pump diaphragm over the spring with the tab on the diaphragm facing **DOWN.** Install the accelerator pump cover over the pump diaphragm with the pump arm extending towards the linkage, as shown.

6- Secure the cover in place with the four Torx-head screws.

7- Install one of the power valves, using a **NEW** gasket. Tighten the valve to a torque value of 8 ft lbs (11 Nm). Install the other power valve in the same manner.

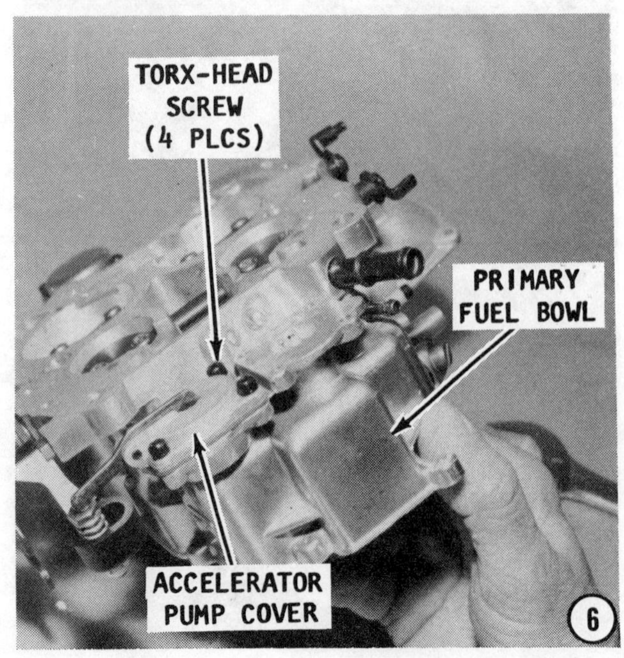

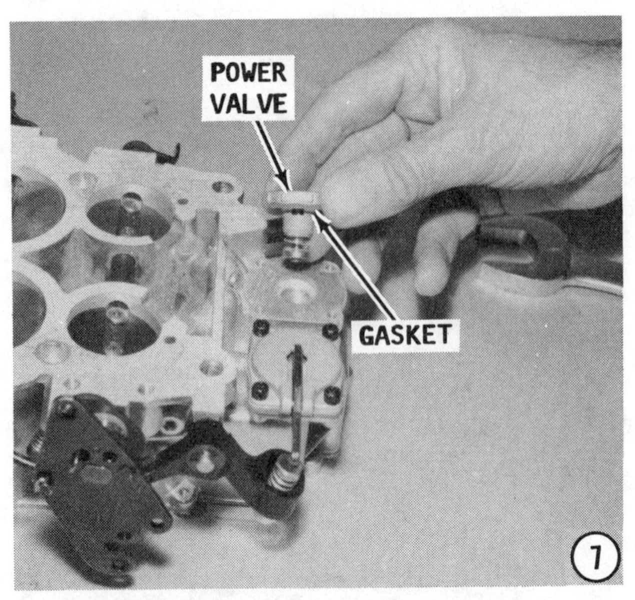

8- Install the power valve covers using **NEW** gaskets. Tighten the Torx-head screws alternately and securely.

9- Slide a spring onto one of the idle adjustment needles and then thread the needle into the side of the primary fuel bowl

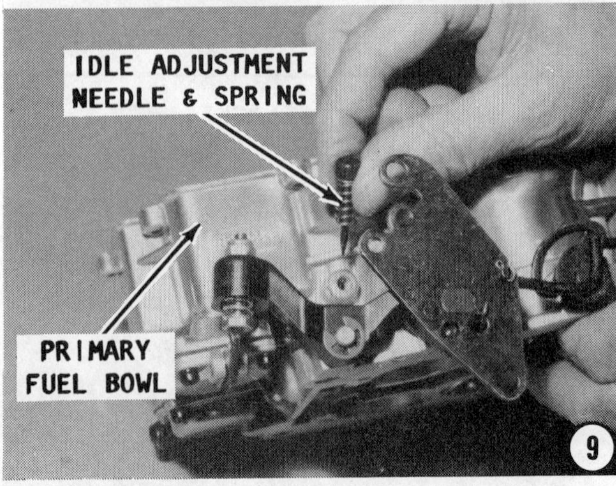

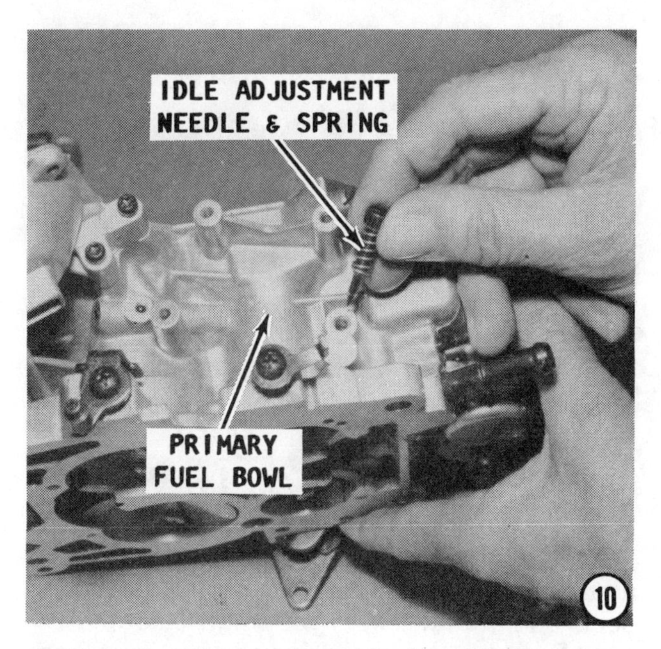

until it barely seats, and then back it out 3/4 turn (counterclockwise) as a preliminary adjustment at this time.

10- Slide a spring onto the other idle adjustment needle; thread it into the other side of the primary fuel bowl until it barely seats; and then back it out 3/4 turn.

11- Drop the check ball into the fifth hole of the secondary diaphragm housing.

12- Insert the secondary diaphragm into the housing. Check to be **SURE** the hole in the diaphragm aligns with the fifth hole in the housing.

13- Position the spring over the diaphragm with the taper facing **UP.** Install the diaphragm cover over the spring with the hole next to one of the cover shoulders aligned with the fifth hole in the housing, as shown in the accompanying illustration. Only one shoulder has a hole next to it.

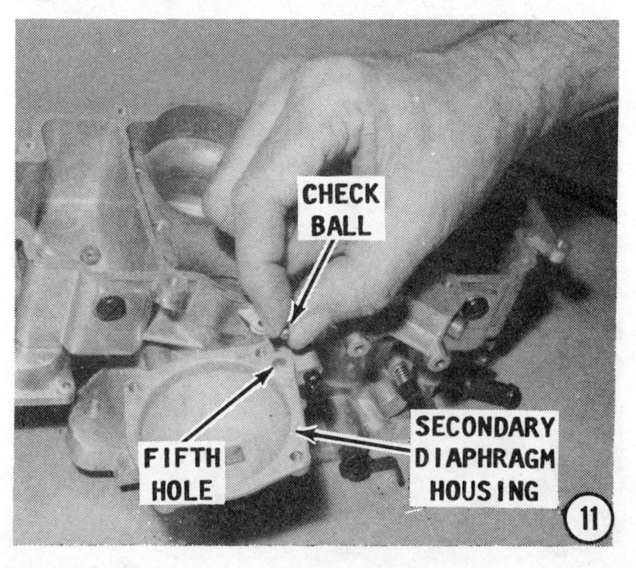

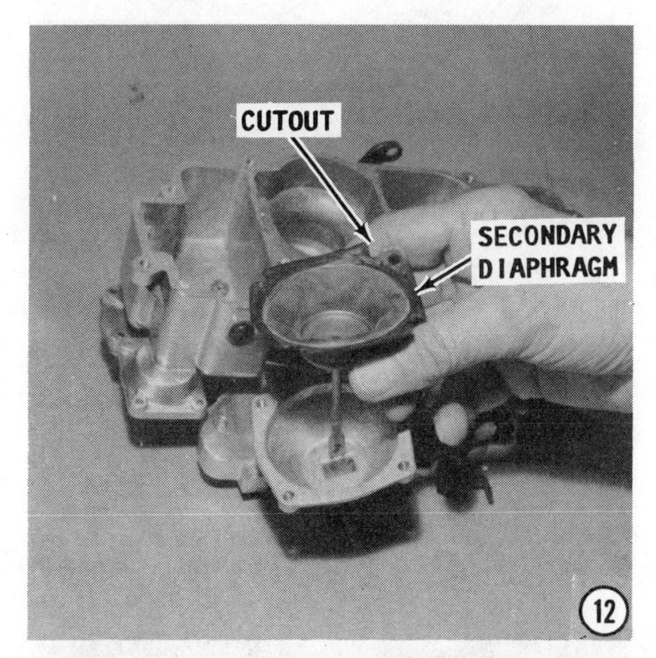

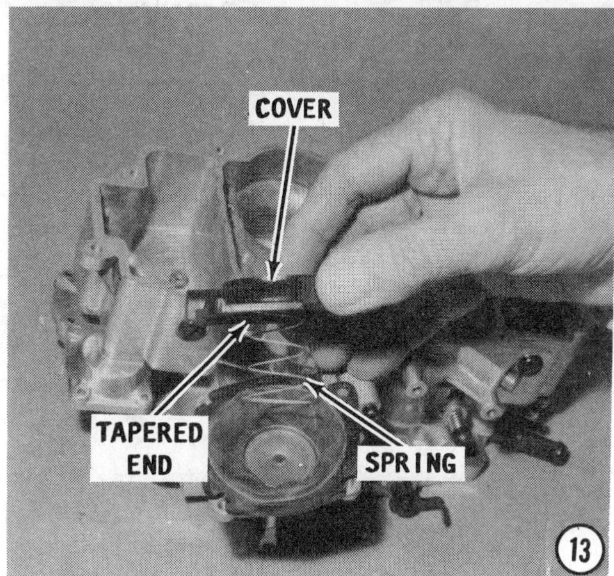

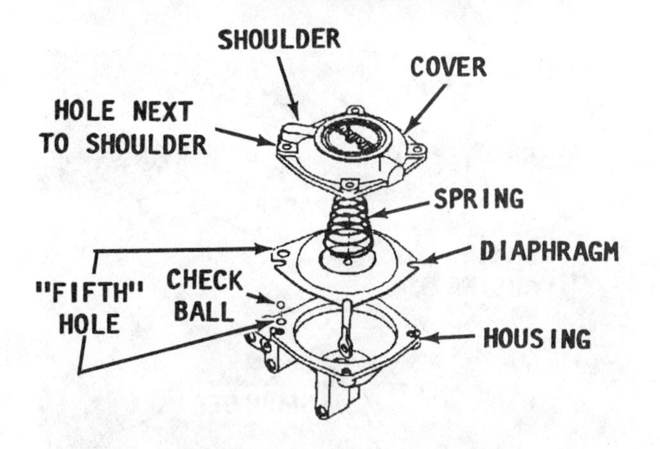

Exploded drawing of the secondary diaphragm housing with major parts including the "fifth" hole identified.

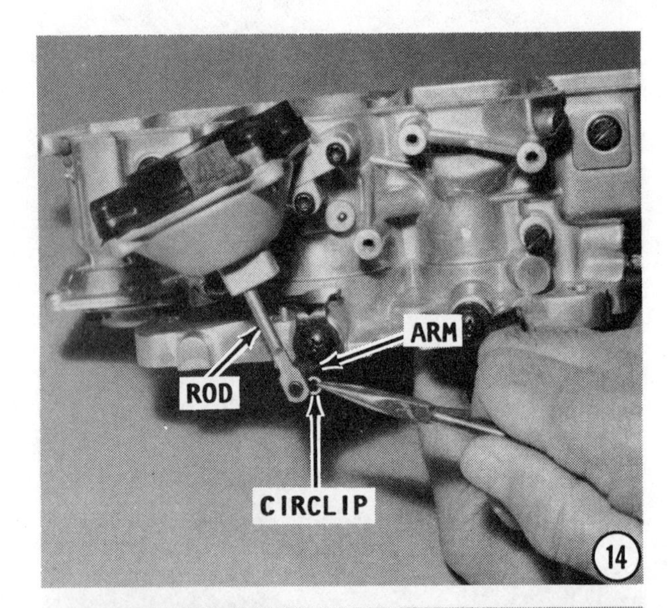

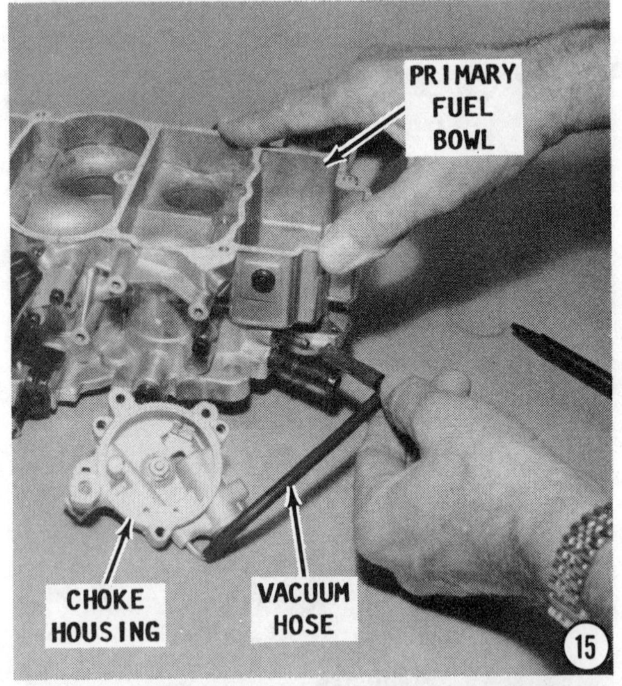

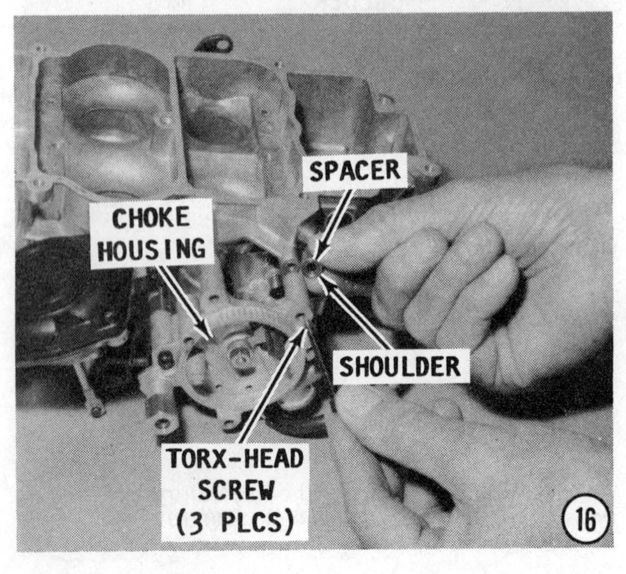

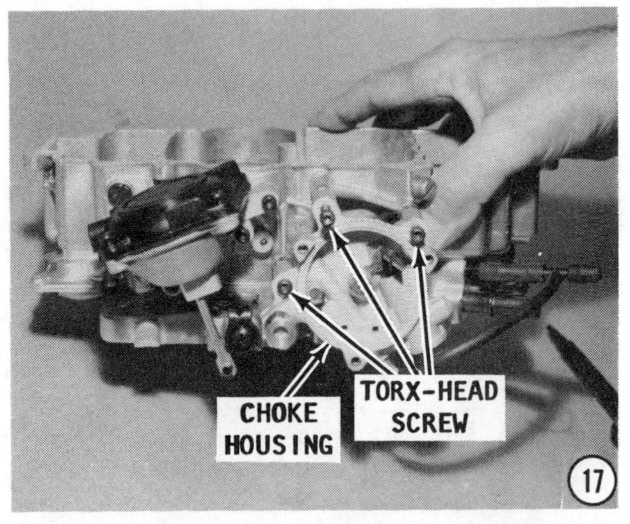

14- Guide the diaphragm rod over the arm on the secondary throttle shaft and secure it in place with the small circlip.

15- Push the choke housing vacuum hose onto the fitting under the primary fuel bowl.

16- Lightly secure the choke housing to the carburetor body with the three long Torx-head screws. A spacer is used at each mounting boss and installed with the shoulder on the spacer facing the choke housing.

17- Check behind the choke housing to be sure the choke actuating rod is above the fast idle cam. Tighten the three Torx-head screws securely.

18- Install the primary and secondary main jets into their original locations. Standard primary jets are embossed with the number **"64"** and standard secondary jets are embossed with the number **"68"**.

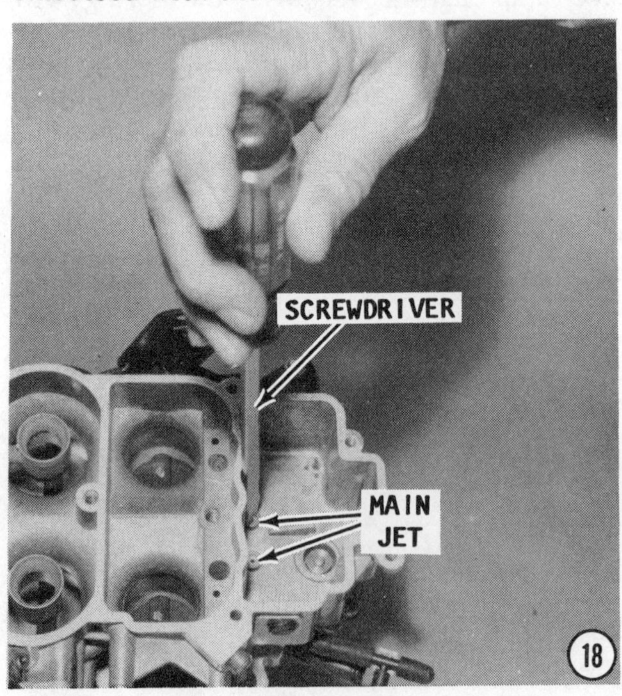

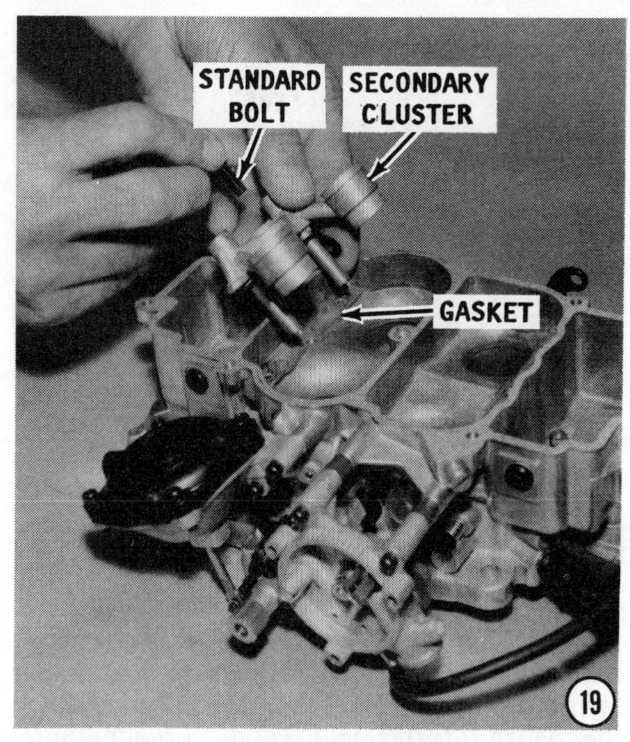

19- Position a **NEW** gasket into the secondary fuel bowl. Lower the secondary cluster into place. The secondary cluster usually has no "shooter" holes in the chamfered edge of the hole for the bolt. The primary cluster will have "shooter" holes in this location. Install and tighten the securing bolt. This bolt is normally black and solid to identify it from the primary bolt which is usually silver and of the "banjo" type - hollow, with holes at the top to allow fuel to pass through the bolt. No check ball and weight rod is used on the secondary side under the cluster.

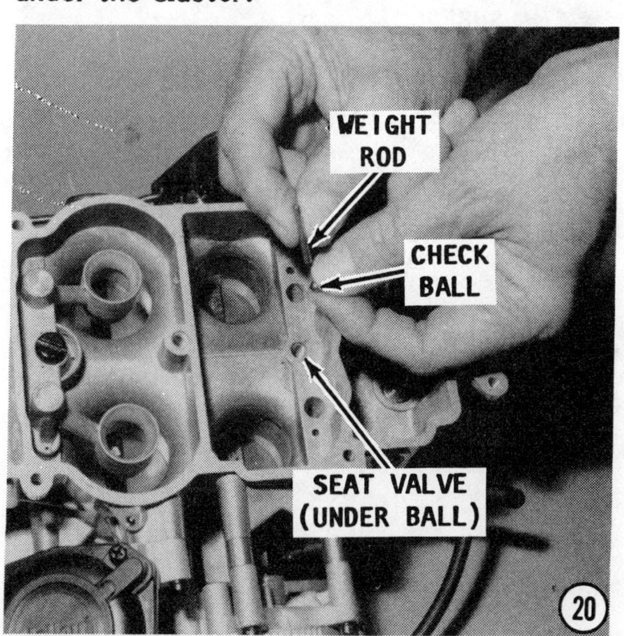

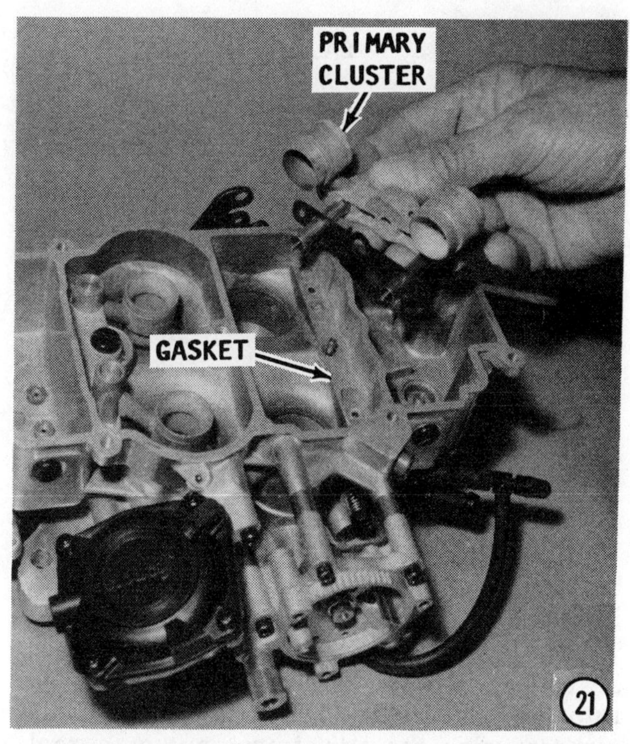

20- Use the proper size screwdriver and install the seat valve for the accelerator pump fuel channel, located at the base of the threaded hole in the center of the primary fuel bowl. Drop the check ball followed by the weight rod down the banjo bolt hole in the primary fuel bowl.

21- Position a **NEW** gasket in the primary fuel bowl. Lower the primary cluster into place over the gasket.

22- Install and tighten the silver banjo bolt securing the primary cluster to the carburetor.

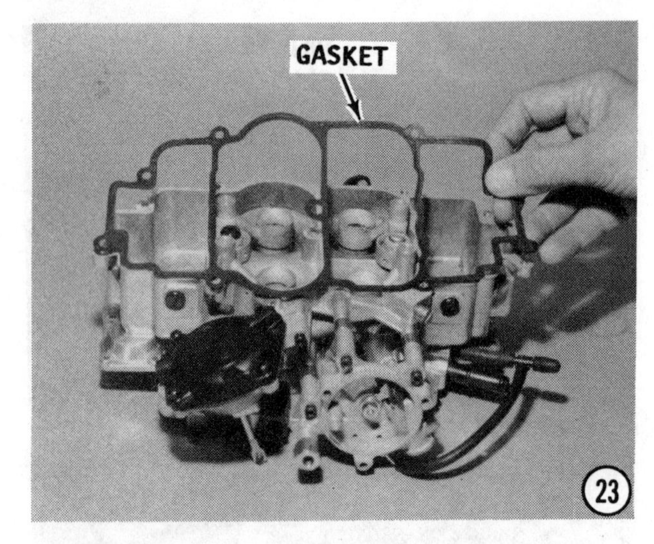

23- Install a **NEW** gasket onto the carburetor body. This gasket can only be installed one way.

24- Lubricate the **NEW O**-ring on each seat valve with engine oil. Install the valves into the primary and secondary sides of the airhorn using the adjustment nut as a tool. The precise depth to which each valve is threaded into the airhorn can only be determined during engine operation. However, a preliminary adjustment is covered in Carburetor Adjustments, Step 2 following the assembling procedures. The seat valve must be temporarily installed to a depth enabling the float - installed in the following step -to hang freely and parallel to the surface of the airhorn.

25- Position one of the floats between the airhorn bosses. Using a pair of needle-nose pliers insert the float pin to secure the float between the two mounting bosses. Center the pin. Install the other float in the same manner.

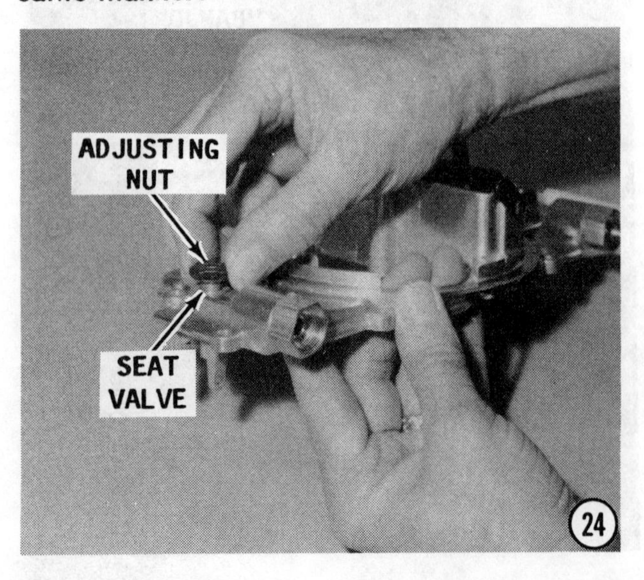

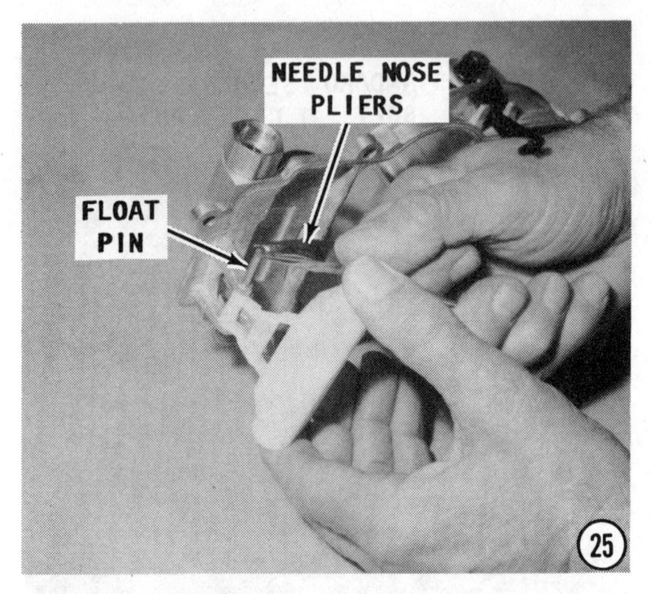

26- Invert the airhorn and allow both floats to hang freely. The upper surface of each float should be parallel to the horizontal surface of the airhorn, as shown. Adjust the depth of the seat valve threaded into the airhorn to obtain the proper alignment for each float.

27- Place **NEW** gaskets above and below the adjustment nut. The larger gasket is

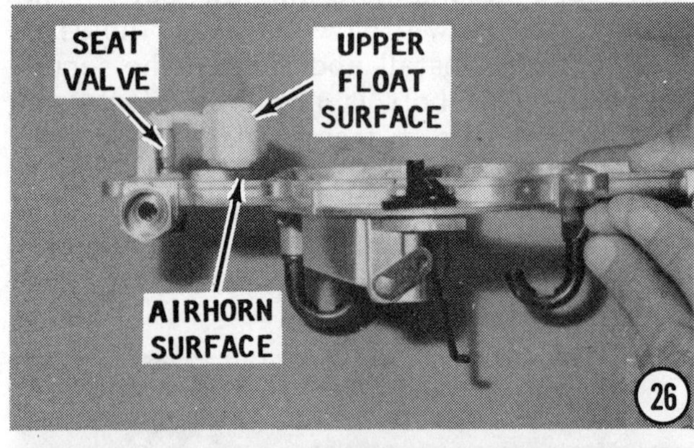

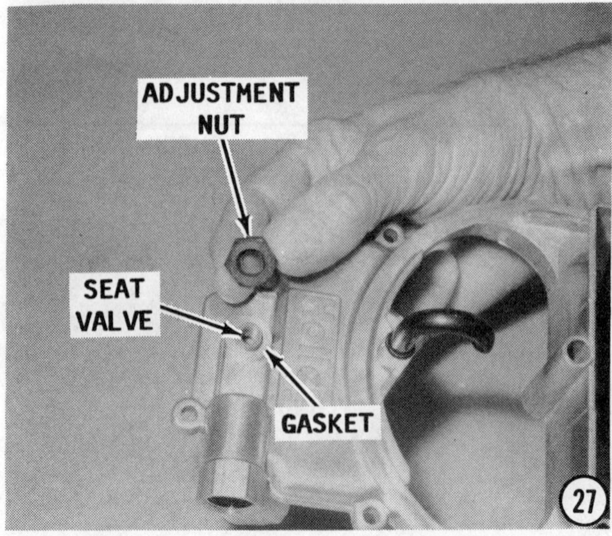

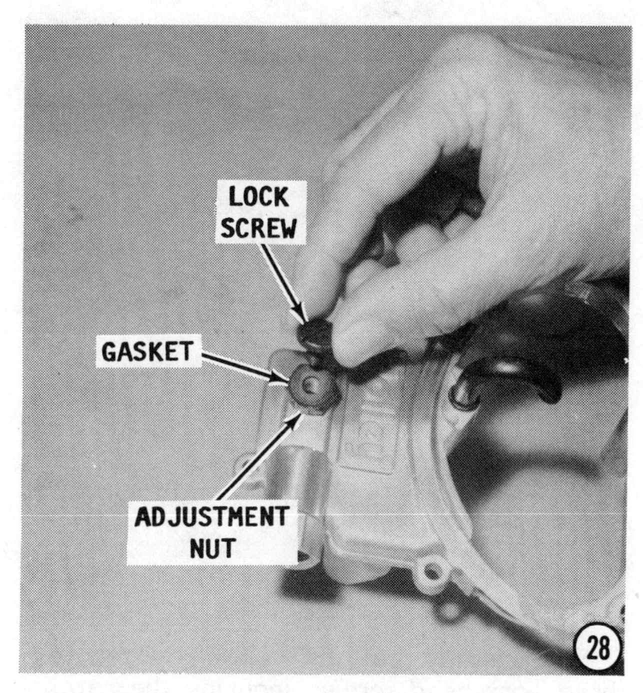

used below the nut. Install the adjustment nuts, one on each seat valve.

28- Use the correct size wrench to prevent the adjustment nut from rotating, and install the lock screw into the seat valve. **TAKE CARE** not to disturb the seat valve setting. Install the other adjustment nut in the same manner. Repeat Step 26 on the previous page, to check the float settings before proceeding with the work.

29- Lower the assembled airhorn down over the fuel bowls. Guide the choke link rod through the hole in the top of the carburetor body and into place close to the choke linkage. Make sure the lower end of the choke rod faces outward, away from the carburetor.

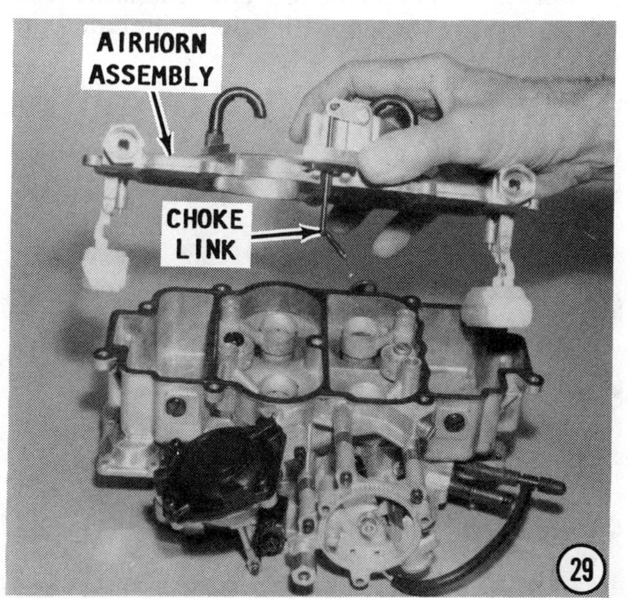

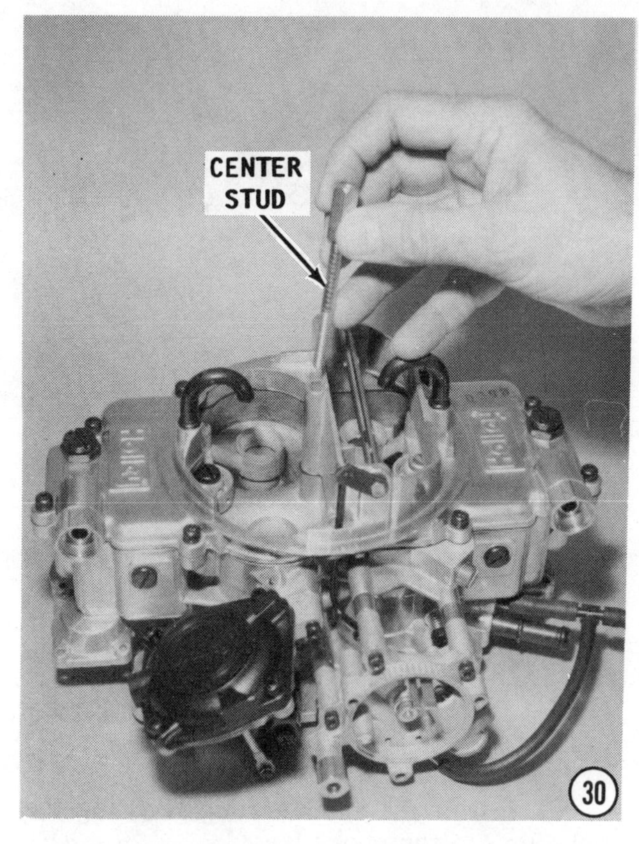

30- Install the center stud through the airhorn and tighten it securely.

31- Using a small screwdriver, guide the lower end of the choke link through the hole in the choke lever.

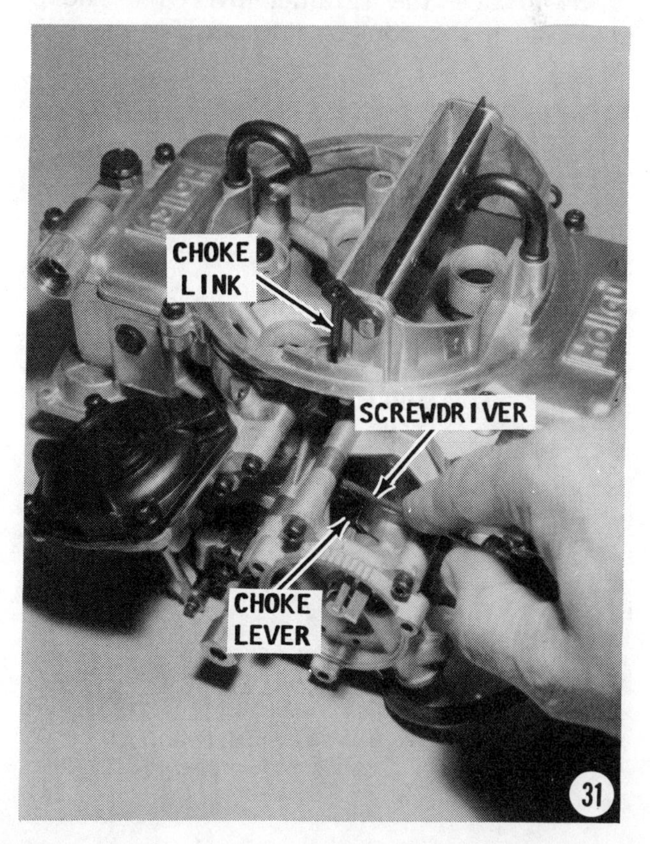

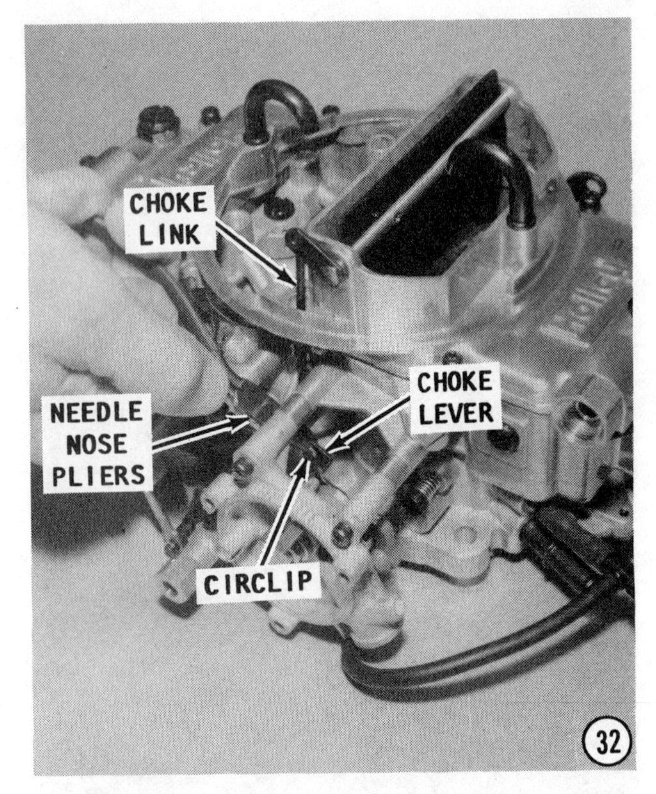

32- Install the small circlip securing the choke link to the choke lever.

33- Install the choke thermostat cover into the choke housing using a **NEW** gasket. The small tang attached to the choke piston must index into the loop on the thermostat.

34- Place the retainer over the choke cover with the convex side against the cover.

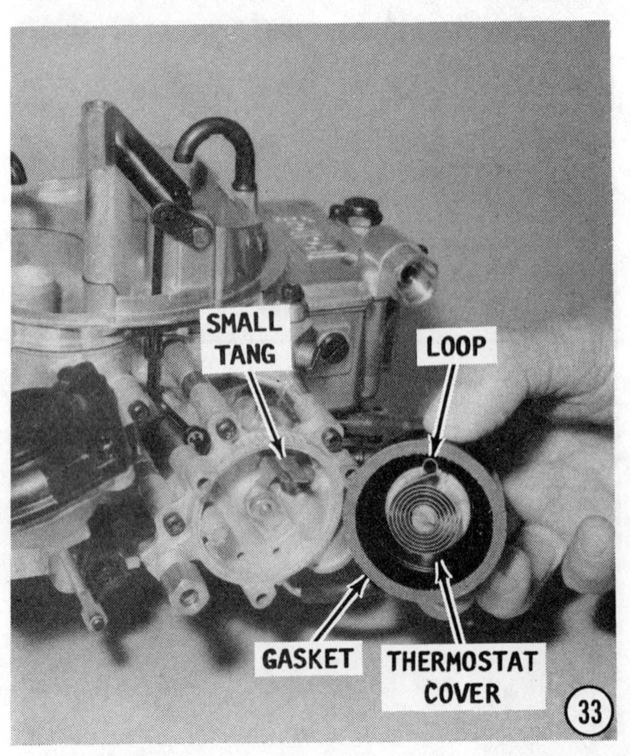

35- Install, but **DO NOT** tighten the three Torx-head screws securing the retainer to the housing, at this time. A choke adjustment follows.

CARBURETOR ADJUSTMENTS MODEL 4011 — FOUR BARREL

CRITICAL WORDS

Carburetor adjustments **MUST** be performed in the sequence outlined in the following steps. Unfortunately, this means the carburetor is installed on the intake manifold; the engine is operated at idle speed while some adjustments are made; the carburetor is removed for other adjustments; and finally, the carburetor is installed a second time.

For best performance and economy results, follow the specifications included in the carburetor repair kit.

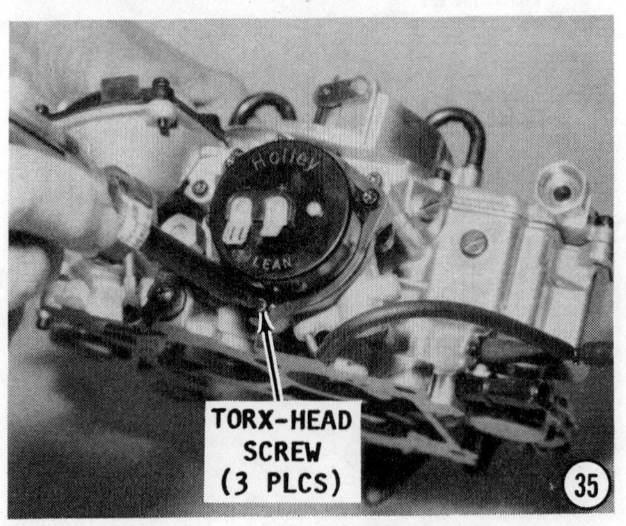

Automatic Choke Adjustment

1- The automatic choke is set at the factory to give maximum performance under all weather conditions. If necessary, the choke can be returned to its original position by aligning the index mark with the proper specification. Loosen the three Torx-head screws around the retainer. Rotate the choke cover to align the mark on the cover with the sixth mark from the left on the choke housing. Tighten the three Torx-head screws.

A richer or leaner mixture can be obtained for the warm-up period, if desired, by rotating the choke cover. **NEVER** set the index mark on the cover more than two marks from the specified setting.

Fuel Level Adjustment

2- Start the engine and allow it to warm to full operating temperature.

CAUTION: Water must circulate through the lower unit to the engine any time the engine is run to prevent damage to the water pump mounted on the engine.

Place a shop towel beneath the level plug to catch any spilled fuel. With the engine approximately level and running at idle speed, remove the level plug on the primary fuel bowl. The fuel level in the bowl must be in line with the threads at the

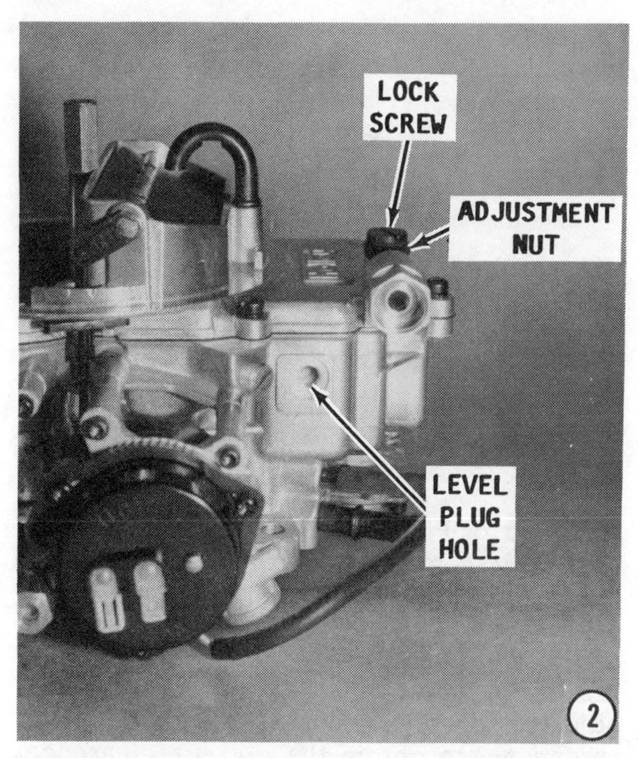

bottom of the plug. To adjust the level: Loosen the lock screw and turn the adjusting nut clockwise to lower the fuel level and counterclockwise to raise the fuel level. A 1/16 turn of the nut will equal approximately 1/16 inch difference in fuel level. After the adjustment has been made, tighten the lock screw and install the level plug. Repeat this procedure for the secondary fuel bowl.

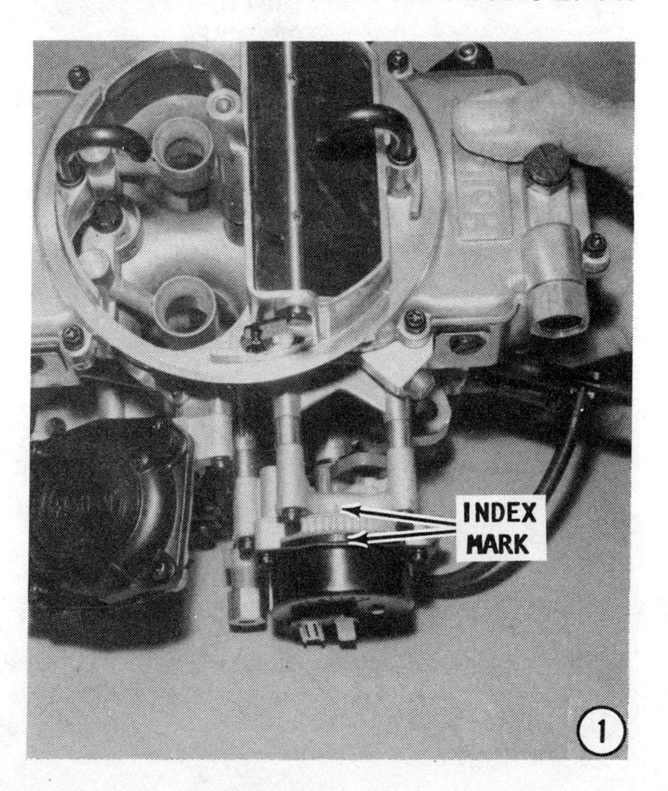

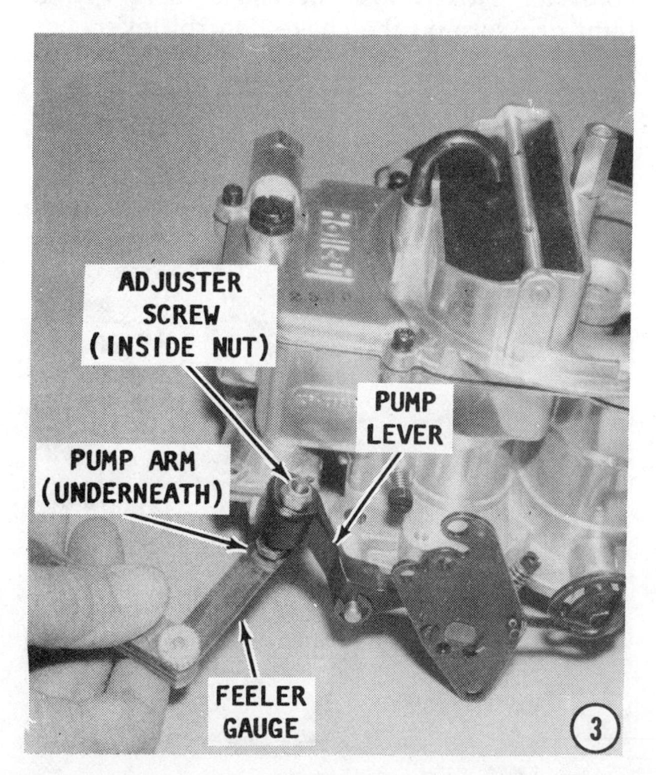

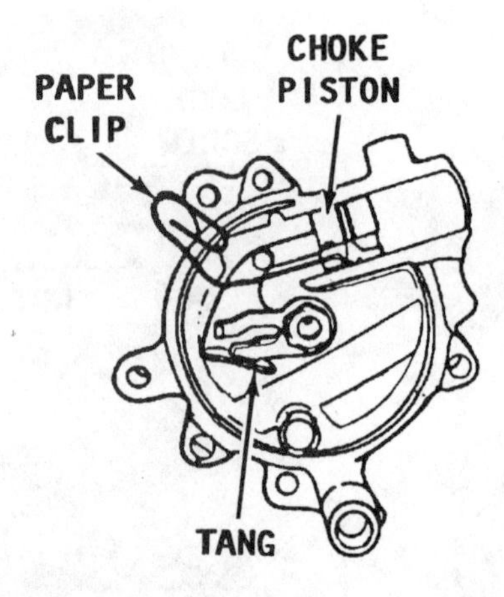

Accelerator Pump Lever Clearance

3- First, move the throttle to the wide-open position. Hold the pump lever down. Insert a 0.015" gauge between the bolt head under the pump lever and the pump arm. Loosen the top locknut and rotate the pump adjuster screw -- a slot cut through the bolt threads -- to obtain the correct clearance. There must be **NO FREE PLAY** with the pump lever at idle.

Choke Control Lever Adjustment

4- Open out a small size paper clip and bend over the last 1/8" (3mm) to form a right angle. Insert the bent end onto the choke piston and hook it onto the piston edge. Pull out the clip and piston as far as possible. Next, close the choke valve with a light pressure on the choke control lever.

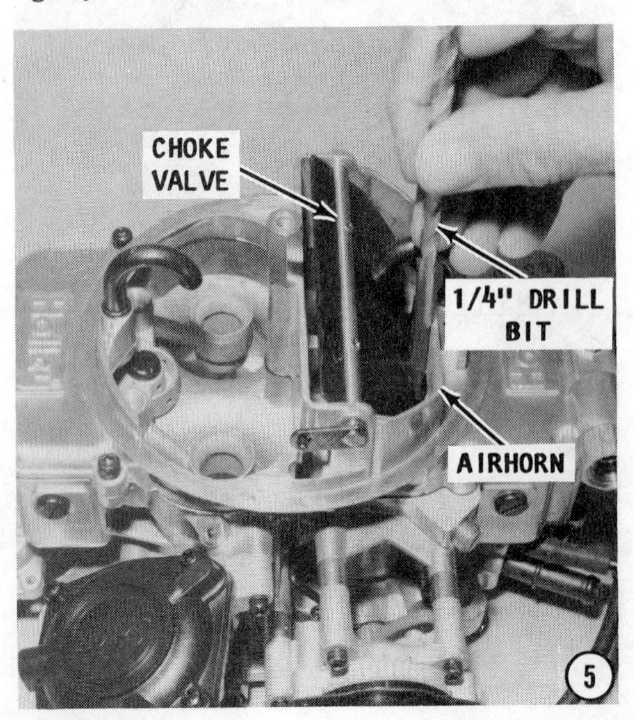

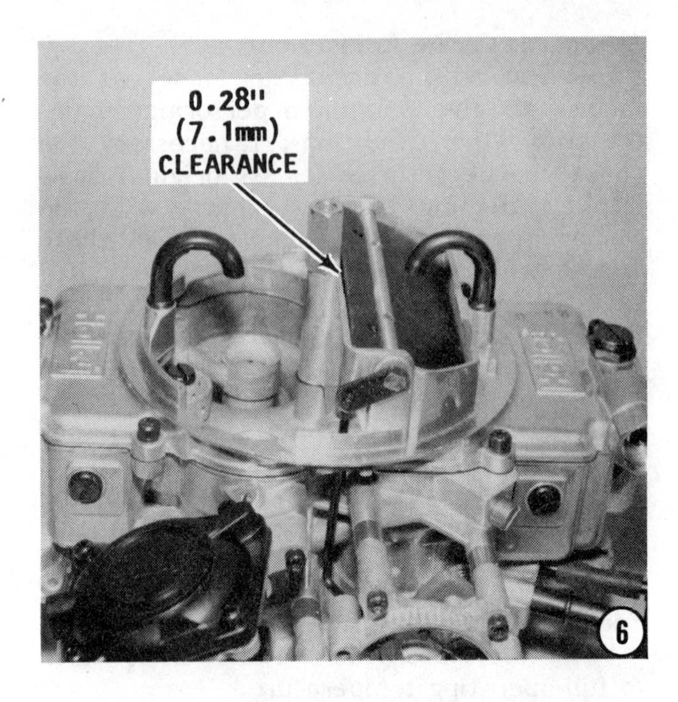

5- Insert a 1/4" (7mm) drill bit between the upper edge of the choke valve and the inner wall of the air horn. **CAREFULLY** bend the indicated tang (shown in illustration No. 4), with needlenose pliers until the required measurement is obtained.

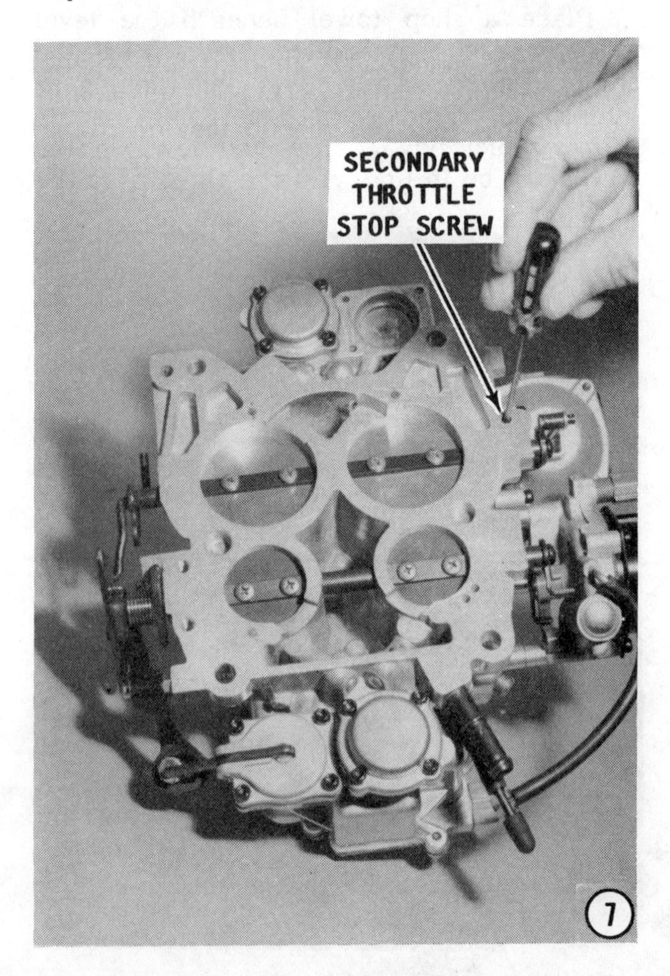

Choke Unloader Adjustment

6- Hold the throttle in the wide-open position and at the same time insert a 1/4" drill bit (7mm) between the upper edge of the choke valve and the inner wall of the air horn. **CAREFULLY** bend the tang, located on the choke lever under the choke housing, using needlenose pliers until the required measurement is obtained.

Secondary Throttle Valve Stop Adjustment

7- This adjustment is made with the carburetor removed from the manifold. First, back out the secondary throttle stop screw until the valves are fully closed. Next, turn the stop screw in until it barely contacts the secondary throttle stop lever. Now, turn the adjusting screw in 1/8 turn.

Secondary Throttle Valve Opening Adjustment

8- Set the secondary lever to the idle position. Measure the clearance, to the right, between the link arm and the oval hole. This clearance shuld be no more than 0.020" (0.5mm). Bend the link arm if necessary.

Idle Speed and Mixture Adjustment

9- Turn each idle adjustment needle - one on each side of the primary fuel bowl -

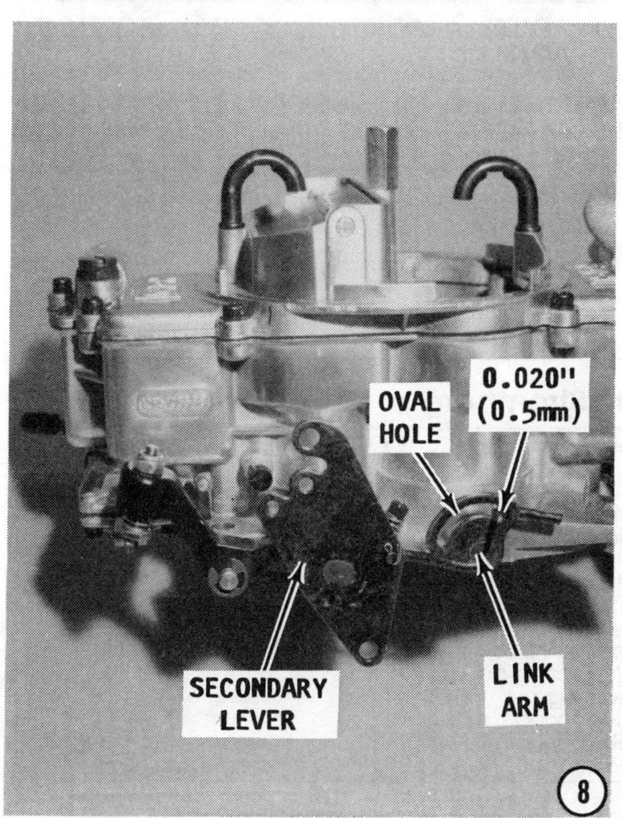

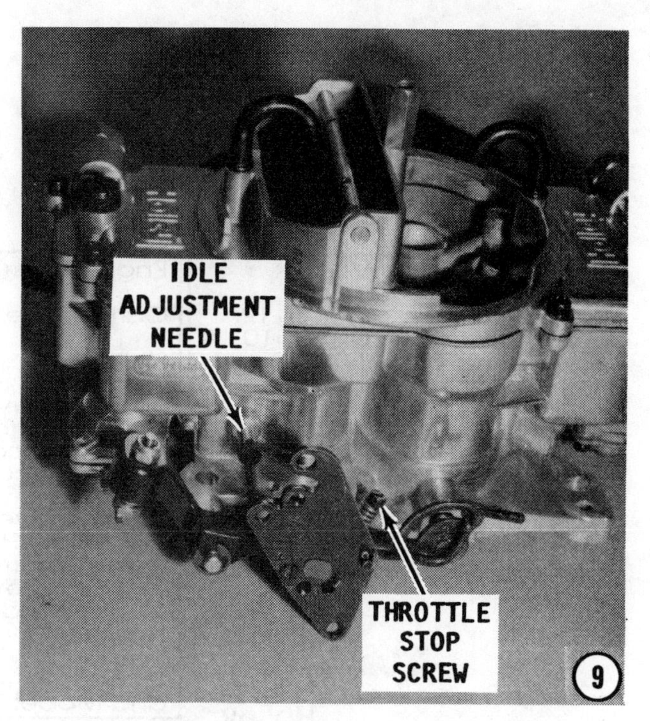

in, until it barely seats, and then back it out 3/4 turn. **TAKE CARE** not to seat the needles tightly, because it would groove the tip of the needle. If the needle should become grooved, a smooth idle cannot be obtained. Start the engine and allow it to warm to full operating temperature.

CAUTION: Water must circulate through the lower unit to the engine any time the engine is run to prevent damage to the water pump mounted on the engine.

Adjust the throttle stop screw on the throttle valve lever until the tachometer indicates 750 rpm.

Holley Model 4011 4-barrel carburetor ready for service.

CARBURETOR SPECIFICATIONS

Holley

Engine	V6
Carb Model	2300C
Holley No.	R-6317-1A
Ford No.	DIFF-9510-LA
Bore Primary	1-1/2"
Venturi Primary	1-3/16"
Main Jet	No. 64
Power Valve	No.50
Cam Hole Pos.	No.2
Pump Override	.015 Min.
Choke Index	3 Ntch. Ln
Choke Pulldown	.140
Dechoke	.300

Holley

Engine	V6 V8*
Carb Model	4150 4160
Holley No.	R-6576A
Ford No.	D2JL-E
Bore Primary	1-9/16"
Bore Secondary	1-9/16"
Venturi Primary	1-1/4"
Venturi Secondary	1-5/16"
Main Jet	No. 64
Power Valve	No. 50
Cam Hole Pos.	No.1
Pump Override	.015 Min.
Choke Index	3 Ntch. Ln
Choke Pulldown	.140
Dechoke	.300

*Except for AQ311, AQ271 and Model 740 (454CID)

Holley

Engine	454 CID
Holley No.	84024
Carb Model	4011
Inlet Seat Pri.	0.110"
Inlet Seat Sec.	0.110"
Pri. Main Jet	No. 64
Sec. Main Jet	No. 68
Power Valve Pri.	6.5"
Power Valve Sec.	6.5"
Pri. Pump Shooter	0.026"
Pri. Cluster	21R-854A
Sec. Cluster	21R-858A
Airhorn Gasket	108-64
Pri. Venturi	1.200"
Sec. Venturi	1.415"
Pri. Bore	1.375"
Sec. Bore	2.000"

Holley

Engine	AQ311 AQ271*	AQ271**
Carb Model	4150	4150
Primary Jet	74	72
Secondary Jet	85	80, 90
Accelerator pump setting	Position 2	Position 2
Choke setting	4 notches to the right	4 notches to the right

* Up to # 4-100-106052
** From # 4-100-106053

Solex

Engine	AQ120B AQ140A	AQ120B AQ125A	AQ145A AQ151A,B	AQ151C,D	AQ125B AQ131A,B AQ145B	AQ131C,D	AQ171A,B	AQ171C,D
Type	Sidedraft	Downdraft	Downdraft	Downdraft	Downdraft	Downdraft	Downdraft	Downdraft
Make	44 PHN	44 PAI-4	44 PAI-4	44 PAI-7	44 PAI-5	44 PAI-7	44 PAI-5-6	44 PAI-7
Venturi	34	34	31	31	34	34	32	32
Main Jet	165	155	145	145	165	165	147	147
Idle Jet	60	62	62	62	65	65	65	60
Air Jet	130	180	185	180	185	185	190	200
Needle Valve	1.7	1.7	1.5	1.7	1.7	2.0	1.7	1.7
Float, Wt. gr.	Double	7.3	7.3	7.3	7.3	7.3	7.3	7.3
Accel. Jet	---	70	60	60	70	70	70	70
Econostat Jet	---	---	---	---	110	110	---	---

Solex

Engine	AQ115A	AQ130C,D	AQ170C,D
Type	Downdraft	Downdraft	Downdraft
Make	44 PAI	44 PAI	44 PAI
Main Jet	145	145	145
Idle Jet	55	70	70
Emulsion Jet	145 E5	190 E5	210 E5
Pump Jet	70	70	70
Needle Seating	2	2	2

Zenith Stromberg

Engine	AQ105A	AQ130A,B	AQ165A
Type	Horizontal	Horizontal	Horizontal
Make	150CD	175CDSE	175CDSE
Metering needle mrkd	8B	3B	2AA
Float level			
at highest point mm	17.5	16	16
at lowest point mm	14.5	11	11
Float needle valve mm	2.0	1.5	1.5
Washer thick. under float needle valve mm	1.0	1.6	1.6
Air valve spring color marking	Red	Blue	No Color
Air valve spring coil thickness mm	1.0	0.9	0.8
Oil for damper	Same as in engine		

Rochester

See Section 4-10 and 4-11 for the Rochester 2GC, 2GV and 4MV specifications.

5
IGNITION

5-1 DESCRIPTION

Three ignition systems are used on the Volvo power plants covered in this manual. One is a conventional distributor with points which is discussed and serviced in the first portion of this chapter.

The second system is an electronic Renix system first used on the AQ171 series double overhead cam (DOHC) engines. This system is described and troubleshooting procedures outlined beginning with Section 5-11, on Page 5-30.

The third system is called the Standard Electronic ignition system and has been used since about 1990 on V6 and V8 powerplants. This system is described and troubleshooting procedures outlined beginning with Section 5-12 on Page 5-35.

Conventional Point System

Engine performance and efficiency are, to a large degree, governed by how fine the engine is tuned to factory specifications as determined by the designers. The service work outlined in this chapter must be performed in the sequence given and to the Specifications listed in the Appendix.

The ignition system consists of a primary and a secondary circuit. The low-voltage current of the ignition system is carried by the primary circuit. Parts of the primary circuit include the ignition switch, ballast resistor, neutral-safety switch, primary winding of the ignition coil, contact points in the distributor, condenser, and the low-tension wiring.

The secondary circuit carries the high-voltage surges from the ignition coil which result in a high-voltage spark between the electrodes of each spark plug. The secondary circuit includes the secondary winding of the coil, coil-to-distributor high-tension lead, distributor rotor and cap, ignition cables, and the spark plugs.

When the contact points are closed and the ignition switch is on, current from the battery or from the alternator flows through the primary winding of the coil, through the contact points to ground. The current flowing through the primary winding of the coil creates a magnetic field around the coil windings and energy is stored in the coil.

Now, when the contact points are opened by rotation of the distributor cam, the primary circuit is broken. The current attempts to surge across the gap as the points begin to open, but the condenser absorbs the current. In so doing, the condenser creates a sharp break in the current flow and a rapid collapse of the magnetic field in the coil. This sudden change in the strength of the magnetic field causes a voltage to be induced in each turn of the secondary windings in the coil.

The ratio of secondary windings to the primary windings in the coil increases the voltage to about 20,000 volts. This high voltage travels through a cable to the center of the distributor cap, through the rotor to an adjacent distributor cap contact point, and then on through one of the ignition wires to a spark plug.

A fully charged battery, filled to the proper level with electrolyte, is the heart of the ignition system. Engine starting and efficient performance can never be obtained if the battery is below its full charge rating.

When the high-voltage surge reaches the spark plug it jumps the gap between the insulated center electrode and the grounded side electrode. This high voltage jump across the electrodes produces the energy required to ignite the compressed air-fuel mixture in the cylinder.

The entire electrical build-up, breakdown, and transfer of voltage is repeated as each lobe of the distributor cam passes the rubbing block on the contact breaker arm, causing the contact points to open and close. At high engine rpm operation, the number of times this sequence of actions take place is staggering.

BALLAST RESISTOR

Beginning at the key switch, current flows to the ballast resistor and then to the positive side of the coil. When the resistor is cold its resistance is approximately one ohm. The resistance increases in proportion to the resistor's rise in temperature.

While the engine is operating at idle or slow speed, the cam on the distributor shaft revolves at a relatively slow rate. Therefore, the breaker points remain closed for a slightly longer period of time. Because the points remain closed longer, more current is allowed to flow and this currrent flow heats the ballast resistor and increases its resistance to cut down on current flow thereby reducing burning of the contact points.

During high rpm engine operation, the reduced current flow allows the resistor to cool enough to reduce resistance, thus increasing the current flow and effectiveness of the ignition system for high-speed performance.

The voltage drops about 25% during engine cranking due to the heavy current demands of the starter. These demands reduce the voltage available for the ignition system. In order to reduce the problem of less voltage, the ballast resistor is by-passed during cranking. This releases full battery voltage to the ignition system.

IGNITION TIMING

In order to obtain the maximum performance from the engine, the timing of the spark must vary to meet operating conditions. For idle, the spark advance should be as low as possible. During high-speed operation, the spark must occur sooner, to give the air-fuel mixture enough time to ignite, burn, and deliver it's power to the piston for the power stroke.

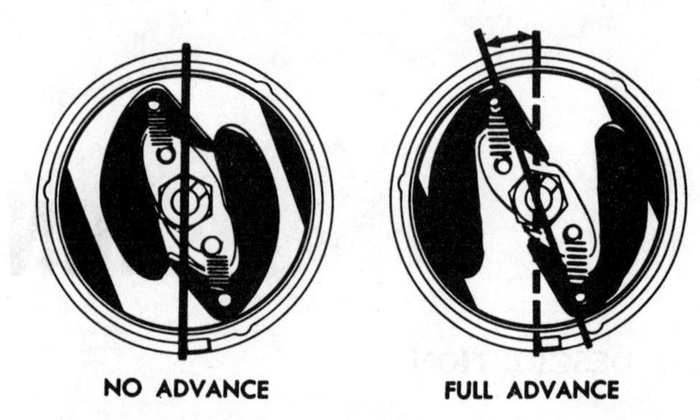

Movement of the centrifugal advance mechanism. All parts must be free to move properly or the ignition timing will not advance wtih engine speed and all phases of performance will suffer, as explained in the text.

Manual setting and centrifugal advance are the two methods of obtaining the constantly changing demands of the engine. The manual setting is made at idle speed. This setting allows the contact points to open at a specified position of the piston before Top Dead Center (TDC). The TDC position for the No. 1 cylinder is indicated by the timing mark on the crankshaft pulley.

The centrifugal advance is controlled entirely by engine speed. A pair of weights, two springs, and a weight base plate comprise the centrifugal advance mechanism.

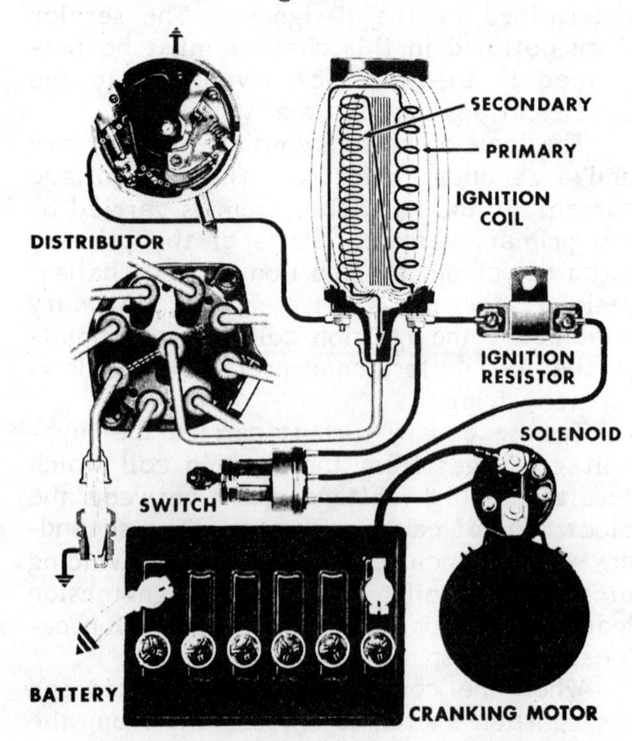

Functional diagram of the ignition system.

As engine speed increases, centrifugal force causes the weights to move outward and push against the cam ramps which in turn rotate the distributor cam ahead of the distributor shaft. This action causes the distributor cam lobes to open the contact points sooner and the timing of the spark is advanced.

Cylinder pressure should not vary between cylinders by more than 15 psi for the engine to run smoothly. Variation between the cylinders is much more important than the actual individual readings.

5-2 IGNITION TROUBLESHOOTING

COMPRESSION

Before spending time and money seeking a problem in the ignition system, a compression check should be made of each cylinder. Without adequate compression, your efforts in the ignition system will not give the desired results.

Remove each spark plug in turn; insert a compression gauge in the hole; crank the engine several times; and note the reading. A variation between the cylinders is more important than the actual reading. A variation of more than 20 psi indicates either a ring or valve problem.

To determine which is defective, squirt about a teaspoonful of oil into the cylinder that has the low reading. Crank the engine a few times to distribute the oil. Now, recheck the compression and note the difference from the first reading. If the pressure increased, the compression loss is past the piston rings; if no change is noted, the loss is past a burned valve.

IGNITION SYSTEM TESTS

Any problem in the ignition system must first be localized to the primary or secondary circuit before the defective part can be identified.

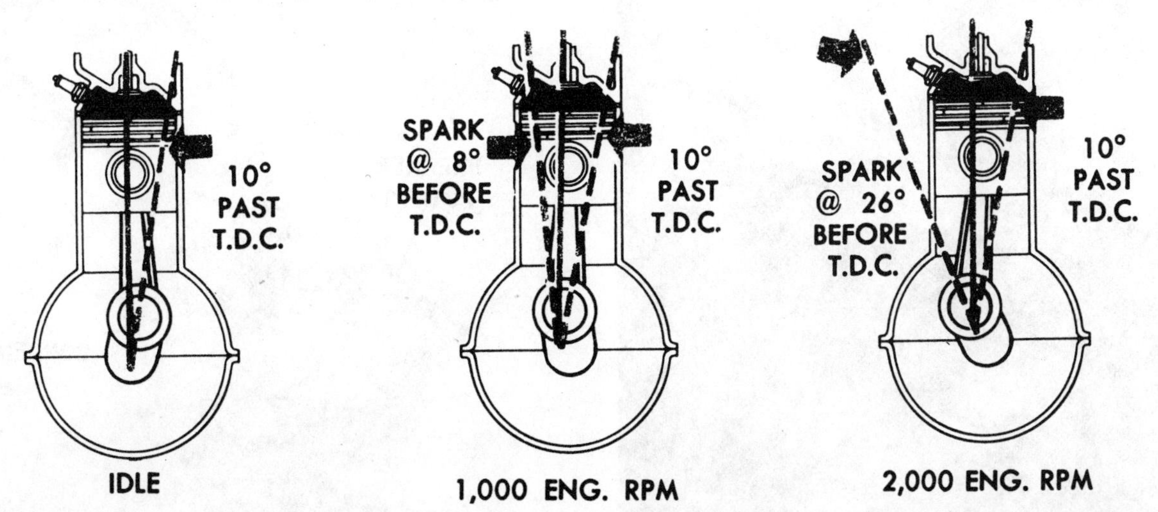

*The mechanical spark advance mechanism advances ignition timing when engine speed increases, as shown in these three drawings. Maximum pressure must occur by 10° **ATDC**, at all speeds. Therefore, the spark must occur at **TDC** for idle speeds (left); at approximately 8° before **TDC** for 1,000 rpm (center); and at 26° before **TDC** for 2,000 rpm (right).*

After taking the first pressure reading, a small amount of oil inserted into a low reading cylinder, followed by a second pressure reading will indicate whether the compression loss is past the rings or the valves.

1- GENERAL TESTS

Disconnect the wire from the center of the distributor cap and hold it about 1/4" from a good ground. Turn the ignition switch to **START**, and crank the engine with the starter. If you observe a good spark, go to Test 5 (Secondary Circuit Test). If you do not have a good spark, go to Test 2 (Primary Circuit Test).

2- PRIMARY CIRCUIT TEST

Remove the distributor cap; lift off the rotor; and then turn the crankshaft until the contact points close. Turn the ignition switch on, and open and close the contact points using a small screwdriver or a non-metallic object. Hold the high-tension coil wire about 1/4" from a good ground. If you observe a good spark jump from the wire to the ground, the primary circuit checks out. Go to Test 5, (Secondary Circuit Test). If there is no spark, go to Test 3, (Contact Point Test).

3- CONTACT POINT TEST

Remove the distributor cap and rotor. Turn the crankshaft until the points are open, and then insert some type of insulator between the points. Now, hold the high-tension coil wire about 1/4" from a good

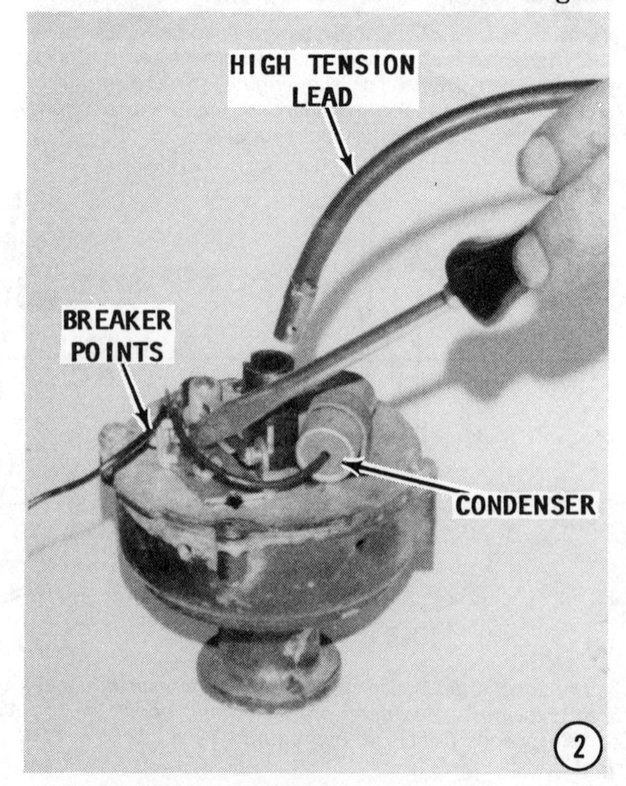

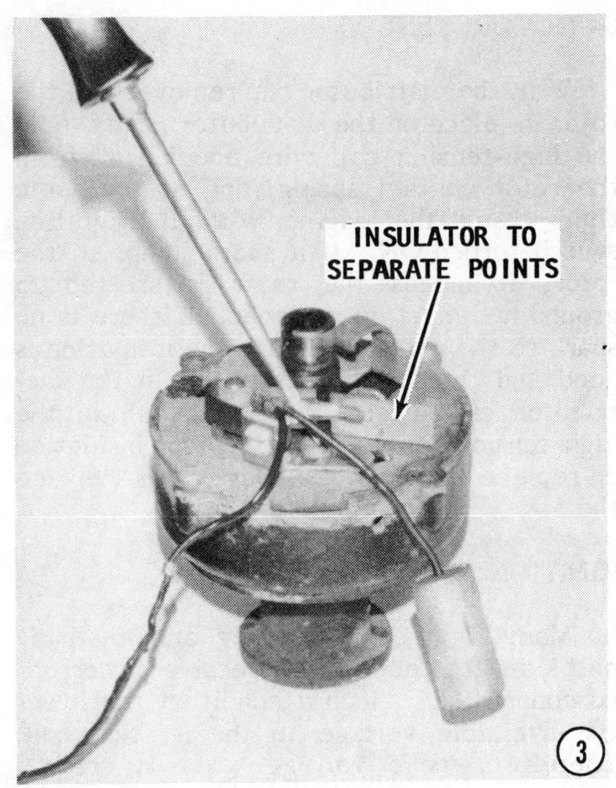

ground, and at the same time move a small screwdriver up and down with the screwdriver shaft touching the moveable point and the tip making intermittent contact with the contact point base plate. In this manner, you are using the screwdriver for a set of contact points. If you get a spark from the high-tension wire to ground, then the problem is in the contact points. Replace the set of points. If there is no spark from the high-tension wire to ground the problem is either a defective coil or condenser. To test the condenser, go to Test 4 (Condenser Test).

4- CONDENSER TEST

Condensers seldom cause a problem. However, there is always the possibility one may short out and ground the primary circuit. In the "old" days it was considered good shop practice to replace the condenser anytime the point set was replaced. Almost all point sets would contain a new condenser. Today, times have changed. Only rarely is the condenser included in the point set package.

Before testing the condenser, check to be sure one of the primary wires or connections inside the distributor has not shorted out to ground.

The most accurate method of testing a condenser is with an instrument manufactured for that purpose. However, seldom is one available, especially during an emergency. Therefore, the following procedure is outlined for emergency troubleshooting the condenser and the primary circuit insulation for a short.

First, remove the condenser from the system. **TAKE CARE** that the metallic case of the condenser does not touch any part of the distributor. Next, insert a piece of insulating material between the contact points. Now, move the blade of a small screwdriver up and down with the shaft of the screwdriver making contact with the moveable contact point and the tip making and breaking contact with the contact point base plate. Observe for a low-tension spark between the tip of the screwdriver and the contact point base plate as you make and break the contact with the screwdriver tip. You should observe a spark during this test and it will prove the primary circuit complete through the neutral-safety switch, the primary side of the ignition coil, the ballast resistor, and the primary wiring inside the distributor. If you have a spark, reconnect the condenser and again make the same test with the screwdriver. If you do not get a spark, the condenser is defective and should be replaced,

If you were unable to get a spark with the condenser disconnected, it means no current is flowing to this point, or there is a

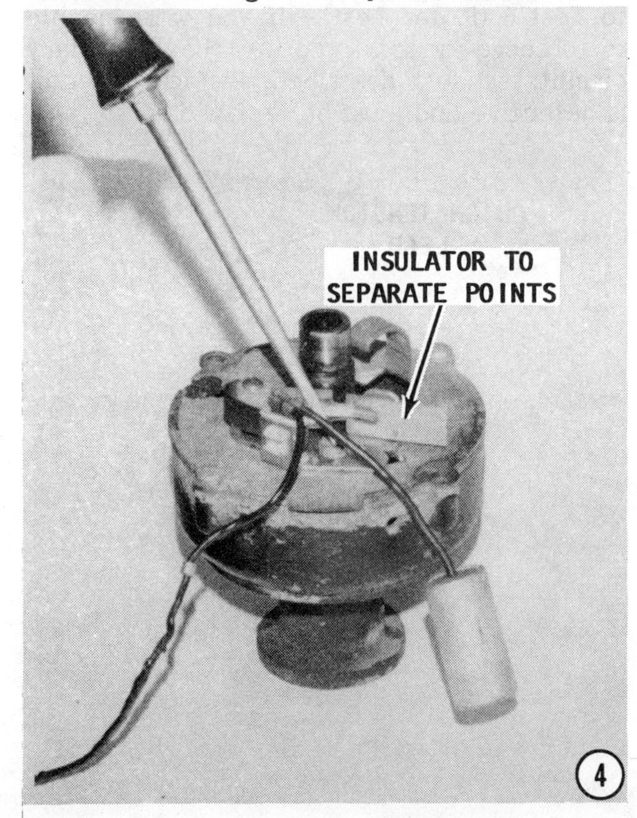

short circuit to ground. Use a continuity tester and check each part in turn to ground in the same manner as you did at the moveable contact point. If you get a spark, indicating current flow, at one terminal of the part, but not at the other, then you have isolated the defective unit.

5- SECONDARY CIRCUIT TEST

The secondary circuit cannot be tested using emergency troubleshooting procedures **UNLESS** the primary circuit has been tested and proven satisfactory, or any problems discovered in the primary cirucuit have been corrected.

If the primary circuit tests are satisfactory, use the same procedures as outlined in Test 2, Primary Circuit Test, to check the secondary circuit. Hold the high-tension wire about 1/4" from a good ground and at the same time, use a small screwdriver to open and close the contact points. A spark at the high-tension wire proves the ignition coil is good. However, if the engine still fails to start and the problem has been traced to the ignition system, then the defective part or the problem must be in the secondary circuit.

The distributor cap, the rotor, high-tension wires, or the plugs may require attention or replacement. To test the rotor, go to Test 6 (Rotor Test). If you were unable to observe a spark during the secondary circuit test just described, the ignition coil is defective and must be replaced.

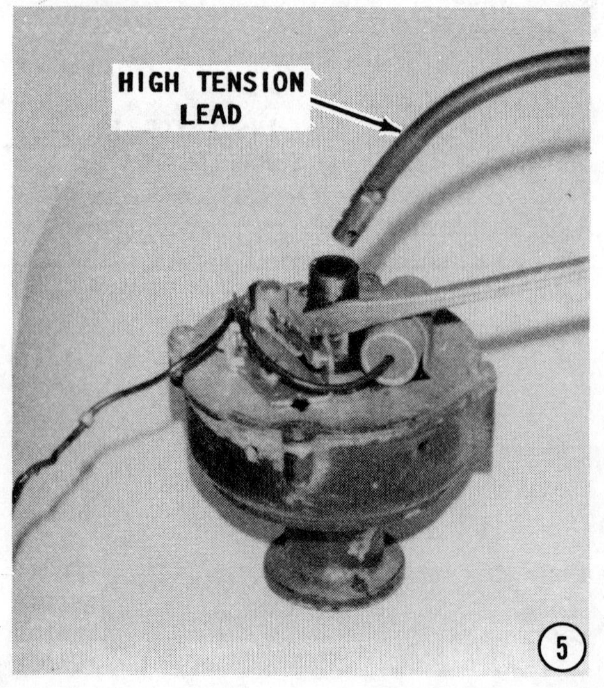

6- ROTOR TEST

With the distributor cap removed and the rotor in place on the distributor shaft, hold the high-tension coil wire about 1/4" from the rotor contact spring, and at the same time crank the engine with the ignition switch turned **ON.** If a spark jumps to the rotor, it means the rotor is shorted to ground and must be replaced. If there is no spark to the rotor, it means the insulation is good and the problem is either in the distributor cap (check it for cracks), in the high-tension wires (check for poor insulation or replace it), or in the spark plugs (replace them).

IGNITION VOLTAGE TESTS

Many times hard starting and misfiring problems are caused by defective or corroded connections. Such a condition can lower the available voltage to the ignition coil. Therefore, make voltage tests at critical points to isolate such a problem. Move the voltmeter test probes from point-to-point in the following order.

TEST 1 Voltage Loss Across Entire Ignition Circuit: Connect a voltmeter between the battery side of the ignition coil and the positive post of the battery, as shown in the Test 1 illustration. Crank the engine until the contact points are closed. Turn the ignition switch to ON. The voltage loss should not exceed 3.2 volts. This figure

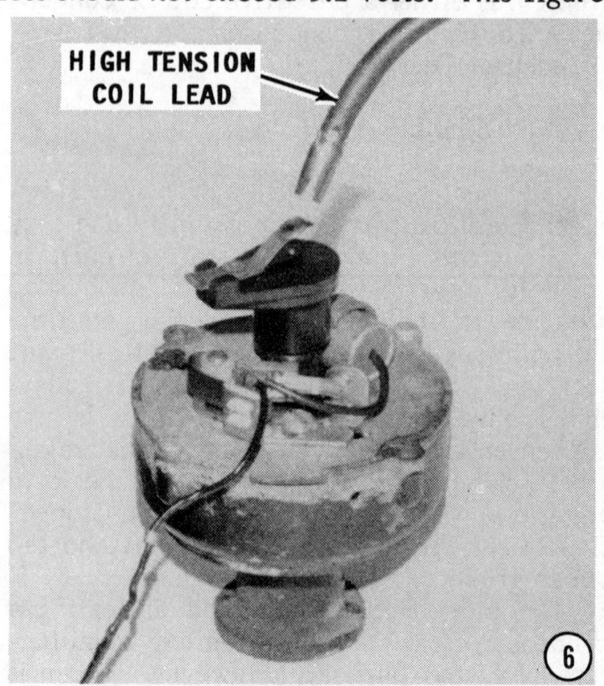

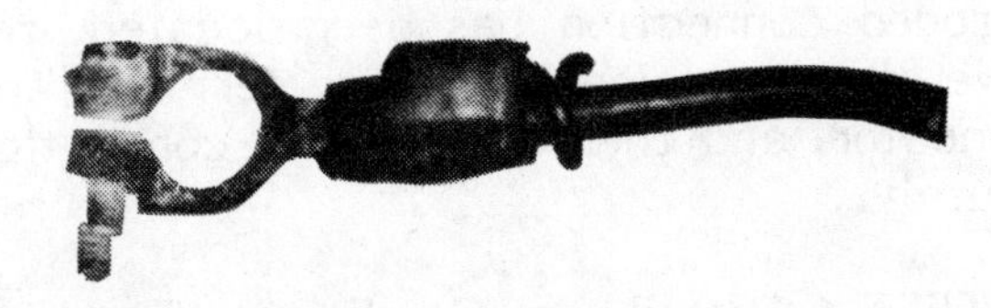

Many ignition system problems can be traced to a high-resistance connection caused by corroded battery posts and cables.

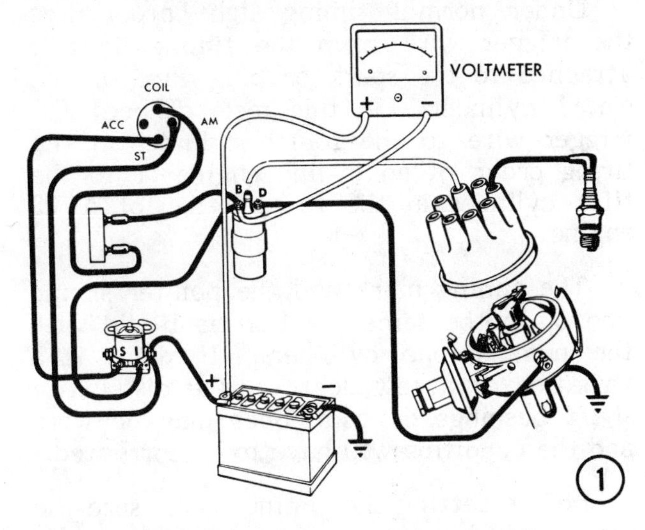

allows for a 0.2 loss across each of the connections in the circuit, plus a calibrated 2.4 volt drop through the ballast resistor. If the total voltage loss exceeds 3.2 volts, then it will be necessary to isolate the corroded connection in the circuit of the key resistor, the wiring, or at the battery.

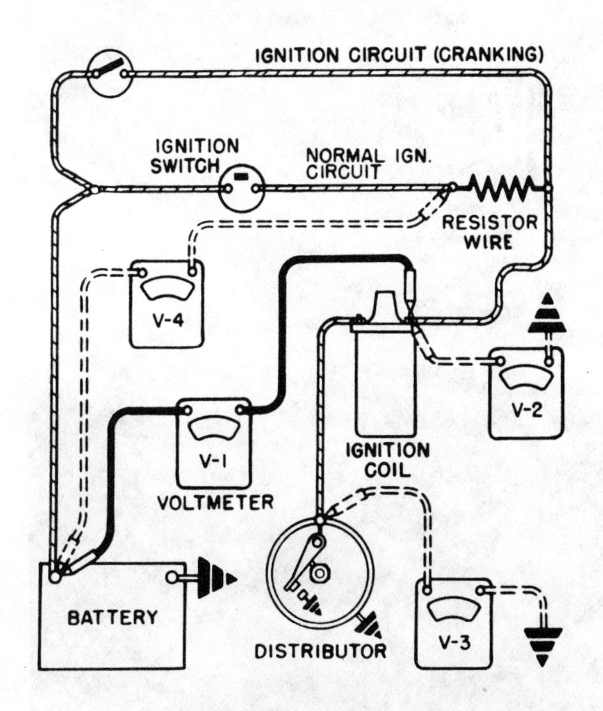

Functional diagram showing voltmeter hookups.

TEST 2 Cranking System: First disconnect the high-tension wire and ground it securely to minimize the danger of sparks and a possible fire. Next, connect a voltmeter between the battery side of the ignition coil and ground, as shown in Test 2 illustration. Now crank the engine and check the voltage. A normal system should have a reading of 8.0 volts. If the voltage is lower, the battery is not fully charged or the starter is drawing too much current.

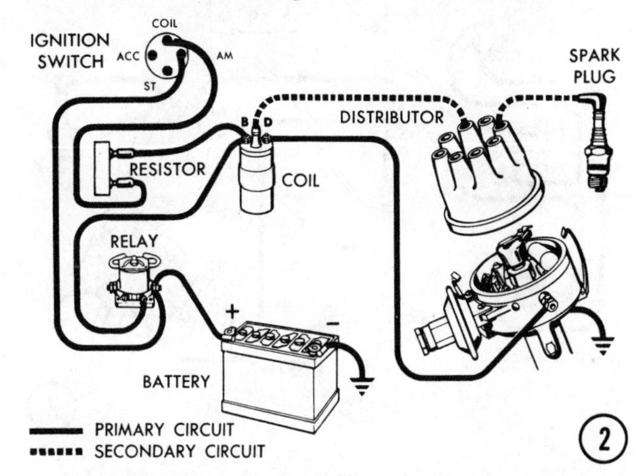

TEST 3 Contact Points and Condenser: Measure the voltage between the distributor primary terminal and ground, as shown in Test 3 illustration. Crank the engine until the contact points are closed. Turn the ignition switch to the ON position. The voltage reading must be less than 0.2 volts. A higher reading indicates the contact points are oxidized and must be replaced. To check the condenser, crank the engine until the contact points are open, and then take a voltage reading. If the reading is not equal to the battery voltage, the condenser is shorted to ground. Check the condenser installation or replace the condenser.

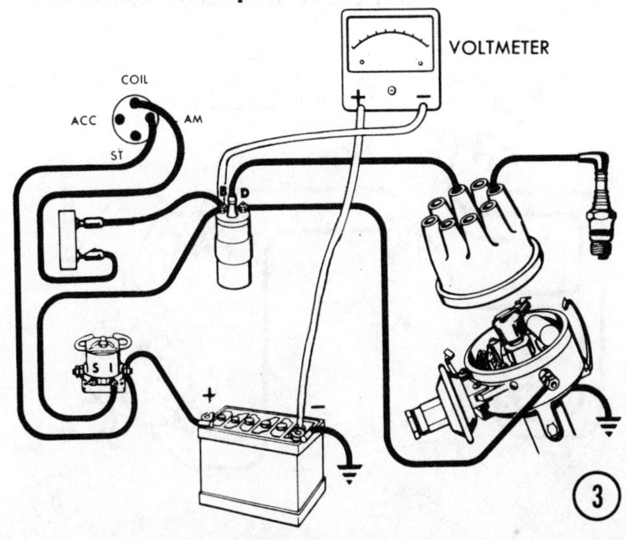

TEST 4 Primary Resistor: Disconnect the battery wire at the primary resistor to prevent damage to the ohmmeter. Connect an ohmmeter across the terminals of the resistor, as shown in Test 4 illustration. The specified resistance is between 1.3 and 1.4 ohms. If the reading does not fall within this range, replace the resistor.

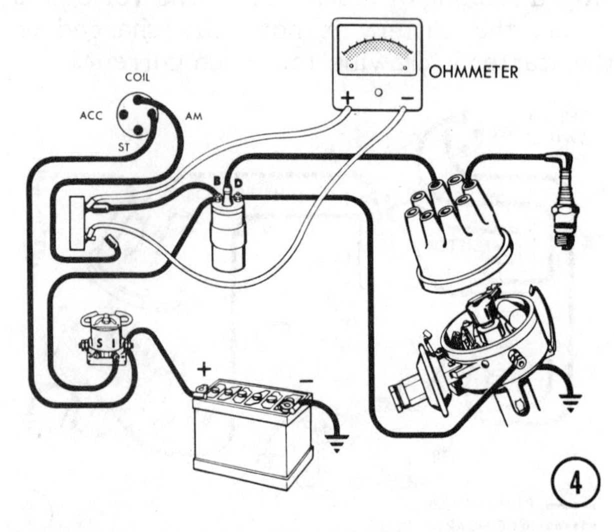

TEST 5 Voltage Loss in the ignition switch, ammeter, and battery cable: Crank the engine until the points are closed. Connect the voltmeter to the battery post (not the cable terminal) and to the load side of the ignition switch, as shown in Test 5 illustration. Now, turn the ignition switch to the ON position and note the voltage reading. The meter reading should not be more than 0.8 volts. A 0.2 volt drop across each of the connections is permitted.

If the voltage drop is more than 0.8 volts, move the test probe to the "hot" side of the ammeter. If the reading is 0.4 volts, the ignition switch is satisfactory. Once the corroded connection has been located, remove the nut, clean the wire terminal and connector, and then tighten the connection securely.

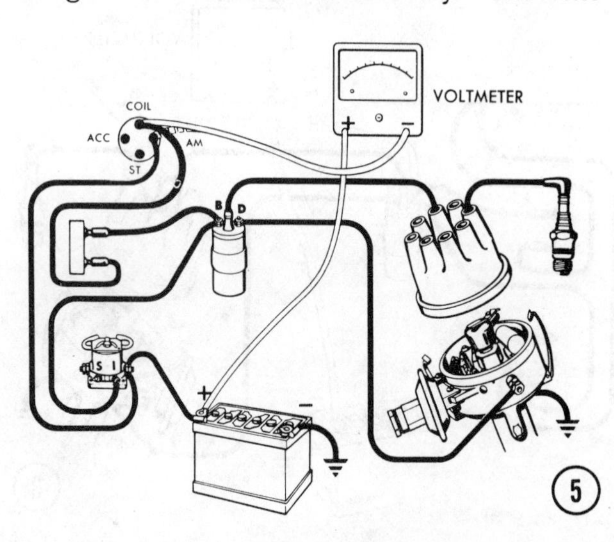

TEST 6 Distributor Condition: The condition of the distributor can be quickly and conveniently checked with a timing light.

Under normal timing light procedures the trigger wire from the timing light is attached to the spark or plug wire of the No. 1 cylinder. In this test, connect the trigger wire to the fourth cylinder in the firing order of an in-line engine or to the fifth cylinder in the firing order of a V8 engine.

The timing mark and the pointer should line up in the same position as it did with the number one cylinder. If there is a variation of a few degrees, the distributor shaft bushings or cam lobes may be worn and the condition will have to be corrected.

Before setting the timing, make sure the point dwell is correct. **TAKE CARE** to aim the timing light straight at the mark. Sighting from an angle may cause an error of two or three degrees.

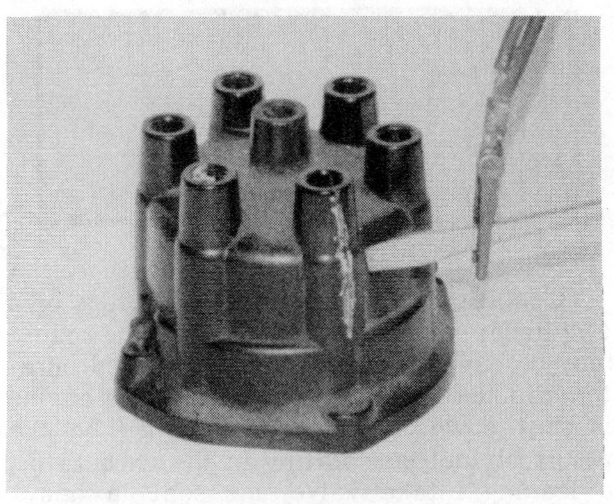

Using a grounded screwdriver blade and spark plug tester to test a cracked distributor cap.

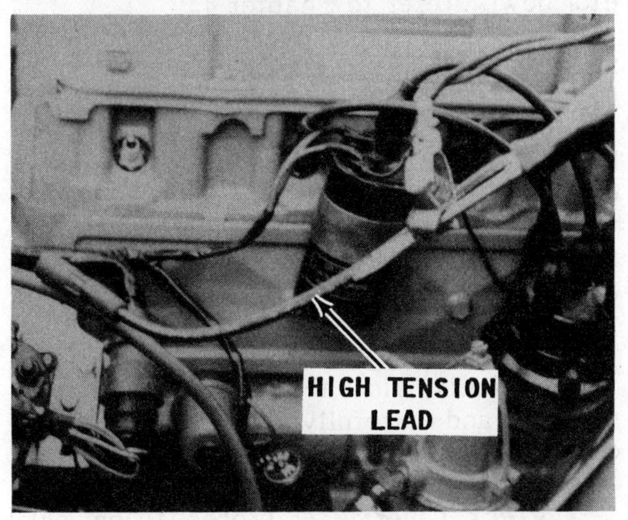

Hook-up for making a voltage check of high-tension spark plug wires. One end of a jumper wire is clipped to the screwdriver; the other end is grounded; and the spark plug wire is disconnected from the plug. As a preventative measure against hard starting, the high tension wires (complete set) should be replaced every two years.

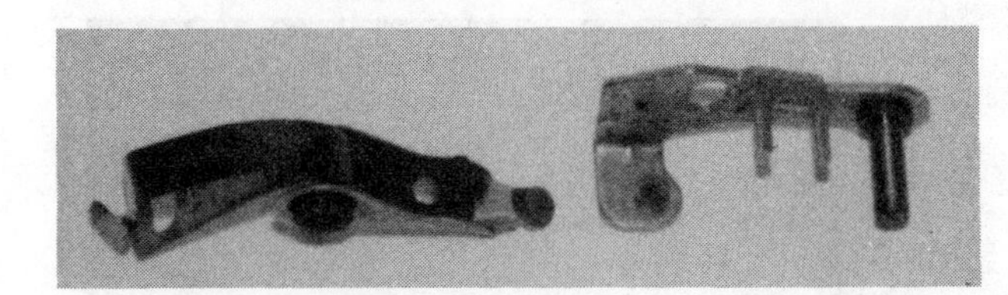

Oxidized G.M. contact points which should be replaced for satisfactory performance.

5-3 SPARK PLUG TROUBLESHOOTING

The following conditions are keyed to the illustrations with the same number. Refer to the illustration for a visual indication of which plugs are involved.

Condition 1: One spark is overheated. Check the firing order. If the burned plug is the second of two adjacent, and consecutive-firing plugs, the overheating may be the result of crossfire. If you found the spark plug of No. 7 cylinder was overheated, and the firing order is 1-8-4-3-6-5-7-2, the crossfire might result because cylinders No. 5 and No. 7 are adjacent to each other physically and No. 7 follows No. 5 in firing order. Separate the high-tension leads to these two plugs and the problem may be corrected.

Condition 2: Four spark plugs are fouled in the unusual pattern shown in the No. 2 illustration. This pattern follows the usual fuel flow in a V8 engine. This condition may indicate one barrel of the carburetor is running too rich.

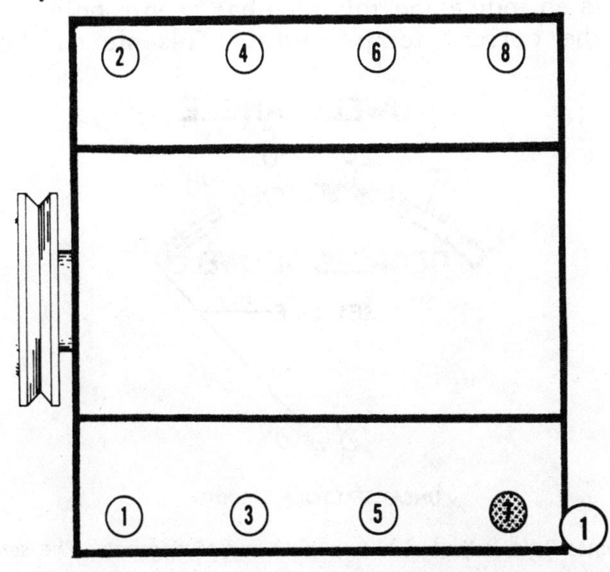

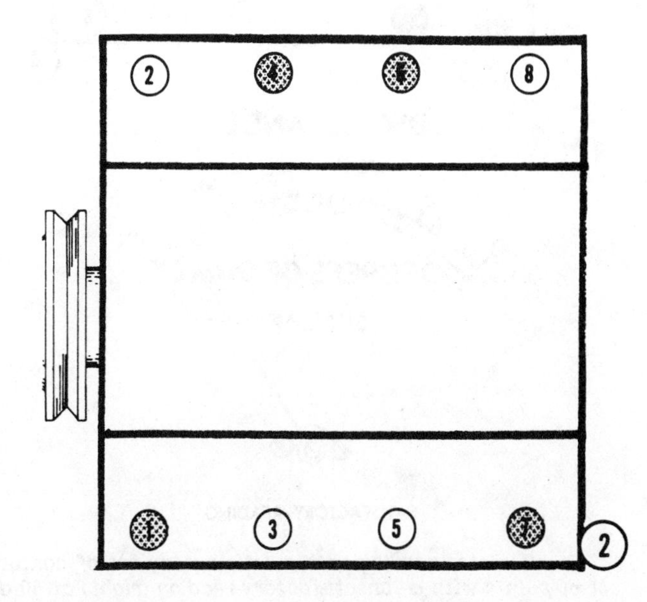

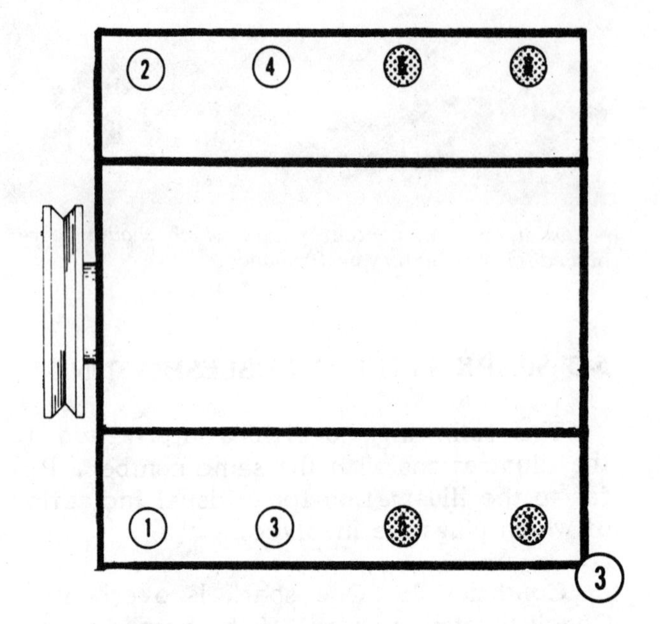

Condition 3: The four rear spark plugs are overheated. This condition indicates a problem in the cooling system. A reverse-flush of the engine may restore circulation to the rear of the cylinder heads.

Condition 4: Two adjacent plugs are fouled. Check the high-tension leads to the spark plugs to be sure they are connected in the proper sequence and leading to the correct plug for firing order. If the high-tension wires are all in good order and connected properly, then check for a blown cylinder head gasket, refer to Chapter 3.

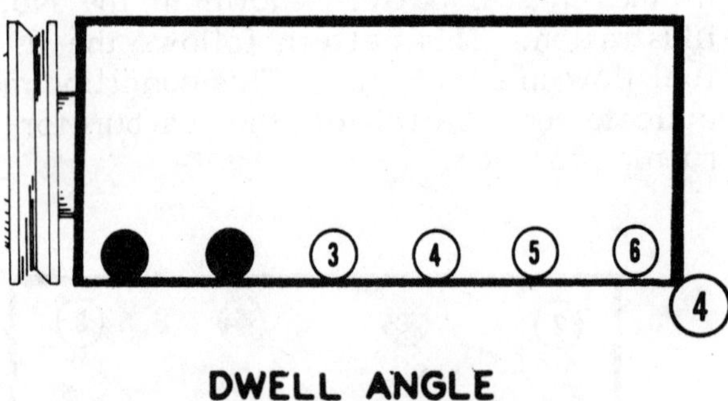

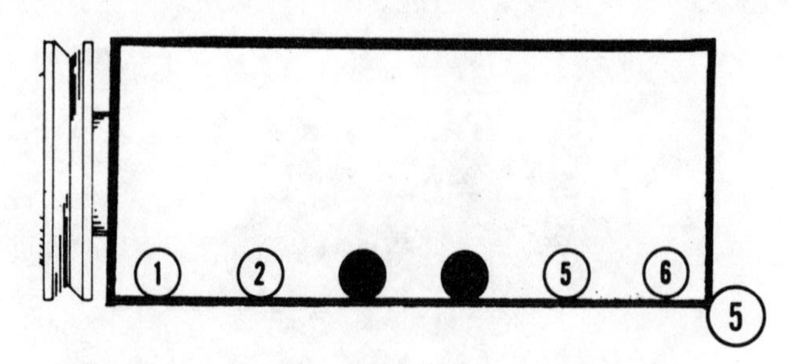

Condition 5: The two center plugs of a 6-cylinder engine are fouled. The cause may be raw fuel boiling out of the carburetor into the intake manifold after the engine is shut down. This condition can be the result of fuel percolating in the carburetor, a leaking intake valve and seat, a heavy float, or a too-high fuel level in the carburetor bowl. Refer to Chapter 4.

SPARK PLUG EVALUATION

Removal: Remove the spark plug wires by pulling and twisting on only the molded cap. **NEVER** pull on the wire or the connection inside the cap may become separated or the boot damaged. Remove the spark plugs and keep them in order. **TAKE CARE** not to tilt the socket as you remove the plug or the insulator may be cracked.

Examine: Line the plugs in order of removal and carefully examine the firing end to determine the firing conditions in each cylinder.

Correct Color: A proper firing plug should be dry and powdery. Hard deposits inside the shell indicate the engine is starting to use some oil, but not enough to cause concern. The most important evidence is the light gray color of the porcelain, which is an indication this plug has been running at the correct temperature. This means the

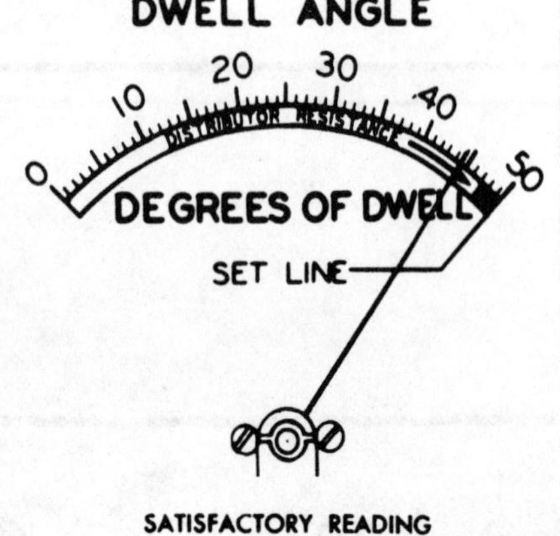

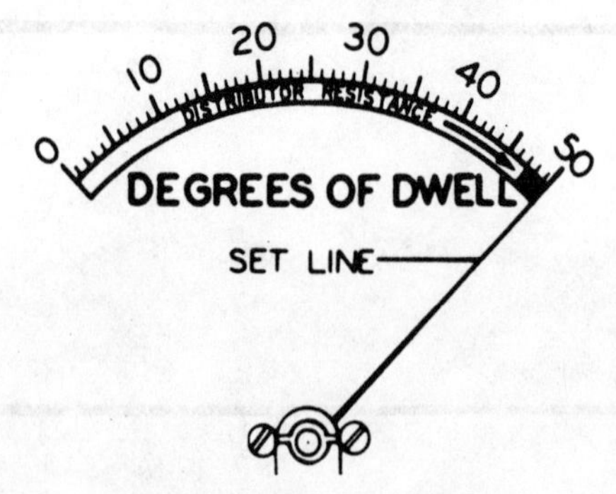

Dwell meter reading with a satisfactory set of contact points (left) adjusted between 44 and 45 degrees. The same set of points with an unsatisfactory reading (right) at 50 degrees.

Example of excessive overheating.

Heavy oil consumption.

plug is one with the correct heat range and also that the air-fuel mixture is correct.

Overheating: A dead white or gray insulator, which is generally blistered, is an indication of overheating and pre-ignition. The electrode gap wear rate will be more than normal and in the case of pre-ignition, will actually cause the electrodes to melt as shown in this illustration. Overheating and pre-ignition are usually caused by overadvanced timing, detonation from using too-low an octane rating fuel, an excessively lean air-fuel mixture, or problems in the cooling system.

Fouled: A foulded spark plug may be caused by the wet oily deposits on the insulator shorting the high-tension current to ground inside the shell. The condition may also be caused by ignition problems which prevent a high-tension pulse to be delivered to the spark plug.

Carbon Deposits: Heavy carbon-like deposits are an indication of excessive oil consumption. This condition may be the result of worn piston rings, worn valve guides, or from a valve seal that is either worn or was incorrectly installed.

Deposits formed only on the shell is an indication the low-speed air-fuel mixture is too rich. At high speeds with the correct mixture, the temperature in the combustion chamber is high enough to burn off the deposits on the insulator.

Too Cool: A dark insulator, with very few deposits, indicates the plug is running too cool. This condition can be caused by low compression or by using a spark plug of an incorrect heat range. If this condition shows on only one plug it is most usually caused by low compression in that cylinder. If all of the plugs have this appearance, then it is probably due to the plugs having a too-low heat range.

Powdery deposits have melted and shorted-out this spark plug.

Red, brown, or yellow deposits are by-products of combustion from fuel and lubricating oil.

Rich Mixture: A black, sooty condition on both the spark plug shell and the porcelain is caused by an excessively rich air-fuel mixture, both at low and high speeds. The rich mixture lowers the combustion temperature so the spark plug does not run hot enough to burn off the deposits.

Electrode Wear: Electrode wear results in a wide gap and if the electrode becomes carbonized it will form a high-resistance path for the spark to jump across. Such a condition will cause the engine to misfire during acceleration. If all of the plugs are in this condition, it can cause an increase in fuel consumption and very poor performance at high-speed operation. The solution is to replace the spark plugs with a rating in the proper heat range and gapped to specification.

Red rust-colored deposits on the entire firing end of a spark plug can be caused by water in the cylinder combustion chamber. This can be the first evidence of water entering the cylinders through the exhaust manifold because of an accumulation of scale or defective exhaust shutter. This condition **MUST** be corrected at the first opportunity. Refer to Chapter 7, Cooling System Service.

POLARITY CHECK

Coil polarity is extremely important for proper battery ignition system operation. If a coil is connected with reverse polarity, the spark plugs may demand from 30 to 40 percent more voltage to fire. Under such demand conditions, in a very short time the coil would be unable to supply enough voltage to fire the plugs. Any one of the following three methods may be used to quickly determine coil polarity.

1- The polarity of the coil can be checked using an ordinary D.C. voltmeter. Connect the positive lead to a good ground.

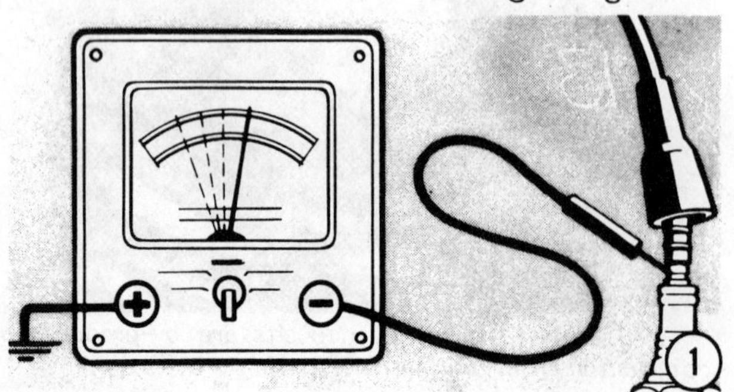

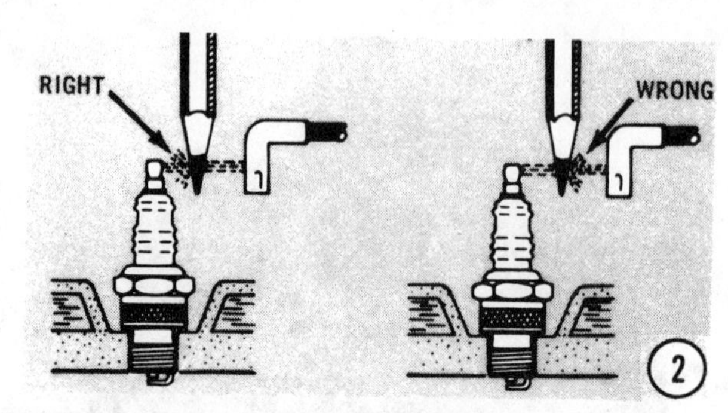

With the engine running, momentarily touch the negative lead to a spark plug terminal. The needle should swing upscale. If the needle swings downscale, the polarity is reversed.

2- If a voltmeter is not available, a pencil may be used in the following manner: Disconnect a spark plug wire and hold the metal connector at the end of the cable about 1/4" from the spark plug terminal. Now, insert an ordinary pencil tip between the terminal and the connector. Crank the engine with the ignition switch ON. If the spark feathers on the plug side and has a slight orange tinge, the polarity is correct. If the spark feathers on the cable connector side, the polarity is reversed.

3- The firing end of a used spark plug can give a clue to coil polarity. If the ground electrode is "dished", it may mean polarity is reversed.

5-4 DISTRIBUTOR SERVICE

The four- and six-cylinder in-line engines all have a Bosch distributor. The V6 and V8 engines may have a Mallory distributor or an Autolite/Prestolite distributor.

Service procedures for these distributors will be given in the above listed order.

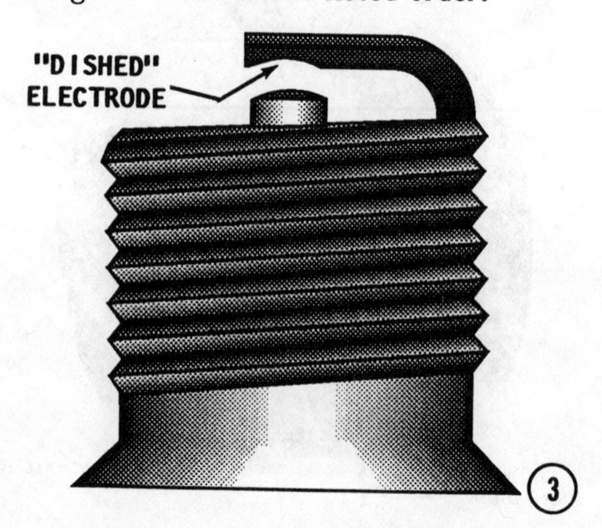

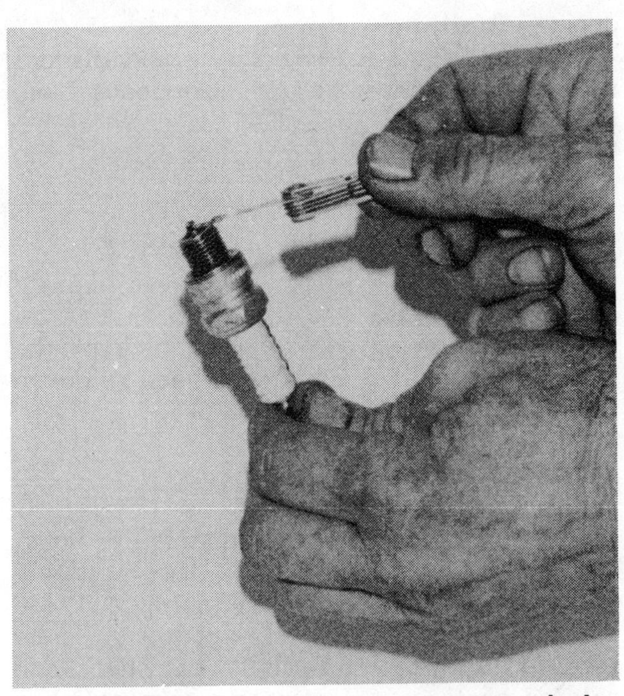

ALWAYS *use a round gauge to measure spark plug gap. A flat gauge will not give an accurate reading.* ***NEVER*** *use a gasket on a spark plug with a tapered seat.*

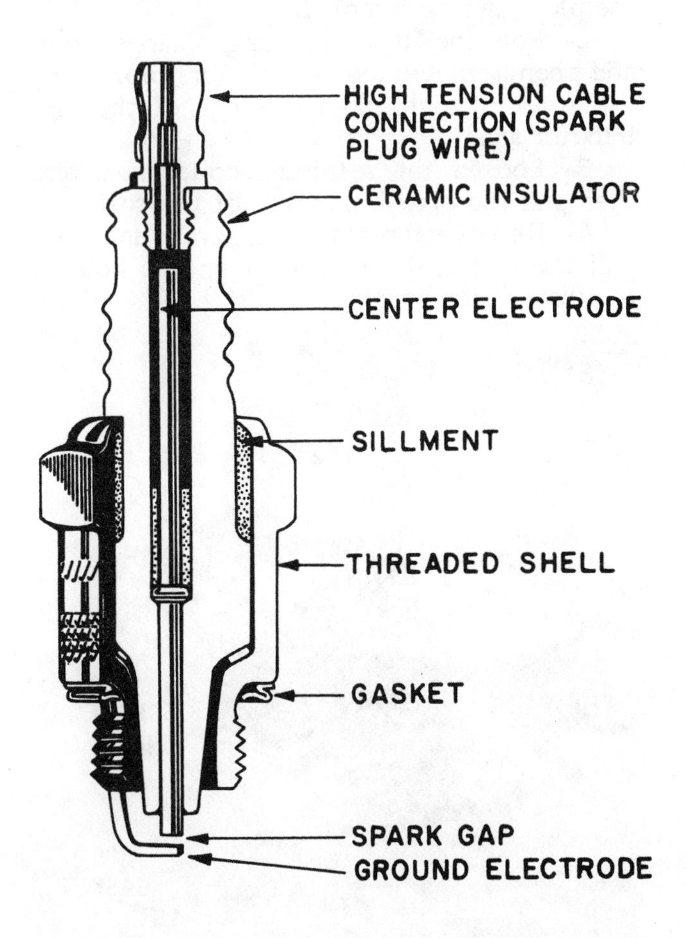

Cross section view of a spark plug.

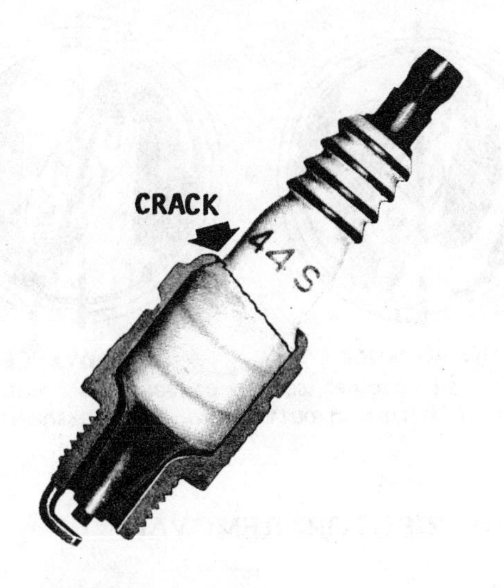

A cracked insulator is always caused by tilting the socket during removal or installation.

The distributor should be removed for any service including the installation and adjustment of the breaker points. New breaker points can be installed with the distributor in place but the position of the distributor usually makes it awkward to do a perfect job of aligning the points and making the point gap setting.

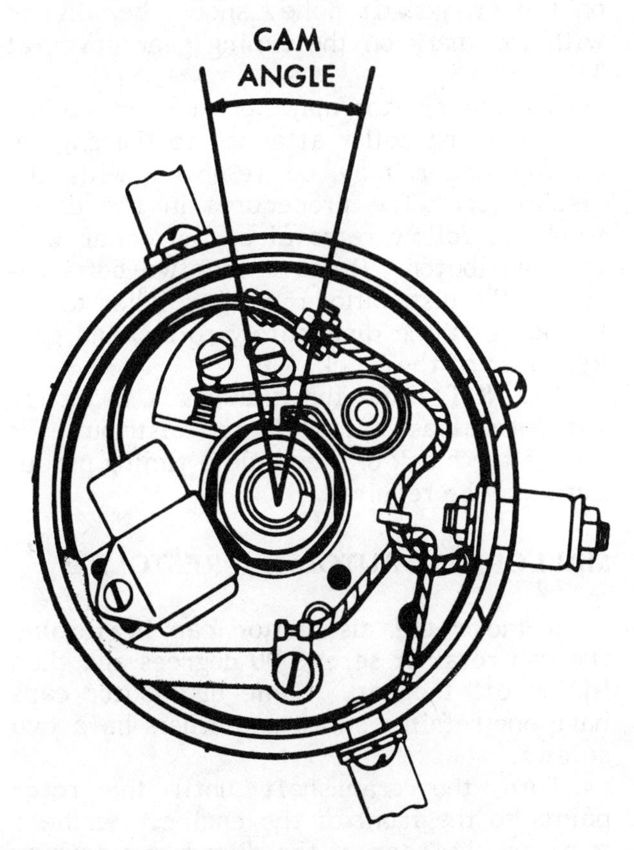

Cam angle of points closed.

Increased engine speed causes the centrifugal weights to be thrown outward to advance the ignition timing.

5-5 DISTRIBUTOR REMOVAL

BOSCH

Identify each opening on the distributor cap to ensure each spark plug wire will be connected back into the same hole from which it was removed.

Pull the primary and secondary wires from the distributor cap. Remove the No. 1 spark plug. Hold a finger over the opening and at the same time have an assistant crank the engine until compression is felt indicating the No. 1 cylinder is approaching top dead center (TDC). The timing marks on the crankshaft pulley should be aligned with the mark on the timing gear cover at TDC.

The distributor may be removed leaving the mounting collar attached to the engine, or the collar may be removed with the distributor. The procedures in the disassembling follow removal of the collar with the distributor. Remove the two bolts securing the distributor retaining collar to the block. Lift the distributor, collar, and gasket clear of the block.

DO NOT allow the crankshaft to be rotated the least bit while the distributor is out of the block or a complete timing procedure may be required.

MALLORY OR AUTOLITE/PRESTOLITE

Remove the distributor cap by turning the cap retainer screws 90 degrees and then lifting off the cap. Some distributor caps have one retainer screw and others have two screws.

Turn the crankshaft until the rotor points to the front of the engine. Scribe a mark on the edge of the distributor housing and a matching one on the block as an aid during installation.

Loosen the retaining screw and disconnect the condenser and primary leads from the terminal. Remove the hold-down clamp bolt, and then remove the distributor from the block.

5-6 BOSCH DISTRIBUTOR SERVICE

SPECIAL WORDS

The following procedures pickup the work after the distributor has been removed from the engine. See Section 5-5.

DISASSEMBLING

1- Scribe a mark on the distributor housing and a matching mark on the mounting collar as an aid during assembling. These marks will ensure the distributor will be installed with the crankshaft, camshaft, and related parts, in the same position as when the distributor was removed. Remove and **DISCARD** the gasket from under the retaining collar. This gasket may have stayed on the block during removal.

2- Pop the two retaining springs free, and then remove the distributor cap. Pull the rotor straight up and clear of the distributor shaft.

3- Loosen the retaining collar nut, and then pull the distributor free of the collar.

4- Remove the retaining screw and then pull the condenser wire through the hole in the distributor housing.

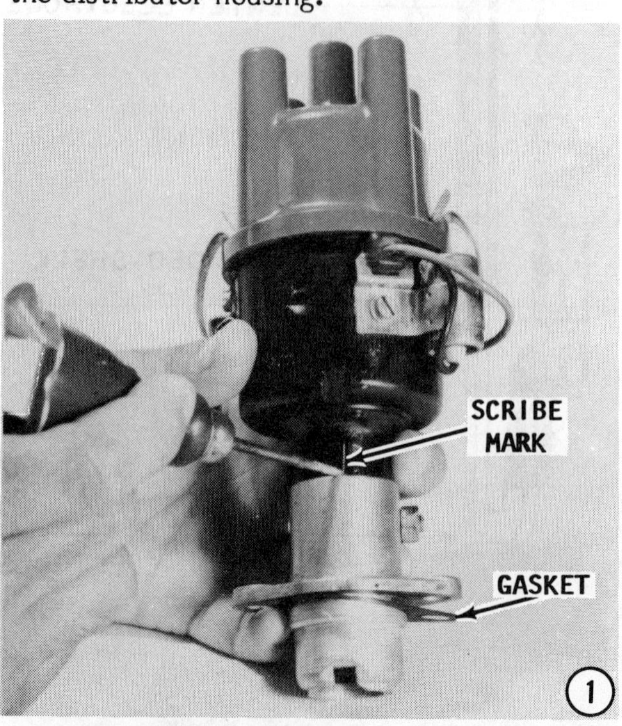

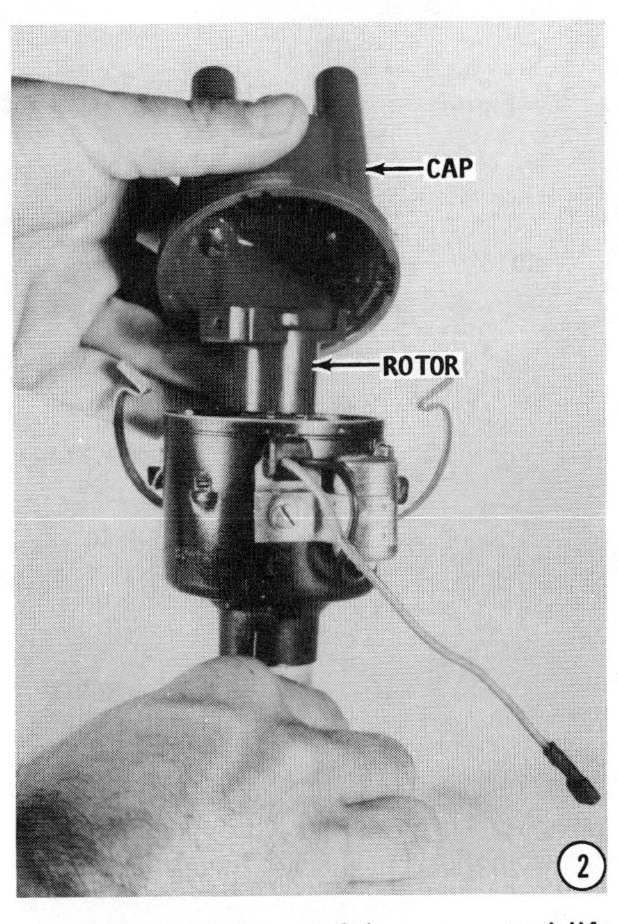

5- Remove the retaining screw and lift the breaker point set out of the distributor.

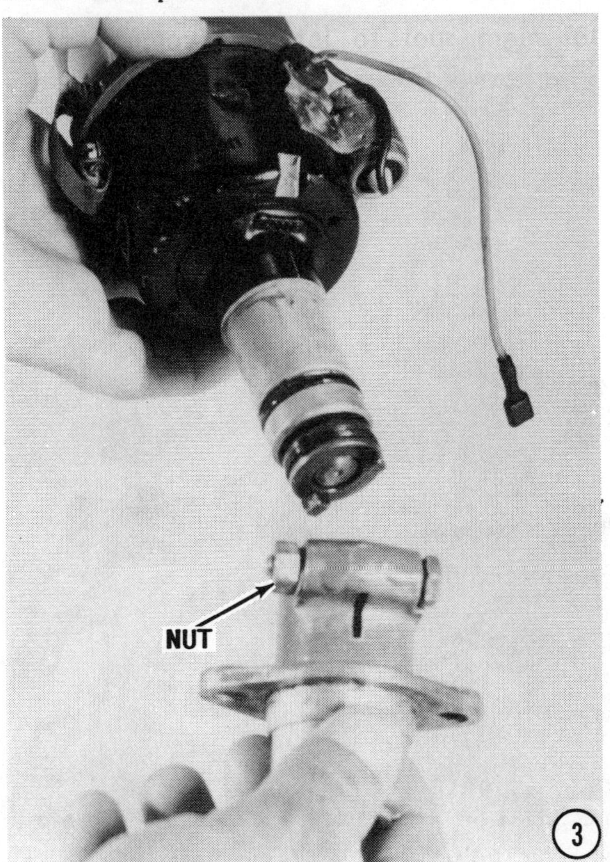

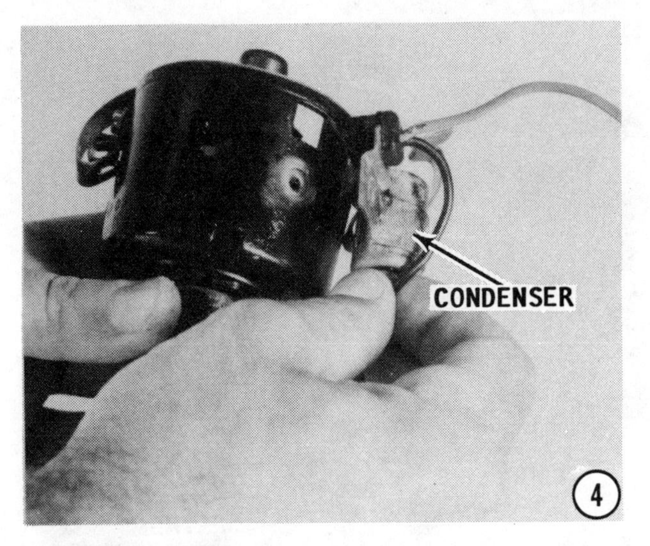

CRITICAL WORDS

Scribe a mark on the breaker plate and a matching mark on the inside of the distributor housing to **ENSURE** the plate will be installed back into the same position from which it was removed.

6- Remove the two screws from the outside of the distributor housing. These two screws hold the two distributor cap retaining springs in place and also the breaker plate. Rotate the plate until the

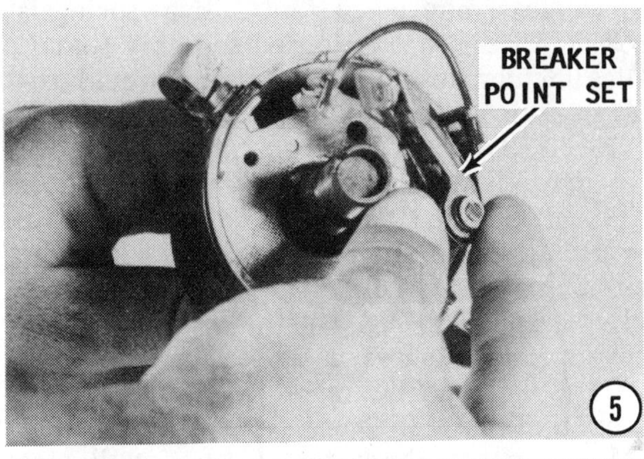

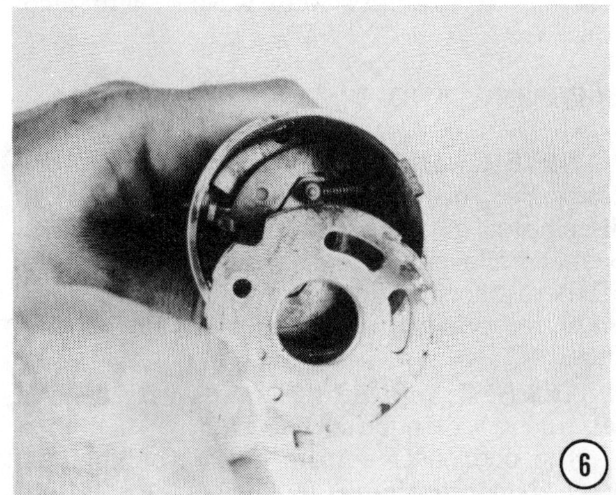

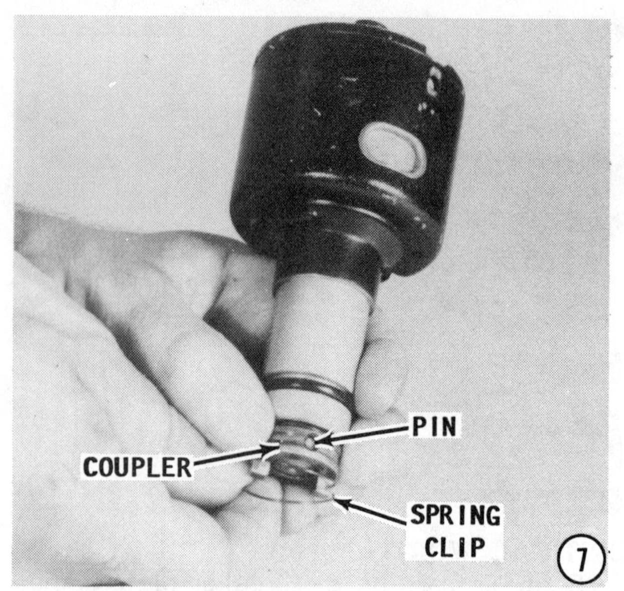

cutouts in the plate are aligned with the humps on the inside surface of the distributor housing. Lift the breaker plate out of the housing.

SPECIAL WORDS

Under normal conditions and only **RARELY** do the centrifugal weights need to be removed. If it is not possible to get any centrifugal advance from the distributor, the weights may be at fault. If a spring is broken, a weight is corroded so badly that it fails to function, then this area would require service.

7- SELDOM is it necessary to remove the drive coupling from the distributor shaft. Remove the spring clip. This clip serves as a retainer for the pin on the distributor drive coupling. Next, drive the pin out, and then remove the coupler. Slide the distributor shaft up out of the housing. The bushing can be replaced, if necessary. When the shaft and bushing are removed, a seal, and a stainless steel washer, will also come free.

CLEANING AND INSPECTING

NEVER wash the distributor cap, rotor, condenser, or breaker plate assembly, of a distributor in any type of cleaning solvent. Such compounds may damage the insulation of these parts or, in the case of the breaker plate assembly, saturate the lubricating felt.

ALWAYS replace the points with a new set during a distributor overhaul.

The condenser seldom gives trouble, but good shop practice a few years ago called

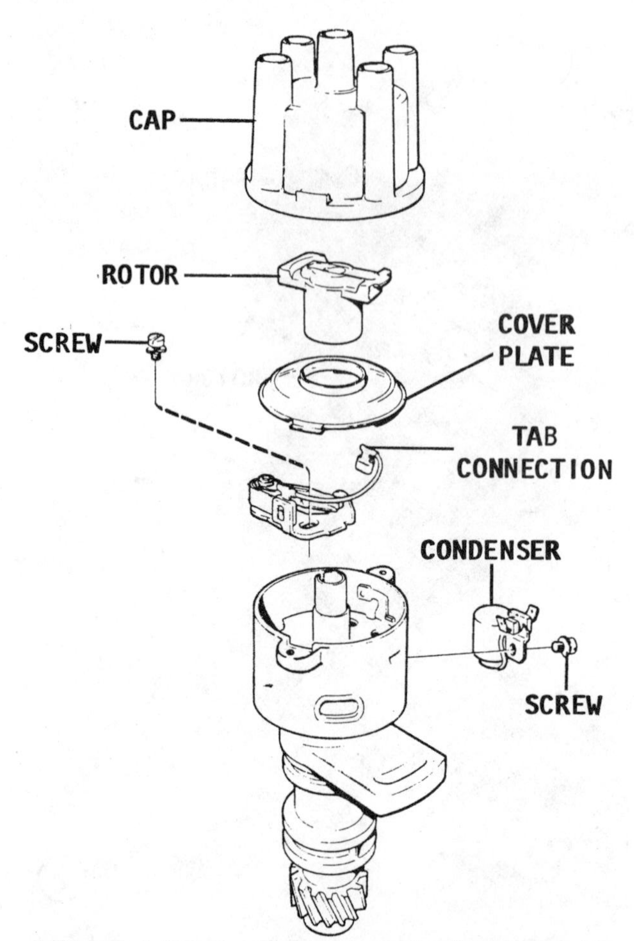

Exploded drawing of the distributor installed on the AQ125 and AQ145 engines, with principle parts identified.

Securing the distributor cap on an AQ145 engine.

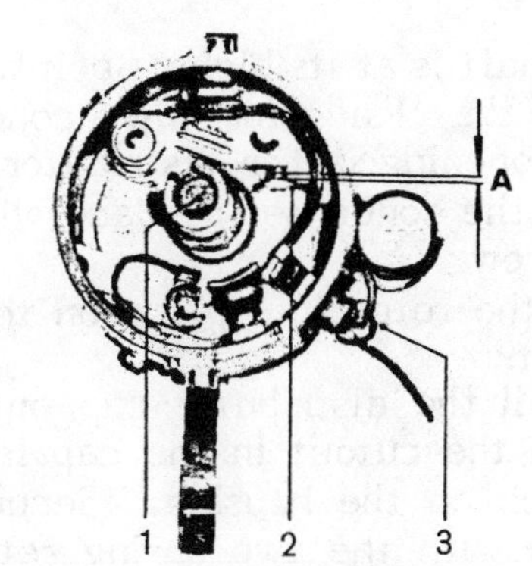

Arrangement of principle parts of a Mallory distributor installed on an AQ190 or AQ240 (Ford V8) engine.

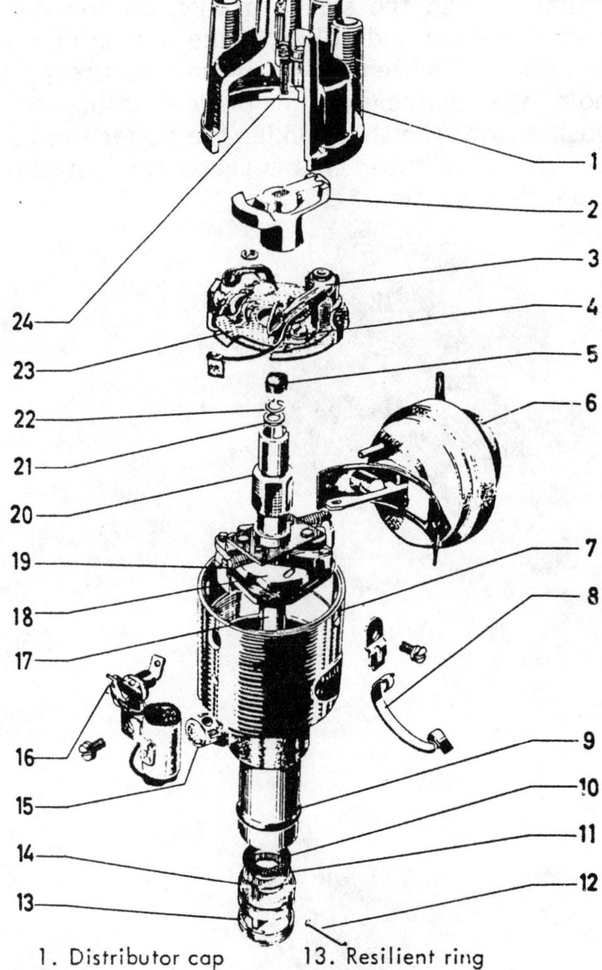

1. Distributor cap
2. Distributor arm
3. Contacts
4. Lock screw for breaker contacts
5. Lubricating felt
6. Vacuum regulator (AQ 165)
7. Distributor housing
8. Cap clasp
9. Rubber seal
10. Fibre washer
11. Steel washer
12. Lock pin
13. Resilient ring
14. Driving collar
15. Lubricator
16. Primary connection
17. Distributor shaft
18. Centrifugal governor weight
19. Centrifugal governor spring
20. Breaker cam
21. Washer
22. Circlip
23. Breaker plate
24. Rod brush (carbon)

Exploded drawing of the Bosch distributor installed on the AQ105, AQ115, AQ130, AQ165, and AQ170 engines.

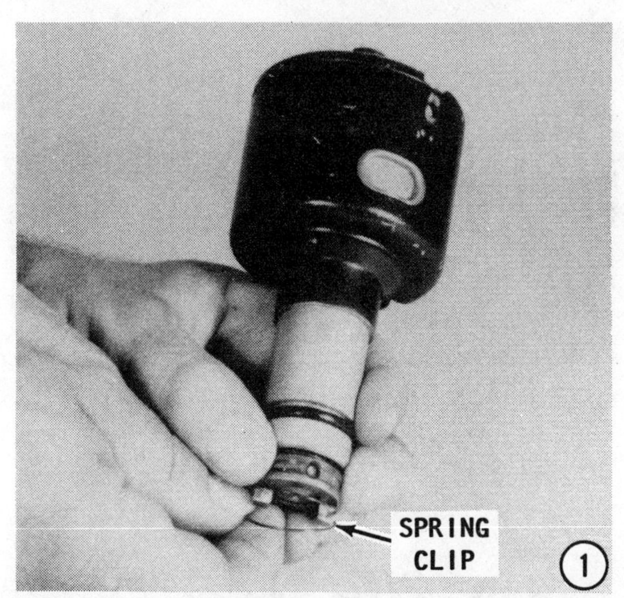

for a new condenser with a new set of points. Some point sets still include a condenser in the package. If you have paid for a new condenser, you might as well install it and be free of concern over that part.

Inspect the distributor cap for cracks or damage. Check the spark plug wires.

If the centrifugal weights were not removed, check their action to be sure they will extend and retract properly.

ASSEMBLING

1- Install the distributor shaft into the housing. Install the bushing, seal, and stainless steel washer onto the lower end of the shaft. Push the pin through, and then secure it with the spring clip.

Check the end play on the shaft. The end play should not be less than 0.004" and not more than 0.010".

2- Slide the breaker plate into the distributor housing with the cutouts in the plate aligned with the humps on the inside surface of the housing. After the plate is seated, rotate it until the scribe mark on the plate is aligned with the scribe mark on the inside surface of the housing. Secure the plate in position with the two screws attaching the two distributor cap retaining spring clips.

3- Install a **NEW** set of breaker points and secure them with the retainer screw.

4- If servicing a 4-cylinder engine, set the breaker points at 0.016" to 0.020". If servicing a 6-cylinder engine, set the points at 0.010" to 0.014". This is the open distance when the high side of the cam on the

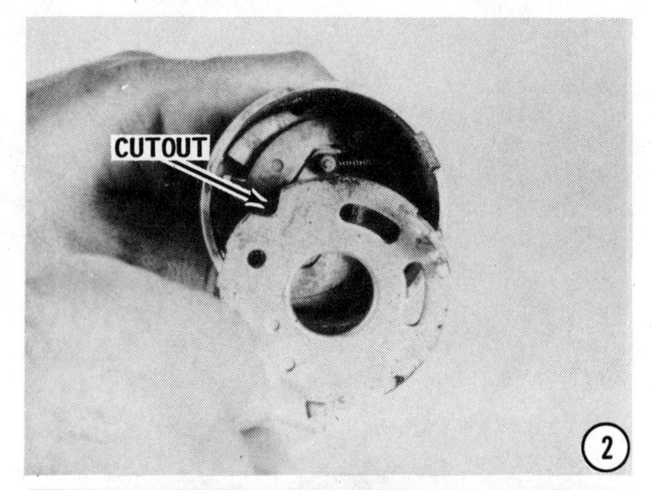

2

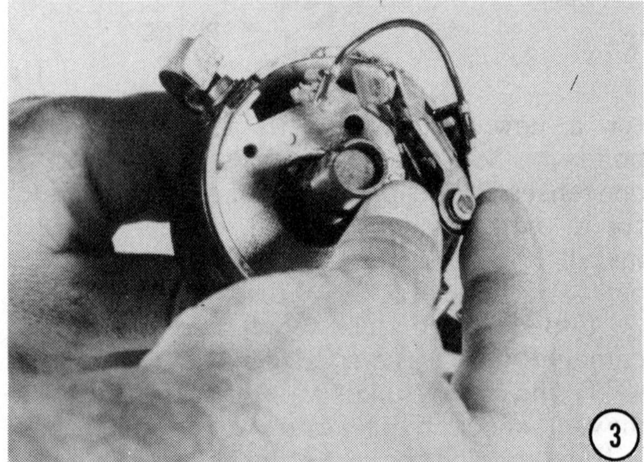
3

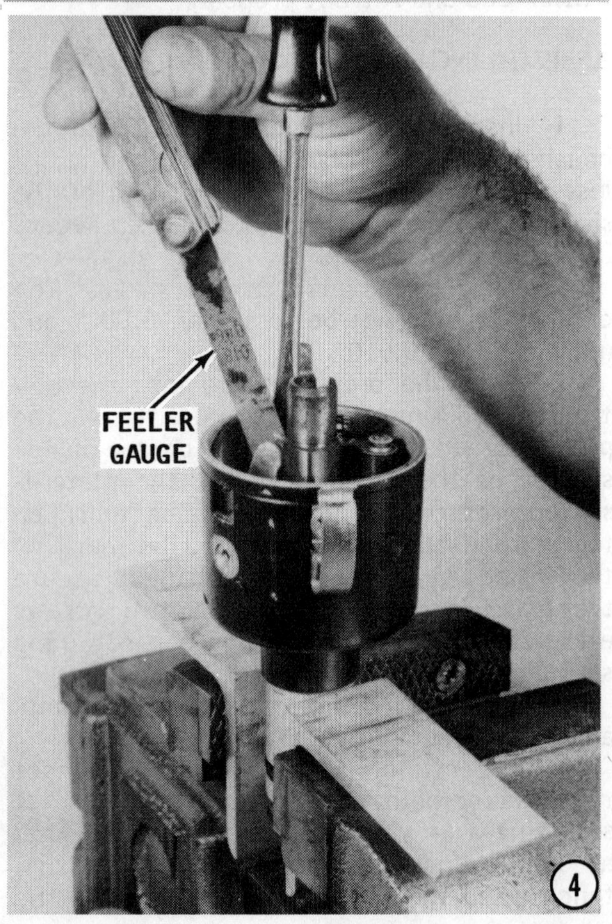

4

distributor shaft is at its highest point.

5- Push the lead for the condenser through the opening in the distributor housing. Secure the condenser in place with the attaching screw.

6- Push the rotor into place on the distributor shaft.

7- Install the distributor cap onto the housing with the cutout in the cap indexed over the knob on the housing. Secure the cap in place with the two spring retaining clips.

8- Slide the distributor shaft through the collar. Align the scribe marks, on the distributor shaft and collar, made during disassembling. Tighten the collar nut securely to hold the distributor in place. Slide the gasket onto the shaft under the collar plate.

The distributor is now ready for installation. See Section 5-10.

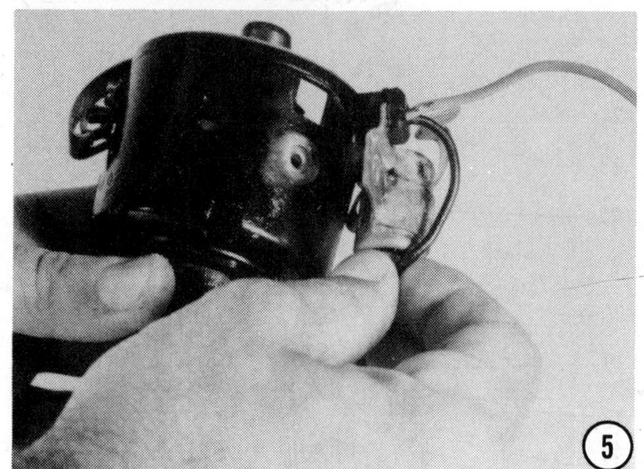
5

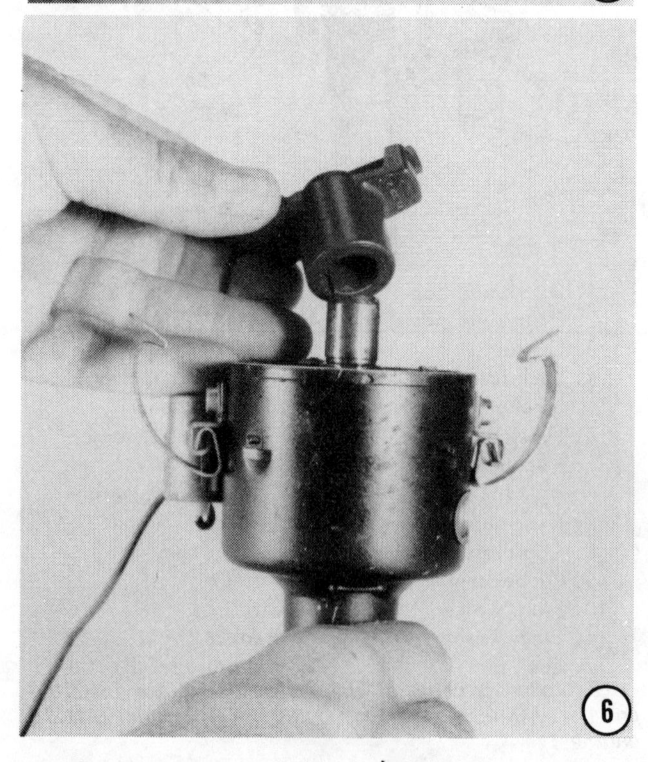
6

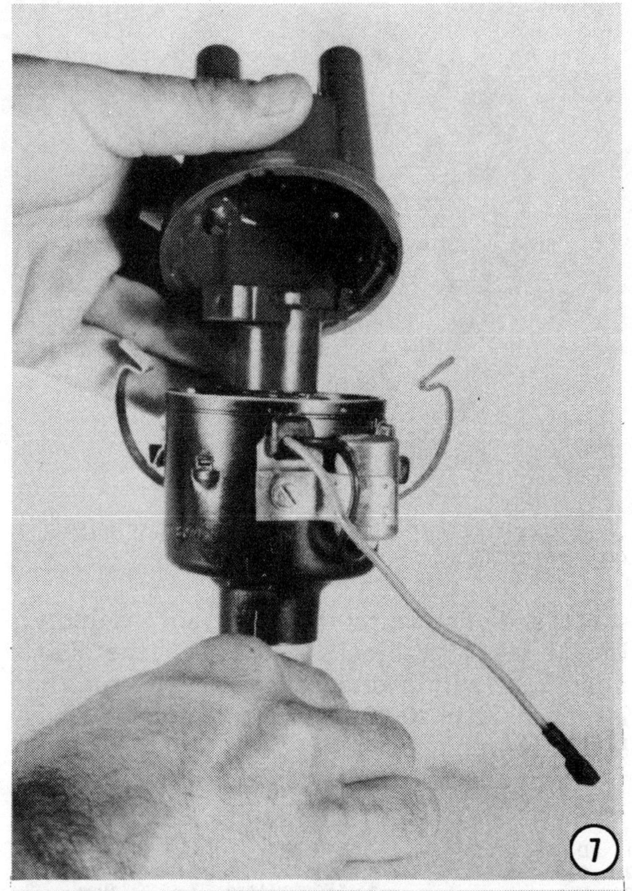

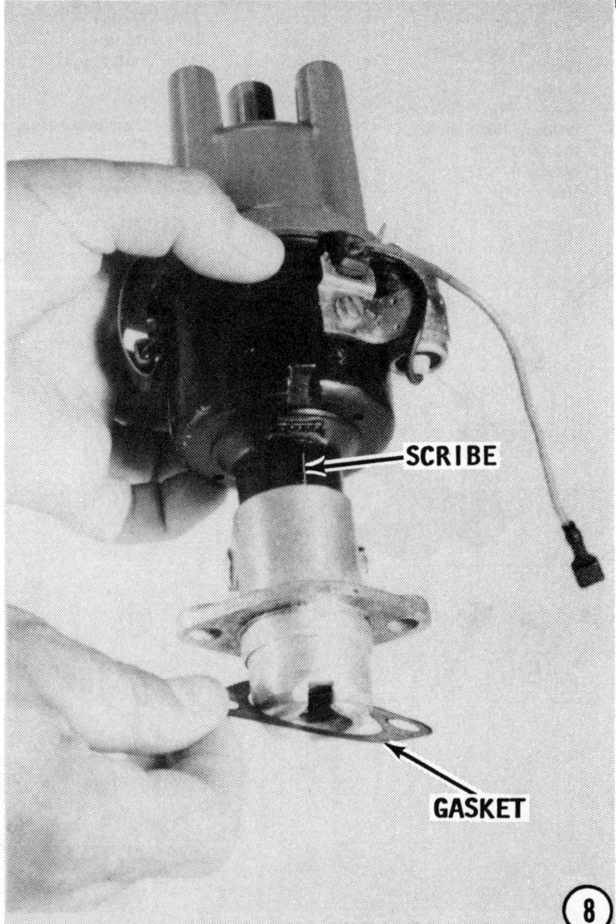

5-7 MALLORY DISTRIBUTOR SERVICE

REMOVAL AND DISASSEMBLY FROM A V8 ENGINE OR A V6 ENGINE 1981-1983

Remove the distributor cap and the cap gasket. Pull the rotor off of the distributor shaft.

Remove the condenser and bracket by first disconnecting the condenser lead from the primary terminal, and then removing the retaining screw. Disconnect the breaker assembly lead from the primary terminal, and then remove the retaining screw and the breaker assembly. Scribe a mark on the housing to indicate the position of the breaker plate as an aid to installation. Remove the plate retaining screws, and then the plate. Remove the oiler wick. The oiler wick is a press fit in the housing and need not be removed.

DO NOT disassemble the distributor further because the oil seal may be damaged.

Drive the retaining pin from the drive gear, and then remove the gear from the shaft. Drive the roll pin out of the distributor shaft collar. Remove the collar, washer, distributor housing, and washer. Remove the vent screen from the housing.

CLEANING AND INSPECTING

NEVER wash the distributor cap, rotor, condenser, or breaker plate assembly of a distributor in any type of cleaning solvent. Such compounds may damage the insulation

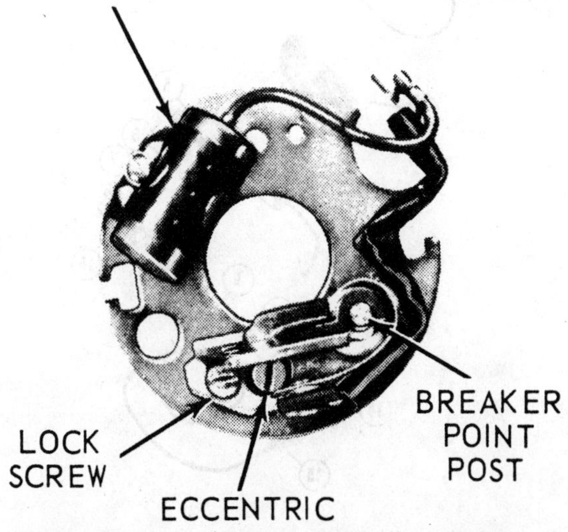

Arrangement of principle parts of a Mallory distributor installed on an AQ190 or 240 (Ford V8) engine.

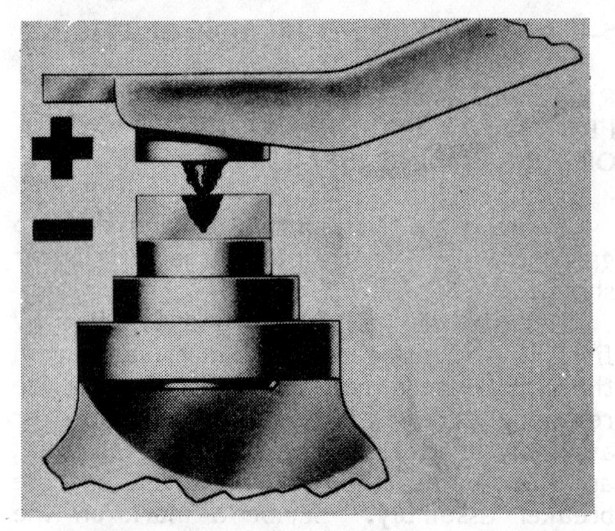

Oxidized contact points which should be replaced for satisfactory performance.

A crack in the distributor cap will cause misfiring and hard starting.

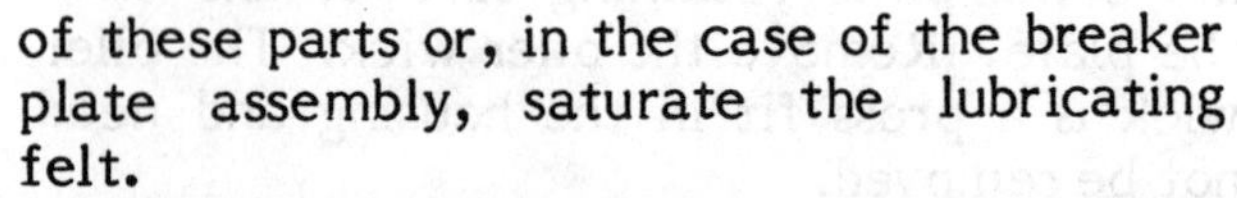

of these parts or, in the case of the breaker plate assembly, saturate the lubricating felt.

Check the shaft for wear and fit in the distributor body bushings. If either the shaft or the bushings are worn, replace the shaft and distributor body as an assembly. Use a set of V-blocks and check the shaft alignment with a dial gauge. If the run-out is more than 0.002", the shaft and body **MUST** be replaced.

Inspect the breaker plate assembly for

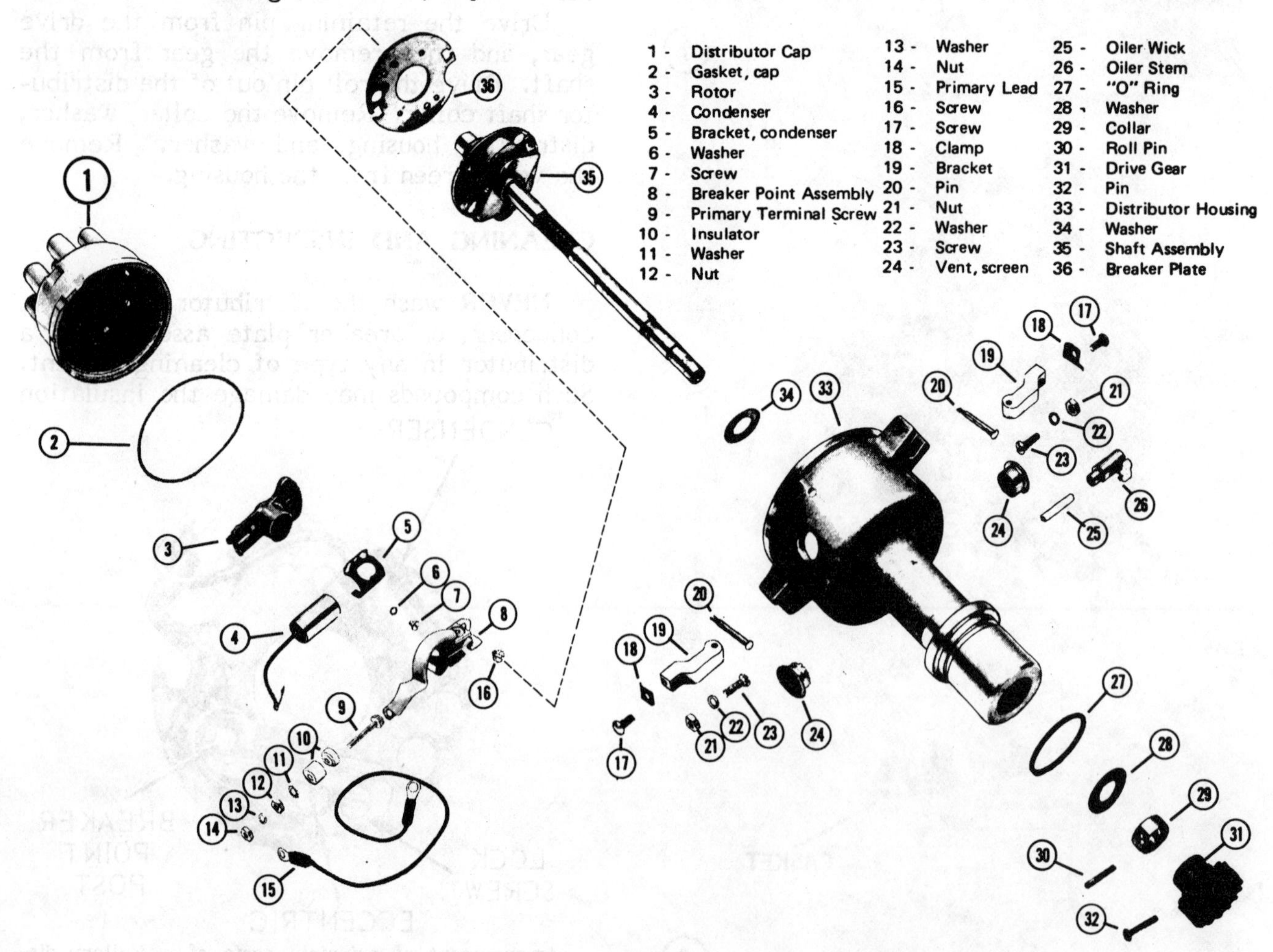

Exploded drawing of a Mallory distributor with principle parts identified.

damage and replace it if there are signs of excessive wear.

Check to be sure the governor weights fit free on their pins and do not have any burrs or signs of excessive wear. Check the cam fit on the end of the shaft. The cam should not fit loose but it should still be free without binding.

ALWAYS replace the points with a new set during a distributor overhaul.

The condenser seldom gives trouble, but good shop practice a few years ago called for a new condenser with a new set of points. Some point sets still include a condenser in the package. If you have paid for a new condenser, you might as well install it and be free of concern over that part.

Inspect the distributor cap for cracks or damage. Check the spark plug wires.

ASSEMBLING THE MALLORY FROM A V8 OR V6 ENGINE

Install the vent screen in the distributor housing vent hole. Secure the screen in place by crimping the inside flange.

Slide a washer onto the distributor shaft, and then install the shaft through the housing. Install the washer and collar on the shaft and secure the collar in place with a **NEW** pin. Check the shaft for end play which should be between 0.008" and 0.010". Install the drive gear on the distributor shaft with the shoulder going on first and secure the gear in place with a **NEW** pin through the shoulder and shaft. Peen both ends of the pin to prevent it from coming out.

Insert the oiler wick into the oiler stem. Install the breaker plate screws, the plate, washers, and nuts. **DO NOT** tighten the nuts at this time.

Install the breaker plate with the marks made during disassembly aligned, and then tighten the screws. Install the breaker assembly and connect the breaker assembly lead to the distributor primary terminal.

Install and secure the condenser with the bracket and attaching screw. Connect the condenser lead to the distributor primary terminal.

Slide the rotor onto the distributor shaft. Install a **NEW** gasket on the distributor cap with the notch in the cap aligned with the

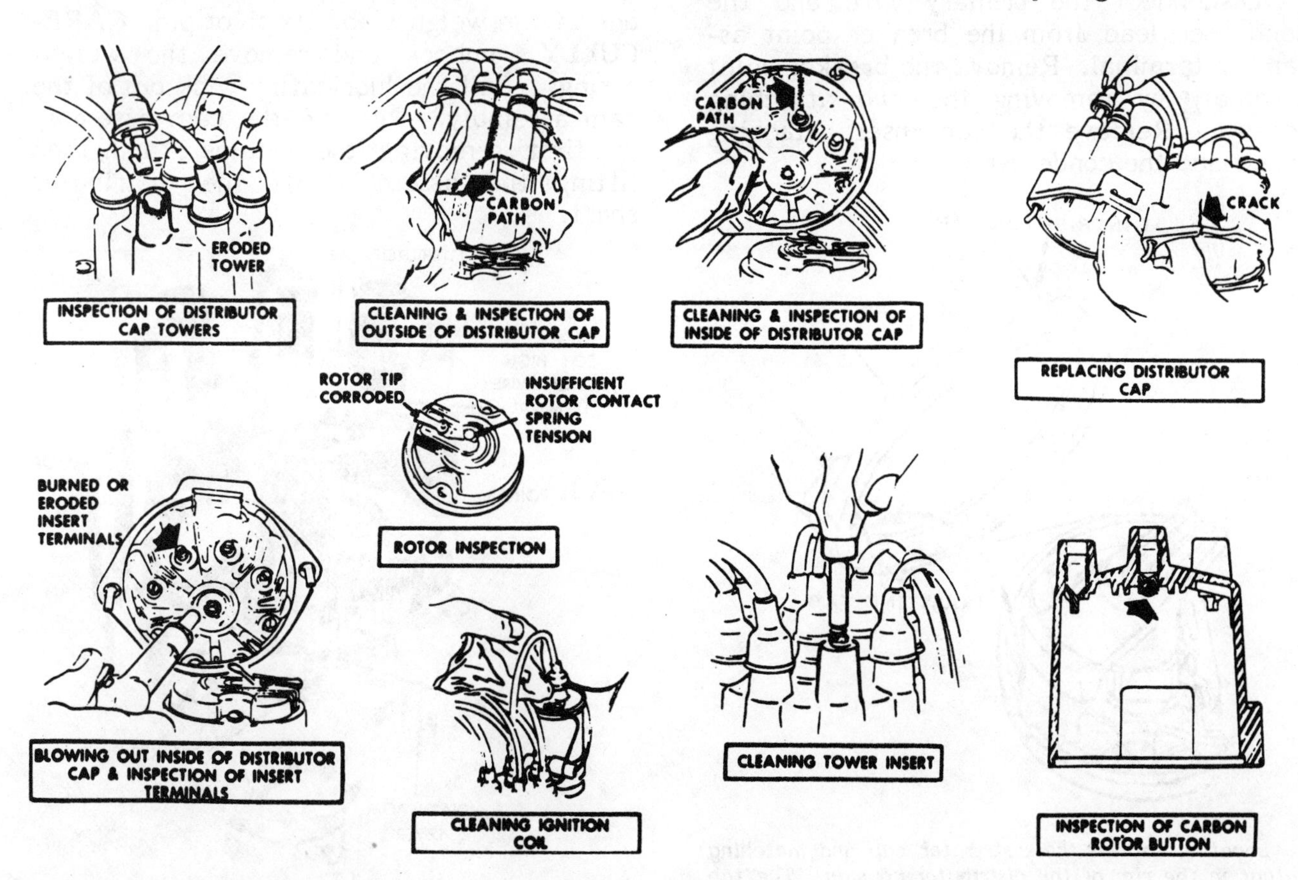

Drawings depicting a thorough cleaning and inspection of the distributor cap.

locating pin on the housing. Install the distributor cap with the tab on the inside rim of the cap aligned with the notch in the housing. If the cap is not properly positioned, as described, the rotor will strike one of the segments in the cap and crack it. Secure the cap in place by tightening the two screws.

GAPPING CONTACT POINTS

After the points have been properly installed, the gap must be accurately set, or the dwell adjusted, and then the ignition timing set. These two adjustments are covered in detail in Section 5-9 and 5-10 of this chapter.

5-8 AUTOLITE/PRESTOLITE DISTRIBUTOR SERVICE

DISASSEMBLING FROM AN AQ190 OR AQ240 (FORD V8) ENGINE

Remove the distributor cap, and then remove the rotor.

Disconnect the primary wire and the condenser lead from the breaker point assembly terminal. Remove the breaker point assembly by removing the two attaching screws. Remove the condenser attaching screw and the condenser.

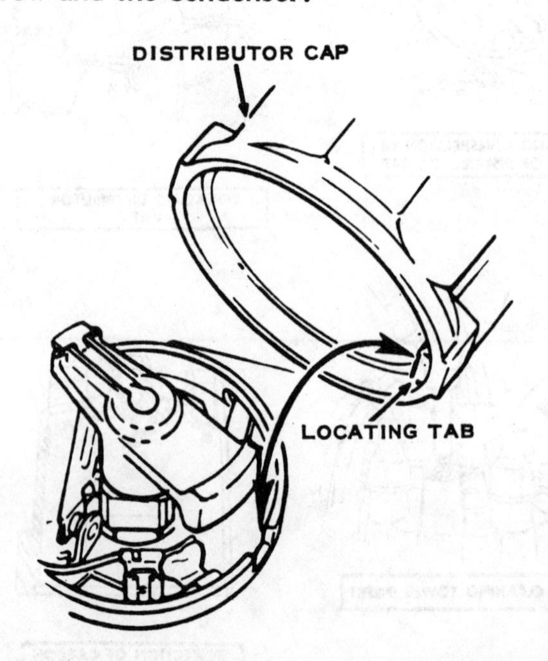

*Locating tab on the distributor cap and matching cutout in the rim of the distributor housing. The tab **MUST** index in the cutout or the tip of the rotor will strike one segment of the cap and crack it.*

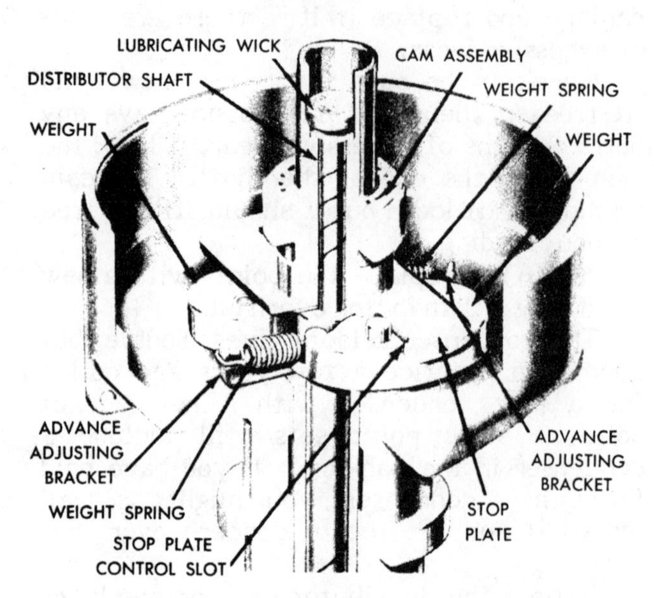

Cutaway view with the breaker plate and associated parts removed. Notice the location of the two weight springs and the relationship of other major parts.

Pull the primary lead through the opening in the housing. Remove the two breaker plate attaching screws, and then remove the breaker plate.

Identify one of the distributor weight springs and its bracket with a mark. Mark one of the weights and its pivot pin. **CAREFULLY** unhook and remove the weight springs. Pull the lubricating wick out of the cam assembly. Remove the cam assembly by first removing the retainer, and then lifting the assembly off the distributor shaft.

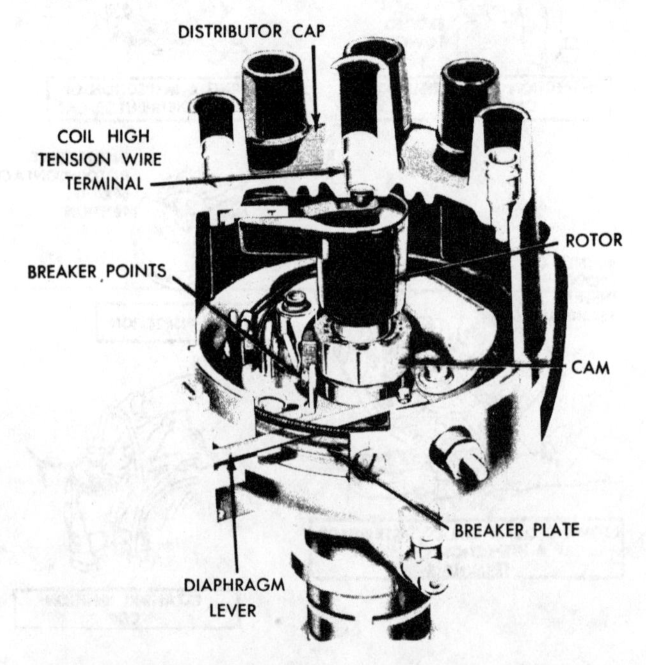

Cutaway view of an Autolite/Prestolite distributor with major parts identified.

Remove the thrust washer installed only on counterclockwise rotating engines.

Remove the weight retainers, and then remove the weights.

Remove the distributor cap clamps. Scribe a mark on the gear and a matching mark on the distributor shaft as an aid in locating the pin holes during assembly. Place the distributor shaft in a V-block, and then use a drift punch to remove the roll pin. Remove the gear from the shaft. Remove the shaft collar roll pin.

CLEANING AND INSPECTING

NEVER wash the distributor cap, rotor, condenser, or breaker plate assembly of a distributor in any type of cleaning solvent. Such compounds may damage the insulation of these parts or, in the case of the breaker plate assembly, saturate the lubricating felt.

Check the shaft for wear and fit in the distributor body bushings. If either the shaft or the bushings are worn, replace the shaft and distributor body as an assembly.

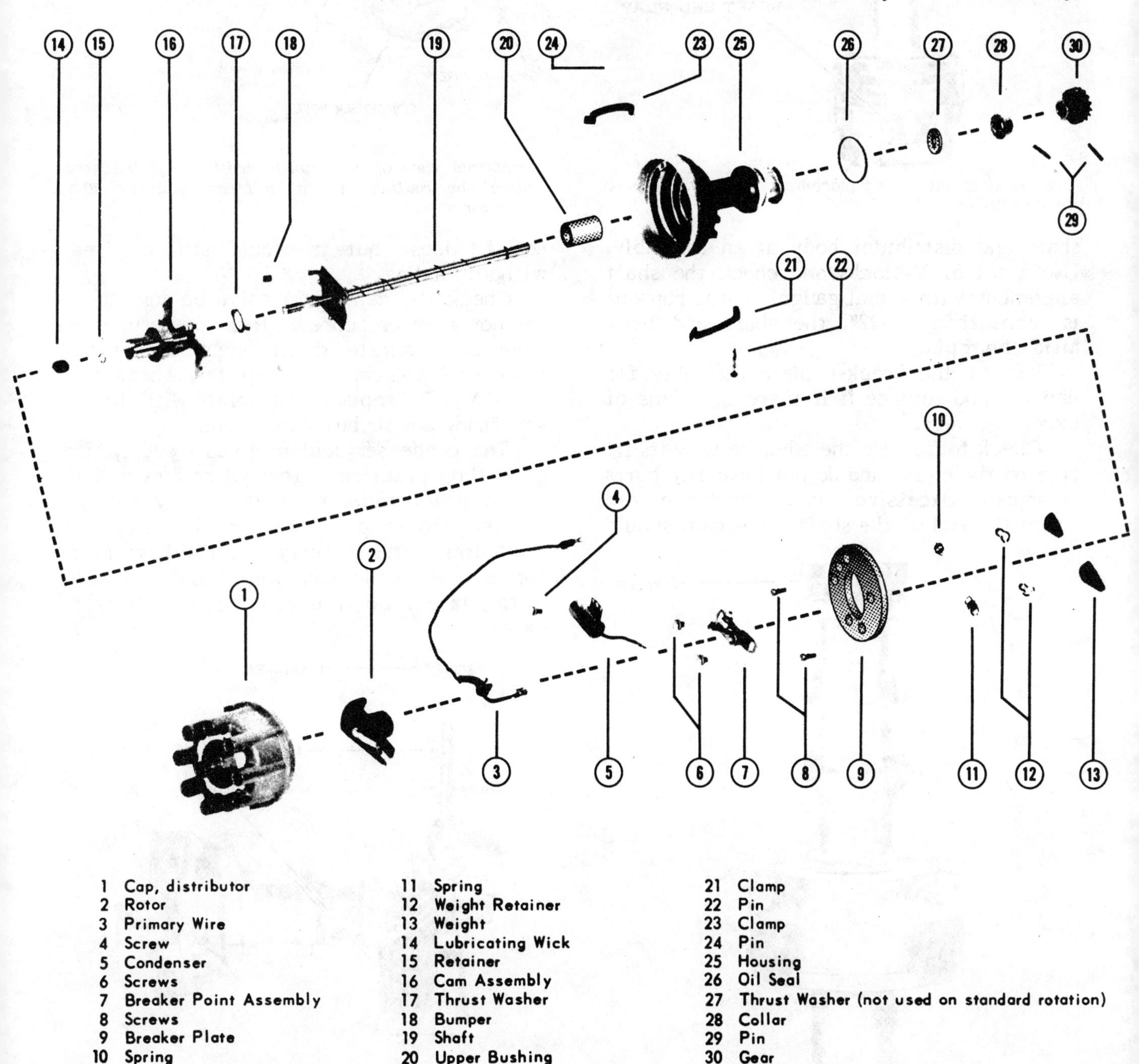

Exploded drawing of an Autolite/Prestolite distributor.

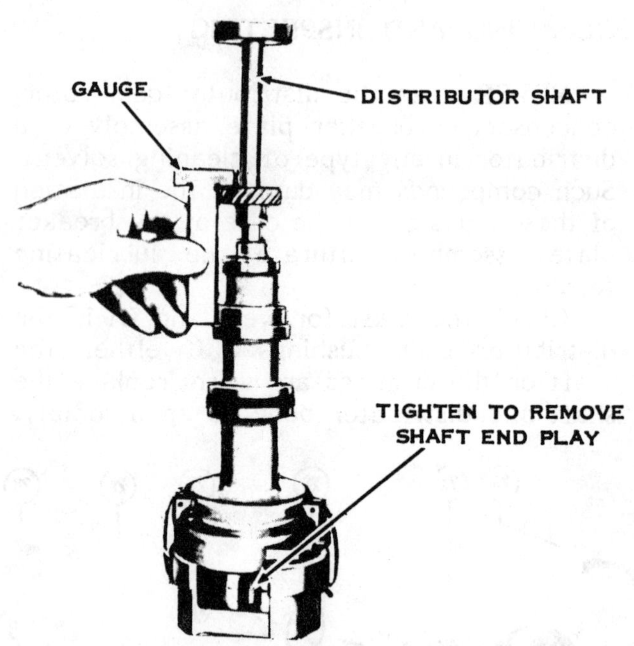

Method of measuring placement of a new gear on the distributor shaft.

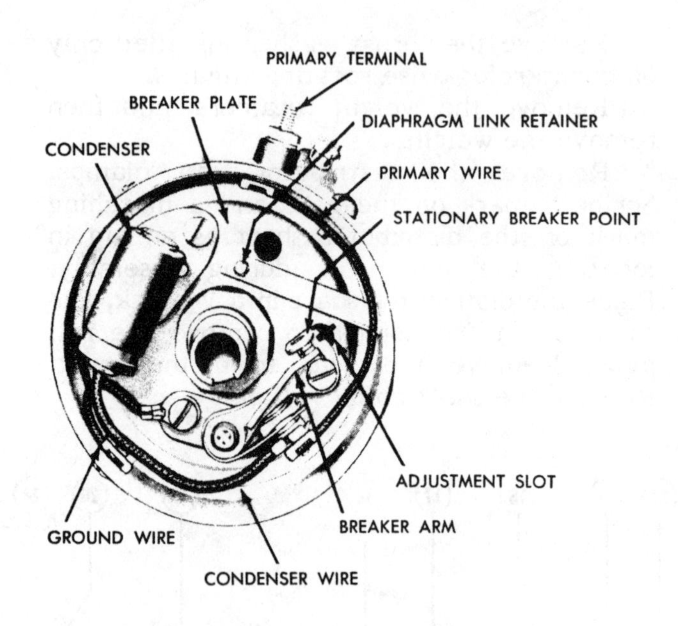

Internal view of an Autolite/Prestolite distributor. Notice the routing of the primary lead and the condenser wire.

shaft and distributor body as an assembly. Use a set of V-blocks and check the shaft alignment with a dial gauge. If the run-out is more than 0.002", the shaft and body **MUST** be replaced.

Inspect the breaker plate assembly for damage and replace it if there are signs of excessive wear.

Check to be sure the advance weights fit free on their pins and do not have any burrs or signs of excessive wear. Check the cam fit on the end of the shaft. The cam should not fit loose but it should still be free without binding.

Check the cam surfaces to be sure they are not worn or scored. Peak performance from an accurate dwell angle cannot be achieved if the cam is excessively worn.

ALWAYS replace the points with a new set during a distributor overhaul.

The condenser seldom gives trouble, but good shop practice a few years ago called for a new condenser with a new set of points. Some point sets still include a condenser in the package. If you have paid for a new condenser, you might as well install it and be free of concern over that part.

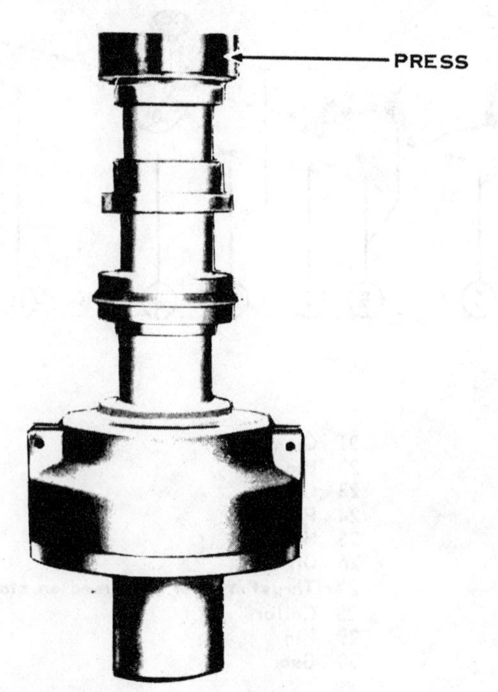

Setup to install a new upper bushing.

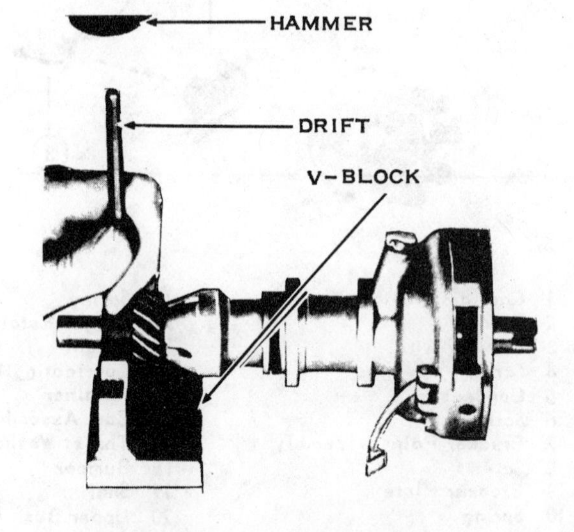

Setup to drive out the roll pin holding the distributor gear in place.

Inspect the distributor cap for cracks or damage. Check the spark plug wires.

ASSEMBLING THE AUTOLITE/PRESTOLITE

Lubricate the distributor shaft with crankcase oil, and then slide it into the distributor body. Slide the collar onto the shaft; align the holes in the collar with the hole in the shaft; and then install a **NEW** pin.

Install the distributor cap clamps. Left-hand rotating engine distributor assemblies have an additional thrust washer between the collar and the base. Use a feeler gauge between the collar and the distributor base to check the shaft end play. The end play should be between 0.024" and 0.035".

Install the gear onto the shaft with the marks on the gear and the shaft you made during disassembly aligned. The holes through the gear and the shaft should be aligned after the gear is installed. Install the gear roll pin.

Fill the grooves in the weight pivot pins with distributor cam lubricant. Position the weights in the distributor with the weight you identified with a mark during disassembly matched with the marked pivot pin. Secure the weights in place with the retainers. Slide the thrust washer onto the shaft. Fill the upper distributor shaft grooves with distributor cam lubricant.

Install the cam assembly with the marked spring bracket near the marked spring bracket on the stop plate. If a new cam assembly is being installed, **TAKE CARE** to be sure the cam is installed with the hypalon-covered stop in the correct cam plate control slot. The proper slot can be determined by measuring the length of the slot used on the old cam, and then using the corresponding slot on the new cam. Some new cams will have the size of the slot stamped in degrees near the slot. If the **WRONG** slot is used, the maximum advance will not be correct.

Coat the distributor cam lobes with a light film of distributor cam lubricant. Install the retainer and wick. Use a few drops of SAE 10W engine oil on the wick. Install the weight springs with the spring and bracket you marked during disassembly matched.

Place the breaker plate in position, and then secure with the attached screws.

Push the primary wire through the opening in the distributor. Place the breaker point assembly and the condenser in position and secure them in place with the attaching screws. Connect the primary wire and the condenser lead to the breaker point primary terminal.

Adjust the point gap and dwell as outlined in the following procedures, then install the rotor and the distributor cap.

5-9 ADJUSTING THE POINT GAP ALL DISTRIBUTORS

A feeler gauge or a dwell meter may be used to adjust the contact points. However, due to the rough surface of used points, a feeler gauge will never provide an accurate setting. The feeler gauge can give satisfactory results if a new set of contact points is being adjusted or if a dwell meter is not available. The feeler gauge is used when the points are adjusted with the distributor out of the engine and the dwell meter when the distributor is installed and the engine is running.

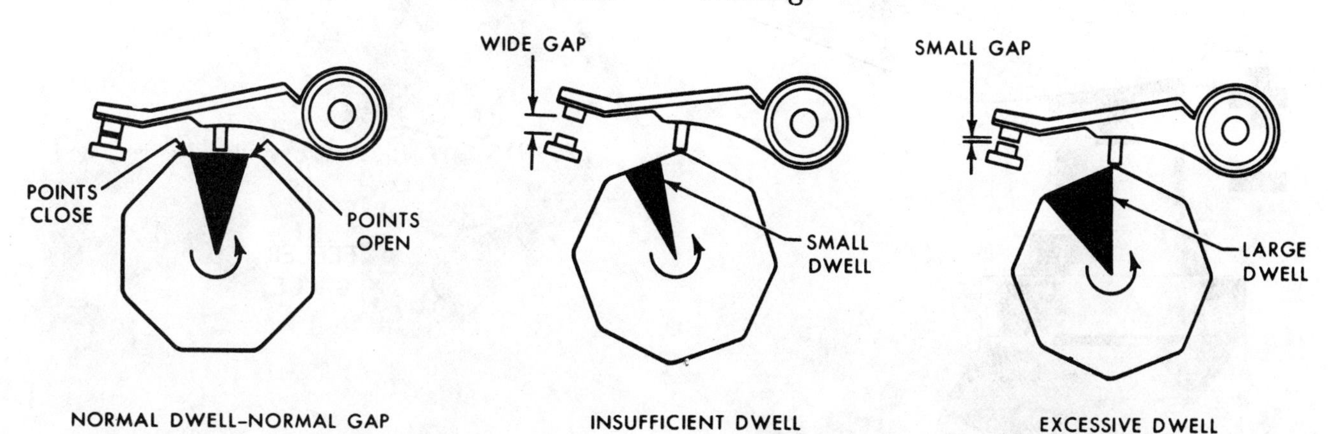

Three views of differing point dwell conditions. At the left is illustrated a normal gap (and dwell angle). The middle picture shows a wide gap, while the right drawing illustrates the result of a small gap. As the gap decreases through wear of the rubbing block, the ignition timing also retards, causing the engine to lose power.

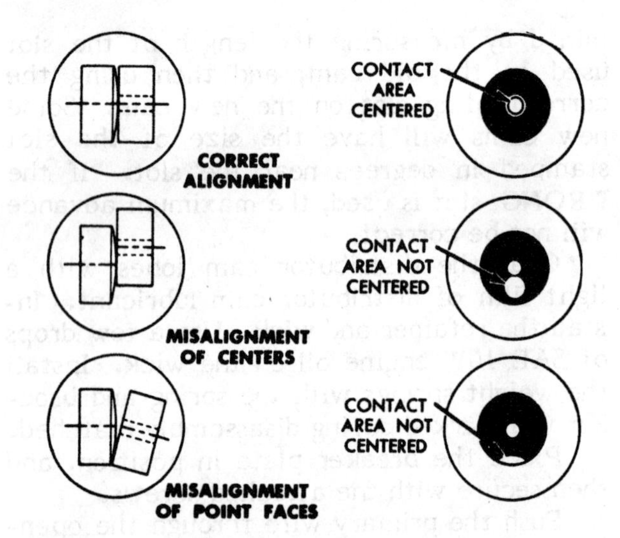

*Before setting the breaker point gap, the points must be properly aligned (top). **ALWAYS** bend the stationary point, **NEVER** the breaker lever. Attempting to adjust an old worn set of points is not practical, because oxidation and pitting of the points will always give a false reading.*

USING A FEELER GAUGE

If the distributor is not installed, rotate the shaft until the fiber rubbing block is on the high point of the cam. If the distributor is installed, crank the engine until the rubbing block is on the high point of the cam. Adjust the gap to the specification given in the Appendix by turning the eccentric adjuster on the stationary contact point or moving the point with a screwdriver in the base slot. Rotate the distributor shaft until the points are closed. Check to see if the points are properly aligned, as shown in the accompanying illustration. If necessary,

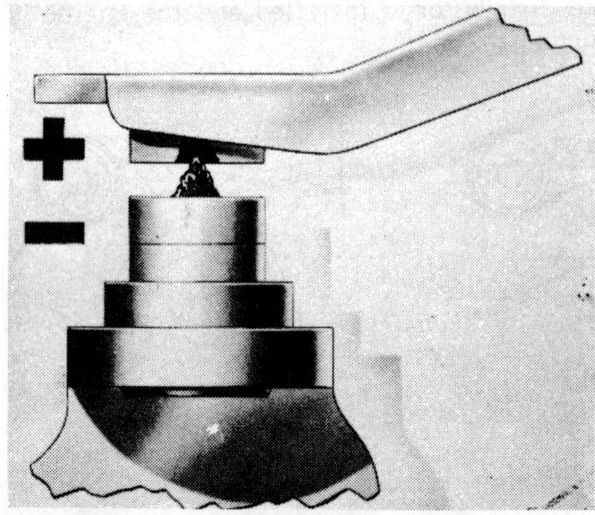

Exaggerated drawing to depict oxidized contact points which should be replaced for satisfactory performance.

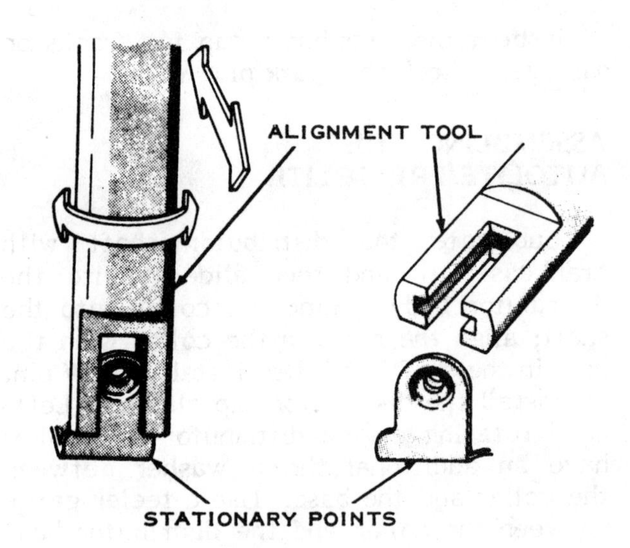

Method of bending the stationary point to align the contact points properly.

use a pair of needle-nose pliers, or a contact point alignment tool, to bend the **STATIONARY** point bracket until the points are aligned, as shown.

ALWAYS use a clean feeler gauge to make the final adjustment or you may leave a thin coating of oil on the points. Any oil on the points will oxidize in a short time and cause problems. **TAKE CARE** when making the gap measurement with the feeler gauge not to twist or cock the gauge. If the guage is not inserted square with the points, you will not get an accurate measurement.

Adjust the point gap about 0.003" wider than the specification to allow for initial rubbing block wear. Keep the contact point retaining screw snug while making the adjustment to keep the gap from changing when the screw is finally tightened.

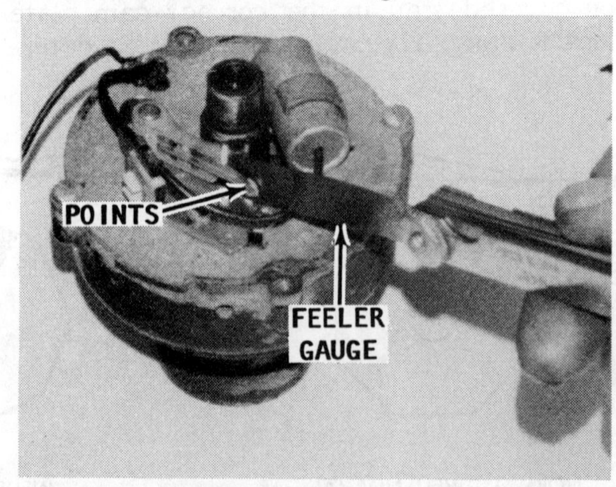

Using a feeler gauge to measure the breaker point gap. Keep the feeler gauge blade clean. The slightest amount of oil film transferred from the blade to the points will cause oxidation and hard starting.

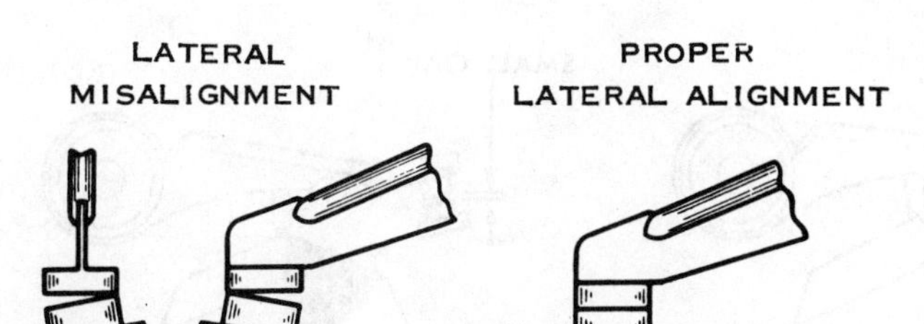

New contact points must be aligned by bending the fixed contact support. *CAUTION: Never bend the breaker lever.*

*Before setting the breaker point gap, the points must be properly aligned (right). Attempting to adjust an old worn set of points is not practical because oxidation and pitting of the point set will always give a false reading. **ALWAYS** bend the stationary point, **NEVER** the breaker lever.*

After the proper gap has been obtained, tighten the retaining screw, and then recheck the gap to be sure the setting has not changed.

Coat the distributor cam with a light film of heavy grease, and then turn the distributor shaft in the normal direction of rotation to wipe the lubricant off against the back of the rubbing block. The lubricant will remain there as a reservoir while the rubbing block wears. Wipe any excess lubricant from the cam.

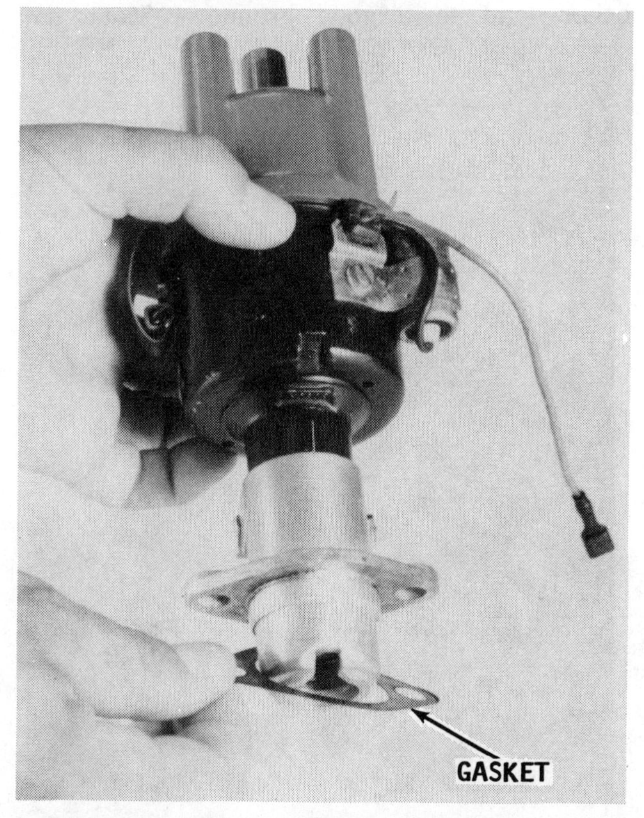

Bosch distributor ready for installation into the block.

5-10 DISTRIBUTOR INSTALLATION

BOSCH

With the gasket in place under the collar flange, slide the distributor shaft into the opening in the block until the flange and gasket are seated.

Secure the distributor and collar with the two bolts through the flange into the block. Tighten the bolts.

Install the primary and secondary wires to the distributor cap in the proper sequence.

MALLORY AND AUTOLITE/PRESTOLITE

During removal you scribed a mark on the distributor housing and a matching one on the block as an aid to installation. Now, slide the distributor shaft into the block with the rotor pointing toward the front of the engine and with the marks you scribed during removal aligned as closely as possible.

If the crankshaft was turned for any reason while the distributor was removed, the timing was lost and it will be necessary to retime the engine.

To time the engine, first remove the rocker arm cover. Next, rotate the crankshaft in the normal direction with a wrench on the harmonic balancer bolt until both valves for No. 1 cylinder are closed and the timing mark on the balancer is aligned with the "O" on the timing indicator. **NEVER** rotate the crankshaft in the opposite direction from the normal or the water pump

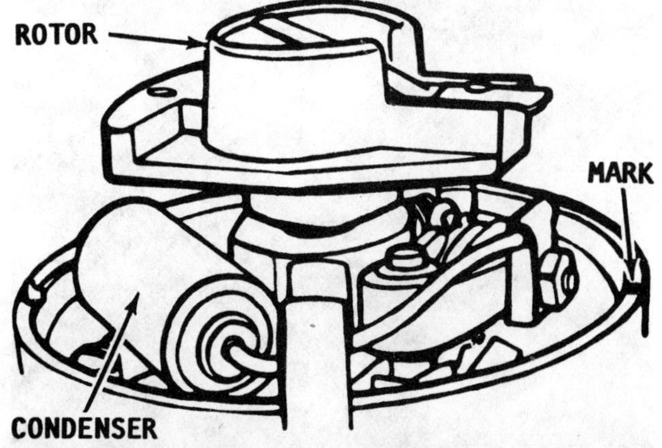

*As the distributor body is pulled up, the rotor will rotate through an arc as the distributor gear slides free of the camshaft gear. A mark should be scribed on the distributor body and a matching one on the block **AFTER** the distributor is removed, but **BEFORE** it is lifted clear. Align these marks during installation. When the distributor body is lowered down into the block and the two gears mesh, the rotor will swing back to its original position.*

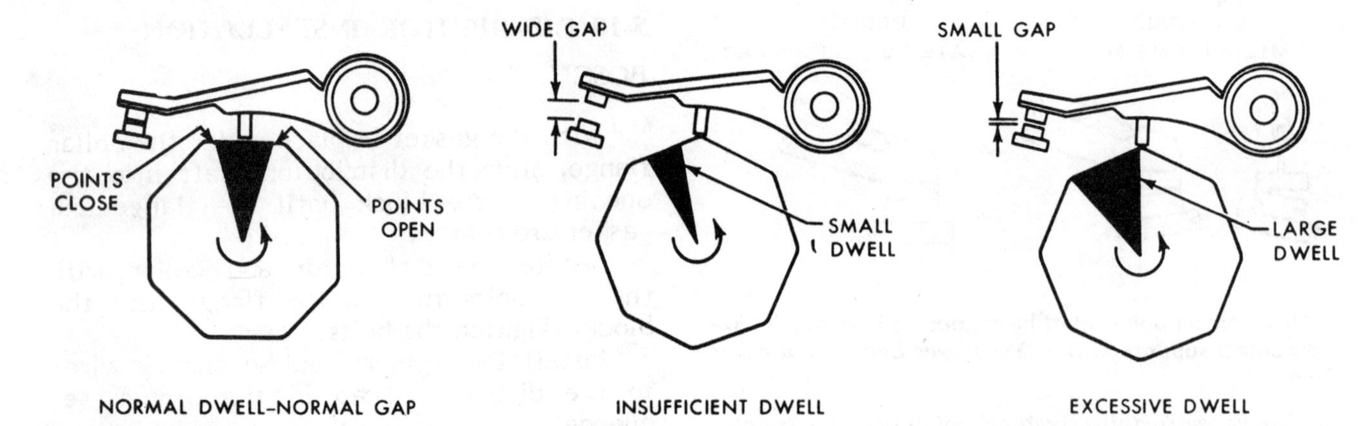

Three different point gap/dwell angle conditions. A normal gap (and dwell angle) is shown in the left view. The center view illustrates an excessive gap with too small a dwell angle. The right illustration depicts too small a gap with the resulting dwell angle too great. If the gap is too small, the ignition timing is retarded, causing loss of power.

mounted on the engine may be damaged. Now, with both valves closed, and the timing mark aligned with the timing indicator, the No. 1 cylinder is in firing position.

Align the rotor with the No. 1 cylinder wire terminal in the distributor cap, and then install the distributor in the block. If the distributor will not seat fully in the block, press down lightly on the housing while a partner turns the crankshaft slowly until the distributor tang snaps into the oil pump shaft slot and the distributor moves into its full seated position. Tighten the distributor hold-down bolt.

Timing light aimed at the timing marks on the damper and on the timing chain cover — AQ200, AQ225, AQ260, AQ280, and AQ290 engines.

Ignition fine-tuning will be accomplished after the engine is running.

Wipe the distributor cap and the coil of any moisture to be sure it does not cause a leakage path.

ADJUSTING THE DWELL

A dwell meter accurately measures the length of time the points are closed, as shown in the accompanying illustration. Connect one lead of the dwell meter to the negative side of the ignition coil and the other lead to a good ground. Start the

Timing light aimed at the timing pointer and the mark on the damper — AQ130 and AQ170 engines.

Small runabout equipped with a 280 stern drive and powered with an AQ130 engine operating in a test tank during final performance adjustments.

Timing pointer and timing notch on the crankshaft pulley -- AQ120, AQ125, AQ140, and AQ145 engines.

engine and adjust the dwell to the Specifications in the Appendix.

CAUTION: Water must circulate through the lower unit to the engine any time the engine is run to prevent damage to the water pump mounted on the engine.

ADJUSTING IGNITION TIMING

The breaker point gap or the dwell must be accurately set before attempting to fine-tune the engine because the point gap directly affects the timing.

Check the timing mark on the balancer or pulley and the lines on the timing tab. If they are hard to see, mark them with paint or chalk. Connect a timing light to an adaptor for No. 1 spark plug. **NEVER** puncture the high-tension wire or you will damage the core.

Timing marks on the upper drive pulley, and belt -- AQ120, AQ125, AQ140, and AQ145 engines.

Timing mark on the accessory drive pulley and on the belt -- AQ120, AQ125, AQ140, and AQ145 engines.

Start the engine and adjust the idle speed to specification.

CAUTION: Water must circulate through the lower unit to the engine any time the engine is run to prevent damage to the water pump mounted on the engine.

Now, aim the timing light at the timing mark on the pulley and the timing tab. **TAKE CARE** to aim the timing light straight at the mark. Sighting from an angle may cause an error. The specified timing mark should align with the pointer. If it does not, loosen the distributor hold-down bolt and rotate the distributor until it is aligned. Tighten the hold-down bolt, and then check the marks again with the light.

Check operation of the centrifugal advance mechanism by accelerating the engine and checking the position of the timing mark with the light. The mark should advance on the pulley if the advance mechanism is operating properly.

ADJUSTING IDLE SPEED AND AIR/FUEL MIXTURE

Detailed procedures to adjust single, double, and triple carburetor installations is presented in Chapter 4, Fuel.

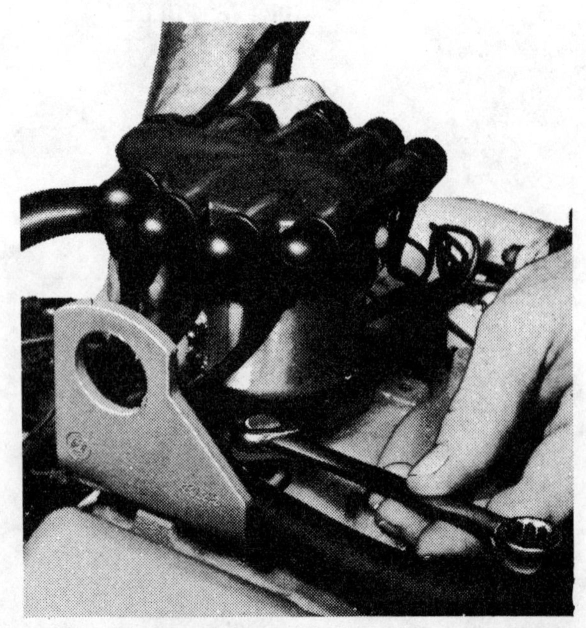

After the distributor is mounted and the engine properly timed, the lock bolt should be tightened securely -- Mallory distributor.

5-11 RENIX ELECTRONIC IGNITION SYSTEM MODEL AQ171 SERIES ENGINES

DESCRIPTION AND OPERATION

The Renix electronic ignition system is installed on the AQ171 series of double overhead cam (DOHC) engines. The system consists of a specially machined flywheel, a flywheel position sender, a control unit, an engine speed limiter, and a distributor.

Flywheel

Forty holes are drilled into the circumference of the flywheel. Two of the holes are larger than the other thirty eight. These two larger holes are used as reference points and are located at exactly 90° before TDC (BTDC) and at 90° before bottom dead center for the No. 1 cylinder.

The flywheel circumference contains two "flat spots" with no holes. These two flat spots are located 180° opposite each other. The ignition sender senses the "flat spots" as they pass. The sender then transmits a signal to the control unit. It is this

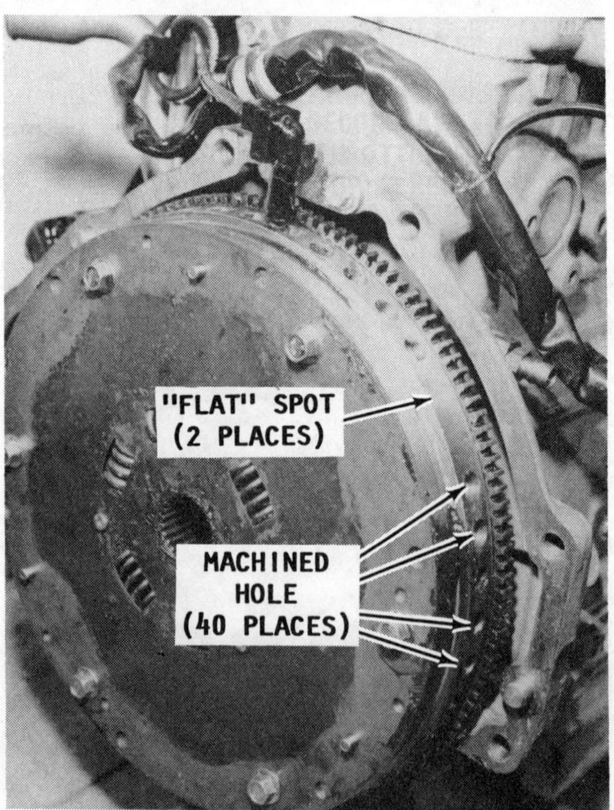

Flywheel of a DOHC engine with some of the machined holes in the perimeter, and one of the "flat" spots showing. The ignition sender is mounted at the top.

signal which results in current being sent to a spark plug. The duration of the signal to the control unit last only during the short interval of time required for the flywheel to rotate from one flat spot through eleven holes to one of the larger holes, approximately 90°.

Position Sender

A flywheel position sender is mounted on the engine block just above the flywheel. This magnetic pickup coil senses each "hole" and the flat spots as they pass and can distinguish the small holes from the larger holes. This means the sender can determine the exact position of the flywheel and therefore the position of the No. 1 cylinder at all times.

Control Unit

The control unit is a transistorized "black box" which gives a more precise control of ignition timing than the conventional contact point ignition systems. The electronic control unit varies the ignition timing to suit changes in engine speed, load, and temperature. The control unit is preprogrammed with 63 engine speed values. These values are the determining factor for engine timing -- advance or retard.

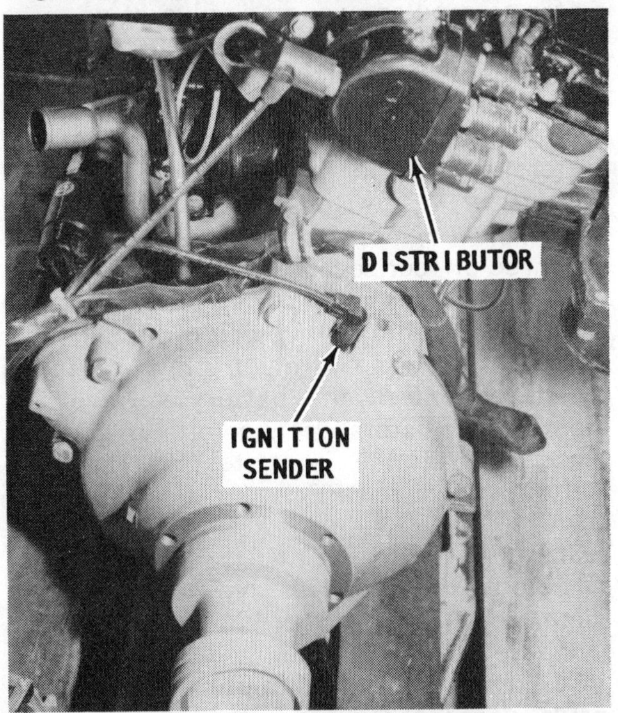

Overall view showing the flywheel cover in place, location of the ignition sender, and the distributor mounted to the cylinder head and indexed with the aft end of the port side camshaft.

Ignition Coil

The ignition coil is a part of the control unit but is replaceable as a separate unit. This coil is the "dry" type, that is, the coil is not filled with oil as the old cylindrical ignition coils.

Distributor

The distributor is mounted on the port side aft to the cylinder head. The distributor is driven by the port side camshaft and its only task is to distribute the high voltage to the correct spark plug. The interior of the distributor is completely empty except for the rotor in the center and a series of contact points around the perimeter. The ignition timing **CANNOT** be adjusted by rotating the distributor.

Speed Limiter

The engine speed limiter is a safety feature to protect the engine from exceeding a designed rpm ("overrun"). The speed limiter is located inside the tachometer. The tachometer is installed on the boat's control panel. When the tachometer needle approaches 6200 rpm, the engine speed limiter sends a signal to the control unit. In turn, the control unit cuts the charging time of the coil. The affect of this action is a reduction in engine rpm.

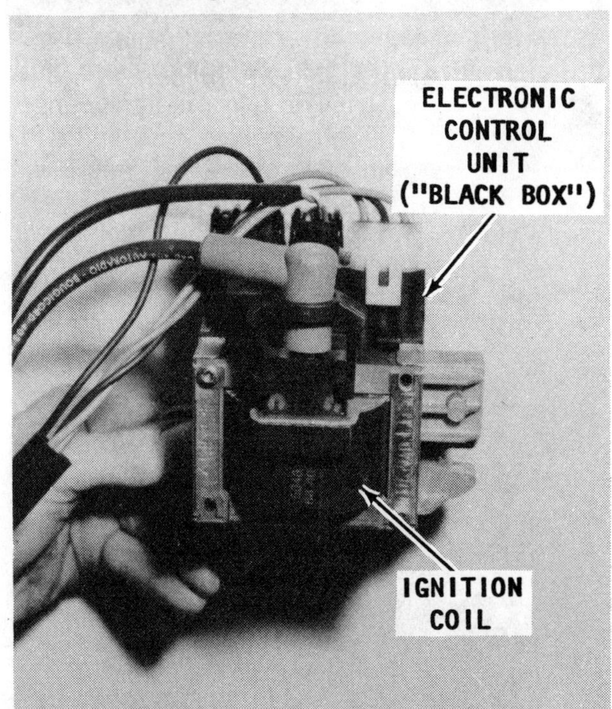

An electronic control unit "black box" for the Renix electronic ignition system for a DOHC engine. The ignition coil attached to the unit is replaceable as a separate item.

CHECKING RENIX ELECTRONIC IGNTION SYSTEM

Start the engine and adjust the speed to 850 rpm.

CAUTION: Water must circulate through the lower unit to the engine any time the engine is run to prevent damage to the water pump mounted on the engine.

Now, aim the timing light at the timing mark on the crankshaft pulley and the timing mark on the timing gear cover. **TAKE CARE** to aim the timing light straight at the mark. Sighting from an angle may cause an error. The pulley timing mark should align with the 10° mark on the timing gear cover.

Next, advance engine speed to 4400 rpm. The timing mark on the pulley should align between the 23° to 25° mark on the timing gear cover.

No adjustment is possible, if the timing marks cannot be aligned at the specified rpm speeds. The manufacturer recommends the electronic "black box" control unit, but not the ignition coil attached to the unit, be replaced.

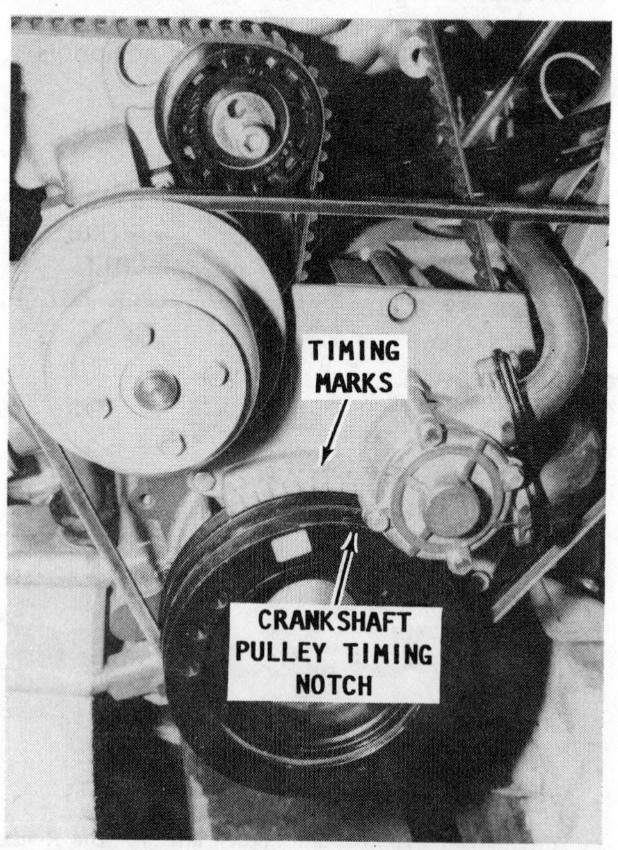

Closeup view of the front of the DOHC engine showing the timing notch on the crankshaft pulley and the timing marks on timing gear cover.

Procedures listed in the next section, Troubleshooting, may help to isolate the problem between the control unit and the ignition coil.

TROUBLESHOOTING RENIX ELECTRONIC IGNITION SYSTEM

If the engine fails to start or starts but operates erratically check the following connections:

The three top connector plugs of the control unit.

The high tension leads at the spark plugs.

The five high tension leads at the distributor.

The ignition position sender at the flywheel.

The cable at the ignition coil (from the ignition sender).

If the engine still fails to start or operate properly, perform the following tests with water circulating because the engine will be cranking and attempting to start.

Test A

Remove the right top connector from the control unit. This connector will have Purple, Black, and Gray leads. Obtain a volt/ohmmeter. Select the scale on the meter to read an expected value of no more than 12 volts.

Connect the Black meter to a suitable ground on the control unit -- a mounting bolt will do. Make contact with the Red meter lead to the inside connector of the Purple wire. Switch on the ignition and crank the engine with the cranking motor for three or four seconds.

The meter should register at least 9.5 volts. If no voltage is registered, look for a short in the Purple wire. If a low voltage is registered, check the battery voltage. If the meter registers just 9.5 volts or greater, proceed with the next test. Turn the ignition switch to **OFF**.

Test B

Select the Rx1000 ohm scale on the ohmmeter. Leave the Black meter lead connected to the ground, as directed in Test A. Connect the Red meter lead to the inside connector of the Black wire. The ohmmeter should register Zero ohms indicating continuity. If the meter registers Zero, proceed with Test C.

If the meter indicates a resistance greater than Zero, recheck the Black wire connection to the connector and repeat the resistance test.

Test C

Leave the top right connector disconnected from the control unit, as in Test B. Remove the center top connector from the control unit. This connector will have White and Red wires. Select the scale on the meter to read an expected value of no more than 300 ohms.

Connect the ohmmeter leads to the inside contacts of the Red and White wires. The meter should indicate between 160 and 280 ohms. If the reading is satisfactory, proceed to Test D.

If the resistance is higher or lower than recommended, remove the ignition position sensor from the flywheel and inspect the sensor for dirt or damage. Inspect the holes in the flywheel for dirt, oil, etc. Clean as required and repeat the resistance test. If the meter reading is still unsatisfactory, replace the ignition position sensor.

Test D

Remove the high tension lead from the top of the ignition coil. Remove the four screws securing the coil to the control unit and remove the coil. Inspect the condition of the two contacts on the ignition coil which index into connectors on the control unit.

Select the Rx1000 scale on the ohmmeter. Measure the resistance between the two connectors on the control unit. The meter should indicate Zero. If the test is satisfactory, proceed to Test E.

If the meter reading is higher than Zero, replace the control unit. Leave the ignition coil for further testing and possible installation.

Test E

Connect the center and right top connectors to the control unit. Obtain a 12 volt test light with a bulb which will withstand at least 4 watts. Make contact with the test light leads to the two connectors on the control unit which receive the ignition coil. Have an assistant switch the ignition switch

Many electrical problems can be traced to poor connections, faulty wiring, or corroded terminals. Wiring at the dashboard should be "neat and tidy". Wires should be routed to permit the making of bundles and then secured with "tie-wraps", as shown. The wires should not be allowed to move as the boat is subjected to violent maneuvers in the water. A properly installed dashboard will permit efficient troubleshooting. Sometimes a malfunction, such as a loose connection, can be detected by visual inspection or simply checking the wire at the terminal.

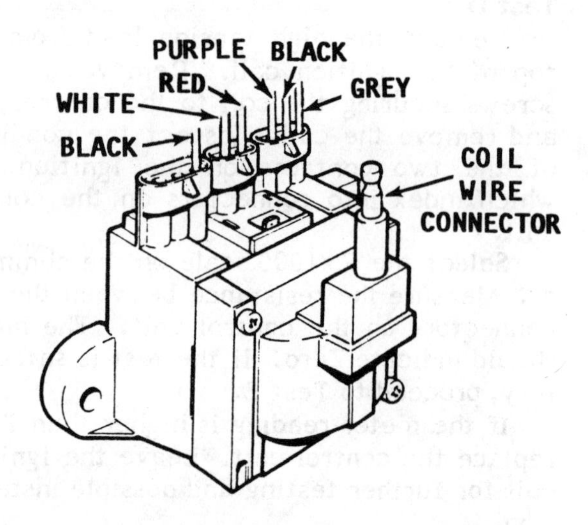

Drawing of the electronic control unit "black box" showing the wire coding for wires connected to the top three connectors.

to **ON** and crank the engine for three or four seconds. The test light should flash on and off, but **NOT** remain on. If the test light flashes as described, proceed to Test F. If the light fails to flash on and off, replace the control unit.

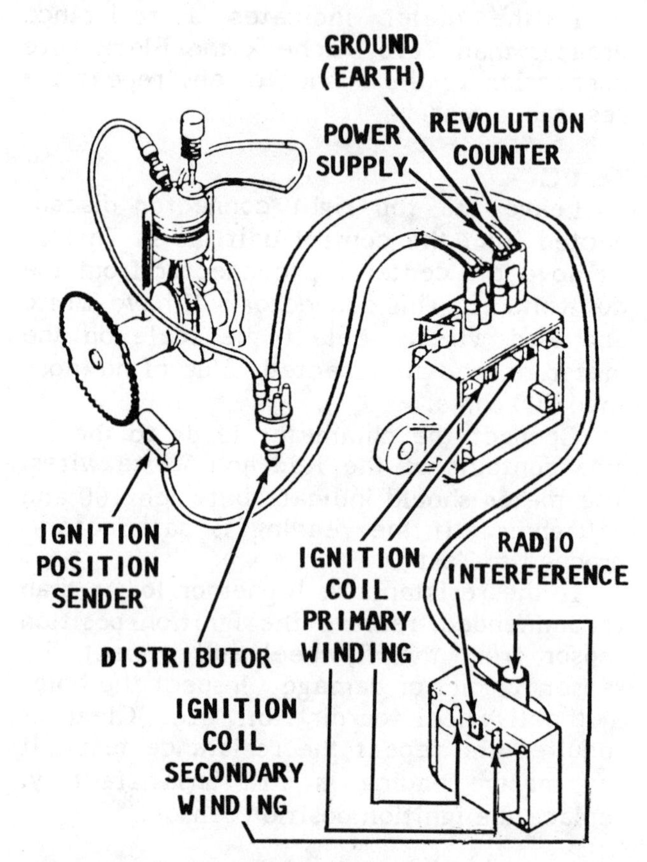

Functional diagram of the Renix electronic ignition system for the DOHC engine. Major parts and wire connections are identified.

Test F

Select the Rx1000 (or the highest scale possible) on the ohmmeter. Connect the Black meter lead to the tall center post of the ignition coil. Connect the Red meter lead to one of the small contacts of the ignition coil. The meter should indicate between 5500 and 2500 ohms. If the reading is satisfactory, proceed to Test G.

If the resistance is significally higher or lower than recommended, replace the ignition coil.

Test G

Select the Rx1 scale on the ohmmeter. Connect the meter leads across the two small contacts of the ignition coil. The meter should indicate between 0.4 and 0.8 ohms. If the reading is satisfactory, proceed to Test H.

If the resistance is significantly higher or lower, replace the ignition coil.

Test H

Install the ignition coil back onto the control unit with the four attaching screws. Connect the high tension lead to the top post of the coil. Disconnect the other end of this lead at the distributor. Hold the disconnected end of the lead about 3/4" (2cm) away from a good ground on the engine block. (The unpainted aft lifting eye bolted to the intake manifold will do.)

CAUTION

NEVER allow a spark to come close to the cover of the electronic control unit during the following test. Such action could seriously damage the unit.

Have an assistant switch the ignition to **ON** and crank the engine for three or four seconds.

A spark should be visible jumping from the high tension lead to engine ground.

If a strong spark is observed, the electronic control unit functions normally. If no spark is observed, replace the electronic control unit.

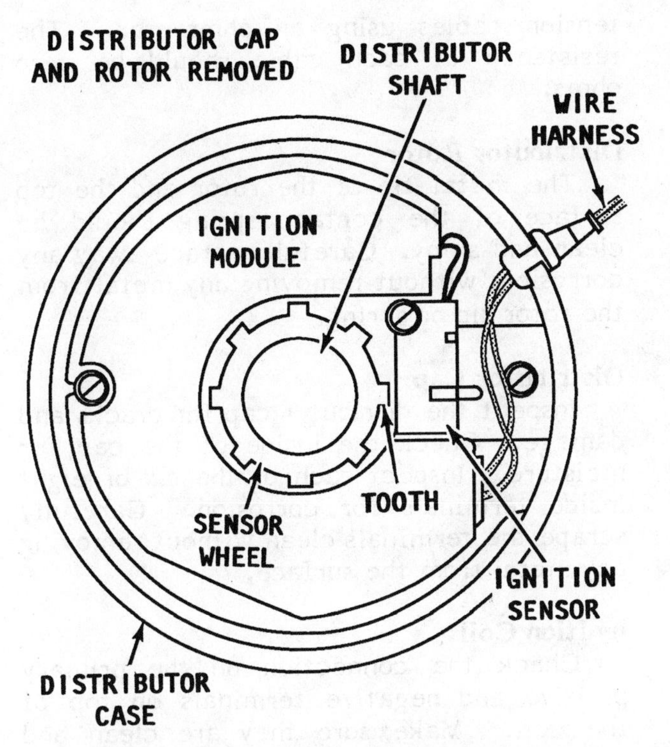

Standard Electronic Ignition distributor for a V8 engine. Note the eight teeth on the sensor wheel. The distributor for a V6 block is identical, except for six teeth on the wheel.

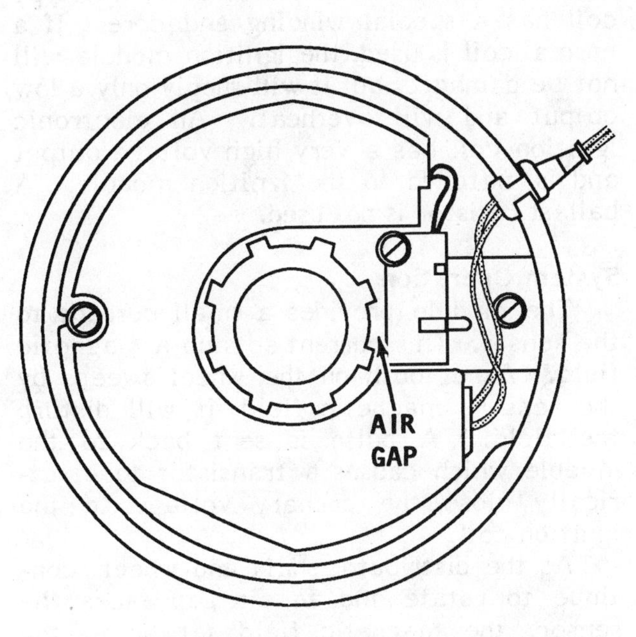

When a tooth on the wheel is aligned with the ignition sensor, as shown in the illustration in the adjacent column, a signal is sent to the module to close primary voltage. This illustration indicates a tooth is not aligned with the sensor and an air gap exists -- the primary voltage is therefore open.

5-12 STANDARD ELECTRONIC IGNITION SYSTEM

System Description

A new standard electronic ignition system was introduced about 1990 on V6 and V8 powerplants. This system is similar to the Renix electronic ignition system used on the AQ171 DOHC engine, but does not use holes drilled in the flywheel as an impulse source.

A standard electronic ignition system does not utilize breaker points but does utilize a mechanical advance mechanism controlled by centrifugal weights in the base of the distributor. It is a breakerless transistorized system, providing accurate and stable ignition timing. It was designed to be almost maintenance free without requirements for periodic adjustment.

In addition to the conventional type distributor shaft, housing, cap and rotor, the system consists of four major components, namely an ignition coil, ignition module, ignition sensor, and a sensor wheel.

Ignition Module

The module is housed within the distributor, in a semi-circular shape - almost encirculing the distributor shaft.

Ignition Sensor and Sensor Wheel

The sensor is attached across the ends of the module with only a small air gap beween it and the rotating teeth of the sensor wheel. The sensor wheel is mounted near the top of the distributor shaft under a conventional rotor. The sensor wheel for a V6 engine has 6 teeth and therefore 6 air gaps, the wheel for a V8 engine has 8 teeth and 8 air gaps. The sensor is used to trigger the ignition. This sensor detects the presence of a tooth as the wheel rotates on the distributor shaft.

The ignition sensor is a device which acts similar to a switch when subjected to a magnetic field. Action of the sensor is not a factor of speed and it will trigger at almost zero rpm. The distributor rotor and cap distribute current to the proper spark plug and do not determine engine timing advance.

The ignition module will withstand a reverse battery connection for no more than a minute, after which time the module will be "fried" — destroyed.

Ignition Coil

The ignition coil is externally mounted on the engine. Externally -- from the outside — the ignition coil appears to be a standard size and shape. However, this type coil has a special winding and core. If a normal coil is used, the ignition module will not be damaged, but it will supply only a low output and will overheat. An electronic ignition coil has a very high voltage output and is matched to the ignition module. A ballast resistor is not used.

System Operation

The module provides a small current to the sensor. This current sets up a magnetic field. As a tooth on the wheel sweeps by the sensor's magnetic field, it will disturb the field. A pulse is sent back to the module which causes a transistor to electrically close the primary voltage to the ignition coil.

As the distributor shaft and wheel continue to rotate and an air gap faces the sensor, the magnetic field set up in the sensor is undisturbed. The module senses this condition and causes the transistor to switch on the primary current to the ignition coil.

Standard tachometers and most synchronizers monitoring ignition impulses at the negative terminal of the coil will operate satisfactorily with the standard electronic ignition system.

Speed Limiter

The engine speed limiter is a safety feature to protect the engine from exceeding a predetermined engine rpm when the propeller leaves the water for a few seconds. The tachometer is installed on the boat's control panel. When the tachometer needle approaches 5700 rpm for the Model AQ311, or 5100 rpm for the Model AQ271, the engine speed limiter sends a signal to the ignition module. In turn the module limits the charging time of the coil. The effect of this action is a reduction in engine rpm.

TROUBLESHOOTING

First make a check of the following areas:

Spark Plugs

See Section 5-3 on Page 5-9, Spark Plug Troubleshooting.

Ignition Cables

Check the resistance of each of the high tension cables using an ohmmeter. The resistance of each cable should be zero ohms.

Distributor Rotor

The metal tip of the rotor and the top surface of the contact spring should be clean and shiny. Carefully scrape away any corrosion, without removing any metal from the rotor tip or spring.

Distributor Cap

Inspect the disributor cap for cracks and damage. Check the inside of the cap for moisture. Inspect each of the six or eight inside terminals for corrosion. Carefully scrape the terminals clean without removing any metal from the surface.

Ignition Coil

Check the connection on the primary positive and negative terminals on top of the coil. Make sure they are clean and tight. Check along the length of each wire for damage to plastic insulation.

Inspect the coil surface for cracks. The entire surface should be clean and dry.

Testing Components

The following testing procedures will identify a defective part. One defective part may affect an otherwise good part. Therefore, if one defective component is identified, **DO NOT STOP** testing. Continue and complete the entire test sequence to ensure the ignition system is in optimum condition.

ENGINE FAILS TO START

1- If the engine fails to start connect a timing light to a 12 volt battery. Connect the pickup around the high tension lead between the distibutor and ignition coil. Obtain the aid of an assistant to crank the engine. If the timing light flashes, the primary circuit is functioning. Proceed to Step 3.

2- If the timing light fails to flash remove the distributor cap and rotor. Observe the sensor wheel and at the same time rotate the crankshaft until one of the teeth aligns squarely with the sensor. Measure the air gap between the tooth and the edge of the sensor closest to the tooth. The distance should be 0.008-0.010" (0.20-0.25mm). Adjust as required.

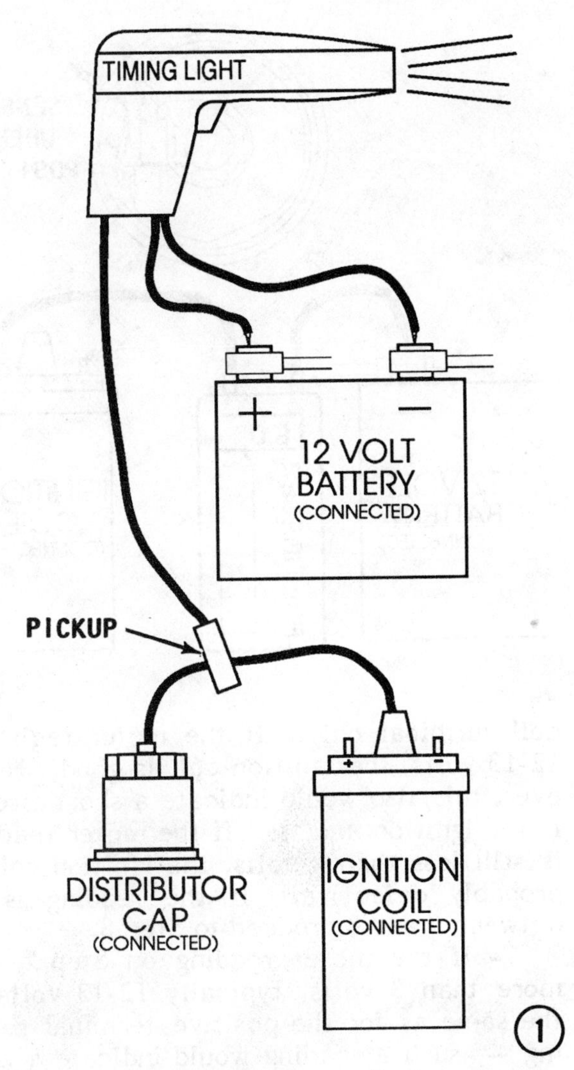

Rotate the crankshaft until the opposite tooth aligns with the sensor edge and measure the air gap again. If there is a significant difference between the two readings, the distributor shaft is bent and must be replaced.

Repeat Step 1. If the timing light still fails to flash, the primary voltage must be measured with a voltmeter.

3- Select a scale on the voltmeter capable of registering up to 15 volts. Rotate the crankshaft until an air gap between the two teeth aligns with the sensor edge. Turn the ignition key to the "ignition" position. The primary circuit is now activated. Measure the voltage across the battery terminals by making contact with the Red meter lead to the positive battery terminal and making contact with the Black meter lead with the negative battery terminal. The meter should register 12-13 volts. If the reading is lower than 12 volts, remove the battery and charge it, see Page 6-2, Battery Service.

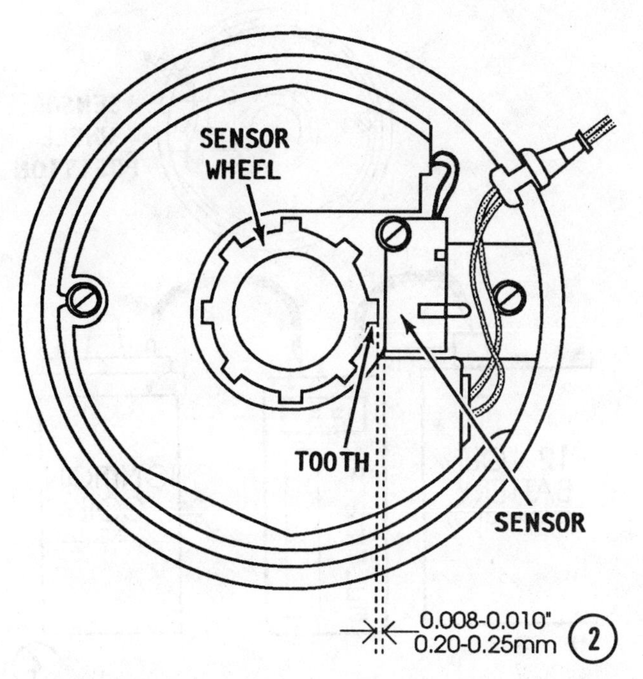

4- Make contact with the Red meter lead to the small positive ignition coil terminal and make contact with the Black meter lead to a suitable ground on the engine. Still with the ignition key in the "ignition" position, the meter should register no more than one volt less than the previous voltage value taken across the battery in Step 3. If the meter reading indicates a voltage drop greater than one volt over the reading taken in Step 3, follow the wiring diagram in the Appendix for the engine being serviced and clean all connections starting with the battery terminals.

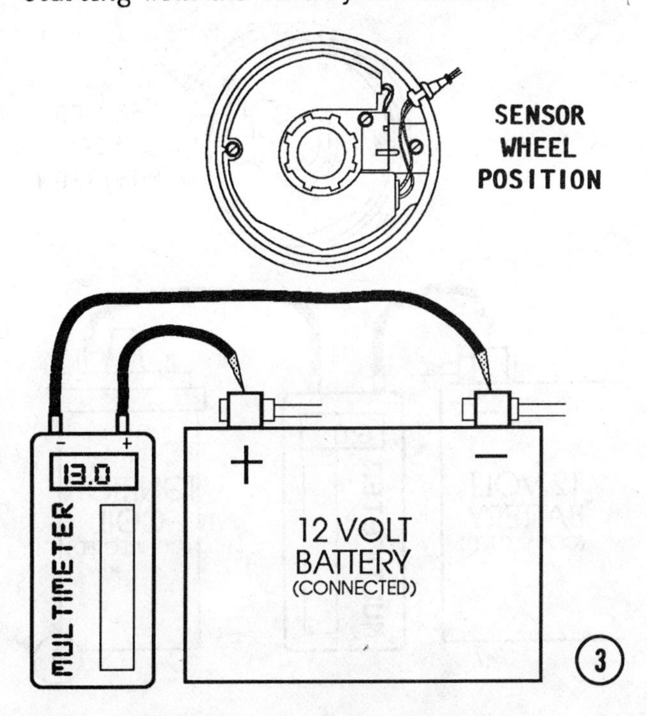

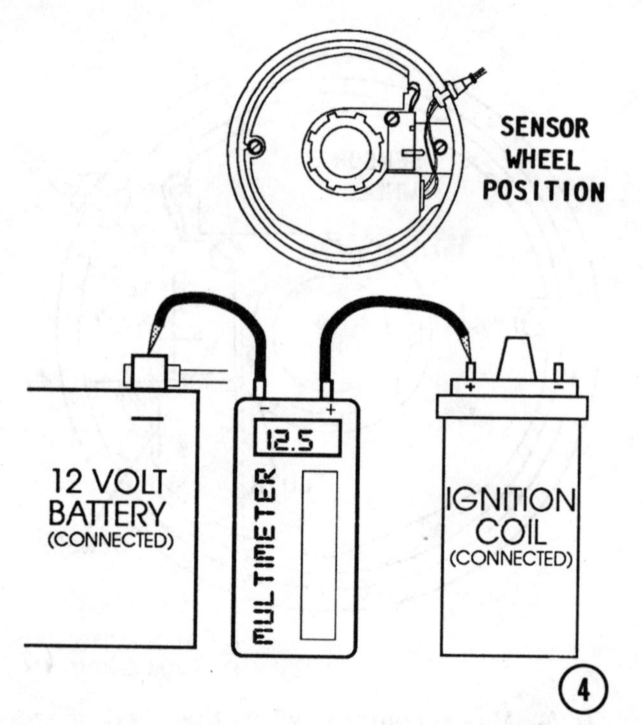

4

If the reading is satisfactory, proceed to the following step.

5- Keep the same connections as in the previous step, but move the Red meter lead to the small negative terminal. The meter reading should be 4-8 volts. If the reading is less than 4 volts proceed to Step 6. If the reading is more than 8 volts proceed to Step 7.

6- Remove the lead from the small negative terminal on the ignition coil. Make the meter connections as before, but this time with the Red meter lead touching the

5

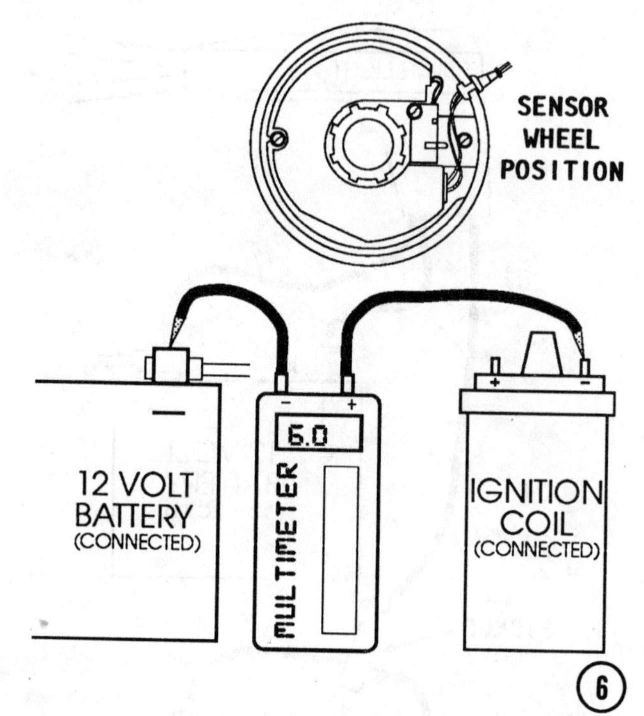

6

coil terminal only. If the meter registers 12-13 volts, the ignition coil is good. However, this also would indicate a short circuit in the ignition module. If the meter reading is still less than 4 volts, the ignition coil is probably defective. If the reading is in between 4 and 8 proceed to Step 8.

7- If the meter reading on Step 5 was more than 8 volts, typically 12-13 volts -- the same as for the positive terminal reading — such a reading would indicate a very poor or no ground connection between the distributor and the block. This situation would be unusual but not impossible. However, if the ground connection is good and

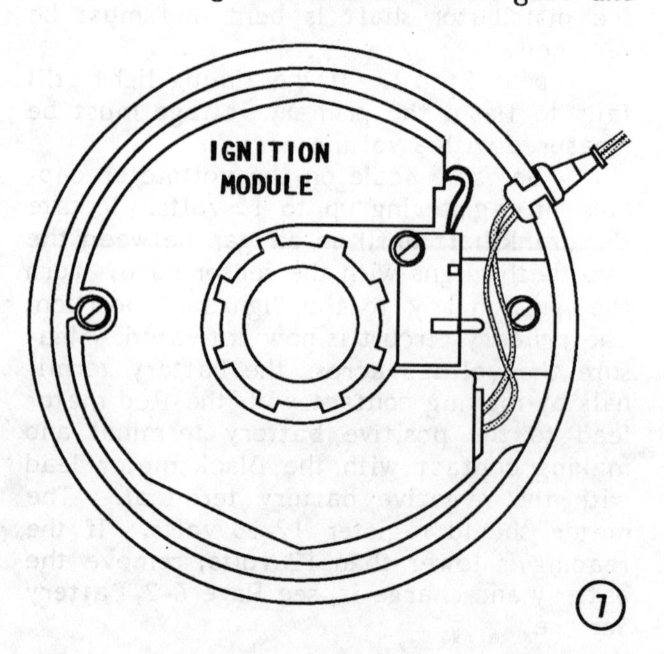

7

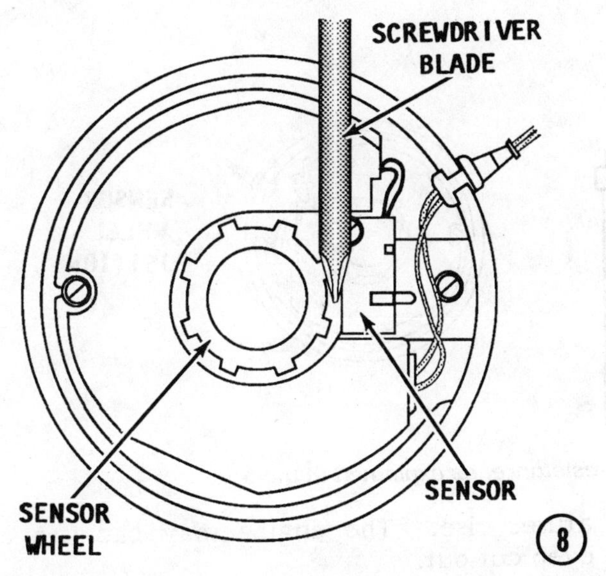

the same readings are obtained, replace the ignition module.

8- Rotate the crankshaft until an air gap faces the edge of the sensor. Insert a flat blade screwdriver into the air gap, gently making contact with the sensor wheel and the sensor. With the same meter connections as before, Black meter lead on negative battery terminal or engine ground and Red meter lead on small negative ignition coil terminal, the meter should register 12-13 volts. If the voltmeter gives a satisfactory reading, but the engine still fails to start, remove all the connections from both small terminals and the high tension lead from the coil tower for the following test.

9- All connections must be removed from the ignition coil terminals for this test. Select a scale on the meter which will register up to 15 ohms resistance. Measure the primary resistance across the two small positive and negative terminals. The resistance should be between 1.25-1.4 ohms at 68°F.

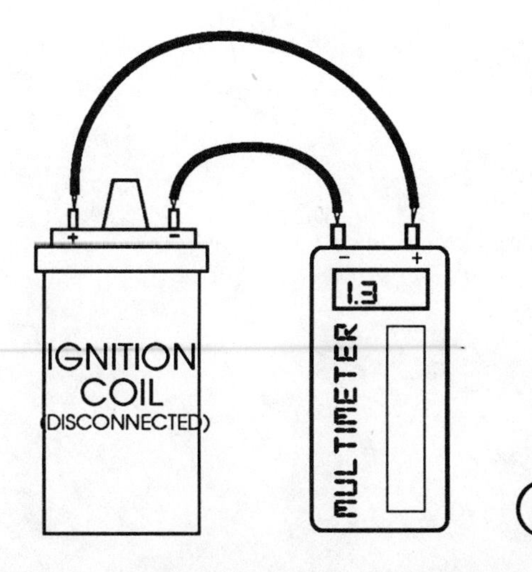

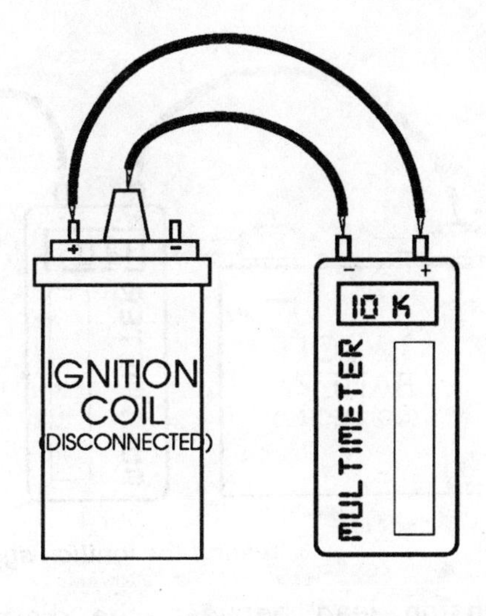

10- Measure the secondary resistance between the coil tower and the small negative coil terminal. The resistance should be 9.4-11.7 ohms at 68°F. If the resistance readings are not as specified, replace the ignition coil.

11- Connect a timing light to a 12 volt battery. Connect the pickup around the

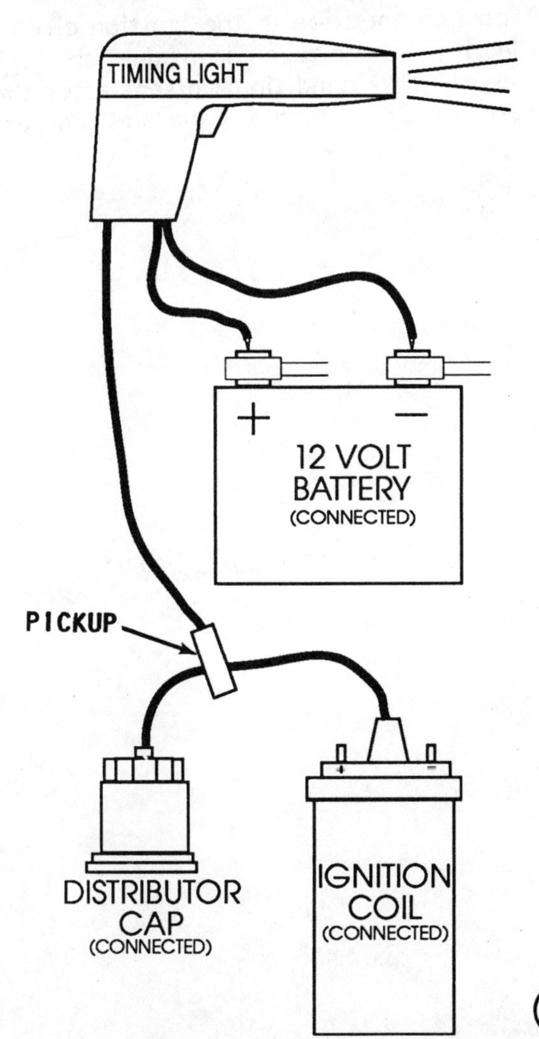

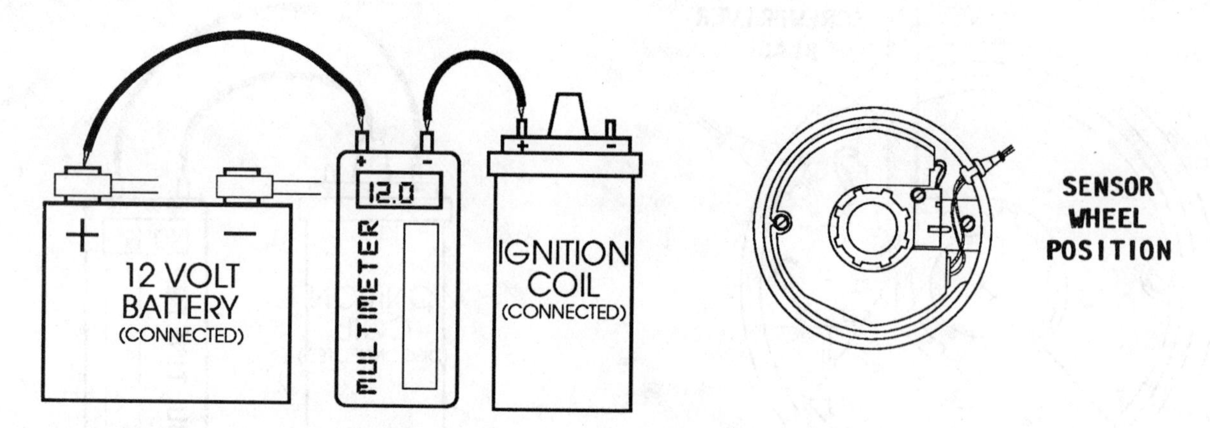

Testing the ignition system for high resistance, as explained in the text.

high tension lead between the distributor and ignition coil. Obtain the aid of an assistant to crank the engine. If the timing light fails to flash the ignition module is defective and must be replaced.

ENGINE STARTS - RUNS ROUGH

A poor connection in the ignition circuit will lead to a voltage drop due to high resistance. This condition worsens after the engine has been running a while and temperatures rise. The engine may backfire or even cut out.

Rotate the crankshaft to align an air gap across from the sensor edge. Make contact with the Red meter lead to the positive battery terminal. Make contact with the Black meter lead to the small positive ignition coil terminal. Turn the ignition key to the **ON** position. The meter should register 11-12 volts. If the reading is less, follow the wiring diagram in the Appendix for the engine being serviced and clean all connections starting with the battery terminals.

6
ELECTRICAL

6-1 INTRODUCTION

The battery, gauges and horns, charging system, and the cranking system are all considered subsystems of the electrical system. Each of these units or subsystems will be covered in detail in this chapter beginning with the battery.

6-2 BATTERIES

The battery is one of the most important parts of the electrical system. In addition to providing electrical power to start the engine, it also provides power for operation of the running lights, radio, electrical accessories, and possibly the pump for a bait tank.

Because of its job and the consequences if it should fail in an emergency, the best advice is to purchase a well-known brand with an extended warranty period from a reputable dealer.

The usual warranty covers a prorated replacement policy which means you would be entitled to a consideration for the time left on the warranty period if the battery should prove defective before its time.

Do not consider a battery of less than 70-ampere hour capacity. If in doubt as to how large your boat requires, make a liberal estimate and then purchase the one with the next higher ampere rating.

MARINE BATTERIES

Because marine batteries are required to perform under much more rigorous conditions than automotive batteries, they are constructed much differently than those used in automobiles or trucks. Therefore, a marine battery should always be the No. 1 unit for the boat and other types of batteries used only in an emergency.

Marine batteries have a much heavier exterior case to withstand the violent pounding and shocks imposed on it as the boat moves through rough water and in extremely tight turns.

The plates in marine batteries are thicker than in automotive batteries and each plate is securely anchored within the battery case to ensure extended life.

The caps of marine batteries are "spill proof" to prevent acid from spilling into the bilges when the boat heels to one side in a tight turn or is moving through rough water.

Because of these features, the marine battery will recover from a low charge condition and give satisfactory service over a much longer period of time than any type of automotive-type unit.

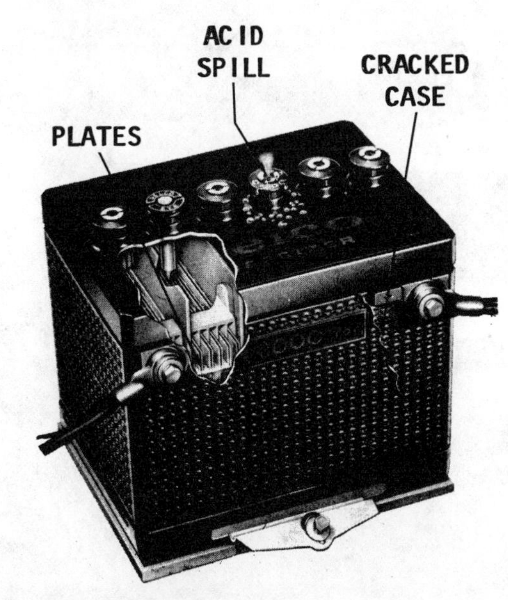

A fully charged battery is the heart of the ignition system. Engine starting and efficient performance can never be obtained if the battery is damaged or below its full charge rating.

BATTERY CONSTRUCTION

A battery consists of a number of positive and negative plates immersed in a solution of dilute sulfuric acid. The plates contain dissimilar active materials and are kept apart by separators. The plates are grouped into what are termed elements. Plate straps on top of each element connect all of the positive plates and all of the negative plates into the groups. The battery is divided into cells which hold a number of the elements apart from the others. The entire arrangement is contained within a hard-rubber case. The top is a one-piece cover and contains the filler caps for each cell. The terminal posts protrude through the top where the battery connections for the boat are made. Each of the cells are connected to each other in a positive-to-negative manner with a heavy strap called the cell connector.

BATTERY RATINGS

Two ratings are used to classify batteries: One is a 20-hour rating at 80°F and the other is a cold rating at 0°F. This second figure indicates the cranking load capacity and is refered to as the Peak Watt Rating of a battery. This Peak Watt Raing (PWR) has been developed as a measure of the batteries cold-cranking ability. The numerical rating is embossed on each battery case at the base and is determined by multiplying the maximum current by the maximum voltage.

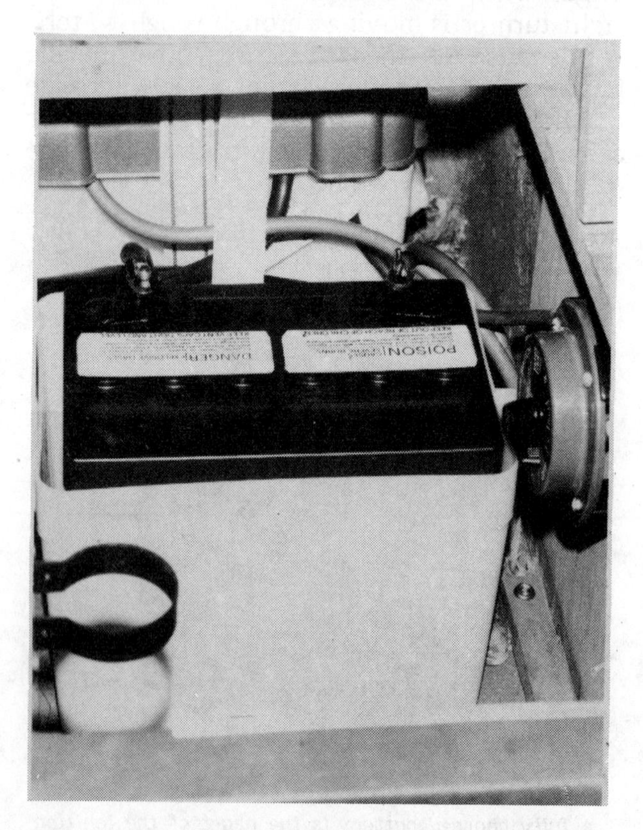

*The battery **MUST** be located near the engine in a well-ventilated area. It must be secured in such a manner that absolutely no movement is possible in any direction under the most violent action of the boat.*

The ampere-hour rating of a battery is its capacity to furnish a given amount of amperes over a period of time at a cell voltage of 1.5. Therefore, a battery with a capacity of maintaining 3 amperes for 20 hours at 1.5 volts would be classified as a 60-ampere hour battery.

Do not confuse the ampere-hour rating with the PWR because they are two unrelated figures used for different purposes.

A replacement battery should have a power rating equal or as close to the old unit as possible.

BATTERY LOCATION

Every battery installed in a boat must be secured in a well-protected ventilated area. If the battery area is not well ventilated, hydrogen gas which is given off during charging could become very explosive if the gas is concentrated and confined. Because of its size, weight, and acid content, the battery must be well-secured. If the battery should break loose during rough boat maneuvers, considerable damage could be done, including damage to the hull.

BATTERY SERVICE

The battery requires periodic servicing and a definite maintenance program will ensure extended life. If the battery should test satisfactorily, but still fail to perform properly, one of several problems could be the cause.

1- An accessory might have accidently been left on overnight or for a long period during the day. Such an oversight would result in a discharged battery.

2- Slow speed engine operation for long periods of time resulting in an undercharged condition.

3- Using more electrical power than the alternator can replace would result in an undercharged condition.

4- A defect in the charging system. A slipping drive belt, a defective voltage regulator, a faulty alternator, or high resistance somewhere in the system could cause the battery to become undercharged.

5- Failure to maintain the battery in good order. This might include a low level of electrolyte in the cells; loose or dirty cable connections at the battery terminals; or possibly an excessive dirty battery top.

Electrolyte Level

The most common practice of checking the electrolyte level in a battery is to remove the cell cap and visually observe the level in the vent well. The bottom of each vent well has a split vent which will cause the surface of the electrolyte to appear distorted when it makes contact. When the distortion first appears at the bottom of the split vent, the electrolyte level is correct.

Some late-model batteries have an electrolyte-level indicator installed which operates in the following manner: A transparent rod extends through the center of one of the cell caps. The lower tip of the rod is immersed in the electrolyte when the level is correct. If the level should drop below normal, the lower tip of the rod is exposed and the upper end glows as a warning to add water. Such a device is only necessary on one cell cap because if the electrolyte is low in one cell it is also low in the other cells. **BE SURE** to replace the cap with the indicator onto the second cell from the positive terminal.

During hot weather and periods of heavy use, the electrolyte level should be checked more often than during normal operation. Add colorless, odorless, drinking water to bring the level of electrolyte in each cell to the proper level. **TAKE CARE** not to overfill because it will cause loss of electrolyte and any loss will result in poor performance, short battery life, and will contribute quickly to corrosion. **NEVER** add electrolyte from another battery. Use only clean pure water.

Cleaning

Dirt and corrosion should be cleaned from the battery just as soon as it is discovered. Any accumulation of acid film or dirt will permit current to flow between the terminals. Such a current flow will drain the battery over a period of time.

Clean the exterior of the battery with a solution of diluted ammonia or a soda solution to neutralize any acid which may be present. Flush the cleaning solution off with clean water. **TAKE CARE** to prevent any of the neutralizing solution from entering the cells by keeping the caps tight.

One of the most effective means of cleaning the battery terminals is by using a wire brush with holder designed for this specific purpose.

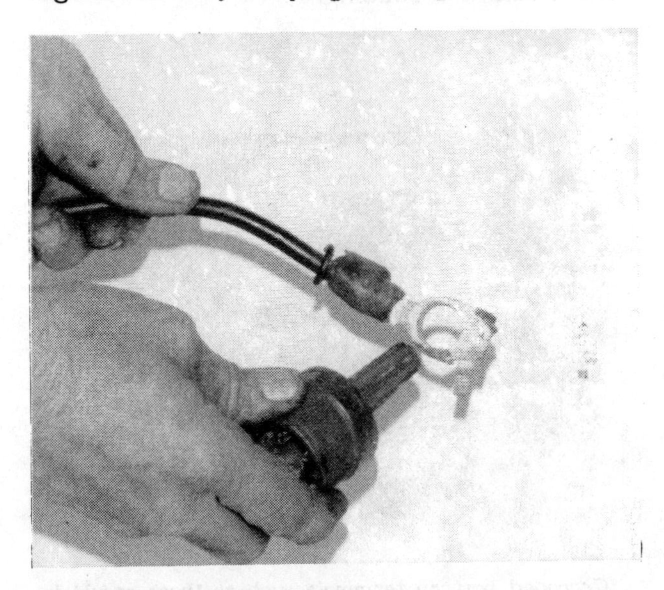

An inexpensive brush can be purchased and used to clean the battery terminals to ensure a proper connection.

A poor contact at the terminals will add resistance to the charging circuit. This resistance will cause the voltage regulator to register a fully charged battery, and thus cut down on the alternator output adding to the low battery charge problem.

Scrape the battery posts clean with a suitable tool or with a stiff wire brush. Clean the inside of the cable clamps to be sure they do not cause any resistance in the circuit.

Battery Testing

A hydrometer is a device to measure the percentage of sulfuric acid in the battery electrolyte in terms of specific gravity. When the condition of the battery drops from fully charged to discharged, the acid leaves the solution and enters the plates, causing the specific gravity of the electrolyte to drop.

The following six points should be observed when using a hydrometer.

1- **NEVER** attempt to take a reading immediately after adding water to the battery. Allow at least 1/4 hour of charging at a high rate to thoroughly mix the electrolyte with the new water and to cause vigorous gassing.

2- **ALWAYS** be sure the hydrometer is clean inside and out as a precaution against contaminating the electrolyte.

3- If a thermometer is an integral part of the hydrometer, draw liquid into it several times to ensure the correct temperature before taking a reading.

Corroded battery terminals such as these result in high resistance at the connections. Such corrosion places a strain on any and all electrically operated devices on the boat and causes hard engine starting.

4- **BE SURE** to hold the hydrometer vertically and suck up liquid only until the float is free and floating.

5- **ALWAYS** hold the hydrometer at eye level and take the reading at the surface of the liquid with the float free and floating.

Disregard the light curvature appearing where the liquid rises against the float stem due to surface tension.

6- **DO NOT** drop any of the battery fluid on the boat or on your clothing, because it is extremely caustic. Use water and baking soda to neutralize any battery liquid that does drop.

After withdrawing electrolyte from the battery cell until the float is barely free, note the level of the liquid inside the hydrometer. If the level is within the green band range, the condition of the battery is satisfactory. If the level is within the white

A check of the electrolyte in the battery should be a regular part of the maintenance schedule on any boat. A hydrometer reading of 1.300, or in the green band, indicates the battery is in satisfactory condition. If the reading is 1.150, or in the red band, the battery needs to be charged. Observe the six points given in the text when using a hydrometer.

band, the battery is in fair condition, and if the level is in the red band, it needs charging badly or is dead and should be replaced. If level fails to rise above the red band after charging, the only answer is to replace the battery.

JUMPER CABLES

If booster batteries are used for starting an engine the jumper cables must be connected correctly and in the proper sequence to prevent damage to either battery, or to the alternator diodes.

ALWAYS connect a cable from the positive terminal of the dead battery to the positive terminal of the good battery **FIRST.** **NEXT,** connect one end of the other cable to the negative terminal of the good battery and the other end to the **ENGINE** for a good ground. By making the ground connection on the engine, if there is an arc when you make the connection it will not be near the battery. An arc near the battery could cause an explosion, destroying the battery and causing serious personal injury.

DISCONNECT the battery ground cable before replacing an alternator or before connecting any type of meter to the alternator.

If it is necessary to use a fast-charger on a dead battery, **ALWAYS** disconnect one of the boat cables from the battery first, to prevent burning out the diodes in the alternator.

NEVER use a fast charger as a booster to start the engine because the diodes in the alternator will be **DAMAGED.**

STORAGE

If the boat is to be laid up for the winter or for more than a few weeks, special attention must be given to the battery to prevent complete discharge or possible damage to the terminals and wiring. Before putting the boat in storage, disconnect and remove the batteries. Clean them thoroughly of any dirt or corrosion, and then charge them to full specific gravity reading. After they are fully charged, store them in a clean cool dry place where they will not be damaged or knocked over.

NEVER store the battery with anything on top of it or cover the battery in such a manner as to prevent air from circulating around the fillercaps. All batteries, both new and old, will discharge during periods of storage, more so if they are hot than if they remain cool. Therefore, the electrolyte level and the specific gravity should be checked at regular intervals. A drop in the specific gravity reading is cause to charge them back to a full reading.

In cold climates, care should be exercised in selecting the battery storage area. A fully-charged battery will freeze at about 60 degrees below zero. A discharged battery, almost dead, will have ice forming at about 19 degrees above zero.

DUAL BATTERY INSTALLATION

Three methods are available for utilizing a dual-battery hook-up.

1- A high-capacity switch can be used to connect the two batteries. The accompanying illustration details the connections for

An explosive hydrogen gas is released from the cells when the caps are removed. This battery exploded when the gas ignited from smoking in the area with the caps removed.

The charging system output can be determined while the engine is running at a fast idle speed by holding an induction-type ammeter over the main wire.

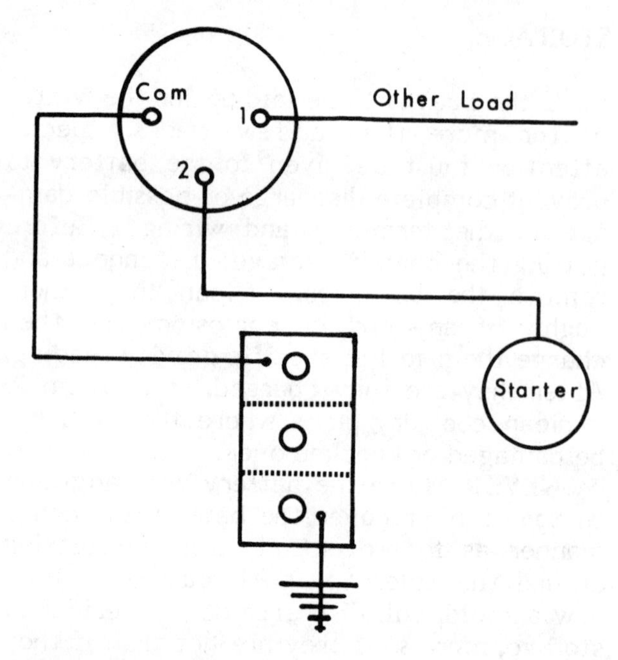

Schematic diagram for a single battery, one engine hookup.

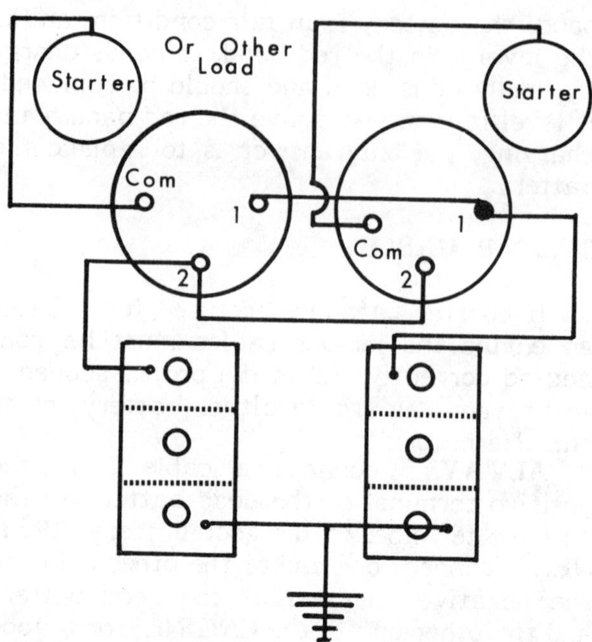

Schematic diagram for a two battery, two engine hookup.

installation of such a switch. This type of switch installation has the advantage of being simple, inexpensive, and easy to mount and hookup. However, if the switch is forgotten in the closed position, it will let the convenience loads run down both batteries and the advantage of the dual installation is lost. However, the switch may be closed intentionally to take advantage of the extra capacity of the two batteries, or it may be temporarily closed to help start the engine under adverse conditions.

2- A relay, can be connected into the ignition circuit to enable both batteries to be automatically put in parallel for charging or to isolate them for ignition use during engine cranking and start. By connecting the relay coil to the ignition terminal of the ignition-starting switch, the relay will close during the start to aid the starting battery. If the second battery is allowed to run down, this arrangement can be a disadvantage since it will draw a load from the starting battery while cranking the engine. One way to avoid such a condition is to connect the relay coil to the ignition switch accessory

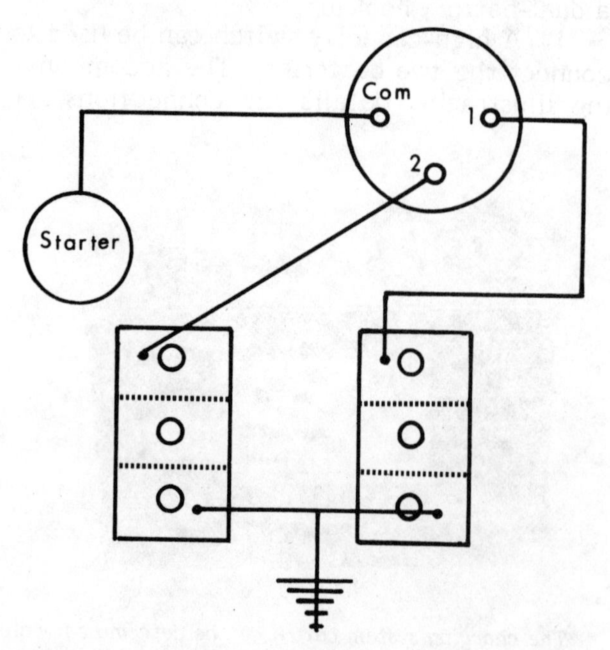

Schematic diagram for a two battery, one engine hookup.

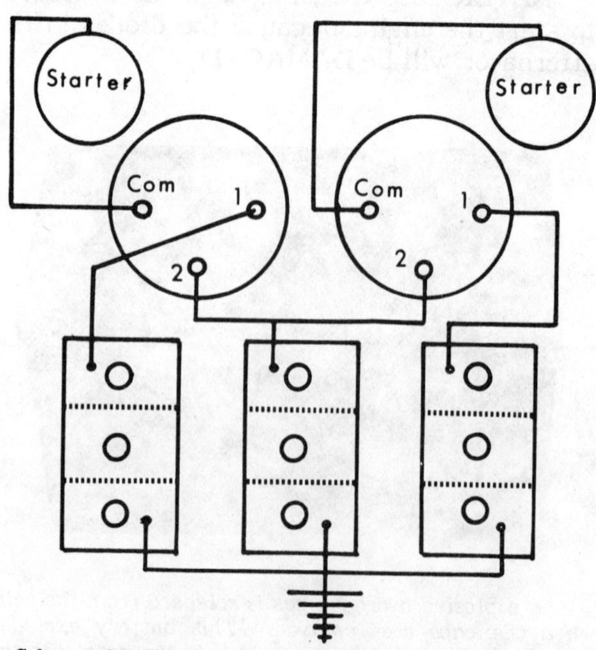

Schematic diagram for a three battery, two engine hookup.

terminal. When connected in this manner, while the engine is being cranked, the relay is open, but when the engine is running with the ignition switch in the normal position, the relay is closed, and the second battery is being charged at the same time as the starting battery.

3- A heavy duty switch installed as close to the batteries as possible can be connected between them. If such an arrangement is used it must meet the standards of the American Boat and Yacht Council, INC. or the Fire Protection Standard for Motor Craft, N.F.P.A. No. 302.

6-3 GAUGES AND LIGHTS

Gauges or lights are installed to warn the operator of a condition in the cooling and lubrication systems that may need attention. The fuel gauge gives an indication of the amount of fuel in the tank. If the engine overheats or the oil pressure drops too low for safety, a gauge or warning light reminds the operator to shut down the engine and check the cause of the warning before serious damage is done.

CONSTANT-VOLTAGE SYSTEM

In order for gauges to register properly, they must be supplied with a steady voltage. The voltage variations produced by the engine charging system would cause erratic gauge operation, too high when the alternator voltage is high and too low when the alternator is not charging. To remedy this problem, a constant-voltage system is used to reduce the 12-14 volts of the electrical system to an average of 5 volts. This steady 5 volts ensures the gauges will read accurately under varying conditions from the electrical system.

SERVICE PROCEDURES

Systems utilizing warning lights do not require a constant-voltage system, therefore, this service is not needed.

Service procedures for checking the gauges and their sending units are detailed in the following sections.

OIL AND TEMPERATURE GAUGES

The body of oil and temperature gauges must be grounded and they must be supplied with 12 volts. Many gauges have a terminal on the mounting bracket for attaching a ground wire. A tang from the mounting bracket makes contact with the gauge. **CHECK** to be sure the tang does make good contact with the gauge.

Ground the wire to the sending unit and the needle of the gauge should move to the full right position indicating the gauge is in serviceable condition.

Check Gauge Sender for a defective motor temperature warning system: Remove the sender from the engine. Connect the sender terminals to an ohmmeter. Submerge the sender in a container of oil with a thermometer. Heat the oil over a fireless heating element. Now, observe the meter and thermometer readings. The meter should read 450 ohms +5% at 220°F.

Check The Sender for a defective oil pressure warning system: Substitute a new sender unit of the correct value. Start the engine with the propeller in the water. Observe the gauge. If the reading is still unsatisfactory, replace the original gauge and test again. If the reading is still unsatisfactory, the problem may be in the engine lubrication system and due to worn bearings. **NEVER** attempt to interchange the sending unit from a system using a gauge with a unit from one using a warning light.

WARNING LIGHTS

If a problem arises on a boat equipped with water, temperature, and oil pressure

The indicator and control panel should be kept clean and protected from water spray, especially when operating in a salt water atmosphere.

lights, the first area to check is the light assembly for loose wires or burned-out bulbs.

When the ignition key is turned on, the light assembly is supplied with 12 volts and grounded through the sending unit mounted on the engine. When the sending unit makes contact because the water temperature is too hot or the oil pressure is too low, the circuit to ground is completed and the lamp should light.

Check The Bulb: Turn the ignition switch on. Disconnect the wire at the engine sending unit, and then ground the wire. The lamp on the dash should light. If it does not light, check for a burned-out bulb or a break in the wiring to the light.

Check The Sender for a defective motor temperature warning system: Remove the sending unit from the engine and connect the terminals to an ohmmeter. Submerge the sender in a container of oil with a thermometer. Heat the oil over a fireless heating element. Observe the thermometer readings. The meter should indicate an open circuit until the temperature reaches 200+5°F. If the circuit does not close at the specified temperature, replace the unit.

Check The Sender for a defective oil pressure warning system: Disconnect the electrical lead at the sending unit. Connect an ohmmeter between the terminal and ground. Turn the ignition switch on and the meter should indicate a complete circuit. Start the engine.

CAUTION: Water must circulate through the lower unit to the engine any time the engine is run to prevent damage to the water pump mounted on the engine.

Increase engine rpm and the meter should then indicate an open circuit. If it does not, replace the sender. If a new sender still fails to open the circuit, the problem may be in the engine lubricating system or because of worn bearings.

THERMOMELT STICKS

Thermomelt sticks are an easy method of determining if the engine is running at the proper temperature. Thermomelt sticks are not expensive and are available at your local marine dealer.

Start the engine with the propeller in the water and run it for about 5 minutes at about 3000 rpm.

CAUTION: Water must circulate through the lower unit to the engine any time the engine is run to prevent damage to the water pump mounted on the engine.

The 140 degree stick should melt when you touch it to the lower thermostat housing. If it does not melt, the thermostat is stuck in the open position and the engine temperature is too low.

Touch the 170 degree stick to the same spot on the lower thermostat housing and it should not melt. If it does, the thermostat is stuck in the closed position or the water pump is not operating properly because the engine is running too hot. For service procedures on the cooling system, see Chapter 7.

FUEL GAUGES

The fuel gauge is intended to indicate the quantity of fuel in the tank. As the experienced boatman has learned, the gauge reading is seldom an accurate report of the fuel available in the tank. The main reason for this false reading is because the boat is rarely on an even keel. A considerable difference in fuel quantity will be indicated by the gauge if the bow or stern is heavy and if the boat has a list to port or starboard.

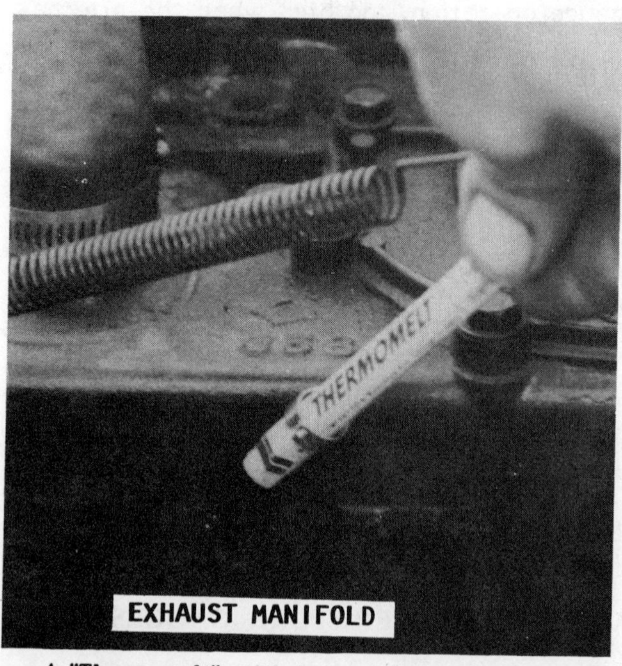

A "Thermomelt" stick is a quick, simple, and fairly accurate method to determine engine operating temperature.

Therefore, the reading is usually low. The amount of fuel drawn from the tank is dependent on the location of the fuel pickup tube in the tank. The engine may cutout while cruising because the pickup tube is clear of the fuel level. Instead of assuming the tank is empty, shift weight in the boat to change the trim and the problem may be solved until you are able to take on more fuel.

FUEL GAUGE HOOKUP

The Boating Industry Association recommends the following color coding be used on all fuel gauge installations:

Black — for all grounded current-carrying conductors.

Pink -- insulated wire for the fuel gauge sending unit to the gauge.

Red -- insulated wire for a connection from the positive side of the battery to any electrical equipment.

Connect one end of a pink insulated wire to the terminal on the gauge marked **TANK** and the other end to the terminal on top of the tank unit.

Connect one end of a black wire to the terminal on the fuel gauge marked **IGN** and the other end to the ignition switch.

Connect one end of a second black wire to the fuel gauge terminal marked **GRD** and the other end to a good ground. It is important for the fuel gauge case to have a good common ground with the tank unit. Aboard an all-metal boat, this ground wire is not necessary. However, if the dashboard is insulated, or made of wood or plastic, a wire **MUST** be run from the gauge ground terminal to one of the bolts securing the sending unit in the fuel tank, and then from there to the **NEGATIVE** side of the battery.

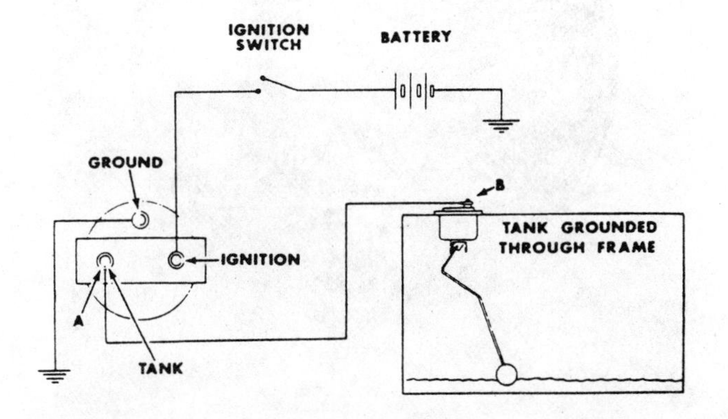

Schematic diagram for safe fuel tank hookup.

FUEL GAUGE TROUBLESHOOTING

In order for the fuel gauge to operate properly the sending unit and the receiving unit must be of the same type and preferably of the same make.

The following symptoms and possible corrective actions will be helpful in restoring a faulty fuel gauge circuit to proper operation.

If you suspect the gauge is not operating properly, the first area to check is all electrical connections from one end to the other. Be sure they are clean and tight.

Next, check the common ground wire between the negative side of the battery, the fuel tank, and the gauge on the dash.

If all wires and connections in the circuit are in good condition, remove the sending unit from the tank. Run a wire from the gauge mounting flange on the tank to the flange of the sending unit. Now, move the float up-and-down to determine if the receiving unit operates. If the sending unit does not appear to operate, move the float to the midway point of its travel and see if the receiving unit indicates half full.

If the pointer does not move from the **EMPTY** position one of four faults could be to blame:

1- The control panel receiving unit is not properly grounded.

2- No voltage at the control panel receiving unit.

3- Negative meter connections are on a positive grounded system.

4- Positive meter connections are on a negative grounded system.

If the pointer fails to move from the FULL position, the problem could be one of three faults.

1- The tank sending unit is not properly grounded.

2- Improper connection between the tank sending unit and the receiving unit on the control panel.

3- The wire from the gauge to the ignition switch is connected at the wrong terminal.

If the pointer remains at the 3/4 full mark, it indicates a six-volt gauge is installed in a 12-volt system.

If the pointer remains at about 3/8 full, it indicates a 12-volt gauge is installed in a six-volt system.

Erratic Fuel Gauge Readings

Inspect all of the wiring in the circuit for possible damage to the insulation or conductor. Carefully check:

1- Ground connections at the receiving unit on the control panel.

2- Harness connector to the control panel unit.

3- Body harness connector to the chassis harness.

4- Ground connection from the fuel tank to the place where it is grounded.

5- Feed wire connection at the tank sending unit.

GAUGE ALWAYS READS FULL, when the ignition switch is **ON:**

1- Check the electrical connections at the receiving unit on the control panel; the body harness connector to chassis harness connector; and the tank unit connector in the tank.

2- Make a continuity check of the ground wire from the tank to the place where it is grounded.

3- Connect a known good tank unit to the tank feed wire and the ground lead. Raise and lower the float and observe the receiving unit on the dash. If the dash unit follows the arm movement, replace the tank sending unit.

GAUGE ALWAYS READS EMPTY, when the ignition switch is **ON:**

Disconnect the tank unit feed wire and do not allow the wire terminal to ground. The gauge on the control panel should read **FULL.**

If Gauge Reads Empty:

1- Connect a spare control panel unit into the control panel unit harness connector and ground the unit. If spare unit reads **FULL,** the original unit is shorted and must be replaced.

2- A reading of **EMPTY** indicates a short in the harness between the tank sending unit and the gauge on the control panel.

If Gauge Reads Full:

1- Connect a known good tank sending unit to the tank feed wire and the ground lead.

2- Raise and lower the float while observing the control panel gauge. If gauge follows movement of the float, replace the tank sending unit.

GAUGE NEVER INDICATES FULL

This test requires shop test equipment.

1- Disconnect the feed wire to the tank unit and connect the wire to ground thru a variable resistor or thru a spare tank unit.

2- Observe the gauge reading. The reading should be **FULL** when resistance is increased to about 90 ohms. This resistance would simulate a full tank.

3- If the check indicates the gauge is operating properly, the trouble is either in the tank sending unit rheostat being shorter, or the float is binding. The arm could be bent, or the tank may be deformed. Inspect and correct the problem.

TACHOMETER

Today, most boats powered by a 4 to 8-cylinder engine and driven by a Vovo Penta stern drive have a tachometer installed on the control panel as standard equipment. The manufacturer recognizes the importance of this instrument as a useful tool in obtaining maximum performance from the boat and its associated equipment.

An accurate tachometer can be installed on any engine. Such an instrument provides an indication of engine speed in revolutions per minute (rpm). This is accomplished by measuring the number of electrical pulses per minute generated in the primary circuit of the ignition system.

Maximum engine performance can only be obtained through proper tuning using a tachometer.

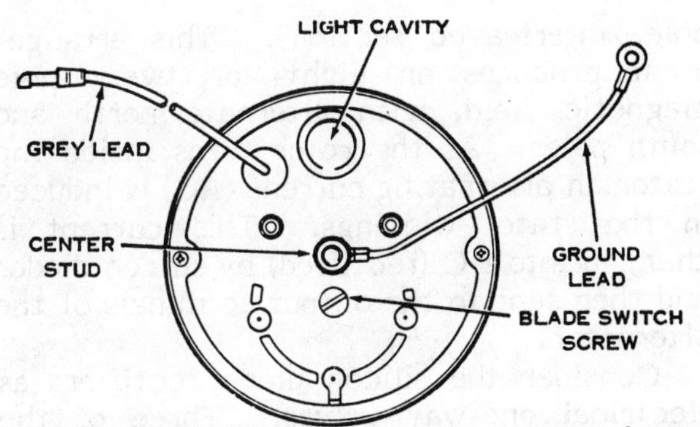

A blade-type tachometer can be installed for use with a four-, six-, or eight-cylinder engine. Simply shift the blade to the connector matching the number of engine cylinders.

6-4 HORNS

The only reason for servicing a horn is because it fails to operate properly or because it is out of tune. In most cases, the problem can be traced to an open circuit in the wiring or to a defective relay.

Cleaning: Crocus cloth and carbon tetrachloride should be used to clean the contact points. **NEVER** force the contacts apart or you will bend the contact spring and change the operating tension.

Check The Relay and Wiring: Connect a wire from the battery to the horn terminal. If the horn operates, the problem is in the relay or in the horn wiring. If both of these appear satisfactory, the horn is defective and needs to be replaced.

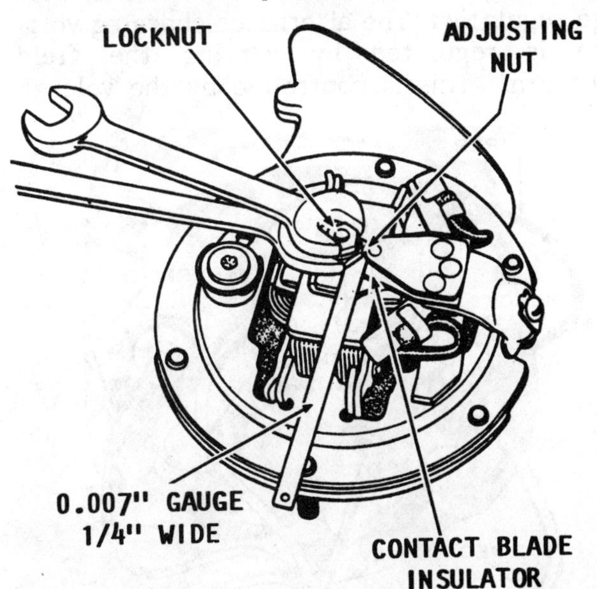

*The tone of a horn can be adjusted with a 0.007" feeler gauge, as described in the text. **TAKE CARE** to prevent the feeler gauge from making contact with the case, or the circuit will be shorted out.*

Before replacing the horn however, connect a second jumper wire from the horn frame to ground to check the ground connection.

Test the winding for an open circuit, faulty insulation, or poor ground. Check the resistor with an ohmmeter, or test the condenser for capacity, ground, and leakage. Inspect the diaphragm for cracks.

Adjust Horn Tone: Loosen the locknut, and then rotate the adjusting screw until the desired tone is reached. On a dual horn installation, disconnect one horn and adjust each one-at-a-time. The contact point adjustment is made by inserting a 0.007" feeler gauge blade between the adjusting nut and the contact blade insulator. **TAKE CARE** not to allow the feeler gauge to touch the metallic parts of the contact points because it would short them out. Now, loosen the locknut and turn the adjusting nut down until the horn fails to sound. Loosen the adjusting nut slowly until the horn barely sounds. The locknut **MUST** be tightened after each test. When the feeler gauge is withdrawn the horn will operate properly and the current draw will be satisfactory.

6-5 CHARGING SYSTEM

The alternator, regulator, battery, ammeter, and the necessary wiring to connect it all together comprise the charging system.

Before the alternator is blamed for battery problems, consider other areas which may be the cause:

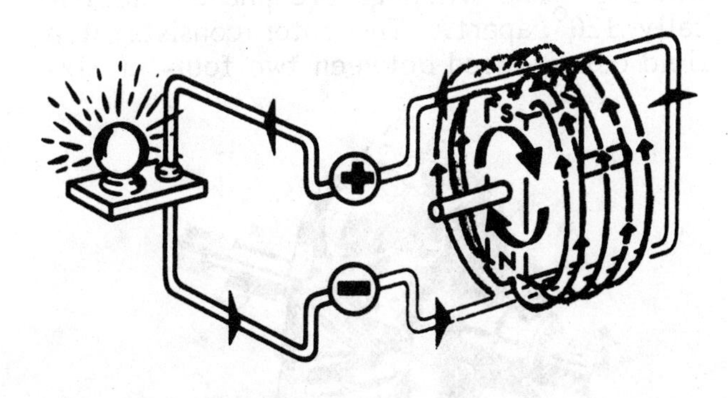

Simplified drawing to illustrate voltage generation in its most basic form. The stator has many coils of wire, shown here as a single loop. The rotor spins; cuts the magnetic field; an electrical voltage is produced in the loop; the voltage moves out of the loop and into the system. Higher revolutions per minute of the loop generates more voltage.

1- Excessive use of lights and accessories while the engine is shut down or operating at low speed for short periods.

2- Voltage losses and the cause.

3- Corroded battery cables, connectors, and terminals.

4- Low electrolyte level in the battery cells.

5- Prolonged disuse of the battery causing a self-discharged condition.

ALTERNATOR DESCRIPTION AND OPERATION

An alternator is an AC generator replacing the conventional DC generator. The alternator has four distinct advantages over the generator:

1- A higher charging rate.
2- A lower cut-in charging speed.
3- A lighter weight.
4- Longer trouble-free service.

The alternator operates on a different principle from the conventional generator. The armature of the alternator is the stationary part and is called the **STATOR;** and the field is the rotating part and is called the **ROTOR.** With this arrangement, the higher current carried by the stator is conducted to the external circuit through fixed leads. This method results in trouble-free operation of a small current supplied to the fields to be conducted through small brushes and rotating slip rings. This is in contrast to the DC generator where the current is carried through a rotating commutator and brushes.

The alternator has a three-phase stator winding. The windings are phased electrically 120° apart. The rotor consists of a field coil encased between two four- or six-pole, interleaved sections. This arrangement produces an eight- or twelve-pole magnetic field, with alternate north and south poles. As the rotor turns inside the stator an alternating current (AC) is induced in the stator windings. This current is changed into DC (rectified) by silicon diodes and then sent to the output terminals of the alternator.

Consider the silicon diode rectifiers as electrical one-way valves. Three of the diodes are polarized one way and are pressed into an aluminum heat sink which is grounded to the slip ring end head. The other three diodes are polarized the opposite way and are pressed into a similar heat sink, which is insulated from the end head and connected to the alternator output terminal. Since a diode has high resistance to the flow of electricity in one direction and passes current with very little resistance in the opposite direction, it is connected in a manner which allows current to flow from the alternator to the battery in a low-resistance direction.

The high resistance in the opposite direction prevents battery current from flowing to the alternator, therefore, no circuit breaker (cutout) is required between the alternator and the battery.

The magnetism left in the rotor field poles is negligible. Thus, the field must be excited by an external source, the battery. The battery is connected to the field winding through the ignition switch and the voltage regulator. The alternator charging voltage is regulated by varying the field strength. This is controlled by the voltage

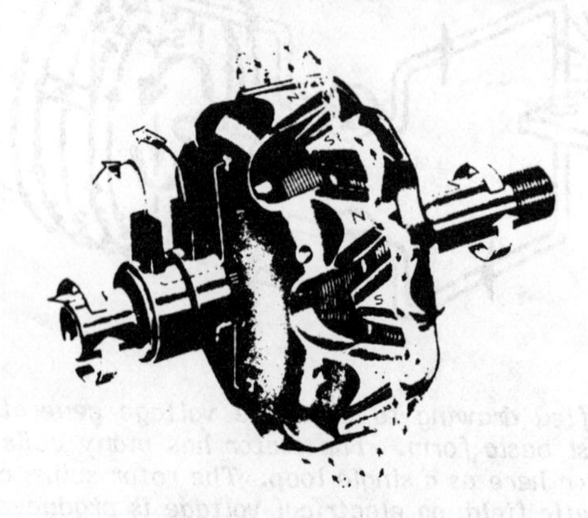

Cutaway view of a modern alternator. Detailed description, operation, and advantages of the alternator are covered in the text.

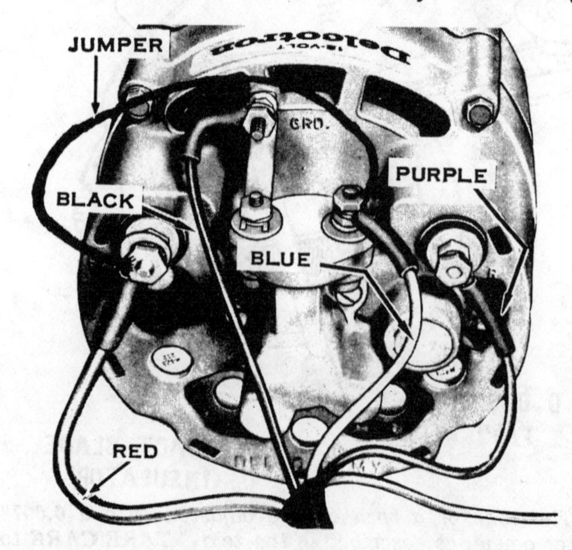

Hookup to check an alternator. Detailed troubleshooting and testing procedures are given in the text.

regulator. No current regulator is required since the alternator has self-limiting current characteristics.

ALTERNATOR PROTECTION DURING SERVICE

The alternator is an important and expensive piece of equipment on the boat. Unless certain precautions are taken before and during servicing, parts of the alternator or the charging circuit may be damaged. **TAKE TIME** to review the following points **BEFORE** serviceing or troubleshooting the charging system.

1- If the battery connections are reversed, the diodes in the alternator will be damaged. Check the battery polarity with a voltmeter before making any connections to be sure the connections correspond to the battery ground polarity.

2- The field circuit between the alternator and the regulator **MUST NEVER BE GROUNDED.** If this circuit is grounded the regulator will be damaged.

3- **NEVER** ground the alternator output terminal while the engine is running or shut down because no circuit breaker is used, and battery current is applied to the alternator output terminal at all times.

4- **NEVER** operate the alternator on an open circuit with the field winding energized, or the unit will be **DAMAGED.**

5- **NEVER** attempt to polarize the alternator because it is never necessary. Any attempt to polarize the alternator will result in **DAMAGE** to the alternator, the regulator, and the wiring.

6- **NEVER** short the bending tool to the regulator base while adjusting the voltage, or the unit will be **DAMAGED.** The bending tool should be **INSULATED** with tape or an insulating plastic sleeve while making the adjustment.

7- **ALWAYS** disconnect the battery ground strap before replacing an alternator or connecting any meter to it.

8- If booster batteries are used for starting an engine the jumper cables must be connected correctly and in the proper sequence to prevent damage to either battery, or to the alternator diodes.

ALWAYS connect a cable from the positive terminal of the dead battery to the positive terminal of the good battery **FIRST. NEXT,** connect one end of the other cable to the negative terminal of the good battery and the other end to the **ENGINE** for a good ground. By making the ground connection on the engine, if there is an arc when you make the connection it will not be near the battery. An arc near the battery could cause an explosion, destroying the battery and causing serious personal injury.

9- If it is necessary to use a fast-charger on a dead battery, **ALWAYS** disconnect one of the boat cables from the battery first, to prevent burning out the diodes in the alternator.

10- **NEVER** use a fast charger as a booster to start the engine because the diodes in the alternator will be **DAMAGED.**

CHARGING SYSTEM TROUBLESHOOTING

The following symptoms and possible corrective actions will be helpful in restoring a faulty charging system to proper operation.

Alternator Fails To Charge

1- Drive belt loose or broken. Replace and/or adjust.

2- Corroded or loose wires or connection in the charging circuit. Inspect, clean, and tighten.

3- Worn brushes or slip rings. Replace as required.

4- Sticking brushes.

5- Open field circuit. Trace and repair.

6- Open charging circuit. Trace and repair.

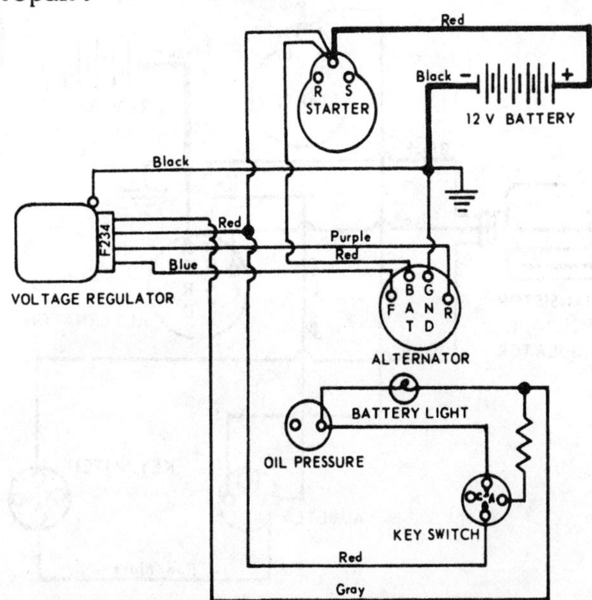

Schematic diagram for a charging system with an indicator light on the dashboard and a relay-type voltage regulator installed.

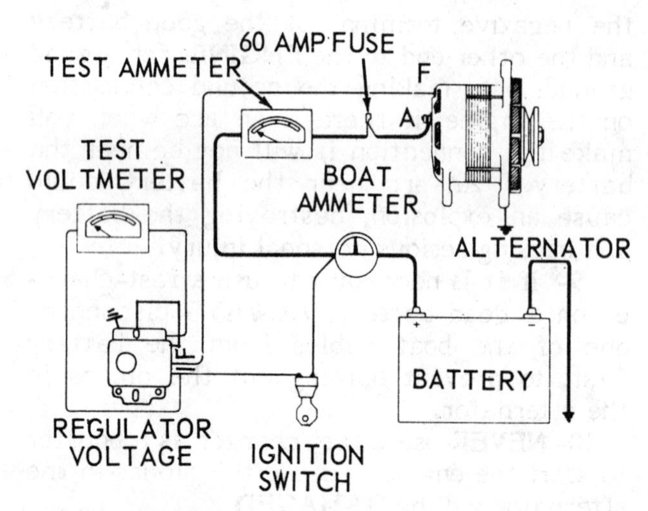

Functional diagram for hookup to test the electrical generating system.

7- Open circuit in the stator windings.
8- Open rectifier. Replace the unit.
9- Defective regulator. Replace the unit.

Alternator Charges Low or Unsteady

1- Drive belt loose or broken. Replace and/or adjust.
2- Battery charge too low. Charge or replace the battery.
3- Defective regulator. Replace the unit.
4- High resistance at the battery terminals. Remove cables, clean connectors and battery posts, replace and tighten.

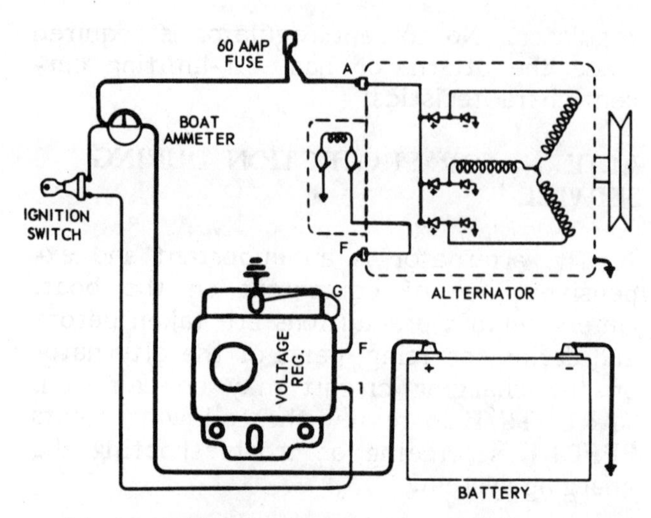

Schematic diagram of a typical Prestolite charging system on Volvo Penta marine installations.

5- High resistance in the charging circuit. Trace and repair.
6- Open stator winding. Replace alternator.
7- High resistance in body-to-ground lead. Trace and repair.

Alternator Output Low and a Low Battery

1- Drive belt slipping. Adjust.
2- High resistance in the charging circuit. Trace and repair.
3- Shorted or open-circuited rectifier. Replace rectifier.
4- Grounded stator windings. Replace alternator.

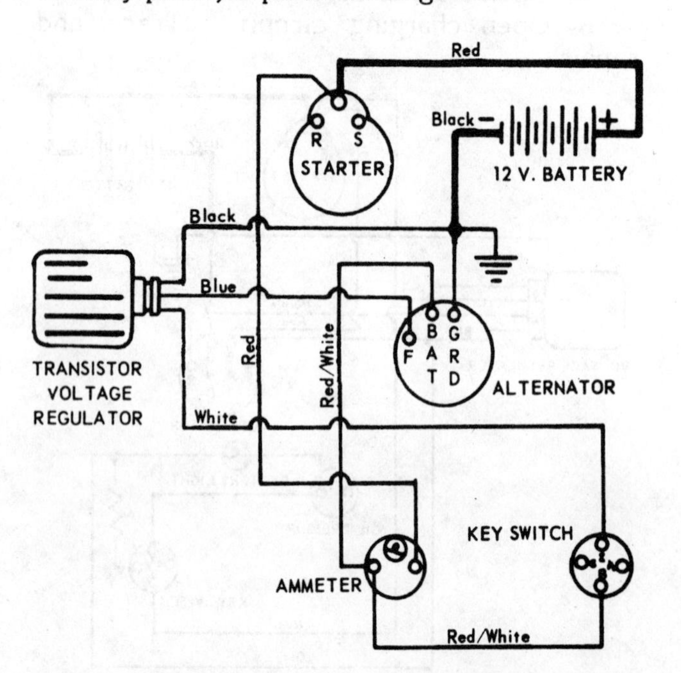

Schematic diagram for a charging system with an ammeter on the dashboard and a relay-type voltage regulator installed.

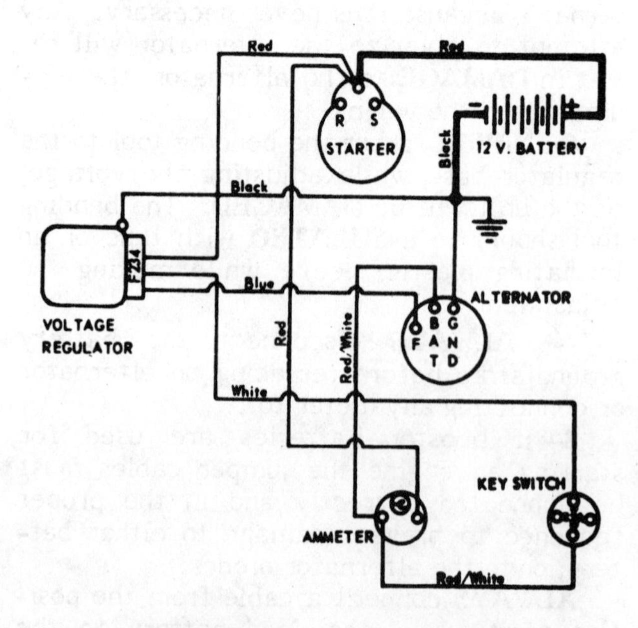

Schematic diagram for a charging system with an ammeter on the dashboard and a transistor-type regulator installed.

Alternator Output Too High — Battery Overcharged

1- Faulty voltage regulator. Replace.

2- Regulator base not grounded properly. Correct condition to make good ground.

3- Faulty ignition switch. Replace.

Alternator Too Noisy

1- Worn, loose, or frayed drive belt. Replace belt and adjust properly.

2- Alternator mounting loose. Tighten all mounting hardware securely.

3- Worn alternator bearings. Replace.

4- Interference between rotor fan and stator leads or rectifier. Check and correct.

5- Rotor or fan damaged. Replace.

6- Open or shorted rectifier. Replace.

7- Open or shorted winding in stator. Replace.

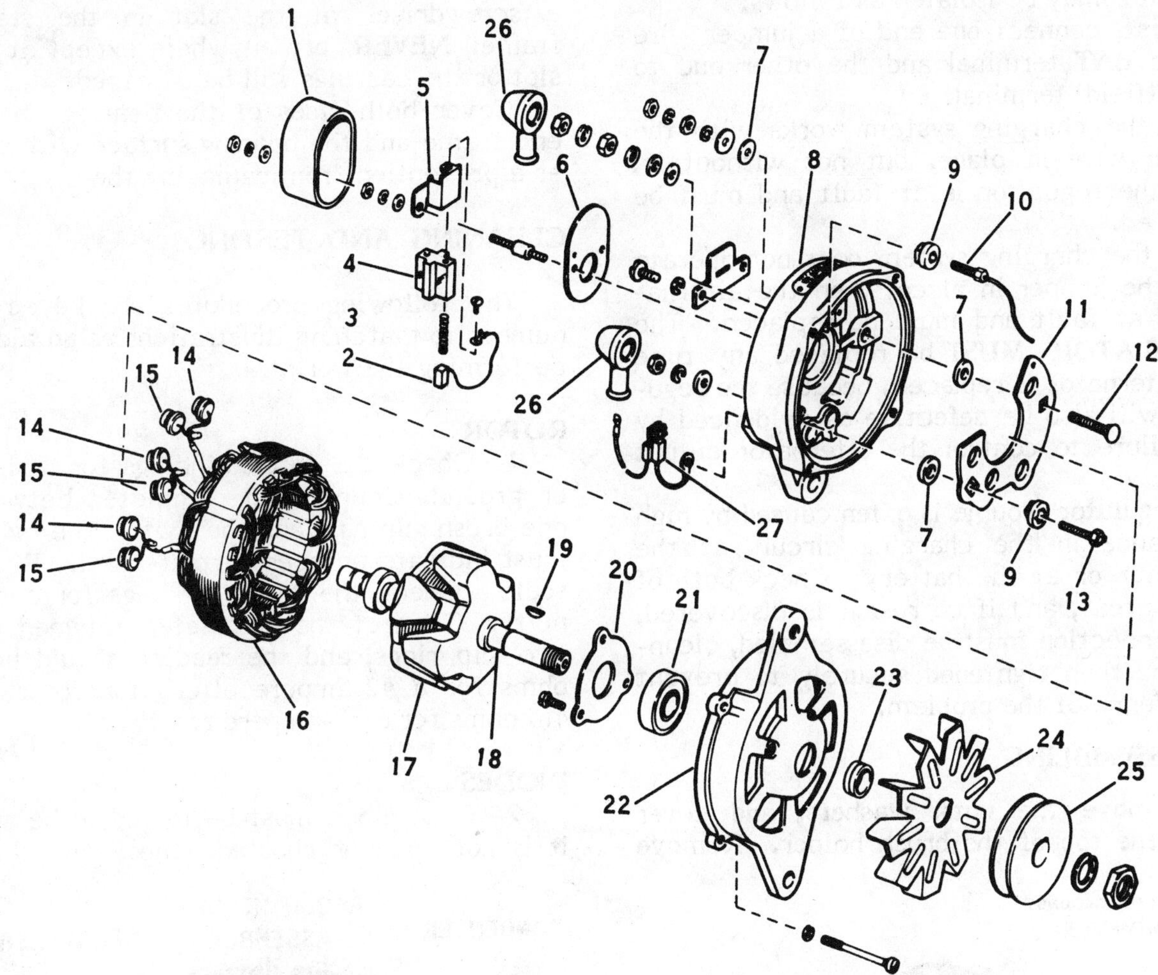

1. **Slip Ring Cover**
2. **Brush Set**
3. **Spring**
4. **Brush Holder**
5. **Brush Holder Cover**
6. **Back-Up Plate**
7. **Insulating Washer**
8. **Slip Ring End Head**
9. **Insulating Bushing**
10. **Field Terminal Wire and Stud**
11. **Rectifier Plate**
12. **Battery Terminal Stud**
13. **Stud**
14. **Negative Rectifiers**
15. **Positive Rectifiers**
16. **Stator**
17. **Rotor and Bearing**
18. **Retaining Ring**
19. **Key**
20. **Retainer Plate**
21. **Bearing**
22. **Drive End Head**
23. **Spacer**
24. **Fan**
25. **Pulley**
26. Cover
27. Fuse and Lead (Shown without Potting Material)

Prestolite alternator with major parts identified.

Ammeter Fluctuates Constantly

High resistance connection in the alternator or voltage regulator circuit. Trace and repair.

6-6 SERVICING CHARGING SYSTEM WITH SEPARATE REGULATOR PRESTOLITE

Trouble between the alternator and the regulator may be isolated as follows:

First, connect one end of a jumper wire to the **BAT** terminal and the other end to the **F** (field) terminal.

If the charging system works with the jumper wire in place, but not without it, then the regulator is at fault and must be replaced.

If the charging system does not operate with the jumper in place, then the alternator is at fault and must be replaced. The **REGULATOR MUST** be replaced any time the alternator is replaced, because the regulator will also be defective as evidenced by its failure to control the alternator output properly.

Regulator trouble is often caused by high resistance in the charging circuit at the ammeter or at the battery. Check both of these areas, and if corrosion is discovered, the connection must be disassembled, cleaned, and then tightened securely to prevent recurrence of the problem.

DISASSEMBLING

Remove the nuts, washers, and cover from the top of the brush holder. Remove the studs and the brush-and-spring assemblies. Remove the screws securing the brush holder to the end frame, and then remove the brush holder.

Scribe a mark on both end frames and matching marks on the stator as an aid to properly assembling the parts.

Remove the four through-bolts. Separate the drive end frame and rotor assembly from the stator by **CAREFULLY** prying with a screwdriver at the slot in the stator frame. **NEVER** pry anywhere except at the slot or the castings will be damaged.

Cover both sides of the bearing on the end frame and the bearing surface with tape as a prevention from damaging them.

CLEANING AND TESTING

The following procedures are keyed by number to matching illustrations as an aid in performing the work.

ROTOR

1- Check the rotor windings for a short or ground: Connect an ohmmeter between one brush slip ring and the shaft. The meter must indicate an open circuit on the R-100 scale. Check the rotor windings for continuity: Connect the ohmmeter between the two slip rings, and the reading should be 6 ohms for a 42-ampere alternator rotor, or 4.5 ohms for a 32-ampere rotor.

DIODES

2- Each diode must be tested to be sure it is not open or shorted. Check the three

(CHECK FOR GROUNDS)
OHMMETER

OHMMETER
(CHECK FOR SHORTS AND OPENS)

1

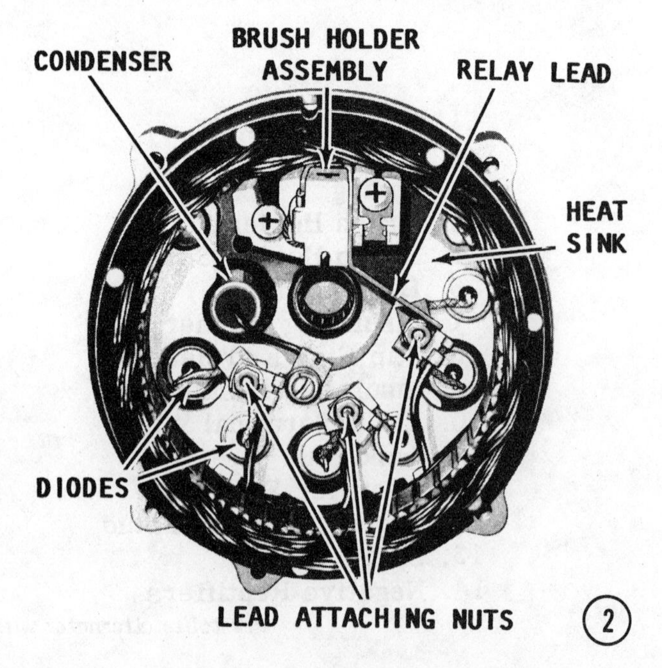

2

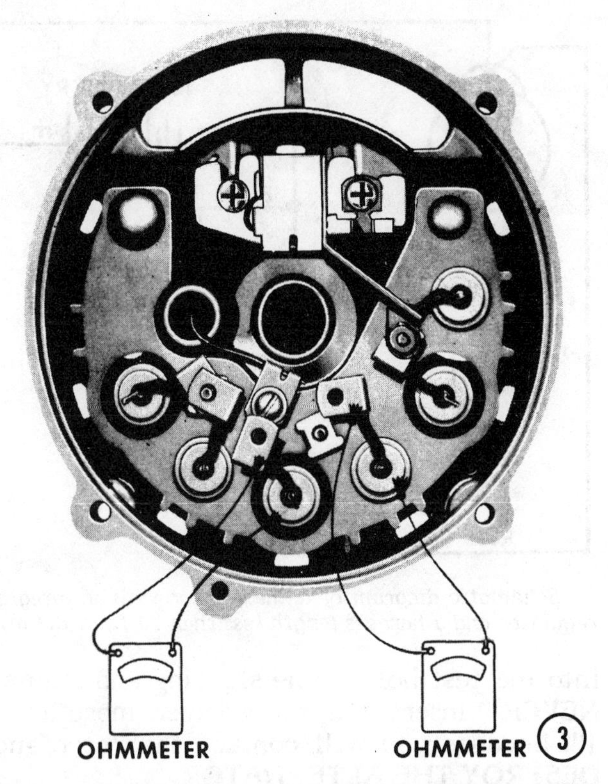

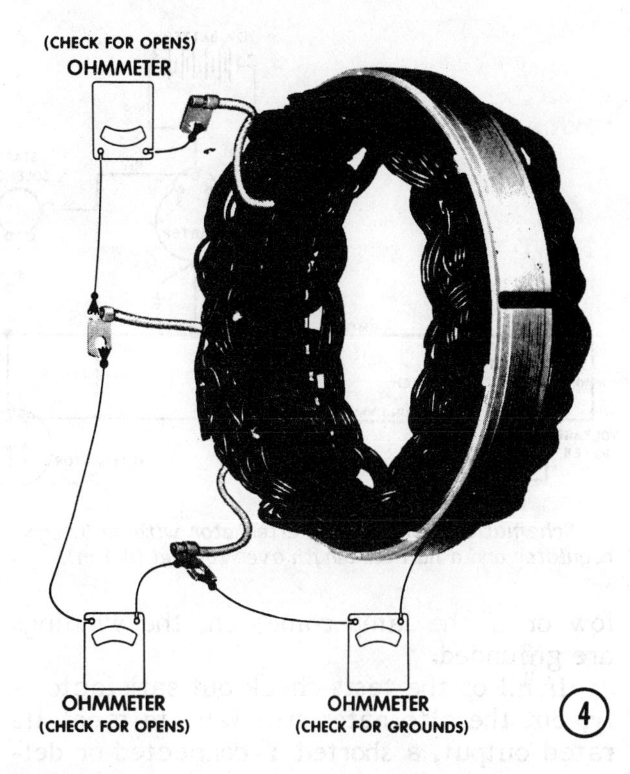

diodes in the heat sink: Disconnect the stator leads and test each diode with one ohmmeter probe on the heat sink and the other on the disconnected diode terminal. Note the reading, and then reverse the test leads and again note the reading. If one reading is high and the other low, the diode is satisfactory. If both readings are zero or infinity, the diode is defective and must be replaced.

3- Check the three diodes in the end frame: Connect the ohmmeter in the same manner as for the previous test EXCEPT connect one test lead to the end frame instead of the heat sink. Note the reading, then reverse the leads and again note the reading. The results should be the same as the previous test.

STATOR

4- The stator is not checked for shorts due to the very low resistance of the windings. Neither are checks of the delta stator for opens because the windings are connected in parallel.

Checks that are made on the stator must be made with all of the diodes disconnected from the stator.

Check the Y-connected stator for open circuits: Connect an ohmmeter or test light across any two pair of terminals. If the ohmmeter reading is high, or if the test light fails to come on, there is an open in the winding.

Both types of stator winding can be checked for grounds by connecting an ohmmeter or test light from either terminal to the stator frame. If the ohmmeter reads

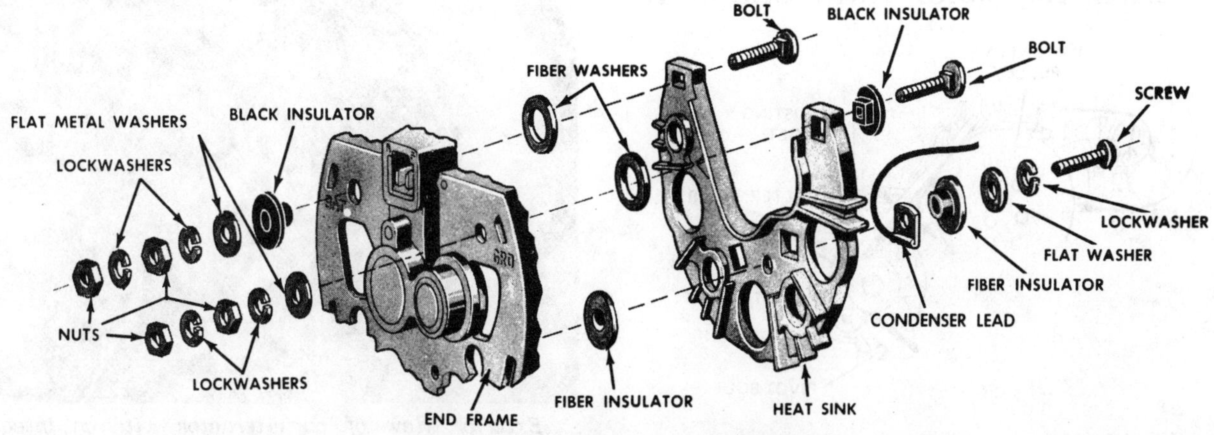

Heat sink with major parts identified.

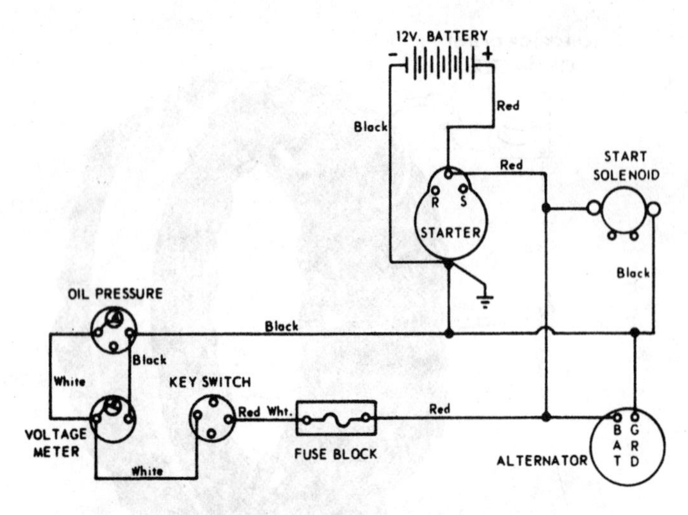

Schematic diagram of an alternator with an integral regulator and a harness length over 20-feet (6.1 m).

low or if the lamp comes on, the windings are grounded.

If all of the tests check out satisfactorily, but the alternator still fails to meet its rated output, a shorted Y-connected or delta stator winding, or an open delta winding, may be the cause.

6-7 SERVICING CHARGING SYSTEM WITH INTEGRAL REGULATOR PRESTOLITE

Alternators with an integral solid-state regulator installed are easily identified by an explosion-resistant screen at each end frame.

Trouble between the alternator and the regulator may be isolated as follows:

Start the engine.

CAUTION: Water must circulate through the lower unit to the engine any time the engine is run to prevent damage to the water pump mounted on the engine.

Now, with the engine running, **CAREFULLY AND SLOWLY** insert a small bladed screwdriver **NO MORE THAN ONE-INCH**

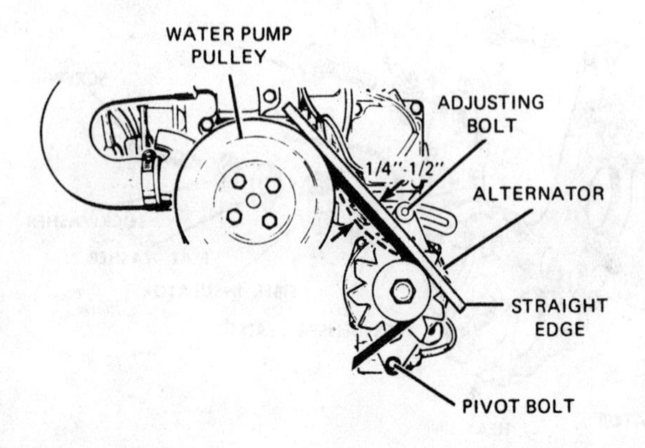

Checking for proper drive belt tension.

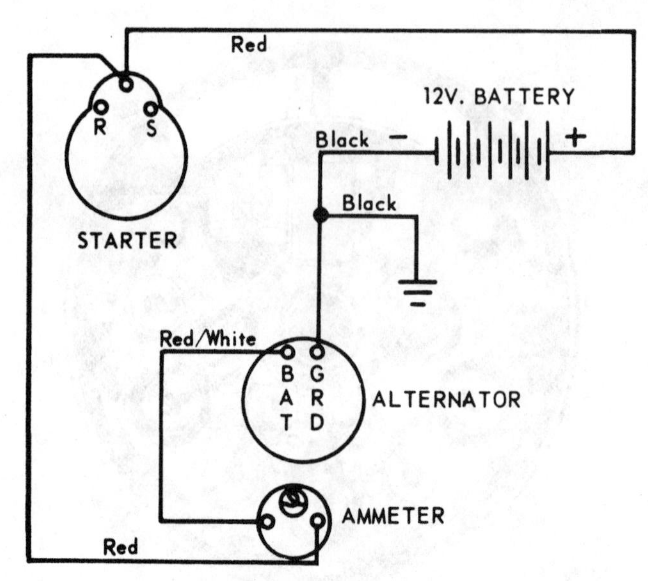

Schematic diagram of an alternator with an integral regulator and a harness length less than 20-feet (6.1 m).

into the test hole in the slip ring end frame. **NEVER** insert the screwdriver more than 1", because you will contact the rotor and **DESTROY THE ALTERNATOR.**

With the screwdriver in place, the field winding is grounded and the regulator is bypassed. If the alternator charges with the screwdriver in place, the regulator is defective and must be replaced. If the alternator does not charge with the screwdriver in place, then the alternator is defective and must be repaired or replaced.

WHENEVER the alternator is repaired or replaced, the regulator must be replaced because it is also defective, as evidenced by its failure to control the alternator output properly.

Exterior view of an alternator with an integral regulator installed on Volvo Penta marine units.

Removing the alternator pulley.

Many times, the cause of a defective regulator is a high resistance connection in the charging circuit, at the ammeter, or at the battery. Each of these areas should be checked and serviced thoroughly to prevent repetition of the trouble. Disassemble the connection, clean, and then tighten it.

DISASSEMBLING

Remove the four through-bolts, and then take off the cover. Secure a 5/16" Allen wrench in a vise, as shown in the accompanying illustration.

Now, slide the alternator shaft over the exposed end of the Allen wrench. Use a 15/16" box-end wrench to loosen, and then remove the end nut, washer, pulley, fan, collar, drive-end frame, and collar from the rotor shaft.

CAREFULLY pull the rotor assembly from the slip-ring end frame, and **TAKE CARE** not to lose the brush springs. Cover the slip-ring end frame with tape to keep dirt out of the bearing. Cover the slip rings with tape also as a precaution against scoring them. **NEVER** use friction tape, which leaves a gummy deposit which is very difficult to remove.

CLEANING AND TESTING

The following procedures are keyed by number to matching illustrations as an aid in performing the work.

ROTOR

1- Check the rotor windings for a short or ground: Connect an ohmmeter between one brush slip ring and the shaft. If the meter does not show an open circuit on the R-100 scale there is a short in the rotor windings. Check the rotor windings for continuity: Connect the ohmmeter between the two slip rings, and a normal reading of 2.6 ohms will indicate the rotor coil is serviceable.

STATOR

2- The stator is not checked for shorts due to the very low resistance of the windings. Neither are checks of the delta stator for opens because the windings are connected in parallel.

Checks that are made on the stator must be made with all of the diodes disconnected from the stator.

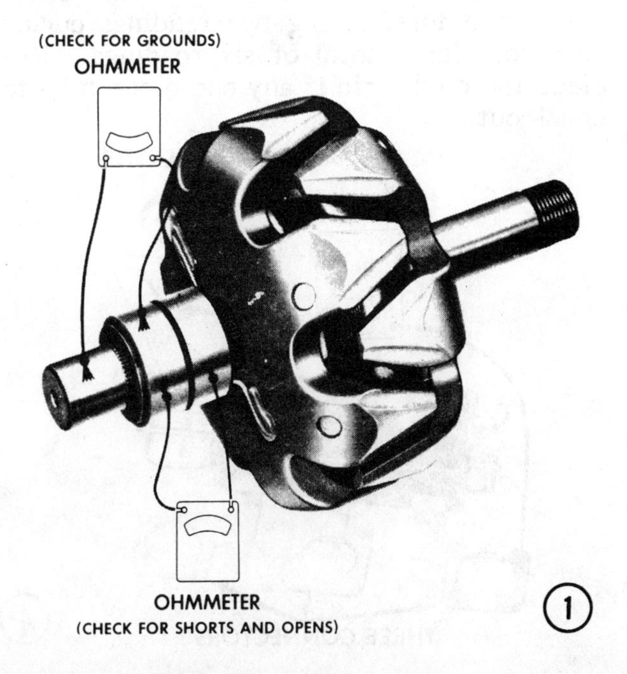

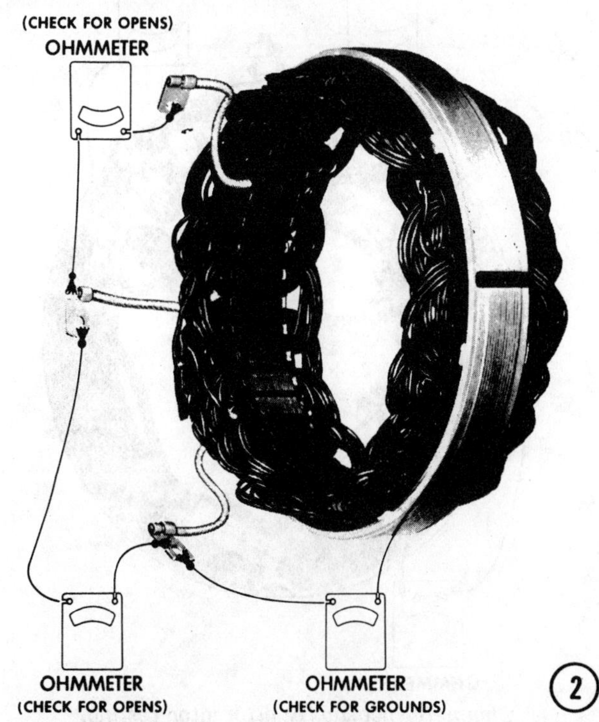

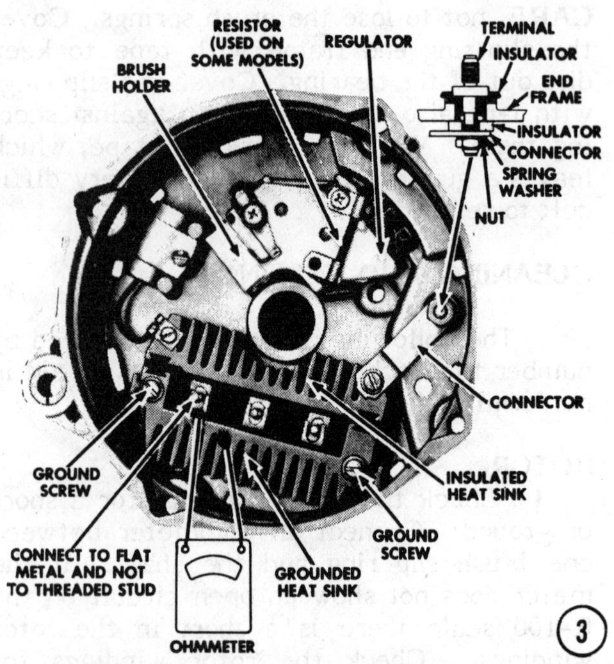

Check the Y-connected stator for open circuits: Connect an ohmmeter or test light across any two pair of terminals. If the ohmmeter reading is high, or if the test light fails to come on, there is an open in the winding.

Both types of stator winding can be checked for grounds by connecting an ohmmeter or test light from either terminal to the stator frame. If the ohmmeter reads low or if the lamp comes on, the windings are grounded.

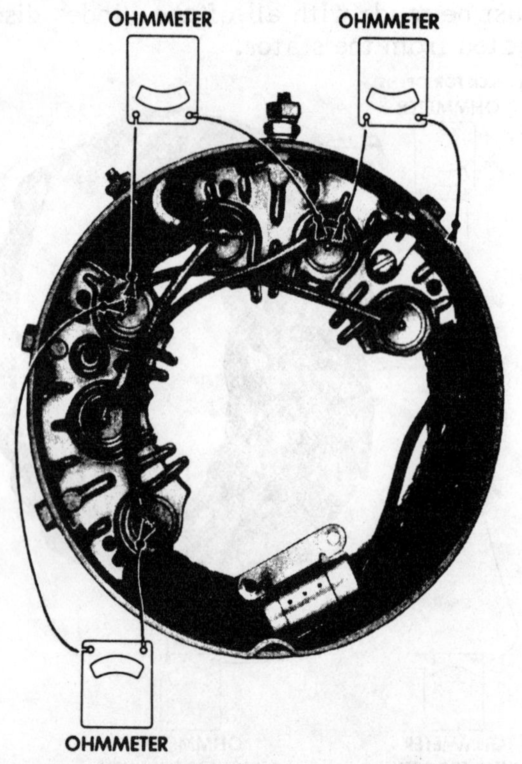

Ohmmeter hookup for alternator testing.

If all of the tests check out satisfactorily, but the alternator still fails to meet its rated output, a shorted Y-connected or delta stator winding, or an open delta winding, may be the cause.

RECTIFIER BRIDGE

3- First, set the ohmmeter scale to R-100. Next, connect one test lead to the grounded heat sink and the other test lead to one of the three terminals. Now, note the reading on the scale. Reverse the test leads, and again note the reading. If both readings are zero or very high, the rectifier bridge is defective. If one of the readings is high and the other low, the rectifier checks out for service.

Repeat the tests between the grounded heat sink and the other two terminals, and then between the insulated heat sink and each of the three terminals. When you are finished, you should have made a total of six tests, with two readings during each test.

DIODE TRIO

4- Remove the diode trio by removing the attaching bolts and nuts. Set the ohmmeter scale to R-100. Check each diode for a short or open by first connecting one test lead to a single connector, of the three connectors, and then noting the reading on the scale. Now, reverse the test leads, and again note the reading. If both readings are zero or very high, the diode is defective. If one reading is low and the other is high, the diode checks out for service.

Repeat each test for each of the other two connectors, taking two readings during each test for a total of six readings. Replace the diode trio if any one diode fails to check out.

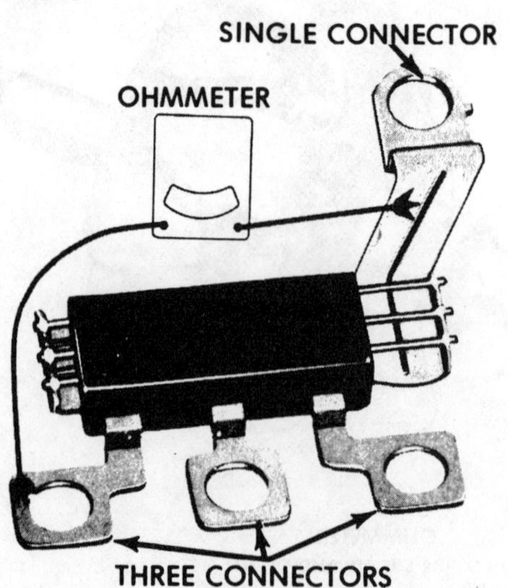

6-8 SERVICING CHARGING SYSTEM WITH SEPARATE REGULATOR SEV MARCHAL ALTERNATOR

INTRODUCTORY WORDS

A 38 amp SEV Marchal alternator was installed on the early AQ130's, AQ170's, AQ200's, AQ225's, AQ250's, AQ260's and AQ290's. A 50 amp Paris Rhone alternator was installed on late model AQ125, AQ145, AQ175, AQ200, AQ225, AQ260, and AQ290. Service procedures and testing of these two alternators are almost identical.

A new Paris Rhone "Valeo" alternator installed on the Model 740 since 1990 is covered in Section 6-11.

The regulator for the SEV MARCHAL alternator is mounted on the back of the unit.

The regulator for the Paris Rhone alternator is an integral part of the alternator.

If the alternator has been submerged; an armature has failed; the windings have become loose, or become burned; the practical solution to restoring generating power is to replace the complete alternator. The regulator for the SEV Marchal alternator can be removed and replaced.

The back of the SEV Marchal alternator contains three diodes. These diodes are also replaceable. However, they are sold **ONLY** as complete sets. The diodes can be checked individually. Detailed instructions and illustrations are included in this section.

The only other service work practical in the field might possibly be to replace the bearings. If the alternator fails to generate current, the most practical remedy is to seek the services of a shop specializing in alternator repair.

REMOVAL

1- Identify, and then disconnect the wires from the terminals on the back of the alternator.

2- Loosen the alternator adjustment bolt and pivot bolt. Slacken the drive belt and then slip if off the alternator pulley. Remove the alternator mounting bolts, and then lift the unit clear of the engine.

BENCH TESTING

3- With the alternator on the work bench: identify the wires from the regulator, and then disconnect them from the alternator. Remove the two attaching screws, and then remove the regulator. Remove the two attaching nuts, and then lift off the diode cover on the back.

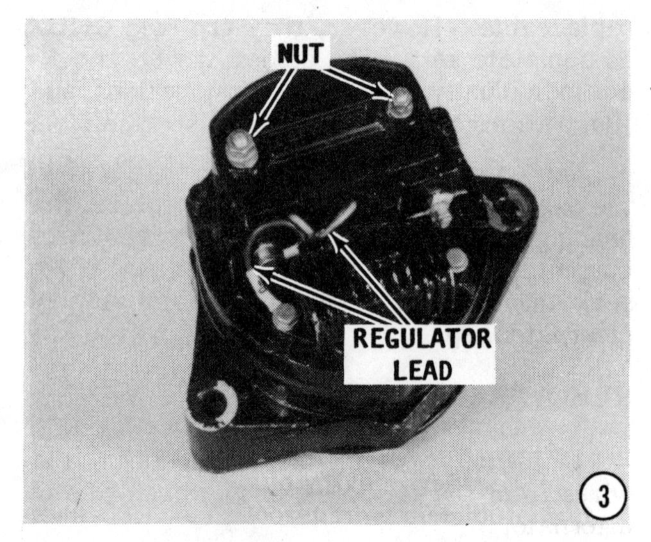

4- Connect one lead of an ohmmeter to a good ground on the alternator. Connect the other lead to the **DF** terminal.

If the resistance is less than 3 ohms or more than 15 ohms, there is a fault in the armature winding and the alternator **MUST** be replaced.

If the resistance is between 3 and 15 ohms, the winding is good.

If a control lamp is used instead of an ohmmeter, make the connections the same as with the ohmmeter. The lamp should glow weakly. If it glows brightly or not at all, there is a fault in the winding and the alternator **MUST** be replaced.

5- Remove the three nuts and the brass plate which attach the main wiring cables to the diode bridge. Remove the three cables and bend the ends upward.

Check the triple outlet by connecting one lead of the ohmmeter to the **61+** and the

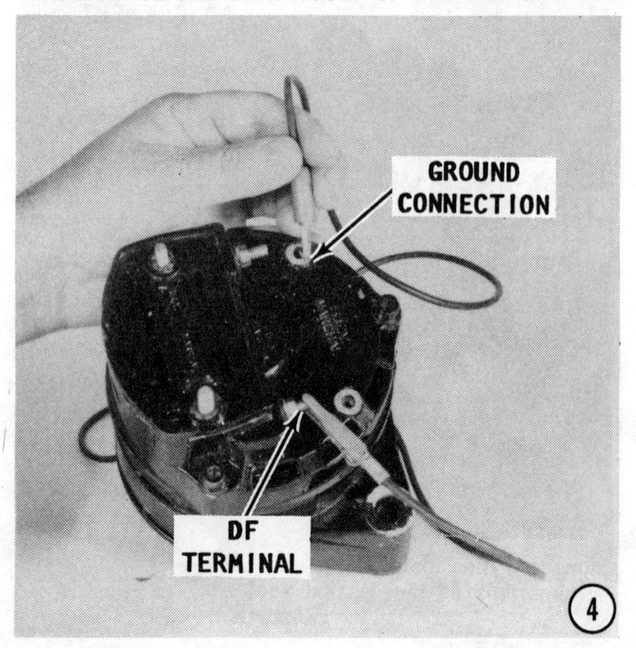

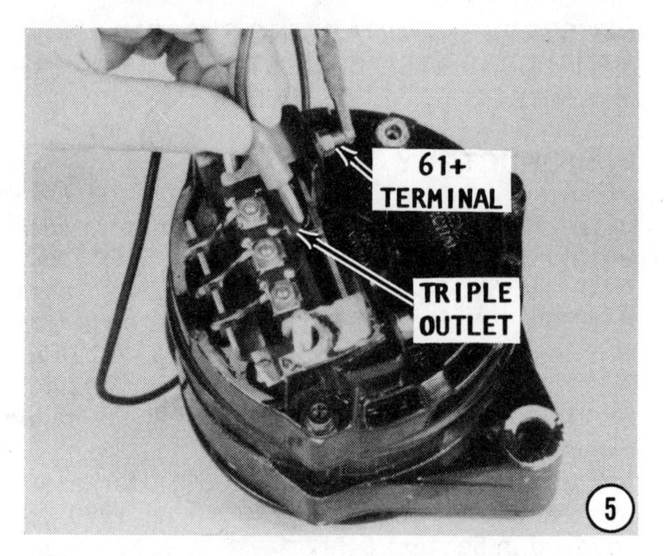

other lead to the common point for the three outlets (the cable terminal on the printed circuit from where the cable originates).

If a current is registered, continue with the testing. If no current is registered, there is a break in the circuit between the triple outlet and the **61+** terminal.

6- The diodes can be checked with the aid of a control lamp and an ohmmeter. Connect one of the test leads to the triple outlet and measure on each soldered point for each of the three diode connections. Change over the test leads.

SPECIAL WORDS

Notice that the diode plate is varnished as a protection against corrosion. The test lead connections **MUST** penetrate through the varnish for an accurate test.

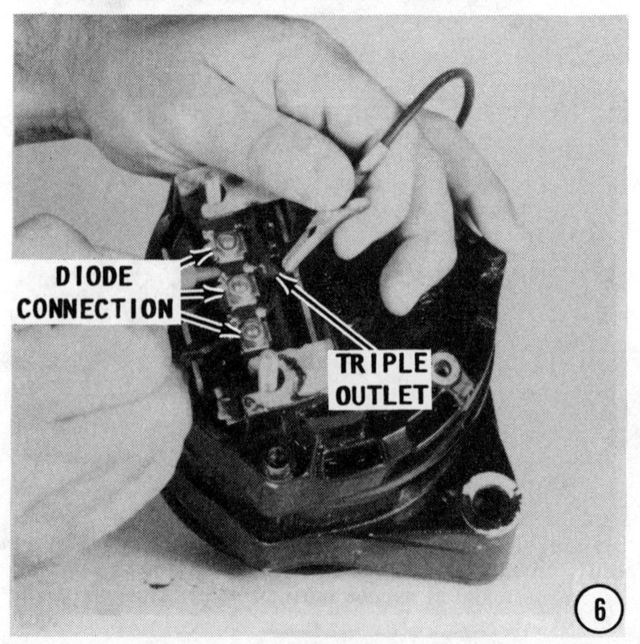

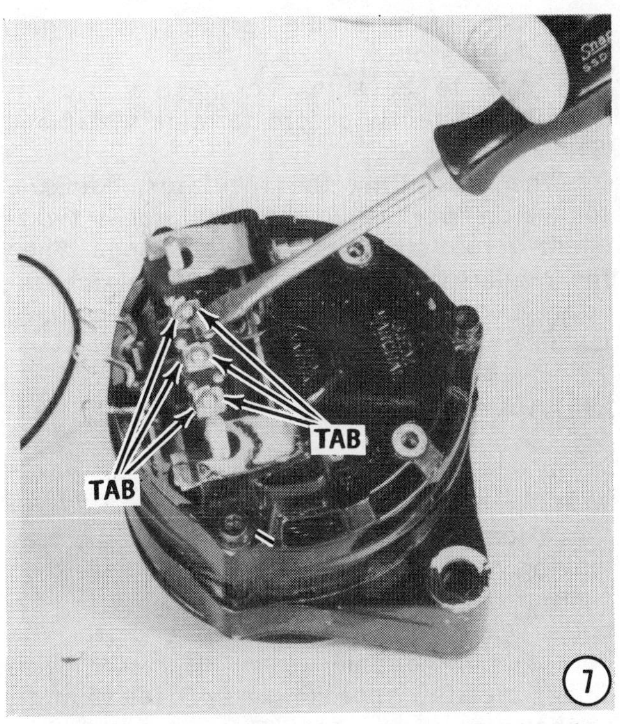

7

If a current is registered in one of the directions only on each of the diodes, then they are serviceable. Continue with the testing.

If there is current in both directions or in neither direction on one or more of the three diodes, the diodes are at fault.

The remedy is to change the plate with the magnetizing diodes. This is accomplished by first removing the nuts and washers on connection **61+** and carefully loosening the tongue (the black cable) under connection **B+**. Remove the rectifier bridge by loosening outlet connection **61+**. Install a new plate.

If a short circuit exists in one or more of the rectifier diodes, the voltage given out by the alternator is considerably lower than normal.

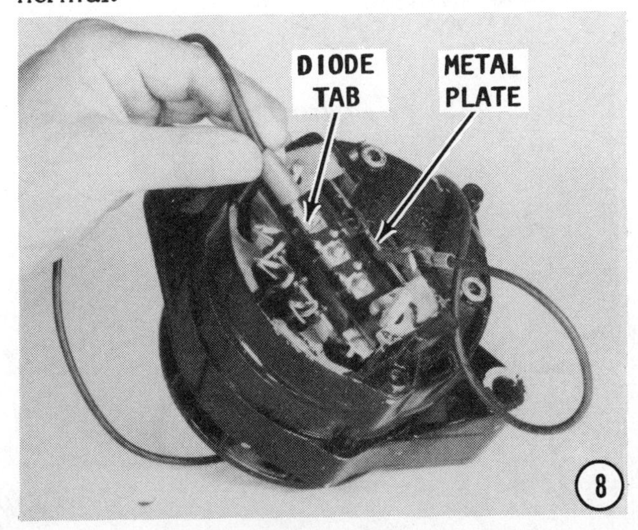

8

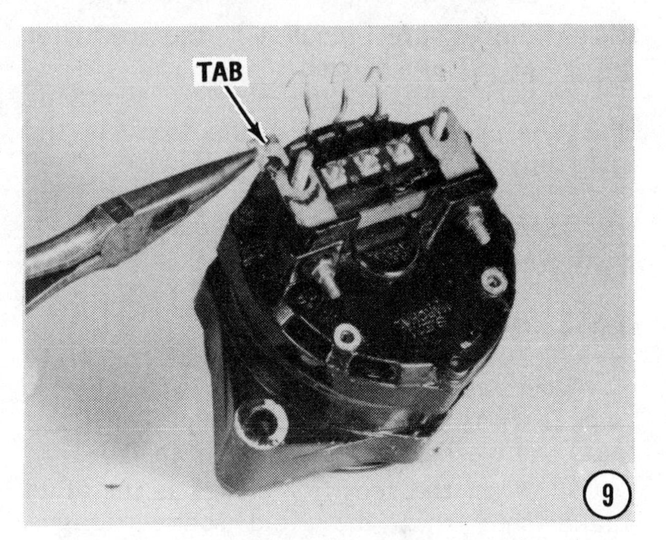

9

7- Use a small screwdriver and bend up the six rectifier diodes connecting tabs.

8- Connect one lead of an ohmmeter or test light, to the metal plate (on the same side as the diodes) and the other test lead to the respective diode.

If current is registered in one direction and not in the other on each diode, the rectifier diodes are seviceable. Continue with the testing.

If a current is registered or not registered in both directions on one or several of the six diodes, three diodes are plus **(+)** and

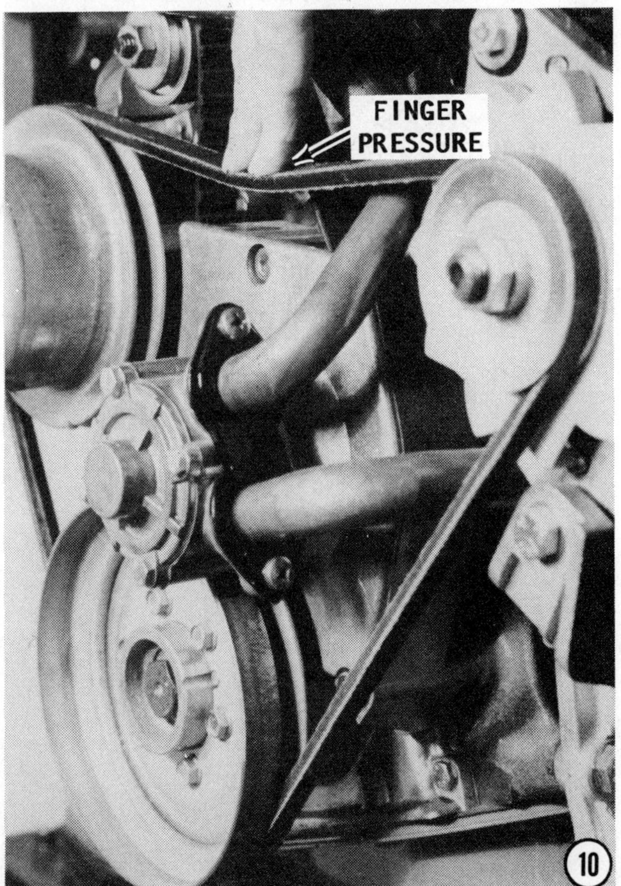

10

three diodes are minus (-), the rectifier bridge **MUST** be changed.

This is accomplished by first removing the nuts and washers on connection **61+** and carefully loosening the tongue (the black cable) under connection **B+**. Remove the rectifier bridge by loosening outlet connection **61+**. Install a new plate.

ASSEMBLING

When assembling socket **"61+"**, hold the socket from the underneath side until the nuts and washers have been tightened.

9- When the rectifier bridge is installed make sure that the field socket tongue (black cable) is pushed in between the upper part of the rectifier bridge and the black insulation on **"B+"**. Also, be sure the cable is pushed into the plastic holder, as shown.

When the main cables are connected, be sure the brass tabs fit into the proper sockets. Push the cables into the plastic holders.

If the thin film of varnish on the diode set and the bridges has been disturbed (damaged) during any of the testing, the varnish **MUST** be restored.

Check to be sure the plastic cover is installed correctly before terminals **"B+"** and **"D-"**are tightened.

When installing the regulator, hold the tongue on **"61+"** while the regulator is tightened. Avoid contact with the casing. When the regulator attaching screws are tightened, a good ground is made. The **"D-"** contact is connected under one of these screws.

INSTALLATION

10- Mount the alternator and secure it with the attaching bolts. Slip the belt over the alternator pulley. Tighten the belt by moving the alternator outboard, and then tighten the alternator in place with the bolts. Check the tension on the belt. The belt should be able to be depressed with finger pressure approximately 3/16" (5 mm), as shown.

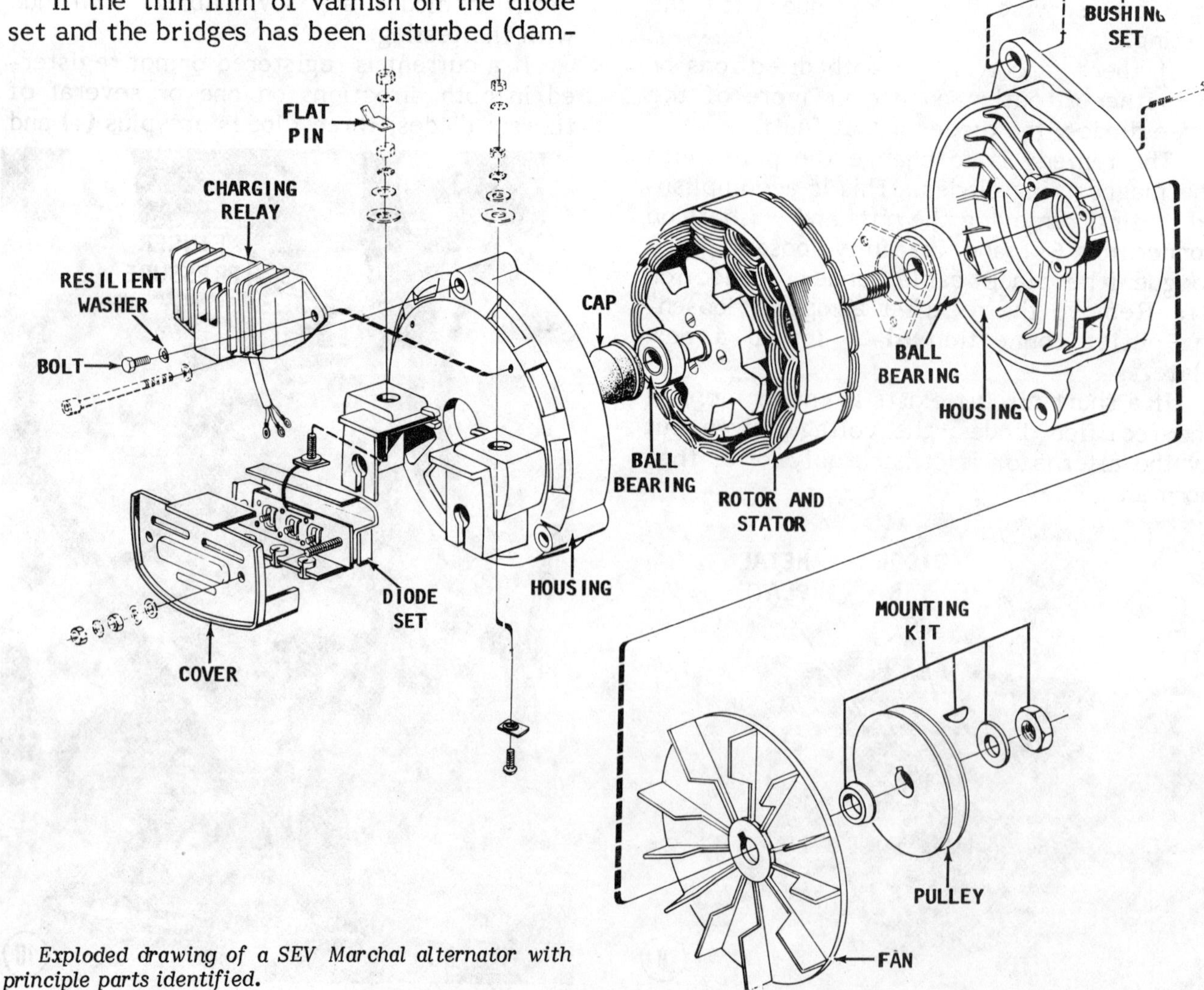

Exploded drawing of a SEV Marchal alternator with principle parts identified.

*After the correct tension is obtained on the belt, tighten the **UPPER** bolt securely first, and then the lower bolt to hold the adjustment.*

6-9 SERVICING DELCOTRON CHARGING SYSTEM

Alternators with an integral solid-state regulator installed, such as the Delcotron, are easily identified by an explosion-resistant screen at each end frame.

Regulator Testing

Trouble between the alternator and the regulator may be isolated as follows:

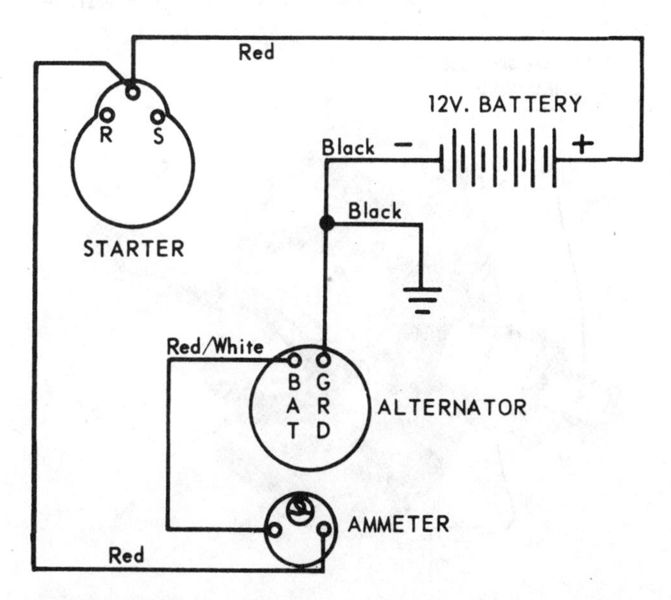

Schematic diagram of an alternator with an integral regulator and a harness length less than 20 feet.

CAUTION

Water must circulate through the lower unit and upper gear housing anytime the engine is operating to prevent damage to the water pump installed in the upper gear housing. Just a few seconds without water will damage the water pump impeller.

Start the engine.

Now, with the engine running, **CAREFULLY AND SLOWLY** insert a small bladed screwdriver **NO MORE THAN ONE INCH** into the test hole, shaped like a backward "D", in the slip ring end frame. **NEVER** insert the screwdriver more than 1", because such action will contact the rotor and **DESTROY THE ALTERNATOR.**

With the screwdriver in place, the field winding is grounded and the regulator is bypassed. If the alternator charges with the screwdriver in place, the regulator is defective and must be replaced. If the alternator does not charge with the screwdriver in place, then the alternator is defective and must be repaired or replaced.

WHENEVER the alternator is repaired or replaced, the regulator must be replaced because it is also defective, as evidenced by its failure to control the alternator output properly.

Many times, the cause of a defective regulator is a high resistance connection in the charging circuit, at the ammeter, or at the battery. Each of these areas should be checked and serviced thoroughly to prevent repetition of the trouble. Disassemble the connection, clean, and then tighten it.

Exterior view of an alternator with an integral regulator.

DISASSEMBLING

Scribe a mark across the two halves of the alternator housing to ensure the electrical connections on the back of the alternator are brought back to their original locations during assembling. Remove the four thru bolts and pry the metal tabs away from the rubber plug. Remove the cover. Pull off the rubber connector from the side of the alternator housing.

Normally the pulley and the fan would remain on the alternator during cleaning and testing. These two items should only be removed if one or the other is so badly bent as to need replacement.

Secure a 5/16" Allen wrench in a vise, as shown in the accompanying illustration.

Now, slide the alternator shaft over the exposed end of the Allen wrench. Use a 15/16" open-end wrench to loosen, and then remove the end nut, washer, pulley, fan, and collar from the rotor shaft.

CAREFULLY pull the rotor assembly from the slip ring end frame, and **TAKE CARE** not to lose the brush springs. Cover the slip ring end frame with tape to keep dirt out of the bearing. Cover the slip rings with tape also as a precaution against scoring them. **NEVER** use friction tape, which leaves a gummy deposit which is very difficult to remove.

Remove the screw securing the diode trio to the regulator. This screw has a special insulating washer and a special coating on the threads, therefore do not substitute this screw with another without these features. Remove the three nuts and washers securing the diode trio to the rectifier bridge, and then remove the diode trio and the stator.

Removing the alternator pulley.

CLEANING AND TESTING

Clean the stator, rotor, brush holder assembly, heat sink and both halves of the housing with a brush or compressed air. Use a soft cloth dipped in a mild solvent to wipe the brushes and slip rings clean, and then blow them dry with compressed air. Inspect the brushes for cracks, grooves and oil saturation.

The original brush set may be reused, if the brushes are 3/16" (4.75 mm) or onger.

NEVER use a grease dissolving solvent to clean electrical parts or bushings because the solvent would damage the insulation and remove the lubricating qualities of the bushings.

The following procedures are keyed by number to matching illustrations as an aid in performing the work.

Rotor

1- Check the rotor windings for a short or ground: Connect an ohmmeter between one brush slip ring and the shaft. If the meter does not show an open circuit on the Rx100 scale there is a short in the rotor windings. Check the rotor windings for continuity: Connect the ohmmeter between the two slip rings, and a normal reading of 2.6 ohms will indicate the rotor coil is serviceable.

Stator

2- The stator is not checked for shorts due to the very low resistance of the windings.

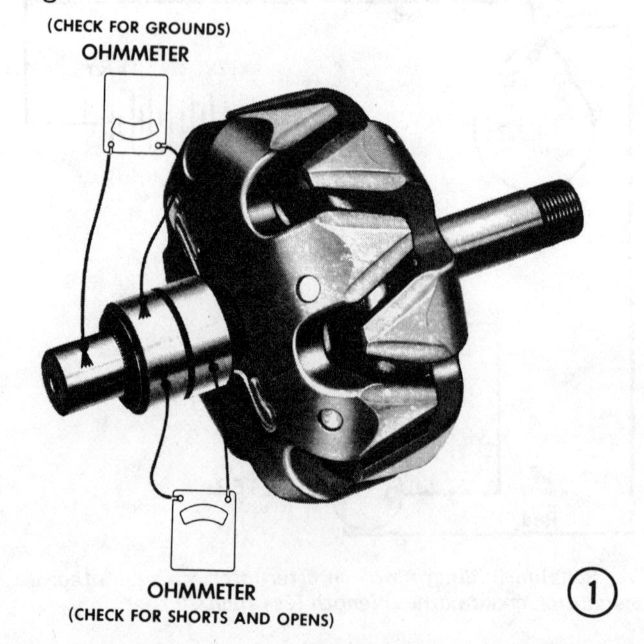

Checks that are made on the stator must be made with all of the diodes disconnected from the stator.

Check the Y-connected stator for open circuits: Connect an ohmmeter or test light across any two pair of terminals. If the ohmmeter reading is high, or if the test light fails to come on, there is an open in the winding.

The stator winding can be checked for grounds by connecting an ohmmeter or test light from either terminal to the stator frame. If the ohmmeter reads low or if the lamp comes on, the windings are grounded.

If all of the tests check out satisfactorily, but the alternator still fails to meet its rated output, a shorted Y-connection may be the cause.

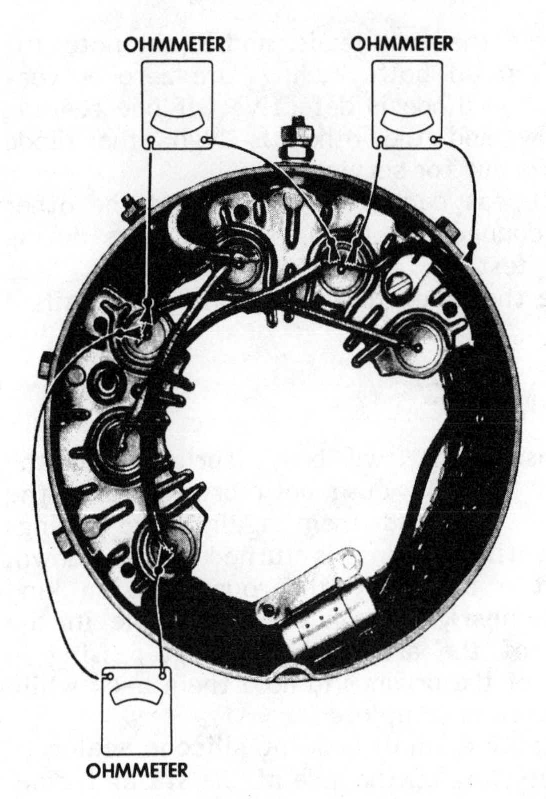

Ohmmeter hookups for testing.

Rectifier Bridge

3- First, set the ohmmeter scale to Rx 100. Next, connect one test lead to the grounded heat sink and the other test lead to one of the three terminals. Now, note the reading on the scale. Reverse the test leads, and again note the reading. If both readings are zero or very high, the rectifier bridge is defective. If one of the readings is high and the other low, the rectifier checks out for service.

Repeat the tests between the grounded heat sink and the other two terminals, and then between the insulated heat sink and each of the three terminals. When complete a total of six tests should have been made, with two readings during each test.

Diode Trio

4- Set the ohmmeter scale to the Rx100 scale. Check each diode for a short or open by first connecting one test lead to a single connector, of the three connectors, and then noting the reading on the scale. Now,

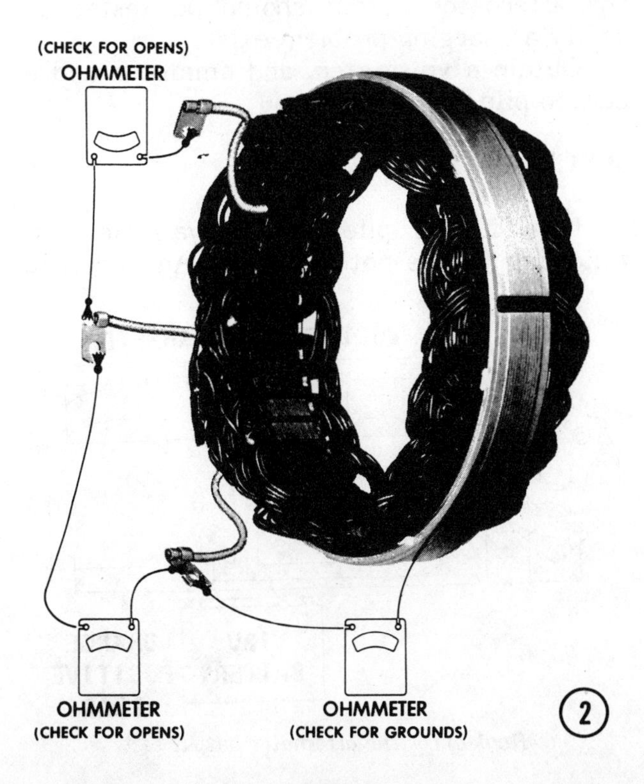

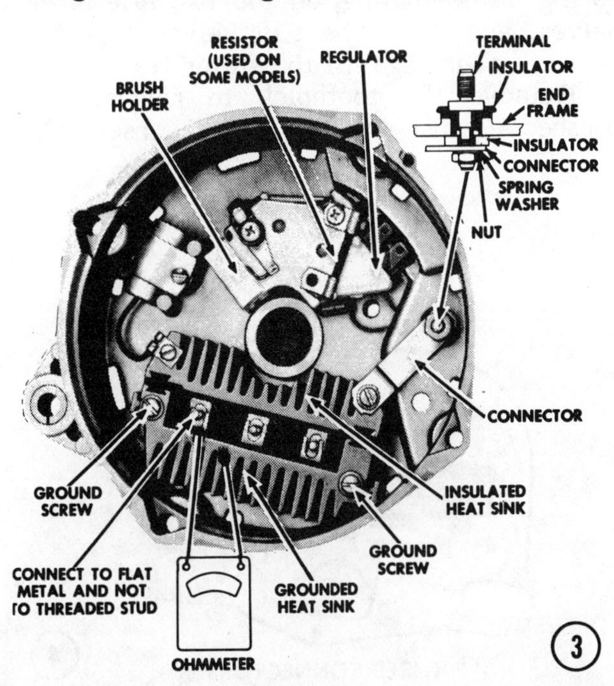

reverse the test leads, and again note the reading. If both readings are zero or very high, the diode is defective. If one reading is low and the other is high, the diode checks out for service.

Repeat each test for each of the other two connectors, taking two readings during each test for a total of six readings. Replace the diode trio if any one diode fails a test.

ASSEMBLING

Insert the two brush springs into the brush holder. Push both brushes into the holder and hold them against the springs while the housing is turned upside down. Insert a toothpick (the round wooden kind works best), through the small hole in the back of the alternator housing, passing in front of the brushes to hold them back while the work is completed.

Apply a small bead of silicone sealer to the opening on the side of the stator frame. This procedure is recommended by the manufacturer as a safety device.

Position the stator in the alternator housing and rotate it until the notches in the stator frame align with the holes in the housing. Install the diode trio and secure it in place with the special insulated screw. Tighten the screw securely.

Hook the stator terminals onto the rectifier bridge. Install the bridge with the three washers and nuts. Align the marks, scribed during disassembling, on the two alternator halves, and bring the two halves together until they seat against the stator frame.

Remove the toothpick to release the brushes against the slip rings. Insert the rubber plug into the voltage regulator terminals and press the plug firmly into place. Position the cover over the back of the alternator and secure it with the four housing screws. Tighten the screws to 40 in lb (4.5 Nm). Push back the metal tabs against the rubber plug.

If either the pulley or the fan were replaced, install the collar, fan, pulley, washer, and end nut onto the rotor shaft. Secure a 5/16" Allen wrench in a vice, slide the alternator shaft over the exposed end of the allen wrench and use a 15/16" open-end wrench to tighten the nut **SECURELY.**

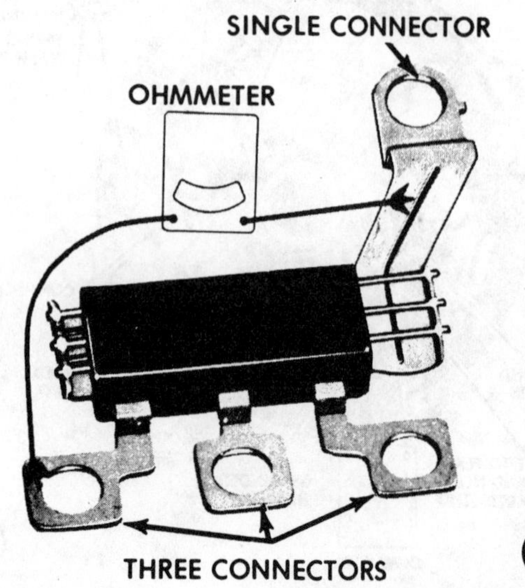

6-10 SERVICING A MOTOROLA ALTERNATOR

TROUBLESHOOTING

Before attempting troubleshooting procedures, first check the alternator belt for correct tension. A loose belt will slip, reducing alternator output. Such a condition will, in time, lead to an undercharged battery. A belt which is overtightened will place a great strain on the bearings and bushings inside the alternator, leading to premature failure of these components.

Alternator Output Test

Before beginning charging system tests, the alternator output should be tested to verify a charging problem exists.

Obtain a voltmeter, and ammeter and a carbon pile.

SPECIAL WORDS

If a carbon pile is not available, the following test is not possible. An accurate

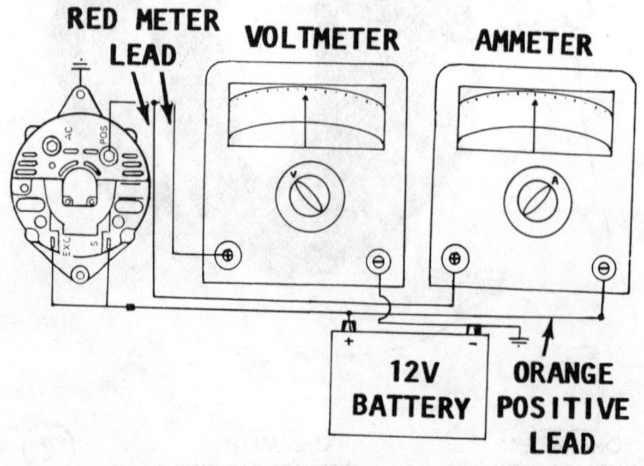

Hookup for the alternator output test.

determination of the alternator output is necessary to isolate the problem area. One of three conditions will exist: overcharging, undercharging, or no charging.

If excessive water is required of the battery, the battery may be continually overcharged. Such a condition will most likely develop from a defective voltage regulator.

Without a carbon pile, it is not possible to distinguish between an undercharging condition and a no charging condition. Therefore, proceed to the next series of tests, which only require a voltmeter and an ammeter.

The following test is conducted with the alternator mounted on the engine and with the boat in a body of water. Disconnect the Orange lead from the terminal marked **POS+** on the back of the alternator. Connect the Red lead of the ammeter to the **POS+** terminal. Connect the Black lead of the ammeter to the Orange lead which was disconnected. Connect the Red lead of the voltmeter to the same alternator terminal, and the Black voltmeter lead to a suitable engine ground. Connect the carbon pile across the battery terminals.

CAUTION

Water must circulate through the lower unit and upper gear housing anytime the engine is operating to prevent damage to the water pump installed in the upper gear housing. Just a few seconds without water will damage the water pump impeller.

Start the enging and allow it to warm to normal operating temperature.

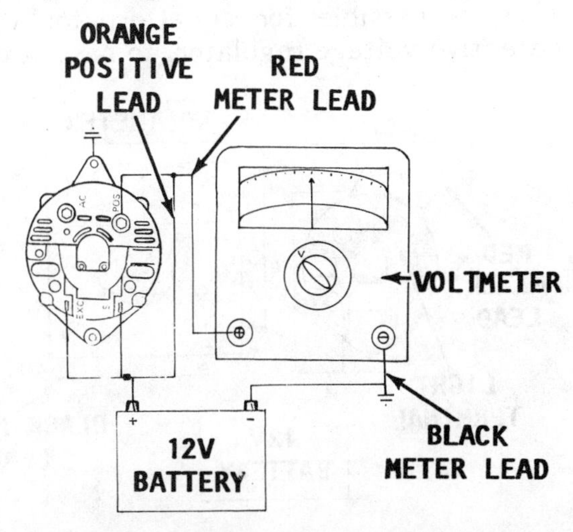

*Hookup for the Orange **POS+** lead test.*

Adjust the carbon pile to obtain maximum output, at the following engine speeds:

650 rpm -- 20 amps minimum
1500 rpm -- 13.0 to 14.7 volts
3000 rpm -- 45 amps minimum

If the test results show a problem with the alternator output, then proceed with the following tests. If the test results are within the given limits, then the battery, or the wiring harness may be at fault. See Section 7-2 for further tests on batteries. Remove the carbon pile from the battery, the meter leads, and reconnect the Orange lead to the **POS +** terminal.

The following tests require a voltmeter and an ammeter only. The alternator remains mounted on the engine and the battery must be fully charged.

In order for the battery to receive sufficient charge from the alternator, the following three wires **MUST** have continuity:

Orange output wire connected to the **POS +** terminal.

Red/Purple excite wire connected to the **EXC** terminal.

Purple sense wire connected to the **S** terminal.

Orange POS+ Lead Test

Connect the Red lead of the voltmeter to the **POS +** terminal, with the Orange wire still attached to the terminal. Connect the Black voltmeter lead to a suitable engine ground. The voltmeter should register close to 12 volts. If the meter indicates less than 11 volts, check the circuit between the Orange wire and the cranking motor solenoid terminal. Repeat the alternator output test. If the output is not within specifications, proceed to the following test.

If the voltmeter registers zero, there is an open in the Orange wire.

Red/Purple EXC Lead Test

Remove the Red/Purple lead from the **EXC** terminal. Connect the Red voltmeter lead to the disconnected wire, and connect the Black voltmeter lead to a suitable engine ground. Turn the key switch to the **ON** position.

The voltmeter should register close to 12 volts. If the meter indicates less than 11 volts, check the circuit between the Red/Purple wire and the primary positive ignition coil terminal. Repeat the alter-

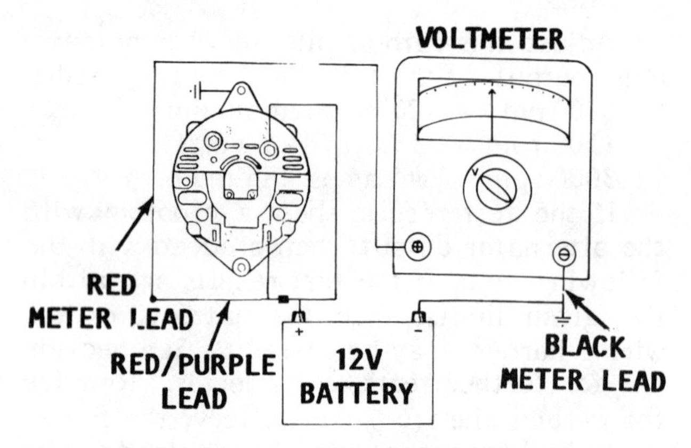

*Hookup for the Red/Purple **EXC** lead test.*

nator output test. If the output is not within specifications, proceed to the following test.

If the voltmeter registers zero, there is an open in the Red/Purple wire. A reading of zero, accompanied with a blown fuse, indicates the Red/Purple wire is **GROUNDED.**

Purple "S" Lead Test

Remove the Purple lead from the **"S"** terminal. Connect the Red voltmeter lead to the disconnected wire, and connect the Black voltmeter lead to a suitable engine ground. Turn the key switch to the on position. The voltmeter should register close to 12 volts. If the meter indicates less than 11 volts, check the circuit between the Purple wire and the main cable connector and the E.S.A. module connector. Repeat the alternator output test. If the output is not within specifications, proceed to the following test.

If the voltmeter registers zero, there is an open in the Purple wire. A reading of zero, accompanied with a blown fuse, indicated the Purple wire is **GROUNDED.**

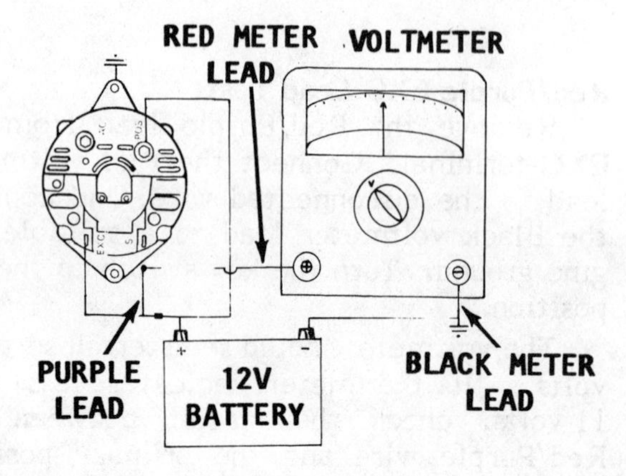

Hookup for the Purple S lead test.

Excite/Sense Circuit Test

Identify the "light" terminal on the back of the alternator. No wires are attached to this terminal. Remove and save the protective cap from the terminal. Connect the Red voltmeter lead to the "light" terminal. Connect the black voltmeter lead to a suitable engine ground. Turn the key to the **ON** position, but do **NOT** start the engine. The voltmeter should register between 1.5 and 3.0 volts. Turn the key switch to the **OFF** position. If this reading is not obtained, replace the voltage regulator. If the reading is within specifications, proceed to the next test.

Diode Trio Test

Connect a jumper wire between the "light" terminal and the **POS+** terminal. This jumper wire will bypass the diode trio inside the alternator. Connect the Red voltmeter lead to the **POS +** terminal, and the Black voltmeter lead to a suitable engine ground.

CAUTION

Water must circulate through the lower unit and upper gear housing anytime the engine is operating to prevent damage to the water pump installed in the upper gear housing. Just a few seconds without water will damage the water pump impeller.

Start the engine and allow it to reach normal operating temperature, at idle speed.

If the voltmeter registers 12 volts or more, the diode trio is defective and should be replaced. Install the protective cap over the "light" terminal.

It is possible for an alternator with a defective voltage regulator, to pass both the

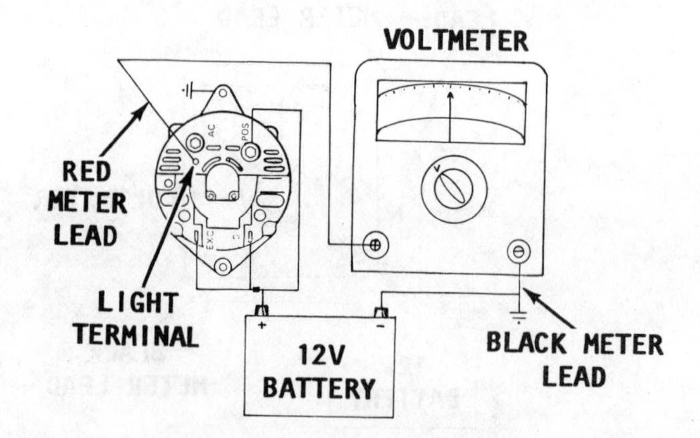

Hookup for the Exite/Sense test.

excite/sense and the diode trio tests. Therefore if the alternator passed the preceding tests, replace the voltage regulator. If the alternator still fails to produce the necessary output, it will have to be disassembled and bench tested component by component.

DISASSEMBLING

Remove the protective cover from the "light" terminal. Remove the nuts and washers from the **AC** and **POS+** terminals. Remove the two securing screws and lift off the back cover, be sure to save the small felt gasket behind the cover. Remove the two retaining locknuts, and then remove the brush holder cover. Loosen, but do **NOT** remove, the nut at the "light" terminal, securing the strap from the brush holder to the "light" terminal. Lift the strap from the brush holder screw.

Slide the brush holder out of the housing, but do **NOT** allow the brushes to fly out of the housing as they move past the slip rings. Remove the two screws, one on either side, securing the voltage regulator, and then remove the regulator. Remove the locknut on the **POS +** terminal, and lift off the condenser lead. Remove the three screws attaching the stator leads to the diode trio. Lift away the small metal tab connecting the stator lead to the **AC** terminal. Remove the four screws retaining the diode trio and the rectifier bridge. Using a pair of needle nose pliers, straighten the strap at the **POS+** terminal and lift off the diode trio.

The rectifier bridge is secured to the alternator with a heat sink compound. With care it may be loosened and pried free.

Scribe a mark across the two halves of the alternator and across the stator frame. This mark will act as a reference during assembling to ensure the two halves are returned to their original locations. Returning the two halves to their original location is most important, as the two halves can be matched in four different positions. Only **ONE** correct position will allow the electrical harness wires to reach their appropriate terminals.

Remove the four thru bolts from the back of the alternator. Separate the two halves from the stator frame. If separation is difficult, insert two screwdrivers, one on either side of the alternator, and simultaneously pry one half of the alternator housing against the stator frame. Do not insert the screwdrivers deeper than 1/16" to avoid causing damage to the stator windings. Lift the stator from the housing.

CLEANING AND TESTING

Clean the stator, brush holder assembly and both halves of the housing with a brush or compressed air. Use a soft cloth dipped in a mild solvent to wipe the brushes and slip rings clean, then blow them dry with compressed air. Inspect the brushes for cracks, grooves, and oil saturation.

The original brush set may be reused, if the brushes are 3/16" (4.75 mm) or longer.

NEVER use a grease dissolving solvent to clean electrical parts or bushings because the solvent would damage the insulation and remove the lubricating qualities of the bushings.

Diode Trio Test

Obtain an ohmmeter and select the Rx1000 scale. Attach one ohmmeter lead to any one of the three small tabs. Attach the

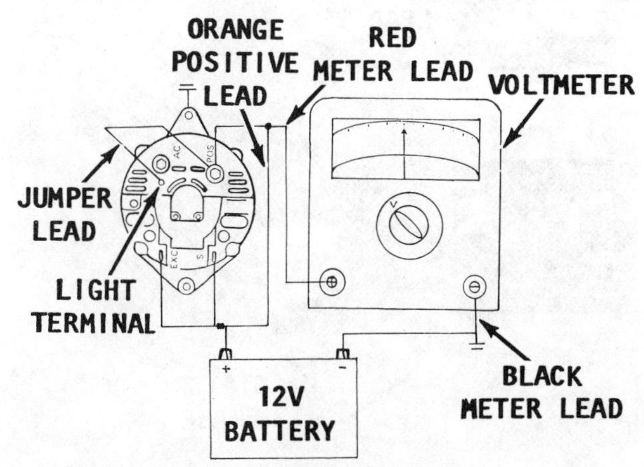

Hookup for the Diode Trio test.

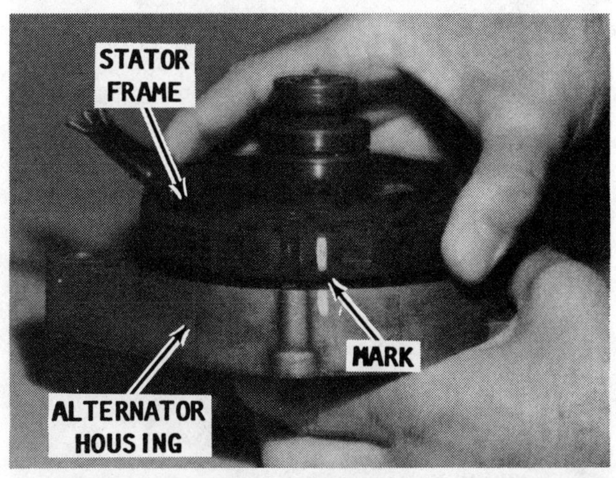

Separating the stator frame from the alternator housing after location identifying marks have been made on both surfaces.

other meter lead to the "light" terminal (the terminal with the metal strap). Observe the meter reading, and then reverse the meter leads and again observe the reading. The ohmmeter should indicate continuity in one direction and an infinite resistance, when the meter leads are reversed. Test the other two small tabs in the same manner. If any one of the three tabs fail the test, the diode trio, as an assembly, must be replaced.

Rectifier Bridge Positive Diode Test

Attach one ohmmeter lead to the **POS +** terminal strap. Attach the other meter lead to one of the three stator attachment terminals. As in the previous test, the ohmmeter should indicate continuity in one direction and an infinite resistance when the meter leads are reversed. Test the other two stator attaching terminals in the same manner. If any one of the three stator attaching terminals fails the test, the entire rectifier bridge assembly must be replaced.

Rectifier Bridge Negative Diode Test

Attach one ohmmeter lead to the rectifier bridge body. Attach the other meter lead to one of the three stator attachment terminals. As in the previous two tests, the ohmmeter should indicate continuity in one direction and an infinite resistance when the meter leads are reversed. Test the other two stator attaching terminals in the same manner. If any one of the three stator attaching terminals fail the test, the entire rectifier bridge assembly must be replaced.

Stator Open Circuit Test

Select the lowest possible scale on the ohmmeter. Make contact with the meter leads to the first and second stator leads. The meter should register approximately 0.1 ohm. Repeat this test with the first and third stator leads, and then the second and third stator leads. In each test, the resistance value should be the same. If the results vary, the stator assembly must be replaced.

Stator Ground Test

Select the Rx1000 scale on the ohmmeter. Make contact with one meter lead to the stator frame. Make contact with the other meter lead to one of the stator leads. The meter must indicate infinity, or no continuity. Repeat this test for the other two stator leads. If any one lead fails the test, the stator assembly must be replaced.

Rotor Open/Short Circuit Test

Select the Rx10 scale on the ohmmeter. Make contact with one meter lead to one slip ring, and the other slip ring with the second meter lead. The meter should indicate 3.5 to 4.7 ohms. A lower reading indicates shorted windings. A higher reading indicates an excessive resistance. An infinite reading indicates an open circuit. If the reading is not within specification, the rotor assembly must be replaced.

Rotor Ground Test

Select the Rx1000 scale on the ohmmeter. Make contact with one meter lead to any one of the two slip rings. Make

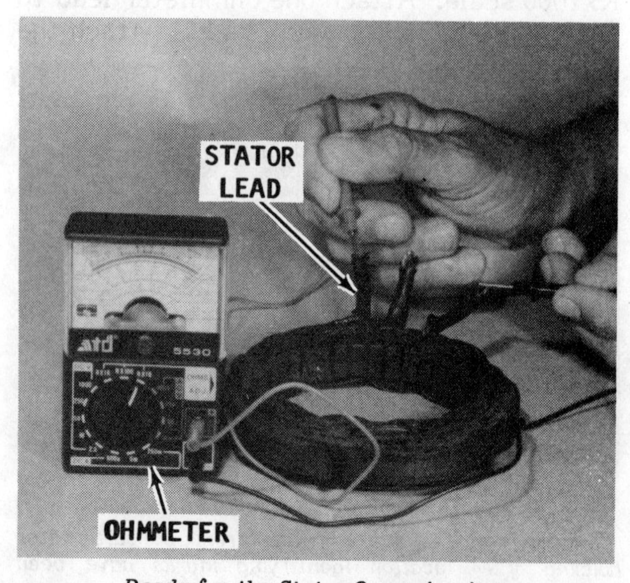

Ready for the Stator Open circuit test.

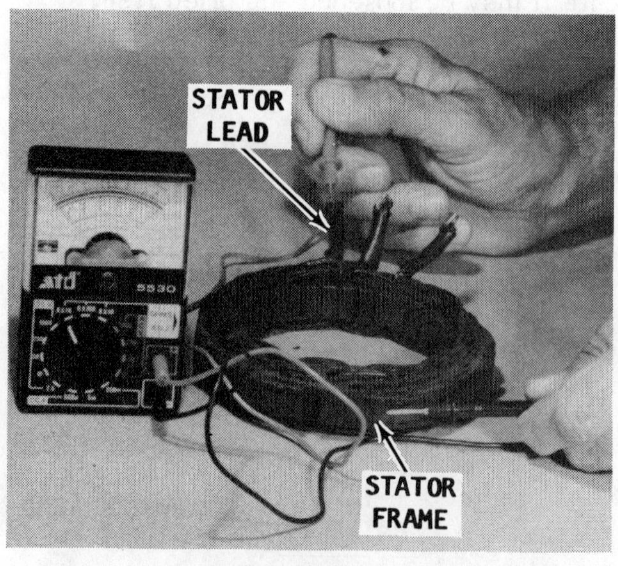

Performing the Stator Ground test.

contact with the other meter lead to the rotor body. The meter should indicate infinity, or no continuity. Any other reading would indicate the rotor assembly is grounded and therefore must be replaced.

ASSEMBLING

SPECIAL WORDS

If a new front housing is being installed, the thru bolt holes will **NOT** be tapped. Therefore, prior to assembling, special thread cutting thru bolts must be obtained from the dealer, and installed with a socket wrench to cut the threads.

Slide the stator frame assembly over the rotor, and align the mark scribed on the frame with the mark scribed on the front alternator housing. In this position, the reliefs in the stator frame should align with the thru bolt mounting holes. Straighten the three stator leads, to stand vertically and evenly spaced. Lower the other half of the alternator housing over the stator frame, allowing the three stator leads to pass through the holes in the housing. Align the mark scribed on the back alternator housing with the mark scribed on the stator frame. The thru bolt holes and the stator frame reliefs should all align in a straight line. Install the thru bolts and tighten them to a torque value of 50 to 60 in lb (5.6 to 6.8 Nm). Spin the pulley and check to be sure the rotor does not bind.

Straighten the **B+** rectifier strap. Apply a thin coat of Heat Sink Compound P/N 322170 to the back of the rectifier bridge and to the installation area in the rear housing. Install the rectifier in the housing. Thread the **B+** strap through the slot in the diode trio body and install the diode trio with the four mounting screws. Slide the **B+** strap over the output terminal, then slide the condenser lead over the same terminal. Thread the locknut over the terminal and tighten the locknut to seat both leads against the terminal base.

Slide a connecting strap over the **AC** terminal and place the other end of the strap between the closest stator lead and the diode trio terminal.

Install and tighten the three screws securing the stator leads to the diode trio.

If the terminal screws slid out of the regulator body, push them back into place. Install the voltage regulator onto the back alternator housing and secure it in place with two screws. Tighten the screws securely.

Position the brush holder with the metal band facing **UPWARD**, away from the voltage regulator. Push the brushes in, to clear the slip rings, and at the same time, slide the holder into the grooves in the housing. Slide the strap from the "light" terminal over the brush/regulator terminal. Install and tighten a locknut at each end of the terminals supporting the strap. Position the felt gasket over the slip ring recess in the rear cover. Install the cover and secure it in place with the two attaching screws. Place an insulator, washer, and locknut on each of the **AC** and **POS +** terminals. Tighten both locknuts securely. Install the protective cap over the "light" terminal.

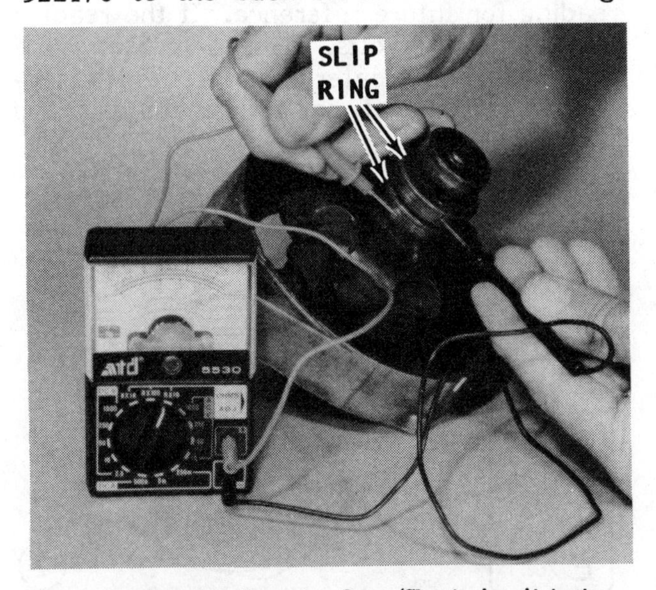

Performing the rotor Open/Short circuit test.

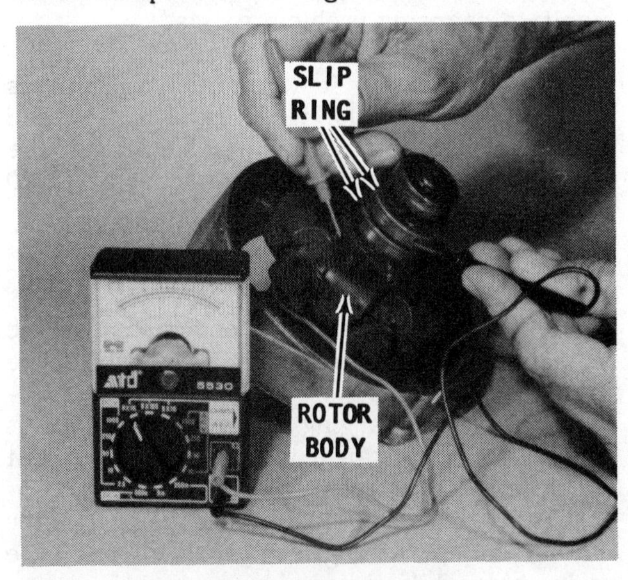

Performing the Rotor Ground test.

6-11 SERVICING VALEO ALTERNATOR

TROUBLESHOOTING

A great many charging system problems may be traced to a defective battery. Before attempting any troubleshooting of the charging system make a thorough check of the battery condition as described in Section 6-1.

Drive Belt Check

Before attempting troubleshooting procedures, first check the alternator belt for correct tension. A loose belt will slip, reducing alternator output. Such a condition will, in time, lead to an undercharged battery. A belt which is overtightened will place a great strain on the bearings and bushings inside the alternator, leading to premature failure of these components.

Replace the drive belt if it is cracked, worn or saturated with oil.

Check the drive belt tension by depressing the belt at a point midway between pulleys. Heavy thumb pressure on the belt should depress it approximately 1/4" (6mm). To adjust tension on the belt, loosen the alternator pivot and mounting bolts, and then tighten them again finger tight. Move the alternator inboard or outboard, until the correct belt tension is obtained. Tighten the mounting bolt first and then the pivot bolt.

Alternator Output Test

Before beginning charging system tests, the alternator output should be tested to verify a charging problem exists.

The alternator must be tested at operating temperature. Some electrical problems only occur when the temperature rises and increased resistance occurs. Operate the engine for approximately 3 minutes at 2,000 rpm before making any tests.

CAUTION

DO NOT disturb any connections for the alternator, regulator, or battery while the engine is operating. Disconnecting any portion of the charging circuit while it is operating, may cause the electrical current to reverse itself and burn out diodes.

Take care to keep measuring meter cables, hair, clothing, etc., well clear of the drive belt while the engine is operating.

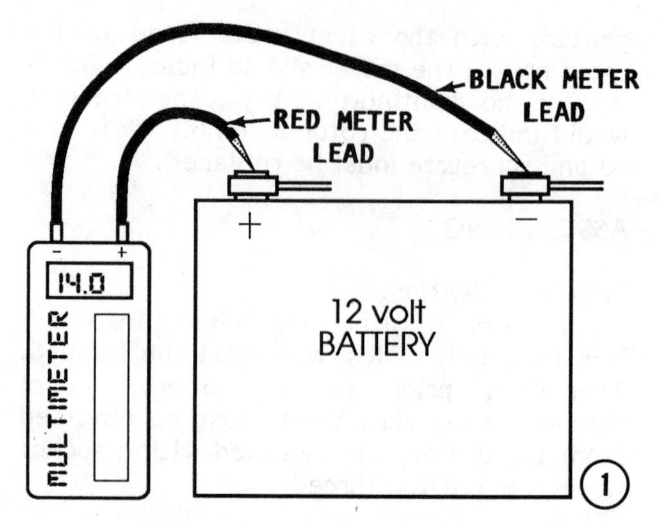

1- Obtain a multimeter, preferably a digital multimeter. This type meter gives a very high degree of accuracy and is easier to read.

CAUTION

Water must circulate through the lower unit and upper gear housing anytime the engine is operating to prevent damage to the water pump installed in the upper gear housing. Just a few seconds without water will damage the water pump impeller.

Start the engine and allow it to reach normal operating temperature at idle speed.

Select a meter scale capable of registering more than 14 volts. Make contact with the positive Red meter lead to the positive battery terminal. Make contact with the negative Black meter lead to the negative battery terminal. The meter should register approximately 14 volts. Make a note of the reading for future reference. If the reading is less than 14 volts, proceed to Step 2. If the reading is greater than 14.4 volts, proceed to Removal and Disassembling and

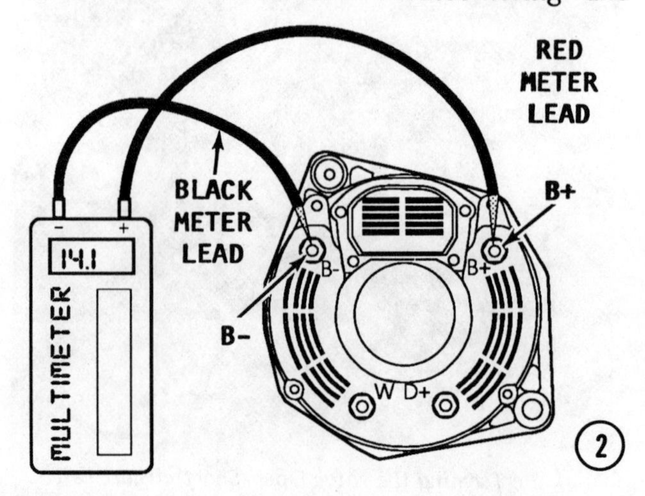

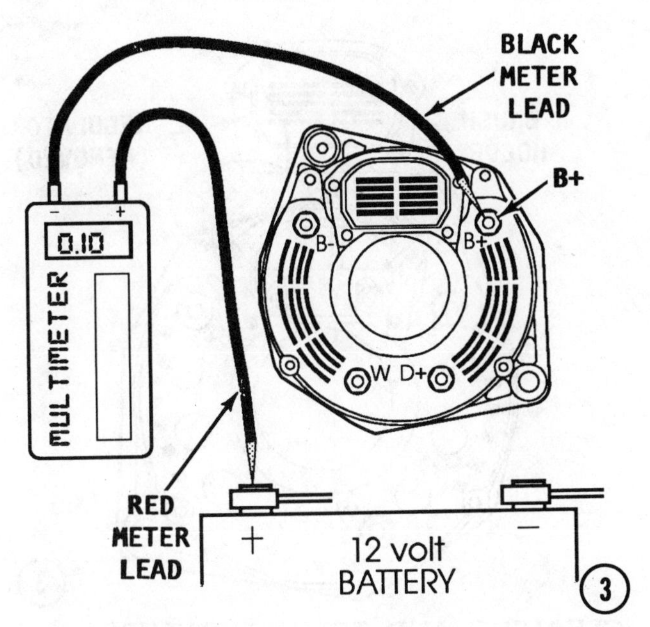

further testing of the internal components of the alternator.

2- Use the same multimeter scale and make contact with the Red meter lead to the **B+** alternator terminal. Make contact with the Black meter lead to the **B-** alternator terminal. Keep the engine idle at 2,000 rpm. The meter should register between 14.0-14.4 volts **AND** only be 0.4 volts less than the reading obtained in the previous step. If the voltage drop is less than 0.2 volts, both the battery cables are satisfactory. If the voltage drop exceeds 0.3 volts, proceed to Step 3 to check the positive battery cable.

3- Select a meter scale capable of registering 0.00 to 0.50 volts. Keep the engine idle at 2,000 rpm.

Make contact with the Red meter lead to the positive battery terminal. Make contact with the Black meter lead to **B+** alternator terminal. The meter should register no more than 0.2 volts. If the meter registers more than 0.2 volts, the positive battery cable has excessive resistance causing too great a voltage drop in the circuit. The positive battery cable should be replaced. If the meter registers less than 0.2 volts, proceed to the following test for the negative battery cable.

4- Keep the same meter scale and engine rpm as in the preceding step. Make contact with the Red meter lead to the negative battery terminal. Make contact with the Black meter lead to the **B-** alternator terminal. The meter should register no more than 0.2 volts. If the meter registers more than 0.2 volts, the positive battery cable has excessive resistance causing too great a voltage drop in the circuit. The positive battery cable should be replaced.

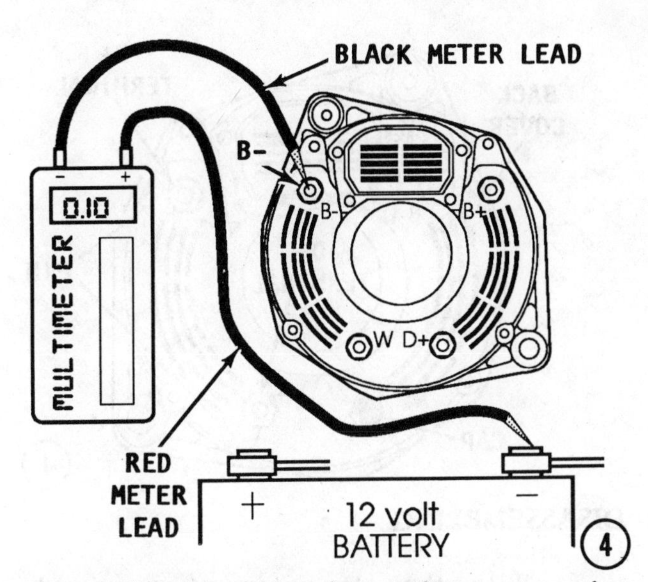

If the meter registers less than 0.2 volts, and has passed the preceding tests and still fails to produce the necessary output; it will have to be disassembled and bench tested component by component.

SPECIAL TESTING WORDS

Accurate bench testing of the regulator can only be accomplished with Volvo Regulator Tester P/N 884892-1. The following tests are limited to those which can be performed using a multimeter with a diode testing function. If the rotor windings, stator windings, and diode bridge are found to be satisfactory, then replace the regulator.

REMOVAL

Disconnect both battery cables from the battery. **REMEMBER,** even with the ignition key in the **OFF** position, the **B+** terminal may still be "hot". Therefore, the battery cables **MUST** be disconnected before starting work on the charging system. Remove the protective boots from the electrical connections on the back of the alternator and then remove the electrical connections from the **B+, B-,** and **D+** alternator terminals. Tag each wire **BEFORE** removal, as an assist in later work.

Loosen the pivot and mounting bolts. Move the alternator inboard and slide the drive belt from the pulley. Remove the bolts, spacers (if used), washers and bolts securing the alternator to the engine. Make a note of the location of any spacers used.

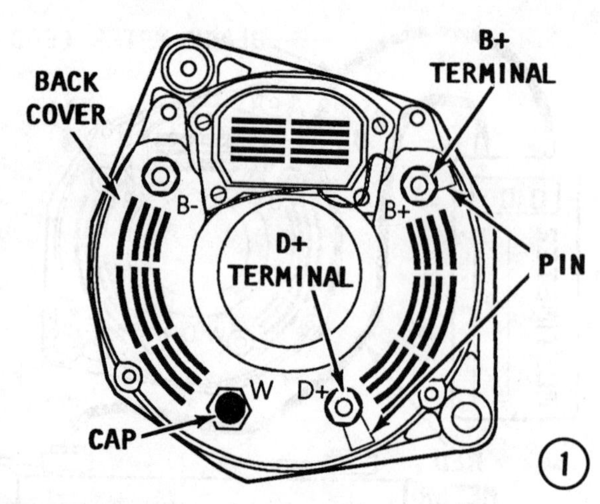

DISASSEMBLING

1- With the alternator on a suitable workbench, remove the nut on the **B+** alternator terminal and slide the lead from the regulator free of the terminal. Remove the connecting pin from the **D+** terminal. Pry the protective plastic cap from the **W** terminal. Remove the back cover.

2- Remove the four securing screws from the regulator. Insert a flat blade screwdriver under the securing lug - one on each side of the regulator, and pry outward to release the regulator.

Use a pair of needlenose pliers to disconnect the two regulator leads from the terminals on the diode bridge. **DO NOT** pull the leads free of the terminals, because the leads may be pulled away from their connectors. The regulator will still be connected to the brush holder. Therefore set the regulator to one side for later work.

3- Remove the two screws securing the brush holder to the alternator. Lift out the regulator and brush holder. To separate the regulator from the brush holder, unsolder the two leads joining the two components.

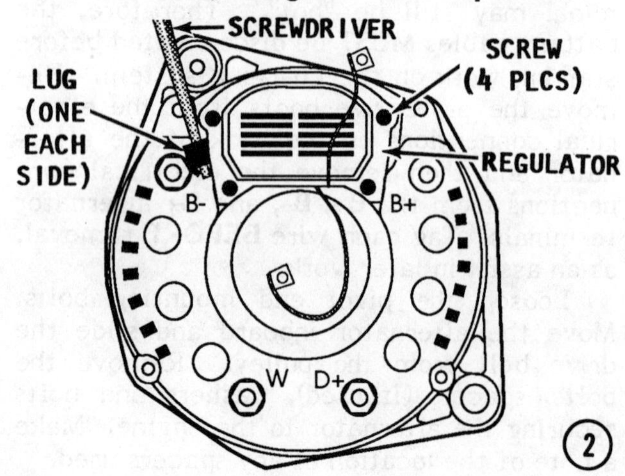

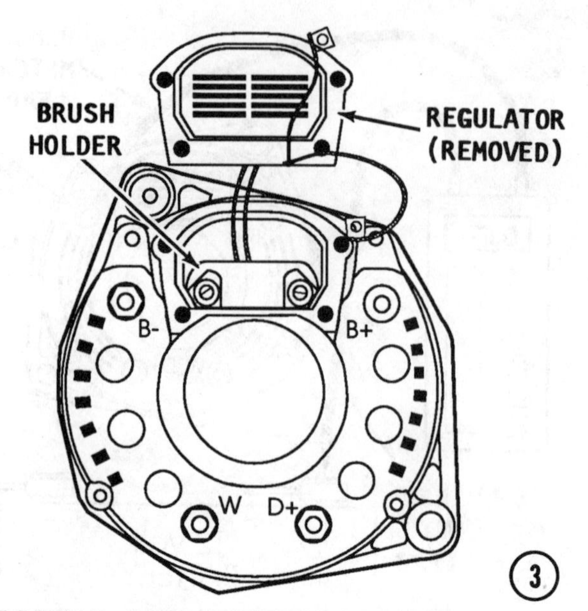

CLEANING AND TESTING BRUSHES

Clean the brush holder assembly with a brush or compressed air. Use a soft cloth dipped in a mild solvent to wipe the brushes clean, then blow them dry with compressed air. Inspect the brushes for cracks, grooves, and oil saturation.

The original brush set may be reused, if the brushes are 5/16" (8mm) or longer. The brushes and holders are replaced as a unit. If the brush holder is to be replaced, unsolder the two regulator leads from the holder. Solder the two leads back onto the new brush holder using acid free solder.

Alternator Testing

1- Select a meter scale capable of registering 10.0 ohms. Connect the meter leads across the two slip rings to test the rotor windings, one meter lead probe on each ring. Make sure each probe makes good contact with the metal ring. The meter should register 4 - 6 ohms. Keep one meter probe on a slip ring. Move the other meter probe to the alternator frame. The meter should register **NO CONTINUITY.** Repeat this test for the other ring. If one ring shows continuity, the other ring will

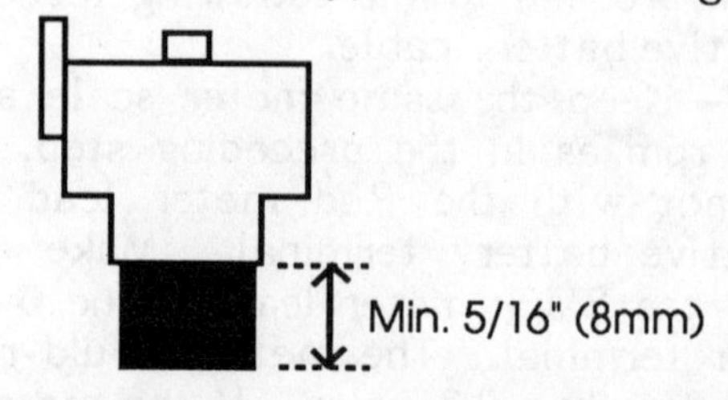

Replace the brush holder if the brushes are worn beyond the specified limit.

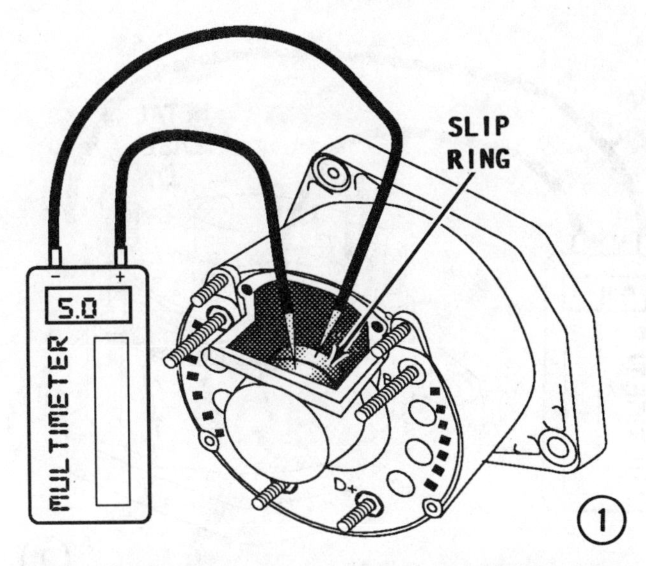

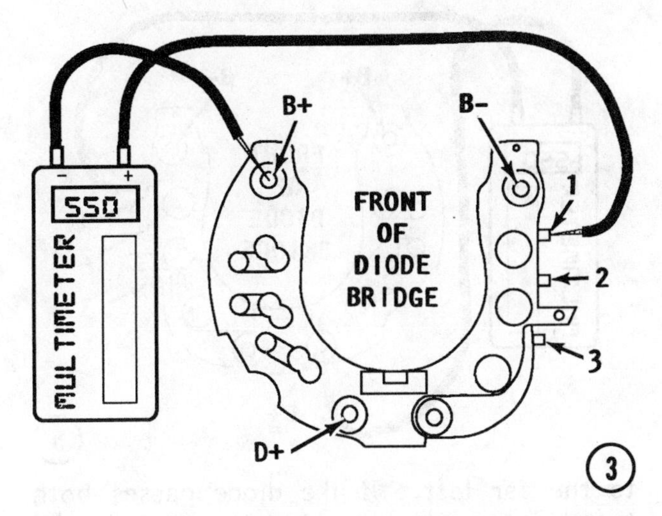

also. If both rings indicate continuity, the alternator must be disassembled and the grounded part located and insulated.

2- The diode bridge must be removed to check the diodes and the stator windings. Carefully unsolder the three stator winding leads. Do not overheat the solder connections, because excessive heat will damage the diodes. Remove the remaining nuts, spacers and washers securing the diode bridge to the alternator.

Testing Diodes

If the reason for testing the charging system is because the battery apparently discharges itself overnight, the cause is probably a bad diode. At the begining of these procedures, it was stated that the **B+** alternator terminal remained "hot" even though the ignition switch was in the **OFF** position. A bad diode will allow the battery to discharge overnight.

The three diodes on the diode bridge can only be tested if the multimeter being used has a diode test function. If the meter is not equipped with this feature, take the diode bridge together with these instructions, to your nearest electronic equipment repair shop.

A diode will only pass current in one direction. The multimeter applies a small voltage to the diode in a specified direction. If the diode passes the voltage in this specified direction, the diode is considered good. The meter leads are then changed around and the voltage is applied in the opposite direction. A good diode will not allow the voltage to pass. A bad diode will allow the voltage to pass in both directions. If any one of the three diodes on the diode bridge fails the test, the bridge, as a unit, must be replaced.

3- Select the diode test function on the multimeter. Make contact with the Red meter lead to the No. 1 stator winding connection, identified in the accompanying illustration. Make contact with the Black meter lead to the **B+** terminal on the diode bridge. The meter should register 450 - 650 mV (milli-volt - 1/1000 volt). Reverse the meter leads. The meter needle should swing

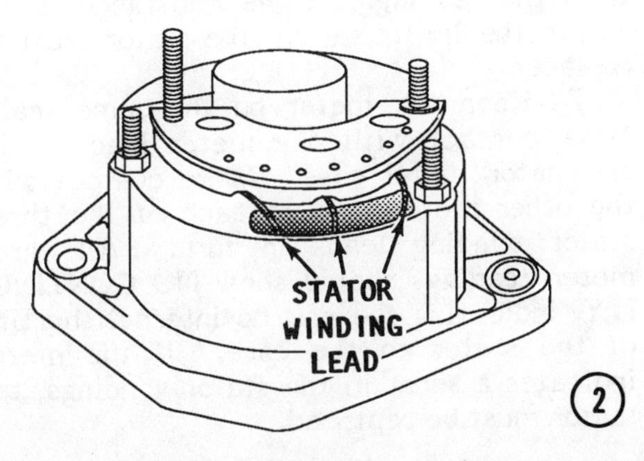

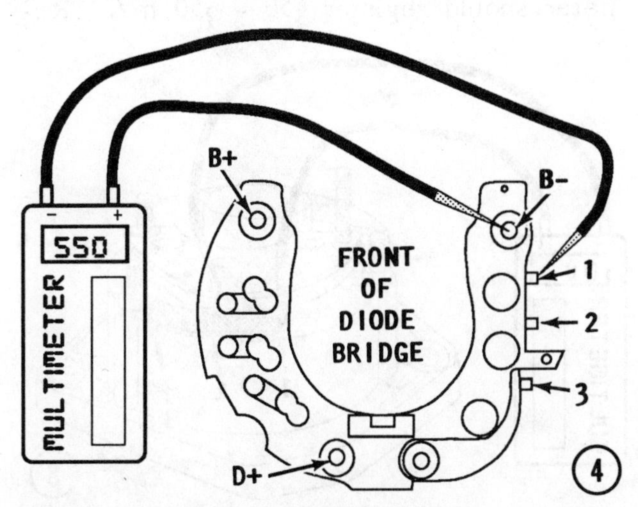

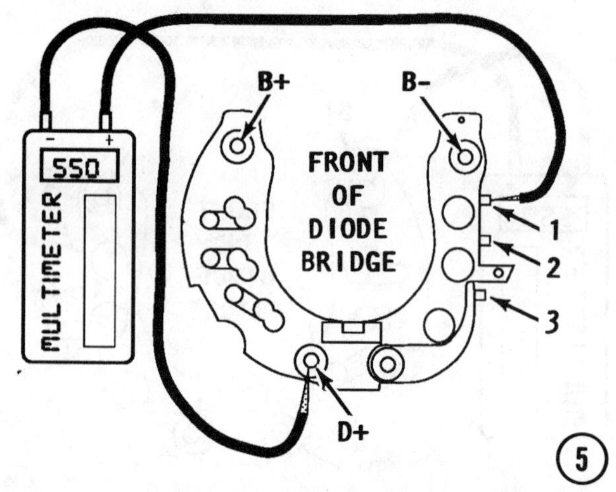

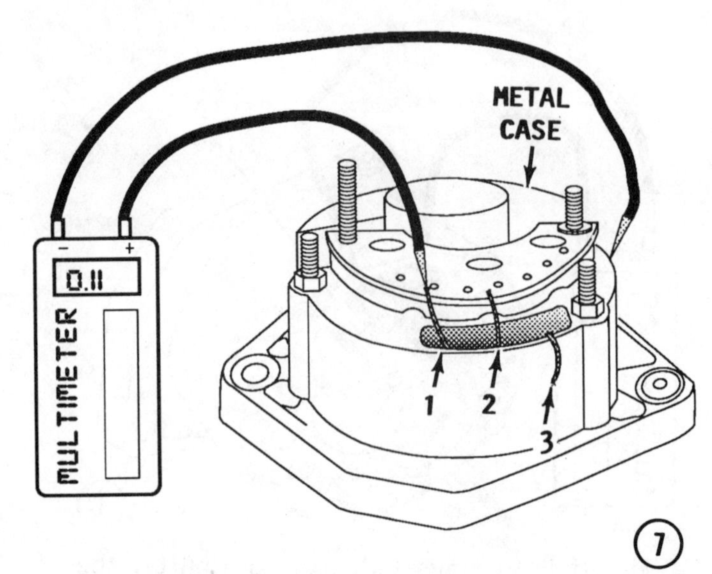

to the far left. If the diode passes both tests, it can be considered good. If the diode fails one or both tests, the entire diode bridge must be replaced. Repeat the test for the No. 2 and No. 3 stator winding connections.

4- Keep the meter on the diode test function. Make contact with the Red meter lead to the No. 1 stator winding connection, identified in the accompanying illustration. Make contact with the Black meter lead to the **B-** terminal on the diode bridge. The meter should register 450 - 650 mV. Reverse the meter leads. The meter needle should swing to the far left. If the diode passes both tests, it can be considered good. If the diode fails one or both tests, the entire diode bridge must be replaced. Repeat the test for the No. 2 and No. 3 stator winding connections.

5- Keep the meter on the diode test function. Make contact with the Red meter lead to the No. 1 stator winding connection, identified in the accompanying illustration. Make contact with the Black meter lead to the D+ terminal on the diode bridge. The meter should register 450 - 650 mV. Reverse the meter leads. The meter needle should swing to the far left. If the diode passes both tests, it can be considered good. If the diode fails one or both tests, the entire diode bridge must be replaced. Repeat the test for the No. 2 and No. 3 stator winding connections.

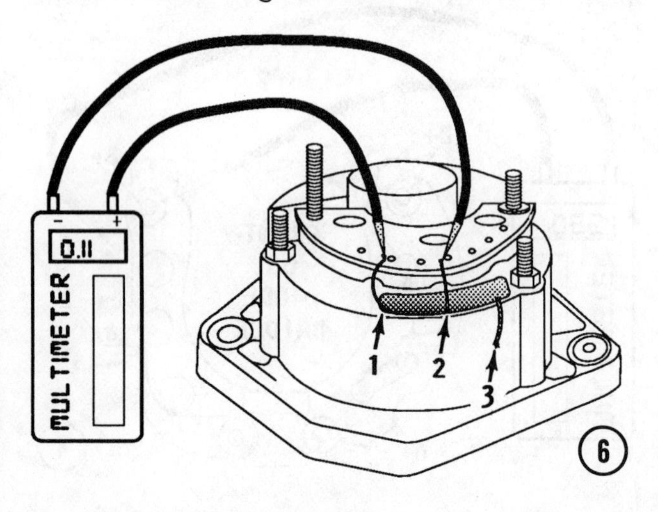

6- Select a meter scale capable of registering 0.00 - 1.00 ohms. Touch both meter leads together and make a note of the resistance of the measuring cables (e.g. 0.10 ohm). This value must be deducted from all test measurements taken in this step.

The three stator leads **MUST** be unsoldered, as directed previously in Step 2, for the test to be valid. Identify the leads as No. 1, No. 2, and No. 3. Determine the resistance as follows:

No. 1 and No. 2
No. 1 and No. 3
No. 2 and No. 3

The meter should register 0.11 - 0.15 ohms for each test **AFTER** the resistance of the measuring cables has been subtracted from the reading. If the resistance is not within the limits shown, the stator must be replaced.

7- Keep the meter on the same scale. Make contact with one meter lead to the alternator metal case. Make contact with the other meter lead to each of the three stator winding leads in turn. All three meter readings should show **NO CONTINUITY**, indicating there is no internal shorting of the stator to the case. If the meter indicates a short in the stator windings, the stator must be replaced.

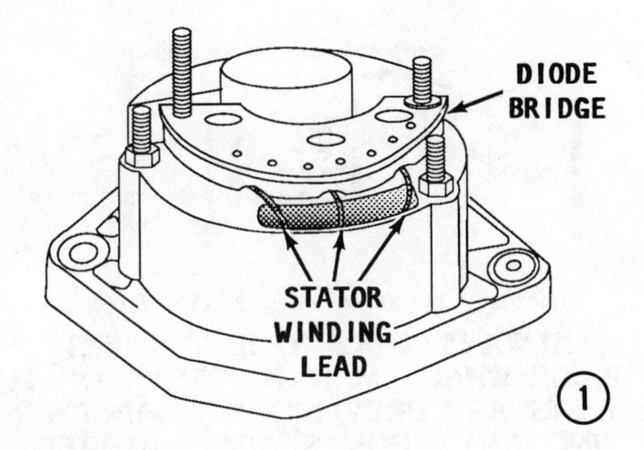

ASSEMBLING

1- Position the diode bridge over the mounting studs on the back of the alternator. Solder the three stator windings to the three terminals on the diode bridge using acid free solder. Secure the bridge with the attaching hardware.

2- Solder the two regulator leads to the brush holder using acid free solder. Insert the brush holder into the alternator and secure it in place with two screws. Rotate the alternator pulley and listen for any noises which may indicate the brushes are not seated correctly. The pulley should rotate smoothly.

3- Place the regulator over the brush holder and snap the mounting lugs into place. Secure the regulator with the four attaching screws. Install the two regulator leads onto their respective terminals on the diode bridge.

4- Install the back cover and secure it with the attaching hardware. Install the plastic protective cap over the **W** terminal.

INSTALLATION

5- Mount the alternator and secure it with the pivot bolt and mounting bolt. Tighten the bolts fingertight. Slip the drive belt over the alternator pulley. Tighten the belt by moving the alternator outboard and first tightening the mounting bolt, then the pivot bolt. Check the drive belt tension by depressing the belt at a point midway between pulleys. The belt should be able to be depressed with thumb pressure approximately 1/4" (6mm). To adjust the tension on the belt, loosen the alternator pivot and mounting bolts, and then tighten them finger-tight. Move the alternator inboard or outboard, until the correct belt tension is obtained. Tighten the mounting bolt first and then the pivot bolt.

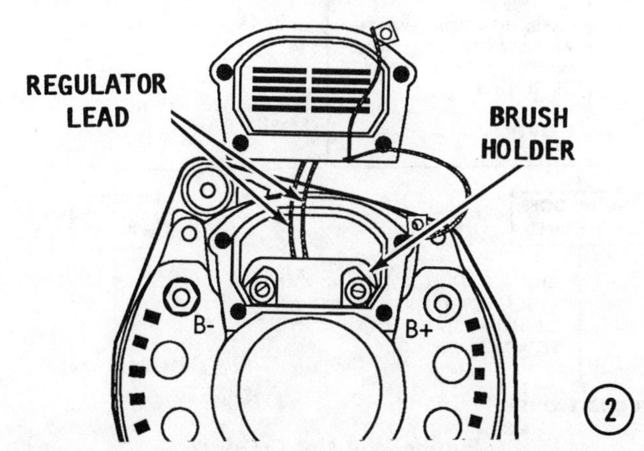

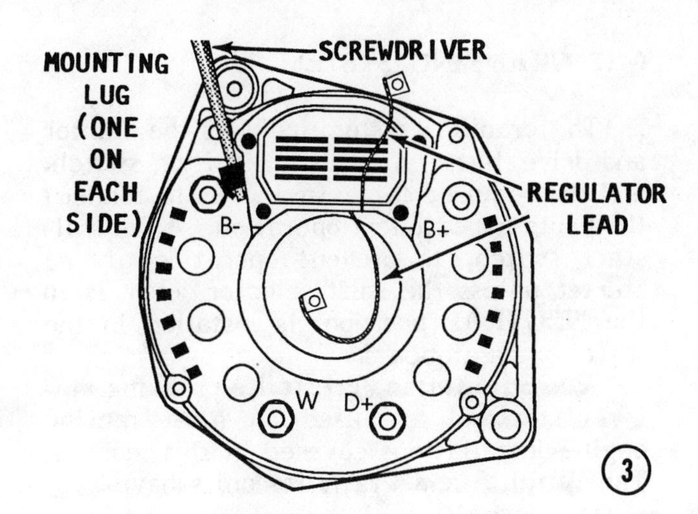

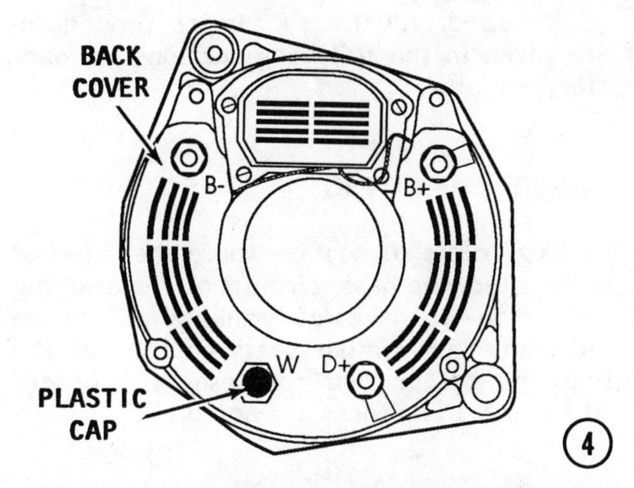

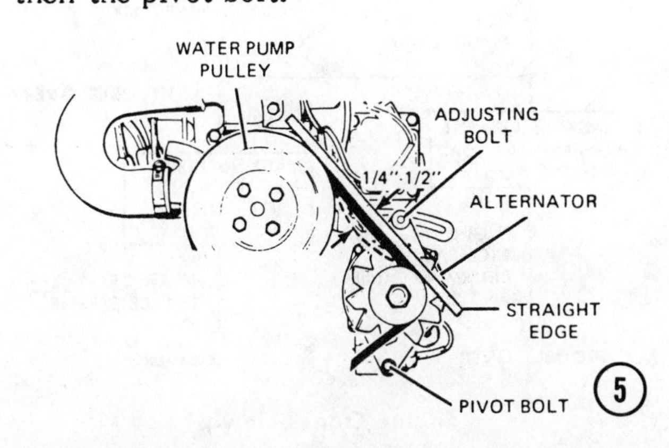

6-12 CRANKING SYSTEM

The cranking system includes the starter and drive, battery, solenoid, ignition switch, and the required cables and wires to connect the parts for efficient operation. A neutral-start switch, to prevent operation of the starter unless the shift selector lever is in the **NEUTRAL** position, is installed in the shift box on all boats.

The Autolite and Prestolite cranking motors (starters) are used on most marine engine installations covered in this manual. The Autolite and Prestolite units have separate solenoids.

Detailed, illustrated service procedures are given in the following sections for each starter motor.

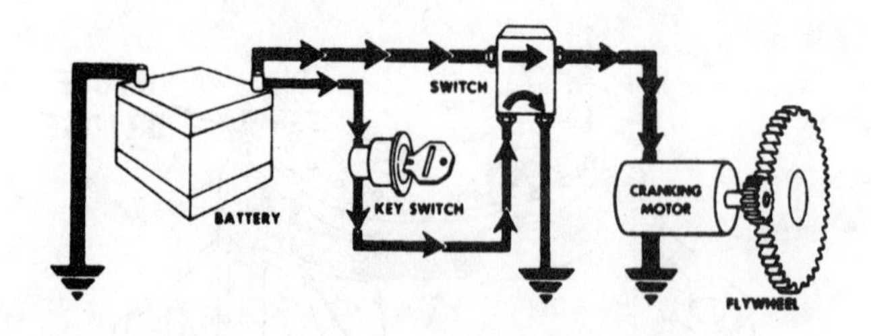

Functional diagram of a typical cranking circuit.

ALWAYS TAKE TIME TO VENT THE BILGE WHEN MAKING ANY OF THE TESTS AS A PREVENTION AGAINST IGNITING ANY FUMES ACCUMULATED IN THAT AREA. AS A FURTHER PRECAUTION, REMOVE THE HIGH-TENSION WIRE FROM THE CENTER OF THE DISTRIBUTOR CAP AND GROUND IT SECURELY TO PREVENT SPARKS.

TROUBLESHOOTING

Regardless of how or where the solenoid is mounted, the basic circuits of the starting system on all makes of cranking motors are the same and similar tests apply. In the following testing and troubleshooting procedures, the differences are noted.

All starter problems fall into one of three problem areas:

1- The starter fails to turn.

2- The starter spins rapidly, but does not crank the engine.

3- The starter cranks the engine, but too slowly.

The following paragraphs provide a logical sequence of tests designed to isolate a problem in the cranking system.

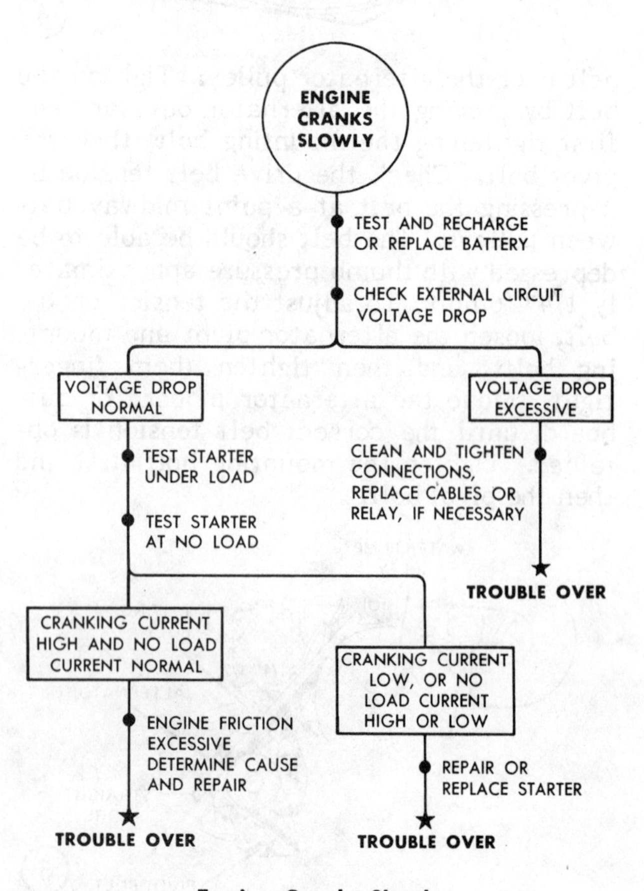

Engine Cranks Slowly

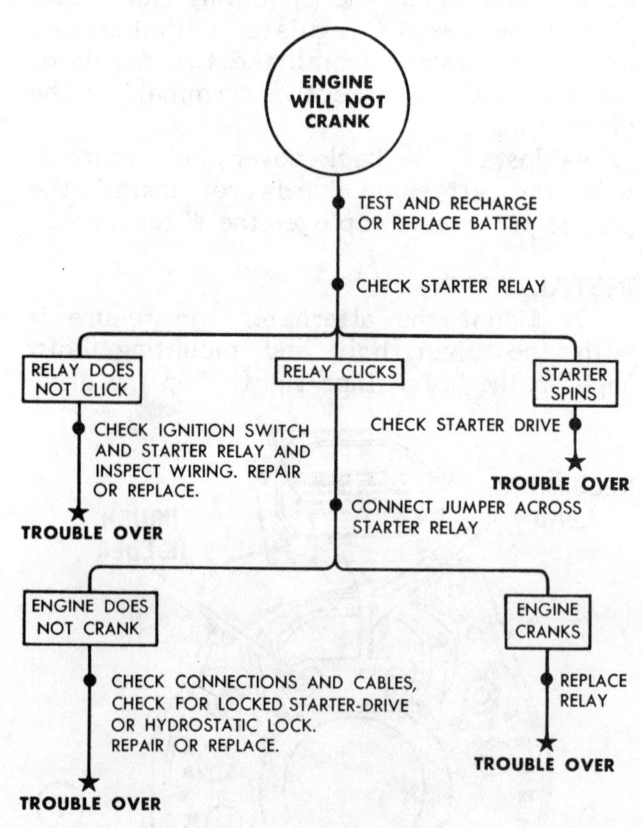

Engine Will Not Crank

Typical installation of Volvo Penta instruments. The tachometer is a critical item to ensure proper timing and performance of the power plant and stern drive.

The procedures are keyed by number to matching numbered illustrations as an aid in performing the work.

BATTERY TEST

1- Turn on several of the cabin lights. Now, turn the ignition switch to the **START** position and note the reaction to the brightness of the lights.

With a normal electrical system, the light will dim slightly and the starter will crank the engine at a reasonable rate. If the lights dim considerably and the engine does not turn over, one of several causes may be at fault.

a- If the lights go out completely, or dim considerably, the battery charge is low or almost dead. The obvious remedy is to charge the battery; switch over to a secondary battery if one is available; or to replace it with a known fully charged one.

b- If the starting relay clicks, sounding similar to a machine gun firing, the battery charge is too low to keep the starting relay engaged when the starter load is brought into the circuit.

c- If the starter spins without cranking the engine, the drive is broken. The starter will have to be removed for repairs.

d- If the lights do not dim, and the starter does not operate, then there is an open circuit. Proceed to Test 2, Cable Connection Test.

CABLE CONNECTION TEST

2- If the starter fails to operate and the lights do not dim when the ignition switch is turned to **START**, the first area to check is the connections at the battery, starting relay, starter, and neutral-safety switch.

First, remove the cables at the battery; clean the connectors and posts; replace the cables; and tighten them securely.

Now, try the starter. If it still fails to crank the engine, try moving the shift box selector lever from **NEUTRAL** to **FORWARD** to determine if the neutral-safety switch is out of adjustment or the electrical connections need attention.

Sometimes, after working the shift lever back-and-forth and perhaps a bit sideways, the neutral-switch connections may be temporarily restored and the engine can be started. Disconnect the leads; clean the connectors and terminals on the switch; replace the leads; and tighten them securely at the first opportunity.

If the starter still fails to crank the engine, move on to Test 3, Solenoid Test.

SOLENOID TEST

3- The solenoid, commonly called the starting relay, is checked by directly bridging between the terminal from the battery (the large heavy one) to the terminal from the ignition switch.

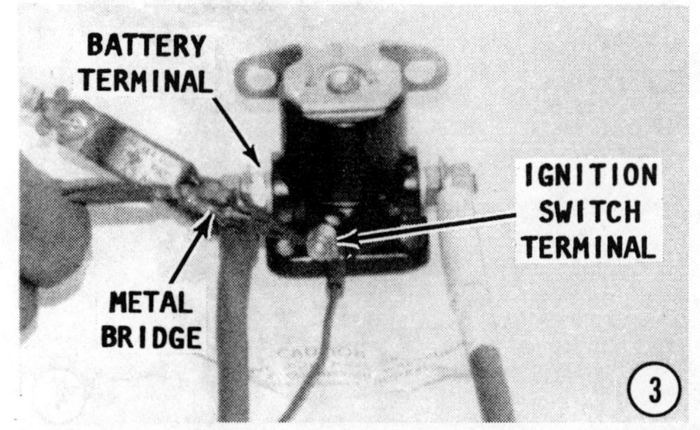

TAKE EVERY PRECAUTION TO ENSURE THERE ARE NO GASOLINE FUMES IN THE BILGE BEFORE MAKING THESE TESTS.

If a bilge blower is installed, operate it for at least five minutes to clear any fumes accumulated in the bilge.

Turn the ignition switch to the **START** position. Now, bridge between the battery lead terminal and the ignition lead terminal with a very heavy piece of wire. If the relay operates, the trouble is in the circuit to the ignition switch. If the starter still fails to operate, continue with Test 4, Current Draw Test.

CURRENT DRAW TEST

4- Lay an amperage gauge on the cable between the battery and the starter. Attempt to crank the engine and note the current draw reading of the amperage gauge under load. The current draw should not exceed 190 amperes.

CHECKING CRANKING CIRCUIT RESISTANCE

If the starter turns very slowly or not at all, or if the solenoid fails to engage the starter with the flywheel, the cause may be

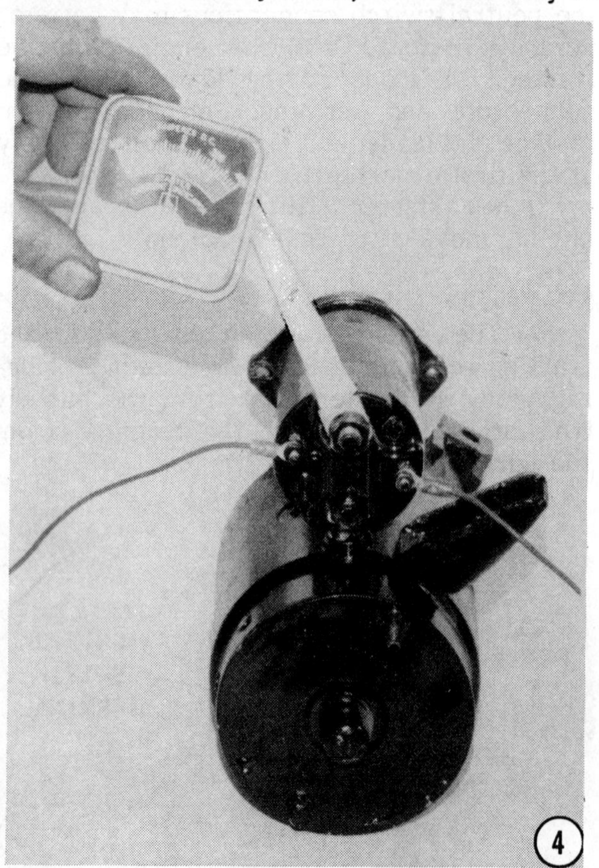

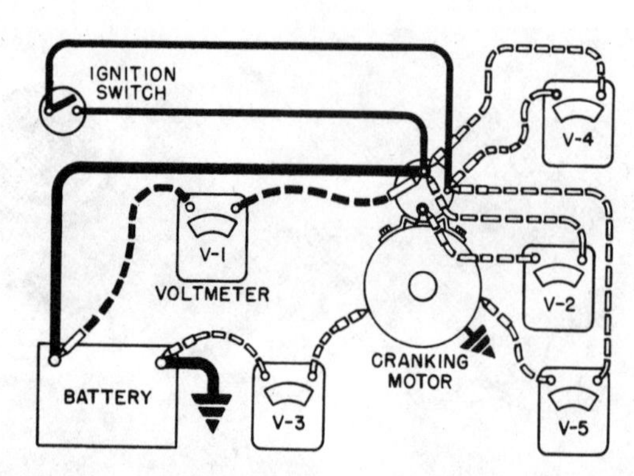

Functional diagram of hookup for various voltage tests outlined in the text.

excessive resistance in the cranking circuit.

The following checks can be performed with the starter installed on the engine in the boat.

1- Test the battery and bring it up to a full charge, if necessary. **GROUND** the distributor primary lead to prevent the engine from firing during the following checks.

2- Measure the voltage drop during cranking between the positive battery post and the battery lead terminal of the solenoid.

3- Measure the voltage drop during cranking between the battery lead terminal of the solenoid and the motor lead terminal of the solenoid.

4- Measure the voltage drop during cranking between the negative battery post and the starter motor frame.

If the voltage drop during any of the previous three tests is more than 0.2 volt, excessive resistance is indicated in the circuit being checked. Trace and correct the cause of the resistance.

If the solenoid fails to pull in, the problem may be due to excessive high voltage drop in the solenoid circuit. To check voltage drop in this circuit, measure the voltage drop during cranking, between the battery terminal of the solenoid and the switch terminal of the solenoid. If the voltage drop is more than 2.5 volts, the resistance is excessive in the solenoid circuit.

If the voltage drop is not more than 2.5 volts and the solenoid does not pull in, measure the voltage available at the switch terminal of the solenoid. The solenoid should pull in with 8.0 volts at temperatures up to 200°F. If it does not pull in, remove the starter motor and test the solenoid.

AUTOLITE AND PRESTOLITE CRANKING SYSTEM VOLTAGE TESTS

Even though the starter cannot be tested accurately while it is mounted on the engine, several tests can be made for excessive resistance in the cranking system circuits. The following four tests are designed to isolate excessive resistance with the system under load.

TAKE EVERY PRECAUTION TO ENSURE THERE ARE NO GASOLINE FUMES IN THE BILGE BEFORE MAKING THESE TESTS.

If a bilge blower is installed, operate it for at least five minutes to clear any fumes accumulated in the bilge.

REMOVE THE HIGH-TENSION WIRE FROM THE CENTER OF THE DISTRIBUTOR AND GROUND IT BEFORE MAKING ANY OF THE FOLLOWING TESTS.

The circled numbers on the accompanying illustration are keyed to the tests and indicate the connection for the voltmeter leads.

TEST 1

Voltage Drop Across the Battery Terminal Cable and Post:

Connect the positive lead of the voltmeter to the positive post of the battery and the negative lead of the voltmeter to the battery terminal of the starter. Have a partner turn the ignition switch to **START** and note the reading of the voltmeter. If a partner is not available, connect a remote starter cable switch between the **S** and battery terminals of the starter relay.

If the reading is more the 0.5 volt, there is a high-resistance connection between the battery post and the cable. The battery post is the first place for corrosion to form due to the high corrosive nature of the battery electrolyte (acid).

Remove both battery cable connectors; clean the connectors and the battery posts thoroughly; replace the connectors; and tighten them securely.

TEST 2

Voltage Drop In the Battery Cable to Starter Relay:

Leave the positive lead of the voltmeter connected to the positive post of the battery. Connect the negative lead of the voltmeter to the battery terminal of the starter relay. Do not connect it to the cable, but directly to the terminal.

Attempt to crank the engine as described in Test 1, and note the voltage drop. If the voltage drops more than 0.1 volt, the starter is drawing too much current because the cable is too thin, or the connection at the relay is corroded. Clean and tighten the connection or replace the cable.

TEST 3

Voltage Drop Across the Contacts Inside the Starter Relay:

Leave the positive lead of the voltmeter connected to the positive battery post. Connect the negative lead of the voltmeter to the starter terminal of the starter relay, as shown. Attempt to crank the engine as described in Test 1 and note the voltage drop. If the voltage drop is more than 0.3 volt, the starter relay must be replaced because the contacts are burned.

TEST 4

Voltage Drop Between the Negative Battery Post and the **Engine Ground:**

Connect the positive lead of the voltmeter to a **GOOD** engine ground. Do not attempt to obtain a good ground on a painted surface. Connect the negative lead of the voltmeter to the negative battery post.

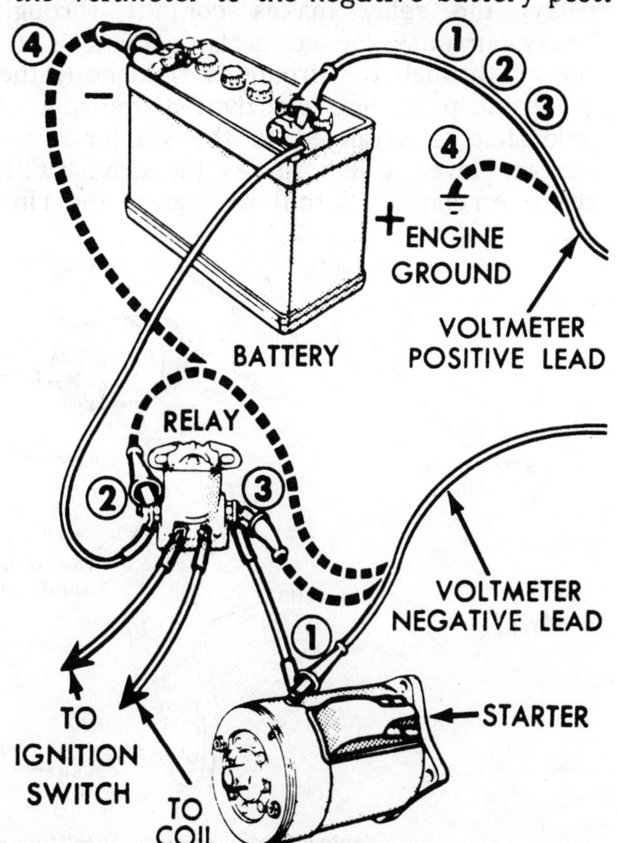

Attempt to crank the engine as described in Test 1 and note the voltage drop. If the voltage drop is more than 0.1 volt, the negative battery post or the cable connector is corroded and must be cleaned and the connector tightened securely. One other possibility is paint between the engine and the cable. Remove the cable, scrape the paint away to bare metal, connect the cable again and repeat the test.

6-13 AUTOLITE CRANKING SYSTEM

The Autolite cranking system consists of the starter, relay, ignition switch, battery, cables, and necessary wiring for efficient operation. The starter has a special moveable pole shoe within the field housing to engage the drive assembly. As on all marine installations, a neutral-start switch is installed in the shift box to permit operation of the starter only when the shift lever is in the **NEUTRAL** position.

AUTOLITE CRANKING MOTORS DESCRIPTION AND OPERATION

The starter has an integral, positive-engagement drive mechanism. The sequence of events in the cranking operation is as follows: The ignition key is turned to the **START** position; current flows to the relay; the relay makes contact through heavy-current type contacts; current is directed through the grounded field coil; the moveable pole shoe is activated; the special pole shoe is attached to the starter drive plunger lever which forces the drive (with the overrunning clutch) to engage the ring gear of the engine flywheel; after the shoe is fully seated, it opens the field coil grounding contacts; and the starter is in normal cranking operation. While the starter is turning the engine flywheel, a holding coil holds the moveable pole shoe in the fully seated position.

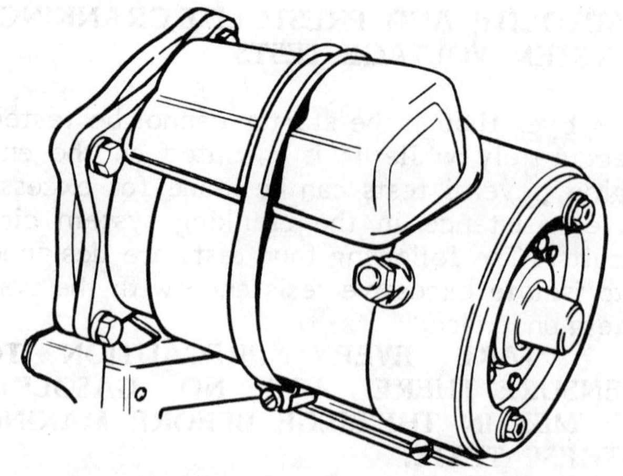

A marine starter motor must have a safety band installed. The starter motor cap being thrown free during engine operation could cause a serious fire.

DISASSEMBLING AN AUTOLITE STARTER

Loosen the retaining screw, and then slide the brush cover band off the starter. Remove the starter drive plunger cover and gasket. Take note of the lead position as an aid to assembling. Remove the commutator brushes from the brush holders.

Remove the long through bolts. Separate the drive end housing from the starter. Remove the plunger return spring. Drive out the pivot pin retaining the starter gear

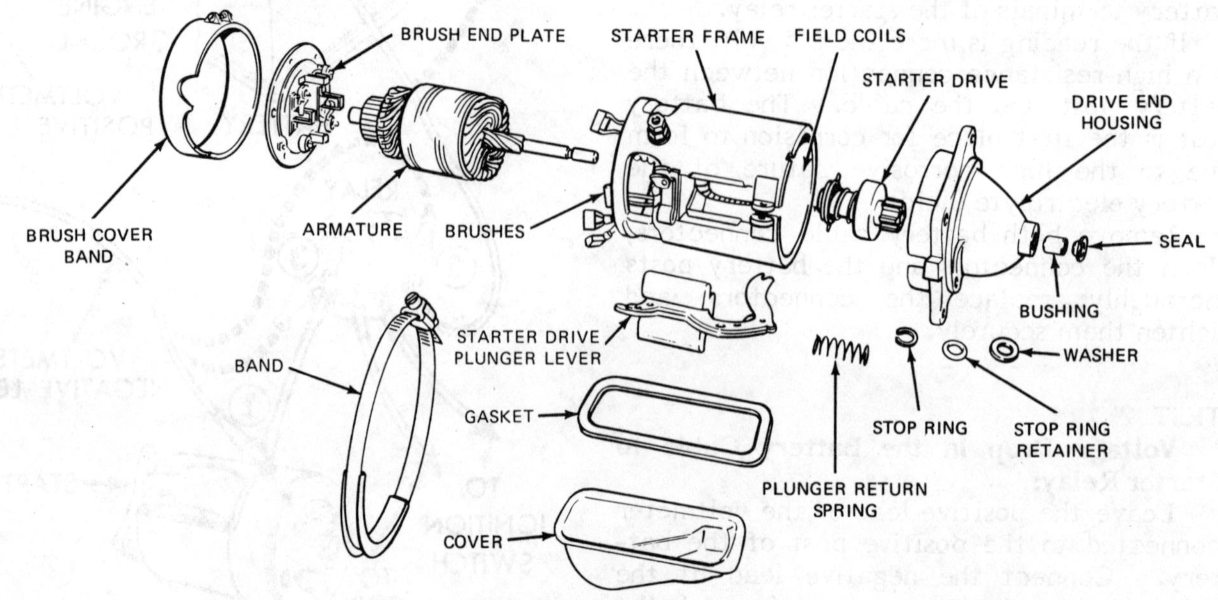

Exploded drawing of an Autolite starter motor with principle parts identified.

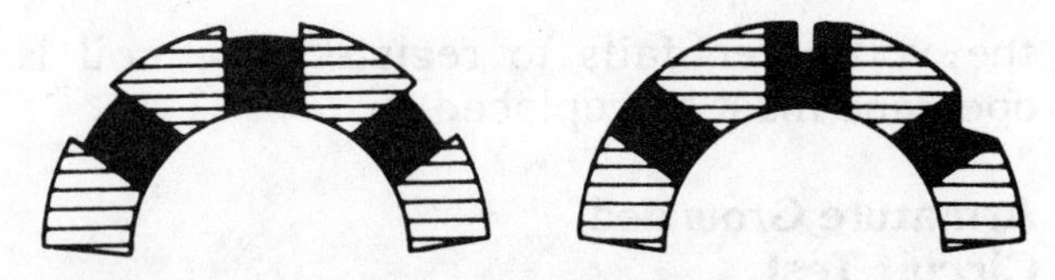

Armature segments properly cleaned (left) and improperly cleaned (right).

plunger lever, and then remove the lever and the armature.

Remove the stop ring retainer from the shaft, and then remove and discard the stop ring. Slide the starter drive gear assembly off the shaft.

Remove the retaining screws, and then the brush end plate. Remove the screws retaining the ground brushes to the frame.

CAREFULLY bend the tab on the field coil retaining sleeve up, and then remove the sleeve.

CLEANING AND INSPECTING

Clean the field coils, armature, commutator, armature shaft, brush-end plate and drive-end housing with a brush or compressed air. Wash all other parts in solvent and blow them dry with compressed air.

Inspect the insulation and the unsoldered connections of the armature windings for breaks or burns.

Perform electrical tests on any part suspected of defect, according to the procedures outlined in the Testing Sections of this Chapter.

Check the commutator for run-out. Inspect the armature shaft and both bearings for scoring.

Turn the commutator in a lathe if it is out-of-round by more than 0.005".

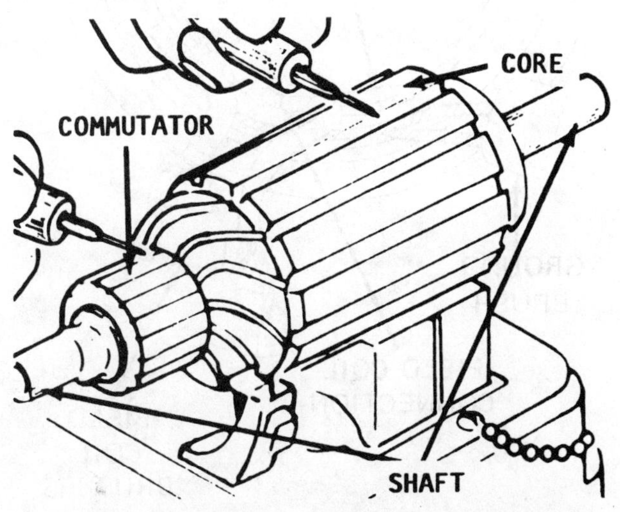

Armature check for a short: one test light lead on each commutator segment, alternately, and the other lead on the armature core. Result -- no continuity.

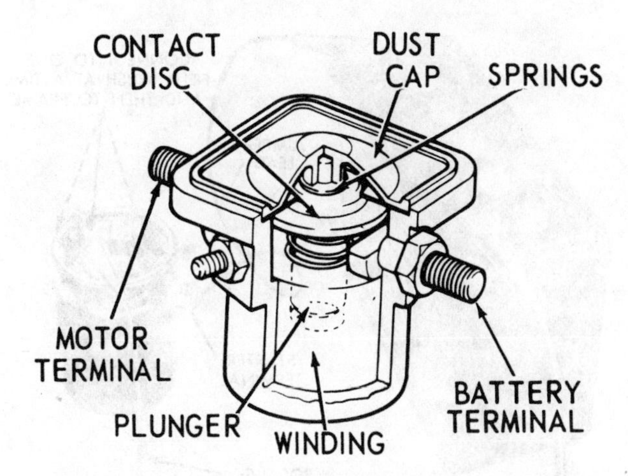

Cutaway view of a starter solenoid showing major parts.

Check the springs in the brush holder to be sure none are broken. Check the spring tension and replace if the tension is not 32-40 ounces. Check the insulated brush holders for shorts to ground. If the brushes are worn down to 1/4" or less, they must be replaced.

Check the field brush connections and lead insulation. A brush kit and a contact kit are available at your local marine dealer, but all other assemblies must be replaced rather than repaired.

BENCH TESTING AUTOLITE STARTERS

The following paragraphs provide a logical sequence of bench tests designed to isolate a defective part in the starter.

The procedures and suggestions are keyed by number to matching numbered illustrations as an aid in performing the work.

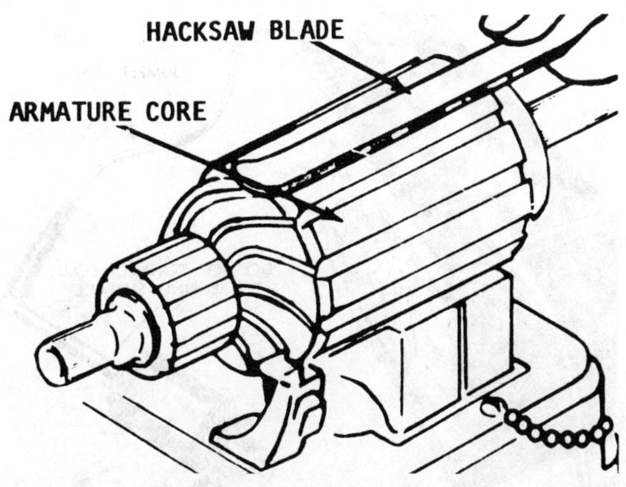

Method of testing the armature for a short circuit using a growler and hacksaw blade. If the blade vibrates, the mica must be cleaned out or the armature replaced.

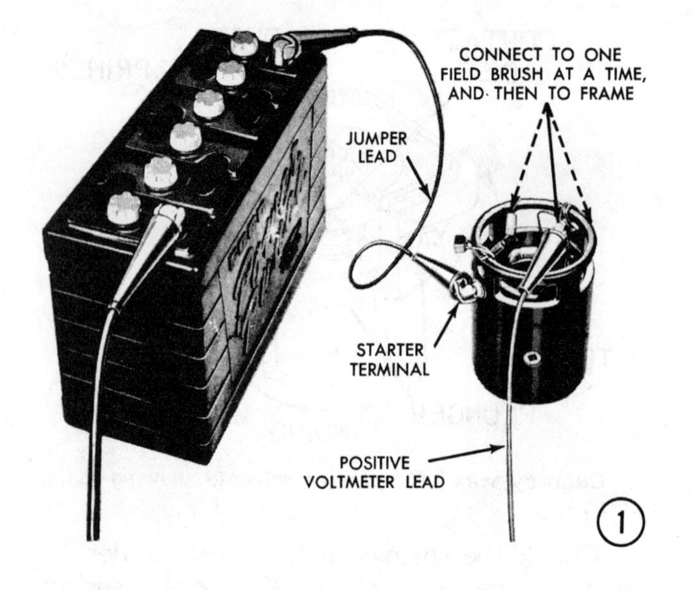

Armature and Field Open Circuit Test

1- Examine the commutator for any evidence of burning. A spot burned on the commutator is caused by an arc formed every time the commutator segment connected to the open-circuit winding passes under a brush.

Connect a jumper wire from the positive terminal of a battery to the starter terminal. Connect the negative lead from a voltmeter to the negative battery terminal. Connect the positive voltmeter lead to one field brush, as shown. Since the starter has three field windings, it will be necesasary to check each of the windings separately. If the voltmeter fails to register, the coil is open and must be replaced.

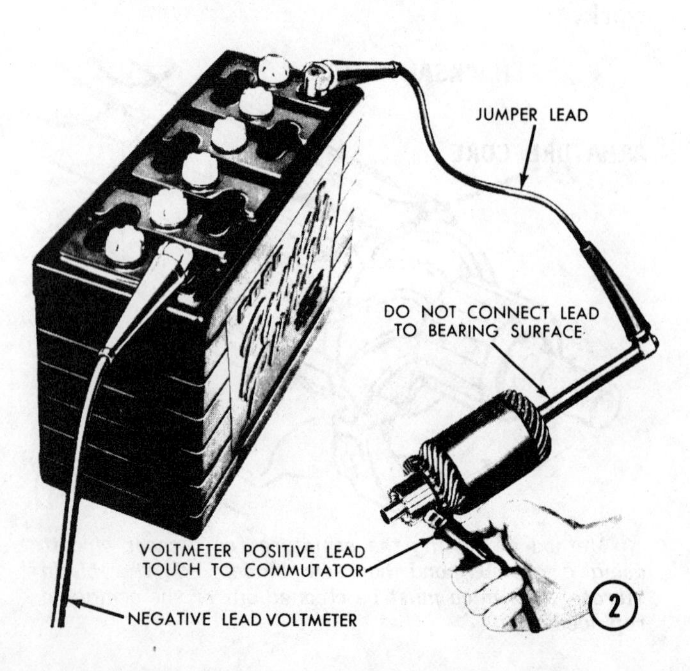

Armature Grounded Circuit Test

2- This test will determine if the winding insulation has failed, permitting a conductor to touch the frame or armature core.

Connect a jumper wire from the positive terminal of a battery to the end of the starter shaft opposite the commutator, as shown. Connect the negative lead from a voltmeter to the negative battery terminal. Make contact with the positive voltmeter lead to the commutator and check for voltage. If the voltmeter fails to register, the windings are grounded and must be replaced.

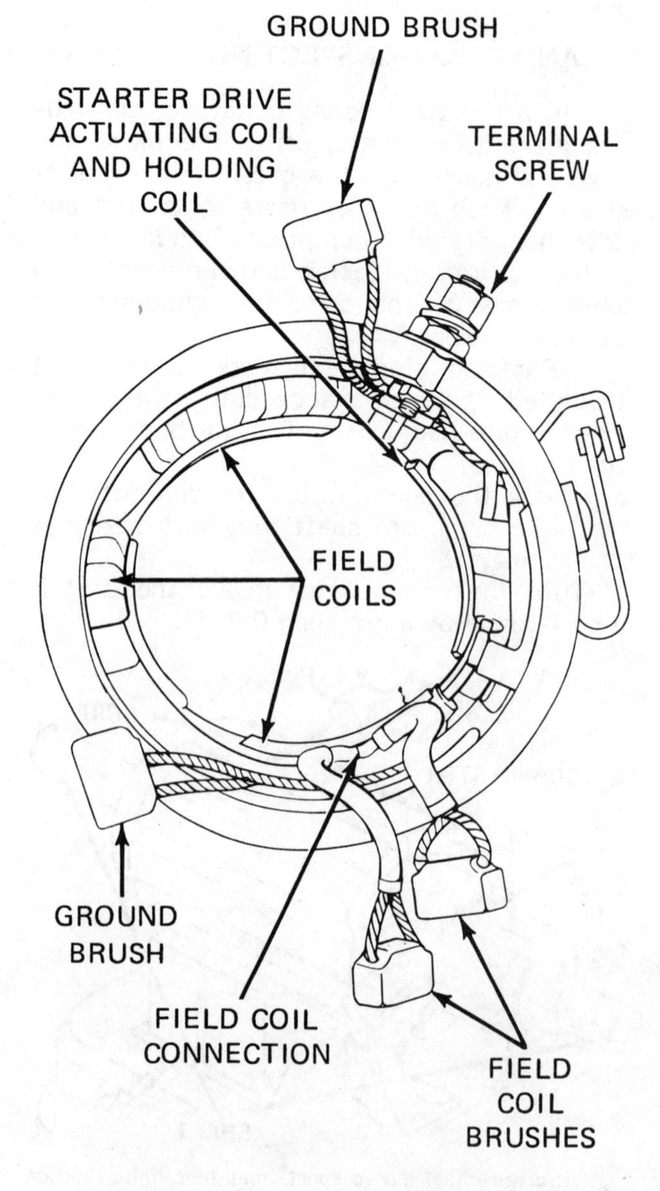

Principle parts of the field coil and brushes.

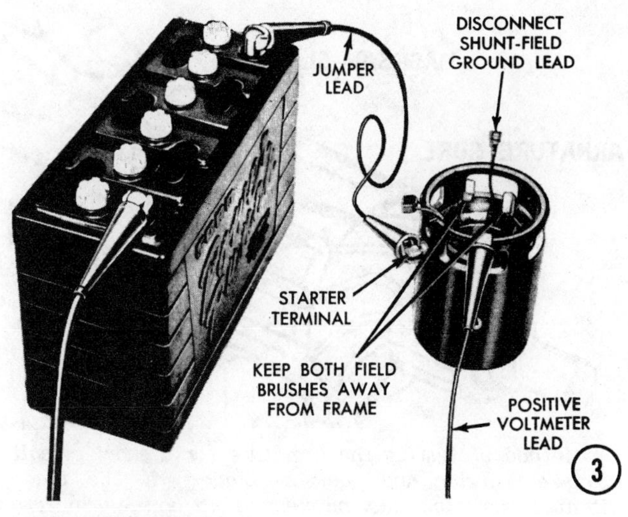

Field Ground Circuit Test

3- Grounded field windings can be detected using a voltmeter and battery. First, **DISCONNECT** the shunt-field ground lead. Next, connect a jumper lead from the positive lead of the battery to the starter terminal. Connect the negative lead from the voltmeter to the negative battery terminal. Now, keep both field brushes away from the starter frame, and make contact with the positive lead from the voltmeter to the starter frame, as shown. If the voltmeter indicates any voltage, the field windings are grounded and must be replaced.

ASSEMBLING AN AUTOLITE STARTER

Position the new insulated field brushes lead onto the field coil terminal and secure it with the clip. To ensure extended service, solder the lead, clip, and terminal together, using resin core solder.

Place the solenoid coil ground terminal over the nearest ground screw hole. Place each ground brush in position to the starter frame and secure them with the retaining screws.

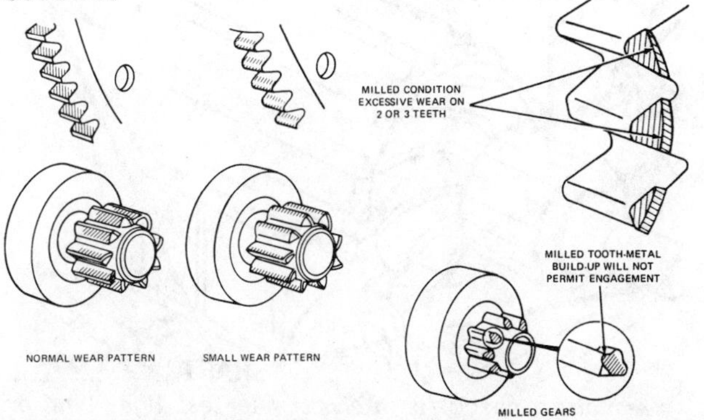

Wear pattern on a starter motor drive gear and on the flywheel ring gear.

Place the brush end plate into position on the starter frame with the boss on the plate indexed in the slot on the frame.

Coat the armature shaft splines with a thin layer of Lubriplate. Slide the starter drive assembly onto the armature shaft, and secure it in place with a **NEW** stop ring and stop-ring retainer.

Slide the fiber thrust washer onto the commutator end of the shaft, then insert the armature shaft into the starter frame. (The thrust washer is not used with molded commutator armatures.)

Place the drive gear plunger lever in position in the frame and the starter drive assembly, and then install the pivot pin.

Fill the drive-end housing bearing bore **ONLY 1/4 FULL** with lubrication. Insert the starter drive plunger lever return spring in position, and then mate the drive-end housing with the starter frame. Install and tighten the through bolts to a torque value of 55-75 in-lbs. **TAKE CARE** not to pinch the brush leads between the brush plate and the frame. **CHECK TO BE SURE** the stop ring retainer is properly seated in the drive housing.

Install the brushes in the brush holder with the **SPRINGS CENTERED** on the brushes.

Check the brush spring tension by pulling on a line with a scale. The line should

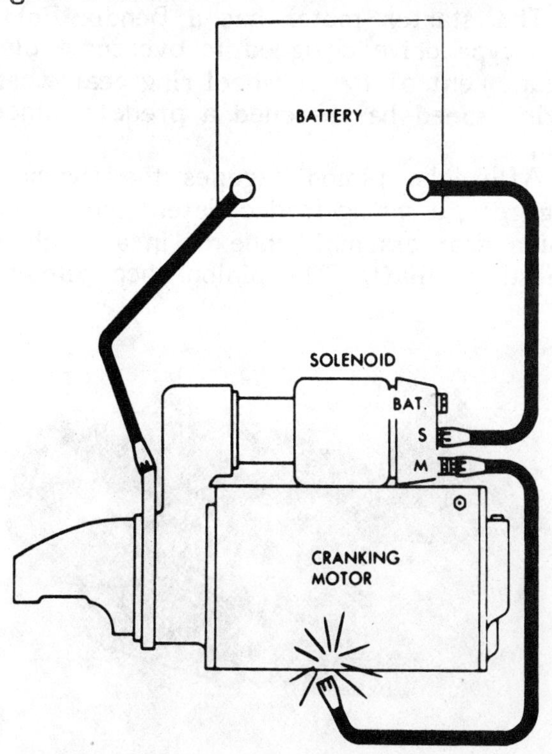

Hookup for bench testing a starter motor.

be hooked under the brush spring near the brush and the pull should be parallel to the face of the brush. Take the reading just as the spring leaves the brush. The reading should be from 32 to 40 ozs.

Finally, position the gasket and drive gear plunger lever cover in place, then slide the brush cover band in place and secure it with the retaining screw.

If the engine should start, but fail to reach the predetermined rpm to disengage the detent pin, a ratchet-type clutch on the screw shaft allows the pinion to over-run the armature shaft and prevent damage to the starter. When this over-run occurs, a light buzzing sound is audible from the clutch ratcheting. When the engine is accelerated, the drive will release and the buzzing sound will stop.

6-14 PRESTOLITE CRANKING SYSTEM

DESCRIPTION AND OPERATION

The Prestolite cranking system includes the starter, relay, ignition switch, battery, cables, wiring, and a special starter relay. As with all marine installations, a safety-start switch is installed in the shift box. This switch prevents operation of the cranking system unless the shift lever is in the **NEUTRAL** position.

The starter motor has a Bendix Folo-Thru type drive designed to overcome disengagement of the flywheel ring gear when engine speed has reached a predetermined rpm.

After the pinion engages the flywheel ring gear, a spring-loaded detent pin in the pinion gear assembly indexes in a notch in the screw shaft. The pinion then remains

Prestolite starter motor used with some American engine installations and a Volvo Penta stern drive unit.

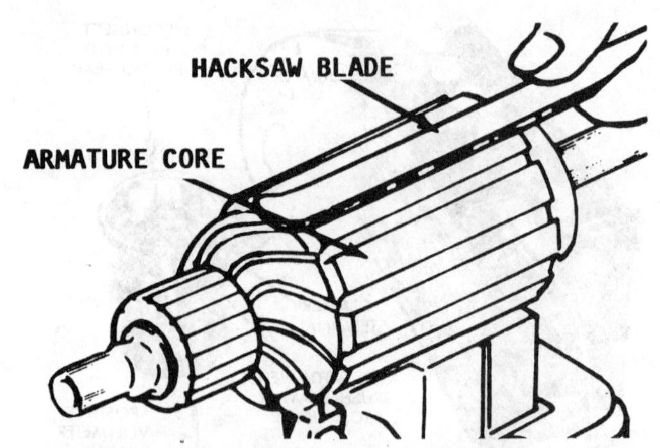

Method of testing the armature for a short circuit using a growler and hacksaw blade. If the blade vibrates, the mica must be cleaned out or the armature replaced.

locked in the engaged position until the engine rpm reaches a predetermined speed. At that point, centrifugal force moves the detent pin out of the notch in the shaft and allows the pinion to disengage the flywheel.

If, during the starting operation, the engine fails to continue running, movement of the pinion in a disengaging direction is prevented by the pin indexed in the screw shaft. For this reason, if the starting motor is re-engaged while the engine kicks back, the starter will not be damaged.

DISASSEMBLING A PRESTOLITE STARTER

First, remove the two through-bolts, and then separate the cover assembly from the commutator end. Next, remove the starter housing from the armature and the end frame.

Now, remove the two screws securing the bearing assembly to the end frame, and then remove the end frame assembly.

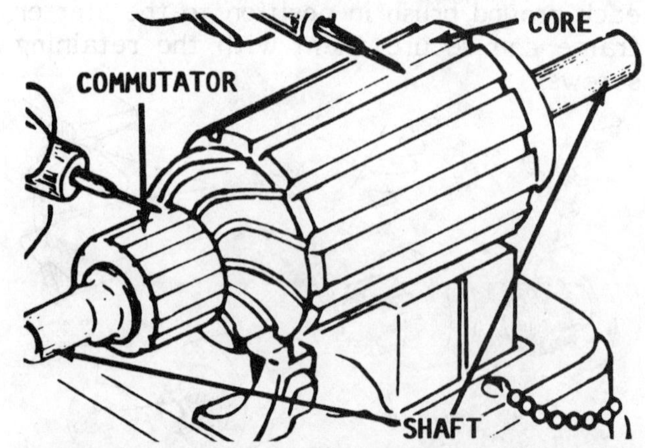

Armature check for a short: one test light lead on each commutator segment, alternately, and the other lead on the armature core. Result -- no continuity.

Remove the pin and the Bendix drive. Remove the bearing assembly from the armature shaft. Remove the three attaching screws, and then the brush plate. Remove the springs and brushes.

If the field is to be removed, disconnect the field wire from the terminal stud; the four screws; field assembly; and pole shoes.

CLEANING AND INSPECTING

Clean the field coils, armature, commutator, brushes, and bushings, with a brush or compressed air. Wash all other parts in solvent and blow them dry with compressed air. **NEVER** use a grease dissolving solvent to clean electrical parts and bushings, because the solvent would damage the insulation and remove the lubricating qualities from the bushings.

Perform electrical tests on any part suspected of defect, according to the procedures outlined in the Testing Sections of this Chapter.

Check the armature shaft to be sure it is not worn or bent. Check the other armature parts: Commutator worn; laminations core

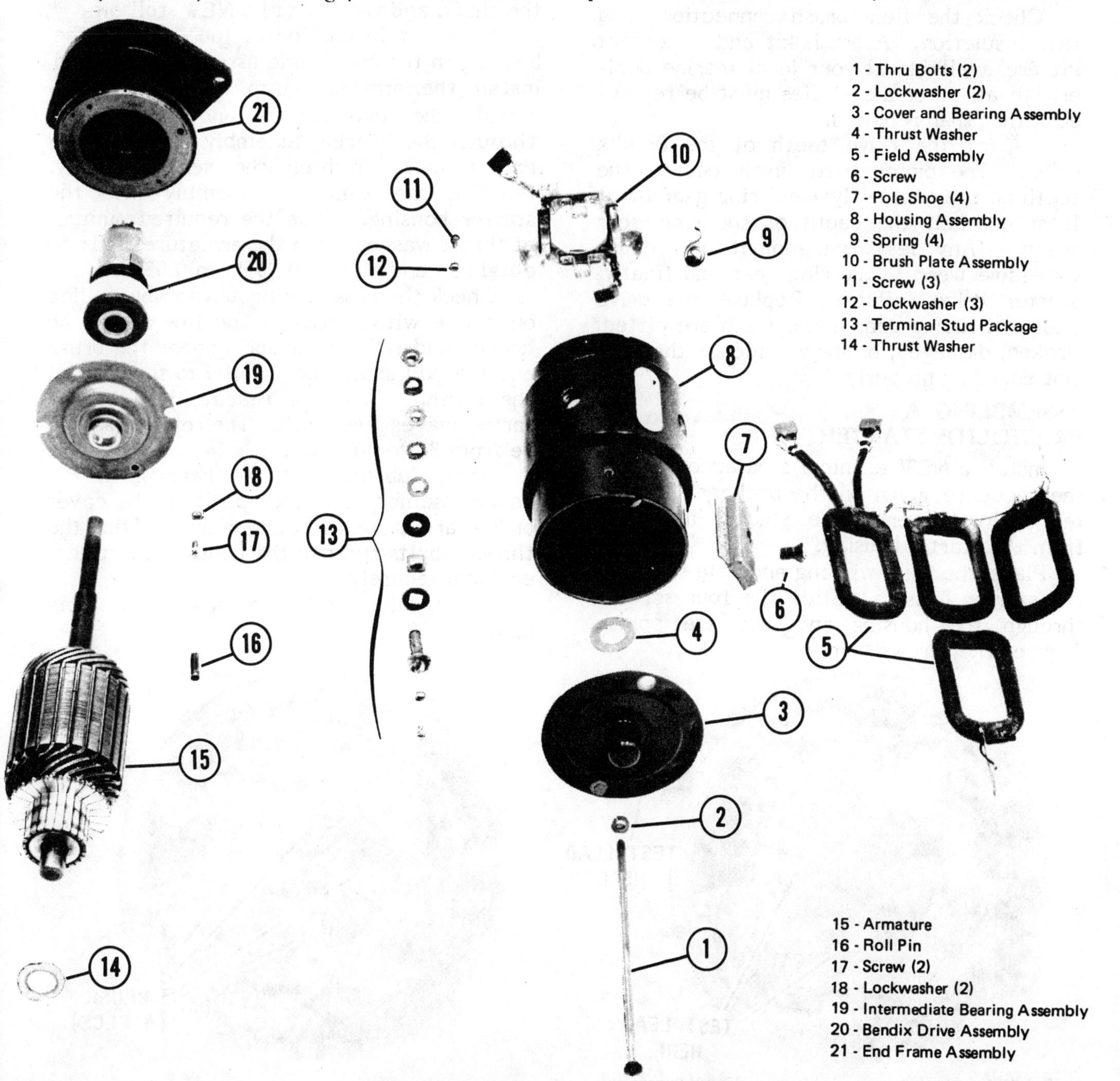

Prestolite starter motor with major parts identified.

scored; or connections requiring attention.

Inspect the armature shaft and bearings for scoring or excessive wear. If the commutator is rough, burned, out-of-round, or has high mica on it, the commutator must be turned on a lathe. The out-of-round limit is 0.005".

Check the springs in the brush holder to be sure none are broken. Check the spring tension and replace if the tension is not 32-40 ounces. Check the insulated brush holders for shorts to ground. If the brushes are worn down to 1/4" or less, they must be replaced.

Check the field brush connections and lead insulation. A brush kit and a contact kit are available at your local marine dealer, but all other assemblies must be replaced rather than repaired.

Inspect the drive teeth of the Bendix drive. The pinion teeth must engage the teeth on the engine flywheel ring gear by at least one-half the depth of the ring gear teeth. Any less engagement will cause excessive wear to the ring gear and finally, starter drive failure. Replace the drive gear or the ring gear if the teeth are pitted, broken, damaged, or show evidence they are not engaging properly.

ASSEMBLING A PRESTOLITE STARTER

Install a **NEW** terminal stud according to the sequence given on the package. Use a test light to verify the stud is insulated from the starter housing.

Place the field winding and pole shoes in the starter housing. Slide the four screws through the housing and into the poles. Tighten the screws firmly.

Position the brush plate assembly in place and install the three screws and lockwashers through the starter housing and into the plate assembly, then tighten them firmly.

Connect the shunt ground wire to the brush plate assembly and install the brush springs.

Apply a thin coating of lubricant to the bearing surface of the bearing assembly, and then slide the assembly into place on the armature shaft. Apply a thin coating of lubricant to the drive-end of the armature shaft. Slide the Bendix drive assembly onto the shaft, and then install a **NEW** roll pin.

Apply a thin coating of lubricant to the bearing in the end frame assembly, and then install the armature into the end frame. Install the two screws and lockwashers through the bearing assembly and into the end frame. Tighten the screws firmly. Position the armature assembly into the starter housing. Slide the required number of thrust washers onto the armature shaft to obtain an end play of 0.005" to 0.030".

Check the brush spring tension by pulling on a line with a scale. The line should be hooked under the brush spring near the brush and the pull should be parallel to the face of the brush. Take the reading just as the spring leaves the brush. The reading should be from 32 to 40 ozs.

Apply lubricant to the bearing in the cover assembly, and then position the cover on the armature shaft and starter. Slide the through-bolts through the housing and tighten them securely.

Check the armature shaft end play again.

Points to attach test leads when testing a Prestolite starter.

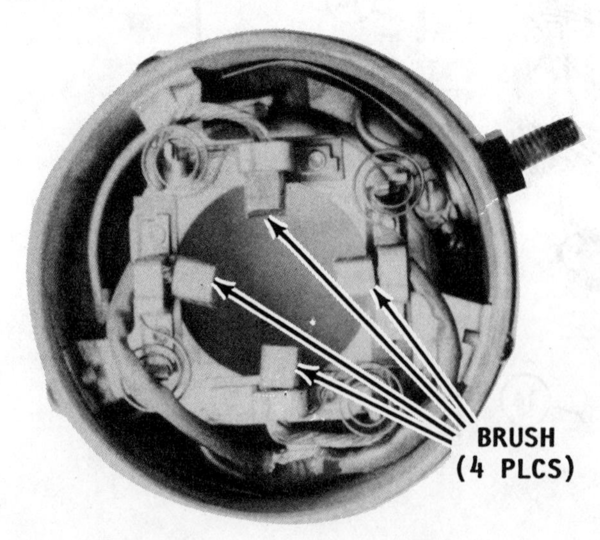

Prestolite starter with the brushes installed.

6-15 BOSCH CRANKING MOTOR

DESCRIPTION AND OPERATION

The Bosch starter motor is used on the 4- and 6-cylinder pushrod engines and on the late model 4-cylinder OHC engines.

The solenoid is an electro-magnet which pushes the bendix drive forward to engage it with the ring gear on the engine flywheel. When the current is released, the bendix mechanism is spring-loaded to the retracted position which disengages it from the flywheel.

The Bosch cranking motor consists of a set of field coils positioned over pole pieces, which are attached to the inside of a heavy iron frame. An armature and an overrunning clutch drive mechanism are housed inside the iron frame.

The armature consists of a series of iron laminations placed over a steel shaft, a commutator, and the armature winding. The windings are heavy copper ribbons assembled into slots in the iron laminations. The ends of the windings are soldered or welded to the commutator bars. These bars are electrically insulated from each other and from the iron shaft.

An overrunning clutch drive arrangement is installed near one end of the motor shaft. This clutch drive assembly contains a pinion gear which is made to move along the shaft by means of a shift lever to engage the engine ring gear for cranking. The relationship between the pinion gear and the ring gear on the engine flywheel provides sufficient gear reduction to meet cranking requirement speed for engine start.

The overrunning clutch drive has a shell and sleeve assembly, which is splined internally to match the spiral splines on the armature shaft. Some starters use helical springs and others use accordian type springs. A collar and spring, located over a sleeve completes the major parts of the clutch mechanism.

When the solenoid is energized and the shift lever operates, it moves the collar endwise along the shaft. The spring assists movement of the pinion into mesh with the ring gear on the flywheel. If the teeth on the pinion fail to mesh for just an instant with the teeth on the ring gear, the spring compresses until the solenoid switch is closed; current flows to the armature; the armature rotates; the spring is still pushing on the pinion; the pinion teeth mesh with the ring gear; and cranking begins.

When the engine starts, the ring gear drives the pinion faster than the armature; the rollers move away from the taper; the pinion overruns the shell; the return spring moves the shift lever back; the solenoid switch is opened; current is cutoff to the armature; the pinion moves out of mesh with the ring gear; and the cranking cycle is completed. The start switch should be opened immediately, when the engine starts, to prevent prolonged overrun.

SERVICING BOSCH STARTER MOTORS

The starter motor is usually installed in an inaccessible location. For this reason the motor drive-end bushing receives little or no lubrication. For lack of lubrication, the drive-end bushing wears; the armature drops down and rubs against the field pole pieces; internal drag is created; and, the output torque of the motor is reduced.

Late model Bosch starter motor.

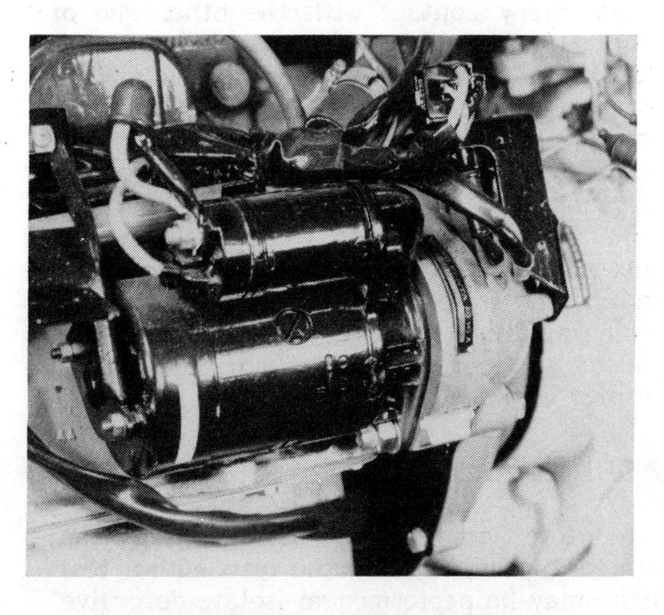

Bosch starter motor installed on an AQ145 engine.

The commutator-end bushing is more accessible and therefore, may receive too much lubrication. The commutator becomes covered with oil. This oil insulates the commutator from the brushes; resistance is increased; and, the efficiency of the cranking system is reduced.

Service of the starter motor consists of replacing defective switches, bushings, brushes, and turning the commutator in a lathe to make it true.

REMOVAL

Disconnect the battery terminals at the battery. Identify, tag, and then disconnect the electrical wiring from the terminals at the starter motor and solenoid. Remove the mounting bolts, and then slide the starter motor forward slightly and free of the engine block.

QUICK BENCH TEST

Secure the complete starter motor lightly in a vise equipped with soft jaws. **DO NOT** overtighten the vise or the starter motor case will be distorted. If a vise is not available, place the starter motor on the floor with the solenoid also against the floor. Hold the motor with one foot while making the following tests.

Connect a 12-volt battery ground cable to the motor case. Connect the positive battery lead to the heavy terminal coming out the top of the solenoid. Connect one end of a jumper wire to this terminal. Make momentary contact with the other end of the jumper wire to terminal No. 50, the knife-type fitting on the solenoid. The starter motor should be energized and spin rapidly.

If the starter motor does **NOT** spin rapidly, make contact with the positive lead from the battery directly to the post below the normal battery connection point on the solenoid. The starter motor should energize and spin rapidly. If it does spin, the solenoid needs service. If it does not, the starter motor is in need of service work.

ADDITIONAL TESTS

Before completely disassembling the starter motor, a few simple tests with a test light may be performed to isolate defective parts.

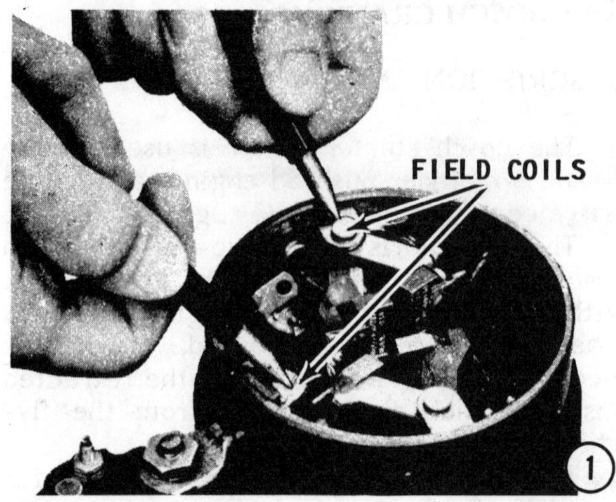

After the starter motor has been removed from the engine, remove the two thru-bolts and then, the commutator end frame. The following tests are keyed by number to illustrations for testing a common starter motor. Your starter motor may appear slightly different, but the tests are valid.

After complete disassembling, the armature can be tested with a growler and hacksaw blade for a short circuit, and with a test light for armature grounded. These tests are presented following the disassembling procedures.

1- Make contact with one probe of a test light on each end of the field coils connected in series. If the test light fails to come on, there is an open in the field coils and repair or replacement is required.

2- Disconnect the shunt coil or coil ground. Make contact with one probe of the test light on the connector strap and on the field frame with the other probe. If the test light comes on, the field coils are grounded and the defective coils must be repaired or replaced.

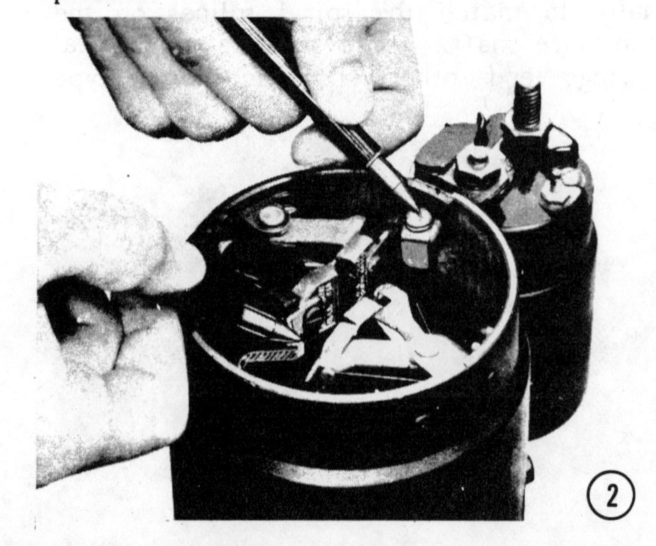

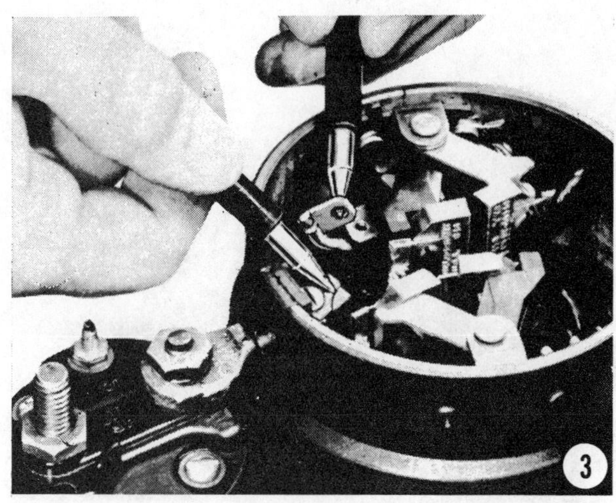

3- Disconnect the shunt coil grounds. Make contact with one probe of the test light on each end of the shunt coil, or coils. If the light fails to come on, the shunt coil is open and must be repaired or replaced.

DISASSEMBLING

The following procedures pick up the work after the starter motor has been removed from the engine. The thru-bolts and the commutator end frame cover are still in place.

1- Clamp the starter motor lightly in a vise equipped with soft jaws. **DO NOT** overtighten the vise or the starter motor case will be distorted. Remove the Bendix arm hinge pin.

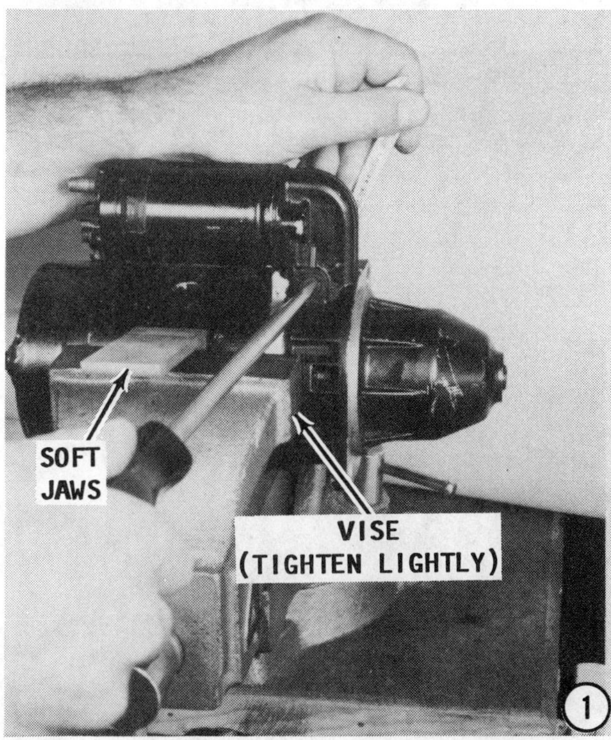

2- Remove the nut from the power cable leading into the starter motor.

3- Remove the two solenoid mounting bolts from the front end of the solenoid.

4- Slide the solenoid back, and at the same time, lift the link arm free of the clutch lever.

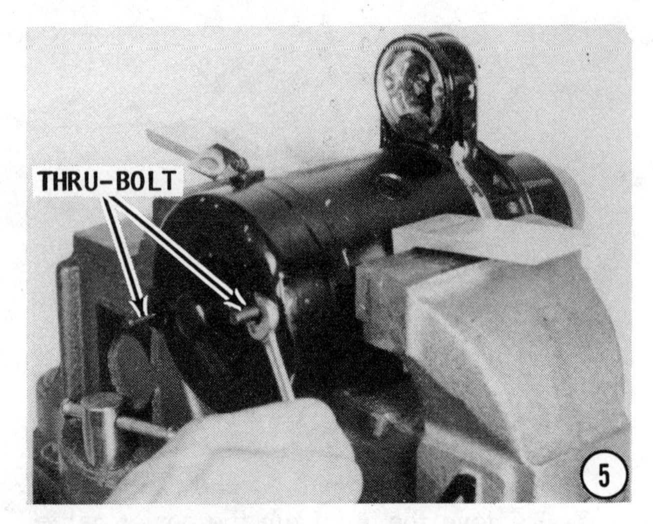

SPECIAL WORDS

If the solenoid has been working off-and-on, and a new solenoid is not available, it is possible to remove the back and clean the contacts. This cleaning may restore the solenoid to satisfactory service.

5- Remove the two thru bolts.

6- Remove the two screws securing the shaft end plate to the end frame cover, and then remove the plate.

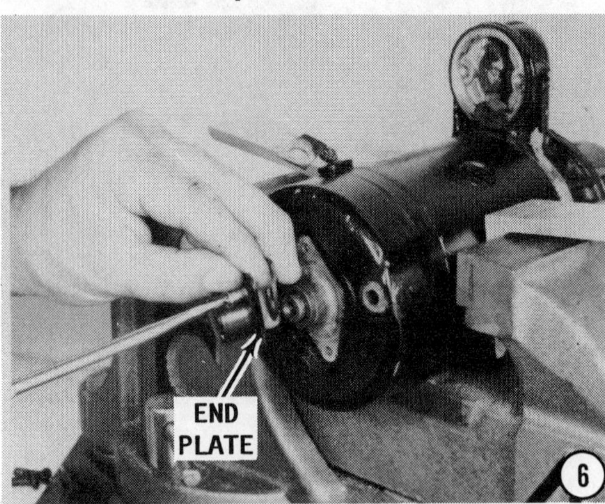

7- Snap the keeper free from the groove in the shaft, and then remove the spacers and washers. Keep the spacers and washers separate and identified to **ENSURE** the same number will be installed during assembling.

8- Slide the end frame cover free of the starter motor case.

9- Remove the two brushes wired to the brush plate.

SPECIAL WORDS

After the brush plate has been removed, the commutator can be inspected. Slight grooves can be cleaned and resurfaced with

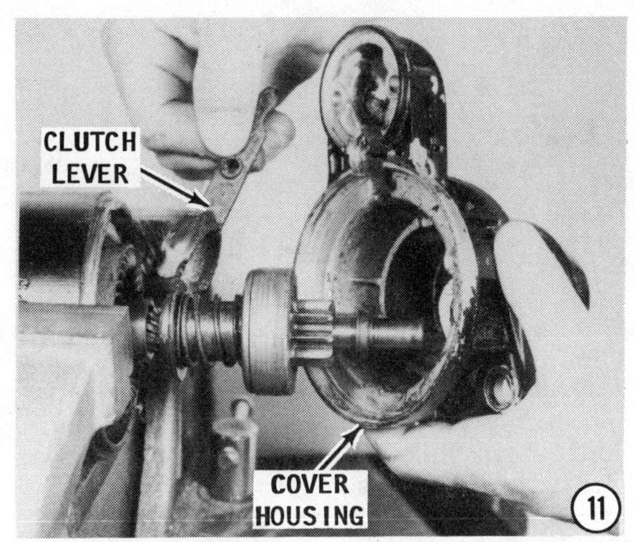

fine grain emory cloth. If the commutator is severely grooved, it must be turned down on a lathe. The most practical remedy is to seek the services of a starter motor/alternator shop specializing in reconditioning these items.

10- The brushes wired to the field may be easily inspected and replaced, as necessary.

11- Separate the bendix spring cover housing from the starter motor case, and at the same time remove the clutch lever.

12- Withdraw the armature from the field and starter motor case.

13- If the bendix spring or the gear teeth are worn or damaged and need to be replaced, first remove the thrust collar. Next,

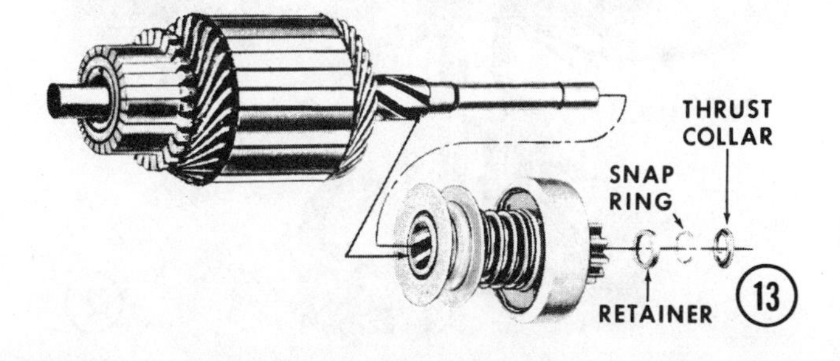

The drive-end bushing should always be replaced during starter motor overhaul.

pop the snap ring inside the retainer out of its groove on the shaft. Work the snap ring off the end of the shaft. Once the snap ring is free, slide the other parts free of the shaft.

CLEANING AND INSPECTING

ALWAYS replace the drive-end bushing during a starter motor overhaul.

True the commutator, if necessary, in a lathe. **NEVER** undercut the mica because the brushes are harder than the insulation.

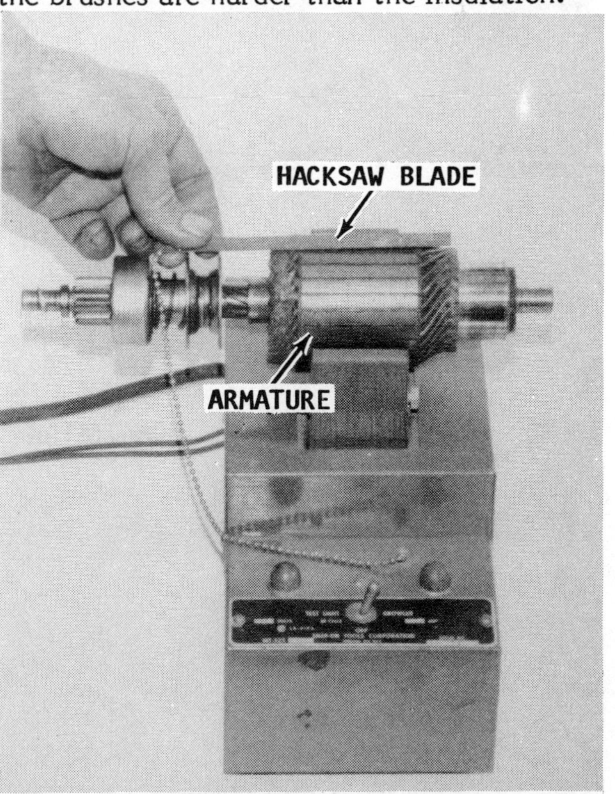

Checking the armature for a short using a hacksaw blade and a growler.

Check the armature for a short circuit by placing it on a growler and holding a hack saw blade over the armature core while the armature is rotated, If the saw blade vibrates, the armature is shorted.

Clean between the commutator bars, and then check again on the growler. If the saw blade still vibrates, the armature **MUST** be replaced.

Check the armature for a ground. Make contact with one probe of a test light on the armature core or shaft and the other probe on the commutator. If the light comes on, the armature is grounded and **MUST** be replaced.

Wash the brush plate in solvent, and then blow it dry with compressed air.

ALWAYS install new brushes whenever the starter motor is serviced.

NEVER wash the clutch in solvent or the lubricant will be dissolved; the clutch will fail; the engine will drive the cranking motor armature at high speed; the windings will be thrown out by centrifugal force; and, the armature will be destroyed.

ASSEMBLING

1- Lubricate the drive end of the armature shaft with a thin coating of silicone

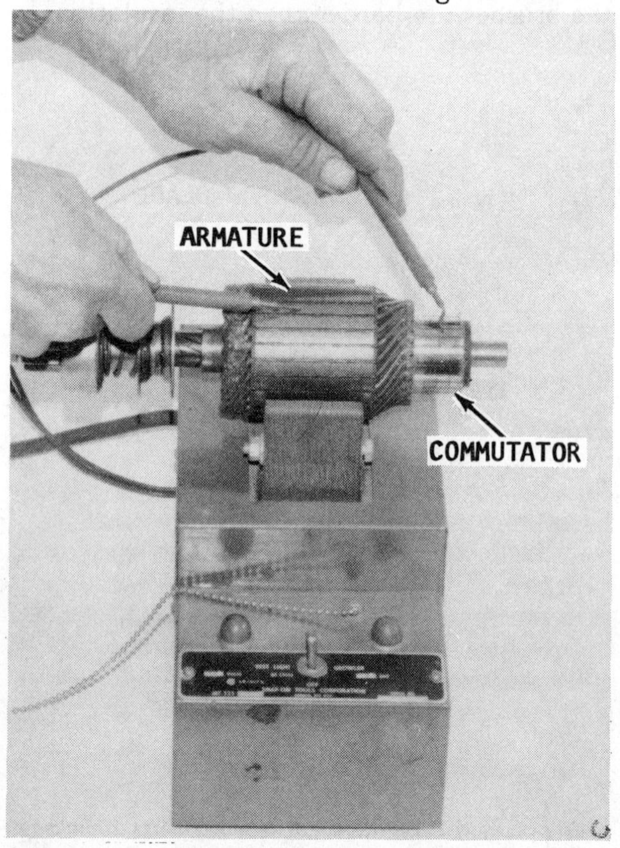

Checking the armature for a ground.

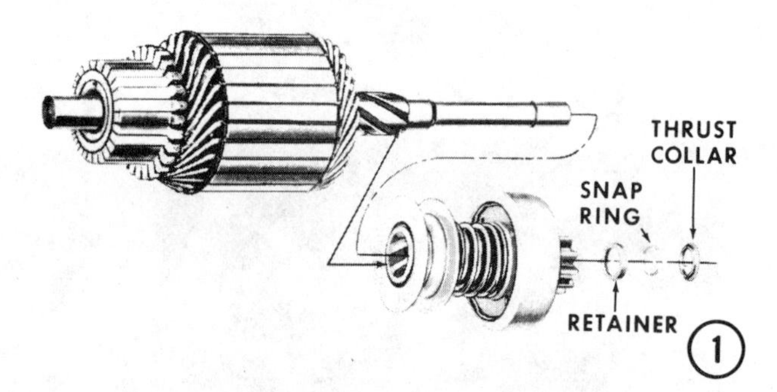

lubricant. Slide the clutch assembly onto the armature shaft with the pinion facing outward. Slide the retainer onto the shaft with the cupped surface facing the end of the shaft (away from the pinion).

2- Stand the armature on end on a wooden surface with the commutator down. Position the snap ring on the upper end of the shaft and hold it in place with a block of wood. Now, tap on the wooden block with a hammer to force the snap ring over the end of the shaft. Work the snap ring down into the groove on the shaft. Assemble the thrust collar onto the shaft with the shoulder next to the snap ring. Position the retainer and thrust collar next to the snap ring. Using two pair of pliers, as shown, grip the retainer and thrust collar and squeeze until the retainer is forced over the snap ring.

Lubricate the drive housing bushing with a thin coating of silicone lubricant. **MAKE SURE** the thrust collar is in place against the snap ring and the retainer.

3- Slide the assembled armture and associated parts into the field and starter motor case.

4- Insert the clutch lever into place, and then slide the cover housing over the pinion gear end of the shaft with the lever moved into the slot in the housing. Align the

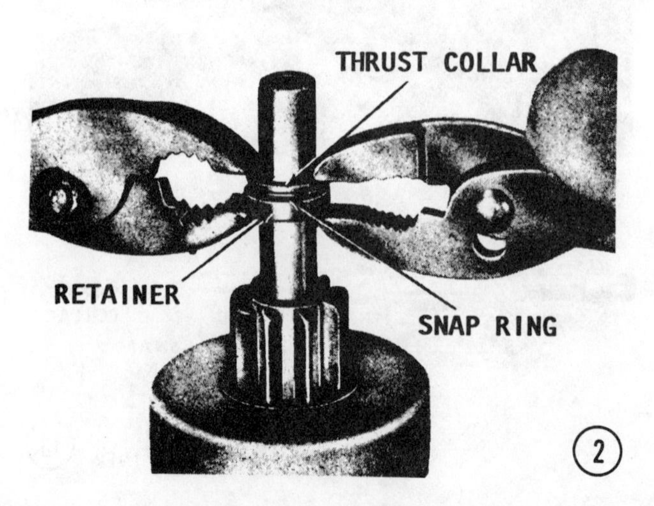

holes in the cover with the holes in the motor case.

5- Assemble two **NEW** brushes into place in the brush plate with the springs snapped over the end of the brush to hold them in position. These two are the ground brushes and the leads are grounded.

6- Connect two new field brushes into place in the motor case.

7- Slide the assembled brush plate into position. Work it on with the brushes over

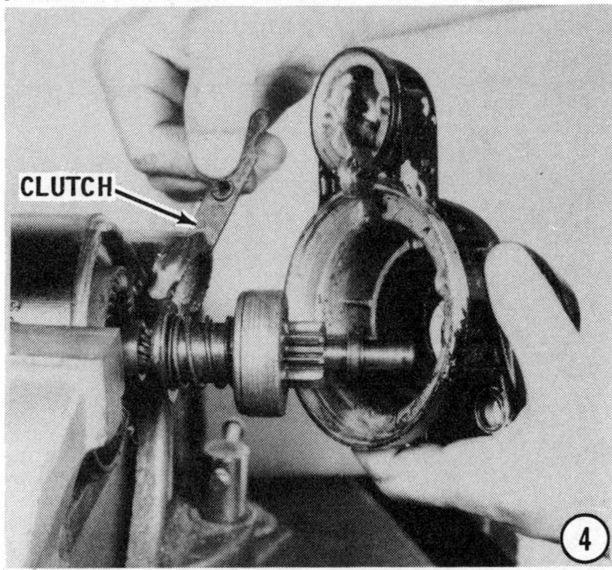

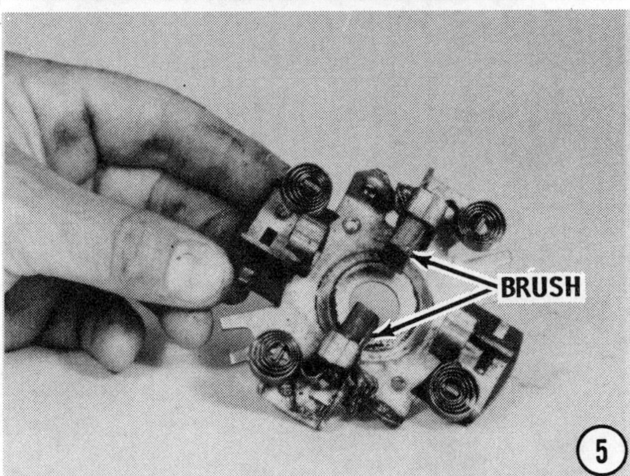

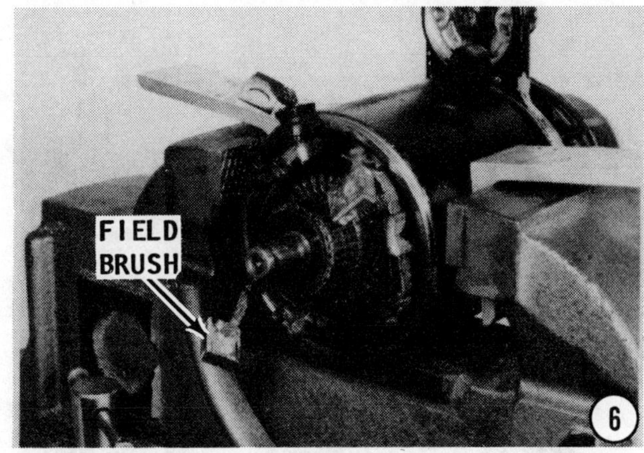

the commutator and the springs behind the brushes. This is a tricky manuever, but with patience it will all go into place.

8- Slide the end frame cover over the end of the shaft and into place with the notch in the cover indexed over the electrical post. The holes for the thru-bolts will then be aligned with the holes in the starter case.

9- Slide the shaft end plate into place and secure it with the two screws.

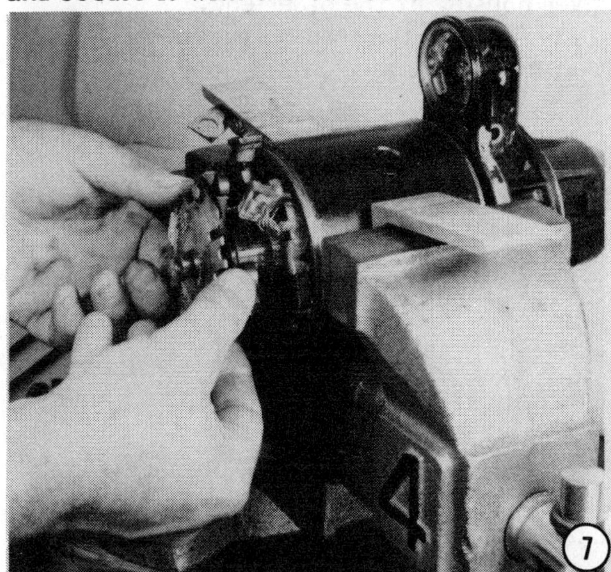

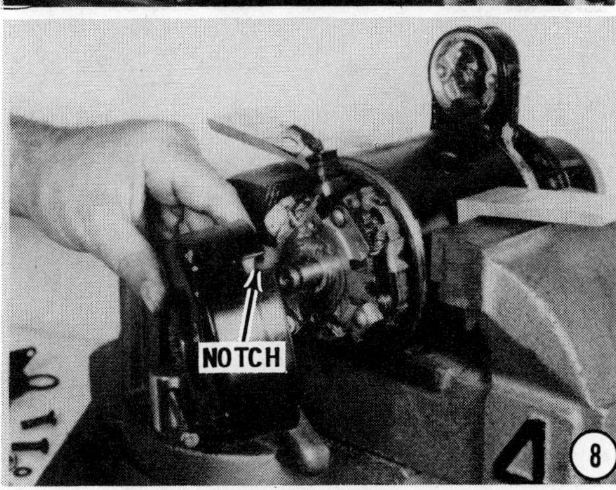

10- Slide the same number of spacers and washers, removed during disassembling, onto the shaft. Secure them in position by snapping the keeper into place

11- Coat the threads of the thru-bolts with Loctite, or similar material. Insert the two thru-bolts through the end frame cover and the motor case. Thread them into the cover housing at the other end.

12- Tighten the thru-bolts securely.

13- Engage the solenoid link arm with the clutch lever, and then move the solenoid into place against the mating surface of the cover housing.

14- Coat the threads of the two solenoid attaching bolts with Loctite, or similar material, and then thread them into place. Tighten the bolts securely.

15- Install the nut onto the power cable leading into the starter motor.

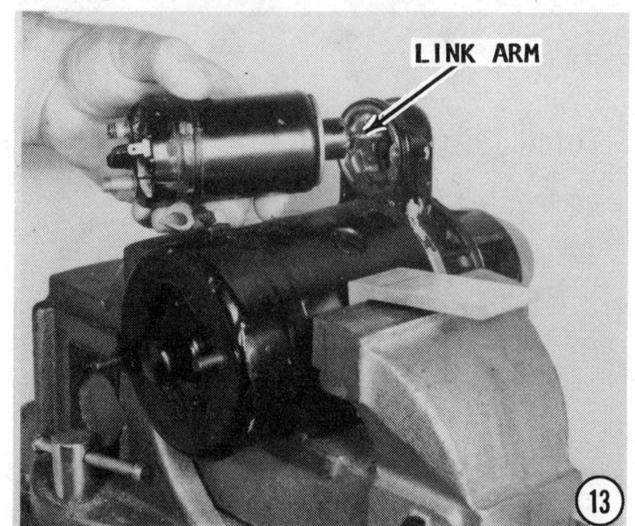

16- Install the pivot pin through the housing and secure it with the nut. An ice pick or similar tool will assist in making the alignment for the pin to pass through. Tighten the nut securely.

TESTING COMPLETED WORK

Before the starter motor is installed, a quick bench test may be performed to verify the work has been satisfactorily performed and the motor is ready for service.

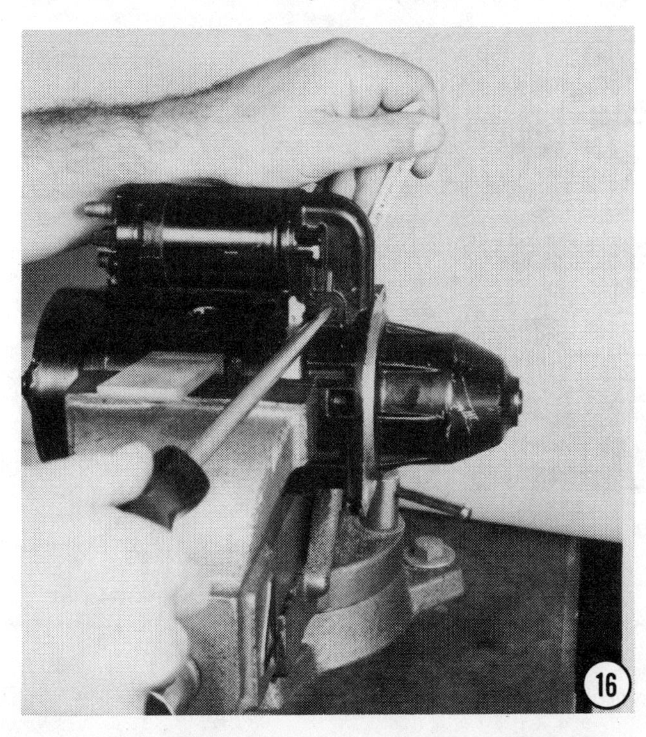

Secure the starter motor lightly in a vise equipped with soft jaws. **DO NOT** overtighten the vise or the starter motor case will be distorted. If a vise is not available, place the starter motor on the floor with the solenoid also against the floor. Hold the motor with one foot while making the following test:

Connect a 12-volt battery ground cable to the motor case. Connect the positive battery lead to the heavy terminal coming out the top of the solenoid. Connect one end of a jumper wire to this terminal. Make momentary contact with the other end of the jumper wire to terminal No. 50, the knife-type fitting on the solenoid. The starter motor should be energized and spin rapidly.

INSTALLATION

Insert the starter motor into the opening in the engine block and secure it in place with the attaching bolts.

Connect the electrical wires. Connect the battery cables. Momentarily activate the starter switch. The starter motor should crank the engine.

6-16 DELCO CRANKING MOTOR

DESCRIPTION AND OPERATION

Delco cranking motors consist of a set of field coils positioned over pole pieces, which are attached to the inside of a heavy iron frame. An armature, an overrunning clutch drive mechanism, and a solenoid are included inside the iron frame.

The armature consists of a series of iron laminations placed over a steel shaft, a commutator, and the armature winding. The windings are heavy copper ribbons assembled into slots in the iron laminations. The ends of the windings are soldered or welded to the commutator bars. These bars are electrically insulated from each other and from the iron shaft.

An overrunning clutch drive arrangement is installed near one end of the starter shaft. This clutch drive assembly contains a pinion which is made to move along the shaft by means of a shift lever to engage the engine ring gear for cranking. The relationship between the pinion gear and the ring gear on the engine flywheel provides sufficient gear reduction to meet cranking requirement speed for starting.

The overrunning clutch drive has a shell and sleeve assembly, which is splined internally to match the spiral splines on the armature shaft. The pinion is located inside the shell. Spring loaded rollers are also inside the shell and they are wedged against the pinion and a taper inside the shell. Some starters use helical springs and others use accordion type springs. Four rollers are used. A collar and spring, located over a sleeve completes the major parts of the clutch mechanism.

When the solenoid is energized and the shift lever operates, it moves the collar endwise along the shaft. The spring assists movement of the pinion into mesh with the ring gear on the flywheel. If the teeth on the pinion fail to mesh for just an instant with the teeth on the ring gear, the spring compresses until the solenoid switch is closed; current flows to the armature; the armature rotates; the spring is still pushing on the pinion; the pinion teeth mesh with the ring gear; and cranking begins.

Torque is transferred from the shell to the pinion by the rollers, which are wedged tightly between the pinion and the taper cut into the inside of the shell. When the engine starts, the ring gear drives the pinion faster than the armature; the rollers move away from the taper; the pinion overruns the shell; the return spring moves the shift lever

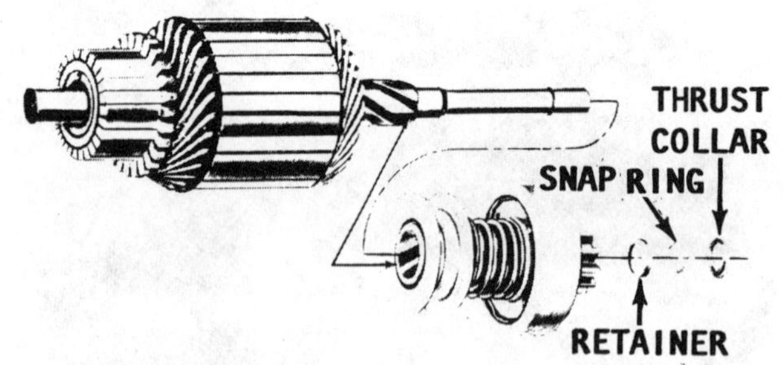

*Principle parts of the starter armature and clutch assembly. **NEVER** wash the clutch assembly in solvent or the lubricant will dissolve causing the unit to seize.*

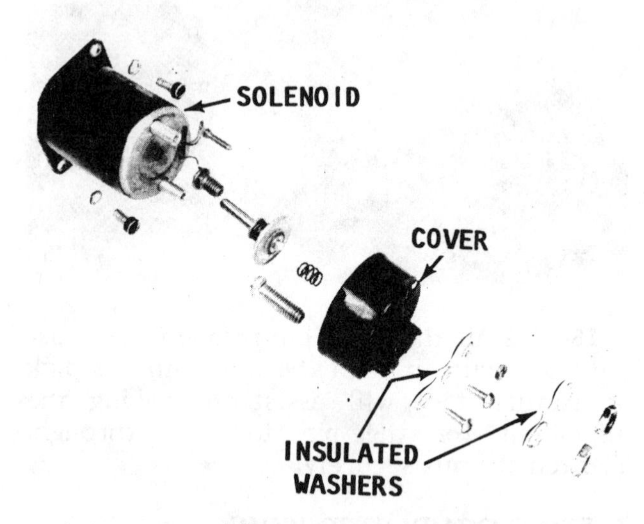

Arrangement of external parts of a Delco solenoid.

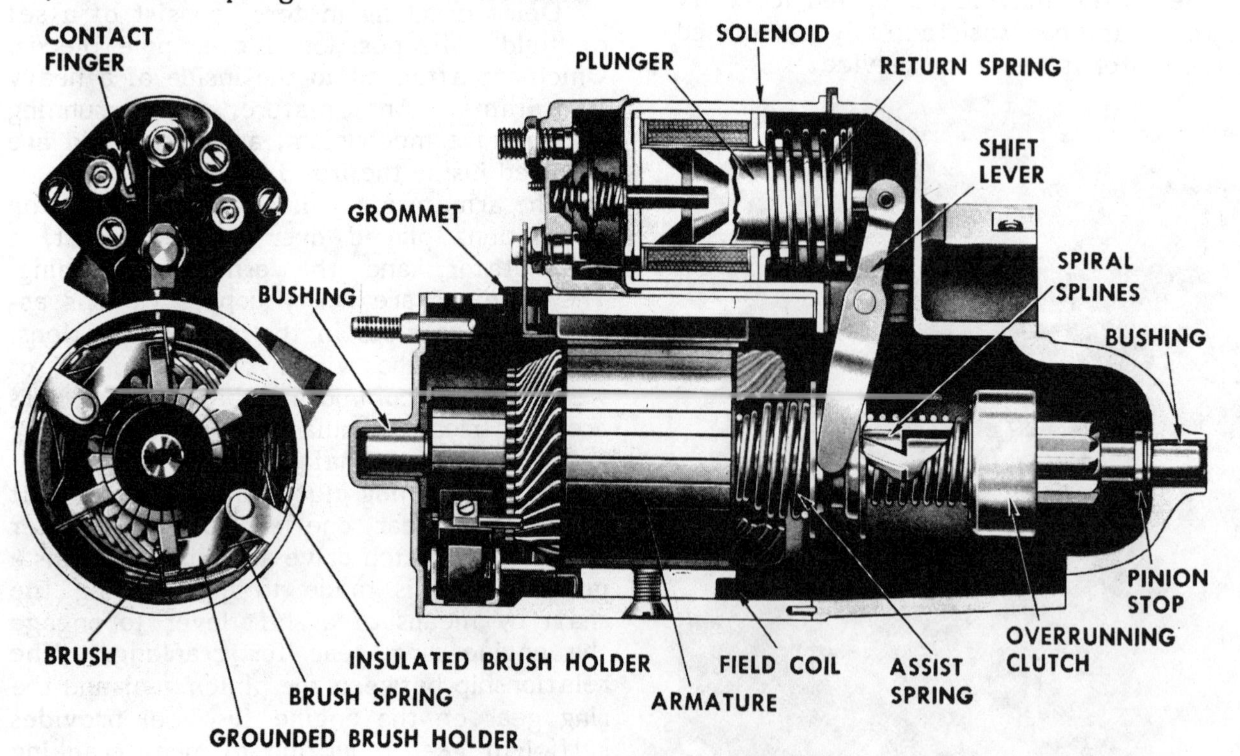

Cross-section view of a cranking motor.

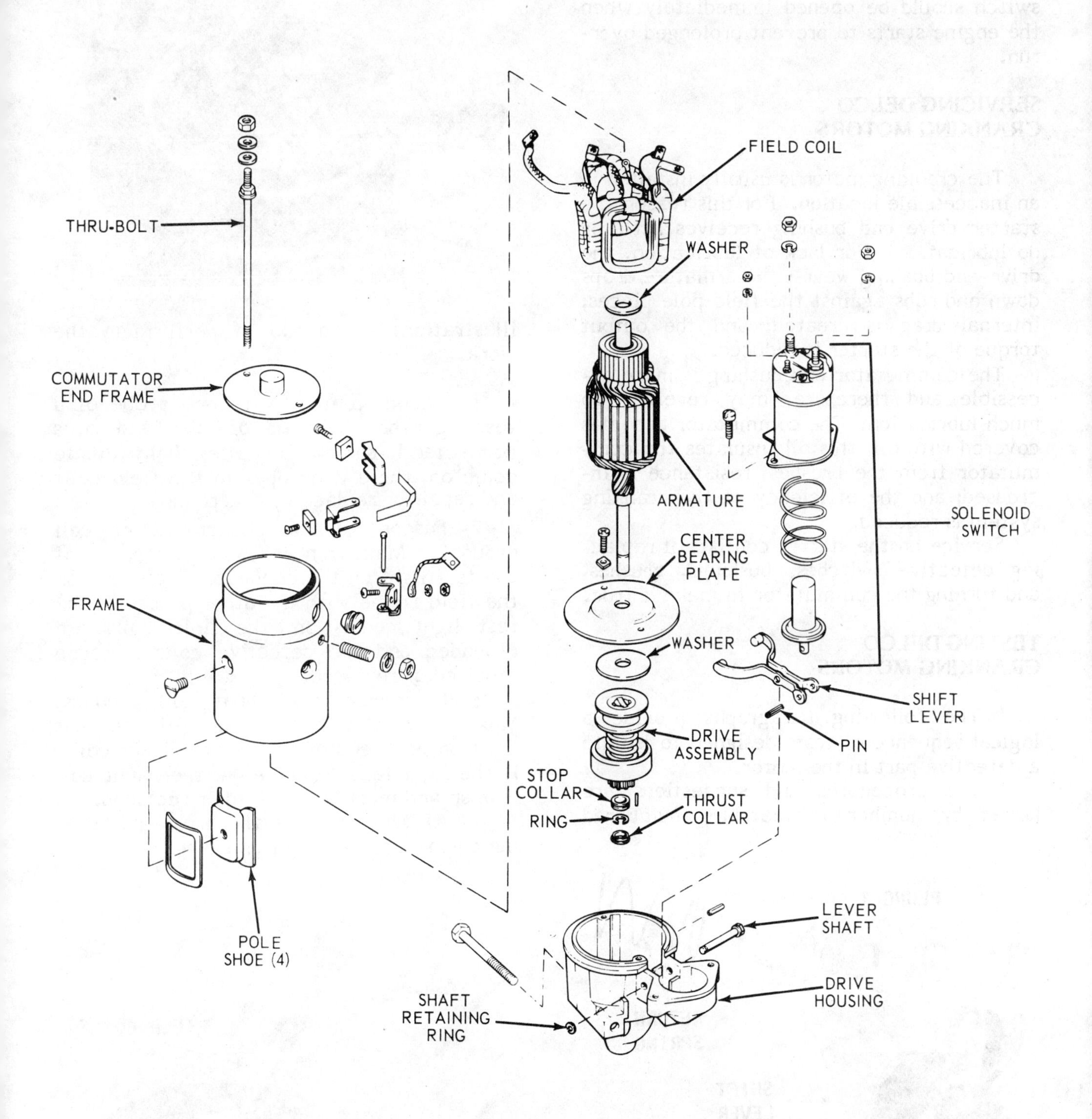

Exploded drawing of the Delco cranking motor, with major parts identified.

back; the solenoid switch is opened; current is cutoff to the armature; the pinion moves out of mesh with the ring gear; and the cranking cycle is completed. The start switch should be opened immediately when the engine starts to prevent prolonged overrun.

SERVICING DELCO CRANKING MOTORS

The cranking motor is usually installed in an inaccessible location. For this reason the starter drive end bushing receives little or no lubrication. For lack of lubrication, the drive-end bushing wears; the armature drops down and rubs against the field pole pieces; internal drag is created; and the output torque of the starter is reduced.

The commutator end bushing is more accessible and therefore, may receive too much lubrication. The commutator becomes covered with oil; this oil insulates the commutator from the brushes; resistance is increased; and the efficiency of the cranking system is reduced.

Service on the starter consists of replacing defective switches, bushings, brushes, and turning the commutator to make it true.

TESTING DELCO CRANKING MOTORS

The following paragraphs provide a logical sequence of tests designed to isolate a defective part in the motor.

The procedures and suggestions are keyed by number to matching numbered illustrations as an aid in performing the work.

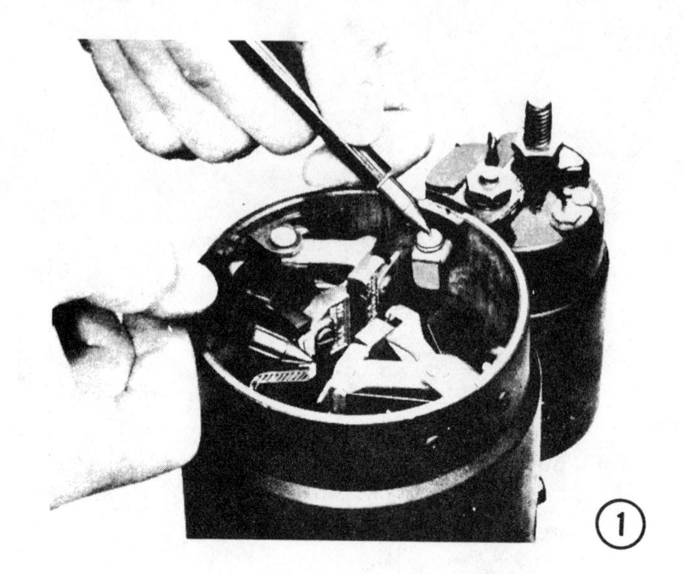

1- Make contact with one probe of a test light on each end of the field coils connected in series. If the test light fails to come on, there is an open in the field coils and repair or replacement is required.

2- Disconnect the shunt coil or coil ground. Make contact with one probe of the test light on the connector strap and on the field frame with the other probe. If the test light comes on, the field coils are grounded and the defective coils must be repaired or replaced.

3- Disconnect the shunt coil grounds. Make contact with one probe of the test light on each end of the shunt coil, or coils. If the light fails to come on, the shunt coil is open and must be repaired or replaced.

4- **ALWAYS** replace the drive end bushing during a starter overhaul.

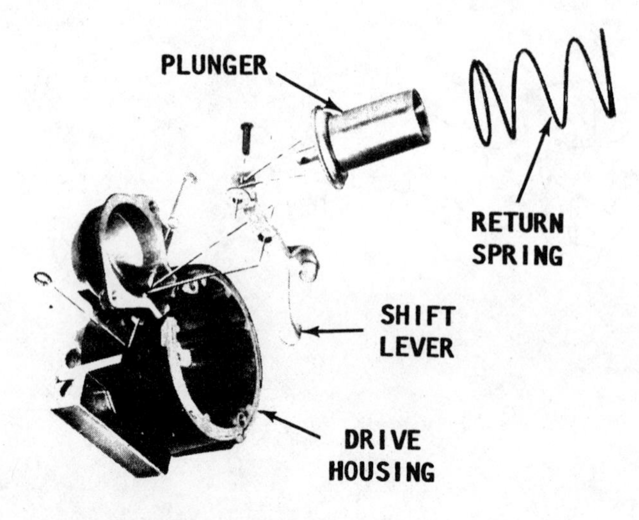

Arrangement of the drive housing and related nose cone parts of a Delco cranking motor.

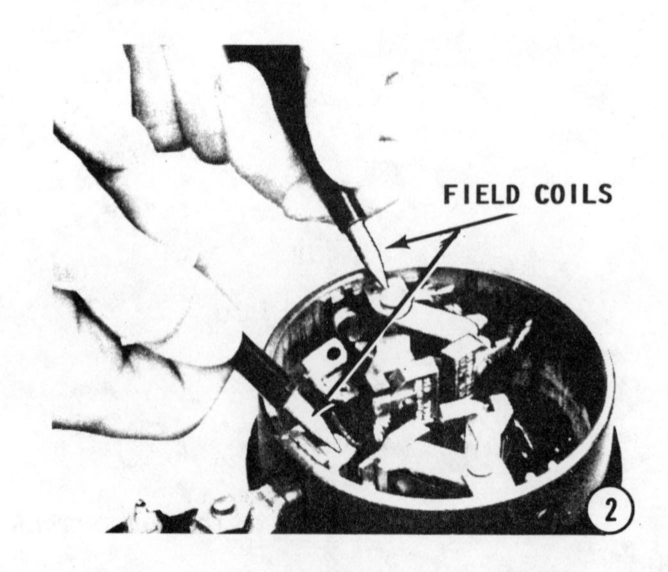

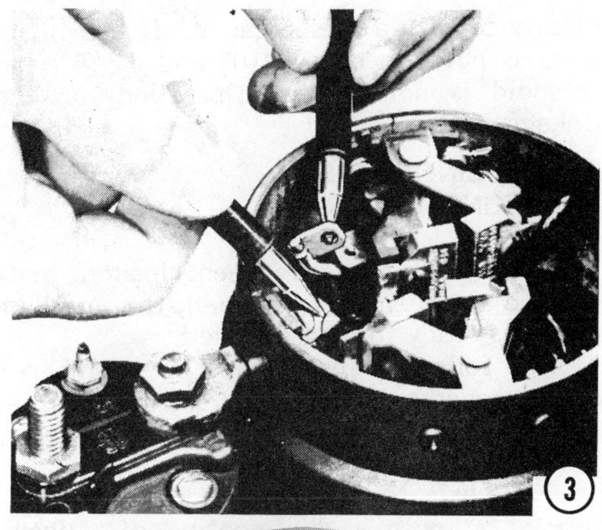

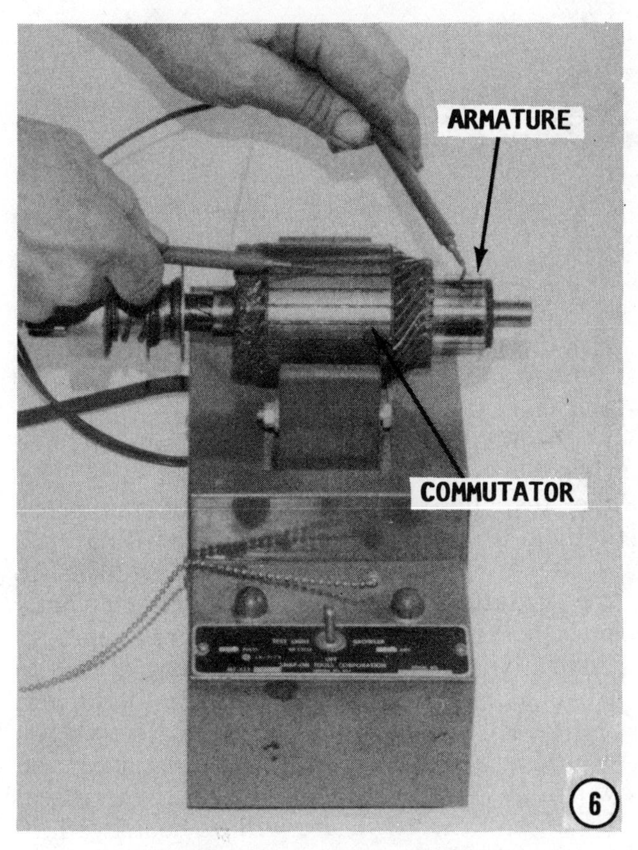

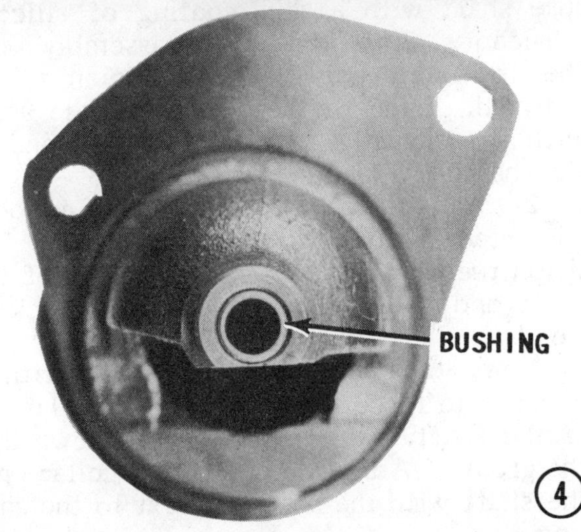

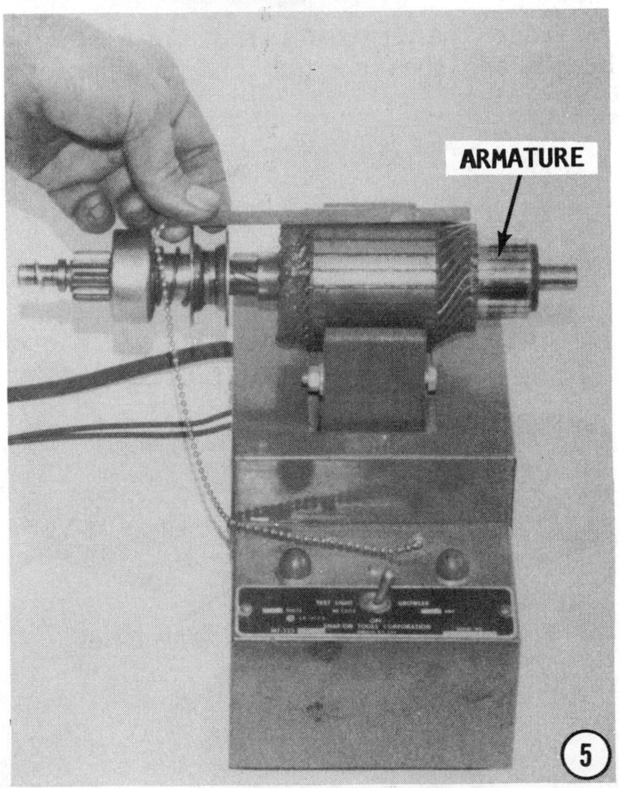

5- True the commutator, if necessary, in a lathe. **NEVER** undercut the mica because the brushes are harder than the insulation. Check the armature for a short circuit by placing it on a growler and holding a hack saw blade over the armature core while the armature is rotated. If the saw blade vibrates, the armature is shorted. Clean between the commutator bars, and then check again on the growler. If the saw blade still vibrates, the armature must be replaced.

6- Make contact with one probe of the test light on the armature core or shaft and

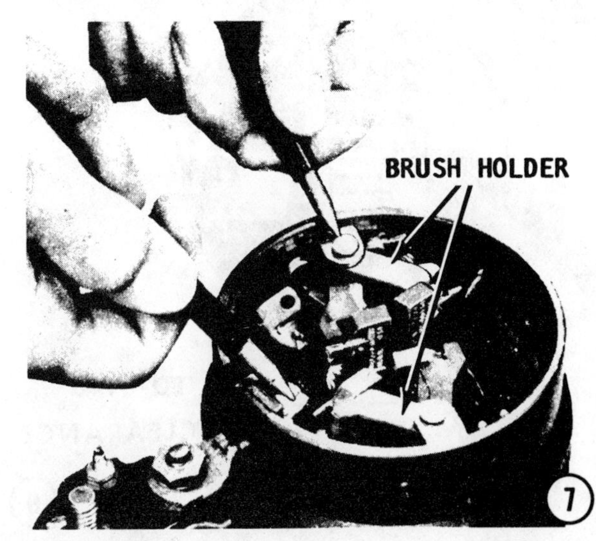

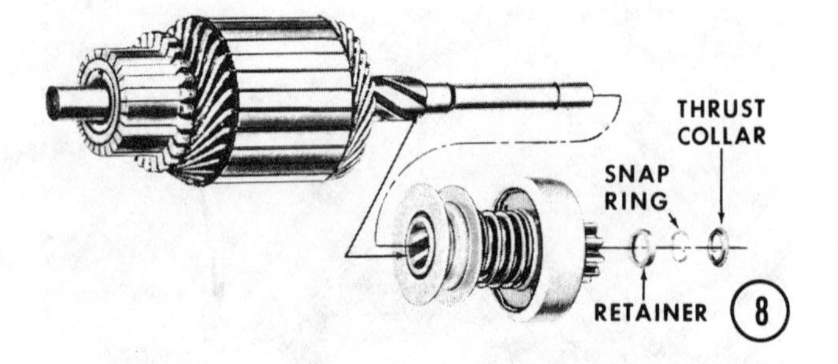

the other probe on the commutator. If the light comes on, the armature is grounded and must be replaced.

7- Wash the brush holder in solvent, and then blow it dry with compressed air. Use the test light to verify two of the brush holders are grounded and two are insulated.

8- The overrunning clutch is secured to the armature by a snap ring. This ring may be removed if the clutch requires replacement. **NEVER** wash an overrunning clutch in solvent or the lubricant will be dissolved; the clutch will fail; the engine will drive the cranking motor armature at high speed; the windings will be thrown out by centrifugal force; and the armature will be destroyed.

9- Check the clearance between the pinion and the retainer. **NEVER** use a 12-volt battery while making this check or the armature will turn. Using a **6-VOLT** battery, energize the solenoid coil; push the pinion back to take up the slack; and measure the clearance with a feeler gauge. If the clearance is not between 0.010" and 0.140", the solenoid is not properly installed, or the linkage is worn.

PRESS ON CLUTCH
AS SHOWN TO
TAKE UP MOVEMENT
PINION
RETAINER
.010" TO .140"
PINION CLEARANCE
FEELER
GAUGE
9

ASSEMBLING A DELCO CRANKING MOTOR

After all parts have been cleaned, tested, and replacements obtained, the cranking motor is ready to be assembled.

The following instructions are numbered and matched to numbered illustrations as an aid in performing the work.

1- Lubricate the drive end of the armature shaft with a thin coating of silicone lubricant. Slide the clutch assembly onto the armature shaft with the pinion facing outward. Slide the retainer onto the shaft with the cupped surface facing the end of the shaft (away from the pinion).

2- Stand the armature on end on a wooden surface with the commutator down. Position the snap ring on the upper end of the shaft and hold it in place with a block of wood.

Now, tap on the wood block with a hammer to force the snap ring over the end of the shaft. Slide the snap ring down into the groove. Assemble the thrust collar onto the shaft with the shoulder next to the snap ring.

Place the armature flat on the work bench, and then position the retainer and

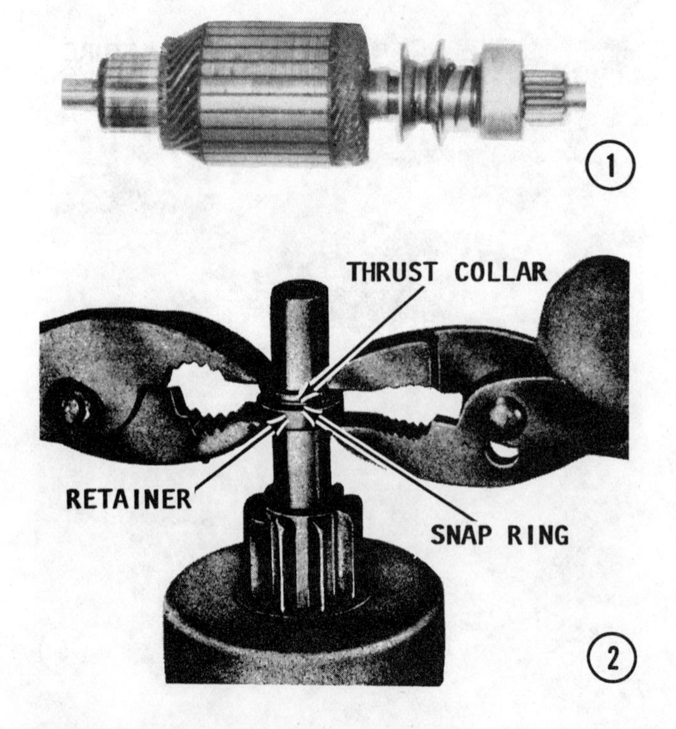

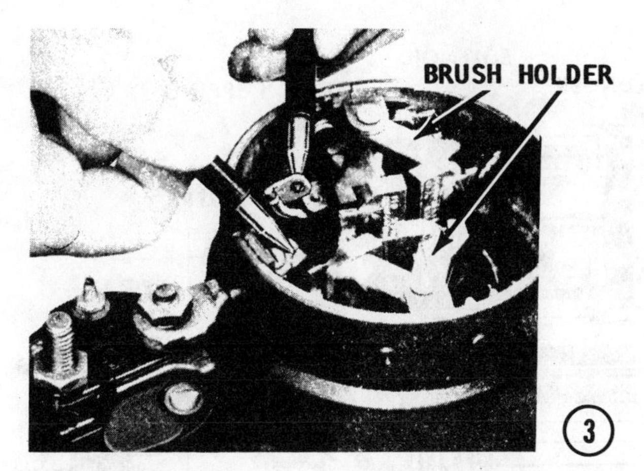

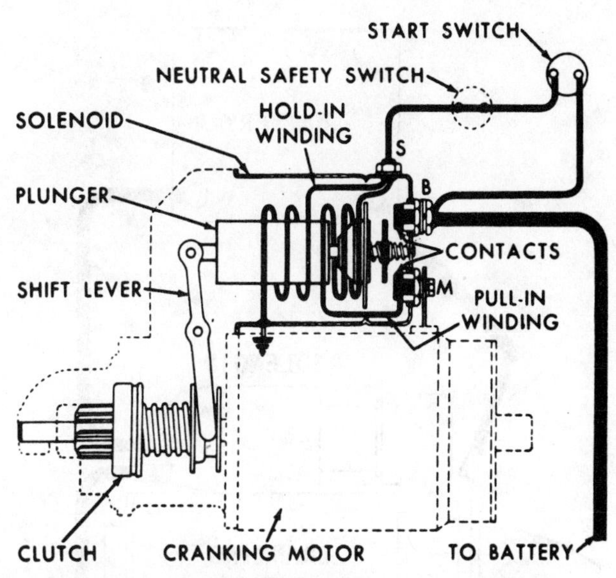

Functional diagram of a Delco solenoid circuit.

thrust collar next to the snap ring. Next, using two pair of pliers at the same time (one pair on each side of the shaft), grip the retainer and thrust collar and squeeze until the snap ring is forced into the retainer.

Lubricate the drive housing bushing with a thin coating of silicone lubricant. **MAKE SURE** the thrust collar is in place against the snap ring and the retainer.

3- Install the brushes into the brush holders. Assemble the insulated and grounded brush holders together with the "V" spring and position them as a unit on the support pin. Push the holders and springs to the bottom of the support, then rotate the springs to engage the "V" in the support.

Attach the ground wire to the grounded brush and the field lead wire to the insulated brush.

4- Slide the armature and clutch assembly into place in the drive housing engaging the shift lever with the clutch.

Position the field frame over the armature and apply a thin coating of liquid neoprene (Gaco) between the frame and the solenoid case. Place the frame in position against the drive housing. **TAKE CARE** not to damage the brushes.

Apply a coating of silicone lubricant to the bushing in the commutator end frame. Place the leather brake washer onto the armature shaft and slide the commutator end frame onto the shaft. Connect the field

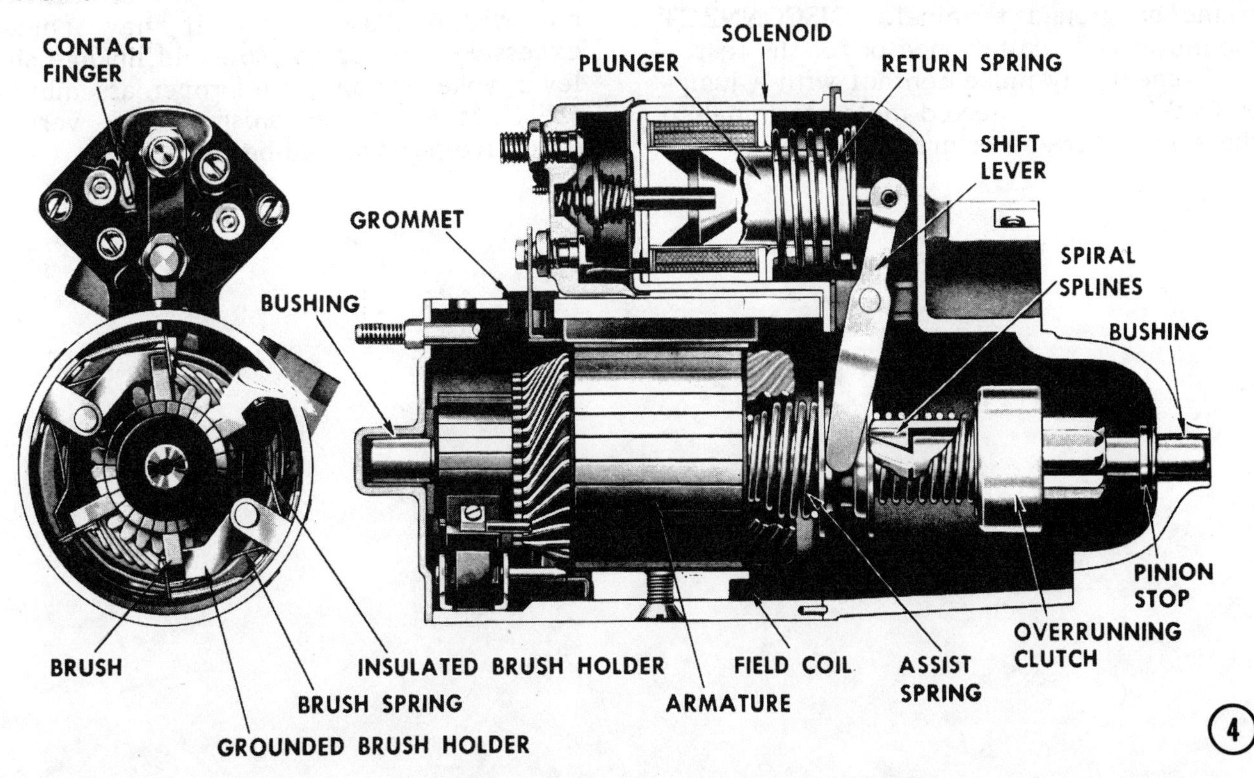

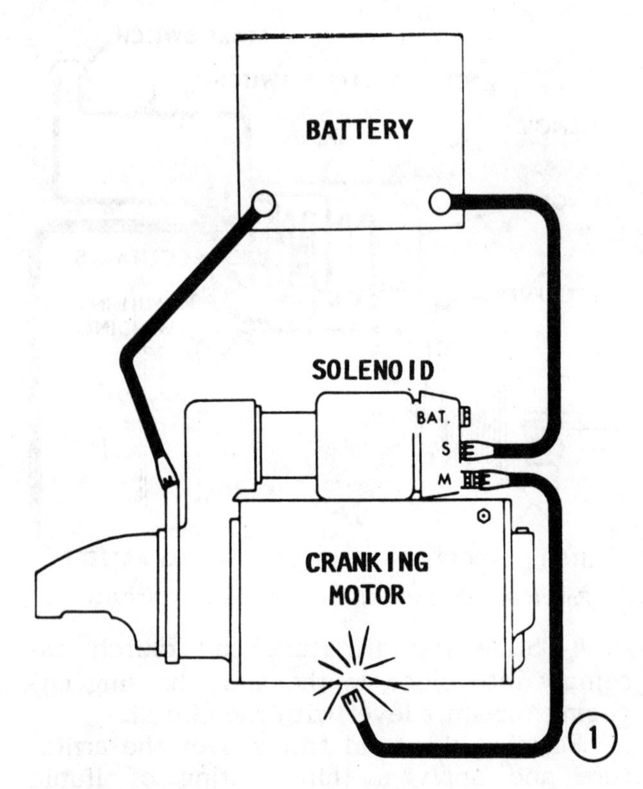

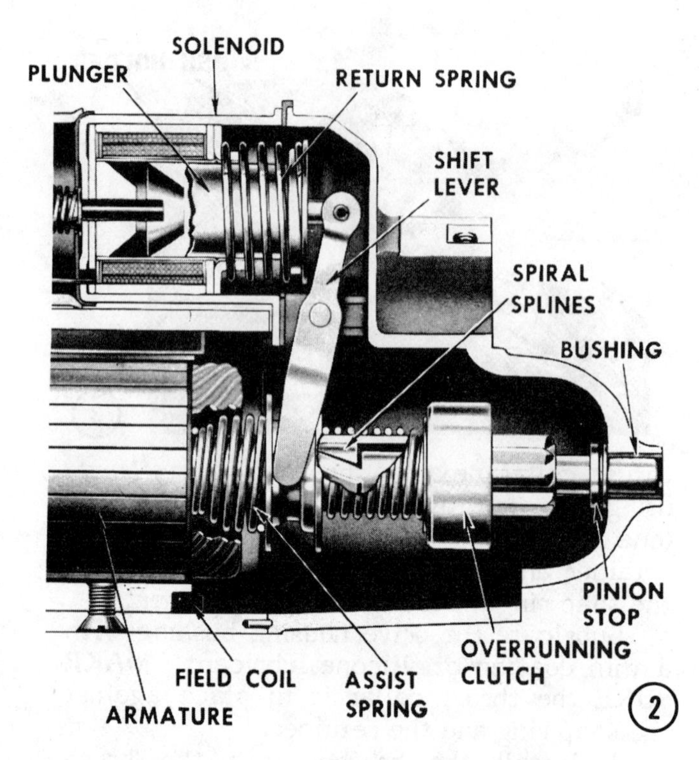

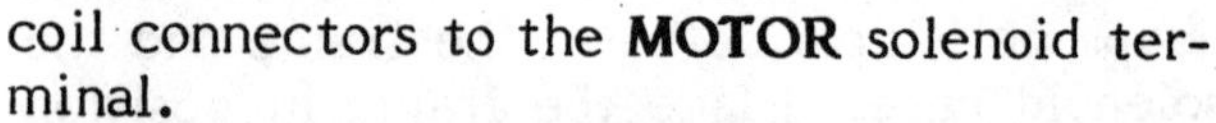

coil connectors to the **MOTOR** solenoid terminal.

Pinion Clearance Check:

After the cranking motor has been assembled, check the pinion clearance as follows:

1- Connect a 12-volt battery across the solenoid switch terminal to the solenoid frame or ground terminal. **DISCONNECT** the motor field coil connector for the test.

Momentarily make contact with a jumper lead from the solenoid motor terminal to the solenoid frame or ground terminal. The pinion will now shift into the cranking position and it will remain there until the battery is disconnected.

2- Push the pinion back towards the commutator end to eliminate any slack movement. Now, measure the distance between the pinion and the pinion stop with a feeler gauge. The clearance should be between 0.10" and 0.040". If the clearance is not within these limits, it may indicate excessive wear of the solenoid linkage shift lever yoke buttons or improper assembly of the shift lever mechanism. Any worn or defective parts should be replaced.

7
COOLING

7-1 DESCRIPTION

Cooling is a critical phase of engine operation. Cool water passing through the cylinder block and head to draw off heat generated by cylinder operation and the friction of parts moving at incredible speed is essential to satisfactory performance.

Sometimes failure to perform the most simple and no cost task can result in tremendous damage and outrageous cost due to unjustified negligence.

A case in point can be graphically illustrated by the picture of the destroyed AQ130 cylinder block on this page. Lack of maintenance and attention to the cooling system directly caused the engine destruction and the expensive replacement cost.

This AQ130 cylinder block was destroyed by negligence of the owner when the "raw" water strainer became clogged and the engine overheated, as explained in the text.

The sequence of events were as follows: The "raw" water filter was not checked at regular intervals and became partially clogged with debris drawn in from outside the boat.

The restriction of adequate "raw" water circulating through the block caused the engine to overheat. The overheated condition was not observed by the boat operator and the head gasket "blew" allowing water to enter the No. 4 cylinder. The engine was operating at high rpm with water skiers in tow behind the boat.

Water will not compress, and very quickly a hydrostatic condition developed forcing the piston rod to bend and soon break. At high rpm the rod was immediately "blown" through the block with the damage shown.

Perhaps this picture and brief explanation will convince the reader of the importance for proper attention to maintenance areas on his boat.

OHC Engine

The cooling system for the 4-cylinder engine actually consists of two water systems. Fresh water is circulated through the cylinder block and head for direct cooling purposes. This fresh water passes through a heat exchanger where heat is drawn off by outside "raw" water circulating through the exchanger.

Both the fresh water and the "raw" water systems have separate reservoirs and pumps to move the water. A thermostat controls the temperature of the water circulating through the engine. Engine lubricating oil is pumped from the crankcase pan under pressure through an oil cooler mounted on the outside of the cylinder block on the starboard side. From the cooler, the oil passes through a full-flow through type fil-

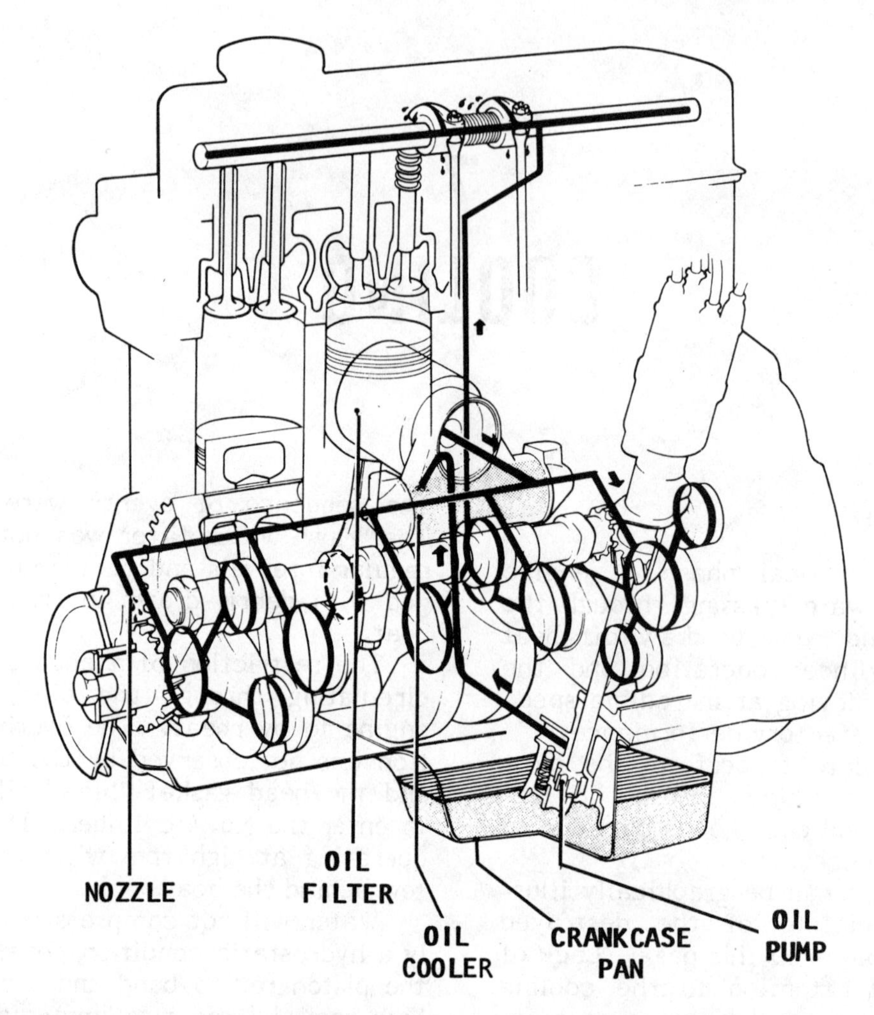

Oil circulation routing for the AQ105, AQ115, AQ130, AQ165, and AQ170 engines. Oil movement in other Volvo Penta engines is similar.

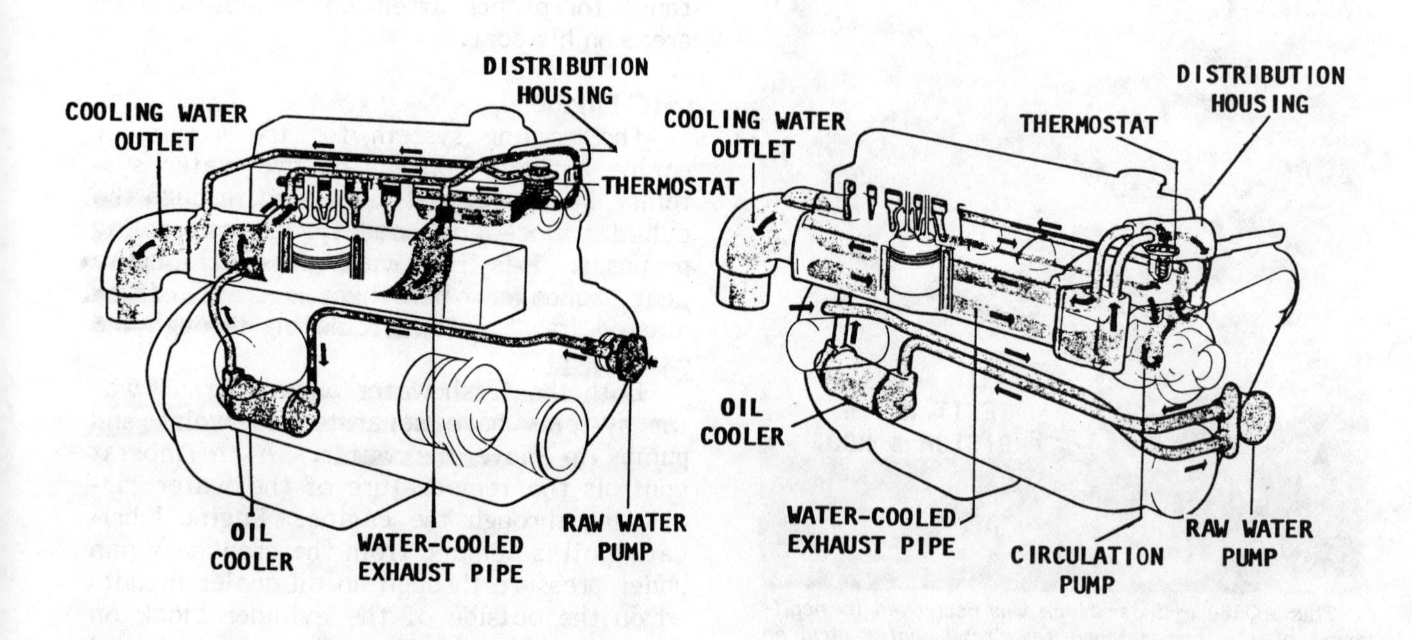

Water circulation routing for the AQ105, AQ115, and AQ130 engines (left), and the AQ165 and AQ170 engines (right).

Cooling equipment installed on early model AQ140 engines. The oil cooler is just aft the oil filter.

Cooling equipment installed on late model AQ145 engines. The oil cooler is visible.

ter and then on to the various lubricating points of the engine. The temperature of the oil is reduced as it circulates through the cooler by "raw" water passing through. The oil cooler is the first area served by the "raw" water drawn in by the "raw" water pump from outside the boat.

On early model engines, the raw water filter was exposed above the reservoir. On late model units, the filter is encased in a housing, as shown in the accompanying illustrations.

In-Line Pushrod Engine

The cooling system, for the in-line pushrod engines, consists of a "raw" water pump to draw water in from outside the boat, a simple screen-type filter, and a thermostat to control the temperature of the water circulating through the cylinder block and head.

To remove the circulating water pump, the fresh water tank must be moved aside; the hoses disconnected; the pump pulley removed; and then the pump removed.

A "raw" water oil cooler is mounted on the side of the cylinder block on the starboard side aft.

Oil cooler installed on in-line pushrod engines. Several items have been removed from this cylinder block.

"Raw" water strainer installed on the starboard side forward on this AQ170 engine.

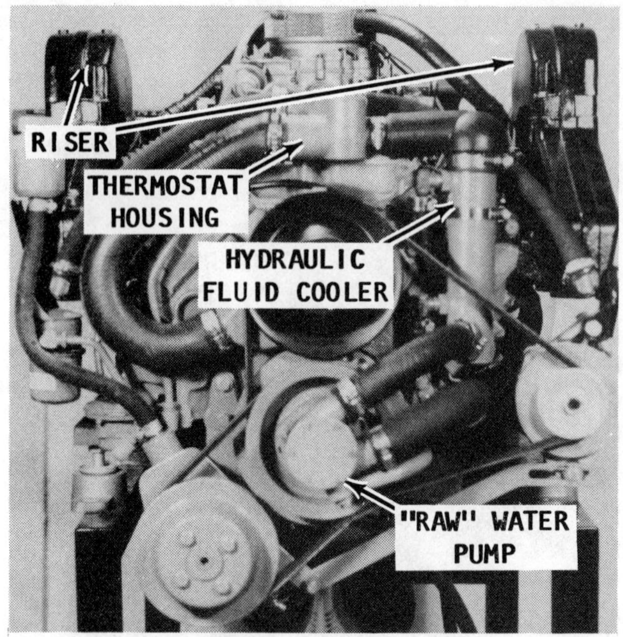

Location of cooling equipment on the AQ290 engine. The oil cooler is installed on the port side forward. The engine cooler is mounted on top of the bell housing.

V6 and V8 Engines

The V6 and V8 engines are cooled in much the same manner as the in-line units. A "raw" water pump pulls water in from outside the boat, and then forces it through a thermostat and distribution housing, where it is circulated through the cylinder block and head by a circulation pump. The heated water is exhausted through the risers, exhaust manifolds, and then overboard through the stern drive.

The AQ290 engine is equipped with a "raw" water oil cooler mounted on the port side forward.

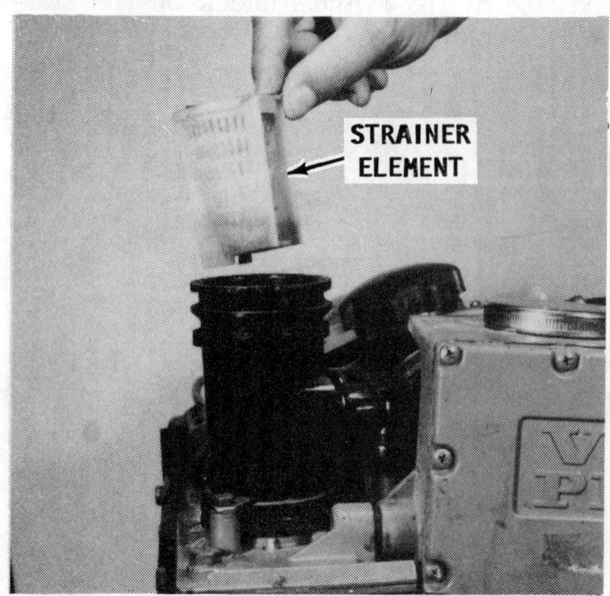

The "raw" water strainer should be checked on a ***DAILY*** *basis when operating in dirty water.*

7-2 "RAW" WATER STRAINER

If the boat is being operated in fairly contaminated water, the strainer should be checked on a daily basis. If the flow of "raw" water is restricted, heating problems will develop very quickly with disastrous results, see the picture on the first page of this chapter.

Strainer with Heat Exchanger

On engines equipped with a heat exchanger, the strainer is located on top of the "raw" water reservoir. To clean the strainer, simply remove the cap and lift the strainer up and out of the housing. On late model units with a heat exchanger a rubber ring is installed over the strainer housing. This rubber ring **MUST** be installed with the word **UP** on the tab facing upward. If the strainer, rubber ring, or the top of the fresh water tank are not installed properly, the cap for the top of the strainer cannot be installed and rotated into the locking position.

The cover for the strainer **MUST** be lubricated at regular intervals. Remove the rubber washer and the sealing disc, see the accompanying exploded drawing on Page 7-5 of the heat exchanger. Coat the inside of the cover evenly with grease. Insert the sealing disc into the cover, and then the rubber washer. Screw the cover into place.

On in-line pushrod engines, beginning about 1973, a "raw" water strainer is installed to filter out sizeable foreign matter from the water drawn in by the "raw" water

The water strainer is installed on the port side forward on this AQ130 engine.

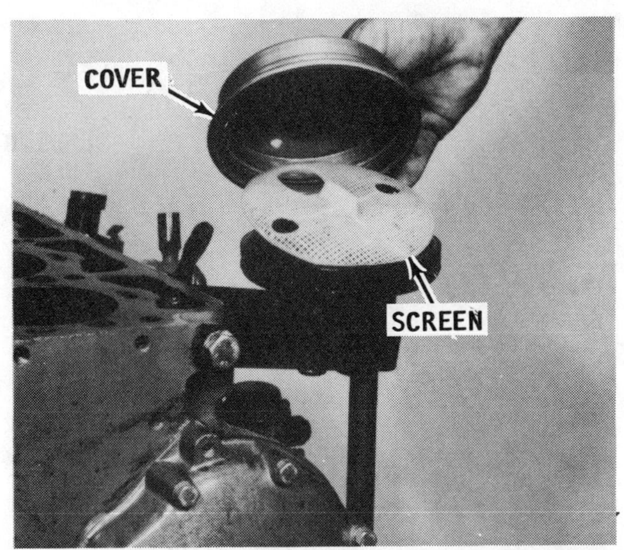

The nuts on the cover should only be fingertight. Pop the cover off by hand and clean the strainer screen.

pump. Water passes through the strainer before reaching the pump. On some units the strainer was installed on the port side forward and on other units on the starboard side forward.

The strainer consists of a simple housing containing a plastic screen-type filter insert. To remove the cover, back off the two nuts on top. These nuts should have been installed just fingertight. Press upward on the cover with finger pressure and the cover should come free. **DO NOT** use a tool in an attempt to pry the cover free.

Preparing to remove the heat exchanger on an AQ140 engine.

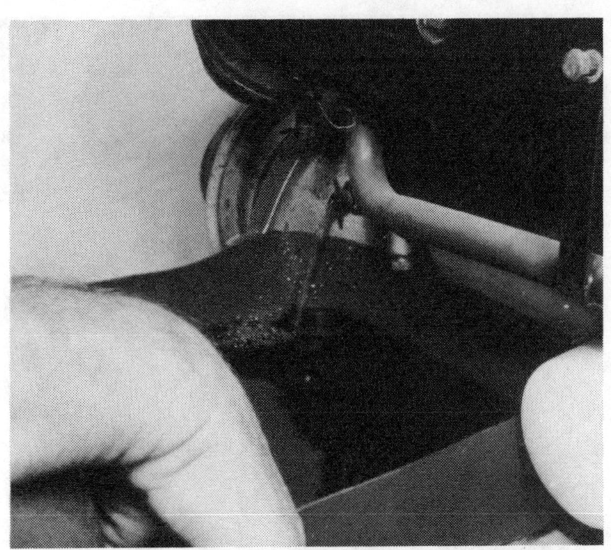

Draining the coolant from the cylinder block in preparation for service of the cooling system.

Lift out the strainer screen and clean it thoroughly. Install the strainer with the tapered projection facing downward, and at the same time press down on the cover. Secure the cover in place with the two nuts and tighten them **ONLY** with your fingers.

7-3 HEAT EXCHANGER

The heat exchanger is a conventional type with a wafer-type filter. Heat from the fresh water cooling solution is drawn off by outside "raw" water passing through the exchanger.

Removing/installing the heat exchanger.

Tightening attaching bolts for the fresh water tank.

The heat exchanger is located in an exposed position making removal for service and installation a comparatively simple matter. No other equipment need be disturbed.

Two plugs are located in the bottom of the heat exchanger. The small plug drains the "raw" water, and the larger plug drains the fresh water from the heat exchanger.

However, these plugs seem difficult to remove after the unit has been in service for some time. Therefore, if they are not released easily, secure a suitable container and drain the coolant from the block by opening the petcock on the starboard side of the block, aft, and just above the crankcase pan line, as shown. Open the fresh water fill cap to allow air to enter for faster draining.

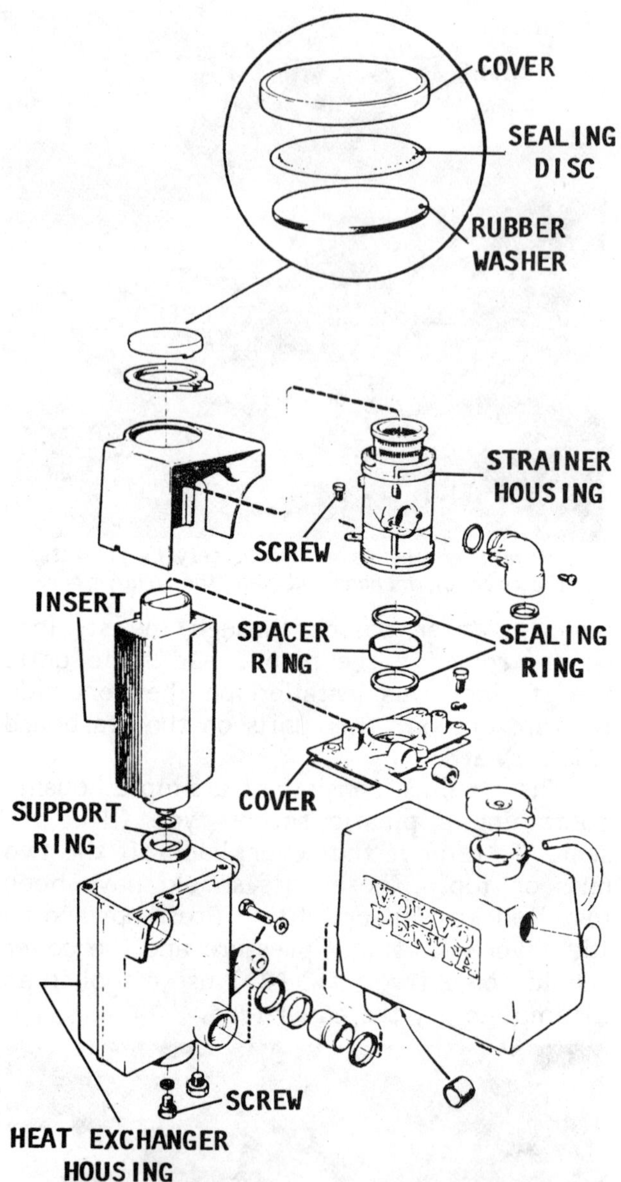

Exploded drawing of the heat exchanger.

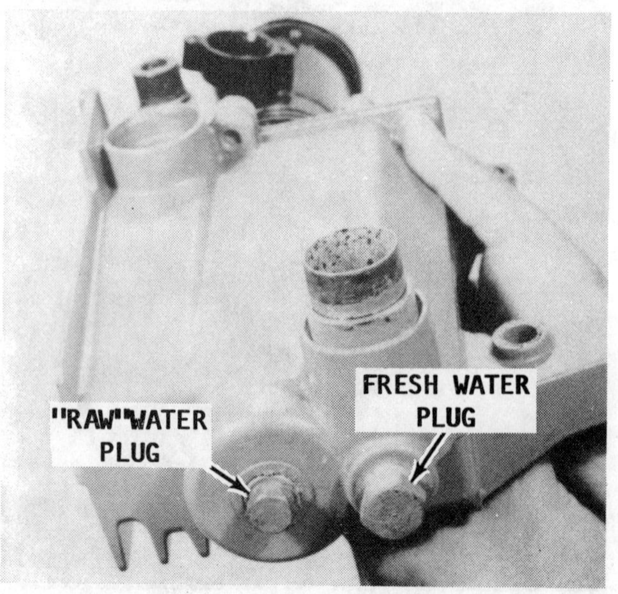

Heat exchanger showing the drain plugs.

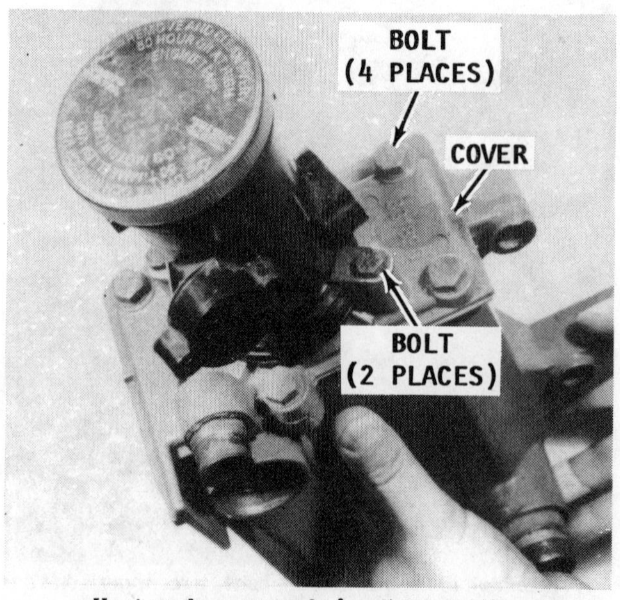

Heat exchanger ready for disassembling.

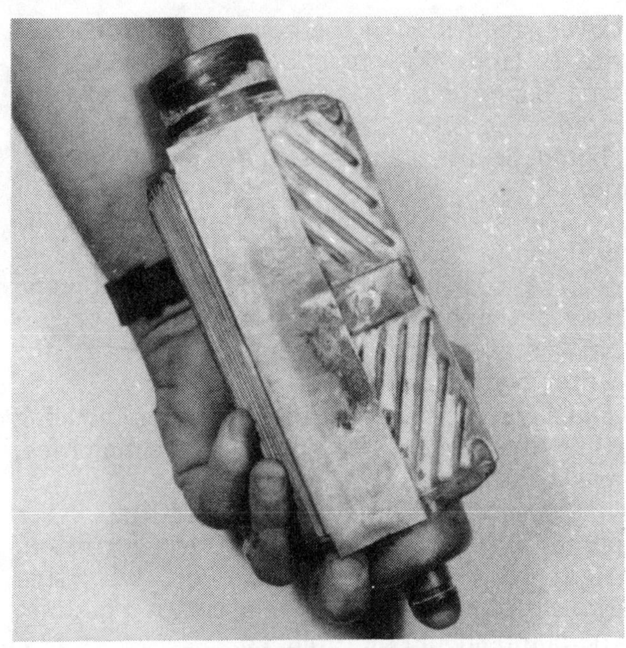

Wafer-type element from the heat exchanger.

The recommended coolant is a mixture of 50% fresh water and 50% anti-freeze.

Remove the bolts attaching the fresh water tank to the cylinder block. Three bolts are threaded into the block, one bolt on the strap for the water pipe, and two bolts to disconnect the water pipe from the strainer. On the other side of the engine, there are two bolts thru the thermostat housing.

Release the strap securing the water pipe in place.

Remove the fresh water tank.

Remove the two screws securing the water pipe elbow to the heat exchanger, and then remove the heat exchanger.

The heat exchanger has a wafer-type element. A clogged element is very difficult to clean thoroughly for optimum service. Therefore, if it cannot be cleaned to your satisfaction, purchase and install a new element and remove this area as a source of problems for some time.

To gain access to the wafer-type element in the heat exchanger, first remove the four bolts securing the top cover to the heat exchanger.

Next, remove the two bolts in the neck of the "raw" water strainer. Remove the top cover and the gasket. Withdraw the element from the heat exchanger.

During installation use **NEW** sealing rings, O-rings, and gasket.

7-4 "RAW" WATER PUMP

The "raw" water pump on all Volvo Penta engines is mounted in an exposed position on the front of the engine and is easily accessible for service.

To remove the impeller, first remove the bolts securing the front cover to the pump. After the cover is free, the impeller may be worked out using a couple screwdrivers, one on either side of the impeller. **TAKE CARE** not to damage the pump body.

Check to be sure the key is in the shaft keyway. Slide the new impeller onto the shaft, with the keyway in the impeller indexed over the key. Continue forcing the impeller into the pump body with hand pressure. It is not important how the blades of the impeller face, because they will flop over into the correct position as soon as the pump rotates. The pump is self-priming, therefore water will be drawn in at engine startup.

Position the front cover onto the pump with a **NEW** gasket and secure it in place with the attaching hardware.

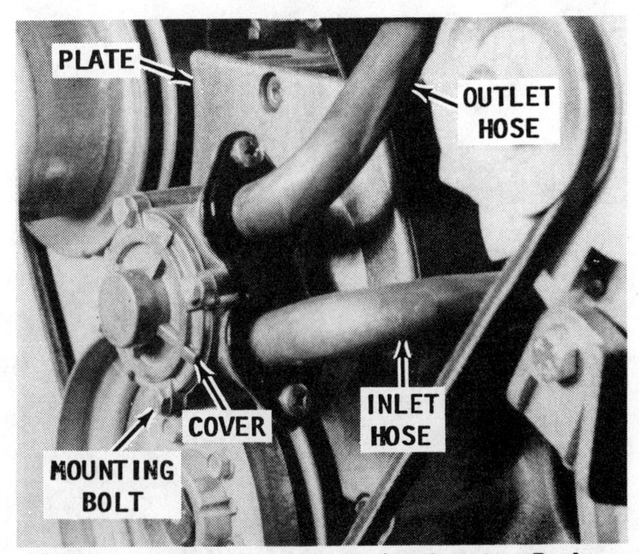

"Raw" water pump and associated parts. Replacement of the impeller is not a difficult task.

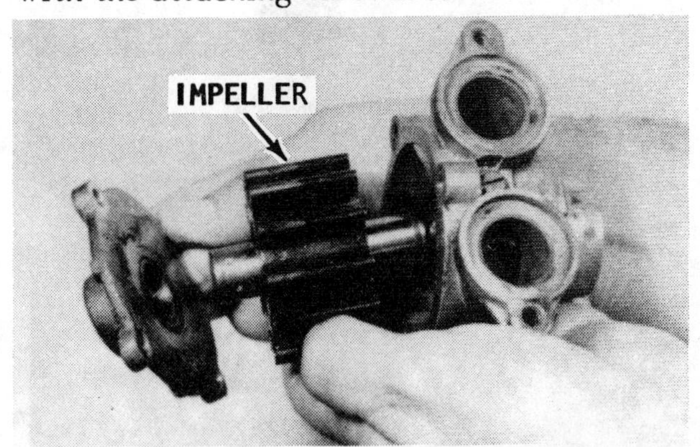

*"Raw" water pump being assembled with a **NEW** gasket, as described in the text.*

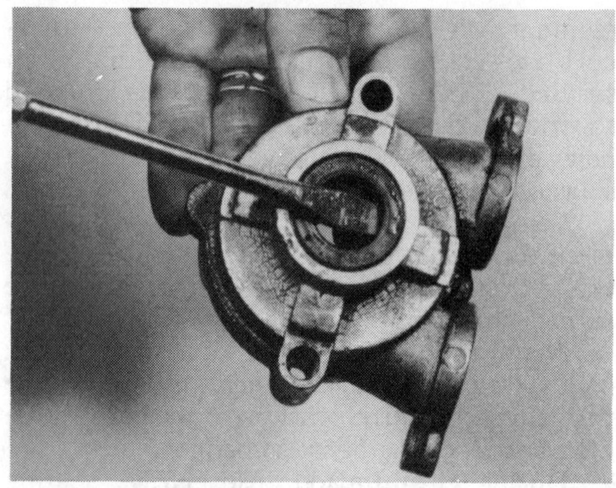

Removing the seal from the back side of the pump, by "popping" it out with a screwdriver.

If the pump needs to be replaced or serviced, beyond replacing the impeller, proceed as follows:

Remove the two bolts through the pump housing and into the plate secured to the cylinder block. Remove the two screws securing the plate holding the inlet and outlet copper tubing to the pump. Remove the pump.

The pump may be disassembled by removing the two screws securing the front cover, then jarring the cover loose to separate it from the pump body. Discard the thin paper gasket, because it can be damaged so easily. Therefore, a **NEW** gasket should be used anytime the pump is serviced.

Installing a new seal into place on the back side of the pump using the correct size socket and a hammer.

The shaft and the impeller may now be pushed out of the body from the back side.

The seal in the back side of the pump may be popped free with a screwdriver. A **NEW** seal should be installed whenever the water pump body is removed.

Drive the new seal into the pump housing using the proper size socket and a hammer, as shown.

Inspect the inside surface of the water pump for scores, damage or corrosion. Light scores can be polished out with fine grain emory cloth. If this is not possible, the pump should be replaced.

Place the key in the shaft keyway. Slide the impeller onto the shaft with the keyway in the impeller indexed over the key. Insert the shaft through the housing with hand pressure. Place a **NEW** gasket in position between the cover and the pump body. Attach the cover to the pump with the four bolts. Tighten the bolts securely.

Move the pump into position with the pump shaft indexed with the driveshaft. Secure the pump in place with the four bolts through the pump body into the plate attached to the cylinder block.

Connect the two copper water tubes to the pump and secure them in place with the plate and two attaching bolts. The plate may be installed in either of two positions.

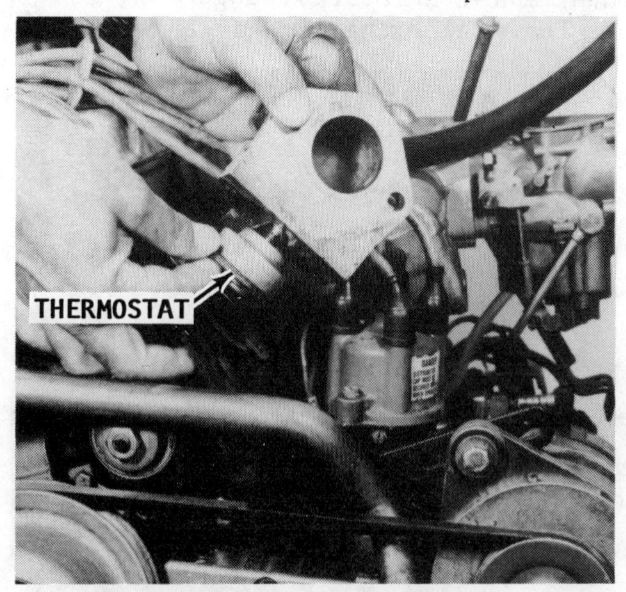

The thermostat can only be installed correctly one way, or the housing will not fit flush against the block.

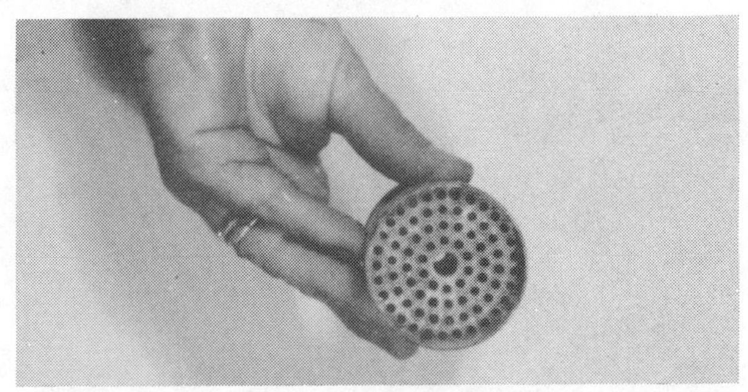

The oil cooler element may be "rodded" out with a piece of welding rod or other suitable material.

7-5 THERMOSTAT

A conventional thermostat is installed on all engines to control the water temperature passing through the cylinder block and head. The thermostat is manufactured to begin opening at 180°F (82°C) and be fully open at 198°F (92°C) If necessary the thermostat can be checked in a pan of hot water with a thermometer.

To remove the thermostat, begin by disconnecting the copper tubing to the thermostat. Remove the two mounting bolts and lift the thermostat housing. The thermostat may come with the housing or it may remain in the cylinder block. The thermostat has a rubber seal. The thermostat **MUST** be installed with the seal against the block. The spring side of the thermostat goes into the block. Actually the thermostat can only be installed properly one way. If it is installed backwards, the housing will not fit down flush against the block.

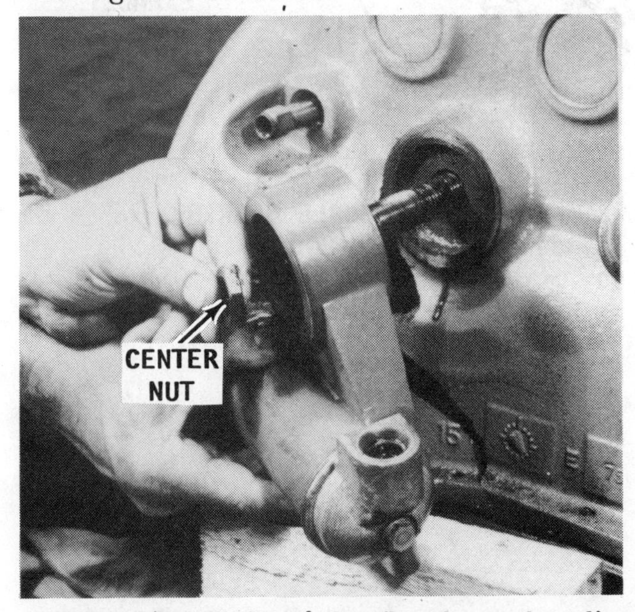

Removal/installation of the oil cooler to the cylinder block. Do not tighten the center nut until the water tubes have been connected.

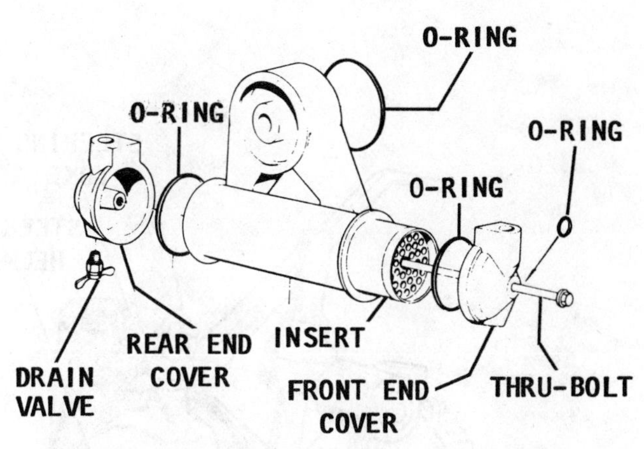

Exploded drawing of the oil cooler. Always use new O-rings and a new sealing ring following service.

7-6 OIL COOLER

On in-line pushrod engines, an oil cooler is installed on the starboard side of the cylinder block just aft of the oil filter. This cooler may be removed for cleaning. On some pushrod engines the cooler is mounted in the horizontal position and on others it is mounted vertically.

On the OHC engine, the cooler is mounted in the vertical position, on the starboard side forward, just aft of the oil filter.

The AQ290 engine has an oil cooler installed on top the bell housing. On units equipped with power steering, a fluid cooler is mounted on the port side forward.

To remove the oil cooler for service on the in-line engine, first, remove the oil filter. Next, disconnect the inlet and outlet tubing to and from the cooler. Now, remove the center nut and pull the cooler free of the cylinder block.

To clean the element remove the thru-bolt from one end, and then separate the two ends from the cooler body. Withdraw the element. The element may be "rodded" out with a stiff piece of welding rod, stiff wire, or other suitable device.

After cleaning, insert the element into the cooler body. Place both ends in position with **NEW** O-rings. Secure the cooler together with a **NEW** O-ring under the head of the thru-bolt. Use a **NEW** O-ring when installing the cooler onto the cylinder block.

DO NOT tighten the nut securing the oil cooler to the cylinder block until after the water tubes to and from the cooler have been connected. It may be necessary to loosen the thru-bolt slightly in order to rotate the top or bottom covers a bit, when connecting the tubes.

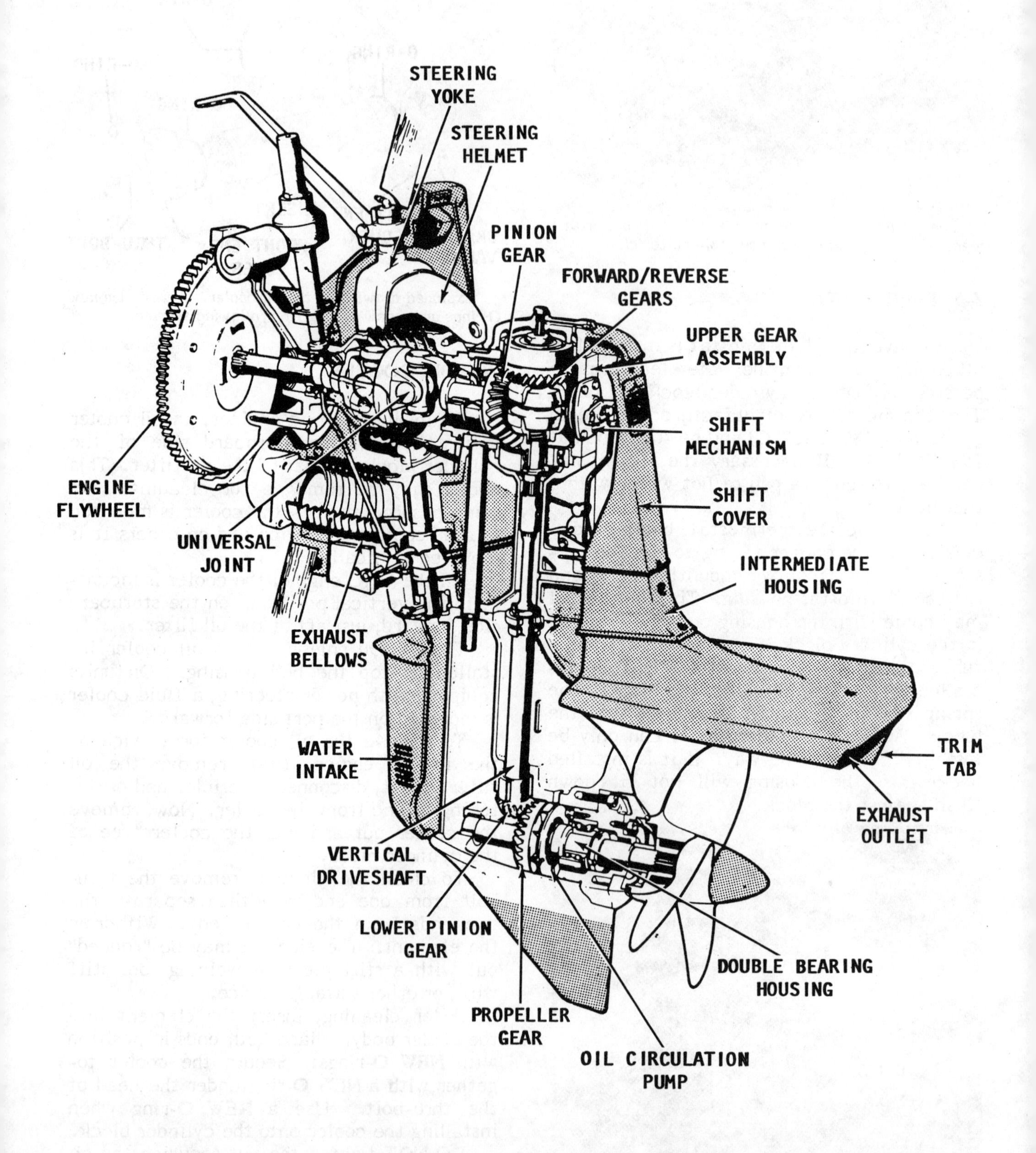
STEERING YOKE
STEERING HELMET
PINION GEAR
FORWARD/REVERSE GEARS
UPPER GEAR ASSEMBLY
SHIFT MECHANISM
SHIFT COVER
INTERMEDIATE HOUSING
ENGINE FLYWHEEL
UNIVERSAL JOINT
EXHAUST BELLOWS
WATER INTAKE
TRIM TAB
EXHAUST OUTLET
VERTICAL DRIVESHAFT
LOWER PINION GEAR
DOUBLE BEARING HOUSING
PROPELLER GEAR
OIL CIRCULATION PUMP

8
STERN DRIVE

8-1 INTRODUCTION

Volvo Penta stern drive units covered in this manual are the 250, 270, 275, 280, and the 290. The model number of each unit is prominently displayed on the shift cover. The detailed procedures in this chapter contain complete instructions for these units.

The information is divided into the following broad areas:

Operation
Troubleshooting
Propeller Removal and Installation
Stern Drive Removal and Installation
Upper Gear Assembly Service
Intermediate Housing Service
Lower Unit Service

Each section is a complete unit with detailed instructions and illustrations to accomplish the entire task of servicing the total unit. In this manner the work can be stopped at any point if further service is not required. Therefore, it is possible to skip unnecessary disassembling and assembling work and jump to the steps involved in returning the unit to full operational capability.

OPERATION

Power from the engine is transferred through a splined universal joint to the upper gear assembly of the stern drive unit. Splines on the forward end of the universal joint shaft mate with matching splines of the intermediate shaft in the engine bell housing.

From the universal joint, power is transferred through a pinion gear to either the forward gear or the reverse gear in the upper gear assembly. These helical gears are manufactured and sold as matched sets. A cone clutch with tapered mated surfaces is used to shift from one gear to the other or to the neutral position.

The forward and reverse gears change the direction of power 90° through a vertical shaft and coupling to the lower unit. A pinion gear on the lower end of the vertical shaft transfers power through a mating gear to the propeller shaft. The pinion gear on the vertical shaft and the driven gear on the propeller shaft may be changed, affecting the gear ratio, to fully utilize the torque developed by different engine installations. These gears are sold as a matching set and **MUST** be changed and installed as a set.

An electrical/mechanical tilt mechanism allows the stern drive to be raised for operation at **IDLE** speed **ONLY** in shallow water and during approach to the beach. This tilt mechanism is also used to raise the stern drive for trailer transporting to and from the water.

An optional hydraulic trim system is available to permit trim while the boat is at cruising speed. This unit is usually installed only on light weight boats. Its affect on operation of a heavy displacement boat would be negligible.

Engine exhaust is routed aft through an exhaust bellows, then down through the stern drive, and finally out through the cavitation plate under water.

The shift mechanism is contained in the intermediate housing between the upper gear assembly and the lower unit. The shift mechanism is a positive mechanical operation.

8-2 TROUBLESHOOTING

Troubleshooting should be done before the unit is removed from the boat to permit isolating the problem to one area.

Check the propeller and the rubber hub. Check to determine if the hub is shredded. If the propeller has been subjected to many strikes with underwater objects, it could slip on its hub. If the hub appears to be damaged, replace it with a **NEW** hub. Replacement of the hub must be done by a propeller rebuilding shop equipped with the proper tools and experience for such work.

Remove the shift cover and check to be sure the linkage is properly connected and the cotter key is in place.

STERN DRIVE NOISES

When the stern drive is positioned for dead-ahead operation, the U-joints will make very little noise. However, as the stern drive is moved to port or starboard, the noise will increase, due to the working of the U-joints. You will be accustomed to the normal noise of the U-joints. During the following tests, be sensitive to "abnormal" noises.

Noise in lower unit

Propeller not installed properly.
Propeller shaft bent.
Worn, damaged, or loose parts.
Metal particles in lubricant.
Gear alignment incorrect.
Incorrect shimming.

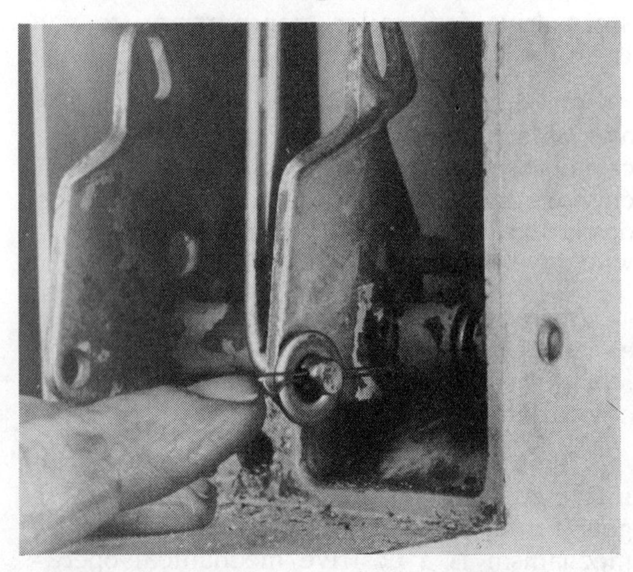

Checking to be sure the shift rod washer and cotter pin are in place.

Noise in upper gear assembly

Lubrication level low.
Worn U-joints.
Worn bearings.
Incorrect shimming.
Gear alignment incorrect.
Engine not aligned properly.
Worn engine coupler.
Boat transom too light for stern drive installation.
Worn, damaged, or loose internal parts.

Connect a flush attachment to the lower unit and turn the water on, or move the boat into a body of water.

CAUTION: Water must circulate through the lower unit to the engine any time the engine is run to prevent damage to the water pump mounted on the engine.

Start the engine and run it at idle speed. Have an assistant turn the stern drive slightly to port and then slightly to starboard, and at the same time you listen at the bellows. An unusual noise is an indication the U-joints are worn or a bearing is defective.

Tilt/Trim System Check

Have an assistant operate the control buttons on the dashboard, and at the same time, check to determine if the stern drive unit moves to the full up and full down positions. If the drive unit fails to move, or

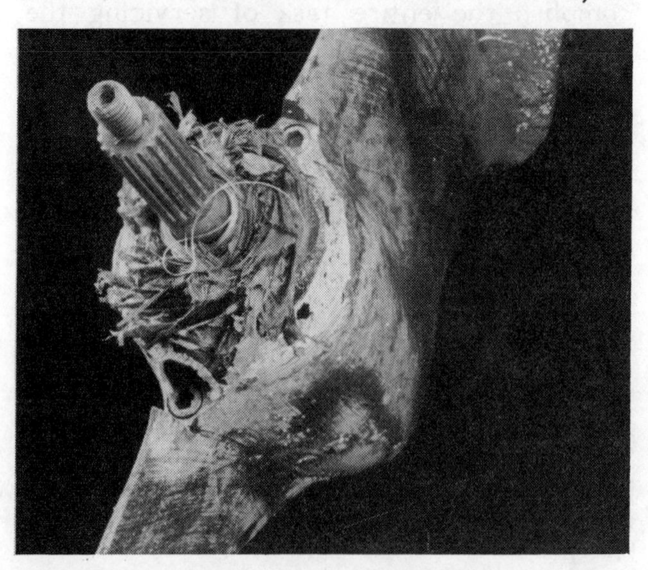

Debris and fish line entangled around the propeller shaft. In this case, the seals were damaged, water entered the lower unit, lubricant escaped, and considerable damage was caused to expensive parts.

Standard installation of the tilt and trim controls.

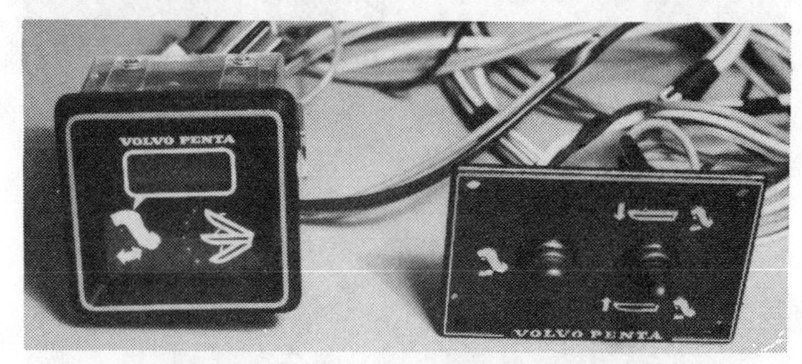

New LCD digital readout dashboard trim unit. The switching mechanism is basically the same as the old style with a push button override for tilt.

if it does not move properly, the tilt/trim mechanism, especially the tilt cylinders, may need service. Damage to the O-rings in the cylinders is the most common cause of problems. See Chapter 9.

SERVICE PRECAUTIONS

The following safety precautions should be observed while servicing the stern drive. The list is presented here as an assist to you in preventing personal injury or expensive damage to parts.

A- Whenever possible, use special tools when they are specified. Always use the proper size common tool when removing or installing attaching parts.

B- Threaded parts are right-hand, unless otherwise indicated.

C- Clean parts in regular cleaning solvent and blow them dry with compressed air. Disassembly and assembly should be done on a clean work bench to prevent contamination (foreign matter, dirt) from entering the hydraulic system, or from becoming attached to parts. Such material can result in false shimming indications and will lead to premature bearing failure.

D- If bearings are removed from assemblies, keep them **TOGETHER**, with the original shims as an aid during assembling. If new parts are used with the assembly, the unit will have to be reshimmed in order to obtain the proper bearing preload or gear lash.

E- Work slowly when removing shims and take time to note the number and location of the shims. Follow shimming instructions to the letter. Gears must be mounted to the correct dial indicator reading and with the correct gear backlash to avoid noisy performance and possibly early failure. Check the specifications for the proper preload. This is the **ONLY** way satisfactory service may be obtained from the unit after the work has been completed. There are **NO** short cuts.

F- Always keep bearing assemblies and spacers together and note the order of removal as an aid during installation.

G- The proper mandrels and supports **MUST** be used to remove pressed-on parts, such as bearings and gears. Therefore, do not use force on the bearing cages, mating surfaces, or edges which could be damaged or cause failure.

H- Lubricate parts with oil or grease to ease installation, retard rust, and prevent corrosion. One exception is the installation of oil seals. These seals must be coated with Loctite "A" to prevent lubrication seepage and to hold the seal in position.

I- When it is necessary to hold a part or housing in a vise, **ALWAYS** use soft brass jaws or wooden blocks to prevent marring or damaging the part.

8-3 PROPELLER SERVICE
STANDARD PROPELLER

REMOVAL

1- If servicing an early model, bend the locking tabs on the lock washer up clear of the slots in the cone. **NEVER** pry on the

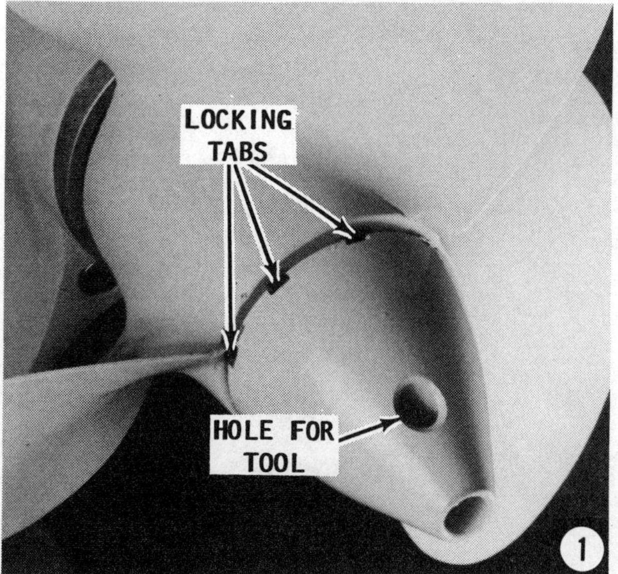

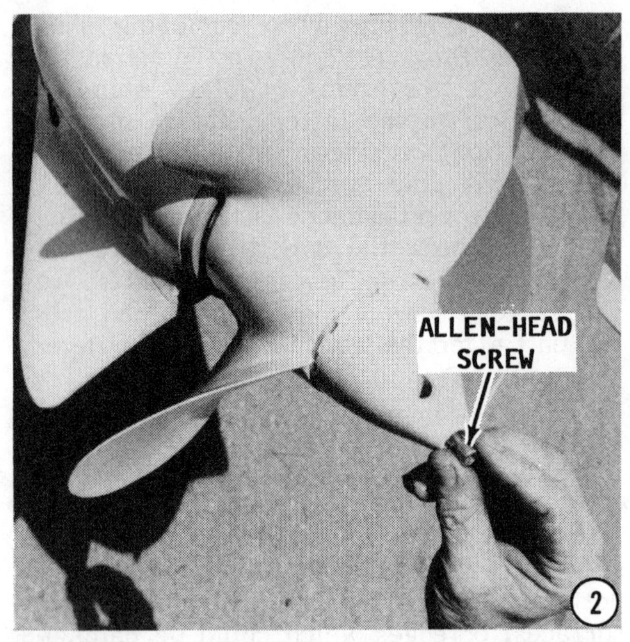

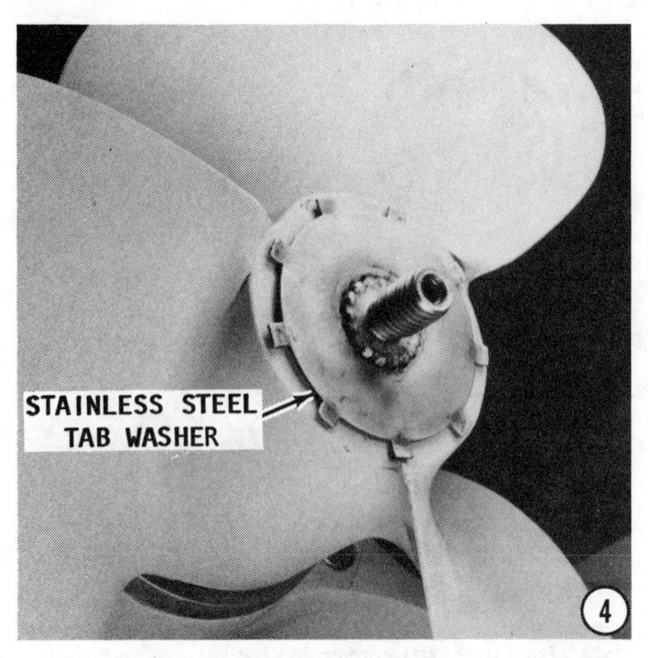

edge of the propeller. Any small distortion will affect propeller performance. Place a block of wood between one blade of the propeller and the anti-cavitation plate to keep the shaft from turning. Insert a stiff round bar, large Phillips screwdriver, or other suitable tool into the hole near the end of the propeller cone and using quick **COUNTERCLOCKWISE** jerks, loosen the cone. Remove the cone from the shaft. Skip Steps No. 2 and 3.

2- On late model propellers, back off the Allen-head screw in the center of the cone.

3- On late model propellers, remove the cone, and then remove the nylon washer. Skip Step No. 4.

4- On early model propellers, remove the stainless steel tab washer.

5- Slide the propeller aft and free of the shaft. Remove the deflector ring.

6- If the zinc ring is to be replaced, remove the Allen-head screws, and then slide the ring back off the propeller shaft.

"FROZEN" PROPELLER

If the propeller is "frozen" to the shaft because an anti-seizing compound was not

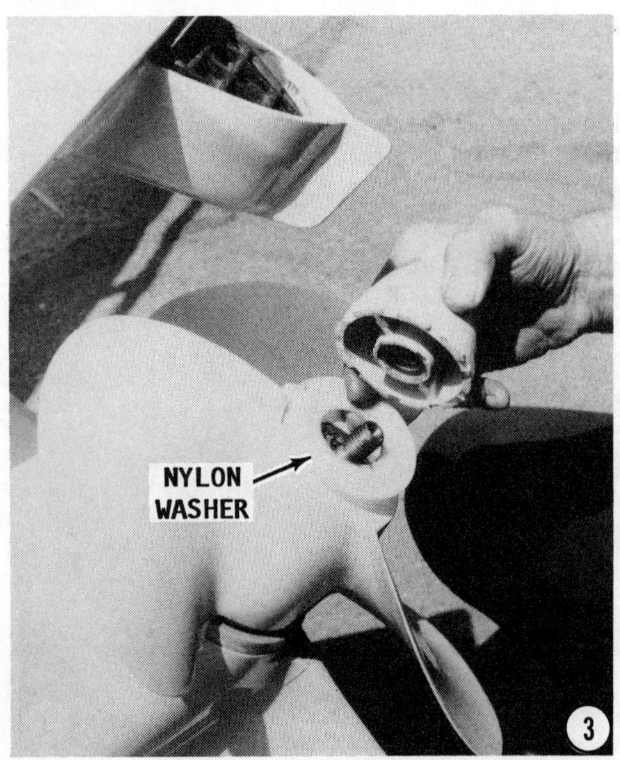

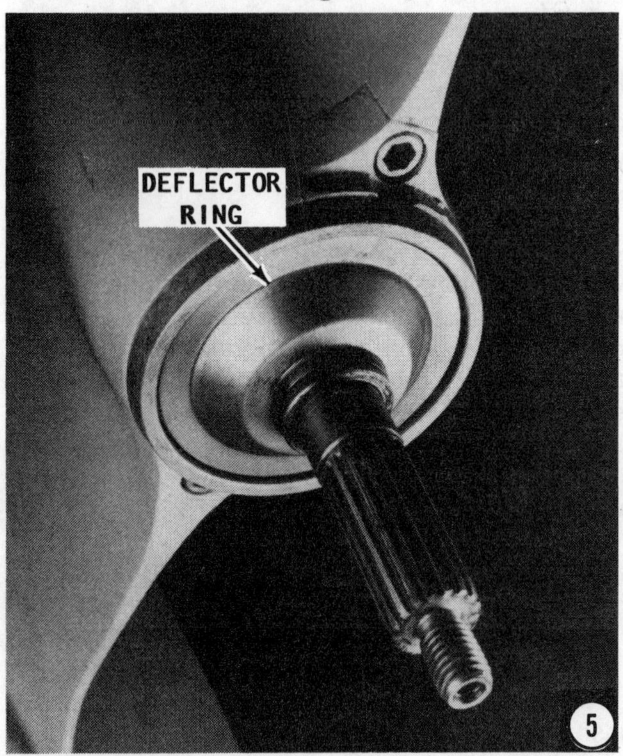

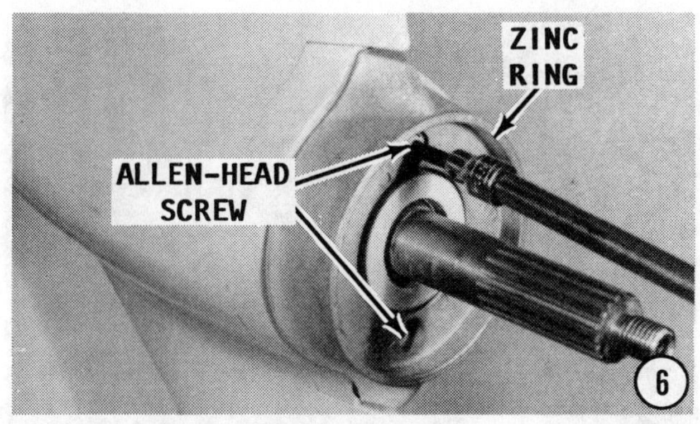

used the last time the propeller was installed, or because of any number of other reasons, heat must be applied and a propeller puller must be used. Apply heat with **CARE** so as not to damage the propeller shaft seals. It would be an excellent idea to replace the seals if heat is applied while removing the propeller.

After the propeller is "pulled" the rubber hub will still be attached to the shaft. Cut it free with a hacksaw or split it free with a suitable tool.

CHANGE PROPELLER ROTATION

The gear arrangement in the upper gear housing is so designed that no modifications are necessary in order to change propeller

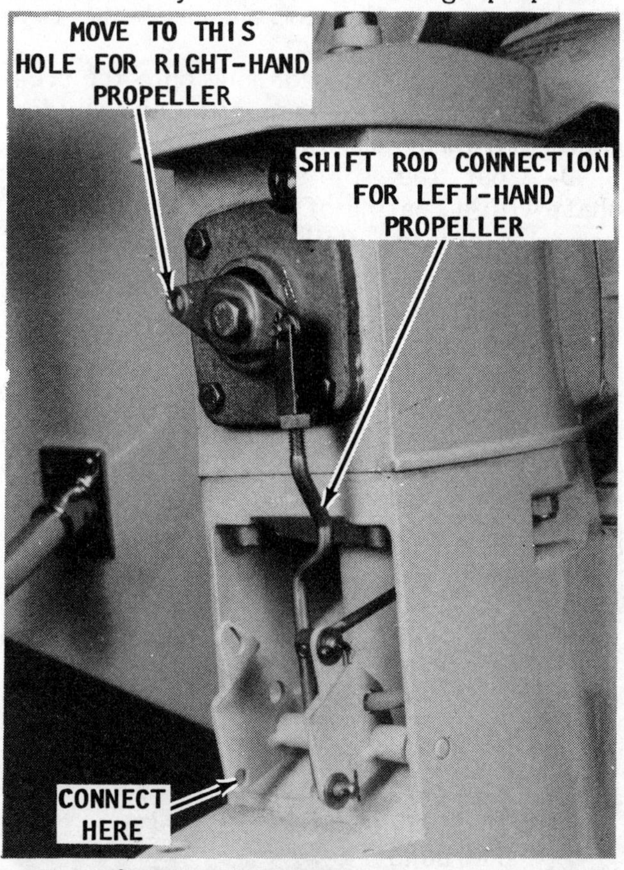

The shift rod can be moved to accommodate a right-hand propeller, as explained in the text.

rotation. Only the linkage needs to be altered and this operation is a very simple task.

With a standard left-hand rotating propeller, the **LOWER** gear operates as the **FORWARD** gear. In this case the shift rod is connected as shown in the accompanying illustration.

With a right-hand rotating propeller, the **UPPER** gear becomes the **FORWARD** gear.

Simply disconnect and move the shift rod to the opposite end of the eccentric piston arm, as indicated. Connect the rod to the opposite hole in the yoke and secure it in place with the cotter key. **DO NOT** rotate the rod when making the new connection. The bend in the rod **MUST** face the same direction for both installations.

If an adjustment is necessary, see next section, Retaining Pawl Adjustment.

RETAINING PAWL ADJUSTMENT

The reference numbers in the following procedures are identified in the accompanying illustration on this page.

1- Remove the protective cover (12). Shift the control lever to the **NEUTRAL** position.

2- Disconnect the shift cable swivel (9) and the fork (11).

3- Release the lock nut for the fork (11). Check to be sure when the fork is connected to the lever, the retaining pawl rod (6) reaches the retaining pawl bracket at **"A"** without pressing against it. Make an adjustment if necessary, and then secure the fork in position with the lock nut.

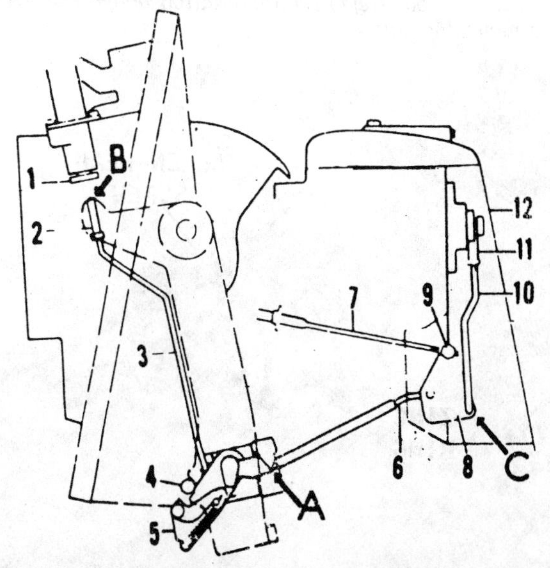

Points identified by number are referenced in the text for making a retaining pawl adjustment.

4- Adjust the swivel (9) to allow it to easily enter the hole in the shift yoke. Shift the control lever to the **FORWARD** position. Check to be sure the corner **"C"** does not make contact with the housing. Install the protective cover.

5- Push the drive forward against the adjusting pin. Check the position of the rod (3). The upper part (2) should extend just a whisker (1/16", 1.6 mm) through the fork at **"B"**. This position will permit the lift (1) to disengage the retaining pawl (5) when the drive is tilted up. Loosen the lock nuts and adjust the upper part of the rod (2) as required.

New zinc ring (right) for installation on the lower unit and zinc bar (left) for installation under the stern drive mounting collar.

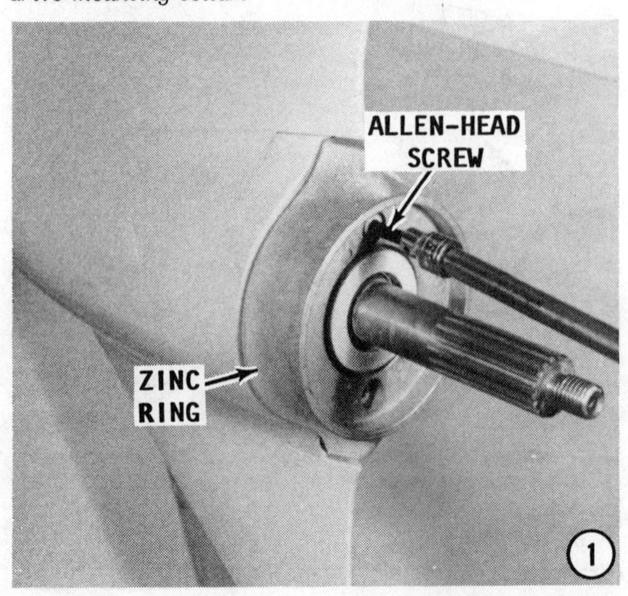

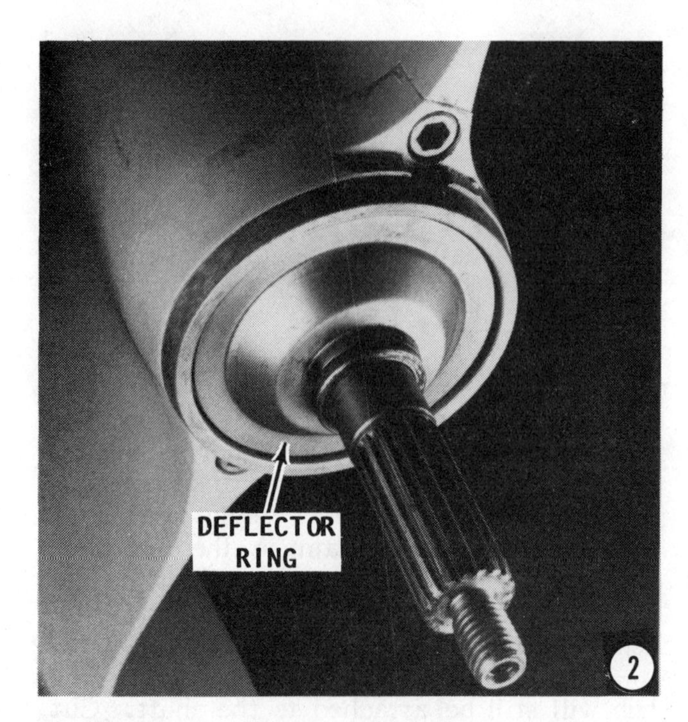

PROPELLER INSTALLATION

1- Check to be sure the splines and threads on the end of the propeller shaft are clean and in good condition. If the zinc ring was removed, slide a **NEW** one onto the shaft and into position against the bearing housing. Secure the zinc ring with the two Allen-head screws.

2- Slide the deflector ring onto the propeller shaft and into place with the shoulder facing outward, as shown.

3- Coat the splines of the propeller shaft with a coating of water-resistant anti-corrosion grease. Failure to apply this lu-

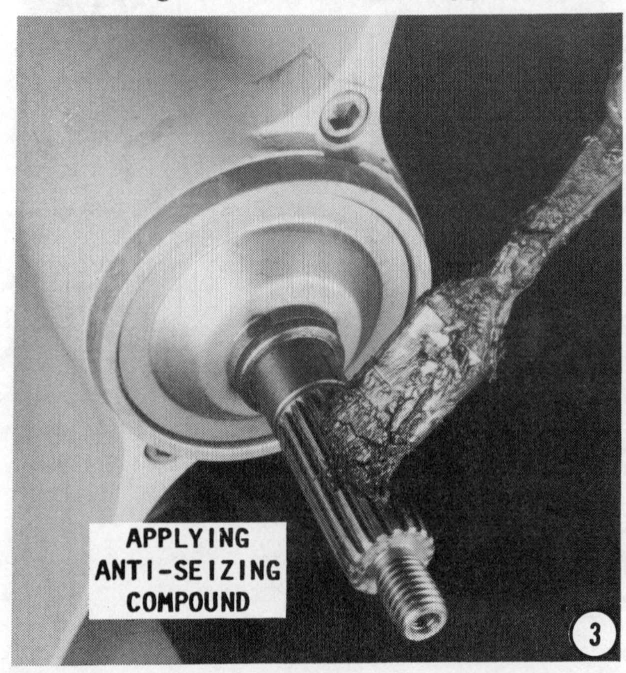

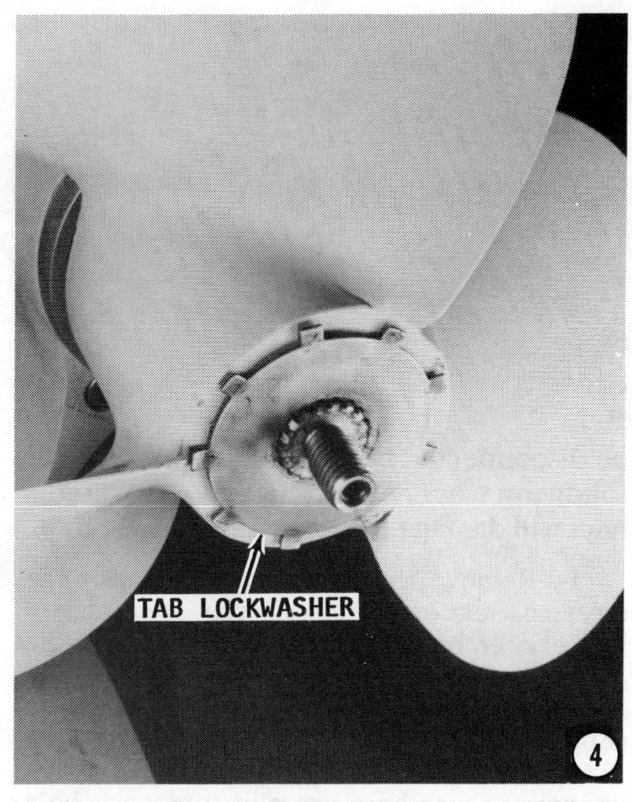

bricant could cause the propeller to "freeze" on the propeller shaft.

4- On an early model propeller installation, slide the tab lockwasher onto the shaft against the propeller.

5- On an early model, thread the cone onto the shaft. Place a block of wood between one of the propeller blades and the anti-cavitation plate to prevent the propeller shaft from rotating. Tighten the cone by inserting a stiff rod, large Phillips screwdriver, or other suitable tool into the hole in the aft end of the cone and apply pressure in a **CLOCKWISE** direction. When the cone is tight and the slots in the forward edge of the cone are aligned with the tabs on the lockwasher, bend the tabs into the slots to secure the cone.

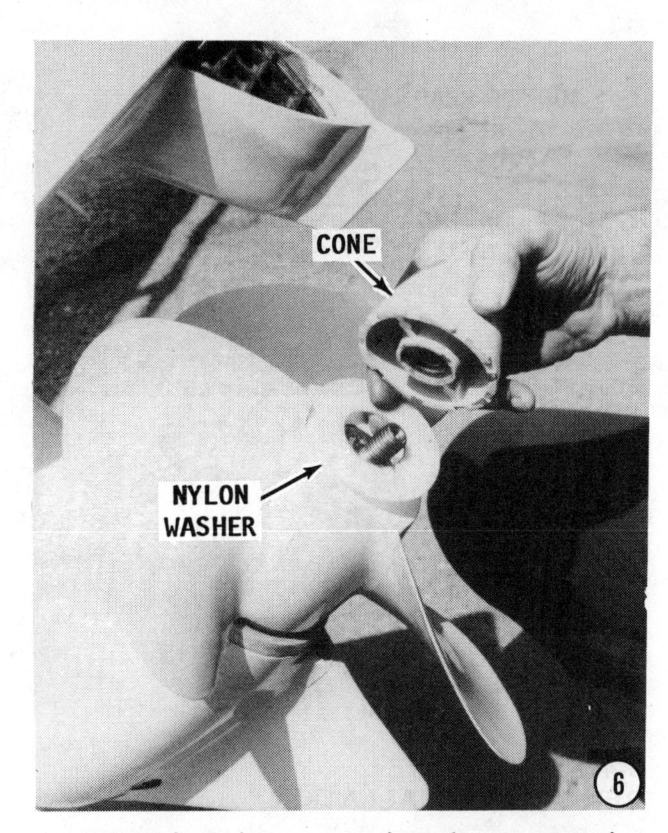

6- On late model propeller installations, slide the nylon washer onto the shaft against the propeller. Thread the cone onto the shaft.

7- Secure the cone in place with the Allen-head screw.

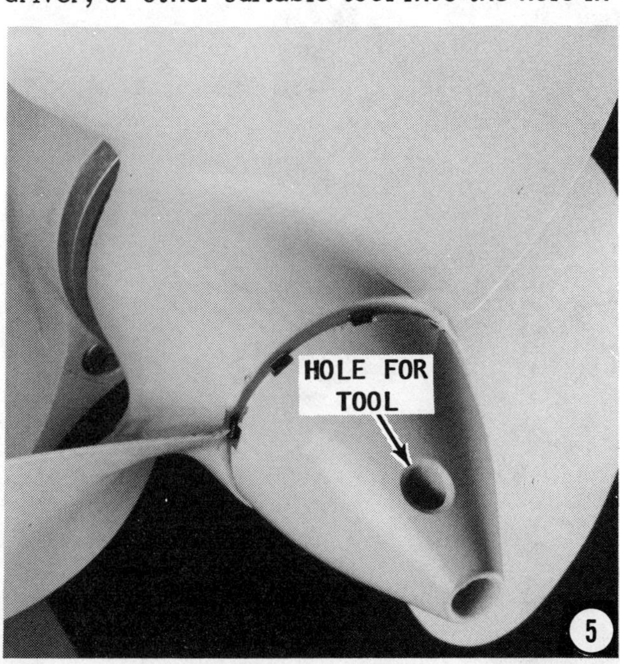

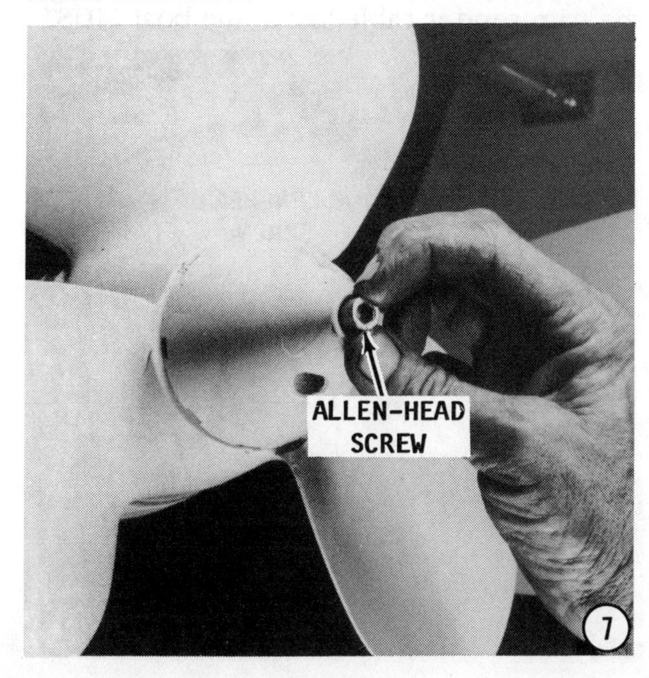

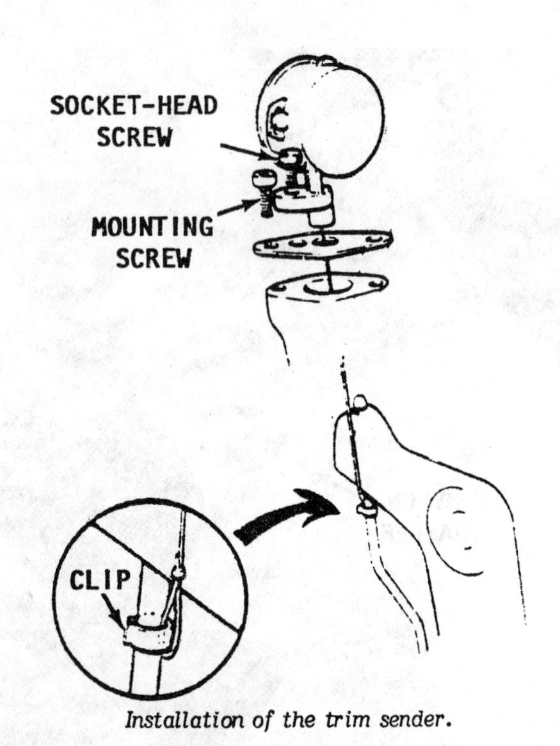

Installation of the trim sender.

8-4 STERN DRIVE REMOVAL/INSTALLATION

Service procedures to remove and install a complete stern drive unit are almost identical for the Model 250, 270, 280, and 290. The 280T does have different hinge pins. These pins are pulled and **NOT** driven out. The procedures identify this area clearly. Differences for the 290 are also noted.

REMOVAL

FIRST, THESE WORDS

On older Volvo units, say prior to 1982, the trim sender cable inside the boat MUST

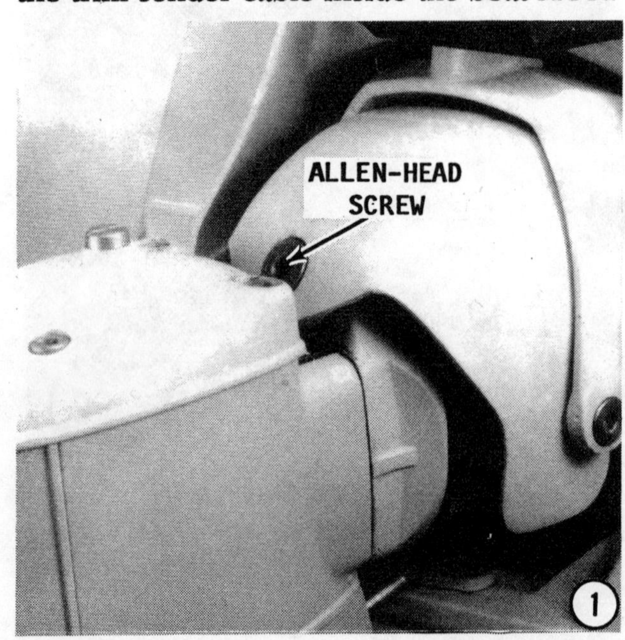

be disconnected before proceeding with the following work. Failure to make the disconnect will damage the expensive sender unit.

1- Remove the attaching bolt out of the steering helmet on top of the stern drive. The earlier model stern drives have a single Allen-head bolt. Later models have two Allen-head bolts plus a sleeve that must be popped out with the aid of two round bars that are bent. These bars resemble an Allen wrench. These bars are slid into each hole of the sleeve. The sleeve is then worked back-and-forth to pull it out of the helmet.

2- Remove the cable clamp for the shift cable located directly underneath the upper gear assembly. Two bolts secure the clamp in place. Both must be removed.

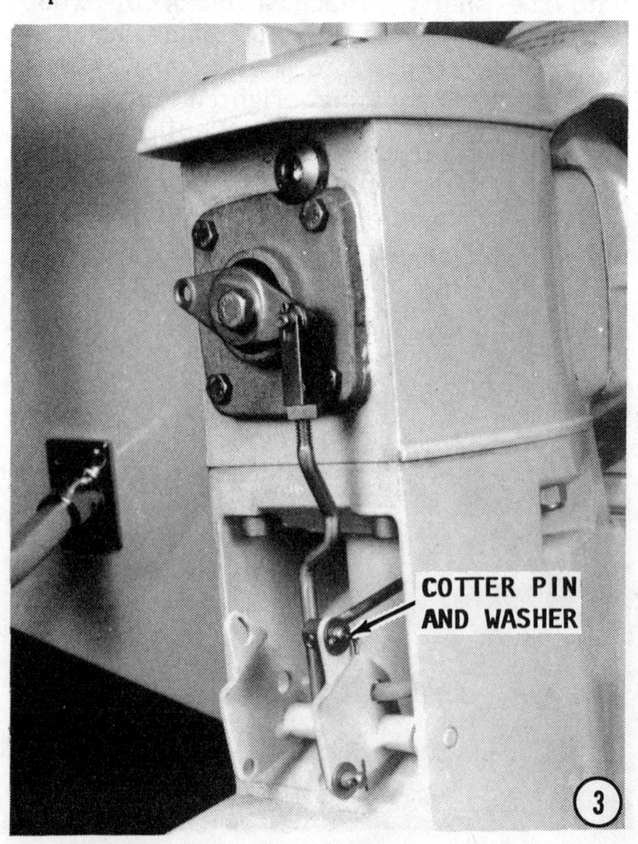

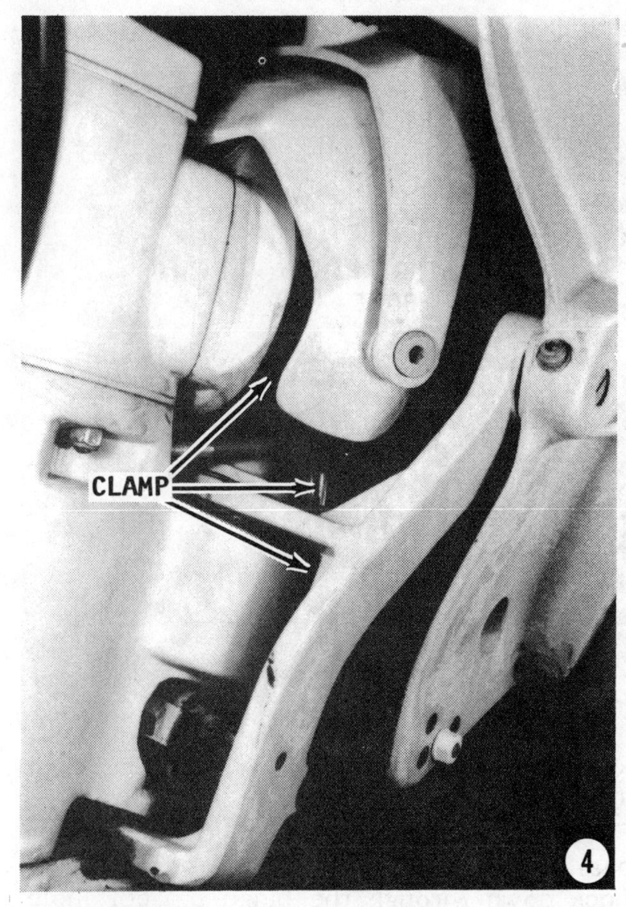

GOOD WORDS

On the Model 290, the bolts securing the shift cable clamp do not have to be removed. The holes in the clamp are slotted. Therefore, the bolts only need to be loosened, and then the plate "rocked" out of the way.

3- Remove the shift cover, and then disconnect the shift cable from the linkage by pulling the cotter pin and making the disconnect.

4- Loosen all of the bellows clamps -- the universal bellows clamp, the water intake hose clamp, and the exhaust hose clamp. After the clamps have been loosened, work the rubber boots free of the stern drive.

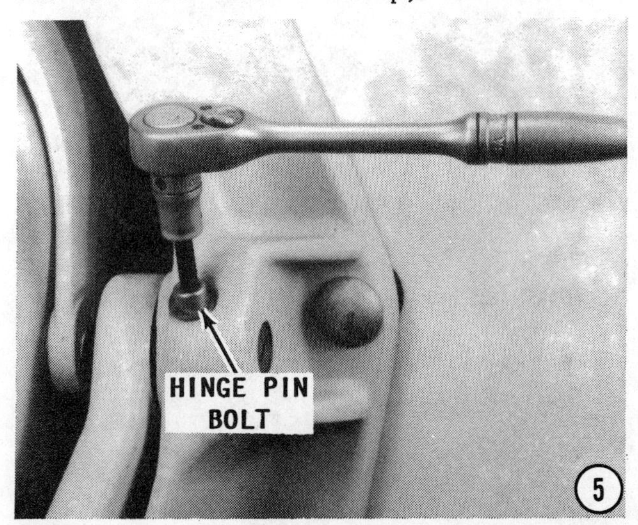

5- Place suitable blocks of wood under the skeg of the stern drive to support its weight. As an aid to accomplishing good support, if the boat is mounted on a trailer, the forward end of the trailer can be raised or lowered until good block support is possible under the stern drive. Remove the hinge pin bolts.

6- Drive the hinge pins free of the transom shield. In order to accomplish this task, first turn the stern drive hard over to port, remove the starboard pin. Next turn the stern drive hard over to starboard and remove the port side pin.

CRITICAL WORDS

If servicing a unit with power trim, the hinge pins are **NOT** driven out, they are actually **PULLED** out of the transom shield. As an aid to pulling these pins, the manufacturer has drilled and tapped the pins with 5/16 x 18 threads. First, thread the proper size bolt into the pin. Next, attach some type of slide hammer or similar tool to the bolt. Operate the slide hammer and **PULL** the pin free of the transom shield. Repeat the procedure for the other pin.

The stern drive is now ready to be removed from the boat. With one hand, reach down and grasp the reverse latch mechanism and pull it down. This action will release the reverse latch mechanism and prevent it from hanging up on the locking pin. At the same time, with the hand under the exhaust housing, pull the stern drive backward. This movement will cause the universal yoke to slide free of the bell housing. Continue to move the stern drive backward away from the transom until it is free. **TAKE CARE** to move the stern drive straight back to allow

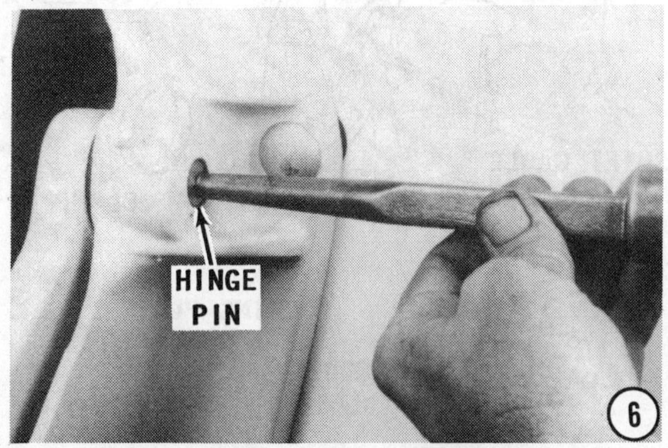

the shift cable to slide through the intermediate housing. If the stern drive is tilted, the cable may be severely bent.

INSTALLATION

1- Clean the hinge pin holes in the transom shield with solvent, and then wipe them dry with a cloth. Smooth the interior surface of the holes with fine grain emory paper. Clean and buff the hinge pins with a wire brush or power wheel. Coat the threads of the hinge pin bolts with water resistant grease, non-hardening Permatex, anti-sieze compound or other suitable material. Check to be sure the nylon or plastic bushings in the steering fork are in good condition. Coat the inside splines of the bell housing with a small amount of water resistant grease. Place a suitable block of wood, preferably the one used during removal, in position under the boat transom for the stern drive skeg to rest upon.

2- Lift the stern drive onto the block of wood placed under the boat transom. Slide the shift cable through the intermediate housing. Lift the universal joint assembly and guide it towards the bell housing. Continue moving forward with the splines of the universal joint indexing with the splines of the intermediate shaft in the bell housing. As soon as the splines begin to index, it is necessary to keep the steering fork square to permit the steering fork to pass into the transom shield. If the fork is not aligned, it is not possible to rotate the fork and align them for the hinge pin to pass through.

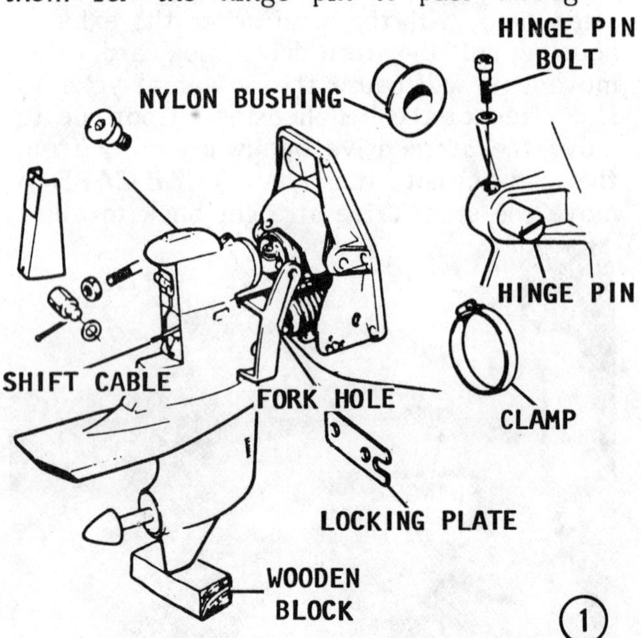

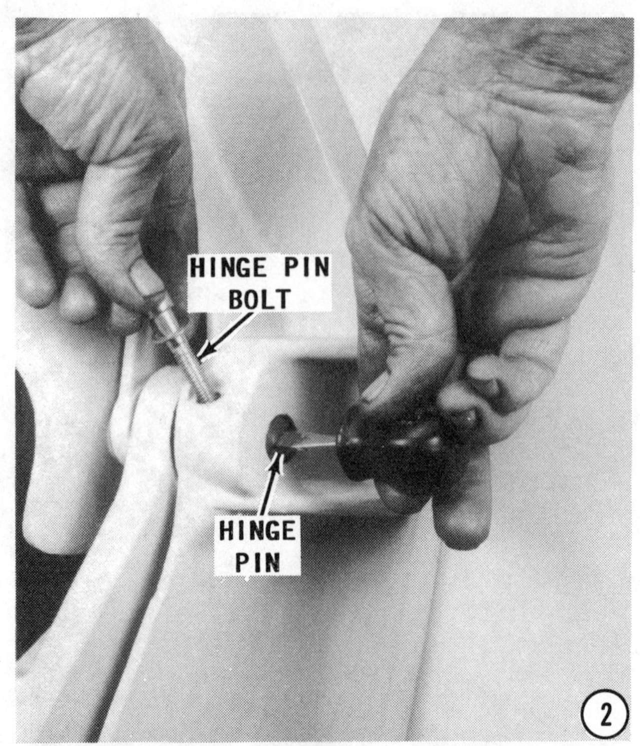

Now, **GENTLY** tap the hinge pins into place with the machined notch in the pin facing **FORWARD** to allow the hinge pin bolt to pass. Continue to tap the pins into place. Look down through the hinge pin bolt hole and at the same time use a large screwdriver to rotate the hinge pin slightly, if necessary, until the bolt will pass through.

3- Work the universal joint bellows into position and secure it in place with a clamp.

CRITICAL WORDS

Always use a socket wrench with a swivel joint from a 1/4" drive set to tighten the

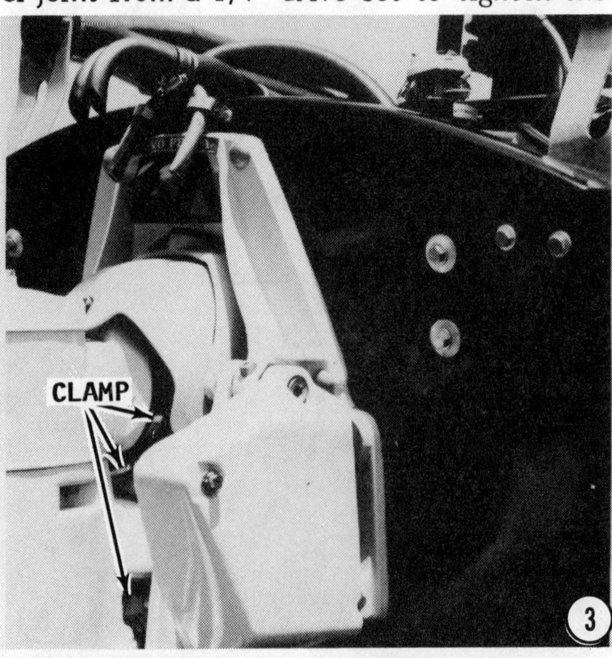

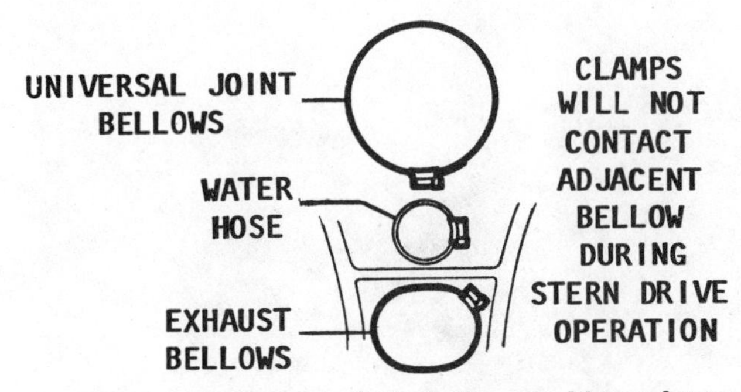

Proper arrangement of the bellows and hose clamps. In this position, the tightening device on one clamp will not chafe against another bellow when the stern drive is in motion.

clamps on the bellows and water hose. If a socket set is not available and a screwdriver is used, exercise extreme **CAUTION** to prevent the screwdriver from slipping off the fastener and puncturing a hole in one of the bellows. **TAKE TIME** to position the clamps as shown in the accompanying illustration to prevent the clamp tightening hardware from coming in contact with another bellow when the stern drive is in motion. Such action could eventually cause the boot to leak, causing extensive damage to expensive parts.

Slide the water intake hose into place with the end marked **"Engine"** towards the transom. Secure the hose with the clamp. Slide the exhaust hose into place and secure it with a clamp.

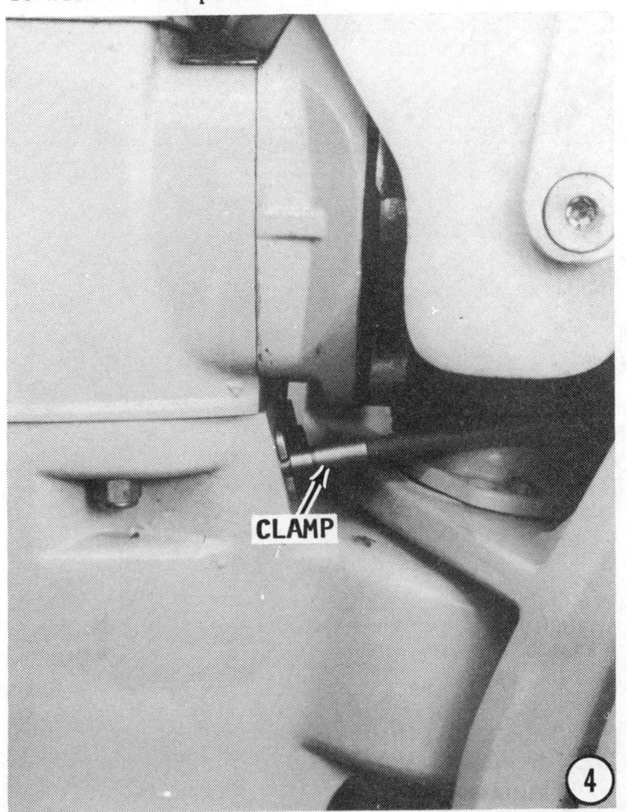

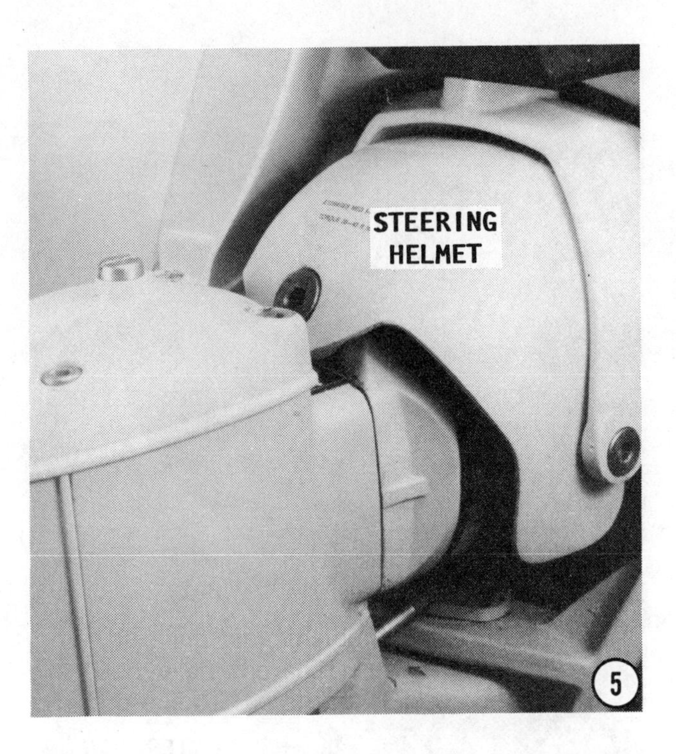

4- Position the shift cable to expose the notch in the cable housing to allow the cable clamp to be installed. Install the clamp with the two attaching bolts.

5- Slide the steering helmet into position and secure it in place with the attaching hardware.

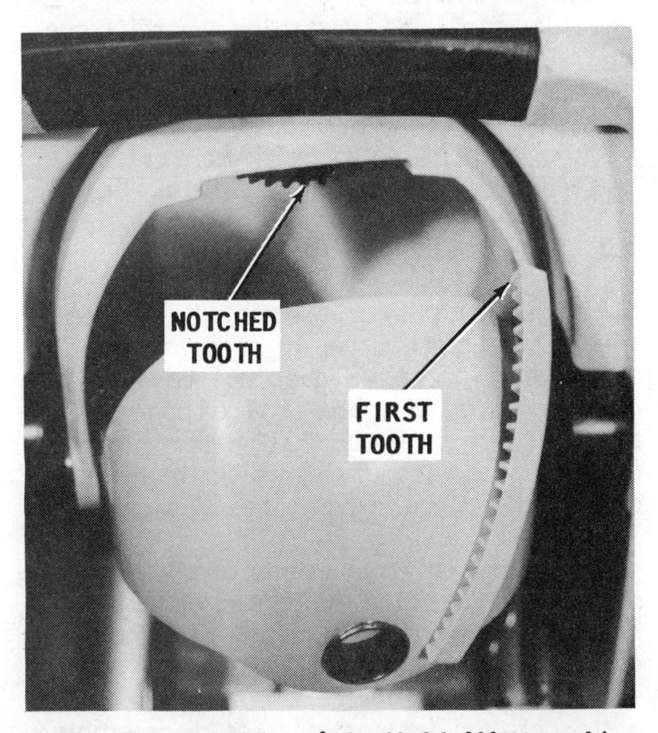

*The rack and pinion of the Model 290 stern drive must be properly mated during installation. The first tooth of the rack **MUST** be indexed with the notched tooth of the pinion on the sender unit.*

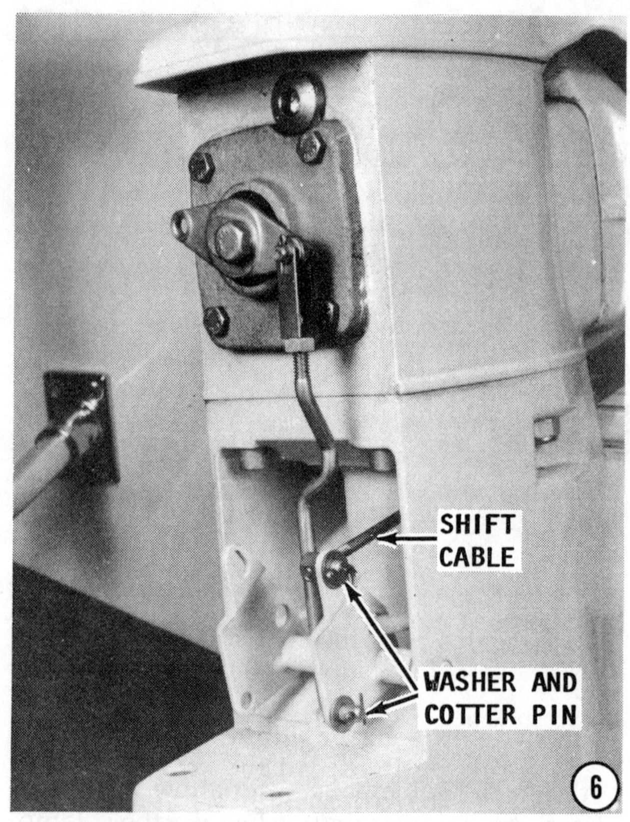

6- Connect the shift cable to the intermediate housing linkage with the clevis, washer, and cotter pin. Install the shift cover.

SPECIAL WORDS

Take care when installing the shift cover. The latch position at the top of the cover **MUST** be engaged into the slot in the cover during installation. Coat the threads of the bottom screw with some type of water resistant grease before installation.

8-5 UPPER GEAR ASSEMBLY

If the upper gear assembly is the only unit of the stern drive to receive service, then the stern drive does not have to be removed from the boat. The upper gear assembly can be "pulled" from the intermediate housing without disturbing the remainder of the stern drive. See Section 8-4, Steps 1, 2, 3, and 4.

If the universal joints or the bellows require service, then the upper gear assembly can be removed leaving the remainder of the stern drive attached to the boat.

The following procedures provide complete detailed instructions for disassembly, assembly, and adjustment, of the upper gear assembly. The procedures take up the work **AFTER** the stern drive has been removed

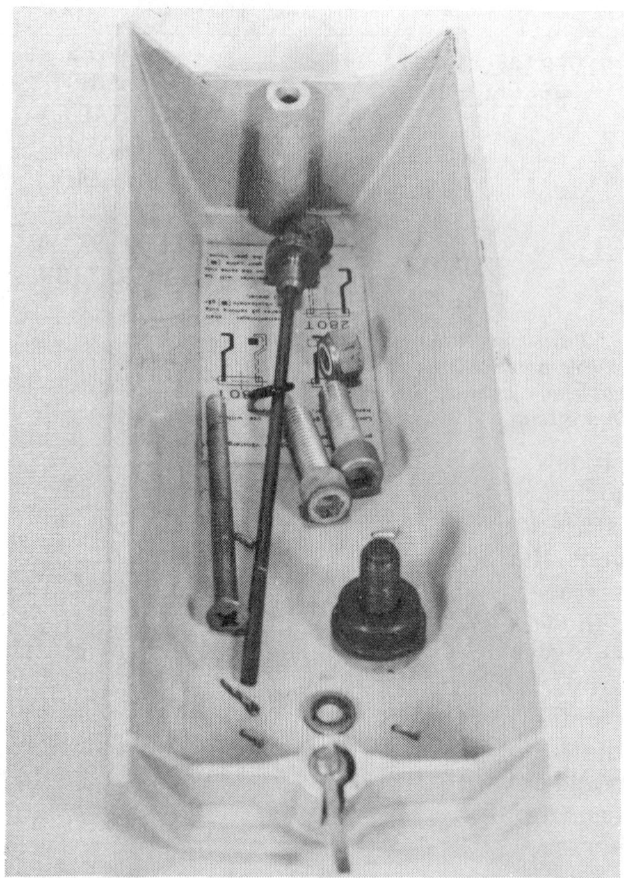

The shift cover makes an excellent container for holding attaching hardware and parts as the work progresses.

from the boat, or after the upper gear assembly has been prepared for removal according to Section 8-4, Steps 1, 2, 3, and 4.

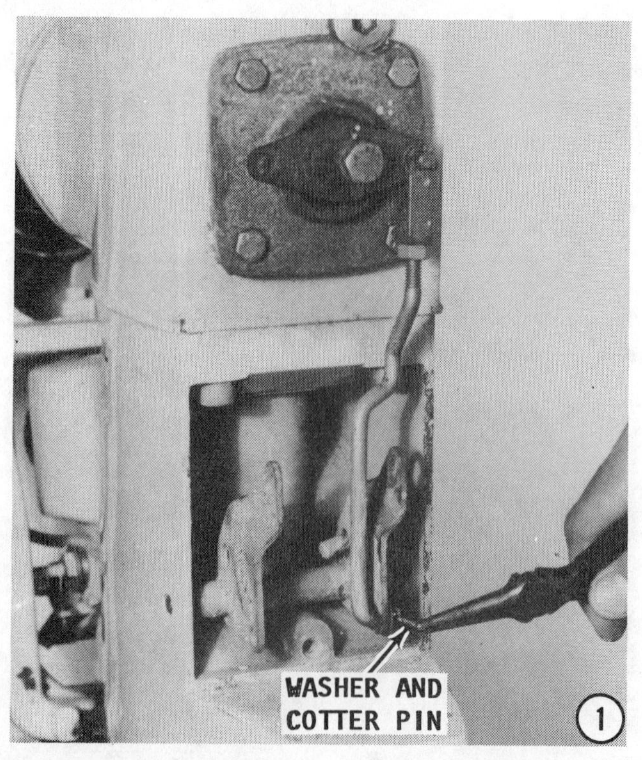

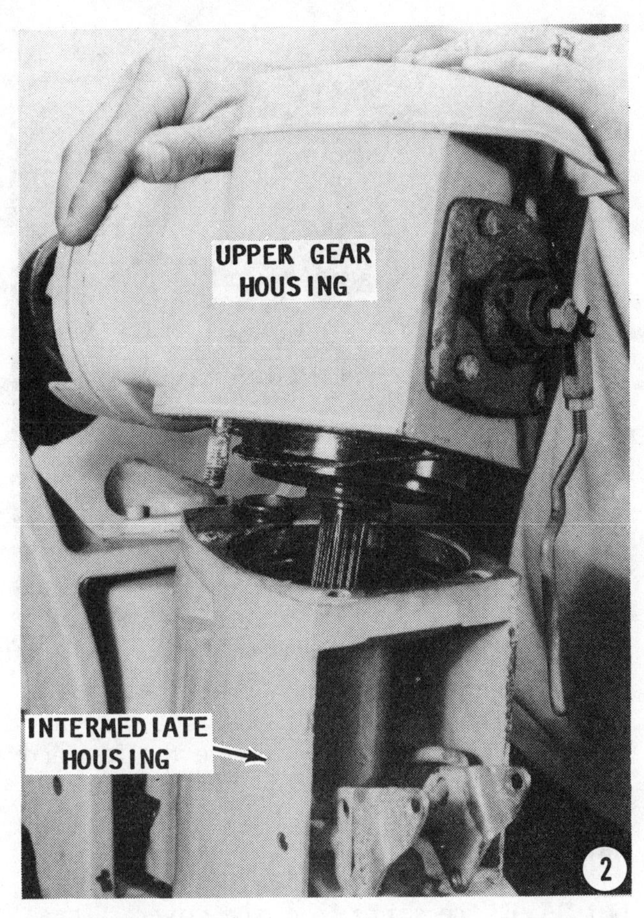

DISASSEMBLING

1- Loosen the dip stick from the upper gear housing. Remove the plug from the bottom of the lower unit and drain the lubricant from the stern drive. Remove the four fasteners securing the upper gear assembly to the intermediate housing. Two of these fasteners are Allen-head screws, and two are nuts, one on either side of the intermediate housing. Pull the cotter pin and washer from the shift linkage rod at the intermediate housing, as shown.

2- Remove the upper gear assembly from the intermediate housing.

3- Remove and **SAVE** the shims from the bearing recess in the intermediate housing. The purpose of these shims is to pre-load the driven gear in the upper gear assembly.

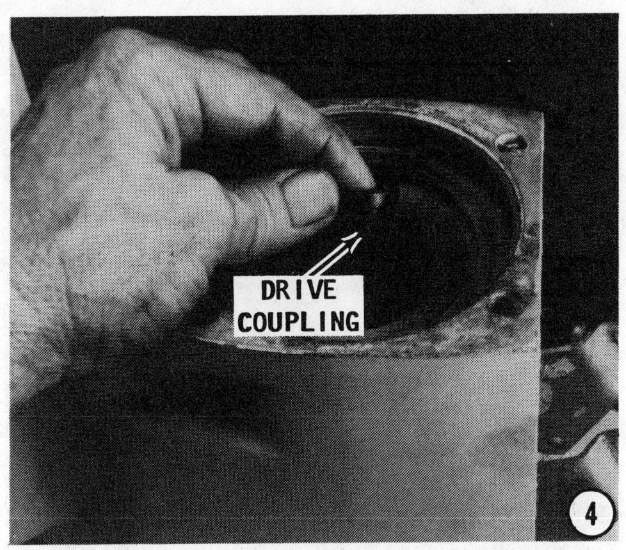

4- Lift the drive coupling out of the intermediate housing. This splined coupling joins the vertical shaft in the upper gear housing with the vertical shaft in the intermediate housing.

5- Remove the four fasteners securing the shift mechanism in the upper gear assembly. Pull the shift mechanism out of the gear housing.

6- Clamp the shift mechanism housing in a vise with the shifting shoe facing upward, as shown. Knock out the lock pin, and then the main pin.

7- Withdraw the piston from the one side of the shift housing and the shifting shoe from the other side. Slide the spring free of the shifting shoe shaft.

8- Pry the O-ring out of the recess on one side of the bearing housing. Work the sealing ring out of the other side.

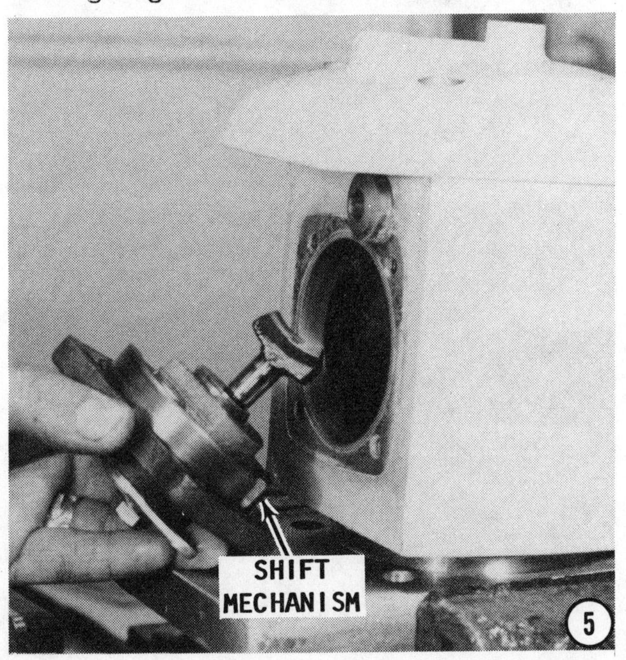

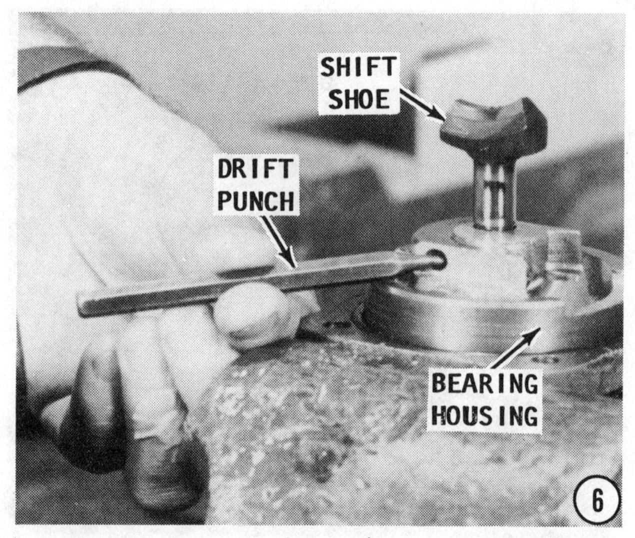

9- Remove the four Allen-head bolts securing the clamp ring to the upper gear housing. Pull the clamp ring free from the upper gear housing.

10- Remove the lubrication dip stick and O-ring from the upper gear assembly.

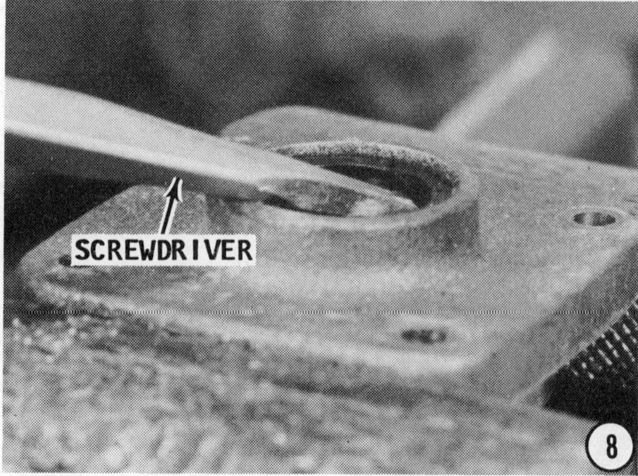

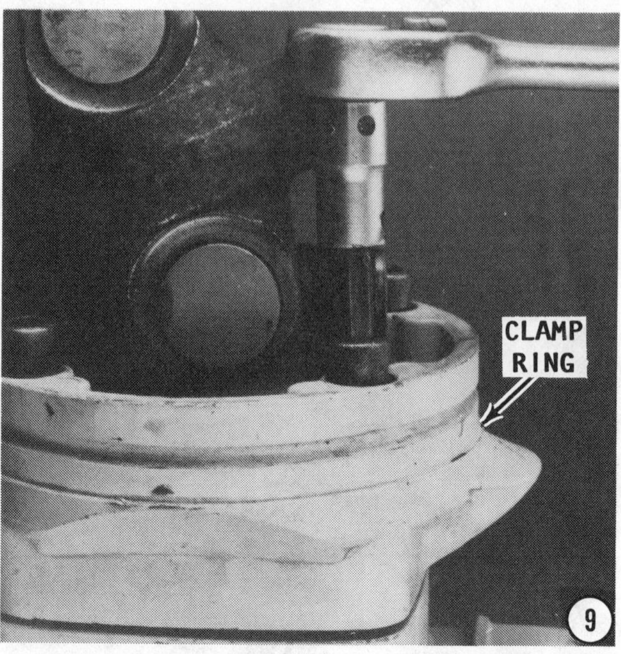

11- Remove the four Allen-head bolts securing the upper gear housing cover, and then lift off the cover. Notice the forward bolt on the starboard side. This special bolt is used to service (fill) the unit with lubricating oil.

12- Remove and **DISCARD** the O-ring, but **SAVE** the shims from the cover. These shims in the cover are used to pre-load the upper driven gear.

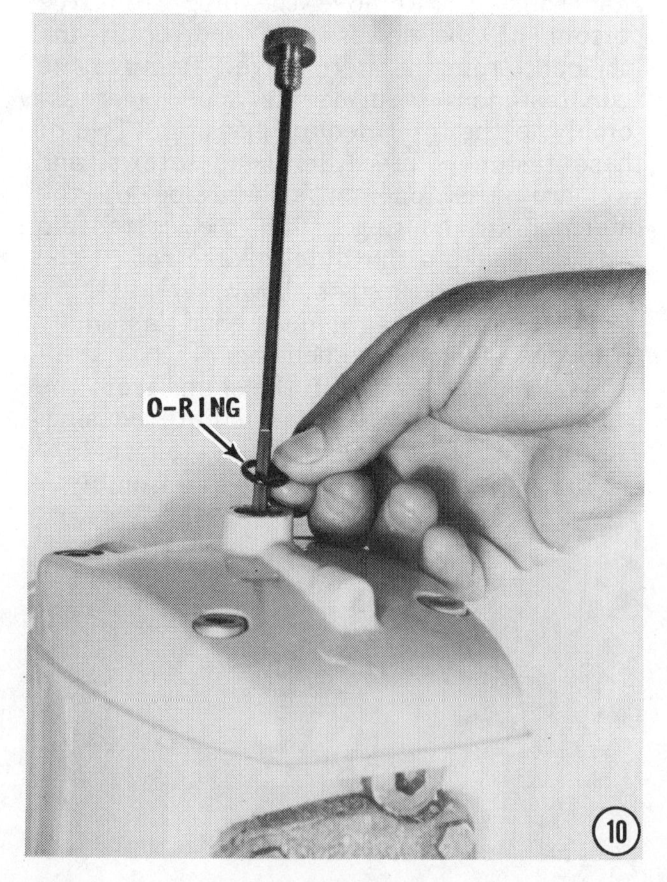

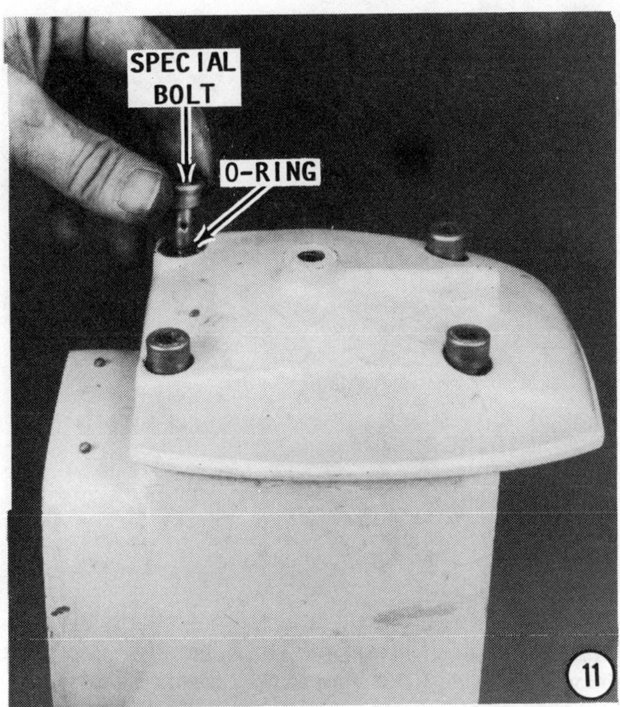

13- Remove the nut from the end of the vertical shaft on top of the gear assembly.

CRITICAL WORD

This nut on the end of the vertical shaft has a **LEFT-HAND** thread. This is the only left-hand thread in the complete Volvo Penta stern drive.

Hold the spline on the bottom of the upper gear assembly with some type of soft jaw clamping device while the nut is turned **CLOCKWISE** for removal. Take **EVERY** precaution to protect the splines. One method would be to clamp the coupling in a soft jaw vise, and then to slide the vertical

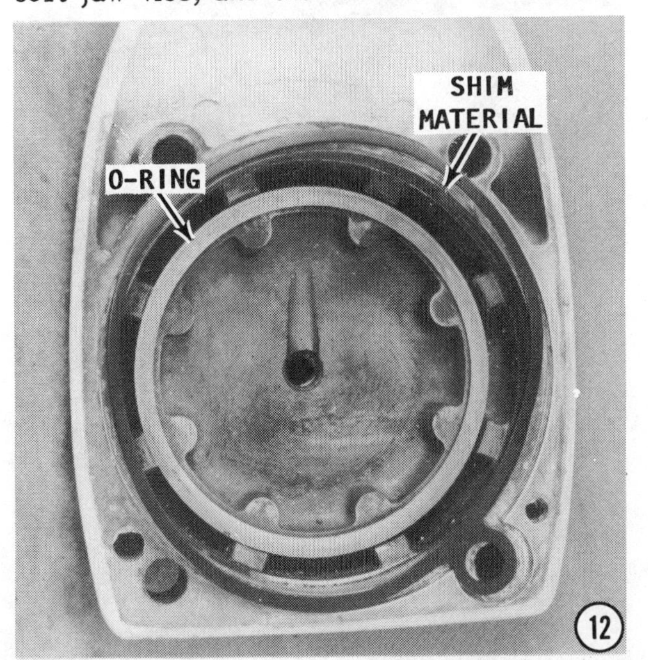

shaft down into the upper gear assembly until the splines index. The nut may be difficult to break loose because it is tightened to a torque value of 92 ft-lbs (125 Nm).

After the nut has been removed, notice the copper wear washer on the bottom of the nut. This nut is available with three different size washers. The purpose of the wear washer is to provide the proper end play for the vertical shaft. A standard nut will have no marking, another will have a **"1"** stamped on it, and the third nut will have a **"2"**. Each nut will have a different size washer. The end play adjustment is explained in detail during the assembling

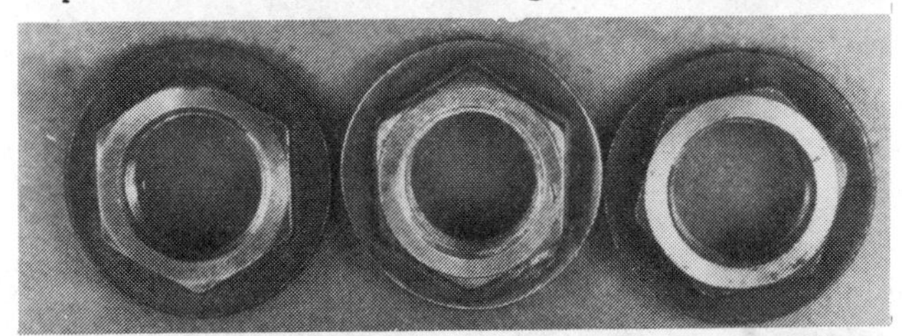

The vertical shaft nut is available with three different washer thicknesses, as explained in the text.

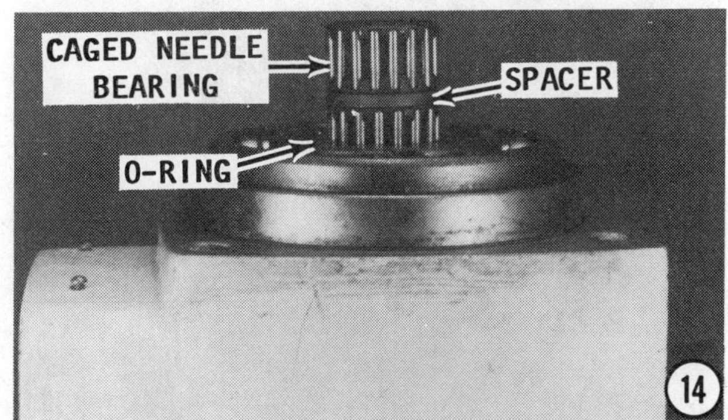

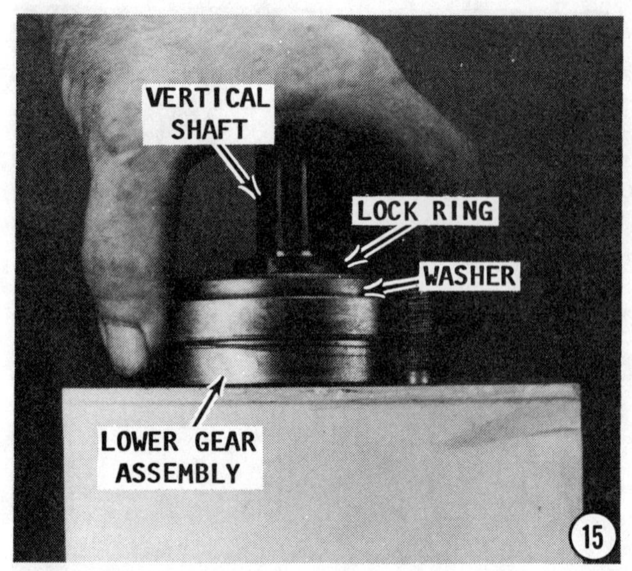

procedures.

14- Remove the upper drive gear, two sets of caged needle bearings, and the spacer between the needle bearings from the shaft. If these needle bearings are to be used again, each set should be installed into the same position from which they were removed. Therefore, make some type of record as to their location.

15- Remove the lower gear assembly along with the vertical shaft and the sliding sleeve assembly from the housing.

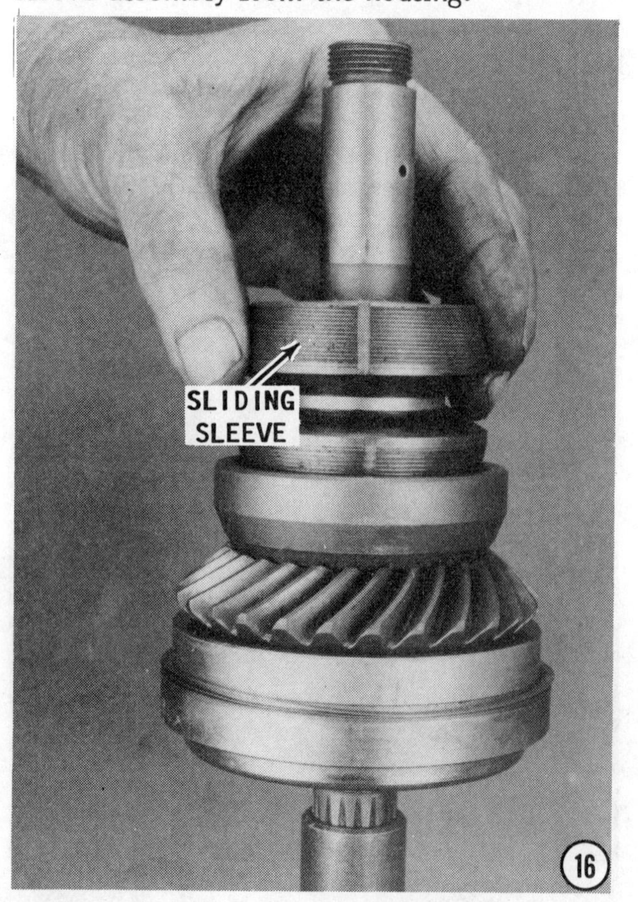

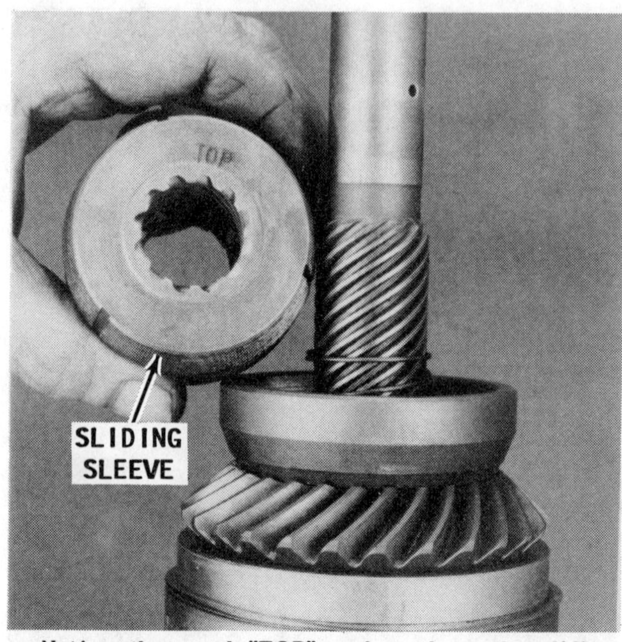

*Notice the word **"TOP"** embossed on the sliding sleeve. This side **MUST** face upward during installation.*

16- Remove the sliding sleeve from the vertical shaft. Notice the word **"TOP"** embossed on the sliding sleeve in the accompanying illustration. During installation this sliding sleeve **MUST** be installed with the word **"TOP"** facing upward.

17- Remove the spring installed under the sliding sleeve. This spring prevents the

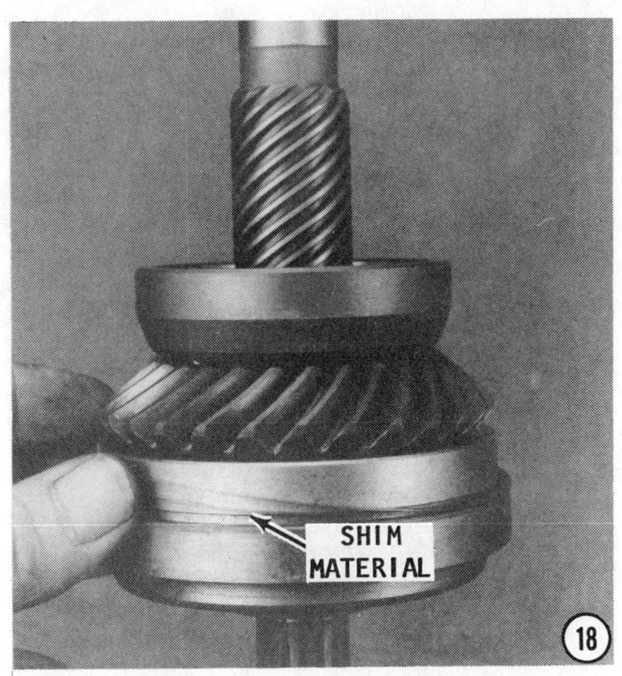

sliding sleeve from dragging on the lower gear when the unit is in the **NEUTRAL** position.

18- Remove the shim material from the housing. These shims are used to determine the depth of the gear and therefore the gear pattern and "backlash". The "backlash" adjustment procedures are given in the assembling instructions.

19- Slide the vertical shaft out of the lower gear assembly, and then slide the collar back. Remove the two halves of the split ring from the shaft.

20- Slide the collar free of the shaft, and then pull the shaft out of the lower gear assembly.

21- If bearings need to be replaced in the upper or lower gear assemblies, press the driven gear out of its bearing housing. Slide or press the bearing free of the housing.

22- Remove the Allen-head bolt from the center of the drive gear. Pull the universal joint free of the double bearing housing.

23- Remove and **DISCARD** the O-ring on the head washer.

WARNING

This next step involves a dangerous procedure and should be executed with care while wearing SAFETY GLASSES. The snap ring is placed under tremendous tension with the Truarc pliers while it is being removed. If it should slip off the Truarc pliers, it will travel with incredible speed causing personal injury if it should strike a person. Therefore, continue to hold the ring and pliers firm after the ring is clear of the double bearing housing. Place the ring on the floor and hold it securely with one foot before releasing the grip on the pliers. An

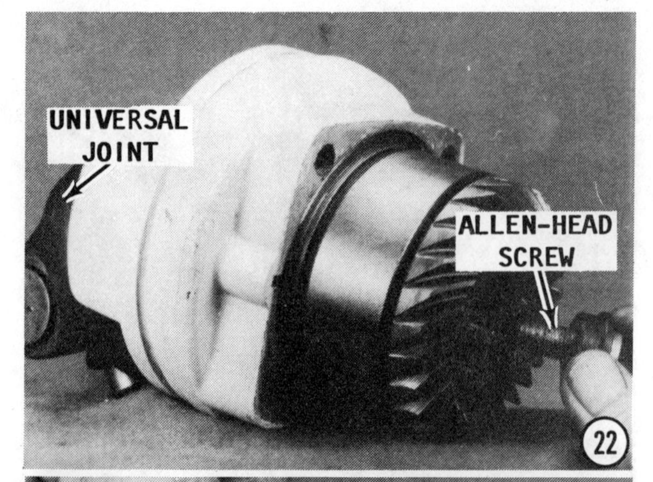

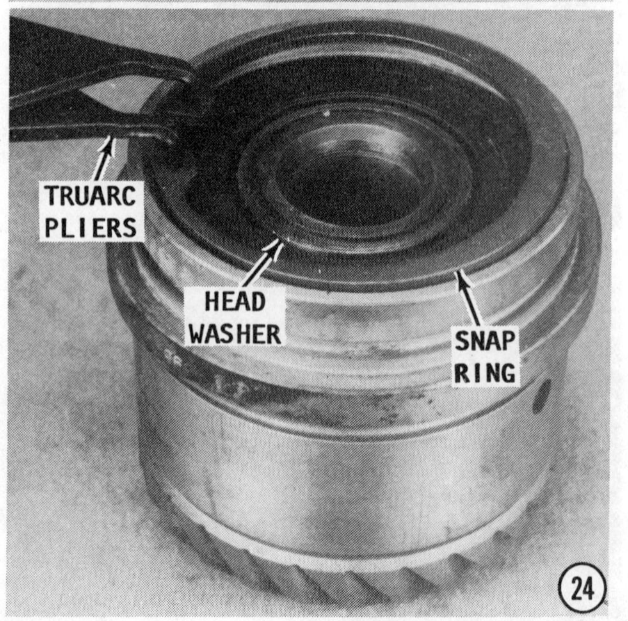

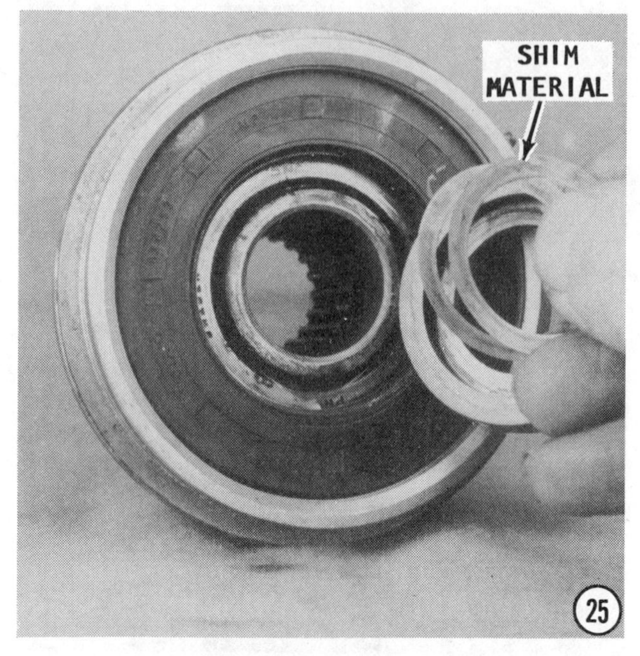

alternate method is to hold the ring inside a trash barrel, or other suitable container, before releasing the pliers.

24- Using a pair of Truarc snap ring pliers, remove the snap ring from the double bearing housing.

25- Pry the head washer out of the double bearing housing. The Model 290A has the head washer but no shim material. Remove the shim material installed under the head washer. **SAVE** and identify these shims because they are used to set the pre-load on the double bearing housing

26- Pry the oil seal loose with a screwdriver. **TAKE CARE** not to damage the bearing cage.

27- Lift out the oil seal.

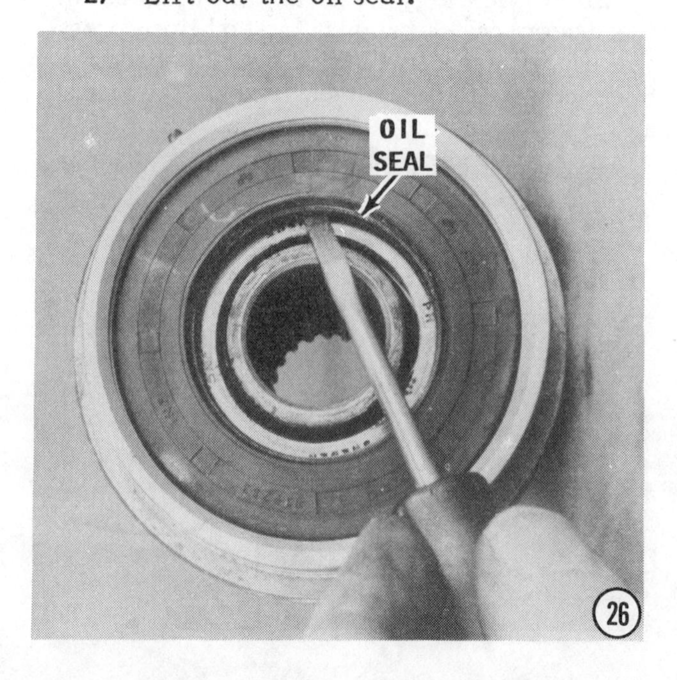

28- To disassemble one side of the universal joint, first, disengage the lock rings securing the needle bearings in place. Secure the main body of the universal joint in a vise. Next, use a hammer and drift punch and drive the needle bearings free of the yoke. Remove the cross. Disassemble the other side in the same manner.

29- Press the drive gear out of the double bearing housing. Lift one of the tapered bearings out. The other tapered bearing will remain on the drive gear. If it is necessary to replace this bearing -- removal will destroy the existing bearing. Therefore, be sure a replacement bearing is available. Use some type of clamping jaw under the bearing, and then press on the back side of the drive gear until the bearing is free.

MODEL 290A ONLY

The Model 290A ustilizes a crushable sleeve between the two tapered roller bear-

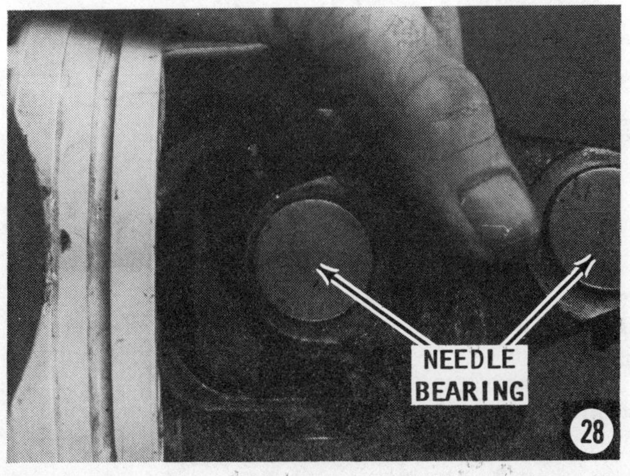

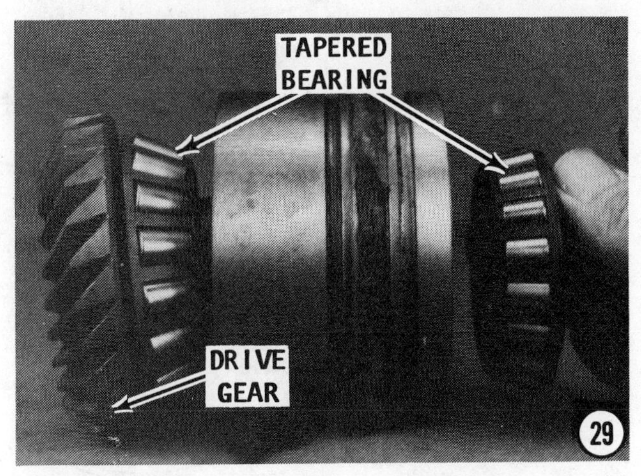

ings. Discard the sleeve. It **CANNOT** be used a second time.

CLEANING AND INSPECTING

Good shop practice dictates that **O**-rings, gaskets, and seals, be discarded and **NEW** ones installed whenever possible.

NEVER dip rubber parts or plastic parts, in cleaning solvent. Clean all upper gear housing parts except the universal joint pieces in solvent and blow them dry with compressed air. **NEVER** spin bearings with compressed air because such action will ruin them.

After all parts are clean and dry, apply a light coating of engine oil to the bright surfaces of all gears, bearings, and shafts, as a prevention against rusting.

Check the gears and gear teeth carefully to be sure they are not chipped, cracked, nicked, or damaged, in any manner. If one gear is damaged, it is necessary to replace all three, the two driven gears and the one drive gear. These gears are available from the local Volvo Penta dealer as a set.

Inspect the bearings and bearing races for pits, grooves, or uneven wear.

Inspect the splines on the shaft, sliding sleeve, and gears, for worn or damaged edges.

Check the spring to be sure it is not distorted nor has lost any of its tension. If in doubt, and if it is possible, check it against a new spring.

If the universal joint was disassembled, check the bearings for looseness ("play"). Inspect the bearing caps for scores or other damage. Check to be sure the needle caps fit snug on the cross. Check to be sure the bearing caps have no "play" in the yoke trunnions. If the cross or needle bearings

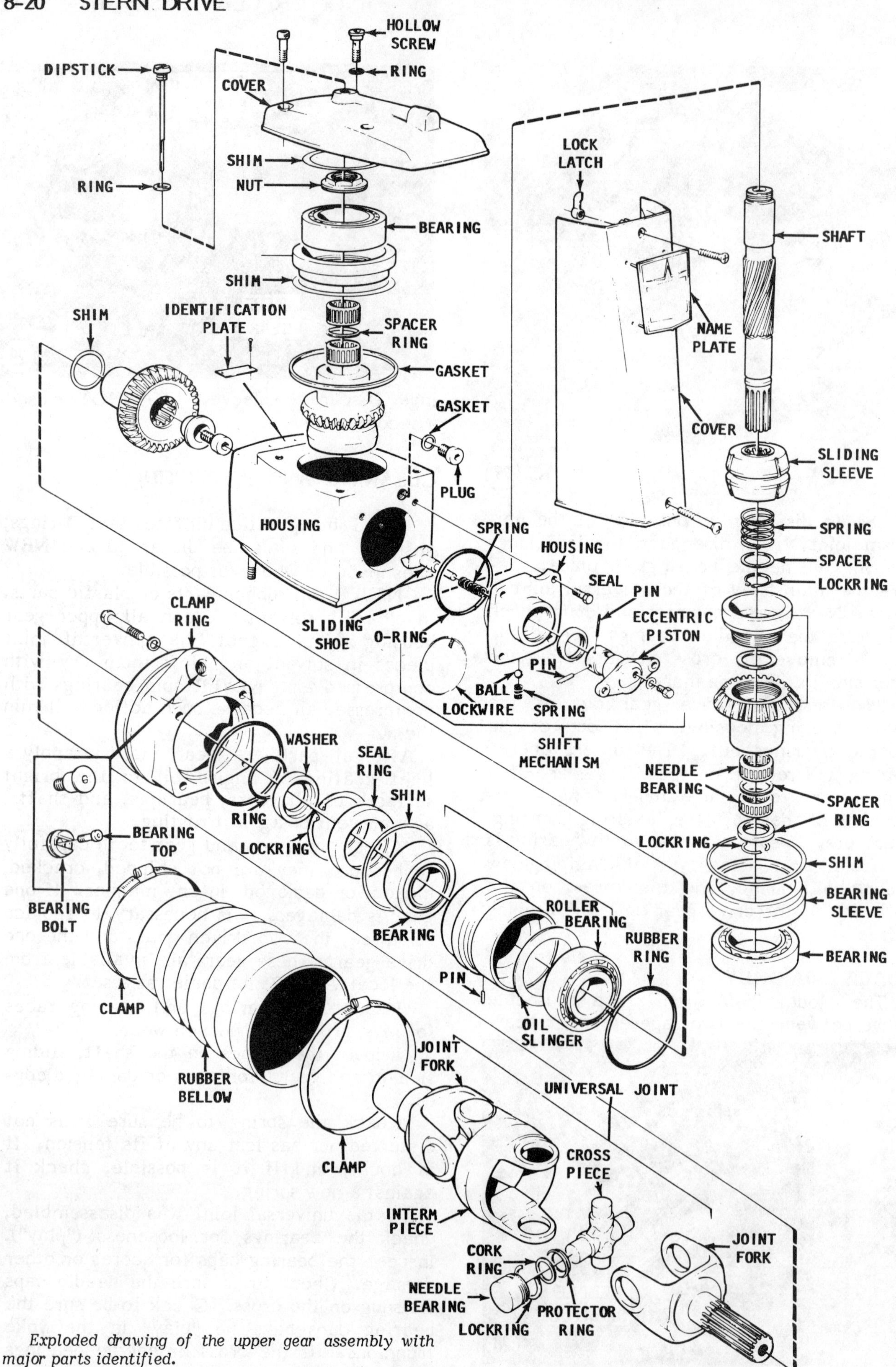

Exploded drawing of the upper gear assembly with major parts identified.

are damaged, replace the cross and the needle bearings as a set. If the bearing caps have "play" in the yoke trunnions, the yokes must be replaced.

Inspect the rubber guard for cracks, cuts, and punctures. Check the clamps for cracks or nicks. Replace the clamps if in doubt as to their condition. **ALWAYS** use Volvo Penta or Volvo-style clamps as a replacement. These clamps are smooth on the clamping surface without slots or holes.

ASSEMBLING

1- If the universal joint was disassembled, first coat the needle bearings with an adequate amount of water-resistant grease for lubrication. **DO NOT** overdo the grease because it will be extremely difficult to compress the needle bearing caps into position. Insert the cross into the yoke. Push the cross far enough in one direction to allow the needle bearing cap to go on. Press the needle bearing cap onto the cross until the snap ring can be seated. Perform the same procedure for the other side.

2- Assemble the double bearing housing, the two sets of tapered bearings, and the drive gear, as shown.

3- Using an arbor press, press the bearings and drive gear into the double bearing housing. Rotate the drive gear a few revolutions to ensure the bearings are properly seated.

SPECIAL WORDS

A preload of 1 to 2 ft.-lbs. must be obtained. If a used bearing is being assembled, the preload desired is 1 ft-lb. If a new bearing is being assembled, the preload should be close to 2 ft-lbs.

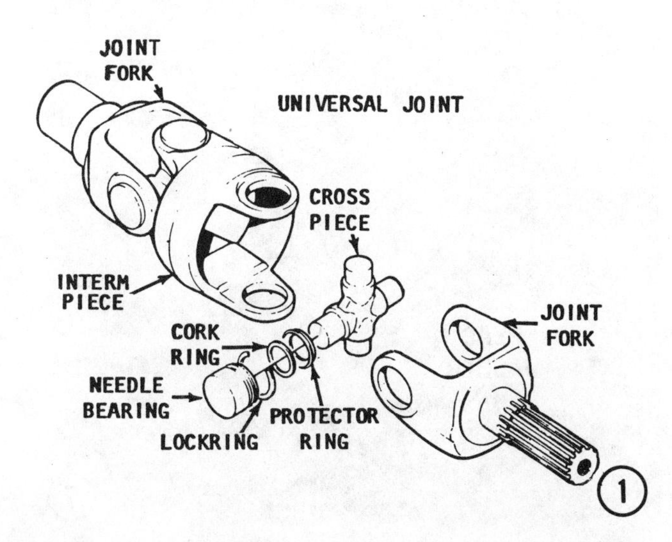

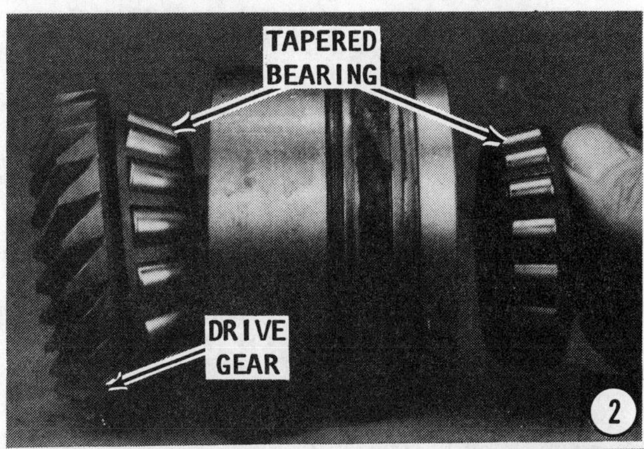

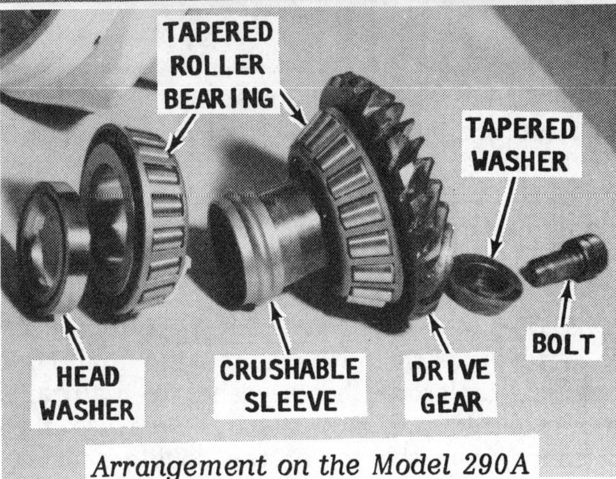

Arrangement on the Model 290A

MODEL 290A ONLY

The Model 290A utilizes a crushable sleeve between the two tapered roller bearings. When pressing the tapered roller bearing into the double bearing housing, press **SLOWLY** and rotate the bearing housing from time to time. As soon as a slight amount of "drag" is felt, cease pressing. Tightening the assembly bolt in Step 10, will provide the remainder of required 1 to 2 ft lbs preload.

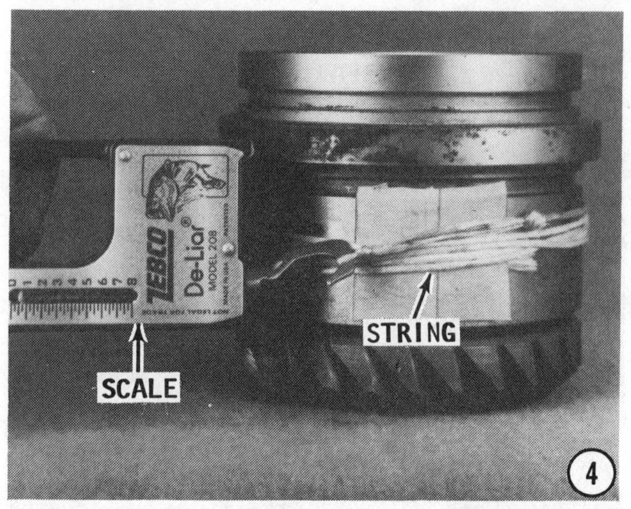

4

4- Measure the preload on the bearing by first wrapping a length of string around the bearing. Next, hook a small scale to the string. Pull on the string and observe the amount of "pull" required to move the bearing.

5- Once the desired preload is obtained, it is then necessary to take measurements to determine the amount of shim material required under the head washer to maintain the required preload. First, measure the distance from the top of the double bearing housing to the backside of the gear, as shown.

6

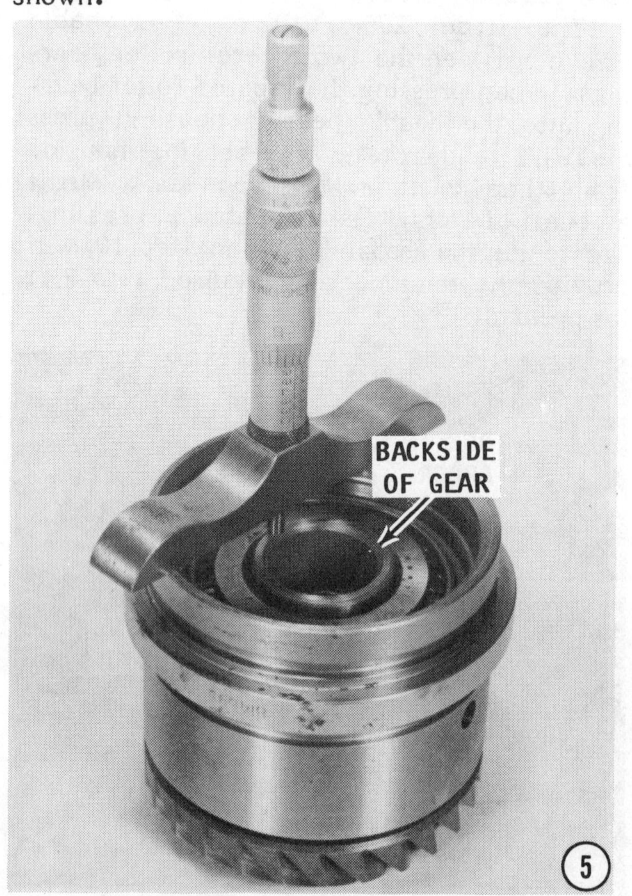

5

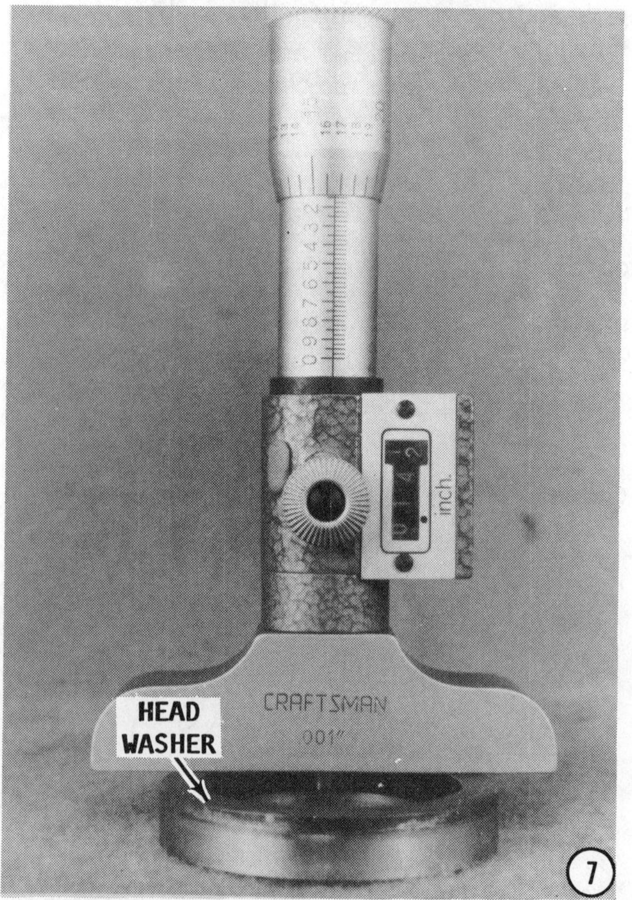

7

6- Measure the distance from the top of the double bearing housing to the inner bearing race. Subtract the measurement obtained in Step 5 from the measurement obtained in this step. The difference between the two measurements will be the height of the shoulder.

7- Measure the head washer for its depth. Subtract the measurement of the shoulder and the head washer. The answer will be the amount of shim material required to hold the double bearing housing at the required pre-load.

8- Measure the shim material removed during disassembling to be sure it agrees with the answer you received after making the measurements in Steps, 5, 6, and 7. Insert the correct amount of shim material in the head washer, and then insert the head washer on the double bearing housing.

9- Strike the back side of the gear with a hard nylon or rubber mallet to loosen the upper gear housing. When the remainder of

this step is accomplished the upper gear housing will come back to the preload condition. Take the splined yoke from the universal assembly and use it as a tool by temporarily installing it into the double bearing housing.

10- Tighten the yoke to the required torque value, using an Allen-head wrench arrangement with socket and torque wrench.

Tighten the Allen head bolt until 1 to 2 ft lb of rolling torque is maintained.

11- Check the preload in the double bearing housing. The preload should remain the same as determined in Step 4. If the desired peload has been maintained, proceed with Step 12. If the preload in the double bearing housing has not been maintained, the shim material under the head washer must be changed. Remove shim material to increase the preload; add shim material to decrease the preload.

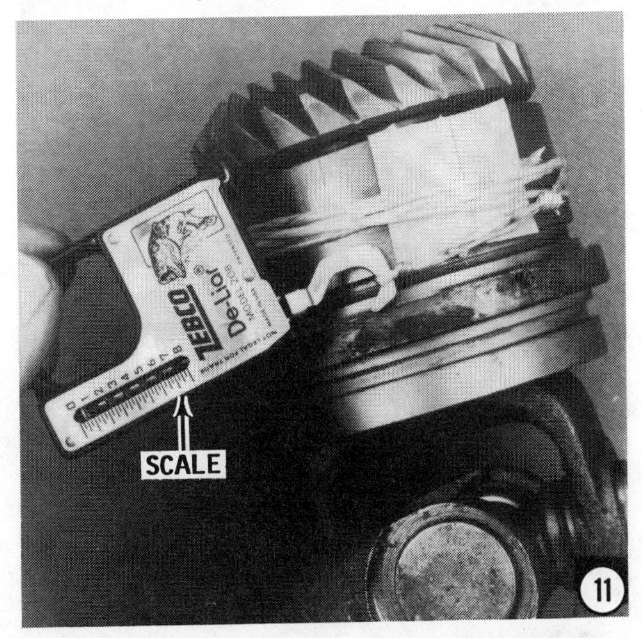

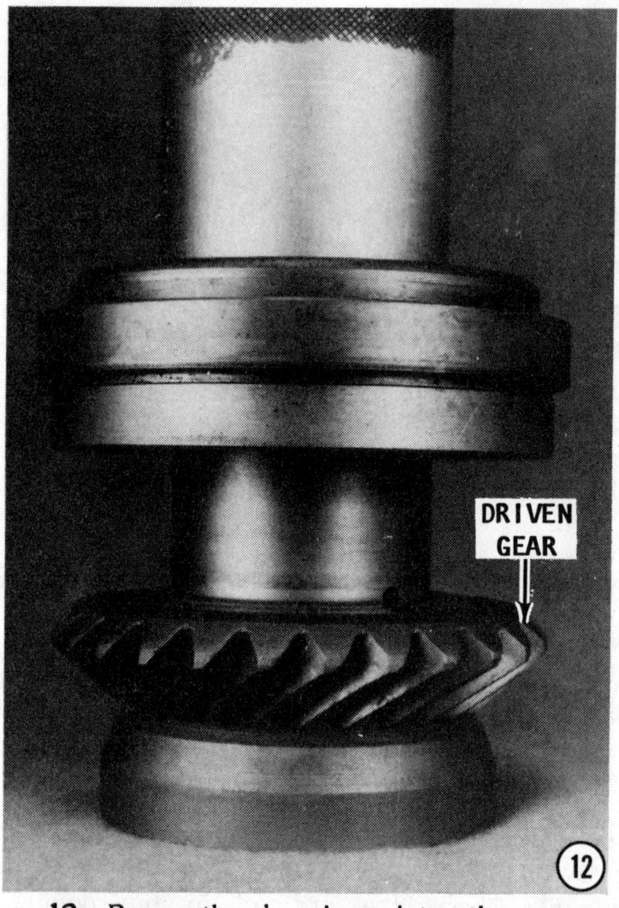

12- Press the bearings into the upper and lower driven gears.

SPECIAL WORDS

In a normal left-hand stern drive, using a left-hand propeller, the lower gear in the upper gear assembly will be the **FORWARD** gear. The upper gear will be the **REVERSE** gear.

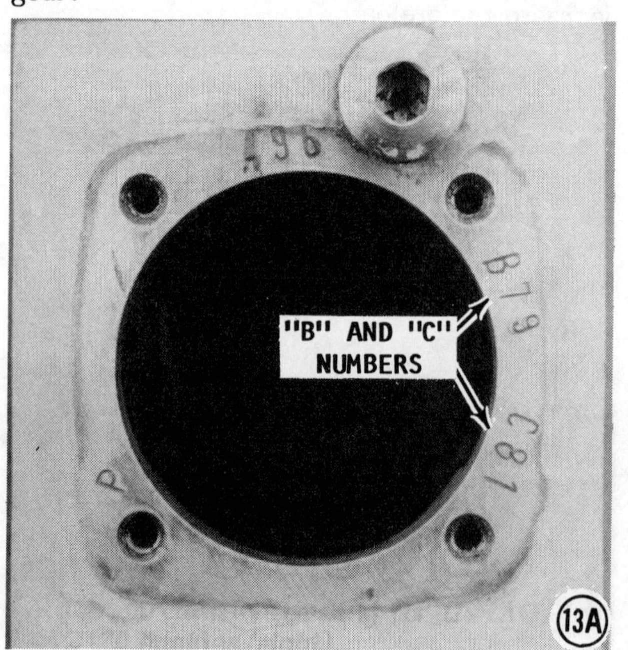

13- Several measurements are necessary in order to determine the amount of shim material required under the **FORWARD** and **REVERSE** gears.

Dimensions required for calculation:

First, the **"A"** dimension which is a fixed dimension, 62.05 mm, for a "perfect" housing.

Next, the **"B"** and **"C"** dimensions which are etched on the upper gear housing in the area where the shift cover is attached. On a standard stern drive, the **"B"** dimension is for the **REVERSE** gear. The **"C"** dimension is for the **FORWARD** gear. These two dimensions are clearly visible in the accompanying illustration,

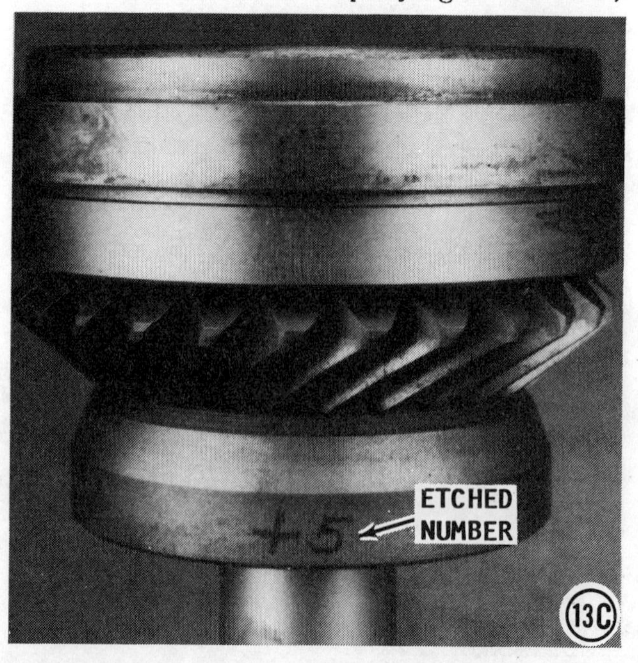

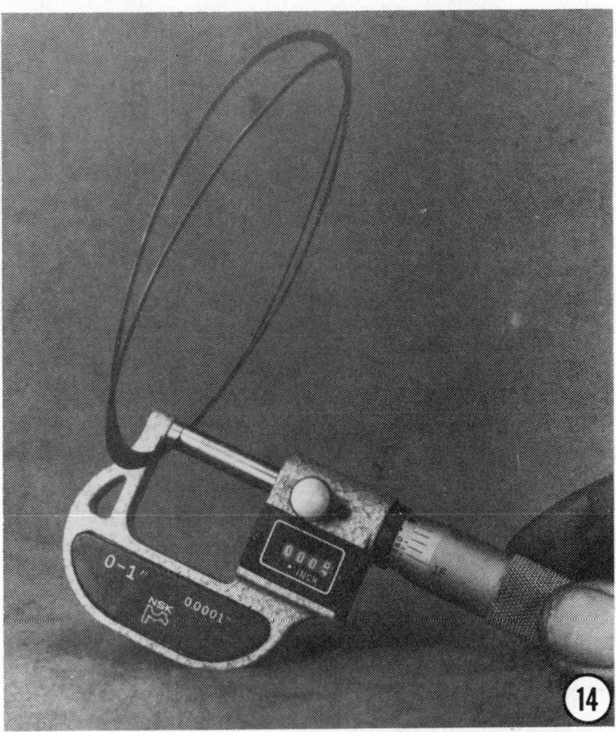

No. 13A. Notice the B79 and C81. It is necessary to assume a 61.79 mm for the "B" dimension and a 61.81 mm for the "C" dimension. The 61 is a set figure provided by the factory. Finally, the marking on each gear. The marking is located on the engaging cone of the gear assembly, as shown. Notice one is a minus (-) 3, illustration **No. 13B** and the other is plus (+) 5, illustration **No. 13C.** Per factory instructions, assume the "3" is actually .03 mm and the "5" is .05 mm.

14- Determine the amount of shim material required for the **FORWARD** gear. Begin with the 62.05 mm for a "perfect" housing. Add the +.05 as stamped on the **FORWARD** gear. The answer is 62.10 mm record this figure. Now, for the "C" dimension. Take the 61.81 mm obtained as the "C" dimension in the previous step, and subtract it from the 62.10 mm. The answer is .29 mm (.012") --the shim material required under the **FORWARD** gear.

Determine the amount of shim material required for the **REVERSE** gear. This procedure is exactly the same as for the **FORWARD** gear. Add or subtract the number etched on the **REVERSE** gear from the 62.05 mm for a "perfect" housing. Record this figure. Next, subtract the 61.81 mm obtained from Step 13. The answer is the amount of shim material required for the **REVERSE** gear.

Determine the amount of shim material required between the clamp ring and the upper gear housing.

15- Measure the depth of the clamp ring from the mounting surface to the recess where the upper gear housing is clamped.

16- Measure the thickness of the shoulder on the double bearing housing and subtract this figure from the measurement obtained as the depth of the recess. Now, add .004" and the answer is the amount of shim material required. The .004" will give that amount of "crush" when the assembly is

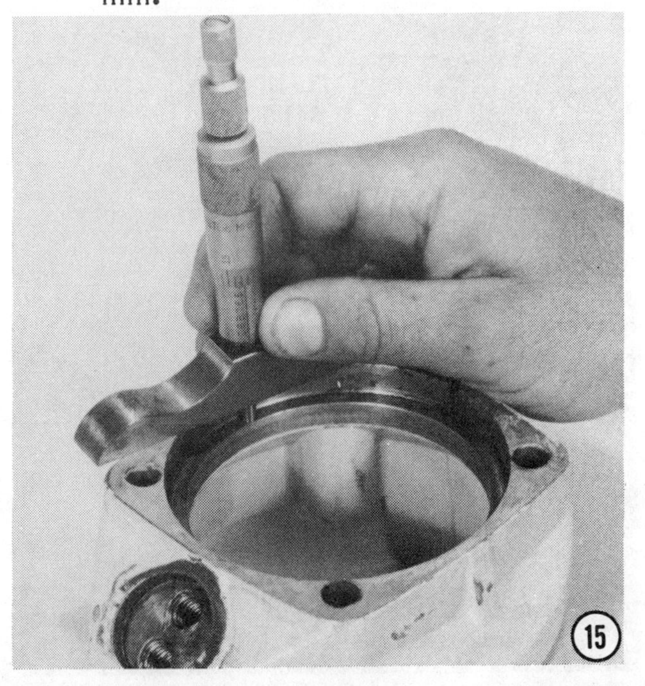

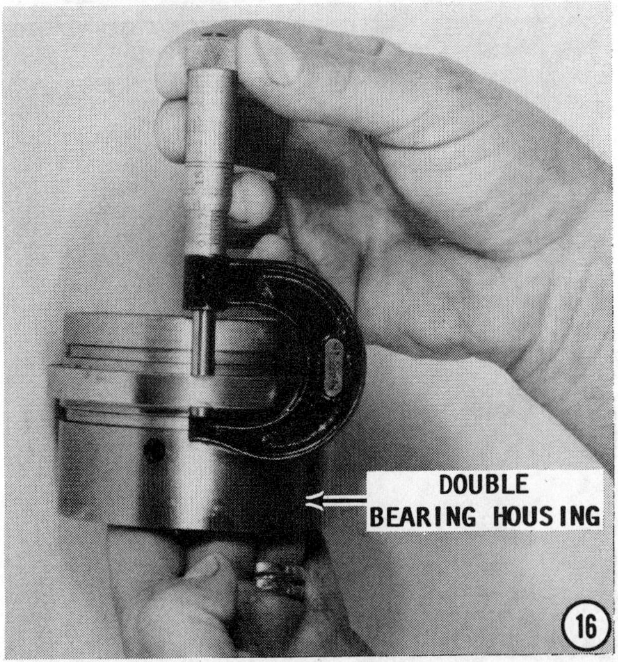

tightened. Divide the total amount of shim material required in **HALF.** Install 1/2 the amount of shim material on each side of the shoulder of the double bearing housing.

17- Check the amount of "backlash" and the gear pattern in the assembled upper gear housing. Begin by taking the vertical shaft and assembling it into the **FORWARD** gear. Install the collar and the two halves of the split ring onto the shaft. It is not necessary to install the sliding sleeve onto the shaft **AT THIS TIME.**

18- Slide the **FORWARD** gear and shaft assembly into the upper gear housing. Slide

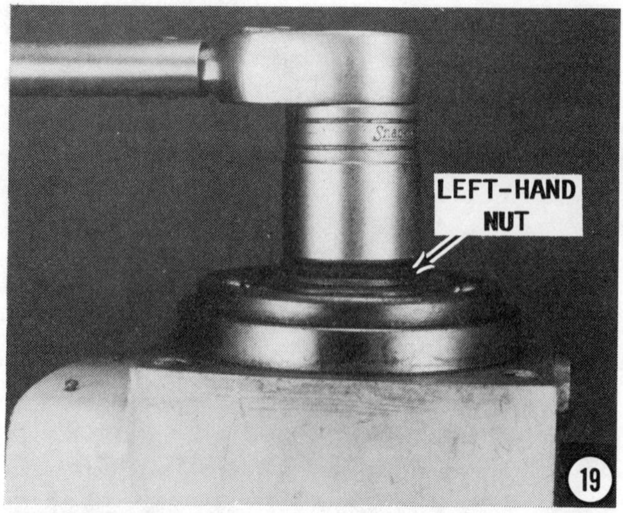

the **REVERSE** gear, with the proper amount of shim material under it, into the upper gear housing from the top.

19- Thread the **LEFT-HAND** nut onto the end of the shaft **COUNTERCLOCKWISE** and tighten it to a torque value of 92 ft-lbs (125 Nm).

20- Slide the clamp ring and the drive gear into the upper gear housing. Secure the ring with the four Allen-head fasteners.

21- Install the upper bearing cap onto the upper gear housing.

22- Obtain Volvo tool No. 884387. Clamp the tool onto the upper gear housing, as shown, to hold the lower gear and bearing assembly in place while checking the "backlash". If this tool is not available, the intermediate housing can be used as a "tool". In this case, clamp the upper gear assembly to the intermediate housing and

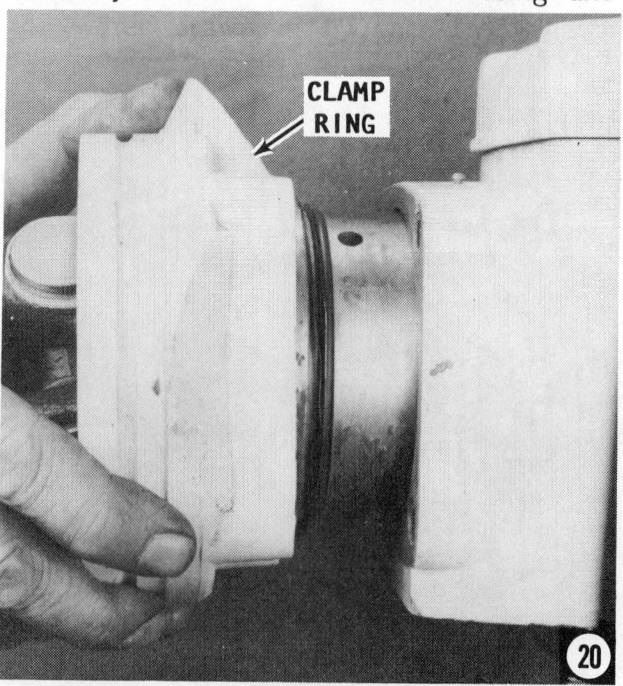

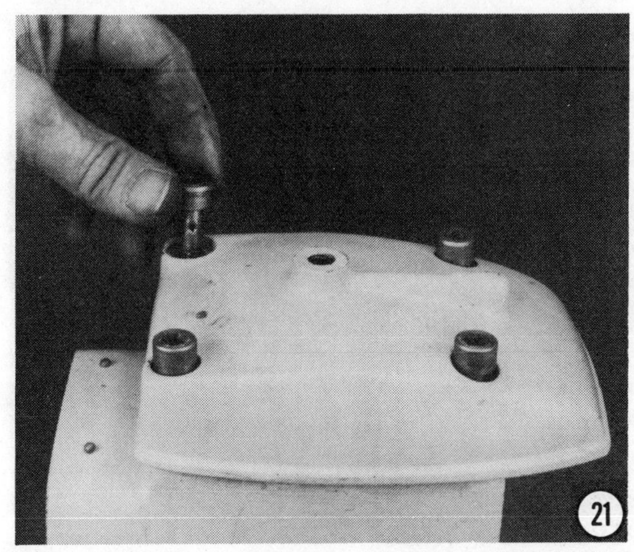
21

then make an attachment for the dial indicator to check the "backlash". Install the dial indicator and check the "backlash". The acceptable amount of "backlash" varies depending on the engine installed with the stern drive unit.

If the unit being serviced has a V8 power plant coupled with a 1.61:1 gear ratio, the acceptable "backlash" is 0.006" to 0.010".

If the unit being serviced has 4-cylinder power plant with a 2.15:1 gear ratio, the acceptable "backlash" is 0.003" to 0.007".

23- Once the proper amount of "backlash" is obtained, check the gear mesh pattern. This is accomplished by first marking the driven side of the gear with some type of material that will leave a pattern on the other gear when they mesh.

22

23

24- Use a screwdriver or hammer handle and apply friction to the driven gear. At the same time, rotate the drive gear at the universal joint yoke in a **CLOCKWISE** direction approximately 6 to 8 complete revolutions. Observe the gear pattern made from the dye material through the shift mechanism opening.

Compare the pattern made with the accompanying illustrations to determine how the two gears mesh. The pattern should almost be oval on the drive side and be positioned about halfway up the gear teeth, but on the small end, as shown in **"A"**. If the pattern resembles that shown in **"B"**, the shim thickness under the double bearing housing should be reduced. This will move the pattern towards the center of the tooth. If the pattern resembles that shown in **"C"**, the shim thickness under the double bearing housing should be increased.

If the pattern appears to be satisfactory, proceed with the assembling. If the pattern does **NOT** appear to be satisfactory, add or remove shims from under the double bearing housing to move the gear pattern towards or away from the center.

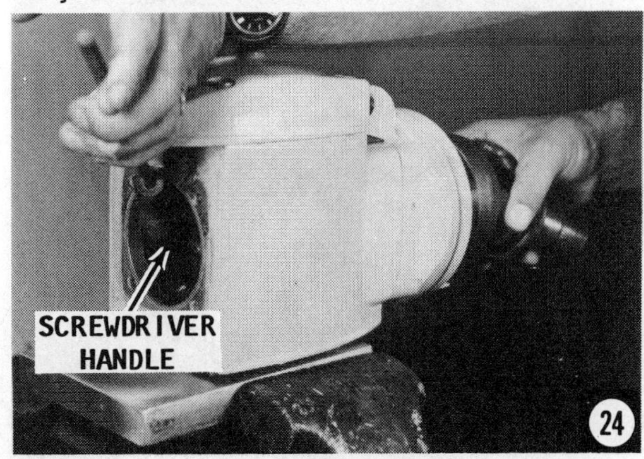

24

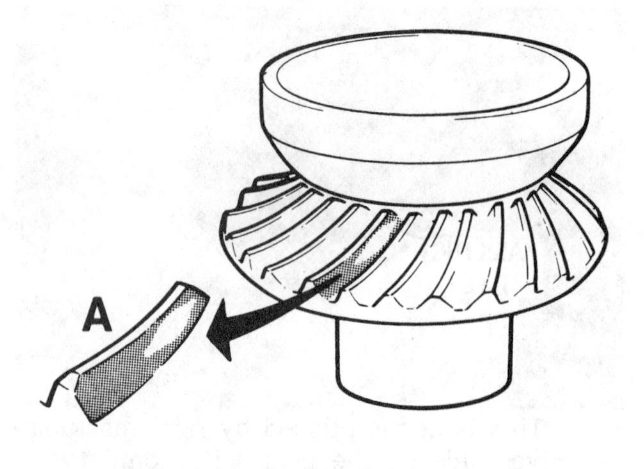

Gear patterns described in the text.

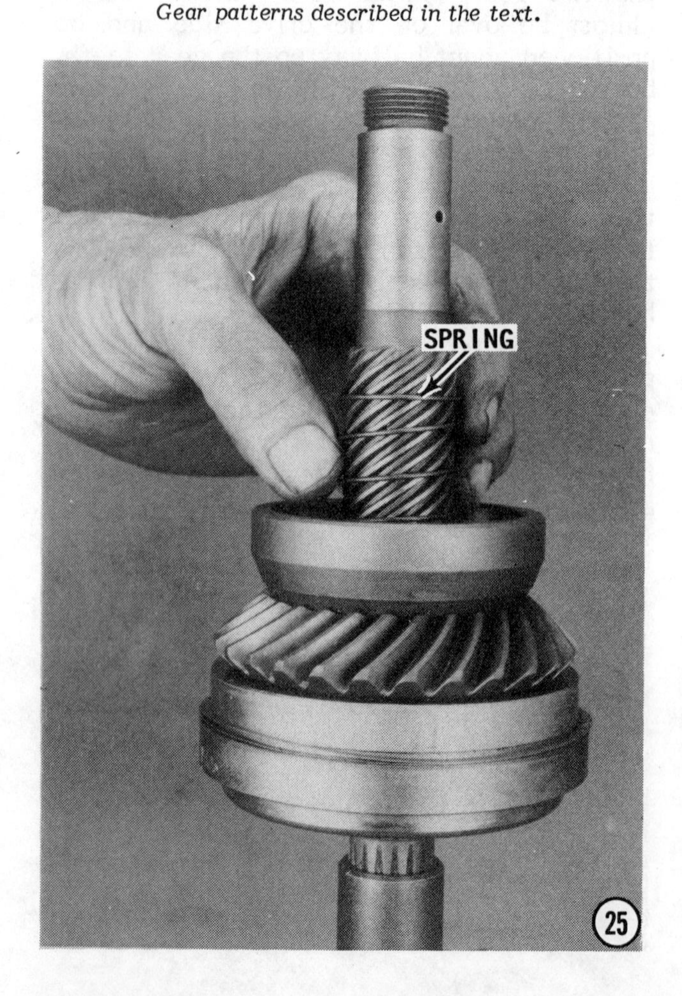

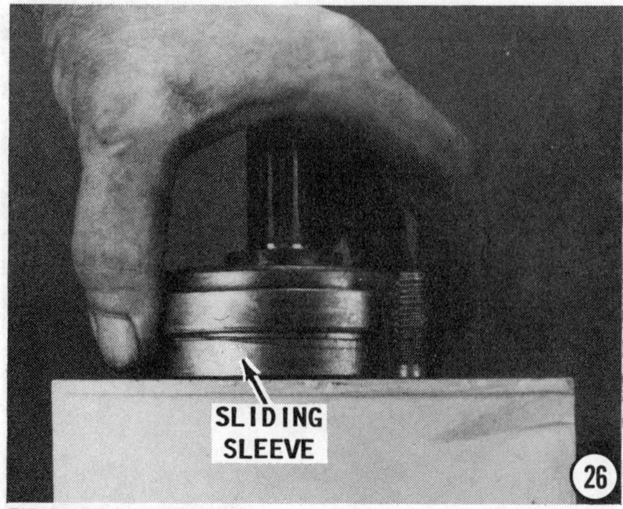

SPECIAL WORDS

After the correct "backlash" and gear pattern have been obtained, the upper gear must be disassembled and cleaned of all marking dye. The bearings and screws should all be oiled before assembling them the final time. New seals and O-rings **MUST** be installed.

Remove the clamp ring and the universal joint assembly. Remove the upper bearing cap. Remove the housing from the tool or off the intermediate housing, whichever was used.

25- Slide the spring onto the shaft. Install the sliding sleeve onto the shaft.

26- Install the lower driven **(FORWARD)** gear and vertical shaft assembly into the upper gear housing.

27- Work the upper driven gear onto the shaft and into the housing.

28- Thread the vertical shaft nut onto the shaft. Tighten the nut to a torque value of 92 ft-lbs (125 Nm).

29- After the nut has been tightened to the proper torque value, check the shaft "end play" under the nut -- the clearance

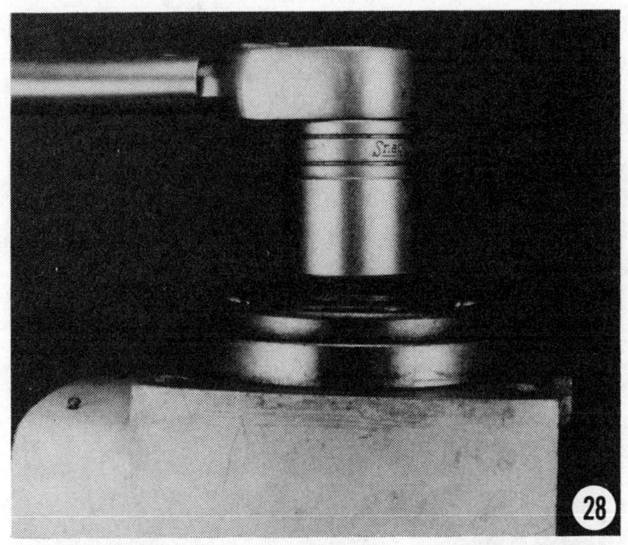

between the nut and the bearing. This clearance should be 0.004" to 0.020" (0.1 to 0.5 mm). If the "end play" is not within the allowable tolerance, then a nut with a thicker or thinner washer may be purchased and installed.

This nut is available with three different size washers, as shown in the accompanying illustration, to provide the proper "end play" for the vertical shaft. A standard nut will have no marking, another will have a **"1"** stamped on it, and the third nut will have a **"2"**. Each nut will have a different size washer.

If the proper "end play" cannot be obtained by using any one of the three nuts

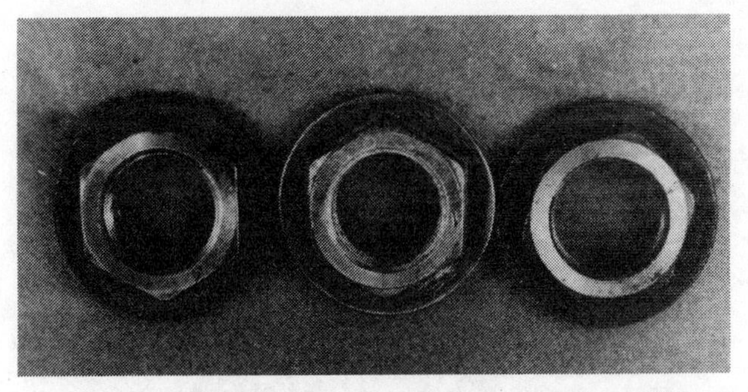

The vertical shaft nut is available with three different washer thicknesses, as explained in the text.

available, the bearing under the nut is most likely worn excessively and no longer fit for service. Another cause for failure to obtain the proper "end play" from one of the three nuts, is excessive wear on the vertical shaft collar. Damage to this collar is usually caused by shifting the unit at high engine rpm.

30- Place a **NEW** O-ring on both sides of the double bearing housing shoulder.

31- Install the oil seal into the double bearing housing.

WARNING

This next step involves a dangerous procedure and should be executed with care while wearing SAFETY GLASSES. The snap ring is placed under tremendous tension with the Truarc pliers while it is being installed. If it should slip off the Truarc pliers, it will

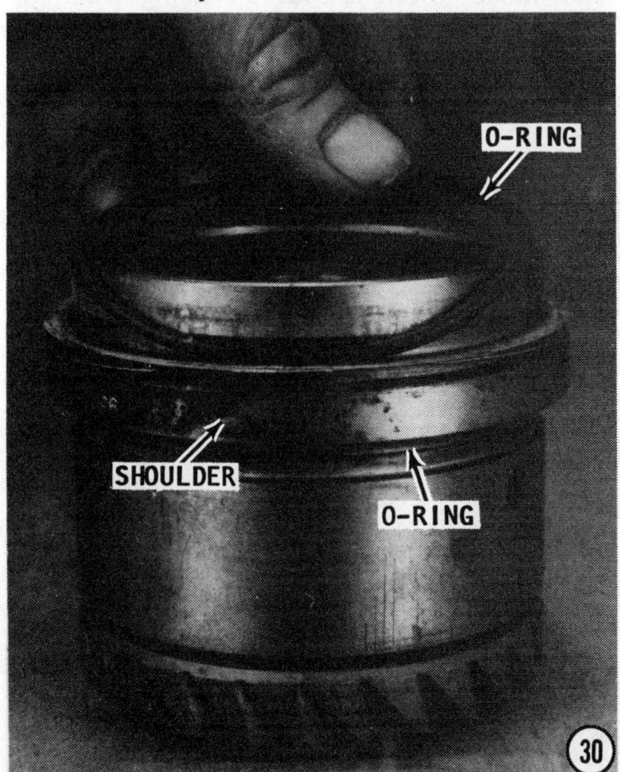

31

travel with incredible speed causing personal injury if it should strike a person.

32- Using a pair of Truarc pliers, **CAREFULLY** install the snap ring.

33- Place the correct amount of shim material inside the head washer. Install the head washer into the double bearing housing. Fit a **NEW** O-ring onto the back side of the head washer.

34- Slide the clamp ring down over the assembled double bearing housing, with the locating pin on the double bearing housing

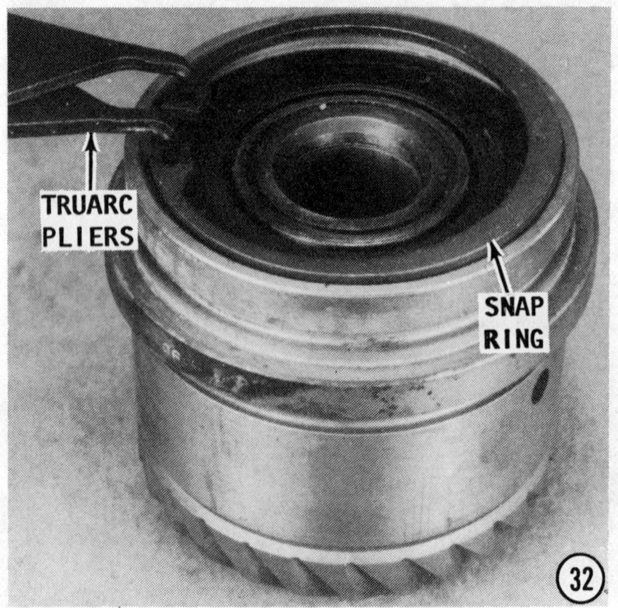

32

33

sliding into the index groove of the clamp ring.

35- Slide the universal joint through the double bearing housing and clamp ring. Rotate the universal joint slightly to allow the splines on the shaft to index with the splines in the drive gear. Coat the threads of the retaining bolt with Loctite or other suitable

34

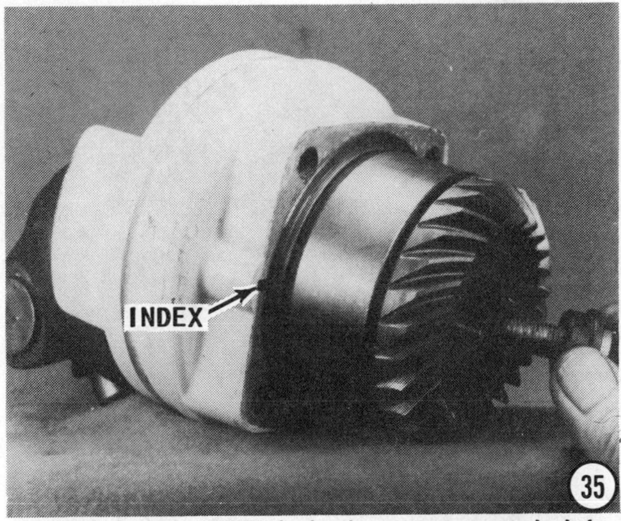

material. A **NEW** bolt is recommended by the factory. Slide the washer onto the bolt and then thread the bolt into the end of the universal joint shaft. Tighten the bolt to a torque value of 62.7 ft-lbs (85 Nm).

36- Slide the assembled double bearing housing and clamp ring into the upper gear housing.

37- Press the clamp ring down into place by hand and at the same time measure the clearance between the clamp ring and the upper bearing housing with a feeler gauge, as shown. This is a check at this time. If the proper amount of shim material has

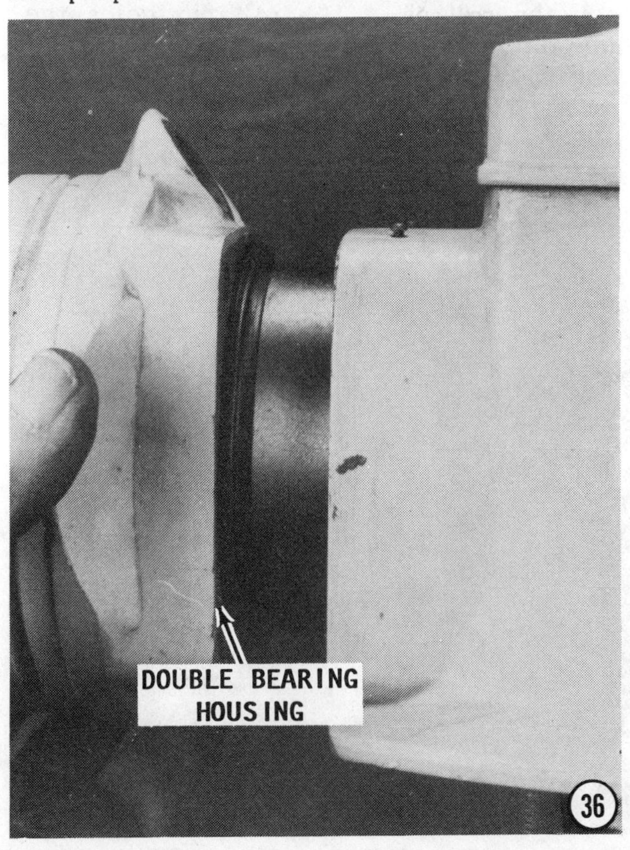

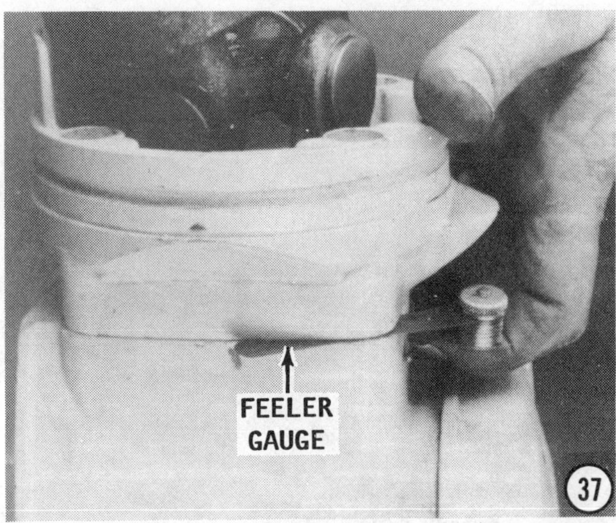

been installed on each side of the double bearing housing, this clearance will be approximately 0.004" (0.1 mm). Remove the assembled double bearing housing and clamp ring.

38- Coat the mating surfaces of the upper gear housing and the clamp ring with Permatex, or other suitable sealing material. Mate the clamp ring to the upper gear housing. The guide pin in the double bearing housing **MUST** be turned downward. Coat the threads of the Allen-head bolts with Loctite, or similar material. Thread the screws into place, and then tighten them to a torque value of 25.9 ft-lbs (35 Nm).

39- Measure the depth of the upper gear housing cap -- from the mating surface of the cap to the recess for the upper bearing, as shown.

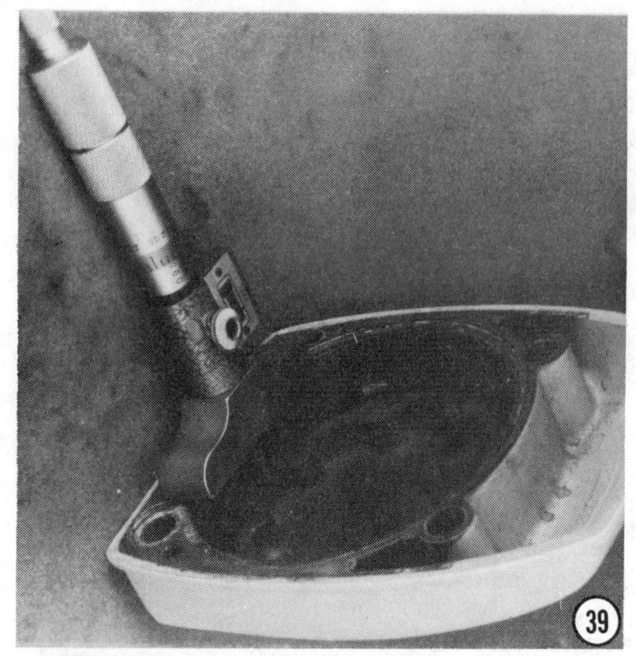
39

40- Measure the distance from the top of the bearing to the mating surface of the upper gear housing. Subtract this measurement from the measurement obtained in the previous step (the depth of the upper gear housing cap). Add 0.004" to the answer for "crush" on the bearing.

41- Place the correct amount of shim material into the upper gear housing cover. Position a **NEW** O-ring into the groove in the cover with the special added O-ring seal fitting over the hole for the special cover bolt, as shown.

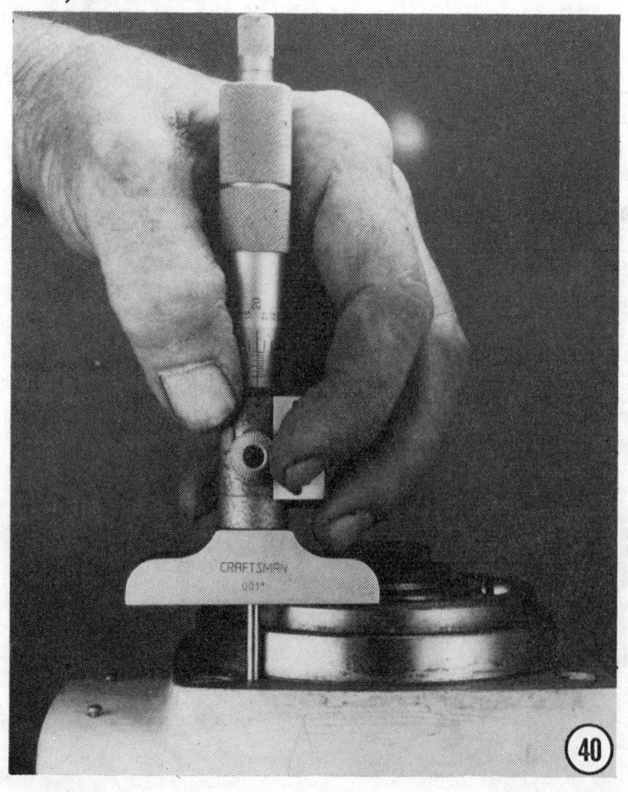

40

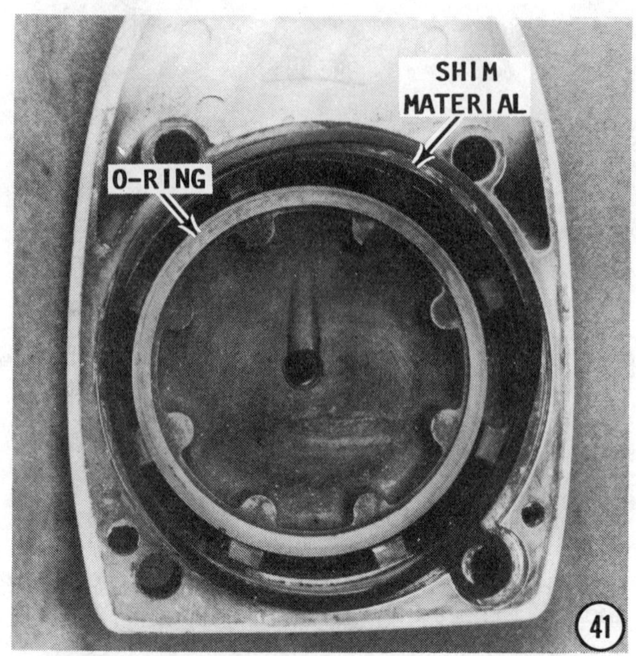

41

42- Install the upper gear housing cap. The forward bolt on the starboard side is a special bolt with an O-ring. This bolt is removed to add lubricant through the upper gear assembly.

43- Drive a new sealing ring into the shift cover with the spring facing inward.

44- Insert the eccentric piston into the shift housing.

45- Insert the main pin into place, and then the roll pin. The roll pin holds the main pin in the proper position.

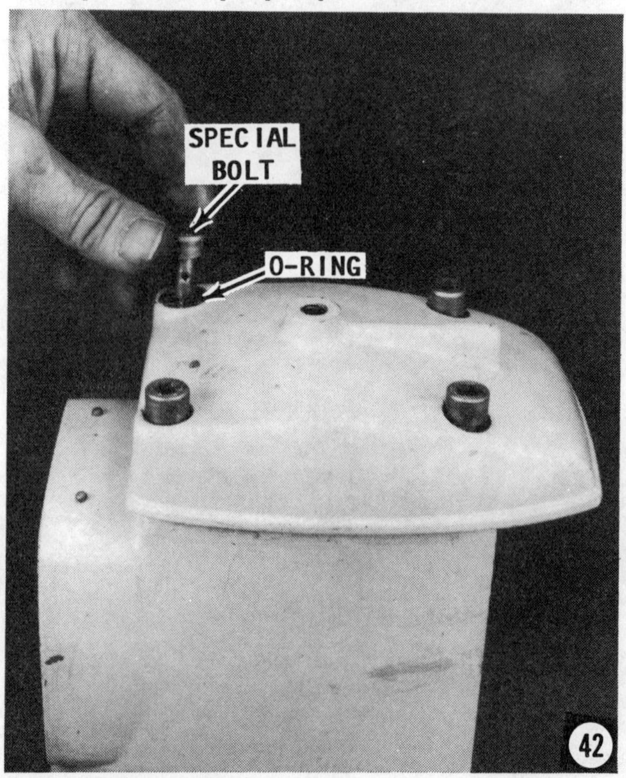

42

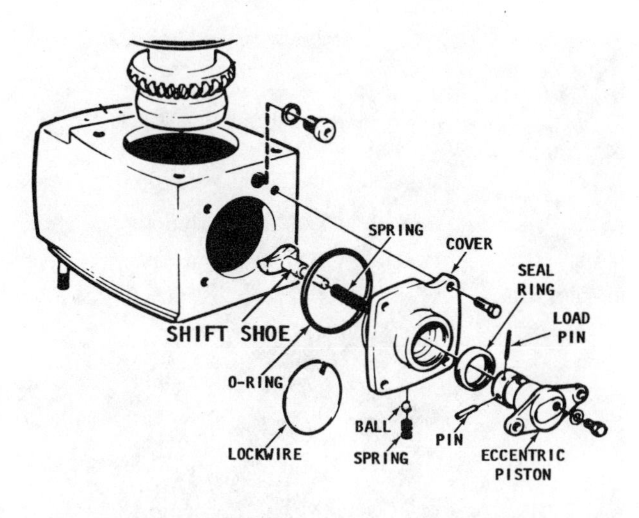

Exploded drawing of the early model upper gear housing. Later models do not have the ball and spring.

46- Slide the spring onto the shift shoe shaft, and then insert it into the eccentric piston.

47- Install the shift shoe tensioning bolt with the same amount of shim material under the head as was removed during disassembling. Install the assembled shift mechanism to the upper gear housing with the shift shoe shaft retainer bolt offset to starboard. When the shift shoe enters the housing, the high side of the shoe **MUST** be on the starboard side. The shift mechanism will not function properly if the shoe is turned the other way. Secure it in place with the attaching bolts.

SPECIAL WORDS

If the amount of shim material was not saved or recorded, or if the upper gear

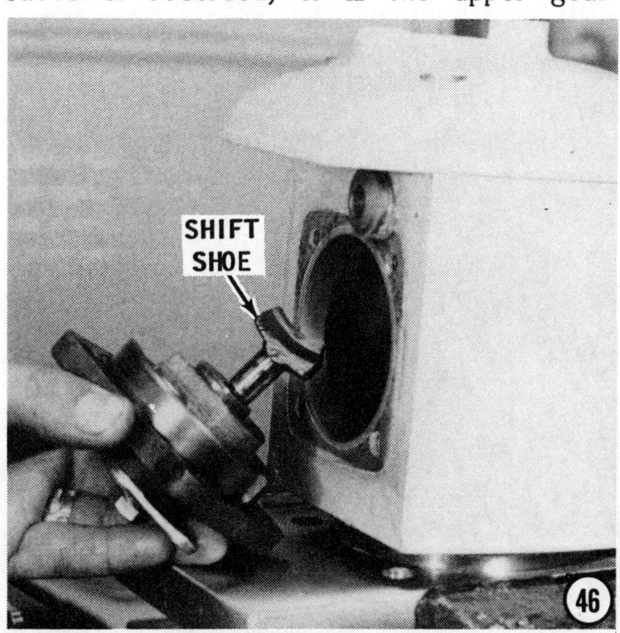

47

assembly has been rebuilt with new parts, an adjustment must be made to the shift mechanism as follows:

With no shim material under the head, thread the shift shoe tensioning bolt into the end of the shaft. Install the assembled shift mechanism to the upper gear housing with the shift shoe shaft retainer bolt offset to starboard. When the shift shoe enters the housing, the high side of the shoe **MUST** be on the starboard side. The shift mechanism will not function properly if the shoe is turned the other way. Secure it in place with the attaching bolts.

Attempt to rotate the vertical shaft in the upper gear housing. The attempt should **FAIL.**

Remove the bolt and place one spacer under the head. Replace the bolt. Again

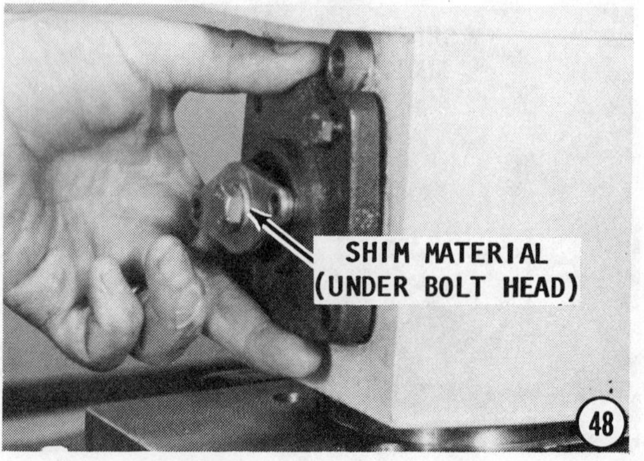

48

49

attempt to rotate the vertical shaft. Continue placing one spacer at-a-time under the head until the shaft can be rotated without resistance.

48- After the adjustment is satisfactory, once again remove the shift mechanism, bolt, and shim material. Coat the bolt and shim material with sealant, and then install and tighten the bolt securely. Coat the mating surfaces of the shift cover and the upper gear housing with Permatex or other suitable sealing material. Install the mechanism bolt, and shim material one final time. Check to be sure the high side of the shift shoe is on the starboard side and the

50

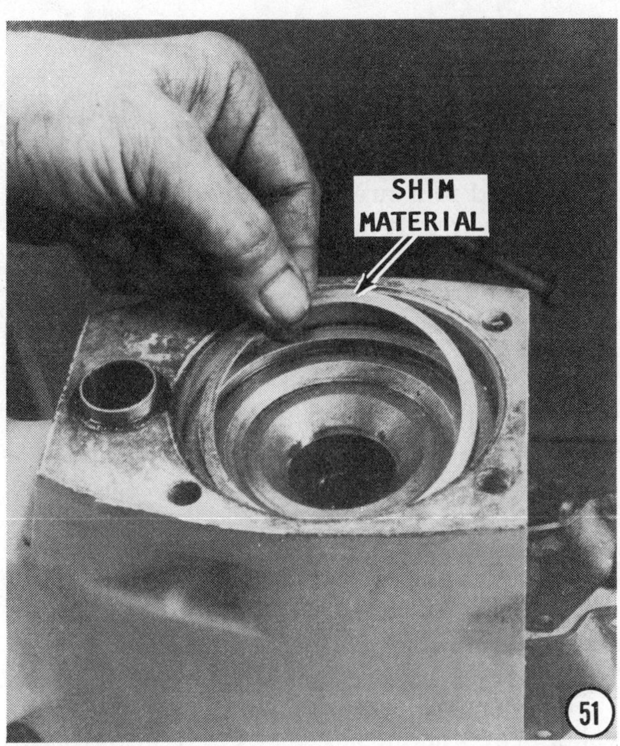

bolt offset to starboard. Tighten all bolts securely.

The upper gear assembly is now ready to be mated with the intermediate housing.

49- Measure the distance from the mating surface of the upper gear housing to the lower driving gear bearing housing, as shown. Record this figure.

50- Measure the distance from the mating surface of the intermediate housing down to the shim recess. Subtract this figure from the measurement obtained in Step No. 49. Add 0.004" to the answer. The sum will be the amount of shim material to be installed in the intermediate housing recess. The 0.004" will provide adequate preload for the lower driven gear when the upper gear housing is mated to the intermediate housing.

51- Install the proper amount of shim material, as determined from Step No. 50, into the recess in the intermediate housing.

52- Slide the coupling into the intermediate housing with the splines of the coupling indexed with the splines of the vertical shaft.

53- Position a **NEW** O-ring over the bearing housing of the upper gear assembly.

HELPFUL HINT

Before the upper gear housing is mated to the intermediate housing, insert a funnel into the oil passage tube, and then pour in

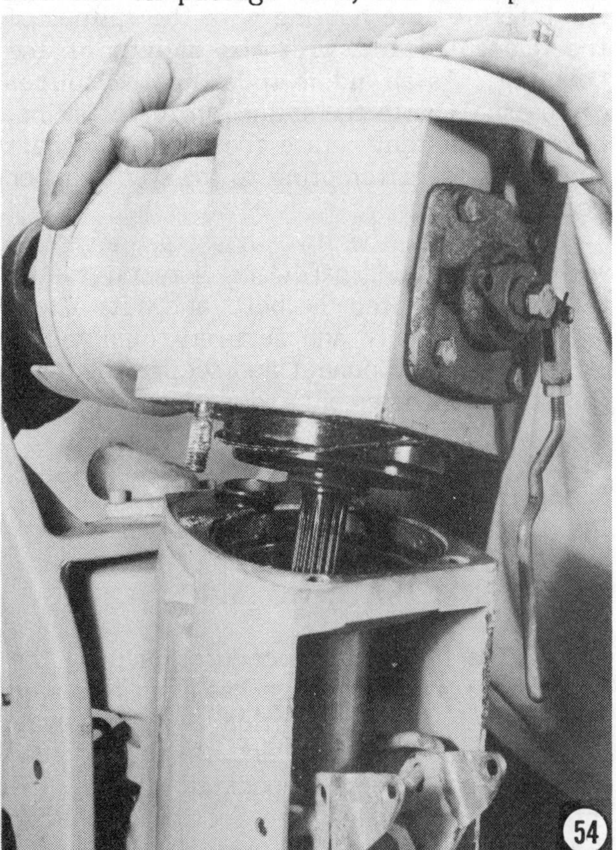

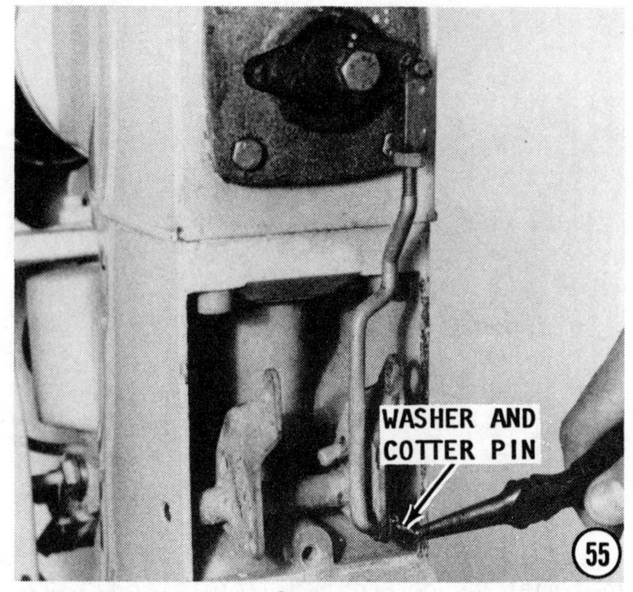

approximately 2-1/4 quarts of lubricating oil. Filling the lower unit at this time is much quicker and easier than adding the full amount of oil after the upper gear housing is secured and the cap is in place.

Check to be **SURE** the drive coupling is installed before the upper gear housing is mated to the intermediate housing.

54- Coat the mating surfaces of the intermediate and upper gear housings with Permatex or other suitable sealing material. Lower the upper gear assembly down onto the intermediate housing with the splines of the shaft indexed with the splines of the coupling. As an aid to indexing the splines of the shaft with the splines in the coupling, have an assistant rotate the propeller shaft slightly while attempting to lower the upper gear housing.

55- Secure the two housings together with the attaching hardware -- two nuts and two bolts. Tighten the bolts and nuts alternately and evenly and securely in a criss-cross pattern. Connect and secure the shift rod with the cotter pin and washer.

The assembled stern drive is now ready for installation to the boat, see Section 8-4, Installation, beginning on Page 8-10.

8-6 INTERMEDIATE HOUSING

The following procedures pick up the work after the stern drive has been removed from the boat, See Section 8-4, Removal, beginning on page 8-8, and the upper gear housing has been separated from the intermediate housing, see Section 8-5, Page 8-13.

DISASSEMBLING

1- Remove the seven attaching bolts securing the intermediate housing to the lower unit. Three bolts are located on top, as shown, and four are located under the flange on the lower unit housing.

2- Separate the intermediate housing from the lower unit.

3- Pull the oil strainer from the lower unit housing. This strainer should be pulled and cleaned any time the intermediate housing is separated from the lower unit for any reason.

4- Remove the two bolts securing the water intake neck to the steering fork. Remove the hose neck. Remove and **DISCARD** the gasket.

5- Start to drive the water tube free of the steering fork, using a suitable driving tool.

6- Remove the water tube from the yoke. This tube also acts as a steering pivot point.

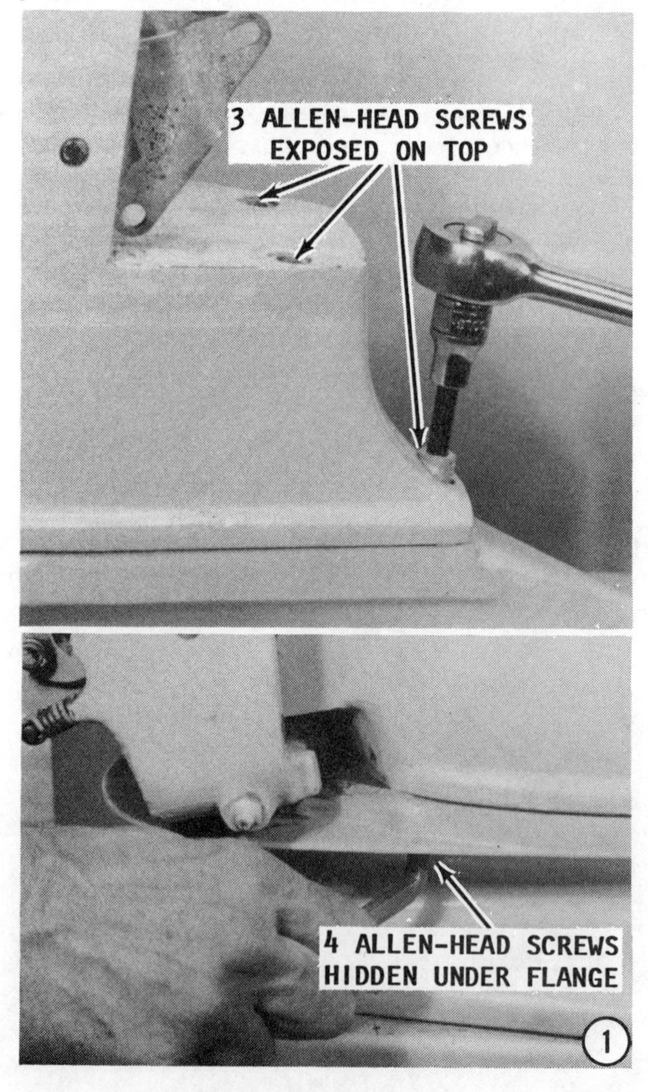

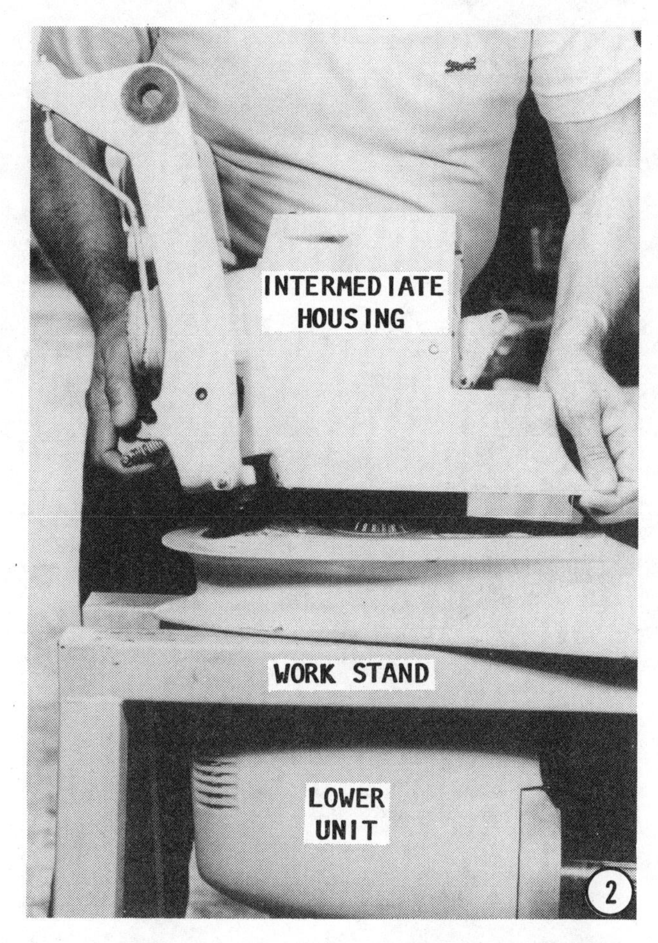

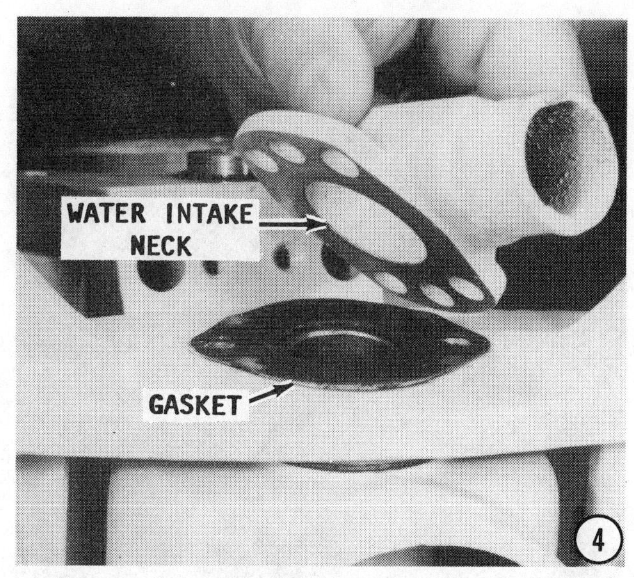

7- Under normal maintenance of the intermediate housing the oil distribution tube is not removed. However, if it needs to be serviced or replaced, simply pull it out. The fit is not a press fit, it is a slide fit.

8- Remove and **DISCARD** the **O**-ring water seal from the lower unit.

9- Use Volvo special tool No. 884259 and No. 9991801, or similar arrangement, and press the sealing rings and the needle bearing housing free of the steering yoke.

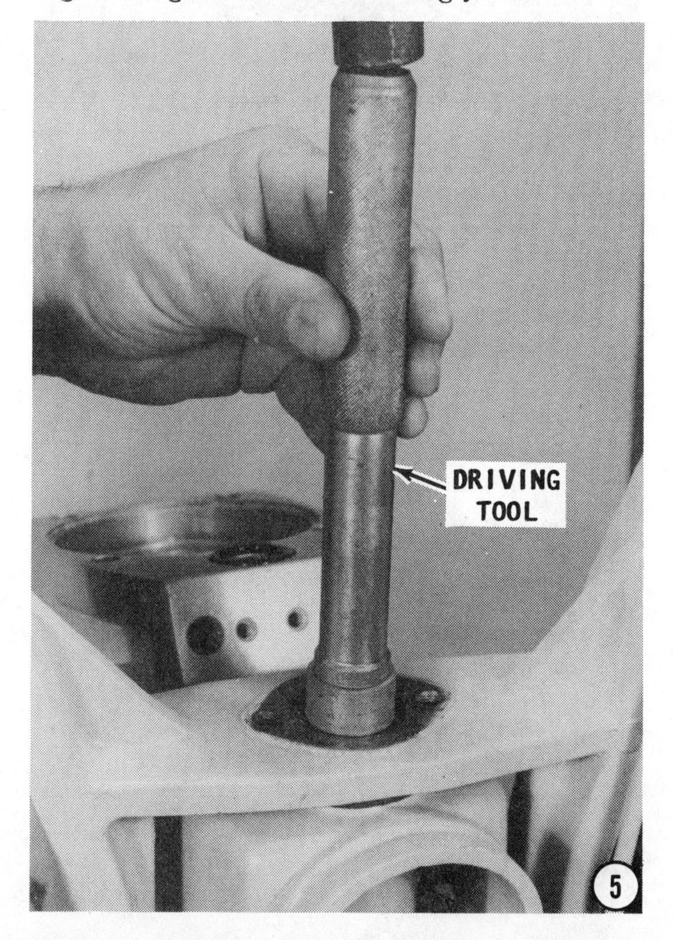

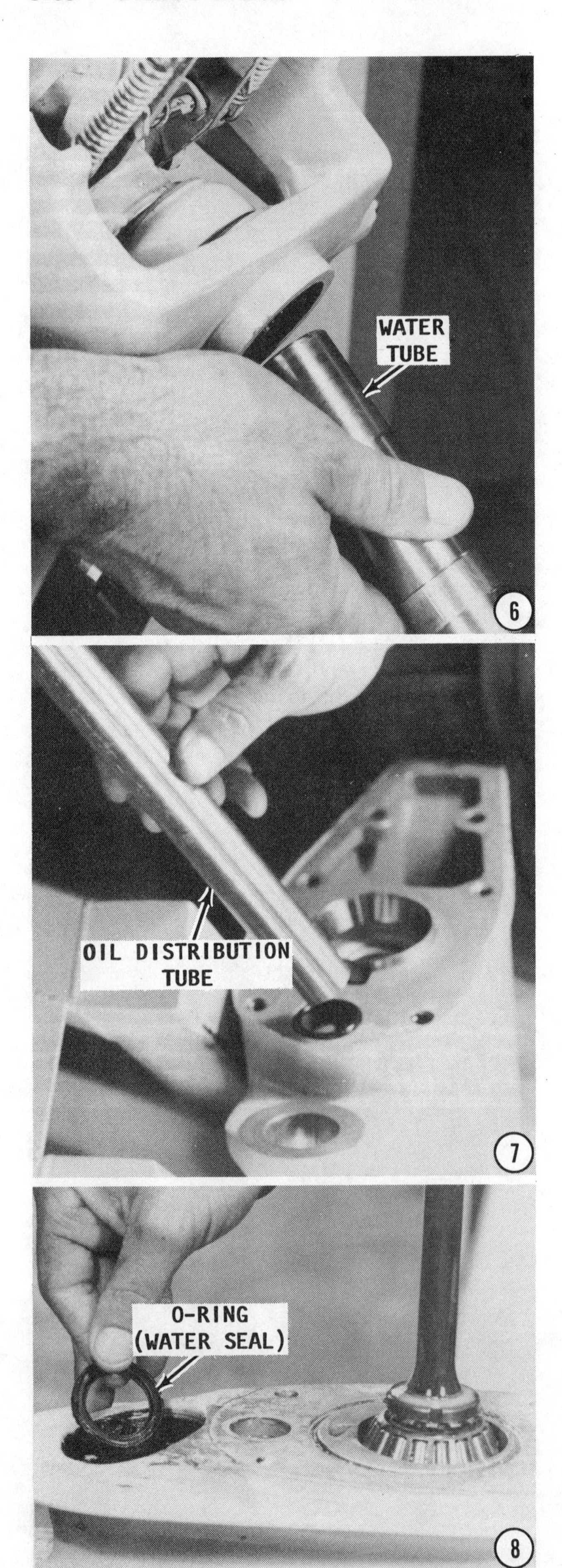
WATER
TUBE
6
OIL DISTRIBUTION
TUBE
7
O-RING
(WATER SEAL)
8

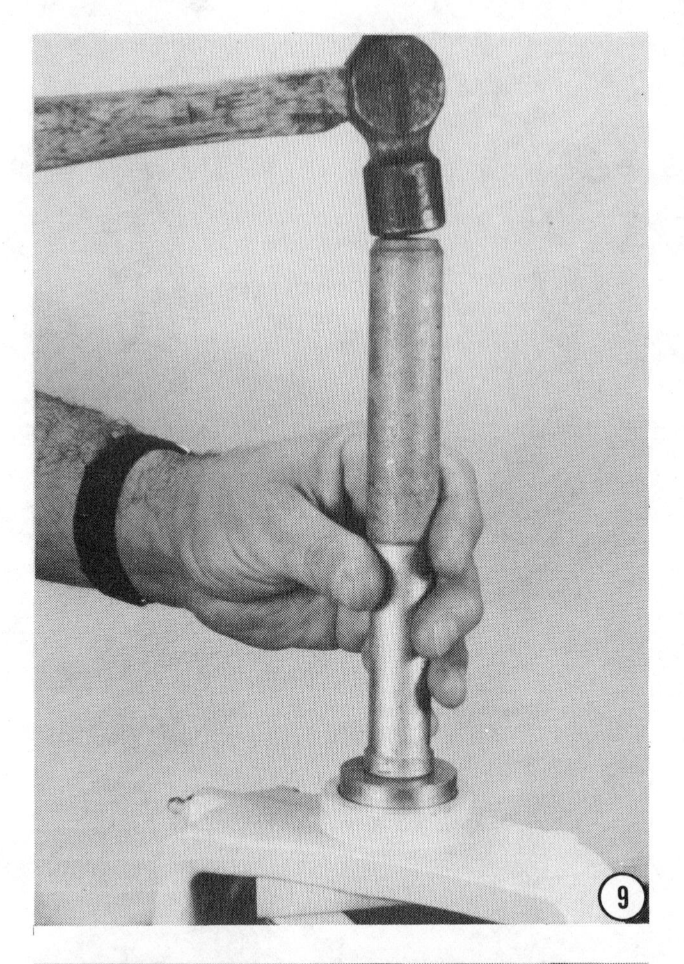
9

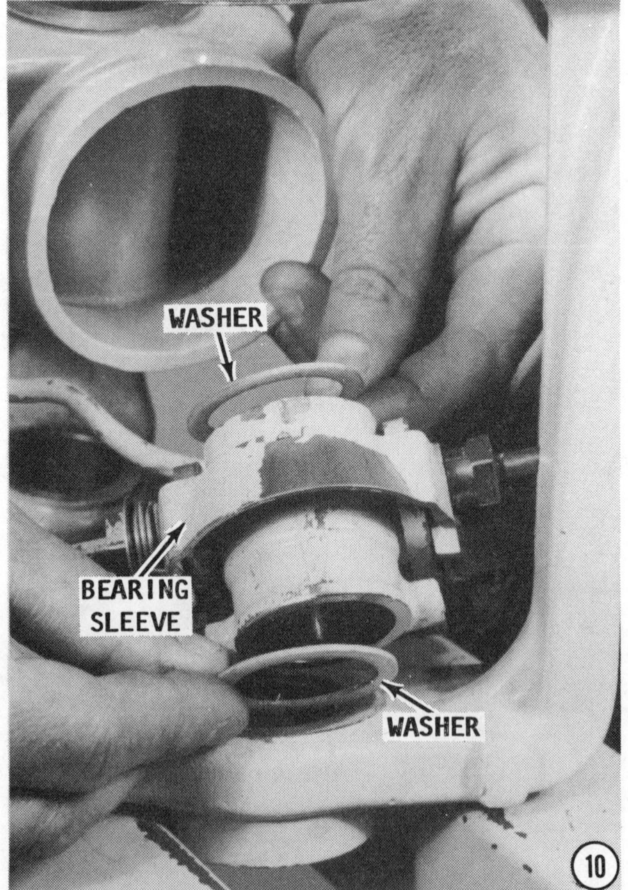
WASHER
BEARING
SLEEVE
WASHER
10

10- Pull the assembled latch mechanism free of the steering yoke. Save the two nylon washers installed on either end of the sleeve.

11- Clamp the reverse latch mechanism in a vise equipped with soft jaws. Pry the cross bar off the lock body. The rod is under spring tension. Exercise **CARE** to prevent the rod from flying loose.

12- Use the proper size open-end wrench

to loosen, and then remove the locating pin from one side of the latch body.

13- Remove the locating pin and spring from the other side of the latch body. Remove the reverse latch locking hoop.

14- Remove the nut from each end of the connecting rod, and then remove the reverse latch pawls.

15- Slide the reverse levers free of the stop shaft. The push rod will come free on the one side.

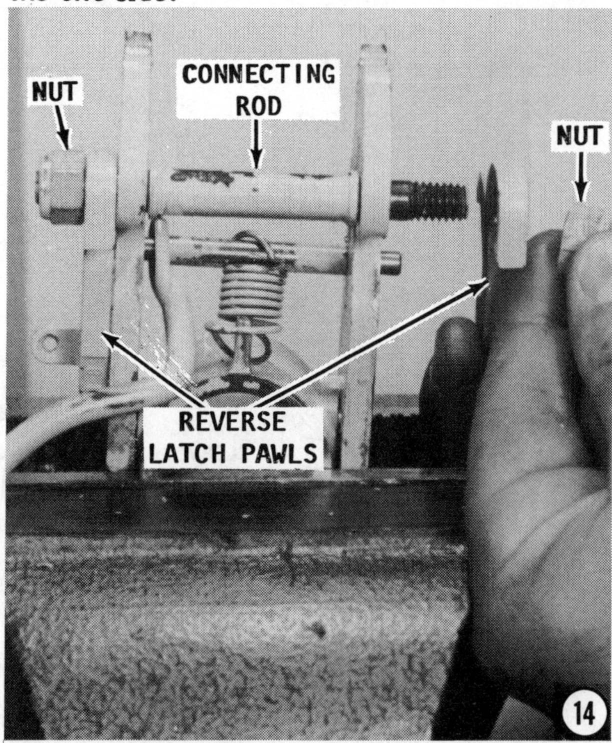

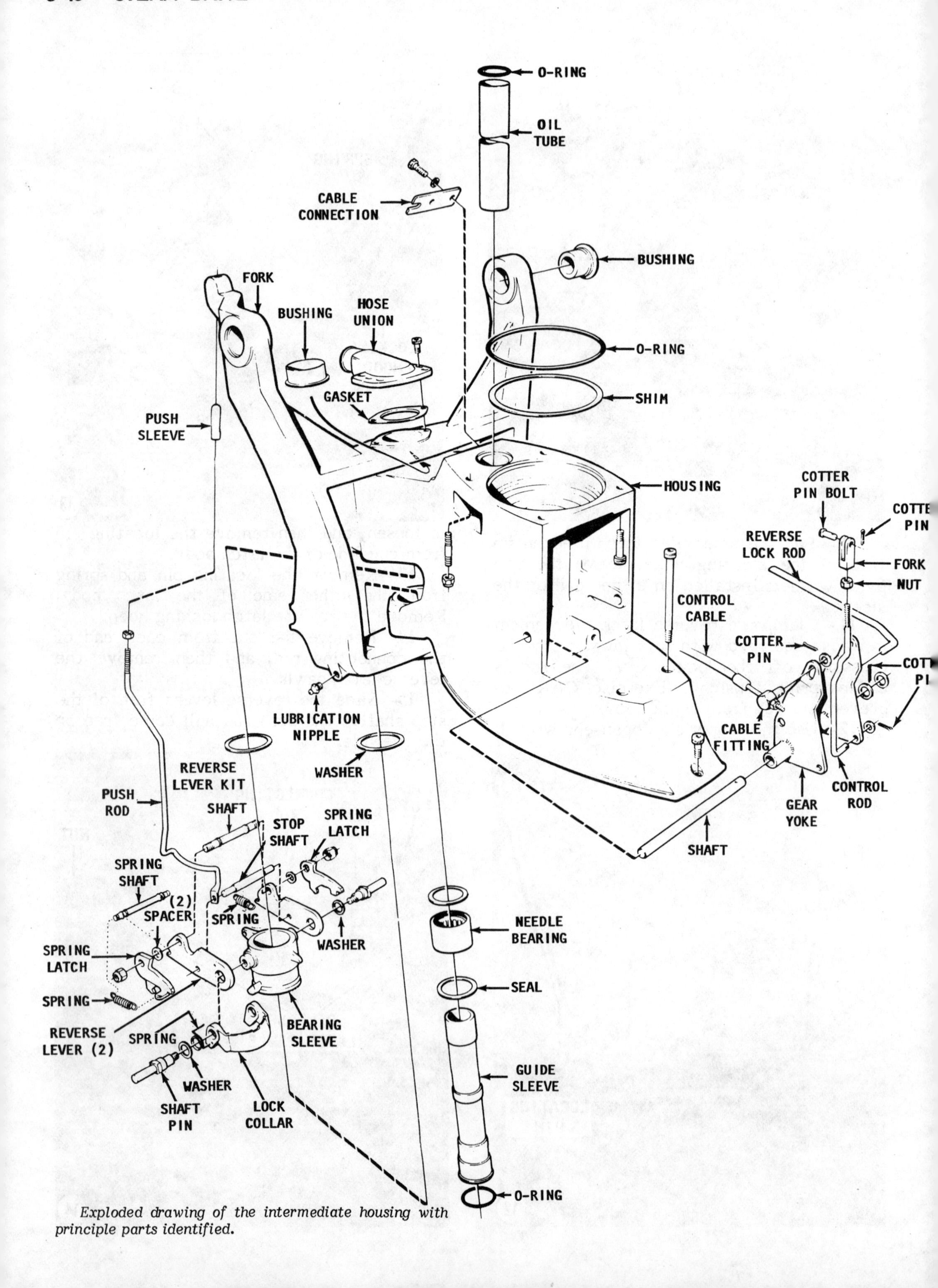

Exploded drawing of the intermediate housing with principle parts identified.

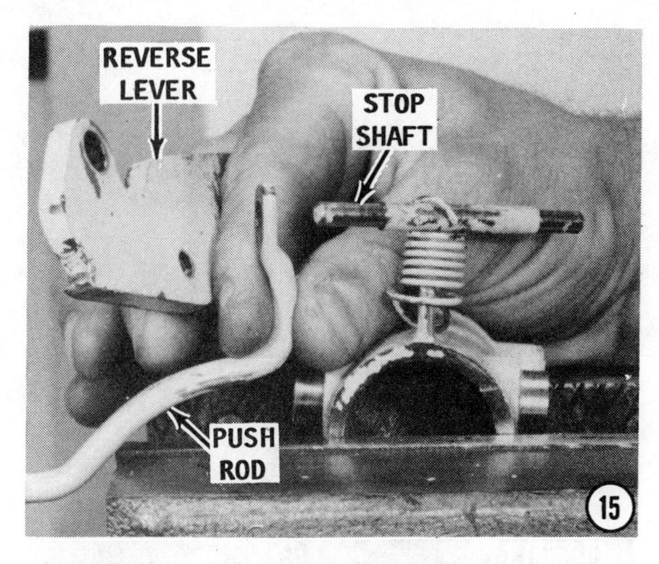

CLEANING AND INSPECTING

Discard all gaskets and O-rings. Replacement of these items with **NEW** ones will help to ensure satisfactory service after the work is completed.

Clean all parts with solvent and blow them dry with compressed air.

Check all mating surfaces to be sure they are clean and free of any old gasket material or sealing compound.

Clean the oil strainer screen thoroughly and blow it dry with compressed air.

Blow all water, oil passageways, and screw holes, clean with compressed air. After all parts are clean and dry, apply a light coating of engine oil to the bright surfaces of all bearings and shafts as a prevention against rusting.

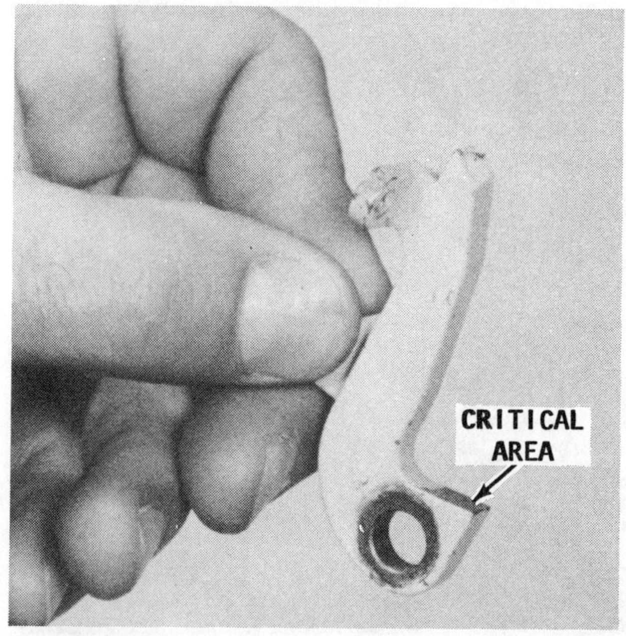

Critical area of the latch pawl to be closely inspected, as explained in the text.

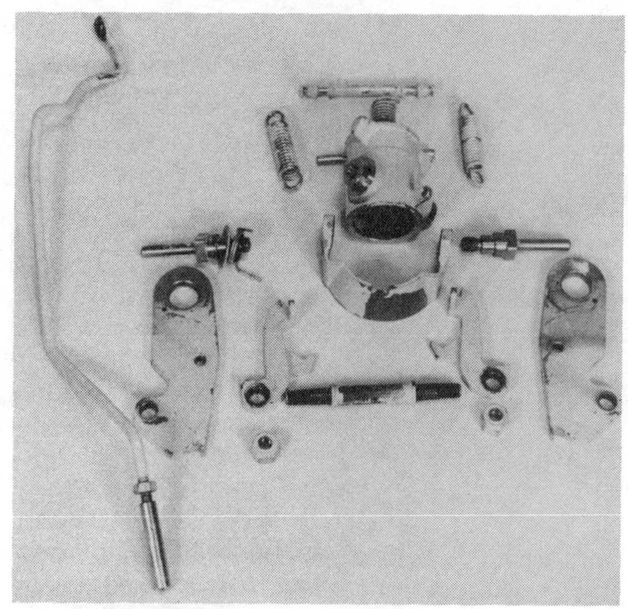

General arrangement of parts for the shift mechanism.

Inspect the latch pawls for excessive wear. Check the area where the pawl makes contact with the tilt rod for any "mushroom" condition or roundness. Excessive wear in this area would not allow the pawl to function properly. The stern drive would kick upward when the unit was shifted into **REVERSE** gear and the throttle advanced.

Check the housing for cracks, or other damage.

Check the needle bearing set for galling, and to be sure they all rotate freely in the cage.

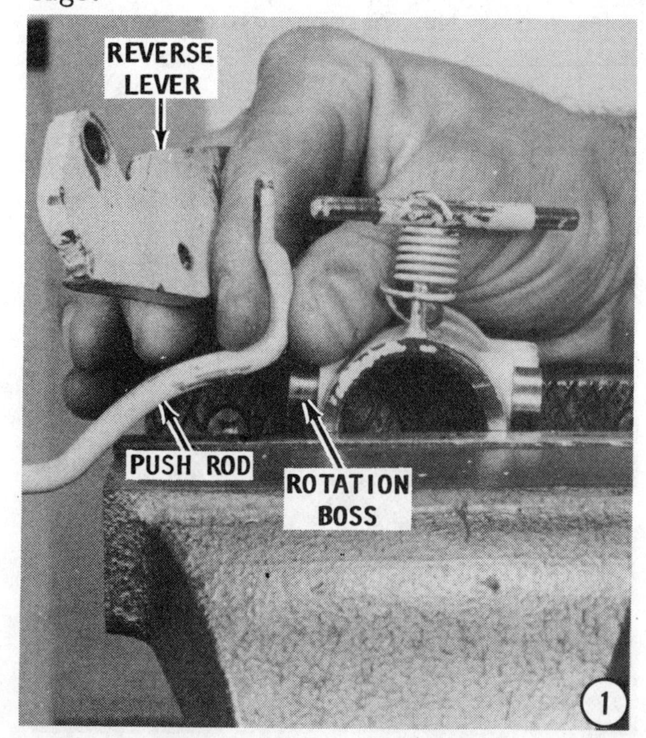

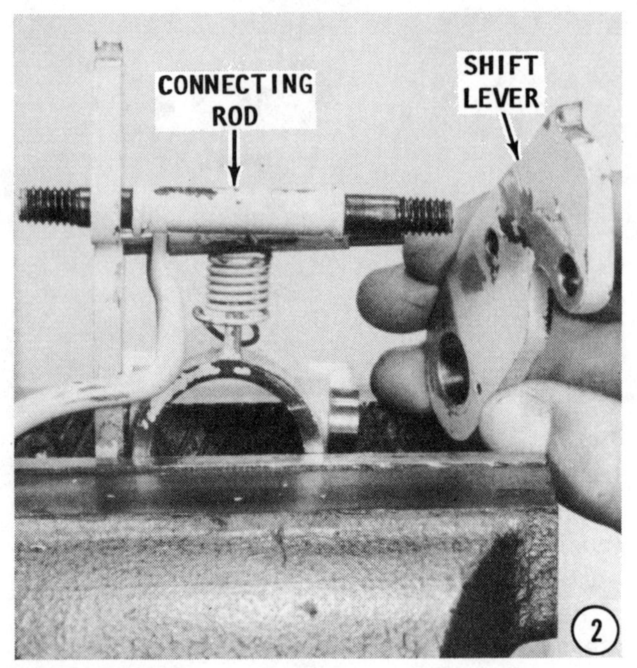

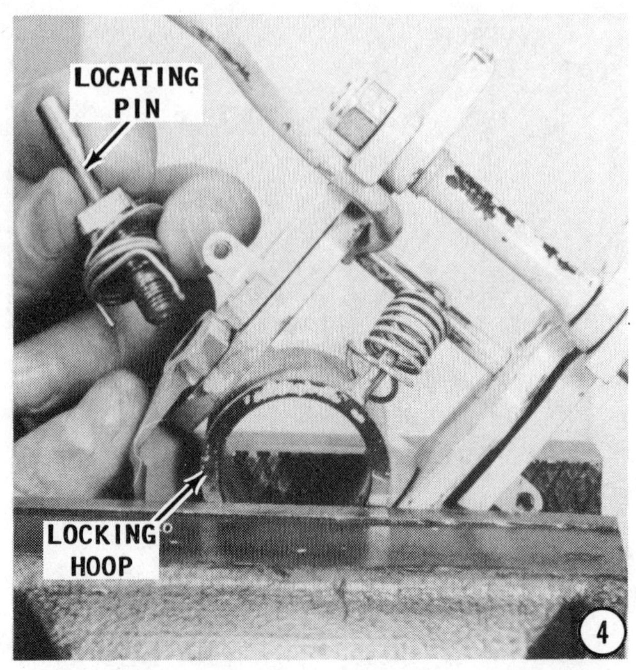

ASSEMBLING

1- Clamp the reverse latch mechanism body in a vise equipped with soft jaws. Slide the push rod onto the stop shaft. Slide the reverse lever onto the stop shaft on the same side and also onto the rotation boss on the bearing sleeve. Check to be sure the push rod passes the shift lever, as shown. Push the connecting rod through the one reverse lever.

2- Slide the other reverse lever onto the other end of the stop shaft with the connecting rod passing through and the lever indexed over the rotation boss on the bearing sleeve.

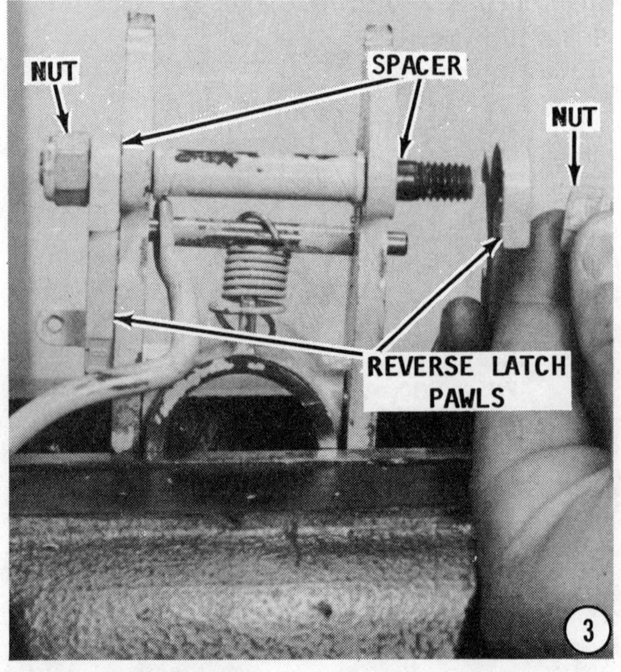

3- Slide a spacer onto the connecting rod outside each reverse lever. Slide the left reverse latch pawl onto the connecting rod and secure it in place with the nut. Slide the other latch pawl onto the connecting rod, and then secure it in place with the nut. Tighten the nut on both ends of the connecting rod just **SNUG**, and then back them off 1/8 turn, or just enough to permit the reverse latch pawl to rotate freely without binding. Check to be sure the levers and pawls are in their proper position and they all rotate without binding.

4- Push the locking hoop into position from the underneath side. This can be accomplished while the assembly is still

Use a pair of needle nose pliers as an aid to installing the return spring for the reverse latch hoop.

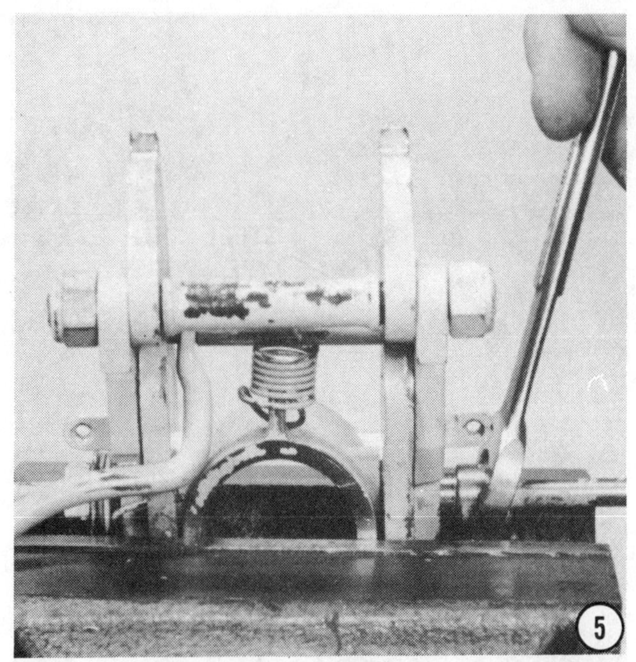

clamped in the vise. Slide the spring onto the inside shaft of the locating pin. Thread the pin into place and then hook the spring behind the reverse pawl in such a manner that the pawl is held against the reverse lever.

5- Thread the other locating pin into place. Tighten the pins snugly with the proper size wrench.

6- Slip the eye of each cross bar spring over the end of the cross bar. Hook the other end of each spring into the hole in the arm on each reverse pawl. Pry the cross bar up over the ends of the reverse levers and into the recess on both levers. A screwdriver under the cross bar with leverage on the connecting rod, as shown, works very well.

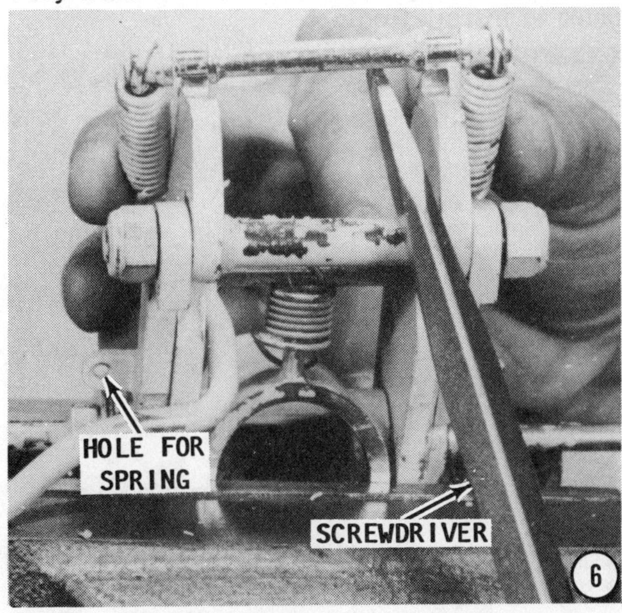

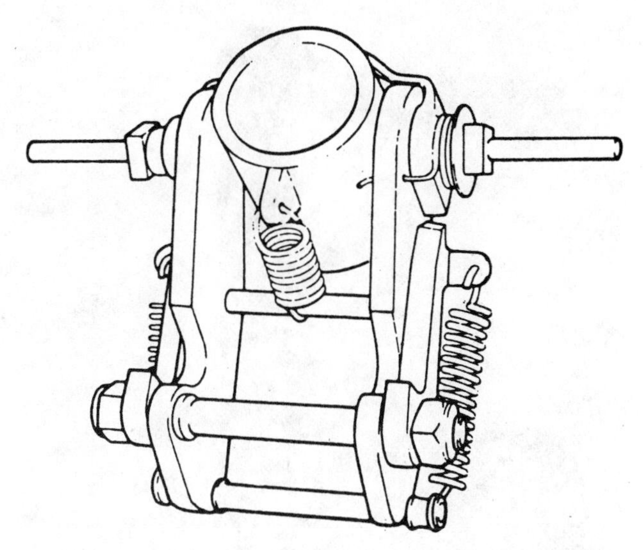

Drawing to depict the latch assembly properly assembled.

SAFETY WORDS

Exercise **CARE** to prevent the springs or rod from coming loose. The cross bar could be thrown with considerable force, causing personal injury.

The accompanying illustration depicts the complete unit properly assembled

7- Coat the needle bearing with a water resistant grease. Press the needle bearing housing into the steering yoke. Use Volvo special tool No. 884259 and No. 9991801 or equivalent pressing device. Press against the side of the bearing housing with the numbers stamped. Leave equal distance in the yoke on both ends of the bearing housing for the sealing rings. **CAREFULLY** tap one of the sealing rings into place from one side

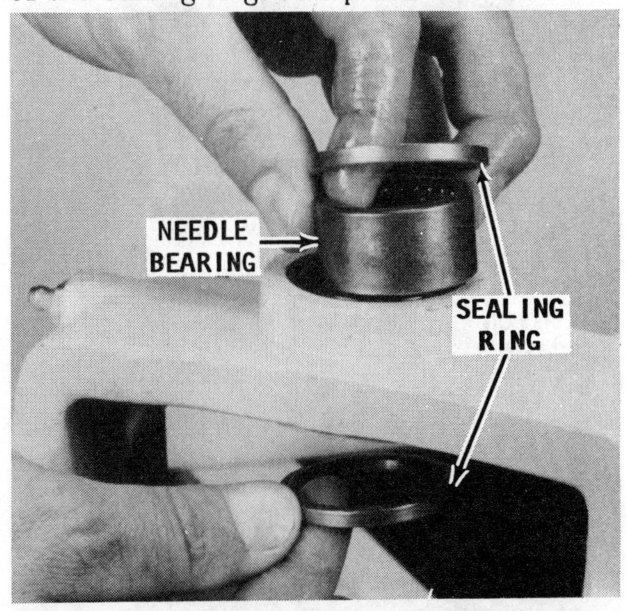

Proper arrangement for the needle bearing and two sealing rings in the steering yoke.

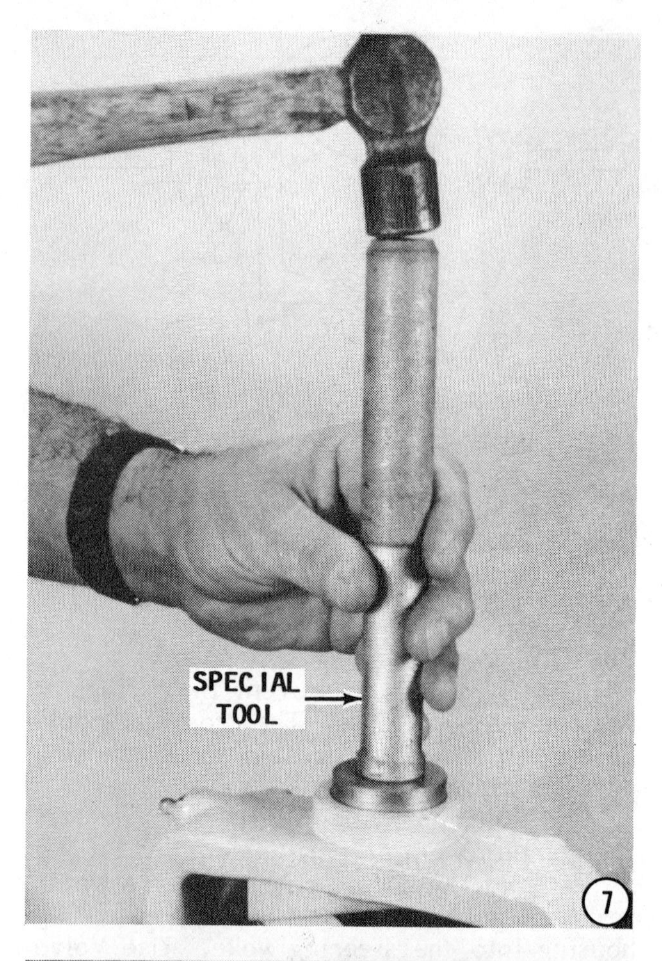

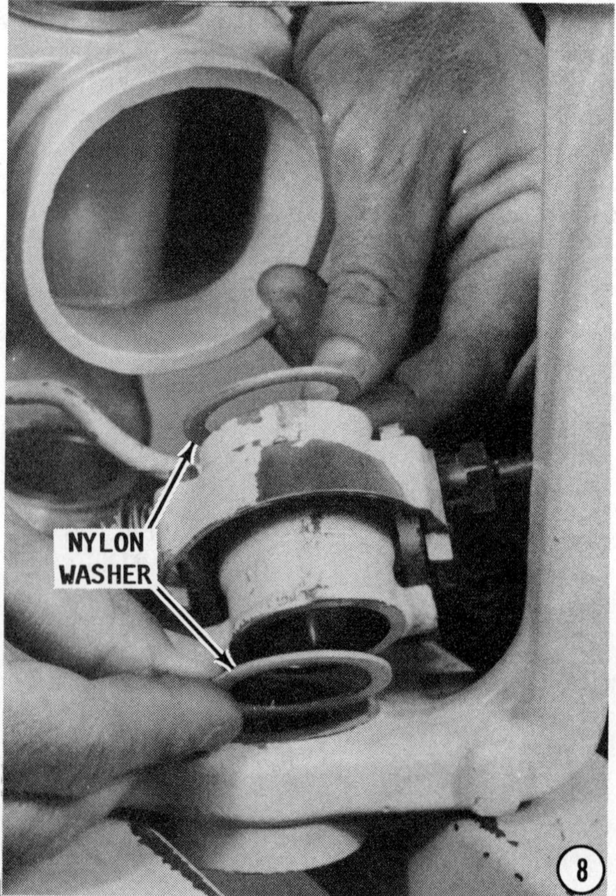

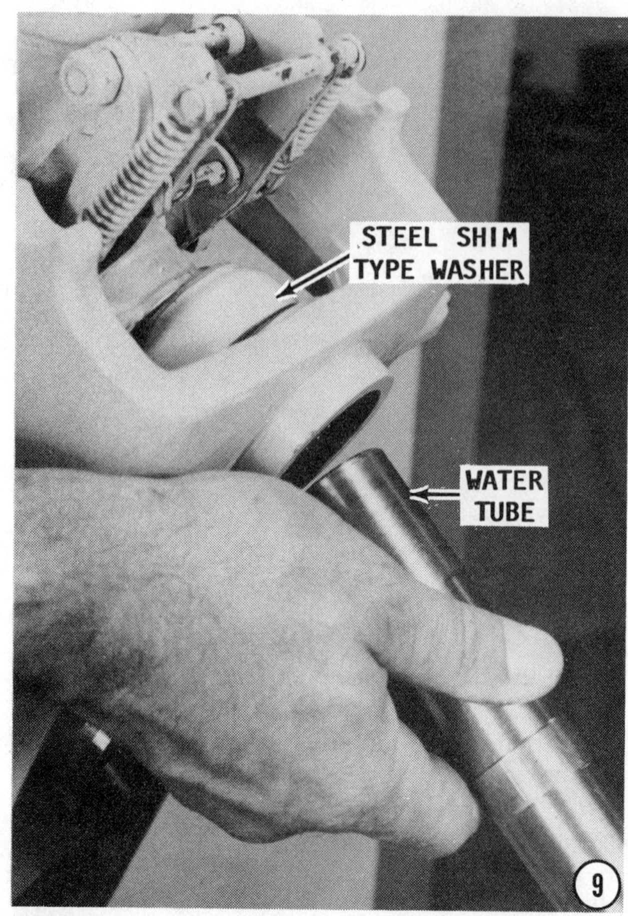

of the yoke with the **FLAT** side going in first. Tap on the outside shell of the sealing ring until the ring is flush with the surface of the yoke. Tap the second sealing ring into place from the other side of the yoke with the **FLAT** going in first. These seals prevent water from entering the needle bearing. The accompanying illustration shows the proper arrangement of the needle bearing housing and the two sealing rings prior to installation.

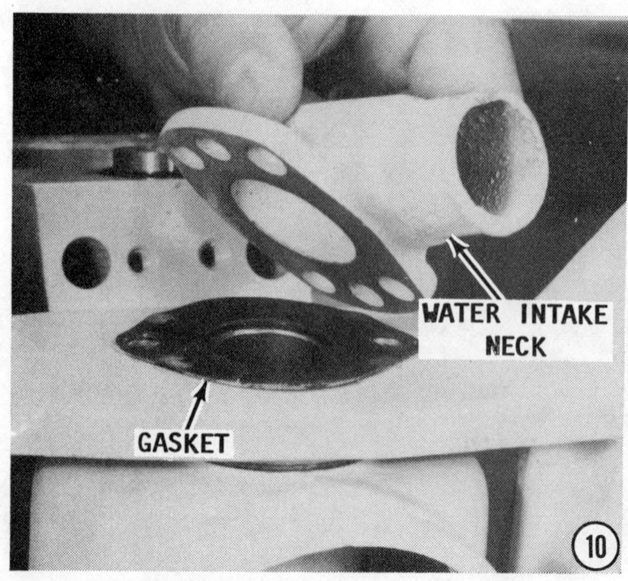

8- Install the assembled reverse latch mechanism into the steering yoke with a nylon washer on both ends, as shown.

9- Push the water tube into the steering yoke. This water tube also serves as a steering spindle for the yoke. As the water tube moves into place, slide the steel shim-type washer into place between the bottom of the steering fork and the pivot boss of the intermediate housing. Continue to push the tube on through the intermediate housing boss. Place a nylon washer between the intermediate housing and the reverse latch mechanism. Continue pushing the tube through the reverse latch mechanism, and then a second nylon washer. Push the tube into the other side of the intermediate housing. At this point, it may not be possible to simply push the tube through. Use a soft-head mallet and **CAREFULLY** drive the tube the remainder of the way into the steering fork. Exercise **CARE** to center the tube in the bushing in the steering yoke. If the tube is not centered, the bushing will be destroyed. Continue to **SLOWLY** drive the tube in until the collar on the tube bottoms out on the yoke.

10- Coat the mating surfaces of the steering yoke and the water intake neck with Permatex or other suitable sealing compound. Place a **NEW** gasket for the water intake neck in position on the steering fork with the holes in the gasket aligned with the holes in the fork. Install the water intake neck onto the steering yoke with the neck facing straight **FORWARD** for installation on the Model 280 stern drive. On the Model 270T stern drive the neck will be turned to port.

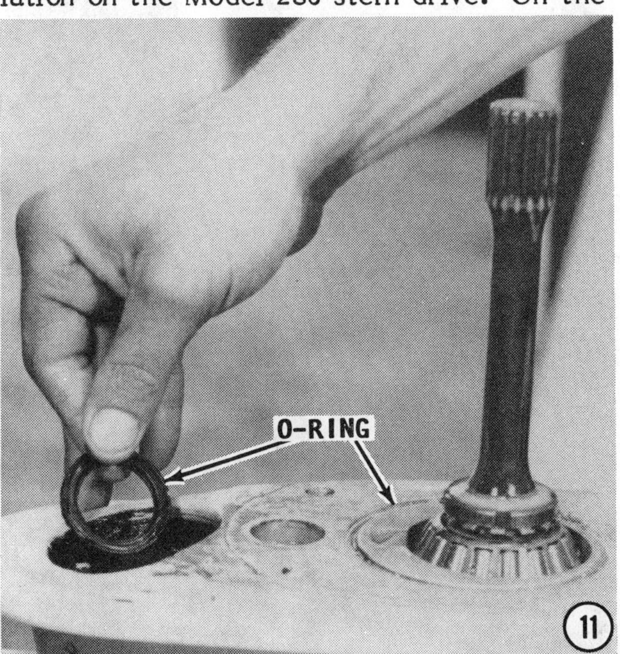

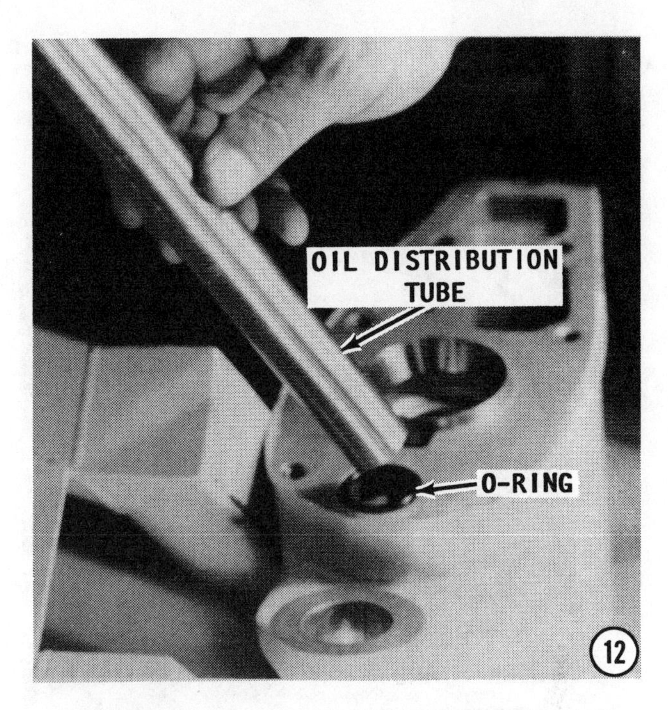

11- Place a **NEW** O-ring into the recess for the water distribution tube. Place a **NEW** O-ring around the vertical shaft thrust bearing.

12- Push the oil distribution tube into the intermediate housing. The fit is not a press fit, simply a push slide fit. The oil distribution tube should extend beyond the intermediate housing an equal amount on both sides. Slide a **NEW** O-ring onto both ends of the tube. Push the O-rings down into the recesses in the intermediate housing.

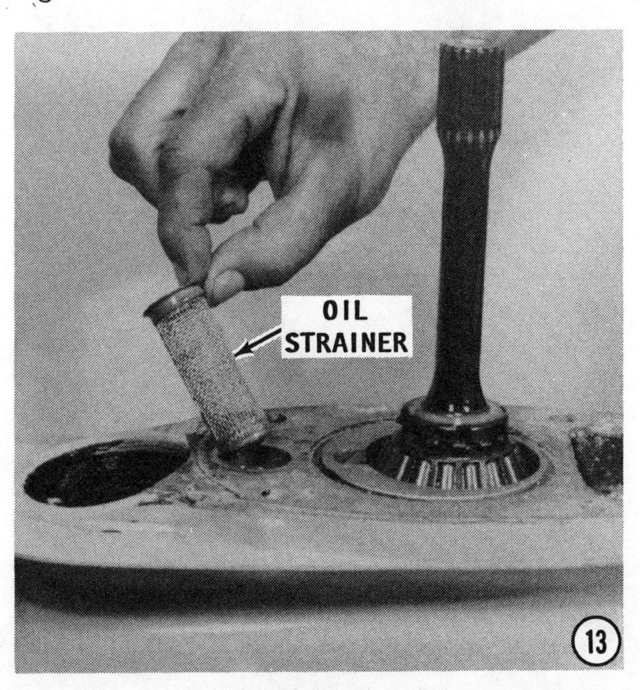

13- Insert a **CLEAN** oil strainer into place in the lower unit.

14- Coat the mating surfaces of the lower unit and the intermediate housing with Permatex or other suitable sealer.

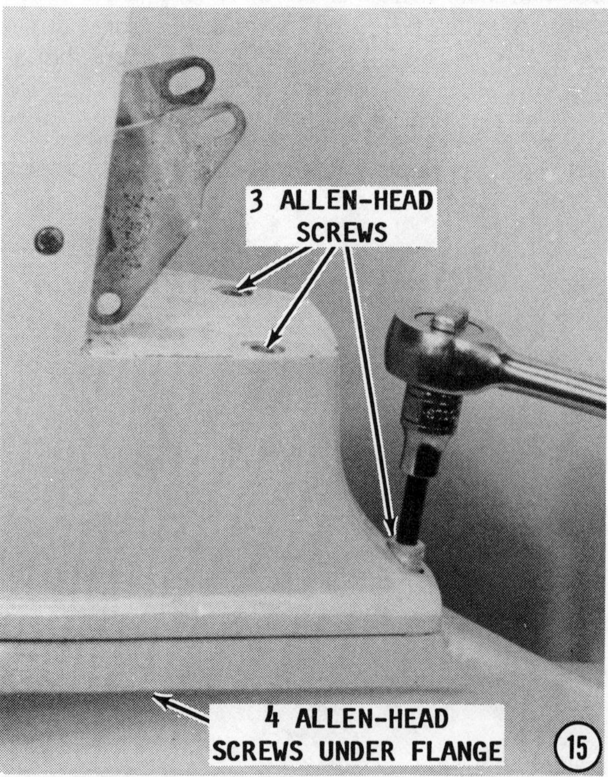

SLOWLY lower the intermediate housing down over the lower unit with the oil distribution tube sliding into the opening in the lower unit. Keep the two surfaces as parallel as possible while lowering the intermediate housing. If the intermediate housing should become "cocked" to one side, the water tube may flip the O-ring out of position. If this should happen, an air leak would develop on the intake side of the cooling system. Such a leak would result in the engine overheating from lack of proper water flow.

15- Secure the intermediate housing with the seven attaching Allen-head screws.

HELPFUL HINT

Before the upper gear housing is mated to the intermediate housing, insert a funnel into the oil passage tube, and then pour in approximately 2-1/4 quarts of lubricating oil. Filling the lower unit at this time is much quicker and easier than adding the full amount of oil after the upper gear housing is secured and the cap is in place.

Check to be **SURE** the drive coupling is installed before the upper gear housing is mated to the intermediate housing.

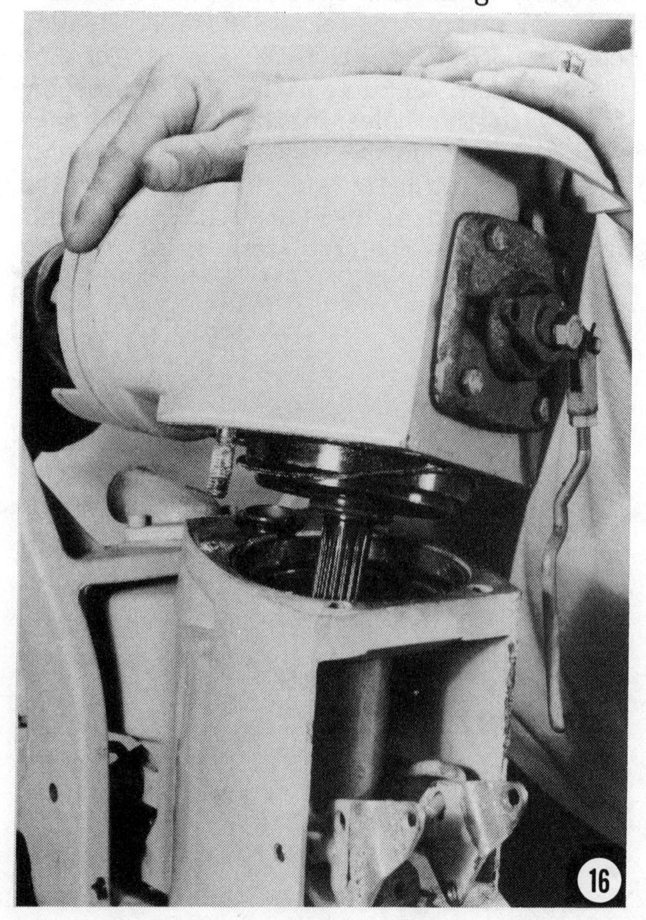

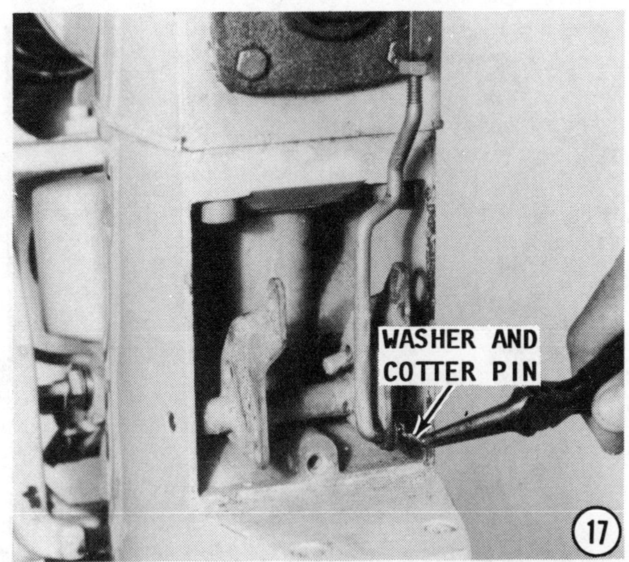

16- Coat the mating surfaces of the intermediate and upper gear housings with Permatex or other suitable sealing material. Lower the upper gear assembly down onto the intermediate housing with the splines of the shaft indexed with the splines of the coupling. As an aid to indexing the splines of the shaft with the splines in the coupling, have an assistant rotate the propeller shaft slightly while attempting to lower the upper gear housing.

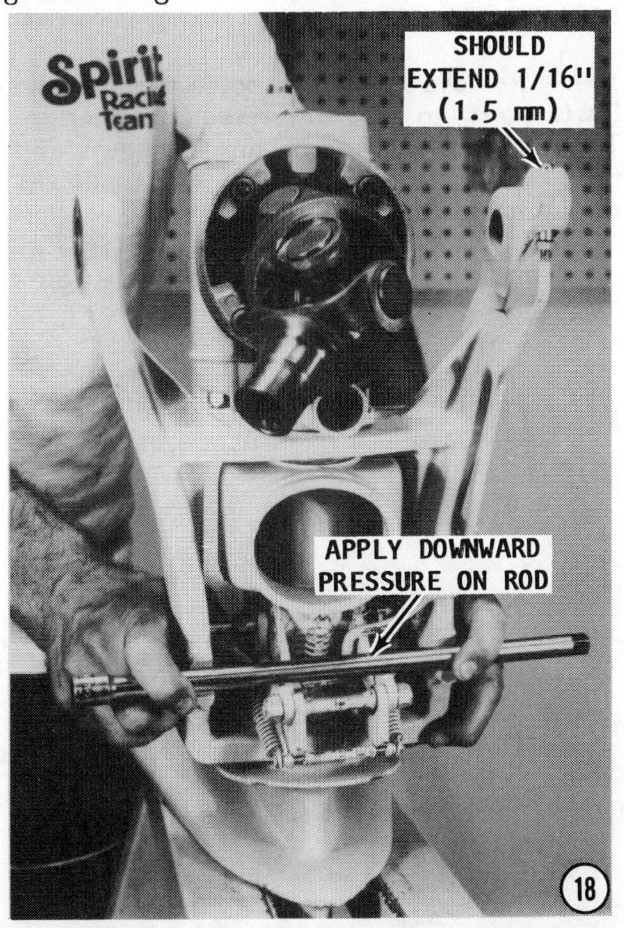

17- Secure the two housings together with the attaching hardware -- two nuts and two bolts. Tighten the bolts and nuts alternately and evenly and securely in a criss-cross pattern. Connect and secure the shift rod with the cotter pin and washer.

18- A check of the completed work may be made by pressing down on both reverse levers at the same time with a stiff rod, as shown. The push rod should extend through the fork approximately 1/16" (1.6 mm).

The assembled stern drive is now ready to be installed on the boat. See Section 8-4, Installation, beginning on Page 8-10.

8-7 LOWER UNIT

PREPARATION

The following procedures pick up the work after the stern drive has been removed from the boat, see Section 8-4, Stern Drive Removal, beginning on Page 8-8.

Loosen the dip stick from the upper gear housing. Remove the plug from the bottom of the lower unit and drain the lubricant from the stern drive. Remove the four

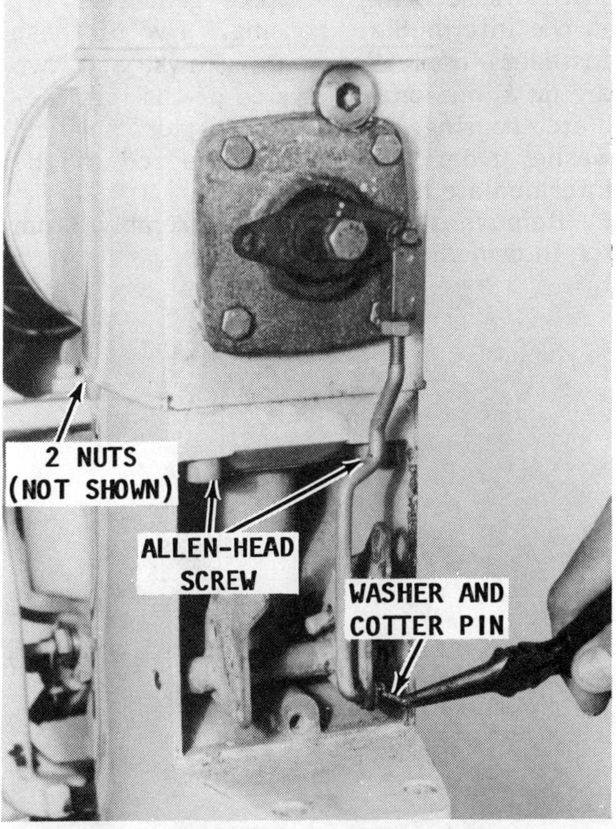

Items to be removed or disconnected prior to separating the upper gear assembly from the intermediate housing.

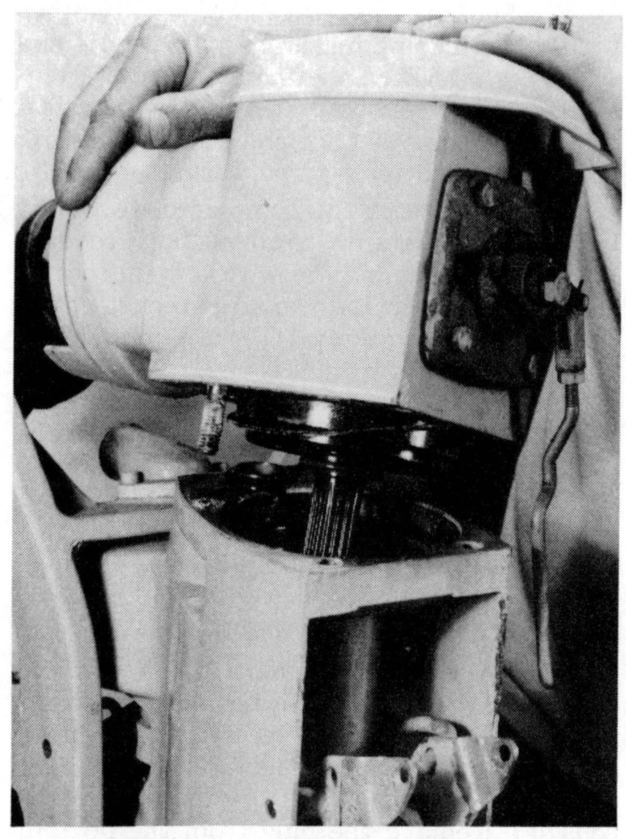

Removing the upper gear assembly from the intermediate housing.

fasteners securing the upper gear assembly to the intermediate housing. Two of these fasteners are Allen-head screws, and two are nuts, one on either side of the intermediate housing. Pull the cotter pin and washer from the shift linkage rod at the intermediate housing, as shown.

Remove the upper gear assembly from the intermediate housing.

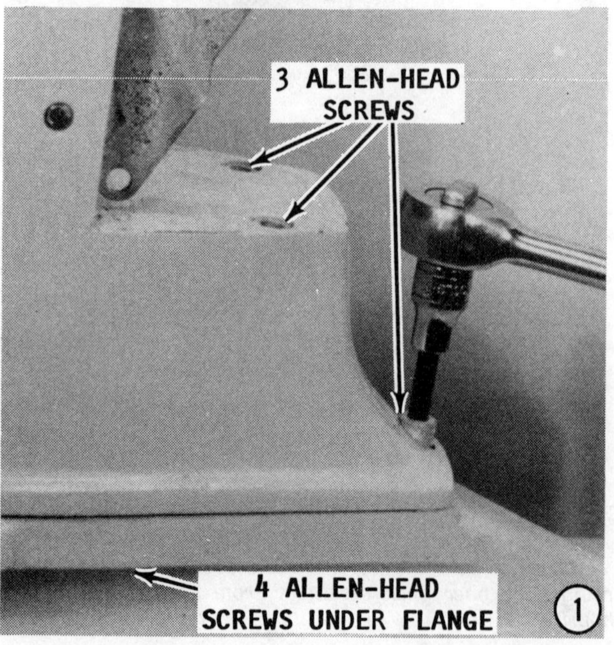

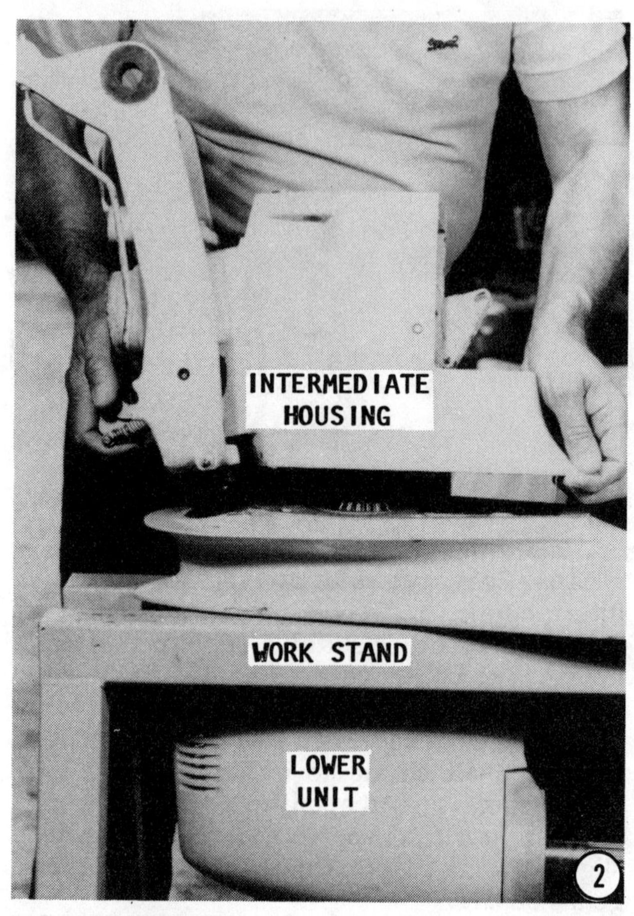

DISASSEMBLING

1- Remove the seven bolts securing the lower unit to the intermediate housing. Three of the bolts are on top, as shown, and four are located on the underneath side of the lower unit flange.

2- Lift the intermediate housing free of the lower unit. Notice the basket oil strain-

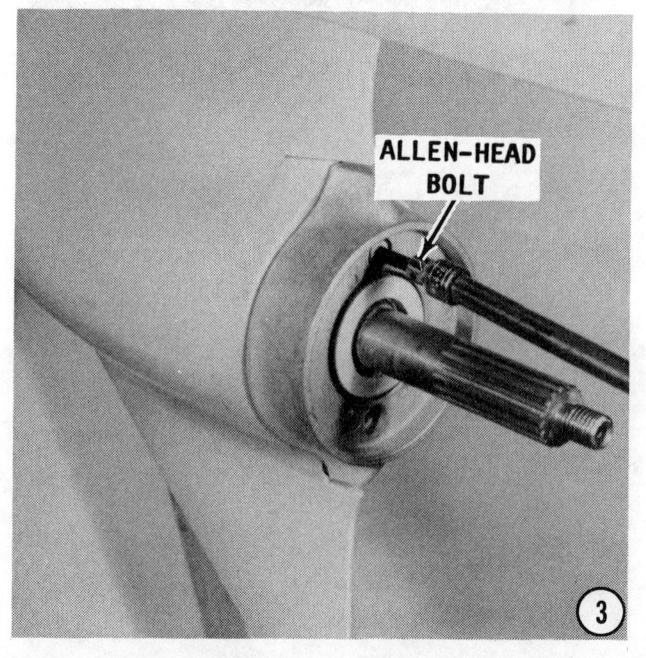

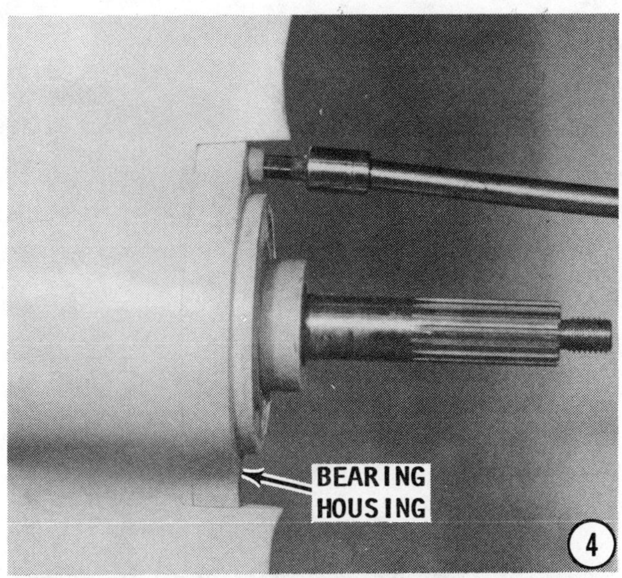

er in the top of the lower unit. This strainer is designed to prevent any debris from circulating in the oil through the lower unit.

3- Back out the two Allen-head bolts securing the zinc ring to the propeller shaft bearing housing. Slide the ring off the propeller shaft.

4- Remove the two bolts securing the propeller bearing housing in the lower unit.

5- Obtain Volvo special slide hammer tool No. 884161. Install and use the slide hammer to remove the propeller shaft and bearing housing from the lower unit.

6- Remove the O-rings from the bearing housing.

7- Remove the six nuts securing the bearing in the bearing housing.

8- Turn the assembly upside down on the work bench. Use a nylon or rubber head hammer and knock the assembly free of the bearing. After the bearing has been removed, remove, save, and identify the shim in the bearing housing. This shim sets the gear depth on the lower gear assembly.

9- Remove the propeller shaft seals

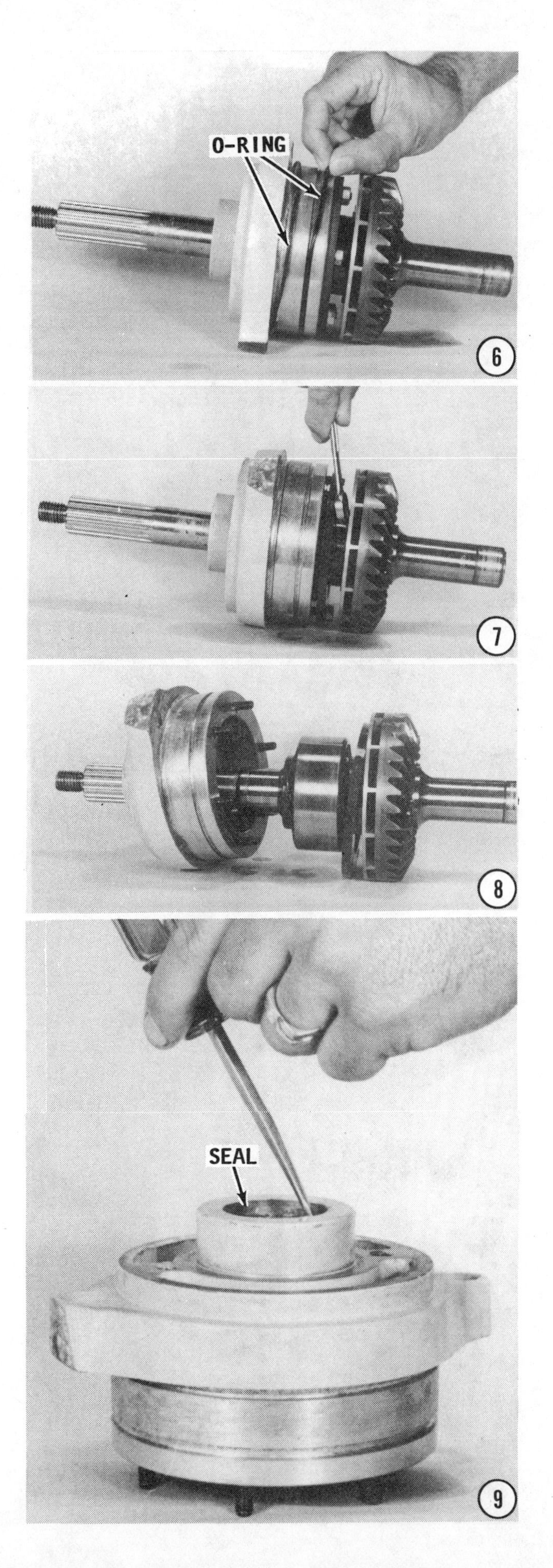

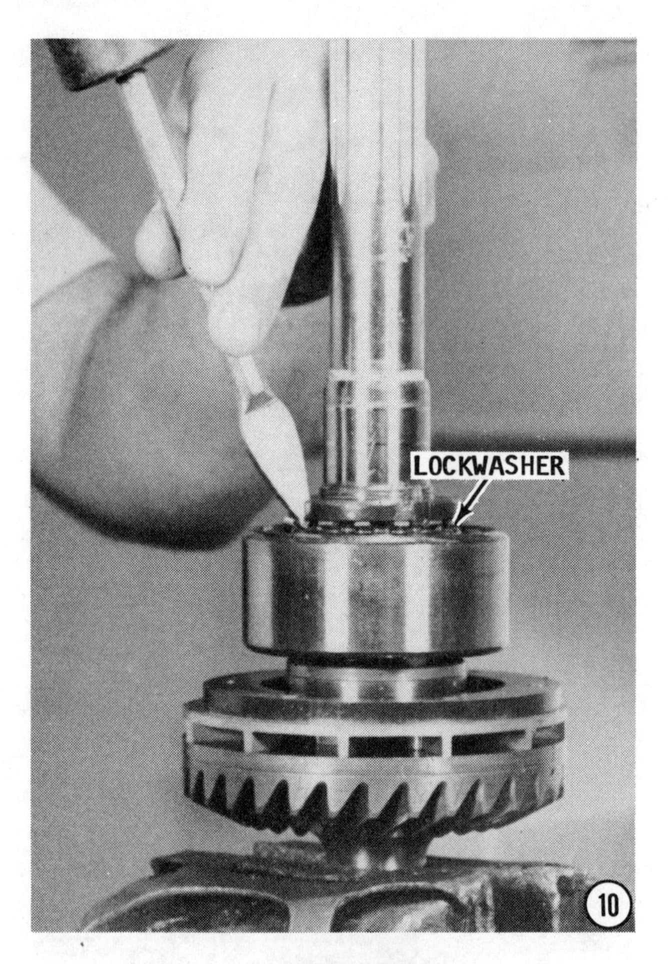
LOCKWASHER
10

ARBOR
PRESS
12

SPANNER
NUT
11

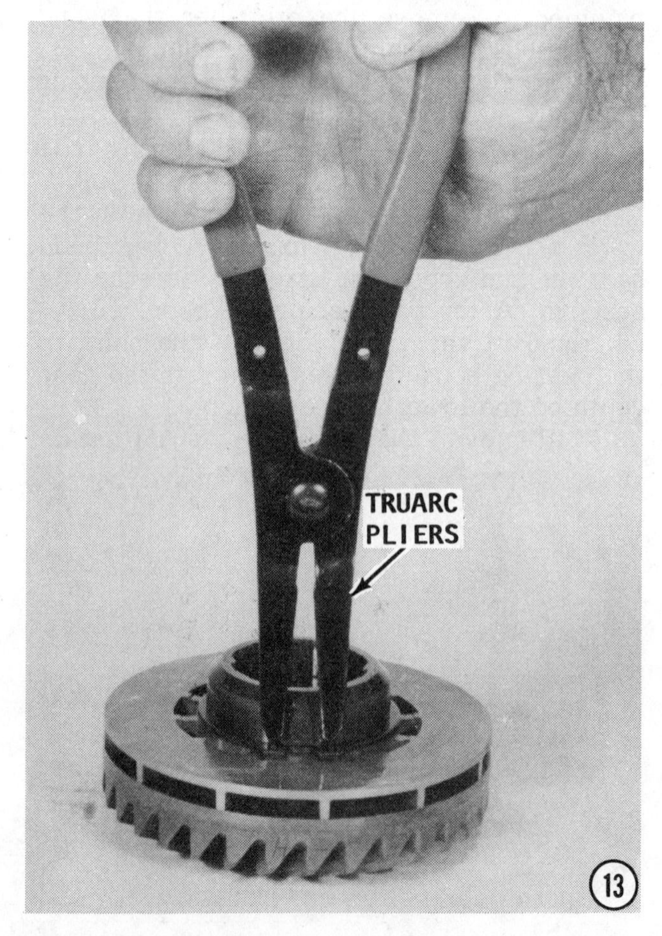
TRUARC
PLIERS
13

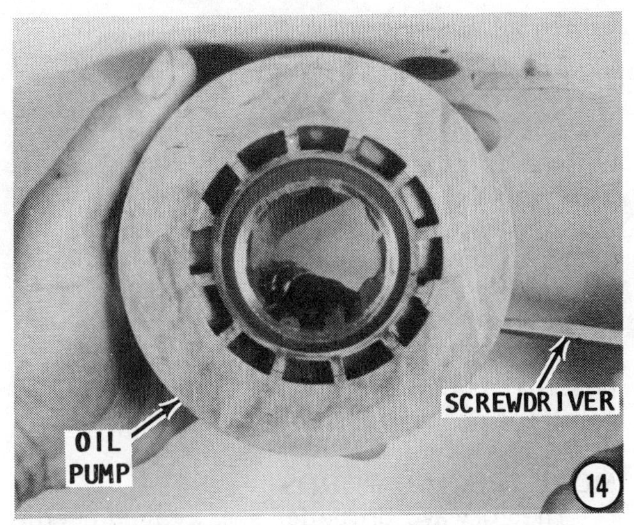

from the housing.

10- To remove the thrust bearing from the propeller shaft, first bend the tab down on the locking washer.

11- Loosen the spanner nut with a spanner wrench or a punch and hammer.

12- Press the bearing, bearing collar, and gear, free of the propeller shaft. This can be accomplished by securing the propeller shaft in a vise equipped with soft jaws, and then working the bearing free of the shaft. **DISCARD** the locking tab. A **NEW** one should **ALWAYS** be used during assembling.

WARNING

The next step involves a dangerous procedure and should be executed with care while wearing SAFETY GLASSES. The snap ring is placed under tremendous tension with the Truarc pliers while it is being removed. If it should slip off the Truarc pliers, it will travel with incredible speed causing personal injury if it should strike a person.

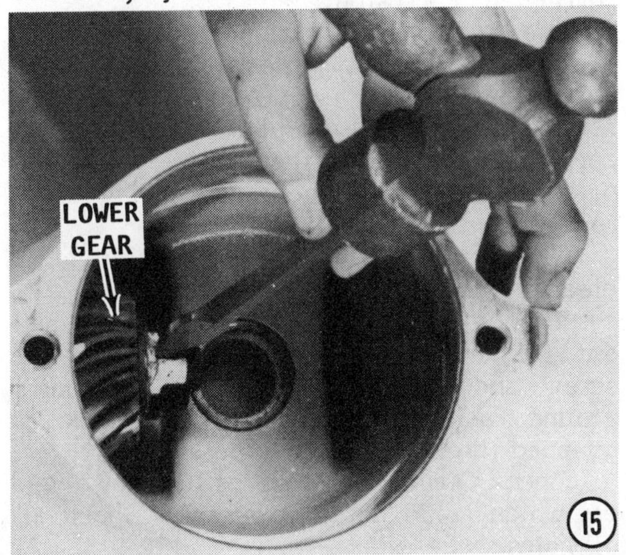

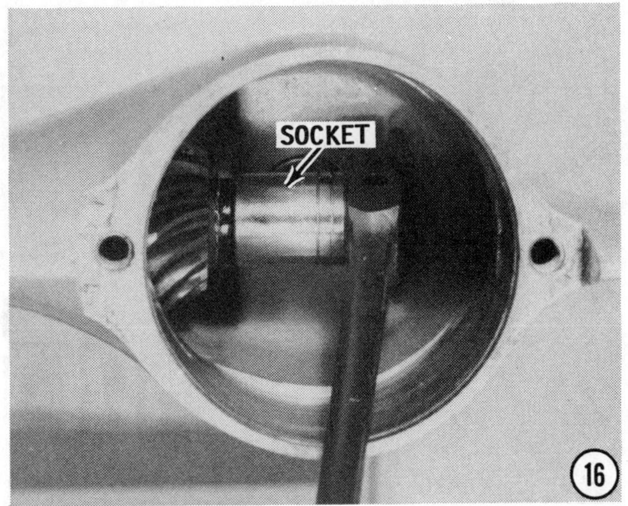

Therefore, continue to hold the ring and pliers firm after the ring is clear of the pump. Place the ring on the floor and hold it securely with one foot before releasing the grip on the pliers. An alternate method is to hold the ring inside a trash barrel, or other suitable container, before releasing the pliers.

13- Use a pair of Truarc snap ring pliers and remove the snap ring from the gear.

14- CAREFULLY pry the oil pump free of the gear.

15- Use a punch and hammer and bend over the tab washer on the lower gear nut on the end of the vertical shaft.

16- Clamp the upper end of the vertical shaft in a vise equipped with soft jaws. Use the proper size socket to loosen the nut on the lower end of the vertical shaft. Put a socket on the nut between the housing and the collar on the nut and turn the shaft. This action will loosen the nut further until the shaft is actually pushed away from the tapered spline in the gear. The gear will

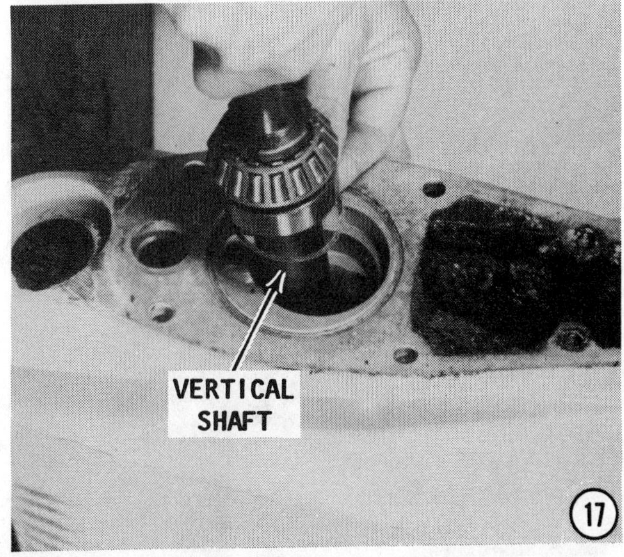

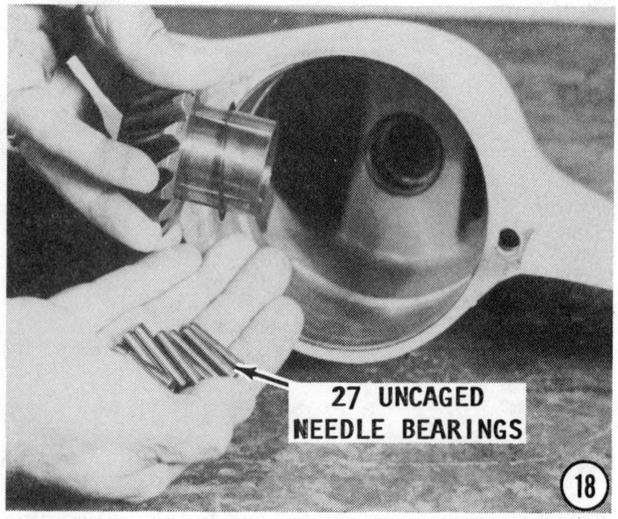

then be loose from the shaft. Remove the nut from the lower end of the vertical shaft.

17- Hold the gear with one hand and with the other hand pull the vertical shaft out of the lower unit.

18- Withdraw the gear and at the same time catch and **SAVE** the twenty-seven (27) uncaged needle bearings.

19- Bend the lock tab down, and then loosen the spanner nut on the vertical drive shaft. Press the bearings from the shaft. Notice the spacer between the two bearings. These bearings need only to be removed from the shaft if they have been damaged or corroded.

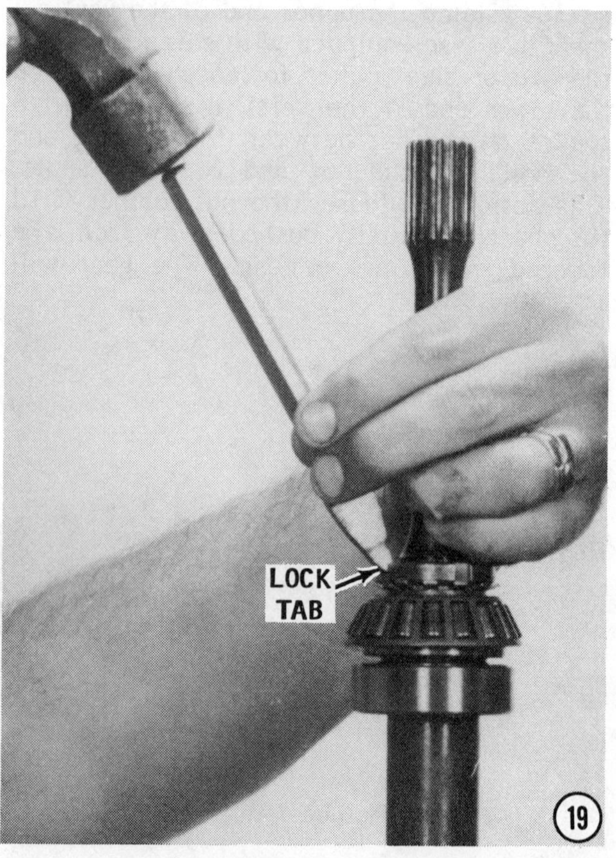

SPECIAL WORDS

The race at the lower end of the vertical shaft in the lower unit housing can be driven out, using a special Volvo tool or making a fixture to grasp the upper lip of the race. This is **ONLY** done if the race has been damaged or corroded. Replacement of the bearing race is **NOT** an easy task. A special puller or a makeshift arrangement is required to "pull" the new race up into the cavity. Again, do **NOT** attempt to remove this race unless it is damaged or corroded so badly as to affect performance of the bearing.

MORE SPECIAL WORDS

The forward propeller shaft caged needle bearing can be removed using a slide hammer with fingers to grip the bearing. If possible, obtain Volvo special tool No. 884316. Do **NOT** attempt to remove this bearing unless there is definite evidence of damage or corrosion. Position the puller in the needle bearing with the fingers engaged onto the back side of the needle-rollers. Screw the extractor so that the fingers are forced apart and the bearing can be removed.

CLEANING AND INSPECTING

Wash all parts in solvent and blow them dry with compressed air. Remove all traces of seal and gasket material from all mating surfaces. Blow all water, oil passageways, and screw holes, clean with compressed air. After cleaning, apply a light coating of engine oil to the bright surfaces of all gears, bearings, and shafts, as a prevention against rusting and corrosion.

Use a fine file to remove burrs. **REPLACE** all **O**-rings, gaskets, and seals, to ensure satisfactory service from the unit, after the work is completed. Clean the corrosion from inside the housing where the propeller shaft and bearing housing were removed.

Check to be sure the water intake is clean and free of any foreign material.

Inspect the lower unit housing inside and out for cracks. Check carefully around screw and shaft holes. Check for burrs around machined faces and holes. Check for stripped threads in screw holes.

Check **O**-ring seal grooves for sharp edges, which could cut a new seal. Check all oil holes.

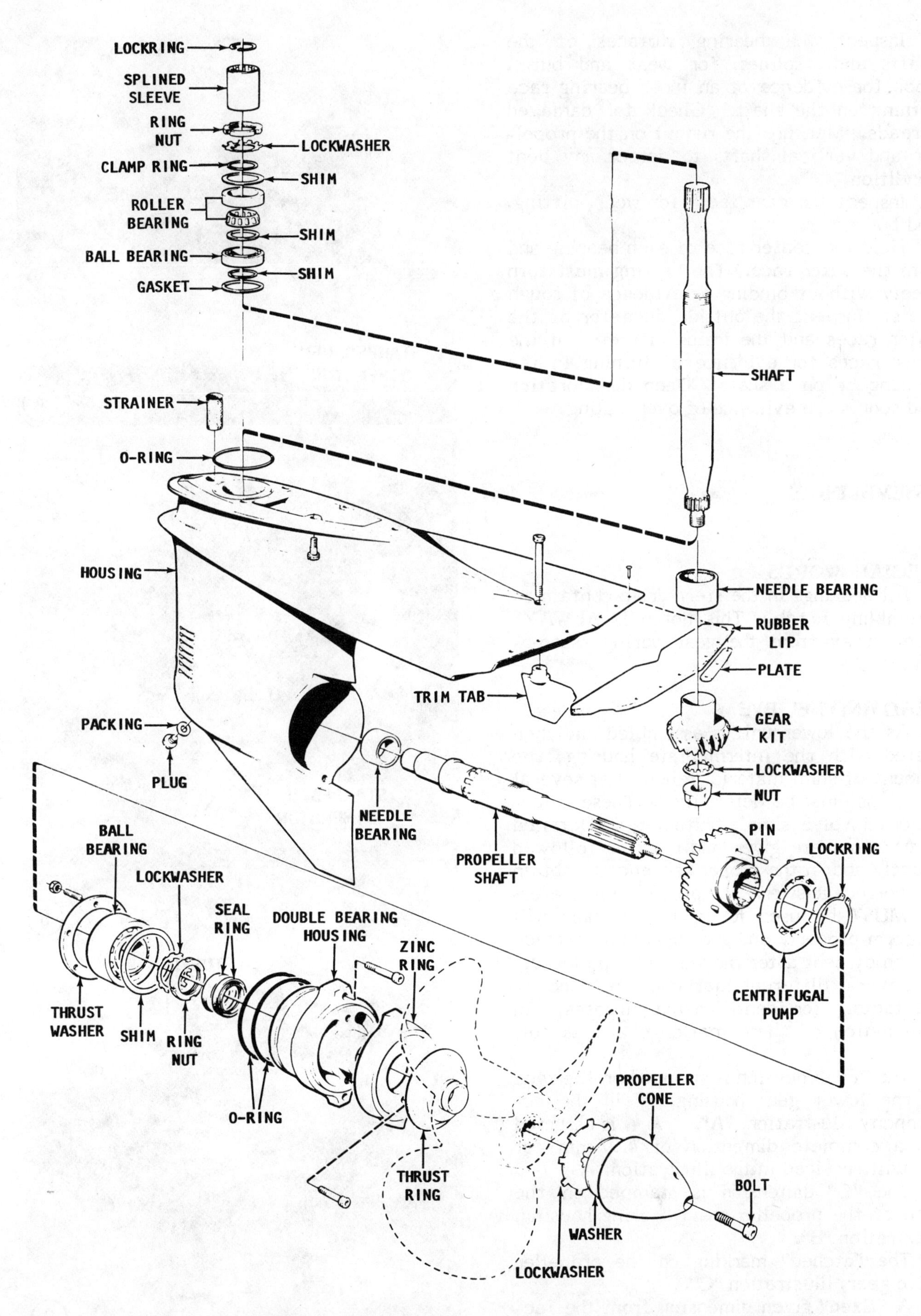

Exploded drawing of the lower unit with principle parts identified.

Inspect the bearing surfaces of the shafts and splines, for wear and burrs. Look for evidence of an inner bearing race turning on the shaft. Check for damaged threads. Measure the runout on the propeller and vertical shafts to detect any bent condition.

Inspect the gear teeth for wear, pitting, and burrs.

Hold the center race of each bearing and turn the outer race. The bearing must turn freely without binding or evidence of rough spots. Inspect the outside diameter of the outer races and the inside diameter of the inner races for evidence of turning in the housing or on a shaft. Deep discoloration and scores are evidence of overheating.

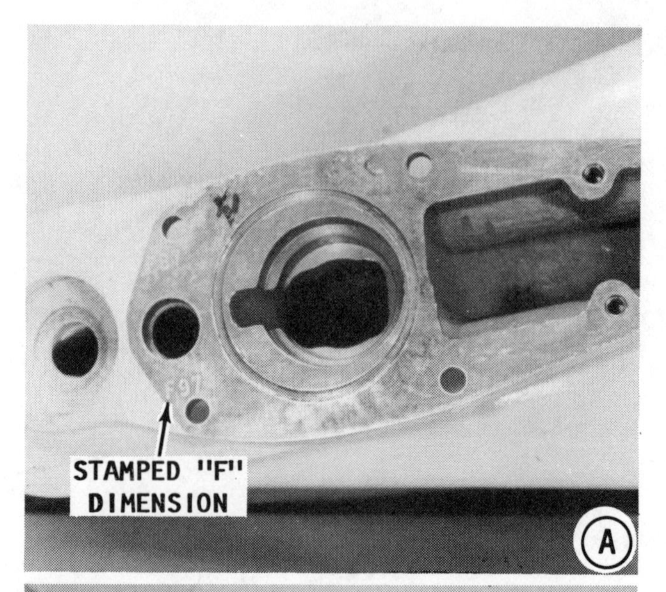

ASSEMBLING

SPECIAL WORDS

All bearings in the stern drive contain an assembling notch. This notch is **ALWAYS** faced away from the gear during assembling.

READ AND BELIEVE

As the lower unit is assembled and then mated with the intermediate housing, the amount of shim material required at several locations must be determined. These calculations involve simple arithmetic. Detailed instructions are given and must be followed exactly and in the proper sequence to obtain the correct answer. The proper shim material **MUST** be used to ensure the unit will perform properly and give maximum service and enjoyment after the work is completed.

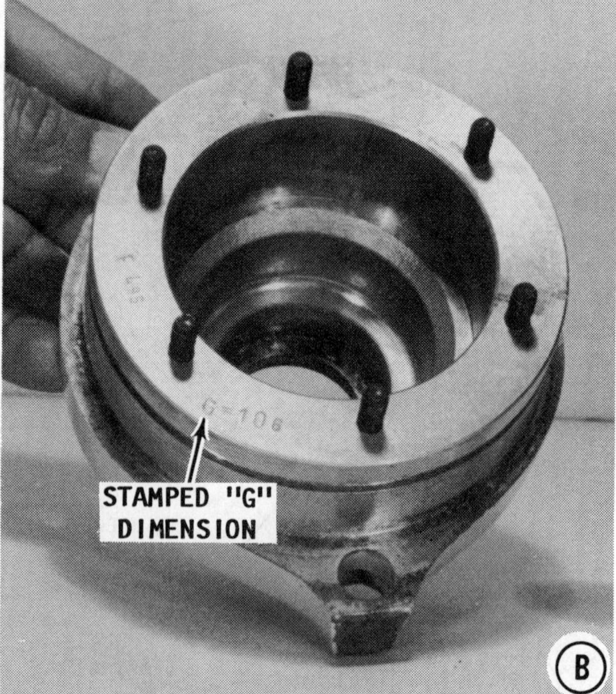

Several different markings are used by the factory for identification purposes. An explanation of these markings are as follows:

The **"F"** dimension is stamped on the face of the lower gear housing, 97 in the accompany illustration, **"A"**. A 4. is assumed for a complete dimension of 4.97 mm for the unit serviced in the illustration.

The **"G"** dimension is stamped on the face of the propeller shaft bearing housing, illustration **"B"**.

The **"etched"** marking on the propeller drive gear, illustration **"C"**.

A **"fixed"** given dimension from the factory, 5.85 mm.

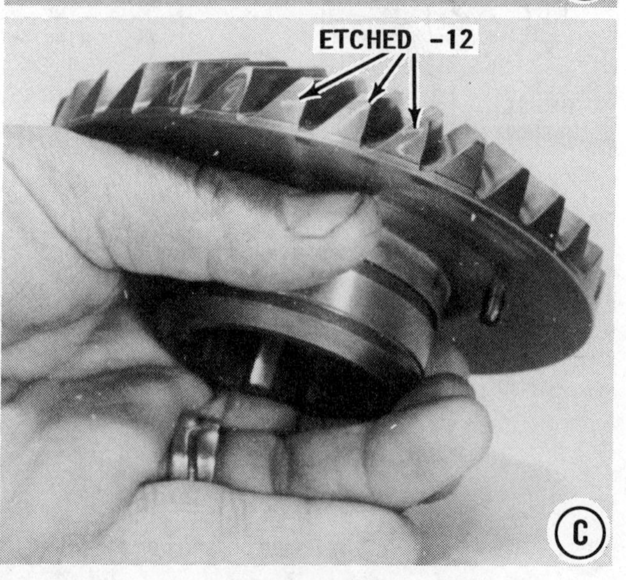

Propeller Bearing Housing Shim Calculation

F = 4.96 mm
G =+1.06 mm
6.02 mm

5.85 mm (Factory figure)
-.12 mm (Etched on prop shaft gear)
5.73 mm

Now 6.02 mm (Answer from above)
-5.73 mm (Answer from above)
= .29 mm (Required shim material)

Conversion = 0.010"

Vertical Driveshaft Shim Calculation

Begin by installing 0.014" shim material under the vertical shaft bearing. A 0.010" paper shim sandwiched between two 0.002" steel shims.

TEMPORARY ASSEMBLING

CRITICAL WORDS

In order to check the shim material required for proper "backlash" and gear pattern, the lower unit is assembled **WITHOUT** the use of O-rings or seals. The unit disassembled, the proper amount of shim material used, and **NEW** seals and O-rings installed as the unit is assembled for service.

Forward Propeller Shaft Needle bearing

1- If the forward propeller shaft caged needle bearing was removed during disassembling obtain Volvo special tool No. 884283 and No. 9991801. If these tools are not available obtain a driver-type tool with a shaft that will fit inside the bearing. Apply a generous coating of lubricant to the needle bearings and then drive the bearing into place. The side of the bearing on which the identification marks are stamped **MUST** face outward (aft).

Lower Vertical Shaft Bearing

2- If the race at the lower end of the vertical shaft in the lower unit housing was driven out during disassembly, obtain Volvo special tool No. 884385 and No. 884241. It is almost impossible to "pull" this bearing into place properly without these two special tools. However, a shaft with threads on

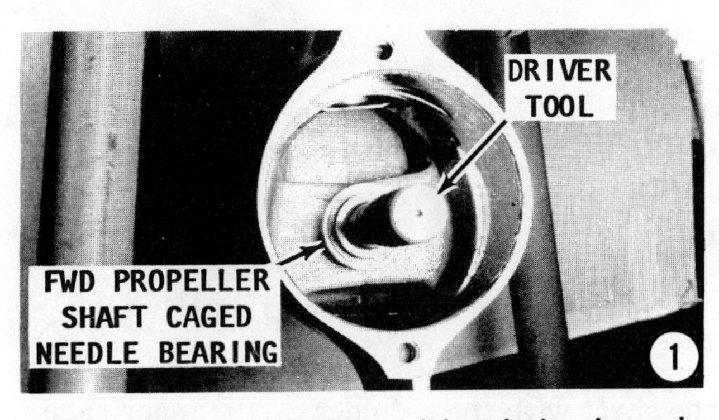

both ends, a small plate with a hole through which the shaft can be passed, and a nut for the lower end, can be used to accomplish this task. Threads on the upper end of the shaft with a plate and nut are also required. Pass the Volvo special tool or the makeshift shaft down through the lower unit and then force the bearing onto the end of the tool or shaft. Place the plate on the shaft and then the nut to hold it all in place. At the upper end, screw special tool No. 884241 onto the Volvo special tool or place the other plate on the shaft and then the nut. Tighten the special tool or the nut on the end of the shaft until the bearing is "pulled" into place. Disassemble and remove the special tools or the makeshift shaft.

3- If the bearings were removed from the vertical shaft, obtain Volvo special tool No. 884266. Press the ball bearing set onto

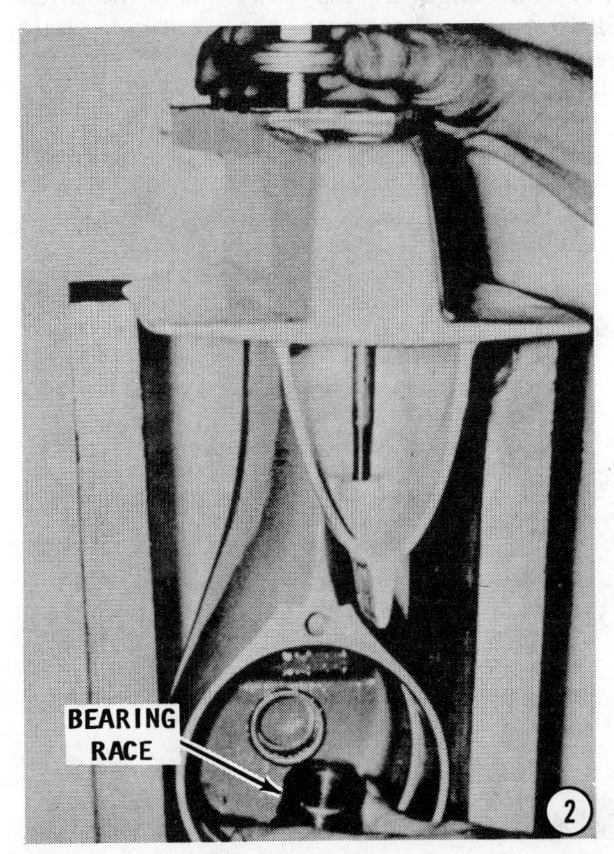

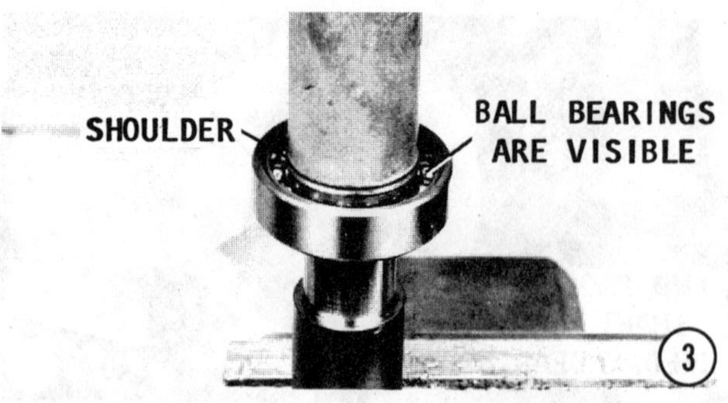

the shaft with the shoulder of the inner race facing **UPWARD.** Slide the spacer insert onto the shaft and into position on the ball bearing. Use Volvo special tool No. 884266 and press the tapered roller bearing onto the shaft with the larger diameter of the bearings facing **DOWNWARD** (toward the ball bearing set). Actually, it is possible to press both bearing sets and the spacer between them onto the shaft in one operation. If this is done, **ENSURE** each bearing set is facing the proper direction as just described. Slide the spacer ring, lockwasher, and round nut onto the vertical shaft. Thread the nut onto the shaft and tighten it snugly, using a spanner-type wrench or a punch and hammer. **DO NOT** bend the locking tab over **AT THIS TIME.**

4- Use a thick bearing grease and insert the 27 loose needle bearings, one at-a-time into the cage in the lower unit housing. Slide the washer onto the vertical shaft gear. This washer serves as a lower race for the needle bearings. **CAREFULLY** and without disturbing the needle bearings, insert the vertical shaft gear inside the lower unit housing.

5- Hold the vertical shaft gear firmly in place with one hand. With the other hand,

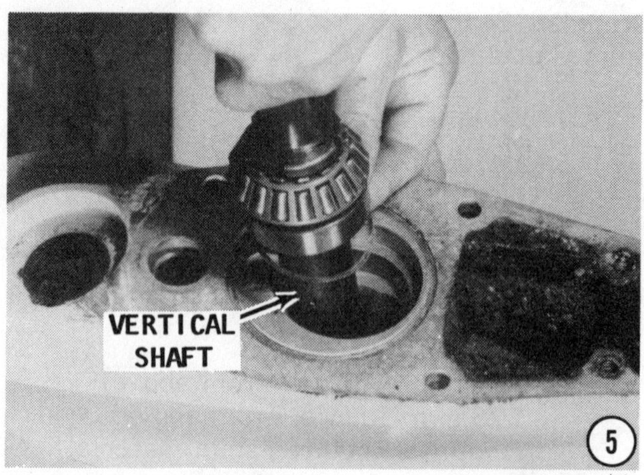

insert the assembled vertical shaft into the lower unit housing, with 0.014" (0.35 mm) thick shim material under the ball bearing cage. The shim material consists of a 0.010" (0.25 mm) paper shim sandwiched between two metal shims, each 0.002" (0.05 mm) thick.

6- Slide the lock tab washer onto the lower end of the vertical shaft, and then thread the nut on. Hold the splined end of the vertical shaft with Volvo special tool No. 884264 or with some type of soft jaw clamping device. Tighten the nut to a torque value of 118 ft-lbs (160 Nm). Do **NOT** bend over the locking tab **AT THIS TIME.**

7- Push the oil pump into place in the propeller shaft gear.

WARNING

This next step can be dangerous. The snap ring is placed under tremendous tension with the Truarc pliers while it is being installed. Therefore, wear **SAFETY GLASSES** and exercise care to prevent the

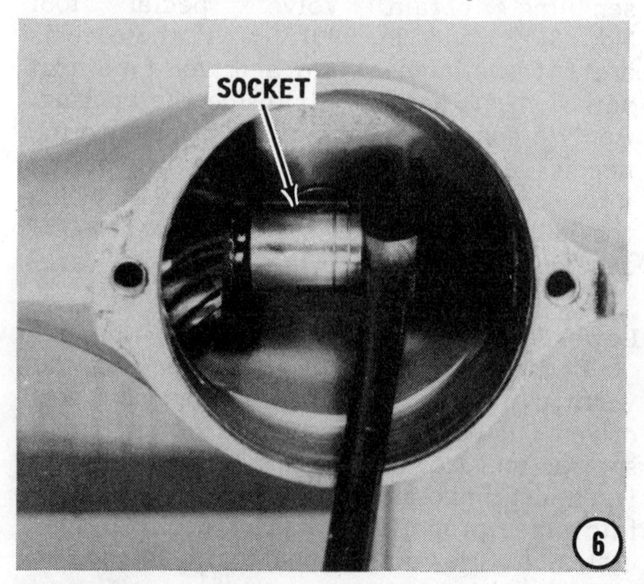

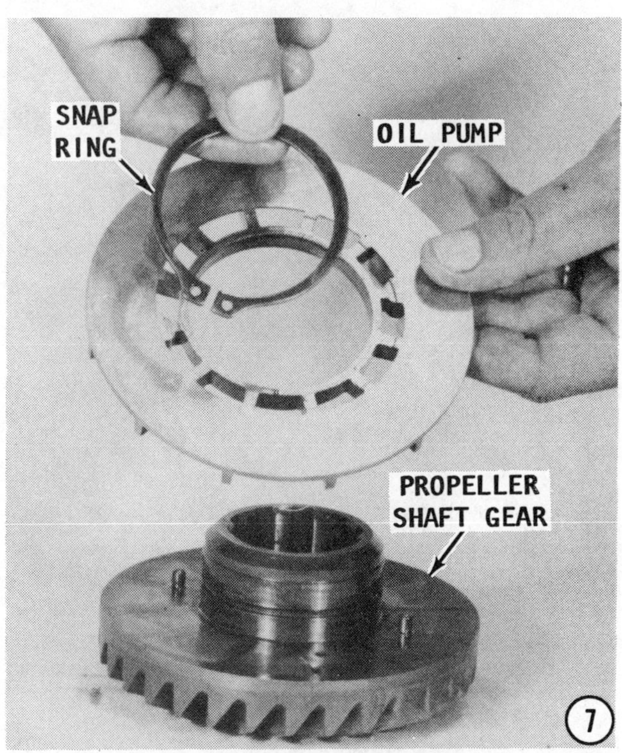

snap ring from slipping out of the pliers. If the snap ring should slip out it would travel with incredible speed and cause personal injury if it struck a person.

8- Use a pair of Truarc pliers and secure the pump in place with the snap ring.

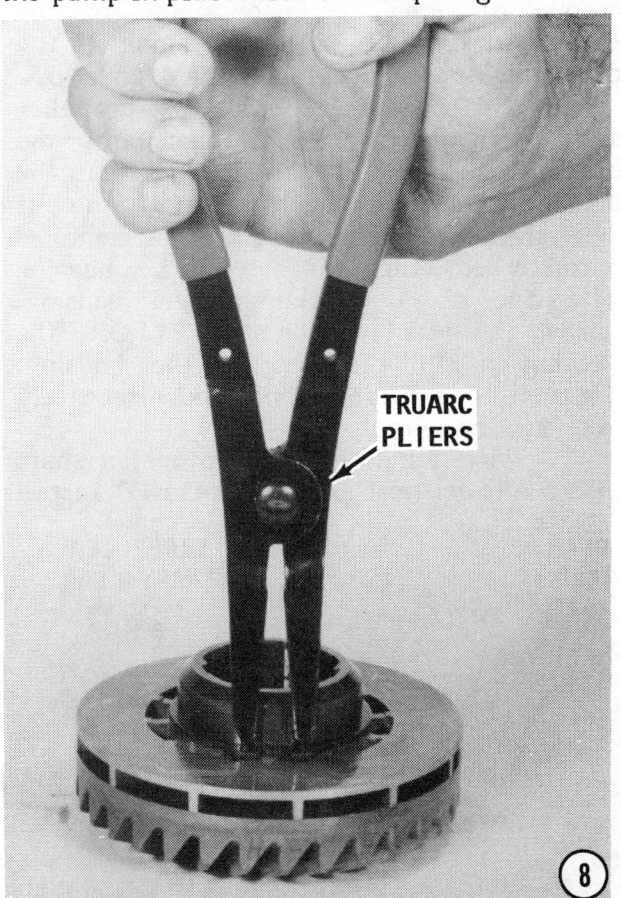

9- Press the bearing race onto the propeller shaft, if it was removed. Slide the propeller shaft gear onto the propeller shaft until the splines in the gear begin to index with the splines on the shaft. Slide the bearing down to the gear. The bearing **MUST** be installed with the cutout in the bearing race with the balls visible **FACING** the propeller end of the shaft. Use Volvo special tools No. 884265 and No. 884263 and press the bearing and gear until the gear bottoms out on the shoulder of the propeller shaft. If these special tools are not available use an Arbor press with the proper size section of heavy wall pipe over the propeller shaft. Slide a **NEW** lockwasher onto the shaft, and then thread the spanner nut into place. Tighten the spanner nut with a spanner wrench or punch and hammer.

10- Bend a tab on the lockwasher up to lock the nut in place.

11- Install the double set of seals into the propeller shaft housing, back-to-back, flat side-to-flat side. Coat the outside perimeter of both seals with sealing compound. Install the first seal with the flat side facing **OUTWARD.** This seal prevents water from entering the lower unit. Coat

the flat side of the first seal with a water repellent grease. Coat the flat side of the second seal with the lubricant. Install the second seal with the flat side **FACING** the flat side of the first seal. The seals are now back-to-back. The second seal prevents lubricant in the lower unit from escaping.

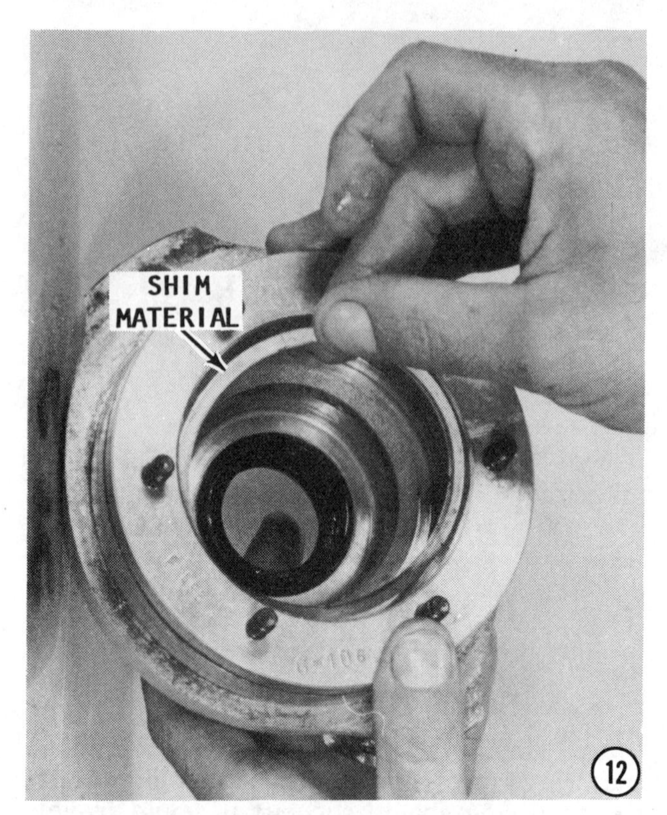

12- Place the same amount of shim material that was removed during disassembling into the bearing housing recess. However, if the gear set was replaced, insert the amount of shim material obtained from the calculations made at the beginning of the assembling procedures, Pages 8-54 and 8-55.

13- CAREFULLY install the propeller shaft with the bearing in place into the bearing housing. Exercise **CARE** that the bearing enters the housing squarely, not at an angle, and that the seals are not damaged or disturbed. Align the holes in the bearing with the studs in the bearing housing. Thread the nuts onto the studs **BEFORE** the bearing is fully seated in the housing. Tighten the six nuts evenly and alternately in a criss-cross manner.

14- Install the assembled propeller shaft into the lower unit housing. **DO NOT** install

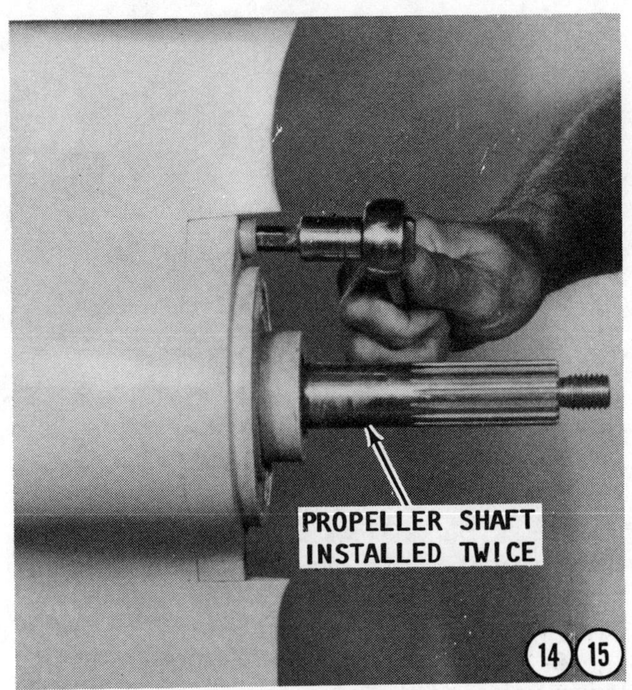

the O-rings on the bearing housing **AT THIS TIME.** It may be necessary to rotate the propeller shaft slightly to allow the gear on the propeller shaft to index with the pinion gear on the lower end of the vertical shaft. Secure it in place with the screws. Tighten the screws to a torque value of 29.5 ft-lbs (40 Nm).

Checking "Backlash"

Secure a dial indicator to the top base of the lower unit housing. Measure the "backlash" at the vertical shaft spline. Hold the propeller shaft firmly with some type of clamping device that will not damage the splines. Now "rock" the vertical shaft back-and-forth and measure the amount of "backlash" with the dial indicator. Approximately 0.002" to 0.004" (0.06 mm to 0.10 mm) clearance at the vertical shaft spline is desired. This figure represents .006" to .010" (0.15 to 0.25 mm) "backlash" at the actual gear set in the lower unit for satisfactory service.

If the "backlash" is too small, increase the amount of shim material under the ball bearing on the vertical shaft. If the "backlash" is too great, reduce the amount of shim material under the ball bearing.

After the proper "backlash" is established, the gear pattern must be checked.

Gear Pattern Check

15- Back out the screws securing the propeller shaft bearing housing to the lower unit. Pull the propeller shaft assembly out of the lower unit. Mark the drive gear on the lower end of the vertical shaft with some type of material that will leave a pattern on the other gear when they mesh. Install the assembled propeller shaft into the lower unit again. Tighten the mounting screws to a torque value of 29.5 ft-lbs (40 Nm). Affix Volvo special tool No. 884264 or similar device that will clamp to the upper end of the vertical shaft without damaging the splines. Apply a heavy breaking force to the propeller shaft. Exercise **CARE** not to damage the splines or threads on the propeller shaft. Now, while applying the braking force to the propeller shaft, rotate the vertical shaft in its normal direction, **CLOCKWISE,** for a left-hand propeller and the opposite direction for a right-hand propeller. Rotate the vertical shaft 6 to 8 complete revolutions. Remove the propeller shaft and check the dye pattern on the gear. Compare the pattern with the accompanying

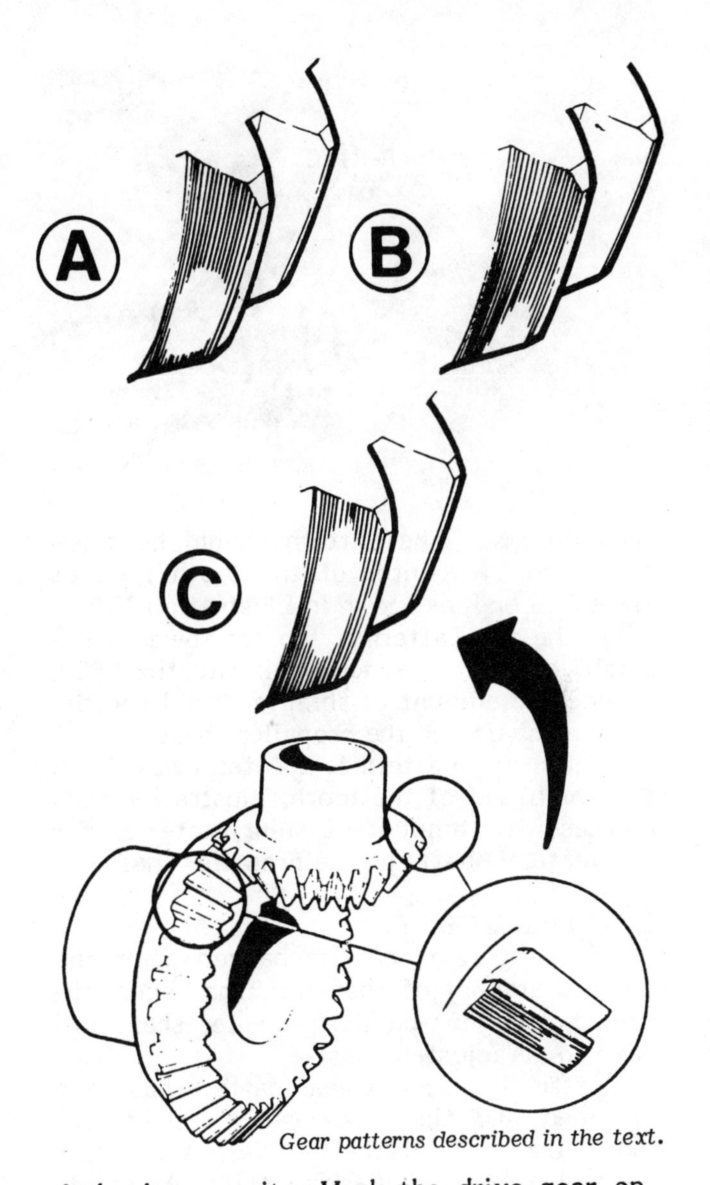

Gear patterns described in the text.

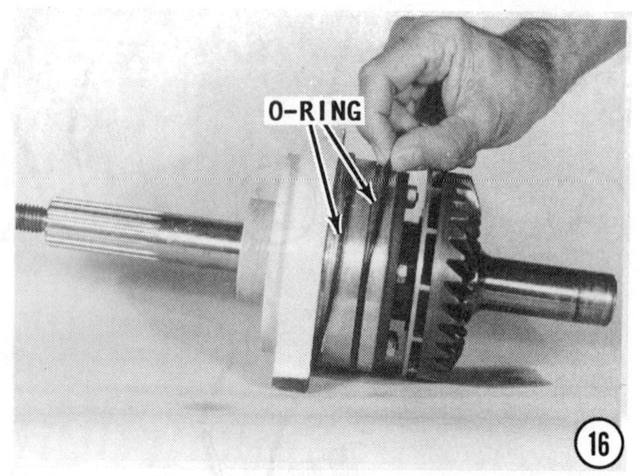

16

illustrations. The pattern should be positioned in the center of the tooth towards the small end, as shown in Illustration "A".

If the dye pattern is too far towards the small end of the tooth, Illustration "B", reduce the amount of shim material for the vertical shaft and the propeller shaft.

If the dye pattern is too far away from the small end of the tooth, Illustration "C", increase the amount of shim material for the vertical shaft and the propeller shaft.

CRITICAL WORDS

If the pinion gear is changed after the correct amount of "backlash" has been obtained, the gear on the propeller shaft will have to be adjusted.

After the correct gear pattern has been obtained and the "backlash" checked, remove the vertical shaft. Wash the pinion gear clean of the marking dye. Assemble the shaft, spacer ring, lockwasher, and nut, again. Tighten the nut to a torque value of 118 ft-lbs (160 Nm). Use Volvo special tool No. 884264 to hold the upper splined end of

17

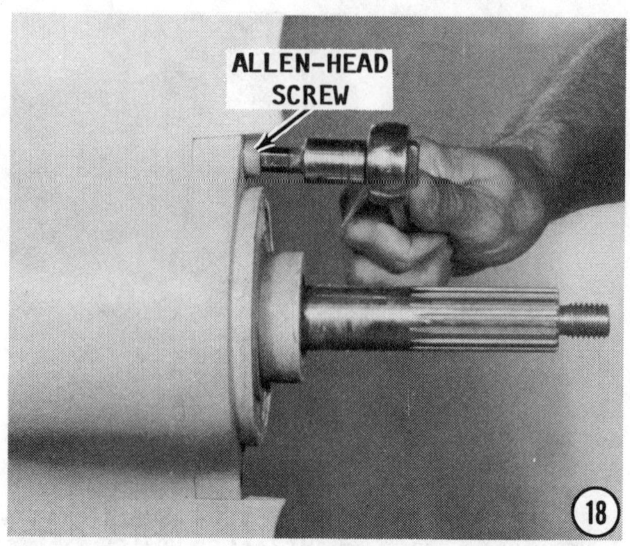

18

the vertical shaft while the nut is being tightened. Bend a tab of the lockwasher down to secure the nut in place.

16- Coat the mating surfaces of the lower unit and the propeller bearing housing with Permatex, or other suitable sealing material. Install the two O-rings into the grooves of the bearing housing.

17- **CAREFULLY** install the assembled propeller shaft into the lower unit. If necessary, rotate the shaft slightly to allow the gear to index with the pinion gear.

18- Install the two attaching screws and tighten them to a torque value of 29.5 ft-lbs (40 Nm).

19- Install the zinc collar and secure it in place with the attaching screws.

The lower unit is now ready to mate with the intermediate housing.

SPECIAL WORDS

If the shim material under the ball bearing on the vertical shaft was changed, it will

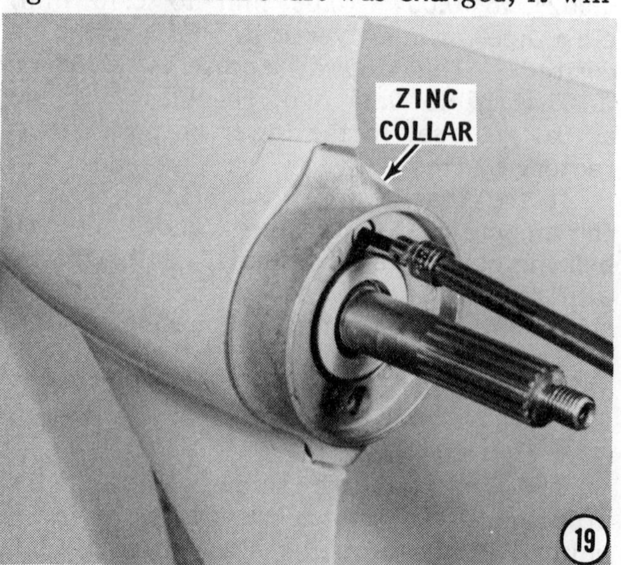

19

be necessary to determine the amount of shim material required to preload the vertical shaft tapered bearing between the lower unit and the intermediate housing.

First, remove the bearing race out of the intermediate housing. Install this bearing race into the lower unit housing on top of the tapered bearing. Measure the distance from the top of the bearing race to the mating surface of the lower unit.

Next, measure the distance from the mating surface of the intermediate housing into the bearing race cavity.

Now, subtract the difference between these two measurements. The "crush" should be 0.001" and the clearance not more than 0.002" which will give 0.003" total tolerance in that area. Approximately 0.001" is factory recommended.

20- Coat the mating surfaces of the lower unit and the intermediate housing with Permatex or other suitable sealer. **SLOWLY** lower the intermediate housing down over the lower unit with the oil distribution tube sliding into the opening in the lower unit. Keep the two surfaces as parallel as possible while lowering the intermediate housing. If the intermediate housing should become "cocked" to one side, the water tube may flip the O-ring out of position. If this should happen, an air leak would develop on the intake side of the cooling system. Such a leak would result in the engine overheating from lack of proper water flow.

21- Secure the intermediate housing with the seven attaching Allen-head screws.

HELPFUL HINT

Before the upper gear housing is mated to the intermediate housing, insert a funnel into the oil passage tube, and then pour in approximately 2-1/4 quarts of lubricating oil. Filling the lower unit at this time is much quicker and easier than adding the full amount of oil after the upper gear housing is secured and the cap is in place.

Check to be **SURE** the drive coupling is installed before the upper gear housing is mated to the intermediate housing.

22- Coat the mating surfaces of the intermediate and upper gear housings with Permatex or other suitable sealing material. Lower the upper gear assembly down onto the intermediate housing with the splines of the shaft indexed with the splines of the coupling. As an aid to indexing the splines

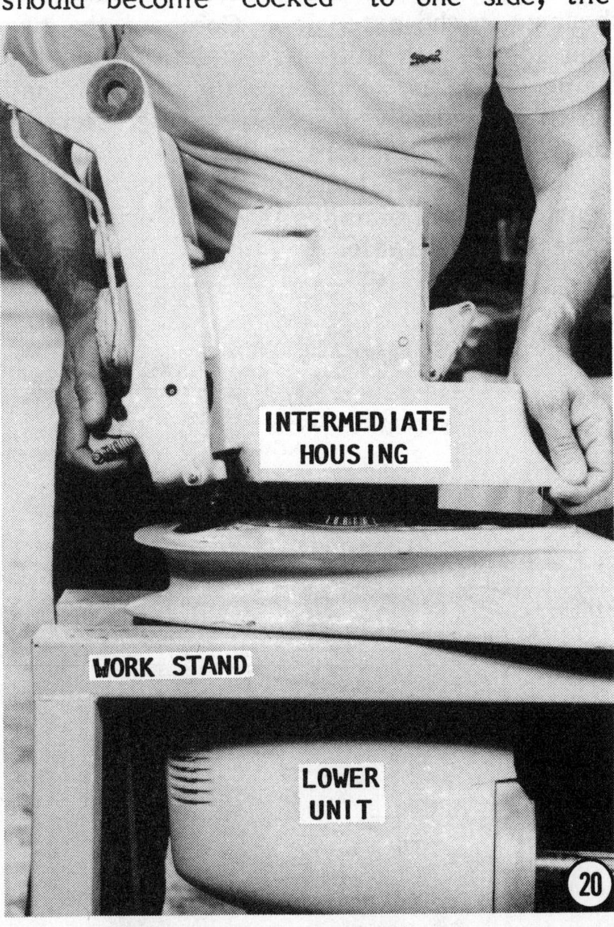

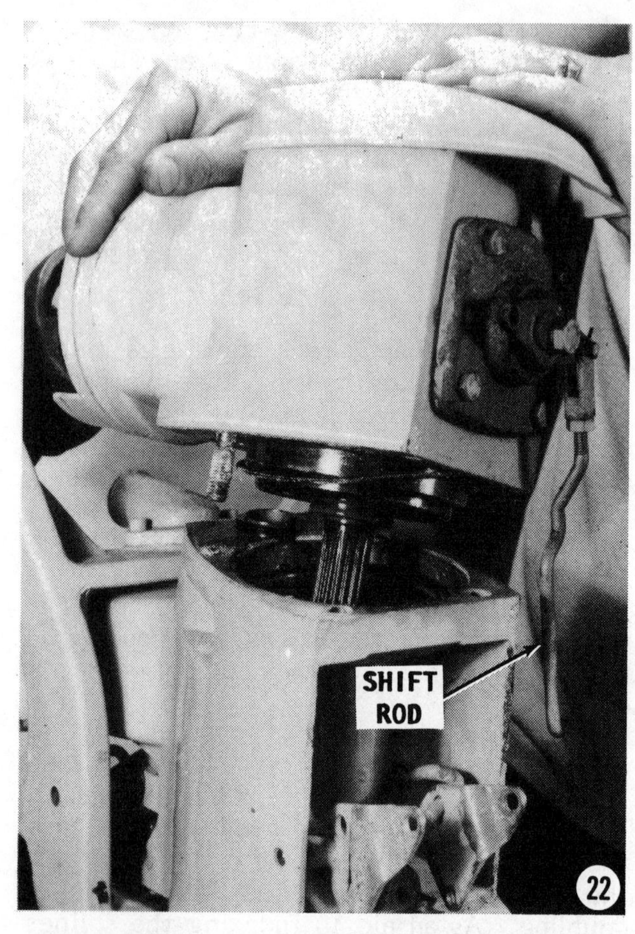

of the shaft with the splines in the coupling, have an assistant rotate the propeller shaft slightly while attempting to lower the upper gear housing.

Secure the two housings together with the attaching hardware -- two nuts and two bolts. Tighten the bolts and nuts alternately and evenly and securely in a criss-cross pattern. Connect and secure the shift rod with the cotter pin and washer.

The assembled stern drive is now ready to be installed on the boat. See Section 8-4, Installation, beginning on Page 8-10.

8-8 EXTENSIONS

Stern drive extensions, including the necessary fittings and attaching hardware, are available in three different sizes, through the local Volvo Penta dealer.

The accompanying exploded drawing of an average extension is almost self explanatory.

One adjustment may be necessary. Measure the depth of the recess in the intermediate housing for the tapered vertical shaft bearing. Measure the recess for the same bearing in the extension. Compare the two and adjust with shim material to obtain the same clearance to provide the same preload on the lower driven gear when the intermediate housing is mated to the extension.

Coat the mating surfaces of the extension and adjoining units with Permatex or other suitable sealer.

STERN DRIVE OIL CAPACITIES and GEAR RATIOS

Model No.	Capacity Quarts	Gear Ratio
100	2	1.66:1
100B	2	1.66:1
250C	2.2	1.89:1
250D	2.2	2.15:1
270C	2.8	1.89:1
270D	2.8	2.15:1
275	2.8	1.61:1
280B	2.8	1.61:1
280C	2.8	1.89:1
280D	2.8	2.15:1
290B	2.8	1.61:1
290D	2.8	2.15:1

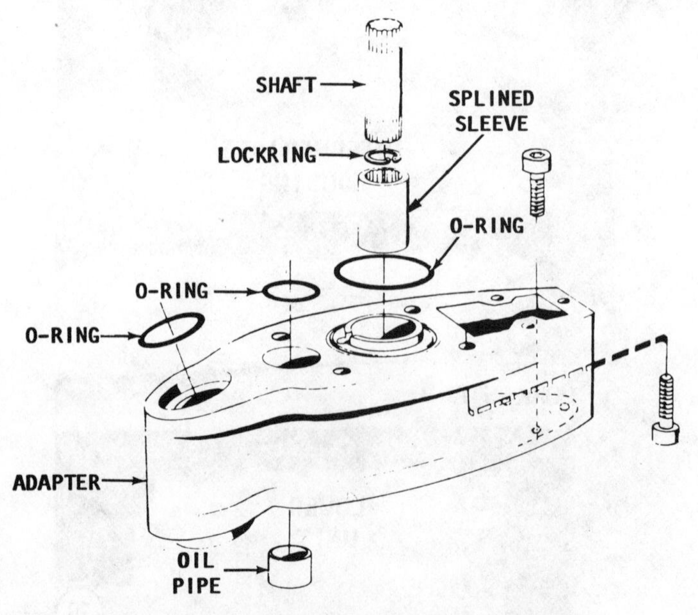

Principle parts of a typical lower unit extension.

8-9 DUOPROP LOWER UNIT

FIRST, THESE WORDS

Extreme care **MUST** be exercised when working with, on, or around the duoprop propeller shaft. If the shaft or the splines for the inner or outer propeller, or any other part is damaged, a complete new shaft, including the forward gear set **MUST** be purchased as a unit. Replacement of these parts could be quite expensive. Therefore, work carefully, move parts on and off the shafts slowly, and use the special tools called out in the procedures.

Propeller installation on a Duoprop lower unit. The propellers are a "matched" set with size and identification number stamped on the side of the hub.

8-10 PROPELLER SERVICE DUOPROP

FIRST, THESE WORDS

The two propellers of the Duoprop unit are a "matched" set and must be kept as a set. The part number and a size code from B1 thru B10 for gas powered units are stamped on the side of the propeller hub. An "A" prefix is used for diesel powered stern drives. Do not attempt to use one propeller from another unit **UNLESS** the prefix letter and numbers match for size and part number.

REMOVAL

CAUTION

Extreme care **MUST** be exercised when working with, on, or around the duoprop propeller shaft. If the shaft or the splines

for the inner or outer propeller, or any other part is damaged, a complete new shaft, including the forward gear set **MUST** be purchased as a unit. Replacement of these parts could be expensive. Therefore, move parts on and off the shafts slowly and with **CARE.**

1- Carefully wedge a block of wood between one blade of the aft propeller and the anti-caviatation plate to prevent the inner propeller shaft from rotating. Remove the bolt and washer from the end of the propeller shaft.

2- Remove the aft propeller nut, and washer, and then remove the block of wood.

3- Carefully remove the aft propeller free of the propeller shaft. Remove the aft fish line protector from the propeller shaft.

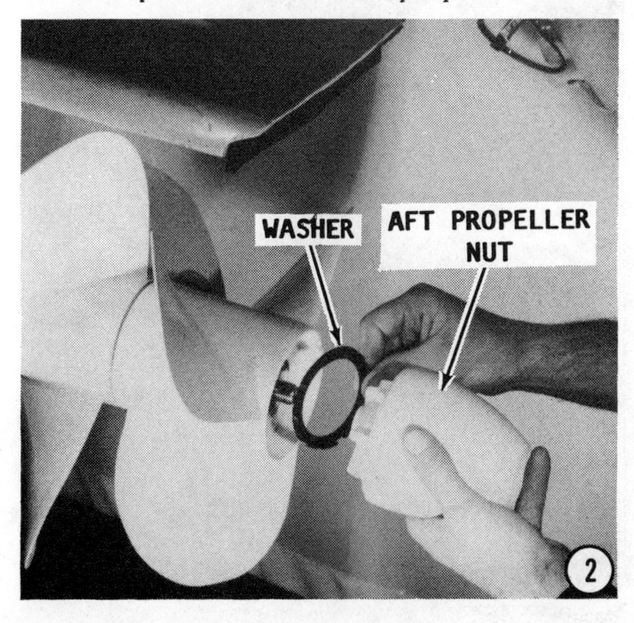

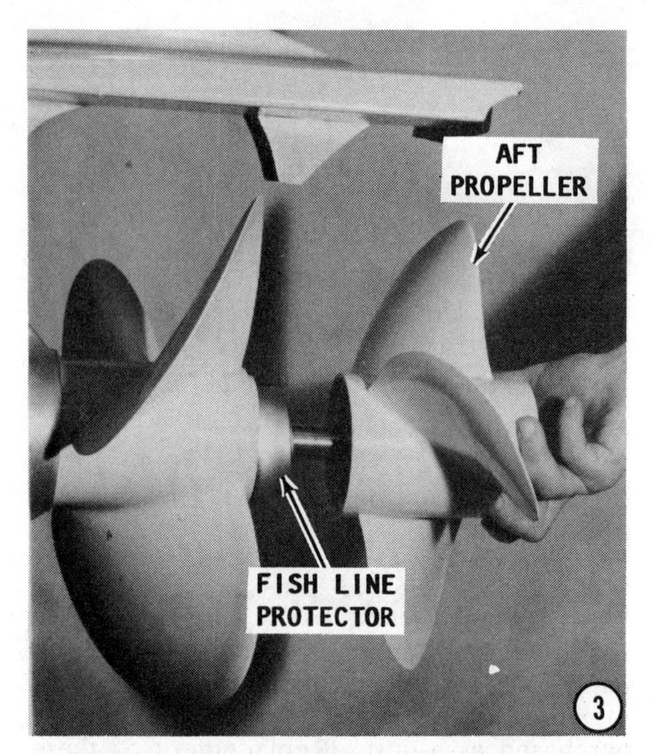

4- Carefully wedge the block of wood between one blade of the forward propeller and the anti-cavitation plate to prevent the outer propeller shaft from rotating. Bend the locking tab or tabs (if more than one was bent down to secure the spanner nut), up to release the spanner nut.

5- Remove the spanner nut, the locking tab washer and a plastic washer.

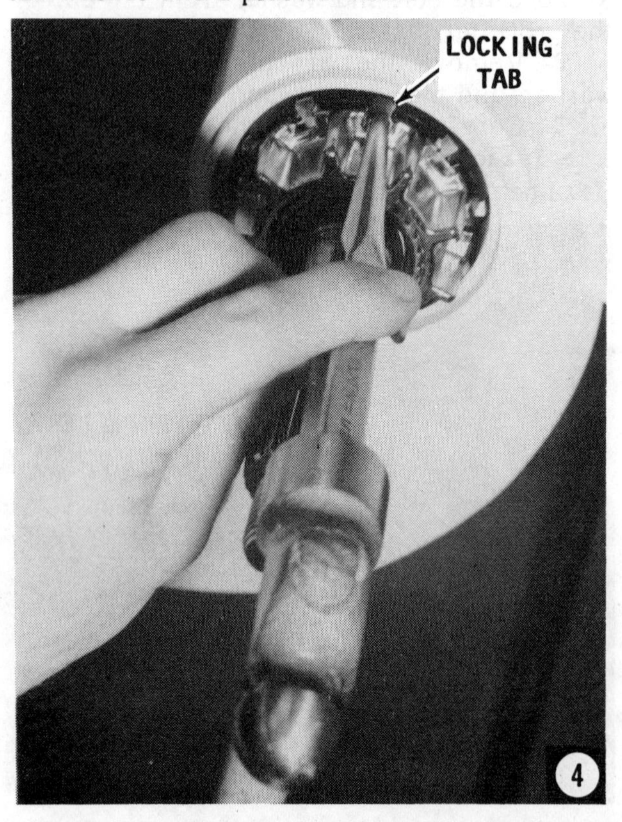

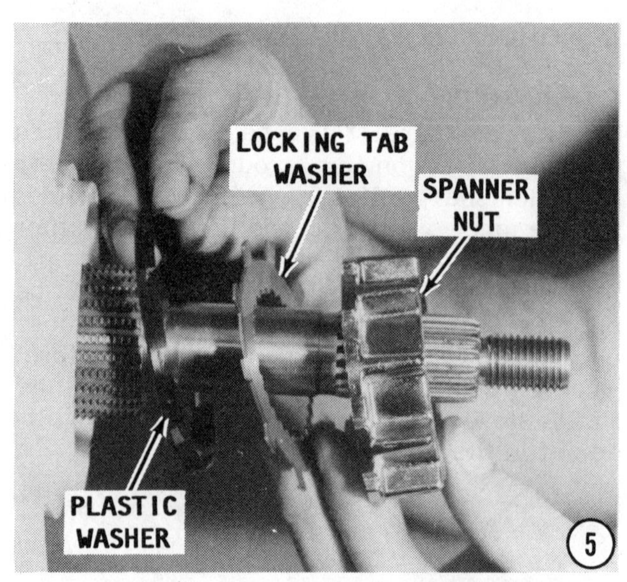

6- Remove the forward propeller from the outer and inner propeller shafts.

7- Slide the forward fish line protector free of both propeller shafts.

"FROZEN" PROPELLER

If the propeller is "frozen" to the shaft because an anti-seizing compound was not used the last time the propeller was installed, or because of any number of other reasons, heat must be applied and a propeller puller must be used. Apply heat with **CARE** so as not to damage the propeller shaft seals. It would be an excellent idea to replace the seals if heat is applied while removing the propeller.

After the propeller is "pulled" the rubber hub will still be attached to the shaft. Cut it free with a hacksaw or split it free with a suitable tool.

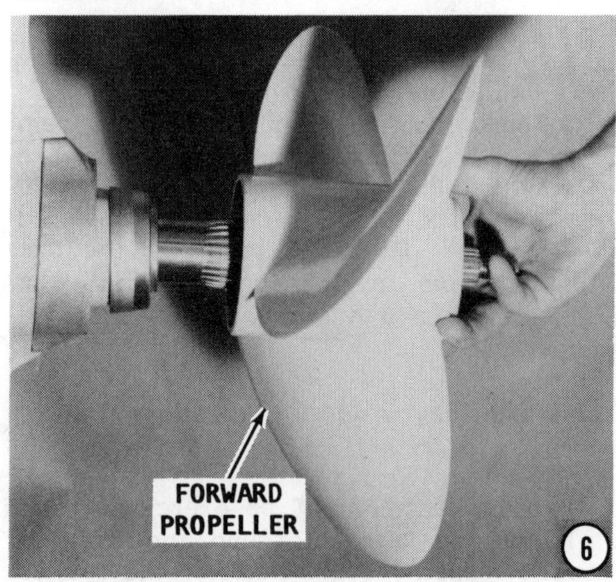

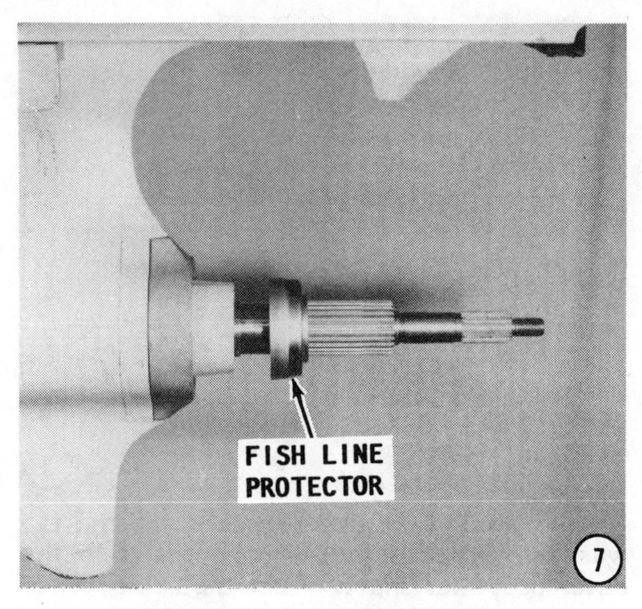

RETAINING PAWL ADJUSTMENT

The reference numbers in the following procedures are identified in the accompanying illustration on this page.

1- Remove the protective cover (12). Shift the control lever to the **NEUTRAL** position.

2- Disconnect the shift cable swivel (9) and the fork (11).

3- Release the lock nut for the fork (11). Check to be sure when the fork is connected to the lever, the retaining pawl rod (6) reaches the retaining pawl bracket at **"A"** without pressing against it. Make an adjustment if necessary, and then secure the fork in position with the lock nut.

4- Adjust the swivel (9) to allow it to easily enter the hole in the shift yoke. Shift the control lever to the **FORWARD** position. Check to be sure the corner **"C"** does not make contact with the housing. Install the protective cover.

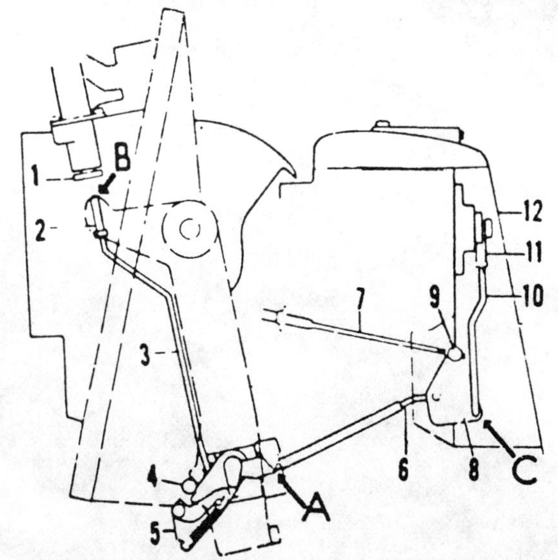

Points identified by number are referenced in the text for making a retaining pawl adjustment.

5- Push the drive forward against the adjusting pin. Check the position of the rod (3). The upper part (2) should extend just a whisker (1/16", 1.6mm) through the fork at **"B"**. This position will permit the lift (1) to disengage the retaining pawl (5) when the drive is tilted up. Loosen the lock nuts and adjust the upper part of the rod (2) as required.

PROPELLER INSTALLATION DUOPROP

FIRST, THESE WORDS

The two propellers of the Duoprop unit are a "matched" set and must be kept as a set. The part number and a size code from B1 thru B10 for gas powered units are stamped on the side of the propeller hub. An "A" prefix is used for diesel powered stern drives. Do not attempt to use one propeller from another unit **UNLESS** the prefix letter and numbers match for size and part number.

CAUTION

Extreme care **MUST** be exercised when working with, on, or around the duoprop propeller shaft. If the shaft or the splines for the inner or outer propeller, or any other part is damaged, a complete new shaft,

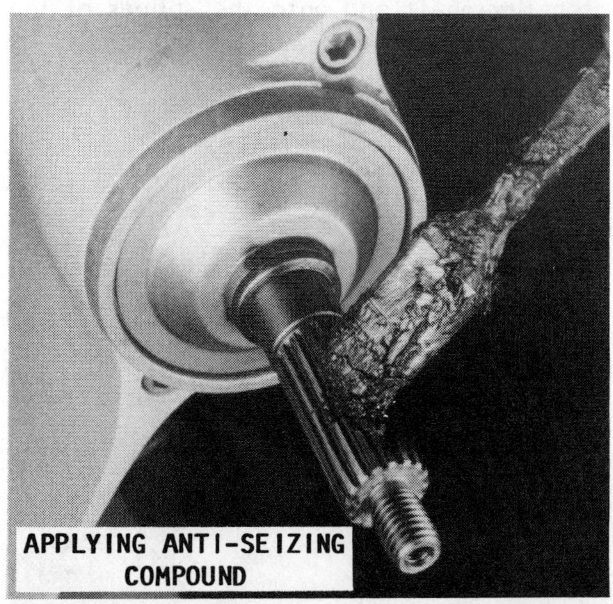

Applying anti-seizing compound to a standard propeller shaft. The same technique applies to a Duoprop installation.

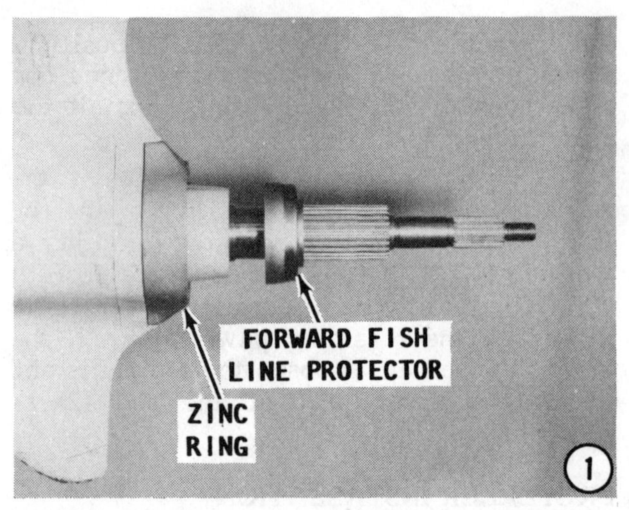

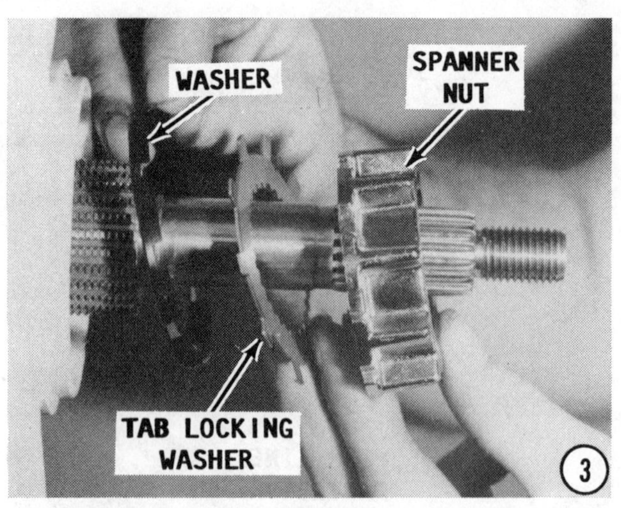

including the forward gear set **MUST** be purchased as a unit. Replacement of these parts could be expensive. Therefore, move parts on and off the shafts slowly and with **CARE.**

1- Check to be sure the splines on the outer propeller shaft are clean and in good condition. If the zinc ring was removed, install **NEW** one in position against the bearing housing. Secure the zinc ring with the two Allen head screws. Install the fish line protector onto the outer propeller shaft with the shoulder facing outward, as shown.

2- Coat the splines of the outer propeller shaft with a good film of water-resistant anti-corrosion grease. Failure to apply this lubricant could cause the propeller to "freeze" on the propeller shaft. Install the forward propeller over the inner propeller shaft and onto the splines of the outer propeller shaft. Move the propeller up against the fish line protector.

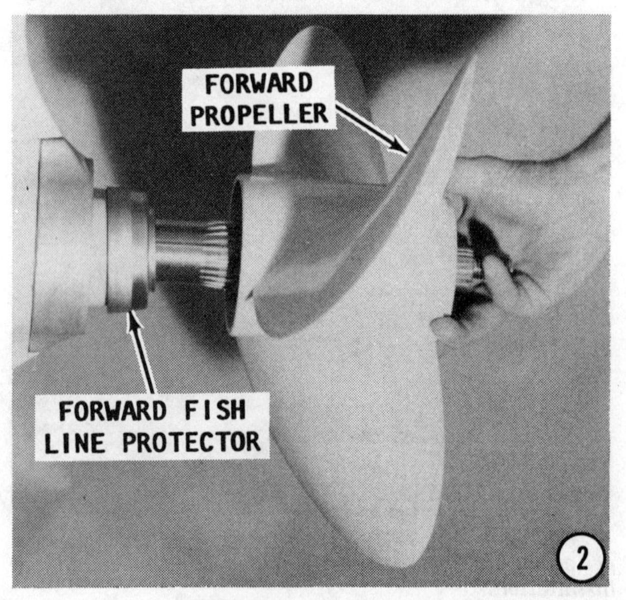

3- Install the plastic washer and the tab washer. Thread the spanner nut onto the outer propeller shaft. Place a block of wood between one of the blades of the forward propeller and the anti-cavitation plate to prevent the outer propeller shaft from rotating. Tighten the spanner nut securely.

4- Bend one, or more if possible, tabs down between the fins of the spanner nut to lock the nut in place.

5- Install the second fish line protector into place on the inner propeller shaft and against the spanner nut. Coat the splines of the inner propeller shaft with a coating of water-resistant anti-corrosion grease. Install the aft propeller onto the inner

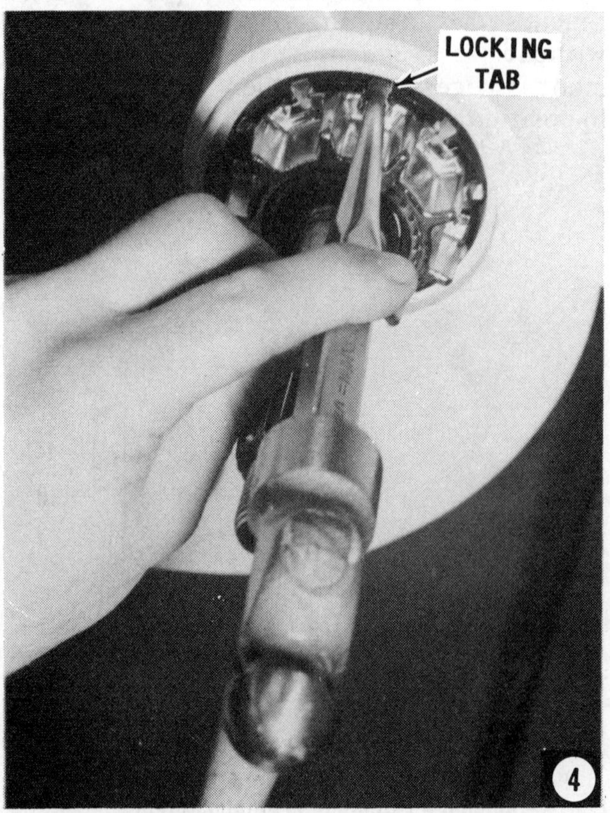

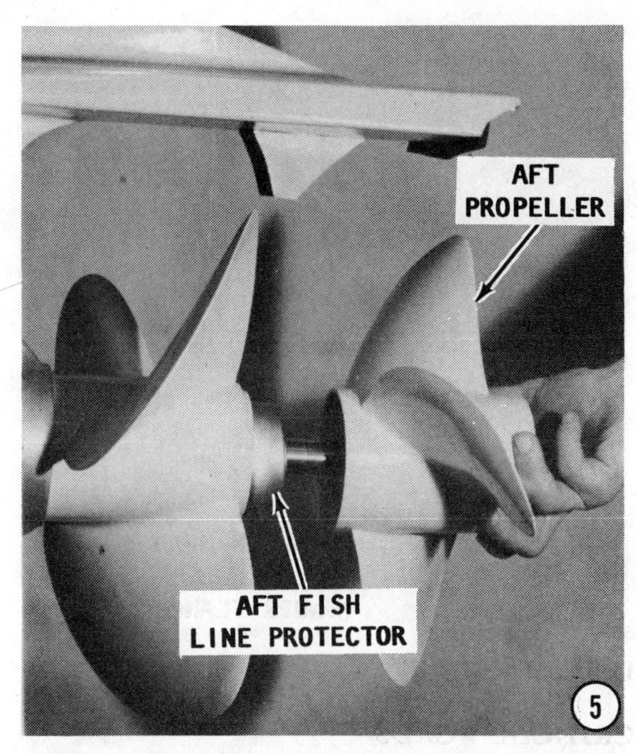

peller shaft and up against the fish line protector.

6- Install the black plastic washer up against the propeller. Thread the propeler nut onto the inner propeller shaft. Move the block of wood to the aft propeller to prevent the inner propeller shaft from rotating. Tighten the propeller nut securely.

7- Slide a washer onto the inner propeller shaft bolt, and then install and tighten the bolt securely. In some cases, the washer may be affixed to the bolt.

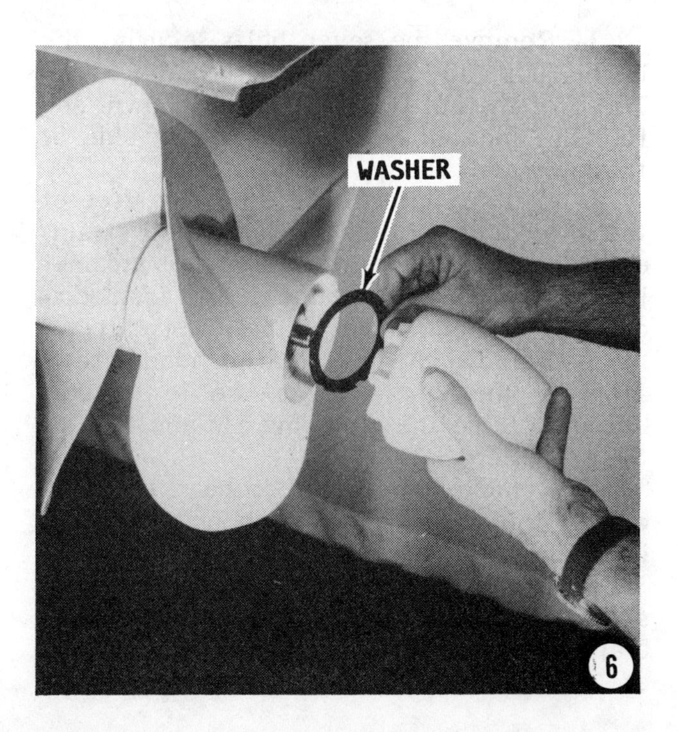

8-11 DUOPROP LOWER UNIT

FIRST, THESE WORDS

Extreme care **MUST** be exercised when working with, on, or around the duoprop propeller shafts. If the shafts or the splines for the inner or outer propeller, or any other part is damaged, a complete new shaft, including the forward gear set **MUST** be purchased as a unit. Replacement of these parts could be quite expensive. Therefore, work carefully, move parts on and off the shafts slowly, and use the special tools called out in the procedures.

PREPARATION

The following procedures pick up the work after the stern drive has been removed from the boat, see Section 8-4, Stern Drive Removal, beginning on Page 8-8.

Loosen the dip stick from the upper gear housing. Remove the plug from the bottom of the lower unit and drain the lubricant from the stern drive. Remove the four fasteners securing the upper gear assembly to the intermediate housing. Two of these fasteners are Allen head screws, and two are nuts, one on either side of the intermediate housing. Pull the cotter pin and washer from the shift linkage rod at the intermediate housing, as shown.

Remove the upper gear assembly from the intermediate housing.

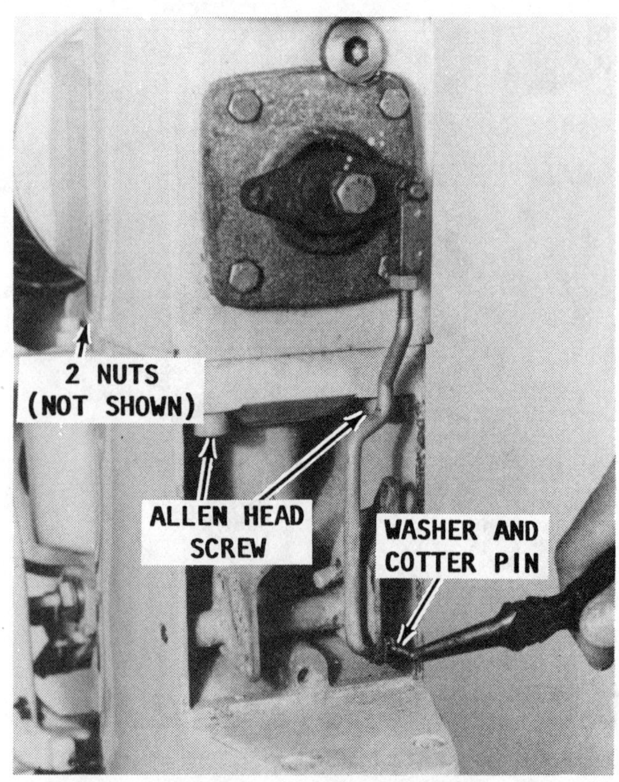

Pulling the cotter pin and washer from the shift linkage rod in preparation to removing the upper gear assembly.

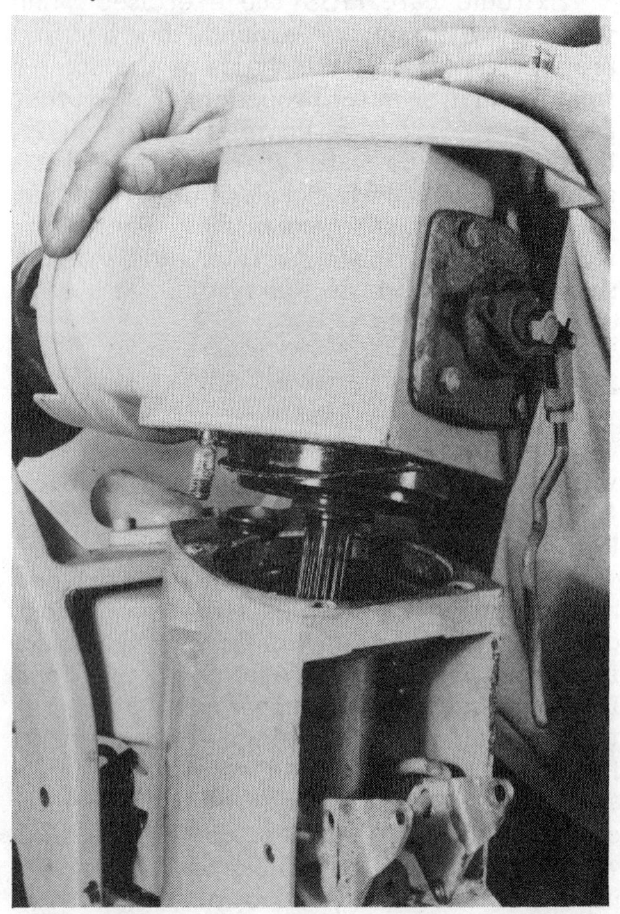

Separating the upper gear assembly from the intermediate housing.

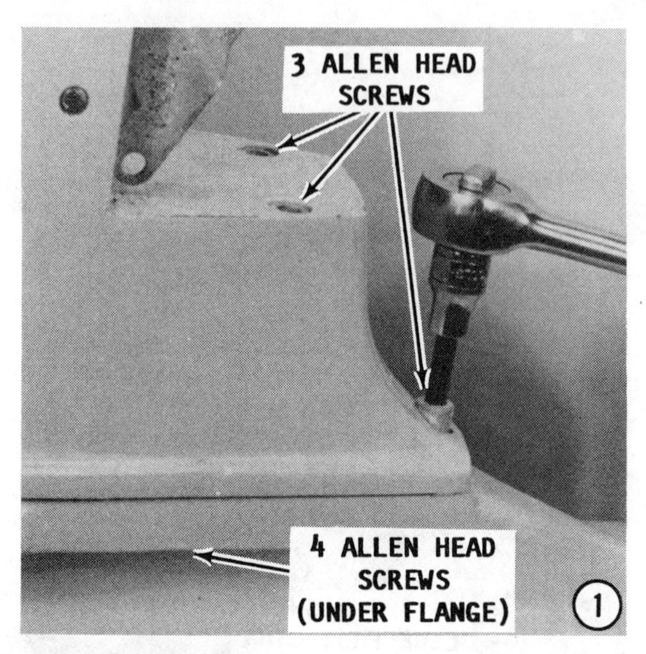

DISASSEMBLING

CRITICAL WORDS

It is most important to record the location of certain bearings, bearing races, and the shim material removed. As an example, the forward gear, bearing, and race is almost identical to the rear gear, bearing, and race. However, they do differ and **MUST** be installed in the correct location.

Considerable time can be saved during assembling, if the exact amount of shim material removed from a certain location can be installed with confidence it is the correct thickness.

1- Remove the seven bolts securing the lower unit to the intermediate housing. Three of the bolts are on top, as shown, and four are located on the underneath side of the lower unit flange.

2- Lift the intermediate housing free of the lower unit. Notice the basket oil strainer in the top of the lower unit. This strainer is designed to prevent any debris from circulating in the oil through the lower unit.

3- Remove the two Phillips head screws securing the zinc ring to the lower unit housing. Slide the zinc ring aft and free of the propeller shafts.

4- Remove the two Allen head bolts securing the double bearing housing to the lower unit.

5- Obtain special tool No. 884789 and slide hammer No. 884161. Secure the special tool to the outer propeller shaft. Operate the slide hammer.

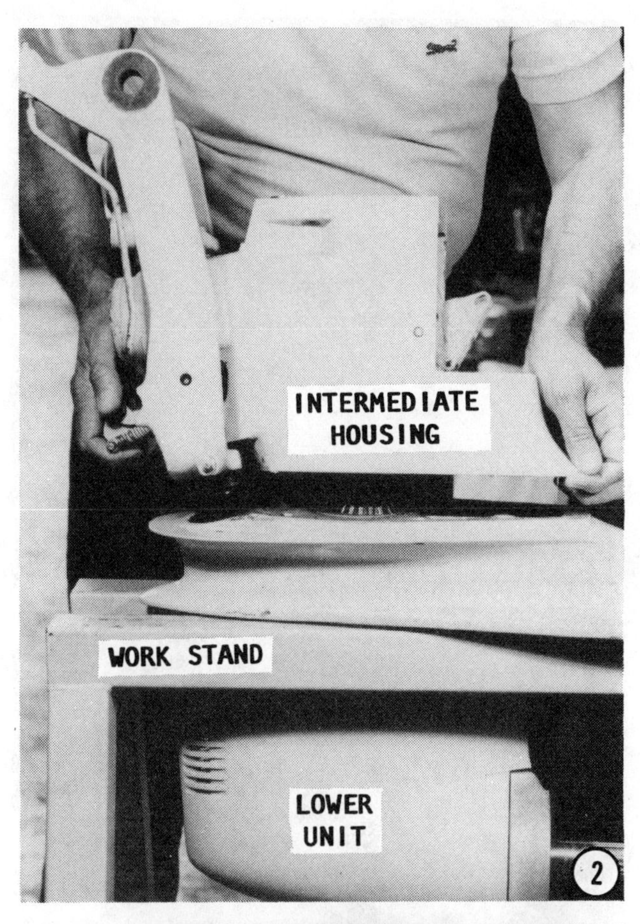

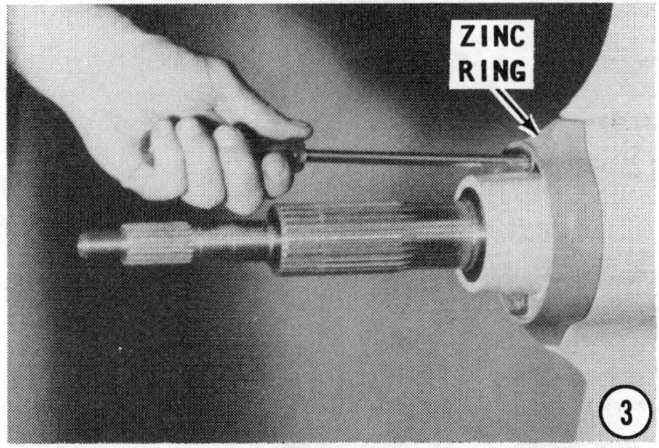

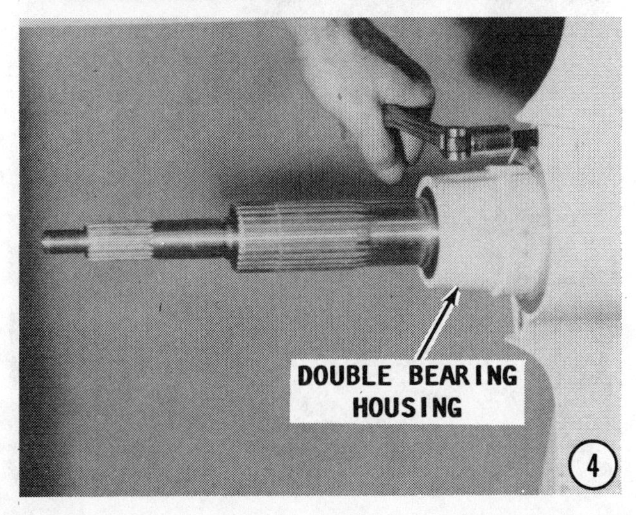

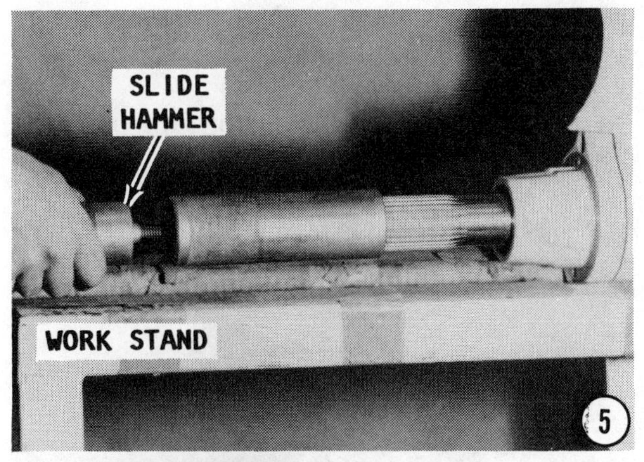

6- Continue operating the slide hammer and pull the outer propeller shaft and the double bearing housing free of the lower unit housing.

7- Obtain special tool No. 884841. Thread the special tool onto the end of the inner propeller shaft.

8- Obtain special tool No. 884802. Insert the tool into the lower unit with the cutout facing up to pass over the pinion gear nut on the lower end of the driveshaft. Install the nut and washer onto the end of the special tool attached to the end of the propeller shaft.

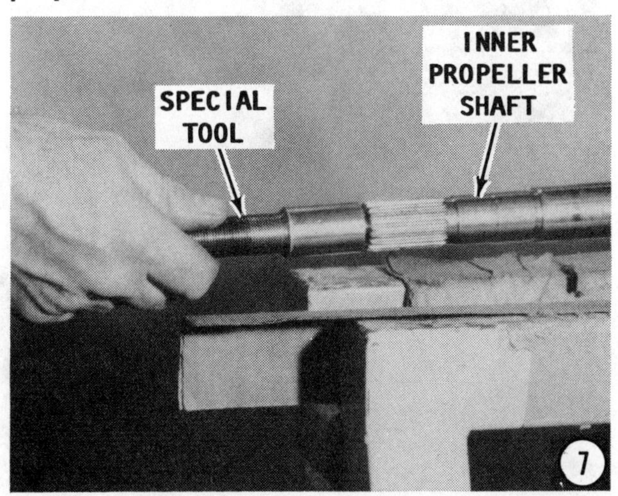

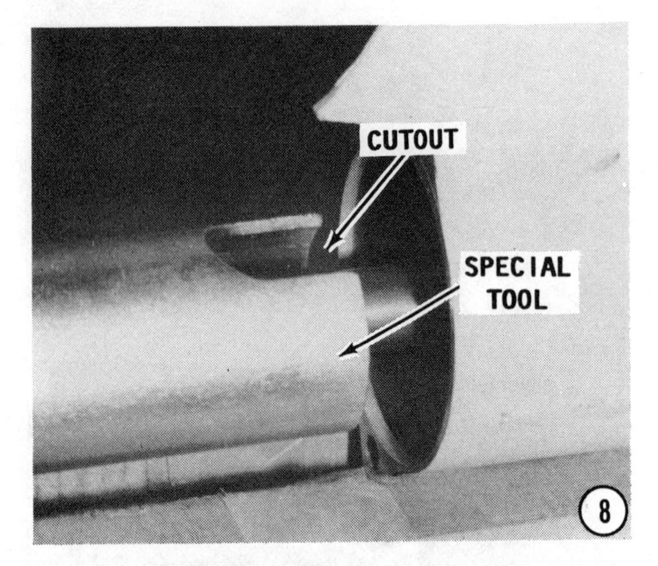

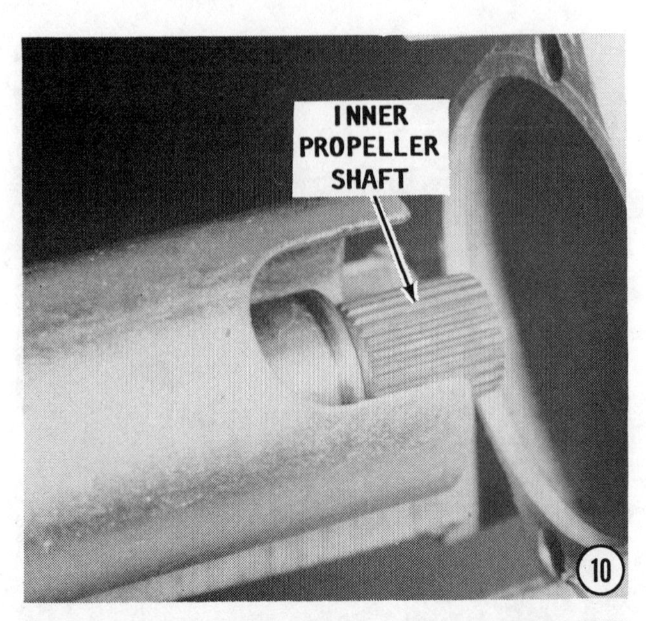

9- Tighten the nut on the end of the propeller shaft. As the nut is tightened, the shaft is moved out of the forward gear. Continue tightening on the nut until the shaft is free.

10- Withdraw the shaft from the lower unit and remove the special tools.

11- Bend down the lock tab on the lock-washer, and then remove the spanner nut. Install the special tool onto the end of the driveshaft.

12- Obtain special tool No. 884267. Install the tool on the driveshaft until it seats on the upper surface of the lower unit. Turn the lower unit upside down and clamp the special tool in a vise equipped with soft jaws.

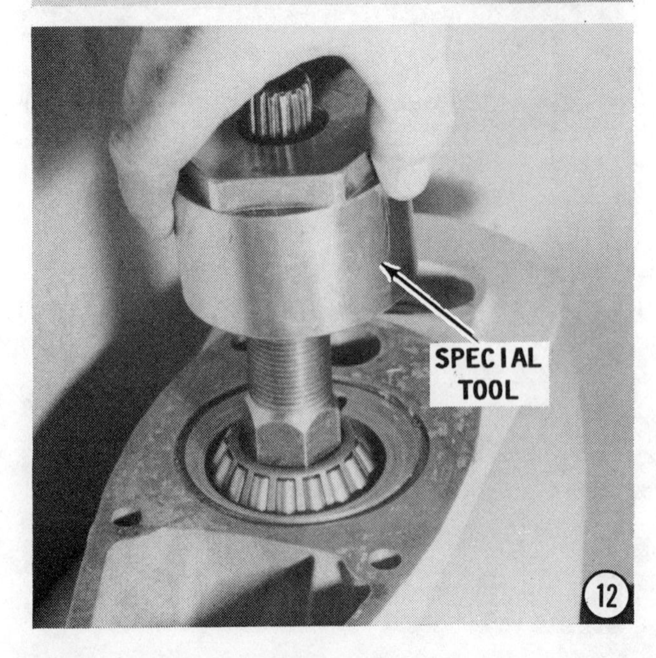

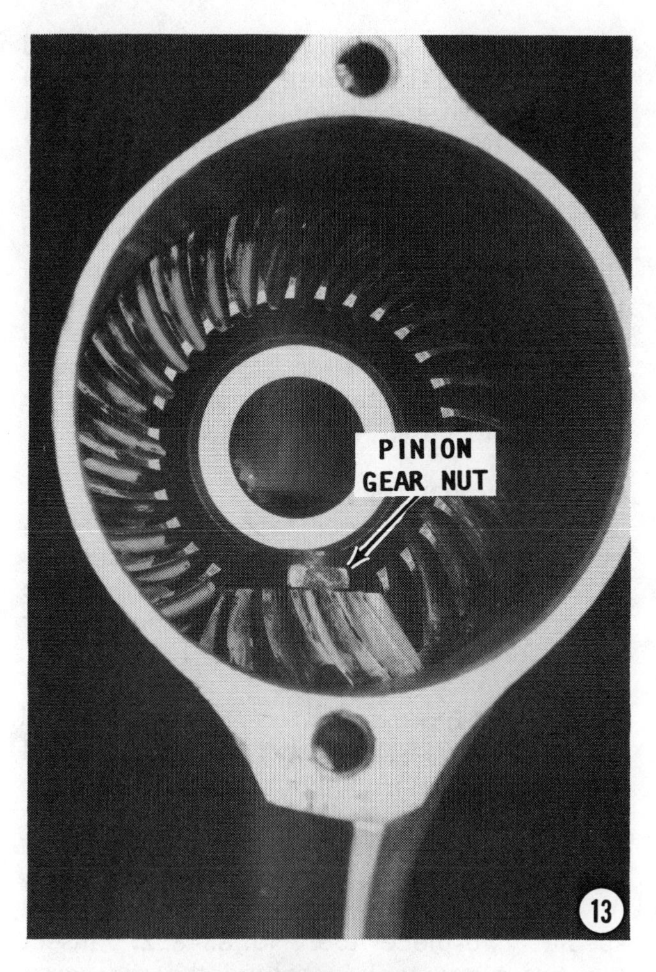

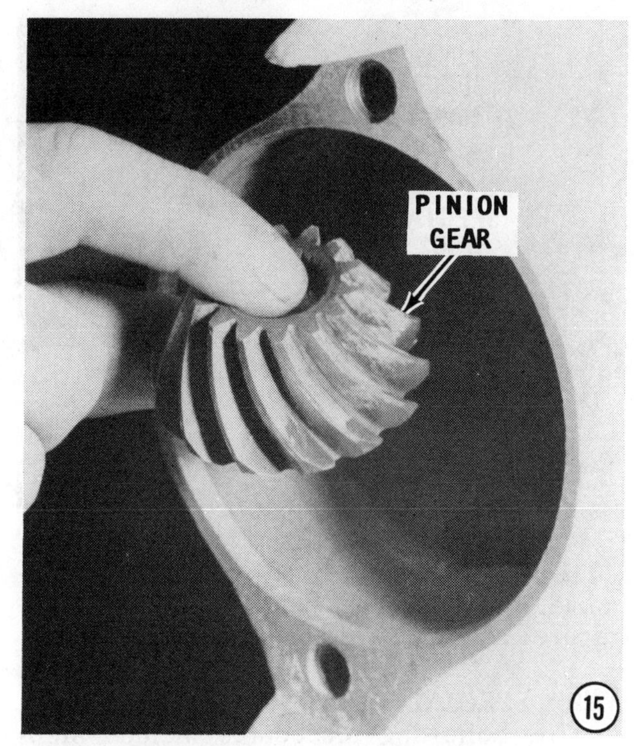

13- Hold the pinion gear nut with the proper size socket. Now, rotate the lower unit **COUNTERCLOCKWISE** until the driveshaft has been released from the pinion gear.

14- Lift the lower unit up and free of the driveshaft. Remove the special tool from the driveshaft. Save any shim material from atop the tool.

15- Remove the pinion gear nut and pinion gear from the lower unit cavity. **DO NOT** discard the pinion gear nut at this time. It will be used later for temporary installation to check prestress, backlash and gear pattern.

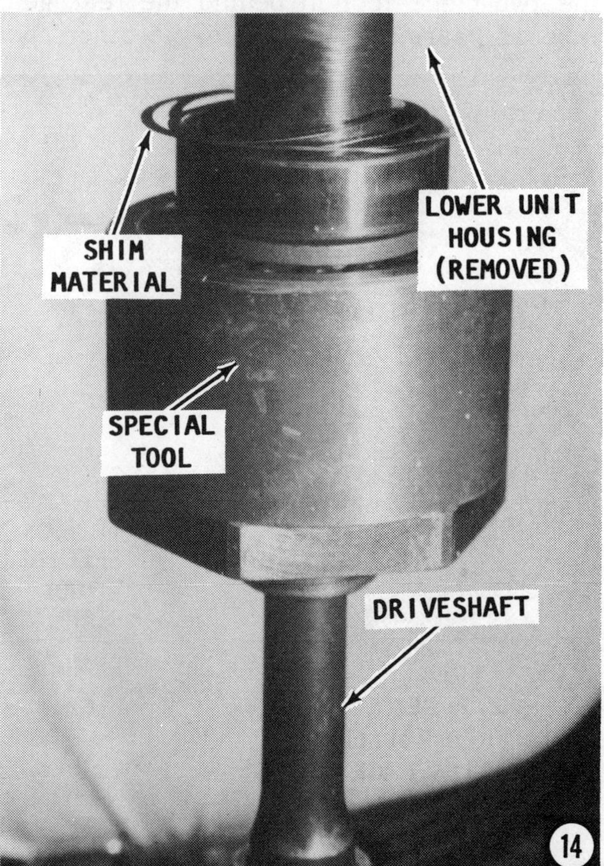

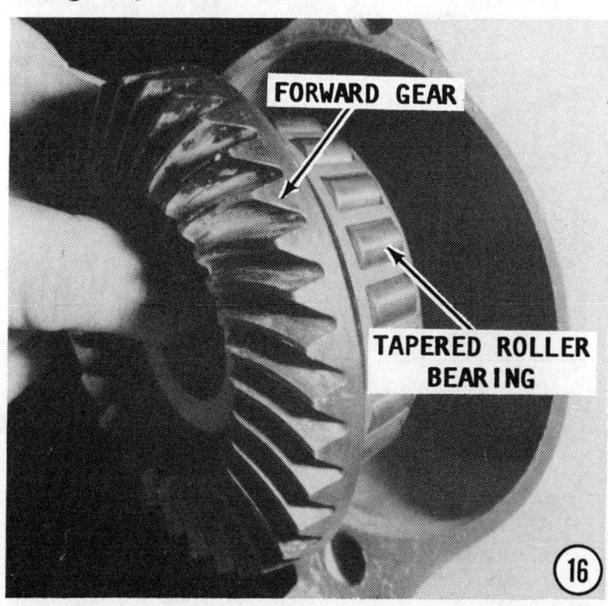

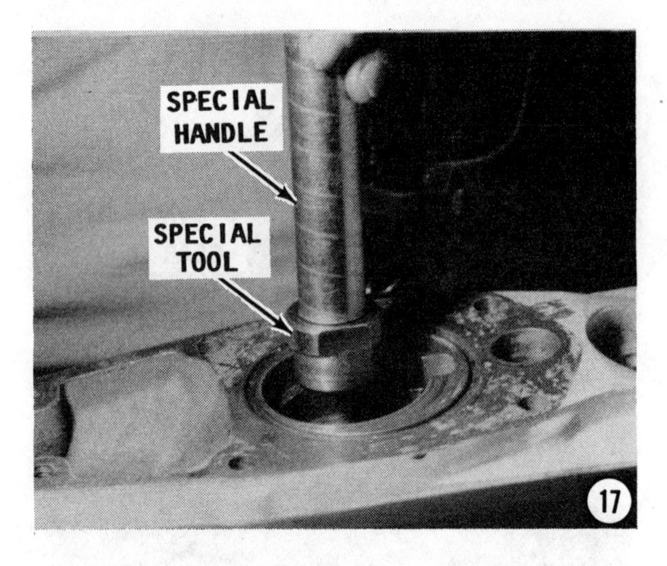

16- Reach into the lower unit and remove the forward gear and the forward gear tapered roller bearing.

VERY SPECIAL WORDS

The following procedures include steps to remove all bearings and races located in the lower unit. All require special tools for removal of the existing part and installation of the replacement bearing or race. **DO NOT** attempt to remove the bearings or races unless the item is determined to be unfit for further service **AND** unless the proper tools called out in the step are available. Any attempt to remove bearings and/or races with makeshift equipment could damage the part being removed/installed or ajacent parts, and lead to frustration in the attempt.

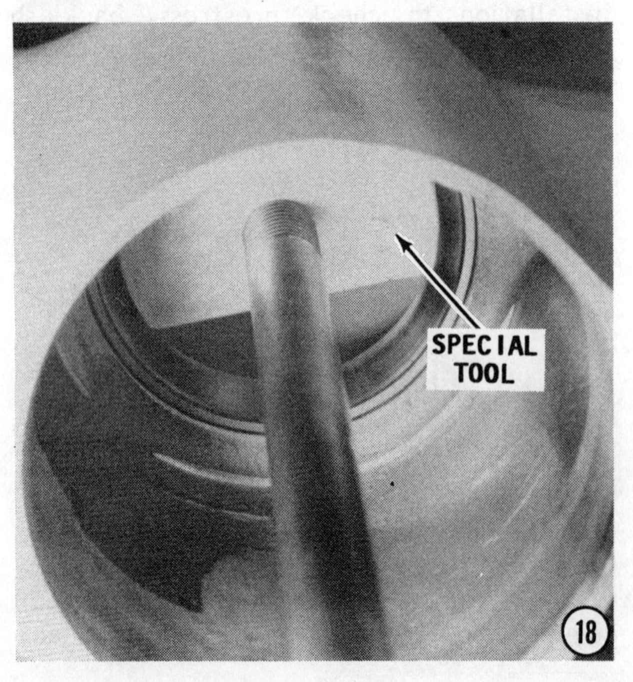

17- Obtain special tool No. 884791 and handle No. 884143. Insert the special tool down through the lower unit until the tool rests on top of the lower driveshaft needle bearing. Drive the bearing down until it falls free in the lower unit cavity.

18- Obtain special tool No. 884794 and slide hammer No. 884161. Insert the special tool in behind the bearing race, and then thread the slide hammer into the special tool. Operate the slide hammer and "pull" the forward gear tapered bearing race free of the lower unit.

CRITICAL WORDS

The front and rear tapered roller bearings and their associated bearing races appear to be quite similar. However, they are different and **MUST** not be exchanged or mixed, one with the other.

19- Obtain special tool No. 884796 and handle No. 991801. Insert the special tool into the double bearing housing and seat the tool against the rear gear tapered roller bearing race. Thread the handle into the tool and drive the race free of the housing.

20- Obtain special tool No. 884140 and special two-piece tool No. 884832. Insert the two-piece tool in behind the rear gear tapered bearing race. Slide the other spe-

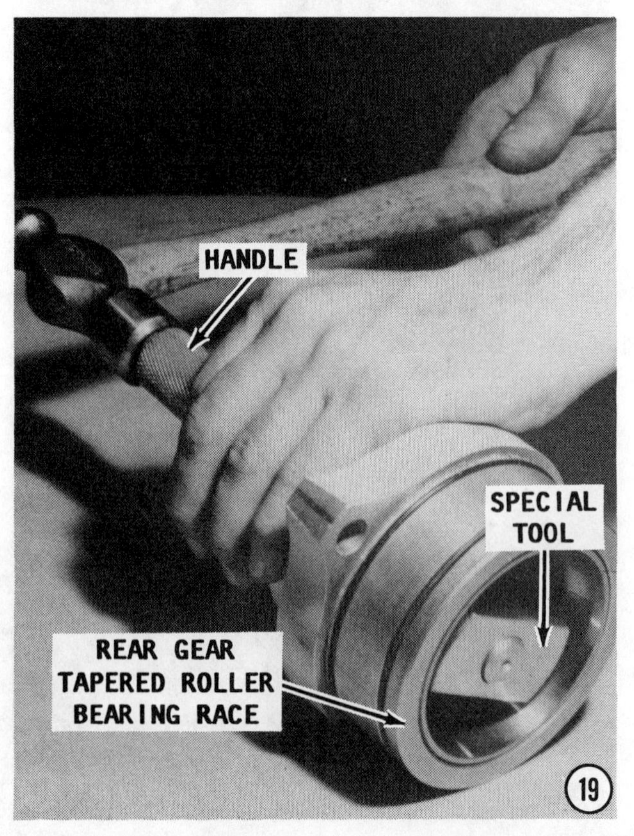

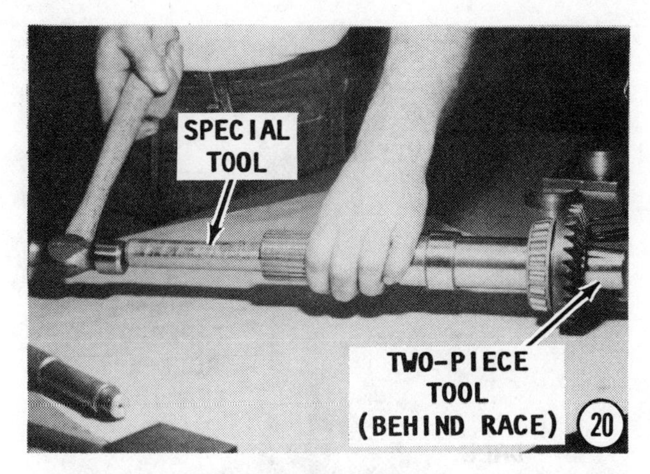

cial tool through the outer propeller shaft with the small end going in first. Push the tool all the way up against the two-piece tool. Now, drive the bearing race free of the rear gear. Exercise care when driving, because the bearing race and shim material will come out with force.

21- **SAVE** any and all shim material removed from behind the bearing race.

22- Obtain special tool No. 884797. Use the special tool and an arbor press to separate the tapered roller bearing from the inner propeller shaft.

23- Obtain Volvo special three-piece tool No. 884831. Slide the plate over the outer drive shaft, and then position the two halves of the collar around the ball bearing race, with the inner recess facing up, as shown. Place the assembled parts in an arbor press. Press down on the end of the propeller shaft and separate the bearing race from the outer propeller shaft.

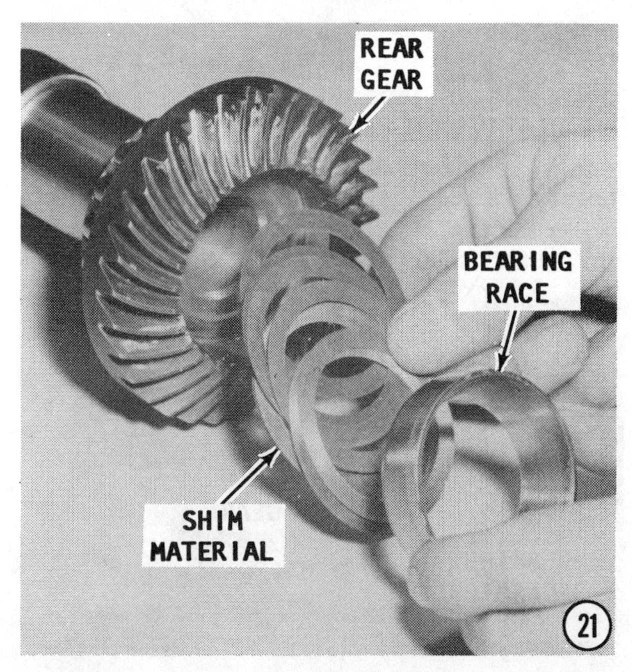

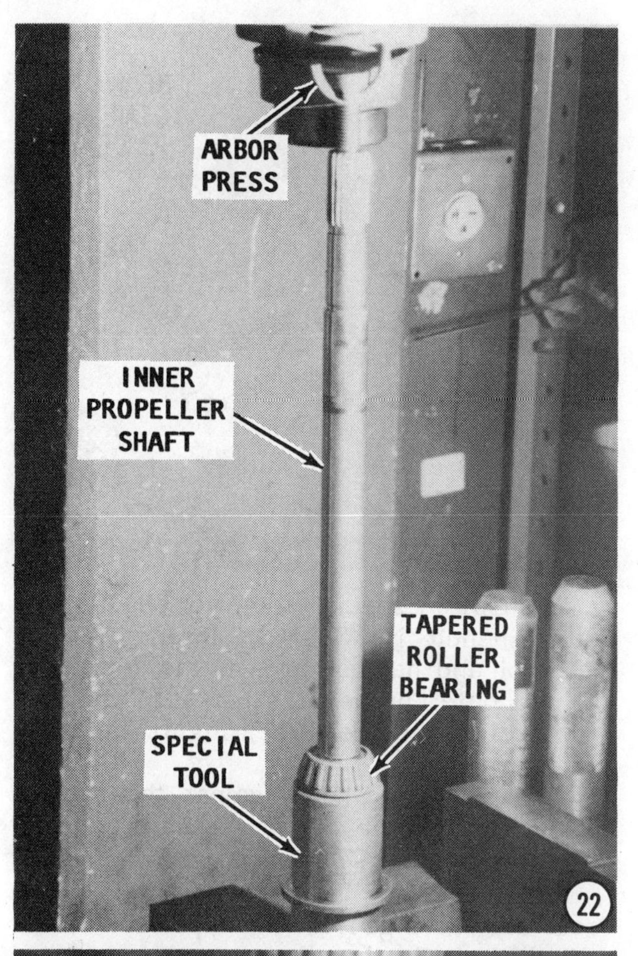

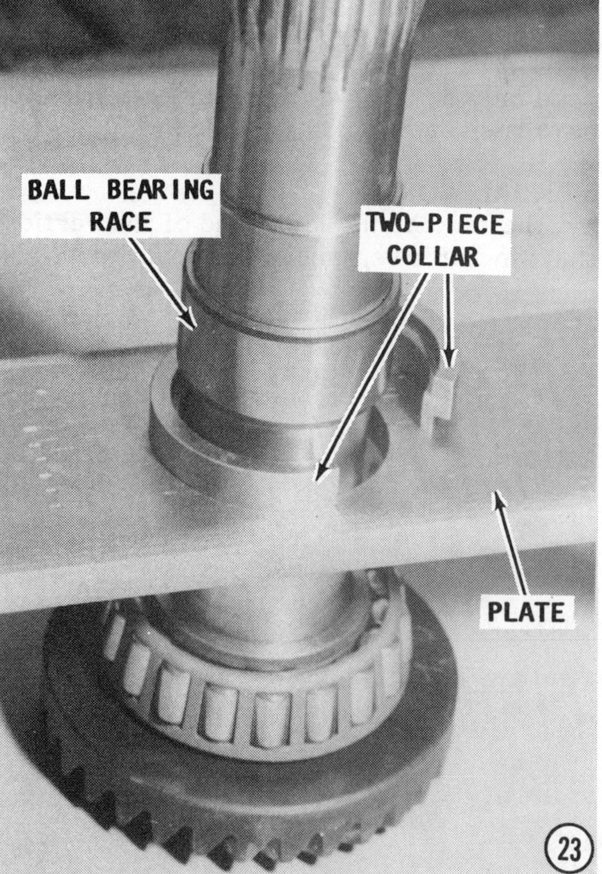

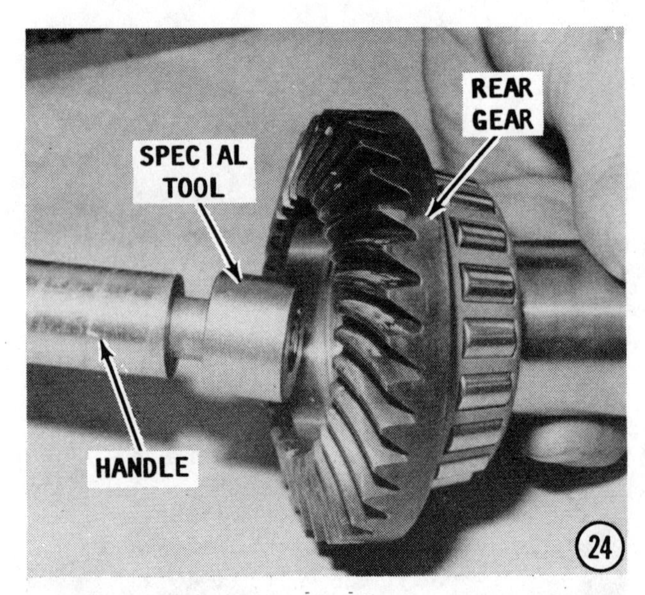

24- Obtain special tool No. 884803 and special handle No. 884143. Insert the special tool and handle through the rear gear and the outer propeller shaft until the tool is up against the needle bearing at the other end of the shaft.

25- Use the handle against the special tool and drive the needle bearing and the two seals free of the outer propeller shaft.

26- Obtain a bearing separator tool. Press the bearings from the upper end of the driveshaft. The bearings may be removed together or one by one. Notice the spacer between the two bearings. These bearings need only be removed from the shaft if they have been damaged or corroded.

SPECIAL WORDS

The race at the lower end of the vertical shaft in the lower unit housing can be driven out, using a special Volvo tool or making a fixture to grasp the upper lip of the race. This is **ONLY** done if the race has been damaged or corroded. Replacement of the bearing race is **NOT** an easy task. A special puller or a makeshift arrangement is required to "pull" the new race up into the cavity. Again, do **NOT** attempt to remove this race unless it is damaged or corroded so badly as to affect performance of the bearing.

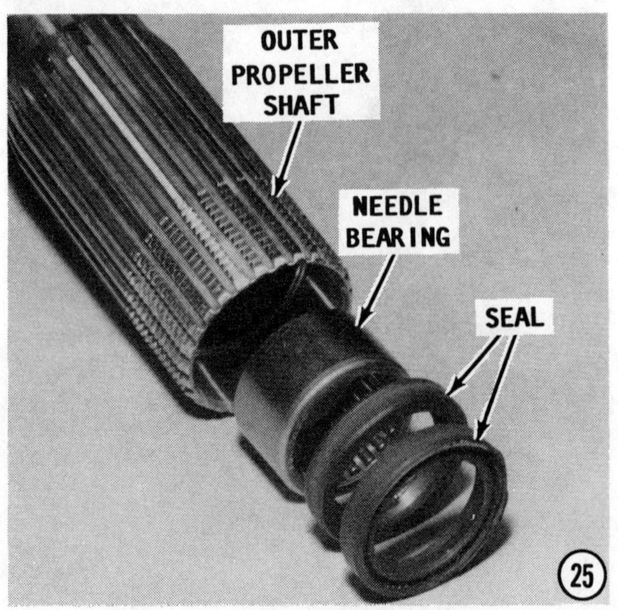

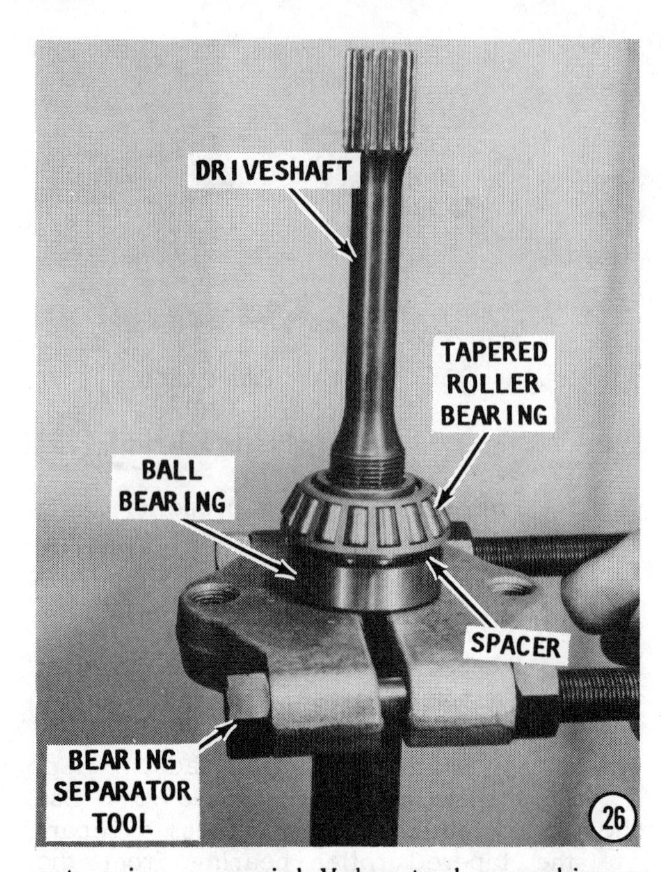

27- Obtain a bearing separator tool. Use tool and an arbor press to separate the forward gear tapered roller bearing from the forward gear.

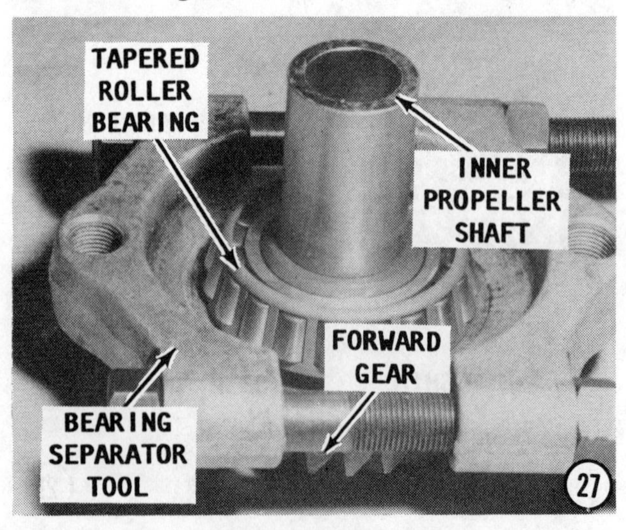

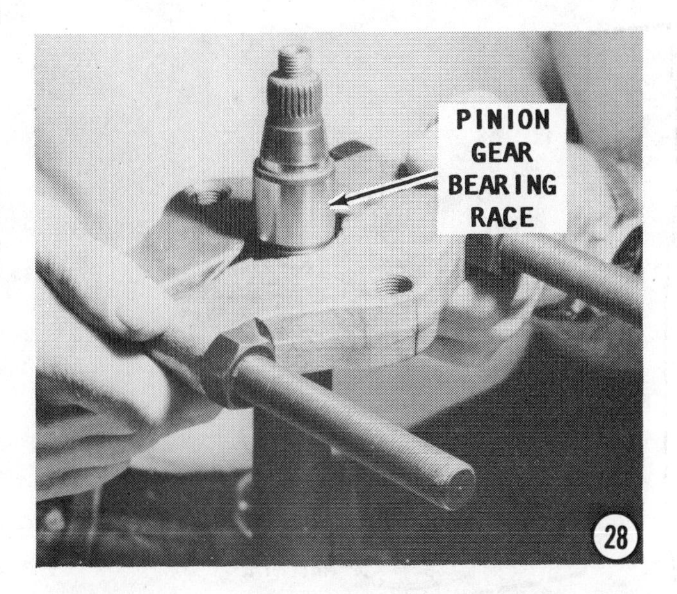

28- Use a bearing separator tool and arbor press to separate the pinion bearing race from the lower end of the vertical driveshaft.

29- Use a bearing separator tool and an arbor press to separate the tapered roller bearing from the rear gear on the outer propeller shaft. **BEFORE** operating the press, check to be sure the lips of the tool are in position under the bearing.

30- Obtain special tool No. 884797. Set up the special tool and the propeller shaft bearing housing in an arbor press. Push the needle bearing and two seals free of the double bearing housing.

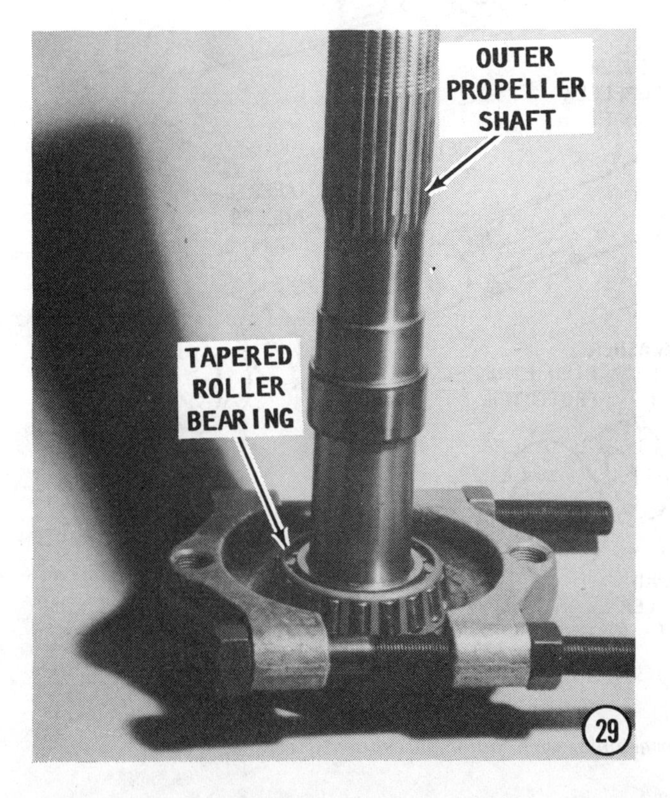

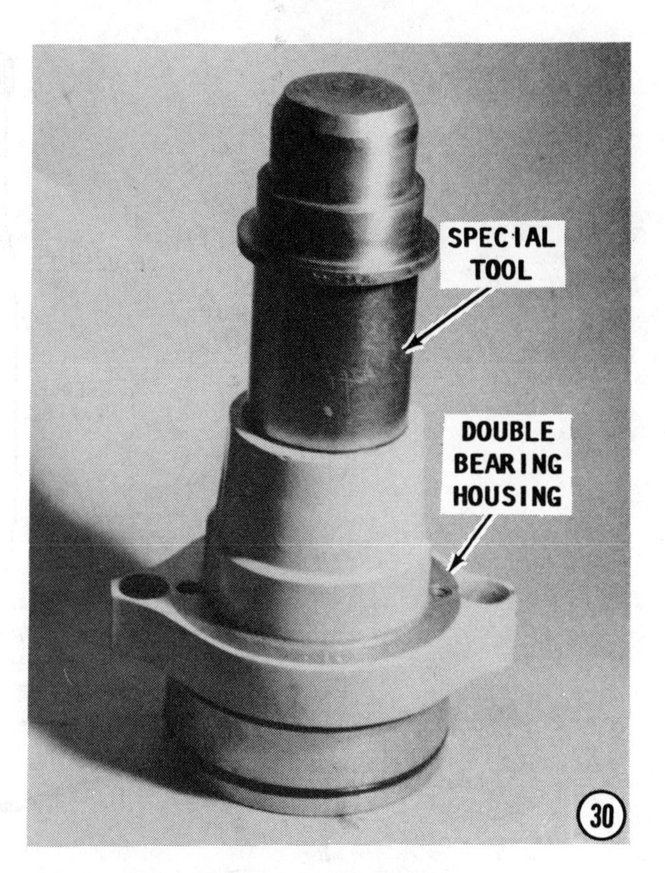

CLEANING AND INSPECTING ALL UNITS

Wash all parts in solvent and blow them dry with compressed air. Remove all traces of seal and gasket material from all mating surfaces. Blow all water, oil passageways, and screw holes, clean with compressed air. After cleaning, apply a light coating of engine oil to the bright surfaces of all gears, bearings, and shafts, as a prevention against rusting and corrosion.

Use a fine file to remove burrs. **REPLACE** all O-rings, gaskets, and seals, to ensure satisfactory service from the unit, after the work is completed. Clean the corrosion from inside the housing where the propeller shafts and bearing housing were removed.

Check to be sure the water intake is clean and free of any foreign material.

Inspect the lower unit housing inside and out for cracks. Check carefully around screw and shaft holes. Check for burrs around machined faces and holes. Check for stripped threads in screw holes.

Check O-ring seal grooves for sharp edges, which could cut a new seal. Check all oil holes.

Inspect the splines on the outer and inner propeller shafts for any sign of damage.

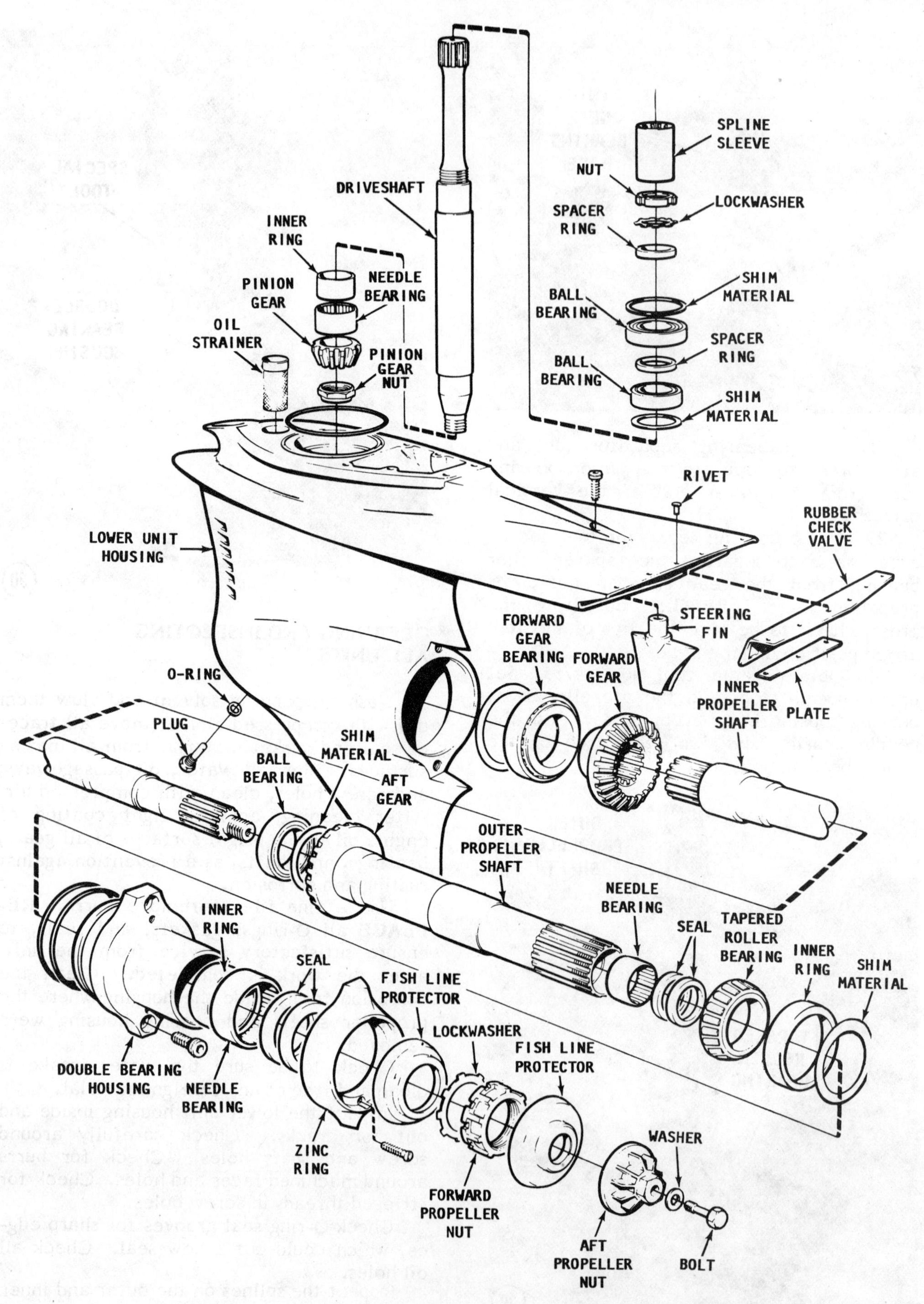

Exploded drawing of a Duoprop lower unit, with major parts identified.

Inspect the bearing surfaces of the shafts and splines, for wear and burrs. Look for evidence of an inner bearing race turning on the shaft. Check for damaged threads. Measure the runout on the propeller shafts and vertical shafts to detect any bent condition.

Inspect the teeth of of pnion gear, forward gear, and aft gear, for wear, pitting, and burrs.

Hold the center race of each bearing and turn the outer race. The bearing must turn freely without binding or evidence of rough spots. Inspect the outside diameter of the outer races and the inside diameter of the inner races for evidence of turning in the housing or on a shaft. Deep discoloration and scores are evidence of overheating.

ASSEMBLING DUOPROP LOWER UNIT

The following procedures pickup the work after all parts have been cleaned, inspected, and replacement parts have been purchased and are available for installation.

The steps outlined in this section include replacement and shimming instructions for all bearings and bearing races. If the particular bearing or bearing race mentioned was not removed, simply skip the step and proceed with the assembling of parts which were removed.

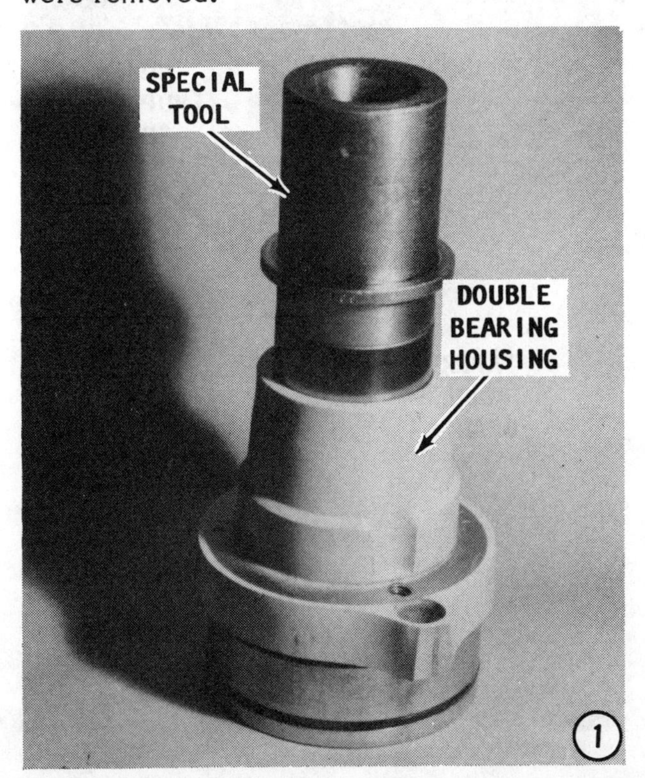

As explained in the disassembling procedures, the proper tools must be used to install bearings, bearing races, and seals. If an attempt is made to fabricate and use makeshift equipment, not only may the installed item be damaged but added expensive adjacent parts as well.

SPECIAL WORDS

The various shimming procedures are to be performed individually and for the complete lower unit **PRIOR** to final installation. Actually, the lower unit is assembled **TWICE.** The first time installation is without seals and O-rings installed to check for prestress, backlash, and gear pattern. The second installation is performed as a final assembly after prestress and gear patterns are satisfactory and all seals and O-rings have been installed.

Propeller Shaft Needle Bearings Into Double Bearing Housing

1- Obtain Volvo special bearing installation tool No. 884797. Insert the needle bearing into the double bearing housing with the writing on the bearing casing facing **UPWARD.** Set the special tool on top of the bearing and press the bearing into place with an arbor press.

DO NOT install the two seals at this time. They will be installed later after backlash and gear pattern have been checked.

Lower Driveshaft Bearing

2- Obtain special Volvo tools No. 884241 and No. 884792. The No. 884241 tool is similar to a long bolt with a washer under the bolt head and a large plate beneath the washer. The 884792 is a bearing installer tool with threads to receive the "bolt".

Coat the bearing location with a good grade of multi-purpose lubricant. By hand, push the bearing up into place into the lower end of the driveshaft opening with the writing on the bearing facing down.

Next, feed the "bolt" part of the special tool down through the driveshaft opening and through the bearing. Reach in with the bearing installer part of the tool and thread in the "bolt". Hold up onto the "bolt" to place the shoulder of the installer up against the bearing. Now, rotate the vertical tool with a wrench. The plate will come down on the lower unit housing. Continue rotating the tool and this action will "pull" the bearing up into place. When the bearing "bottoms out" against a shoulder in the lower unit, cease rotating and remove the special tool.

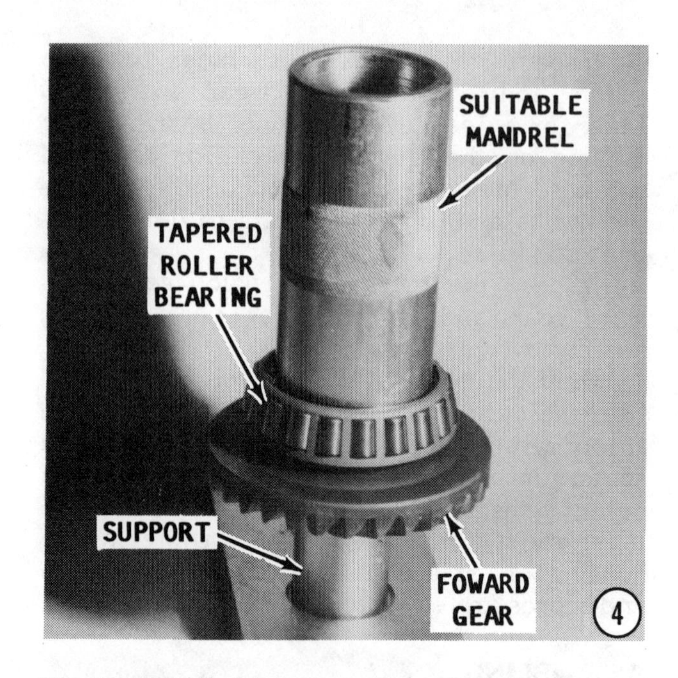

Tapered Roller Bearing Onto Rear Gear

3- Obtain Volvo special tool No. 884801. Place the bearing in position on the rear gear with the smaller tapered end facing **UP**, as shown. Carefully lower the special tool down over the outer driveshaft until it is seated on top of the tapered roller bearing. Force the bearing into place using an arbor press.

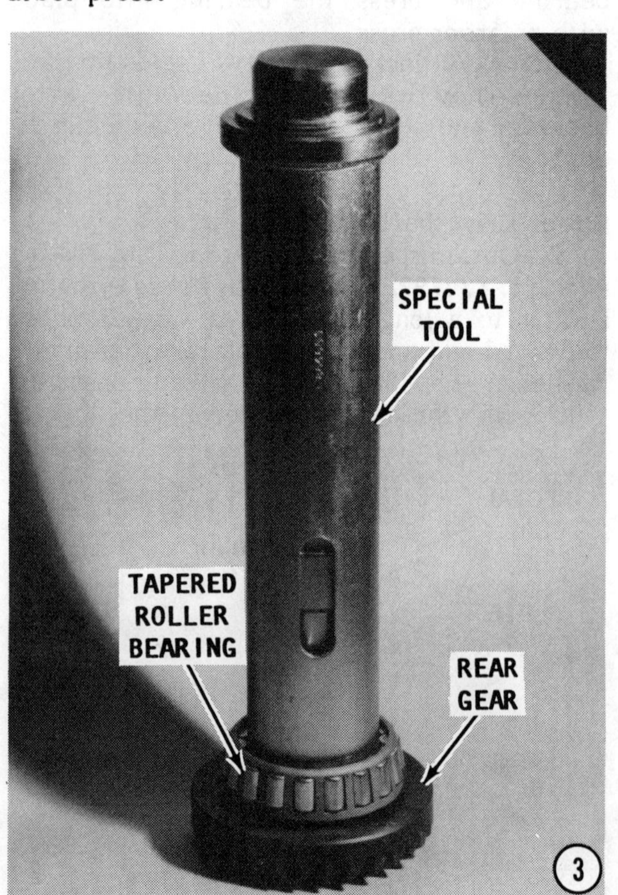

Tapered Roller Bearing Onto Forward Gear

4- Place the bearing in position on the forward gear with the smaller tapered end facing **UP**, as shown. Support the forward gear with a plate and suitable tool to prevent any pressure on the gear teeth. Place a suitable mandrel on top of the bearing. Move the setup into an arbor press and force the bearing into place.

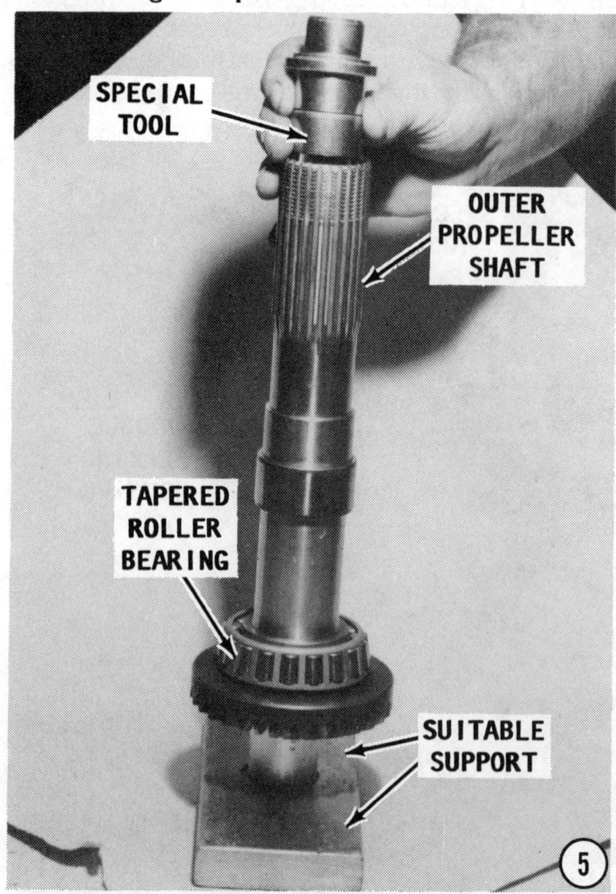

Roller Bearing
Onto the Outer Shaft

5- Obtain Volvo special tool No. 884806. Place the bearing in position on the forward gear, with the gear teeth facing down, and the writing on the bearing facing **UP.** Support the forward gear with suitable items to prevent any pressure on the gear teeth. Place the special tool on the upper end of the outer propeller shaft. Move the setup into an arbor press and force the bearing into the forward gear.

DO NOT install the sealing rings at this time. They will be installed later after backlash and a gear contact pattern has been determined.

Lower Driveshaft Bearing Race

6- Obtain Volvo special tool No. 884793. Set the bearing race in position on the lower end of the vertical driveshaft. Place the special tool on top of the race. Move the setup into an arbor press and force the race onto the end of the driveshaft. The race will "bottom out" against a shoulder of the driveshaft.

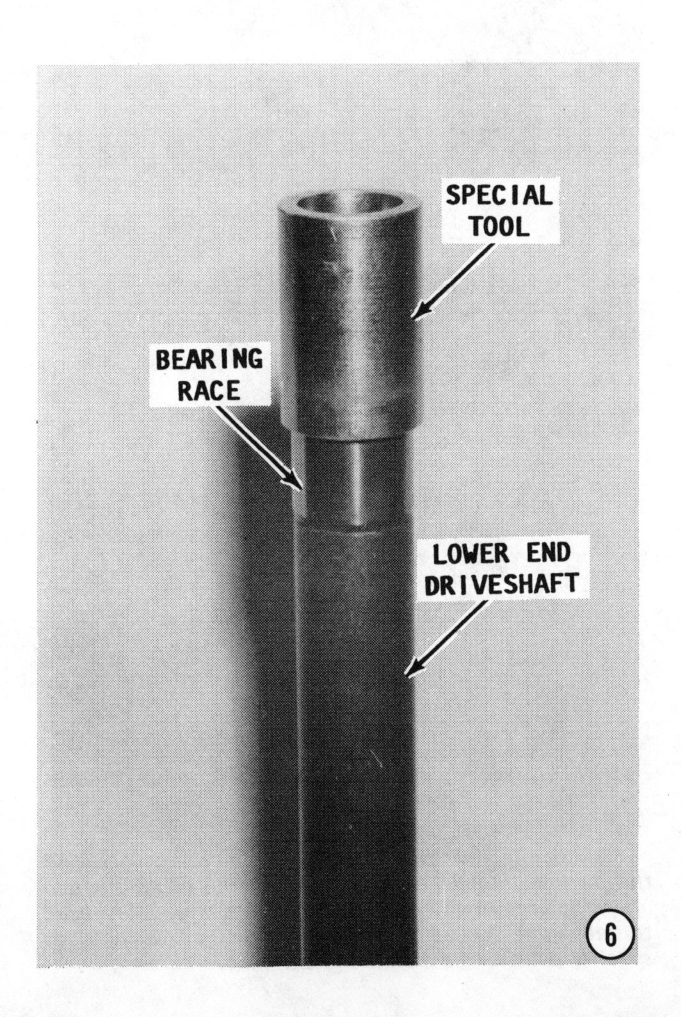

Upper Driveshaft Bearings

7- Obtain Volvo special tool No. 884266. Press the ball bearing set onto the shaft with the large shoulder of the inner race facing **UPWARD.** Slide the thin spacer insert onto the shaft and into position on the ball bearing. Use special tool No. 884266 and press the tapered roller bearing onto the shaft with the smaller diameter of the tapered bearing facing **UPWARD.** Actually, it is possible to press both bearings and the spacer between them onto the shaft in one operation. If this is done, **ENSURE** each bearing set is facing the proper direction as just described. Slide the thick spacer ring, lockwasher, and then thread the round nut onto the vertical shaft.

8- Thread the nut onto the shaft with the flat facing **UP.** Tighten it snugly, using a spanner-type wrench or a punch and hammer. **DO NOT** bend the locking tab over at this time.

STOP

At this point, the assembly work is interrupted to make a series of simple calculations (arithmetic -- addition and subtraction only), to determine the correct amount of shim material required at several points.

The numbers stamped on various parts, in most cases, represent digits following the decimal in **MILLIMETERS.** Therefore, it would become quite awkward and almost

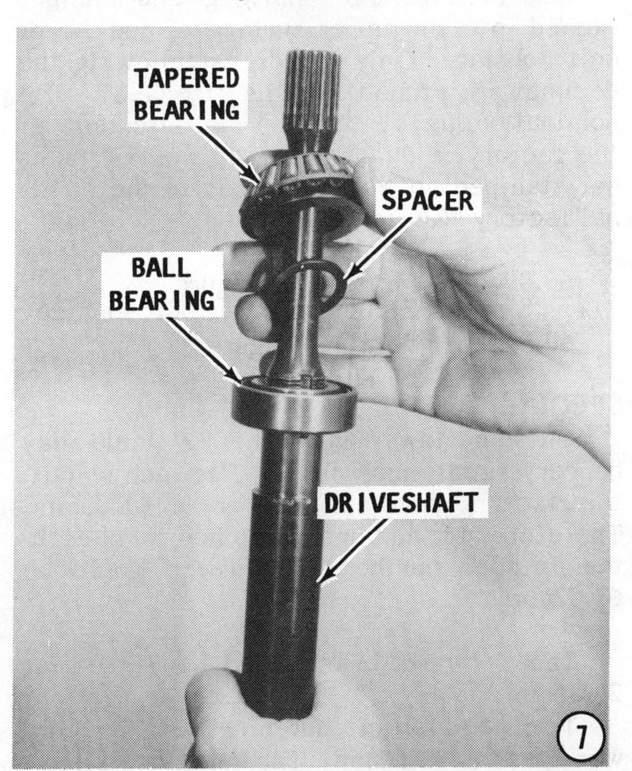

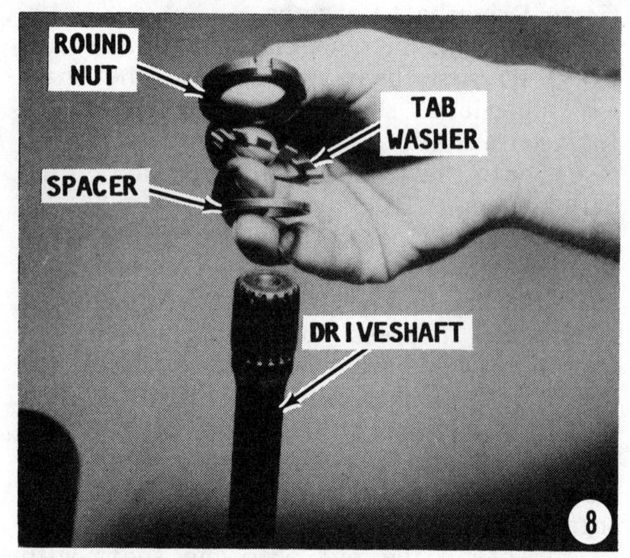

impossible to make actual measurements using portions of an inch. As a result, the measurements, calculations, and amount of shim material required will all be given using only the metric system.

CRITICAL WORDS

The front and rear tapered roller bearings and their associated bearing races appear to be quite similar. However, they are different and **MUST** not be exchanged or mixed, one with the other.

Shim Material
Front Bearing Race

Observe the **"G"** marking and number etched into the upper surface of the lower unit housing. Only the digits following the decimal are etched. In this case **G=89.** The nominal value of the **"G"** measurement at the factory is 60.00mm. Add the **"G"** number stamped on the lower unit to the nominal factory **"G"** figure, thus:

60.00mm	Factory **"G"** figure
+ .89mm	Etched on lower unit surface
60.89mm	**Total "G" figure**

SPECIAL WORDS

On some lower units, the **"G"** value may be very small, such as 02. For such a unit, the factory nominal number is 61.00mm. Therefore, by adding the etched number to the factory number the answer would be **61.02mm.**

The forward bearing height is set at 20.85mm.

The nominal measurement of the forward gear is 39.50mm. The tolerance allowed, either plus or minus is etched on the back side of the forward gear. Again, only the digits following the decimal are etched.

Observe the number etched on the back side of the forward gear. In this case, the number following the decimal is **"0".**

CRITICAL WORDS

The **"9"** etched on the back side of the forward gear before the decimal, is the number to indicate a **MATCHED** set with the rear gear and pinion gear. The rear gear **AND** pinion gear will also have a **"9"** etched on the back surface. **NEVER** attempt to install a forward gear, rear gear, and pinion gear with unmatched numbers. Such action would only lead to extensive rapid gear wear because the proper amount of shim material could not be achieved.

If a plus or minus would have been etched before the decimal, the figure following the decimal would be added or substracted to the nominal figure of 39.50mm.

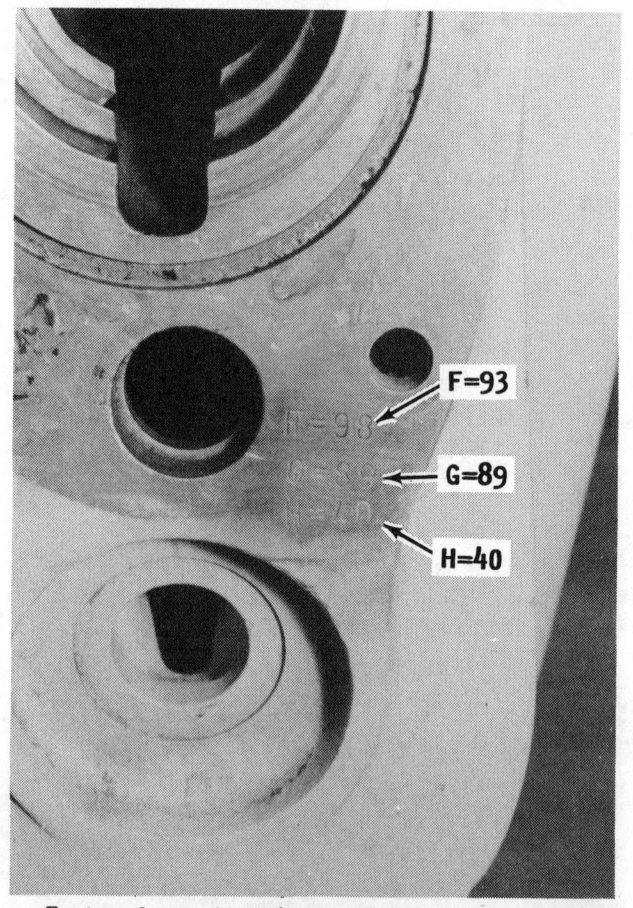

Factory letter identification and numbers following the decimal etched or stamped into the upper surface of the lower unit. These letters and numbers are used to determine the amount of shim material required at specific points, as explained in the text.

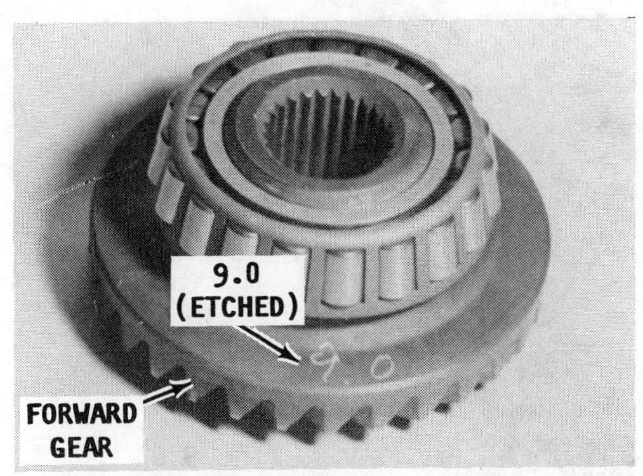

Numbers etched or stamped onto the back side of the forward gear for calculating amount of shim material to be used behind the forward gear bearing.

Calculation

Add the height of the front bearing to the forward gear figure.

20.85mm	Front bearing height
+39.50mm	Forward gear figure
60.35mm	**SUM**

Subtract this answer from the total "G" measurement.

60.89mm	Total "G" figure
-60.35mm	(Answer just above)
.54mm	**Fwd. bearing race shim material required.**

Shim Material
Upper Driveshaft Bearings

Observe the "H" marking and number etched in the upper surface of the lower unit housing. Only the digits following the decimal are etched. In this case **H=40.** The nominal figure from the factory is 277mm. Add the figure etched on the housing to the nominal figure, thus:

277.00mm	(Factory nominal figure)
+ .40mm	(Etched on lower unit)
277.40mm	**Total "H" figure**

Next, observe the number etched on the back side of the pinion gear. In this case, + 20. The factory nominal measurement of the pinion gear is 60.00mm. Add the figure etched on the pinion gear to the nominal figure, thus:

60.00mm	Factory nominal figure
+ .20	Etched on pinion gear
60.20mm	**Total pinion gear figure**

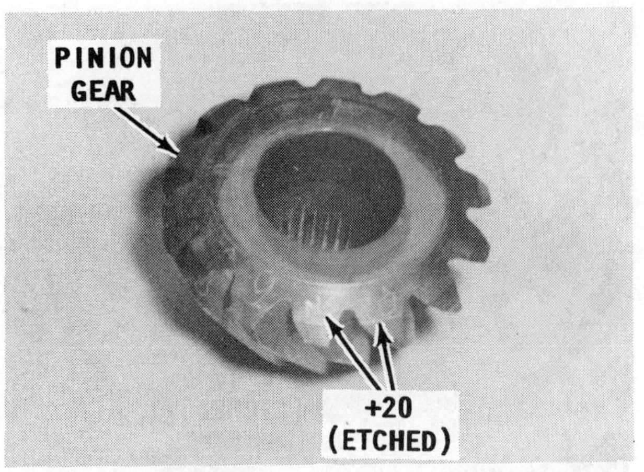

Numbers etched or stamped on the back side of the pinion gear for calculating amount of shim material to be used on the driveshaft.

The factory nominal measurement of the vertical driveshaft is 217.75mm. Add the total pinion gear figure to this number, thus:

217.75mm	Factory nominal figure
+ 60.20mm	Pinion gear figure
277.95mm	**Total driveshaft figure**

Now, subtract the total "H" figure from the total driveshaft figure, thus:

277.95	Total driveshaft figure
-277.40	Total "H" figure
.55	**Driveshaft shim material req'd.**

Shim Material
Double Bearing Housing

Observe the "F" marking and number etched into the upper surface of the lower unit housing. Only the digits following the decimal are used. In this case **F=93.** The nominal factory measurement is 80.00 ±0.10mm.

Now, read slowly and follow closely. If the figures etched into the housing are between 90 and 99, add the etched figure to 79.00mm. If the figures etched into the housing are between 00 and 10, add the etched figures to 80.00mm.

In our case the etched numbers are 93. Therefore we add the etched numbers to 79.00mm, thus:

79.00mm	Factory nominal figure
+ .93	Etched figure on housing
79.93mm	**Total "F" figure**

Next, observe the "C" marking and number etched into the double bearing housing

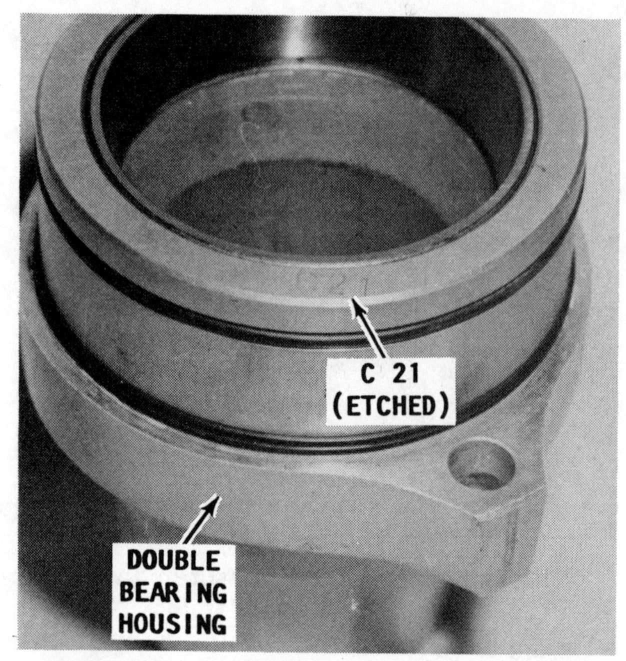

Letter and number etched or stamped on the forward face of the double bearing housing for calculating amount of shim material to be used behind the rear gear.

Number etched or stamped on the back side of the rear gear for calculating amount of shim material to be used behind the rear gear.

collar. Only the digits following the decimal are etched. In this case **C=21.** The nominal factory measurement is 19.00mm. Add the figure etched on the double bearing housing to the nominal figure, thus:

19.00mm	Factory nominal figure
+ .21	Etched on dbl. bearing house
19.21mm	Total **"C" figure**

Subtract the total **"C"** figure from the **"F"** figure, thus:

79.93mm	Total **"F"** figure
-19.21mm	Total **"C"** figure
60.72mm	Difference of **"F"** and **"C"**

Observe the etched number on the back side of the rear gear. Only the digits following the decimal are etched. In this case we have the number "0".

CRITICAL WORDS

The "9" etched on the back side of the forward gear before the decimal, is the number to indicate a **MATCHED** set with the rear gear and pinion gear. The rear gear **AND** pinion gear will also have a "9" etched on the back surface. **NEVER** attempt to install a forward gear, rear gear, and pinion gear with unmatched numbers. Such action would only lead to extensive rapid gear wear because the proper amount of shim material could not be achieved.

If a plus or minus would have been etched before the decimal, the figure following the decimal would be added or subtracted to the nominal figure.

The nominal measurement for the rear gear is 39.50. The etched number on the back of the rear gear after the decimal is "0". Add the etched number to the nominal factory number, thus:

39.50mm	Factory nominal figure
.00mm	Etched number on rear gear
39.50mm	**Total rear gear figure**

CRITICAL WORDS

The forward gear bearing and the rear gear bearing appear quite similar, but they are different. **ENSURE** the proper bearing is installed in the rear location by verifying the letters and numbers BK-358X are stamped on the inner ring and K-354X on the outer ring (the bearing race).

The height of the rear gear tapered roller bearing is given as 20.75mm. Add the height of the bearing to the total rear bearing figure, thus:

39.50	Total rear bearing figure
+20.75	Height of rear bearing
60.25mm	**Answer**

Now, subtract this answer from the answer obtained when "C" was subtracted from "F", thus:

60.72mm	Difference of "C" from "F"
-60.25mm	Answer just obtained
.47mm	**Rear Bearing Shim Material Req'd.**

Shim Material
Intermediate Bearing

Begin by adding the "G" figure to the "F" figure, thus:

60.89mm	"G" Figure
+79.93mm	"F" Figure
140.82mm	**Sum**

From the answer obtained, subtract the "C" figure, thus:

140.82mm	Sum from above
- 19.21mm	Total "C" Figure
121.61mm	**Sum**

Now, add the amount of shim material required for the front and rear bearings, thus:

.60mm	Fwd. Bearing Shim Material
+ .47mm	Rear Bearing Shim Material
1.07mm	**Sum**

To this figure, add a factory constant 120.00, thus:

1.07mm	Sum just obtained
+120.00mm	Factory constant
121.07mm	**Sum**

Finally, subtract the sum, just obtained from the sum, when "C" was subtracted from sum of "G" and "F" above, thus:

121.61mm	Answer after subtracting "C" from the sum of "G" and "F", above
-121.07mm	Answer after adding the two previous amounts
.64mm	**Shim material req'd for intermediate bearing.**

End Shim Material Calculations

ASSEMBLYING LOWER UNIT CONTINUES

Forward Gear Bearing Race

9- Obtain Volvo special tool No. 884795. Lubricate the area in the lower unit where the forward gear bearing race will seat. Place the amount of shim material calculated under Shim Material -- Forward Gear Bearing Race, onto the shelf in the lower unit housing. Carefully position the bearing race in the lower unit. Insert the special tool in the lower unit housing and on top of the bearing race. Now, slowly drive the race into position until it "bottoms out". A distinct change in sound of the hammer striking the tool will be noticed when the race is properly seated.

Tapered Bearing Race
In Double Bearing Housing

10- Lubricate the area in the double bearing housing where the tapered roller bearing race will seat. Obtain Volvo special tool No. 884795. Place the calculated amount of shim material for the intermediate bearing onto the shelf in the double bearing housing. If possible, place thin shim material pieces between thicker pieces of shim material. Position the bearing race squarely on the double bearing housing and drive it (or use an arbor press), into place with the special tool. If driving, when the bearing race "bottoms out" a distinct change in sound of the hammer striking the special tool will be heard.

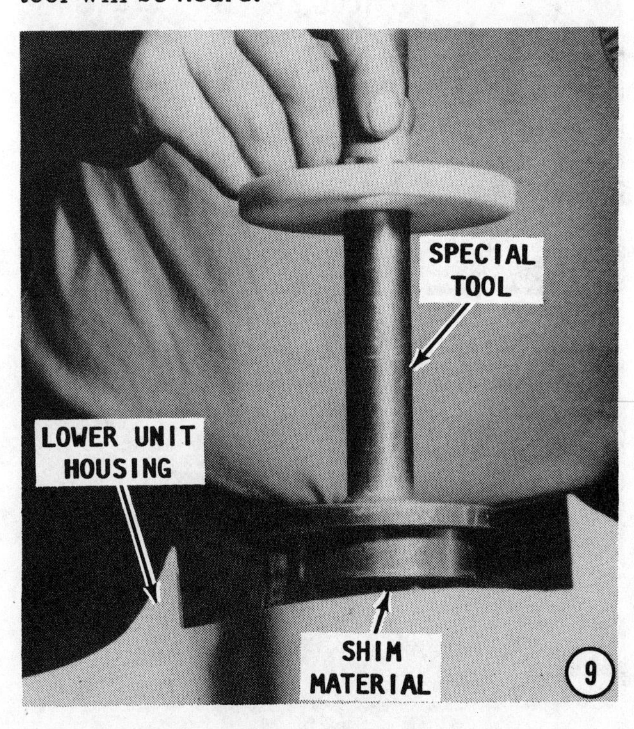

Tapered Bearing Race Into Rear Gear

11- Lubricate the area of the rear gear where the tapered roller bearing race will seat. Obtain Volvo special tool No. 884797. Place the calculated amount of shim material onto the shelf in the rear gear. If possible, place thin shim material between thicker pieces of shim material. Place the bearing squarely in position, and then drive it (or use an arbor press), into place with the special tool. If driving, when the bearing race "bottoms out" a distinct change in sound of the hammer striking the special tool will be heard.

TEMPORARY ASSEMBLING

EXTRA SPECIAL WORDS

The lower unit is to be assembled **TWICE**. The first time without seals and O-rings to establish and check the prestress, backlash, and gear mesh pattern. If the gears and the shim material used to adjust them have not been disturbed, then only a final installation need be performed.

If such is the case, a **NEW** pinion nut must be used and the **NEW** seals in the aft end of the outer propeller shaft and the seals in the double bearing housing must be installed prior to final assembling.

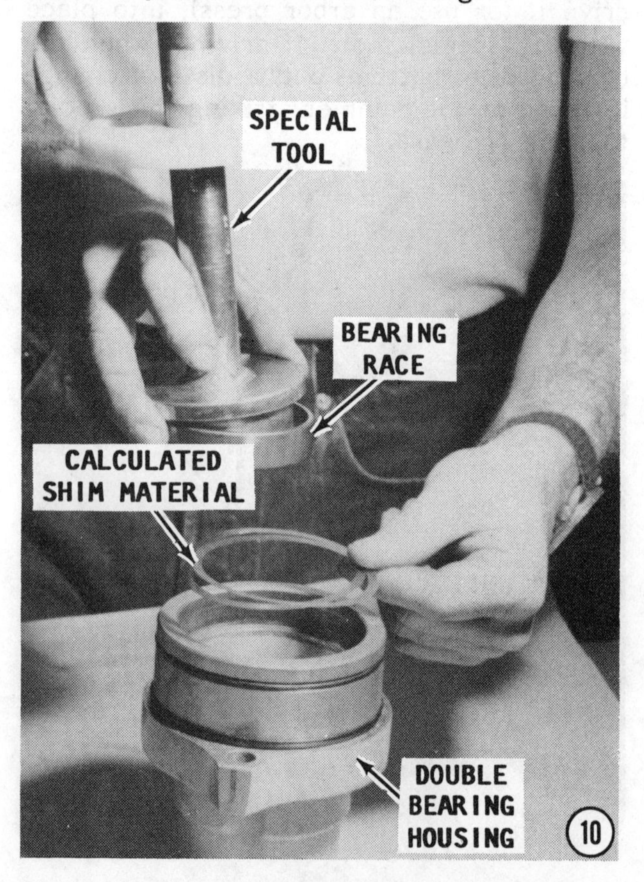

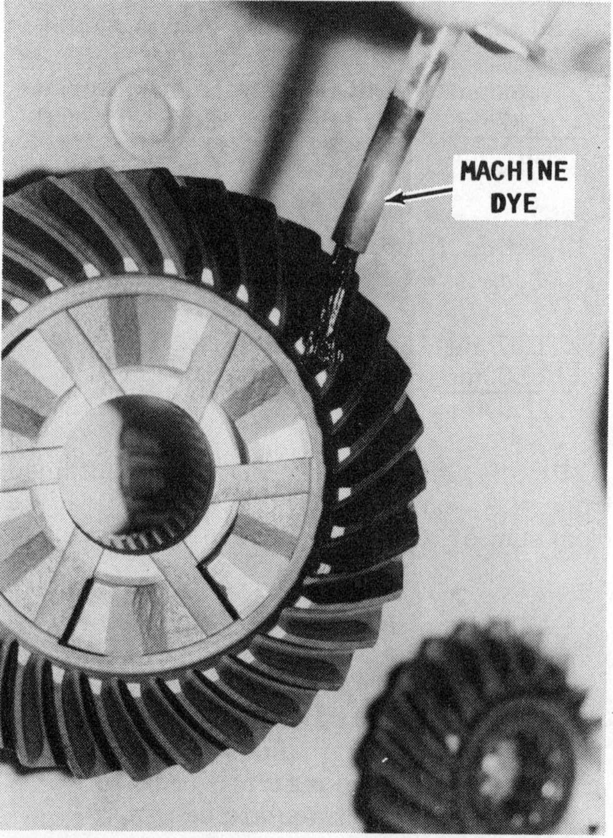

Machine dye being applied to a typical gear in preparation to obtaining a gear mesh pattern.

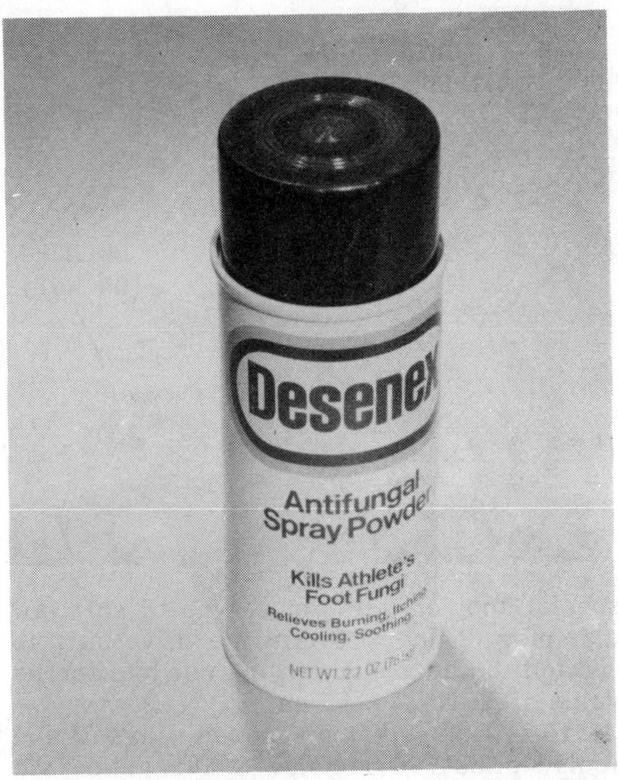

Can of Desenex foot spray powder. The powder may be applied to a gear prior to assembly to obtain a gear mesh pattern, as explained in the text.

PLAN AHEAD

Obtain a suitable substance which can be used to indicate a wear pattern on the gears in the lower unit as they mesh, one with the other. Machine dye may be used and if this material is not available, Desenex Foot Powder, or equivalent, may be substituted. Desenex is a white powder pressurized in an aerosol container, available at the local Drug Store/Pharmacy.

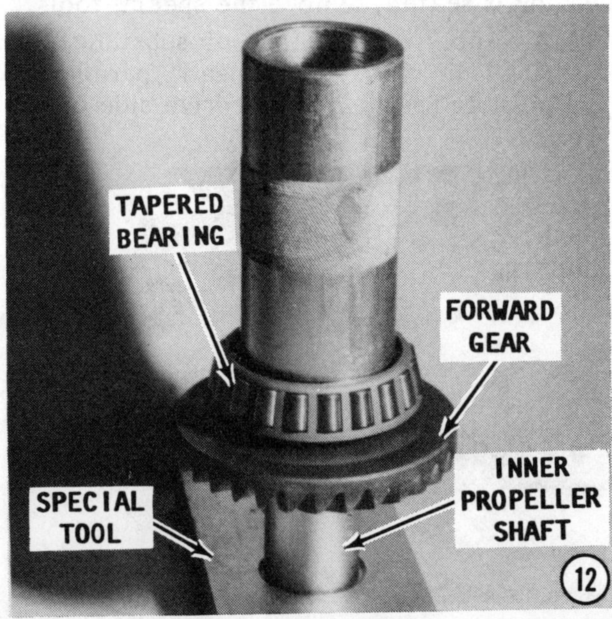

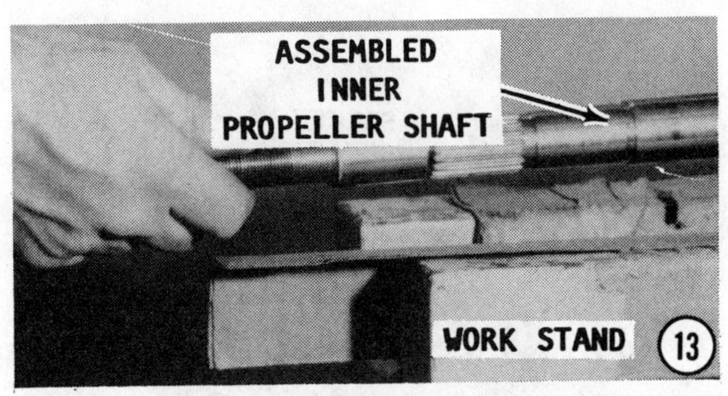

Before assembling the forward gear, rear gear, or the pinion gear, apply a light film of the dye, Desenex, or equivalent, to the drive side of the gear. After the gears are assembled in the lower unit and rotated several times, they will be disassembled and the wear pattern can be examined. The substance will be removed from the gears prior to final assembly.

12- Install the inner propeller shaft with the splines on the end of the shaft indexing with the splines of the forward gear. Press the forward gear onto the shaft using Volvo special tool No. 884263.

13- Lubricate the forward tapered roller bearing with grease. Apply a light film of substance, as described in the "Plan Ahead" paragraphs above, to the drive side of the forward gear. Insert the assembled inner propeller shaft with the forward gear and tapered roller bearing in place, into the lower unit. Push in on the propeller shaft to move the bearing into the bearing race.

14- Insert the amount of shim material calculated for the pinion gear onto the shelf in the lower unit housing. Apply a light film of substance, as described in the "Plan Ahead" paragraphs just before Step 12, to the drive side of the pinion gear. Insert the pinion gear and the pinion gear nut up into place in the lower unit housing. The groove

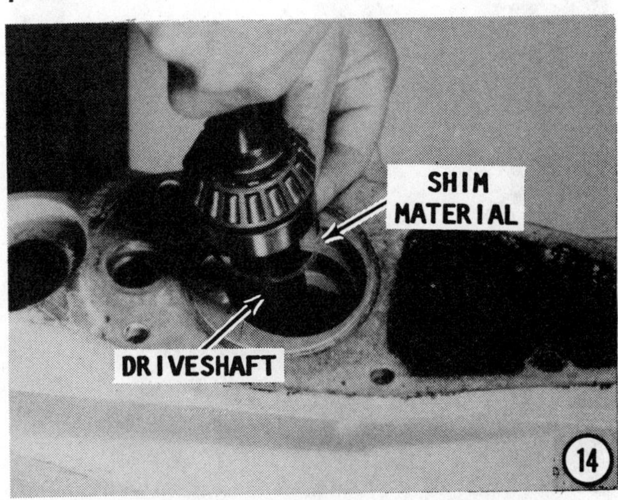

on the nut must face the pinion gear. Hold the pinion gear and nut with one hand, and then with the other hand insert the driveshaft down through the lower unit housing, through the pinion gear. Start the pinion gear nut onto the lower end of the driveshaft.

15- Obtain Volvo special holding tool No. 884264. After the nut has been started on the driveshaft, hold the nut with a socket and the upper end of the driveshaft with the

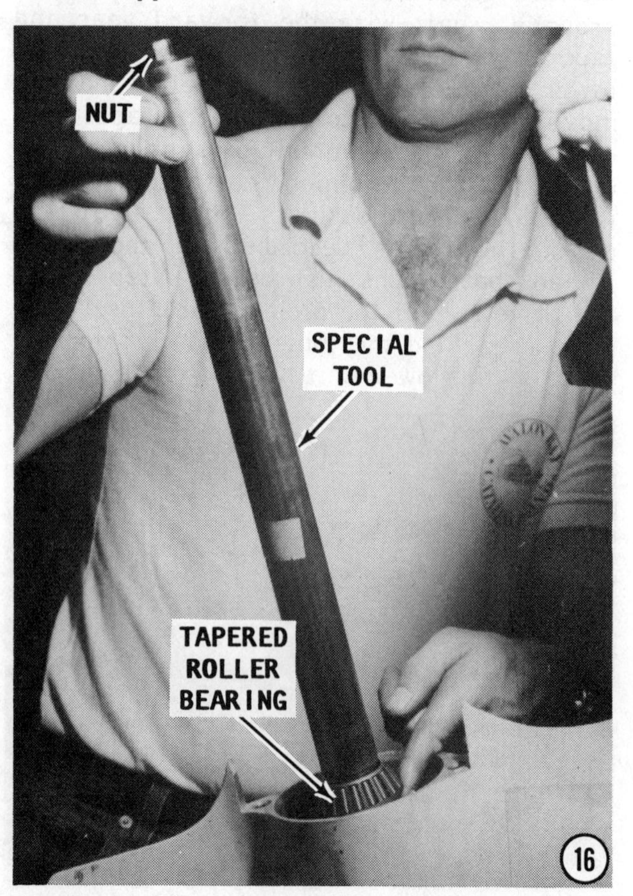

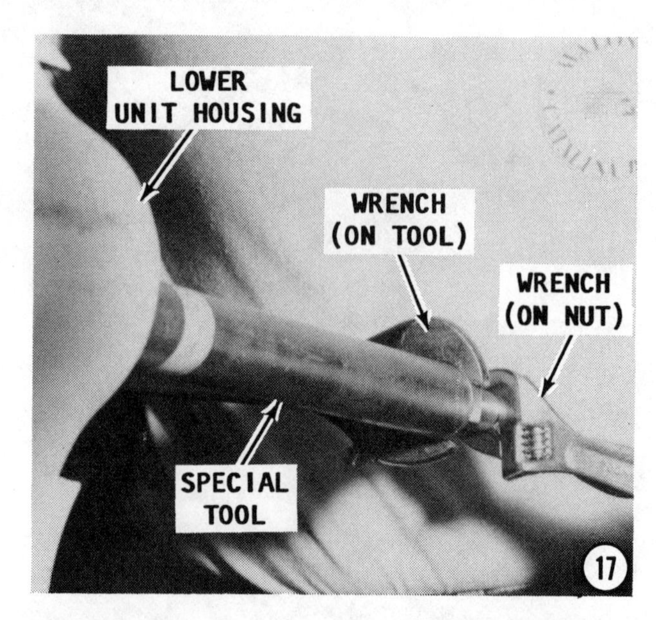

special tool or with some type of soft jaw clamping device. Rotate the driveshaft to tighten the nut. Tighten the nut to a torque value of 81 ft lbs (110Nm).

16- Obtain Volvo special installation tools No. 884798 and No. 884841. Slide the intermediate tapered roller bearing onto the propeller shaft. Next, install the special tool No. 884841 on the end of the propeller shaft. Now, install special tool No. 884798 down over the propeller shaft and against the bearing.

17- Turn the lower unit upside down, and then clamp the driveshaft in a vise equipped with soft jaws. Hold the one tool with a wrench and tighten the nut with another wrench to press the bearing onto the shaft. Tighten the nut until the bearing "bottoms out" on the propeller shaft. After the bearing is seated, remove the special tools.

18- Apply a light film of substance, as described in the "Plan Ahead" paragraphs just prior to Step 12, to the drive side of the aft gear.

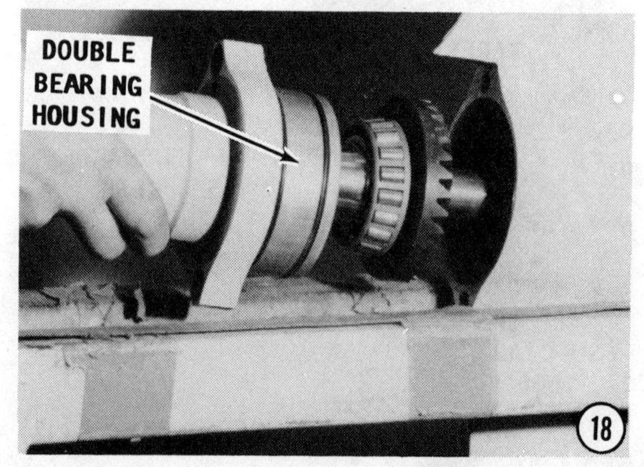

SPECIAL WORDS

DO NOT install the double seals into the aft end of the outer propeller shaft, at this time. They will be installed later after a gear pattern has been established and checked. Insert the aft gear and outer propeller shaft into the lower unit housing.

Install the double bearing housing. **DO NOT** install the double seals in the bearing housing at this time. They will be installed later after a gear pattern has been established and checked. Secure the double bearing housing with the two Allen head bolts. Tighten the bolts to a torque value of 30 ft lbs (40Nm).

CHECKING BACKLASH

19- Obtain Volvo Penta special tools No. 884140 and No. 884143. Insert the expander special tool into the intermediate housing tapered bearing race. Use the drift special tool and drive the bearing race free. Position this bearing race over the roller bearing on the lower gear housing. Install spanner special tool No. 884348. If necessary, insert shim material between the bearing race and the special tool to prevent the driveshaft from moving sideways. Secure the special tool in place with the two bolts and nuts.

20- Secure the lower gear housing in a vise equipped with soft jaws. Lock the driveshaft. Install the propeller nut onto the outer propeller shaft and the nut onto the inner propeller shaft. Set up a dial indicator gauge with the shaft of the gauge out on the wing of the inner propeller nut as far as possible.

Check the forward gear backlash by rotating the inner propeller shaft clockwise and counterclockwise by hand. The "play"

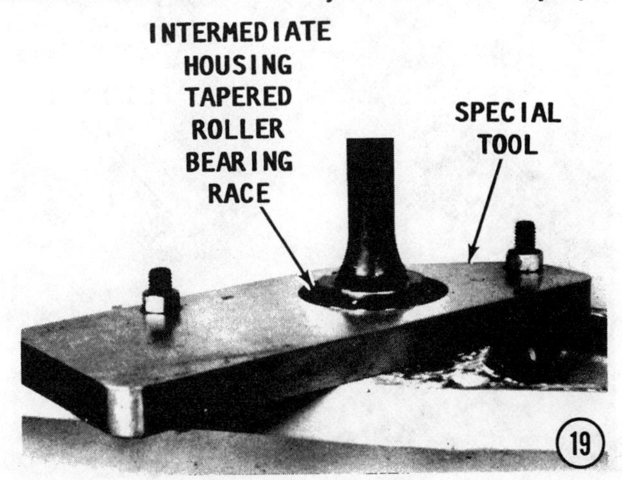

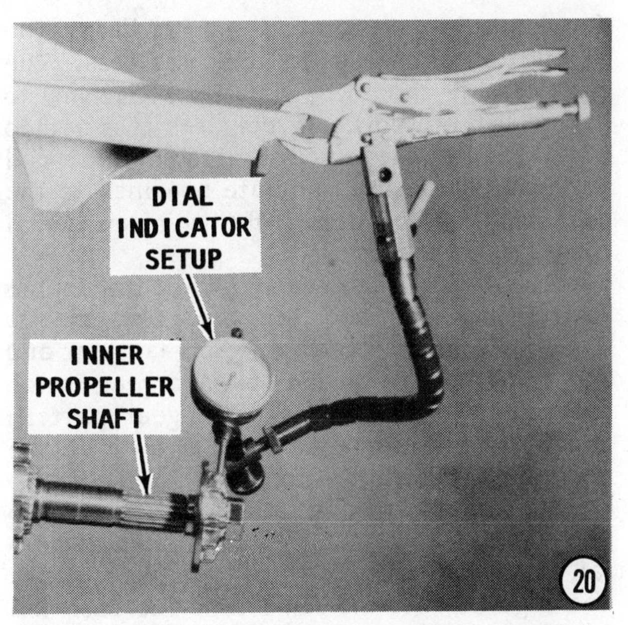

at the wing of the nut should be 0.003 to 0.007" (0.08 to 0.17mm). This amount of "play" will indicate a backlash at the forward gear of 0.006 to 0.013" (0.15 to 0.30mm).

21- Move the dial indicator gauge setup to the outer propeller nut. Check the aft gear backlash in the same manner as for the forward gear described in the previous step. The "play" at the outer propeller nut should be 0.004 to 0.010" (0.15 to 0.27mm). This amount of "play" will result in a backlash at the aft gear of 0.008 to 0.014" (0.20 to 0.35mm).

22- Obtain Volvo Penta special tool No. 884830. Attach the special tool to the upper end of the driveshaft. Use a small torque wrench and check the amount of

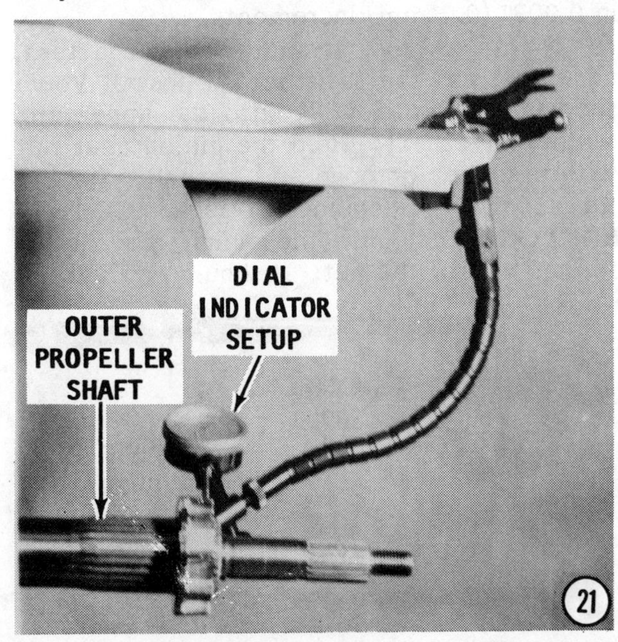

force required to rotate the driveshaft. The amount of force is called the prestress. The prestressing should be 0.9 ft lbs (1.2Nm) to 1.7 ft lbs (2.3Nm). If the prestress is too high, reduce the amount of shim material underneath the intermediate bearing. If the prestress is too light, the shim material must be increased.

23- Remove the spanner tool, the double bearing housing and the propeller shafts. Check the gear pattern on the rear gear and the gear pattern on the forward gear inside the lower unit housing. The gear pattern should favor the big end as shown in the accompanying illustration "A".

If the pattern favors the small end, as in "B", remove 0.004" (0.10mm) of shim material from behind the front and rear roller bearings (1). At the same time increase the amount of shim material under the intermediate bearing (3) by the same amount **AND** reduce the shim material under the upper driveshaft bearing (2) by 0.008" (0.20mm). Assemble the lower unit and check the gear pattern again. If more shim material must be removed, reduce the amount in increments of 0.004" (0.10mm).

If the pattern favors the large end, as in "C", add 0.002" (0.05mm) shim material under the front and rear roller bearings (1). Reduce the amount of shim material under the intermediate bearing (3) by the same amount. Increase the shim material under the upper driveshaft bearing (2) by 0.004" (0.10mm). Assemble the lower unit and check the gear pattern again. If the shim material should be increased, add material in 0.002" (0.05mm) increments.

24- After the backlash, gear pattern, and prestress are satisfactory, install Volvo special tool No. 884264 onto the upper end of the driveshaft. Hold the pinion gear nut with a socket or open end wrench. Rotate the special tool on the driveshaft **COUNTERCLOCKWISE** until the pinion gear nut is free. Discard the nut. It should not be used

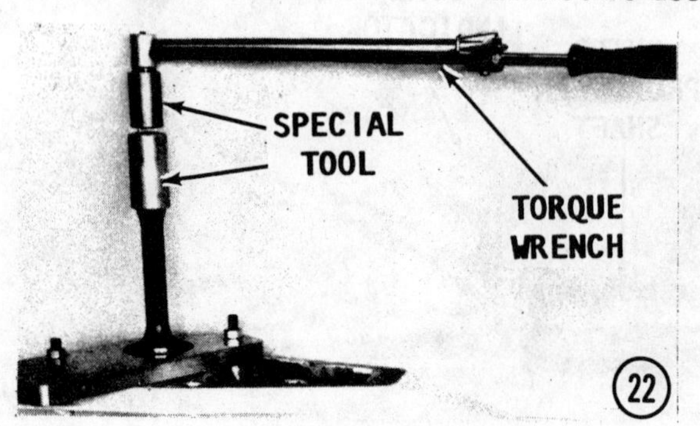

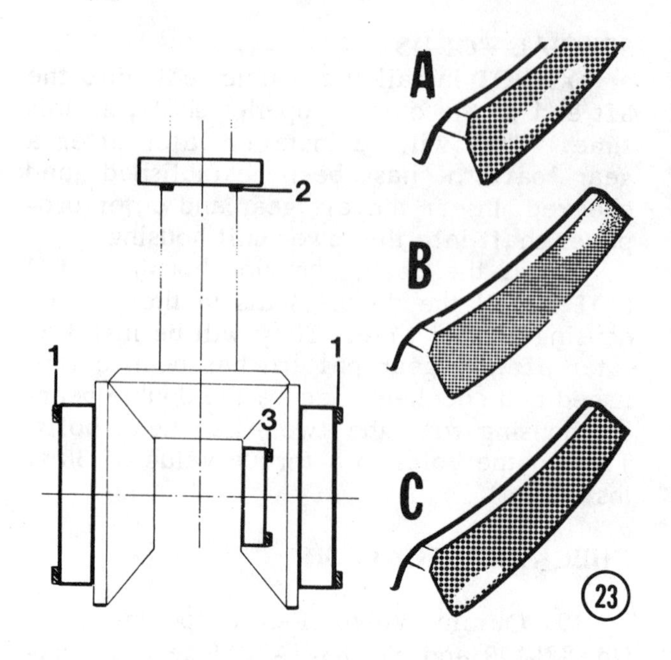

a second time. Thread a **NEW** pinion gear nut onto the lower end of the driveshaft. Hold the nut with a socket and torque wrench. Rotate the tool, and driveshaft **CLOCKWISE** until the nut is tightened to a torque value of 81 ft lbs (110Nm).

25- Bend one of the locking tabs up into one of the grooves of the nut to secure the nut in place.

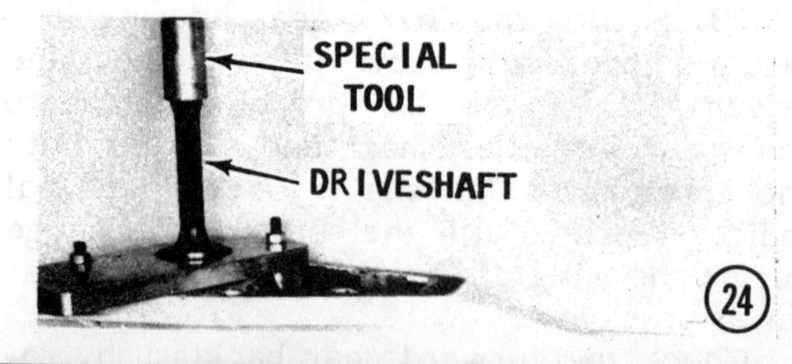

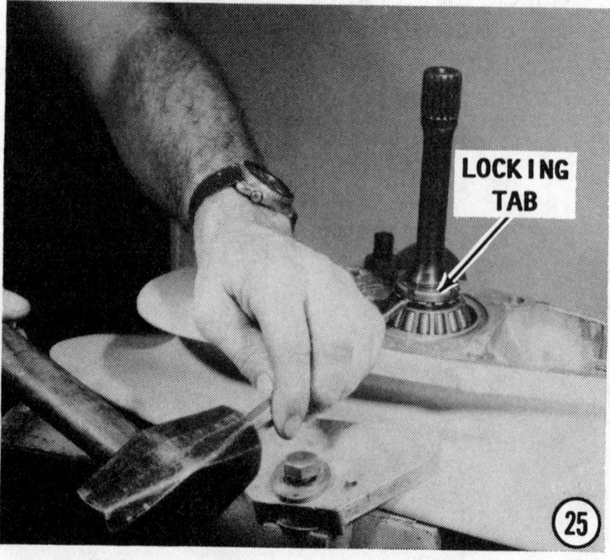

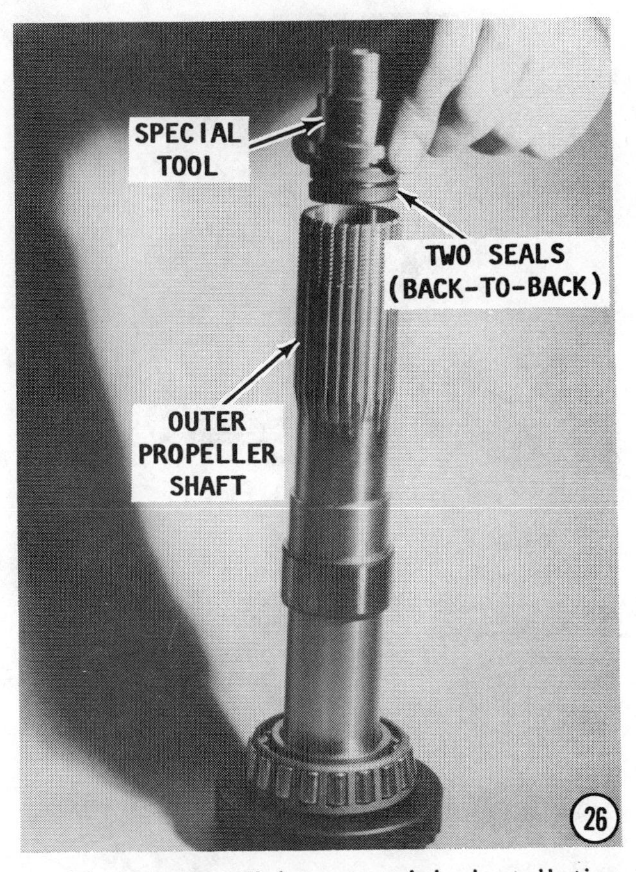

26- Obtain Volvo special installation tool No. 884806. Coat the outside perimeter and backs of the two outer propeller shaft seals with multipurpose lubricant. Carefully drive the two seals, **BACK-TO-BACK** (flat side-to-flat side), into the end of the outer propeller shaft, using the special tool and a soft head mallet. The outer seal will prevent water from entering the lower unit and the inner seal will prevent lubricant in the lower unit from escaping.

27- Obtain Volvo special installation tool No. 884801. Coat the outside perimeter and backs of the two double bearing housing seals with multipurpose lubricant. Carefully drive the two seals, **BACK-TO-BACK** (flat side-to-flat side), into the double bearing housing with a soft head mallet. The outer seal will prevent water from entering the lower unit and the inner seal will prevent lubricant in the lower unit from escaping. Install the sealing ring into the other end of the double bearing housing.

28- Install two **NEW O**-rings into the grooves of the double bearing housing.

29- Obtain Volvo special installation tool No. 884807. Push the special protective ring tool into the aft end of the double bearing housing. This tool is most important to protect the two seals from being damaged by the splines on the outer propeller shaft as the bearing housing is being installed.

30- Coat the mating surfaces of the lower unit and the double bearing housing with Permatex or other suitable sealing material. Install the outer propeller shaft into the lower unit housing, and then **CAREFULLY** move the double bearing housing forward over the inner and outer propeller shafts and up against the lower unit.

31- Secure the bearing housing in place with the two Allen head bolts. After the bearing housing is secure, remove the special protective ring tool.

32- Install the zinc ring and secure it with the two Phillips head screws.

SPECIAL WORDS

If the shim material under the ball bearing on the vertical shaft was changed, it will be necessary to determine the amount of shim material required to preload the vertical shaft tapered bearing between the lower unit and the intermediate housing.

First, remove the bearing race out of the intermediate housing, if the race was not removed in Step 19. Install this bearing race into the lower unit housing on top of the tapered bearing. Measure the distance from the top of the bearing race to the mating surface of the lower unit.

Next, measure the distance from the mating surface of the intermediate housing into the bearing race cavity.

Now, subtract the difference between these two measurements. The "crush" should be 0.001" and the clearance not more than 0.002" which will give 0.003" total tolerance in that area. Approximately 0.001" is factory recommended.

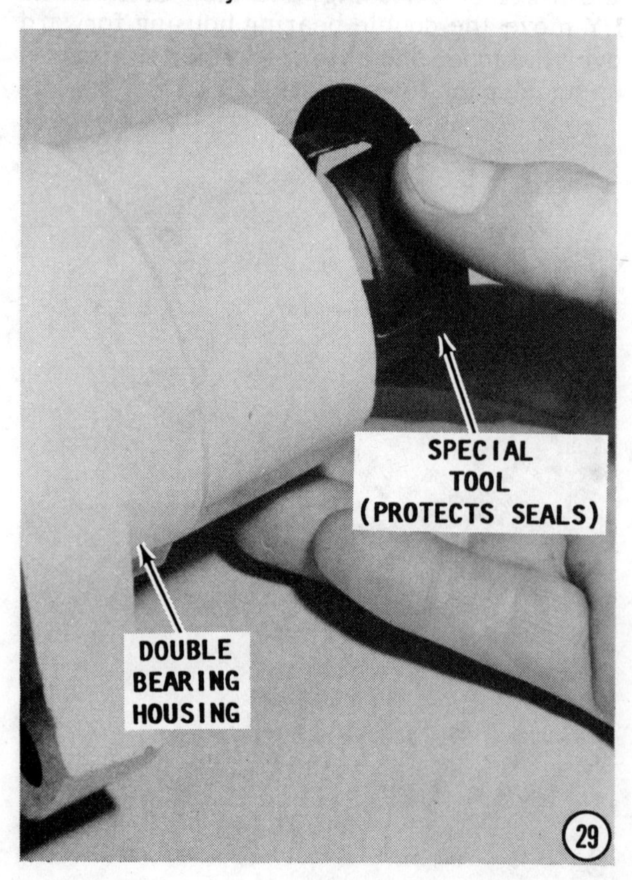

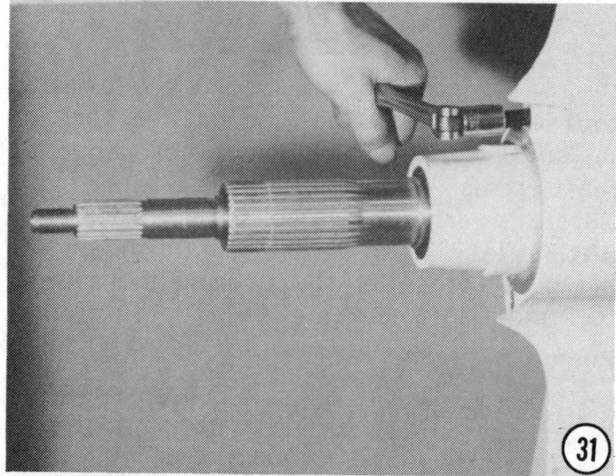

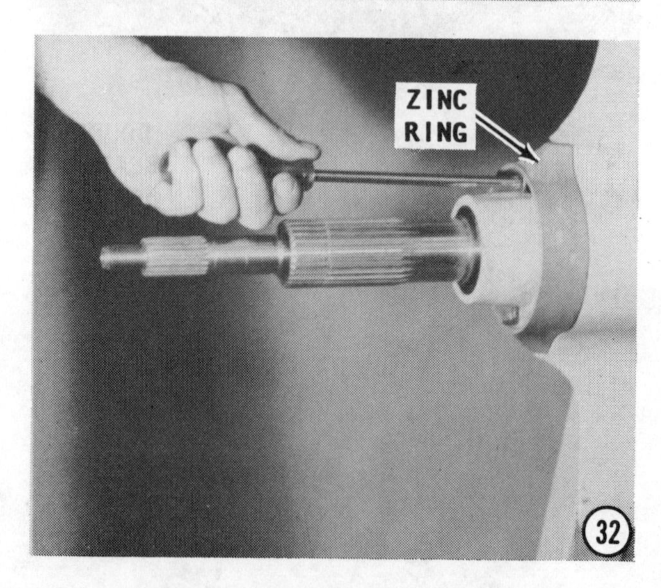

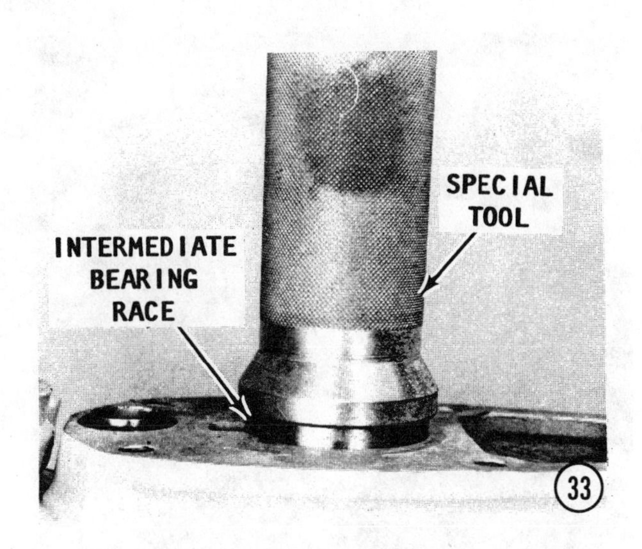

33- Obtain Volvo special tool No. 884168 and install the intermediate housing tapered roller bearing race which was removed in Step 19.

34- Insert the oil strainer into the oil channel of the lower unit. Coat two new O-rings with Permatex, or similar material and place them in position in the recesses of the lower unit.

35- Coat the mating surfaces of the lower unit and the intermediate housing with Permatex or other suitable sealer. **SLOWLY** lower the intermediate housing down over the lower unit with the oil distribution tube sliding into the opening in the lower unit. Keep the two surfaces as parallel as possible while lowering the intermediate housing. If the intermediate housing should become "cocked" to one side, the water tube may flip the O-ring out of position. If this should happen, an air leak would develop on the intake side of the cooling system. Such a leak would result in the engine overheating from lack of proper water flow.

36- Secure the intermediate housing with the seven attaching Allen head screws.

HELPFUL HINT

Before the upper gear housing is mated to the intermediate housing, insert a funnel into the oil passage tube, and then pour in approximately 2-1/4 quarts of lubricating oil. The manufacturer recommends a gear

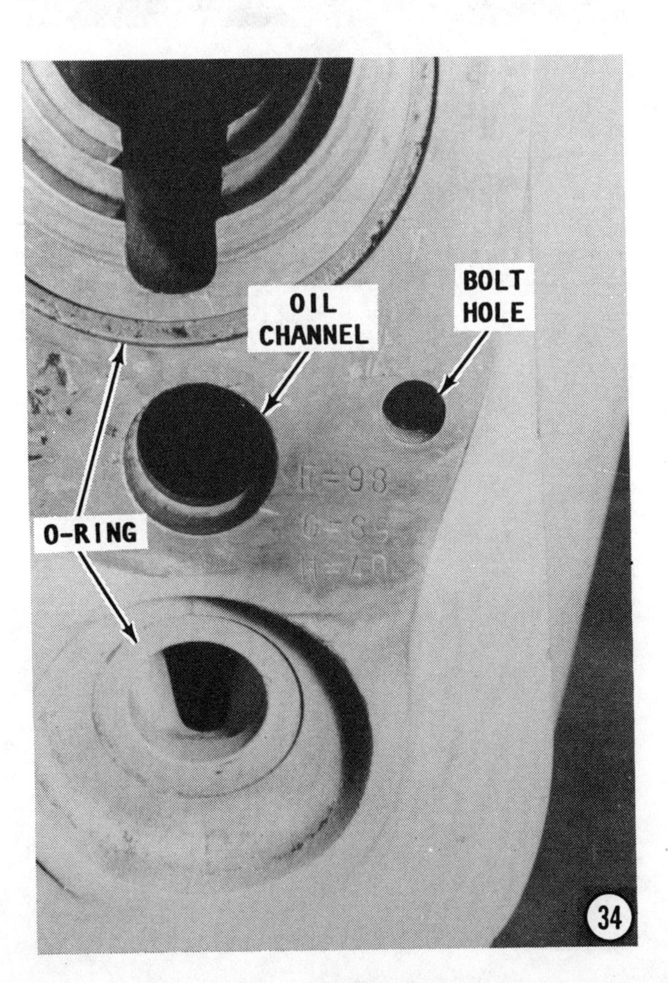

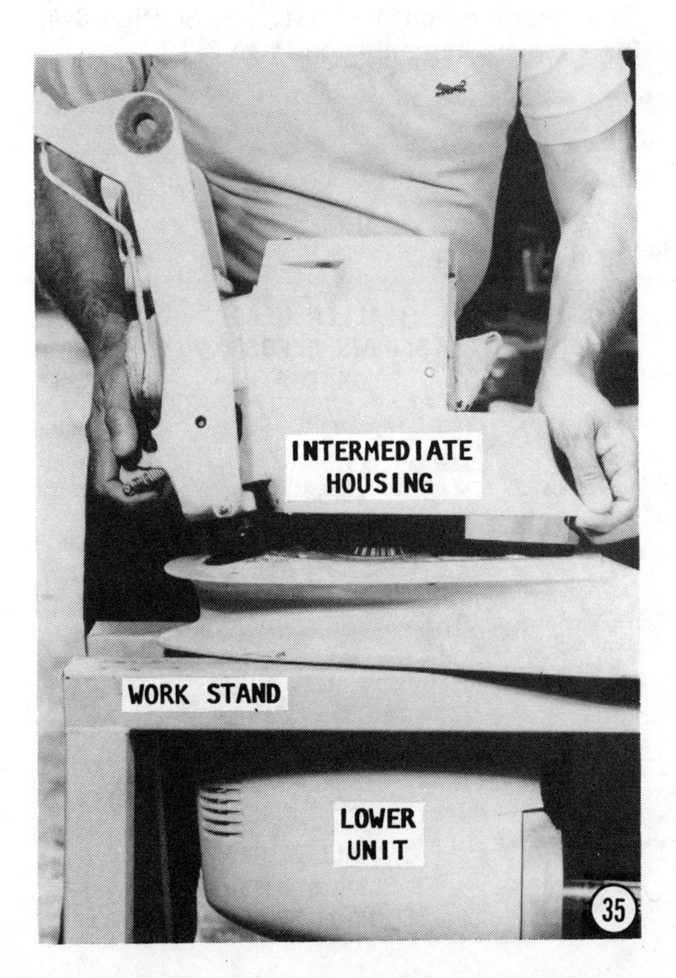

oil with a "GL-5" rating. Filling the lower unit at this time is much quicker and easier than adding the full amount of oil after the upper gear housing is secured and the cap is in place.

Check to be **SURE** the drive coupling is installed before the upper gear housing is mated to the intermediate housing.

37- Coat the mating surfaces of the intermediate and upper gear housings with Permatex or other suitable sealing material. Lower the upper gear assembly down onto the intermediate housing with the splines of the shaft indexed with the splines of the coupling. As an aid to indexing the splines of the shaft with the splines in the coupling, have an assistant rotate the propeller shaft slightly while attempting to lower the upper gear housing.

Secure the two housings together with the attaching hardware -- two nuts and two bolts. Tighten the bolts and nuts alternately, evenly, and securely in a criss-cross pattern.

38- Connect and secure the shift rod with the cotter pin and washer.

The assembled stern drive is now ready to be installed on the boat. See Section 8-4, Installation, beginning on Page 8-10.

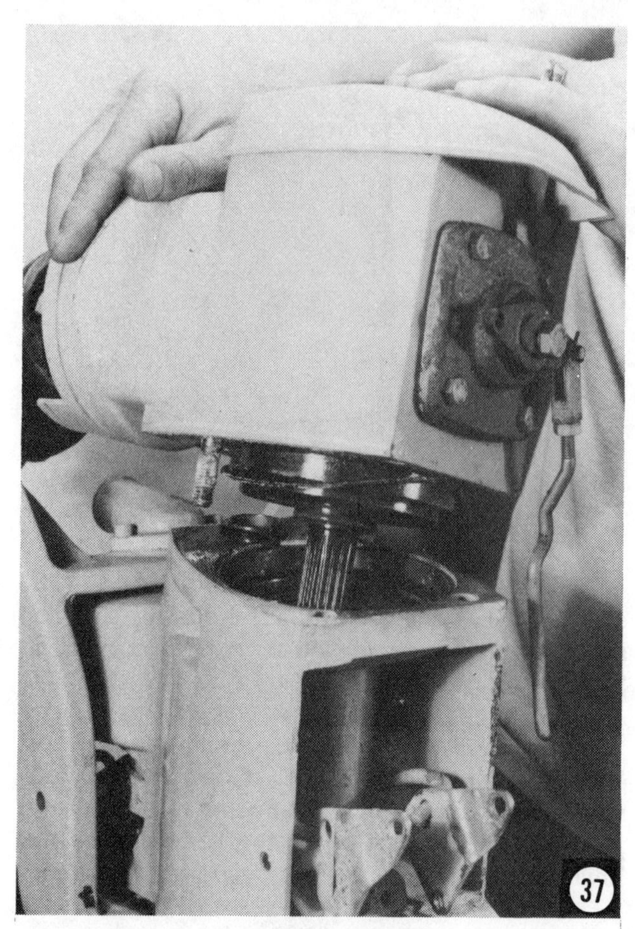

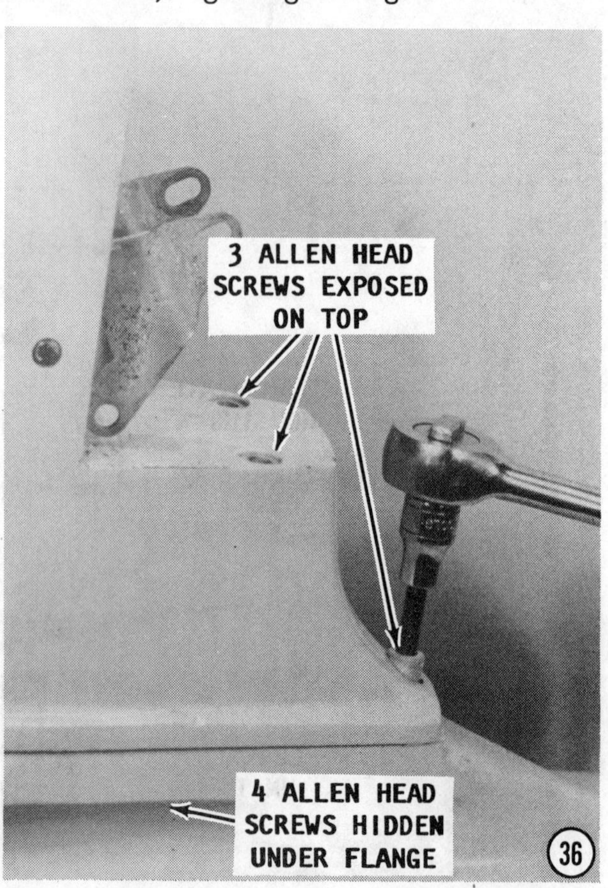

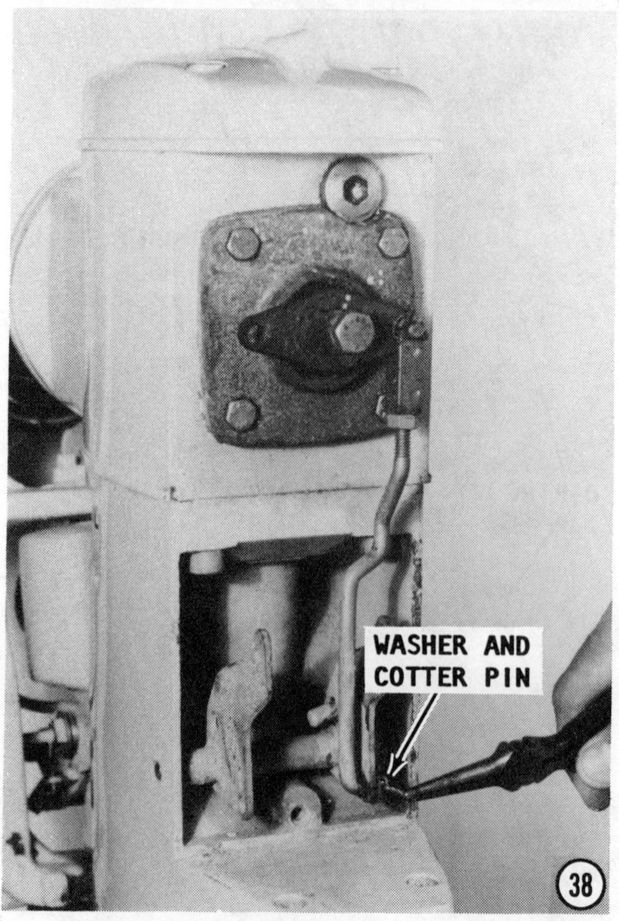

9
TILT/TRIM

9-1 INTRODUCTION

TILT

The tilt system is an electro-mechanical mechanism that allows the stern drive to be raised when moving through shallow water at **IDLE** speed; when approaching the beach, again at **IDLE** speed; and for trailering the boat to and from the water.

An electric motor, controlled by a toggle switch and two relays, rotates a worm screw and gear set to move a spindle screw shaft to raise and lower the stern drive. A control switch is mounted on the dashboard and activated by the helmsperson. When the switch is activated, either the up relay or the down relay is energized. Current to the motor activates the motor in one of two directions to raise or lower the stern drive. When the stern drive reaches its full down position, a limit switch cuts power to the relays and the motor stops. A built-in circuit breaker protects the operating parts from overload. A warning light illuminates on the dashboard while the stern drive is in the **TILT** position. When the drive is down and locked, the light goes out.

TRIM

The trim system is mounted in most cases on light displacement high-speed boats. Attempting to control the position of the bow on a heavy displacement boat would be neglible with a trim system.

The trim system consists of an electric motor and pump connected to a set of hydraulic cylinders mounted port and starboard in the transom shield. Hydraulic fluid is routed from a reservoir through the necessary hoses and fittings to the cylinders and back to the reservoir. The hydraulic fluid level in the reservoir remains constant during trim operation.

A control switch, mounted on or near the dashboard controls movement of the stern drive. An indicator dial installed near the switch provides a visual indication of the relative position of the stern drive. The dial has three plainly marked positions: **TRIM**, **BEACH**, and **TILT**.

The stern drive can be adjusted within the **TRIM** range, while the boat is underway at normal operating range in order to maintain the very best bow position for maximum speed and efficiency.

Operation in the **BEACH** range may be done while the boat is underway at **LOW** or **IDLE** speed. After the stern drive has been set to the desired position, **FORWARD** or **REVERSE** shifting may be done while the boat is underway. Speed may also be increased or decreased while in the **BEACH** range.

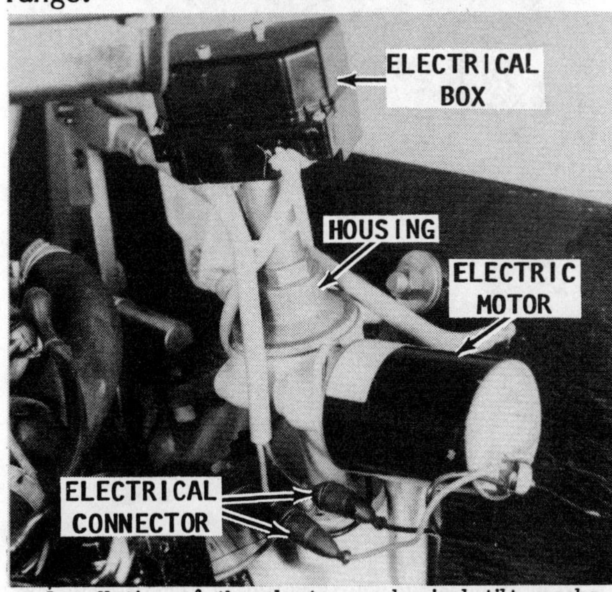

Installation of the electro-mechanical tilt mechanism with the major exterior parts identified.

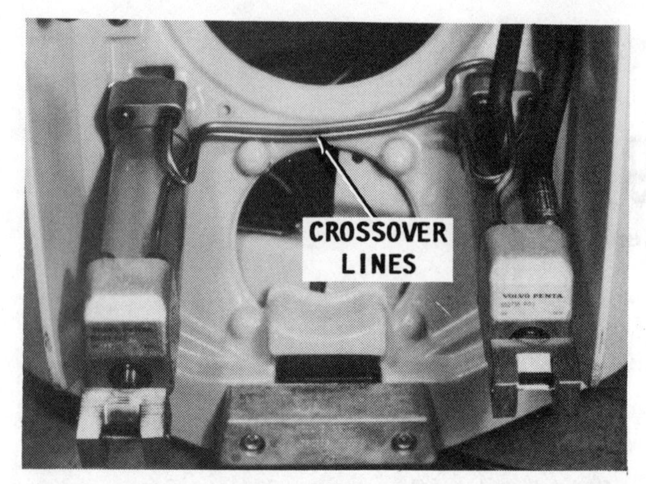

The Model 290 stern drive is equipped with a power trim system as standard equipment. The design of the pump is slightly different than the Model 280 and Model 270 but the function is the same. Instead of four hydraulic lines passing through the transom shield there are now only two with two stainless steel lines crossing over on the outside.

When the **TILT** range is to be used, the engine **MUST** be shut down and **NOT** be started or operated. This position is to be used **ONLY** while the boat is anchored or moored in very shallow water or during movement on a trailer to and from the water.

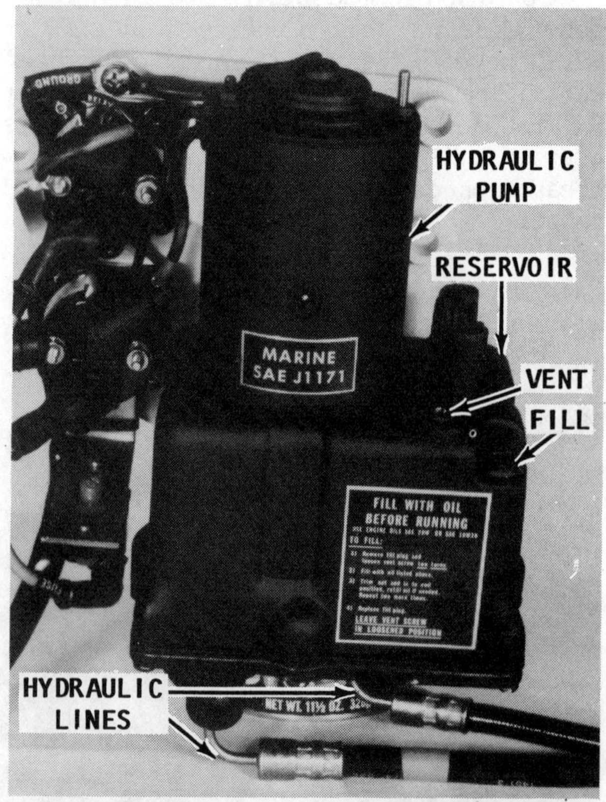

Hydraulic pump and reservoir of the tilt/trim system installed on light-weight boats. This system permits the stern drive to be adjusted to raise and lower the bow of the boat for maximum speed and efficiency.

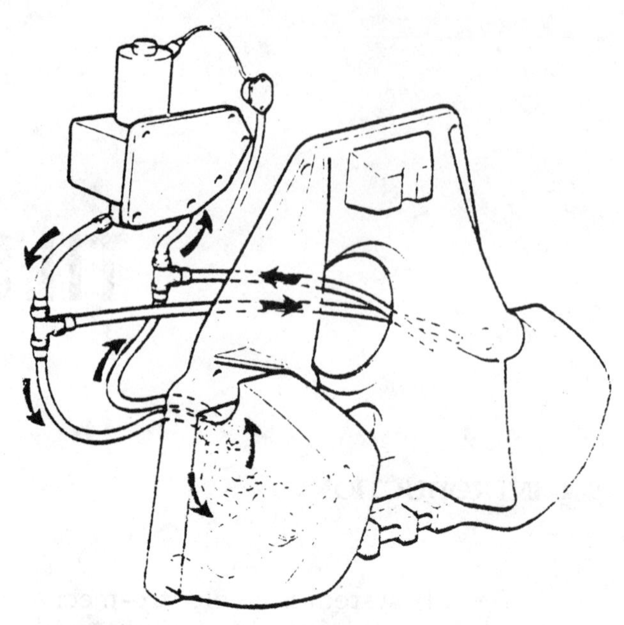

Simple drawing to depict hydraulic flow to and from the cylinders.

9-2 TROUBLESHOOTING

Troubleshooting must be a well thought-out procedure. To be successful, you must start by accurately determing the problem; then use a logical approach to arrive at the proper solution. The common phrase "shotgun approach" only leads to wasted time, money, and frustration. Obviously, if the instructions are to be of maximum benefit as a guide, they must be fully understood and followed.

TILT SYSTEM

If the stern drive fails to rise, or to lower, a simple check is to unplug the electric motor from the wiring harness. With a set of jumper wires, operate the electric motor independent of the electrical box. By switching the wires it is possible to operate the motor to raise or lower the stern drive. If there is no problem with the electric motor then the problem is in the electrical box, most likely one of the relays.

If the stern drive fails to lower, then the **DOWN** relay or the limit switch is at fault. Use the jumper wires and bypass the limit switch to determine if it is at fault. If the stern drive fails to rise, then the **UP** relay is at fault.

If there is evidence water has entered the electrical box, then there is a strong possibility both relays and the limit switch should be replaced.

If the tilt system has been operated up and down several times in a very short time, the electric motor may overheat. If this happens a built-in overheating protection cutout switch may open the circuit when the temperature reaches a predetermined amount. Under this condition, the electric motor **MUST** be allowed to cool for approximately 1/2 hour before the stern drive can be moved up or down.

If the tilt motor is driven too long after the drive has reached its uppermost position, the electric motor may overheat and burn. To eliminate this problem, late model units have an overload cutout switch installed as standard equipment. If the unit being serviced does not have the overload switch, one should be purchased and installed at the first opportunity.

To install the overload switch, first secure the switch to the transom near the tilt mechanism. Next, loosen the tilt mechanism ground wire (blue) on the flywheel cover and strip the cable's loom back far enough that the ground wire can be fed through the overload cutout switch and back to the flywheel cover. Cut the wire and connect the two wire connectors. Secure the ground wire to the flywheel cover.

If the brake nut (clutch) seems to be slipping when the stern drive approaches the full up position, the housing may need to be tighten a bit to add tension on the spring. If the motor appears to be laboring, the spring tension may be too great. In this case, loosen the housing cover just a bit to relieve some of the spring tension.

If the motor seems to operate, but the stern drive does not raise and lower properly, internal parts of the tilt mechanism may be corroded or damaged, requiring removal, disassembly, and replacement of affected parts.

The only electric motor parts available from the local Volvo Penta dealer are brush sets. The back cover plate of the motor can be removed and an inspection of the brushes made.

If the windings or internal parts are damaged, it would be best to seek the services of a speciality shop with experienced personnel and equipment available for such repair.

TILT/TRIM SYSTEM

The following paragraphs list the most common faults in the tilt/trim system. Most of the faults may be caused by more than one problem. Therefore, the least expensive and less time consuming solutions are given first. Only after eliminating these areas as the cause, should the system be opened and disassembled.

SPECIAL WORDS

When adding hydraulic fluid to the pump reservoir, the vent screw must be loosened and the fill screw removed. The vent allows air in the reservoir to escape while fluid is being added. Add fluid until the level is up to the bottom of the fill hole when the reservoir is approximately level.

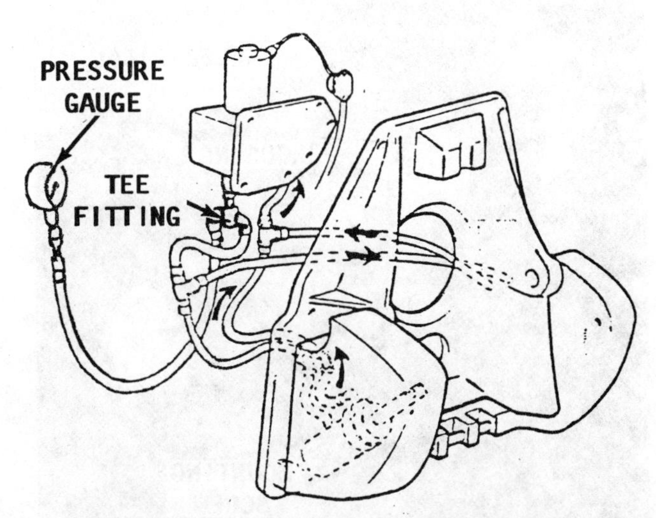

Hookup to test the hydraulic pump using a tee fitting and gauge, as described in the text.

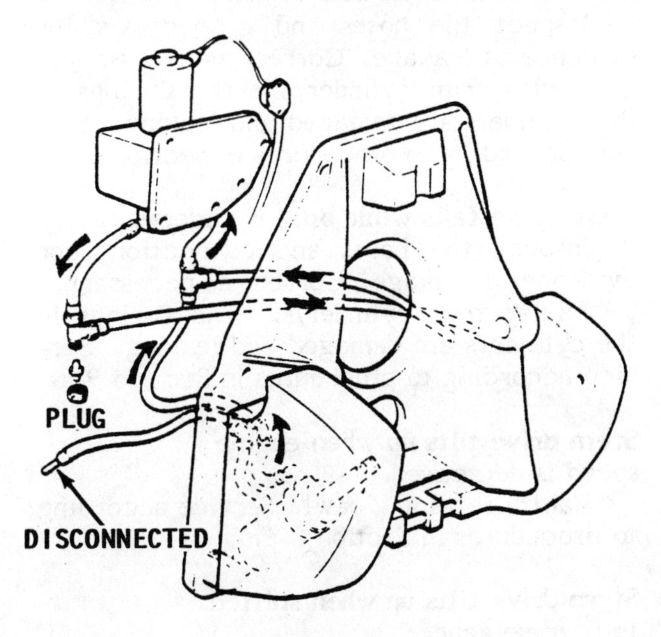

Hookup to test the cylinders using the tee fitting and a plug, as described in the text.

Drive does not lift.

Check the oil in the hydraulic tank and fill with the same quality as is already used in the hydraulic pump. Air vent the system.

Inspect the hoses and connections for evidence of leakage. Correct as necessary.

Check all electrical connections for corrosion or breakage. If necessary, replace the rotor in the hydraulic pump.

Faulty pump. Check the pump by connecting a dial indicator to the delivery side of the pump using a tee fitting, as shown. When the drive reaches maximum travel position against the rubber buffer a minimum pressure of 22 psi (150 kp/cm^2) should be obtained. Replace the hydraulic pump if the pressure is low.

Faulty trim cylinder/s. The O-rings in the cylinders are damaged and leaking. Service according to procedures in Section 9-4.

Check valve in the pump valve body is defective.

Stern drive lifts partially or hesitantly.

Check the oil in the hydraulic tank and fill with the same quality as it already used in the hydraulic pump. Air vent the system.

Inspect the hoses and connections for evidence of leakage. Correct as necessary.

Faulty trim cylinder/s. The O-rings in the cylinders are damaged and leaking. Service according to procedures in Section 9-4.

Check valve in the pump valve body is defective.

Stern drive will not hold in raised position.

Inspect the hoses and connections for evidence of leakage. Correct as necessary.

Faulty trim cylinder/s. The O-rings in the cylinders are damaged and leaking. Service according to procedures in Section 9-4.

Stern drive falls while boat is underway.

Inspect the hoses and connections for evidence of leakage. Correct as necessary.

Faulty trim cylinder/s. The O-rings in the cylinders are damaged and leaking. Service according to procedures in Section 9-4.

Stern drive tilts up when engine speed is decreased.

Faulty retaining pawl. Service according to procedures in Section 8-6.

Stern drive tilts up when shifted to reverse gear.

Faulty retaining pawl. Service according to procedures in Section 8-6.

Stern drive lowers partially or hestitantly.

Check the oil in the hydraulic tank and fill with the same quality as is already used in the hydraulic pump. Air vent the system.

Inspect the hoses and connections for evidence of leakage. Correct as necessary.

Faulty pump. Check the pump by connecting a pressure gauge to the delivery side of the pump using a tee fitting, as shown. When the drive reaches maximum travel position against the rubber buffer a minimum pressure of 22 psi (150 kp/cm^2) should be obtained. Replace the hydraulic pump if the pressure is low.

Faulty trim cylinder/s. The O-rings in the cylinders are damaged and leaking. Service according to procedures in Section 9-4.

Check valve in the pump valve body is defective.

9-3 TILT SYSTEM SERVICE

REMOVAL AND DISASSEMBLING

1- Move the stern drive to the full down position. This action will move the screw spindle all the way back into the housing. Disconnect the tilt system wiring harness from the boat wiring harness by separating the Cannon-type connector. Cut any tie wraps securing the wiring to the transom. Remove the two Allen-head screws securing the mounting flange to the transom shield. Pull the unit up and free of the transom shield. If a new gasket is not available, save the gasket from under the flange. Remove the two thru-bolts securing the electrical

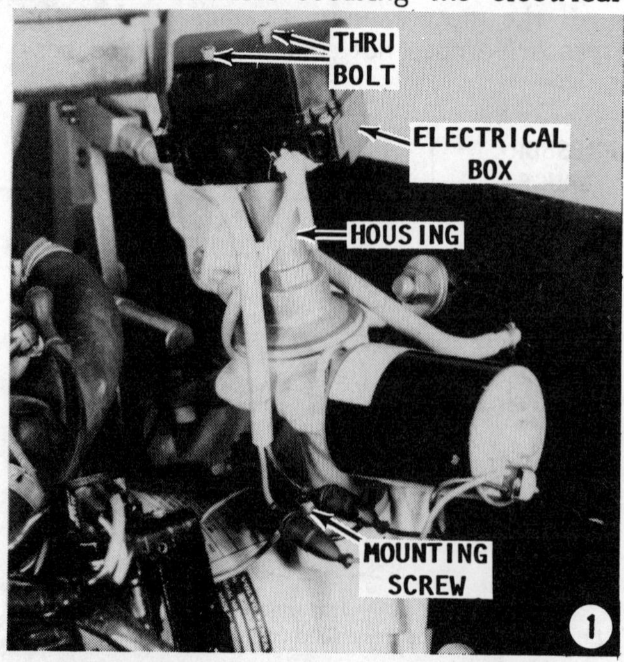

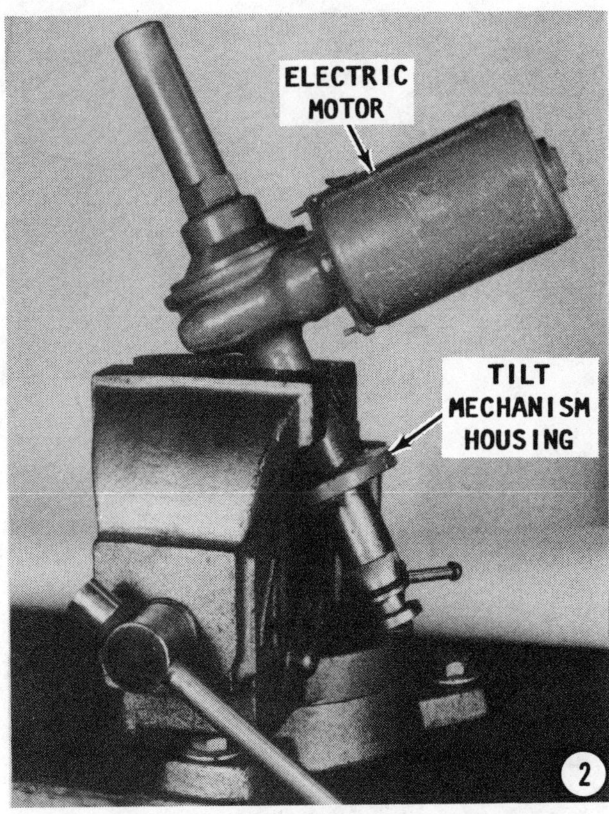

box to the housing. Disconnect the electrical wires between the electric motor and the electrical box. Slide the electrical box up and free of the housing cover.

2- Clamp the unit in a vise at a slight angle, as shown. Do not overtighten the vise to prevent distorting the housing.

3- Remove the two nuts securing the electric motor to the housing.

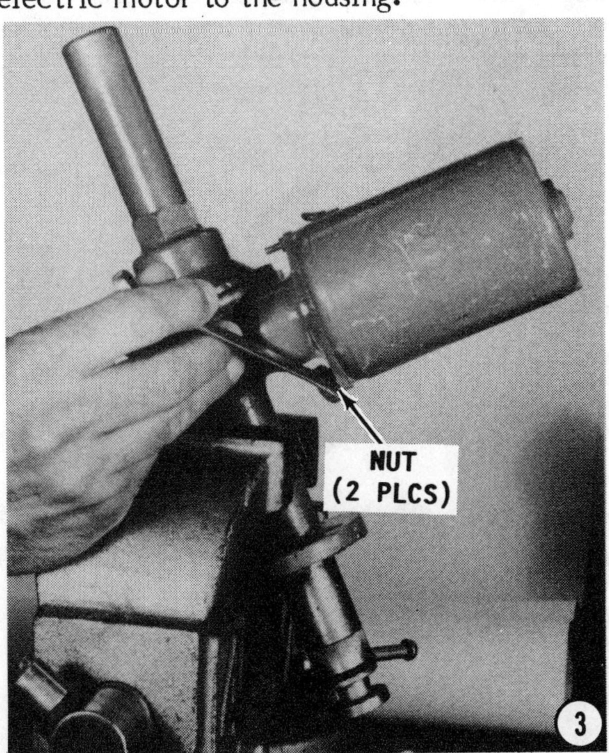

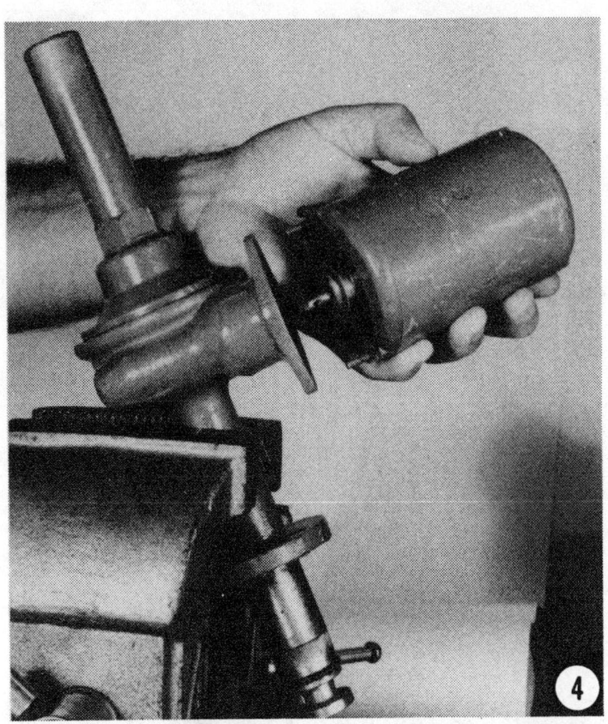

4- Lift the electric motor free of the tilt mechanism housing.

5- **TAKE NOTE** of the distance between the housing cover and the housing. The cover may not be down tight against the housing. The amount the cover is tightened sets the tension on the spring and therefore on the brake nut. Record the distance between the cover and the housing as an aid during installation. Loosen the housing cover using the proper size wrench on the collar.

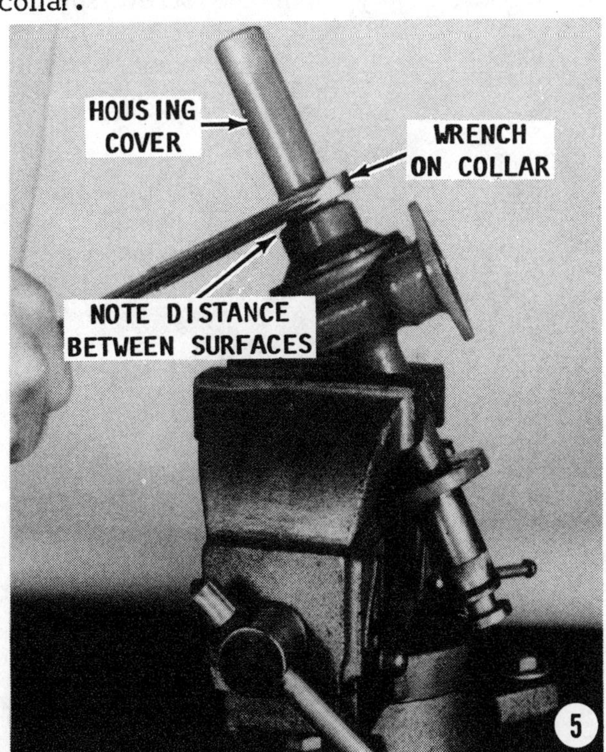

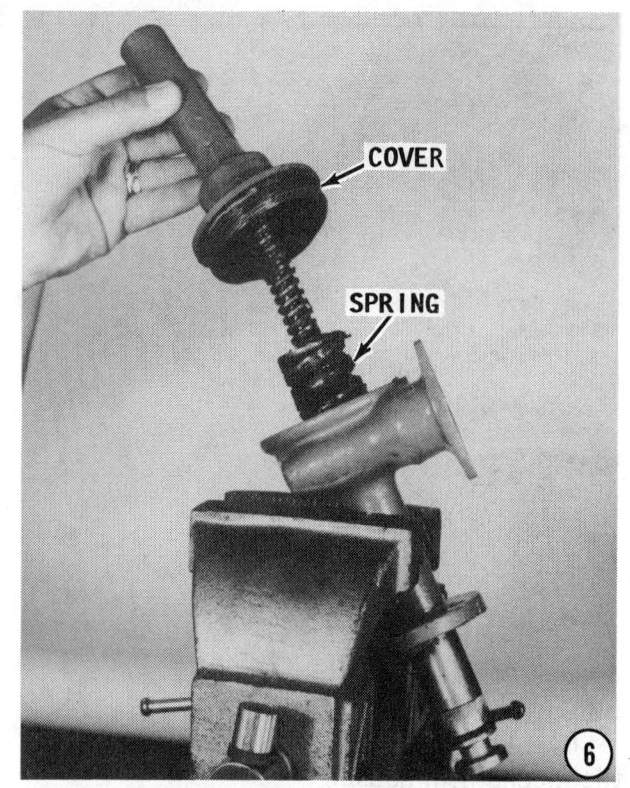

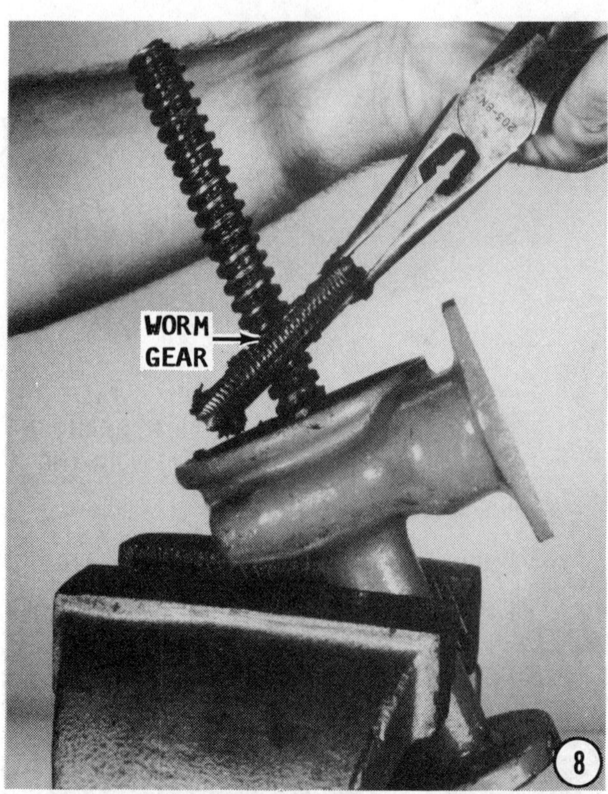

6- Lift the cover off the housing. Remove the large spring.

7- Slide the upper bearing set off the screw spindle. Back off and remove the brake nut.

8- Reach into the housing and grasp the circular part of the worm gear set with a pair of needle nose pliers. Work the gear up and off the screw spindle. Remove and save the Woodruff key from the screw spindle shaft.

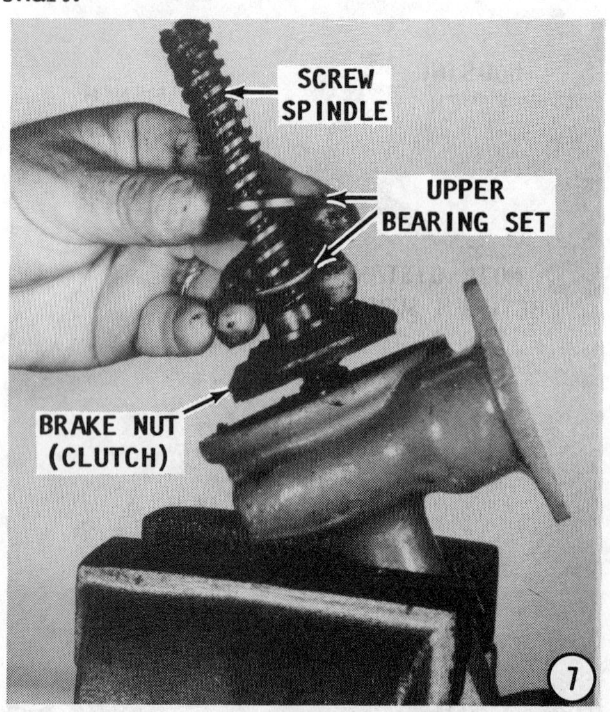

9- After the worm gear is removed, lift the lower bearing out of the housing. The lower bearing race will remain in the housing. If it is not damaged, it is not necessary to remove it. If it is damaged, it can be removed with a sharp pointed tool.

10- Bend the locking tab over on the lower end of the screw spindle shaft.

11- Unscrew the pressure plate from the end of the shaft using a pair of water pump or channel-lock pliers. Remove the lockwasher.

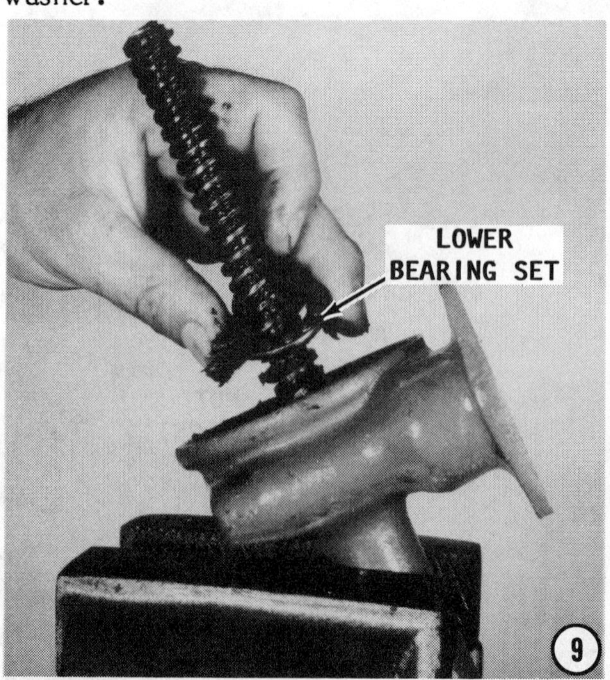

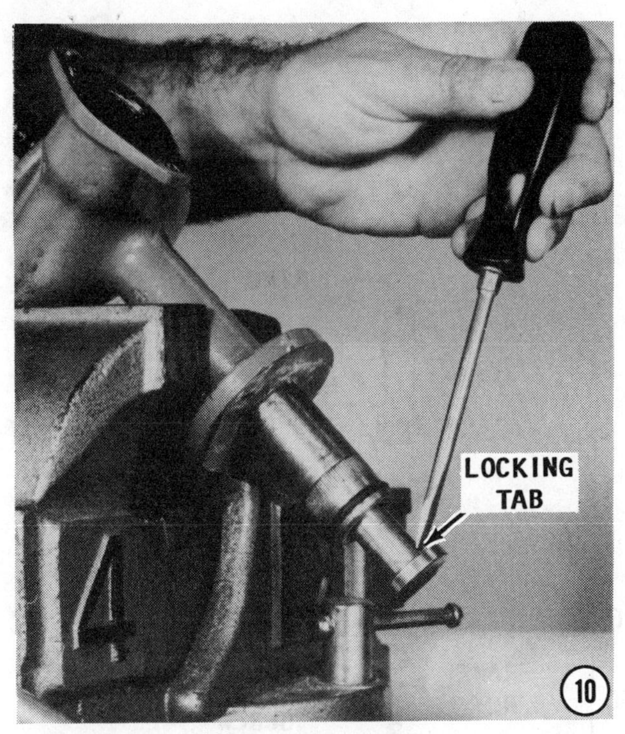

12- Remove the sealing nut.

13- Two O-rings are installed in the sealing nut. One on the outside, which seals the housing to the transom shield, and another inside the nut. This second O-ring prevents water from entering the lift mechanism. These two O-rings should be replaced anytime the lift mechanism is disassembled.

14- Withdraw the screw spindle from the housing.

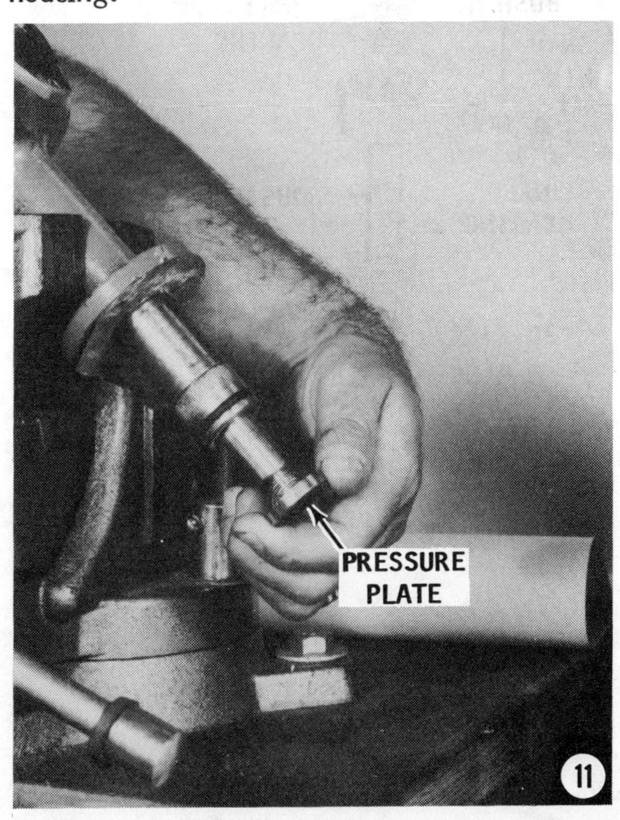

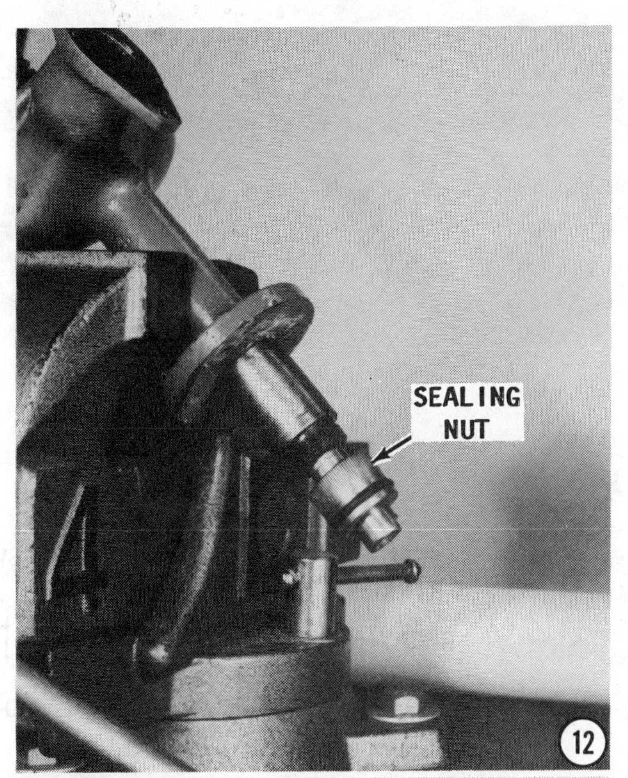

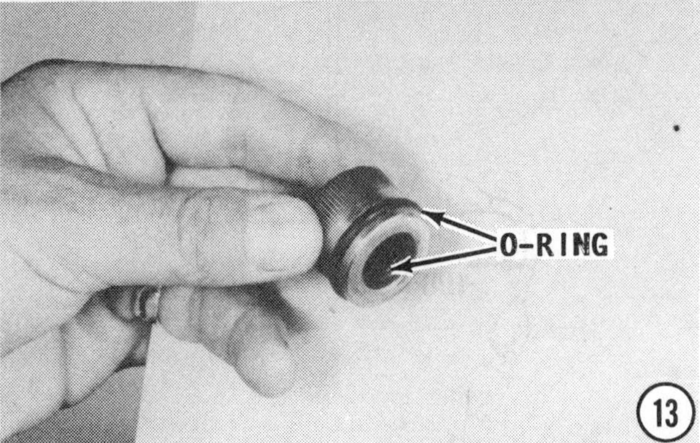

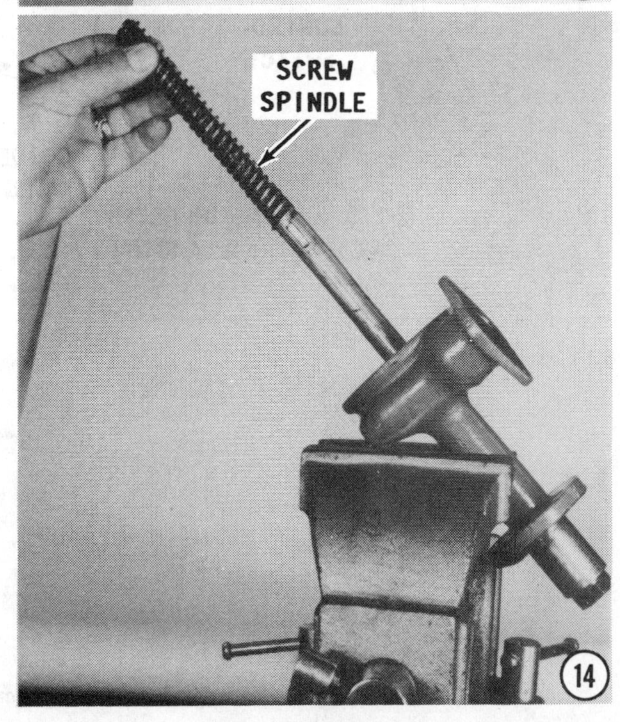

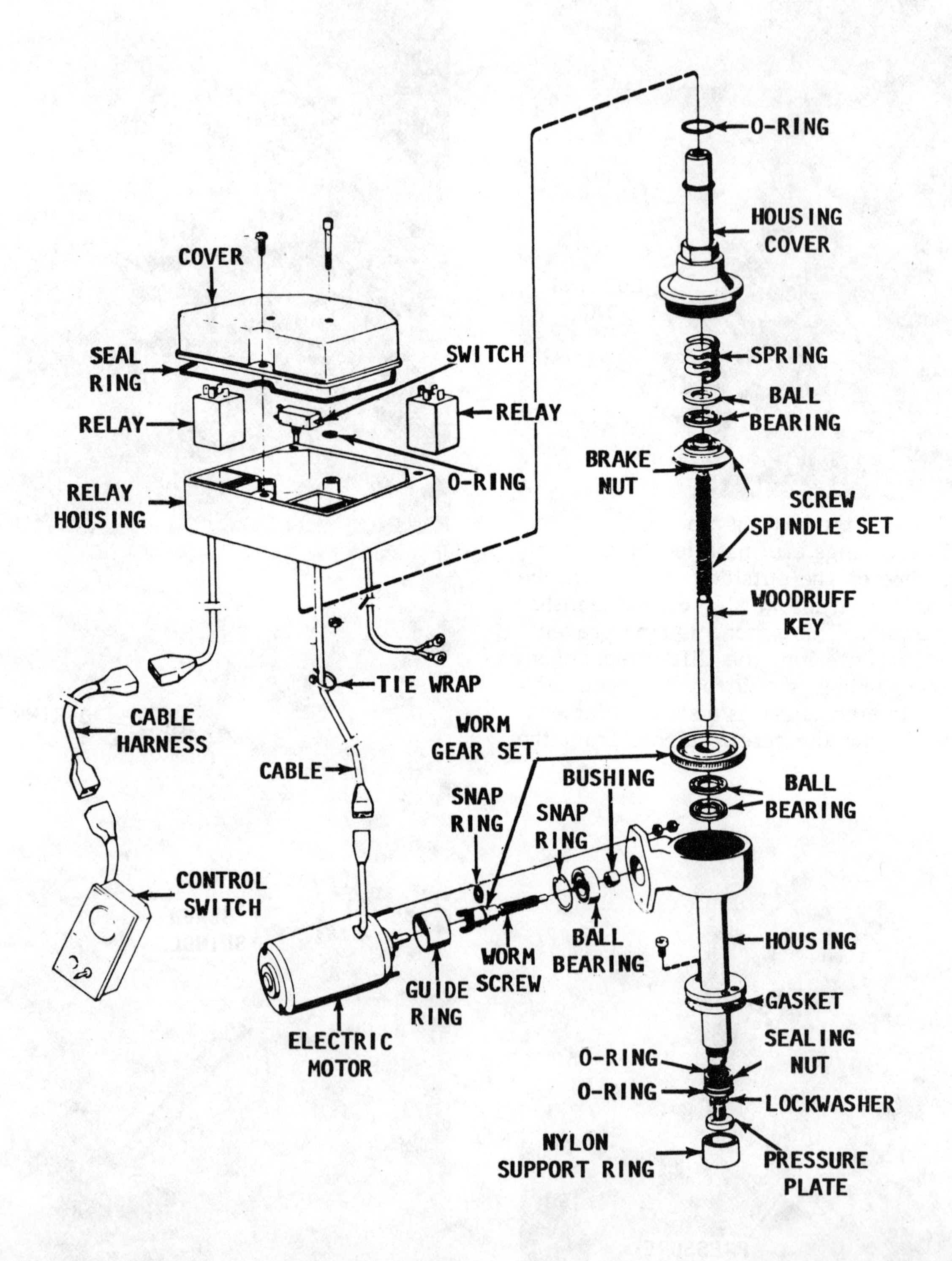

Exploded drawing of the tilt mechanism with principle parts identified.

15- The worm gear left in the housing should only be removed if there is evidence that it is damaged. If the gear turns freely and the bearings do not appear to bind, the worm gear does not have to be removed.

16- The worm gear, ball bearing, and bushing may be removed from the electric motor opening by first removing the snap ring on the worm gear shaft, and then removing the worm gear. Reach in with a pair of Truarc pliers and remove the snap ring from the housing. The ball bearing set may then be removed. As previously stated in Step 15, this worm gear and bearing set are **ONLY** removed if there is evidence they are damaged.

CLEANING AND INSPECTING

Clean the parts thoroughly in solvent, and then blow them dry with compressed air. Check for corrosion and damaged parts.

Inspect the teeth of the part of the worm gear removed and also the worm gear remaining in the housing. Check the bearings for scores, galling, pitting, or other damage. If the bearings are damaged, replace them as a set -- the bearing and the race.

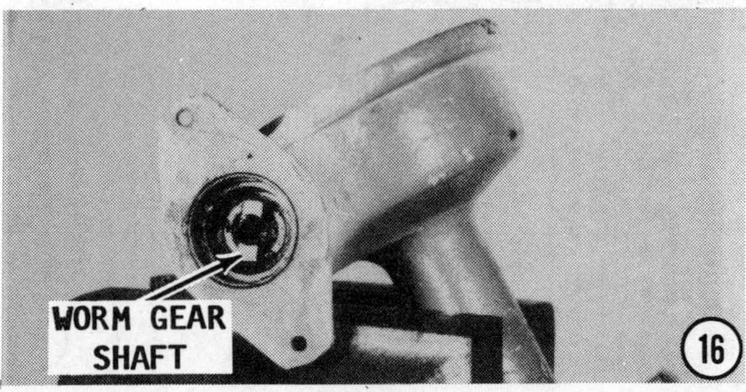

Install **NEW** gaskets and O-rings.

If the stern drive failed to raise to the full up position with an indication the tilt mechanism was slipping, the clutch assembly may be glazed. Use fine grade valve lapping compound and seat the clutch mechanism.

ASSEMBLING AND INSTALLATION

GOOD WORDS

During the assembling process, coat all parts and pack the bearings with water resistant grease.

1- Coat the O-ring grooves of the sealing nut with water resistant grease, and then install **NEW** O-rings into the grooves. The O-ring in the inside groove prevents water from entering the tilt mechanism. The second O-ring on the outside seals the housing to the transom shield. Set the sealing nut aside for later installation.

2- Coat the worm gear in the housing and the inside surface of the housing with water repellent grease.

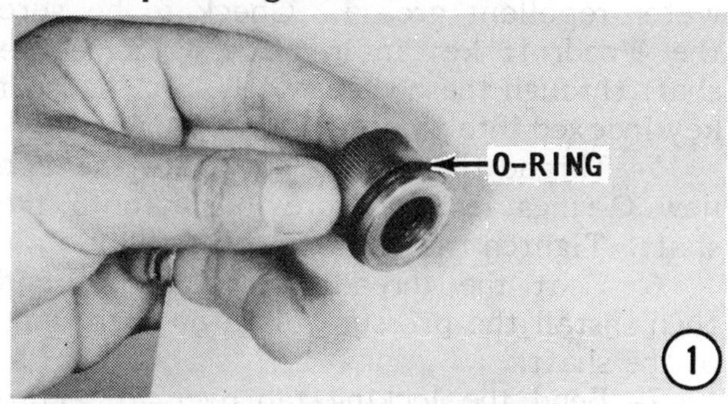

3- If the small worm gear and associated parts were removed, coat them all with water resistant grease. As an aid to installing these parts, refer to the exploded drawing of the tilt mechanism on Page 9-8. Install the bushing into the housing by driving it in until it is fully seated. Install a new bearing onto the worm gear shaft. Secure the bearing in place with the snap ring. Slide the worm gear and bearing into the housing. Secure it in place with another snap ring. Push the guide ring into the housing until the outer surface is flush with the housing.

4- Coat the screw spindle and shaft with water repellent grease. Check to be sure the Woodruff key in in place. Slide the shaft through the housing with the Woodruff key indexed into the slot in the housing.

5- Thread the sealing nut, with the two new O-rings installed previously, onto the shaft. Tighten the sealing nut securely.

6- Coat the threads with Loctite, and then install the pressure plate onto the end of the shaft.

7- Bend the locking tab over to secure the pressure plate.

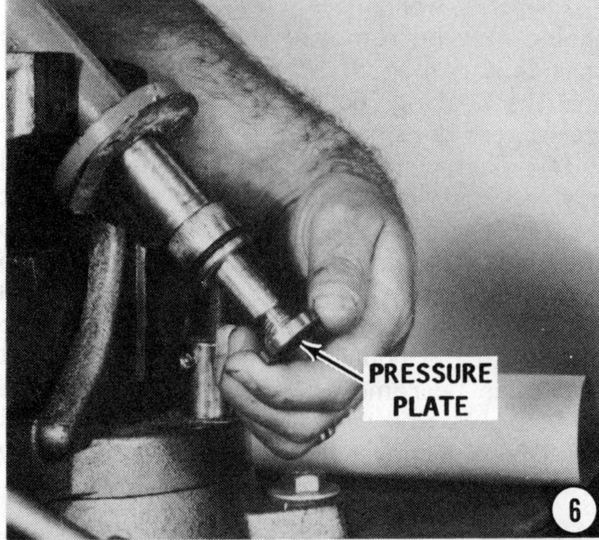

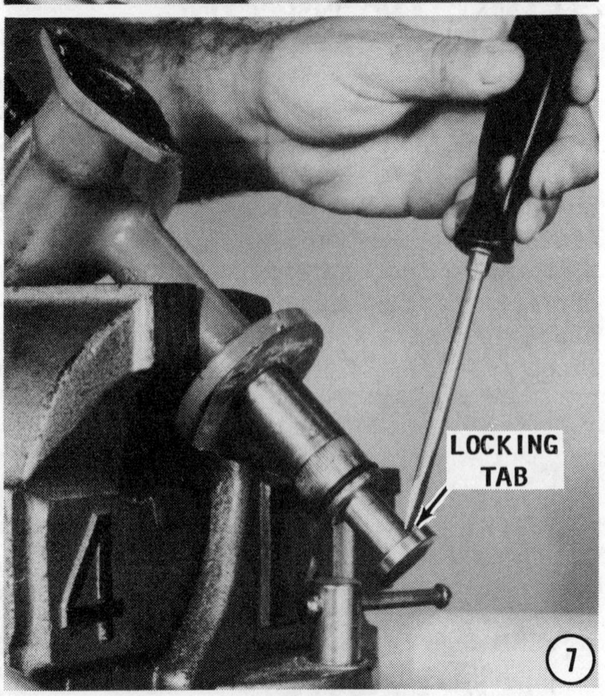

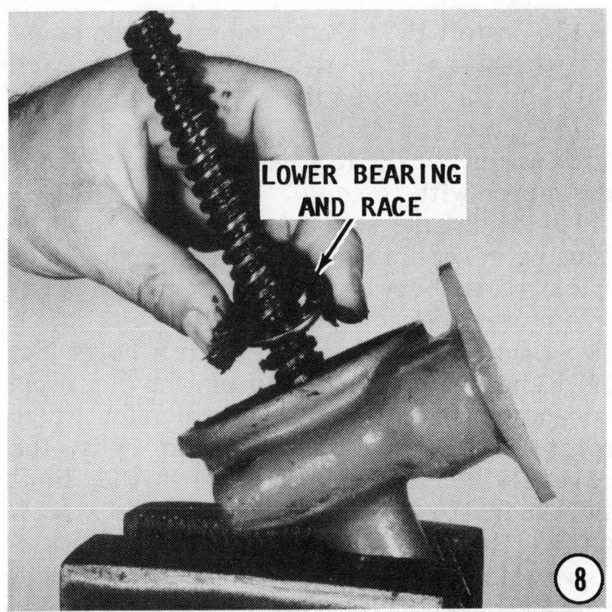

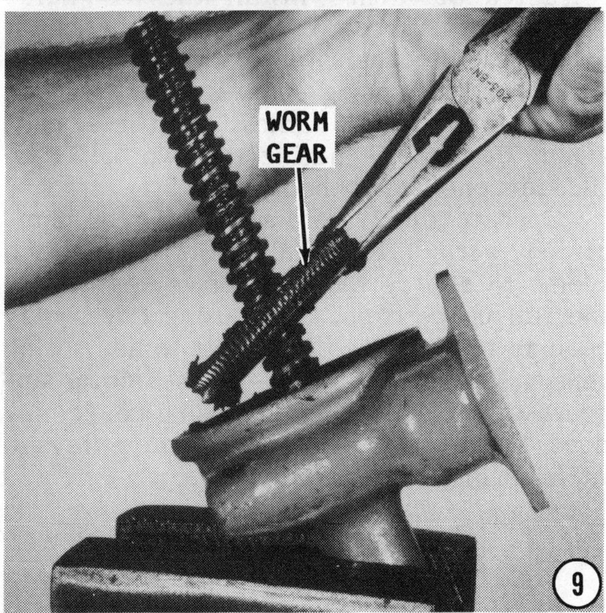

8- If the lower bearing bearing race was removed during disassembling, coat it with water repellent grease, and then force it into the housing until it is fully seated. Slide the lower ball bearing set down the screw spindle into place against the race.

9- Lower the worm gear down the screw spindle.

10- Install the ball bearing set and race onto the screw spindle.

11- Coat the threads of the housing cover with water repellent grease, and then work the housing cover down over the screw spindle and spring. The tension on the spring is determined by the amount the housing is tightened down.

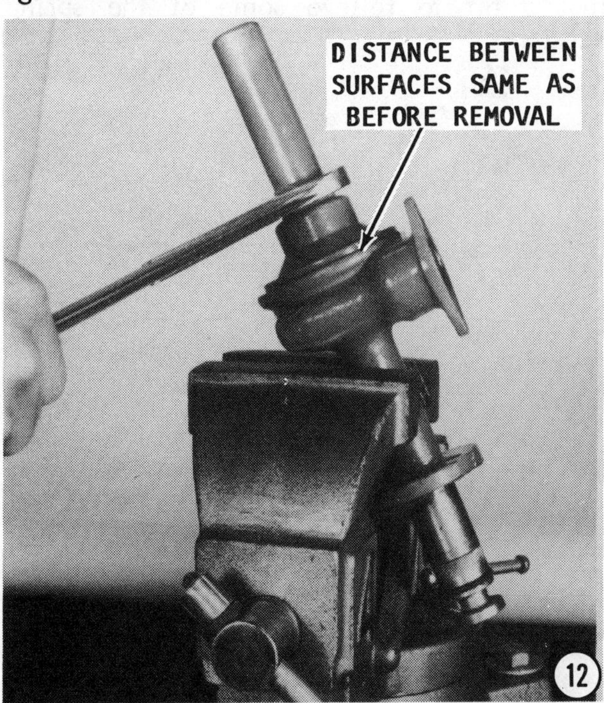

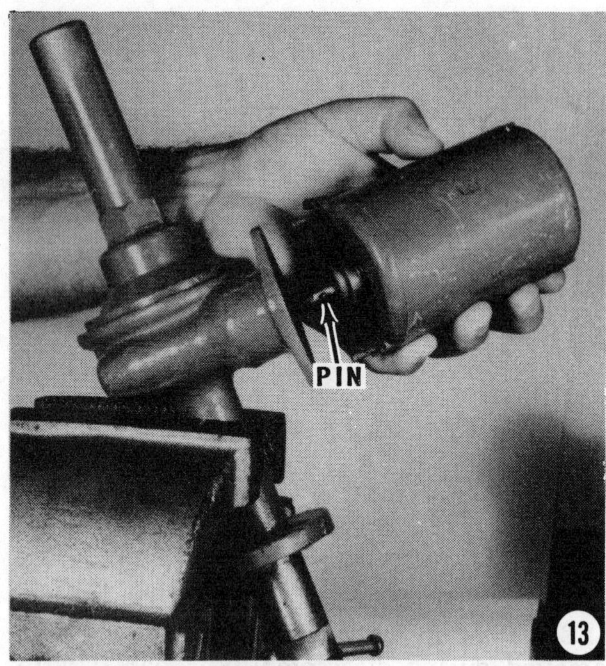

12- Tighten the housing until the distance from the housing mating surface is the same as recorded during disassembling, Step No. 5.

SPECIAL WORDS

If, after installation of the complete unit to the transom shield, the brake nut (clutch) seems to be slipping when the stern drive is raised to the full up position, the housing may need to be tighten a bit to add tension on the spring. If the motor appears to be laboring, the spring tension may be too great. In this case, loosen the housing cover just a bit to relieve some of the spring tension.

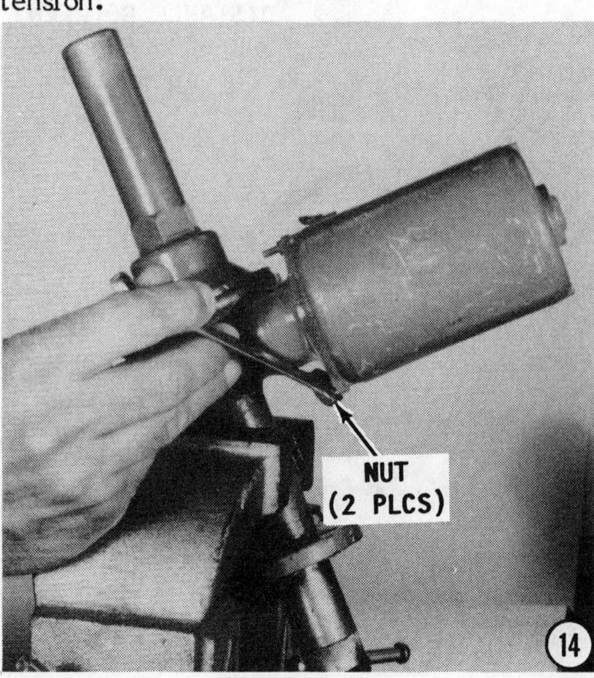

13- Install the electric motor to the housing with the pins on the motor shaft indexed between the forks of the worm gear.

14- Coat the threads of the motor studs with water repellent grease, and then secure the motor with the nuts.

15- Slide the electrical box down the housing cover into place. Connect the electrical wires between the box and the electric motor. Secure the box in place with the two thru-bolts. Tighten the thru bolts just **SNUG** to prevent the box from moving in an up-and-down direction, but still allow it to rotate. If the bolts are too tight, the cover will make contact with the limit switch in the electrical box. Slide a **NEW** gasket onto the lower end of the tilt housing. Move the screw spindle to the full retracted position. Install the assembled tilt mechanism onto the transom shield. Secure the mechanism in place with the two Allen-head mounting screws. Connect the tilt system wiring harness with the boat wiring harness by mating the two halves of the Cannon-type connector.

Operate the system and check the completed work. **REMEMBER** if the clutch seems to slip as the stern drive approaches the full up position, the housing cover may need to be tightened just a bit to add spring tension onto the clutch. If the motor appears to be laboring, the housing cover may need to be loosened just a bit to relieve a little spring tension on the clutch.

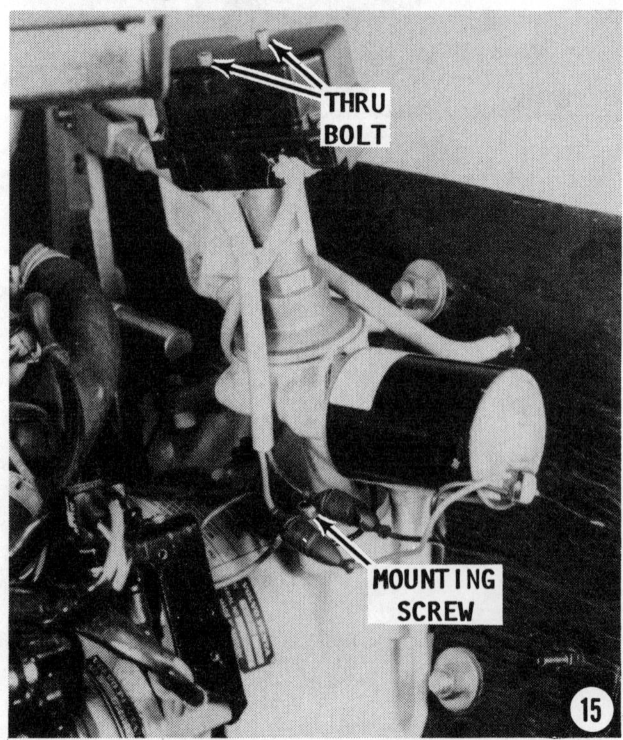

9-4 TRIM SYSTEM SERVICE

REMOVAL AND DISASSEMBLING

For **PHOTOGRAPHIC** clarity, the pictures for removal were taken with the transom shield off the boat.

1- Raise the stern drive up approximately half way through its travel. Block the stern drive in this position with a box, empty five-gallon container upside down, sturdy chair, or other suitable device placed under the skeg of the lower unit. Disconnect the two hydraulic hoses from the transom shield.

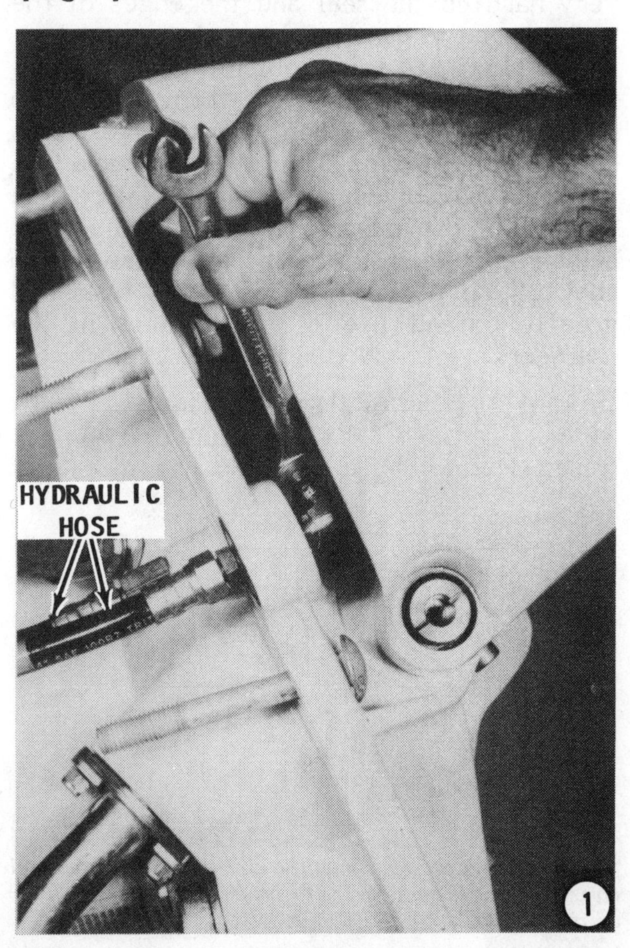

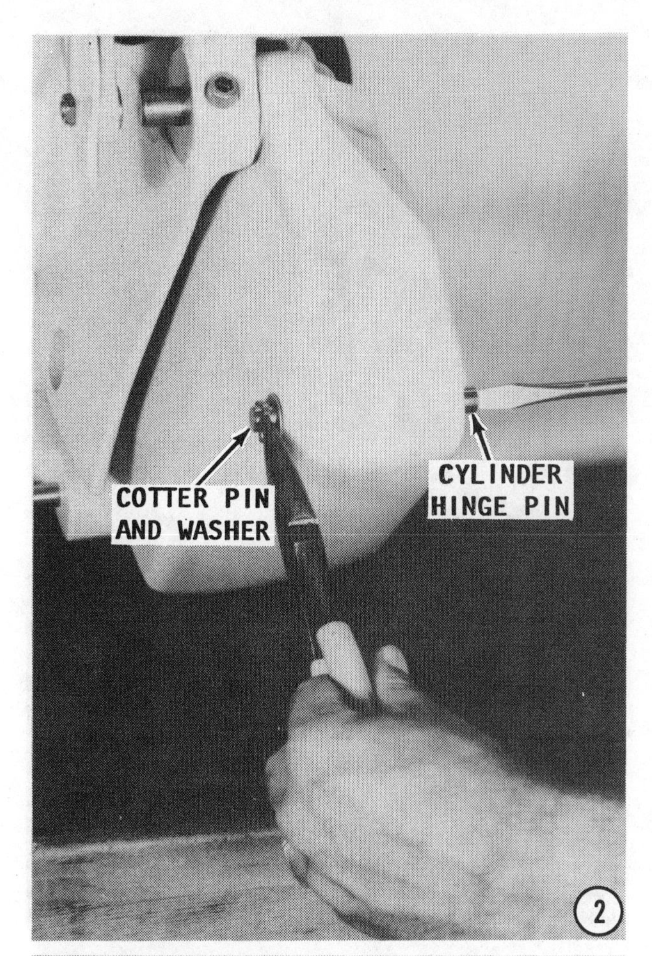

2- Hold one of the cylinder hinge pins firm with a screwdriver, and at the same time pull the cotter pin. Remove the washer, and then push the pin free. Remove the second cylinder hinge pin in the same manner. Remove the pin from the piston end of each cylinder. Lift the cylinders free of the transom shield.

3- Clamp the cylinder in a vise in the upright position, as shown.

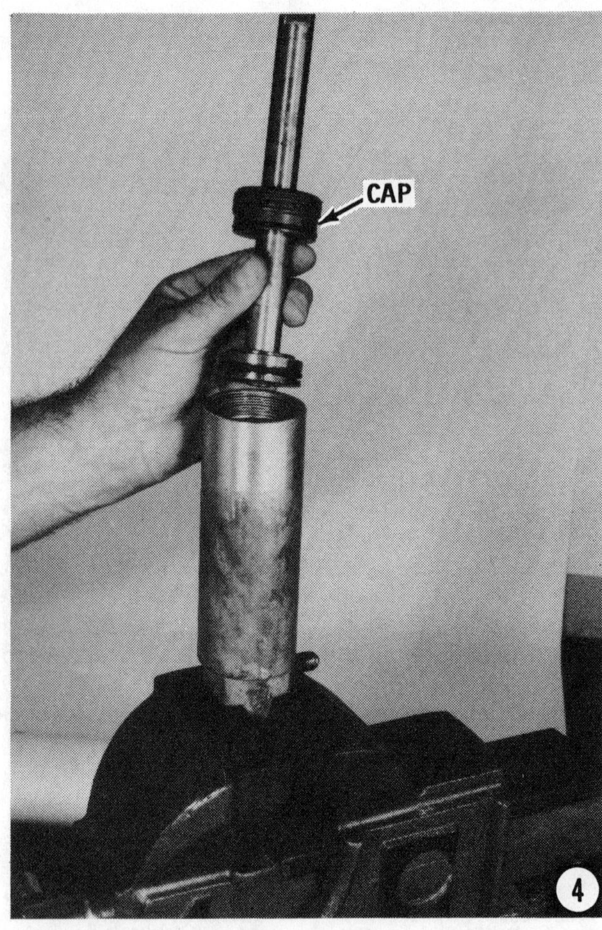

4- Back off the cylinder cap, and then pull the piston free.

5- Remove and discard the O-ring. If the O-ring is damaged, as the one shown in the accompanying illustration, the cylinder would not hold its position, because hydraulic fluid would be able to pass the O-ring.

CLEANING AND INSPECTING

Install a **NEW** O-ring on the lower end of the piston whenever the cylinder is disassembled for service. Volvo does not sell the O-ring as a separate item. However, this same O-ring is used on many items in the

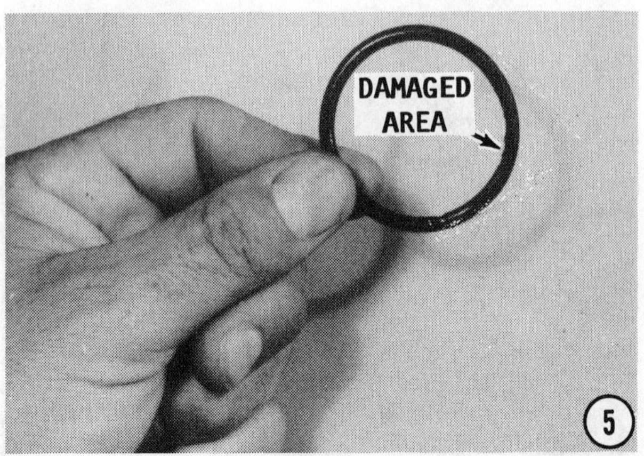

Check the interior surface of the cylinder and remove any light scores by polishing them out with crocus cloth.

marine industry and is available from an OMC marine dealer. The part number for the O-ring is OMC No. 321446.

Check the edges of the O-ring groove. Sharp edges could cut and damage the O-ring. Therefore, take the sharp edge off by making a very slight radius with crocus cloth.

Inspect the O-ring in the cap. If it is not damaged, it may be used again. The O-ring serves as a seal for the cap.

VERY GOOD WORDS

Many times the cylinder O-ring is damaged because the boat operator raises the stern drive up hard against the rubber cushion on the transom shield. This action is very hard on the seal and the edges of the groove tend to damage the seal.

To eliminate this problem, have an assistant watch as you raise the stern drive with the tilt control. When the stern drive is about 1" (2.54 cm) from the pad have the assistant give you a signal. Notice the position of the indicator, and then adjust the sender unit to indicate the stern drive is in the full up position. This little task will greatly extend life of the O-rings in the cylinders.

The O-ring groove should be carefully inspected and any sharp edges removed with crocus cloth, as explained in the text.

Inspect the interior of the cylinder carefully. Check for signs of scoring, pitting, or other damage. If the cylinder wall is lightly scored, it can be returned to serviceable condition by polishing the score out with fine grain crocus cloth. If the wall is deeply scored, the trim cylinder **MUST** be replaced. The housing is not sold separately, therefore, a complete trim cylinder assembly must be purchased.

ASSEMBLING AND INSTALLATION

1- Install a new O-ring into the groove in the lower end of the piston. Coat the cylinder wall and the O-ring with hydraulic fluid. Apply a light coating of Loctite or other suitable material to the threads of the cap. Insert the piston into the cylinder and secure it by threading the cap into the cylinder. Tighten the cap securely.

2- Position the trim cylinder into the transom shield with the hole in the housing flange aligned with the hole in the transom shield. Insert the hinge pin through the transom shield, cylinder, and out through the hole on the other side. Push the pin through until it bottoms out against the washer.

3- Secure the hinge in place with the washer and cotter pin. A slot in the hinge pin permits the use of screwdriver on that end to hold the pin firmly while the cotter pin is being installed and opened. Install the pin in the piston end of each cylinder. Secure the pins with the washers and cotter pins. Connect the hydraulic hoses to the cylinder.

4- Add fluid to the hydraulic pump reservoir. Loosen the vent screw and remove the fill screw. The vent **MUST** be opened to allow air in the reservoir to escape while fluid is being added. Fill the reservoir until

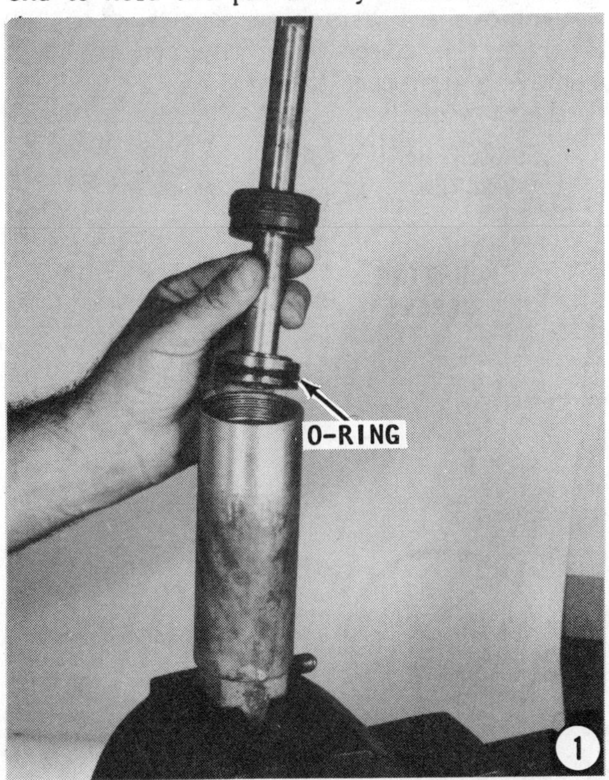

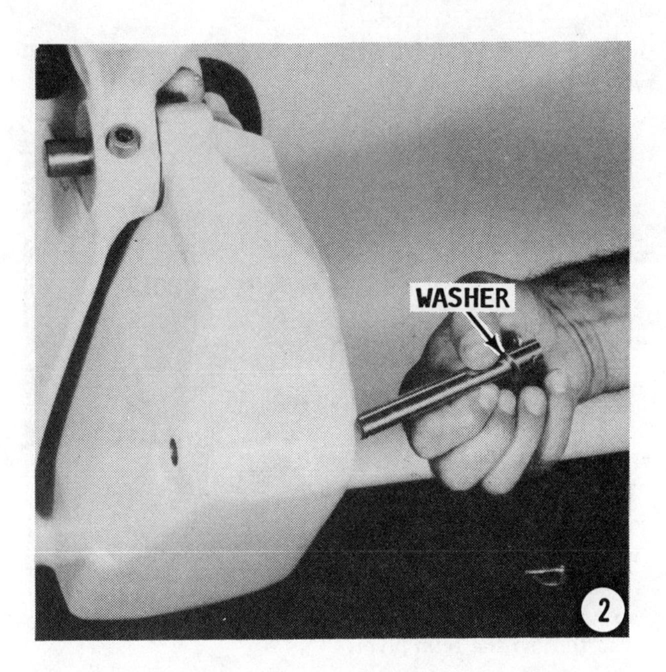

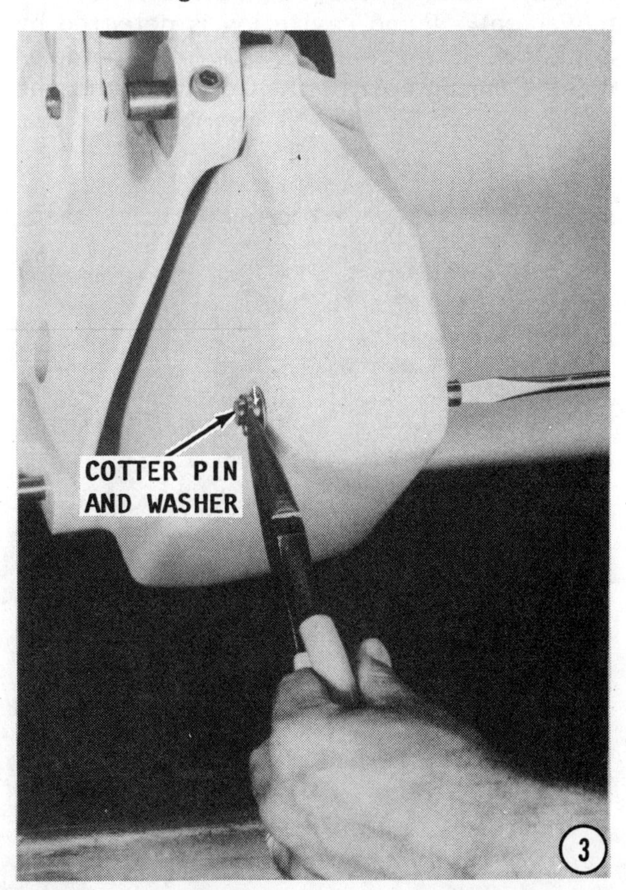

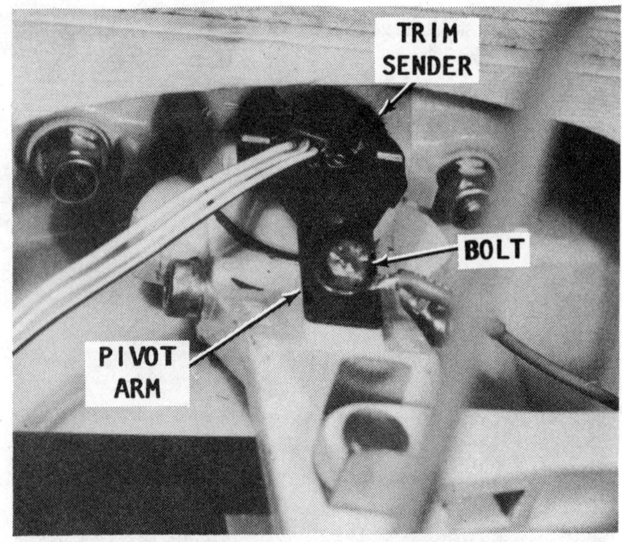

The power trim sender unit is attached to the tiller arm and can be adjusted by loosening the clamp bolt and turning the trim pivot.

the fluid is up to the bottom of the fill hole when the reservoir is approximately level. Install the fill screw, but leave the vent screw open just a wee bit to allow trapped air in the system to escape while the pump is operating. Operate the system through several cycles. Operate the switch momentarily during these cycles in order to vent the air from the system. The pump will cavitate and it will be necessary to stop movement. Pump cavitation is detected by a marked change in the sound the pump is making during operation. Continue with the

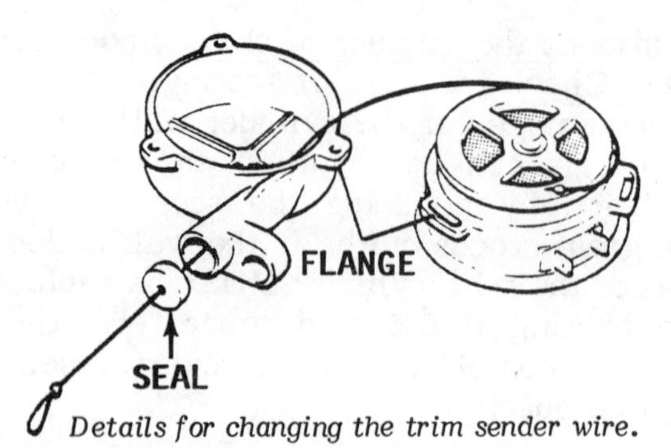

Details for changing the trim sender wire.

cycles, check the reservoir level, and add fluid, if necessary. When the pump is operating steady with no cavitation, all air has been removed from the system. Tighten the vent screw. The system is now ready for normal operation.

9-5 TRIM SENDER SERVICE

REPLACE WIRE

SPECIAL WORDS

Refer to the two accompanying drawings while performing the following steps. Parts mentioned in the procedures are identified.

Disconnect the wire from the reverse release rod.

Identify, mark, and then disconnect the electrical wires from the sender unit. Loosen, and then lift up the socket-head screw securing the sender. Pry out the seal, and then remove the sender.

Remove the attaching screws, and then separate the cover from the sender body. Remove and discard the wire.

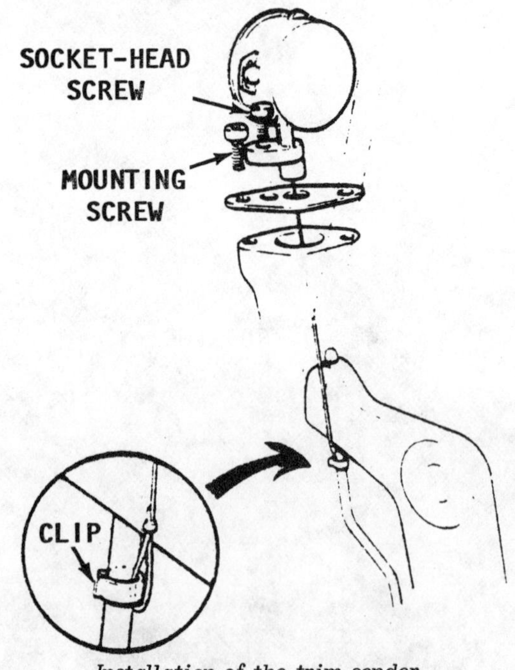

Installation of the trim sender.

Insert the new wire into the cover. If the hole is too small, bore it out a bit. Feed the wire into the groove around the ring.

Bring the two halves together with the flanges of the body and cover matched. Secure the havles together with the attaching screws.

Check to be sure the wire runs freely. Raise and lower the sender with the wire.

Carefully press the wire eyelet through the seal and push the seal into the recess. If there is evidence water has entered the sender install a special washer into the recess before the seal. This special washer is available from the local Volvo Penta dealer.

Connect the electric wires to the trim gauge: yellow/green to the **G** terminal and the black wire to the **"-"**, negative terminal.

Install the sender and secure it in place with the socket-head screw.

Pull the wire down and feel if the spring wants to pull the wire back into the sender. Secure the wire to the reverse release rod with the clip.

ADJUST THE SENDER

The sender is adjusted very similar to adjusting an electric choke on a carburetor. The sender rotates on a spring.

Lower the stern drive to the full **DOWN** position. Turn the key switch on the dashboard to the **ON** position. Loosen the three screws securing the two halves together. Have an assistant observe the dial indicator on or near the dashboard. Rotate the sender until the dial indicator needle in on the first dot in the green section, indicating the stern drive is all the way down. Tighten the screws.

Raise the stern drive until it is about an 1" (2.54 cm) from the rubber cushion mounted on the transom shield. Make a note or a mark where the trim indicator is pointing. Now, each time the stern drive is raised, stop the movement when the trim indicator reaches the mark and the stern drive will not come in contact with the rubber cushion. Taking time to determine this position on the trim indicator will help to extend the life of the seals in the cylinders.

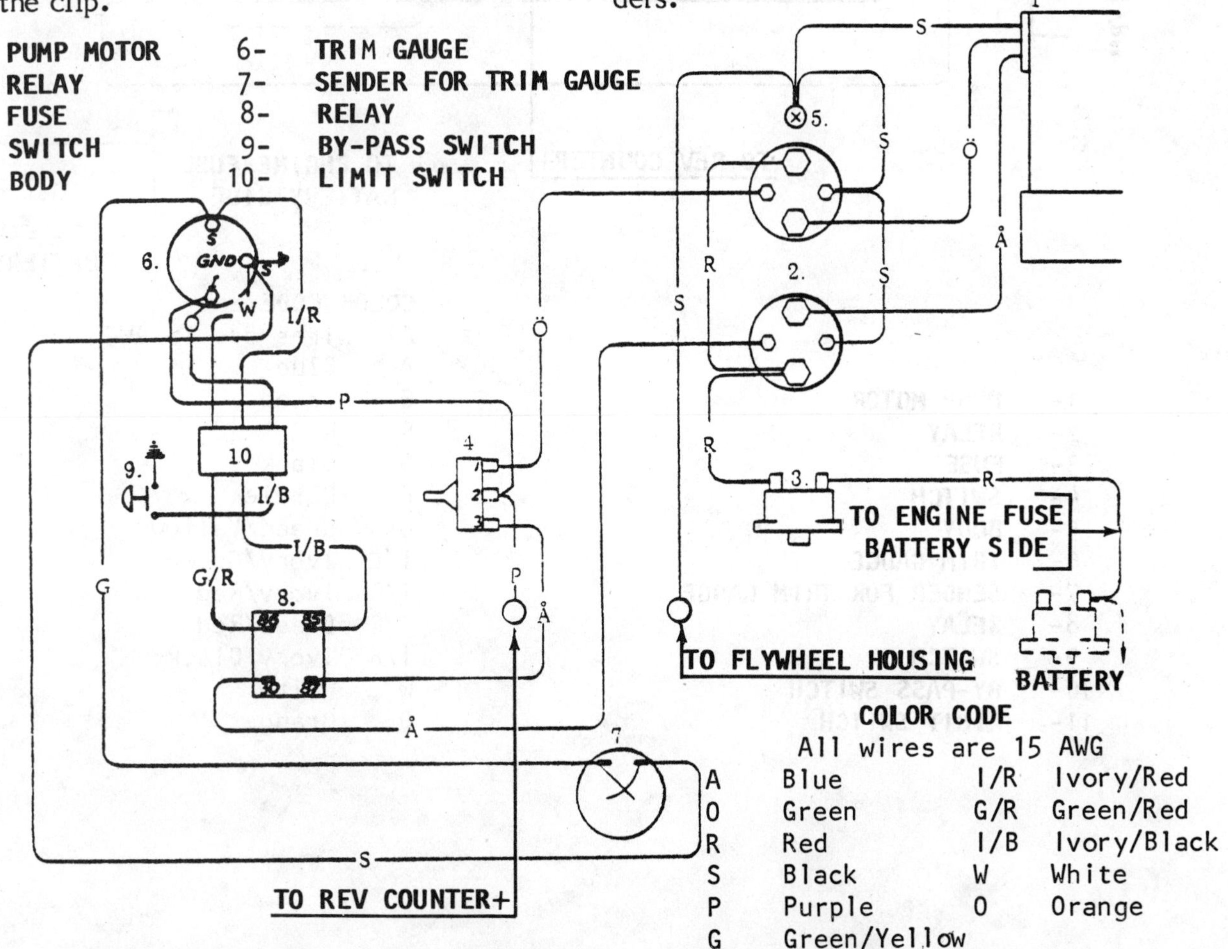

Wiring diagram for the trim system with wire and components identified.

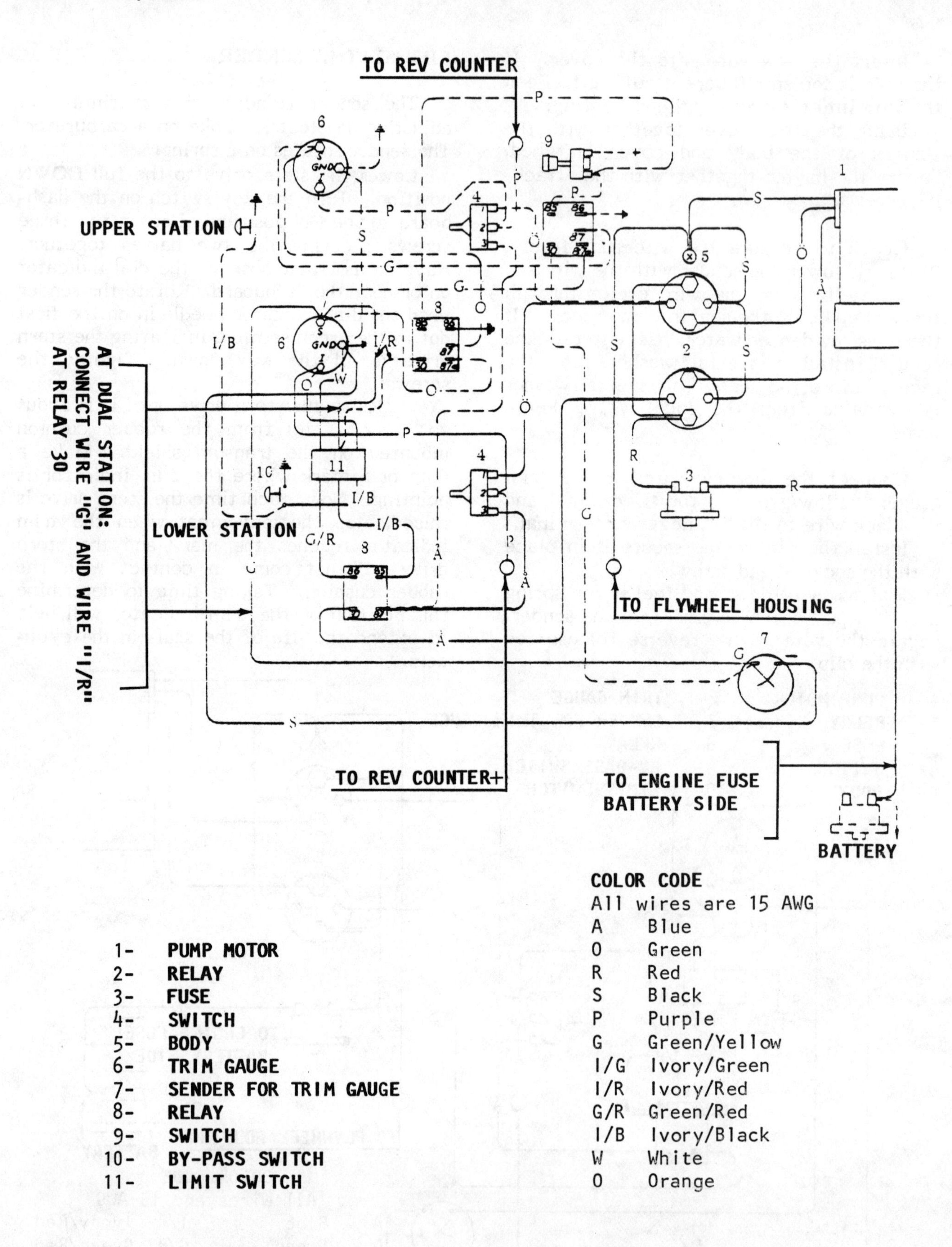

Wiring diagram for the trim system and a dual station installation with wire and components identified.

10 STEERING

10-1 DESCRIPTION

Three types of steering systems are used on Volvo Penta installations: a leverage arm steering arrangement, a direct cable connection without a lever arm, and a power steering system.

The leverage arm arrangement is a standard factory installed item on all Volvo Penta installations. This system incorporates a lever arm as part of two link arms. One link arm is connected to the steering cable and the other connects the lever arm to a short "tiller" arm.

The second steering system was installed on very early Volvo Penta units and consisted of a cable mounted in a fixture on the transom and connected directly to the "tiller" arm. This system did not have a lever arm and therefore, steering was much more "stiff". The lever arm installed on later units was incorporated as an assist to overcome the torque developed by the higher horsepower engines.

The third steering method is a power steering system which may be added to a V6 or V8 power installation as optional equipment. A hydraulic pump is mounted to the front of the engine and is driven by a belt off the crankshaft pulley. Hydraulic lines connect the pump to a servo valve and actuating cylinder mounted over the engine bell housing. A hydraulic reservoir tank is mounted to the engine above the pump. An oil cooler is connected to the system to prevent the hydraulic fluid from overheating.

Adding Hydraulic Fluid

Fluid is added to the system as follows:

Remove the cap on top of the tank. Start the engine and allow it to warm to operating temperature.

Late model manual steering installation.

CAUTION: Water must circulate through the lower unit to the engine any time the engine is run to prevent damage to the water pump mounted on the engine.

Operate the engine at idle speed and do not turn the steering wheel. Add hydraulic fluid (ATF oil type F) until the fluid level stabilizes at the upper mark.

After the fluid level stabilizes, stop the engine and install the cap.

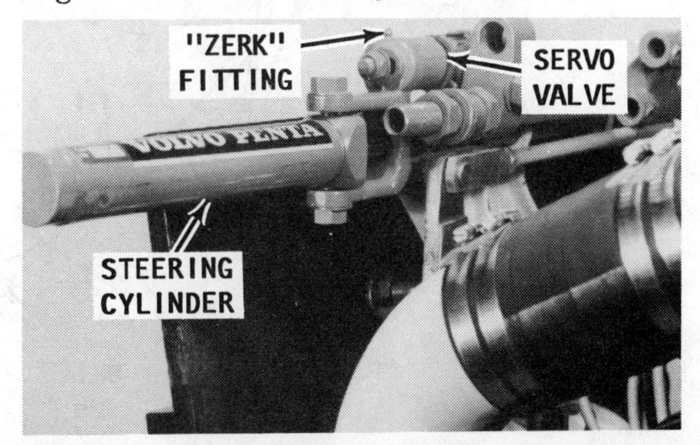

Power steering installation without the steering cable connected.

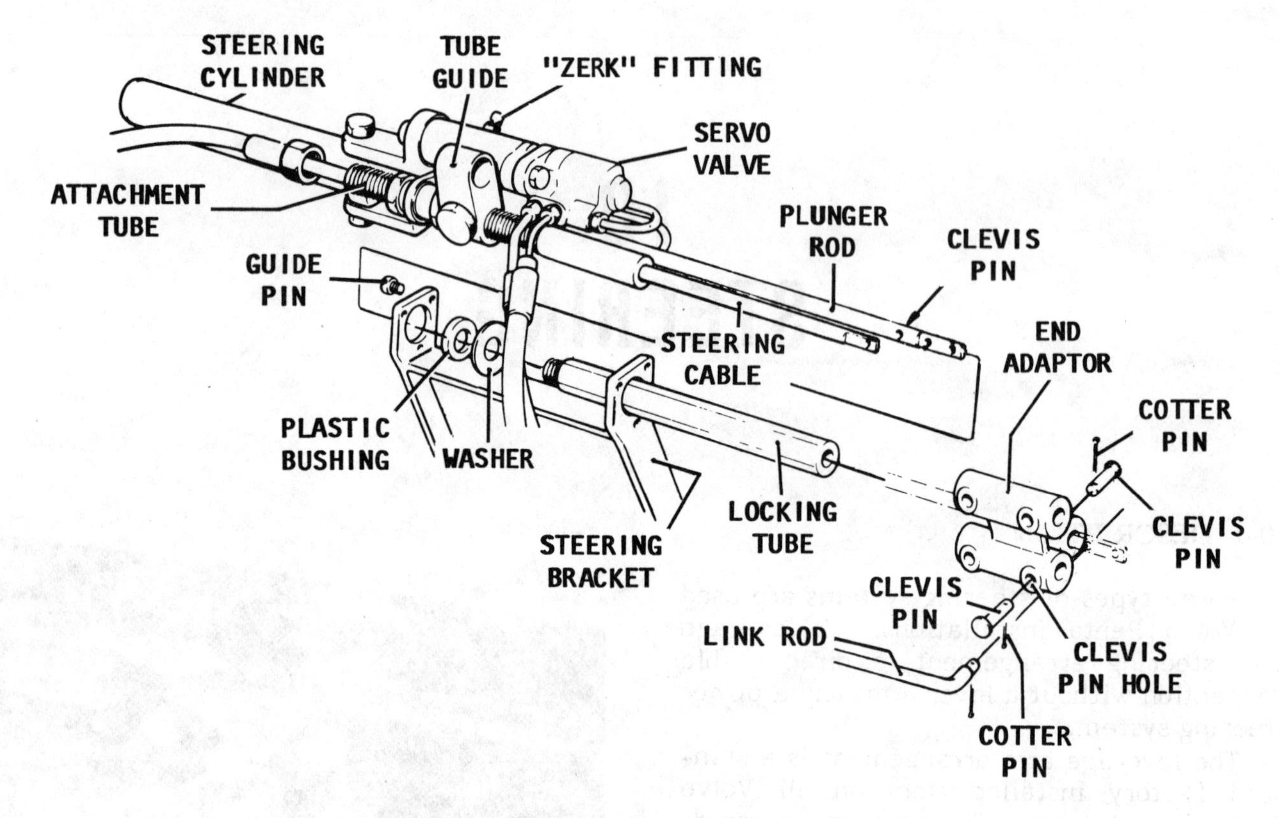

Details of a power steering cylinder installation.

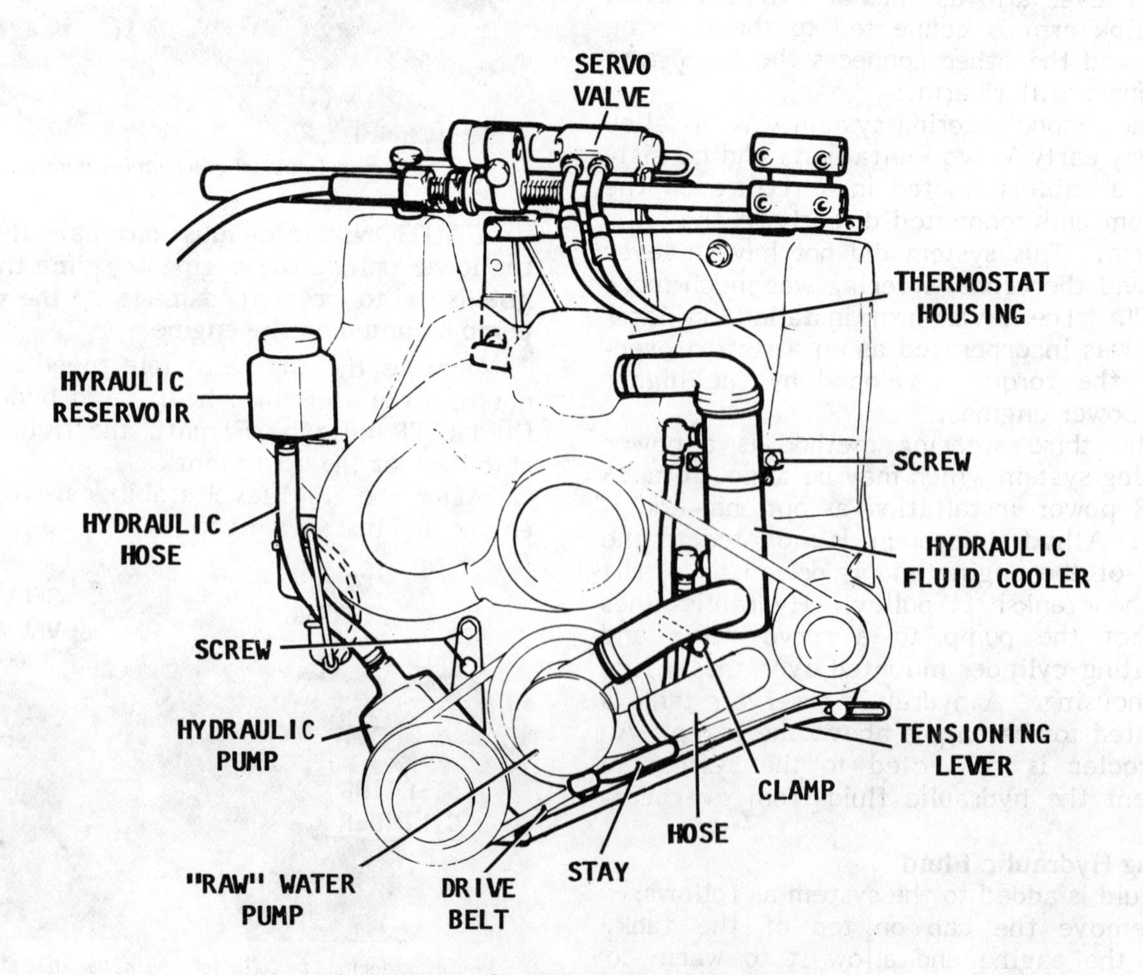

Arrangement of power steering components.

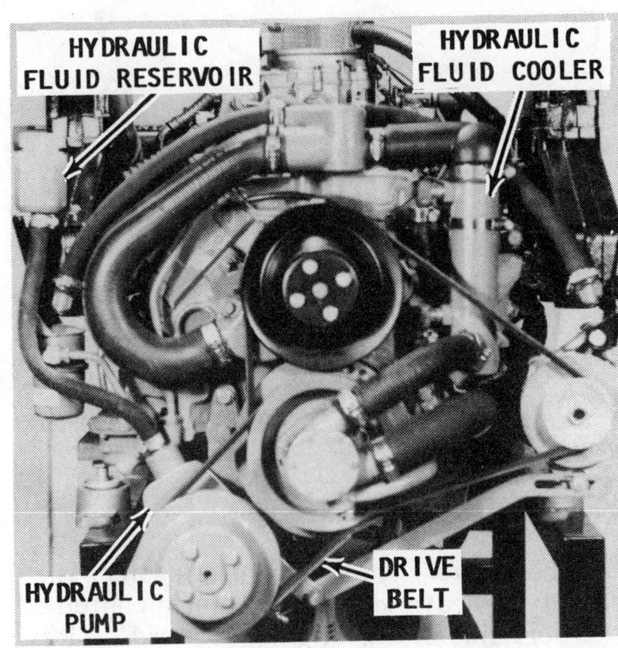

Front view of a V8 engine with the power steering system attached, ready for installation into the boat.

SYSTEM MAINTENANCE

The system requires very little attention and seldom does it require major maintenance. An exposed "Zerk" fitting on the servo valve provides the necessary lubrication for the valve. This fitting should be serviced a couple times each season.

The pump cannot be repaired. Therefore, if a malfunction should develop and the determination is made finding the pump at fault, it must be replaced with a new unit.

The hydraulic fluid cooler is located in an exposed position and can be removed and "rodded" out with a stiff piece of welding rod or similar tool.

The servo valve can be adjusted to correct any adverse steering "drag" to either port or starboard.

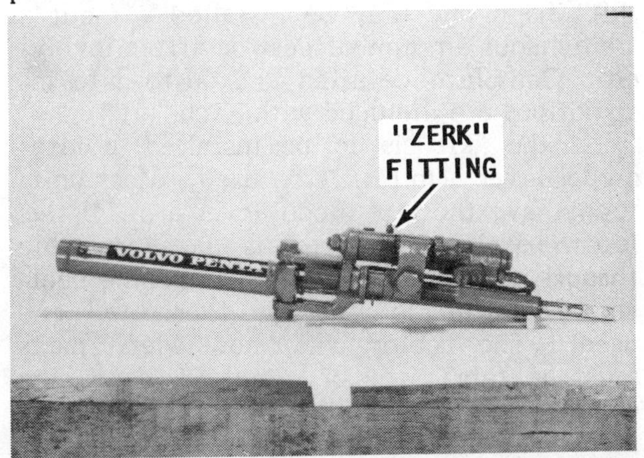

Power steering cylinder and servo valve with the lubrication "Zerk" fitting exposed.

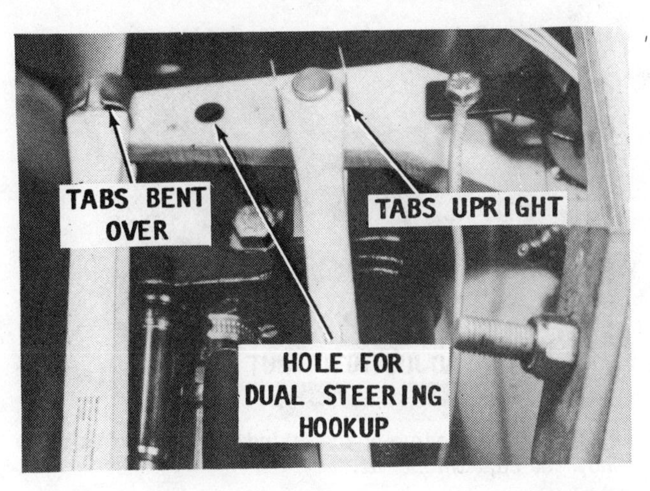

New steering attachment arms for the Model 290 stern drive. There are no measurements or adjustments to make. The power steering unit is bolted in place, the cable screwed in and the cable pins installed. The tabs on the arms are bent over as a safety feature. Only one power steering unit is used with a dual engine installation. A tie rod is installed to the steering arm of the second engine.

10-2 SERVO VALVE ADJUSTMENT

WARNING

Exercise care when making an adjustment to the servo valve because the cable end adaptor will move while the adjustment is being made.

First, check the hydraulic fluid level in the reservoir and fill as described in the previous section. If the fluid level was low, the problem may be corrected. If not, disconnect the steering control cable and the link rod from the end adaptor of the cylinder.

Next, remove the protection cap from the end of the servo valve.

Now, start the engine and allow it to warm to operating temperature. With the engine running at a normal idle rpm, **SLOW-**

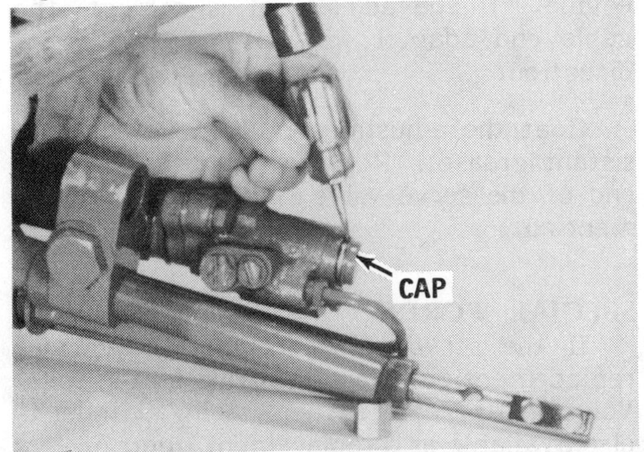

Removing the end cap from the servo valve in preparation to making an adjustment.

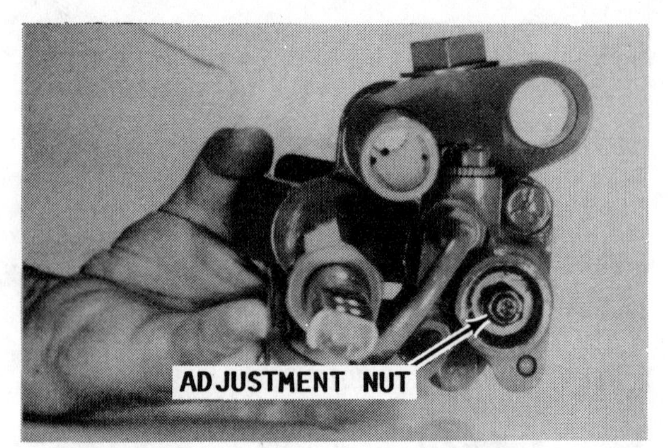

Closeup end view of the cylinder and servo valve to show the adjustment nut.

End view of the hydraulic fluid cooler. The tubes may be "rodded" out with a stiff piece of welding rod.

LY rotate the adjusting nut **CLOCKWISE** until the steering cylinder begins to move by itself. As soon as it begins to move, **SLOWLY** rotate the nut **COUNTERCLOCKWISE** counting the number of revolutions, until the cylinder just begins to move in the opposite direction.

At this point, rotate the nut again **CLOCKWISE** exactly 1/2 the number of turns counted as the nut was being turned counterclockwise.

This position neutralizes the servo valve in the center.

CRITICAL WORDS

Do not turn the adjusting nut any more than necessary. Excessive rotation of the nut can cause the nut to lose its locking effect.

After the adjusting nut has been properly positioned, it should be possible to push the plunger rod in and out with hand pressure.

Shut the engine down and connect the steering cable and link rod. Again, start the engine. If the adjustment is correct, the cable end adaptor will not move in either direction.

Coat the adjustment nut with water resistant grease. Replace the cap over the end of the servo valve covering the adjustment nut.

SPECIAL WORDS

If the servo valve should be damaged, replacement parts are available at the local Volvo Penta dealer. The steering cylinder is also available as a replacement item.

10-3 POWER STEERING COOLER

The power steering fluid cooler is a tube-type heat exchanger cooler mounted on the port side of the engine, forward. Raw water from outside the boat is drawn in by the "raw" water pump and circulated through the cooler. Hydraulic fluid heat is drawn off by the "raw" water passing through the heat exchanger and into the engine cooling system. Therefore, if the cooler should become plugged, the engine could overheat.

The cooler may be easily removed and the tubes rodded out with a heavy piece of welding rod or similar tool. Take care not to use a sharp pointed piece of material which could cause damage to the tubes.

To remove the cooler, first disconnect the inlet and outlet "raw" water hoses. Next, disconnect the inlet and outlet hydraulic lines. Remove the attaching hardware and the cooler will come free.

10-4 POWER STEERING KIT

The power steering system is available in "kit" form and may be installed without a tremendous amount of cost or effort involved. Complete detailed and up-to-date instructions are included with each "kit".

If the "kit" is to be installed, a late-model lever arm **MUST** be used. Most units today have the late model lever arm. If this late model lever arm is not available through the local Volvo Penta marine dealer, special parts must be fabricated and used. The dealer will know which parts must be fabricated or modified, depending on the type and model of manual steering installed on the boat being serviced.

11 MAINTENANCE

11-1 IN-SEASON MAINTENANCE

In order to obtain maximum performance and enjoyment from your boat, power unit, and stern drive, the following maintenance tasks should be performed as indicated.

Daily Before Starting — First Time

Check the engine oil level.

Check the "raw" water strainer, if one is used.

Every Two Weeks During Operation

Check the lubrication level in the stern drive.

High performance units, such as the "Spirit of Western Engine", have crews with but one thought in mind, "maintain every part to the max. and keep it going".

Check the "zincs" -- anti-corrosion plate and ring. If more than 50% is eaten away, replace the plate or ring.

Check and clean the fuel filter.

Check the electrolyte level in the battery.

Check the hyraulic fluid level in the power steering reservoir, if installed.

Check the drive belt tension for each belt. Tighten as required.

Every 50 Hours Operation

Change the engine oil in the crankcase.

Lubricate the driveshaft bearing grease cap on the bell housing. Lubricate steering

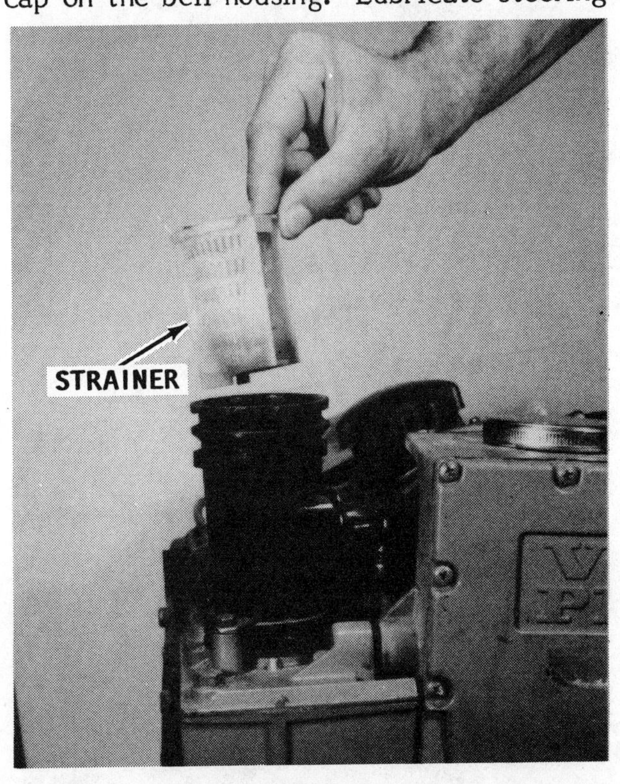

*Checking the "raw" water strainer on a regular basis should be a **MUST** to prevent the engine from overheating. The picture on Page 7-1 is very convincing.*

The battery should be kept clean and secured in a well-ventilated area close to the engine with the electrolyte level kept to the full mark, and the battery fully charged, even when in storage.

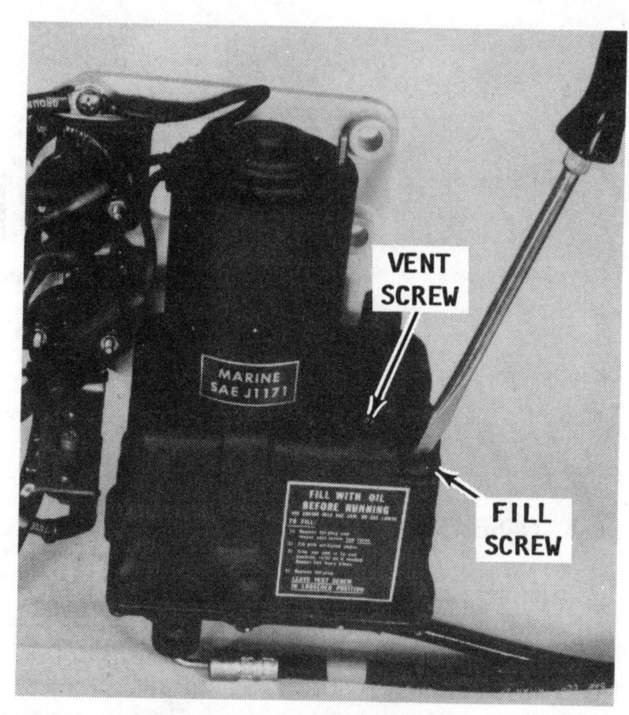

If a power trim unit is installed on the boat, the hydraulic pump fluid level should be checked every 100 hours.

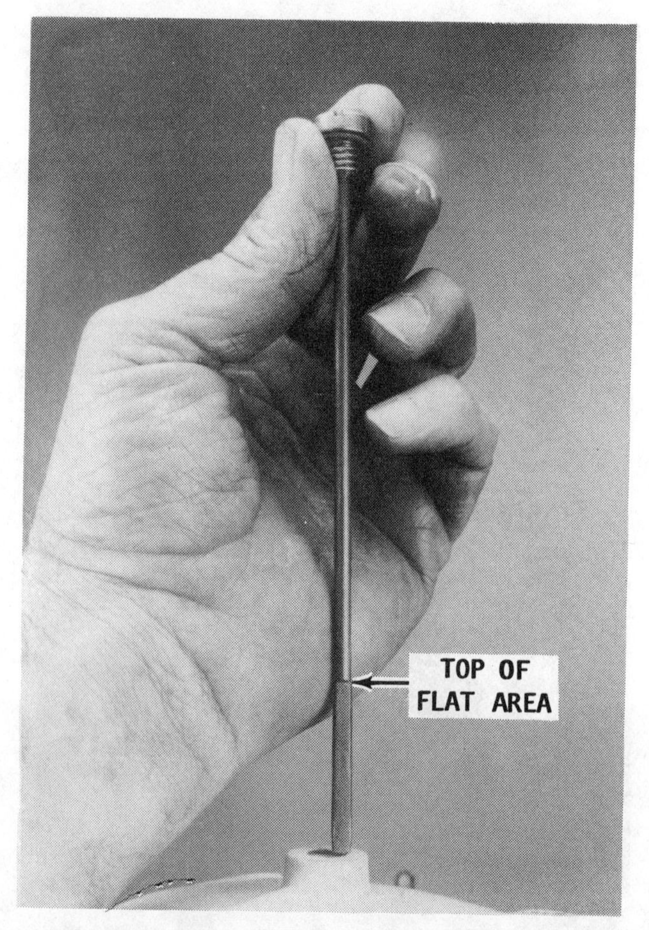

The lubrication level in the stern drive should be within the flat area but ***NEVER*** *above the flat area. During a level check, the stern drive must be down.*

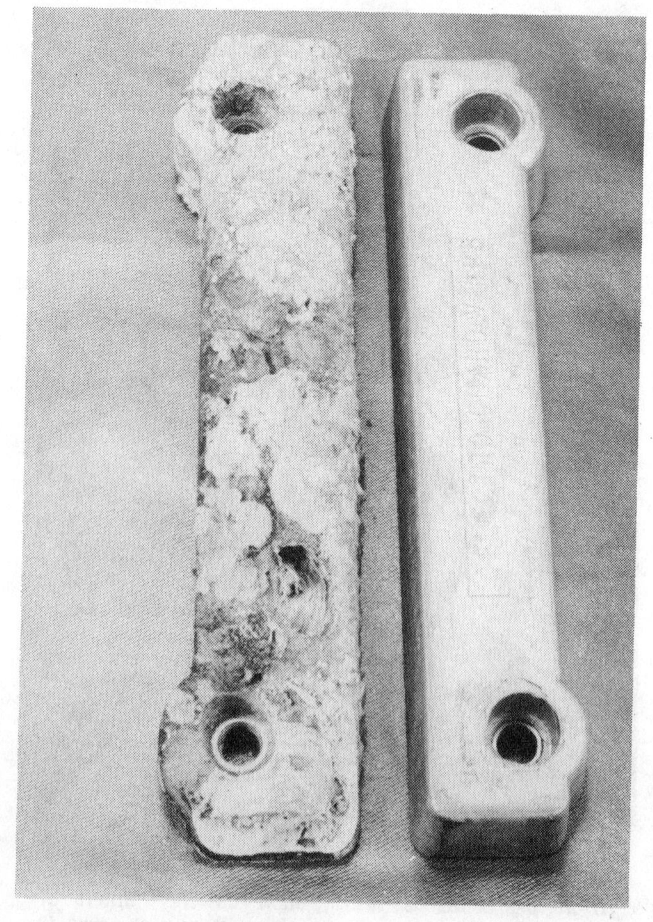

A used zinc alongside a new one for comparison. This zinc has corroded to the point where it should be replaced.

The fluid level in the power steering reservoir should be checked every two weeks.

Applying "finger pressure" to the alternator drive belt as a regular check during the season.

Grease cap and cup installed on the bell housing. This lubrication fitting should be checked and filled each 50 hours of operation.

bearings at the "Zerk" fittings, inside and outside the transom. Early models have one fitting inside and later models have two inside. Both have one outside.

Check the spark plugs and replace the set if any one shows signs of excessive wear.

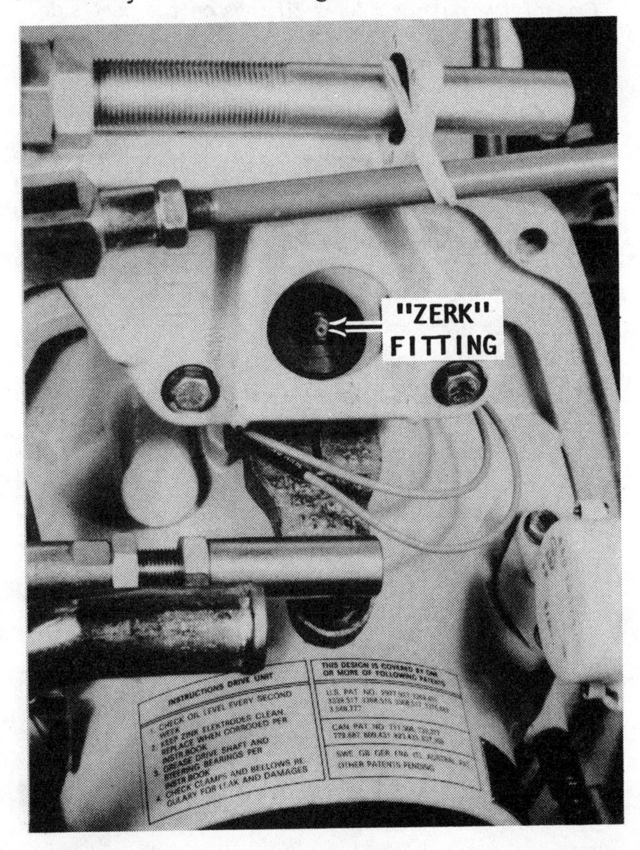

"Zerk" fitting in the transom shield inside the boat.

One of the outside "Zerk" fittings on the port side.

A compression check of each cylinder can be considered a maintenance task. A drastic variation between cylinders indicates serious work is required.

Draining the lubricant from the stern drive should be accomplished after every 200 hours of operation.

Every 100 Hours Operation Or Once Each Season

Change the oil filter.

Change the lubrication in the stern drive (every 200 hours operation).

Check the drive belts for wear and tension.

Check the ignition system, see Chapter 5.

Check and adjust the carburetors, see Chapter 4.

Check the cooling system, "raw" water pump impeller, thermostat, strainer, and the water intake. Check the heat exchanger, and the power steering hydraulic fluid cooler, see Chapter 7.

Check the electrical system, see Chapter 6.

Check the adjustment of the retaining pawl rod in relation to the retaining pawl, see next section.

Check the fluid level in the hydraulic reservoir for the power trim, if installed.

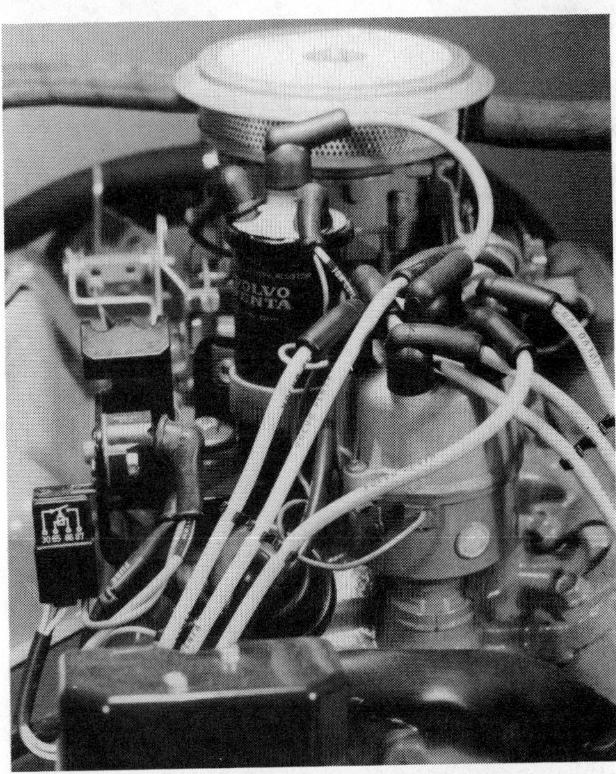

A thorough check of the ignition system as a pre-season task will reward the owner with proper performance the first time the boat is put into operation.

The impeller in the "raw" water pump can be easily replaced. The heat exchanger can be removed without a great deal of effort. These tasks will ensure adequate cooling of water and oil circulating through the engine.

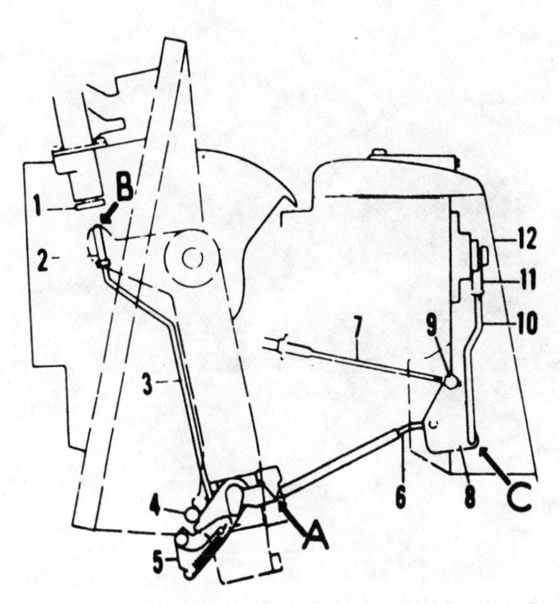

Points identified by number are referenced in the text for making a retaining pawl adjustment.

RETAINING PAWL ADJUSTMENT

The reference numbers in the following procedures are identified in the accompanying illustration on this page.

1- Remove the protective cover (12). Shift the control lever to the **NEUTRAL** position.

2- Disconnect the shift cable swivel (9) and the fork (11).

3- Release the lock nut for the fork (11). Check to be sure when the fork is connected to the lever, the retaining pawl rod (6) reaches the retaining pawl bracket at **"A"** without pressing against it. Make an adjustment if necessary, and then secure the fork in position with the lock nut.

4- Adjust the swivel (9) to allow it to easily enter the hole in the shift yoke. Shift the control lever to the **FORWARD** position. Check to be sure the corner **"C"** does not make contact with the housing. Install the protective cover.

5- Push the drive forward against the adjusting pin. Check the position of the rod (3). The upper part (2) should extend just a whisker (1/16", 1.6 mm) through the fork at **"B"**. This position will permit the lift (1) to disengage the retaining pawl (5) when the drive is tilted up. Loosenen the lock nuts and adjust the upper part of the rod (2) as required.

11-2- OFF-SEASON STORAGE

1- Start the engine and allow it to warm to normal operating temperature.

The oil cooler can be easily removed and the tubes rodded out. This task should be accomplished after each 100 hours of operation.

CAUTION: Water must circulate through the lower unit to the engine any time the engine is run to prevent damage to the water pump mounted on the engine.

If the engine is equipped with a closed-circuit cooling system, check the anti-freeze. Add rust inhibitor if the anti-freeze has been used for more than one season. If the solution is contaminated, flush and replace with a new mixture of 50% ethylene glycol and 50% water. Stop the engine, drain the oil in the crankcase, and remove the oil filter. Install a **NEW** filter element, and then fill the crankcase with the prescribed weight and amount of oil. Start the engine and allow it to run at a high idle for a few minutes.

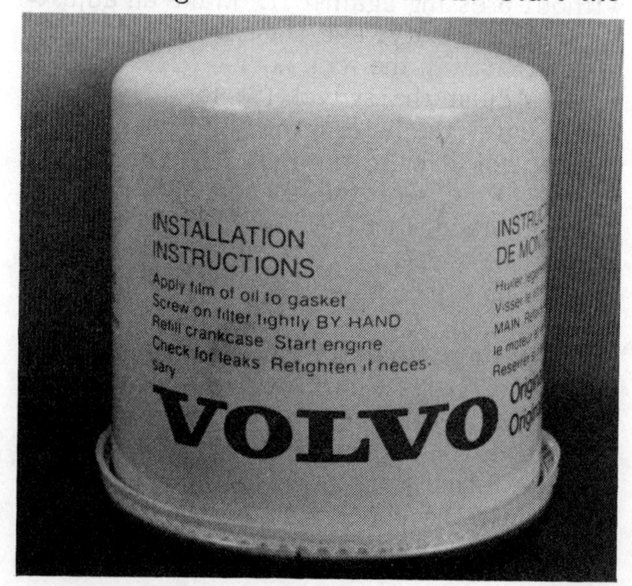

Use authorized Volvo Penta parts and lubricants whenever possible. This habit will help prolong the life of your unit and add to your enjoyment with very little added cost.

CAUTION: Water must circulate through the lower unit to the engine any time the engine is run to prevent damage to the water pump mounted on the engine.

Check the oil level as indicated on the dip stick. Some amount of oil may remain in the engine without draining down into the crankcase. A slightly lower reading may, therefore, be indicated on the dip stick. Add only enough oil to bring the reading into the safe running range above the **ADD** mark to allow for the amount of oil held in various areas of the engine.

2- Shut off the gasoline supply at the fuel tank. Disconnect the fuel line between the valve and the fuel pump. Drain the fuel from the line. Insert the end of the line disconnected from the fuel pump into a can containing several ounces of fuel mixed with a rust inhibitor. Start the engine and run it at a fast idle until it stalls from lack of fuel.

CAUTION: Water must circulate through the lower unit to the engine any time the engine is run to prevent damage to the water pump mounted on the engine.

3- Remove the flame arrestor and **SLOWLY** pour about a pint of rust-preven-

Removing the flame arrestor and air filter in preparation to pouring about a pint of rust-preventative oil into the carburetor, while the engine is operating at a fast idle.

tive oil into the carburetor air intakes while running the engine at a fast idle. This can be done at the same time as Step 2, while using up the fuel in the fuel lines, fuel pump, and in the carburetor. Clean the fuel filter and sediment bowl. Install the bowl with a **NEW** gasket and reconnect the fuel line. Clean the flame arrestor in solvent, and then blow it dry with compressed air. Lubricate the alternator, starter, distributor, and control linkage.

4- Remove all spark plugs and squirt about a teaspoonful of rust-preventive oil into each cylinder. Crank the engine through several revolutions to allow the oil to coat the cylinder walls. Remove any excess oil from around the spark plug holes, and then install the spark plugs.

Overhead Valve Engines

5- Remove the rocker arm covers and inspect the valve train mechanism for worn or damaged parts. Clean the inside of the covers. Apply a liberal coating of crankcase oil to the valve mechanism and onto the inside of the covers. Install the covers with **NEW** gaskets. Using new gaskets will ensure a good seal with the valve covers. Clean the outside of the engine, and then wipe it down with an oily rag. Cover the engine with a protective cover, but **ALLOW** for air circulation.

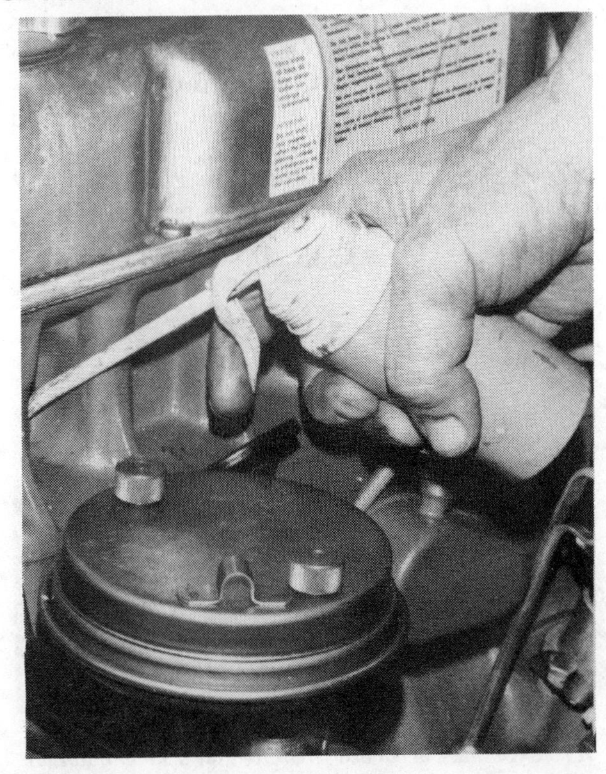

Inserting a little rust-preventative oil into each cylinder, as part of the off-season storage procedure.

6- Flush the cooling system with fresh water. Allow the water to circulate for at least 5 minutes. If the engine is equipped with a raw-water cooling system, open all engine and manifold water jacket drains. Allow the water to drain completely. **ALWAYS** have the stern drive in a horizontal position when draining the system to ensure all of the water is able to leave the system. If the stern drive is not horizontal, water will be trapped inside. If this water should freeze, serious damage to expensive parts may result.

Leave all the drains open. Disconnect the water hoses at the low end and allow them to drain. Remove the drain plugs from the water pumps and allow them to drain.

Remove the propeller. Clean and lubricate the propeller shaft. Install the propeller.

Wipe the outside of the stern drive unit with an oily rag. Check all the steering connections. Lubricate all joints and pulleys.

7- Check the oil level in the stern drive. Add lubricant, as required. Lubricate all the fittings on the steering and stern drive.

8- Seal off all openings to the carburetor and exhaust system to prevent dust, water, and insects from entering the engine.

9- Remove the batteries from the boat

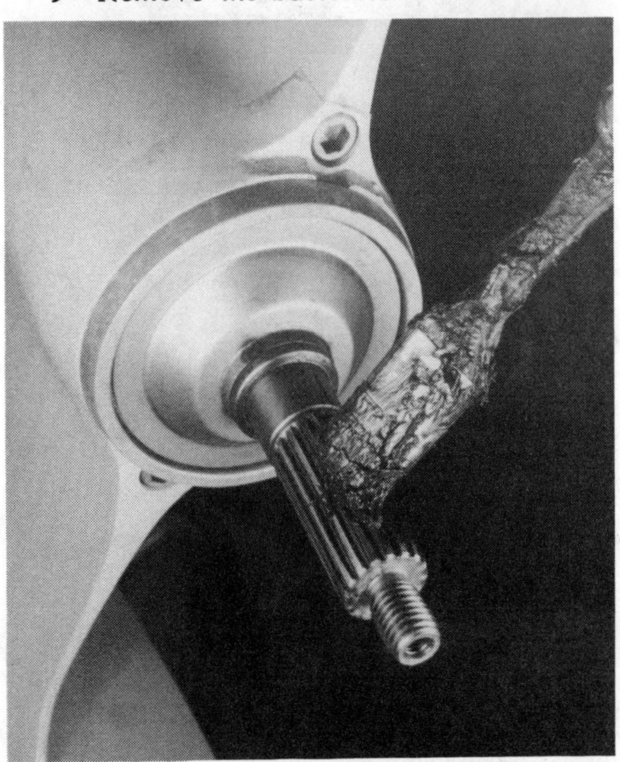

The propeller should be removed and the shaft coated with anti-sieze compound before installing the propeller prior to off-season storage.

The flame arrestor should be kept clean for maximum engine performance. It should be clean and in place during winter storage.

and keep them charged during the storage period. Clean the batteries thoroughly of any dirt or corrosion, and then charge them to full specific gravity reading. After they are fully charged, store them in a clean cool dry place where they will not be damaged or knocked over.

NEVER store the battery with anything on top of it or cover the battery in such a manner as to prevent air from circulating around the filler caps. All batteries, both new and old, will discharge during periods of storage, more so if they are hot than if they remain cool. Therefore, the electrolyte level and the specific gravity should be checked at regular intervals. A drop in the specific gravity reading is cause to charge them back to a full reading.

In cold climates, care should be exercised in selecting the battery storage area. A fully-charged battery will freeze at about 60°F below zero. A discharged battery, almost dead, will have ice forming at about 19°F above zero.

ALWAYS remove the drain plug and position the boat with the bow higher than the stern. This will allow any rain water and melted snow to drain from the boat and prevent "trailer sinking". This term is used to describe a boat that has filled with rain water and ruined the interior including the engine because the plug was not removed or the bow was not high enough to allow the water to drain properly.

A Volvo Penta spare parts kit includes a raw water pump impeller, set of spark plugs, condenser, V-belt for the alternator, and a fuse. Such a kit, sold at modest cost, could prove most valuable in an emergency situation. A zinc bar and ring (lower right), and a spare oil filter (left), might be added to the kit by the prudent boat owner.

11-3- PRE-SEASON PREPARATION

1- Lubricate the engine according to the manufacturer's recommendations. Remove, clean, inspect, adjust, and install the spark plugs with new gaskets if they require gaskets. Make a thorough check of the ignition system. This check should include: the points, coil, condenser, condition of the wiring, and the battery electrolyte level and charge.

2- Take time to check the gasoline tank and all the fuel lines, fittings, couplings,

A four-barrel carburetor installed on a V8 engine. If adequate off-season maintenance was performed, the unit will be ready for operation after several months in storage.

valves, flexible tank fill, vent, and fuel lines. Turn on the gasoline supply valve at the tank. If the gas was not drained at the end of the previous season, make a careful inspection for gum formation. When gasoline is allowed to stand for long periods of time, particularly in the presence of copper, gummy deposits form. This gum can clog the filters, lines, and passageways in the carburetor. See Chapter 4, Fuel System Service.

3- All marine engines **MUST** be equipped with an effective means of backfire flame control. This can be accomplished by one of two methods. The first and most popular is through installation of a Coast Guard approved flame arrestor on the carburetor. The second method is by ducting the air intakes outside the engine compartment to the atmosphere.

Clean and inspect the flame arrestors. Check and adjust the alternator belt tension, and replace them if they are worn or frayed. Check the oil level in the crankcase. The oil should have been changed prior to storage after the previous season. If the oil was not changed, do so and install a **NEW** oil filter.

4- Close all the water drains. Check and replace any defective water hoses. Connect and check to be sure the connections do not leak. Replace any spring-type hose clamps with band-type clamps, if they have lost their tension or if they have distorted the water hose. Check the sea cocks of the cooling system. Check to be sure the through-hull fittings are in good condition. If the engine is equipped with a closed-circuit cooling system, check the level of water. See Chapter 7, Cooling System.

5- The engine can be run with the lower unit of the stern drive submerged in water to flush it. If this is not practical, a Flush-it attachment may be used with a garden hose. Attach a garden hose, turn on the water, allow the water to flow into the engine for awhile, and then run the engine.

CAUTION: Water must circulate through the lower unit to the engine any time the engine is run to prevent damage to the water pump mounted on the engine.

Check the exhaust outlet for water discharge. Check for leaks. Check operation of the thermostat. After the engine has reached operating temperature, tighten the cylinder head bolts to the torque value given in the Specifications in the Appendix.

6- Check the electrolyte level in the batteries and the voltage for a full charge. Clean and inspect the battery terminals and cable connections. **TAKE TIME** to check the polarity, if a new battery is being instal-

Operating the engine with the boat in a test tank during the pre-season preparation work.

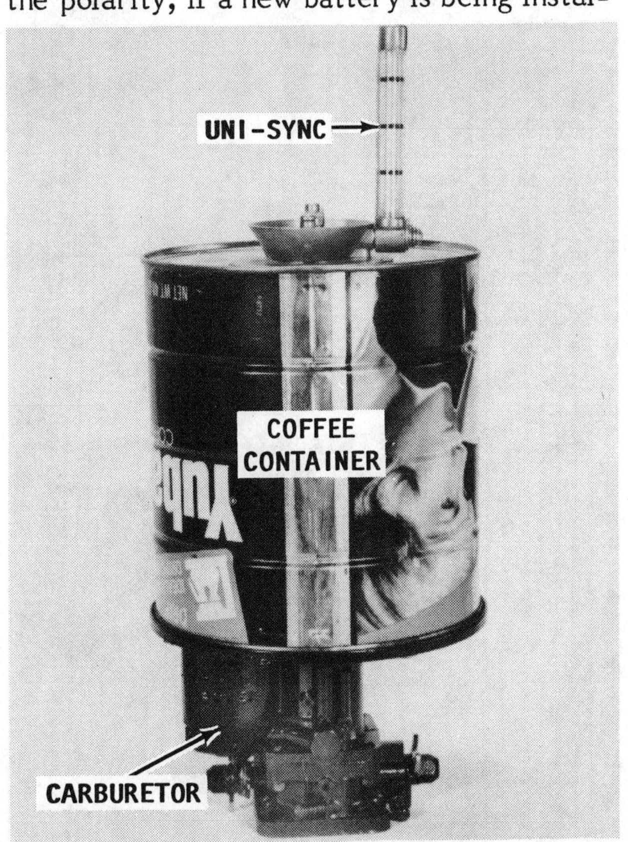

A coffee container and Uni-Sync vacuum measuring device used when synchronizing multiple carburetors.

led. Cover the cable connections with grease or special protective compound as a prevention against corrosion forming. Check all electrical wiring and grounding circuits.

7- Check the engine compartment for proper ventilation. **THERE MUST** be adequate means for removing combustible vapors from the boat. Coast Guard standards include specially designed hardware to do the best job of preventing any vapors from accumulating in the engine compartment or in the bilge.

8- Check the tension of the engine drive belt to ensure proper operation of the engine water pump and alternator. If the belt can be depressed more than 1/4" midway between the water pump and alternator, loosen the alternator and make the proper adjustment to the belt.

9- Check all electrical parts in the engine compartment and lower portions of the hull to be sure they are not of a type that could cause ignition of an explosive atmosphere. Rubber caps help keep spark insulators clean and reduce the possibility of arcing. Starters, distributors, alternators,

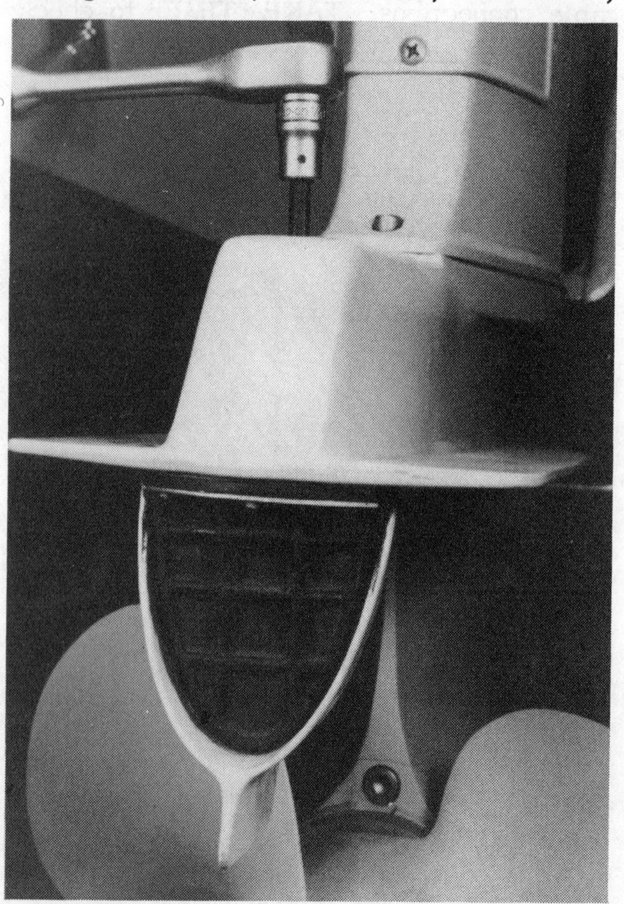

Making an adjustment to the trim tab, as described in the text.

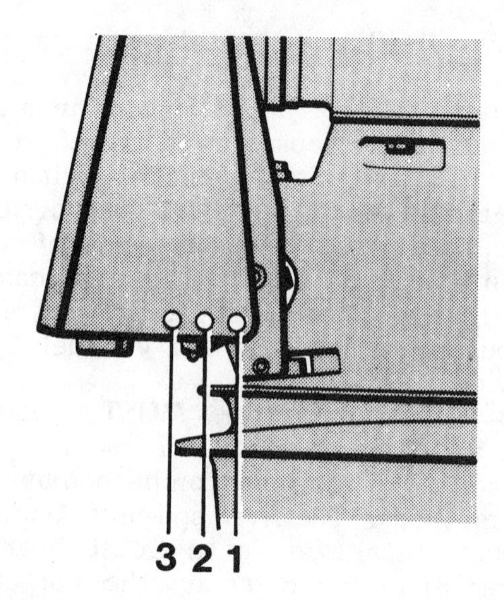

The three pin positions for adjusting boat trim, as described in the text.

electric fuel pumps, voltage regulators, and high-tension wiring harnesses should be of a marine type that cannot cause an explosive mixture to ignite.

11-4 ADJUSTING BOAT TRIM

An adjustment pin in the transom shield may be shifted to obtain proper boat trim, boat up or bow down for proper planing.

If the boat has a tendancy to over plane, bow seems to be down, shift the pin to the No. 1 position in the accompanying illustration.

If the stern seems to be down and fails to rise for proper planing, shift the pin to the No. 3 position.

The No. 2 position is the normal operating position.

11-5 ADJUSTING TRIM TAB

The boat should hold a fairly straight course when the wheel is released and the boat is planing properly as described in the previous section.

If the boat seems to veer to port or starboard when the wheel is released, adjust the trim tab by shifting the aft edge of the tab slightly toward the side the boat is veering towards when underway.

11-6 FIBERGLASS HULLS

Fiberglass-reinforced plastic hulls are tough, durable, and highly resistant to impact. However, like any other material they can be damaged. One of the advantages of this type of construction is the relative ease with which it may be repaired. Because of its break characteristics, and the simple techniques used in restoration, these hulls have gained popularity throughout the world. From the most congested urban marina, to isolated lakes in wilderness areas, to the severe cold of far off northern seas, and in sunny tropic remote rivers of primative islands or continents, fiberglass boats can be found performing their daily task with a minimum of maintenance.

A fiberglass hull has almost no internal stresses. Therefore, when the hull is broken or stove-in, it retains its true form. It will not dent to take an out-of-shape set. When the hull sustains a severe blow, the impact will be either absorbed by deflection of the laminated panel or the blow will result in a definite, localized break. In addition to hull damage, bulkheads, stringers, and other stiffening structures attached to the hull may also be affected and therefore, should be checked. Repairs are usually confined to the general area of the rupture.

11-7 BELOW WATERLINE SERVICE

A foul bottom can seriously affect boat performance. This is one reason why racers, large and small, both powerboat and sail, are constantly giving attention to the condition of the hull below the waterline.

In areas where marine growth is prevalent, a coating of vinyl, anti-fouling bottom paint should be applied. If growth has developed on the bottom, it can be removed with a solution of muriatic acid applied with a brush or swab and then rinsed with clear water. **ALWAYS** use rubber gloves when working with muriatic acid and **TAKE EXTRA CARE** to keep it away from your face and hands. The **FUMES ARE TOXIC.** Therefore, work in a well-ventilated area, or if outside, keep your face on the windward side of the work.

A badly corroded stern drive and boat bottom. Lack of attention to proper maintenance can cause expensive damage.

Barnacles have a nasty habit of making their home on the bottom of boats which have not been treated with anti-fouling paint. Actually they will not harm the fiberglass hull, but can develop into a major nuisance.

If barnacles or other crustaceans have attached themselves to the hull, extra work will be required to bring the bottom back to a satisfactory condition. First, if practical, put the boat into a body of fresh water and allow it to remain for a few days. A large percentage of the growth can be removed in this manner. If this remedy is not possible, wash the bottom thoroughly with a high-pressure fresh water source and use a scraper. Small particles of hard shell may still hold fast. These can be removed with sandpaper.

ONE FINAL WORD

Before putting the boat in the water, **TAKE TIME** to check to be sure the drain plugs are installed. Countless number of boating excursions have had a very sad beginning because the boat was eased into the water only to have water begin filling the inside.

Joan & Clarence

Clean boat, engine and stern drive ready for a day of fun on the water or ready to be laid-up for off-season storage. Compare this boat with the one shown in the adjoining column covered with marine growth and corrosion.

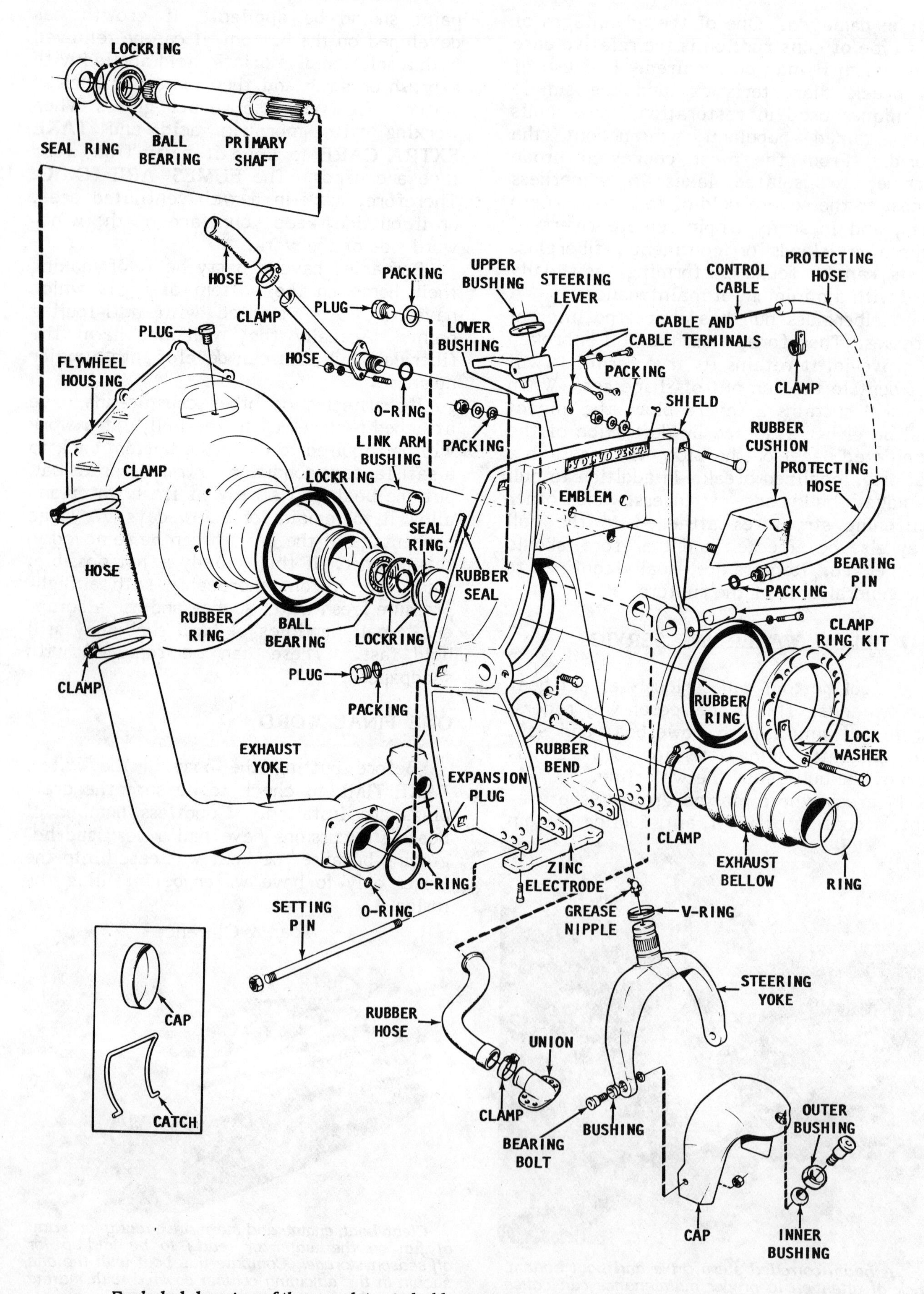

Exploded drawing of the complete gimbal housing arrangement, with major parts identified.

APPENDIX

METRIC CONVERSION CHART

LINEAR

inches	X 25.4	= millimetres (mm)
feet	X 0.3048	= metres (m)
yards	X 0.9144	= metres (m)
miles	X 1.6093	= kilometres (km)
inches	X 2.54	= centimetres (cm)

AREA

$inches^2$	X 645.16	= $millimetres^2$ (mm^2)
$inches^2$	X 6.452	= $centimetres^2$ (cm^2)
$feet^2$	X 0.0929	= $metres^2$ (m^2)
$yards^2$	X 0.8361	= $metres^2$ (m^2)
acres	X 0.4047	= hectares (10^4 m^2) (ha)
$miles^2$	X 2.590	= $kilometres^2$ (km^2)

VOLUME

$inches^3$	X 16387	= $millimetres^3$ (mm^3)
$inches^3$	X 16.387	= $centimetres^3$ (cm^3)
$inches^3$	X 0.01639	= litres (l)
quarts	X 0.94635	= litres (l)
gallons	X 3.7854	= litres (l)
$feet^3$	X 28.317	= litres (l)
$feet^3$	X 0.02832	= $metres^3$ (m^3)
fluid oz	X 23.57	= millilitres (ml)
$yards^3$	X 0.7646	= $metres^3$ (m^3)

MASS

ounces (av)	X 28.35	= grams (g)
pounds (av)	X 0.4536	= kilograms (kg)
tons (2000 lb)	X 907.18	= kilograms (kg)
tons (2000 lb)	X 0.90718	= metric tons (t)

FORCE

ounces - f (av)	X 0.278	= newtons (N)
pounds - f (av)	X 4.448	= newtons (N)
kilograms - f	X 9.807	= newtons (N)

ACCELERATION

$feet/sec^2$	X 0.3048	= $metres/sec^2$ (m/S^2)
$inches/sec^2$	X 0.0254	= $metres/sec^2$ (m/s^2)

ENERGY OR WORK (watt-second - joule - newton-metre)

foot-pounds	X 1.3558	= joules (j)
calories	X 4.187	= joules (j)
Btu	X 1055	= joules (j)
watt-hours	X 3500	= joules (j)
kilowatt - hrs	X 3.600	= megajoules (MJ)

FUEL ECONOMY AND FUEL CONSUMPTION

miles/gal	X 0.42514	= kilometres/litre (km/l)

Note:
235.2/(mi/gal) = litres/100km
235.2/(litres/100 km) = mi/gal

LIGHT

footcandles	X 10.76	= lumens/$metre^2$ (lm/m^2)

PRESSURE OR STRESS (newton/sq metre - pascal)

inches HG (60°F)	X 3.377	= kilopascals (kPa)
pounds/sq in	X 6.895	= kilopascals (kPa)
inches H_2O (60°F)	X 0.2488	= kilopascals (kPa)
bars	X 100	= kilopascals (kPa)
pounds/sq ft	X 47.88	= pascals (Pa)

POWER

horsepower	X 0.746	= kilowatts (kW)
ft-lbf/min	X 0.0226	= watts (W)

TORQUE

pound-inches	X 0.11299	= newton-metres (N·m)
pound-feet	X 1.3558	= newton-metres (N·m)

VELOCITY

miles/hour	X 1.6093	= kilometres/hour (km/h)
feet/sec	X 0.3048	= metres/sec (m/s)
kilometres/hr	X 0.27778	= metres/sec (m/s)
miles/hour	X 0.4470	= metres/sec (m/s)

TEMPERATURE

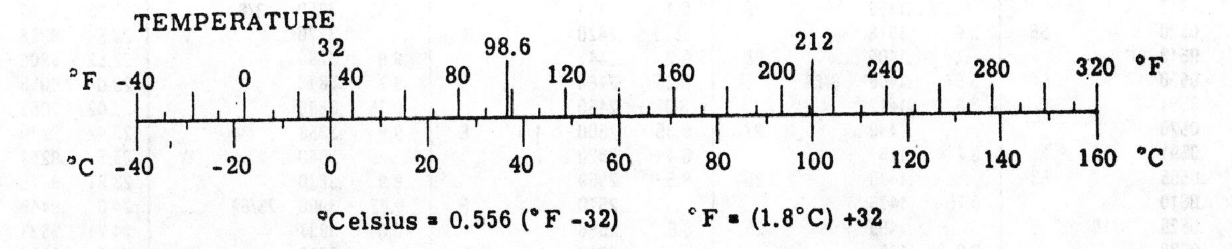

°Celsius = 0.556 (°F -32) °F = (1.8°C) +32

DRILL SIZE CONVERSION CHART

SHOWING MILLIMETER SIZES, FRACTIONAL AND DECIMAL INCH SIZES AND NUMBER DRILL SIZES

Milli-Meter	Dec. Equiv.	Frac-tional	Num-ber
.1	.0039		
.15	.0059		
.2	.0079		
.25	.0098		
.3	.0118		
....	.0135		80
.35	.0138		
....	.0415		79
.39	.0156	1/64	....
.4	.0157		
....	.0160		78
.45	.0177		
....	.0180		77
.5	.0197		
....	.0200		76
....	.0210		75
.55	.0217		
....	.0225		74
.6	.0236		
....	.0240		73
....	.0250		72
.65	.0256		
....	.0260		71
....	.0280		70
.7	.0276		
....	.0292		69
.75	.0295		
....	.0310		68
.79	.0312	1/32	....
.8	.0315		
....	.0320		67
....	.0330		66
.85	.0335		
....	.0350		65
.9	.0354		
....	.0360		64
....	.0370		63
.95	.0374		
....	.0380		62
....	.0390		61
1.0	.0394		
....	.0400		60
....	.0410		59
1.05	.0413		
....	.0420		58
....	.0430		57
1.1	.0433		
1.15	.0452		
....	.0465		56
1.19	.0469	3/64	...
1.2	.0472		
1.25	.0492		
1.3	.0512		
....	.0520		55
1.35	.0513		
....	.0550		54
1.4	.0551		
1.45	.0570		
1.5	.0591		
....	.0595		53
1.55	.0610		
1.59	.0625	1/16	....
1.6	.0629		
....	.0635		52
1.65	.0649		
1.7	.0669		
....	.0670		51

Milli-Meter	Dec. Equiv.	Frac-tional	Num-ber
1.75	.0689		
....	.0700		50
1.8	.0709		
1.85	.0728		
....	.0730		49
1.9	.0748		
....	.0760		48
1.95	.0767		
1.98	.0781	5/64	....
....	.0785		47
2.0	.0787		
2.05	.0807		
....	.0810		46
....	.0820		45
2.1	.0827		
2.15	.0846		
....	.0860		44
2.2	.0866		
2.25	.0855		
....	.0890		43
2.3	.0905		
2.35	.0925		
....	.0935		42
2.38	.0937	3/32	....
2.4	.0945		
....	.0960		41
2.45	.0964		
....	.0980		40
2.5	.0984		
....	.0995		39
....	.1015		38
2.6	.1024		
....	.1040		37
2.7	.1063		
....	.1065		36
2.75	.1082		
2.78	.1094	7/64	....
....	.1100		35
2.8	.1102		
....	.1110		34
....	.1130		33
2.9	.1141		
....	.1160		32
3.0	.1181		
....	.1200		31
3.1	.1220		
3.18	.1250	1/8	....
3.2	.1260		
3.25	.1279		
....	.1285		30
3.3	.1299		
3.4	.1338		
....	.1360		29
3.5	.1378		
....	.1405		28
3.57	.1406	9/64	....
3.6	.1417		
....	.1440		27
3.7	.1457		
....	.1470		26
3.75	.1476		
....	.1495		25
3.8	.1496		
....	.1520		24
6.9	.1535		
....	.1540		23
3.97	.1562	5/32	....

Milli-Meter	Dec. Equiv.	Frac-tional	Num-ber
....	.1570		22
4.0	.1575		
....	.1590		21
....	.1610		20
4.1	.1614		
4.2	.1654		
....	.1660		19
4.25	.1673		
4.3	.1693		
....	.1695		18
4.37	.1719	11/64	....
....	.1730		17
4.4	.1732		
....	.1770		16
4.5	.1771		
....	.1800		15
4.6	.1811		
....	.1820		14
4.7	.1850		13
4.75	.1870		
4.76	.1875	3/16	....
4.8	.1890		12
....	.1910		11
4.9	.1929		
....	.1935		10
....	.1960		9
5.0	.1968		
....	.1990		8
5.1	.2008		
....	.2010		7
5.16	.2031	13/64	....
....	.2040		6
5.2	.2047		
....	.2055		5
5.25	.2067		
5.3	.2086		
....	.2090		4
5.4	.2126		
....	.2130		3
5.5	.2165		
5.56	.2187	7/32	....
5.6	.2205		
....	.2210		2
5.7	.2244		
5.75	.2263		
....	.2280		1
5.8	.2283		
5.9	.2323		
....	.2340		A
5.95	.2344	15/64	....
6.0	.2362		
....	.2380		B
6.1	.2401		
....	.2420		C
6.2	.2441		
6.25	.2460		D
6.3	.2480		
6.35	.2500	1/4	E
6.4	.2520		
6.5	.2559		
....	.2570		F
6.6	.2598		
....	.2610		G
6.7	.2638		
6.75	.2657	16/64	....
6.75	.2657		
....	.2660		H

Milli-Meter	Dec. Equiv.	Frac-tional	Num-ber
6.8	.2677		
6.9	.2716		
....	.2720		I
7.0	.2756		
....	.2770		J
7.1	.2795		
....	.2811		K
7.14	.2812	9/32	....
7.2	.2835		
7.25	.2854		
7.3	.2874		
....	.2900		L
7.4	.2913		
....	.2950		M
7.5	.2953		
7.54	.2968	19/64	....
7.6	.2992		
....	.3020		N
7.7	.3031		
7.75	.3051		
7.8	.3071		
7.9	.3110		
7.94	.3125	5/16	....
8.0	.3150		
....	.3160		O
8.1	.3189		
8.2	.3228		
....	.3230		P
8.25	.3248		
8.3	.3268		
8.33	.3281	21/64	....
8.4	.3307		
....	.3320		Q
8.5	.3346		
8.6	.3386		
....	.3390		R
8.7	.3425		
8.73	.3437	11/32	....
8.75	.3445		
8.8	.3465		
....	.3480		S
8.9	.3504		
9.0	.3543		
....	.3580		T
9.1	.3583		
9.13	.3594	23/64	....
9.2	.3622		
9.25	.3641		
9.3	.3661		
....	.3680		U
9.4	.3701		
9.5	.3740		
9.53	.3750	3/8	....
....	.3770		V
9.6	.3780		
9.7	.3819		
9.75	.3838		
9.8	.3858		
....	.3860		W
9.9	.3839		
9.92	.3906	25/64	....
10.0	.3937		
....	.3970		X
....	.4040		Y
10.32	.4062	13/32	....
....	.4130		Z
10.5	.4134		

Milli-Meter	Dec. Equiv.	Frac-tional
10.72	.4219	27/64
11.0	.4330	
11.11	.4375	7/16
11.5	.4528	
11.51	.4531	29/64
11.91	.4687	15/32
12.0	.4724	
12.30	.4843	31/64
12.5	.4921	
12.7	.5000	1/2
13.0	.5118	
13.10	.5156	33/64
13.49	.5312	17/32
13.5	.5315	
13.89	.5469	35/64
14.0	.5512	
14.29	.5624	9/16
14.5	.5709	
14.68	.5781	37/64
15.0	.5906	
15.08	.5937	19/32
15.48	.6094	39/64
15.5	.6102	
15.88	.6250	5/8
16.0	.6299	
16.27	.6406	41/64
16.5	.6496	
16.67	.6562	21/32
17.0	.6693	
17.06	.6719	43/64
17.46	.6875	11/16
17.5	.6890	
17.86	.7031	45/64
18.0	.7087	
18.26	.7187	23/32
18.5	.7283	
18.65	.7344	47/64
19.0	.7480	
19.05	.7500	3/4
19.45	.7656	49/64
19.5	.7677	
19.84	.7812	25/32
20.0	.7874	
20.24	.7969	51/64
20.5	.8071	
20.64	.8125	13/16
21.0	.8268	
21.04	.8218	53/64
21.43	.8437	27/32
21.5	.8465	
21.83	.8594	55/64
22.0	.8661	
22.23	.8750	7/8
22.5	.8858	
22.62	.8906	57/64
23.0	.9055	
23.02	.9062	29/32
23.42	.9219	59/64
23.5	.9252	
23.81	.9375	15/16
24.0	.9449	
24.21	.9531	61/64
24.5	.9646	
24.61	.9687	31/32
25.0	.9843	
25.03	.9844	63/64
25.4	1.0000	1

REPLACEMENT PISTONS AVAILABLE
AQ125 AND AQ145 ENGINES

Measure the pistons with a micrometer at right angles to the gudgeon pin hole and 6 mm (0.25") from the lower edge.

Piston diameter standard. AQ145
(C-marked): 95.940–95.950 mm (3.7776–3.7780 in.)
(D-marked): 95.950–95.960 mm (3.7787–3.7791 in.)
(E-marked): 95.960–95.970 mm (3.7780–3.7783 in.)
(G-marked): 95.980–95.990 mm (3.7787–3.7791 in.)
Oversize I: (0.3–96.245): 96.237–96.252
(0.0118-3.7892): 3.7889–3.7894"
II: (0.6–96.545): 96.537–96.552
(0.0236–3.8010): 3.8007–3.8013"

Piston diameter standard. AQ125
(D-marked): 91.990–92.000 mm (3.6216–3.6220 in.)
(G-marked): 92.020–92.030 mm (3.6228–3.6232 in.)
Oversize I: (0.5–92.480): 92.474–92.492
(0.0196–3.6409): 3.6407–3.6414 in.)
II: (1.0–92.980): 92.977–92.922
(0.394–3.6606): 3.6605–3.6611 in.)

VALVE ADJUSTING KITS

Kit No. 884516 for AQ125A and AQ145A
Kit No. 884516-4 for AQ131 and AQ151

Both kits have the same shim part numbers, but the quantities differ.

Part No.	Quantity 884516	Quantity 884516-4	Designation	Thickness In./mm
463551	2	6	Shim Material	0.140 (3.55)
463552	3	6	"	0.142 (3.60)
463553	6	6	"	0.144 (3.65)
463554	8	6	"	0.146 (3.70)
463555	12	6	"	0.148 (3.75)
463556	12	12	"	0.150 (3.80)
463557	12	12	"	0.152 (3.85)
463558	12	12	"	0.154 (3.90)
463559	12	12	"	0.156 (3.95)
463560	12	6	"	0.158 (4.00)
463561	8	6	"	0.160 (4.05)
463562	6	6	"	0.162 (4.10)
463563	3	6	"	0.164 (4.15)
463564	2	6	"	0.166 (4.20)

VALVE ADJUSTING KIT FOR AQ131 AND AQ151
KIT NO. 884516-4

TUNE-UP ADJUSTMENTS

DESIGNATION	ENGINE MFG.	VALVE ACTION	H.P.	DISPL. CU. IN.	FIRING ORDER	CARB. TYPE
AQ105A	Volvo	OHV	105	121	1-3-4-2	1V
AQ115A & B	Volvo	OHV	115	121	1-3-4-2	1V
AQ120B	Volvo	OHC	110	130	1-3-4-2	1V
AQ125A	Volvo	OHC	110	130	1-3-4-2	1V
AQ125B	Volvo	OHC	120	141	1-3-4-2	1V
AQ130A, B, C	Volvo	OHV	130	121	1-3-4-2	2x1V
AQ130D	Volvo	OHV	130	121	1-3-4-2	2x1V
AQ131A,B,C,D	Volvo	OHC	120	141	1-3-4-2	2x1V
AQ140A	Volvo	OHC	125	130	1-3-4-2	2x1V
AQ145A	Volvo	OHC	138	140	1-3-4-2	2x1V
AQ145B	Volvo	OHC	139	141	1-3-4-2	2x1V
AQ151A	Volvo	OHC	146	152	1-3-4-2	2x1V
AQ151B	Volvo	OHC	138	140	1-3-4-2	2x1V
AQ171A	Volvo	DOHC	167	152	1-3-4-2	2x1V
AQ165A	Volvo	OHV	165	182	1-5-3-6-2-4	3x1V
AQ170A,B,C	olvo	OHV	170	182	1-5-3-6-2-4	3x1V
AQ175A	GMC	OHV	170	229	1-2-3-4-5-6	3x1V
AQ205	GMC	OHV	205	262	1-6-5-4-3-2	4V
AQ190A	Ford	OHV	182	302	1-5-4-2-6-3-7-8	2V
AQ200A & B	GMC	OHV	200	307	1-8-4-3-6-5-7-2	2V
AQ200C	GMC	OHV	200	305	1-8-4-3-6-5-7-2	2V
AQ200D	GMC	OHV	200	305	1-8-4-3-6-5-7-2	2V
AQ211A	GMC	OHV	209	305	1-8-4-3-6-5-7-2	2V
AQ225A & B	GMC	OHV	225	307	1-8-4-3-6-5-7-2	4V
AQ225C	GMC	OHV	225	305	1-8-4-3-6-5-7-2	4V

TUNE-UP ADJUSTMENTS

IGN. TIME. BASIC	IGN. TIME. STROBE	DWELL	POINT GAP	PLUG GAP	IDLE RPM	MAX. RPM
---	22-24°@2000	62°+3	.016-.020	.028	900-1000	5100
9°BTDC	24-26°@2000	62°+3	.016-.020	.028-.032	900-1000	5100
10°BTDC	36-38°@2800	62°+3	.016	.028-.032	900	44-4800
10°BTDC	36-38°@2800	62°+3	.016	.028-.032	900	44-4800
6°BTDC	32-36°@4200	62°+3	.016	.028	900	47-5000
12°BTDC	27-29°@2000	62°+3	.016-.020	.028-.032	900-1000	5100
8°BTDC	25-27°@3000	62°+3	.016-.020	.028-.032	900	5100
6°BTDC	32-36°@4200	62°+3	.016	.028	900	47-5000
6-10°BTDC	36-38°@2800	62°+3	.016	.028	900	5100
4°BTDC	30°@4000	62°+3	.016	.028	900	45-5100
6°BTDC	32-35°@4200	62°+3	.016	.028	900	4800-5500
6°BTDC	32-36°@4200	62°+3	.016	.028	900	48-5500
6°BTDC	32-36°@4200	62°+3	.016	.028	900	48-5500
10°BTDC	23-25°@4400	Note 1	---	.028	900	50-5700
12°BTDC	18-20°@1500	40°+3	.010-.014	.028-.032	900-1000	45-5000
12°BTDC	25-27°@2000 Note 2	40°+3	.010-.014	.028-.032	900-1000	45-5000
10°BTDC	34°@4200	35°+2	.017-.022	.028	700	4800
8°BTDC	16° @2500	Note 1	---	.035	750	4800
10°BTDC	37°@4200	31°	.014-.019	.032-.036	600-650	4200
4°BTDC	---	30°	.018	.028	600-650	40-4400
8°BTDC	---	31°+3	.018	.028	600-650	40-4400
8°BTDC	30°@4200	31°+3	.014-.019	.028	600-650	40-4400
6°BTDC	30°@3000	31°+3	.014-.019	.028	750	40-4400
4°BTDC	---	30°	.018	.028	600-650	40-4400
8°BTDC	---	31°+3	.018	.028	600-650	40-4400

TUNE-UP ADJUSTMENTS

DESIGNATION	ENGINE MFG	VALVE ACTION	H.P	DISPL. CU. IN.	FIRING ORDER	CARB. TYPE
AQ225D	GMC	OHV	225	305	1-8-4-3-6-5-7-2	4V
AQ225E & F	GMC	OHV	225	305	1-8-4-3-6-5-7-2	4V
AQ231	GMC	OHV	229	305	1-8-4-3-6-5-7-2	4V
AQ240A	FORD	OHV	231	351	1-3-7-2-6-5-4-8	4V
AQ255A	GMC	OHV	255	350	1-8-4-3-6-5-7-2	4V
AQ255B	GMC	OHV	255	350	1-8-4-3-6-5-7-2	4V
AQ260A & D	GMC	OHV	260	350	1-8-4-3-6-5-7-2	4V
AQ271 A & C	GMC	OHV	270	350	1-8-4-3-6-5-7-2	4V
AQ280A	GMC	OHV	280	350	1-8-4-3-6-5-7-2	4V
AQ290A	GMC	OHV	290	350	1-8-4-3-6-5-7-2	4V
AQ311A & B	GMC	OHV	307	350	1-8-4-3-6-5-7-2	4V
740	GMC	OHV	330	454	1-8-4-3-6-5-7-2	4V
430	GMC	OHV	175	262	1-6-5-4-3-2	2V
431	GMC	OHV	205	262	1-6-5-4-3-2	4V

GENERAL NOTES

1- Engine identification given on manufacturer's plate.
2- OHV = Overhead valve -- push rod action.
3- OHC = Single overhead cam engine.
4- DOHC = Double overhead cam engine.
5- BTDC = Before top dead center.
6- Carburetor type, V = number of throats (barrels). 2x1V = 2 one barrel carbs.

TUNE-UP ADJUSTMENTS

IGN.TIME. BASIC	IGN. TIME STROBE	DWELL	POINT GAP	PLUG GAP	IDLE RPM	MAX. RPM
8°BTDC	30°@4200	31°+3	.014-.019	0.028	600-650	40-4400
8°BTDC	30°@3000	31°+3	.014-.019	0.028	750	40-4400
6°BTDC	30°@3000	31°+3	.014-.019	0.028	750	4600
10°BTDC	37°@4200	31°	.014-.019	.032-.036	600-650	4200
8°BTDC	—	31°+3	0.018	0.028	600-650	40-4400
8°BTDC	30°@4200	31°+3	.014-.019	0.028	600-650	40-4400
8°BTDC	30°@4200	31°+3	.014-.019	0.028	600-650	40-4400
Note 3	30°@3000	Note 1	—	0.028	750	40-4600
8°BTDC	30°@4200	31°+3	.014-.019	0.028	600-650	47-5200
8°BTDC	30°@4200	30.5°+1.5	.010-.014	0.028	600-650	47-5200
8°BTDC	30°@3000	Note 1	—	0.028	750	46-5200
8°BTDC	25°@2500	—	—	0.035	750	40-4400
8°BTDC	16°@2500	—	—	0.035	750	40-4400
8°BTDC	16°@2500	—	—	0.035	750	40-4400

SPECIAL NOTES

1- Distributor sensor gap is 0.0078-0.0098". Early models with conventional distributor with points -- dwell angle is 31° +3 and point gap is 0.014-0.019".
2- Ignition timing, strobe for 170C engine is 32-34° @ 3000 rpm.
3- Basic ignition timing for AQ271A -- 8° BTDC -- for AQ271C -- 6° BTDC.

ENGINE SPECIFICATIONS -- 4-CYL. OHC

GENERAL			
Type Designation	AQ120B AQ125A	AQ140A	AQ145A
Type of Operation	4-Stroke OHC	4-Stroke OHC	4-Stroke OHC
Horsepower	115 hp	125 hp	138 hp
Maximum Speed	4800 rpm	5500 rpm	5500 rpm
Maximum Cruising Speed	4500 rpm	5200 rpm	5200 rpm
Compression Ratio	9.3:1	9.3:1	9.7:1
Compression press. @ Str. Mtr. Sp.	146-174 psi	145-174 psi	146-174 psi
No. Cylinder	4 In-line	4 In-line	4 In-Line
Bore	3.622"	3.622"	3.7795"
Stroke	3.150"	3.150"	3.150"
Displacement	129.798 CID	129.798 CID	141.271 CID
Weight Incl. S/D W/O oil & wtr.	529 lbs	540 lbs	529 lbs
Idle Speed	900 rpm	900 rpm	900 rpm
CYLINDER BLOCK			
Material	Cast Iron	Cast Iron	Cast Iron
Bore Standard	3.6224-3.6228"	3.6224-3.6228"	3.7795-3.7897"
Bore Oversize 1	3.6417"	3.6417"	3.7913"
Bore Oversize 2	3.6614"	3.6614"	3.8031"
PISTONS			
Material	Light Alloy	Light Alloy	Light Alloy
Height Total	2.795"	2.795"	3.165"
Hgt: Piston pin ctr. to piston top	1.811"	1.811"	1.8268"
Piston Clearance	.0004-.0012"	.0004-.0012"	.0020-.0028"
Piston Std. Size	3.6217-3.6220"	3.6217-3.6220	3.7772-3.7791"
Piston Oversize 1	3.6406-3.6412"	3.6406-3.6412"	3.7889-3.7894"
Piston Oversize 2	3.6603-3.6605"	3.6603-3.6605	3.8007-3.8013"
PISTON RINGS			
Piston Ring Gap (Oil Rings)	.0098-.0157"	.0098-.0157"	.0118-.0236"
Piston Ring Gap (Compress. Rings)	.0138-.0217"	.0138-.0217"	.0157-.0256"
Piston Ring Oversize 1	.0197"	.0197"	.0118"
Piston Ring Oversize 2	.0394"	.0394"	.0236"

ENGINE SPECIFICATIONS -- 4-CYL. OHC

Type Designation	AQ120B AQ125A	AQ140A	AQ145A
PISTON RINGS (Continued)			
Top Compression Ring	Chrome	Chrome	Chrome
Lower Ring Marked	"TOP"	"TOP"	"TOP"
Number Ring, ea. Piston	2	2	2
Height, Upper	.0779-.0783"	.0779-.0783"	.0680-.0685"
Height, Lower	.0779-.0783"	.0779-.0783"	.0779-.0783"
Piston Ring Clearance, Upper	.0016-.0028"	.0016-.0028"	.0024-.0036"
Piston Ring Clearance, Lower	.0016-.0028"	.0016-.0028"	.0016-.0028"
OIL RINGS			
Number ea. piston	1	1	1
Height	.1566-.1571"	.1566-.1571"	.1565-.1571"
Piston Ring Clearance, in groove	.0012-.0024"	.0012-.0024"	.0012-.0026"
PISTON PINS			
Fully Floating. Circlips @ both ends.			
Connecting Rod Fit	Push	Push	Push
Piston Fit	Close Running	Close Running	Close Running
Diameter, Standard	.9449"	.9449"	.9449"
Dimension,Oversize	.9469"	.9469"	.9469"
CRANKSHAFT			
Crankshaft End Play	.0015-.0058"	.0015-.0058"	.0015-.0058"
Main Bearings, Radial Clearance	.0011-.0033"	.0011-.0033"	.0011-.0033"
Big Bearings, Radial Clearance	.0009-.0028"	.0009-.0028"	.0009-.0028"
MAIN BEARINGS			
Main Bearing Journals, Dia. Std.	2.4980-2.4984"	2.4980-2.4984"	2.4980-2.4984"
.0100" Undersize	2.4882-2.4886"	2.4882-2.4984"	2.4882-2.4886"
.0200" Undersize	2.4783-2.4787"	2.4783-2.4787"	2.4783-2.4787"
Main Bearings			
Diameter Standard	2.4996-2.5008"	2.4996-2.5008"	2.4996-2.5008"
.0100" Undersize	2.4898-2.4909"	2.4898-2.4909"	2.4898-2.4909"
.0200" Undersize	2.4799-2.4811"	2.4799-2.4811"	2.4799-2.4811"

ENGINE SPECIFICATIONS -- 4-CYL. OHC

Type Designation	AQ120B AQ125A	AQ140A	AQ145A
BIG END BEARINGS			
Big End Bearing Journals			
Width of Bearing Recess	.9744-.9783"	.9744-.9783"	.9744-.9783"
Diameter, Standard	2.1255-2.1260"	2.1255-2.1260"	2.1255-2.1260"
.0100" Undersize	2.1157-2.1161"	2.1157-2.1161"	2.1157-2.1161"
.0200" Undersize	2.1059-2.1063"	2.1059-2.1063"	2.1059-2.063"
Big End Bearing Shells			
Thickness, Standard	.0783"	.0783"	.0783"
.0100" Undersize	.0833"	.0833"	.0833"
.0200" Undersize	.0883"	.0883"	.0883"
CONNECTING RODS			
End Float on Crankshaft	.0059-.0138"	.0059-.0138"	.0059-.0138"
Length, Center - Center	5.705-5.717"	5.705-5.717"	5.705-5.717"
CAMSHAFT			
Number of Bearings	5	5	5
Bearing Journals, Diameter	1.1791-1.1799"	1.1791-1.1799"	1.1791-1.1799"
Radial Clearance, Bearing Journals	.0012-.0028"	.0012-.0028"	.0012-.0028"
Axial Clearance	.004-.016"	.004-.016"	.004-.016"
CAMSHAFT BEARINGS			
Camshaft Bearings, Diameter	1.1811-1.1819"	1.1811-1.1819"	1.1811-1.1819"
Transmission			
Number of Teeth, Crankshaft Gear	19	19	19
Intermediate Shaft Gear	38	38	38
Camshaft Gear	38	38	38
Timing Belt	123	123	123
Intermediate Shaft			
Number of Bearings	3	3	3
Diameter Bearing Journal, Forward	1.8494-1.8504"	1.8494-1.8504"	1.8494-1.8504"
Diameter Bearing Journal, Center	1.6939-1.6949"	1.6939-1.6949"	1.6939-1.6949"
Diameter Bearing Journal, Rear	1.6900-1.6909"	1.6900-1.6909"	1.6900-1.6909"
Radial Clearance	.0008-.0029"	.0008-.0029"	.0008-.0029"
Axial Clearance	.0079-.0181"	.0079-.0181"	.0079-.0181"
Diameter Intermediate Shaft Bearings			
Front	1.8511-1.8524"	1.8511-1.8524"	1.8511-1.8524"
Center	1.6957-1.6969"	1.6957-1.6969"	1.6957-1.6969"
Rear	1.6917-1.6929"	1.6917-1.6929"	1.6917-1.6929"

ENGINE SPECIFICATIONS -- 4-CYL. OHC

Type Designation	AQ120B AQ120A	AQ140A	AQ145A
VALVES			
Intake			
Valve Head Diameter	1.7323"	1.7323"	1.7323"
Stem Diameter	.3130-.3138"	.3130-.3138"	.3130-.3138"
Valve Seat Angle	45.5°	45.5°	45.5°
Seat Angle in Cylinder Head	44.75°	44.75°	44.75°
Seat Width in Cylinder Head	.0669"	.0669"	.0669"
Clearance, When Checking, Intake or Exhaust			
Cold Engine	.010-.018"	.020-.083"	.010-.018"
Hot Engine	.012-.020"	.012-.002"	.012-.020"
Clearance, When Adjusting, Intake or Exhaust			
Cold Engine	.014-.016"	.014-.016"	.014-.016"
Hot Engine	.016-.018"	.016-.018"	.016-.018"
Exhaust			
Valve Head Diameter	1.3780"	1.3780"	1.3780"
Stem Diameter	.3118-.3126"	.3118-.3126"	.3118-.3126"
Seat Angle in Cylinder Head	44.75°	44.75°	44.75°
Seat Angle, Cylinder Head	44.75°	44.75°	44.75°
Seat Width in Cylinder Head	.091"	.091"	.091"
VALVE GUIDES			
Length	2.047"	2.047"	2.047"
Inner Diameter	.3150-.3157"	.3150-.3157"	.3150-.3157"
Clearance, Valve Stem Guide			
Intake Valve	.0012-.0024"	.0012-.0024"	.0012-.0024"
Exhaust Valve	.0024-.0035"	.0024-.0035"	.0024-.0035"
Valve Springs			
Length, Unloaded, Approximately	1.77"	1.77"	1.77"
Length with 63-72 lb. Load	1.50"	1.50"	1.50"
Length with 160-178 lb. Load	1.06"	1.06"	1.06"

ENGINE SPECIFICATIONS -- 4-CYL. OHC

Type Designation		AQ120B AQ125A	AQ140A	AQ145A
LUBRICATING SYSTEM				
Oil Change Excluding Filter		9.5 U.S. Pints	10.6 U.S. Pints	9.5 U.S. Pints
Oil Change Including Filter		11.0 U.S. Pints	12.0 U.S. Pints	11.0 U.S. Pints
Oil Press. 2 max. speed, warm eng.		36-87 psi	36-87 psi	36-87 psi
Oil Viscosity		SAE 10W/40	SAE 10W/40	SAE 10W/40
Oil Pump				
Type		Gear Driven	Gear Driven	Gear Driven
Number of Teeth, ea. Gear		9	9	9
End Play		.0008-.0047"	.0008-.0047"	.0008-.0047"
Relief Valves Spring				
Length, Unloaded		1.8740"	1.5433"	1.8740"
Length, with 8.81-10.58 lb. Load		1.2598"	---	1.2598"
Length, with 9.92-11.68 lb. Load			1.0335"	
Length, with 12.12-14.76 lb. Load		1.0236"	---	1.0236"
Length, with 13.22-16.75 lb. Load			.8268"	
FUEL SYSTEM				
Fuel Pump				
Type		Diaphragm	Diaphragm	Diaphragm
Make		Carter	Pierburg APG	Carter
Feed Pressure		2-4 psi	2-4 psi	2-4 psi
Carburetor	AQ120B			
Type	Sidedraft	Downdraft	Sidedraft	Downdraft
Make	Solex 44PHN	Solex 44PAI-4	Solex 44PHN	Solex 44PAI-4
Venturi	K34	34	K34	31
Main Jet	Gg 165	155	Gg 165	145
Idle Jet	gf 60	62	gf 60	62
Air Jet	A 130	180	A 130	185
Needle Valve	1.7	1.7	1.7	1.5
Float, Weight gr	Double	7.3	Double	7.3
Accelerator Jet	---	70	---	60

ENGINE SPECIFICATIONS -- 4-CYL. OHC

Type Designation	AQ120B AQ125A	AQ140A	AQ145A
ELECTRICAL SYSTEM			
Battery			
Grounded	Negative	Negative	Negative
Voltage	12V	12V	12V
Capacity	60 Ah	60 Ah	60 Ah
Specific Gravity			
Fully Charged, Lbs/cu. in.	.0460-.0464	.0460-.0464	.0460-.0464
Discharged, Lbs./cu. in.	.0444	.0444	.0444
IGNITION SYSTEM			
No. 4 Cylinder -- Closest to Flywheel			
Spark Plugs, Bosch (or equivalent)	W6DC	W200T30	W6DC
Spark Plug Gap	.028"	.028"	.028"
Distributor			
Type, Bosch	JF4	JF4	JF4
Basic Point Setting @750 rpm	10° BTDC	6°BTDC	4°BTDC
Strobe Setting @4200 rpm	36°BTDC	36-38°BTDC	29°BTDC
Point Gap	.016"	.016"	.016"
Dwell Angle ±3°	62°	62°	62°
COOLING SYSTEM			
Thermostat			
Start Opening @	180° F	180° F	180° F
Fully Open @	198° F	198° F	198° F

TIGHTENING TORQUE VALUES

Cylinder Head Screws	66 ft-lbs	66 ft-lbs	80 ft-lbs
Big End Bearing Screws	46 ft-lbs	46 ft-lbs	46 ft-lbs
Main Bearing Bolts	90 ft-lbs	80 ft-lbs	90 ft-lbs
Camshaft Nut	35 ft-lbs	35 ft-lbs	35 ft-lbs
Flywheel Screw	53 ft-lbs	53 ft-lbs	53 ft-lbs
Spark Plugs	18.5 ft-lbs	18.5 ft-lbs	18.5 ft-lbs
Intermediate Shaft, Forward	35 ft-lbs	35 ft-lbs	35 ft-lbs
Camshaft Bearings	15 ft-lbs	15 ft-lbs	15 ft-lbs
Crankshaft, Forward	122 ft-lbs	122 ft-lbs	122 ft-lbs

ENGINE SPECIFICATIONS -- 4-CYL. OHC & DOHC

GENERAL			
Type Designation	**AQ125B AQ131A AQ145B**	**AQ151A**	**AQ171A**
Type of Operation	4-Stroke OHC	4-Stroke OHC	4-Stroke DOHC
Horsepower	Note 1	146 hp	167 hp
Maximum Speed	4700-5000 rpm	4800-5500 rpm	5000-5700 rpm
Maximum Cruising Speed	Note 2	Note 2	Note 2
Compression Ratio	9.7:1	9.7:1	9.7:1
Compression press. @ Str. Mtr. Sp.	142-170 psi	142-170 psi	142-170 psi
No. Cylinder	4 In-line	4 In-line	4 In-lin
Bore	3.78"	3.78"	3.78"
Stroke	3.150"	3.386"	3.386"
Displacement	141 CID	152 CID	152 CID
Weight Incl. S/D W/O oil & wtr.	529 lbs	551 lbs	637 lbs
Idle Speed	900 rpm	900 rpm	900 rpm
CYLINDER BLOCK			
Material	Cast Iron	Cast Iron	Cast Iron
Bore Standard	3.780-3.781"	3.780-3.781"	3.780-3781"
Bore Oversize 1	3.791"	3.791"	3.791"
Bore Oversize 2	3.803"	3.803"	3.803"
PISTONS			
Material	Note 3	Note 3	Note 3
Height Total	2.547"	2.429"	2.429"
Hgt: Piston pin ctr. to piston top	1.562"	1.445"	1.445"
Piston Clearance	0.0004-0.0012"	0.0004-0.0012"	0.0004-0.0012"
Piston Std. Size	3.779-3.780"	3.779-3.780"	3.779-3.780"
Piston Oversize 1	3.790-3.791"	3.790-3.791"	3.790-3.791"
Piston Oversize 2	3.802-3.803"	3.802-3.803"	3.802-3.803"

Note 1 The AQ125B and AQ131A have a 120hp rating, the AQ145B is rated at 139hp.

Note 2 200 rpm lower than maximum speed.

Note 3 Light alloy. The maximum weight difference between pistons of the same engine is 0.056 oz (16 grams).

ENGINE SPECIFICATIONS -- 4-CYL. OHC & DOHC

TYPE DESIGNATION	AQ125B AQ131A AQ145B	AQ151A	AQ171A
PISTON RINGS			
Ring Gap (Oil Rings)	0.0118-0.0236"	0.0118-0.0236"	0.0118-0.0236"
Ring Gap (Compress. Rings)	0.0118-0.0217"	0.0118-0.0217"	0.0118-0.0217"
Ring Oversize 1	0.0118"	0.0118"	0.0118"
Ring Oversize 2	0.0236"	0.0236"	0.0236"
Compression Rings			
Top Compression Ring	Chrome	Chrome	Chrome
Lower Ring Marked "Top"	Faces Up	Faces Up	Faces Up
Number Ring, ea. Piston	2	2	2
Height, Upper	0.068-0.069"	0.068-0.069"	0.068-0.069"
Height, Lower	0.068-0.069"	0.068-0.069"	0.068-0.069"
Ring Clearance, Upper	0.0016-0.0028"	0.0016-0.0028"	0.0016-0.0028"
Piston Ring Clearance, Lower	0.0016-0.0028"	0.0016-0.0028"	0.0016-0.0028"
Oil Rings			
Number ea. Piston	1	1	1
Height	0.1368-0.1374"	0.1368-0.1374"	0.1368-0.1374"
Ring Clearance, in Groove	0.0012-0.0024"	0.0012-0.0024"	0.0012-0.0024"
PISTON PINS			
Fully Floating Circlips	Both Ends	Both Ends	Both Ends
Connecting Rod Fit	Push	Push	Push
Diameter, Standard	0.906"	0.906"	0.906"
Dimension, Oversize	0.907"	0.907"	0.907"
CRANKSHAFT			
Crankshaft End Play	0.0031-0.0106"	0.0031-0.0106"	0.0031-0.0106"
Main Bearings, Radial Clearance	0.00094-0.00283"*	0.00094-0.00283"	0.00094-0.00283"
Rod Bearings, Radial Clearance	0.00091-0.00264"	0.00091-0.00264"	0.00091-0.00264"
MAIN BEARINGS			
Bearing Journals, Dia. Std.	2.1648-2.1654"**	2.1648-2.1654"	2.1648-2.1654"
.0100" Undersize	2.1550-2.1555"#	2.1550-2.1555"	2.1550-2.1555"
.0200" Undersize	2.1452-2.1457"##	2.1452-2.1457"	2.1452-2.1457"

*AQ131C & D 0.009-0.0025" #AQ131C & D 2.4700-2.4705"
** AQ131C & D 2.4798-24803" ##AQ131C & D 2.4001-2.4606"

ENGINE SPECIFICATIONS -- 4-CYL. OHC & DOHC

Type Designation	AQ125B AQ131A AQ145B	AQ151A	AQ171A
CONNECTING RODS			
Rod Bearing Journals			
Width of Bearing Recess	0.949-1.0276"	0.949-1.0276"	0.949-1.0276"
Diameter, Standard	1.9285-1.9293"	1.9285-1.9293"	1.9285-1.9293"
.0100" Undersize	1.9187-1.9195"	1.9187-1.9195"	1.9187-1.9195"
.0200" Undersize	1.9088-1.9096"	1.9088-1.9096"	1.9088-1.9096"
Rod End Bearing Shell			
End Float on Crankshaft	0.0098-0.0177"	0.0098-0.0177"	0.0098-0.0177"
Length, Center - Center	5.9843"	5.9843"	5.9843"
CAMSHAFT			
Number of Bearings	5	5	5
Bearing Journals, Diameter	1.179-1.180"	1.179-1.180"	1.179-1.180"
Radial Clearance, Brg. Journals	0.0012-0.0028"	0.0012-0.0028"	0.0012-0.0028"
Axial Clearance	0.0039-0.0157"	0.0039-0.0157"	0.0039-0.0157"
CAMSHAFT BEARINGS			
Bearing Diameter	1.1811-1.1819"	1.1811-1.1819"	1.1811-1.1819"
TRANSMISSION			
Number of Teeth, Crankshaft Gear	19	19	19
Intermediate Shaft Gear	38	38	38
Camshaft Gear	38	38	38
Timing Belt, No. Teeth	123	123	123
INTERMEDIATE SHAFT			
Number of Bearings	3	3	3
Diameter Bearing Journal, Fwd.	1.849-1.850"	1.849-1.850"	1.849-1.850"
Diameter Bearing Journal, Ctr.	1.694-1.695"	1.694-1.695"	1.694-1.695"
Diameter Bearing Journal, Rear	1.690-1.691"	1.690-1.691"	1.690-1.691"
Radial Clearance	0.0008-0.0030"	0.0008-0.0030"	0.0008-0.0030"
Axial Clearance	0.0079-0.0181"	0.0079-0.0181"	0.0079-0.0181"
Diameter Intermediate Shaft Bearings			
Front	1.851-1.852"	1.851-1.852"	1.851-1.852"
Center	1.696-1.697"	1.696-1.697"	1.696-1.697"
Rear	1.692-1.693"	1.692-1.693"	1.692-1.693"

ENGINE SPECIFICATIONS -- 4-CYL. OHC & DOHC

Type Designation	AQ125B AQ131A AQ145B	AQ151A	AQ171A
INTAKE VALVES			
Valve Head Diameter	1.732"	1.732"	1.358"
Stem Diameter	0.313-0.314"	0.313-0.314"	0.274-0.275"
Valve Seat Angle	44.5°	44.5°	44.5°
Seat Angle in Cylinder Head	45°	45°	45°
Seat Width in Cylinder Head	0.0512-0.0748"	0.0512-0.0748"	0.0512-0.0748"
EXHAUST VALVES			
Valve Head Diameter	1.378"	1.378"	1.240"
Stem Diameter	0.3128-0.3134"	0.3128-0.3134"	0.2730-0.2740"
Seat Valve Angle	44.5°	44.5°	44.5°
Seat Angle in Cylinder Head	45°	45°	45°
Seat Width in Cylinder Head	0.0669-0.0906"	0.0669-0.0906"	0.0669-0.0906"
Clearance, When Checking, Intake or Exhaust			
Cold Engine	0.0118-0.0157"	0.0118-0.0157"	0.0118-0.0157"
Hot Engine	0.0138-0.0177"	0.0138-0.0177"	0.0138-0.0177"
Clearance, When Adjusting, Intake or Exhaust			
Cold Engine	0.0138-0.0157"	0.0138-0.0157"	0.0138-0.0157"
Hot Engine	0.0157-0.0177"	0.0157-0.0177"	0.0157-0.0177"
VALVE GUIDES			
Length	2.047"	2.047"	2.165"
Inner Diameter	0.315-0.316"	0.315-0.316"	0.275-0.276"
Clearance, Valve Stem Guide			
Intake Valve	0.0012-0.0024"	0.0012-0.0024"	0.0012-0.0024"
Exhaust Valve	0.0024-0.0035"	0.0024-0.0035"	0.0016-0.0028"
VALVE SPRINGS			
Length, Unloaded, Approximately	1.772"	1.772"	1.772"
Length with 63-72 lb. Load	1.496"	1.496"	1.496"
Length with 160-178 lb. Load	1.063"	1.063"	1.063"

ENGINE SPECIFICATIONS -- 4-CYL. OHC & DOHC

Type Designation	AQ125B AQ131A AQ145B	AQ151A	AQ171A
LUBRICATING SYSTEM			
Oil Change Excluding Filter	1.06 US Gal.	1.06 US Gal.	1.06 US Gal.
Oil Change Including Filter	1.38 US Gal.	1.38 US Gal.	1.38 US Gal.
Oil Press., max. speed	36-87 psi	36-87 psi	36-87 psi
Oil Viscosity	SAE 10W/30	SAE 10W/30	SAE 10W/30
OIL PUMP			
Type	Gear Driven	Gear Driven	Gear Driven
End Play	0.0008-0.0047"	0.0008-0.0047"	0.0008-0.0047"
Radial Clearance	0.0008-0.0035"	0.0008-0.0035"	0.0008-0.0035"
Relief Valve Spring			
Length, Unloaded	1.874"	1.874"	1.874"
Length, 8.8-10.6 lb. Load	1.260"	1.260"	1.260"
Length, 12.1-14.8 lb. Load	1.024"	1.024"	1.024"
FUEL SYSTEM			
Fuel Pump			
Type	Diaphragm	Diaphragm	Diaphragm
Make	Carter	Carter	Carter
Feed Pressure	2-4 psi	2-4 psi	2-4 psi

CARBURETOR	AQ131C & D	AQ125B & AQ131A & B	AQ145B	AQ151A	AQ171A
Type	Downdraft	Downdraft	Downdraft	Downdraft	Downdraft
Make	Solex 44 PAI-7	Solex 44 PAI-5	Solex 44 PAI-5	Solex 44 PAI-4	Solex PAI-5-6
Venturi	34	34	34	31	32
Main Jet	165	165	165	145	147
Idle Jet	65	65	65	62	70
Air Jet	185	185	185	185	190
Needle Valve	2.0	1.7	1.7	1.5	1.7
Float Weight gr.	7.3	7.3	7.3	7.3	7.3
Accelerator Jet	70	70	70	60	80
Econostat Jet	110	110	—	—	—

ENGINE SPECIFICATIONS -- 4-CYL. OHC & DOHC

Type Designation	AQ125B AQ131A AQ145B	AQ151A	AQ171A
ELECTRICAL SYSTEM			
Battery			
Grounded	Negative	Negative	Negative
Voltage	12 Volt	12 Volt	12 Volt
Capacity	60 Ah	60 Ah	60Ah
Specific Gravity			
Fully Charged, Lbs./cu. in.	0.0460-0.0464	0.0460-0.0464	0.0460-0.0464
Discharged, Lbs./cu. in.	0.0444	0.0444	0.0444
IGNITION SYSTEM			
No. 4 Cylinder -- Closest to Flywheel			
Spark Plugs, Bosch (or equiv.)	W6DC	W6DC	W6DC
Spark Plug Gap	0.028"	0.028"	0.028"
Distributor			
Type, Bosch	JF4	JF4	Electronic
Basic Ign. Setting @850 rpm	6°BTDC	6°BTDC	N/A
Strobe Setting @4200 rpm	32-36°BTDC	32-36°BTDC	N/A
Point Gap	0.016"	0.016"	N/A
Dwell Angle ±3°	62°	62°	N/A
COOLING SYSTEM			
Thermostat			
Start Opening @	82°	82°	82°
Fully Open @	92°	92°	92°

ENGINE SPECIFICATIONS -- IN-LINE 4 & 6

GENERAL

Aquamatic, engine version	AQ105A	AQ115A	AQ130A	AQ130B	AQ130C,D	AQ165A	AQ170A,B,C
Max. output, h.p. (SAE)	105	115	130	115	130	165	170
Max. engine speed r.p.m.	5100 1)	5100 1)	5100 1)	5100 1)	5100 1)	5000	5000
Compression ratio	9.5:1	9.5:1	9.5:1	8.4:1	9.5:1	9.2:1	9.5:1
Compression pressure (at starter motor speed 200 r.p.m.) kg/cm^2 (lb./sq.in.)	11-13(155-185)	12-14(170-200)	12-14(170-200)	10-12(145-170)	12-14(170-200)	10-12(145-170)	12-14(170-200)
Number of cylinders	4	4	4	4	4	6	6
Bore, mm (in.)	88.90(3.50)	88.90(3.50)	88.90(3.50)	88.90(3.50)	88.90(3.50)	88.90(3.50)	88.90(3.50)
Stroke, mm (in.)	80.0(3.15)	80.0(3.15)	80.0(3.15)	80.0(3.15)	80.0(3.15)	80.0(3.15)	80.0(3.15)
Capacity, liters (cu.in.)	1.986(121)	1.986(121)	1.986(121)	1.986(121)	1.986(121)	2.979(182)	2.979(182)
Weight, incl. elec. equipment, carburetor, approx. kg (lb.)	180(396)	180(396)	180(396)	180(396)	180(396)	220(485)	220(485)
Idling speed, r.p.m.	900 - 1000	900 - 1000	900 - 1000	900 - 1000	900 - 1000	800 - 900	900 - 1000

CYLINDER BLOCK

	All versions
Material	Special alloy cast iron
Bore, nominal, standard mm (in.)	88.90(3.50) 88.92(3.508) 2)
Bore, 0.030" oversize mm (in.)	89.66(3.529) 89.68(3.530) 2)

PISTONS

	All versions
Material	Light-alloy
Permissible weight deviation between pistons in the same engine, g (oz.)	10 (0.35)
Height, overall, mm (in.)	71.0 (2.79)
Weight standard, g (oz.)	495-505 (19.48-19.88), 502-512 (19.74-20.14) 4)
Height, piston pin centre - piston crown, mm (in.)	46 (1.81)
Piston clearance	0.02-0.04 (0.0008-0.0016) 0.04-0.06 (0.0016-0.0024) 2)

PISTON RINGS

	All versions
Piston ring gap, mm (in.)	0.40-0.55 (0.0158-0.0217)
Piston ring oversizes	0.030"

COMPRESSION RINGS

	All versions
Marked "TOP". Top ring chromed.	
Number on each piston	2
Height, mm (in.)	1.98 (0.078)
Piston ring clearance in groove, mm (in.)	0.045-0.072 (0.0018-0.0028)

OIL SCRAPER RINGS

	All versions
Number on each piston	1
Height, mm (in.)	4.74 (0.186)
Piston ring clearance in groove, mm (in.)	0.045-0.072 (0.0018-0.0028)

PISTON PINS

	All versions
Floating fit. Circlips at both ends in piston.	
Fit:	
In connecting rod	Close running fit
In piston	Push fit
Diameter, standard, mm (in.)	22.00 (0.866)
0.05 mm (0.002") oversize, mm (in.)	22.05 (0.868)

CYLINDER HEAD

	All versions
Height, measured from cylinder head contact face to face for bolt heads, mm (in.)	86.7 (3.413)
Cylinder-head gasket, thin mm (in.)	0.8 (0.0315) 3)
Cylinder-head gasket, thick 5), mm (in.)	4-cyl.: 2.5 (0.098) 5); 6-cyl.: 2.5 (0.098) 5)

1) In light boats with a speed exceeding 30 knots (35 miles) the maximum speed of the 4-cyl. engine may reach 5500 r.p.m.
2) AQ170A from No. 520.
3) Not AQ130B.
4) AQ105A-130A from No. 5929; AQ165A " " 1379; AQ170A " " all
5) Engines equiped with thick cylinder-head gasket is only marketed in countries where there is no premium petrol (gasoline) with an octane rating of at least 97 (Research Method).

ENGINE SPECIFICATIONS -- IN-LINE 4 & 6

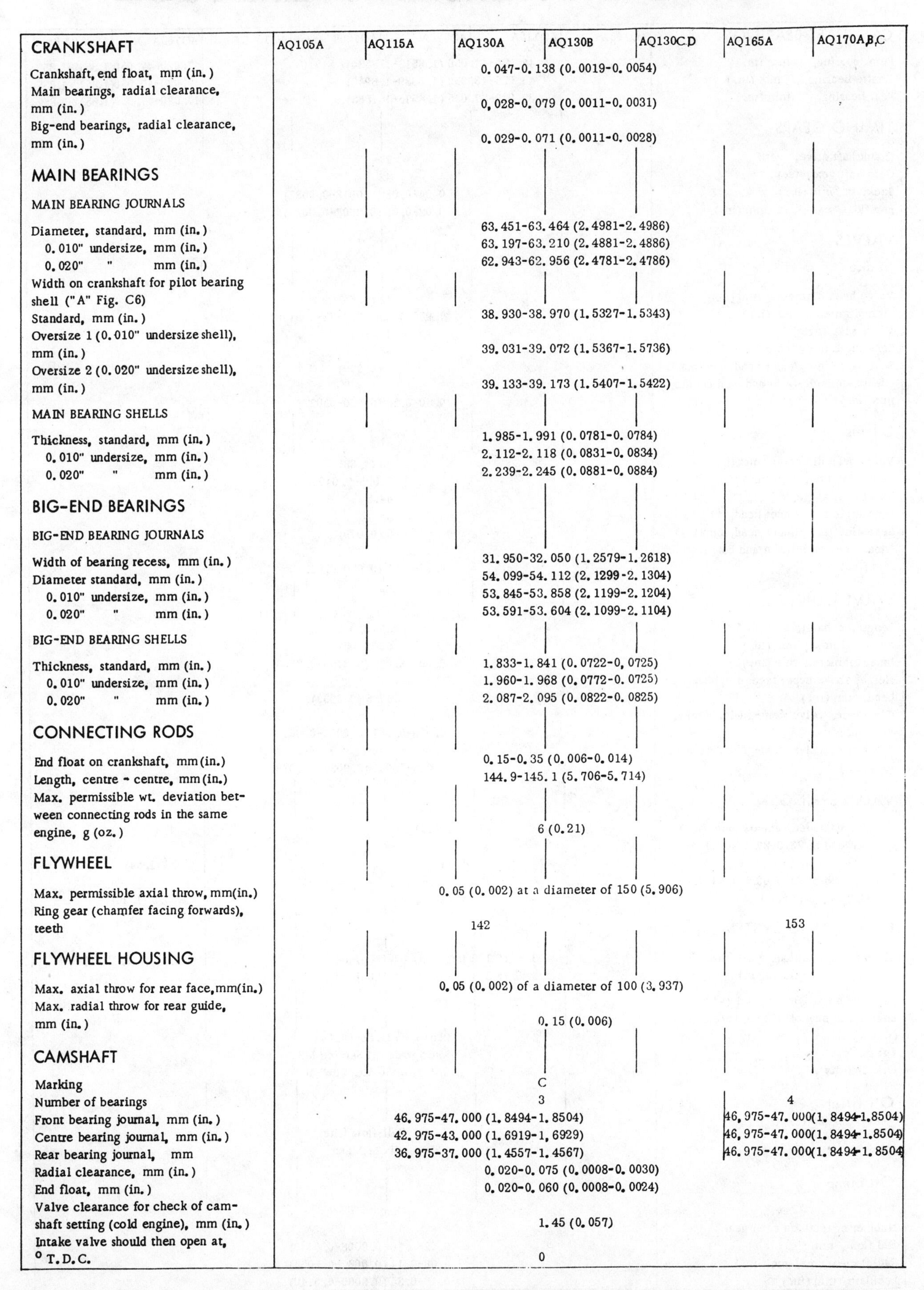

CRANKSHAFT	AQ105A	AQ115A	AQ130A	AQ130B	AQ130C,D	AQ165A	AQ170A,B,C
Crankshaft, end float, mm (in.)	0.047-0.138 (0.0019-0.0054)						
Main bearings, radial clearance, mm (in.)	0.028-0.079 (0.0011-0.0031)						
Big-end bearings, radial clearance, mm (in.)	0.029-0.071 (0.0011-0.0028)						
MAIN BEARINGS							
MAIN BEARING JOURNALS							
Diameter, standard, mm (in.)	63.451-63.464 (2.4981-2.4986)						
0.010" undersize, mm (in.)	63.197-63.210 (2.4881-2.4886)						
0.020" " mm (in.)	62.943-62.956 (2.4781-2.4786)						
Width on crankshaft for pilot bearing shell ("A" Fig. C6)							
Standard, mm (in.)	38.930-38.970 (1.5327-1.5343)						
Oversize 1 (0.010" undersize shell), mm (in.)	39.031-39.072 (1.5367-1.5736)						
Oversize 2 (0.020" undersize shell), mm (in.)	39.133-39.173 (1.5407-1.5422)						
MAIN BEARING SHELLS							
Thickness, standard, mm (in.)	1.985-1.991 (0.0781-0.0784)						
0.010" undersize, mm (in.)	2.112-2.118 (0.0831-0.0834)						
0.020" " mm (in.)	2.239-2.245 (0.0881-0.0884)						
BIG-END BEARINGS							
BIG-END BEARING JOURNALS							
Width of bearing recess, mm (in.)	31.950-32.050 (1.2579-1.2618)						
Diameter standard, mm (in.)	54.099-54.112 (2.1299-2.1304)						
0.010" undersize, mm (in.)	53.845-53.858 (2.1199-2.1204)						
0.020" " mm (in.)	53.591-53.604 (2.1099-2.1104)						
BIG-END BEARING SHELLS							
Thickness, standard, mm (in.)	1.833-1.841 (0.0722-0,0725)						
0.010" undersize, mm (in.)	1.960-1.968 (0.0772-0.0725)						
0.020" " mm (in.)	2.087-2.095 (0.0822-0.0825)						
CONNECTING RODS							
End float on crankshaft, mm (in.)	0.15-0.35 (0.006-0.014)						
Length, centre - centre, mm (in.)	144.9-145.1 (5.706-5.714)						
Max. permissible wt. deviation between connecting rods in the same engine, g (oz.)	6 (0.21)						
FLYWHEEL							
Max. permissible axial throw, mm(in.)	0.05 (0.002) at a diameter of 150 (5.906)						
Ring gear (chamfer facing forwards), teeth	142					153	
FLYWHEEL HOUSING							
Max. axial throw for rear face, mm(in.)	0.05 (0.002) of a diameter of 100 (3.937)						
Max. radial throw for rear guide, mm (in.)	0.15 (0.006)						
CAMSHAFT							
Marking	C						
Number of bearings	3					4	
Front bearing journal, mm (in.)	46.975-47.000 (1.8494-1.8504)					46,975-47.000(1.8494-1.8504)	
Centre bearing journal, mm (in.)	42.975-43.000 (1.6919-1,6929)					46,975-47.000(1.8494-1.8504)	
Rear bearing journal, mm	36.975-37.000 (1.4557-1.4567)					46.975-47.000(1.8494-1.8504)	
Radial clearance, mm (in.)	0.020-0.075 (0.0008-0.0030)						
End float, mm (in.)	0.020-0.060 (0.0008-0.0024)						
Valve clearance for check of camshaft setting (cold engine), mm (in.)	1.45 (0.057)						
Intake valve should then open at, ° T.D.C.	0						

ENGINE SPECIFICATIONS -- IN-LINE 4 & 6

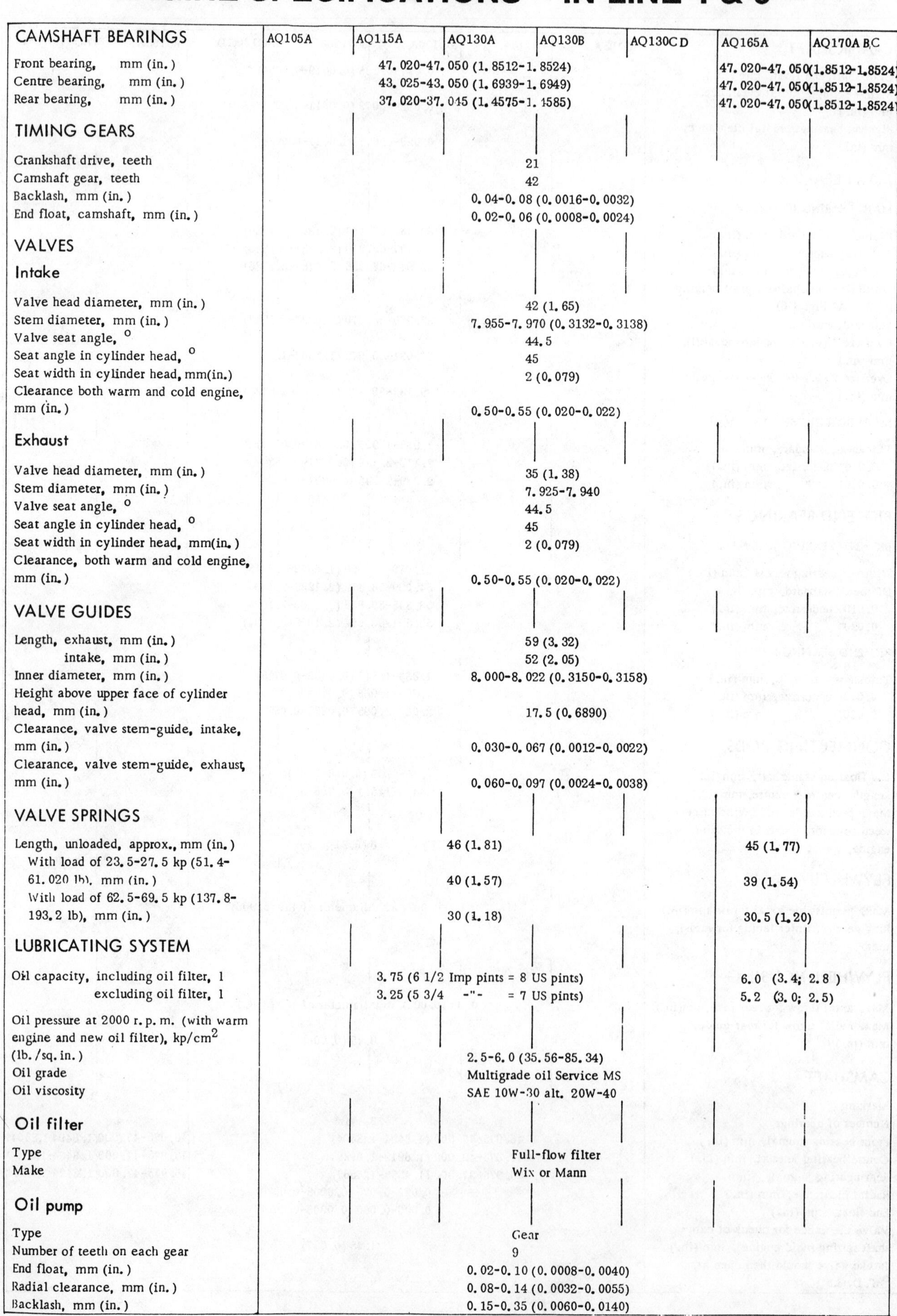

CAMSHAFT BEARINGS	AQ105A	AQ115A	AQ130A	AQ130B	AQ130C D	AQ165A	AQ170A B C
Front bearing, mm (in.)			47.020-47.050 (1.8512-1.8524)			47.020-47.050 (1.8512-1.8524)	
Centre bearing, mm (in.)			43.025-43.050 (1.6939-1.6949)			47.020-47.050 (1.8512-1.8524)	
Rear bearing, mm (in.)			37.020-37.045 (1.4575-1.4585)			47.020-47.050 (1.8512-1.8524)	
TIMING GEARS							
Crankshaft drive, teeth				21			
Camshaft gear, teeth				42			
Backlash, mm (in.)				0.04-0.08 (0.0016-0.0032)			
End float, camshaft, mm (in.)				0.02-0.06 (0.0008-0.0024)			
VALVES							
Intake							
Valve head diameter, mm (in.)				42 (1.65)			
Stem diameter, mm (in.)				7.955-7.970 (0.3132-0.3138)			
Valve seat angle, °				44.5			
Seat angle in cylinder head, °				45			
Seat width in cylinder head, mm(in.)				2 (0.079)			
Clearance both warm and cold engine, mm (in.)				0.50-0.55 (0.020-0.022)			
Exhaust							
Valve head diameter, mm (in.)				35 (1.38)			
Stem diameter, mm (in.)				7.925-7.940			
Valve seat angle, °				44.5			
Seat angle in cylinder head, °				45			
Seat width in cylinder head, mm(in.)				2 (0.079)			
Clearance, both warm and cold engine, mm (in.)				0.50-0.55 (0.020-0.022)			
VALVE GUIDES							
Length, exhaust, mm (in.)				59 (3.32)			
intake, mm (in.)				52 (2.05)			
Inner diameter, mm (in.)				8.000-8.022 (0.3150-0.3158)			
Height above upper face of cylinder head, mm (in.)				17.5 (0.6890)			
Clearance, valve stem-guide, intake, mm (in.)				0.030-0.067 (0.0012-0.0022)			
Clearance, valve stem-guide, exhaust, mm (in.)				0.060-0.097 (0.0024-0.0038)			
VALVE SPRINGS							
Length, unloaded, approx., mm (in.)			46 (1.81)			45 (1.77)	
With load of 23.5-27.5 kp (51.4-61.020 lb), mm (in.)			40 (1.57)			39 (1.54)	
With load of 62.5-69.5 kp (137.8-193.2 lb), mm (in.)			30 (1.18)			30.5 (1.20)	
LUBRICATING SYSTEM							
Oil capacity, including oil filter, l			3.75 (6 1/2 Imp pints = 8 US pints)			6.0 (3.4; 2.8)	
excluding oil filter, l			3.25 (5 3/4 -"- = 7 US pints)			5.2 (3.0; 2.5)	
Oil pressure at 2000 r.p.m. (with warm engine and new oil filter), kp/cm² (lb./sq.in.)				2.5-6.0 (35.56-85.34)			
Oil grade				Multigrade oil Service MS			
Oil viscosity				SAE 10W-30 alt. 20W-40			
Oil filter							
Type				Full-flow filter			
Make				Wix or Mann			
Oil pump							
Type				Gear			
Number of teeth on each gear				9			
End float, mm (in.)				0.02-0.10 (0.0008-0.0040)			
Radial clearance, mm (in.)				0.08-0.14 (0.0032-0.0055)			
Backlash, mm (in.)				0.15-0.35 (0.0060-0.0140)			

ENGINE SPECIFICATIONS -- IN-LINE 4 & 6

	AQ105A	AQ115A	AQ130A	AQ130B	AQ130CD	AQ165A	AQ170ABC
Relief valve spring (in oil pump)							
Length, unloaded, mm (in.)			39.0 (1.54)				
Length, loaded with 4.6-5.4 kp (10.1-11.9 lb.), mm (in.)			26.3 (1.04)				
Length, loaded with 6.2-7.8 kp (13.6-17.2 lb.), mm (in.)			21.0 (0.83)				
FUEL SYSTEM							
Fuel pump							
Type			Diaphragm pump				
Make			Pierburg: PE15572				
Fuel pressure, measured at same level as pump kp/cm² (p.s.i.)			0.22 (3.2)				
Carburetors							
No. of carburetors	1	1	2	2	2	3	3
AQ115A, AQ130C, AQ170A							
Type			Down-draught carburetors				
Make and designation			Solex 44 PAI				
Main jet			145				
Idling jet		55			70		70
Emulsion jet		145 E5			190 E5		210 E5
Pump jet			70				
Needle seating			2				
AQ105A, AQ130A-B, AQ165A							
Type			Horizontal carburetors				
Make and designation Zenith-Stromberg	150CD		175CDSE			175CDSE	
Metering needle marked	8B		3D			2AA	
Float level ("A" Fig. F5), mm (in.)	17-18 (0.67-0.71)			15-17 (0.59-0.67)			
Float level ("B" Fig. F5), mm (in.)	14,5 (0.5716)			11.0 (0.4331)			
Float needle valve, mm (in.)	2 (0.08)				1.5 (0.06)		
Washer thickness under float needle valve, mm (in.)	1.0 (0.04)				1.6 (0.06)		
Air valve spring colour marking	Red		Blue			Uncoloured	
Air valve spring coil thickness, mm (in.)	1.0 (0.040)		0.9 (0.036)			0.8 (0.032)	
Oil for damper			Same as in engine				
BATTERY							
Earthed			Negative terminal				
Voltage, V			12				
Battery capacity, standard, Ah			60				
Specific gravity of electrolyte:							
Fully charged battery, g/cm³			1.275-1.285				
When charging is necessary, g/cm³			1.230				
Recommended charging current, A			4.5				
DYNAMO/ALTERNATOR							
Type	Bosch LJ/6EH 90/12 1800 FR20		S.E.V. Motorola				
Voltage, V			12				
Rated output, W	90				450		
Max. current, A	7.5				38		
Earthed			Negative terminal				
Direction of rotation			Clockwise				
STARTER MOTOR							
Type			Bosch 0 001 311 032				
Voltage, V			12				
Earthed			Negative terminal				
Direction of rotation			Clockwise				
Output, hp			Approx. 1				

ENGINE SPECIFICATIONS -- IN-LINE 4 & 6

IGNITION SYSTEM	AQ105A	AQ115A	AQ130A	AQ130B	AQ130CD	AQ165A	AQ170ABC
Order of firing			1-3-4-2			1-5-3-6-2-4	
Spark plug type			Bosch W 225T35 or corresponding				
Spark plug gap, mm (in.)				0.7 (0.028)			
DISTRIBUTOR							
Type Bosch			0 231 153 012				0 231 152 060
Dwell angle, °			59-65				37-43
Breaker points, gap, mm (in.)			0.40-0.50 (0.016-0.020)			0.25-0.35 (0.010-0.014)	
Basic setting°, BTDC	9			12			12, 15 4)
Stroboscope setting)2000 r.p.m.), ° BTDC	22-24	24-26	27-29	26-28	27-29	22-24 3), 28-30 3)4)	
COOLING SYSTEM							
Thermostat							
Type			Bellows thermostat				
Marked			54			55	
Begins opening at, °C (°F)			51-56 (124-133)			53-56 (127-133)	
Fully open at, °C (°F)			67 (153)			66-70 (151-158)	
WEAR TOLERANCES							
Cylinders							
To be rebored when wear amounts to (if engine has abnormal oil consumption, mm (in.)				0.25 (0.010)			
Crankshaft							
Max. permissible out-of-round on main bearing journals, mm (in.)				0.05 (0.0020)			
Max. permissible out-of-round on big-end bearing journals, mm (in.)				0.07 (0.0028)			
Max. crankshaft end float, mm (in.)				0.15 (0.0060)			
Valves							
Max. permissible clearance between valve stem and valve guide, mm (in.)				0.15 (0.0060)			
Max. permissible wear, valve stem, mm (in.)				0.02 (0.0008)			
Camshaft							
Permissible out-of-round (with new bearings), max., mm (in.)				0.07 (0.0028)			
Bearings, permissible wear, max., mm (in.)				0.02 (0.0008)			
Timing Gears							
Permissible backlash, max., mm (in.)				0.12 (0.0048)			

TIGHTENING TORQUES	kpm	(ft. lb.)
Cylinder head	9	(65)
Main bearings	12.5	(90)
Big-end bearings	5.5	(40)
Flywheel	5	(35)
Spark plugs	4	(30)
Camshaft nut	13-15	(95-108)
Bolt for crankshaft belt pulley	7-8	(51-58)
Nipple for oil filter	4.5-5.5	(38-40)
Oil sump bolts	1.0-1.2	(7.3-8.7)
Tensioning ring - flywheel casing	3.5	(25)
Nut, oil cooler	3.0-3.5	(21.7-25.3)
Centre bolt, oil cooler	1.2-1.4	(8.7-10.1)
V-belt A.C. generator approx.	1.5 1), 1.7 2)	(10.85 1); 12.30 2))
V-belt D.C. generator approx.	0.9	(6.51)

1) AQ130A-C
2) AQ165A-170A
3) AQ165A, with vacuum governor uncoupled.
4) AQ165-170A, engines equiped with thick cylinder-head gasket.

ENGINE SPECIFICATIONS -- GMC V6

GENERAL

Type designation	AQ175A
Maximum flywheel output[1], kW (hp)	125 (170)
Max. speed of rotation (r/min)	4800
Compression ratio	8.6.1
Compression pressure at starter motor speed	10–11 kp/cm^2 (142–156 p.s.i.)
Number of cylinders	6
Cylinder diameter, mm (in)	94.87–94.96 (3.735–3.7385)
Cylinder bore, oversize 0.76 mm (0.030 in)	95.7 (3.77)
Max. out-of-round bore mm (in)	0.05 (0.002)
Max. conicity of bore mm (in)	0.025 (0.001)
Max. piston clearance mm (in)	0.069 (0.0027)
Stroke, mm (in)	88.4 (3.48)
Displacement dm^3 (cu.in)	3.8 (229)
Weight with drive model 280, without oil and water approx kg (lbs)	342
Idling speed (r/min)	700
Direction of rotation, seen from front	Clockwise

[1] According to SAE J607

PISTON RINGS

Piston ring gap measured in the opening of the compression ring,	
upper	0.25–0.76 mm (0.01–0.03 in)
lower	0.25–0.89 mm (0.010–0.035 in)
oil ring	0.38–0.65 mm (0.01496–0.02559 in)
Piston ring oversize	0.76 mm (0.030 in)

COMPRESSION RINGS

Top ring chromed	
Number on each piston	2
Piston ring clearance in groove "upper"	0.03–0.11 mm (0.0012–0.0042 in)
Piston ring clearance in groove "lower"	0.03–0.11 mm (0.0012–0.0042 in)

OIL RINGS

Number on each piston	1
Piston ring clearance in groove	0.05–0.20 mm (0.002–0.008 in)

PISTON PINS

Press fit	
Fit: In connecting rod, interference	0.02–0.04 mm (0.0008–0.0016 in)
Diameter, standard	23.50–23,55 mm (0.9270–0.9273 in)
Piston pin clearance in piston, max.	0.025 mm (0.001 in)

CRANKSHAFT

Crankshaft, end float	0.05–0.15 mm (0.002–0.006 in.)

MAIN BEARINGS

Main bearings, journals, No (from front end)	1	2.3	4
Diameter, standard	62.189–62.21 mm (2.4484–2.4493 in)	62.18–62.20 mm (2.4481–2.4490 in)	62.177–62.199 mm (2.4479–2.4488 in)
Bearing clearance	0.025–0.038 mm (0.001–0.0015 in)	0.025–0.064 mm (0.001–0.0025 in)	0.064–0.089 mm (0.0025–0.0035 in)
Out-of-round, max.		0.025 mm (0.001 in)	
Conicity, max.		0.025 mm (0.001 in)	

ENGINE SPECIFICATIONS -- GMC V6

Main bearing shells

Undersize 1st	0.25 mm (0.010 in)
Undersize 2nd	0.50 mm (0.020 in)

BIG-END BEARING JOURNALS

Diameter, standard	53.31–53.33 mm (2.0988–2.0998 in)
Big-end bearing clearance	0.033–0.076 mm (0.0013–0.0030 in)
Out-of-round, max.	0.025 mm (0.001 in)
Conicity, max.	0.025 mm (0.001 in)

BIG-END BEARING SHELLS

Undersize, 1st	0.25 mm (0.010 in)
Undersize, 2nd	0.50 mm (0.020 in)

CONNECTING RODS

End float on crankshaft	0.2–0.36 mm (0.008–0.014 in)

CAMSHAFT

Number of bearings	4
Bearing journal, diameter	47.45–47.48 mm (1.8682–1.8692 in)
Bearing journal, out-of-round, max.	0.025 mm (0.001 in)
Camshaft straightness, max. throw	0.038 mm (0.0015 in)
Axial clearance	0.1–0.3 mm (0.004–0.012 in)
The lobe lift	6,83 mm (0.269")
Tolerance on lift	±0.051 mm (±0.002 in)

CYLINDER HEAD

VALVES

Intake

Valve head diameter	46,74 mm (1.84 in)
Stem diameter	8.64 mm (0.341 in)
Valve seat angle	45°
Seat angle in cylinder head	46°
Seat width in cylinder head	0.794–1.588 mm (0.031–0.062 in)

Exhaust

Valve head diameter	38,10 mm (1.50 in)
Stem diameter	8.64 mm (0.341 in)
Valve seat angle	45°
Seat angle in cylinder head	46°
Seat width in cylinder head	1.588–2.381 mm (0.062–0.093 in)

VALVE GUIDES

Clearance, valve stem-guide, intake valve	0.025–0.094 mm (0.001–0.0037 in)
Clearance, valve stem-guide, exhaust valve	0.025–0.120 mm (0.001–0.0047 in)

VALVE SPRINGS

Length, unloaded approx.	51.5 mm (2.03 in)
Length with load of 334–370 N	44.0 mm (1.70 in)
Length with load of 853–906 N	31.7 mm (1.25 in.)

LUBRICATING SYSTEM

Oil capacity excluding oil filter dm^3	3.8
Oil capacity including oil filter dm^3	4.1
Oil pressure at idling speed, warm engine	0.7 kp/cm^2 (10 p.s.i.)
Oil pressure at max. speed, warm engine	2.5–3.16 kp/cm^2 (35–45 p.s.i.)
Oil grade (API)	Motor oil SE
Viscosity	Multigrade oil SAE 10W–30, 20W–40, 10W–40[1)]

1) Volvo Penta oil has viscosity SAE 10W–40.

ENGINE SPECIFICATIONS -- GMC V6

FUEL SYSTEM

Octane rating, min.	91 octane (RON)

FUEL PUMP

Feed pressure	0.31–0.42 kp/cm^2 (4.5–6 p.s.i.)

CARBURETTOR

Make	Rochester
Type	2GE

ELECTRICAL SYSTEM

BATTERY

Earth	Negative (–)
Voltage	12 V
Capacity	60 Ah
Density of battery electrolyte:	
fully charged battery	1.275–1.285 g/cm^3 (0.460–0.464 lb/cu.in.)
battery to be recharged at	1.230 g/cm^3 (0.444 lb/cu.in.)

GENERATOR

Type	Alternator
Output max. A (W)	50 (50 x 14)

STARTER MOTOR

Starter motor output kW (HP)	0.96 (1.3)

IGNITION SYSTEM

Spark plug	AC43MT
Spark plug gap	0.7 mm (0.028")
Firing sequence	1-6-5-4-3-2
Basic setting	10° B.T.D.C.
Stroboscopic setting 4200 r/m	34° B.T.D.C.
Gap mm (in)	0.43–0.56 mm (0.017–0.022")
Cam angle	33–37°

COOLING SYSTEM

THERMOSTAT

Starts opening at	62°C (154°F)
Fully open at	72°C (160°F)

TIGHTENING TORQUES

	Nm	kpm	lb/ft	Lubricant
Cylinder head screws, 1st stage	50	5	36	Permatex
2nd stage	90	9	65	
Main bearing screws	108	11	70	Molykote
Connecting rod bolt 3/8"	61	6.2	45	oil
Flywheel screws	82	8.2	60	oil
Flywheel housing bolts	41	4.1	30	oil
Centre bolt, crankshaft, forward	82	8.2	60	oil
Intake manifold bolts	41	4.1	30	oil
Exhaust manifold bolts	35	3.5	25	Permatex
Riser, bolts	25	2.5	18	Permatex
Spark plugs	20	2	15	Dry
Oil pump	90	9	65	oil
Oil filter	35	3.5	25	oil
Oil pan drain plug	27	2.7	20	oil
Circulating pump (cooling water)	41	4.1	30	Permatex

ENGINE SPECIFICATIONS -- GMC V6

General

Type designation	**431 (AQ205A)**
Output, see sales literature	
Compression ratio	9.3:1
Compression pressure at starter motor speed	10—11 kp/cm² (142—156 lb/in²)
Number of cylinders	6
Cylinder diameter mm (in)	101.59—101.66 (3.9995—4.0025)
Cylinder bore, oversize 0.76 mm (0.030 in)	102.4 (4.03)
Max cylinder ovality mm (in)	.05 (.0020)
Max cylinder conicity mm (in)	.025 (.0010)
Stroke mm (in)	88.4 (3.480)
Swept volume, dm³ approx (cu.in.)	4.293 (262)
Weight with drive without oil and water approx. kg (lbs)	DP355 (782) SP351 (773)
Idling speed r/s (r/min)	12.5 (750)
Direction of rotation (viewed from front)	clockwise

Piston

Max piston clearance mm (in)	0.069 (.0027)
Piston diameter mm (in)	101.595—101.608 (3.9998—4.0003) 101.608—101.620 (4.003—4.008)
Piston diameter, oversize	0.762 mm (.030'') 102.362 (4.03'')

Piston rings

Piston ring gap in bore,	
compression ring upper	0.25—0.76 mm (.010—.030 in)
lower	0.25—0.89 mm (.010—.035 in)
Oil control ring	0.38—1.65 mm (.015—.065 in)
Oversize piston rings	0.76 mm (.030 in)

Compression rings

Upper ring chromium plated, marked side up	
Number on each piston	2
Height	1.98 mm (.078 in)
Piston ring clearance in groove, upper	0.03—0.11 mm (.0012—.0043 in)
Piston ring clearance in groove, 2nd	0.03—0.11 mm (.0012—.0043 in)

Oil control rings

Number on each piston	1
Height	4.74 mm (.187 in)
Piston ring clearance in groove	0.05—0.20 mm (.002—.008 in)

Wrist pins

Grip fit	
Fit: in connecting rod, negative clearance	0.02—0.04 mm (.0008—.0016 in)
Diameter, standard	23.50 mm (.925 in)
Wrist pin clearance in piston, max.	0.025 mm (.0010 in)

Crankshaft

Crankshaft end play	0.05—0.15 mm (.002—.006 in)

ENGINE SPECIFICATIONS -- GMC V6

Main bearings

Main bearing journals,

journal No. (from front)	1	2, 3,	4
Diameter, standard	62.179—62.212 mm (2.4484—2.4493 in)	62.181—62.204 mm (2.4481—2.4490 in)	62.176—62.199 mm (2.4479—2.4488 in)
Bearing clearance	0.025—0.038 mm (.0010—.0015 in)	0.025—0.064 mm (.0010—.0025 in)	0.064—0.089 mm (.0025—.0035 in)
Ovality, max		0.025 mm (.0010 in)	
Conicity, max		0.025 mm (.0010 in)	

Main bearing shells

1 st undersize	0.254 mm (.010 in)
2nd undersize	0,508 mm (0.020 in)

Crank pin bearing journals

Diameter standard	57.116—57.144 mm (2.2487—2.2498)
Crank pin bearing clearance	0.033—0.076 mm (.0013—.0030 in)
Ovality, max	0.025 mm (.001 in)
Conicity, max	0.025 mm (.001 in)

Crank pin bearing shells

1st undersize	0.254 mm (.010 in)
2nd undersize	0.508 mm (.020 in)

Connecting rods

Side clearance at crankshaft	0.15—0.36 mm (.006—.014 in)

Camshaft

Number of bearings	4	
Bearing journal, diameter	47.452—47.477 mm (1.8682—1.8692 in)	
Bearing journal ovality max	0.0254 mm (.0010 in)	
Camshaft's straightness, max throw	0.051 mm (.002 in)	
End play	0.1—0.3 mm (.004—.012 in)	
	Inlet	Exhaust
Cam lift height	5.94 mm (.234 in)	6.52 mm (.257 in)
Lift height tolerance	±0.051 mm (.002 in)	

Cylinder head

Valves

Intake

Valve disc diameter	49.149—49.403 mm (1.935—1.945 in)
Valve stem diameter	8.64 mm (.34 in)
Oversize .015 in	9.01 mm (.355 in)
Oversize .030 in	9.40 mm (.370 in)
Valve seat angle	45°
Seat width in cylinder head	0.80—1.60 mm (.031—.063 in)

Exhaust

Valve disc diameter	37.97—38.23 mm (1.495—1.505 in)
Valve stem diameter	9.45 mm (.372 in)
Oversize .015 in	9.01 mm (.355 in)
Oversize .030 in	9.40 mm (.370 in)
Valve seat angle	45°
Seat angle in cylinder head	46°
Seat width in cylinder head	1.60—2.40 mm (.063—.094 in)

ENGINE SPECIFICATIONS -- GMC V6

Valve guides

Clearance, valve stem-guide, intake valve	.025—.094 mm (.0010—.0037 in)
Clearance, valve stem-guide, exhaust valve	.025—.120 mm (.0010—.0047 in)

Valve springs

	Intake, Exhaust
Length without load	51.6 mm (2.03 in)
Length with load 339—374 N (34.5—38.1 kp/76—84 lb	43.2 mm (1.7 in)
Length with load 863—916 N (88—93.5 kp/194—206 lb)	31.8 mm (1.252 in)

Valve lash

Intake, exhaust	3/4 turn down from zero lash

Lubricating system

Oil quality	Service SG
Viscosity	SAE 20W/50 (15W/50)
Oil capacity excl. oil filter, litre (Imp.gals/US gals)	3.8 (.83/1.00)
Oil capacity incl. oil filter, litre (Imp.gals/US gals)	4.2 (.92/1.11)
Oil pressure at full speed, hot engine	2.5—3.16 kp/cm^2 (35.5—45.0 lbf/in^2
Oil pressure at idle, hot engine	.7 kp/cm^2 (9.96 lbf/in^2)

Fuel system

Oil quality, min	91 octane (RON). (USA 87 R + M/2)
The engine can be run on unleaded fuel	

Fuel pump

Feed pressure	.42—.56 kp/cm^2 (5.97—7.96 lb/in^2)

Electrical system

Battery

Earth connection	Negative (—)
System voltage	12 V
Capacity	60 Ah
Density of battery electrolyte:	
fully charged battery	1.275—1.285 g/cm^3
battery to be recharged at	1.230 g/cm^3

Alternator

Type	AC
Output max	700 W (50A)

ENGINE SPECIFICATIONS -- GMC V6

Starter motor

Starter motor output kW (hp) .96 (1.3)

Ignition system

Spark plugs	Volvo part No. 876046-4, AC MR 43T or similar
Spark plug gap	.9 mm (.035 in)
Firing order	1-6-5-4-3-2
Stroboscope setting, 2500 r/min	16° B.T.D.C.
Basic setting (idling), 750 r/min	8° B.T.D.C.
Distributor, sensor gap	.20—.25 mm (.0078—.0098 in)

Cooling system (sea water cooled)

Thermostat

Starts opening at	62°C (144°F)
Fully open at	72°C (162°F)

Fresh water cooling system (accessorie)

Thermostat

Starts opening at	68°C (155°F)
Fully open at	83°C (182°F)

Tightening torques

	Nm	Kpm	Lbf.ft	Lubrication
Cylinder head bolts, 1st tightening	50	5	36	Permatex
2nd tightening	90	9	66	
Main bearing bolts	108	11	80	Molykote
Big-end bearing caps	61	6.2	45	oil
Flywheel bolts	82	8.2	60	oil
Flywheel housing bolts	41	4.1	30	oil
Center bolt, crankshaft, front	82	8.2	60	oil
Bolts for camshaft gear	24	2.4	17.5	oil
Intake manifold bolts	41	4.1	30	oil
Exhaust manifold bolts	35	3.5	26	Permatex
Riser, bolts	25	2.5	18.5	Permatex
Spark plugs	20	2	14.5	Dry
Bolts for oil pump	90	9	66	oil
Oil pan bolts	11	1.1	8	oil
Oil pan nuts	22	2.2	16	oil
Timing gear casing bolts	14	1.4	10.3	oil
Valve cover bolts	6	.6	4.5	oil
Oil drain plug	27	2.7	19.8	oil
Circulation pump bolts (coolant)	41	4.1	30	Permatex
Distributor bracket bolts	46	4.6	34	oil
Carburetor bolts	13	1.3	9.5	oil

ENGINE SPECIFICATIONS -- GMC V8

Check engine tune-up chart on Pages A4 and A6 for engine identification by CID

GENERAL				
Type Designation	**305 CID**	**307 CID**	**350 CID**	**454 CID**
Compression Ratio	8.5:1	8.5:1	9.0:1	8.0:1
Compression pressure	142-156	142-156	142-156	142-156
No. Cylinder	8	8	8	8
Bore	3.736"	3.875"	4.000"	4.250"
Stroke	3.48"	3.25"	3.48"	4.00"
Displacement	305	307	350	454
CYLINDER BLOCK				
Material	Cast Iron	Cast Iron	Cast Iron	Cast Iron
Bore Standard	3.7350-3.7385"	3.8745-3.8775"	3.9995-4.0025"	4.2495-4.2525"
Oversize (0.030")	3.766"	3.905"	4.030"	4.2800"
Oversize (0.060")	3.796"	3.935"	4.060"	Not Recom.
Out-of-round, max.	0.001"	0.001"	0.001"	0.002"
Taper, thrust side, max.	0.0005"	0.0005"	0.0005"	0.0005"
Taper, relief side, max.	0.001"	0.001"	0.001"	0.001"
PISTONS				
Piston Clearance	0.0007-0.0017"	0.0007-0.0017"	0.0007-0.0017"	0.005"
Piston Std. Size	3.7343-3.7353	3.8743-3.8733"	3.9983-3.9993"	4.245"
0.030" Oversize	3.7643-3.7653"	3.9043-3.9033"	4.0283-4.0293"	4.275"
0.060" Oversize	3.7943-3.7953"	3.9343-3.9333"	4.0583-4.0593"	Not Avail.
PISTON RINGS				
Oil Ring Gap	0.015-0.055"	0.015-0.055"	0.015-0.055"	0.015-0.065"
Compression Ring Gap	0.010-0.025"	0.010-0.025"	0.010-0.025"	0.010-0.030"
Ring, 0.030" Oversize	Available	Available	Available	Available
Ring, 0.060" Oversize	Available	Available	Available	Not Avail.

ENGINE SPECIFICATIONS -- GMC V8

Type Designation	305 CID	307 CID	350 CID	454 CID
PISTON RINGS (Continued)				
COMPRESSION RINGS				
Each Piston	2	2	2	2
Height, Upper	0.078"	0.078"	0.078"	No Info.
Height, Lower	0.078"	0.078"	0.078"	No Info.
Upper Ring Clear.	0.0012-0.0042"	0.0012-0.0042"	0.0012-0.0042"	0.0017-0.0042"
Lower Ring Clear.	0.0012-0.0042"	0.0012-0.0042"	0.0012-0.0042"	0.0017-0.0042"
OIL RINGS				
Number ea. piston	1	1	1	1
Height	0.1866"	0.1866"	0.1866"	0.1866"
Ring Groove Clear.	0.002-0.008"	0.002-0.008"	0.002-0.008"	0.005-0.008"
PISTON PINS				
Circlips	Both ends	Both ends	Both ends	Both ends
Rod Fit	Press fit	Press fit	Press fit	Press fit
Clearance	0.001"	0.001"	0.001"	0.001"
Diameter, Std.	0.9270-0.9273"	0.9270-0.9273"	0.9270-0.9273"	0.9895-0.9898"
CRANKSHAFT				
End Play	0.002-0.006"	0.002-0.006"	0.002-0.006"	0.006-0.010"
Rod Bearings				
Radial Clear.	0.0013-0.0030"	0.0013-0.0030"	0.0013-0.0030"	0.0009-0.0030"
Out-of-round, Max.	0.001"	0.001"	0.001"	0.001"
Taper, Max.	0.001"	0.001"	0.001"	0.001"
Rod Journal Dia.	2.0988-2.0998"	2.0988-2.0998"	2.0988-2.0998"	2.1990-2.2000"
Brgs. avail. if ground 0.010" & 0.020" under	Yes	Yes	Yes	Yes

ENGINE SPECIFICATIONS -- GMC V8

Type Designation	305 CID	307 CID	350 CID	454 CID
MAIN BEARINGS				
All Main Brgs.				
Radial Clear.				
No. 1, 2, 3, 4	0.0010-0.0015"	0.0010-0.0015"	0.0010-0.0015"	0.0010-0.0025"
No. 5	0.0025-0.0035"	0.0025-0.0035"	0.0025-0.0035"	0.0025-0.0035"
Journals, Dia. Std.				
No.1	2.4484-2.4493"	2.4484-2.4493"	2.4484-2.4493"	2.74810-2.7490"
No. 2, 3, 4	2.4481-2.4490"	2.4481-2.4490"	2.4481-2.4490"	2.7481-2.7490"
No. 5	2.4479-2.4488"	2.4479-2.4488"	2.4479-2.4488"	2.7476-2.7486"
.010" Undersize	Available	Available	Available	Available
.020" Undersize	Available	Available	Available	Available
CAMSHAFT				
No. of Bearings	5	5	5	5
Brg. Journal, Dia.	1.8682-1.8692"	1.8682-1.8692"	1.8682-1.8692"	1.9482-1.9492"
Runout, Max.	0.0015"	0.0015"	0.0015"	0.0015"
End "Play"	0.001-0.005"	0.001-0.005"	0.001-0.005"	0.001-0.006"
VALVES				
Lifter	Hydraulic	Hydraulic	Hydraulic	Hydraulic
Intake				
Stem Dia.	0.341"	0.341"	0.341"	Note 1
Valve Seat Angle	45°	45°	45°	45°
Seat Angle - Cyl. Head	46°	46°	46°	46°
Seat Width - Cyl. Head	0.0313-0.0625"	0.0313-0.0625"	0.0313-0.0625"	0.0313-0.0625"
Seat Runout, Max.	0.002"	0.002"	0.002"	0.002"
Exhaust				
Stem Dia., Early	0.341"	0.341"	0.341"	---
Stem Dia., Late	0.372"	0.372"	0.372"	Note 2
Seat Valve Angle	45°	45°	45°	45°
Seat Angle - Cyl. Head	46°	46°	46°	46°
Seat Width - Cyl. Head	0.0625-0.0938"	0.0625-0.0938"	0.0625-0.0938"	0.0625-0.0938"
Seat Runout, Max.	0.002"	0.002"	0.002"	0.002"
VALVE GUIDES				
Stem to Guide Clear.				
Intake Valve	0.0010-0.0027"	0.0010-0.0027"	0.0010-0.0027"	0.0010-0.0037"
Exhaust Valve	0.0012-0.0029"	0.0012-0.0029"	0.0012-0.0029"	0.0012-0.0049"

ENGINE SPECIFICATIONS -- GMC V8

Type Designation	305 CID	307 CID	350 CID	454 CID
Valve Springs				
Unloaded Length				2.12"
One lavender stripe	2.03"	2.03"	2.03"	No Color
Two green stripes	1.91"	1.91"	1.91"	No Color
With 76-84 lb. Load	1.61"	1.61"	1.61"	2.12"
With 183-195 lb. Load	1.20"	1.20"	1.20"	1.80"
Installed height	1-23/32"	1-5/8"	1-23/32"	1-51/64"
Damper Spring,				
Free length	1.86"	1.86"	1.86"	1.86"
Number of Coils	4	4	4	4
Valve Lash Adjust.				
Intake & Exhaust	One turn down from zero lash.			3/4 Turn dn.
LUBRICATING SYSTEM				
Oil Change Excl. Filter	5 qts	5qts	5qts	---
Oil Change Incl. Filter	5.5 qts	5.5 qts	5.5qts	Note 3
Oil Press. max. rpm	35-45 psi	35-45 psi	35-45 psi	40-60 psi
Oil Grade	SE	SE	SE	SG
Oil Viscosity	10W40	10W40	10W40	20W50
FUEL SYSTEM				
Octane Rating, min .	91	91	91	91
Fuel Pressure	6-8 psi	6-8 psi	6-8 psi	7-8.5 psi
ELECTRICAL SYSTEM				
Battery				
Grounded	Negative	Negative	Negative	Negative
Voltage	12V	12V	12V	12V
Capacity	60AH	60AH	60AH	60AH
Specific Gravity				
Fully Charged,				
Lbs./cu. in.	0.460-0.464	0.460-0.464	0.460-0.464	0.460-0.464
Discharged,				
Lbs./cu. in.	0.044	0.044	0.044	0.044

ENGINE SPECIFICATIONS -- GMC V8

Type Designation	305 CID	307 CID	350 CID	454 CID
IGNITION SYSTEM				
Spark Plugs, Bosch	HR8A	HR8A	HR8A	Note 4
Spark Plug Gap	0.028"	0.028"	0.028"	0.035"
Distributor				
Basic Ignition Timing	See Tune-up Adjustment Chart Page A4 & A6			
Strobe Setting	See Tune-up Adjustment Chart Page A4 & A6			
Point Gap, New	0.016"	0.016"	0.016"	N/A
Point Gap, Used	0.019"	0.019"	0.019"	N/A
Dwell Angle $\pm 3^{o}$	31°	31°	31°	N/A
COOLING SYSTEM				
Thermostat				
Start Opening @	154°	154°	154°	Note 5
Fully Open @	160°	160°	160°	Note 6

SPECIAL NOTES GIVEN IN ENGINE SPECIFICATIONS

Note 1 Standard intake valve stem diameter is 0.3714-0.3722".
Oversize (0.015") intake valve stem diameter is 0.3872-0.3864".
Oversize (0.030") intake valve stem diameter is 0.4022-0.4014".

Note 2 Standard exhaust valve stem diameter is 0.3712-0.3720".
Oversize (0.015") exhaust valve stem diameter is 0.3870-0.3862".
Oversize (0.030") exhaust valve stem diameter is 0.4020-0.4012".

Note 3 Oil capacity including filter and cooler -- 12 quarts US.

Note 4 Manufacturer recommends Volvo Penta P/N 876047-2 or AC MR43.

Note 5 Sea water cooled thermostat starts to open at 149°F; fresh water at 158°F.

Note 6 Sea water cooled thermostat fully open at 162°F; fresh water at 181°F.

TIGHTENING TORQUE VALUES
All Values Given in Ft. Lbs.

	All Except 454	454 CID	Thread Lube.
Cylinder Head Bolts			
1st stage	36	36	Permatex
2nd stage	65	81	Permatex
Connecting Rod Bearing Bolts	45	48	Engine Oil
Main Bearing Bolts	80	110	Molykote
Flywheel Screw	60	65	Engine Oil
Spark Plugs	18.5	22	Dry
Crankshaft, Timing Gear Nut	60	20	Engine Oil
Intake Manifold Bolts	30	30	Engine Oil
Exhaust Manifold Bolts	25	40	Permatex

ENGINE SPECIFICATIONS -- FORD V8

General

Type designation	AQ190A	AQ240A
Number of cylinder (2 banks, 90° V-from)	8	8
Max output at 70 rev/sec 4200 rev/min	182 hk, (134 kW)	231 hk. (170 kW)
Bore, mm (in.)	101, 6 (4.0)	
Stroke, mm (in.)	76.2 (3.0)	88.9 (3.5)
Displacement, dm^3 (cu. in)	4.95 (302)	5.75 (351)
Compression ratio	7.9:1	8.1:1
Compr. press (starter motor MPa (kp/cm^2)	100-110 (10-11) (142-156 p.s.i.)	
Max speed rev/sec	70 (4200 rev/min)	
Idling speed, rev/sec	10.0-10.8 (600-650 rev/min)	
Direction of rotation, viewed from front	Clockwise	
Total weight, engine, rises and outboard drive, approx kg	405 (890 lb)	410 (900 lb)

Cooling system

Thermostat limit values °C (°F)	66-80 (119-143) DI MA-8575

Lubricating system

Oil quality	Service SE
Oil viscosity	Multigrade Oil SAE 10W-30 or 10W-40
Oil capacity including oil filter, litres (Imp. qts./US qts.) appr.	5.2 (4.6/5.5)
Oil capacity excluding cleaner, litres (Imp. qts./US qts.) appr.	4.8 (4.2/5.1)
Oil capcity between max and min marks approx litre (Imp qts./US qts.)	1.0 (1.0 qt)

Ignition system

Firing order	1-5-4-2-6-3-7-8	1-3-7-2-6-5-4-8
Ignition distributor, type PRESTOLITE	IBM 7007S	IBM 7008S
Basic setting (0-90 r/m)	10° B.T.D.C.	
Ignition distributor, contact gap, mm	0.36-0.48 (0.014-0.019 in)	
Dwell angle	31°	
Spark plugs, type AUTOLITE	ARF-32M	
Spark plugs, electrode gap, mm	0.8-0.9 (0.032-0.036 in)	
Stroboscope setting 70 r/s (4200 r/m)	37°	

Electrical system

Voltage V	12V Negative Earth
Battery capacity, Standard Ah	60
Battery electrolyte specific gravity:	
Fully charged battery	1.275-1.285
When re-charging is necessary	1.230
Alternator	Ford Autolite (DOFF-10300-k)
Output, max	450 W (38A)
Starter motor, type Ford Motorcraft	D5FF 11001 AA
Output, kW	0.8

Fuel quality

Leaded fuel min	94 octane R.O.N.

Fuel system

Fuel pump, Carter, type	EM D2JL 9350A (M-66 96S)	
Carburetor, Holley, type	D3JL 9510S	D2JL 9510E

ENGINE SPECIFICATIONS -- FORD V8

GENERAL SPECIFICATIONS

Engine	Bore and Stroke	Compression Pressure PSI (Sea Level) @ Cranking Speed	Oil Pressure-Hot @ 2000 RPM	Firing Order	Drive Belt Tensions (lbs.) (5)			
					Size	Newly Installed (1)	Used: Up to 10 Min.	Used: Over 10 Min.
AQ190	4.00 x 3.00	The lowest reading must be	40-60	1-5-4-2-6-3-7-8	1/4"	50-80	40-80 (2)	40-60 (2)
AQ240	4.00 x 3.50	within 75% of the highest	40-65	1-3-7-2-6-5-4-8	Exc. 1/4"	120-160	90-160 (3)	75-120 (4)

(1) Read tension immediately after belt is installed, before it stretches or seats in pulley grooves.
(2) If less than 40 pounds, readjust to 40-60 pounds.
(3) If less than 90 pounds, readjust to 90-120 pounds.
(4) If less than 75 pounds, readjust to 90-120 pounds.
(5) A used belt is any in operation for 10 minutes.

CYLINDER HEAD

Engine	Combustion Chamber Volume	Valve Guide Bore Diameter (Standard Intake and Exhaust)	Valve Seat Width		Valve Seat Angle	Valve Seat Runout (Maximum)	Valve Arrangement (Front to Rear)	Rocker Arm Stud Bore Dia. Std.	Gasket Surface Flatness ①
			Intake	Exhaust					
AQ190	56.7-59.7	0.3433-0.3443	0.060-0.080	0.060-0.080	45°	0.0020	Right I-E-I-E-I-E-I-E	0.3680-0.3695	0.003 inch In Any 6 Inches
AQ240	58.9-61.9			0.060-0.090			Left E-I-E-I-E-I-E-I		0.006 Overall

① Head Gasket Surface Finish R.M.S. 60-150

VALVE ROCKER ARMS, PUSH RODS AND TAPPETS

Engine	Rocker Arm Lift Ratio	Valve Push Rod (Maximum Runout)	Valve Tappet Or Lifter			Collapsed Tappet Gap	
			Standard Diameter	Clearance To Bore ①	Hydraulic Lifter Leakdown Rate	Allowable	Desired
AQ190	1.61:1	0.015	0.8740-0.8745	0.0007-0.0027	5-50 Seconds Maximum — Measured at 1/16 Inch plunger travel	0.090-0.190	0.115-0.130
AQ240						0.106-0.206	0.131-0.181

① Wear Limit - 0.005

VALVE SPRINGS

Engine		Valve Spring Pressure Lbs @ Specified Length		Valve Spring Free Length Approximate	Valve Spring Assembled Height Pad to Retainer	Valve Spring Out-of-Square (Maximum)
		Pressure	Wear Limit			
AQ190	Intake	76-84 @ 1.69 190-210 @ 1.31	10% Pressure loss @ Specified Length	1.94	1-43/64-1-45/64	5/64 (0.078)
	Exhaust	76-84 @ 1.60 190-210 @ 1.22		1.85	1-19/32-1-39/64	
AQ240		71-79 @ 1.790 190-210 @ 1.340		2.06	1-49/64-1-13/16	

VALVES

Engine	Valve Stem To Valve Guide Clearance ①		Valve Head Diameter		Valve Face Angle ②
	Intake	Exhaust	Intake	Exhaust	
AQ190 AQ240	0.0010-0.0027	0.0015-0.0032	1.773-1.791	1.442-1.460	44°

① Wear Limit . 0.005 ② Valve Face Runout – Maximum 0.002.

VALVES (Continued)

Engine	Valve Stem Diameter							
	Standard		0.003 Oversize		0.015 Oversize		0.030 Oversize	
	Intake	Exhaust	Intake	Exhaust	Intake	Exhaust	Intake	Exhaust
AQ190 AQ240	0.3416-0.3423	0.3411-0.3418	0.3446-0.3453	0.3441-0.3448	0.3566-0.3573	0.3561-0.3568	0.3716-0.3723	0.3711-0.3718

ENGINE SPECIFICATIONS -- FORD V8

Engine	Lobe Lift ①		Theoretical Valve Lift		Camshaft End Play ②	Camshaft Journal to Bearing Clearance Clearance ③	Timing Chain Deflection (Maximum)
	Intake	Exhaust	Intake	Exhaust			
AQ190	0.2303	0.2375	0.3707	0.3823	0.001-0.007	0.001-0.003	0.005
AQ240	0.2600	0.2780	0.4180	0.4480			

① Maximum allowable lobe lift loss (All engines) ...... 0.005 ② Wear Limit – 0.009 ③ Wear Limit – 0.006.

CAMSHAFT (Continued)

Item	Bearing	AQ190 AQ240	Item	Bearing	AQ190 AQ240
Camshaft Journal Diameter–Standard ①	(No 1)	2.0805-2.0815	Camshaft Bearings Inside Diameter	(No 1)	2.0825-2.0835
	(No 2)	2.0655-2.0665		(No 2)	2.0675-2.0685
	(No 3)	2.0505-2.0515		(No 3)	2.0525-2.0535
	(No 4)	2.0355-2.0365		(No 4)	2.0375-2.0385
	(No 5)	2.0205-2.0215		(No 5)	2.0225-2.0235
Camshaft Bearing Location ②	(No 1)	0.0050-0.0200			

① Camshaft journal maximum runout 0.005
Camshaft journal maximum out-of-round 0.0005
② Distance in inches that the front edge of the bearing is installed towards the rear from the front face of the cylinder block.

APPROXIMATE OIL PAN CAPACITIES ①

Engine	U.S. Measure	Imperial Measure
AQ190 AQ240	3-1/2 Quarts	4 Quarts

① Add 1/2 quart with filter replacement

CYLINDER BLOCK

Engine	Cylinder Bore Diameter ① ③	Lifter Bore Diameter	Main Bearing Bore Diameter	Cylinder Block Distributor Shaft Bearing Bore Diameter	Head Gasket Surface Flatness ②	Crankshaft To Rear Face Of Block Runout TIR Max
AQ190	4.0004-4.0052	0.8752-0.8767	2.4412-2.4420	0.4525-0.4541	0.003 inch in any 6 inches or 0.006 inch overall	0.010
AQ240	4.0000-4.0048		3.1922-3.1930	0.5155-0.5171		

① Maximum out-of-round 0.0015
Wear Limit 0.005
Cylinder Bore Surface Finish RMS 18-38
② Head Gasket Surface Finish RMS 60-150
③ Maximum Taper 0.001
Wear Limit 0.010

CRANKSHAFT AND FLYWHEEL

Engine	Main Bearing Journal Diameter ①	Main Bearing Journal Runout-Maximum ②	Main Bearing Journal Thrust Face Runout	Main Bearing Journal Taper Max	Thrust Bearing Journal Length	Main Bearing Surface Finish RMS Maximum	
						Journal	Thrust Face
AQ190	2.2482-2.2490	0.002	0.001	0.0006 Per Inch	1.137-1.139	12	35 Front 25 Rear
AQ240	2.9994-3.0002						

① Main bearing journal out-of-round maximum 0.0004 (all engines) ② Wear Limit – 0.005

Engine	Connecting Rod Journal Diameter ②	Connecting Rod Bearing Journal Maximum Taper	Crankshaft Free End Play ①	Flywheel Clutch Face Runout	Flywheel Ring Gear Lateral Runout Transmission	
					Standard	Automatic
AQ190	2.1228-2.1236	0.0006 Per Inch	0.004-0.008	0.010	0.030	0.060
AQ240	2.3103-2.3111			–	–	

① Wear Limit – 0.012
② Connecting rod journal out-of-round maximum 0.0006 (all engines).

OIL PUMP

Engine	Rotor-Type Oil Pump Relief Valve Spring Tension Lbs @ Specified Length	Drive Shaft To Housing Bearing Clearance	Relief Valve Clearance	Rotor Assembly End Clearance	Outer Race To Housing (Radial Clearance)
AQ190	10.6-12.2 @ 1.704	0.0015-0.0029	0.0015-0.0029	0.001-0.004	0.001-0.013
AQ240	18.2–20.2 @ 2.49				

ENGINE SPECIFICATIONS -- FORD V8

CRANKSHAFT BEARINGS

Engine	Connecting Rod Bearings			Main Bearings		
	To Crankshaft Clearance		Wall Thickness	To Crankshaft Clearance		Wall Thickness
	Desired	Allowable	Standard ①	Desired	Allowable	Standard ②
AQ190	0.0008-0.0015	0.0008-0.0026	0.0572-0.0577	No. 1 Bearing 0.0001-0.0005 All Others 0.0005-0.0015	No. 1 Bearing 0.0001-0.0020 All Others 0.0005-0.0024	No. 1 Bearing 0.0961-0.0966 All Others 0.0957-0.0962
AQ240				0.0008-0.0015	0.0008-0.0026	0.0957-0.0960

① 0.002 Undersize Thickness .... Add 0.0010 to Standard Thickness.

② 0.002 Undersize Thickness .... Add 0.0010 to Standard Thickness.

CONNECTING ROD

Engine	Piston Pin Bore Or Bushing ID	Connecting Rod Bearing Bore Diameter ①	Connecting Rod Length Center To Center	Connecting Rod Alignment Maximum Total Difference ②		Connecting Rod Assembly (Assembled To Crankshaft)	
				Twist	Bend	Side Clearance	Wear Limit
AQ190	0.9104-0.9112	2.2390-2.2398	5.0885-5.0915	0.024	0.012	0.010-0.020	0.023
AQ240			5.9545-5.9575				

① Connecting rod bearing bore maximum out-of-round and taper (All Engines) 0.0004

② Pin bushing and crankshaft bearing bore must be parallel and in the same vertical plane within the specified total difference at ends of 8-inch long bar measured 4 inches on each side of rod.

PISTON

Engine	Diameter ①				Piston To Cylinder Bore Clearance	Piston Pin Bore Diameter	Ring Groove Width
	Coded Red	Coded Blue	0.003 Oversize	Coded Yellow			
AQ190	3.9984-3.9990	3.9996-4.0002	4.008-4.0014	4.0020-4.0026	0.0018-0.0026	0.9122-0.9126	Upper Compression Ring 0.080-0.081 Lower Compression Ring 0.080-0.081
AQ240	3.9973-3.9984	3.9990-3.9996	4.0002-4.0008	4.0014-4.0020	0.0018-0.0026	0.9124-0.9127	Oil Ring 0.1880-0.1890

① Measured at the piston pin bore centerline at 90° to the pin bore

PISTON PIN

Engine	Length	Diameter		To Piston Clearance	To Connecting Rod Clearance
		Standard	0.001 Oversize		
AQ190	3.010-3.040	0.9120-0.9123	0.9130-0.9133	0.0002-0.0004	Interference Fit
AQ240				0.003-0.0005	

PISTON RINGS

Engine	Ring Width		Side Clearance			Ring Gap Width		
	Compression Ring		Compression Ring ①		Oil Ring	Compression Ring		Oil Ring ②
	Top	Bottom	Top	Bottom		Top	Bottom	
AQ190 AQ240	0.077-0.078	0.077-0.078	0.002-0.004	0.002-0.004	Snug	0.010-0.020	0.010-0.020	0.015-0.055

① Wear Limit .... 0.006 ② Steel Rail

HOLLEY CARB. SPECIFICATIONS

Engine	V6
Carb Model	2300C
Holley No.	R-6317-1A
Ford No.	DIFF-9510-LA
Bore Primary	1-1/2"
Venturi Primary	1-3/16"
Main Jet	No. 64
Power Valve	No.50
Cam Hole Pos.	No.2
Pump Override	.015 Min.
Choke Index	3 Ntch. Ln
Choke Pulldown	.140
Dechoke	.300

Engine	V6 V8*
Carb Model	4150 4160
Holley No.	R-6576A
Ford No.	D2JL-E
Bore Primary	1-9/16"
Bore Secondary	1-9/16"
Venturi Primary	1-1/4"
Venturi Secondary	1-5/16"
Main Jet	No. 64
Power Valve	No. 50
Cam Hole Pos.	No.1
Pump Override	.015 Min.
Choke Index	3 Ntch. Ln
Choke Pulldown	.140
Dechoke	.300

*Except for AQ311, AQ271 and Model 740 (454CID)

Engine	454 CID
Holley No.	84024
Carb Model	4011
Inlet Seat Pri.	0.110"
Inlet Seat Sec.	0.110"
Pri. Main Jet	No. 64
Sec. Main Jet	No. 68
Power Valve Pri.	6.5"
Power Valve Sec.	6.5"
Pri. Pump Shooter	0.026"
Pri. Cluster	21R-854A
Sec. Cluster	21R-858A
Airhorn Gasket	108-64
Pri. Venturi	1.200"
Sec. Venturi	1.415"
Pri. Bore	1.375"
Sec. Bore	2.000"

Engine	AQ311 AQ271*	AQ271**
Carb Model	4150	4150
Primary Jet	74	72
Secondary Jet	85	80, 90
Accelerator pump setting	Position 2	Position 2
Choke setting	4 notches to the right	4 notches to the right

* Up to # 4-100-106052
** From # 4-100-106053

TORQUE SPECIFICATIONS

TORQUE LIMITS – FT-LBS

Item	Engine AQ190	Engine AQ240
Cylinder Head Bolts Step 1	55-65	95-105
Step 2	65-72	105-112
Oil Pan to Cylinder Block	9-11 (5/16 x 18) 7-9 (1/4 x 20)	
Intake Manifold Bolts	23-25	
Exhaust Manifold Bolts	18-24	
Distributor Vacuum Control Valve	15-18	
Flywheel to Crankshaft	75-85	
Main Bearing Cap Bolts	60-70	95-105
Oil Drain Plug	15-25	
EGR Valve to Carburetor Spacer	12-18	
Oil Inlet Tube to Oil Pump	10-15	
Oil Filter Insert to Block	20-30	20-30
Oil Filter to Insert (Cartridge Type)	With grease on gasket surface, hand-tighten until gasket contacts adapter face then tighten 1/2 turn more.	
Camshaft Sprocket to Camshaft	40-45	
Camshaft Thrust Plate to Block	9-12	
Vibration Damper to Crankshaft	70-90	
Crankshaft Pulley to Vibration Damper	35-50	
Connecting Rod Nuts	19-24	40-45

Item	Engine AQ190	Engine AQ240
Valve Rocker Arm Cover	3-5	
Fuel Pump to Cylinder Front Cover	19-27	
Rocker Arm Stud Nut	17-23 Ft-Lbs After Nut Contacts Shoulder.	
Alternator Pivot Bolt	45-57	

MISCELLANEOUS TORQUE VALUES

Item	Torque
Alternator Bracket to Cylinder Block	15-20
Alternator Pivot Bolt	45-57
Alternator Adjusting Arm to Cylinder Block	15-20
Alternator Adjusting Arm to Alternator	24-40
Carburetor Mounting Nuts	12-15
Carburetor Mounting Stud to Intake Manifold	15 max.
Fuel Filter to Carburetor	80-100 in-lb

TORQUE LIMITS FOR VARIOUS SIZE BOLTS–FT-LBS

CAUTION: If any of the torque limits listed in this table disagree with any of those listed in the preceding tables, the limits in the preceding tables prevail.

Size (Inches)	1/4–20	5/16–18	3/8–16	7/16–14	1/2–13	9/16–18
Torque (Ft-Lbs)	6–9	12–18	22–32	40–55	55–80	85–120

Note: Oil threads with engine oil unless the threads require oil resistant or water resistant sealer.

Legend

1. Key switch with starter contact
2. Instrument lighting switch
3. Temperature gauge
4. Warning lamp for "low oil pressure"
5. Revolution counter
6. Charging control lamp
7. Extra switch
8. Connection terminal
9. Cable harness
10. Battery
11. Master switch
12. Starter motor
13. Charging regulator
14. Alternator (or D.C. generator)
15. Fuse
16. Oil pressure sender
17. Temperature sender
18. Advance engaging resistor
19. Ignition coil
20. Distributor
21. Connector
22. Cable harness

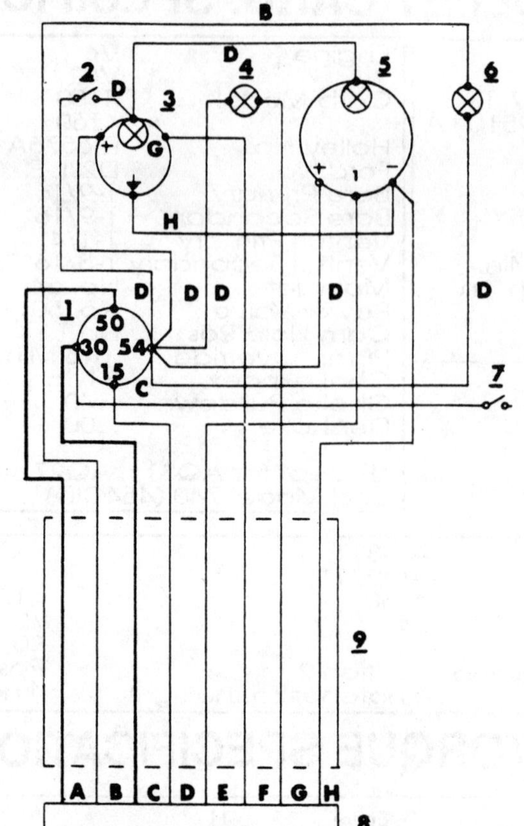

Instrument Panel -- all engines.

Cable markings

Mark	Colour	sq. mm	AV
A	Ivory	6	[illegible]
B	Black	1.5	15
B´	Black	0.6	19
B^+	Black	35	1
C	Red	6	9
C^+	Red	35	1
C^{++}	Red	4	11
C"	Red	0.6	19
D	Green	1.5	15
D^+	Green	0.6	19
E	Grey	1.5	15
F	Yellow	1.5	15
G	Brown	1.5	15
H	Blue	1.5	15
H´	Blue	4	11

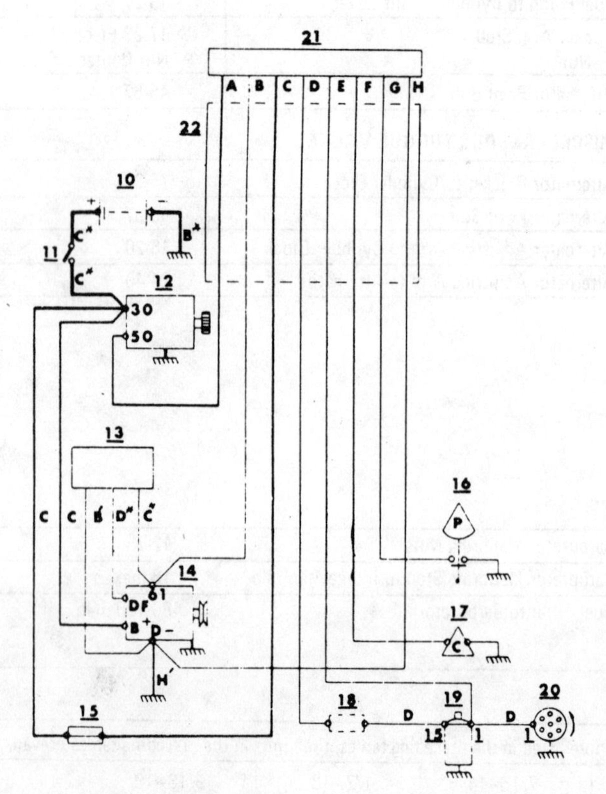

4-cylinder engine with DC generator.

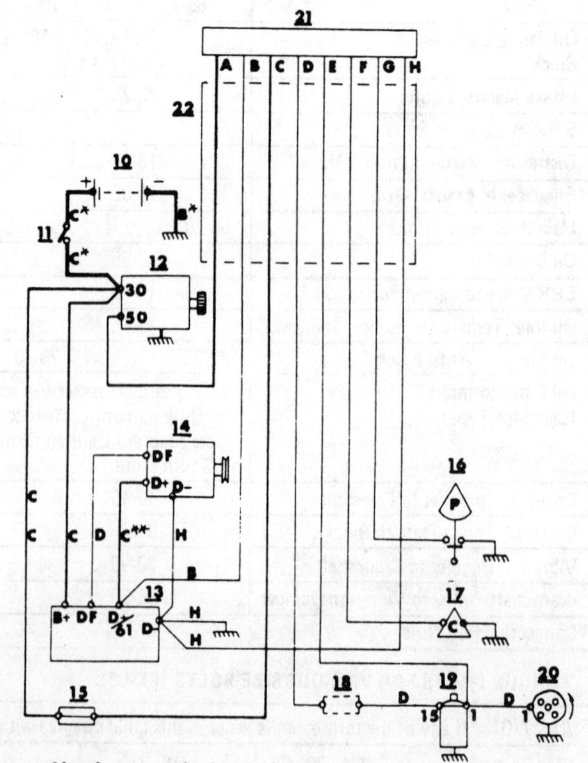

4-cylinder and 6-cylinder engine with alternator.

Wire Identification — AQ105A, AQ115A, AQ130A, B, C, and D, AQ165A, & AQ170A.

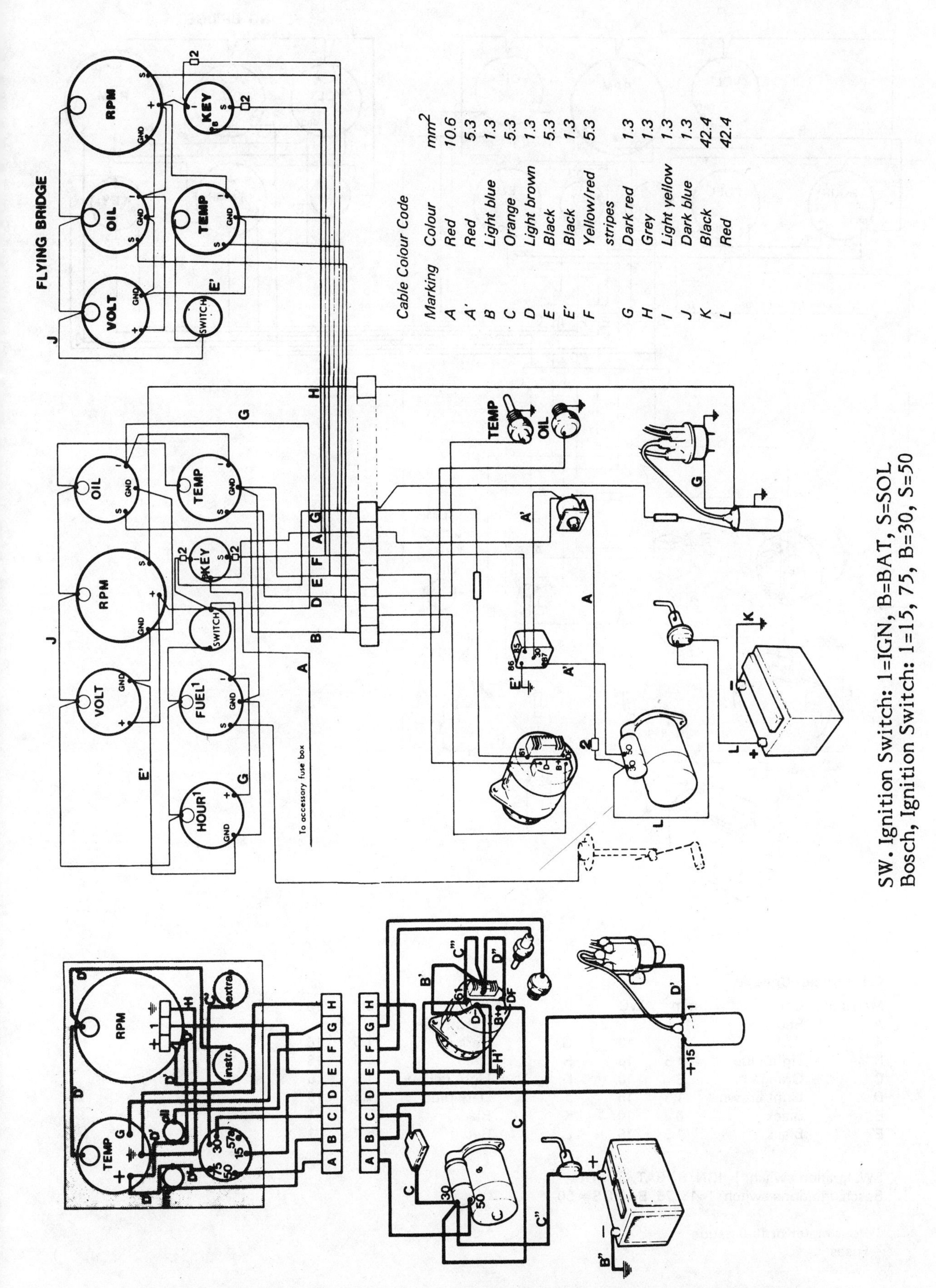

Cable Colour Code

Marking	Colour	mm^2
A	Red	10.6
A'	Red	5.3
B	Light blue	1.3
C	Orange	5.3
D	Light brown	1.3
E	Black	5.3
E'	Black	1.3
F	Yellow/red stripes	5.3
G	Dark red	1.3
H	Grey	1.3
I	Light yellow	1.3
J	Dark blue	1.3
K	Black	42.4
L	Red	42.4

Wire Identification -- AQ120B, AQ125A & B, AQ140A, AQ145A & B.

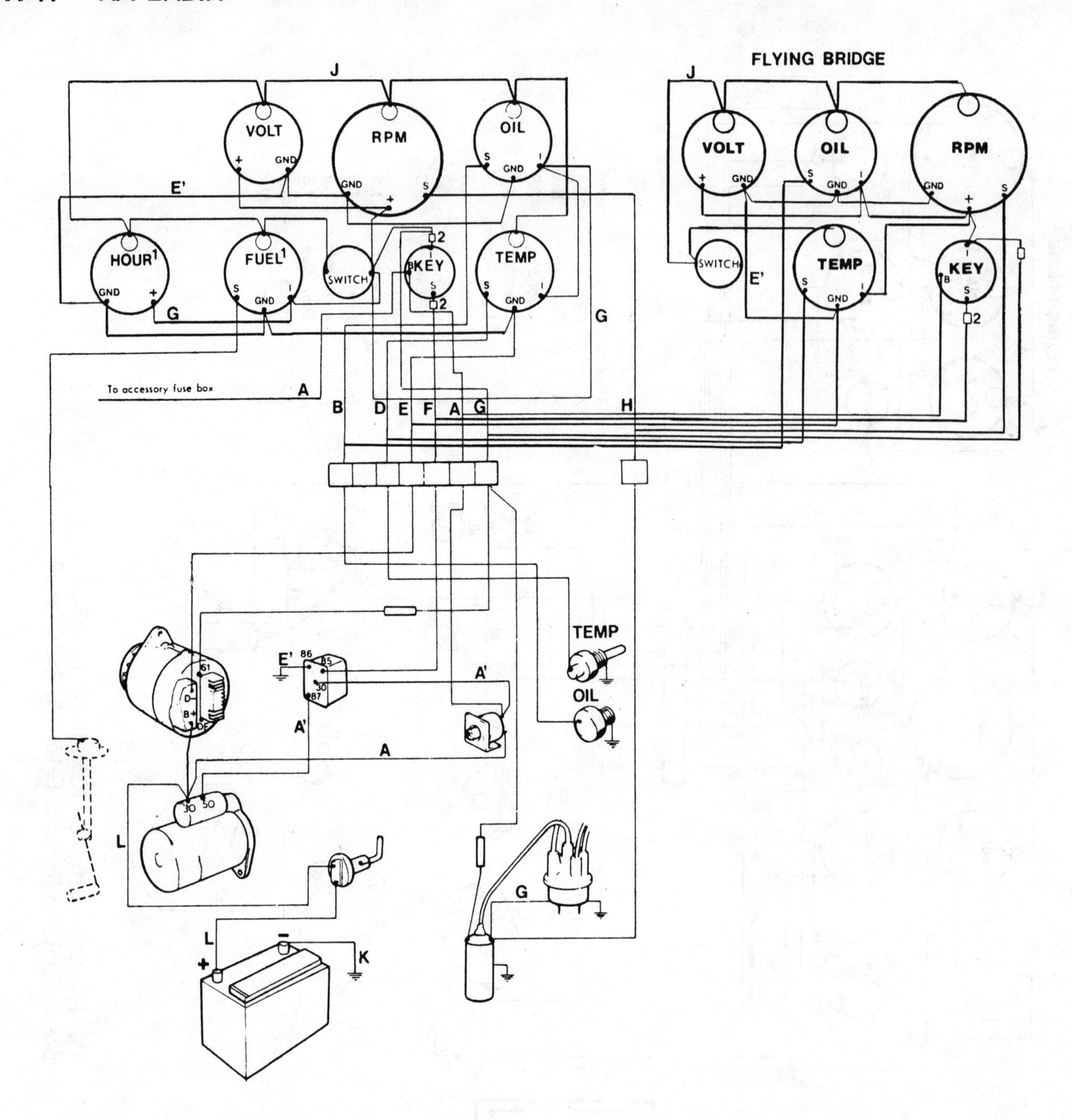

Cable Colour Code

Marking	Colour	mm²	AWG
A	Red	10	8
A	Red	6	10
B	Light blue	1.5	16
C	Orange	6	10
D	Light brown	1.5	16
E	Black	6	10
E′	Black	1.5	16
F	Yellow/ red stripes	6	10
G	Dark red	1.5	16
H	Grey	1.5	16
I	Light yellow	1.5	16
J	Dark blue	1.5	16
K	Black	35	1
L	Red	35	1

SW, Ignition switch: I=IGN, B=BAT, S=SOL
Bosch, Ignitions switch: I=15, 75, B=30, S= 50

1) Hour meter or fuel gauge
2) Fuses

Wire Identification--- AQ131A,B,C,D, AQ151A & B.

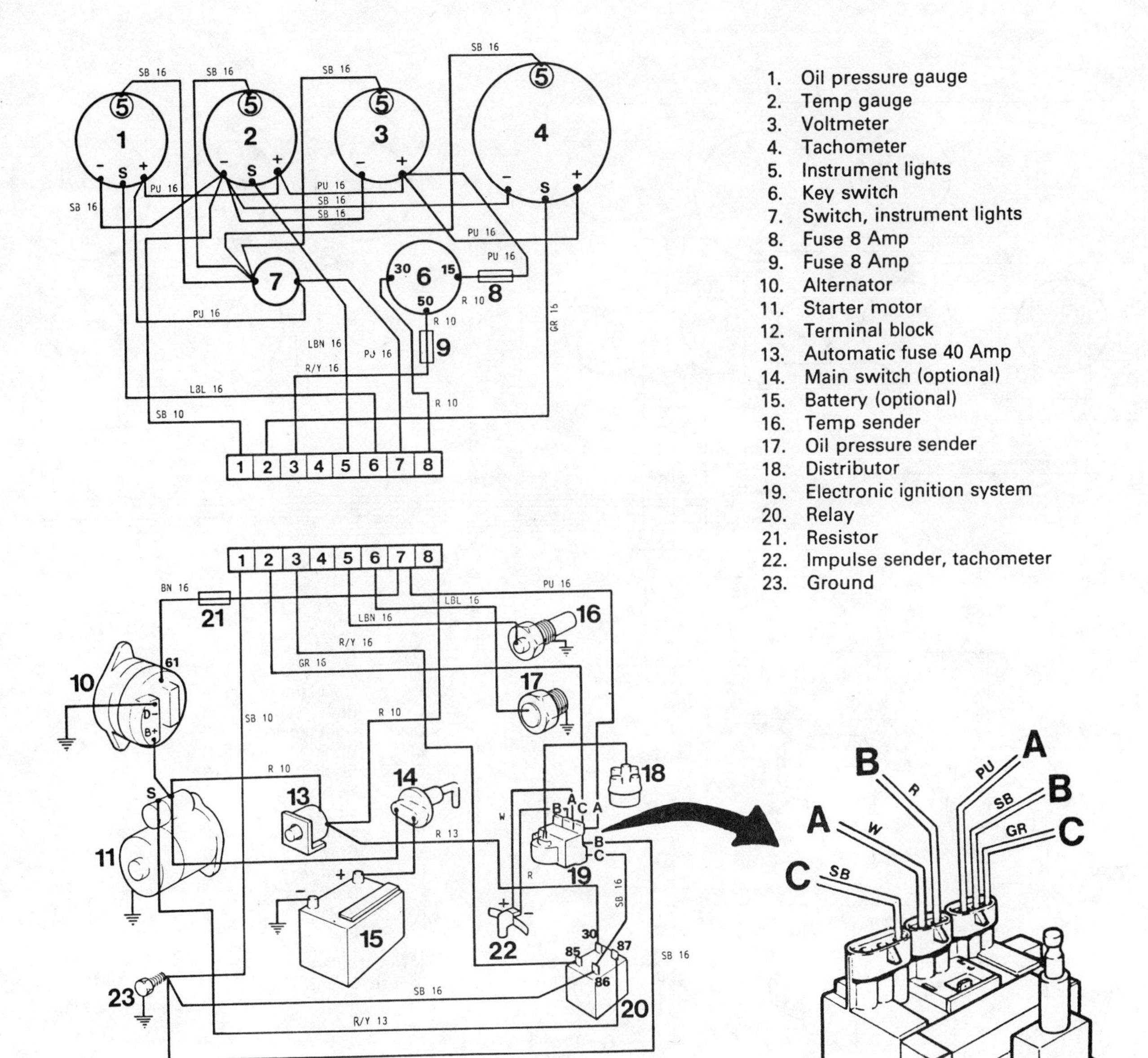

Cable colour Code

SB = Black
PU = Purple
LBN = Light brown
R = Red
GR = Grey
LBL = Light blue
R/Y = Red/Yellow
BN = Brown
W = White

Wire Identification -- AQ171A.

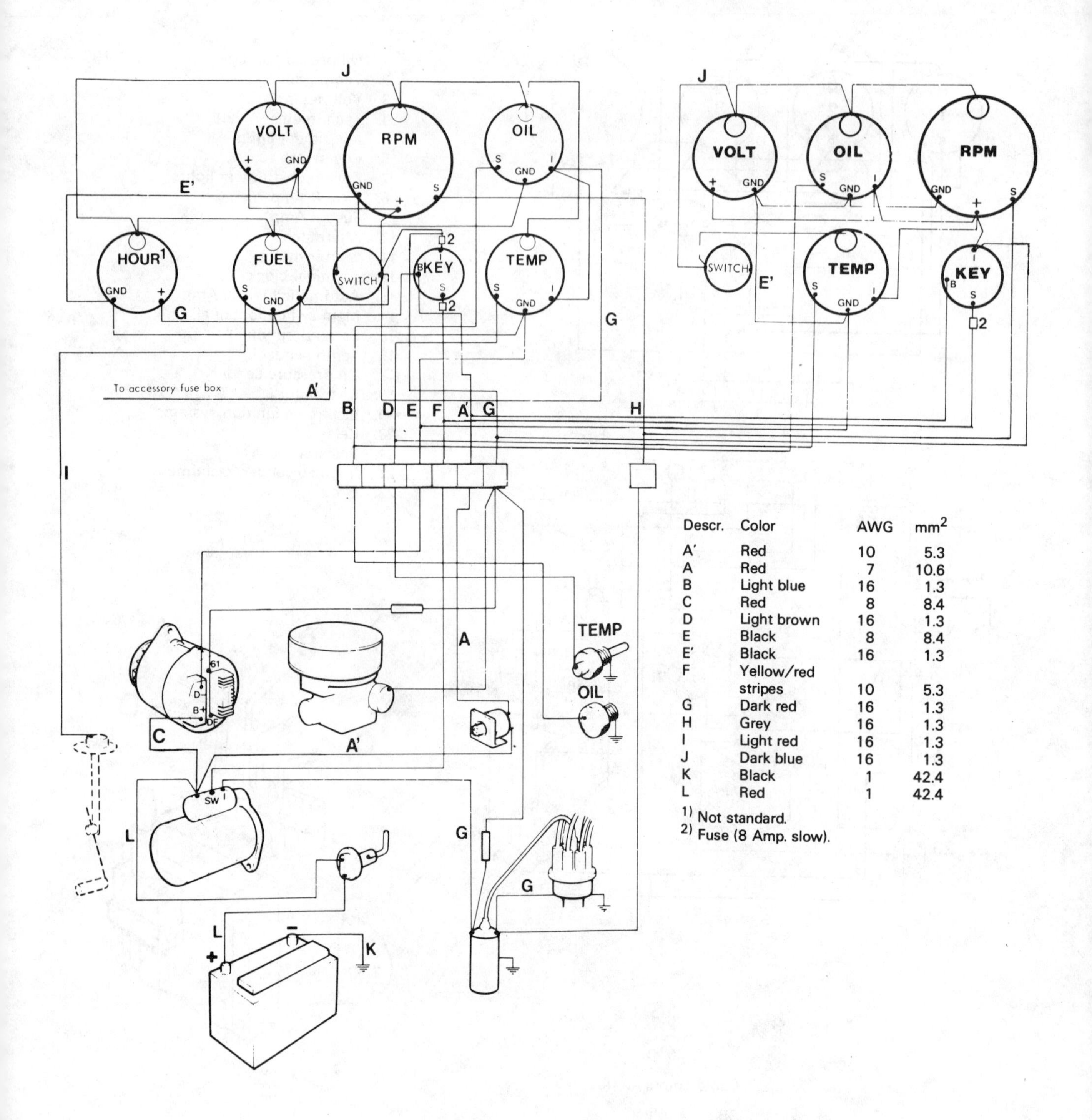

Descr.	Color	AWG	mm²
A'	Red	10	5.3
A	Red	7	10.6
B	Light blue	16	1.3
C	Red	8	8.4
D	Light brown	16	1.3
E	Black	8	8.4
E'	Black	16	1.3
F	Yellow/red stripes	10	5.3
G	Dark red	16	1.3
H	Grey	16	1.3
I	Light red	16	1.3
J	Dark blue	16	1.3
K	Black	1	42.4
L	Red	1	42.4

1) Not standard.
2) Fuse (8 Amp. slow).

Wire Identification -- AQ175A.

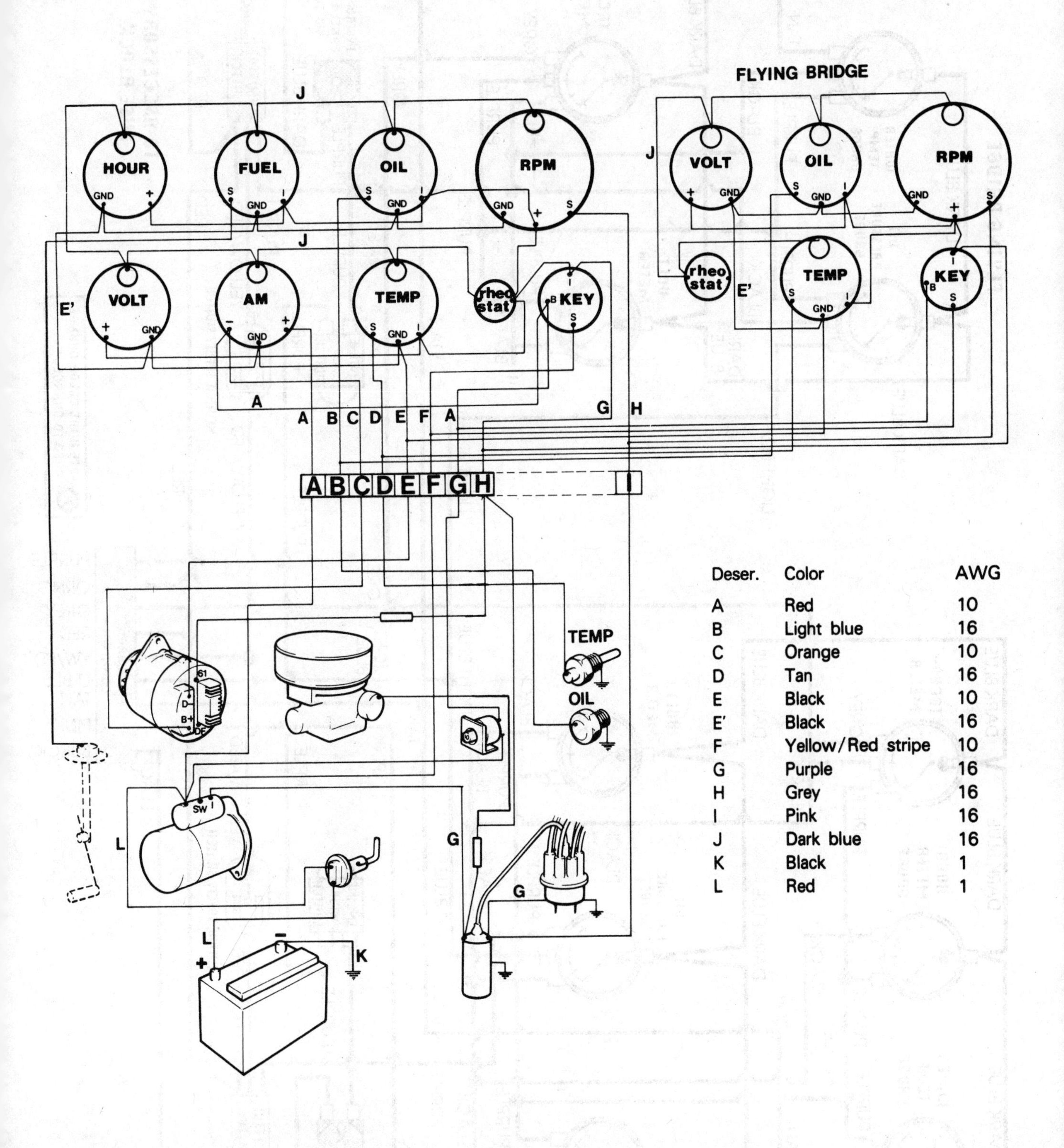

Deser.	Color	AWG
A	Red	10
B	Light blue	16
C	Orange	10
D	Tan	16
E	Black	10
E'	Black	16
F	Yellow/Red stripe	10
G	Purple	16
H	Grey	16
I	Pink	16
J	Dark blue	16
K	Black	1
L	Red	1

Wire Identification -- Early model AQ200D, AQ225D, and AQ255D.

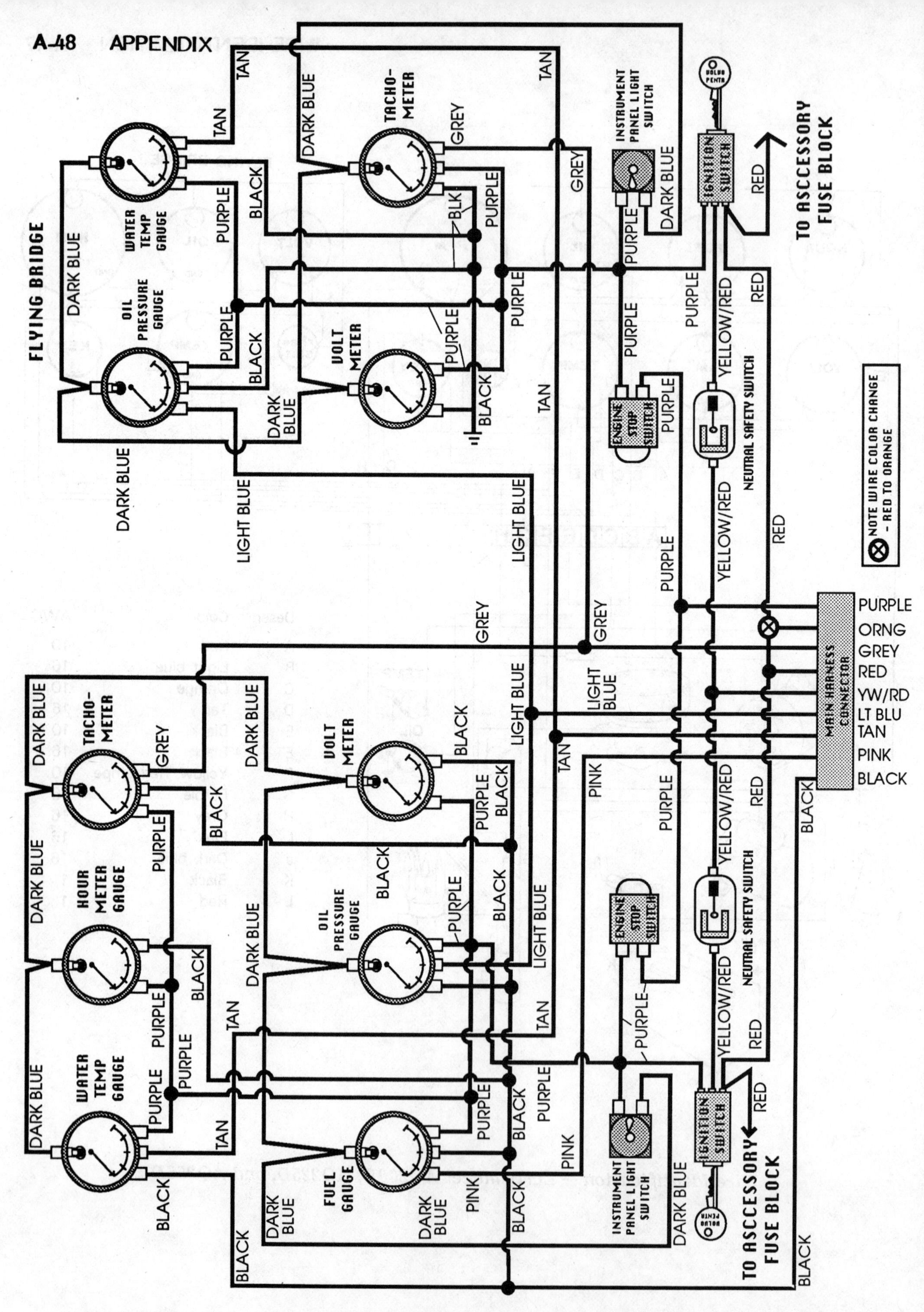

Wire Identification -- Instrument panels -- Model AQ190A and AQ240A with voltmeter.

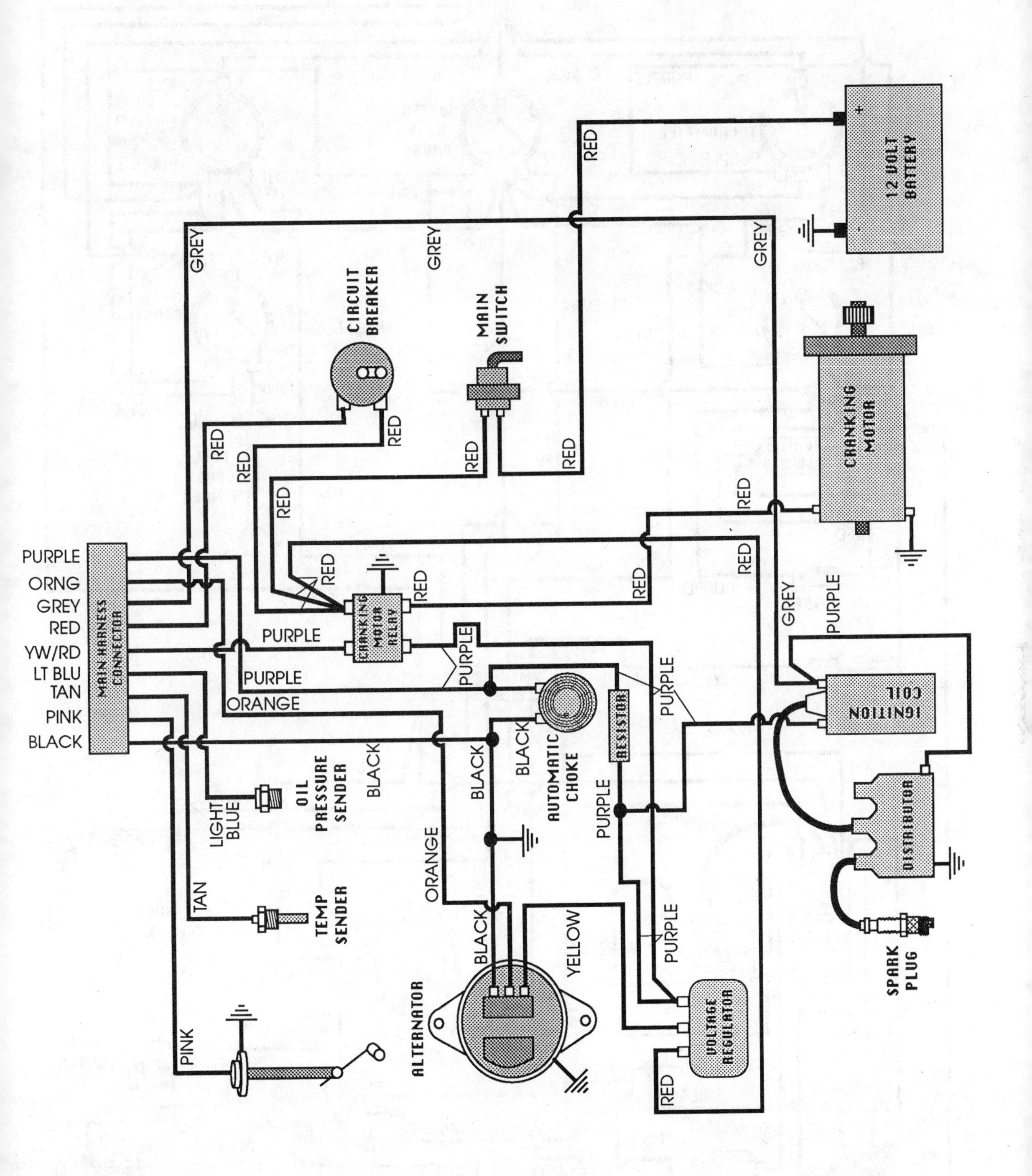

Wire Identification -- Engine compartment -- Model AQ190A and AQ240A.

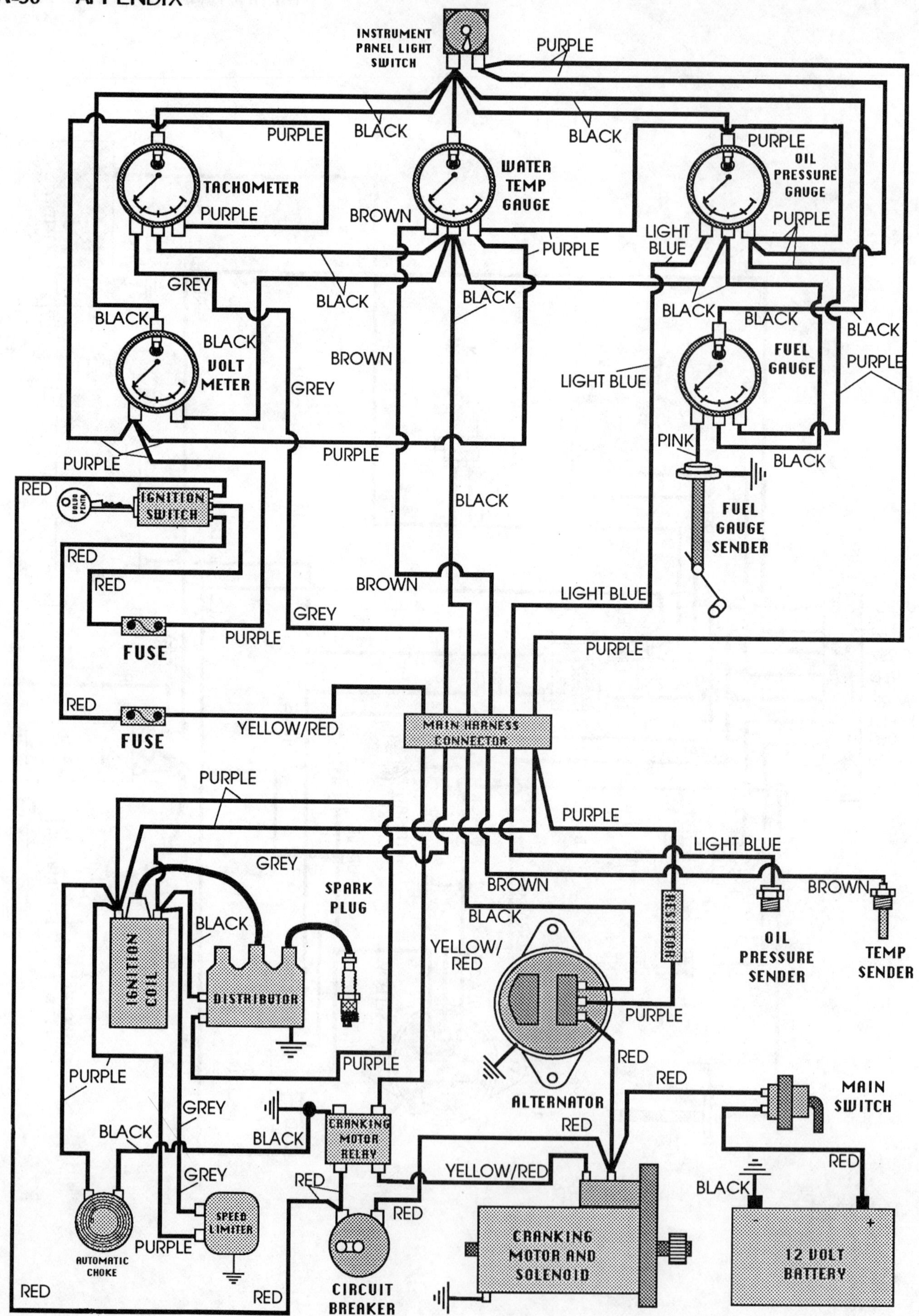

Wire Identification -- Model AQ205, AQ211A, AQ231, AQ271C, and AQ311A & B.

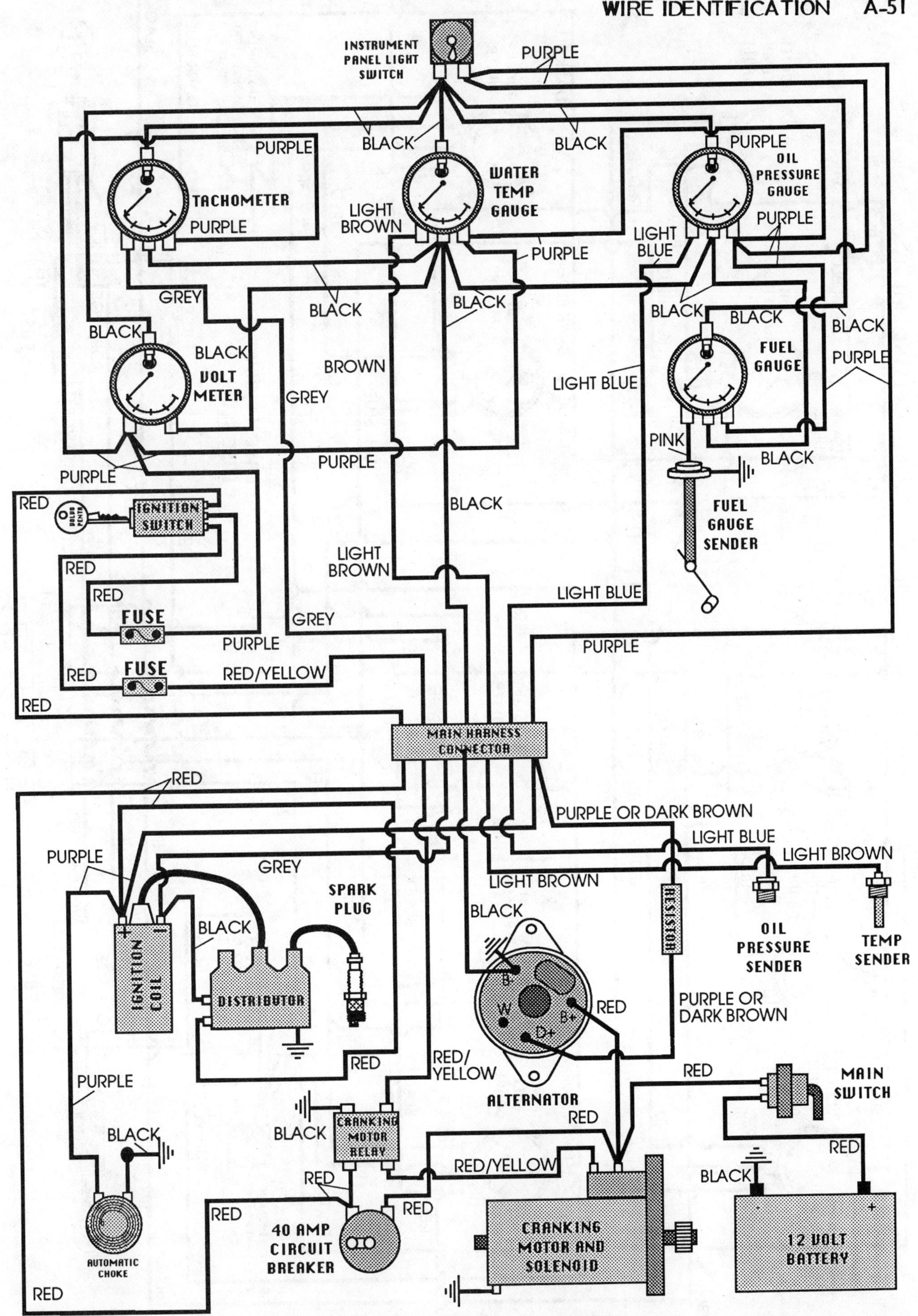

Wire Identification -- Model 431.

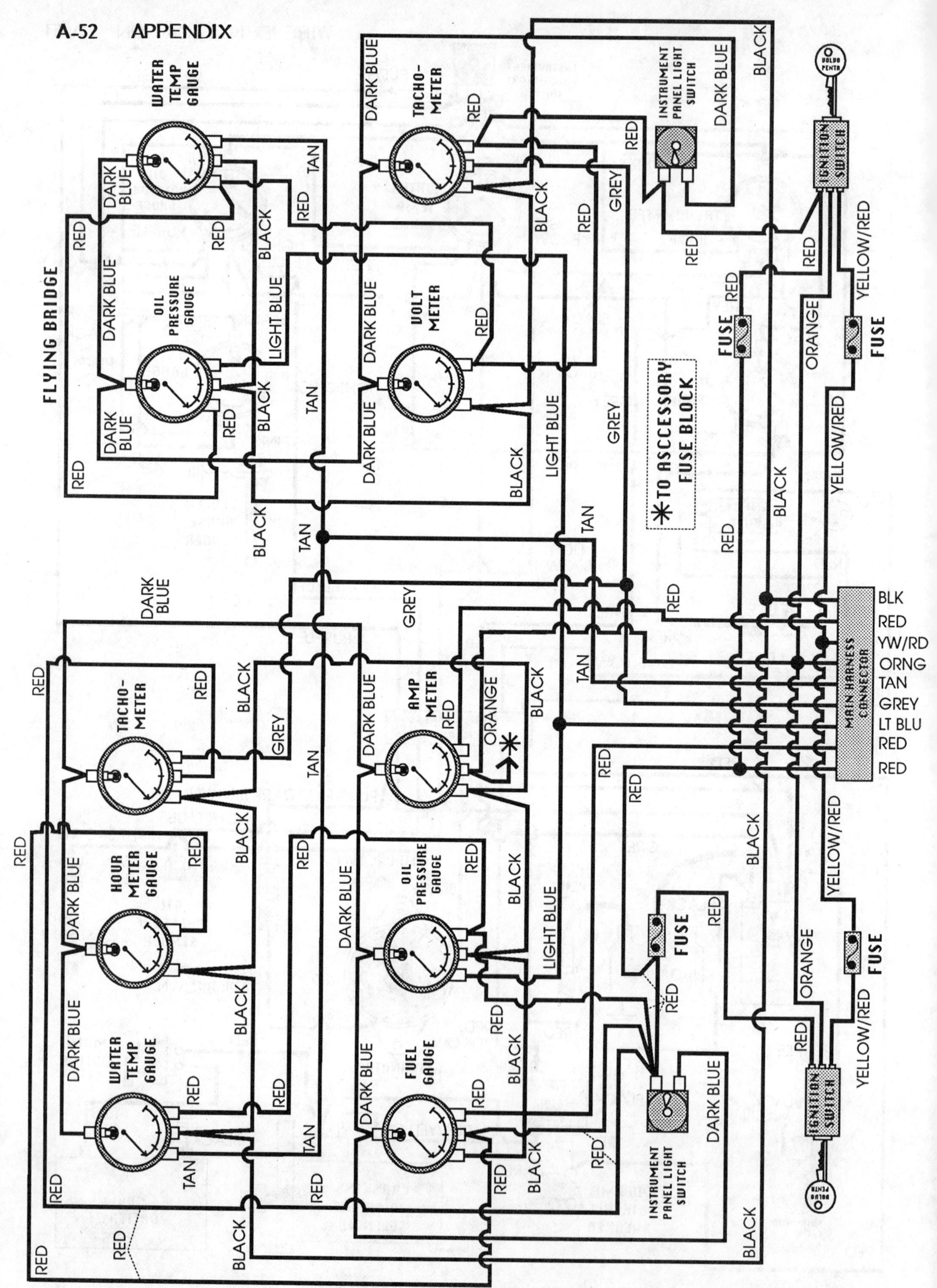

Wire Identification -- Instrument panels -- Model AQ200AB and AQ290A with ammeter.

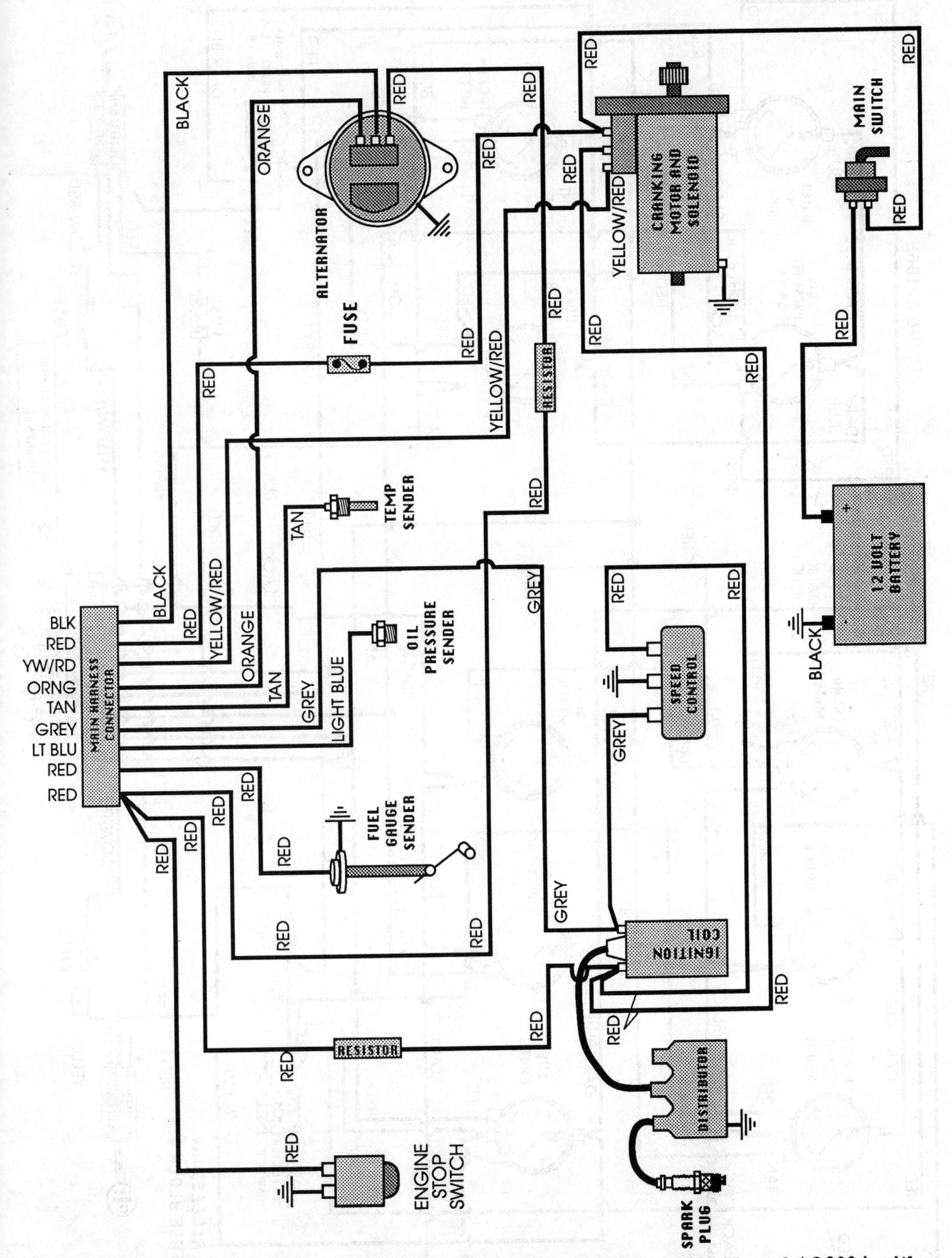

Wire Identification -- Engine compartment -- Model AQ200AB, AQ225AB, and AQ290A with ammeter.

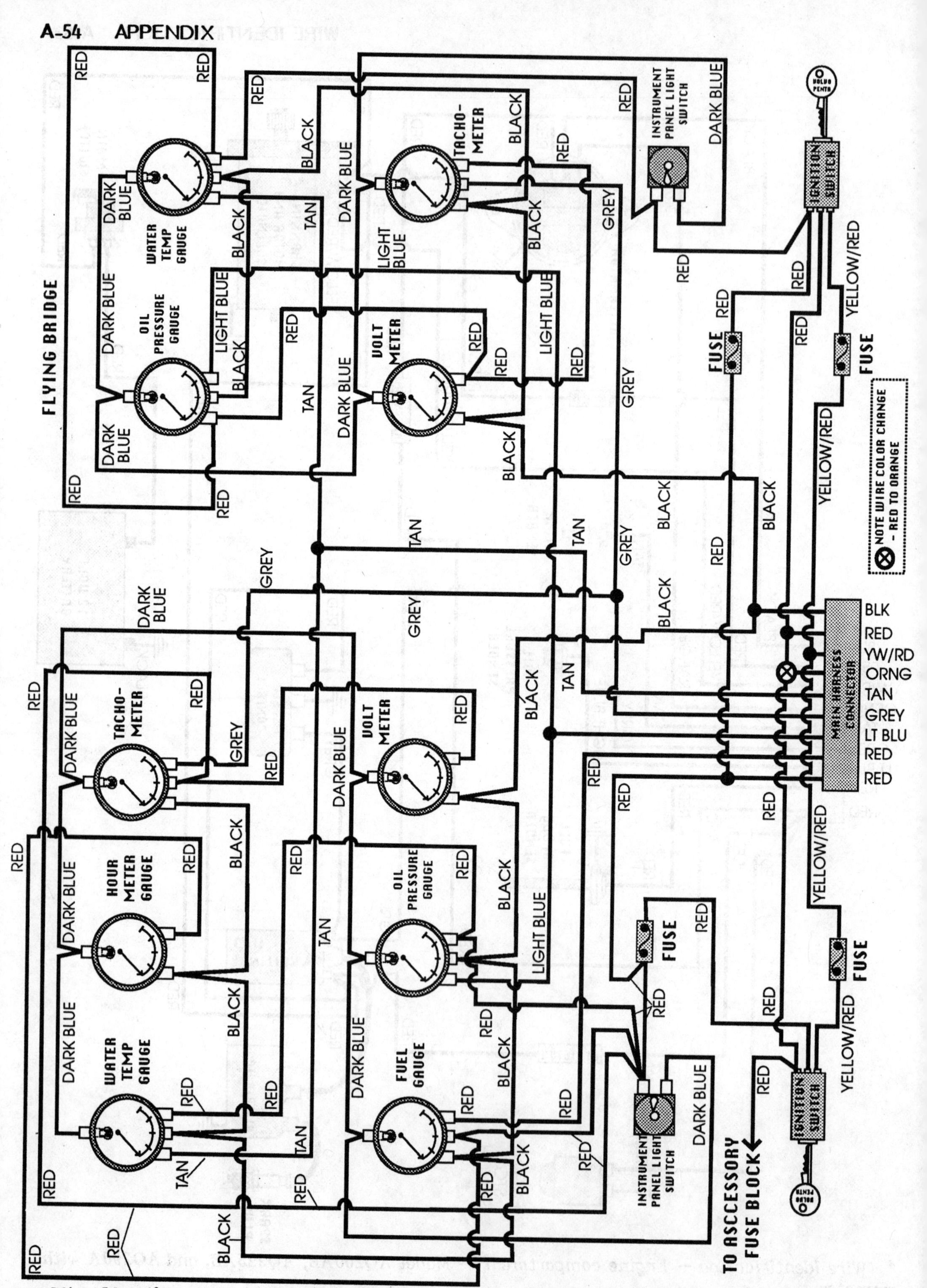

Wire Identification -- Instrument panels -- Model AQ200AB, AQ225AB, and AQ290A with voltmeter.

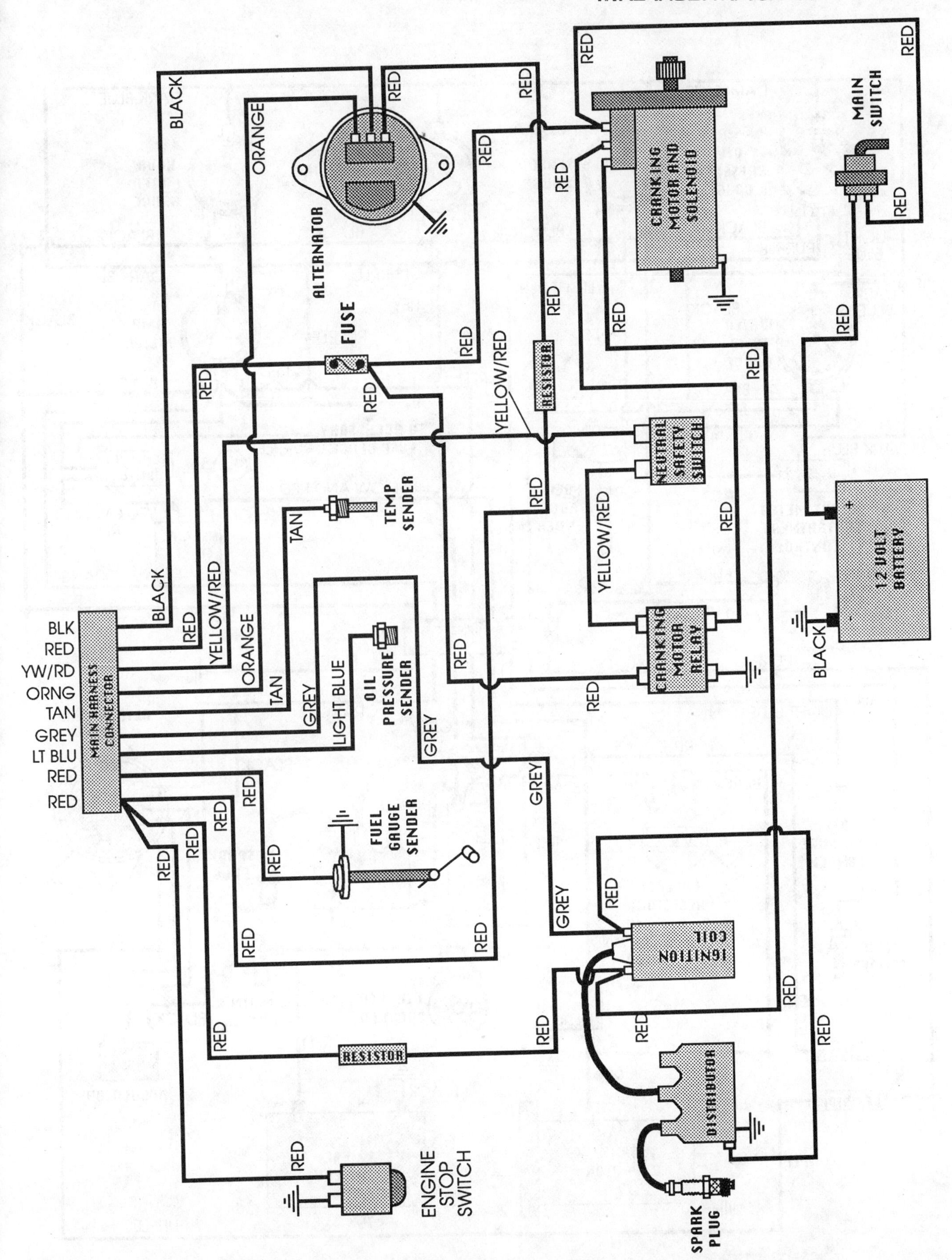

Wire Identification -- Engine compartment -- Model AQ200AB, AQ225AB, and AQ290A with voltmenter.

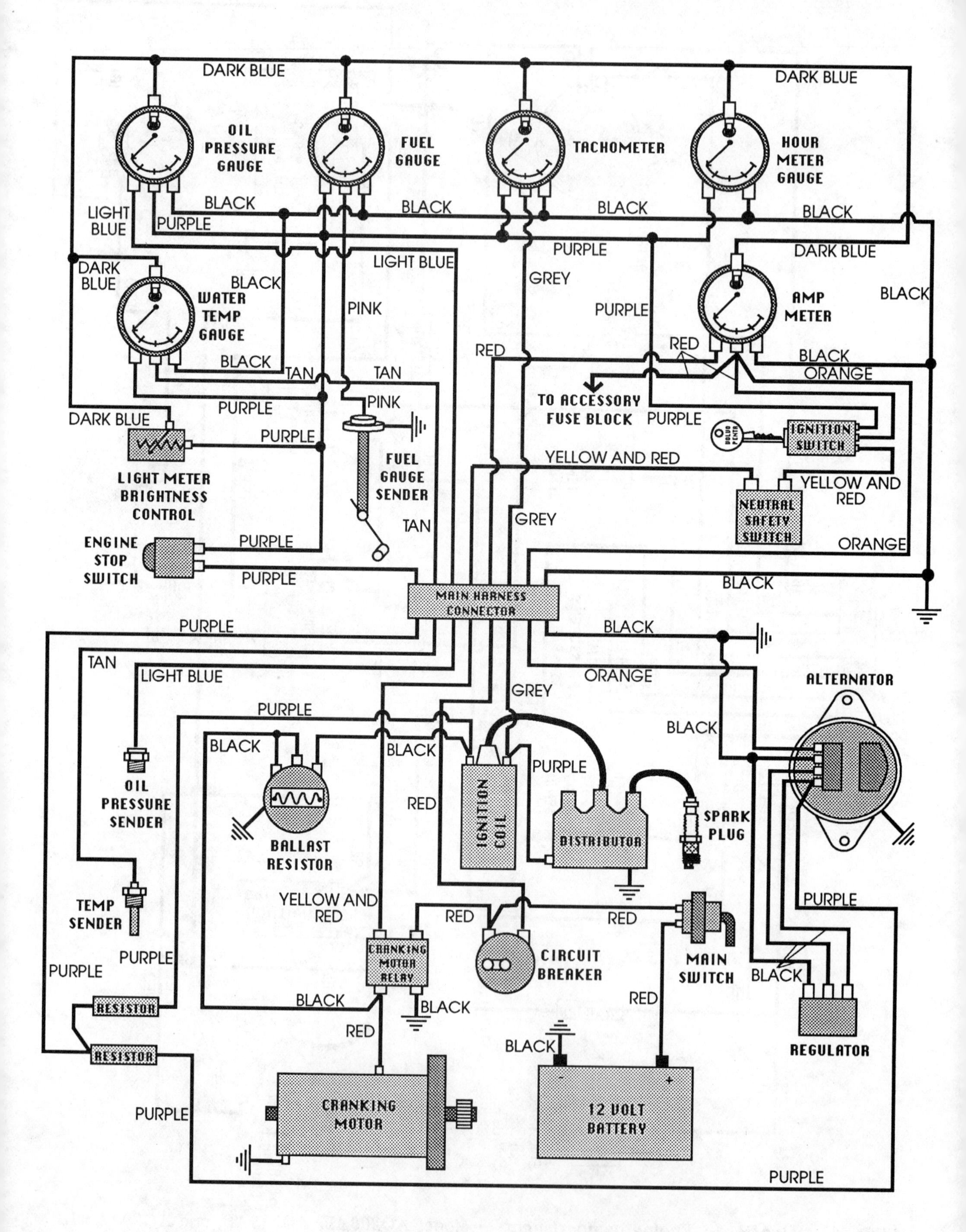

Wire Identification -- Model AQ200C, AQ225C, and AQ225A with ammeter.

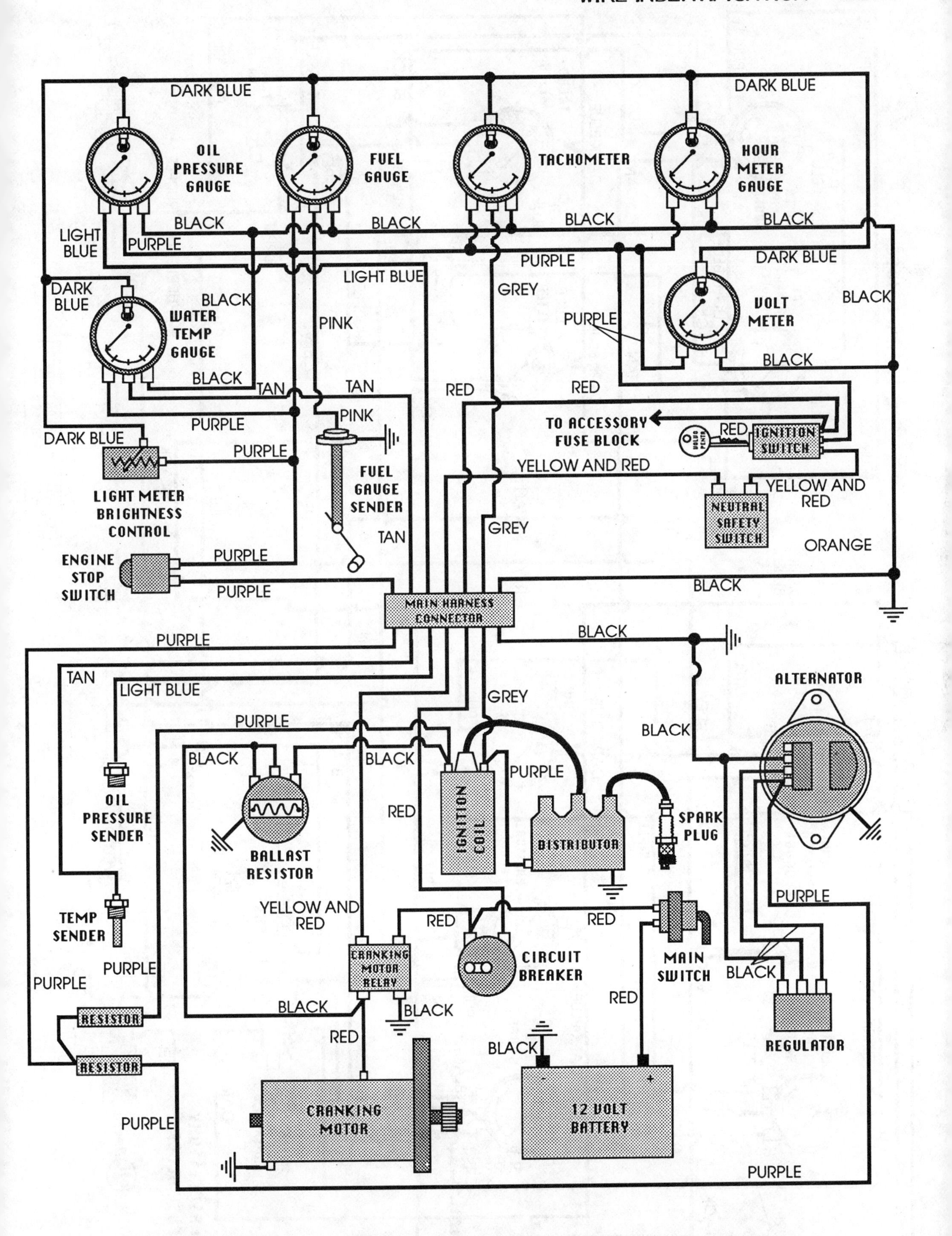

Wire Identification -- Model AQ200C, AQ225C, and AQ225A with voltmeter.

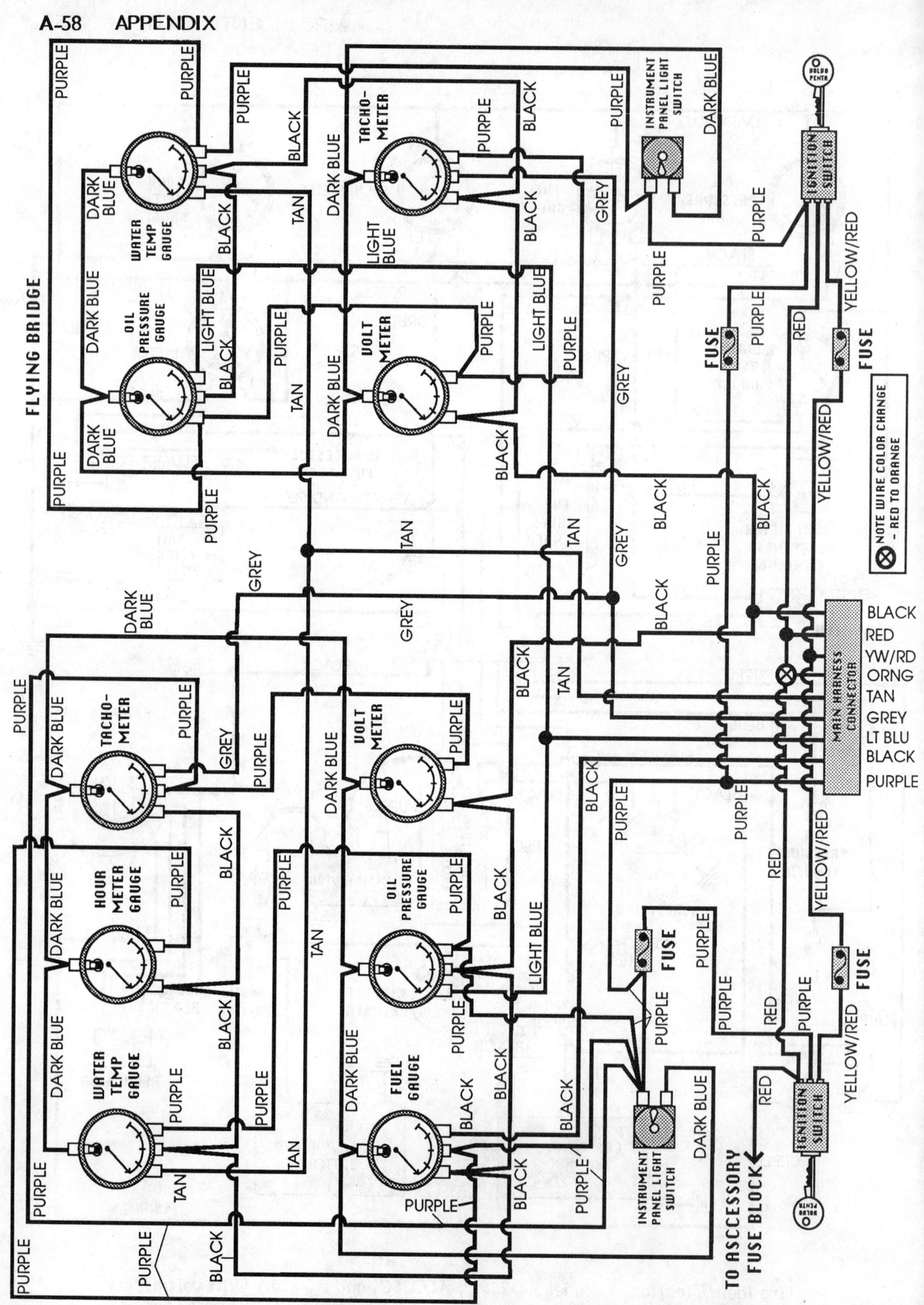

Wire Identification -- Instrument panel -- Model AQ280A and AQ225A.

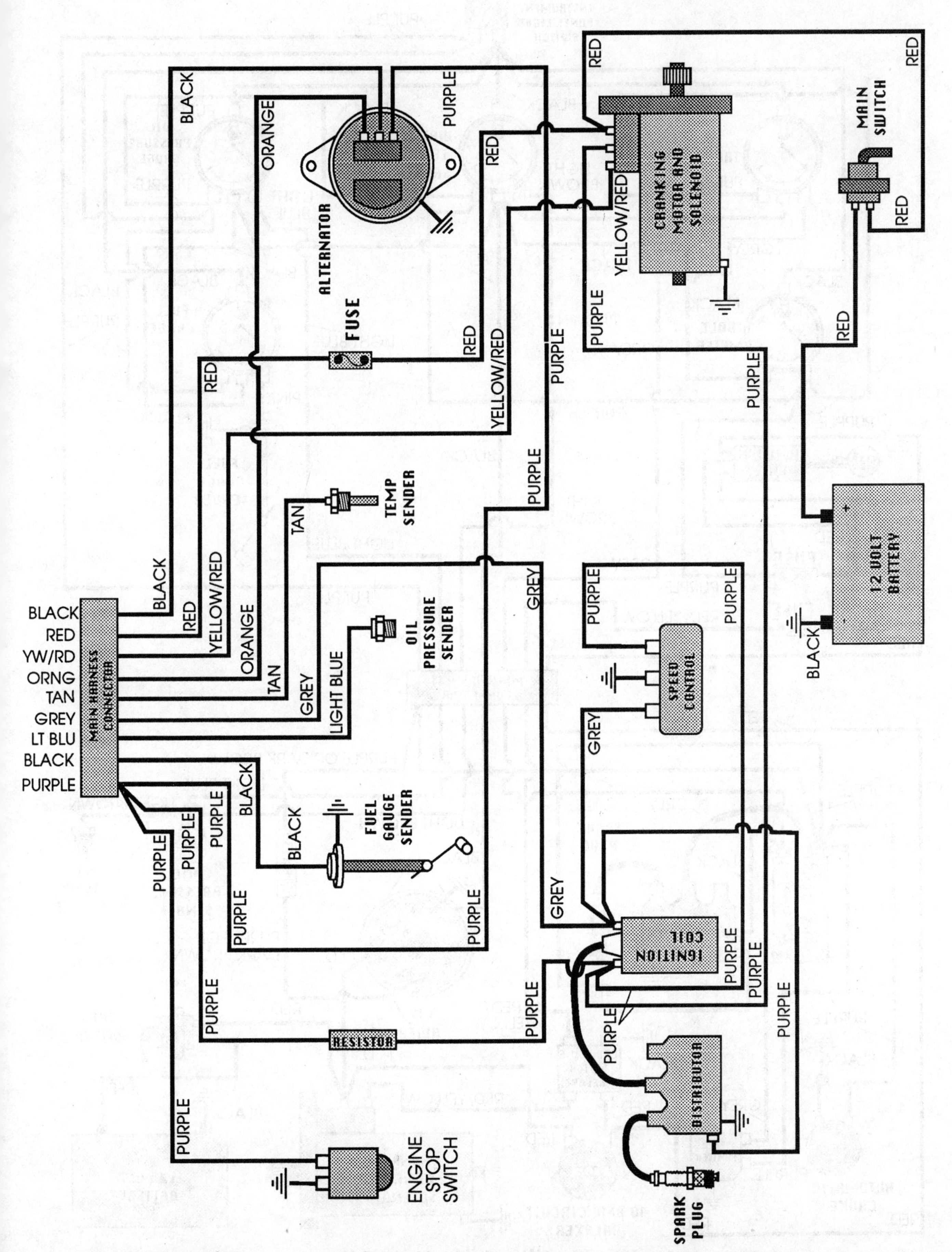

Wire Identification -- Engine compartment -- Model AQ280A and AQ225A.

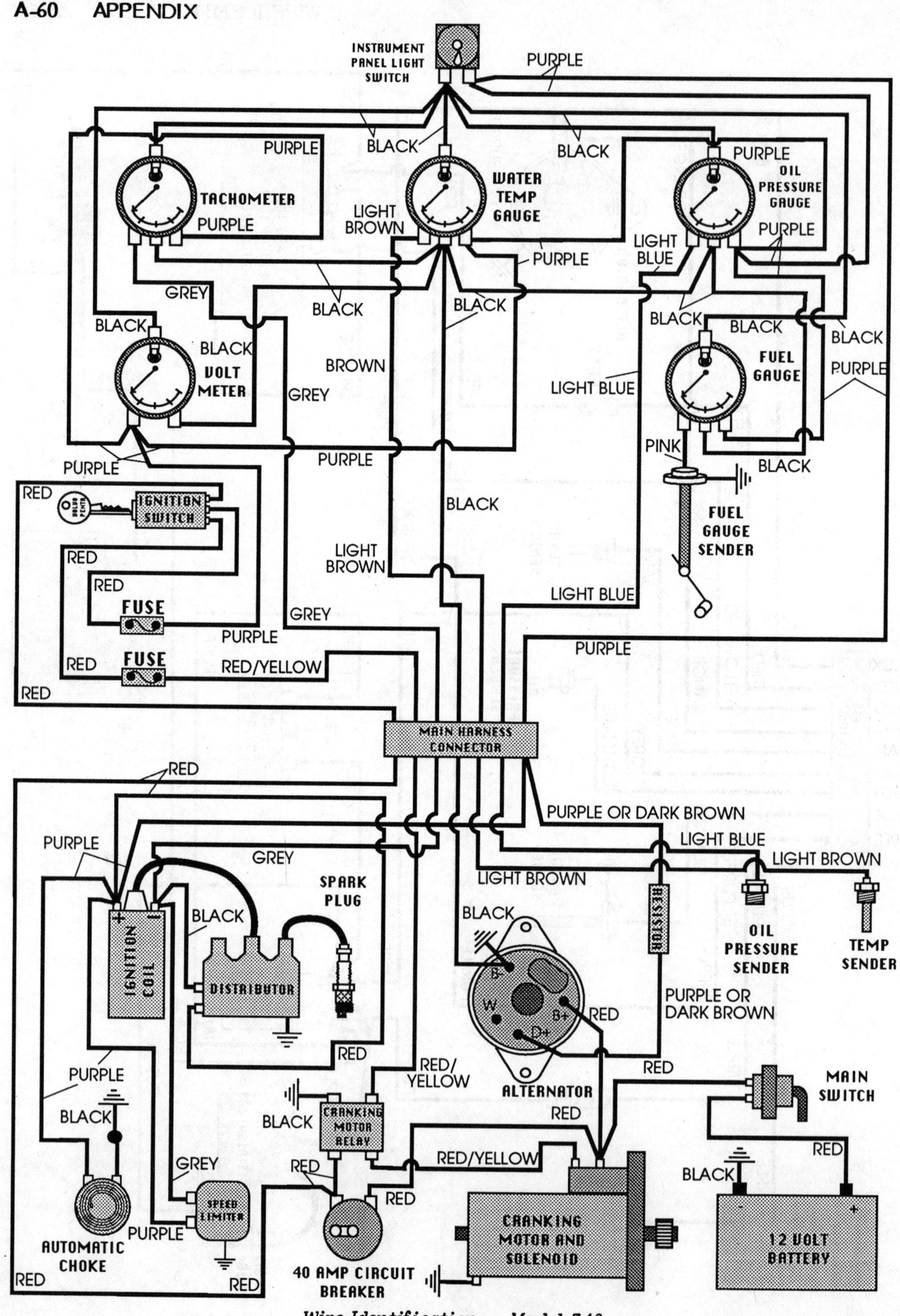

Wire Identification -- Model 740.

Instrument panel with ammeter

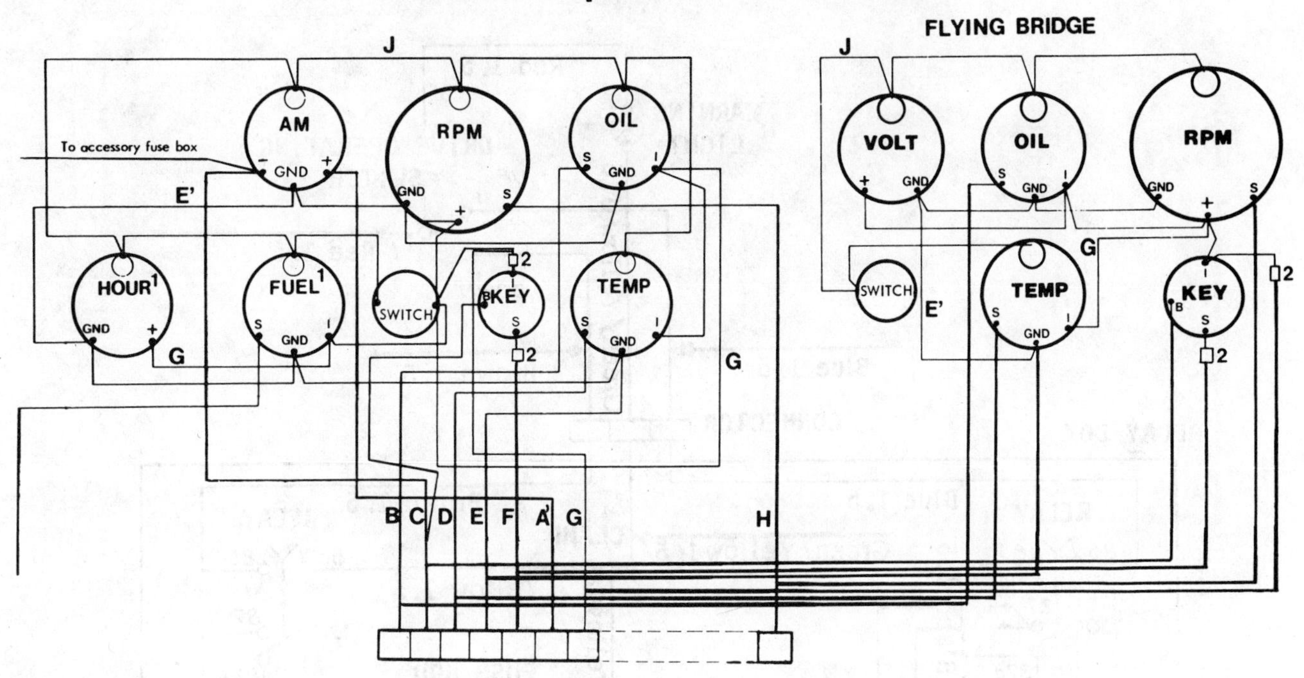

Instrument panel with voltmeter

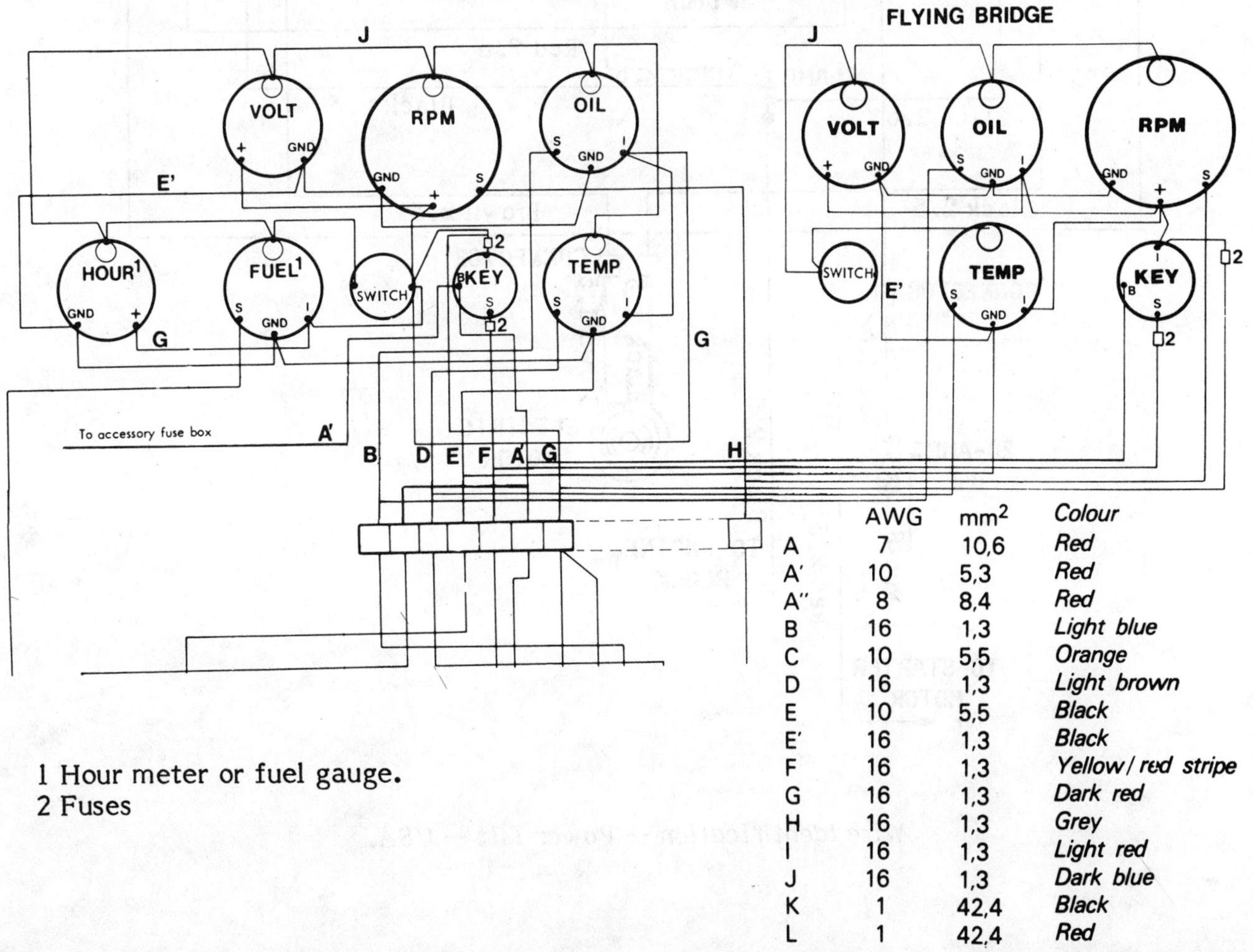

	AWG	mm²	Colour
A	7	10,6	*Red*
A'	10	5,3	*Red*
A''	8	8,4	*Red*
B	16	1,3	*Light blue*
C	10	5,5	*Orange*
D	16	1,3	*Light brown*
E	10	5,5	*Black*
E'	16	1,3	*Black*
F	16	1,3	*Yellow/red stripe*
G	16	1,3	*Dark red*
H	16	1,3	*Grey*
I	16	1,3	*Light red*
J	16	1,3	*Dark blue*
K	1	42,4	*Black*
L	1	42,4	*Red*

1 Hour meter or fuel gauge.
2 Fuses

Wire Identification -- Late Model AQ200D, AQ225D, AQ260A & D, AQ290A, and all other GMC V6, V8, and Ford V8 powerplants.

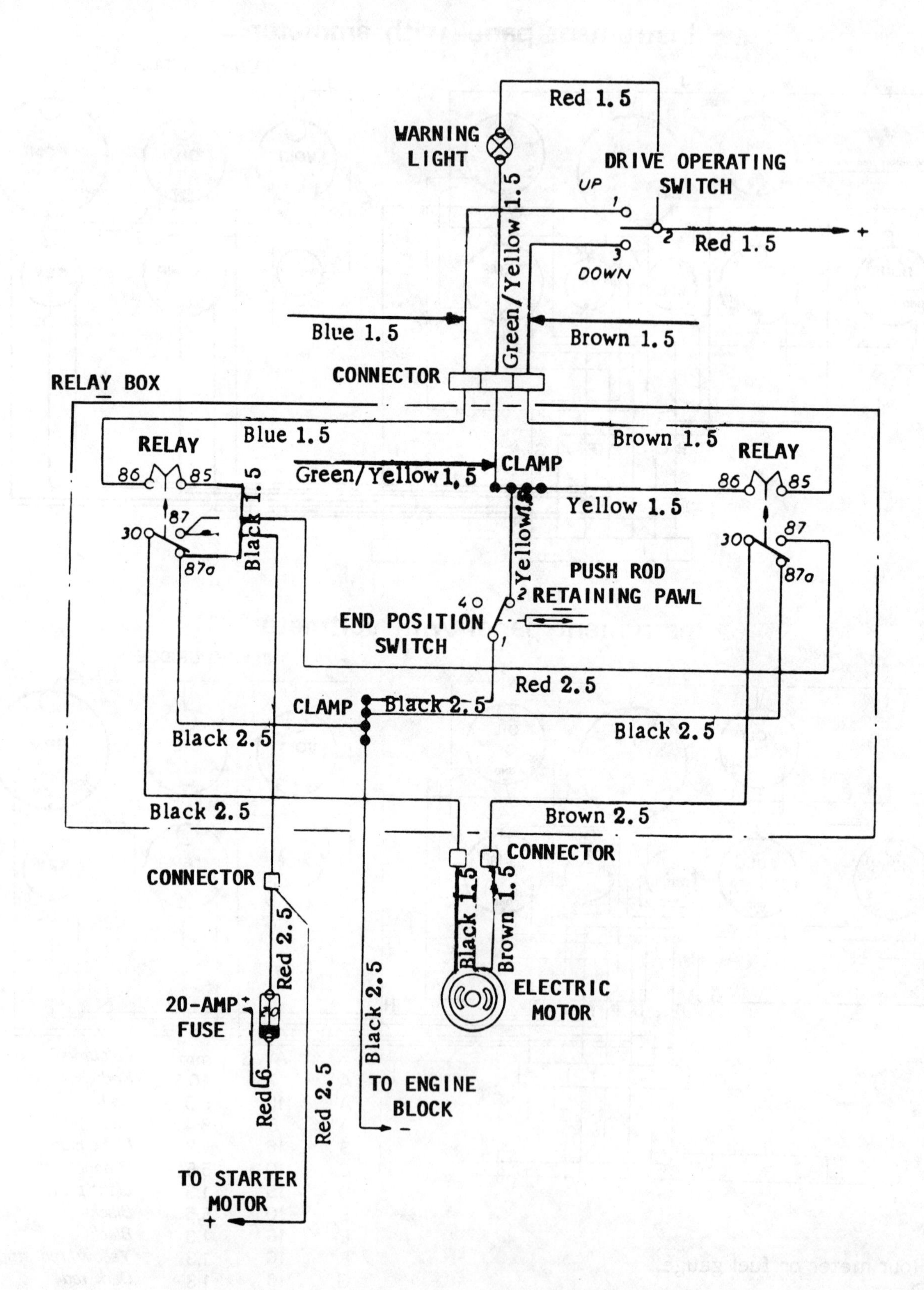

Wire Identification -- Power Tilt -- USA.

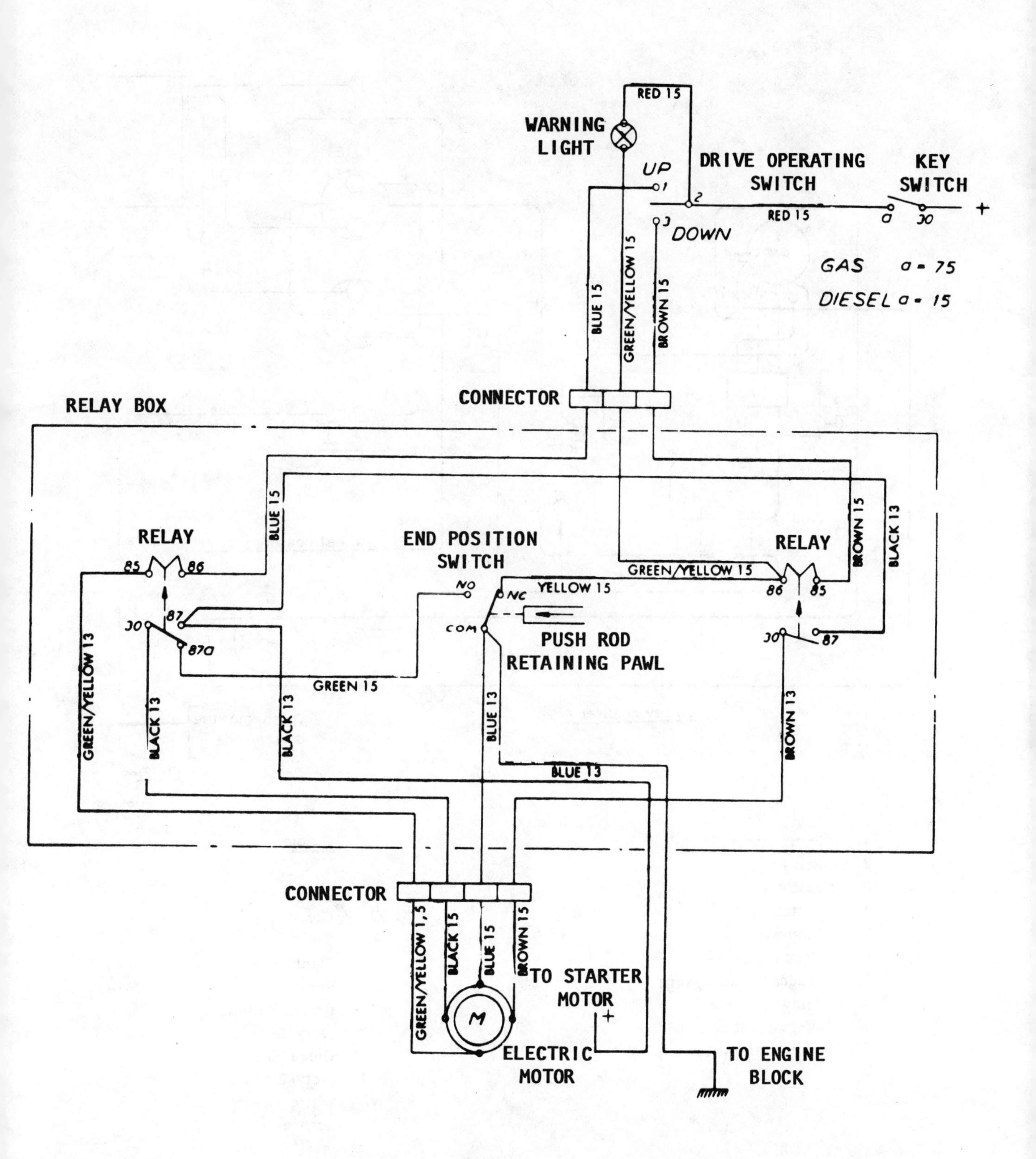

Wire Identification -- Power Tilt -- all except USA.

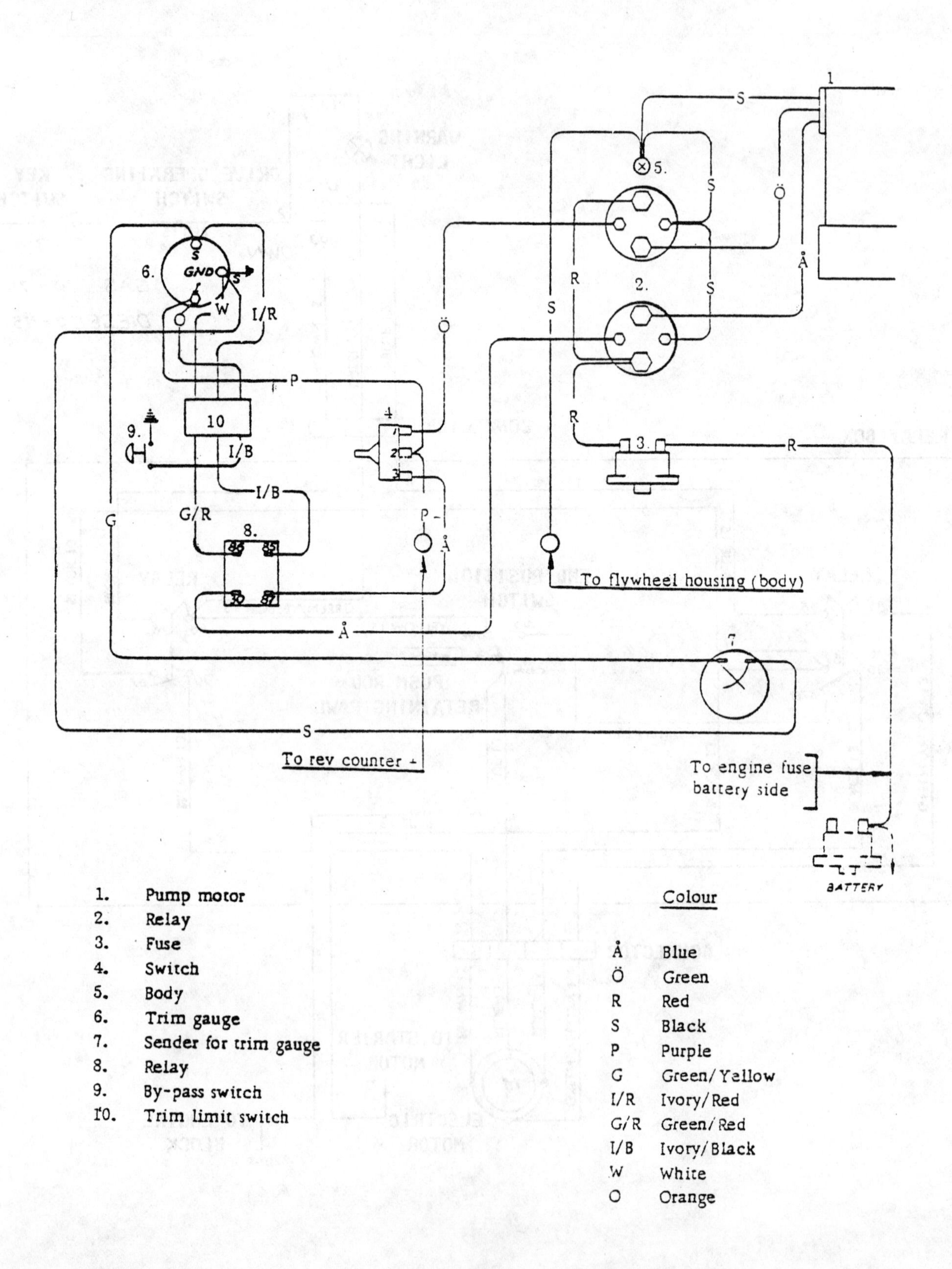

Wire Identification -- Power Trim -- single station with trim gauge and trim limit switch.

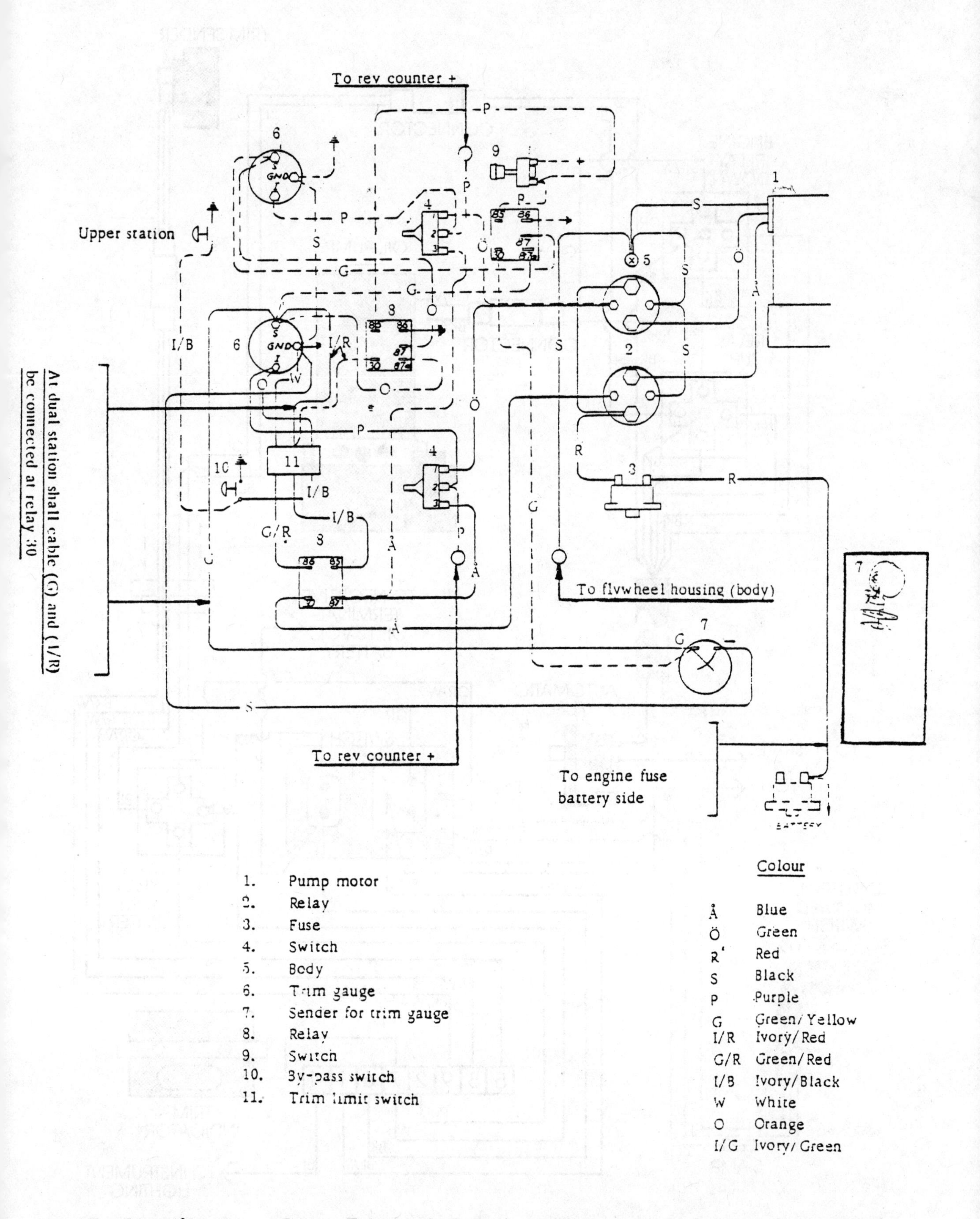

1. Pump motor
2. Relay
3. Fuse
4. Switch
5. Body
6. Trim gauge
7. Sender for trim gauge
8. Relay
9. Switch
10. By-pass switch
11. Trim limit switch

Colour

Å	Blue
Ö	Green
R	Red
S	Black
P	Purple
G	Green/Yellow
I/R	Ivory/Red
G/R	Green/Red
I/B	Ivory/Black
W	White
O	Orange
I/G	Ivory/Green

Wire Identification -- Power Trim -- dual station with trim gauge and trim limit switch.

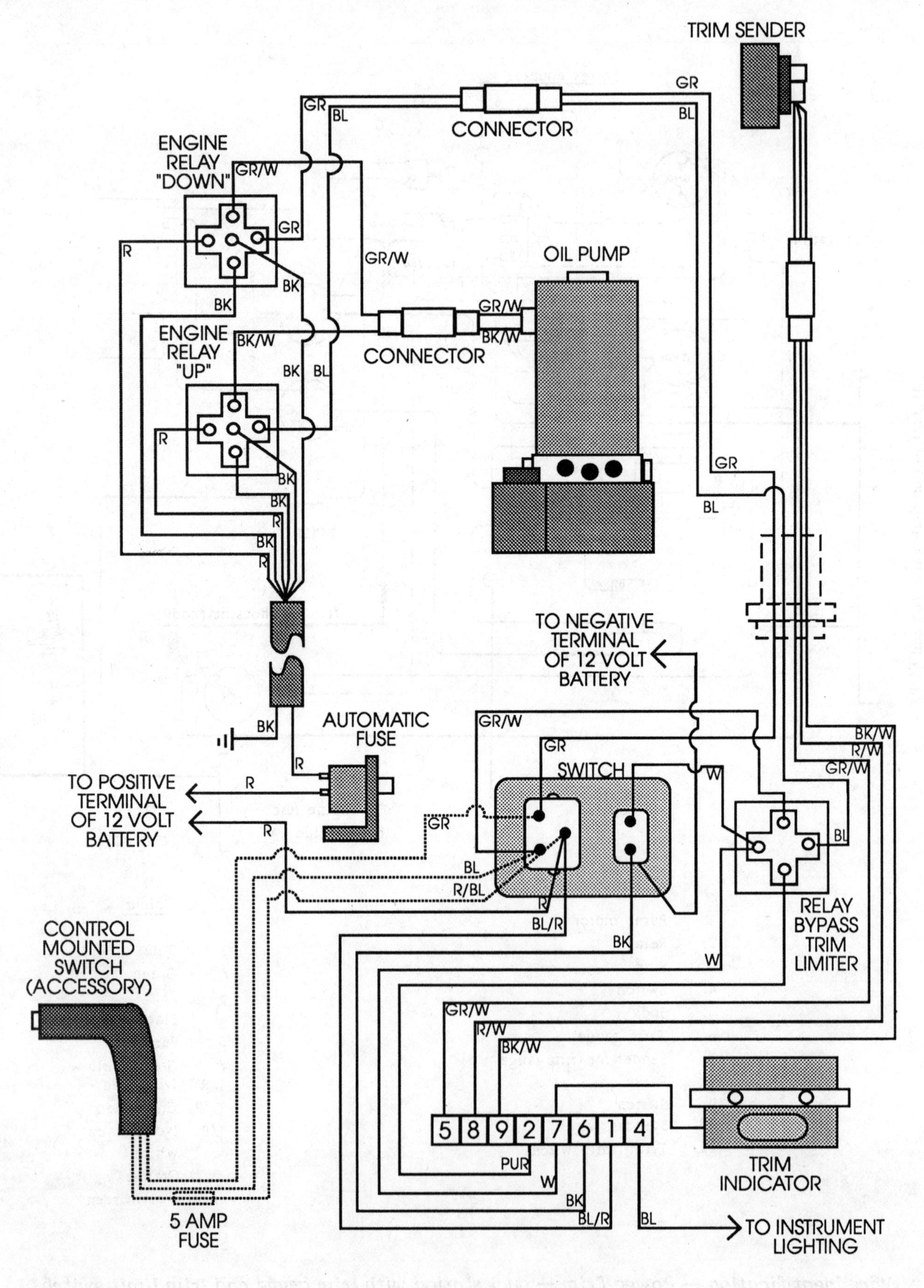

Wire identification -- Power Trim -- late type -- single station with trim gauge and trim limit switch.

NOTES & NUMBERS

New titles are constantly being produced and the updating work on existing manuals never ceases. All manuals contain complete detailed instructions, specifications, and wiring diagrams.